Studies in Surface Science and Catalysis 119

NATURAL GAS CONVERSION V

Studies in Surface Science and Catalysis

Advisory Editors: B. Delmon and J.T. Yates

Vol. 119

NATURAL GAS CONVERSION V

Proceedings of the 5th International Natural Gas Conversion Symposium,
Giardini Naxos - Taormina, Italy, September 20-25, 1998

Editors

A. Parmaliana
University of Messina, Italy

D. Sanfilippo
Snamprogetti SpA, Milan, Italy

F. Frusteri
Istituto CNR-TAE, Messina, Italy

A. Vaccari
University of Bologna, Italy

F. Arena
University of Messina, Italy

1998
ELSEVIER
Amsterdam — Lausanne — New York — Oxford — Shannon — Singapore — Tokyo

ELSEVIER SCIENCE B.V.
Sara Burgerhartstraat 25
P.O. Box 211, 1000 AE Amsterdam, The Netherlands

Library of Congress Cataloging in Publication Data.
A catalog record from the Library of Congress has been applied for.

ISBN 0-444-82967-9

∞The paper used in this publication meets the requirements of ANSI/NISO Z39.48-1992 (Permanence of Paper)

Transferred to digital printing 2006

Printed and bound by CPI Antony Rowe, Eastbourne

PREFACE

Many words have been used to give the more appropriate idea of the scientific, economic and technological impact of the Natural Gas Conversion on energy production, chemical and petrochemical industry as well as on the economy of the countries possessing large reserves. NG Conversion has been considered a challenging topic for the modern catalysis, now, at the eve of the third millennium, it is one of the greatest and proven scientific achievement of the last decade which will imply significant change in the current technology related to fuel, gasoline , intermediates and chemicals production. A rapid look at the volumes collecting the papers presented at the NG Conversion Symposia allows to experience the growing interest devoted to the NG Conversion along the years as well the consolidated trend to lessen the attention towards topics which in spite of their potential importance are quite far from the industrial exploitation and to focus all the efforts towards research subjects which deserve a greater technological interest being then economically rewarding.

Along the years the number of papers aimed to present technological issues and economical evaluation of the gas to liquid (GTL) processes is growing, the Fischer-Tropsch chemistry, catalysis and technology is currently revisited, new approaches for syngas production are currently pursued. However, even if such topics constitute the driving forces for attracting more and more interest towards the NG Conversion we must consider that innovative research approaches for the NG Conversion involving the use of membrane reactor and/or electrochemical devices original methods for the direct conversion of natural gas to formaldehyde and methanol as well oxygenates of higher added value obtained trough two-step or cross-coupling reaction systems are presently pursued by many academic and industrial research groups worldwide. On other hand, such a great scientific and technological interest posed in the NG Conversion, apart the reasons above outlined, arises from the fact that on January 1988 the ascertained and economically accessible reserves of NG amounted worldwide to over 144,000 billion cubic meters, corresponding to 124 billion tons of oil equivalents (comparable with the liquid oil reserves estimated to 138 billion TOE). It is hypothesized that the volume of NG reserve will continue to grow at the same rate of the last decade. Forecasts on production indicate a potential increase from about 2,000 billion cubic meters of 1990 to not more than 3,300 billion cubic meters in 2010, even in a high economic development scenario. NG consumption represents only one half of oil one: 1.9 billion TOE/y as compared with 3.5 of oil. As a consequence in the future gas will exceed oil as carbon atom source.

All these aspects indicate that in the future the potential for getting energetic vectors or petrochemicals from NG will continue to grow.

The first need is to transport NG from production sites to the consumption markets. Current technologies for marking available this "remote gas" are basically as CNG via pipelines (on-shore – off-shore) or LNG via ocean shipping in dedicated tankers. The delivered cost is relevant to the distance and over 1,000 – 2,000 kilometers LNG becomes competitive with CNG. The value at which this remote gas is made available in the developed markets represents the break-even price or the economic baseline for any alternative uses.

The presence of light paraffins (C_2-C_4) in the NG can be a key factor in promoting further exploitation of the NG conversion.

Indeed, light NG paraffins, apart from their use in steam cracking, have had some additional exploitation: maleic anhydride from butane and the selective production of olefins (propylene and isobutylene, butadiene) via dehydrogenation are the most significant examples. On this account, the changing scenario of the chemical commodities producer countries due to the increasing tendency of developing countries to better exploit their internal resources, and not only for captive utilization, have led to the development of technologies aimed at transforming NG components into more valuable or transportable products.

During the last twenty years, a network of new and old technologies aimed at making available wider possibilities of economically attractive transformation of natural gas to higher valued chemicals or liquid fuels has been growing in a more or less co-ordinated effort of technological innovation taking into consideration the presence of C_2-C_4 hydrocarbons together with methane.

In the last 3-4 years information on the NG conversion has overcome the limit of the scientific or technological literature and has entered the financial news world, meaning that the attention of market operators is addressed to this opportunity.

It is in this context that we present this volume collecting the Proceedings of the Fifth Natural Gas Conversion Symposium which will be held in Giardini Naxos-Taormina the 20-25 September 1998. The Symposium continues the tradition set by four previous meetings held in Auckland (New Zealand, 1987), Oslo (Norway, 1990), Sydney (Australia, 1993) and Kruger National Park (South Africa,1995).

The scientific programme consists of invited plenary and key-note lectures, oral and poster contributions. The papers cover the following area topics:

Catalytic combustion; Integrated production of Chemicals and Energy from Natural Gas, Fischer-Tropsch Synthesis of Hydrocarbons; Innovative Approaches for the Catalytic Conversion of Natural Gas and Novel Aspects of Oxidative Coupling; Natural Gas Conversion via Membrane based Catalytic Systems; Synthesis of Oxygenates from Syngas,

Partial Oxidation of Methane and Light Paraffins to Oxygenates; Catalytic Conversion of light Paraffins; Production of Syngas (Oxyreforming, Steam Reforming and Dry Reforming); Natural Gas Conversion-Industrial Processes and Economics.

The topics of the Symposium witness the large global R&D effort to look for new and economic ways of NG exploitation, ranging from the direct conversion of methane and light paraffins to the indirect conversion through synthesis gas to fuels and chemicals. Particularly underlined and visible will be the technologies already commercially viable.

The 5th NGCS is therefore a way of showing the increasing role of NG a source of value creation for companies and as a perspective clean raw material for answering to the environmental societal concerns.

The interest raised by the Symposium has been overwhelming as accounted by the large number of papers presented and delegates. The countries participating in the congress and contributing to the Proceedings reported here are:

- Algeria
- Argentina
- Australia
- Canada
- China
- Denmark
- England
- Finland
- France
- Germany
- Greece
- Hungary
- Ireland
- Italy
- Japan
- Korea
- Malaysia
- Mexico
- Norway
- Poland
- Portugal
- Russia
- Saudi Arabia
- Slovakia
- South Africa
- Spain
- The Netherlands
- U.S.A.
- Venezuela

The Organising Committee is grateful to the International Scientific Committee for having given to the Italian Chemical Community the chance and the honour to handle the organisation of such international scientific event as well for the scientific co-operation in the choice of the congress topics and paper selection.

The 5th Natural Gas Conversion Symposium is supported by the Division of the Industrial Chemistry and Catalysis Group of the Italian Chemical Society, the Institute CNR-TAE and the University of Messina.

The Italian Catalysis Community is particularly keen to gather in Italy all the Scientists active in this strategic area. We feel that this event marks also the active role played by the Italian Scientific Community in developing original and viable routes for the NG Conversion.

We are confident and the content of this volume proves this view, that mature and technologically feasible processes for the natural gas conversion are already available and that new and improved catalytic approaches are currently developing and we hope that their validity and feasibility are soon documented. This is an exciting area of the modern catalysis which certainly will open novel and rewarding perspectives for the chemical, energy and petrochemical industries. With this optimism we address the Symposium to all the participants, to all the scientists active in the area.

It is a pleasure to acknowledge the generous support given by the Sponsors which have greatly contributed the success of the event, the assistance of the members of the International Scientific Committee, the hard work of the Organising Committee and the many student assistants and all who have contributed to the success of the Symposium through presentation, discussion, chairing of Sessions and refereeing of manuscripts.

Messina 25 June 1998

Adolfo Parmaliana
Domenico Sanfilippo
Francesco Frusteri
Angelo Vaccari
Francesco Arena

ORGANIZERS

The symposium has been organized by:

- ***Division of Industrial Chemistry and Catalysis Group of the Italian Chemical Society***
- ***Institute CNR-TAE (Messina)***
- ***University of Messina***

ORGANIZING COMMITTEE

A. Parmaliana	*University of Messina*	*Italy*
D. Sanfilippo	*Snamprogetti SpA, Milano*	*Italy*
F. Frusteri	*Istituto CNR-TAE, Messina*	*Italy*
F. Arena	*University of Messina*	*Italy*
G. Cacciola	*Istituto CNR-TAE*	*Italy*
G. Deganello	*University of Palermo*	*Italy*
P. Garibaldi	*Euron SpA, Milano*	*Italy*
R. Maggiore	*University of Catania*	*Italy*
G. Petrini	*Enichem SpA, Milano*	*Italy*
A. Vaccari	*University of Bologna*	*Italy*

SCIENTIFIC COMMITTEE

C. Apesteguia	*Argentina*
M. Baerns	*Germany*
T.H. Fleisch	*USA*
A. Holmen	*Norway*
G. Hutchings	*UK*
E. Iglesia	*USA*
B. Jager	*South Africa*
E. Kikuchi	*Japan*
W. Li	*China*
J. Lunsford	*USA*
I. Maxwell	*The Netherlands*
C. Mirodatos	*France*
J. Ross	*Ireland*
J. Rostrup-Nielsen	*Denmark*
D. Sanfilippo	*Italy*
L.D. Schmidt	*USA*
D. Trimm	*Australia*

5th Natural Gas Conversion Symposium , 20-25 September 1998, Giardini Naxos - Taormina

FINANCIAL SUPPORT

The organising committee would like to thank the following Organisations for the financial support sponsorship:

LIST of SPONSORS

- **Azienda Autonoma per l'Incremento Turistico della Provincia di Messina**
- **AKZO NOBEL CHEMICALS S.p.A.**
- **AMOCO Corp.**
- **BANCO di SICILIA**
- **ENGELHARD**
- **ENI S.p.A.**
- **ENITECNOLOGIE S.p.A.**
- **EURO SUPPORT MANUFACTURING B.V.**
- **EURON S.p.A.**
- **HALDOR TOPSØE A.S.**
- **K.T.I. S.p.A.**

❖LIBRERIA BONANZINGA

❖MICROMERITICS S.r.l.

❖PHILIPS S.p.A.

❖Presidenza dell' Assemblea della Regione Siciliana

❖Presidenza del Governo della Regione Siciliana

❖PROVINCIA REGIONALE di MESSINA

❖REPSOL PETROLEO S.A.

❖SASOL

❖SHELL

❖SNAM S.p.A.

❖SNAMPROGETTI S.p.A.

❖STATOIL A.S.

❖SÜD-CHEMIE A.G.

❖THERMOQUEST – C.E. Instruments

TABLE of CONTENTS

Plenary Lectures

Topic C

Catalytic Combustion; Integrated Production of Chemicals and Energy from Natural Gas

Topic H

Fischer-Tropsch Synthesis of Hydrocarbons

Topic I

Innovative Approaches for the Catalytic Conversion of Natural Gas and Novel Aspects of Oxidative Coupling

Topic M

Natural Gas Conversion via Membrane based Catalytic Systems

Topic O

Synthesis of Oxygenates from Syngas; Partial Oxidation of Methane and Light Paraffins to Oxygenates

Topic P

Catalytic Conversion of Light Paraffins (C2-C5)

Topic S

Production of Syngas

Keynote Lectures

Topic T

Natural Gas Conversion: Industrial Processes and Economics

Keynote Lectures

NATURAL GAS CONVERSION V
Studies in Surface Science and Catalysis, Vol. 119
A. Parmaliana et al. (Editors)

Natural Gas as raw material for clean fuels and chemicals in the next decades

by M. Colitti

EniChem, Milan, Italy

Modern industry originates in a change of feedstock, the result of a never-ending quest for a better, cheaper raw material. In the origin, there was coal, a source of both energy and industrial feedstock, the so-called synthesis gas. Then, coal was replaced by liquid hydrocarbons, coming from a refinery or from wells of crude oil and gas. The passage from solid to liquids was part and parcel of a structural change which has produced an extraordinary acceleration of economic growth. We might hope that the same will happen when natural gas will comes in to displace the liquids.

New feedstocks do not come in without a fight. It is not only a matter of price, but also of technology, and of the natural tendency of industries to protect their own investments in plants which are all of a sudden made to look old. Rather than repeat for the nth time the list of technologies which can turn natural gas into a basic feedstock for oil and petrochemical industries, I will try to discuss how will companies decide upon this matter. What are the main elements of such a decision? Predictably, its main element is a comparison between costs and prices. However, this is not a simple matter, to be decided on a back-of-an-envelope calculation: it is, rather, a differential decision, based on a comparison between the situation in which we are now, and a future one, by itself uncertain.

The first element is the price of the new feedstock per ton of the product we want, which is deeply influenced by the cost of new technological processes and therefore requires a technological assessment of the variable as well as the fixed costs, and of the direct costs as well as the amortisation of the capital invested. These costs will be compared with the price of the products we want, which, as history will teach us, might change together with the feedstock. Let's try to deal with these elements, briefly but, if possible, clearly, although in a purely descriptive way.

Who makes the price of gas? Methane is in great demand as fuel and it is said that it will fully displace oil in such uses. Therefore, its value in any area of the world tends to be what can be netted back from sales to Europe. In the rich markets of the Old Continent, gas is sold as fuel in competition with delivered light gasoil, at prices which leave to the seller a good part of its competitive advantage (with the exception, of course, of the ecological improvement, which is collective). It is therefore too highly priced for it to become a raw material. A large industrial conversion plant could not conceivably pay the same price that can be extracted from a household consumer for gas delivered inside his house.

This means that gas is available at industrial prices only if produced in areas which are too far from Europe, and do not have a great fuel demand themselves. In these areas the industrial transformation of gas into liquids should take full advantage of the lower cost of transporting a liquid.
The other big factor in defining the long term gas price is the large reserves of natural gas in the world, which should tend to keep its price down, at least in certain areas, but this is by no means certain: gas sources which cannot enter the rich markets do not seem to influence the price at all.
The second element, technology, determines the capital to be invested to obtain a ton of the product we desire from the new feedstock. This number should be quite certain, based on rock-bottom certainty of the engineers' calculations. However, first generation plants do require more capital per ton of product than second or third generation ones. Not only the scale of the plant increases with experience; also, technological change starts with high-pressure high-temperature plants and moves into low-pressure, low-temperature ones, which cost much less to build and maintain. So the comparison, to be honest, has to be done taking into account future things like the experience curve, the acceleration of reactions produced by accumulated know how, in short, the overall technological trends.
Let's talk of the third element, the price of the products that can be obtained. The tendency to manufacture the most valuable product possible has to be balanced by the fact that highly-priced products are often small-volumes specialities. To combine the high volumes which come from world-scale plants with high-price products, a sometimes impossible operation, could perhaps be performed by aiming at the market for ecological components, that is, products so fine that they can be used to upgrade low-cost base products. This case is however partly clouded by the uncertainty on the future trends of the environmental legislation, which seems already bound to swallow every product in an ever lengthening list of baddies. Alternatively, one can imagine plants which combine productions, for example, of liquids (methanol) with that of electricity. Trouble is, the areas which do not offer a great market for gas are not hungry for other energy sources either, and to sell there large volumes of electricity might be as difficult, if not more, than to transport that gas to the nearest high-price market.
The choice is therefore complex, and the qualitative elements we have just briefly listed do become figures only after assumptions which do not always reduce the uncertainty, but sometimes increase it. Different companies will react differently to this challenge, the majority of them falling into one or the other of the following categories.
The innovator, who runs the risk of investing in new technologies or in old ones revisited and adapted to new productions. He may be motivated by the lure of large innovation profits to be obtained either by producing more cheaply something already in the market, or by marketing a new product. Paradoxically, this decision may be justified in two opposite ways: by saying either that you have more investment capital that investment opportunities in proven technologies; or that, having invested and found gas, you cannot allow that sunk capital simply to lie fallow, not producing anything.
The follower, who tends to avoid risk, and therefore leaves to the innovators not only the capital risk, but also the job to improve the technology and to develop the know-how.

He may hope that he will be able to obtain both from one of the innovators at a reasonable price, which will work out to be lower that the cost of the risk; or he might be developing his own process, which may not be ready yet, etcetera. He enters in the action later, and possibly not alone, to distribute the risk.

The laggard, who is content of the profit he is making and moves much later than both the innovator and the follower, and only when he considers it really unavoidable; that is, if he identifies the new technologies as a menace to his market position and his current profits.

The non-player, who does not want to run any risk, possibly because he does not believe in the new opportunity (and in some cases it might be right): or because he does not have the finance or the management to exploit. He would therefore exit from that area of products, rather than participate in the new developments.

The companies listening to me can easily classify themselves, a function which I would not dare to do for them.

Bear in mind that the risk is not necessarily limited to the technology. Liquids may be obtained from gas using old, or in any case well proven, process, like, for example, the production of methanol. In this case the risk is predominantly a market one, because the people who run, say, the power stations might resist the use of a new fuel, not thoroughly proven in all its aspects, technical as well as environmental. This kind of risk seems to me of a lower level than the technological one.

However, even in the most sophisticated projects we are talking about a technology which, if I am not wrong, dates back to pre-World War Two, when it was applied to obtain liquid from coal. The basic process operates, as of yesteryear, on Synthesis gas, and then goes through Hydrocracking or Dewaxing to obtain a mix of oil products: in some configurations Naphtha, Jet Fuel, Gasoil and Lube bases. The level of purity (zero sulphur, zero aromatics, zero metals) of these products qualifies them as ecological additives to oil products normally obtained by a refinery, and also qualifies them to prices which might be some 30% higher that the normal product. This has already been seen on the market when the price of MTBE was set between 1.2 and 1.5 times that of premium gasoline.

All this means that the differential evaluation has to take into account the alternative to obtain the full slate of oil products at a acceptable level of purity: in fact, traditional desulphurisation cannot reach the zero point, and the lower is the sulphur level, the higher is the cost of reaching it.

Perhaps it might pay to have a quick review of the products obtainable with the different technologies, starting from the more obvious.

The first, one could say traditional, way of obtaining a liquid from gas is to produce methanol. This idea has already been applied by gas-rich areas which could not supply the high price European market: the South Chilean Cape Horn plant, the Caraibic Coast Venezuelan ones, and of course the large methanol producing capacity in the Arab-Persian Gulf and especially in the Kingdom of Saudi Arabia are all examples of this strategy. A new technologies seems to offer the opportunity of going beyond the accepted maximum scale of two thousand tons per day, but only, as we have already

said, by going through a large electricity output, which can create some marketing problems.
All the plants we have quoted produce methanol as a chemical intermediate, but methanol might be more flexible that that. It can certainly be used in modern power stations, where it could largely improve both production efficiency and the environmental impact. Or it could be used to go to olefins, something apparently quite interesting. There is a general tendency to side-step the cracker, a plant that, when fed with Virgin Naphtha, produces such a large stream of different products that it creates some embarrassment for their final utilisation. The success of the Dehydro concept, which produces butadiene from normal butane, isobutilene from isobutane, and propylene from propane - an interesting case of an old technology revisited with great success- goes exactly in this direction.
Second, the oil products we have quoted before. In this case, one could say that we have a clear-cut case of substitution. What you could obtain in a refinery you will produce in a different plant, using a different technology and feedstock. If that was true, one could object that refining capacity is quite large today both in Europe and in the US, and that it is not very profitable, and also that demand of oil products does not increase much. However, from 1995 and 2010 demand for Virgin Naphtha is supposed to increase, for example in Europe, at 3,3% per year, while the other oil products are expected to grow at lower rate, about 1.8%, so a gas-to-liquids plant which would produce a number of oil products would see its demand grow at something between 2% and 3%, which is not bad at all. However, this calculation is by far on the over-conservative side. A gas-to-liquid plant would produce lubricant bases, whose demand seems to increase at rates near to ten per cent; and it would produce more gas oil than gasoline following the present market trend. Moreover, the products, as we have seen, would not be the same, and one would expect high purity components - because this is what they would be - of oil products to grow at a much higher rate.
All this means that the substitution of a feedstock is not a mere technological change, which would leave more or less the rest as it was. It is a structural operation which not only offers to change the way the products we utilise now are obtained, but also to change the products themselves. The more we move forwards towards the next millennium, the more we can expect that the environmental premium to grow higher that it is now. If we don't want our atmosphere to grow worse, the strictness of the discipline needed to protect the environment must increase at least at the same rate of increase of the volumes of the products utilised. If the market works, we can expect the environmental premium to increase more or less at the same rate. It might be that the innovators will turn out to be the real risk-averse ones, as the risk of doing nothing increasingly seems to be deadlier than that of making a mistake.

NATURAL GAS CONVERSION V
Studies in Surface Science and Catalysis, Vol. 119
A. Parmaliana et al. (Editors)

Promotion of Steam Reforming Catalysts

I. Alstrup, B.S. Clausen, C. Olsen, R.H.H. Smits, J.R. Rostrup-Nielsen*

Haldor Topsøe A/S, Nymøllevej 55, DK-2800 Lyngby, Denmark

ABSTRACT

The use of more economic reforming conditions is limited by the requirement for carbon-free operation. This constraint can be weakened by promotion of the catalyst. The principal mechanisms of avoiding carbon formation are analysed and the experimental evidence discussed on the basis of new data on spill-over of adsorbed water, the role of alkali and ensemble effects by alloying and by decoration with surface oxides.

1 INTRODUCTION

A key to improving the process for steam or CO_2 reforming of hydrocarbons is to expand the room for carbon-free operation [1]. The selection of operating parameters as well as the design of the reforming catalyst are dictated by the need for carbon-free operation. With improved catalysts it is possible to design for lower steam-to-carbon ratios and higher preheat temperatures and to achieve higher feedstock flexibility [2].

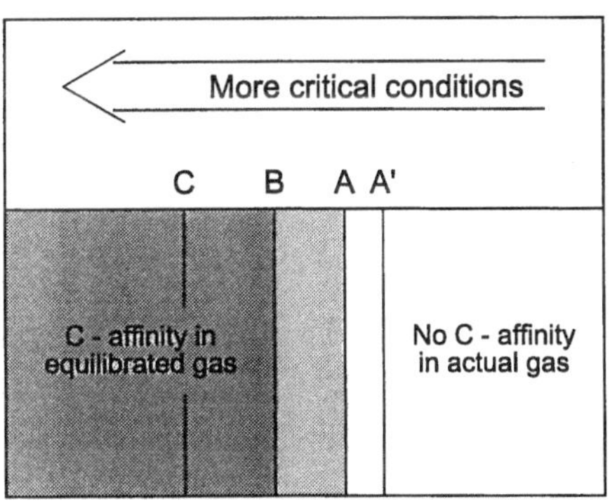

Fig. 1 Carbon Limits
A' no affinity for actual gas
A real carbon limit
B principle of equilibrated gas
C sulphur passivation, noble metals

At a given temperature and for a given hydrocarbon feed carbon will be formed below a critical steam-to-carbon ratio (carbon limit A in *Fig. 1*). This critical steam-to-carbon ratio increases with temperature. By promotion of the catalyst, it is possible to push this limit towards the thermodynamic limit B reflecting the principle of the *equilibrated* gas [3]:

Carbon formation is to be expected on nickel catalysts if the gas shows affinity for carbon after the establishment of the methane reforming and the shift equilibria.

By use of noble metals or sulphur passivation, it is possible to push the limit

* Corresponding author

beyond limit B to Limit C. A safe design criteria is to require that the *actual* gas shows no affinity for carbon formation. This results in carbon limit A'. For higher hydrocarbons, for which the carbon reactions are irreversible carbon limit A' applies.

Whether carbon-free operation is possible depends on the kinetic balance as illustrated in the simplified two-step mechanism [3] shown in *Fig. 2*. For a nickel catalyst carbon is normally formed by the whisker mechanism [1,4]. Adsorbed carbon atoms that do not react to gaseous molecules are dissolved in the nickel crystal and carbon whiskers nucleate from the nickel support interface of the crystal. Carbon formation is avoided when the concentration of carbon dissolved in the nickel crystal is smaller than that at equilibrium, in other words, when the steady state activity of carbon is smaller than one. In terms of the sequence in Fig. 2, the steady state activity is proportional with [C*] which can be expressed by [3]:

$$CH_4 + * \xrightarrow{k_1} CH_x *$$

$$CH_x \xrightarrow{k_2} C * \longleftrightarrow [C, Ni]_{bulk} \rightarrow \text{whisker carbon}$$

$$CH_x * + OH_y * \xrightarrow{k_3} \text{gas}$$

$$C * + OH_y * \xrightarrow{k_4} \text{gas}$$

Fig. 2 Methane Reforming. Simplified Reaction Sequence [1]
* represents nickel site disregarding the ensemble size.

$$a_c^s \sim [C*] \sim \frac{k_1 k_2}{k_3 k_4} \cdot \frac{1}{[OH_y *]^2} \qquad (1)$$

Hence, the steady state carbon activity can be decreased by:

- enhancing the adsorption of steam or CO_2
- enhancing the rate of the surface reaction
- decreasing the rate and degree of methane activation and dissociation.

The whisker mechanism may also be blocked by use of noble metal catalysts because these metals do not dissolve carbon [1,5,6].

This paper will focus on attempts to achieve these effects.

2 SPILL-OVER OF STEAM

The impact of alkali and active magnesia on carbon-free steam reforming of higher hydrocarbons is well known [1]. Kinetic studies indicated that the adsorption of steam was enhanced by "active" magnesia and alkali and that spill-over of adsorbed steam to the metal surface may play a role [1]. This was reflected by negative reaction orders with respect to steam [1,7]. Similar effects of La_2O_3 and Ce_2O_3 on CO_2 and steam reforming have been observed [8-11]. However, little fundamental work has been done to clarify the detailed role of enhanced adsorption of steam and CO_2 on the catalyst.

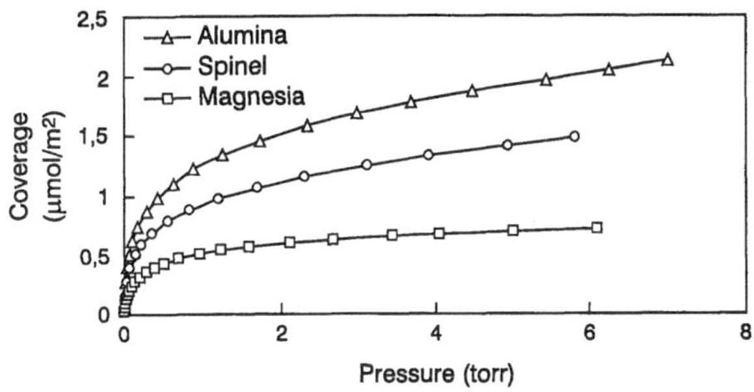

Fig. 3 Coverage per unit area as a function of steam pressure for three commonly used supports for steam reforming catalyst: Al_2O_3 ($\theta+\alpha$), $MgAl_2O_4$ and MgO

In order to achieve a better understanding of the phenomenon, steam adsorption on various supports commonly used for nickel-based steam reforming catalysts was studied by micro-calorimetry [12]. Some results are shown in *Fig. 3* [13].

In contrast to what would be expected from the simple model sketched above, the magnesia support showed the lowest amount of adsorption of steam, followed by the spinel and the alumina, respectively. A similar unexpected order was found in the heats of adsorption of steam on the supports: at a given coverage, steam was found to adsorb the strongest on alumina. The difference between spinel and magnesia was small, magnesia showing slightly lower heats of adsorption than spinel did.

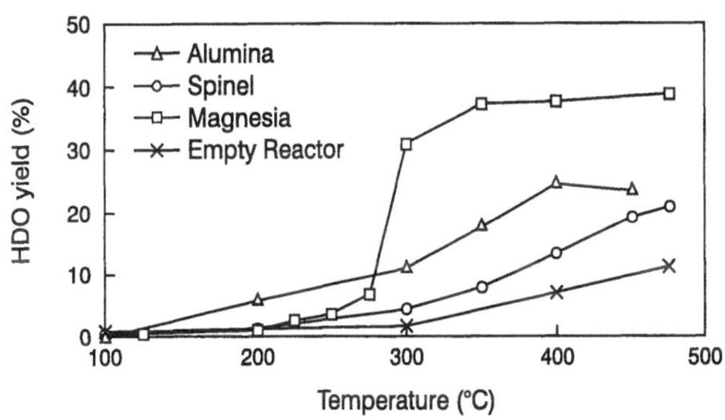

Fig. 4 Yield of HDO from a mixture of 1% H_2 and 2.8% D_2O in He on various supports as a function of temperature

However, by isotope exchange studies [13] it was shown that the magnesia based catalyst is more active to dissociate the adsorbed steam as illustrated in *Fig. 4*. In this figure, the results of H/D isotope exchange experiments between H_2 and D_2O are shown. The yield of HDO from the spinel support is hardly more than that formed without any material in the reactor. The alumina support shows a somewhat higher activity, but the magnesia support is a very active catalyst for the H_2/D_2O isotope exchange reaction: the statistical H-D distribution is reached at temperatures where the other supports have just started to show activity. The sudden increase in activity between 275°C and 300°C is remarkable. One may speculate that this is related to the fact that these temperatures are not far above the temperature range where $Mg(OH)_2$ is stable; under the conditions used the bulk phase transformation from $Mg(OH)_2$ to MgO is calculated to occur at 175°C.

The two sets of experimental results described above indicate that the improved adsorption of steam on magnesia supports resulting in improved resistance to carbon formation is not a static but a dynamic effect. As discussed earlier [1] enhanced steam adsorption cannot reflect a true equilibrium constant. This would violate the principle of microscopic reversibility, because steam is also adsorbed directly on the nickel surface. The following reaction scheme, which is a modification of the one proposed in [1], illustrates this:

$$H_2O + *_{sup} = H_2O *_{sup} \quad (2)$$

$$H_2O *_{sup} + *_{sup} = OH*_{sup} + H *_{sup} \quad (3)$$

$$OH*_{sup} + *_{Ni} = OH*_{Ni} + *_{sup} \quad (4)$$

$$H_2O + 2*_{Ni} = OH*_{Ni} + H*_{Ni} \quad (5)$$

(Whether $OH*_{Ni}$ in eqs.(4) and (5) is further dissociated to $O*_{Ni}$ before reaction with CH_x is an open question.)

The above experimental results show that the quantity and strength of steam adsorption on magnesia are lower than on non-promoting supports. The improved resistance to carbon formation of magnesia-supported nickel catalysts is thus not caused by an increased adsorption equilibrium constant of steam on the support (reaction (2)). Instead, the *rate* of dissociation of water on magnesia (reaction (3)) is much higher than on non-promoting supports. As a result, the amount of OH_y species present on the nickel is increased on magnesia-supported nickel, thereby enhancing the removal of CH_x and reducing the full dehydrogenation of CH_x to C*.

The above means that the spill-over of steam as suggested in the literature probably involves OH species instead of molecular water. This is in agreement with many recently published results. Bradford and Vannice [14] developed a kinetic model for Ni/MgO and Ni/TiO_2 and concluded that surface OH groups possibly situated on the support react with CH_x intermediates absorbed on the nickel. Work by Efstathio et al. [15] indicated that spill-over of lattice oxygen from yttrium stabilised zirconia was involved in the reforming reaction. Bitter et al. [16] found for CO_2 reforming of methane on Pt-ZrO_2 catalysts that the rate of reaction was proportional to the length of the metal-support perimeter. They suggested that reaction takes place between CH_4 activated on the metal and CO_2 activated in the form of carbonate on the support, without the need for spill-over of an oxygen species from the support to the metal or for adsorption of oxidants from the gas phase onto the metal.

3 THE FUNCTION OF ALKALI

Apart from the enhanced steam adsorption on alkali promoted catalyst [1], it is well known that the addition of alkali to steam reforming catalysts results in a decrease of the reforming rate [1,7] sometimes by more than one order of magnitude. The effect has been observed on a number of different group VIII metals and on a variety of supports. The decrease in reaction rate is reflected by lower preexponential factors whereas the activation energies are almost unchanged [7]. In contrast, the enhancement of steam activation on magnesia based catalyst has no impact on the preexponential factor. It is remarkable that the decline in activity when promoting with alkali is also observed when testing the catalyst for hydrogenolysis of ethane [1], i.e. without the presence of steam.

The impact of alkali is stronger on less acidic supports which suggests that the alkali partial pressure over the catalyst is important [1]. A less acidic support has a weaker bonding of alkali resulting in easier transport (via the gas phase) from the support to the metal. This effect of alkali on the activity of nickel is not fully understood.

The influence of alkali on the chemisorption of a number of different molecules on transition metal surfaces has been explained as the result of electrostatic interactions [17]. Coadsorption of steam and alkali has been studied on Ru(001), Pt(111), and Ni(111) single crystal surfaces. Most of the studies have been reviewed in [18]. More recently Kuch et al. [19,20] have studied the influence of preadsorbed potassium on the adsorption of water molecules on Ni(111). The simple electrostatic model cannot explain all the results obtained for the different metals. Some of the results depend sensitively on the metal in question. Thus it has been concluded that the H_2O molecule on Li and Na precovered Ru(001) is adsorbed with the oxygen atom pointing towards the surface, while for K precovered Pt(111) it was found that the molecular H-O-H plane was tilted 160° with the hydrogen atoms pointing towards the surface. In contrast angle-resolved photoemission measurements on Ni(111) by Bornemann et al. [21] indicated that the H_2O molecule was tilted *in* the H-O-H plane. The direction could not be determined. However, in all cases it was found that above a critical alkali coverage in the range 0.04 - 0.15 ML part of the adsorbed H_2O is dissociated into OH and H. The maximum OH coverage is equal to the alkali coverage. Both adsorbed H_2O and OH are strongly stabilised by the presence of alkali.

Ceyer et al. [22] stated on the basis of methane beam studies that preadsorbed potassium does not influence the chemisorption of methane on Ni(111). Due to the nonpolar nature of the methane molecule this result seems to be in accordance with the above-mentioned electrostatic theory of alkali promotion.

It has been speculated that the role of alkali in steam reforming is associated with the structure sensitivity of methane chemisorption on nickel. Beebe et al. [23] found that the sticking probability of methane is significantly smaller on the close packed Ni(111) surface than on the more open Ni(100) and Ni(110) surfaces. It is well known that alkali induces reconstructions of the Ni(110) surface, thereby creating (111) facets [7,24]. However, it remains to be shown that such restructuring takes place on the real catalyst. It is not very likely that the open surface planes constitute a significant part of the surface of the nickel particles of the working catalyst.

In order to achieve a better understanding of the influence of alkali, the impact of preadsorbed potassium on the chemisorption of methane on Ni(100) and Ni(111) surfaces was studied by Alstrup et al. [25]. Measurements of the chemisorption of methane at 475 and 500K for a range of potassium coverages showed that the initial sticking probability is influenced significantly by preadsorbed potassium on both surfaces as shown in *Fig. 5*. It is seen that the influence of potassium adatoms is much stronger on the (111) than on the (100) surface. However, for both surfaces the K-coverage dependence at low coverages is too strong to be explained by a simple ensemble blocking effect. These results seem to be in conflict with the above-mentioned electrostatic model of the influence of alkali. However, accurate density functional theory (DFT) calculations of the chemisorption of methane on Ni(111) show that during the course of the dissociative chemisorption event the methane molecule

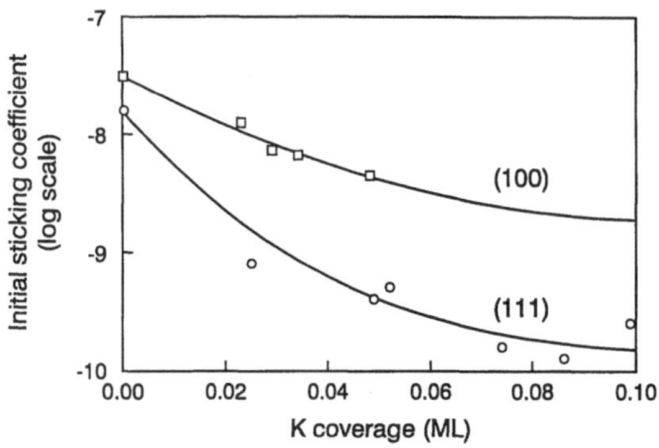

Fig. 5 Logarithmic plots of the initial sticking probabilities of CH_4 on Ni(100) and Ni(111) determined for a number of K coverages at 500K. (The dashed curves are guide to the eye.)

acquires a significant dipole moment in the transition state [26]. Therefore the electrostatic model may also be able to explain the new results. Similar calculations have unfortunately not been carried out for methane chemisorption on Ni(100), so it is not yet possible to explain the difference between the two surfaces. Neither is it clear why Ceyer et al. [22] did not observe any influence of potassium on the chemisorption of methane on Ni(111). It may be suggested that the difference between the results of the two studies is related to the fact that Ceyer et al. [22] used a molecular beam with far higher molecular energies than the main part of the methane molecules in the experiments of Alstrup et al. [25], in which the surface is bombarded by molecules approximately in thermal equilibrium with the surface.

In conclusion, the promoting effect of alkali inhibiting the formation of carbon may be related to these effects as well as to the spill-over of dissociated water.

4 DISSOCIATION OF METHANE

A number of recent studies have dealt with the impact of changing the catalyst composition on the activation of methane. Osaki et al. [27] studied the degree of dehydrogenation of CH_x-species on various catalysts and found indirectly x to be larger for nickel than for cobalt and larger for MgO supported catalyst than for those based on SiO_2. Aparicio [28] also observed a smaller degree of methane dehydrogenation on a Ni/MgO catalyst compared to an $MgAl_2O_4$ supported one. The catalysts had roughly the same nickel surface area and showed similar activities for steam reforming of methane, but the Ni/$MgAl_2O_4$ catalyst was significantly more active for CH_4/D_2 exchange meaning that methane was dissociated to a smaller degree on the Ni/MgO catalyst. Hence, the promoting effect of magnesia may be related to this effect on methane activation as well as enhanced adsorption and dissociation of steam [3].

Zhang and Verykios [8] claimed a similar double effect (i.e., methane activation as well as enhanced adsorption of CO_2) to be responsible for the promoting effect when using La_2O_3 as support for a nickel catalyst. Other investigations have shown similar promoter activity of Ce_2O_3-containing catalysts for steam reforming of butane [3,29,30]. Borowiecki et al. [,31] have reported retarding effects of Mo and W on the coking rate. Later work by these authors suggests that it is Mo oxide which is the species causing the reduced rate of carbon formation [32].

More studies are required to explain these promoting effects of various oxides and to clarify whether the promoters are acting by decorating the nickel surfaces. Promotion was demonstrated by Bradford and Vannice [33,34] who studied Pt-TiO_x and Pt-ZrO_2 catalysts for CO_2-reforming. The Pt-TiO_x catalyst showed much higher activity than did a pure Pt catalyst which was ascribed to creation of special sites at the metal/support interface similar to the ideas of Bitter et al. [16]. There was also strong evidence for TiO_x-layers on the Pt surface suppressing carbon deposition, probably by ensemble control.

A direct blockage of surface nickel atoms with resulting *ensemble* control was observed over partly sulphur poisoned nickel catalysts [35]. By controlling the sulphur content in the feed, it is possible to establish a situation on the nickel surface with ensembles available for the dissociation of methane but not for the dissolution of carbon atoms into the nickel crystal and the nucleation of the whisker carbon. This way of obtaining carbon-free operation was brought into practice in the SPARG process [36]. It is the result of a dynamic situation since methane may well decompose over a passivated catalyst in the absence of steam. However, this results in carbon whiskers with another structure ("octopus" carbon). Trimm has suggested a similar mechanism for the promoting effect he found for Bi addition to Ni [6].

Alloying nickel with copper [37,38] can also decrease the rate of carbon formation, but it is not possible to achieve the required high surface coverage of copper atoms as with sulphur atoms to eliminate carbon formation. A very surprising result of these studies was that the rate of carbon formation was even enhanced by low additions of copper. An electronic effect revealed by density functional theory (DFT) calculations of the influence of various alloying elements on the chemisorption of methane on Ni(111) [39] may explain this result. They showed as illustrated in Table 1 that the activation energy of methane chemisorption on a nickel atom in the Ni(111) surface is significantly smaller if the neighbour atoms are copper atoms than if they are nickel atoms.

Table 1

Change of energy barrier for the dissociation of CH_4 on a Ni atom with 1 or 2 Au or 6 Cu neighbour atoms on a Ni(111) surface [26,39]

Neighbour Atoms	*Change of Energy Barrier (kJ/mol)*
6 Ni	0
6 Cu	-5
1 Au; 5 Ni	16
2 Au; 4 Ni	38

While Ni and Cu form a stable random alloy, this is not the case for the Ni-Au system. Ni and Au do not mix in the bulk but may form stable alloys in the outermost layer [40]. DFT calculations (Table 1) predict that one Au neighbour increases the activation barrier for the methane dissociation over a Ni atom by 16 kJ/mole and two Au neighbours increase it by 38 kJ/mol. The suggestion by these DFT calculations that Au impedes methane dissociation was verified by molecular beam scattering experiments on well defined Ni(111) surfaces [41]. The DFT calculations also suggested that the stability of adsorbed carbon on the Ni(111) surface is drastically reduced in the vicinity of an Au atom resulting in a lower carbon coverage. Since the probability of the nucleation of whiskers is determined indirectly by the coverage of carbon (see above), Au may also in this way reduce the tendency for whisker formation on nickel catalysts [42].

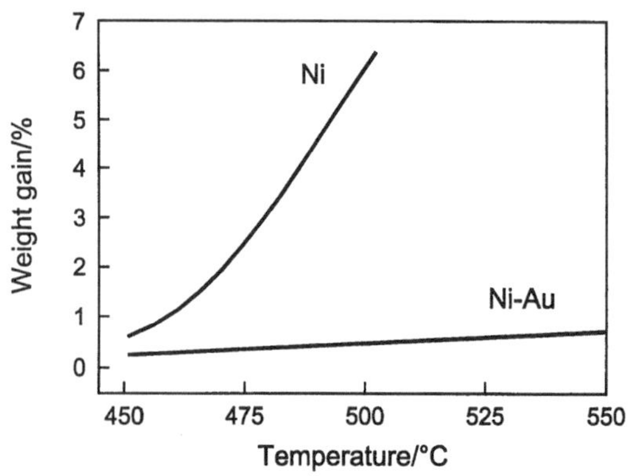

Fig. 6 The weight increase measured by TGA of a Ni catalyst compared to that of an Au-Ni (1.85% Au) catalyst during steam reforming of butane. Gas composition: 3.8% butane: 22.9%

The higher resistance to carbon formation of an Au-Ni surface alloy compared to that of pure Ni as suggested by the surface science work and theory was verified in TGA measurements for steam reforming of butane on a high surface area Au-Ni catalyst [43]. In contrast to the pure Ni catalysts, also Au-Ni catalyst appears to be resistant to carbon formation, as illustrated in *Fig. 6.* The activity for the reforming reaction was found to be reduced by only 40% compared to the pure Ni catalyst.

5 NOBLE METALS

A number of recent papers [44] have dealt with the use of noble metals to eliminate carbon formation. This effect has been described mainly in relation to CO_2 reforming on rhodium, ruthenium and platinum catalysts. As stated above, carbon formation on noble metals is probably prevented because carbon is not dissolved in these metals, thus preventing the diffusion of carbon through the metal to form whisker carbon [1,6]. Palladium is the only noble metal that still forms carbon, probably because of the formation of a carbide [5]. The superior carbon resistance of noble metal catalysts [3] has been demonstrated for CO_2-reforming of methane as well as steam reforming of higher hydrocarbons. CO_2-reforming of natural gas is practised with a noble metal catalyst [45] at conditions for which the principle of equilibrated gas would predict carbon formation (see Fig. 1).

6 CONCLUSIONS

Promotion of reforming catalysts may allow operation at more economic conditions such as low steam-to-carbon ratio and high preheat temperature. Moreover, increased carbon resistance means higher flexibility to feedstock composition. The promotion may be related to enhanced steam adsorption coupled with spill-over of OH species to the nickel surface as well as to a reduced degree of dissociation of the adsorbed methane.

Almost 30 years ago, Andrew [46] claimed in a discussion of the promotion of steam reforming catalysts for naphtha that "it seems unreasonable to expect that one immobile solid (refractory oxide) could effectively catalyse the oxidation of another immobile solid (carbon) on the surface of a third solid (nickel)". Today, surface science has provided a better understanding of phenomena like spill-over and ensemble control.

There is still a need for more fundamental studies of these effects.

ACKNOWLEDGEMENTS

The Danish Research Councils through the Center for Surface Reactivity supported part of the work.

REFERENCES

1. J. R. Rostrup-Nielsen, "Catalytic Steam Reforming", in J. R. Anderson and M. Boudart (Editors), Catalysis, Science and Technology, Vol. 5, Springer, Berlin, 1983, p. 1.
2. J. R. Rostrup-Nielsen and I. Dybkjær, Proc. 1st European Conf. on Chemical Engineering (ECCE), Firenze, May 4-7, 1997.
3. J. R. Rostrup-Nielsen, J.-H. Bak Hansen and L. M. Aparicio, J. Jap. Petr. Inst., 40 (1997) 366.
4. I. Alstrup, J. Catal., 109 (1988) 241
5. J. R. Rostrup-Nielsen and J.-H. Bak Hansen, J. Catal., 144 (1993) 38.
6. D. L. Trimm, Catal.Today, 37 (1997) 233.
7. J. R. Rostrup-Nielsen and L. J. Christiansen, Appl.Catal. A., 126 (1995) 381.
8. Z. Zhang and X. E. Verykios, Catal. Lett., 38 (1996) 175.
9. T. Horiuchi, K. Sakuma, T. Fukui, Y.Kubo, T. Osaki and T. Mori, Appl. Catal. A, 144 (1996) 111.
10. K. Seshan, H. W. ten Barge, W. Hally, A. N. J. van Keulen and J. H. R. Ross, Stud. Surf. Sci. Catal., 81 (1994) 285.
11. L. Basini and D. Sanfilippo, J. Catal. 157 (1995) 162.
12. J. A. Dumesic, private communication.
13. R. H. H. Smits, to be published.
14. M. C. J. Bradford and A. M. Vannice, Appl. Catal. A, 142 (1996) 97.
15. A. M. Efstathio, A. Kladi, V. A. Tsipouriari and X. E. Verykios, J. Catal., 158 (1996) 64.
16. J. H. Bitter, K. Seshan and J. A. Lercher, J. Catal., 171 (1997) 279.
17. J. K. Nørskov, in D. A. King and D. P. Woodruff (Editors), The Chemical Physics of Solid Surfaces, Vol. 6, Elsevier, Amsterdam, 1993, p. 1.
18. H. P. Bonzel and G. Pirug, in D. A. King and D. P. Woodruff (Editors), The Chemical Physics of Solid Surfaces, Vol. 6, Elsevier, Amsterdam, 1993, p. 51.
19. W. Kuch, M. Schulze, W. Schnurnberger and K. Bolwin, Surf. Sci., 287/288 (1993) 600.
20. W. Kuch, W. Schnurnberger, M. Schulze and K. Bolwin, J. Chem. Phys., 101 (1994) 1687.
21. T. Bornemann, H.-P. Steinrück, W. Huber, K. Eberle, M. Glanz and D. Menzel, Surf. Sci., 254 (1991) 105.
22. S. T. Ceyer, Q. Y. Yang, M. B. Lee, J. D. Beckerle and A. D. Johnson, Stud. Surf. Sci. Catal., 36 (1988) 51.
23. T. P. Beebe, Jr, D. W. Goodman, B. D. Kay and J. T. Yates, Jr., J. Chem. Phys., 87 (1987) 2305.

24. R. J. Behm, D. K. Flynn, K. D. Jamison, G. Ertl and P. A. Thiel, Phys. Rev., B36 (1987) 9267.
25. I. Alstrup, I. Chorkendorff and S. Ullmann, to be published.
26. P. Kratzer, private communication.
27. T. Osaki, H. Masuda , T. Horiuchi and T. Mori, Catal. Lett., 34 (1995) 59.
28. L. M. Aparicio, unpublished results.
29. T. Inui, K. Saigo, Y. Fujii and K. Fujioka, Catal. Today, 26 (1995) 295.
30. Z. Cheng, Q. Wu, J. Li and Q. Zhu, Catal. Today, 30 (1996) 147.
31. T. Borowiecki and A. Golebiowski, Catal. Lett., 25 (1994) 309.
32. T. Borowiecki, A. Golebiowski and B. Stasinska, Appl. Catal. A, 159 (1997) 141.
33. M. C. J. Bradford and M. A. Vannice, J. Catal., 173 (1998) 157.
34. M. C. J. Bradford and M. A. Vannice, Catal. Lett., 48 (1997) 31.
35. J. R. Rostrup-Nielsen, J. Catal., 85 (1984) 31.
36. N. R. Udengaard, J.-H. Bak Hansen, D. C. Hanson and J. A. Stal, Oil Gas J., 90 (1992) 62.
37. C. A. Bernardo, I. Alstrup and J. R. Rostrup-Nielsen, J. Catal., 96 (1985) 517.
38. I. Alstrup and M. T. Tavares, J. Catal., 139 (1993) 513.
39. P. Kratzer, B. Hammer and J. K. Nørskov, J. Chem. Phys., 105 (1996) 5595.
40. L. Pleth Nielsen, F. Besenbacher,I. Stensgaard, E. Lægsgaard, C. Engdahl, P. Stoltze, K. W. Jacobsen and J. K. Nørskov, Phys. Rev. Lett., 71 (1993) 754.
41. P. M. Holmblad, J. Hvolbæk Larsen, I. Chorkendorff, L. Pleth Nielsen, F. Besenbacher, I. Stensgaard, E. Lægsgaard, P. Kratzer, B. Hammer and J. K. Nørskov, Catal. Lett., 40 (1996) 131.
42. F. Besenbacher, I. Chorkendorff, B. S. Clausen, B. Hammer, A. M. Molenbroek, J. K. Nørskov and I. Stensgaard, Science, in press.
43. J. K. Nørskov, J. E. Hyldtoft and B. S. Clausen, Patent Appl. No. 0683/97, 1997.
44. S. Wang, G. Q. Lu and G. J. Millar, Energy & Fuels, 10 (1996) 896.
45. S. E. L. Winter, J.-H. Bak Hansen, and J. R. Rostrup-Nielsen, paper at AIChE National Spring Meeting, March 9-13, 1997, Houston.
46. S. P. S. Andrew, Ind. Eng. Chem. Prod. Res. Develop., 8 (1969) 321.

NATURAL GAS CONVERSION V
Studies in Surface Science and Catalysis, Vol. 119
A. Parmaliana et al. (Editors)

Reductive activation of oxygen for partial oxidation of light alkanes

K. Otsuka, I. Yamanaka and Ye Wang*

Department of Chemical Engineering, Tokyo Institute of Technology,
Ookayama, Meguro-ku, Tokyo 152-8552, Japan

The topics in this paper are (1) the selective oxidation of methane and ethane into their oxygenates by using a gas mixture of H_2 and O_2, (2) the partial oxidation of light alkanes (CH_4, C_2H_6, C_3H_8) with a catalytic system of $EuCl_3$ / Zn powder / CF_3CO_2H, and (3) the reductive activation of oxygen and partial oxidation of alkanes ($\geq C_3$) at the cathode by applying [H_2 | H_3PO_4 | O_2] cell reactors.

(1) Methane was selectively converted to methanol by a mixture of H_2 and O_2 at > 600K and atmospheric pressure over $FePO_4$ catalyst, while, in the absence of H_2, the conversion of methane required temperatures higher than 700K and formaldehyde was the initial product at a low methane conversion. The *in situ* FT-IR spectroscopy indicated the absorption band due to a peroxide species on $Fe_{0.5}Al_{0.5}PO_4$, a model catalyst of $FePO_4$, in the presence of H_2 and O_2. The reaction of methane with this peroxide at ≥ 473K generated methoxide and OH group, suggesting that the adsorbed peroxide could be the active oxygen species for the formation of methanol. The structure of catalytic active site and the reaction mechanism for the oxidation of methane to methanol were discussed.

(2) The catalytic system made up from Eu salts or complexes, CF_3CO_2H and Zn powder without organic solvents caused the oxidations of methane, ethane and propane into their corresponding oxygenates at 313K. The turnover number based on $EuCl_3$ for the formation of methanol was 4.0 (0.8% yield) in 1h at the reaction conditions; $EuCl_3$ (30 μmol), CF_3CO_2H (4 ml), Zn (1 g), O_2 (0.4 MPa), CH_4 (1.0 MPa). Other rare earth metal chlorides and transition metal chlorides did not show catalytic activities for the oxidation of methane. The unique catalysis of Eu salts was ascribed to a good matching of the redox potentials of Eu(III) / Eu(II) with that of Zn(II) / Zn(0). The reductively actived oxygen by zinc powder through the redox of Eu(III) / Eu(II) was responsible for the partial oxidations of light alkanes at room temperature.

(3) Oxygen is reduced at the cathode of H_2-O_2 fuel cell, generating reductively activated oxygen species which enables partial oxidations of aromatics and alkanes at the cathode. Carbon fiber and carbon whisker were good host carbon materials for the cathode. Addition of $VO(acac)_2$ and Pd black into the carbon fiber enhanced the oxidation of propane, producting acetone as the main oxygenates at room temperature. The oxidation of propane was

---

*present address; Institute for Chemical Reaction Science,
Tohoku University, Katahira 2 chome, Aoba-ku, Sendai 980-8577, Japan

hypothesized to be initiated by OH radical released from the cathode. Methane and ethane were also oxidized at room temperature, though the main product was CO_2.

1. INTRODUCTION

One of the challenges in catalytic reactions is to develop new catalytic systems for the direct oxidation of light alkanes such as CH_4, C_2H_6 and C_3H_8 into their corresponding oxygenates. Among these light alkanes, the partial oxidation of CH_4, the major component of natural gas, has long been expected in industry and will be so in the next century. At present, most of research in the catalytic partial oxidation of CH_4 with O_2 focuses on the high temperature oxidation using metal oxide catalysts. Although HCHO could be produced with a limited yield ($\leq$ 4%), it has been unsuccessful to obtain CH_3OH at < 0.1 MPa with solid catalysts [1-6]. On the other hand, monooxygenase and its mimic systems are often applied for the oxygenations of light alkanes under mild conditions, using a reductant such as ascorbic acid, NADPH, $NaBH_4$ or zinc which enables the reductive activation of O_2 [7-12]. Although the active sites and the mechanisms for the activation of O_2 are complicated and the active oxygen species for the monooxygenations are quite different, the over-all activation of O_2 in acidic media and its addition to alkanes can be represented by equation 1 and 2.

$$O_2 + 2H^+ + 2e^- \longrightarrow O^* + H_2O \quad (1)$$
$$RH + O^* \longrightarrow ROH, RO, H_2O \quad (2)$$

where e^- is provided from the reductants being present or added to the catalytic systems and O* is the reductively activated oxygen species responsible for the oxygenations of light alkanes.

1. 1.

Methane monooxygenase (MMO) catalyzes the oxidation of CH_4 to CH_3OH with O_2 under ambient conditions [13]. O_2 is reductively activated on the iron centers of MMO by e^- and H^+ supplied from a reductant such as NADH or NADPH, generating active oxygen species on the iron site which directly convert CH_4 to CH_3OH [14-16]. We expect that the heterogeneous oxidation of CH_4 to CH_3OH on solid catalysts may also be realized if a reductant is co-fed with oxygen in the reaction system. In this case, H_2, a cheap and easy handling gaseous reductant, is most desirable.

1. 2.

We have reported that the catalytic system composed of Eu salts / Zn / CH_3CO_2H / CH_2Cl_2 enables the partial oxidation of cyclohexane [17,18] and epoxidations of hexenes [19] and propene [20]. For this catalytic system, we have chosen zinc powder as a reductant because it is most easily handled and does not evolve H_2 in a weak acid medium. By using zinc powder as a reductant as well as an electron conducting medium and acetic acid as a proton conducting medium, Eu cations are assumed to work as catalysts for the reductive activation of O_2 with H^+ and e^- as schematically demonstrated in Figure1. The active oxygen generated on Eu cations oxygenates alkanes and alkenes into alcohols, ketones and epoxides. The second purpose in this work is to apply similar catalytic systems to the partial oxidation of CH_4, C_2H_6 and C_3H_8.

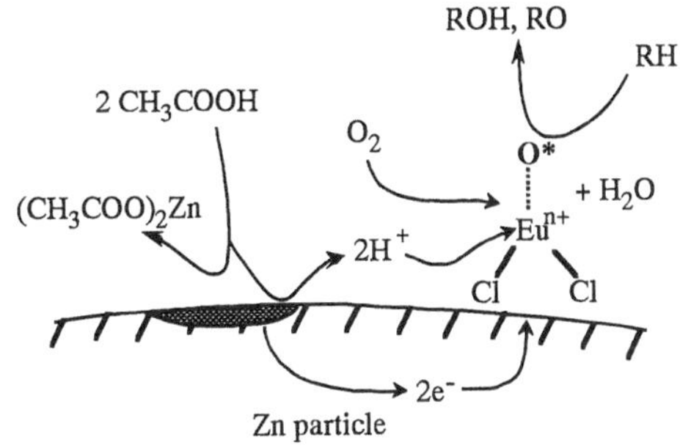

Figure 1. Conceptual model of the $EuCl_3$-catalytic system for reductive activation of dioxygen in light alkane oxidation.

1. 3.

When an acidic electrolyte is used in H_2-O_2 fuel cell, the stoichiometric reactions at the anode and the cathode are written simply as,

$$\text{(Anode)}\ H_2 \longrightarrow 2H^+ + 2e^- \qquad (3)$$
$$\text{(Cathode)}\ 1/2O_2 + 2H^+ + 2e^- \longrightarrow H_2O \qquad (4)$$

where the reduction of O_2 at the cathode may proceed stepwise as follows:

$$\begin{array}{ccccccccc} O_2 & \longrightarrow & O_2^- & \longrightarrow & O_2^{2-} & \longrightarrow & O_2^{3-} & \longrightarrow & 2H_2O \\ & & (HO_2\cdot) & & \left(\begin{array}{c} H_2O_2 \\ 2HO\cdot \\ MO_2H \end{array}\right) & & \left(\begin{array}{c} O^-, H_2O \\ HO\cdot, H_2O \\ M\text{-}O\cdot, H_2O \end{array}\right) & & \end{array} \qquad (5)$$

If the reduced oxygen intermediates, including the protonated ones and metal oxo species, have a finite lifetime in the presence of a suitable catalyst(M), we expect that these reduced oxygen species might activate alkanes and aromatics at the cathode side, resulting in their oxygenation during H_2-O_2 cell reactions. On the basis of this idea, we have developed a simple method for the reductive activation of O_2 at the cathode of $[H_2 \mid H_3PO_4 \mid O_2]$ cell systems, which realized selective oxygenations of alkanes and aromatics at room temperature [21-23]. The third purpose in this work is to apply the similar cell system for the activation and oxygenation of light alkanes.

2. EXPERIMENTAL

2. 1. Partial oxidation of CH_4 and C_2H_6 with a gas mixture of H_2 and O_2

The $FePO_4$ catalyst used was prepared from a mixed solution of $Fe(NO_3)_3$ and $NH_4H_2PO_4$. After the solution was dried at 363K for 12h, the resultant was calcined at 823K for 5h in air. The BET surface area of the $FePO_4$ powder was 8.5 m^2g^{-1}. The iron aluminum phosphate ($Fe_{0.5}Al_{0.5}PO_4$) used for FTIR studies was prepared by sol-gel method from aqueous solutions of $FeCl_3$, $AlCl_3$ and $NH_4H_2PO_4$ (moler ratio, 0.50 : 0.50 : 1.00) added with propylene oxide at 273K. The gel was calcined at 823K in a flow of O_2. The BET surface area of the $Fe_{0.5}Al_{0.5}PO_4$

was 275 m^2g^{-1}.

The steady-state catalytic activities of each catalyst for CH_4 and C_2H_6 oxidations in the absence and presence of H_2 were measured using a conventional fixed bed flow reactor at atmospheric pressure. When H_2 was cofed with CH_4 (or C_2H_6) and O_2, special caution should be taken to prevent explosion. The entire reaction system was barricaded with acrylic planks, and most experiments were carried out beyond explosion limits.

The $Fe_{0.5}Al_{0.5}PO_4$ catalyst used for *in situ* FTIR-transmission measurements was pressed into a self-supporting wafer. The catalyst wafer could be heated to 1000K at the center of the quartz-made IR cell. The IR spectra were usually recorded at ambient temperature after the sample had been contacted with H_2, $H_2 + O_2$, N_2O or CH_4 at higher temperatures > 473K.

2. 2. Partial oxidation with Eu(III) / Zn / RCO_2H catalytic systems

The partial oxidation of CH_4, C_2H_6 and C_3H_8 were performed as follows. $EuCl_3 \cdot 6H_2O$ (30 μmol) was dissolved into CH_3CO_2H or CF_3CO_2H (4 ml) in a glass tube holder in an autoclave. After Zn powder (1.0 g) was added to the solution, oxygen (0.4 MPa) and light alkane (CH_4, C_2H_6 or C_3H_8, 0.1~1.0 MPa) were introduced to the autoclave. The oxidation of alkanes was continued for 1h by stirring the solution with a magnetic spin-bar at 273~313K.

2. 3. Partial oxidation applying a [H_2 | H_3PO_4 | O_2] cell reactor

The H_2-O_2 cell reactor and the principle of the method for the oxidation of light alkanes are demonstrated in Figure 2. A detailed description of the cell setup has been given elsewhere [22]. A silica-wool disk (2.0 mm thickness, 26 mm diameter) impregnated with aqueous H_3PO_4 (1 M, 1 ml) as an electrolyte separates the anode and the cathode compartments. The anode was made from a mixture of Pt-black, graphite and Teflon powder by a hot-press method. The cathodes were prepared by the same method from a mixture of carbon fiber (VGCF, Vapor Grow Carbon Fiber, obtained from Showa Denko Co.) with various metal blacks and metal salts. Usually, the contents of metal blacks and metal salts were 0.5 and 1.0 mol% of carbon (50 mg), respectively. Superficial area of the electrode wafers was ca. 3.1 cm^2.

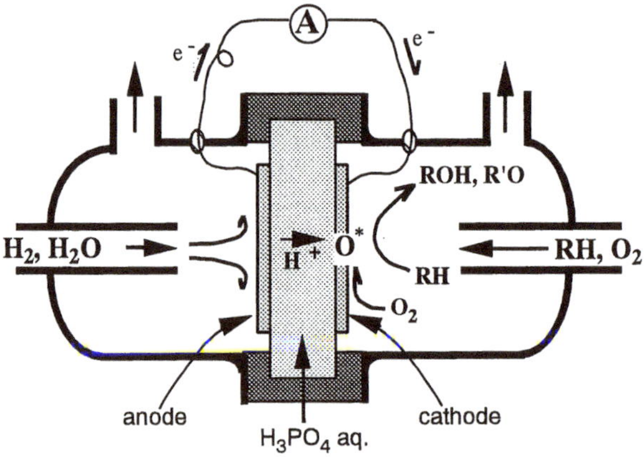

Figure 2. Diagram of the H_2-O_2 cell for oxidation of light alkanes.

The oxidation of light alkanes was carried out by passing a gas mixture of alkanes (50 kPa) and O_2 (51 kPa) in the cathode compartment. H_2(49 kPa) and H_2O vapor (4 kPa, to keep the electrolyte always wet) were passed through the anode compartment. The reaction was started by shorting the circuit at 300K. The rate of formation of products was recorded after the steady state rate was obtained. The products dissolved in the electrolyte were analyzed by extracting the solutes with water after the reaction.

3. RESULTS AND DISCUSSION

3. 1. Partial oxidation of CH_4 by H_2-O_2 gas mixture

We have tested various solid catalysts for the catalytic conversion of CH_4 to CH_3OH using H_2 as a reductive activator of O_2. Among a wide variety of catalysts tested, some iron-containing catalysts showed an enhancing effect of H_2 on the conversion of CH_4. Particularly, $FePO_4$ showed a very unique property for the selective synthesis of CH_3OH in the presence of H_2. The cofeed of H_2 with O_2 remarkably increased the conversion of CH_4 as well as the selectivity to CH_3OH [24,25]. Kinetic studies have suggested that a new oxygen species generated on $FePO_4$ in the presence of H_2-O_2 gas mixture is responsible for this selective formation of CH_3OH [24,25].

The catalytic performance of the $Fe_{0.5}Al_{0.5}PO_4$ catalyst in a gas mixture of H_2 and O_2 was quite similar to that of $FePO_4$ except for a larger catalytic activity per weight of catalyst due to higher specific surface area [26]. Thus, we used this $Fe_{0.5}Al_{0.5}PO_4$ as a model catalyst of $FePO_4$ for investigating the catalytic active sites and the active oxygen species responsible for the specific conversion of CH_4 to CH_3OH in the presence of H_2-O_2 gas mixture.

Characterization of the $Fe_{0.5}Al_{0.5}PO_4$ catalyst by XPS before and after the reaction suggested the redox of Fe(III) / Fe(II) on the catalyst surface during the oxidation of CH_4 with a H_2-O_2 gas mixture.

The adsorbed oxygen species generated on the catalyst in a H_2-O_2 gas mixture and its reactivity with CH_4 were studied by *in situ* FT-IR spectroscopy. The absorption band at 895 cm^{-1} was observed in the presence of H_2 and O_2 when the temperature was raised above 573K. The isotope substitution of $^{16}O_2$ with $^{18}O_2$ shifted the absorption band at 895 cm^{-1} to 849 cm^{-1}. Three absorption bands at 895, 870 and 849 cm^{-1} were observed when a mixture of $^{16}O_2$, $^{16}O^{18}O$ and $^{18}O_2$ with H_2 was contacted with the catalyst at ≥ 573K. These observations strongly suggest that the band at 895 cm^{-1} is ascribed to a peroxide species adsorbed on the iron site of the catalyst. The intensity of the band due to this peroxide species decreased with reaction time when CH_4 was added at ≥ 473K as a result of the oxidation of CH_4 with the peroxide species. The new bands ascribed to the stretching vibrations of CH_3 groups and of C-O of methoxide species appeared simultaneously. The absorption band due to the stretching vibration of the adsorbed OH groups increased in parallel to that of methoxide species. These results suggest that the adsorbed peroxide anions activate CH_4 at > 473K, producing CH_3O and OH groups as reaction intermediates for the formation of CH_3OH [26]. The role of H_2 added is to reduce Fe(III) to Fe(II) at the catalyst surface. O_2 must be reductively activated on the Fe(II) site by accepting electrons, initially forming O_2^- species, then further reduced into O_2^{2-} species by electrons trapped in the neighboring Fe(II) sites.

The catalytic functions observed for the three model catalysts demonstrated in Figure 3 suggest that the tetrahedrally coordinated iron site isolated from each other by phosphate groups (model C) is the active site for the selective oxidation of CH_4 to CH_3OH with H_2-O_2 gas mixture. The comparison of the results for $FePO_4$, $FeAsO_4$ and $FeSbO_4$ has indicated that the Brönsted acidity of the surrounding groups of iron site plays an important role as the acceptor and donor of protons and thus enhances the formation of O_2^{2-} which is responsible for the activation and the partial oxidation of CH_4 [27].

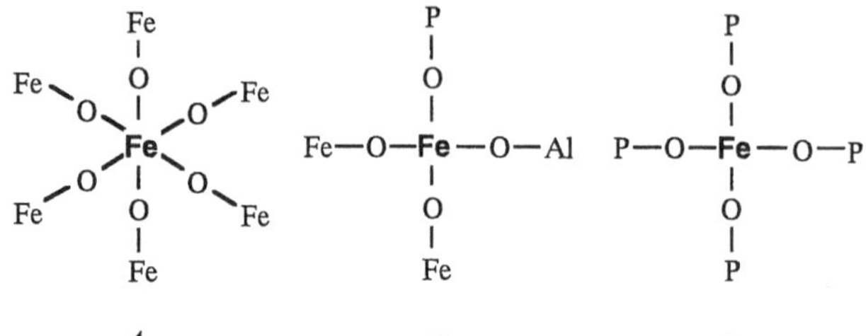

Figure 3. Three model catalysts for oxidation of CH_4.

We propose the mechanism in Figure 4 on the bases of the results described above.

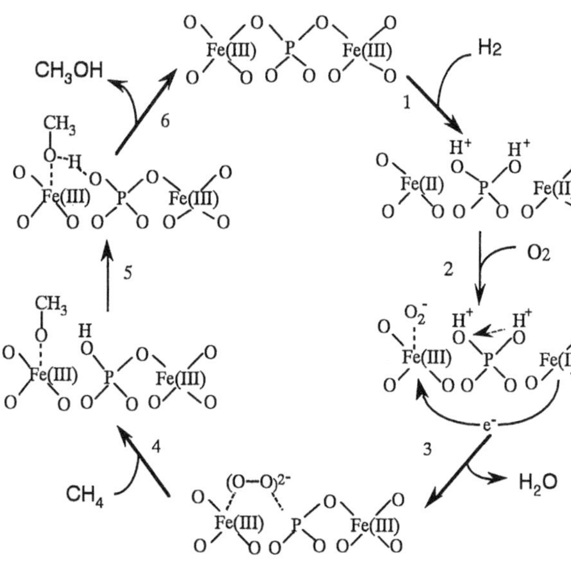

Figure 4. Reaction mechanism of the partial oxidation of CH_4 to CH_3OH.

3. 2. Oxidation of light alkanes with Eu(III) catalytic systems

The catalytic system composed of $EuCl_3$ / Zn powder / CH_3CO_2H / CH_2Cl_2 (as solvent) was not appropriate for the oxidation of CH_4 to CH_3OH because CH_3OH and CO_2 were produced mainly from CH_3CO_2H. We found that the use of CF_3CO_2H in place of CH_3CO_2H and without using any solvent realized the oxidation of CH_4 to CH_3OH [28].

Figure 5 shows the results of oxidation of CH_4, C_2H_6 and C_3H_8 with O_2 catalyzed by the $EuCl_3$ / Zn powder / CF_3CO_2H (or CH_3CO_2H) catalytic systems at 313K. In the case of

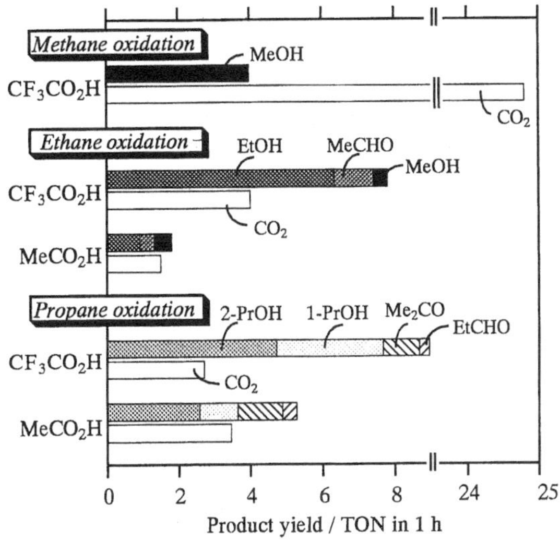

Figure 5. Oxidations of CH_4 (10 atm), C_2H_6 (10 atm) and C_3H_8 (8 atm) using $EuCl_3$-O_2-Zn-CF_3CO_2H (or $MeCO_2H$)-system at 313K.

oxidation of CH_4, the production of CH_3OH was obvious, but accompanied by a large formation of CO_2. In the absence of CH_4, larger formation of CO_2 (TON =39.5) was observed, but CH_3OH was not produced at all. These observations indicate that CH_3OH is produced only from CH_4, but CO_2 mainly from CF_3CO_2H. While, CO_2 was a minor product in the oxidations of C_2H_6 and C_3H_8. It should be noted that CH_3CO_2H is also applicable as a proton donor for the oxidations of C_2H_6 and C_3H_8. Therefore the partial oxidations of C_2H_6 and C_3H_8 into their corresponding oxygenates can be performed effectively by using either of CF_3CO_2H or CH_3CO_2H as an element in the $EuCl_3$ catalytic systems.

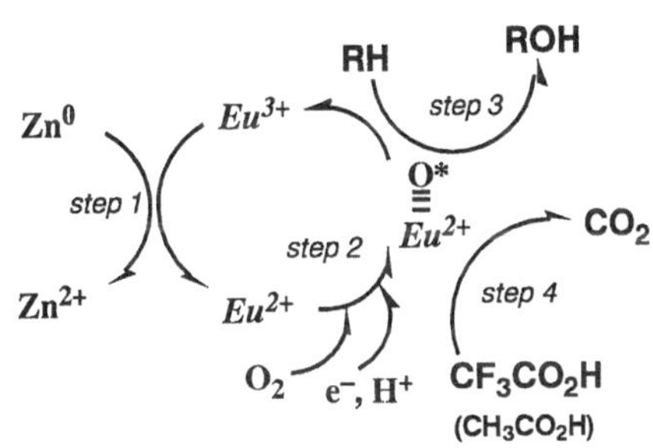

Scheme 1. Reaction mechanism for the oxidation of CH_4 with the $EuCl_3$-catalytic system.

The informations obtained from the studies on the stoichiometric reactions among Eu(II), O_2 and alkanes by using UV-visible spectroscopy and electrochemical voltammetry suggested that the redox coupling among Zn(II) / Zn(0), Eu(III) / Eu(II) and O_2 / O_2^- (H^+) would generate a powerful active oxygen species for the oxidation of CH_4 to CH_3OH (Scheme 1). The addition of titanium compounds ($TiO(acac)_2$ and TiO_2) enhanced the conversion of CH_4 considerably into both CH_3OH and CO_2 [29]. The redox of Ti(IV) / Ti(III) or Ti (III) / Ti(II) must accelerate the formation of active oxygen species because these standard redox potentials are in between those of Zn(II) / Zn(0) and Eu(III) / Eu(II).

3. 3. [H_2 | H_3PO_4 | O_2] cell for the oxidation of light alkanes

The cathode electrocatalysts prepared from the mixtures of Pd black, Fe-compounds

(Fe_2O_3, Fe_3O_4, $FeCl_3$, or $Fe_2(SO_4)_3$) and carbon whisker showed relatively high catalytic performance in the hydroxylation of benzene [30]. Coexistence of Pd black and Fe-compounds showed a marked synergism for the formation of phenol. In this case, OH radical was suggested to be the active oxygen species responsible for the hydroxylation of benzene.

The effects of additives to the host carbon fiber (VGCF) have been examined for the oxidation of C_3H_8 by the method illustrated in Figure 2. Among various additives tested (Fe_2O_3, $CuSO_4$, TiO_2, V_2O_5, $VO(acac)_2$, NH_4VO_3, ammonium molybdate and tangstate, with and without Pd black), the combination of Pd black and $VO(acac)_2$ showed the best results for the partial oxidation of propane to acetone at the moment. The selectivities to acetone and acetic acid were 48 and 12%, respectively. The results for the cathodes of carbon fiber alone, with $VO(acac)_2$ or Pd black, and with both $VO(acac)_2$ and Pd black are shown in Figure 6. The results in this figure indicate the synergism of Pd black and $VO(acac)_2$ on the conversion of propane and the formation of acetone. The synergism of Pd black and $VO(acac)_2$ may be explained in the terms of the cooperative actions of the two additives. (i) Addition of Pd black increases the rate of reduction of O_2, thus the current enhanced remarkably compared to that for VGCF alone in Figure 6. (ii) The vanadium compound improves the formation of active oxygen, probably OH radical, through the Fenton type reaction,

$$VO^{2+} + 2H^+ + e^- \longrightarrow V^{3+} + H_2O \quad (6)$$
$$H_2O_2 + V^{3+} \longrightarrow \cdot OH + H^+ + VO^{2+} \quad (7)$$

where the formation of H_2O_2 from O_2 must be enhanced by Pd black. Thus, the coaddition of Pd black and $VO(acac)_2$ shows the synergism for the oxidation of propane. The OH radical generated on the cathode might be released into the gas phase, which causes gas phase autooxidation, and the formations of acetone, acetic acid and CO_2.

The (Pd black + $VO(acac)_2$) / VGCF cathode described above was also used for the oxidations of C_2H_6 and CH_4. In the case of C_2H_6 oxidation, acetic acid was the partial oxidation product (selectivity 22%). However, the oxidation of CH_4 produced only CO_2. The rate of alkane oxidation decreased as C_3H_8 > C_2H_6 >

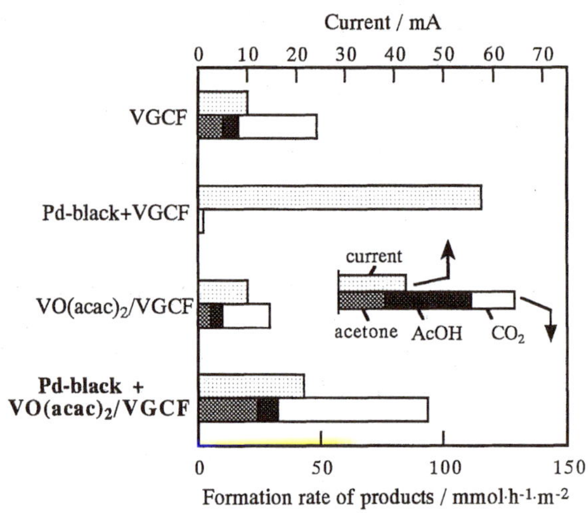

Figure 6. Oxidation of propane using H_2 | H_3PO_4 | O_2 cell: Synergism of Pd-black and $VO(acac)_2$ on the reaction. Content of additives : Pd-black (0.5 mol%), $VO(acac)_2$ (1 mol%).

CH_4 as we expected. It should be noted that, although CO_2 is the main product, CH_4 and C_2H_6 can be oxidized at 303K. In contrast with the results of CH_4 and C_2H_6, the selectivity to the useful oxygenates (acetone + acetic acid) in the oxidation of propane exceeded 60% on the basis of the C_3H_8 reacted.

REFERENCES

1. J. M. Fox, Catal. Rev. -Sci. Eng., 35 (1993) 169.
2. R. Pitchai and K. Klier, Catal. Rev. -Sci. Eng., 28 (1986) 13.
3. M. J. Brown and N. D. Parkyns, Catal. Today, 8 (1991) 305.
4. N. D. Parkyns, C. I. Warburton and J. D. Wilson, Catal. Today, 18(1993) 385.
5. O. V. Krylov, Catal. Today, 18 (1993) 209.
6. T. J. Hall, J. S. J. Hargreaves, G. J. Hutchings, R. W. Joyner and S. H. Taylor, Fuel. Proc. Tech., 42 (1995) 151.
7. J. T. Groves, T. E. Nemo and R. S. Myers, J. Am Chem. Soc., 101 (1979) 1032.
8. Ortiz de Montellano (ed.), "Cytochrome P-450, Structure, Mechanism and Biochemistry", Plenum press, New York, 1986.
9. F. Montanari and L. Casella (eds.), "Metalloporphyrins Catalyzed Oxidations", Kluwer Acad. Pub., Dordrecht, 1994.
10. D. H. R. Barton, M. J. Gastiger and W. B. Motherwell, J. Chem. Soc., Chem. Commun., (1983) 41; D. H. R. Barton et al., J. Chem. Soc., Perkin Trans. I. (1986) 947.
11. N. Kitajima, H. Fukui and Y. Moro-oka, J. Chem. Soc., Chem. Commun., (1988) 485.
12. H. Dalton and J. Green, J. Biol. Chem., 264 (1989) 17698; J. Colby, K. I. Stirling and H. Dalton, Biochem. J., 165 (1977) 395.
13. J. Colby, D. I. Stirling and H. Dalton, Biochem. J., 165 (1977) 395.
14. H. Dalton and J. Green, J. Biol. Chem., 264 (1989) 17698.
15. S. -K. Lee, J. C. Nesheim and J. D. Lipscomb, J. Biol. Chem., 268 (1993) 21569.
16. M. J. Ratai, J. E. Kauth and M. I. Donnelly, J. Biol. Chem., 166 (1991) 18684.
17. I. Yamanaka and K. Otsuka, J. Mol. Catal. A, 95 (1995) 115; J. Mol. Catal., 83 (1993) L15.
18. I. Yamanaka, T. Akimoto, K. Nakagaki and K. Otsuka, Chem. Lett. (1994) 1717; I. Yamanaka, T. Akimoto and K. Otsuka, Chem. Lett. (1994) 1514.
19. I. Yamanaka, T. Akimoto, K. Nakagaki and K. Otsuka, J. Mol. Catal. A, 110 (1996) 119.
20. I. Yamanaka, K. Nakagaki and K. Otsuka, J. Chem. Soc., Chem. Commun., (1995) 1185.
21. K. Otsuka, I. Yamanaka and K. Hosokawa, Nature, 345 (1990) 697.
22. I. Yamanaka and K. Otsuka, J. Chem. Soc., Faraday Trans., 89 (1993) 1791.
23. I. Yamanaka and K. Otsuka, J. Chem. Soc., Faraday Trans., 90 (1994) 451.
24. Y. Wang and K. Otsuka, J. Chem. Soc., Chem. Commun., (1994) 1893.
25. Y. Wang and K. Otsuka, J. Catal., 155 (1995) 256.
26. Y. Wang and K. Otsuka, Stud. Surf. Sci. Catal., 101 (1996) 397.
27. Y. Wang and K. Otsuka, J. Mol. Catal. A, 111 (1996) 341.

28. I. Yamanaka, M. Soma and K. Otsuka, J. Chem. Soc., Chem. Commun., (1995) 2235.
29. I. Yamanaka, M. Soma and K. Otsuka, Chem. Lett., (1996) 565.
30. K. Otsuka, M. Kunieda and H. Yamagata, J. Electrochem. Soc., 139 (1992) 2381; K. Otsuka, M. Kunieda and I. Yamanaka, Stud. Surf. Sci. Catal., 82 (1994) 703.

NATURAL GAS CONVERSION V
Studies in Surface Science and Catalysis, Vol. 119
A. Parmaliana et al. (Editors)

Developments in Fischer-Tropsch technology

B Jager

Sasol Technology Research & Development, PO Box 1, Sasolburg, 9570, Republic of South Africa

1. INTRODUCTION

To convert natural gas to hydrocarbons, three major steps are involved. First the natural gas is converted to syngas, a mixture of hydrogen and carbon monoxide, which in the second step is converted to hydrocarbons. In the third step these primary hydrocarbons in the form of a syncrude, have to be worked up to final products.

The production of syngas is achieved by reforming the natural gas in steam reformers, autothermal reformers or partial oxidation reformers. Alternatively, coal may be used as a source of syngas through gasification. This is done commercially in South Africa but is more complex and more expensive than starting with natural gas. Both reforming and gasification are well established processes.

Conversion of syngas to hydrocarbons can be achieved by means of the Fischer-Tropsch (FT) process. The Fischer-Tropsch process can be operated at low temperatures (LTFT) to produce a syncrude with a large fraction of heavy, waxy hydrocarbons or it can be operated at higher temperatures (HTFT) to produce a light syncrude and olefins

With HTFT the primary products can be refined to environmentally friendly gasoline and diesel, solvents and olefins. With LTFT, the heavy hydrocarbons can be refined to speciality waxes or if hydrocracked and/or isomerised, to produce excellent diesel, base stock for lube oils and a naphtha that is ideal feedstock for cracking to light olefins. The work up of FT syncrude, although somewhat different from that of normal crude, falls well within the scope of established refinery processes and operations.

2. SYNGAS PRODUCTION

The production of syngas is obtained by reforming natural gas with either steam or carbon dioxide, or by partial oxidation or by a combination of the three. When water is present, which is normally always the case, the water gas shift reaction also plays a role.

Reforming by:

Steam	$CH_4 + H_2O$	$\rightarrow$	$CO + 3H2$	$\Delta H = 206,3$ kJ/mole
CO_2	$CH_4 + CO_2$	$\rightarrow$	$2CO + 2H_2$	$\Delta H = 246,9$ kJ/mole
Partial oxidation	$CH_4 + \frac{1}{2} O_2$	$\rightarrow$	$CO + 2H_2$	$\Delta H = -35,6$ kJ/mole
Water gas shift reaction:	$CO + H_2O$	$\rightarrow$	$CO_2 + H_2$	$\Delta H = -40,6$ kJ/mole

Whereas Fischer-Tropsch reactions require an H_2/CO ratio of about 2, these different reforming reactions give syngases with H2/CO ratios ranging from less than 1 to more than 3. A combination of these reactions in the presence of a catalyst is normally practised to obtain a suitable H_2/CO ratio. This may be done in a tubular catalytic steam reformer with a furnace to provide external heat, in an autothermal reformer where partial oxidation is combined with adiabatic catalytic steam reforming, or in a partial oxidation reformer followed by a shift reactor. When reforming is followed by FT, it is possible to recycle FT tail gas to the feed of reforming to provide some CO_2 as another means to obtain a suitable H_2/CO ratio.

3. THE FISCHER-TROPSCH PROCESS

Frans Fischer and Hans Tropsch of the Kaiser Wilhelm Institute developed the FT process for Coal Research in Mülheim in 1923. They showed that carbon monoxide and hydrogen in the presence of iron, cobalt or nickel catalyst at 180 - 250°C and atmospheric pressure, produce a mixture of straight chain hydrocarbons ranging from methane to waxes of high molecular weight and smaller amounts of oxygenates [Ref 1]. The FT reaction can be written as

$$CO + 2H_2 \quad \rightarrow \quad [CH_2] + H_2O \qquad\qquad \Delta H = -167{,}4 \text{ kJ/mole CO}$$

where $[CH_2]$ is the basic unit building block of the hydrocarbon molecules. The reaction is highly exothermic which makes heat transfer a major issue in the design of FT reactors.

Based on these discoveries, in 1929 commercial production of synthetic oil began in Germany and during World War II an annual output of over half a million tons of syncrude was achieved. The best catalyst was found to be based on cobalt and this catalyst was used at pressures up to about 10 bar. Later a switch was made to the more economic iron based catalyst.

Originally LTFT was practised in fixed bed reactors, later in the form of tubular fixed bed reactors. These reactors are still used by Sasol in South Africa, by Shell in Malaysia and also in Russia. They typically operate between 180 and 250^{0}C at pressures ranging from 10 - 45 bar.

In South Africa Sasol has also operated a high temperature Fischer-Tropsch process since 1955. This HTFT process, called Synthol, operates at about 25 bar and 330 - 350^{0}C.

The products from the FT synthesis may vary depending on the catalyst formulation and process conditions. Typical product distributions for LTFT and HTFT are shown in Table 1. The product slates follow the Schulz-Flory distributions with, for LTFT, alpha values ranging up to

Table 1
Selectivity (carbon basis) of Sasol processes

Product	LTFT	HTFT
CH_4	4	7
C_2 to C_4 olefins	4	24
C_2 to C_4 paraffins	4	6
Gasoline	18	36
Middle distillate	19	12
Heavy oils and waxes	48	9
Water soluble oxygenates	3	6

0,95 or higher. As mentioned, some oxygenated hydrocarbons are also formed. The lighter water soluble oxygenates dissolved in the aqueous phase, can be recovered (e.g. methanol, ethanol, propanol, aceetic and propionic acids) or they are biologically destroyed. Heavier hydrocarbons are normally hydrogenated in hydrotreating units during work up.

The fuels produced by FT processes are environmentally superior to conventional crude oil derived fuels; they have virtually zero sulphur content. The LTFT fuels have also very low aromaticity. The LTFT derived diesel has a high cetane number (>70) which together with the low aromatics and sulphur levels, results in superior combustion characteristics

4. FISCHER - TROPSCH REACTOR DEVELOPMENTS

4.1. The Sasol Advanced Synthol (SAS) Reactor for HTFT

A total of 19 Synthol-Circulating Fluidised Bed (CFB) reactors were used commercially by Sasol from 1955 to the present. The reactor system is complex and needs a complex support system to cope with the circulating catalyst loads and temperature differences. This makes it expensive. The large tonnage of catalyst circulated, cause relatively high-pressure drops across the reactor system. These and other disadvantages are eliminated when using a Sasol Advanced Synthol (SAS) reactor that makes use of conventional solid-gas fluidisation. The SAS reactor is shown in figure 1.

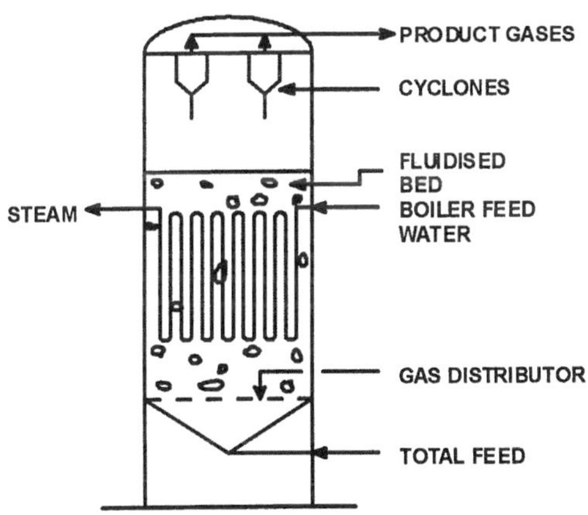

Figure 1. Sasol Advanced Synthol (SAS) Reactor for HTFT

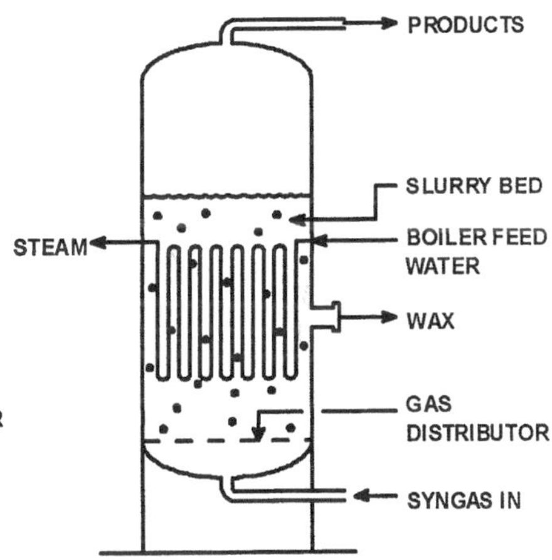

Figure 2. Sasol Slurry Phase Distillate (SPD) Reactor for LTFT

The SAS reactor vessel contains a fluidised bed consisting of reduced, fused iron oxide catalyst [Ref 2]. Syngas is bubbled through the bed where it is catalytically converted to hydrocarbons that are in the vapour phase at the process conditions of about 340°C and 25 bar. The products and unconverted gases leave the reactor through internal cyclones. The process conditions in the SAS reactor are such that the cyclones can retain the catalyst very effectively. Unlike with the CFB reactors, scrubber towers are therefore not needed to remove the last traces of catalyst before the product stream is passed to the condensing train. As part of the development of the SAS reactor, considerable development was done on the design, operation and control of the cyclones used in these reactors.

The major advantages of the SAS reactor over the CFB reactor are its simplicity, ease of operation and lower operating cost due to elimination of the catalyst recycle. Catalyst consumption is reduced to about 40% and maintenance to about 15% of that of the CFB systems. In general higher conversions are obtained at higher gas loads. The latter together with the fact that more cooling coils can be installed in the SAS reactor and more heat can be removed, allows for larger capacity equipment, which translates into advantages of economy of scale. Maximum capacities of up to 20 000 bbl/day are feasible, well beyond the 7 500 bbl/day possible for the CFB reactors. They also have thermal efficiencies four percentage points higher than CFB reactor systems.

A 5 m diameter reactor with a capacity of 3 400 bbl/day was successfully operated at Sasolburg during 1989 - 1992 after which it was converted to the Slurry Phase reactor to be referred to below. An 8 m diameter, 11 000 bbl/day SAS reactor was successfully commissioned during June 1995 at Secunda and has been running smoothly ever since at availabilities in excess of 98%. The cost of these reactors is about 40% of that of equivalent CFB reactor systems. At present Sasol is in the process of replacing 16 Synthol-CFB reactors with eight SAS reactors, four with capacities of 11 000 bbl/day and four with capacities of 20 000 bbl/day.

Apart from gasoline and diesel produced by the Sasol's HTFT plants, olefins are also recovered and ethylene, propylene, 1-pentene and 1-hexene are produced more cheaply than by conventional processes. The production of octenes is planned and that for higher olefins is also being considered.

4.2. The Sasol Slurry Phase Distillate Process (SPD) Reactor for LTFT

The tubular fixed bed (TFB) reactor is complex and expensive. The scale-up of the reactor is mechanically difficult and is complicated by the fact that the design has to provide periodic replacement of the iron-based catalyst. The replacement is cumbersome and maintenance and labour intensive [Ref 3].

Because of the exothermic nature of the Fischer-Tropsch reaction, axial and radial temperature profiles exist in the tubes. Maximum average temperature is required for maximum conversion. This is, however, well below the maximum allowable temperature peak that may not be exceeded in order to prevent carbon formation on the catalyst. Carbon formation causes break-up of the catalyst, which in turn causes blockages and a need to replace the catalyst. Too high peak temperatures may also negatively affect product selectivities.

To overcome the problems and limitations associated with tubular fixed bed reactors, the slurry phase reactor was developed as part of the Sasol Slurry Phase Distillate (SPD) process. It was successfully commissioned in 1993 and has been operated commercially by Sasol ever since at a capacity of about 2 500 bbl/day. As shown diagrammatically in figure 2 it consists of a vessel containing slurry of process derived wax with catalyst dispersed in it. Syngas is bubbled through this slurry bed and is converted to hydrocarbons. The heat generated is passed from the slurry to the cooling coils inside the reactor to generate steam.

The heavier liquid hydrocarbons are mixed into the slurry and removed from it in a proprietary solid separation process developed by Sasol. The development of this separation step was crucial to the viability of the process and considerable optimisation of the physical properties of the rather weak and fragile iron catalyst was done. Excellent results are obtained by this separation step.

The Sasol SPD reactor is much simpler in construction than the tubular fixed bed reactor and it lends itself much better to scale up. Maximum capacities of 14 000 bbl/day are envisaged for the SPD reactor as compared to about 1 500 bbl/day for the tubular fixed bed reactors. The most significant advantage for the Sasol SPD reactor is that it is well mixed and can operate isothermally which allows much higher average operating temperatures and higher reaction rates. The yields per reactor volume are higher, and the catalyst consumption per ton of product is only 20 to 30% of that of the tubular fixed bed reactor. The cost of a single 10 000 bbl/day Sasol SPD reactor train is about 25% of that of a tubular fixed bed reactor system with the same capacity.

Another advantage of the Slurry Phase reactor is that it allows for on-line catalyst removal and addition, which is not feasible with the TFB reactor. This is especially important for iron catalyst, which must be replaced periodically. It also reduces maintenance costs. Where cobalt catalyst is used which has a longer life, this is obviously less important.

The pressure drop across the Slurry Phase reactor is less than 1 bar, as compared to 3-7 bar for TFB reactors. With relatively high recycle flows, this gives rise to considerable savings on recompression costs.

Because of the isothermal nature of the reactor and the much smaller pressure drops across the reactor, the control of the reactor is much simpler and operating costs are much reduced. The easier control of average catalyst life through regular catalyst renewal, allows for easier control of the process selectivities and hence the quality of the primary products.

Since commissioning the commercial Slurry Phase reactor, it has performed very well. Its availability has been very high, of the order of 98%. Experimentation on the commercial scale unit has shown that with minor modifications, the capacity can be increased to 120% of design.

The Slurry Phase reactor system is still new technology and further improvements are being developed and implemented; i.e. the technology benefits from being still on the learning curve. The technology using iron-based catalyst has proven to be robust. Using cobalt catalyst would make the operation of the reactor easier as the physical properties of cobalt catalyst are better suited for slurry bed operations.

5. CAPACITIES OF COMMERCIAL FT REACTORS

The present capacities of the different Sasol FT reactor trains are given in the Table 2 below.

Table 2
Sasol Fischer-Tropsch Commercial Reactors - Capacities (bbl/day)

	CFB	SAS	TFB	SPD
Total installed capacity	110 000	11 000	3 200	2 500
Capacity per reactor	6 500	11 000	500-700	2 500
Potential per reactor	7 500	20 000	3 000	20 000

6. FISCHER-TROPSCH CATALYST

In the earlier FT plants, the catalyst was based on cobalt. The FT plants, which started to operate in 1955 in South Africa, used iron-based catalysts. For LTFT a precipitated iron catalyst is used and for HTFT a fused catalyst. In practice only iron-based catalyst is considered for HTFT, as cobalt catalyst at the higher temperatures would produce mainly methane.

Iron catalyst is relatively cheap but has a life of only weeks or months. It can operate over a wide range of temperatures and pressures as shown by it being used in both low and high temperature Fischer-Tropsch processes. It is sensitive to sulphur poisoning and removal of sulphur from syngas to less than 0,1 ppm is required.

At lower conversions the activities of iron and cobalt based catalysts are similar. The activity of iron based catalyst is however negatively affected by water vapour. Since water is a product of the reaction, it is not possible to obtain high per pass conversions with iron catalyst and high conversions can only be achieved by recycling unconverted syngas after removal of the water and products in a condensing train.

6.1. HTFT Catalyst

Catalyst used in the Synthol-CFB and the SAS reactors has to be strong to prevent attrition during the fluidisation process and for that reason a fused iron catalyst is used. In the early days catalyst was prepared from Allanwood ore but subsequently millscale has been used. The iron oxide has promoters added, is fused and cooled and finally milled to the proper particle size distribution. The catalyst is then reduced with hydrogen before it is charged to the HTFT reactors.

In the Synthol reactor the catalyst is conditioned by slowly taking up the process conditions to final values. During this period the catalyst is partly re-oxidised and it is also carbided. The carbides and the metallic iron are the active ingredients catalysing the FT reaction. The iron oxides catalyse the water gas shift reaction.

At reaction conditions, during the normal life of the iron catalyst, carbon is deposited on and in the catalyst, which in the extreme, breaks it up. To some extent carbon make can be controlled by the addition of suitable promoters. In any case, carbon affects the density of the catalyst and this ultimately affects the fluidisation in both the Synthol-CFB and SAS reactors. HTFT catalyst therefore has to be replenished from time to time with fresh catalyst.

Sulphur seriously poisons Fe-based Fischer-Tropsch catalyst. Where poisoning is excessive, the catalyst is permanently deactivated. If only the surface of the catalyst particles is poisoned, the catalyst tends to regenerate itself in time by the spalling of the surface through carbon deposition just below the surface of the catalyst particles.

Since the gas-solid interaction in the Synthol-CFB and SAS reactors are very similar, it is found that the same catalyst can be used for the two processes. Because the hydrodynamics in the two systems is different, the physical properties of the catalyst can be optimised for the respective processes.

6.2. LTFT Catalyst

Iron Based Catalyst

For the production of LTFT iron based catalyst, iron metal is first dissolved in nitric acid, it is then precipitated with alkalis, filtered, slurried with promoters and filtered once more [Ref 4]. The catalyst to be used in the TFB reactors is extruded and calcined. For Slurry Bed operations the catalyst is spray dried. The oxides are then reduced in an atmosphere of hydrogen before they are charged to the Fischer-Tropsch reactors. As with HTFT catalyst, the catalyst is conditioned by slowly taking it up to process conditions.

The gas-liquid-solid interaction in the Sasol SPD reactor is quite different from that in the TFB reactor. Originally milled or ground TFBR catalyst was used in the Sasol SPD reactor.

Considerable adjustments had to be made to the catalyst manufacturing process to obtain a catalyst with the right chemical and physical properties for slurry bed conditions. Precipitated iron based catalyst tends to be weak. It is very important that the catalyst is strong enough to prevent break up which would make the liquid/solid separation more difficult.

The product spectra industrially obtained for LTFT and HTFT processes using iron based catalyst for C-numbers ranging from C5 to C18, are shown in Table 3.

Table 3
Typical product composition for LTFT and HTFT reactors

	TFB Reactor		SP Reactor		HTFT	
	C_5-C_{18}	C_{13}-C_{18}	C_5-C_{12}	C_{13}-C_{18}	C_5-C_{10}	C_{11}-C_{14}
% Paraffins	53	65	29	44	13	15
% Olefins	40	28	64	50	70	60
% Aromatics	0	0	0	0	5	15
% Oxygenates	7	7	7	6	12	10
	100	100	100	100	100	100
% Olefins as n-Alpha	95	93	96	95	55	60

Only about 5% of the hydrocarbons obtained from LTFT are branched while about half of those from HTFT are branched. Currently the olefins from the Sasol SPD process are hydrogenated to straight chain paraffins. Although the fraction of olefins obtained from the newer low temperature Sasol SPD process is smaller than that obtained from HTFT, it consists mainly of straight chain alpha olefins which are potentially very valuable products. The higher olefinicity obtained with the Sasol SPD and SAS processes are thought to be due to the smaller catalyst particles used, which allow a larger portion of the primary FT product, olefins, to escape from the particles before being hydrogenated to paraffins [Ref 5]. It is envisaged that for large grass root plants using the commercially proven iron catalyst based Sasol SPD process, there is considerable incentive to recover these valuable olefins.

Cobalt Based Catalyst

Cobalt catalyst was used in the original German plants and it is used in the Russian plants, which have been operating since the 1950's. It is also used in the plant built more recently by Shell in Malaysia. The modern cobalt catalysts are similar to those prepared by Fischer in 1932 in that they have promoted cobalt precipitated on metal oxide supports, which are then reduced [Ref 4].

Cobalt based catalyst is not negatively affected by water produced by the reaction. For that reason cobalt based catalyst allows for high conversions of syngas in Fischer-Tropsch reactions. It has stronger hydrogenating characteristics than iron based catalyst and in a LTFT slurry bed reactor will only produce about half the olefins that will be obtained with iron based catalyst.

Cobalt based catalyst is much more sensitive to temperature than iron based catalyst. In general cobalt catalyst allows much less flexibility than iron based catalyst with respect to process conditions and hence to the flexibility of plant design.

The catalyst is expensive compared to iron based catalyst and needs a long life for it to be commercially useful. As it is very sensitive to sulphur poisoning, sulphur in the syngas must be reduced to even lower levels than that used for iron based catalyst.

7. NATURAL GAS CONVERSION USING THE SLURRY PHASE DISTILLATE PROCESS

When integrating Reforming with the SPD process, ideally the overall reaction would be

$$CH_4 + \frac{1}{2} O_2 \rightarrow [CH_2] + H_2O \qquad \Delta H = -203{,}0 \text{ kJ/mol}$$

To provide the O_2 from air for this reaction, it is estimated that 23,3 kJ/mol of O_2 is required, so that the exothermic heat liberated for the overall reaction would be 190 kJ/atom of carbon in the final product. Although this analysis does not consider exergy, it is clear that the integrated process is a net producer of considerable energy.

In an integrated Reformer-SPD plant, reforming tends to use a combination of steam reforming, CO_2 reforming and partial oxidation reforming to provide for optimal H_2/CO syngas. Optimality is determined by the relative cost of the synthesis and reforming steps of which the reforming step is the more expensive.

For smaller SPD plants, up to say 5000 bbl/day, syngas can be obtained through reforming natural gas by steam reforming. For larger plants, where the cost of oxygen is reduced by economy of scale, partial oxidation or autothermal reforming is more economical.

When optimised to obtain the best combination of high conversion with acceptable cost, thermal efficiencies based on lower heating value of natural gas fed and saleable products is about 60%. Because the overall process is exothermic, the complex is energy self sufficient by recovering heat from the processes and off gases. Where such opportunities exist, excess heat may be recovered as electrical power and sold. This increases the thermal efficiency to about 63%.

Although the Slurry Phase reactor has been used by Sasol to produce waxes and chemicals, the Sasol SPD process is designed to convert the primary product to predominantly middle distillate fuels such as gasoil or diesel. These fuels are compatible with existing crude oil derived fuels and present distribution infrastructure and engines can be used without modifications.

As mentioned this synthetic diesel has several environmental advantages over conventional fuels. It has superior combustion characteristics due to its high cetane number (>70), and it is free of sulphur and aromatic compounds. This leads to considerable reduction in emissions. The emissions from engines operating on standard, "reformulated" and Fischer-Tropsch diesels have been measured by South West Research Institute, a large independent fuels and lubricants testing organisation in the USA. The tests showed that Fischer-Tropsch diesel results in considerably reduced exhaust emissions (particulates, nitrogen oxides, carbon monoxide and unburned hydrocarbons) compared to both the standard and the "reformulated" diesels. The diesel fuel met all the 1998 CARB specifications and the tests indicated that with engines tuned for this diesel it could be expected also to meet the ULEV (2004) specifications.

Existing refiners often have difficulty in meeting increasingly stringent specifications due to either equipment or feedstock limitations. Blending with an essentially sulphur and aromatic free product, provides an attractive option requiring no extra capital investment for refining.

Economics

Economic studies for integrated Reformer - SPD plants were done for grass root plants. A single module of the SPD process was considered which will convert 110 000 m_n^3/h or 100 MMSCFD of natural gas into 10 000 bbl/day or 425 000 tons per year of liquid transportation fuels. Using US Gulf Coast costs, the capital investment required for such a module is of the order of up to US $300 million, depending on the location and the infrastructure available at the site. This gives an investment of $30 000 per daily barrel. This represents the total cost of the process, utility and infrastructure units for a grassroots plant erected on a green field site in a standard location. It does not include the cost associated with extraordinary infrastructure that may be required for a remote site. All products are finished fuels. The advantages of the modular approach are that the initial investment is limited, but that at the same time the opportunity for expansion exists by the addition of further modules.

The processes would fit in best where the cost of natural gas is low or where it has a negative value as a by product. This would be the case with remote gas fields or with natural gas associated with crude oil, for which there is no ready market. At a gas price of US $0,50/GJ, the feedstock cost is equivalent to about $5/bbl of product. Other fixed and variable costs total approximately another $5/bbl of product, resulting in a direct cash cost of production of about $10/bbl.

The products will fetch at least conventional fuel prices, and potentially also some premium due to their environmental advantages. At crude oil prices in the $16-18/bbl range, the product prices are expected to lie in the $22-25/bbl range. At these prices, the pre-tax return on investment is in the range of 12-15%. For multiple modules, using economy of scale, the capital cost is expected to be reduced considerably and the return on investment increased. The return on investment is bound to be further improved as the process proceeds along the learning curve and a target to reduce the capital cost from $30 000 per daily barrel to $25 000 and eventually $20 000 per daily barrel seems feasible.

The major cost of the Sasol SPD process is in the production of syngas. A major cost element in this is the cost of oxygen, which is sensitive to economy of scale. An alternative would be to use air instead of oxygen. The diluting nitrogen would obviously make the synthesis step much more expensive and it can probably only be considered for small scale operations. A longer term but potentially more important development is the use of membranes to separate oxygen from air at reformer process conditions in the reformer reactor. This could reduce the cost of the process for both small scale and large-scale operations.

Where associated gas is available in remote areas from which crude oil is pumped over large distances, the gas can be converted by the Sasol SPD process to a primary product that blends well with crude oil and can be pumped away with it. If it had been converted to methanol, a separate transport system, e.g. another pipeline, would have to be installed.

By the nature of the Sasol SPD process, it can be applied in special situations where small scale is a requirement and it is conceivable to skid mount a small Sasol SPD plant on a few trucks. This would be useful in remote inaccessible regions where gas is available but transport fuels can only be imported at great cost. Legislation on flaring gas is becoming stricter and this affects the exploitability of small oil fields with associated gas where this gas needs to be reinjected, which is not always economic. In offshore situations where piping natural gas is not economic, it appears quite feasible to put an SPD plant on a ship. The primary products from such a Sasol SPD plant could be blended and transported with crude oil. Several interested parties are investigating this approach.

The co-production of some chemicals is possible with the Sasol SPD process. The products of the low temperature Fischer-Tropsch reaction are primarily a mixture of linear paraffins and olefins with a wide range of carbon chain lengths. It is feasible to recover linear paraffins that can be used for the manufacture of detergent alkylates and other chemical uses. The separation of linear olefins for the production of detergent feedstock is also possible. Especially attractive is the use of isomerisation and dewaxing of the SPD reactor waxes with existing processes to produce excellent base stock for the higher grades of lube oils. Plants producing these high added value products obviously will be more profitable than those producing only fuels and will be less affected by ruling crude oil prices. The earlier plants may have the benefits of producing these products. The market for these products is however limited, whereas the market for diesel fuel is by comparison virtually limitless.

8. THE USE OF HIGH TEMPERATURE FT PLANTS

In the foregoing the emphasis has been on LTFT and its application. On small scale LTFT is more attractive than HTFT, which is, more costly because of the work up needed for the large amounts of lighter products produced. However these products contain considerable fractions of olefins and oxygenates. On larger scale the advantages obtained from these chemicals outweigh the disadvantages of a greater need for work up. At present there is general interest in LTFT because of the fit between a need to use remote natural gas and the easy placing of products derived from it. It is conceivable that with a greater interest in petrochemicals there may be a growing interest in HTFT as well; especially when an HTFT plant can be incorporated in an existing refinery.

9. CONCLUSION

Recent developments in FT technology and optimisation of the integration of FT technology with reforming of natural gas, have considerably reduced the capital and operating costs associated with the production of liquid fuels from natural gas. Where cheap natural gas is available, this technology can provide an attractive option for efficient use of this gas. Co-production of chemicals is also a possibility.

LITERATURE CITED

1 F. Fischer and H. Tropsch, German Patent 484337(1925).
2 B. Jager, M.E. Dry, T. Shingles, and A.P. Steynberg, Experience with a New Type of Reactor for Fischer-Tropsch Synthesis, Catalysis Letters 7, 1990, 293-302.
3 B. Jager, R. K. Kelfkens and A.P. Steynberg, A Slurry Bed Reactor for Low Temperature Fischer-Tropsch, Third International Natural Gas Conversion Symposium, Sydney, July 1993, Elsevier Science B.V.,1994, 419-425.
4 B. Jager and R. Espinoza, Advances in Low Temperature Fischer-Tropsch Synthesis, Catalysis Today 23, 1995, 17-28.
5 E. Iglesia, S.C. Reyes and S.L. Soled, Reaction-Transport Selectivity Models and the Design of Fischer-Tropsch Catalyst, Computer-Aided Design of Catalysis, edited by E.R. Becker and C.J. Pereira, Marcel Dekker, Inc., New York, 1993, 199-257.

NATURAL GAS CONVERSION V
Studies in Surface Science and Catalysis, Vol. 119
A. Parmaliana et al. (Editors)

Economics of Gas to Liquids Manufacture

Michael J. Gradassi

Gas Transportation & Upgrading Division, Exploration and Production Technology Group, Amoco Production Company, 501 Westlake Park Boulevard, Houston, Texas 77079, USA.

Abstract

The last year has seen a great deal in the literature about the rebirth of gas to liquids processes, most notably, Fischer-Tropsch processes. This renewed interest has been brought about by a technology that is said to have been so improved that it is now a commercially attractive option for natural gas monetization. No one single reason can be cited for this positive economic change. Rather, it is the result of several technological improvements that together have cut the capital cost of Fischer-Tropsch gas to liquids projects in half. Among these technological improvements are lower cost syngas preparation and lower cost gas to liquids reactors. This paper examines the economics of Fischer-Tropsch gas to liquids manufacture, using recent literature articles to develop process capital costs, operating expenses, liquid product value parameters, and other economic factors, to paint a general picture of the technology's current economic status. While manufacturing economics are reviewed, the answer to the question of gas to liquids project profitability is left to the individual investor whose economic thresholds must, in the final analysis, be met.

1. Background

Many articles have appeared in the literature of late describing the recent advances in gas to liquids technologies, with a primary focus on Fischer-Tropsch technologies. Articles have been written by technology providers, investment houses, academicians, inventors, and myriad more individuals with varying levels of interest in the technology. While they each have their own viewpoint, a single consensus appears to have emerged: Fischer-Tropsch technology has broken through the profitability barrier. How this has come about is explained as being the result of a collection of technological improvements in syngas generation, Fisher-Tropsch reactor design, catalysts, and overall process scaleablity.

The purpose of this paper is to present representative economic information relating to Fischer-Tropsch gas to liquids manufacture based solely on a selection of recent literature articles. No separate engineering design or costing was carried out. Economics were calculated using a cash flow analysis incorporating time value of money, revenue generation, capital spending patterns, depreciation, escalation, working capital and taxes. In addition to the Base Case economics, a sensitivity analysis is presented because there is no single case that alone can determine the economic viability of a project. Although this paper focuses only on Fischer-Tropsch technology, it should be recognized that it is only one of a family of gas to liquids technologies including methanol, gas to gasoline, DME, and several others that will benefit from recent technological advances.

A simplified process block flow diagram is shown in Figure 1. Detailed process descriptions can be found in the literature and will not be discussed here.

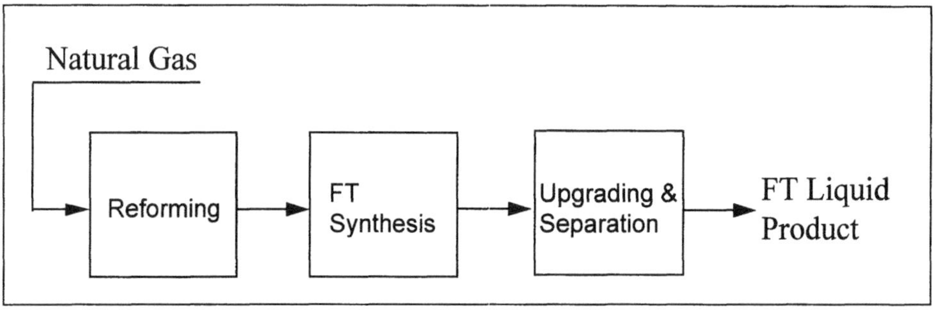

Figure 1. Gas to Liquids Block Flow Diagram

2. Economics Methodology

The economics methodology used in this paper follows that of Stermole, et. al. [1] for the calculation of after-tax cash flows shown in Table 1.

From the cash flow calculation, the Internal Rate of Return (IRR) and Net Present Value at 10% (NPV10) were calculated to establish a set of Base Case economic parameters as well as sensitivities to the Base Case which is explained later.

The calculated economics include the effect of the following parameters on the Base Case IRR and NPV10:

- Capital Cost of the Gas to Liquids Manufacturing Plant
- Natural Gas Feedstock Price
- Crude Oil Market Price
- Liquid Product Market Value

Because there is no one way to evaluate economics of any project, IRR calculations were performed for IRR's of 8%, 10%, 13%, and 15%, with the resultant NPV10 calculated for each IRR level, as well. In deciding the IRR ranges to present in this paper, it was judged that IRR's in the 10% to 13% range define a band representative of the minimum internal rate of return range of many project investors. Furthermore, an IRR of less than 8% was judged unlikely to be found economically attractive to but a few. Correspondingly, an IRR of more than 15% or better was judged likely to be found economically attractive by many investors.

Table 1
After-Tax Cash Flow Calculation

Revenue
- Natural Gas Feedstock Expense
- Operating Expense
- Freight Expense
- Depreciation
Taxable Income
- Income Tax
Net Income
+ Depreciation
- Plant Capital Cost
After-Tax Cash Flow

Table 2
Base Case Gas to Liquids Plant Assumptions Summary

Parameter	Value
Gas to Liquids Plant Capacity	50,000 barrels per stream day
Plant Capital Cost	$26,200 per daily barrel
Gas Consumption	9,500 scf per barrel of liquid product
Non-Gas Operating Expense	$5.00 per barrel of liquid product
Liquid Product Value	143% Crude Oil Price
By-product Power	None

3. Economic Results

For the assumed Base Case, the IRR was calculated at just below 15%, with an NPV10 of $600 MM. Therefore, a gas to liquids manufacturing plant that follows the given Base Case assumptions will provide its investors a 15% return on their investment over the plant's lifetime. Furthermore, such a plant will return $600 MM over and above the net cash outflows and inflows discounted at 10%. The balance of this section addresses manufacturing costs, economic sensitivities to the Base Case plant assumptions, and the implications of the economic cash flows.

Table 3
Economic Assumptions

Parameter	Value
Manufacturing Plant Life	25 years
Depreciation Schedule	10 year, Straight Line
Plant Construction Period	3 years
Plant Construction Capital Spending Profile	25%, 50%, 25%
Owner's Equity	100%
General Inflation	3% per annum
Escalation above general inflation	None
Federal + State Income Taxes	35%
Plant On-Stream Factor	95%
Plant Stream Day Production Profile	50% year 1, 100% Year 2-25
Working Capital Model	15 Day's Liquid Product Inventory
Crude Oil Price	\$18.00 per barrel
Natural Gas Price	\$0.50 per Mscf
Product Shipping Expense	\$0.91 per barrel

3.1. Gas to Liquids Product Manufacturing Cost

Liquid product manufacturing cost is shown in Figure 2 for IRR's of 8%, 10%, 13% and 15%. What the figure illustrates is that for the assumed Base Case, liquid product plant gate manufacturing cost (cash cost) is about \$10.00 per barrel regardless of the IRR. The figure further illustrates that with the addition of freight expense and capital investment (not return on capital), manufacturing costs subtotal about \$14.00 per barrel.

Therefore, regardless of the IRR required to meet investment requirements, the minimum manufacturing cost of gas to liquids products likely will not be materially less than \$14.00 per barrel. This \$14.00 per barrel figure also can be viewed as the break-even cost for gas to liquids manufacture. That is, if the market pays at least \$14.00 per barrel for the liquid product, an investor can expect to break-even on the investment. It does not mean, however, that at this product price level, an investor is likely to make a profit, which is the driving force for investment in the first place.

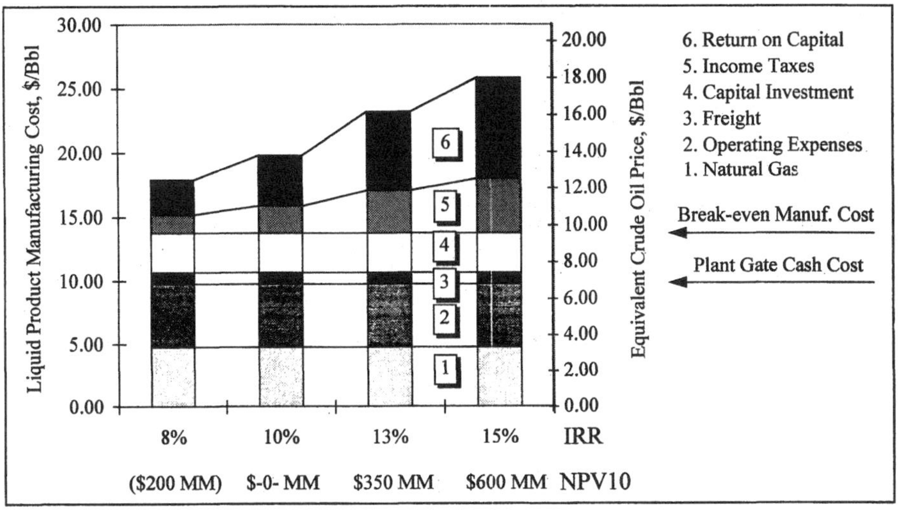

Figure 2. Gas to Liquids Liquid Product Manufacturing Cost

When Return on Capital (profit) and taxes are considered, however, the manufacturing cost picture changes dramatically. Manufacturing costs at 15% IRR rise to about $26.00 per barrel, adding an additional $12.00 per barrel to the expected minimum, or break-even, manufacturing cost. As further illustrated, this 15% IRR likely is achievable in an $18.00 per barrel crude oil market. At an 8% IRR, manufacturing costs rise to only $18.00 per barrel. However, as the illustration shows, even in a $14.00 per barrel crude oil market environment, an IRR of at least 10% should be feasible. Thus, an investment in a Fischer-Tropsch gas to liquids project is judged likely for some.

3.2. Economic Sensitivities

The effect the Base Case parameters of Plant Capacity, Gas Price, Crude Oil Price, and Liquid Product Value have on the overall Gas to Liquids economics is summarized in the Base Case Tornado Diagram, Figure 3.

What Figure 3 shows is that for the illustrated range of each Base Case parameter, Gas Price has the single greatest effect on the IRR of gas to liquids manufacture. This effect is followed in order of sensitivity by Plant Capital Cost (expressed as a percentage of its Base Case value), Crude Oil Price, and the Liquid Product Value (expressed as a percentage of the Crude Oil Price).

The sensitivity of the Base Case economics to Gas Price is especially important for it shows that with even modestly priced gas at $1.00 per Mscf, the IRR of gas to liquids manufacture drops rapidly. In this case, it drops from 15% with $0.50 gas, to 11% with $1.00 gas. While a 15% IRR is judged likely to be attractive to many gas to liquids investors, an 11% IRR is judged likely to be only within the threshold range of economic attractiveness to some. Thus, for a profitable gas to liquids technology application, it is highly important to secure an attractive gas price over the manufacturing plant's lifetime.

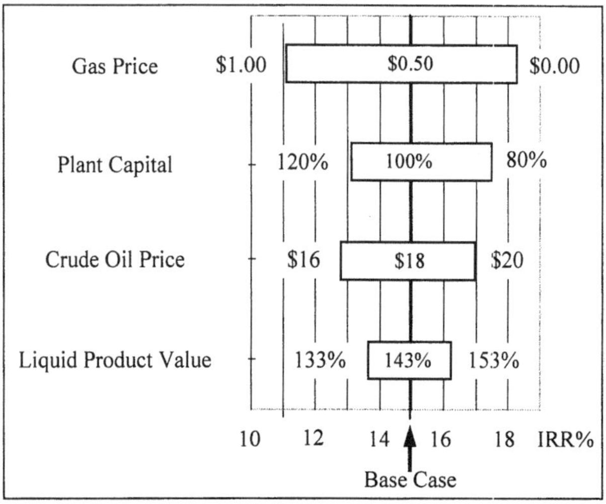

Figure 3. The Base Case Tornado Diagram

To further illustrate the economic sensitivities of the assumed Base Case parameters, the effect of Gas Price versus Capital Cost, and Crude Oil Price versus Liquid Product Value are illustrated in Figures 4 and 5.

The Gas price versus Capital Cost figure shows the sensitivity of project economics to Capital Cost for any given gas price in the range of $0.00 to $1.00 per Mscf. At the Base Case Gas Price of $0.50 per Mscf, Figure 4 shows that with as little as a 20% increase in plant Capital Cost, a project can lose as much as 2% IRR, resulting in an IRR of 13%. Such an increase in Capital Cost over the Base Case apparently is not unlikely when the manufacturing plant is built at a remote site where construction costs can be as much as 20% greater than those of the U.S. Gulf Coast [14]. Under such circumstances, the corresponding loss in NPV10 is $250 MM dollars. In other words, a 20% increase in capital costs versus the Base Case would destroy over $250 MM in the investor's value in the plant. A similar loss in project IRR and NPV10 would result if the Gas price were to rise by only $0.25 per Mscf from $0.50 per Mscf in the Base Case to just $0.75 per Mscf.

The Crude Oil Price versus Liquid Product Value Figure 5 shows the sensitivity of project economics to Product Value for any given Crude Oil Price in the range of $12.00 to $22.00 per barrel. For the Base Case Liquid Product Value of 143%, an increase in the crude oil price for the entire project life would raise the gas to liquids project IRR attractively well beyond a 15%. However, if crude oil prices were to decline significantly, much as they had during early 1998, project economics would suffer. With only a $2.00 per barrel drop in crude oil price, the project economics drop from 15% IRR to 13% IRR. Another $2.00 per barrel drop would move the economics to about 11% IRR.

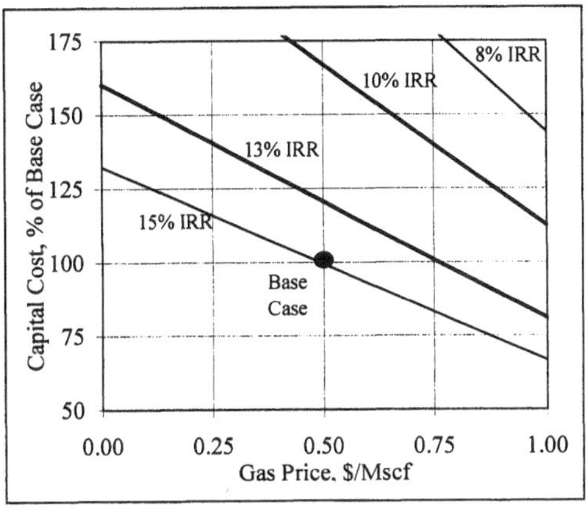

Figure 4. Gas Price versus Capital Cost

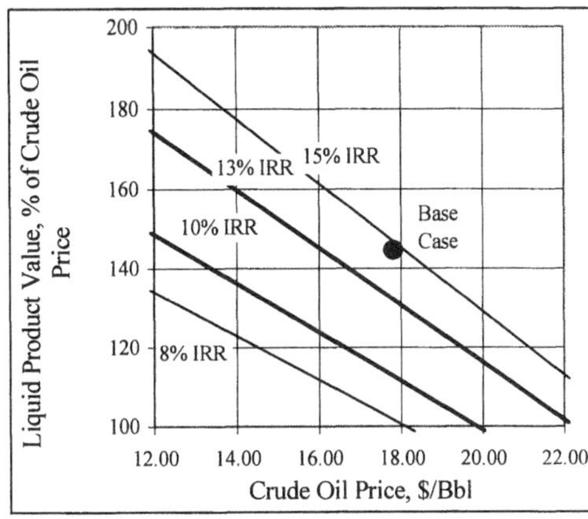

Figure 5. Crude Oil Price versus Liquid Product Value

With respect to the Liquid Product Value, it is judged unlikely for it to drop below the average conventional product value premium versus crude oil of 130%. Thus, in a worst case, a $4.00 per barrel drop in crude oil price should result in no worse than an 8% to 9% IRR for a gas to liquids project. However, values of IRR this low may discourage many investors.

3.3. Project Cash Flows

Base Case annual cash flows are shown in Figure 6. As the illustration shows, significant negative cash flows are experienced during the assumed construction period of 3 years. Following construction, however, the trend reverses, and annual cash flows in the $200 MM to $400 MM range can be expected if market conditions support the Base Case assumptions. Thus, gas to liquids manufacturing projects are expected to be a tremendous source of cash over their lifetime.

A project's cash flow is not without risk for a considerable time, however, and this is illustrated in Figure 7. This figure shows the cumulative cash flows for the duration of a gas to liquids project, from the first year of construction through to the project's final year. By the end of the 3-year construction period, the negative cash flows total in excess of $1,300 MM, a significant exposure for any investor. It is at this point that an investor is most vulnerable to changes in the market place which must generate the anticipated positive future cash flows.

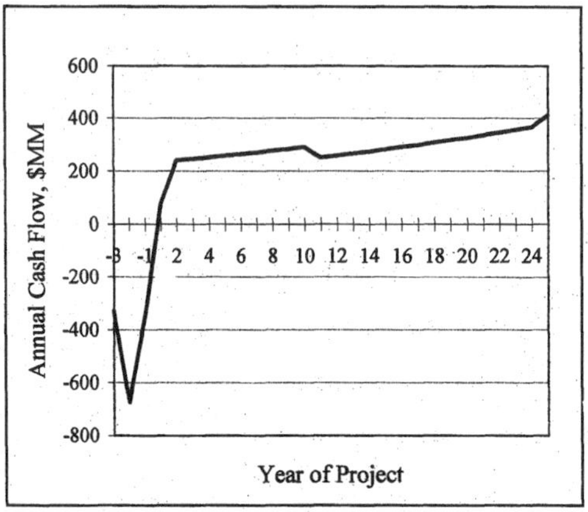

Figure 6. Annual Cash Flow

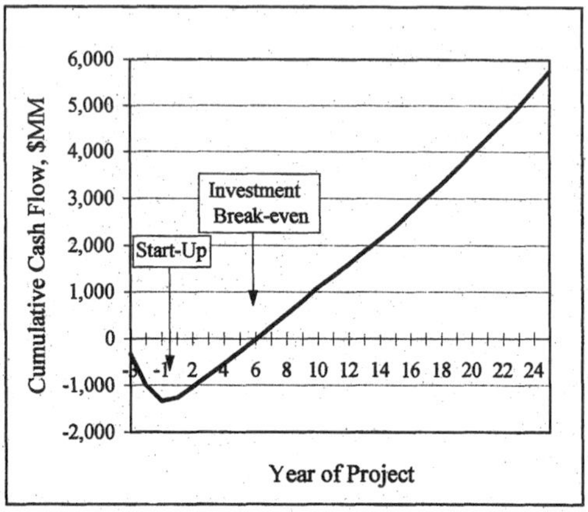

Figure 7. Cumulative Cash Flow

For the Base Case, a positive cash flow is illustrated, ultimately reaching an accumulated total cash flow of nearly $6,000 MM. Note, however, that the invested capital remains at risk for at least 6 years from the start up of the manufacturing plant. That is, during the first 6 years of operation the generated after tax revenues are returning only the initial plant investment, plant gas and operating expenses, and liquid product freight to market. It is not until year 7 following plant start up that an investor will begin to see any pay off. When the 3-year plant construction period is taken into account, it becomes apparent that it can be a full 10 years before any real positive return is realized.

4. Selection of Base Case Gas to Liquids Plant Parameters

To develop the economics discussed in this paper, the cited literature was reviewed for capital cost, operating expense, gas consumption, and product value quotations from which to draw consensus figures for use in the cash flow analysis.

4.1. Gas to Liquids Plant Capacity

A plant liquids capacity of 50,000 barrels per stream day was selected for the Base Case. It is assumed all salable liquids produced by the plant are included in this capacity figure.

4.2. Plant Capital Cost

Capital cost figures ranged from $24,000 to $30,000 per daily barrel for plant sizes ranging from 2,500 to 50,000 barrels per day [2,3,4,5]. For a 50,000 barrel per stream day plant capacity, a regression of these figures resulted in a capital cost of $26,200 per daily barrel.

4.3. Gas Consumption Rates

Gas consumption per barrel of liquid product is quoted from 8,000 standard cubic feet per barrel (scf/Bbl) to as much as 11,400 scf/Bbl [2,3,4,5,6,7,8,9], averaging 9,500 scf/Bbl. No clear reason can be given to the wide range of figures which varied plus or minus 15 to 20 percent from this average. However, it is likely safe to assume the lower gas consumption rates are indicative of more efficient plant design configurations and the higher indicative of the less efficient.

4.4. Plant Operating Expenses

Plant non-gas feedstock operating expenses quoted in the literature varied from $3.00 to $6.00 per barrel [4,5,6,7,8,10]. These expense figures averaged $5.00 per barrel and cover expenses for plant labor, plant maintenance, local taxes, insurance and the like.

4.5. Liquid Product Value

Fischer-Tropsch liquids are well known for their being virtually sulfur, nitrogen, and heteroatom free, and they are said to carry a product premium in excess of conventional crude oil derived liquid products such as diesels and gas oils. Only a few articles in the referenced literature quantified Fischer-Tropsch product value [4,5,9]. Two references quoted Fischer-Tropsch liquids having an average value of 143% of crude oil versus conventionally derived liquids having a value of 130% [11]. A third reference [5] quoted an average product value of 189%, but this value was judged excessive given an in-house study [12] which confirmed the 143% figure as more representative.

5. Summary

Natural gas to liquids technology appears to have reached the threshold of economic attractiveness when gas is priced at about $0.50 per Mscf. For the assumed Base Case, minimum returns on investment are judged to be likely when crude oil is priced between $14.00 to $16.00 per barrel, and more attractive returns are judged to be likely when crude oil is priced at a minimum of $18.00 per barrel. Gas to liquids projects require a significant investment, and at a plant capacity of 50,000 barrels per day, can be expected to require about $1,300 MM. In addition, this capital investment is expected to be at risk for a likely minimum of 6 years following the manufacturing plant startup when revenues generated from product sales begin to pay back the invested capital. Revenues generated from product sales are expected to be quite attractive and range from \$200 MM to \$400 MM annually over the lifetime of the project. Break-even manufacturing costs, including the initial capital investment, are expected to be a minimum of \$14.00 per barrel with gas priced at \$0.50 per Mscf.

References

1. Franklin J. Stermole, et. al., Economic Evaluation and Investment Decision Methods, 8th Edition, Golden: Investment Evaluations Corporation, 1993.
2. Syntroleum, Texaco, Brown & Root Announce Plans for \$75-million Gas-to-Liquids Barge-Mounted Plant, Remote Gas Strategies, Vol. II, No. 1, January, 1998.
3. Gerald Parkinson, Fischer-Tropsch Comes Back, ACHEMA, http://www.che.com/acema/html/ad1p12s1.htm, America Online, February 1998.
4. Ben Jager, The Status of Fischer-Tropsch Technology: A Competitor for LNG?, 20th World Gas Conference, Copenhagen, 1997, http://www.wpc.org/roundtable/commf/103/, America Online, February, 1998.
5. Douglas Terreson, Monthly Perspectives - September 1997: Answering the Gas to Liquids Question, Morgan Stanley Dean Witter, September 9, 1997.
6. Joseph M. Fox, et. al., "An Evaluation of Direct Methane Conversion Processes", Chemical Engineering Progress, April, 1990, 42.
7. Ray Swanepoel, Case Study: Bateman's GTL Initiatives (Mossgas and Syntroleum), Remote Gas Strategies Conference, Monetizing Stranded Gas Reserves, Houston, TX, December 10 -1 12, 1997.
8. J. Jacometti, Economic Perspectives of Fischer-Tropsch Based Gas Conversion: The Shell MDS Perspective, Proceedings 20th World Gas Conference, Copenhagen, 1997, http://www.wpg.org/roundtable/commf/102/, America Online, February 1998.
9. Gerald N. Choi, et. al., Design and Economics of a Fischer-Tropsch Plant for Converting Natural Gas to Liquid Transportation Fuels, Clean Fuels Symposium, American Chemical Society Meeting, San Francisco, CA, April 13 - 17, 1997.
10. The Syntroleum Process, Natural Gas to Synthetic Oil, Syntroleum, 1997.
11. Average IEA CIF Crude Cost and Spot Crude and Product Prices, Table 8, Energy Information Agency [Online], ftp://eia.doe.gov., America Online, February 1998.
12. Amoco Petroleum Products, Private Study, 1996.
13. Mark A. Agee, The Syntroleum Perspective, Remote Gas Strategies Conference, Monetizing Stranded Gas Reserves, Houston. TX, December 10 - 12, 1997.
14. Michael J. Gradassi, et.al., Economics of Natural Gas Conversion Processes, Fuel Processing Technology 42, Elsevier, Amsterdam, (1995) 65-83.
15. "Gas to Oil: A Gusher for the Millennium?", Business Week, May 19, 1997, 130.

NATURAL GAS CONVERSION V
Studies in Surface Science and Catalysis, Vol. 119
A. Parmaliana et al. (Editors)

Catalytic methane combustion on La-based perovskite type catalysts

F. Martinez Ortega[a]., C. Batiot[a], J. Barrault[a], M. Ganne[b] and J.M. Tatibouët[a, *]

[a] Laboratoire de Catalyse en Chimie Organique, UMR CNRS 6503
Ecole Supérieure d'Ingénieurs de Poitiers
40, avenue du Recteur Pineau, 86022 Poitiers cedex (France)

[b] Institut des Matériaux de Nantes, UMR CNRS 110
2, rue de la Houssinière, 44072 Nantes (France)

The catalytic behavior of two series of lanthanum-based perovskite catalysts ($La_{1-x}Sr_xFeO_3$ and $La_{1-x}Sr_xCoO_3$) has been investigated in isothermal conditions close to those existing in a catalytic burner, at 900°C and in the presence of water and carbon dioxide in the gas phase. The study of the influence of the lanthanum partial substitution by strontium has shown that the presence of strontium increases the specific surface area of the catalysts, and accordingly the catalytic activity, the best performances being observed for the $La_{0.8}Sr_{0.2}MO_3$ samples (M=Fe or Co). Moreover, the presence of strontium seems to limit the sintering of the catalysts, even after a drastic aging under conditions modeling the reaction environment.

1. INTRODUCTION

One of the main problems raised by the use of natural gas for heat production by flame combustion in air is the release of a large quantity of nitrogen oxides (NO_x) due to the combination of nitrogen and oxygen in the heatest part of the flame where the temperature can reach a value as high as 1800°C.

The use of a catalyst allows to control the rate of the reaction and then to avoid the hot spots where the NO_x are formed. Moreover, the use of a catalyst leads to a more complete conversion of the natural gas than in a flame and then contribute to reduce both the NO_x and the unburned hydrocarbons emission.

Although the temperature in a catalytic combustor is lower than in a flame, a value in the range 1000-1200°C is commonly reached. At a so high temperature, the noble metal based catalysts deactivate rapidly due to the sintering of the metal particles, to the vaporization of the active phase or its thermal reduction.

Mixed oxide catalysts represent an attractive alternative to noble metals catalysts in view of their thermal stability. Among them, mixed oxides with a perovskite structure ABO_3 appear to be very promising since they are thermally stable and are active in total oxidation reaction. Moreover, the large possibility of partial substitution of the cations A or B, allows us to expect a control of both the catalytic activity and the thermal stability.

* Corresponding author

Perovskite-type mixed oxides catalysts have been extensively studied in hydrocarbon or methane total oxidation [1], but only few data on the kinetic of the reaction, obtained at high temperature, have been yet published [2, 3].

The goal of the present work is to study the catalytic behavior of two series of the lanthanum based perovskites $La_{1-x}Sr_xFeO_3$ and $La_{1-x}Sr_xCoO_3$, at 900°C, in isothermal conditions close to those existing in a catalytic burner, i. e., high temperature and the presence of CO_2 and H_2O in the gas phase. The catalyst deactivation with time on stream was also investigated.

2. EXPERIMENTAL

2.1. Catalyst preparation

The perovskite type oxides were prepared by addition of glycine ($H_2NCH_2CO_2H$) to an aqueous solution of the metal nitrates in order to have a ratio $NO_3^-/NH_2=1$. The resulting solution was slowly evaporated until a vitreous material was obtained, and then calcined at 250°C for one hour. During this calcination, a fast exothermic reaction occurs, yielding to the formation of a pulverulent precursor still containing carbonaceous species. A calcination at 700°C for one hour eliminates all the remaining carbon. The catalysts were subsequently calcined at 900°C before characterization.

2.2. Characterization

The B.E.T. surface area was determined by N_2 adsorption with a Micromeritics, Flowsorb 2300. The XPS analysis were performed with a Leybold apparatus. The La ($3d^{5/2}$), Sr (3d), O (1s), Fe and Co ($2p^{3/2}$) signals were used for the surface composition measurement.

2.3. Catalytic measurements

In order to obtain reliable kinetic data, it is necessary to avoid an increase of the catalyst bed temperature due to the high exothermicity of the total methane oxidation:

$CH_4 + 2\,O_2 \rightarrow CO_2 + 2\,H_2O \quad \Delta H\ (\text{at } 298\ K) = -802.3\ kJ/mol$

The isothermal conditions were obtained by dilution of the catalyst with pumice and by the use of a low methane partial pressure (≈ 1 Torr). This low partial pressure was obtained by the catalytic conversion of about 90% of an usual reactant mixture ($CH_4/O_2/He=1/10/89$ mol%) in a first reactor. The exiting gases are then the reactant mixture used for the kinetic and deactivation measurements which is performed in a second reactor where the isothermal conditions are expected to be realized. A quartz stick was used to fill the void space after the catalytic bed, to avoid non catalytic reactions. In these conditions, the reaction in the absence of catalyst was negligible. The main advantage to use this kind of apparatus is to simply obtain a reactant mixture containing carbon dioxide and water in the stoichiometric proportions of the combustion reaction, thus modeling the conditions existing in a catalytic burner.

The studies of the deactivation and the determination of the kinetic parameters were carried out with 5mg of catalyst diluted by 20 mg of pumice. The products were analyzed at the entry and at the end of the second reactor by a on line gas chromatograph.

To calculate the kinetic parameters the reactor was considered as a plug flow one (VVH=$6.10^5 h^{-1}$). By assuming that the reaction is zero order relatively to the oxygen and first order relatively to the methane, the reaction rate constant can be expressed by the relation: $k=-\ln(1-x).R/(X.pCH_4)$, where X is the methane conversion, pCH_4 the methane partial pressure and R the experimental rate of the methane consumption.
The methane conversion was mainly comprised between 35 and 60% in order to minimize the deviation from the ideal plug flow reactor (conversion ≤60%) and to ensure a good accuracy of the CH_4 conversion measurement (conversion ≥35%). A series of experiments carried out at various temperatures showed that mass transport limitations did not influence the reaction kinetics.

The deactivation of the catalysts was characterized by the factor $D=(k_2-k_{22})/k_2$ where k_X is the rate constant after x hours of reaction.

3. RESULTS

The surface composition of the catalysts, determined by XPS analysis, before and after the reaction, are given in the Table 1. An EDX analysis showed that the composition of all the catalysts is very close to the nominal composition.

Before reaction, the surface composition appears to be significantly different from the bulk. A surface enrichment in lanthanum and in strontium is clearly visible for all the samples. For the iron containing samples, the enrichment in La+Sr is maximum for the $La_{0.8}Sr_{0.2}FeO_3$ sample, whereas it remains almost constant for the Co containing samples.

After reaction (900°C, 24h) the XPS analysis of the surface composition of the catalysts remains different from the bulk composition (Table 1), but during the reaction course, a different behavior has occuring, depending on the presence of iron or cobalt. For the iron containing perovskites, the La+Sr surface enrichment has increased, whereas for the Co containing samples, the La+Sr content has decreased until to obtain an excess in Co for the Sr containing samples (Table 1). However, it should be noticed that more than 70% of the detected surface species are formed by oxygen.

Table 1
XPS analysis before and after reaction

catalyst	before reaction			after reaction		
	Surface composition	La/Sr	La+Sr	Surface composition	La/Sr	La+Sr
$LaFeO_3$	$La_{1.2}FeO_{3+x}$	-	1.2	$La_{2.2}FeO_{3+x}$	-	2.2
$La_{0.8}Sr_{0.2}FeO_3$	$La_{2.4}Sr_{1.2}FeO_{3+x}$	2.0	3.6	$La_{2.4}Sr_{2.3}FeO_3+x$	1	4.7
$La_{0.5}Sr_{0.5}FeO_3$	$La_{0.8}Sr_2FeO_{3+x}$	0.4	2.8	$La_{1.4}Sr_{2.6}FeO_3+x$	0.5	4
$LaCoO_3$	$La_{2.8}CoO_{3+x}$	-	2.8	$La_{1.7}CoO_{3+x}$	-	1.7
$La_{0.8}Sr_{0.2}CoO_3$	$La_{2.0}Sr_{0.9}CoO_{3+x}$	2.2	2.9	$La_{0.32}Sr_{0.24}CoO_{3+x}$	1.6	0.6
$La_{0.75}Sr_{0.25}CoO_3$	$La_{1.73}Sr_{0.67}CoO_{3+x}$	2.6	2.4	$La_{0.26}Sr_{0.08}CoO_{3+x}$	3.2	0.34

The catalytic results are summarized in the Table 2. The deactivation with time on stream is shown in the Figures 1a and 1b.

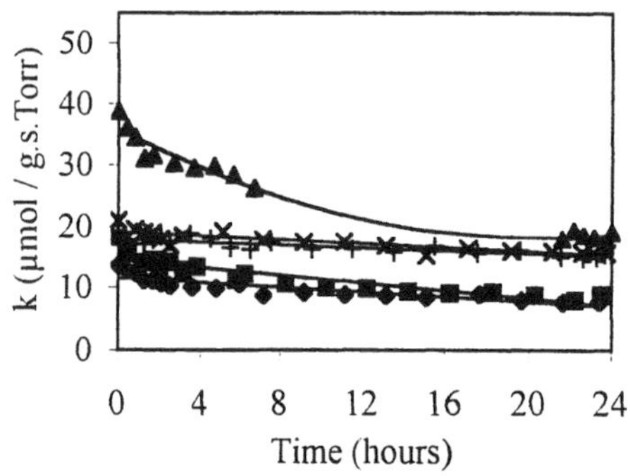

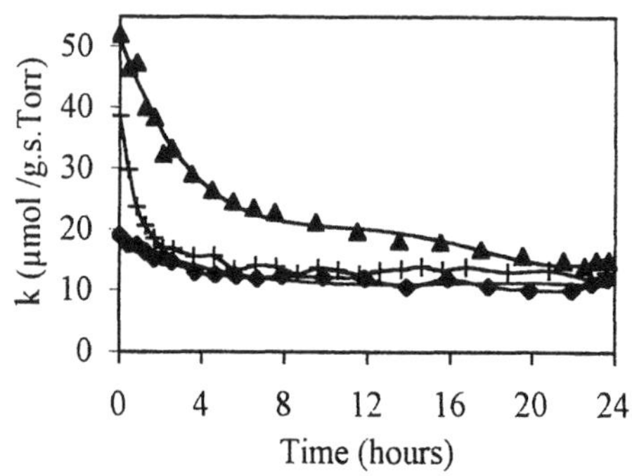

Variation of the rate constants in function of the time on stream (T_R=900°C)

Figure 1a: $La_{1-x}Sr_xFeO_3$
♦: x=0; ■: x=0.1; ▲: x=0.2; ×: x=0.4; +: x=0.5

Figure 1b: $La_{1-x}Sr_xCoO_3$
♦: x=0; +: x=0.15; ▲: x=0.2

Table 2
Rate constants at 900°C and BET surface areas after 2 and 22 hours of reaction.
mcata.= 5mg, pumice : 20mg; pCH_4=1 torr, pH_2O=13.2 torr, pCO_2=6.6 torr, pO_2=62.8 torr

Catalyst	BET surface area (±0.2m^2/g)		Δ=(A-B)/A	k_2 (μmol/g.s.torr)	k_{22} (μmol/g.s.torr)	D=(k_2k_{22})/k_2
	fresh A	aged B				
$LaFeO_3$	3.3	2.1	0.36	10.74	7.72	0.281
$La_{0.9}Sr_{0.1}FeO_3$	5.5	5.0	0.09	12.50	8.73	0.302
$La_{0.8}Sr_{0.2}FeO_3$	6.7	5.6	0.16	30.32	18.05	0.405
$La_{0.7}Sr_{0.3}FeO_3$	2.6	2.1	0.19	20.50	15.26	0.256
$La_{0.6}Sr_{0.4}FeO_3$	2.6	2.6	0	18.02	15.89	0.118
$La_{0.5}Sr_{0.5}FeO_3$	2.3	1.9	0.17	18.37	14.97	0.185
$LaCoO_3$	1.2	0.8	0.33	15.42	10.18	0.340
$La_{0.9}Sr_{0.1}CoO_3$	1.4	1.4	0	14.86	12.78	0.140
$La_{0.8}Sr_{0.2}CoO_3$	4.5	3.8	0.16	32.51	14.90	0.542
$La_{0.75}Sr_{0.25}CoO_3$	1.6	1.3	0.19	14.14	12.14	0.141

The surface area of the iron based catalyst are always higher than those of the cobalt containing ones. Whatever the series, the highest BET surface area is obtained for the $La_{0.8}Sr_{0.2}MO_3$samples (M=Fe or Co).

A decrease in the surface area of the samples is generally observed after a treatment modeling the catalytic reaction. This treatment consists to pass through a larger catalyst bed (100 mg) than for the kinetic studies (5 mg), at 900°C for 6h, a gas mixture ($H_2O/CO_2/O_2/He$) resulting to the total methane conversion of the $CH_4/O_2/He$=3/10/87 (mol %) reactant feed in the first reactor. The presence of Sr seems to partly inhibits the catalyst sintering.

The values of the rate constants (measured after 2 and 22 hours of reaction) indicate that the catalytic activity is maximum for the $La_{0.8}Sr_{0.2}FeO_3$and $La_{0.8}Sr_{0.2}CoO_3$ samples which possess the highest surface area. Nevertheless, their deactivation is important as shown in Figures 1a and 1b. It appears that after 22 hours of reaction the cobalt containing perovskites have almost the same catalytic activity whatever the Sr content.

The influence of strontium content on the catalytic activity for the fresh and aged catalysts is shown in Figure 2. It appears clearly that for the fresh catalyst the rate constant of the Co and Fe based catalysts increases with x up to x=0.2 but decreases sharply for x>0.2. After 22 hours of reaction the promoter effect of strontium content is less visible.

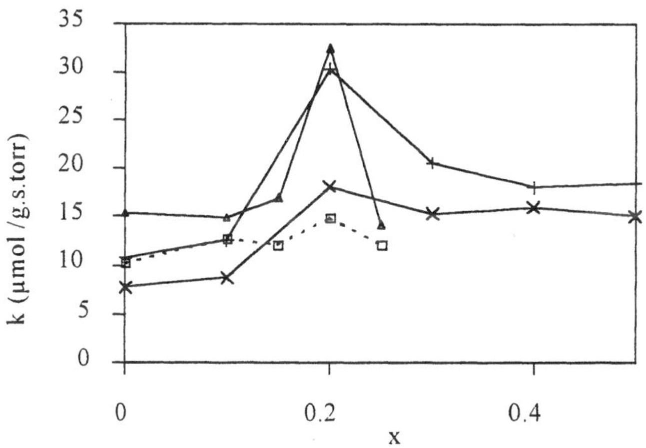

Figure 2:
Variation of the rate constants of $La_{1-x}Sr_xFeO_3$ and $La_{1-x}Sr_xCoO_3$ as a function of the Sr content: $La_{1-x}Sr_xFeO_3$, t=2h: +, t=22h ×; $La_{1-x}Sr_xCoO_3$, t=2 h:Δ, t=22h: □

The influence of the partial pressure of water and carbon dioxide on the deactivation of $La_{0.8}Sr_{0.2}CoO_3$ has been studied at 900°C. The catalyst was previously treated, in situ, at 900°C for 6h, in a $H_2O/CO_2/O_2/He$ feed resulting to the methane total conversion in the first reactor of the reactant feed containing 1 or 3 mol % of CH_4, followed by the usual kinetic study. The results are shown in Figure 3. It appears that the deactivation is more important after the treatment with the highest water and carbon dioxide partial pressure, but whatever the conditions, the rate constants seem to converge to the same value, reached for the treatment performed with the highest water and carbon dioxide partial pressures.

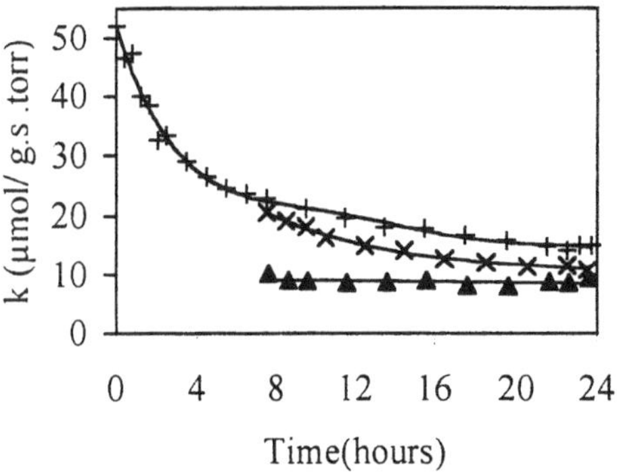

Figure 3 : Variation of the rate constants of $La_{0.8}Sr_{0.2}CoO_3$ with time on stream.
mcata=5mg, T=900°C, pumice=21mg
pCH_4=1torr, pH_2O=13.2torr, pCO_2=6.6torr, pO_2=62.8torr:
(+): without treatment.
(×) after a 6h in-situ treatment under $H_2O/CO_2/O_2/He$=2/1/8/89 mol% , at 900°C.
(▲) after a 6h in-situ treatment under $H_2O/CO_2/O_2/He$=6/3/4/87 mol% , at 900°C

4. DISCUSSION

The variation of the values of the rate constants seem to follow the same trend as the specific surface area of the catalysts, suggesting that this parameter is paramount to control the catalytic activity. Nevertheless, a more accurate examination of the results shows that for the iron containing series, the $La_{0.9}Sr_{0.1}FeO_3$ catalyst is about twice less catalytically active than the $La_{0.7}Sr_{0.3}CoO_3$ catalyst although the specific surface area are 5.5 and 2.6m^2/g, respectively. The same behavior is also observed for the $LaFeO_3$ and $La_{0.7}Sr_{0.3}FeO_3$ catalysts. For these samples, the rate constants differ by a factor of two, whereas the specific surface areas are similar. We can then conclude that the chemical composition of the catalysts plays also a direct role in the catalytic activity, independantly of the value of the specific surface area.

We have then investigated the role of the La substitution by Sr on the $LaFeO_3$ and $LaCoO_3$ based perovskites. A kinetic study has been performed at 900°C in the presence of water and carbon dioxide, modeling the conditions existing in a catalytic burner, and then allowing to follow the catalyst deactivation in function of time on stream. We have shown that $LaCoO_3$ was more active than $LaFeO_3$, in good agreement with the results of McCarty et al. [2] and Arai et al. [4]. A partial substitution of La^{3+} by Sr^{2+}increases the catalytic activity, the best results being obtained with $La_{0.8}Sr_{0.2}FeO_3$ and $La_{0.8}Sr_{0.2}CoO_3$ samples. The Sr substitution was expected to lead to the formation of electrophilic oxygen species (O^-, O_2^-), very reactive towards C-H bonds [5]. The charge compensation could also be achieved either by the formation of tetravalent Co or positive holes [1,6-8]. The significant decrease of the initial catalytic activity when the Sr loading is greater than 0.2 could be explained either by the decrease of the reoxidation rate of the sample or by the drop of the capability of oxygen dissociation on the surface by increasing of Sr^{2+} substitution [9].

The surface composition, determined by XPS, indicates a large enrichment in La and Sr, suggesting that our catalysts could be considered as supported catalysts on the corresponding perovskite with the nominal composition (as probed by EDX and XRD analysis), the active phase being likely formed by amorphous surface oxides. The higher catalytic activity of Sr containing catalysts could be attributed to the large surface enrichment in lanthanum and strontium. The catalysts behave as a lanthanum strontium mixed oxide, the presence of strontium enhancing the catalytic activity as already shown for the reaction of the oxidative coupling of methane [10].

The deactivation of the catalysts has been studied at 900°C, either in the reaction conditions, the reactant feed being formed by a $CH_4/O_2/He=1/10/89$ (mol%) gas mixture previously converted at 90% in a first reactor ($P_{CH_4}=1$ Torr, $P_{H_2O}=13.2$ Torr, $P_{CO_2}=6.6$ Torr) or only in the presence of water and carbon dioxide corresponding to the complete conversion of the reactant feed containing from 1 to 3 mol% of methane and 10 mol% of oxygen diluted by He. In these conditions, the reaction was zero order in oxygen and first order in methane [3]. We have shown that the deactivation depends only on the initial methane partial pressure, i. e., to the water and carbon dioxide partial pressures on the catalyst.

As shown in Fig. 1 and in Table 2, for all the catalysts, the catalytic activity decreases with the time on stream. Simultaneously, we have observed a decrease in the specific surface area of the catalysts due to the sintering of the samples in the reaction conditions. The comparison between the percentage of the decrease of the specific surface area (Δ) and of the rate constants (D) clearly indicate that a simple correlation cannot be found between the sintering of the catalysts and the variation of the catalytic activity. Indeed, the $La_{0.8}Sr_{0.2}FeO_3$ and $La_{0.8}Sr_{0.2}CoO_3$ samples exhibit a decrease in the rate constant of 54% and 41%, respectively, whereas the specific surface areas decrease only by 16% for both catalysts. The surface composition change during the reaction is in opposite trend, depending on the presence of iron or cobalt, (enrichment in lanthanum and strontium for the iron containing samples, disappearance for the cobalt containing samples). It seems then difficult to correlate this behavior with the decrease of the catalytic activity with time on stream, but the large amount of oxygen present on the surface after reaction could indicate, either the extensive formation of surface hydroxyl groups, or more likely an important formation of surface carbonates species, as already mentioned for La_2O_3 and Sr/La_2O_3 catalysts in the reaction of methane oxidative coupling [11]. Lombardo et al have shown that the very stable $La_2(CO_3)_3$ was produced at the surface of $LaFeO_3$ and $LaCoO_3$[12,13]. These carbonate species could inhibit the reaction by blocking the methane adsorption sites. In this point of view, the higher the surface carbonate stability, the lower the catalytic activity. In fact, two parameters acting in an opposite way should control the catalytic activity, i) the specific surface area of the catalysts, or the ability to preserve this surface in the reaction conditions, and ii) the stability of the surface carbonate species formed during the reaction. These two parameters depend roughly on the chemical composition of the perovskite, but the former is controlled by the bulk composition whereas the latter should depends mainly on the surface composition.

5. CONCLUSION

The catalytic behavior of two series of lanthanum-based perovskite catalysts ($La_{1-x}Sr_xFeO_3$ and $La_{1-x}Sr_xCoO_3$) has been investigated in isothermal conditions, modeling the conditions existing in the terminal part of a catalytic burner, i. e. high temperature (900°C), low methane pressure and the presence of water and carbon dioxide in the gas phase. The partial substitution of lanthanum by strontium induces an increase in the specific surface area of the samples and a better resistance to the sintering than the unsubstituted perovskite. The best catalytic performance is obtained for the samples where 20% of the lanthanum is substituted by strontium.

The variations of the surface composition of the catalysts which occur under the reaction conditions suggest that the deactivation with time on stream, observed for all the catalysts, could be due to the concomitantly formation of highly stable carbonate species on the surface, and to the decrease in the specific surface area of the catalysts.

Acknowledgments

The French government and Gaz de France are kindly acknowledged for financial support.

6. REFERENCES

1. T. Seiyama, Catal. Rev.-Sci. Eng., 34 (1992) 281.
2. J.G. Mc Carty and H. Wise, Catal. Today, 8 (1990) 231.
3. D. Klvana, J. Vaillancourt, J. Kirchnerova and J. Chaouki, Appl. Catal., 109 (1994) 181.
4. H. Arai, T. Yamada, K. Eguchi and T. Seiyama, Appl. Catal., 26 (1986) 265.
5. J.H. Lunsford, Langmuir, 5 (1989) 12.
6. A.C.C. Tseung and H.L. Bevan, J. Mater. Sci., 5 (1970) 604
7. J. Kirchnerova and D. Klvana, "Hydrogen Energy Progress IX" Vol. 1 Proc. 9th World Hydrogen Energy Conf.; Paris, june 1992, p. 485-493
8. J.M. Herrmann "Les Techniques Physiques d'Etude des Catalyseurs"; B. Imelik and J.C. Vedrine Eds., Ed. Technip, Paris 1988, ch. 22
9. T. Nitadori and M. Misono, J. Catal., 93 (1985) 459
10 T. Le Van, M. Che and J. M. Tatibouët, Catal. Letters, 14 (1992) 321
11. T. Le Van, M. Che; J.M.Tatibouët and M. Kermarec, J. Catal., 142 (1993) 18
12. E.A. Lombardo, K. Tanaka and I. Toyoshima, J. Catal., 80 (1983) 340
13. V.G. Milt, R. Spretz, M.A. Ulla and E.A. Lombardo, Catal. Letters, 42 (1996) 57

NATURAL GAS CONVERSION V
Studies in Surface Science and Catalysis, Vol. 119
A. Parmaliana et al. (Editors)

THE USE OF METHANE IN MOLTEN CARBONATE FUEL CELLS

S. Freni, P. Staiti, G. Calogero, M. Minutoli

Istituto CNR-TAE, via Salita S.Lucia sopra Contesse 5, 98126 Santa Lucia, Messina, Italy.

The present paper is a summary of the most interesting results obtained during a wide investigation on the use of natural gas as fuel for molten carbonate fuel cells. This research concerns a theoretical evaluation about the feasibility and convenience of systems based on Molten Carbonate Fuel Cells (MCFC) with configurations of direct and indirect internal reforming. Furthermore, some considerations about the open problems on these systems have been made. At last, other two alternatives on the use of methane for MCFC have been considered: the partial oxidation and the autothermal reforming process. The conclusions report what future developments can be expected by these applications.

1. INTRODUCTION

The molten carbonate fuel cell (MCFC) systems represent one of the most flexible systems [1] for in-situ and stationary electrical power plants, because of their high electrical efficiency and possibility to supply hydrogen produced by different raw fuels. Anyway, the possibility to develop methane fuelled MCFC systems is a very attractive aim due to the several advantages correlated to the use of this fuel. In fact, it is known that the conversion of methane to synthesis gas (H_2/CO) by the catalytic steam reforming process is already well established [2]. In particular, the external and internal steam reforming of methane to synthesis gas are the main catalytic processes which produce fuel (hydrogen) for molten carbonate fuel cells, when natural gas is utilised [3].

In the first system, the methane is reformed in a reactor, separated from the MCFC station, that feeds the anode compartment of the cell. In the second process, the methane is directly reformed into synthesis gas in the anode compartment of the cell (IR-MCFC). Generally, the IR-MCFC is considered a more attractive system respect to the MCFC with external reforming of methane, nevertheless, some severe restrictions are present in the former. Recently, much attention has been paid to the study of catalytic methane partial oxidation to synthesis gas (CPOX) and on the use of autothermal reforming (ATR) applied to MCFC systems. These different ways to use natural gas as a fuel for MCFC present a multiplicity of factors that make difficult to ascertain which of them is the most convenient. The aim of the present paper is a description of the most interesting results obtained by a wide theoretical research carried out to investigate the peculiar aspects of the application of the different configurations and processes finalised to the use of methane in MCFC systems.

2. METHANE REFORMING CONFIGURATIONS FOR MCFC

In general, each type of fuel cells has to be supplied by a hydrogen rich gas mixture and requires a preliminary treatment of the raw fuel. At same way, the MCFC's anodes require hydrogen that will be converted following the half-reaction:

$$H_2 + CO_3^{-2} \leftrightarrow H_2O + CO_2 + 2e \tag{1}$$

In the case of methane, the plant has to be supplied by a proper section for the steam reforming of the fuel, with a subsequent enhancement of cost and complexity. Nevertheless, the MCFC operate at such a level of temperature (923 K) that is compatible with that of the steam reforming process. This peculiarity allows to remove the external reformer device and promises an improvement of the thermal balance of the system due to the direct use of the electrochemical heat released by the cell to sustain the steam reforming reaction. The general scheme of a MCFC with internal reforming foresees two different configurations, defined as direct internal reforming (DIR-MCFC) and indirect internal reforming (IIR-MCFC).

2.1 Direct internal reforming configuration

In the configuration of direct internal reforming, the fuel consists of a mixture of steam and methane, and it is directly supplied to the anode compartment that is formed by two adjacent zones: one provided by a proper catalyst for the steam reforming of the methane and the other containing anode for the electrochemical oxidation of the produced hydrogen. Between these two zones there is not any barrier to separate the gases, thus the hydrogen produced in the reformer section is in the meantime oxidised by the anode cell reaction and this shifts of the thermodynamic equilibrium towards further production of hydrogen. The overall process is represented by the following reactions:

$$CH_4 + H_2O \leftrightarrow CO + 3H_2 \qquad \Delta H_{923K} = +225.17 \text{ kJ mol}^{-1} \tag{2}$$

$$CO + H_2O \leftrightarrow CO_2 + H_2 \tag{3}$$

$$H_2 + CO_{2(c)} + \frac{1}{2}O_2 \leftrightarrow H_2O + CO_{2(a)} \qquad \Delta H_{923K} = -247.58 \text{ kJ mol}^{-1} \tag{4}$$

It is evident that in this configuration a complete thermal and chemical integration between the methane reforming and the cell reaction is realised. In fact, it is not necessary to supply heat to the steam reforming reaction, from an external source, because it will be sustained by the heat produced by the electrochemical reaction. The thermodynamic equilibrium of the involved reactions, is also influenced by the fact that hydrogen and steam participate or are produced in the reactions (2), (3) and (4).

This influence is evident on the mass and energy balances of the system. Several studies have been carried out to evaluate the influence of the operative parameters on the characteristics of the systems [4, 5], for a DIR-MCFC and IIR-MCFC, at working conditions reported in Tab.1.

Table 1
Input data for the mass/energy balance calculations

Parameters	Units	Values
Cell Temperature	K	923
Pressure	atm	1
Current Density	mA/cm^2	150
Fuel utilisation	%	60
Oxidant utilisation	%	36
Molar ratio steam to carbon (S/C)	mol/mol	2
Molar ratio O_2/CO_2	mol/mol	0.1672/0.2
Cell internal resistance	Ωcm^2	0.7

As reported in Fig.1, it is evident that an increase of pressure from 1 atm to 5 atm moves the methane conversion from 94.8% to 70.9%, if the cell releases a current density of 50 mA/cm^2, while the conversion remains almost constant if the current density is 200 mA/cm^2. The content of hydrogen produced by this process is reported in the diagram of Fig.2. It appears evident that when the cell is in open circuit condition the hydrogen production is independent by the configuration and it only depends by the pressure. On the other hand, when the current density is high, the influence of pressure on hydrogen production is less. For this reason, the hydrogen produced by this configuration is capable to sustain current densities so high as 220 mA/cm^2. The advantages that the DIR-MCFC configuration offers are so interesting that a lot of researches have addressed their activities to the development of such a system: nevertheless, some severe restrictions exist. The restriction concerns the tendency of the electrolyte to react and poison the internal reforming catalyst [6]. Another critical aspect concerns the difficulty to design bipolar plates without negative or positive peaks of temperature due to the local excessive preponderance of the reactions 2) or 4).

2.2 Indirect internal reforming configuration

In the configuration of indirect internal reforming the MCFC is supplied by a mixture of steam and methane, as for DIR-MCFC. The main difference is that the fuel is reformed on a catalyst placed in a section inside the cell hardware but physically separated by the anode compartment. After the reforming process, the hydrogen rich mixture produced is supplied to the anode compartment, where the hydrogen is oxidised. It is possible with this system to obtain the full thermal integration between the reforming and the cell reactions. It is settled that the maximum overall methane conversion is dictated by the thermodynamic equilibrium of the reactions (2) and (3), because there is not any removal of hydrogen during the evolution of the process. At standard MCFC conditions, the methane conversion obtainable corresponds to 81.6%, but this value decreases substantially when the pressure increases. As it is shown in Fig.2, the methane conversion lowers to 50% for pressure equal to 5 atm and it is not dependent by the current density.

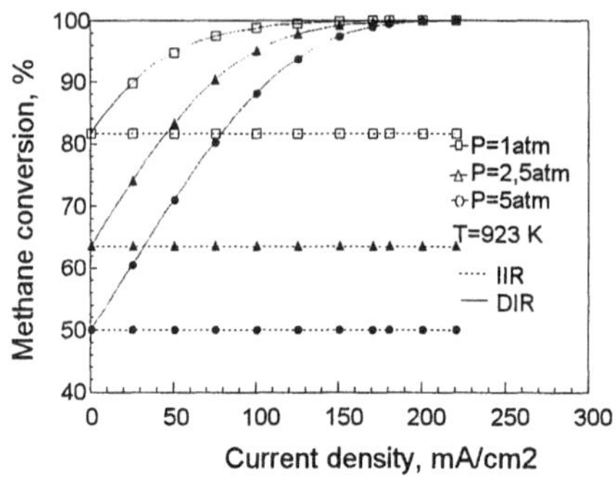

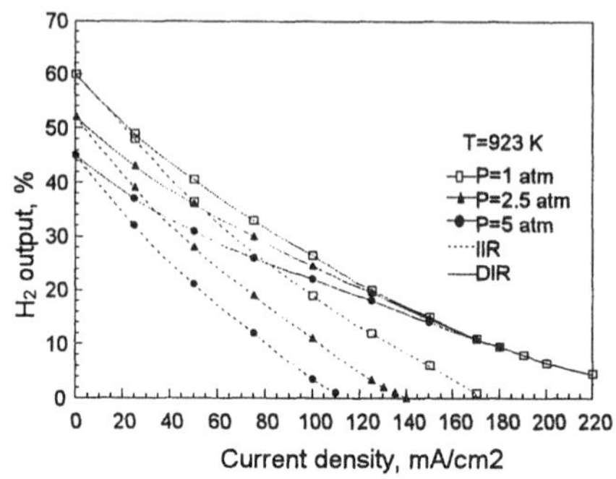

Figure 1 - CH_4 conversion Vs current density, at P=1-5 atm and for DIR and IIR.

Figure 2 - H_2 content Vs current density, at P=1-5 atm and for DIR and IIR.

Furthermore, in this configuration, the influence of the pressure on the methane conversion produces a restriction to the production of hydrogen (see Fig.2) and then to the available limit current. To compensate the limit to the current that can be realise by the cell, some attractive advantages exist like, safer operative conditions for the reforming catalyst, a simplest geometry and an easier management of the process gases with respect to DIR-MCFC.

3. CATALYTIC PARTIAL OXIDATION OF METHANE IN MCFC

An alternative route to the production of hydrogen is the catalytic partial oxidation of methane (CPOX) process. This process presents a series of advantages because it is slightly exothermic and the very fast reaction allows to operate at GHSV values up to one or two orders of magnitude higher than that experienced in the methane steam reforming (SRM). Furthermore, it is characterised by a low specific feed consumption, corresponding to 0.33 Nm^3 CH_4/Nm^3 H_2, against a requirement of 0.45 Nm^3 CH_4/Nm^3 H_2 for SRM.

The operative conditions are characterised by a stoichiometric defect of oxygen, therefore, the main reaction is the combustion of a fraction of the methane with a complete consumption of the oxygen and a high adiabatic increase of temperature, by the reaction:

$$CH_4 + 2O_2 \leftrightarrow CO_2 + 2H_2O \qquad \Delta H_{923K} = -886.16 \text{ kJ mol}^{-1} \qquad (5)$$

Then, the remaining quantity of CH_4 is reformed by both H_2O and CO_2 according to the endothermic reactions:

$$CH_4 + H_2O \leftrightarrow CO + 3H_2 \qquad \Delta H_{923K} = +225.17 \text{ kJ mol}^{-1} \qquad (2)$$

$$CH_4 + CO_2 \leftrightarrow 2CO + 2H_2 \qquad \Delta H_{923K} = +261.00 \text{ kJ mol}^{-1} \qquad (6)$$

The main limit to the oxyreforming gas phase process is the need to operate at high temperatures and pressures. Recently, [7] efforts have been concentrated to optimise a catalytic process, developing an appropriate oxyrefoming catalyst operating at atmospheric pressure and temperature above 973 K. On the other hand, theoretical calculations have been carried out, by the use of mathematical models, to predict the feasibility of this process integrated to a small size DIR-MCFC electricity generator (10 kW). It has been found that temperature, O_2/CH_4 ratio and electrical power released by the cell are important factors that influence the process.

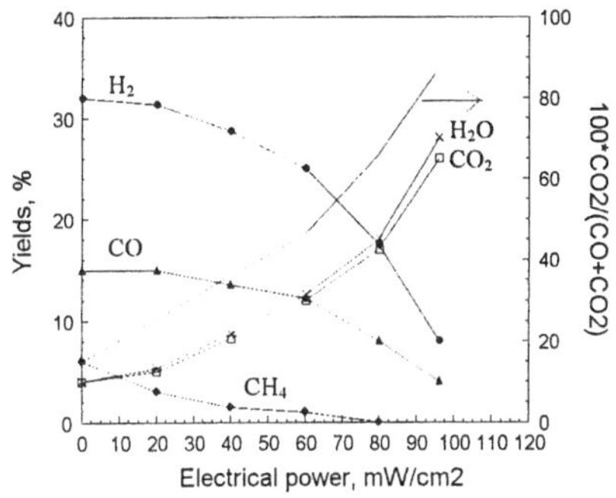

Fig 3 - Gas composition Vs Electr. power, calc. at P=1 atm, T=923 K, and O_2/CH_4=0.5.

In Fig.3, the diagram representing the CO_2 selectivity and the molar fractions of the exhaust anode gas as a function of the electrical power, for an integrated system CPOX-DIRMCFC, is reported. It is evident that the methane conversion is very high for any released electrical power level. In fact, in OCV conditions (W=0) the conversion is about 75% and it reaches the full conversions for W=80 mW/cm^2. This is confirmed also by the corresponding trend of the CO_2 selectivity that reaches the value of 87.3% at W = 97 mW/cm^2. The percentage of unconverted hydrogen is equal to 32% for W=0 but it drops when W increases.

4. AUTOTHERMAL REFORMING OF METHANE IN MCFC

Another process for the hydrogen production could be given by the integration of the steam and oxygen reforming reactions. In this case, the overall process, defined as autothermal reforming (ATR) is catalysed and the final products and energy balance are influenced by both reactions. As for the oxyreforming, the ATR operative conditions are characterised by a stoichiometric defect of oxygen and the composition of the products is controlled by both reactions 2), 3) and 5). The overall process becomes almost isothermal and thermodynamically suitable at MCFC's conditions. Following these considerations, a mathematical model [8] has been developed for carrying out a wide investigation to evaluate the theoretical convenience and feasibility to supply MCFC with ATR process. In particular, the study evidenced that the mass balance of the ATR process changes significantly when change the operative parameters like pressure, H_2O/CH_4 ratio, O_2/CH_4 ratio, temperature and cell current density. Substantially, the faster reaction is that of substoichiometric oxidation of the methane and thus the composition of the process products depends, at first, from the flow rate of inlet oxygen and then from the flow rate of steam. However, the sensitivity analyses carried out to determinate the influence of the inlet flow rates of oxygen and steam on the composition of the process products gave very interesting results. In the diagram shown in Fig. 4, we reported the curves representing the rate of hydrogen produced as a function of the H_2O/CH_4 molar ratio, calculated at different values of the O_2/CH_4 molar ratio for a determined flow rate of methane.

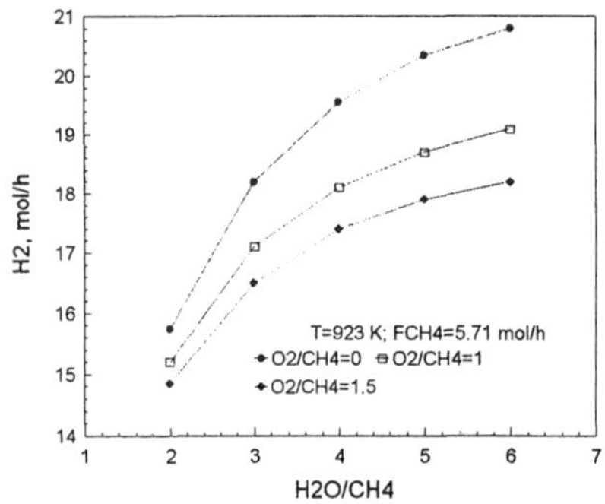

Fig 4 - H_2 flow Vs H_2O/CH_4 and O_2/CH_4 ratios.

It is evident that there is a larger production of hydrogen at a high steam/carbon ratio (for $H_2O/CH_4 \leq 6$) even if the energy process balance becomes endothermic and needs heat from an external source, as evidenced by the energy balance. An enhancement of the inlet oxygen flow rate will promote a further oxidation of methane with a corresponding lowering of hydrogen produced. But, in this case, the energy balance of the system becomes exothermic and excess of heat can be expected. As regard to the influence of the pressure and of the cell current density, it has been seen that both influence the system with mechanisms similar to that of the processes above described.

5. CONCLUSIONS

The investigations demonstrated that the methane can be considered as a valid fuel for molten carbonate fuel cells. Actually, different typologies of configurations and processes have been identified and studied. The traditional methane steam reforming seems to be successfully adaptable to the DIR-MCFC configurations, mainly if an alkali resistant catalyst will be developed. The IIR-MCFC configuration, also, shows very interesting features, as simple plant configuration and no problem of interaction between electrolyte and I.R. catalyst. The CPOX and ATR processes are under investigation, also. Between them, the ATR is very promising because it represents a good compromise, between the CPOX and SRM processes, and can be well integrated to the MCFC concept.

REFERENCES

1. M. A. Rosen, Int. J. Hydrogen Energy, 15, 267-274 (1990).
2. J. R. Rostrup-Nielsen, Catalysis, Science and Technology (Eds. Anderson J.R., Roudart M.), 5, 1-117, Springen, Berlin, (1984).
3) K. Kinoshita, F.R.McLarson and E.J. Cairns, "Fuel Cells, A Handbook", U.S. Dept. of Energy - Office of fossil Energy, (1988).
4) S. Freni, S. Cavallaro, M. Aquino and N. Giordano, J. of Power Sources, 39, 2, 203-214, (1992).
5) S. Freni, G. Maggio, Int. J. of Energy Research, 21, 253-264, (1997).
6) S. Cavallaro, S. Freni, R. Cannistraci, M. Aquino and N. Giordano, Int. journal of hydrogen, 17, 1, 181-186, (1992).
7) S.C. Tsang, J.B. Claridge, M.L.H. Green, Catalysis Today 1995,23,3.
8) S. Freni, S. Cavallaro, CNR-TAE internal report, Messina 1998.

NATURAL GAS CONVERSION V
Studies in Surface Science and Catalysis, Vol. 119
A. Parmaliana et al. (Editors)

LSM-YSZ catalysts as anodes for CH_4 conversion in SOFC reactor

Shizhong Wang, Yi Jiang, Yahong Zhang and Wenzhao Li[1]

Dalian Institute of Chemical Physics, Chinese Academy of Sciences, Dalian 116023, P.R.China

1. INTRODUCTION

Solid Oxide Fuel Cells (SOFC) is one of the most promising fuel cell systems for generating electricity in an efficient and enviroment friendly way[1]. Most of the present Fuel Cell Systems use hydrogen as fuel, which is the product of natural gas through endothermic steam reforming reaction. SOFC can directly run on methane and can, in principle, co-generate both electricity and useful chemicals[1-2].

Nickel cermet has been widely employed as fuel electrode materials for SOFC. However, the electrode activity shows a steady decrease due to the volatility and sintering of nickel and the effect of carbon deposition. Perovskite oxides, such as doped $LaMnO_3$ (LSM), are known for its high catalytic and electrocatalytic activity for the reduction of oxygen [3] and oxidation of hydrocarbons. On most of this kind of perovskite oxides, methane inclines to be oxidized to H_2O and CO_2, which can generate more electricity. They also can be modified by some kinds of additives to cogenerate electricity and useful partial oxidation products. The present paper is focused on the oxidation of methane on catalysts consisting of LSM and Yttria Stabilized Zirconia (YSZ) as anode electrodes in an SOFC reactor.

2. EXPERIMENTAL

The $La_{0.8}Sr_{0.2}MnO_3$ electrode material was prepared as reported previously[3]. The mixture of LSM and YSZ was used as the materials for both cathode and anode. The composite LSM-YSZ electrodes with various YSZ content were deposited on one side of YSZ slabs as anodes by screen printing method (80 mesh) and sintered at 1673 K for 2.5 h. For the counter and reference electrodes, 20 wt% YSZ+LSM and Pt were deposited on the other side. The areas of cathode, anode and reference electrode were 1.2, 1.0, and 0.1 cm^2 respectively.

The PEN structure YSZ slab was sealed to an open end YSZ tube (length = 16cm, diameter = 2cm) with glass seal. The YSZ+LSM anode was exposed to CH_4 or H_2, while the

[1] To whom correspondence should be addressed

cathode and counter electrode were exposed to ambient air. Pt grids were attached to the three electrodes and pressed by spring as current collectors. Three Pt wires were connected to the working, counter and reference electrode and led to a potentiostat/galvonostat. The effluent gas was analyzed with an on-line TCD type gas-chromatography after water-ice trapping.

The electrochemical measurements were made with a potentiostat/galvanostat (EG&G173) equipped with 376 interface and an EG&G175 universal programmer. AC impedance spectroscopy was carried out with an EG&G 5204 lock-in amplifier and a GFG-8019G function generator. The frequency range used in all AC experiments was 2-60 kHz, and the amplitude of the input sinuous signal was 7.4 mV.

3. RESULTS AND DISCUSSION

3.1. Power output and anodic polarization behavior of various LSM-YSZ anodes

The LSM-YSZ anodes were active for the electrocatalytic oxidation of methane. The current-voltage curves of various LSM-YSZ anodes showed that the composite electrode activity increased with increasing amount of LSM in the electrode in terms of maximum power output and the short circuit current(Fig. 1). The short circuit current of the 30 wt% YSZ+LSM anode was about 0.15 $A.cm^{-2}$ and the maximum power density was about 0.05 $W.cm^{-2}$. The open circuit potential (OCP) of various LSM-YSZ anodes were about 0.95 V, less than the theoretical value (≈1.3 V). This might be due to the high oxygen activity on LSM-YSZ electrodes.

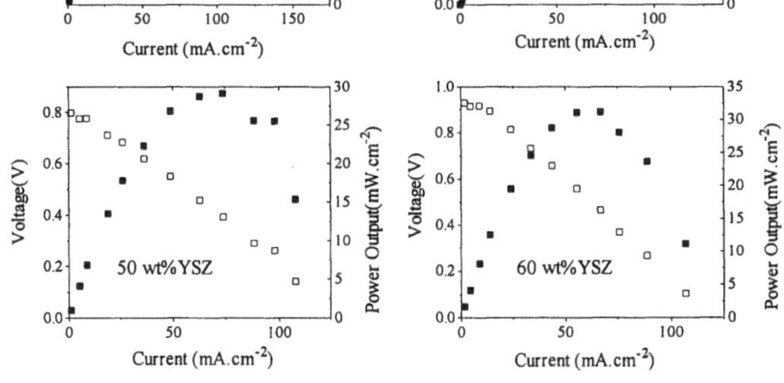

Figure 1. Power output and current-Voltage performance of various LSM-YSZ anodes at 1223 K running on methane at 60 $ml.min^{-1}$.

To further investigate the performance of various LSM-YSZ anodes, the anodic polarization experiments were carried out in methane and helium respectively. The anode performance increased with the increase of LSM content in the anode. Polarization resistances were calculated from the slopes of these curves in high potential region, as shown in Table 1.

Table 1
The polarization resistances, ohmic resistance, electrochemical resistance of various LSM-YSZ anodes at 1223 K

Samples	30YSZ-LSM	40YSZ-LSM	50YSZ-LSM	60YSZ-LSM
R_{He} (ohm)	4.2	4.7	8.6	31
R_{0He} (ohm)	3.1	2.9	6.0	9.0
R_{eHe} (ohm)	1.1	1.8	2.6	22
R_{CH4} (ohm)	4.1	4.9	5.5	9.9
R_{eCH4} (ohm)	3.3	4.1	4.7	9.0

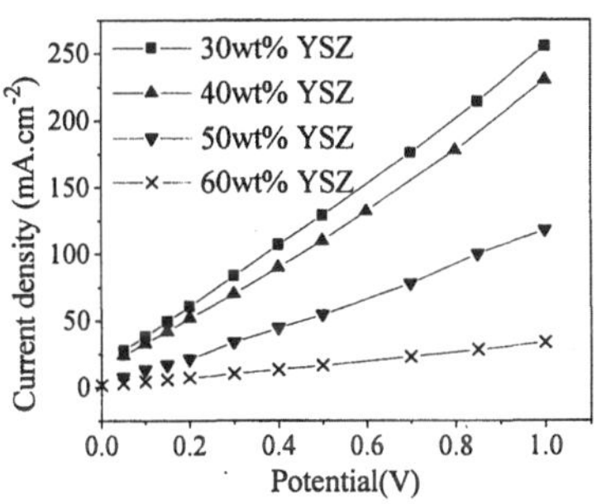

Figure 2. Current-potential curves of various LSM-YSZ anodes in He at 1223K.

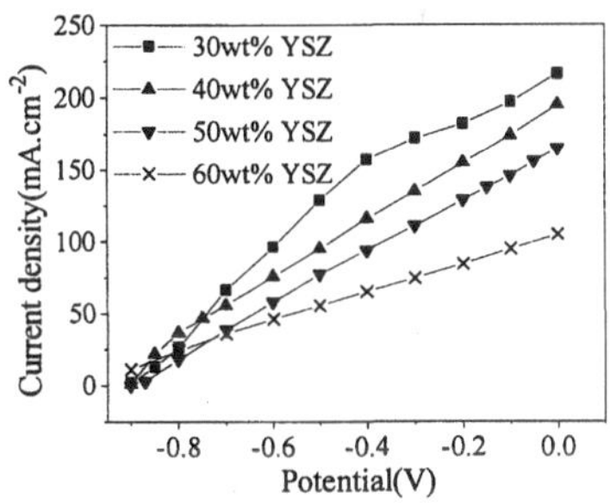

Figure 3. Current-potential curves of various LSM-YSZ anodes in methane at 1223K.

AC impedance experiments were performed to determine the ohmic resistances (the electrolyte and interfacial resistances) of various anodes at 1223K in helium and methane respectively. While exposed to helium, the ohmic resistances increased with increasing amount of YSZ in anodes as shown in Table 1. The electrolyte and interfacial resistances of varoius LSM-YSZ anodes were all about 0.85 ohm while exposed to methane under open circuit condition for 1 hour, which means that the carbon deposition on the electrode determines this resistance.

R_{He}, R_{0He} , R_{eHe}, stand for the total polarization resistance, the electrolyte and interfacial resistance, electrochemical resistance in helium respectively, and R_{CH4}, R_{eCH4} for the total resistance and electrochemical resistance in methane. It can be seen from table 1 that the electrochemical resistance in both helium and methane increased with increasing amount of YSZ in LSM -YSZ electrodes, which demonstrated that the LSM might be the active components for oxygen reduction and methane oxidation. The electrochemical resistances of LSM-YSZ electrodes in methane were higher than that in helium. This might be due to the reduction of methane, which changed the oxygen evolution performance of LSM-YSZ electrode.

3.2. Methane oxidation on 40wt%YSZ+LSM anode

The performance of 40 wt%YSZ-LSM anode under various methane flow rates and methane partial pressures were investigated at 1223K. The performance was positively dependent on the decrease of the flow rate (150-30 $ml.min^{-1}$) or the increase of P_{CH4} (10%-100%) as shown in Fig. 4, Fig. 5. These characteristics have great significance for practical application, since the low feed rate and high feed concentration can reduce the excess heat wasted by the effluent gas and increase the fuel utilization and energy conversion efficiency.

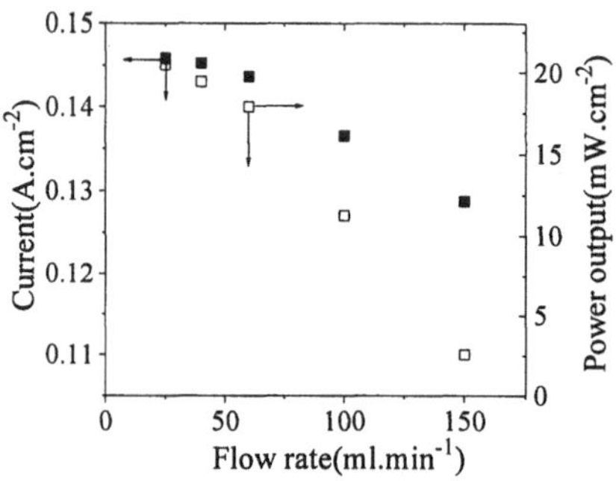

Figure 4. Current-voltage performance of 40 wt%YSZ+LSM anode under various methane flow rates at 1223 K.

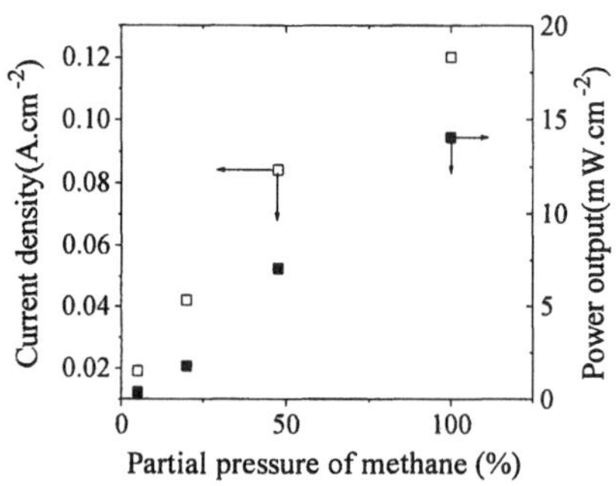

Figure 5. Current-voltage performance of 40 wt%YSZ+LSM anode under various methane partial pressures at 1223 K.

The CO and CO_2 selectivity were 30 and 70 % respectively. This result showed that this electrode inclines to oxidize CH_4 to CO_2 and H_2O, which can generate more electricity. When the reactant mixture (20 $ml.min^{-1}$ O_2 and 40 $ml.min^{-1}$ CH_4) was introduced into SOFC reactor, the production distribution was similar to that for pure methane, thus the avtive oxygen species might related to adsorbed atomic oxygen.

Carbon deposition is the main factor affecting the performance of SOFC anodes. To this end, we investigated the effect of coking on the performance of 40 wt%YSZ+LSM anode. Under open circuit condition, methane was introduced to the anode for various times, then the anode was subjected to polarization at zero volt in helium and methane respectively.

The current transients in helium after various coking times, showed a sharp decrease (Fig. 6 (a)), the longer the coking time, the higher the current transients, which mean that these current transients corresponded to the electrochemical combustion of deposited carbon. The current transients in methane showed some difference from those in helium as shown in Fig. 6

(b). When the coking time was less than 5 minutes, the current transients increased quickly to a stable level during the initial several minutes. Yet the current transients after coking for 15 and 45 minutes showed the same pattern as those in helium, i.e. a sharp decrease first, then a steady decrease to a limiting current.

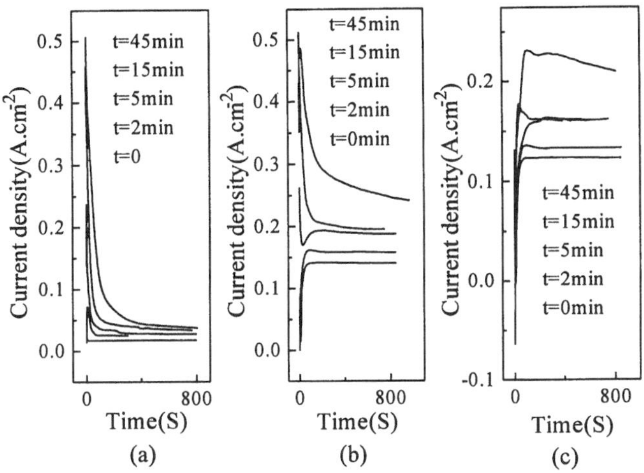

Figure 6. Current transients after coking for various time in helium (a), Methane (b), and the difference between them (c).

The difference between the current transients in methane and helium corresponded to the current due to methane oxidation. It can be seen that the longer the coking time, the better the anode performance, which demonstrated that the carbon deposition on LSM-YSZ composite electrodes can promote methane oxidation significantly. From Fig. 6 (c), the total resistance after coking for various time can be estimated by the ratio of open circuit potential and short circuit current. The results calculated assuming that E_{OCP} was about 0.95 V were listed in Table 2, which showed that the total resistances decreased with increaing coking time. The ohmic resistance difference between samples in helium and samples coking for 1 hr was 2.1 ohm. Though the ohmic resistance difference between samples coking for various time should be less than 2.1 ohm, the decrease of ohmic resistance still may be one of the main reasons leading to the improvement of the performance with increasing coking time. The resistance difference between samples without pre-coking and that coking for 45 minutes was almost 3.3 ohm, far higher than 2.1 ohm. This result demonstrated that the improvement of the anode performance after carbon deposition was not only due to the decrease of interfacial resistance, but also to the new active sites for methane oxidation formed after carbon deposition, i.e. the carbon deposited on LSM-YSZ anodes may be the active sites for methane oxidation.

Table 2
The total resistance after coking for various time

coking time (min)	0	2	5	15	45
resistance (ohm)	7.7	7.2	5.8	5.9	4.2

3.3. The performance in H_2

The performance of LSM-YSZ anodes in H_2 was not so high as that in methane. The performance increased with increasing amount of LSM in LSM-YSZ anodes, due to the decrease of ohmic resistance.

The maximum power output and short circuit current had strong dependency on hydrogen partial pressure and the path to reach it as shown in Figure 7. The power output for the decreasing P_{H2} direction is lower than that for the increasing direction. The ohmic resistance had the same hysteresis pattern as that in Figure 7, which demonstrated that the hysteresis of the performance was due to the hysteresis of interfacial structure. The hysteresis pattern also showed that the current through YSZ can stabilize the LSM phase.

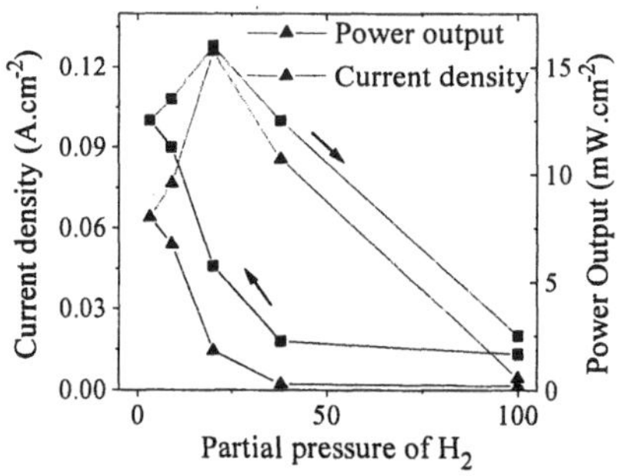

Figure 7. Current density and power output under various partial pressure of hydrogen at 1223K.

4. CONCLUSION

1. LSM-YSZ composite anodes were active for the electrocatalytic oxidation of methane to CO, H_2, CO_2 with simultaneous generation of electricity. The electrode performance increased with the amount of LSM in LSM-YSZ composite electrodes.
2. The increase of the performance of LSM-YSZ composite electrode after carbon deposition was attributed to the decrease of both the ohmic resistance and electrochemical resistance.
3. The performance of LSM-YSZ anode under various P_{H2} showed significant hysteresis, which was due to the hysteresis of interfacial strucure under reducing conditions.

REFERENCES

1. H. Uchida, N. Mochizuki and M. Watanabe, J. Electrochem. Soc., 143 (1996), 1700.
2. Y. Hiei, T. Ishihara and Y. Takita, Solid State Ionics, 86-88 (1996), 1267.
3. S. Wang, Y. Jiang, Y. Zhang, J. Yan, W. Li, J. Electrochem. Soc., 145(6) (1998), 1932.

NATURAL GAS CONVERSION V
Studies in Surface Science and Catalysis, Vol. 119
A. Parmaliana et al. (Editors)

Catalytic combustion of methane over transition metal oxides.

S. Arnone[a], G. Bagnasco[a], G. Busca[b], L. Lisi[c], G. Russo[c], M. Turco[a].

[a] Dipartimento di Ingegneria Chimica, Università "Federico II", Napoli, Italy.

[b]Istituto di Chimica, Facoltà di Ingegneria, Università di Genova, Genoa, Italy.

[c] Istituto di Ricerche sulla Combustione, CNR, Napoli, Italy.

Simple and mixed metal oxides containing Co, Mn, Cr and Fe have been investigated as catalysts for the combustion of methane in the temperature range 300-600°C under diluted conditions. The effect of the catalyst composition on the catalytic performances and on the redox properties has been evaluated. Single metal oxides containing Cr, Co and Mn show comparable activities and were found more active than Fe_2O_3. Mixing of Co with Cr oxide and Fe oxide with Mg and Zn to give spinels, improves the catalytic activity with respect to pure compounds. Temperature programmed reduction (TPR) shows that redox properties are strongly dependent on the catalyst composition. Fe based mixed oxides are more hardly reducible than the other catalysts, this effect being related to the dilution of Fe with bivalent cations. The comparison of the kinetic parameters, evaluated on the base of a first order rate equation, gave evidence of a correlation between the activation energy values and the ease of the reduction showing that the oxides reducible at lower temperature give rise to a reaction mechanism with a lower activation energy.

1. INTRODUCTION

Catalytic combustion has been proposed as an alternative technique to the homogeneous combustion for several applications like as gas turbines, boilers, aircrafts, afterburners, domestic heaters, VOC removal. The process can be carried out in a wide range of fuel/air ratio at low temperature thus leading to a marked reduction of NO_x emission levels [1]. Different systems are reported in literature as active catalysts in methane combustion such as noble metals [2] and perovskite oxides [3, 4]. Nevertheless, less attention has been paid to transition metal oxides with different structures such as spinels, mostly because some of them are instable at high temperature. Nevertheless, some spinels are very active and the study of the factors affecting their activity can give light on the mechanism involved in total combustion catalysis [5]. In this paper different transition metal oxides have been studied as methane combustion catalysts. The effect of the partial substitution of the metal cation was also investigated. TPR technique was employed for the characterization of the catalysts with the aim to define a correlation between redox properties and catalytic activity.

2. EXPERIMENTAL

Single and mixed transition metal oxides were prepared by precipitation, except Zn and Mg ferrites that were prepared as aerogels and Co_3O_4 and γ-Fe_2O_3 that were commercial materials.

XRD analysis was performed using a Philips PW 1710 diffractometer. Specific surface area of catalysts was determined by N_2 adsorption at 77K according to the BET method using a Carlo Erba 1900 Sorptomatic apparatus. Temperature programmed reduction (TPR) experiments were carried out in a Micromeritics 2900 TPD/TPR flow system equipped with a TCD. After treatment in air flow at 600°C samples were reduced with a 2% H_2/Ar mixture (25 cm^3 min^{-1}) at heating rate of 10°C min^{-1} up to 600°C.

Catalytic combustion of methane was studied in a fixed bed quartz micro-reactor. The catalyst (particle size = 300-400μm), diluted 1:10 in quartz powder, was placed on a frit disk. Quartz pellets upside the catalytic bed and a narrowing of the reactor section both in the post-catalytic and pre-catalytic zone reduced the homogeneous volume. A thermocouple placed in the catalytic bed, allowing the monitoring of the temperature during the reaction, showed that the maximum temperature gradient was lower than 5°C.

The feed composition was 0.4% CH_4 and 10% O_2 in a balance of N_2. A constant space velocity of 40000 cm^3 h^{-1} g^{-1} was ensured by Brooks 5850 TR Series mass flow controllers. Catalytic tests were carried out in the temperature range 300-600°C.

The concentration of reactants and products was measured using a Hewlett Packard 6890 gas-chromatograph equipped with two capillary columns (a poraplot Q and a molecular sieve 5A) and thermal conductivity and flame ionization detectors. Carbon balance was verified within ± 5 %.

3. RESULTS AND DISCUSSION

3.1. Physico-chemical characterization

The crystalline phases and the values of the specific surface area of simple and mixed oxides are reported in Table 1.

Table 1
XRD phases, specific surface areas and thermal behaviour of the catalysts.

Catalyst	XRD phase	Specific surface area (m^2 g^{-1})	Thermal behaviuor
Co_3O_4	normal spinel	15	to CoO at T>450°C
Mn_3O_4	random spinel	24	to a-Mn_2O_3 at T=400°C to Mn_3O_4 at T=970°C
Cr_2O_3	corundum	18	thermodynamically stable
γ-Fe_2O_3	non stoichiometric spinel	22	to a-Fe_2O_3 at T>650°C
α-Fe_2O_3	corundum	102	thermodynamically stable
$CoCr_2O_4$	normal spinel	110	thermodynamically stable
$ZnFe_2O_4$	normal spinel	27	thermodynamically stable
$Mg_{0.5}Zn_{0.5}Fe_2O_4$	random spinel	37	thermodynamically stable
$MgFe_2O_4$	inverted spinel	56	thermodynamically stable

All samples show the spinel structure except Cr_2O_3 and one of the ferric oxides that crystallize in the corumdum structure. Simples oxides have a surface area of about 20 m^2 g^{-1}, except α-Fe_2O_3, whereas higher values of surface area are shown by the mixed oxides. For Fe based oxides, the dilution with Mg leads to a marked enhancement of the surface area of the simple oxide, a lower increase is related to the introduction of the Zn cation. TG/DTA experiments show that all samples are structurally stable up to 1000°C except Co_3O_4, Mn_3O_4 and γ-Fe_2O_3 (Table 1).

TPR profiles of simple oxides are reported in Figure 1. TPR curve of Cr_2O_3 sample has not been reported due to the very low H_2 uptake compared to that of the other simple oxides.

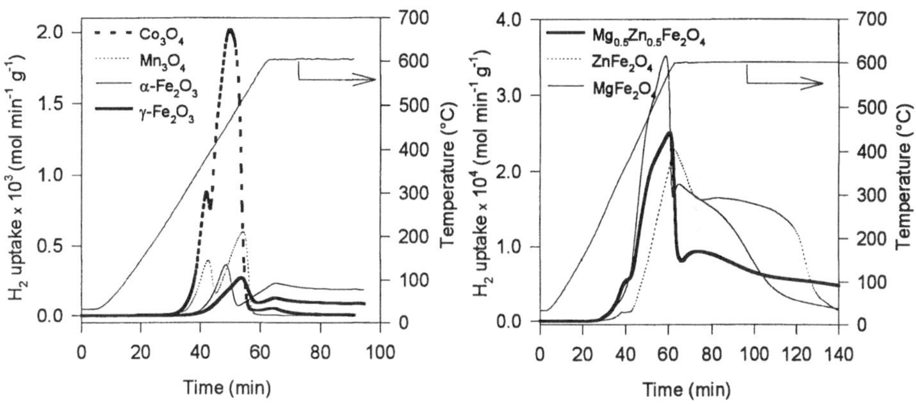

Figure 1. TPR profiles of simple and mixed oxides.

In Table 2 the H_2 consumed in the TPR experiments, the onset temperature (T_{onset}) and the temperature corresponding to the maximum uptake (T_{max}) are reported for all samples.

Table 2
Results of TPR experiments.

Catalyst	H_2 uptake (mol H_2 mol^{-1} M*)	T_{onset} (°C)	T_{max} (°C)
Co_3O_4	1.2	264	382, 470
Mn_3O_4	0.5	187	385, 520
Cr_2O_3	0.025	182	295, 470
γ-Fe_2O_3	0.3	373	448, 535
α-Fe_2O_3	0.3	336	460, 520
$CoCr_2O_4$	0.023	185	255
$ZnFe_2O_4$	1.0	284	-
$Mg_{0.5}Zn_{0.5}Fe_2O_4$	0.6	236	-
$MgFe_2O_4$	0.8	248	-

* M in mixed oxides refers to the total metal content.

The reduction occurs in two or more steps for all oxides starting at quite low temperature for Co_3O_4, Mn_3O_4 and Cr_2O_3 simple oxides and is complete within 600°C. For the two Fe_2O_3 samples the shift of the baseline at 600°C suggests that the the reduction is still continuing isothermally at this temperature. The extent of the reduction is markedly affected by the metal cation only cobalt undergoing a deep reduction. The values of H_2/M ratio suggest that Co cations in Co_3O_4 undergoes the complete reduction to Co^0. The presence of metallic Co was confirmed by XRD analysis carried out after the TPR experiment. Thus, the first peak could correspond to the reduction from the average oxidation state 2.7+ to 2+ and the second one to the reduction from 2+ to metallic oxidation state. The XRD spectra taken on the Mn_3O_4 after TPR experiment show the signal of MnO phase, suggesting that manganese is reduced to 2+ oxidation state. Taking into account this result an average Mn initial oxidation state of 3+ can be evaluated from H_2 consumption, then higher than that expected from the stoichiometry of the compound. This suggests that the oxidation to Mn_2O_3 can occur during the pretreatment, in agreement with literature data reporting that Mn_3O_4 undergoes the transition to Mn_2O_3 in oxidizing atmosphere at about 600°C [6]. A further confirmation was also obtained by XRD analysis effected on the sample after the first TPR peak, showing the signals of Mn_3O_4 phase. For Cr_2O_3 sample the very low extent of the reduction make uncertain the determination of the exact stoichiometry of the final compound. Finally, both Fe_2O_3 samples are reduced to Fe_3O_4 as suggested by the value of H_2/M ratio. In $CoCr_2O_4$ sample the H_2 uptake is strongly reduced with respect to Co_3O_4 sample and is very close to that observed for Cr_2O_3. Moreover, a shift of T_{max} in compared to the that of pure compound was observed. A different behaviour was shown by Fe based mixed oxides that need higher temperatures to activate the reduction. As shown by TPR profiles reported in Figure 1 the reduction shows the maximum rate at temperatures approaching 600°C, therefore higher than the other catalysts, and continues isothermally at this temperature. This suggests that the dilution of Fe with lower valence cations makes the mixed oxides more hardly reducible even if the extent of the Fe reduction increases compared to the simple Fe oxides. It can be supposed that Mg and Zn, being stable in 2+ oxidation state, do not undergo reduction therefore the H_2 uptake can be due to the reduction of Fe cation only.

After TPR experiments the samples were treated in air flow at 600°C and reduced again under the same conditions of the first experiments. The reduction-oxidation process was found reversible for all oxides except for Fe based sample. In this case a shift of T_{onset} and T_{max} and a modification of the intensity of the signals were observed.

3.2. Catalytic activity tests

Preliminary tests, performed under the same conditions of the catalytic tests, but without catalyst, showed that homogeneous reactions are negligible under the experimental conditions investigated. The results of the catalytic activity tests are reported in Figure 2.

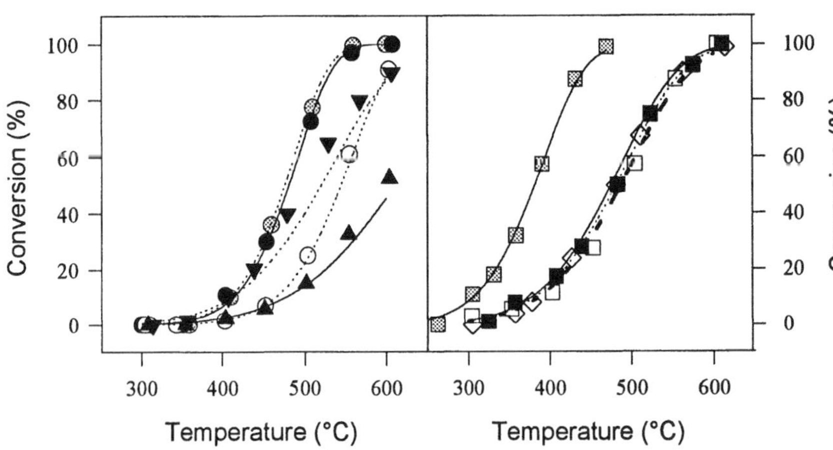

Figure 2. CH_4 conversion as a function of temperature for γ-Fe_2O_3 (▲), α-Fe_2O_3 (▼), $MgFe_2O_4$ (●), $ZnFe_2O_4$ (○), $Mg_{0.5}Zn_{0.5}Fe_2O_4$ (◍), Mn_3O_4 (◆), Cr_2O_3 (□),Co_3O_4 (■), $CoCr_2O_4$ (▩).

All catalysts, except γ-Fe_2O_3, give complete conversion of methane within 600°C with 100% selectivity to CO_2. The catalysts are able to activate the reaction in a temperature range lower than that of perovskite oxides [3] and comparable to that of noble metals [7]. Cr_2O_3, Co_3O_4 and Mn_3O_4 show a comparable activity. Fe_2O_3 is the less active when it crystallizes in a non stoichiometric spinel structure. The substitution of the metal cation enhances the activity of both Fe and Cr based catalysts. In the Fe based oxides, the mixing with Mg oxide gives rise to a larger effect compared to that due to mixing with Zn. The ternary system, obtained by the partial substitution of Zn with Mg, has an activity higher than $ZnFe_2O_4$ but comparable to that of $MgFe_2O_4$.

After a first cycle of tests all catalysts were cooled down to room temperature and a new cycle of experiments was performed. The results of the second cycle were the same of the first one for all catalysts except for Fe based samples that gave rise to some loss of activity suggesting that these oxides undergo a deactivation under the reaction conditions.

Catalytic activity data were elaborated assuming a methane first order rate equation [8] and a plug flow integral reactor. CH_4 conversions ranging from 10 to 90% were used to evaluate the values of activation energy and preexponential factor reported in Table 3.

The activation energy is about 20 Kcal mol^{-1} for Mn, Co, Cr and Fe single oxides and for $CoCr_2O_4$, however, a higher value of the activation energy was evaluated for Fe mixed oxides. This result suggests that the dilution of Fe^{3+} with a bivalent cation can modify the mechanism of methane activation. The higher activity of $CoCr_2O_4$ catalyst (Figure 2) can thus be due to the greater value of the surface area shown by this sample as can be demonstrated by the value of preexponential factor referred to the catalyst specific surface comparable to that of Mn, Co and Cr simple oxides. The comparison of the preexponential factors of simple oxides suggests that Fe_2O_3 oxides exhibit the lowest surface sites concentration. Likewise, the best catalytic performances of α-Fe_2O_3 in respect with γ-Fe_2O_3 could be associated to the higher surface area of our corundum type sample more than to an effect of the different structure of this oxide. Fe mixed oxides show the highest activation energy value despite of their catalytic activity is comparable to that of other catalysts, and significantly higher than that of γ-Fe_2O_3. This effect is due to the higher values of preexponential factors referred to surface area induced by the Fe dilution with Mg or Zn..

Table 3
Activation energy (Ea) and preexponential factor (A).

Catalyst	Ea (Kcal mol^{-1})	A x 10^{-8} (l h^{-1} g^{-1})	A x 10^{-8} (l h^{-1} m^{-2})
Co_3O_4	20	0.4	0.029
Mn_3O_4	20	0.4	0.017
Cr_2O_3	20	0.5	0.027
γ-Fe_2O_3	20	0.07	0.003
α-Fe_2O_3	20	0.2	0.002
$CoCr_2O_4$	20	2.9	0.027
$ZnFe_2O_4$	30	80	3
$Mg_{0.5}Zn_{0.5}Fe_2O_4$	30	395	11
$MgFe_2O_4$	30	350	6

On the base of the above results a correlation between the catalytic activity and the redox behaviour can be drawn. If a relationship between the extent of the reduction seems to be excluded, a correlation between the ease of reducibility and the activation energy appears quite reasonable. Catalysts which are reduced within 600°C show the same value of activation energy. By contrast, mixed Fe based catalysts whose reduction is delayed, as the maximum H_2 uptake occurs at temperatures approaching 600°C and the process continues isothermally, show the same activation energy value, higher then that of the previous materials. This suggests that the availability of the surface lattice oxygen significantly affects the catalytic properties in activating methane oxidation. It is reported that the catalytic activity in the total oxidation of methane is strongly related to the oxidation properties of the catalysts, the surface oxygen being involved in the reaction mechanism [1]. Moreover, the disactivation observed in both reduction with H_2 and CH_4 oxidation processes for Fe based oxides gives a further confirmation of the correlation between redox and catalytic properties, suggesting that the reversibility of the reduction process is an important feature for catalysts that could be employed in the catalytic combustion of methane.

4. CONCLUSIONS

Simple and mixed oxides activate the oxidation of methane in a temperature range comparable to that of noble metals and lower than of perovskite oxides ensuring a 100% selectivity towards the total oxidation products.

All simple oxides catalyse the methane oxidation activating the same reaction mechanism not depending on the nature of the transition metal. They show a comparable density of active sites except Fe_2O_3 oxides which have a lower concentration of surface sites. The partial substitution of Cr^{3+} with Co^{3+} leads to an increase of catalytic activity attributed to the enhancement of the specific surface area. On the contrary, the dilution of Fe with bivalent cations results in a different reaction mechamism and, at the same time, in an increase of surface sites concentration.

The evaluation of the redox properties by TPR analysis showed a close correlation between the range of temperature in which the reduction occurs and the activation energy of methane oxidation estimated for the metal oxides catalysts.

REFERENCES

1. M. F. M. Zwinkels, S. G. Järas, P. G. Menon and T. A. Griffin, Catal. Rev. Sci. Eng., 35 (1993) 319.
2. R.Prasad, L.A. Kennedy and E. Ruckenstein, Catal. Rev. Sci. Eng., 26(1) (1984) 1.
3. L.G. Tejuca, J.L.G. Fierro and J.M.D. Tascon, Adv. Catal., 36 (1989) 37.
4. P. Ciambelli, L. Lisi, G. Minelli, I. Pettiti, P. Porta, G. Russo and M. Turco, Proceedings of 3rd World Congress on Oxidation Catalysis, San Diego, 1997.
5. R.Prasad, L.A. Kennedy and E. Ruckenstein, Comb. Sci. Tec., 22 (1980) 271.
6. M. Baldi, E. Finocchio, F. Milella and G. Busca, Appl. Catal.: B. Environ., in press.
7. R. Burch and P. K. Loader, Appl. Catal. B: Environ., 5 (1994) 149.
8. H. Arai, T. Yamada, K. Eguchi and T. Seiyama, Appl. Catal., 26 (1986) 265.

NATURAL GAS CONVERSION V
Studies in Surface Science and Catalysis, Vol. 119
A. Parmaliana et al. (Editors)
1998 Elsevier Science B.V.

High temperature combustion of methane over hexaaluminate-supported Pd catalysts.

G. Groppi[1], C. Cristiani[1], P. Forzatti[1], F. Berti[2] and S. Malloggi[3]

[1]Dipartimento di Chimica Industriale e Ingegneria Chimica "G. Natta" - Politecnico di Milano, Piazza Leonardo da Vinci 32 - 20133 Milano - Italy
[2]ENEL-CRAM, Via A. Volta 1, 20093 Cologno Monzese (MI) - Italy
[3]ENEL Ricerca Polo Termico, Via A. Pisano 120, 50122 Pisa - Italy

The methane combustion properties of barium hexaaluminate supported Pd catalysts are compared with those of an alumina supported one. Combustion tests over catalysts calcined at 1000°C show that hexaaluminate-supported systems posses lower activity than the alumina-supported one. Tests performed upon treatment under reaction conditions show that marked deactivation occurs to the hexaaluminate based materials.

XRD characterization indicate that the lower combustion activity of barium hexaaluminate-based catalysts is associated with the lower dispersion of Pd species on these supports.

1. INTRODUCTION

Catalytic combustors for gas turbines [1] have recently reached near-commercial development as the most effective method for simultaneous reduction of NO_x, CO and unburned hydrocarbon emissions [2]. In these devices the use of Pd-based catalysts is mandatory in view of the following reasons: i) Pd catalysts exhibit the highest activity in CH_4 combustion [3], so that they posses a unique ability to ignite natural gas at low inlet temperatures and short residence times typical of gas turbine operations; ii) all the relevant Pd species (metal, oxide and hydroxides) in the reaction medium exhibit negligible volatility below 1000°C [4]; iii) reversible PdO (active) $\Leftrightarrow$ Pd° (less active) transformation [5] results in a self-regulation of the catalyst temperature that is useful to reduce the thermal stresses of the catalyst [6].

The choice of the support can greatly affect the behaviour of the Pd based catalysts. For instance the use of zirconia based material has been reported to provide better control of the catalyst temperature with respect to stabilised alumina [7]. As a general matter, supports for noble metal catalysts should exhibit stable surface area and phase composition to limit encapsulation and sintering of the active phase. With respect to this the use of hexaaluminate-type materials as supports for Pd catalysts has been proposed in the literature [8]. Indeed these materials exhibit excellent thermal stability properties that make them promising for this scope. Thermal stability is related to their peculiar layered structure originating from the stacking of spinel blocks containing Al^{3+} separated by mirror planes in which large earth alkaline (Ba and Sr) and rare earth (La) cations are located. This structure, that is stable up to

1600°C, is able to suppress diffusion of large cations along the stacking direction thus preventing high temperature sintering. Surface areas of 15-20 m^2/g upon calcination at 1300°C have been obtained [9].

In this work the methane combustion activity of hexaaluminate-supported Pd catalysts ($Pd/BaAl_{12}O_{19}$ and $Pd/BaMnAl_{11}O_{19}$) has been investigated in comparison with that of an alumina-supported Pd catalyst, aiming at assessing the potential of hexaaluminate-type materials as Pd supports. Characterization by DTA-TG and XRD measurements has also been performed to rationalize the observed catalytic properties.

2. EXPERIMENTAL

2.1 Preparation.

$BaAl_{12}O_{19}$ and $BaMnAl_{11}O_{19}$ have been prepared by coprecipitation in water using $(NH_4)_2CO_3$, as precipitating agent and the nitrates of the components as precursor materials [10]. The final supports have been obtained upon calcination of the dried precursors at 1300°C for 10 h. Both these supports present a monophasic composition consisting of a $BaAl_{12}O_{19}$-type phase with a Ba-β-Al_2O_3 structure [10] and a surface area of 15 and 17 m^2/g for $BaAl_{12}O_{19}$ and $BaMnAl_{11}O_{19}$ respectively.

Al_2O_3 support has been prepared by calcination of a pseudobohemite precursor of 250 m^2/g at 1000 °C for 10 hours. A final material consisting of a mixture of θ- and α-Al_2O_3 phases with a surface area of 100 m^2/g has been obtained.

Deposition of palladium has been performed starting from a $PdCl_2$ precursor according to a wet procedure described elsewhere [11]. This method was demonstrated to allow for direct elimination of chlorides from the surface during the deposition reaction that occurs via the surface hydrolisis of the aquo-chloro Pd complexes.

All the catalysts have been calcined at 1000°C for 20 h after Pd deposition before activity tests and characterization analyses.

2.2 Characterization

XRD analyses have been performed using a Philips PW 1050-70 instrument with a Ni-filtered Cu-Kα radiation. Mean crystallite dimensions of PdO [JCPDS 6-515] have been calculated by Sherrer equation [12] from full width at half maximum (FWHM) of the (101) reflection, evaluated by profile fitting routine.

Surface area measurements have been obtained by a Fison Sorptomatic 1900 instrument using the N_2 adsorption technique.

DTA-TG measurements have been performed by a simultaneous TG-DTA 6300 Seiko instrument. The following experimental parameters have been used: atmosphere air (200 ml/min), heating and cooling rate 10°C/min. Three cycles of heating and cooling in the 450-1000°C have been performed.

2.3 Activity tests

Combustion tests have been performed over powder catalysts with small particle size (d_p=0.1 mm). Quartz powder with the same particle size has been added for dilution (V_{cat}/V_{dil}=2/1). Small particle size and dilution allowed for the suppression of external and internal heat and mass transfer limitations. A catalyst amount of 0.45g has been loaded in a microreactor equipped with a sliding thermocouple. The reactor has been fed with 1% CH_4 in air at GHSV=54,000 Ncc/g_{ctz}h. Analysis of products and reagents has been performed by on line GC. Further details are reported elsewhere [10].

3. RESULTS

3.1 CH_4 combustion activity

Activity tests have been performed over $BaAl_{12}O_{19}$, $BaMnAl_{11}O_{19}$ and Al_2O_3 loaded with 0.9% (w/w) of Pd. These samples will be referred in the following as 0.9Pd-BHA, 0.9Pd-BMHA and 0.9Pd-A respectively.

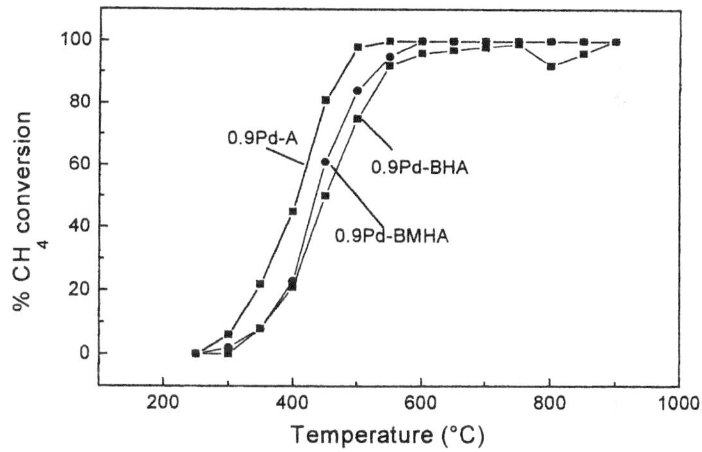

Figure 1. Results of methane combustion tests over calcined catalysts

The experimental conversion curves are compared in Figure 1. Except for a slight enhancement of conversion in favor of 0.9Pd-BMHA, the two samples with hexaaluminate-type supports provide similar performances below 700°C. On the other hand the alumina supported sample shows higher conversion at any temperature below 500°C. As an index for low temperature activity, $T_{10\%}$ (temperature at which 10% conversion is obtained under the adopted experimental conditions) of 315°C, 355°C and 360°C have been obtained for 0.9Pd-A, 0.9Pd-BHA and 0.9Pd-BMHA respectively. For all the investigated samples very similar apparent activation energies of 18-19 kcal/mole have been calculated under the assumptions of first order kinetics and isothermal plug-flow behavior of the reactor.

In the high temperature region the presence of a conversion minimum at 800°C is evident for the 0.9Pd-BHA sample whereas no appreciable deviation from 100% conversion are observed over both 0.9Pd-BMHA and 0.9Pd-A.

In order to investigate the effect of the reaction medium on the catalyst performances the activity tests have been repeated upon treatment under reaction atmosphere at 900°C for 4h.

The results reported in Figure 2 show that marked differences arise from modification of catalyst behavior upon the treatment under reaction conditions. Both the hexaaluminate-supported samples deactivate and show markedly lower conversion than those exhibited by the untreated catalysts. On the other hand 0.9Pd-A presents a slight activation upon the treatment, and, consequently, much higher activity than the hexaaluminate-supported samples.

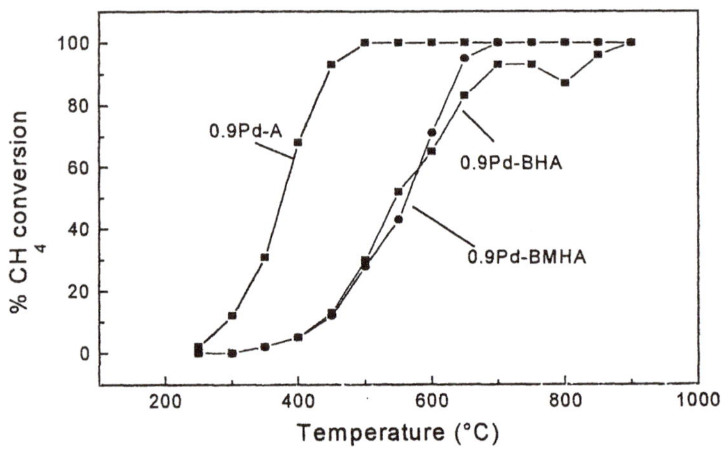

Figure 2. Results of methane combustion tests after treatment under reaction conditions

3.2 Characterization

XRD and DTA-TG measurements have been performed in order to rationalize the observed catalytic behavior. Characterization has been accomplished on $BaAl_{12}O_{19}$ and Al_2O_3 supports loaded with relatively high amount of Pd (2.5% w/w of Pd°) in order to magnify the investigated features. The characterized samples will be referred in the following as 2.5Pd-BHA and 2.5Pd-A.

In Figure 3 are reported the thermograms of 2.5Pd-BHA and 2.5Pd-A obtained during three heating and cooling cycles between 450°C and 1000°C. The two samples show a very similar behavior that resembles well literature indication for PdO ⇔ Pd° reversible transformation in Pd-based systems [5]. During the heating ramp just above 800°C a progressive weight loss starts, that is completed slightly below 900°C. During the cooling ramp the weight is constant down to 660-680°C. Below this temperature a weight increase is observed, that is completed at about 580°C and exactly corresponds to the weight loss previously observed. The weight variations associated to the hystereses compare well with the theoretical value of total PdO ⇔ Pd° reversible transformation (exp. Δw%: 0.36 for 2.5Pd-BHA and 0.43 for 2.5Pd-A. vs th. Δw%: 0.38 for both the samples)

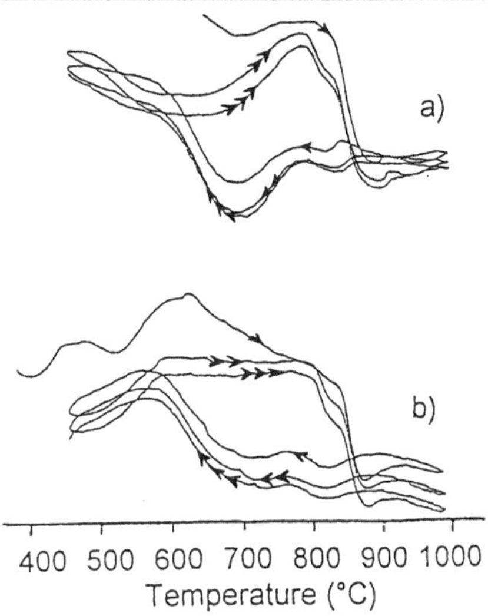

Figure 3: DTA-TG: a) 2.5Pd-A; b) 2.5Pd-BHA arrows indicate heating and cooling ramps ‣ 1st cycle; ‣ ‣ 2nd cycle; ‣ ‣ ‣ 3rd cycle

At 0.9% w/w of Pd loading the interpretation of DTA-TG analyses is difficult due to the small weight variations associated with the relevant phenomena. However Al_2O_3 samples loaded with 5 and 10% w/w of Pd exhibit similar behavior to that loaded with 2.5% w/w of Pd.

In the XRD spectra of both the catalysts in addition to the features of the support phases also the reflections of PdO are observed. The presence of crystalline Pd° has not been detected, likely due to the complete reoxidation of Pd during cooling after calcination. Crystallite dimensions of PdO calculated from FWHM are reported in Table 1. Markedly larger crystallites are observed on the hexaaluminate supported sample with respect to the alumina supported one. This indicates that Ba-hexaaluminate provides a lower dispersion of Pd species, possibly due to its lower surface area. Data in Table 1 also indicates that treatment under reaction conditions results in a further growth of PdO crystallites on the Ba-hexaaluminate surface. Activity tests performed on 2.5Pd-BHA indicate that also this sample deactivates upon treatment under reaction conditions at 900°C.

Table 1
Mean crystallite dimensions of PdO

	2.5Pd-A	2.5Pd-BHA	2.5Pd-BHA after test
mean crystallite dimension	150 Å	350 Å	550 Å

4. DISCUSSION

Activity tests in CH_4 combustion evidence that Pd-catalysts supported over both unsubstituted and Mn-substituted barium hexaaluminates provide worse performances than the alumina supported ones. The former systems show lower CH_4 combustion activity upon calcination at 1000°C and, differently from 0.9Pd-A, markedly deactivate upon treatment under reaction condition at 900°C.

According to the XRD characterization data this behaviour is related to the lower ability of hexaaluminates to disperse Pd oxide species, likely due to the relatively low surface area of these materials. This results in the large dimension of the PdO crystallites that is likely responsible for the lower ignition activity.

Concerning the high temperature behaviour the data of the $BaAl_{12}O_{19}$ -supported sample are in line with previous literature reports on CH_4 combustion over Pd-based systems. Indeed, a typical activity decrease occurs at high temperature, being associated with reduction of very active PdO_x species into less active Pd° as evidenced by DTA-TG measurements.

The absence of the conversion minimum for the $BaMnAl_{11}O_{19}$-supported catalyst could be related to the activity of the Mn-substituted hexaaluminates that has been investigated in a previous work [10]. The activity of the support at high temperature can partially compensate for the deactivation associated with PdO_x → Pd° reduction. Similar data were reported in the literature [8] for Pd supported on $Sr_{0.8}La_{0.2}MnAl_{11}O_{19}$. It is worth stressing that high temperature activity of the support could likely interfere with the mechanisms of temperature self-regulation of Pd.

DTA-TG have shown that PdO → Pd° reduction also occurs over the alumina supported sample with similar features to those observed for 2.5Pd-BAH. The absence of conversion minimum for 0.9Pd-A under the investigated GHSV conditions, is likely related to the higher residual activity of more dispersed Pd species.

5. CONCLUSIONS

The combustion activity data collected in this work have evidenced that hexaaluminate-type materials are not effective as Pd supports. Lower activity and stability have been observed with respect to alumina supported Pd catalysts. According to XRD characterization the lower ability to provide dispersion of Pd species is likely responsible for such worse catalytic performances. On the other hand DTA-TG measurements have shown that with respect to PdO ⇔ Pd° reversible transformation, barium hexaaluminate and pure alumina supports behave very similarly.

Acknowledgments
financial support for this work has been provided by Enel Spa and CNR.

REFERENCES

1. L.D. Pfefferle and W. C. Pfefferle, Catal. Rev. Sci. Eng., 29 (1987) 219
2. J.C. Schlatter et al., ASME paper 97-GT-57, 1997
3. M.F.M. Zwinkels et al., Catal. Rev. - Sci. Eng., 35 (1993) 319.
4. J. McCarty, Proceedings of EUROPACAT III, Krakow (Poland) 1997, p. 90
5. R.J. Farrauto et al, Applied Catalysis A: General, 81 (1992) 227.
6. R.A. Dalla Betta et al., US Patent 5183401, 1993
7. R.A. Dalla Betta et al., US patent 5405260, 1995.
8. K. Sekizawa et al. Journal of Catalysis, 142 (1993) 655.
9. G. Groppi et. Al., Catalysis, 13 (1997) 85
10. G. Groppi et al., Appl. Catal. A: General, 104 (1993) 101.
11. C. Cristiani et al., Catalysis Today in press
12. H.P. Klug and L.E. Alexander, in "X-Ray Diffraction Procedures" (Wiley, 1974).

NATURAL GAS CONVERSION V
Studies in Surface Science and Catalysis, Vol. 119
A. Parmaliana et al. (Editors)

Combustion of methane over palladium catalysts supported on metallic foil

A. Gervasini, C.L. Bianchi, and V. Ragaini

Dipartimento di Chimica Fisica ed Elettrochimica, Università degli Studi di Milano, via C. Golgi 19, I-20133 Milano, Italy.

Fecralloy foils washcoated with dispersed palladium oxide on alumina have been tested in the complete oxidation of methane. Activity and stability have been investigated before and after heat-treatments of the foil catalysts performed at 650 and 800°C. High Pd-loaded catalyst was less active and more stable than that low-Pd loaded. Thermal treatments at 800°C led to an increase of activity but the stability decreased compared with fresh catalysts. Surface composition of the differently treated catalysts was studied by surface spectroscopy (XPS) and electron microscopy (SEM-EDS) to find relations between property and activity.

1. INTRODUCTION

The combustion of natural gas, in particular methane, represents a widespread technology for energy production, either for industrial or for household appliances [1,2,3]. Catalytic combustors can provide high steady combustion activity over temperature range lower than that of traditional thermal combustors exceeding 1000°C, avoiding the problems associated with nitrogen oxide (NO_x) emissions from gas-exhausts [4-5].

The superiority of precious metal based catalysts, in particular palladium, for methane oxidation is well known [1, 6-8]. Supported palladium catalysts show complex behaviors for the combustion of methane over 500-900°C temperature range because of the formation and decomposition of palladium oxide. The decomposition of PdO to metallic Pd leads to reconstruction of palladium oxide crystallites creating palladium-oxygen species dispersed on bulk palladium metal, designated as PdO_x/Pd [7, 9-12]. These transformations lead to large hysteresis in the rates of methane combustion [7, 9, 10, 13].

Most commonly, palladium is distributed in a washcoat which contains various materials to improve reactivity and thermal stability [1,4]. Moreover, the washcoat should maintain its surface area under working conditions. Important loss of surface, due to sintering, can cause encapsulation of the active component leading to a loss of activity [1, 14-16].

Usually for industrial applications, because of the requirements of low pressure drop, the catalysts are cast in the form of monolith which can be ceramic or metallic [17]. Metallic monolith is commonly made of thin foil sheets of alloys. They have high thermal conductivity and the heat generated by combustion can be rapidly removed from the catalyst bed and transferred to suitable devices for warm water or steam production [18].
In this work, thin Fecralloy foils washcoated with dispersed palladium oxide on alumina have been tested in the complete oxidation of methane. Activity and stability of the foil catalysts have been investigated before ad after heat-treatments (650 and 800°C). Surface composition and physical properties have been studied by surface spectroscopy (XPS) and electron microscopy (SEM-EDS) in order to find relations between activity and properties.

2. EXPERIMENTAL

Washcoated $PdO/\gamma-Al_2O_3$ catalysts on corrugated Fecralloy (Fe, Cr, Al) foils (50 μm) were used in the investigation. The two catalysts differ in Pd content (1.3 and 2.7 g of Pd per 100g of washcoat for Pd/F-1 and Pd/F-2, respectively) and in materials introduced in the washcoat. Lanthana and baria were in the catalyst at lower Pd loading, Pd/F-1, and ceria and neodymia in that at higher Pd loading, Pd/F-2. The catalysts were studied either as received or after thermal treatments performed in air atmosphere at 650 and 800°C for 18 h.

Catalytic tests were performed in a laboratory scale reactor system operating at atmospheric pressure. The foil catalyst, cut to 25 mm height and 77 mm length (corresponding to 2.3 and 6.4 mg of Pd for Pd/F-1 and Pd/F-2, respectively) was wrapped around a special support put inside the steel reactor which was clamped vertically inside a tube furnace electrically heated. Reactant mixture (1% CH_4, 30% air and 69% N_2) passed down along both sides of the foil. The contact time during the experiments was between 1000 ad 2000 $g_{Pd}\cdot s/mol_{CH4}$. The analysis of the reactant and the gas-effluent mixture was performed by a total organic carbon apparatus (TOC, from NIRA Instruments, Italy). Activity was measured either as a function of temperature (from 400 to 750°C) and at constant temperature (650 and 750°C) for the durability tests.

Surface properties were analyzed by XPS spectroscopy using a M-Probe Science Surface Instruments with monochromatic Al Kα X-rays (1486.7 eV) equipped with an ion gun (Ar^+) for eroding the surface at a rate of 1 Å/s. The quantitative analyses were performed with the sensitivity factors given by Scofield [19] from the intensities of Al 2s, Ba $3d_{5/2}$, Ce $3d_{5/2}$, La $3d_{5/2}$, Pd $3d_{5/2}$, Yb $4d_{5/2}$. SEM-EDS analysis was performed on a Cambridge Scan 150 Instruments (40 kV) coating the samples with gold. Physical adsorption isotherms of N_2 and Kr and chemical adsorption isotherms of H_2 were collected with automatic Instruments (Thermoquest, Italy). Temperature of 30°C and pressure not greater than 20 Torr were utilized for the H_2 adsorption to avoid the formation of Pd-β hydride. Prior to adsorption, the samples were treated in H_2 flowing (50 ml/min) at 300°C for 3 h and then degassed at the same temperature for 16 h.

3. EXPERIMENTAL RESULTS

3.1. Surface characterization

The two Pd-based catalysts, that contained different amount of Pd (Table 1) were very different from morphological point of view. SEM analysis revealed that Pd/F-2 had a more homogeneous surface than Pd/F-1, that showed many agglomerates and fractures, as evidenced from the images at 1000 magnifications. EDS analysis well evidenced the presence of Pd, at about 3 keV, on both the catalysts. An intensity peak at 1.5 keV, attributed to Al of the alumina which constituted the washcoat, was observed on both Pd/F-1 and Pd/F-2. A broad peak at 4.4-5.4 keV attributed to little amount of Ba and La was present on Pd/F-1, while Pd/F-2 showed distinct peaks in the range 4.8-6 keV typical of the Ce and Nd presence.

The microstructure of the two foil catalysts was also studied by physical adsorption of gas. Because of the low surfaces expected for the samples, the surface area analyses were carried out with N_2 and Kr, the latter gas is a more suitable adorbate for low-surface samples. The results obtained in the two cases are very close; BET surfaces of 63.3 and 64.4 $m^2/g_{washcoat}$ were calculated by using N_2 and Kr, respectively, on Pd/F-2. The low Pd-loaded catalyst, Pd/F-1, has light higher surface and wider pore radius than Pd/F-2 (Table 1).

Table 1
Composition and characteristics of the metallic foil catalysts

Catalyst	Pd content [a] wt(%)	BET surface ($m^2/g_{washcoat}$)	r_{av} [b] (Å)	Pd dispersion (%)	Pd surface (m^2/g_{Pd})	d_{av} [c] (Å)
Pd/F-1	1.3	70.6	83	17.8	79	63
Pd/F-2	2.7	63.3	57	37.2	165	30

[a] amount of Pd per 100 g of washcoat; [b] average pore radius determined at P/P_o = 0.98; [c] average Pd particle size.

Pd/F-1 has lower amount of Pd either in the bulk and on the surface than Pd/F-2, as determined by H_2 adsorption. The analyses were performed by the dual isotherm method (adsorption/outgassing/readsorption, [20]). From the value of H_2 adsorbed at monolayer coverage, the parameters of Pd dispersion and metallic surface were determined (Table 1). The Pd dispersion of Pd/F-2 was not greatly affected by the heat treatment performed at 800°C in air; a slight decrease of about 16% was determined (Pd dispersion, 30.9%). This evidence indicated that Pd was well anchored on the washcoat and it was not subjected to pronounced sintering phenomena by thermal treatments.

XPS analyses also confirmed that the surface of Pd/F-2 contained more Pd than Pd/F-1. In particular, Pd was present in both the catalysts in one only oxidation state (Pd $3d_{5/2}$, BE = 337 eV) characteristic of particulate PdO [21]. Al, Ba, and La at BE of 118, 780, and 837 eV, respectively, were the other elements identified on the surface of Pd/F-1. On Pd/F-2 surface, only Al and Ce were identified at BE of 118 and 883 eV, respectively.

A deeper spectroscopic analysis of the first layers of the two foil catalysts was performed by eroding the surface down to 5000 Å in order to study the distribution profile of Pd and of the other materials of the washcoat. The collected data indicated that Pd was principally present on the surface. Starting from 1000 Å, the Pd amount was constant down to 5000 Å, corresponding to 10 and 20% of that present on the surface for Pd/F-2 and Pd/F-1, respectively. The elements present within the first 5000 Å of Pd/F-1 were the same of those identified on its surface. In the case of Pd/F-2, at 2000 Å of deepness from the surface, a peak at BE of 182 eV appeared indicating the presence of Yb. Nd, which was identified by EDS analysis, was not observed by XPS, suggesting that it was present only in the deep layers of the washcoat.

Table 2
Atomic distribution of the different elements within the first layers of the metallic foil catalysts

Catalyst	Atomic composition (%)							
Element	Surface	500 Å	1000 Å	1500 Å	2500 Å	3500 Å	4500 Å	5000 Å
Pd/F-1								
Ba	6.12	3.46	3.18	3.00	2.82	2.97	2.61	2.68
La	1.07	1.58	1.67	1.88	1.82	1.54	2.12	2.26
Pd	5.22	1.95	1.64	1.60	1.40	1.36	1.25	1.12
Al	87.59	93.01	93.51	93.52	93.74	94.13	94.01	93.94
Pd/F-2								
Ce	18.73	18.27	12.65	9.34	6.88	6.32	3.95	5.48
Pd	6.51	2.30	0.87	1.32	0.64	1.64	0.12	0.74
Yb	-	-	-	-	1.90	2.22	1.57	2.27
Al	74.76	79.43	86.48	89.34	90.57	89.93	94.36	91.52

A summary of the collected data is compiled in Table 2, which reports for the two samples the distribution of the identified elements for different deepness. The treatment of Pd/F-2 at 650°C did not cause important surface modification. On the opposite, the thermal treatment at 800°C led to redistribution of the surface in terms of atomic composition. The amount of Ce remarkably decreased and Yb appeared. The surface amount of Pd decreased of about 20%, lightly more than what observed from the H_2 adsorption measurements.

3.2. Activity and stability

The catalytic combustion activities of Pd/F-1 and Pd/F-2 were determined as a function of temperature from 400 to 750°C at different contact times. It was found that Pd/F-1 (low Pd-loaded catalyst) was more active than Pd/F-2. For example, at contact time of 942 g_{Pd}·s/mol_{CH_4} and at 650°C, Pd/F-2 converted 36.6% of methane while Pd/F-1 converted 68.8%.

Thermal treatment at 650°C did not greatly affect the activity of Pd/F-2, probably as at this temperature the decomposition of PdO is not yet active. On the contrary, heat-treatment at 800°C led to an increased activity of the methane combustion, particularly remarkable at low temperatures (Figure 1). It is known that at 800°C, the decomposition of PdO to Pd(0) occurs [7], the subsequent cooling leads to redispersed stable phases of PdO_x-Pd/Al_2O_3 and PdO/Al_2O_3. These phases could be responsible of the observed enhanced activity.

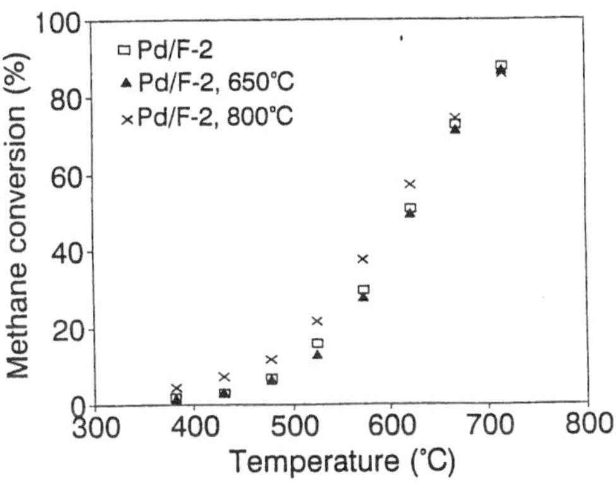

Figure 1. Catalytic combustion of methane over Pd/F-2 after different thermal treatments.

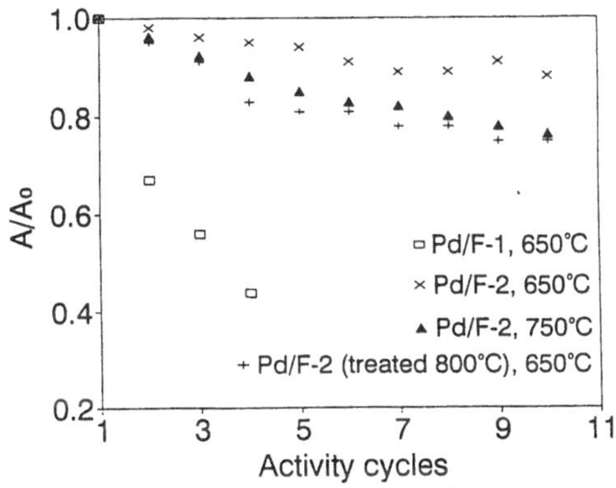

Figure 2. Durability tests of methane combustion over Pd/F-1 and Pd/F-2 at 650 and 750°C.

Although the low Pd-loaded catalyst had greater activity than that at high amount of Pd, the durability tests showed that Pd/F-2 was more stable than Pd/F-1. The tests were carried out performing discontinuous cycles of reactivity at reaction temperatures of 650 and 750°C and maintaining the two catalysts at contact times that corresponded to a methane conversion of about 60%. The catalysts were maintained under the reactant mixture at the reaction temperature for 6 h per day up to 10 days (total of 60 h of activity). Stable values of methane conversion were obtained within each day, therefore, a mean value of activity for each day, i.e., for each cycle of activity, was considered. The comparative results between Pd/F-1 and Pd/F-2 are shown in Figure 2. For a better comparison, the results have been written as ratio between the activity of a given cycle with respect to the initial activity of the fresh catalyst, A/Ao. At

650°C, Pd/F-2 showed 5% of activity decay while Pd/F-1 showed 56% of decay after 4 cycles of activity (24 h). At reaction temperature of 750°C, the stability of Pd/F-2 markedly decreased. Similar activity decay was observed on Pd/F-2 treated at 800°C.

4. CONCLUSION

The better activity of Pd/F-1 having lower amount of surface Pd than that present on Pd/F-2 could be due to the materials present in the washcoat. Moreover, the better stability of Pd/F-2 could be due to high amount of surface Pd, in fact the Pd crystallites could be subjected to reconstruction in order to maintain good activity during the time.

REFERENCES

1. M.F.M. Zwinkels, S.G.Järås and P.G. Menon, Catal. Rev.-Sci. Eng., 35 (1993) 319.
2. A. Nishino, Catal. Today, 10 (1991) 107.
3. D.L. Trimm, Appl. Catal., 7 (1983) 249.
4. R. Prasad, L.A. Kennedy and E. Ruckenstein, Catal. Rev.-Sci. Eng., 26 (1984) 1.
5. L.D. Pfefferle and W.C. Pfefferle, Catal. Rev.-Sci. Eng., 29 (1987) 219.
6. R.B. Anderson, K.C. Stein, J.J. Feenan and L.E.J. Hofer, Ind. Eng. Chem., 53 (1961) 809.
7. R.J. Farrauto, M.C. Hobson, T. Kennelly and E.M. Waterman, Appl. Catal. A: General, 81 (1992) 227.
8. J.G. McCarty, in "Int. Workshop on Catalytic Combustion", H. Arai, Ed. (Catalysis Society of Japan, Tokyo, 1994) p. 108.
9. K. Sekizawa, M. Machuda, K. Eguchi and H. Arai, J. Catal., 142 (1993) 655.
10. P. Salomonsson, S. Johansson and B. Kasemo, Catal. Lett., 33 (1995) 1.
11. T.R. Baldwin and R. Burch, Appl. Catal., 66 (1990) 359.
12. R.J. Farrauto, J.K. Lampert, M.C.Hobson and E.M. Waterman, Appl. Catal. B: Environmental, 6 (1995) 263.
13. J. G. McCarty, Catal. Today, 26 (1995) 283.
14. R.F. Hicks, H. Qi, M.L. Young and R.G. Lee, J. Catal., 122 (1990) 280.
15. R.F. Hicks, H. Qi, M.L. Young and R.G. Lee, J. Catal., 122 (1990) 295.
16. K. Sekizawa, K. Eguchi, H. Widjaja, M. Machida and H. Arai, Catal. Today, 28 (1996) 245.
17. S. Irandoust and B. Andersson, Catal. Rev.-Sci. Eng., 30 (1988) 341.
18. D.L. Trimm, Catal. Today, 26 (1995) 231.
19. J.H. Scofield, J. Elect. Spect. Relat. Phenom., 8 (1976) 129
20. J.E. Benson, H.S. Hwang and M. Boudart, J. Catal., 30 (1973) 146.
21. L.P. Haack and K. Otto, Catal. Lett., 34 (1995) 31.

NATURAL GAS CONVERSION V
Studies in Surface Science and Catalysis, Vol. 119
A. Parmaliana et al. (Editors)

Preparation and study of thermally and mechanically stable ceramic fiber based catalysts for gas combustion.

Z.R. Ismagilov, R.A.Shkrabina, N.V.Shikina, T.V.Chistyachenko, V.A.Ushakov, N.A.Rudina.

Boreskov Institute of Catalysis, 630090 Novosibirsk, pr. Ak.Lavrentieva,5, Russia

Abstract

The effective stable fiber based catalyst with a good permeability is prepared and studied. It is shown that proposed preparation method of catalyst with low concentration of active components provides stable activity and durability in hydrocarbon combustion.

INTRODUCTION

It is known that fibrous catalysts on the base of γ-Al_2O_3 are the high active catalysts for combustion processes, but these catalysts have some disadvantages. The main of which are the low thermal stability of fibrous supports, which does not allow to use these catalysts at high temperature, and the high fragility of γ-Al_2O_3 fibers, that leads to the damage of catalyst pad and formation of catalyst dust during exploitation of catalysts.

Therefore, for high temperature combustion processes the catalysts on the base of silica-alumina fibrous ceramics have been used [1, 2]. But dense fibrous ceramics have rather low specific surface area and total volume of pores and it limits direct application of silica-alumina fibers for the preparation of catalysts with required properties, activity and stability.

The catalytic gas heaters based on oxide catalysts supported on silica-alumina fibrous ceramic carriers have been developed a few years ago at the Boreskov Institute of Catalysis. These heaters are intended for space heating in industry and household. The developed fibrous catalyst with iron oxide has a high stability and efficiency. It consists of iron oxide, with a specific surface area ca.5 m^2/g; the catalyst activity is provided by iron oxide and additionally by low concentration of Pt. Nevertheless, the catalysts due to high concentration of Fe_2O_3 has insufficient permeability, consequently the combustion process on the surface of catalyst is not uniform enough.

This work was devoted to the development of the preparation method and study of thermally and mechanically stable ceramic fiber based catalysts with a high permeability .

EXPERIMENTAL

As shown in [3], top layers on the basis of composition of γ-Al_2O_3 and ZrO_2 are good washcoating materials for combustion catalysts. This composition was chosen for washcoating of ceramic fiber support in preparation of fibrous combustion catalysts in this study. The influence of the following parameters on the formation of mechanically stable washcoating layer with a high permeability of fiber pad has been studied:
- type of alumina used (hydroxide/oxide)
- concentration of anhydrous alumina in the sol;
- amount of ZrO_2 doped into sol;
- dipping time;
- number of dippings;
- sequence of the active components introduction;
- drying and calcination duration.

For preparation of washcoated fiber support the suspension of sol aluminium hydroxide with pseudoboehmite structure have been used. Fine powder of ZrO_2 was introduced into sol. It was found that the following composition of anhydrous alumina and ZrO_2 in sol: (25wt% Al_2O_3 + 75wt% ZrO_2) provides high attrition resistance of washcoated layer and high permeability of washcoated fiber pad.

Iron oxide was introduced into sol as nitric salt and Pt was impregnated as aqueous solution of H_2PtCl_6 after calcination of fibrous pad at 550^0C. Pt containing catalyst was calcinated at 550^0C during 3 hours in air atmosphere.

Catalysts prepared were tested in the reaction of methane oxidation. Catalyst activity is characterized as temperature of 50% conversion of methane in the reaction of methane oxidation in a flow set-up. Testing conditions are gas flow rate 20ml/min; sample weight 0,2 g; gas mixture composition 1 vol%CH_4 in air.

Catalyst permeability, life time and mechanical stability were tested in a model catalytic combustion heater and catalyst samples were characterized by temperature of maximum conversion of propane-butane mixture at their oxidation in the heater. Testing conditions are gas flow rate 10L/hour; sample weight 14-16 g; sample size - 10x10cm; stoichiometric gas mixture composition (3 vol% i-C_4H_{10} + 8,4 vol.% C_2H_6 + 11,1 vol.% C_4H_{10} + 77,2 vol.% C_3H_8) in air.

RESULTS AND DISCUSSION

X-ray data show that the active component in the standard catalyst has the composition of α-Fe_2O_3 and Fe_3O_4. Dispersity of these phases is ca. 30nm. Correlation between intensity of X-ray diffraction for α-Fe_2O_3 ($2\theta = 24^0$) and Fe_3O_4 ($2\theta = 30^0$) shows that ratio between two phases of iron oxides is equal to ca. 1:1. For the new catalyst X-ray data show that the active component is presented only by α-Fe_2O_3 phase with dispersity of ca. 15 nm. The phase of Fe_3O_4 or high dispersed solid solution of iron cation in γ-Al_2O_3 (having maximal diffraction intensity also at $2\theta = 30^0$) can be present in this catalyst as was observed in [4], but they were not detected, probably due to their high dispersion < 5 nm.

Figures 1-3 present the SEM photos of the fiber support, new and standard catalysts. It is seen that the distribution of active components on the washcoating layer is more uniform and has less fiber pad structure density in comparison with its distribution on the standard catalyst.

The activity of the new catalyst in methane oxidation is higher than the activity of the standard catalyst at all temperatures as seen in Fig. 4.

Fig. 1. SEM photo of the support surface.

Fig. 2. SEM photo of the new catalyst surface.

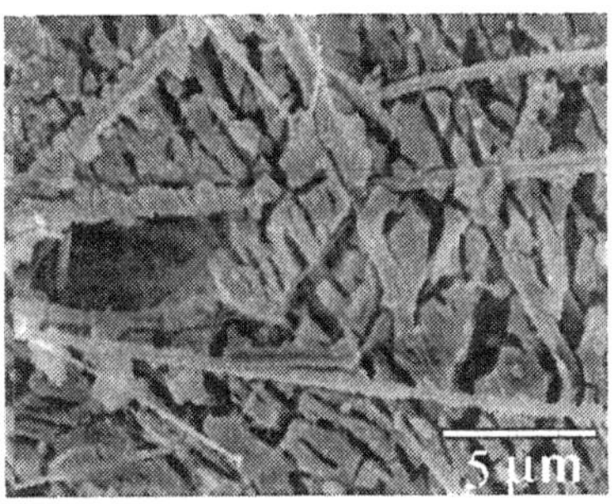

Fig. 3. SEM photo of the standard catalyst surface.

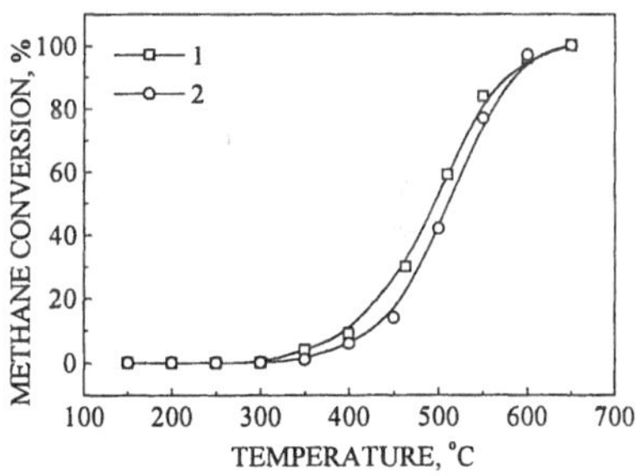

Fig. 4. Methane oxidation on catalysts: 1 - new catalyst; 2 - standard catalyst.

The Table shows the properties of the prepared catalyst in comparison with the properties of the standard catalyst (without washcoating layer) developed and studied in [1,2]. The stable activity of the new catalysts is maintained after long operation in combustion of propane in the heater during 150 hours. The better gas permeability and high dispersion of Fe_2O_3 supported on the washcoated fiber (mixture of alumina and zirconia) allow to reach the high efficiency of the catalyst at a low concentration of active components.

Table.
The properties of developed catalyst

Type of catalyst	Washcoating layer compos.	Active comp. (chem.analys.)	Activity				Permeab.
			CH_4 convers. T_{50}°C	Max.convers. C_3H_8, % (T°C)	Max.convers. C_3H_8, % (T°C) after 150 hrs	220hrs	
new	Al_2O_3+ZrO_2	4%Fe_2O_3 +0,08%Pt	490	100 (330)	100 (330)	80(330)	+++
standard	-	21%Fe_2O_3 +0,08%Pt	515	93,6 (330)	93,6 (330)	80(330)	++

REFERENCES

1. D.A.Arendarskii, Z.R.Ismagilov, I.Zh.Zainieva, T.V.Chistyachenko. //Book of Abstr.3rd Intern.Workshop on Catalytic Combustion, Sept.23-25, 1996, Amsterdam, The Netherlands, p.4.1.
2.Z.R.Ismagilov, I.Zh.Zainieva, D.A.Arendarskii, V.A.Ushakov, T.V.Chistyachenko.// Proceed. 1st World. Conf. "Environmental Catalysis for a Better World and Life", May 1-5, 1995, Pisa, Italy, p.651.
3.Z.R.Ismagilov, R.A.Shkrabina, N.A.Koryabkina, N.V.Shikina, D.A.Arendarskii //Proceed. Capoc4, 1997, April 9-11,Brussels, Belgium, V.2, p,255-258.
4.O.A.Kirichenko, V.A.Ushakov, E.M.Moroz, M.P.Vorob'eva. //Kinet. katal.V.34, N 4, 1993, P.739 (in Russian).

NATURAL GAS CONVERSION V
Studies in Surface Science and Catalysis, Vol. 119
A. Parmaliana et al. (Editors)

Reactivity and Characterization of Pd-containing Ceria-Zirconia Catalysts for Methane Combustion

Alessandra Primavera[a], Alessandro Trovarelli[a], Carla de Leitenburg[a] Giuliano Dolcetti[a], and Jordi Llorca[b]

[a]Dipartimento di Scienze e Tecnologie Chimiche, Università di Udine, via Cotonificio 108, 33100 Udine, Italy.

[b]Departament de Química Inorgànica, Universitat de Barcelona, Diagonal 647, 08028, Barcelona, Spain

The reactivity and characterization of Pd-containing ceria-zirconia catalysts in the combustion of methane is investigated. It is shown that the properties of the mixed-oxide phase play an important role in the behavior of the catalysts in the high-temperature region of the light-off curve. This can be related to the oxygen diffusion features of CeO_2-ZrO_2 which can help transfer of oxygen to Pd and reduce the detrimental effects of PdO→Pd transformation.

1. INTRODUCTION

Catalytic combustion of methane and other light hydrocarbons is receiving a considerable attention in these years with the aim of providing new solutions for reducing emissions of air pollutants. In particular, the interest in new materials for high-temperature applications [1] and for exhaust gas treatments, especially VOC [2], has boosted forward research in this area.

Among several types of catalysts/supports that have been developed at various stages, particular attention is being paid to the preparation and characterization of rare-earth containing catalysts with a specific focus towards materials based on CeO_2 [3]. We have recently reported that mixed oxides of composition 80%CeO_2-20%MO_2 (with M=Zr or Hf) behave as efficient and stable catalysts for the total oxidation of CH_4 [4], and several reports appeared on the use of these materials as a base for catalysts having a high oxygen storage/transport capacity [5], unusual redox properties [6] and remarkable activity in CO oxidation reaction [7]. In addition, the stabilisation of textural properties against sintering, by introduction of ZrO_2 into CeO_2 lattice was also reported [8,9]. The reason for this behavior can be found in the efficiency of the Ce^{4+}-Ce^{3+} redox couple which is strongly enhanced in CeO_2-ZrO_2 solid solutions due to the introduction of the smaller Zr^{4+} cation into the fluorite lattice of CeO_2. This generates defects throughout the crystal, which in turn, brings to an increase of the oxygen mobility and diffusion in the lattice. Therefore it seems that the modification of the redox parameters induced by the presence of a mixed oxide phase could allow the range of catalytic activity and stability to be widened enough to meet more demanding conditions, such as for example those encountered in high-temperature catalytic combustion.

In this paper we describe the characterization and reactivity of Pd-containing CeO_2-ZrO_2 catalysts for the low and high-temperature combustion of methane. Pd-containing catalysts are in fact receiving a considerable interest due to their high activity coupled with their peculiar behavior in high-temperature methane oxidation [10].

2. EXPERIMENTAL

Pd-containing catalysts were prepared by coprecipitation of the mixed oxide with the appropriate amount of $Ce(NO_3)_3$, $ZrO(NO_3)_2$, (Aldrich) and $Pd(NO_3)_2$ (Johnson Matthey) using NaOH as precipitating agent. These materials were compared with catalysts prepared by wet impregnation of $Pd(NO_3)_2$ over the preformed support followed by drying and calcination. Typical compositions of $Ce_{1-x}Zr_{x-z}Pd_zO_{2-y}$ (with x and z in the range 0.2-0.5 and 0.01-0.05) were prepared and the catalysts were characterized by x-ray diffraction, surface area measurements, electron microscopy analysis, XPS and temperature programmed techniques. The x-ray diffraction profiles were collected with a Siemens D-500 instrument, using a graphite monochromator and a Cu target. The x-ray diffractometer was equipped with a Paar HTK 10 AP high-temperature device. In order to calculate the lattice parameters the four main reflections corresponding to (111), (200), (220) and (311) crystallographic planes of a cubic fluorite lattice have been considered. High-resolution transmission electron microscopy combined with energy dispersive x-ray microanalysis were performed using a Philips CM-30 electron microscope working at 300 kV with a 0.2 nm point to point resolution. Redox properties of the solid solutions were studied by quantitative temperature programmed reduction (TPR). Combustion of methane was carried out in a tubular microreactor operating at atmospheric pressure under the following conditions: GHSV of 60000 h^{-1}, total flow of 100 ml/min (STP), and a CH_4:O_2:He ratio of 1:4:95.

3. RESULTS AND DISCUSSION

Table 1 summarizes the composition and textural features of the representative samples examined in our study. The surface area of the materials is dependent on the calcination temperature and typical values range from 100

m^2/g after calcination at 723K down to less than 1 m^2/g after calcination at 1473K, with the most significant drop of surface area observed above 1273K.

Table1
Surface areas (S.A.) and composition of catalysts

	surface area (m^2/g) *a*			Pd content wt%
sample	723K	1073K	1473K	calc. (found)
$Ce_{0.76}Zr_{0.19}Pd_{0.05}O_{2-y}$	110	63	<1	3.2 (3.0)
$Ce_{0.74}Zr_{0.19}Pd_{0.07}O_{2-y}$	108	62	<1	4.4 (3.9)
$Ce_{0.79}Zr_{0.20}Pd_{0.01}O_{2-y}$	110	46	1	- (-)
$Ce_{0.57}Zr_{0.38}Pd_{0.05}O_{2-y}$	111	47	2	3.4 (3.2)
$Pd/Ce_{0.8}Zr_{0.2}O_2$	98	44	<1	- (-)
$Ce_{0.8}Zr_{0.2}O_2$	96	42	<1	- (-)

a: S.A. were measured after calcination at the temperature indicated for 6 h.

The presence of Pd does not seem to greatly influence either the surface area of the fresh catalysts or the textural properties after calcination at high temperature. The formation of a mixed oxide phase is evidenced from the structural parameters calculated from x-ray diffraction analysis. All the samples investigated crystallize in a cubic structure with a shrinking of the cell parameter if compared with pure ceria. This is in agreement with the results reported in the case of CeO_2-ZrO_2 [11], where cubic solid solutions are formed in this composition range. No signals originating from PdO were detected by x-ray diffraction, even at the higher Pd loading (7 mol %), which strongly support the presence of Pd^{2+} dissolved within the CeO_2-ZrO_2 lattice. This is also in agreement with EDX analysis performed on individual crystallites using an electron beam with a diameter between 5 and 8 nm. There are no visible domains enriched in any of the components alone, and in all cases the signals due to Pd, Ce and Zr maintain the same relative abundance.

The stabilities of mixed oxides were investigated by following the structural modifications after treatment under air at increasing temperature. In situ x-ray diffraction experiments were performed with the sample placed in a sealed chamber filled with air. In Figure 1 is shown the sequence of XRD profiles collected when $Ce_{0.76}Zr_{0.19}Pd_{0.05}O_{2-y}$ is subjected to a given temperature cycle (298K-1473K-298K). From the figure it can be concluded that there is a segregation of metallic Pd particles starting at a temperature of 1273K, which are clearly detected with the diffraction peak of Pd(111) located at $2\theta = 40.1°$ (the signal of Pt originates from the sample holder). When the temperature is lowered it is clear that the segregation is not reversible, and it persists at room temperature. Only traces of PdO could be detected from their diffraction intensities collected by conventional x-ray over longer aquisition times. Similar results are obtained with all the coprecipitated samples. Confirming evidence come from analysis of the Pd $3d_{5/2}$ XPS spectra (Figure 2).

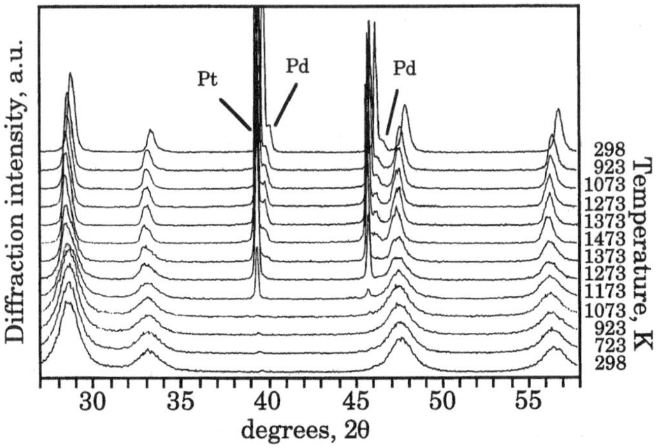

Figure 1: In situ x-ray diffraction profiles of $Ce_{0.76}Zr_{0.19}Pd_{0.05}O_{2-y}$. Sample was hold for 1 h at each temperature before measurements.

Increasing the temperature of calcination, the intensity of the signal originating from Pd increases (which indicates a segregation of the metal from the solid solution) with a strong signal component belonging to Pd in a lower oxidation state (340.8 *vs* 339.3 eV).

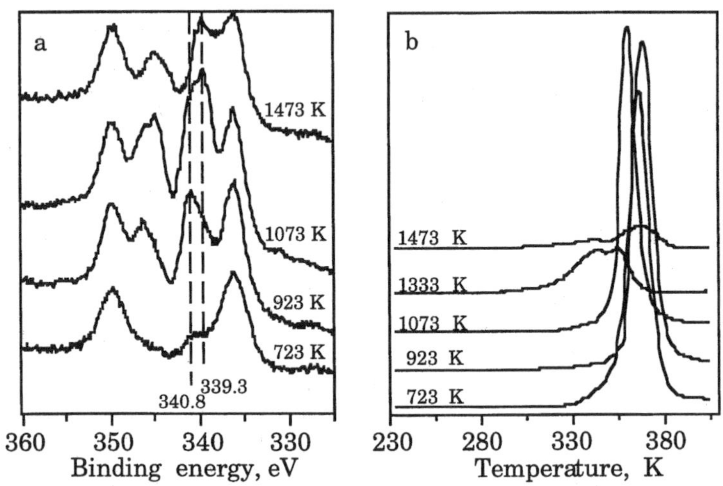

Figure 2: Effect of calcination temperature in the (a) XPS spectra and (b) low temperature TPR profile of $Ce_{0.76}Zr_{0.19}Pd_{0.05}O_{2-y}$.

According to this picture, the temperature programmed reduction profiles of samples calcined at T>1200K give a negligible reduction signal due to PdO → Pd transformation, indicating that the metal is already in a reduced form. The morphology studied by HRTEM suggests that when the temperature is raised the Pd atoms contained in the solid solution segregate and coalesce to form discrete small crystallites at the surface of the resulting Ce-Zr-O solid solution, which is now essentailly free from Pd. These Pd crystallites are heterogeneous in size, ranging from 8 to 28 nm, with most of them centered around 15 nm. The structural parameters of the oxide were also calculated (using both XRD and ED) before and after the segregation. The following values were obtained with $Ce_{0.76}Zr_{0.19}Pd_{0.05}O_{2-y}$ for the *a* cell parameter: 0.5381(6) nm and 0.5365(2) nm. This indicates that Pd is responsible for the slight cell enlargement. Calculations for the ionic radious of Pd^{2+} in a VIII fold coordination [12] (such as that found for Ce^{4+} and Zr^{4+} in CeO_2-ZrO_2 solid solutions) indicates that it is slightly greater than the ionic radius of Ce^{4+} in the same coordination environment (0.95Å vs 0.93Å). This could explain the slight modification of the cell parameter after the segregation of Pd.

As can be seen from Figure 3a, the catalytic activity of the Pd containing catalysts is dependent strongly on the calcination temperature. With samples calcined at the lower temperature (T<1073K) the conversion steadily increases up to a value close to 100%, while for the other samples (T> 1073K) the conversion increases up to ca. 950K and then decreases reaching a minimum at around 1000K. For all samples examined the higher is the calcination temperature the larger is the drop in catalytic activity indicating that the combustion of methane should be highly sensitive to the Pd metal initially present on the support.

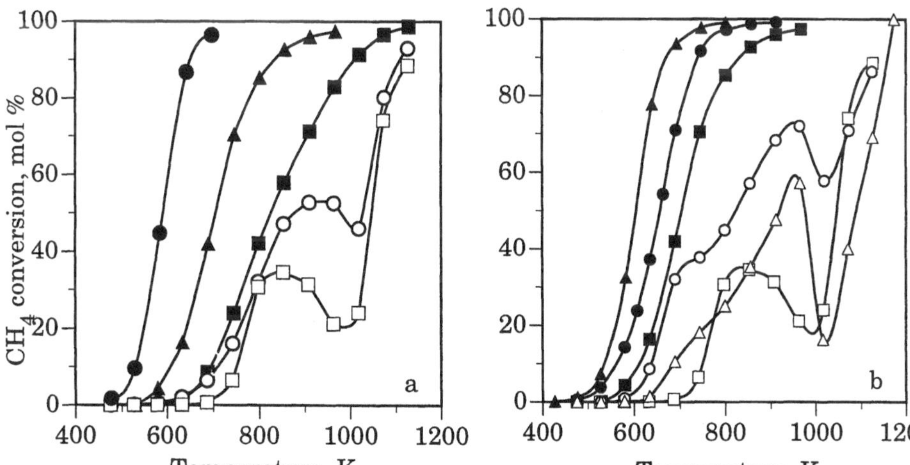

Figure 3: (a) Catalytic combustion of CH_4 over $Ce_{0.76}Zr_{0.19}Pd_{0.05}O_{2-y}$. Calcination temperature: 723K (●), 923K (▲), 1073K (■), 1333K (O) and 1473K (□); (b) Catalytic combustion of CH_4 over $Ce_{0.76}Zr_{0.19}Pd_{0.05}O_{2-y}$ (▲, △), $Ce_{0.57}Zr_{0.38}Pd_{0.05}O_{2-y}$ (●, O), Pd(5%mol)/$Ce_{0.8}Zr_{0.2}O_2$ (■, □). Calcination temperature: filled symbols 923K, open symbols 1473K.

Samples calcined at temperature lower than 1073K, which do not have significant Pd on the surface does not show any inflection in the catalytic activity. A comparison between coprecipitated and supported catalysts (Figure 3b) shows that the latter behaves better at the lower reaction temperature, while the former are better catalysts at higher temperature. The high activity of the supported samples at low reaction temperatures can be related to the their higher surface Pd content compared to the coprecipitated catalysts, where only a few Pd atoms are exposed. The drop of activity observed at high temperature is lower for $Ce_{0.57}Zr_{0.38}Pd_{0.05}O_{2-y}$ than for Pd/$Ce_{0.80}Zr_{0.20}O_2$ at similar value of conversions. The reason for the better behavior could be connected to the faster oxygen transfer rates from the support to the metal in the Zr-rich sample, which could even be enhanced by the presence of residual Pd dissolved in the solid. This oxygen transfer could help the combustion of methane over surface Pd. The high oxygen availability is also related to the faster oxygen diffusion in catalysts containing 50-60 mol% of ceria compared to other compositions [5]. Therefore it seems that the design of a support having high oxygen transfer capacity can be effective in reducing or compensating for the drop in activity. Similarly, it was reported that ceria-zirconia helps the transfer of oxygen lattice from the bulk to the surface of the support, thus increasing the activity of Rh in CO oxidation [7]. In the present case transfer of oxygen from the support to the metal can help maintaining the Pd in a more active PdO_x state.

REFERENCES

1. M. F. M. Zwinkels, S. V. Jaras, P. G. Menon and T. A. Griffin, Catal. Rev.-Sci. Eng., 35(3) (1993) 319.
2. J. J. Spivey, Ind. Eng. Chem. Res., 26 (1987) 2165.
3. A. Trovarelli, Catal. Rev.-Sci. Eng., 38(4) (1996) 439.
4. F. Zamar, A. Trovarelli, C. de Leitenburg and G. Dolcetti, J. Chem. Soc. Chem. Commun., (1995) 965.
5. P. Fornasiero, R. Di Monte, G. R. Rao, J. Kaspar, S. Meriani, A. Trovarelli and M. Graziani, J. Catal., 151 (1995) 168.
6. G. Balducci, P. Fornasiero, R. D. Monte, J. Kaspar, S. Meriani and M. Graziani, Catal. Lett., 33 (1995) 193.
7. T. Bunluesin, R. J. Gorte, and G. W. Graham, Appl. Catal. B: Environ., 15 (1998) 107, and refs. therein.
8. M. Pijolat, M. Prin, M. Soustelle, O. Touret, and P. Nortier, J. Chem. Soc. Faraday Trans., 91 (1995) 3941.
9. C. de Leitenburg, A. Trovarelli, G. Bini, F. Cavani and J. Llorca, Appl. Catal. A: General, 139 (1996) 161.
10. R. J. Farrauto, M. C. Hobson, T. Kennelly and E. M. Waterman, Appl. Catal. A: General, 81 (1992) 227.
11. F. Zamar, A. Trovarelli, C. de Leitenburg and G. Dolcetti, Stud. Surf. Sci. Catal., 101 (1996) 1283.
12. A. Brese and L. O'Keefe, Acta Cryst. B, 47 (1991) 192.

NATURAL GAS CONVERSION V
Studies in Surface Science and Catalysis, Vol. 119
A. Parmaliana et al. (Editors)

Methane catalytic and electrocatalytic combustion over perovskite type oxides deposited on YSZ.

S. Douvartzidis, G. Dimoulas and P. Tsiakaras

Department of Mechanical and Industrial Engineering, University of Thessaly,
Pedion Areos, 383 34 Volos, Greece. E-mail: tsiak@uth.gr

The catalytic and the electrocatalytic behavior of $La_{0.6}Sr_{0.4}Co_{0.8}Fe_{0.2}O_3$ Perovskite type oxide deposited on Yttria Stabilized Zirconia (YSZ) in the form of thin porous films, was studied during the reaction of methane combustion. It was found that oxygen electrochemically supplied through the solid electrolyte (YSZ), affects drastically the rate of methane combustion.It was also found that the observed phenomena are purely Faradaic and the above catalytic system can operate in a wide range of experimental conditions maintaining long time catalytic activity and thermal stability.

1. INTRODUCTION

Today, over 90% of methane is consumed as heating fuel. Because transportation of natural gas from remote sites is costly, it has been often suggested that natural gas, namely methane, should be converted to more easily transported liquid fuel [1].The most common use of methane that is widely applied is it's combustion for energy production, which has also a number of potentially important and practical applications like improved flame stability, generation of low temperature process heat and reduced pollutant emissions. On the other hand methane catalytic combustion is the most common chemical reaction in solid oxide electrochemical devices for direct conversion of chemical to electrical energy. Solid Oxide Fuel Cells (SOFCs) have attracted considerable interest during the past decade as highly effective systems with efficiencies of 50 to 65% and enviromentally acceptable sources of electrical energy production. The most common SOFC feed fuels are hydrogen, carbon monoxide, methane, or other hydrocarbons [2-4]. Moreover, an important advantage of SOFCs is that they can operate at temperatures at which reaction rates attain values of practical interest [5] and they offer the feasibility of chemical co-generation, i.e. the simoultaneous generation of electrical energy and usefull chemicals [6]. In addition, perovskite type oxides have been extensively studied in various applications including catalysis [7-9], electrodes for fuel cells, superconductors, and photocatalytic dissociation of water [10-13].

In the present work the catalytic and electrocatalytic behavior of Perovskite oxide $La_{0.6}Sr_{0.4}Co_{0.8}Fe_{0.2}O_3$ was examined in conditions of total methane oxidation. Perovskite-type oxides containing transition metals can be included as important catalyst candidates for a great number of combustion applications due to their low cost and their reported catalytic properties [3-5].

2. EXPERIMENTAL

2.1 Experimental Setup

The equipment used in the present study is depicted in Figure 1. A *Shimadzu GC-14B* Gas Chromatograph and a *Hartmann and Braun Advance Optima IR* analyzer (CH_4, CO and CO_2) were used for the analysis of reactants and products.

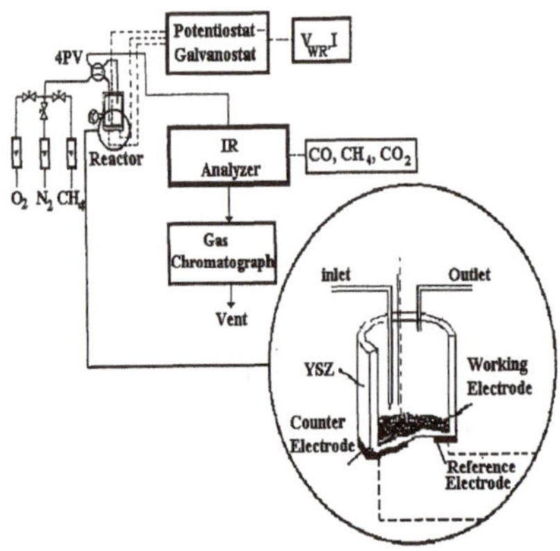

Figure 1. Schematic representation of the experimental set-up and of the electrode system.

Two *Bargragh D.M.M* Units, a differential Voltmeter and an *AMEL System 5000* were used to measure potential and impose currents through the electrochemical cell, respectively.

The YSZ (8 mol % Y_2O_3 in ZrO_2) atmospheric pressure continuous flow well-mixed (CSTR) reactor has a volume of 30 cm^3 and has been described previously [14]. Reactants were certified standards of CH_4-N_2 and O_2-N_2 mixtures (99.99%). They could be further diluted in pure N_2 (99.99%). The results reported here are typically obtained with total flow-rates of 0.5-3 cm^3 STP/sec.

2.2 Catalyst Preparation

The powder of the perovskite was prepared by applying the EDTA method. Nitrates of the constituent metals (Merck, p.a.quality) were dissolved separately in fixed quantity of water and their concentrations were determined by titration with EDTA. Stoichiometric amounts of the solutions were taken and mixed. A solution of ammonia/EDTA was added to the mixed metal solution, the final concentration of EDTA being 1.5 times the total metal cation concentration. The pH of the solution was adjusted to 8-9 by using ammonia. The solution was mixed for several hours under moderate heating to complete complexation and then pyrolysed in an oven at 500K. The resulting powder was milled for several hours in a planetary mill in acetone and after drying calcined at 1200K in stagnant air for 12 hours. Calcination at lower temperatures also resulted in perovskite formation but in these cases XRD revealed traces of $SrCO_3$.

After the powder preparation, the perovskite was mixed with ethyl glycol (50 mg of LSCF and 20 ml of glycol) and the mixture was heated until more than the half of its volume was evaporated. A portion of the perovskite viscous suspension was then deposited on the inside bottom of the Zirconia tube in order to prepare the working (catalyst) electrode, by using painting techniques. The tube was then heated to 1000°C, with a heating rate of about 200°C/h. The thickness of the LSCF film prepared in this way was of the order of 5-20 μm and its superficial area was about 2 cm^2.

Platinum was then used for the preparation of the counter and the reference electrodes. Two porous Pt films were deposited, by applying a thin coating of Engelhard A1121 platinum paste, on the outside wall of stabilized Zirconia tube which was exposed to ambient air.

The thin coating was followed by drying and calcining in air first for 2h at 450°C and then for 30 min at 900°C. In the course of the experiments both galvanostatic and potentiostatic operation was used and both gave similar results. In the galvanostatic mode of operation a constant current is applied between the catalyst (working) and counter electrodes while monitoring the ohmic – drop – free [15] potential air/Pt electrode. In the potentiostatic mode a constant potential V_{WR} is applied between the catalyst and the current I between the catalyst and the counter electrode.

3. RESULTS AND DISCUSSION

3.1. Current-Potential Behavior

The fundamental electrochemical reaction, taking place at the LSCF/YSZ interface, is the oxygen exchange between the gas phase and the electrolyte as is shown in Figure 2.

The above is the determining reaction of the overall electrochemical phenomenon that causes the development of an overpotential η on the LSCF/YSZ interface. This is defined as:

$$\eta = V_{WR} - V^{o}_{WR}$$

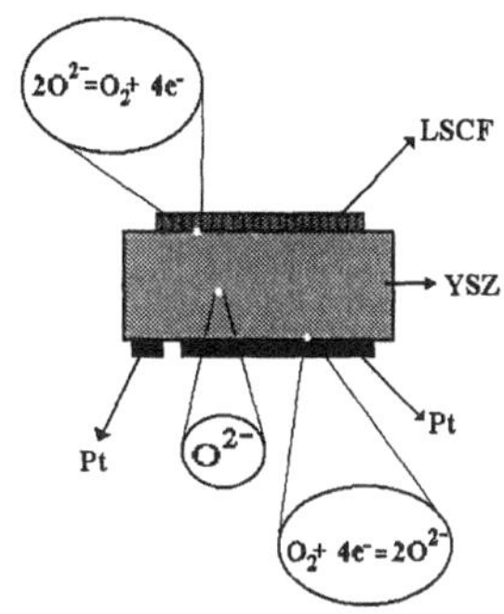

Figure 2. Schematic survey of fluxes and reaction in the LSCF/YSZ/Pt interfaces.

where V^{o}_{WR} is the open circuit catalyst potential relative to the reference electrode. The overpotential η appears due to the delay of the charge movement through the interface which results to charge accumulation. The overall oxygen exchange reaction can be broken down into a number of sequential reaction steps. Each of these reaction steps contributes to the overall electrode polarization behavior. For multi-step reaction, the electrode kinetics can be modified into a form of the Butler-Volmer equation [16]:

$$I = I_o [\exp(\alpha_a F\eta / RT) - \exp(\alpha_c F\eta/RT)]$$

where α_a and α_c are the anodic and the cathodic charge transfer coefficients, respectively, F is the Faraday's constant, R is the ideal gas constant and I_o is the exchange current density. The Butler-Volmer equation is the general equation of electrochemical kinetics that approximates the real current density-overpotential relation for every case that arises depending on the values of the parameters involved.

In Figure 3, is depicted the dependence of the logarithm of the exchange current density on the inverse temperature. The measured apparent activation energy values was about 22 kcal/mol in agreement to similar values that were reported in literature for the case of Ag electrodes under reaction conditions [17] as well as in the presence of pure oxygen .

Experiments under open circuit conditions ($I = 0$, $r = r_o$) were carried out in the reaction apparatus shown in Figure 1 in a temperature range between 500 and 900°C at atmospheric total pressure using a thin LSCF film in the role of catalyst (working electrode). Total combustion products were mainly detected, i.e CO_2 and H_2O. In all cases, the amounts of CO and C_2 hydrocarbons that were also detected during the product analysis were found to be extremely small.

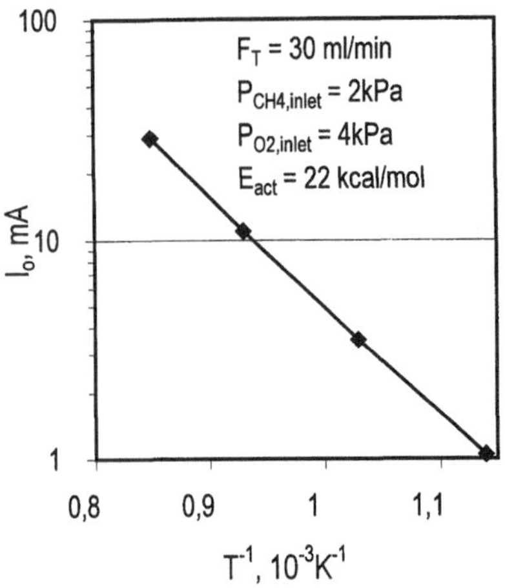

Figure 3. The effect of temperature on the exchange current density.

3.2. Open-circuit measurements

Figure 4a shows the dependence of the rates of formation of CO_2 and H_2O on the partial pressure of oxygen.

The inlet pressure of methane and the temperature T were kept constant at 8 kPa and 750°C, respectively. Moreover, in Figure 4b is depicted the dependence of the above mentioned rates on the partial pressure of methane for T = 750°C and P_{O_2} = 8 kPa. The results obtained revealed that LSCF perovskite exhibits good catalytic activity towards total oxidation.

3.3. Effect of temperature

The effect of temperature on the methane conversion for both stoichiometric and rich oxidation conditions, is reported in Figure 5.

The onset of the combustion reaction (conversion > 1%) started for the stoichiometric case at 250°C and for the rich oxidation conditions at 300°C. In both cases the maximum conversion is attained at about 800°C with a value of about 100% and 30%, respectively.

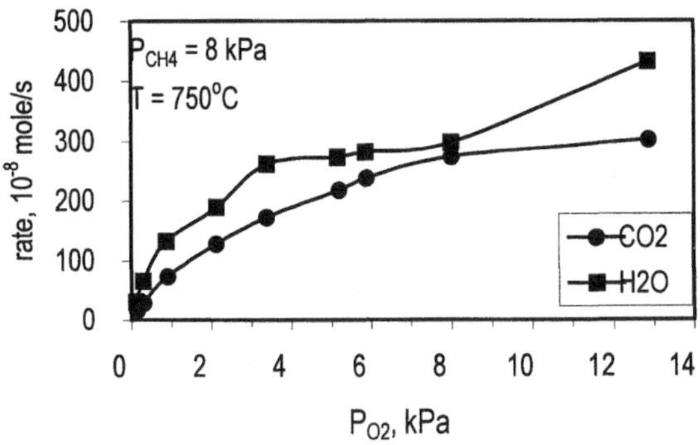

Figure 4 a-. Effect of P_{O2} on the reaction rates at T = 750°C and P_{CH4} = 8kPa

3.4. Closed-circuit measurements

Closed circuit measurements (I ≠ 0) were carried out in the electrochemical device shown in Figure 1 in the presence of gaseous O_2 in the feedstream. Several previous studies with such solid electrolyte cells reported that under certain conditions, the increase in the rate of oxygen

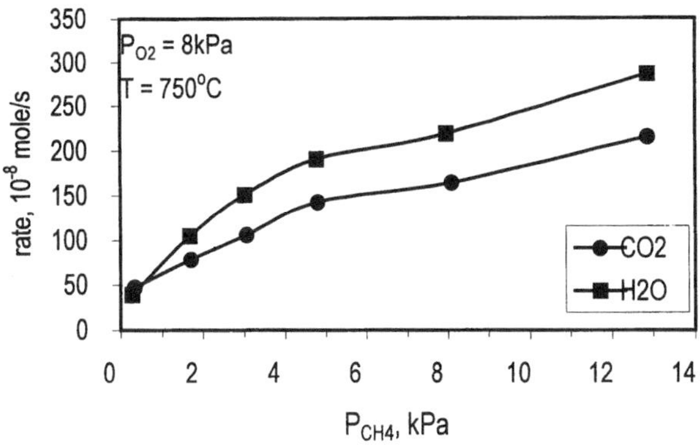

Figure 4 b-. Effect of P_{O2} on the reaction rates at T = 750°C and P_{CH4} = 8kPa

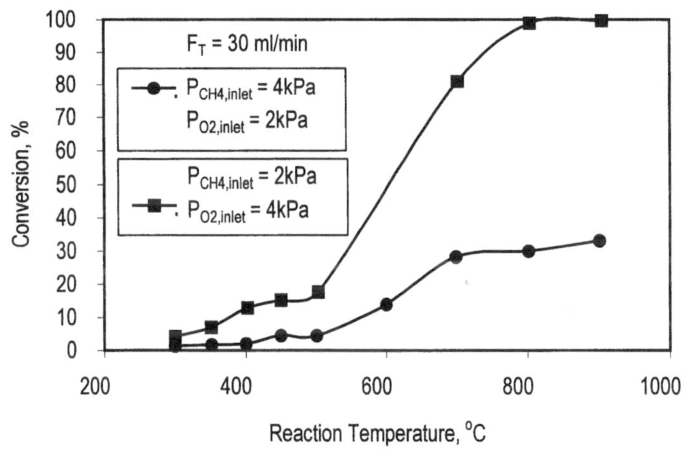

Figure 5. Effect of reaction temperature on the total conversion of methane for stoichiometric and rich oxidation conditions.

consumption is not equal the rate of O^{--} transport through the electrolyte [18, 19]. Vayenas and co-workers [19] defined the dimensionless rate enhancement factor Λ as:

$\Lambda = \Delta r / (I / 2F)$

where Δr (=r-r_o) is the increase in the catalytic rate of oxygen consumption and I/2F is the reported flux of O^{--} through the electrolyte. Figure 6 shows the Δr to I/2F relation of CO_2 formation that was obtained for constant feed composition ($P_{CH_4} = P_{O_2}$ = 5 kPa) at 880°C. In the present case Λ can be defined as: $\Lambda = \Delta r_{CO_2} / (I / 2F)$ taking values between 1 and 2.

This means that the observed phenomena are purely Faradaic compared to other catalytic oxidation reactions that are reported taking values of the order of 10^5 [14, 19].

4. SUMMARY

The results obtained in the present investigation show that the $La_{0.6}Sr_{0.4}Co_{0.8}Fe_{0.2}O_3$ Perovskite behave as an effective catalyst for methane combustion and as a conductive electrode that may be used on both catalytic and electrocatalytic applications. Only traces of CO and C_2's have been detected. It is suggested that LSCF catalyst can be used in a wide range of temperatures with practically constant thermal stability and catalytic activity.

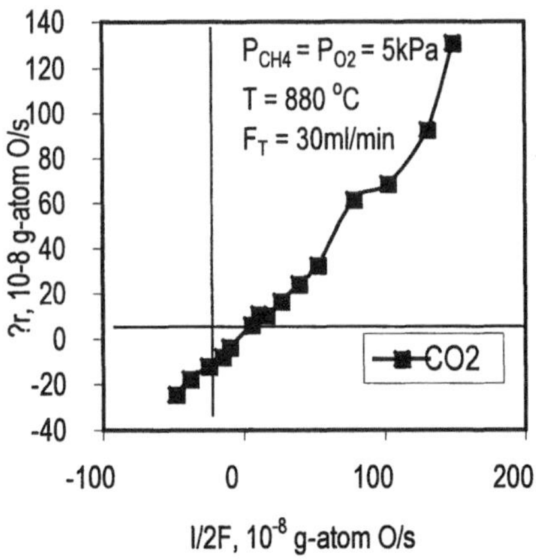

Figure 6. Effect of electrochemical oxygen pumping on the rate of formation of CO_2.

REFERENCES

1. Wolf, E. E., ''Methane Conversion by Oxidative Processes'', Van Norstrand Reinhold, New York (1992).
2. Etsell, T. H., and Flengas, S. N., J. Electoch. Soc., 118, (1971), 1890.
3. Subbarao, E. C., in ''Solid Electrolytes and their Application'', Plenum Press, New York (1980).
4. Steele, B. C. H., in ''Electrolyte Processes in Solid State Ionics'', M. Kleitz and J. Duphy, Editors, p. 307, Reidel, Dordrecht (1976).
5. Stoukides, M., Ind. Eng. Chem. Res. 27, (1988), 1745.
6. Vayenas, C. G., Solid State Ionics, 28, 1521, (1988).
7. Kung, H. H., ''Transition Metal Oxides: Surface Chemistry and Catalysis'', Studies in Surface Science and Catalysis, Vol. 45, Elsevier, Amsterdam, (1988).
8. Voorhove, R. J. H., Remeika, J. P., Freeland, P. E., and Mathias, B. T., Science, 177, (1972), 353.
9. Shimizu, T., Catal. Rev. Sci. Eng., 34, 355, (1992).
10. Libby, W. F., Science (Washington), 171, 499, (1971).
11. Meadowcraft, D. B., Nature (London), 266, 847, (1970).
12. Tejuca, L. J., Fierro, J. G., and Tascon, J. M. D., Adv. Catal., 36, 237, (1989).
13. Avudaithai, M., and Kutty, T. R. N., Mater. Res. Bull., 22, 641, (1987).
14. Vayenas, C. G., Bebelis, S., Yentekakis, I. V., and Lintz, H. G., Catal. Today, 11(3), 303, (1992).
15. Tsiakaras, P., and Vayenas, C. G., Materials Science Forum 76, 179-182, (1991).
16. Bockris, J. O. M., and Reddy, A. K. N., Modern Electrochemistry (Plenum Press, New York, 1977).
17. Tsiakaras, P., and Vayenas, C. G., J. Catal., 144, 333, (1993).
18. Stoukides, M., J. Appl. Electroch., 25, 899, (1995).
19. Vayenas, C. G., Jaksic, M. M., Bebelis, S. I., and Neophytides, S. G., in ''Modern Aspects in Electrochemistry'', Vol 29, eds. J. O. M. Bockris et al, (Plenum Press, New York, 1996), ch. 2.

NATURAL GAS CONVERSION V
Studies in Surface Science and Catalysis, Vol. 119
A. Parmaliana et al. (Editors)

How transient kinetics may unravel methane activation mechanisms.

C. Mirodatos

Institut de Recherches sur la Catalyse, CNRS
2 Avenue Albert Einstein, Villeurbanne Cédex, France.

Novel routes for direct conversion of methane into higher hydrocarbons and oxygenates were investigated during the last decades, in parallel with a renewed interest for the indirect routes through syngas formation. For the latter, despite a large research effort on a less heat-consuming process such as partial oxidation, engineering problems (heat and mass transfer control) have still to be solved and catalysts to be improved. For this, a precise knowledge of the related mechanisms and micro-kinetics is required.

In a study confined to the steady-state kinetics, only lumped kinetic parameters can be estimated from a regression of steady-state kinetic data. By generating transients input signals, one observes signals out of the steady-state, the relaxation of which may be related to the time scale on which the corresponding global reaction occurs at the steady-state. The steady-state isotopic transient kinetic analysis (SSITKA) consists in generating isotopic step functions which let the overall surface occupancy unperturbed and therefore gives access to true time constants of the reactive intermediates and to their concentration. It can also be combined with *in situ* diffuse reflectance infrared spectroscopy (DRIFT) [1]. In the temporal analysis of product (TAP) reactor transient kinetics are generated by pulsing small and known amounts of reactants through a catalyst bed maintained under vacuum and analysing the pulse expansion and relaxation at the reactor exit with a time resolution below the millisecond [2].

Recent reviews demonstrate the unique power of these techniques to investigate catalytic phenomena (non uniform surface, combination of kinetic, thermal and hydrodynamic parameters) [3-6]. This paper aims at illustrating some recent application of these techniques for unravelling the complex mechanisms of methane reforming and methanation.

1. SYNGAS FORMATION BY METHANE REFORMING

The addition of CO_2 in the feedstock of methane steam reforming has the advantage to adjust the CO/H_2 ratio to the oxosynthese requirements and the possible use of natural gas originally mixed with CO_2. However, the main drawback is the coke formation which necessarily derives from lower H/C ratios in the feedstock. Dry reforming is catalysed by most of the group VIII metals [7-10]. Like for the steam reforming, there is still a large debate on the reaction mechanism. A dissociative CH_4 and CO_2 adsorption is proposed by Rostrup-Nielsen and Bak-Hansen [7] (different metals) and by Kroll et al. [10] (Ni/SiO_2). The rate determining step is then the recombination of carbon and oxygen on the surface. Over Ir and Ru on alumina, Mark and Maier [8] propose a mechanism where the dissociative CH_4 adsorption is rate limiting, followed by the direct reaction of CO_2 with adsorbed carbon. They observed no support effect for Rh and Ir. Erdöhelyi et al. 11] observed as well no support effect over Rh, but over Pd. An active role of the support has been proposed over Ni/La_2O_3 [12]. In a recent paper [13], Aparicio developed a microkinetic approach based on transient experiments related to the main steps of the methane reforming over Ni catalysts. Methane dehydrogenation and water adsorption/dissociation were studied by isotopic transient CH_4/H_2-to-CD_4/H_2 and H_2-to-H_2O/D_2 experiments, respectively. The formation and the cleavage of C-O bonds was investigated by non isotopic transient methanation experiments. Rate constants of elementary steps were obtained by fitting the response curves to microkinetic models. No single step actually was found to determine the rate of methane reforming, the surface oxygen playing a key role in that process. This interesting study shows that precise mechanistic data are required for differentiating microkinetic models describing the same transient data.

In a similar approach, we performed SSITKA and TAP experiments over Ni/SiO_2, Ni/Al_2O_3, Ru/SiO_2 and Ru/Al_2O_3 [10,14]. We present here TAP data, completed by recent modelling results.

Alternating pulse experiments ($^{12}CH_4/^{13}CO_2$) are carried out in the TAP-2 reactor [14,21]. Table 1 shows that the catalysts present close steady-state turn-over frequencies, making Ru more active than Ni since these catalytic tests are performed at 550 and 600°C, respectively. The TAP methane conversion is important for all catalysts (>50%) but large differences are observed for H_2 selectivity. An effect of support is also noted, the alumina tending to decrease the H_2 selectivity by comparison with silica.

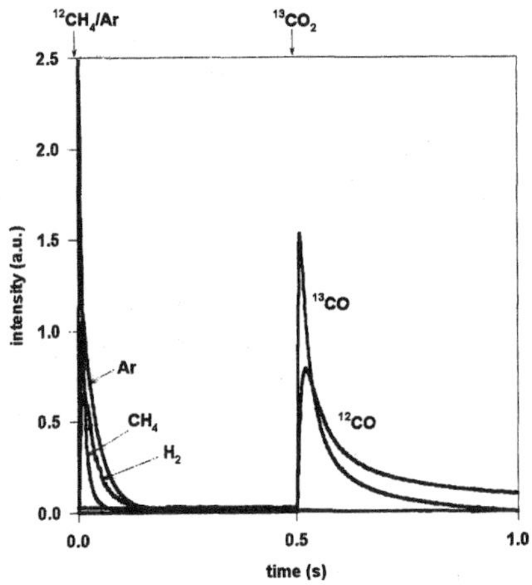

A : Ni/SiO_2

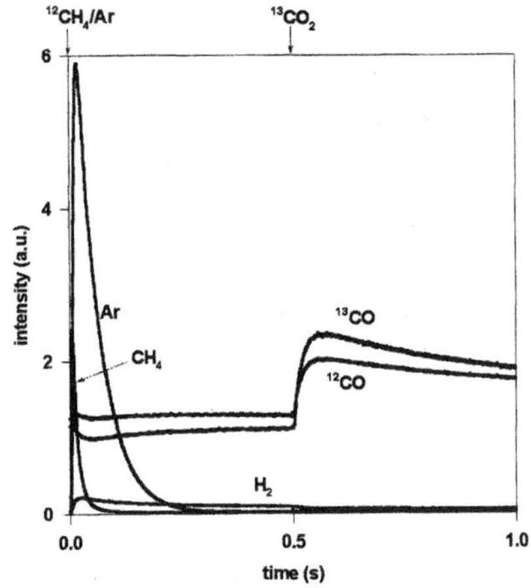

B : $Ni/\alpha Al_2O_3$

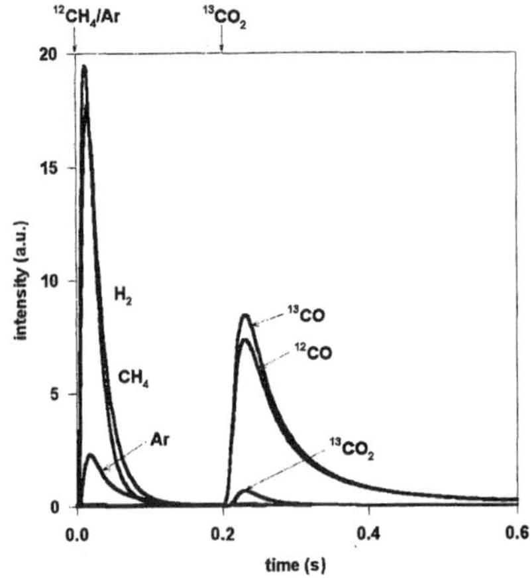

C : Ru/SiO_2

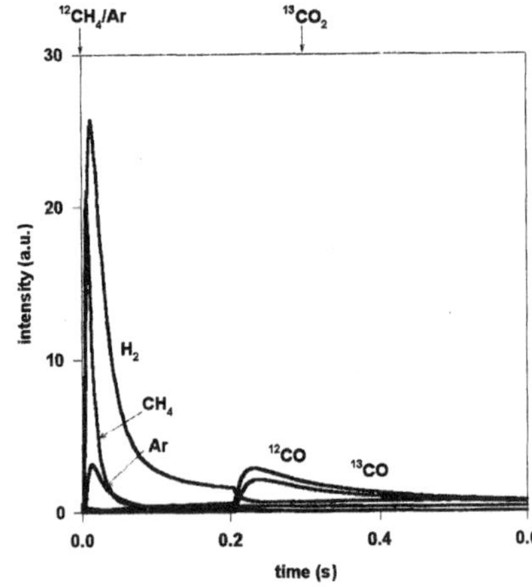

D : $Ru/\gamma Al_2O_3$

Fig. 1 : TAP alternating pulse experiments, $^{12}CH_4/Ar\ (9/1)/^{13}CO_2$ at 600 (A,B) and 550°C (C,D)

Table 1 *Characterization and catalytic data: i) Turn-over frequency at steady-state and ii) CH_4 conversion and H_2 selectivity in the TAP $^{12}CH_4/^{13}CO_2$ alternating pulse experiments.*

catalyst	metal loading (wt%)	metal dispersion %	reduction temperature (°C)	degree of reduction (%)	Turn-over frequency (s^{-1})	TAP experiments	
						X_{CH4}	S_{H2}
Ni/SiO_2	4.0	10	700	97	5.0^a	0.87	0.10
$Ni/\alpha\text{-}Al_2O_3$	4.3	17	750	62	7.2^a	0.99	0.01
Ru/SiO_2	0.7	13	400	nd*	3.5^b	0.50	0.95
$Ru/\gamma\text{-}Al_2O_3$	0.6	51	400	nd*	1.0^b	0.75	0.67

* not determined, [a] measured at 600°C, [b] measured at 550°C

Ni/SiO_2. Figure 1A shows the responses on a pulse of $^{12}CH_4/Ar$ followed 0.5 s later by a pulse of $^{13}CO_2$. H_2 gas is released on the methane pulse. On the $^{13}CO_2$ pulse, ^{13}CO is formed instantly but ^{12}CO response is delayed and tailing.

$Ni/\alpha\text{-}Al_2O_3$. Broader responses of both H_2 and CO are observed (Figure 1B). The CO response has become a continuous signal. The H_2 response decreases suddenly on the $^{13}CO_2$ pulse. Moreover, the H_2 selectivity is much lower than over Ni/SiO_2 (Table 1).

Ru/SiO_2. A different behaviour is observed over Ru by comparison with Ni (Figure 1C). Again the H_2 production is observed upon the introduction of CH_4 with a very high selectivity (Table 1). On the $^{13}CO_2$ pulse, the ^{12}CO and ^{13}CO responses are identical.

$Ru/\gamma\text{-}Al_2O_3$. H_2 is produced on the CH_4 pulse and its response tails considerably, until the introduction of $^{13}CO_2$, upon which a sudden H_2 consumption appears (Figure 1C). The CH_4 conversion is higher with a lower H_2 selectivity as compared to the Ru/SiO_2 catalyst. On the $^{12}CH_4$ pulse the ^{12}CO production already starts, with a superposition of the transient production on the $^{13}CO_2$ pulse. The ^{12}CO and ^{13}CO responses corresponding to the $^{13}CO_2$ pulse are identical but significantly broader than those observed over Ru/SiO_2. Larger $^{13}CO_2$ pulses lead to a higher $^{13}CO/^{12}CO$ ratio.

Role of the metal. Whatever the metal, the H_2 production of hydrogen takes place on the methane pulse. For Ni/SiO_2, it can be written as follows :

$$^{12}CH_4 + Ni \rightleftarrows Ni^{12}C + 2H_2 \qquad (1)$$

Step (1) is assumed to be reversible since fast H/D exchange reactions are observed when switching from CH_4 to CD_4 under steady-state conditions [10]. The intermediate formation of hydrogenated species CH_x is therefore likely. However, step (1) also leads to the accumulation of stable surface carbide, as demonstrated by TPH and XPS [10, 16].

The distinct responses for unlabeled and labelled CO suggest two different routes for the production of CO over Ni/SiO_2 :

$$^{13}CO_2 + Ni \rightleftarrows NiO + {}^{13}CO \qquad (2)$$
$$NiO + Ni^{12}C \rightarrow 2Ni + {}^{12}CO \qquad (3)$$

In step (2), the dissociative adsorption of CO_2 leads to surface oxygen and gaseous CO, since no O_2 release is observed. In contrast, CO is immediately released into the gas phase, as indicated by the absence of tailing for the corresponding TAP response. Step (2), strongly displaced towards the CO_2 dissociation under TAP conditions is equilibrated under steady-state conditions (similar isotopic composition between CO and CO_2 [10]).

Step (3) corresponds to the reaction of surface carbon (arising from step (1)) with surface oxygen atoms (arising from step (2)) into CO. This step 3 can be proposed as rate limiting over Ni/SiO_2 due to the tailing of the related TAP response. This is consistent with the absence of H/D isotopic effect, indicating that the RDS of the process does not involve C-H bond cleavage or formation [10].

Surface oxygen produced upon CO_2 activation (step 2) is stable enough to react with surface carbon since a large part of the hydrogen formed from methane cracking in step (1) is readily oxidized into H_2O (Table 1) though no more gaseous CO_2 crosses the catalytic bed. Thus, the following step must be added to the mechanism :

$$H_2 + NiO \rightleftarrows H_2O + Ni \qquad (4)$$

Like step (2), step (4), strongly displaced towards H_2O formation under TAP conditions, is at equilibrium under steady-state conditions. The combination of steps (2) and (4) forms the WGS equilibrium, generally achieved under reforming conditions [13, 17].

Over Ru/SiO_2 (Figure 1C), the same step (1) of methane cracking is observed, but leading to a much larger H_2 production (Table 1).

$$^{12}CH_4 + Ru \rightarrow Ru^{12}C + 2H_2 \qquad (5)$$

At variance with Ni/SiO_2, the two identical CO responses indicate a unique step:

$$^{13}CO_2 + Ru^{12}C \rightleftarrows {}^{13}CO + {}^{12}CO + Ru \qquad (6)$$

This implies that no surface oxygen accumulates on the Ru surface, which rules out a step equivalent to step (4) for Ru, in agreement with the high H_2 selectivity (Table1). The reaction step (6) is also proposed by Mark and Maier [8] over Ru and Ir/Al_2O_3 catalysts. The CH_4 activation, step (5), is for this case rate determining and step (6) is fast. This conclusion agrees with the fact that Ru is known to maintain carbonaceous residue for a longer period than Ni, which favours the chain growth for Fischer-Tropsch synthesis [18].

Role of the support. Much broader responses of both CO and H_2 products are observed on alumina as compared to silica supported Ni and Ru catalysts (Figures 1B and 1D). For both catalysts ^{13}CO and ^{12}CO are almost continuously produced along the $^{12}CH_4/^{13}CO_2$ pulse experiments. Similarly, a large tailing of hydrogen is observed, even if the H_2 production remains far more larger on Ru than on Ni, as discussed earlier. An important effect to be stressed is also the marked decrease of this continuous H_2 production upon the CO_2 pulse.

γ-Al_2O_3 is well known to act as a reservoir of hydroxyl groups, which favours the spill-over of OH groups (or adsorbed water) to the metal surface. In turn, these metal hydroxyls react fast with surface carbon into CO and H_2 as follows :

$$OH_{AL2O3} + Me \rightarrow MeOH \qquad (7)$$
$$MeOH + MeC \rightarrow CO + MeH + Me \qquad (8)$$
$$2\,MeH \rightarrow H_2 + 2\,Me \qquad (9)$$

The consumed alumina OH groups are regenerated by water formed during the process (through the reverse WGS reaction). A close process was recently proposed for the partial oxidation of methane over Rh/Al_2O_3 [19].

Steps (7-9) thus provide a residual and continuous flow of H_2 and CO (in addition to the production related to the CO_2 pulse), as experimentally observed. This supply of oxygen from the support to the metal may explain the coking resistance of the alumina-supported samples in comparison with the silica-supported ones. The continuous production of CO (labelled and unlabeled) is also found much larger than the one of hydrogen over Ni/Al_2O_3 (Figure 1B). This means that for this case, the alumina also acts as a reservoir of reversible CO_2 and/or CO (formate and/or carbonate groups) :

$$O^{2-} + CO_2 \rightleftarrows CO_3^{2-} \quad (10)$$
$$-OH + {}^{12,13}CO \leftrightarrows -OCHO \quad (11)$$

The latter type of equilibrium was also reported by Amenomiya [20].

The sudden drop of the H_2 response on the CO_2 pulse in Figures 1B and D confirms the reverse WGS reaction, favoured on alumina via the above spill-over (steps 7-9) and CO/formate equilibrium (step 11).

Thus, the above TAP study permitted to discriminate the main elementary steps of CH_4 and CO_2 activation and to get direct information about the origin of the formed CO for various catalysts. The H_2 selectivity was straightforwardly related to the secondary oxidation of H_2 into H_2O, deriving from the presence (Ni) or not (Ru) of oxygen on the metal. Unlike the silica support, the alumina support was also shown to play an intricate role in the mechanism and the side reaction of coke deposition, due to participation of its acid/base groups.

Modelling of TAP data. Based on the above conclusions, a quantitative modelling of TAP dry reforming curves was attempted over Ni/SiO_2 and Ru/SiO_2 [to be published].

The diffusion and the kinetic processes in the catalyst zone are described by the following continuity equations, based on the steps (12) and (13), which represent the reversible dissociative activation and an unknown irreversible step, respectively :

$$CH_4 + 2^* \rightleftarrows CH_3^* + H^* \quad (12)$$
$$CH_3^* \longrightarrow \quad (13)$$

$$\varepsilon \frac{\delta C_{CH4}}{\delta t} = D_{CH4} \frac{\delta^2 C_{CH4}}{\delta z^2} - a_s S_v (1-\varepsilon)(k_a C_{CH4}(1-\theta_{CH3*}-\theta_{H*})^2 - k_d \theta_{CH3*}\theta_{H*}) \quad (14)$$

$$\frac{\delta \theta_{CH3*}}{\delta t} = k_a C_{CH4}(1-\theta_{CH3*}-\theta_{H*})^2 - k_d \theta_{CH3*}\theta_{H*} - k_r \theta_{CH3*} \quad (15)$$

The continuity equations for the quartz zones contain only the accumulation and diffusion term and are omitted here. The initial and boundary conditions which apply in the TAP reactor are reported in [21].

The obtained set of partial differential equations is integrated numerically, by approaching the right-hand side of the equations by finite differences. A variable time-step algorithm is used to calculate properly the steep transient concentration gradients. Optimisation of the parameters is carried out by applying Marquardt's algorithm [22].

Table 2 Kinetic parameters determined from TAP curves modelling

pre-exponential factors	Ni	Ru	energies ($kJ\ mol^{-1}$)	Ni	Ru
adsorption ($m^3\ mol^{-1}\ s^{-1}$)	$1.7\ 10^3$	$5.4\ 10^3$	adsorption	34 ± 3	42 ± 4
desorption (s^{-1})	$1.0\ 10^9$	0	desorption	70 ± 3	0
reaction (s^{-1})	$1.9\ 10^4$	n.d.	reaction	51 ± 6	n.d.

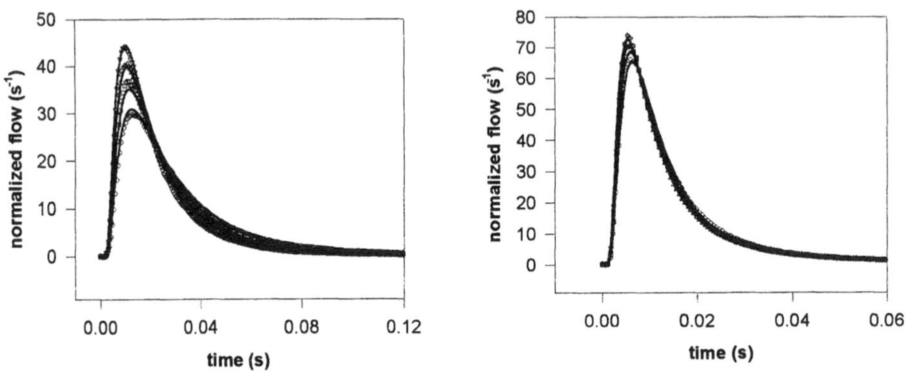

A - Ni/SiO_2 B - Ru/SiO_2

Figure 2: CH_4 TAP data (experimental and calculated) at 773, 823, 873, and 923 K

The transient responses of methane over both the nickel and the ruthenium, shown in Figures 2 A and B, were modelled according to reaction steps (12) and (13). All adsorption sites were assumed to be available for the CH_4 chemisorption, although a slight carburation and/or oxidation of the surface could have taken place during previous pulses. Due to a strong correlation between the active site concentration and the adsorption rate constant, the former was fixed at the value calculated from the dispersion measurements. All other diffusion and kinetic parameters were estimated by a non-linear regression of the transient responses. Initial values for the diffusion coefficients were calculated from independent experiments. Initial values for the adsorption, desorption and reaction rates were obtained by regression of the data at separate temperatures. Initial values for the activation energies were then calculated from the corresponding Arrhenius plots. 95% confidence intervals were found on the estimated parameters statistically significant with a 95% confidence interval of approximately 10%. The correlation coefficient were lower than 0.82.

The CH_4 activation over Ni could only be described satisfactorily if a reversible CH_4 chemisorption and a further CH_4 dissociation was taken into account. In the case of Ru no difference in the model fit was observed between reversible and irreversible CH_4 activation, indicating a negligible CH_4 desorption. These results correspond well with the previous TAP analysis, indicating reversible and irreversible activation of CH_4 on Ni and Ru, respectively.

The activation energies for the methane chemisorption (Table 2) falls in the range of values reported for methane activation over transition metal surfaces [23-25]. For both metals a sticking coefficient of 10^{-6} at 873 K was calculated from the chemisorption rate constant, which compares well with the values measured by Chorkendorff et al. [26] for the chemisorption of CH_4 on Ni(100). The value of the pre-exponential factor for CH_4 desorption from Ni indicates, according to the transition- state theory, a mobile adsorbate and mobile transition state with a large degree of rotational freedom [6]. The difference between the activation energies for adsorption and desorption is equal to the heat of adsorption, which amounts to -36 kJ mol^{-1}. This value is close to the value of -42 kJ mol^{-1} reported in [13].

No meaningful evaluation of the reaction rate parameters for Ni can be done, since this step (13) takes only into account the disappearance of the CH_3* groups, but has no real physical basis (e.g., other slow steps could exist later on in the process). To do this, the development of a full model, involving all reacting and products species, is in progress.

2. SYNGAS METHANATION

The CO hydrogenation into methane and higher hydrocarbons is a particularly favourable example for applying the transient techniques since it combines fast steps such as H_2 and CO activation, and surface and bulk accumulation of reactive and less or unreactive species [27, 28]. In addition, slow processes such as metal sintering may also superimpose to the reaction kinetics as shown by Agnelli et al. [29].

Several mechanisms have been proposed up to now. Alstrup [27] proposed recently a micro-kinetic model to account for the literature data obtained on Ni single crystal surfaces [30]. The coverage of active surface carbon, θ_C, is assumed to be constant for the range of temperature considered (0.05 monolayer, ML). However, much larger carbon accumulation values were found under steady-state conditions (θ_C = 0.38 ML, as shown by Agnelli et al. [28]), or as large as several monolayers as reported by Biloen et al. [31]. Thus, an utmost important question remains to evaluate the exact concentration of "active" carbon and its possible link with "inactive" or "less active" species on the working surface.

An identical surface site and stoichiometry for any adsorbed intermediate species - CO, H, C, CH_x, O, OH - was considered in [27]. Differently, Aparicio [13] assumed two sites for CO adsorption, based on changes in sticking coefficient and adsorption energy observed on single crystals. In addition, the hydrogenation of adsorbed CO into CHO intermediate was found to involve five Ni sites. The "ensemble model" proposed by Dalmon and Martin in [32] assumed that the methanation rate was proportional to $(1 - \theta_{CO})^W$ where θ_{CO} is the CO coverage and W the number of Ni atoms forming an active site. The W value, however, could not be calculated since θ_{CO} was not known under steady-state reacting conditions [32].

By various isotopic transient techniques, we measured directly the steady-state coverage of adsorbed CO and of active intermediates CH_x as detailed in [28, 33]. As shown in Figure 3, the coverage in adsorbed CO decreases and the one in carbonaceous intermediate (CH_x or CD_x) increases with temperature and tends to the limit of one monolayer of surface carbide Ni_3C. The differences observed between hydrogenation and deuteration may directly be related to the reverse isotopic effect generally observed for the methanation reaction [33].

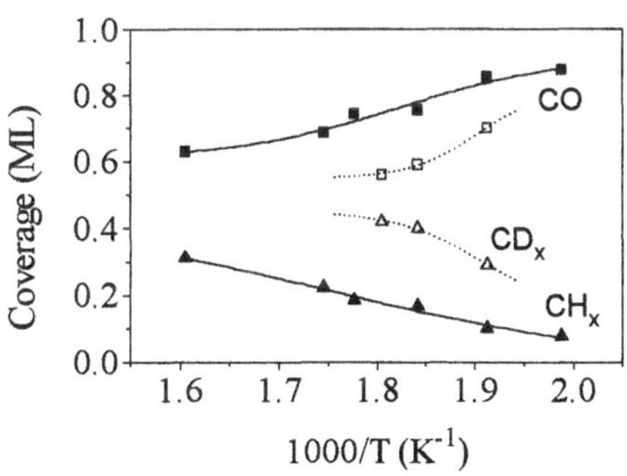

Figure 3. CO, CH_x and CD_x steady-state coverages under $CO+2H_2$ (full symbols) and $CO+2D_2$ (open symbols) reactions over Ni/SiO_2.

From these θ_{CO} and θ_{Chx} values, the following rate equation is proposed :

$$r = k_o \exp(-E_o/RT)\, P_{H2}\, (1-\theta_{CO})^W \qquad (24)$$

with $k_0 = 3\ 10^6$ molec s^{-1} $site^{-1}$ bar^{-1}, which gives a sticking coefficient of $4\ 10^{-3}$, E_0 = 62 kJ and W = 1.2. This rate equation expresses directly that the rate of methanation is controlled by the probability of a H_2 molecule to collide with a site formed of 1-2 nickel atoms free from adsorbed CO. Further isotopic transient evidence presented in the present issue [33] allowed us to propose that the most abundant CH_x intermediates were CH species, indicating that the elementary RDS was the hydrogenation of CH species into CH_2. Thus a large accumulation of carbide species was shown to exists both under methanation and reforming conditions.

3. CONCLUSION

These examples of recent kinetic studies of the methane reforming and methanation demonstrate the power of the transient methods to unravel mechanisms and provide rapidly reliable kinetic data. The limits of such methods are also outlined by observing that distinct and possibly wrong models may account for experimental kinetic data, when the number of parameters to be determined increases due to the reaction complexity.

References

1 Mirodatos, C., Catalysis Today 9 (1991) 83.
2 Gleaves, J.T., Ebner, J.R. and Kuechler, T.C., Catal. Rev.-Sci. Eng. 30 (1988) 49.
3 "Dynamics of surface and reaction kinetics in heterogeneous catalysis", G.F. Froment et al, Eds., Studies in Surface Science and Catalysis, Elsevier, Amsterdam, 109 (1997).
4 "Transient Kinetics", P. Szedlacsek and L. Guczi, Eds., Appl. Catal., 151 (1997).
5 "Kinetic Methods in Heterogeneous Catalysis", G.B. Marin, R.A. van Santen, Eds., Appl. Catal., 160 (1997).
6 "The Microkinetics of Heterogeneous Catalysis", J.A. Dumesic, D.F. Rudd, L.M. Aparicio, J.E. Rekoske and A.A. Trevino, ACS, Washington, DC, 1993.
7 Rostrup-Nielsen, J. R., and Bak-Hansen, J. H., J., Catal. 144 (1993) 38.
8 Mark, M. F., and Maier, W. F., J. Catal. 164 (1996) 122.
9 Kroll, V.C.H., Swaan, H.M., and Mirodatos, C., J. Catal. 161 (1996) 409.
10 Kroll, V.C.H., Swaan, H.M., Lacombe, S., and Mirodatos, C., J. Catal. 164 (1996) 387.
11 Erdöhelyi, A., Cserenyi, J., andSolymosi, F., J. Catal. 141 (1993) 287.
12 Slagtern, A., Schuurman, Y., Leclercq, C., Verykios, X., and Mirodatos, C., J. Catal. 172 (1997) 118.
13 Aparicio, L.M., J. Catal. 165 (1997) 262.
14 Schuurman, Y., Kroll, V.C.H., Ferreira-Aparicio, P., and Mirodatos, C., Catal. Today 38 (1997) 129
15 Schuurman, Y., and Mirodatos, C., Appl. Catal. 151 (1) (1997) 305-331.
16 Kroll, V.C.H., Delichère, P., and Mirodatos, Kinetics and Catalysis, 37 (5) (1996) 698.
17 Swaan, H.M., Kroll, V.C.H., Martin, G.A., and Mirodatos, C., Catalysis Today 21 (1994) 571.
18 de Pontes, M., Yokomiso, G.H., and Bell, A.T., J. Catal., 104 (1987) 147.
19 Wang, D., Dewaele, O., De Groote, A.M., and Froment, G., J. Catal., 159 (1996) 418.
20 Amenomiya, Y., J. Catal. 55 (1978) 205; Amenomiya, Y., and Pleizer, G., J. Catal. 76 (1982) 345.
21 Gleaves, J.T., Yablonskii, G.S., Phanawadee, P., and Schuurman, Y., Appl. Catal. 160 (1997) 55.
22 Marquardt, D.W., J. Soc. Indust. Appl. Math. 11 (1963) 431.
23 Kuijpers, E.G.M., Jansen, J.W., van Dillen, A.J., and Geus, J.W., J. Catal. 72 (1981) 75.
24 Mallens, E. P. J., Hoebink, J.H.B.J., and Marin, G.B., J. Catal. 167 (1997) 43.
25 Beebe Jr., T.P., Goodman, D.W., Kay, B.D., andYates Jr., J.T., J. Chem. Phys. 87 (1987) 2305.
26 Chorkendorff, I., Alstrup, I., and Ullman, S., Surf. Sci. 227 (1990) 291.
27 Alstrup, I., J. Catal. 151 (1995) 216.
28 Agnelli, M., Swaan, H.M., Marquez-Alvarez, C., Martin, G.A., and Mirodatos, C., J. Catal. 174 (1998) 117.
29 Agnelli, M., Kolb, M., and Mirodatos, C., J. Catal. 148 (1994) 9.
30 Goodman, D.W., Kelley, R.D., Madey, T.E., and White, J.M., J. Catal. 64 (1980) 479.
31 Biloen, P., J. Mol. Catal. 21, 17 (1983); Biloen, P., Helle, J.N., Van den Berg, F.G.A. and Sachtler, W.H.M., J. Catal. 81 (1983) 450.
32 Dalmon, J.A., and Martin, G.A., J. Catal. 84 (1983) 45.
33 Márquez-Alvarez, C., Martin, G.A., and Mirodatos, C., to be published in this issue.

NATURAL GAS CONVERSION V
Studies in Surface Science and Catalysis, Vol. 119
A. Parmaliana et al. (Editors)

Modified alumina supports for cobalt Fischer-Tropsch catalysts

F. Rohr[b], A. Holmen[a,b], K.K. Barbo[a], P. Warloe[a], E.A. Blekkan[a,b]

[a]Department of Industrial Chemistry, Norwegian University of Science and Technology (NTNU), N-7037 Trondheim, Norway

[b]SINTEF Applied Chemistry, N-7037 Trondheim, Norway*

The effect of promotion with zirconia on alumina supported cobalt FT-catalysts has been studied. The promoted catalysts show an increase in activity by up to 50 % compared to the unpromoted systems in CO hydrogenation at 5 bar total pressure. The zirconia is introduced as an addition to the support before impregnating with the active metal. For a constant zirconia loading the activity increases with increasing cobalt loading. The increase is asymptotic towards high cobalt loadings as opposed to the unpromoted catalysts which show a linear dependency with cobalt amount. No increase in the cobalt reducibility is observed with TPR-experiments. Neither does hydrogen chemisorption detect any increase in the cobalt dispersion. SSITKA data recorded at low pressure reveals that the overall gain in the catalytic activity is due to an increased coverage with reactive intermediates, and is not caused by an increase of the intrinsic turnover frequency.

1. Introduction

Supported cobalt catalysts are the preferred catalysts for the Fischer-Tropsch synthesis of long chain paraffins from syngas made from natural gas. Alumina is much used as a support for cobalt catalysts due to favourable mechanical properties, but cobalt supported on alumina has a limited reducibility due to a strong interaction between the support and the cobalt oxides [1-3]. This can to a certain extent be overcome by promotion with easily reducible metal promoters, e.g. Pt [2,4]. The Fischer-Tropsch synthesis is generally believed to be structure insensitive and the specific activities at practical conditions are reported to be independent of the support [5].

Catalytic modifiers or promoters can influence the activity or the selectivity of catalysts, either through structural modification or through chemical or electrostatic interaction with the catalytically active material. Zr is reported to be a promoter for Fischer-Tropsch synthesis over Co/SiO_2 [6], where Zr leads to improved activity. In the present work we investigate the possibility of changing the support interaction between cobalt and alumina by adding zirconia to the alumina support and studying the changes in the catalytic properties of cobalt supported on this new material. The use of SSITKA (Steady State Isotopic Transient Kinetic Analysis) [7-9] allows the measurement of intrinsic turnover frequencies as well as the surface coverage with

* The authors are grateful for the financial support for this work from the EU through contract no. JOF3 CT950016

reaction intermediates at steady state conditions. This allows the determination of the origin of an effect leading to a change in the activity of the catalyst.

2. Experimental

The catalyst samples were prepared by means of sequential impregnation of the alumina support with solutions of the cobalt and zirconia precursors respectively. The zironia impregnation was always carried out first. The samples were dried (80°C for 24 hrs in air) and calcined (350°C for 2 hrs in air) after each impregnation step. Different Co and zirconia loadings were achieved by varying the concentration of the precursor solutions. In the case of the samples with the highest zirconia loading the zirconia impregnation had to be carried out twice due to limited solubility of the zirconia precursor. As the starting material for the support commercial Al_2O_3 (boehemite-phase) from Kaiser Chemicals was used. An organic solution of Zr^{IV}-isopropoxid was chosen for the zirconia impregnation. The solvent used was a mixture of toluene, acetylacetone and isopropanol (1:1:1), all p.a. qualities. Cobalt impregnation was accomplished with an aqueous solution of Co^{II}-nitrate. To specify the particular samples the following notation wil be used: A catalyst containing 12 weight percent cobalt and 10 weight percent zirconia supported on alumina will be denoted 12Co10Zr. The results will be compared to unpromoted reference catalysts of the type x weight% cobalt / Al_2O_3. These catalysts will be denoted xCo (e.g. 12Co). The reference catalysts were prepared using basically the same procedure as the one outlined above but with the promotion step omitted.

Testing of CO hydrogenation activity was carried out at methanation conditions in a fixed-bed microreactor (H_2:CO 9:1; T=245°C; 5 bar total pressure; catalyst loading: 100 mg calcined catalyst; feed-rate: 250 ml/min). Prior to the reaction the catalysts were reduced in flowing hydrogen at 350°C and 1 bar pressure. Product analysis was performed online with a HP 5880 GC equipped with TCD and FID detection, allowing for the analysis of CO, CO_2, CH_4 and hydrocarbons up to C_9.

Different analytical techniques were used to characterise the catalysts. The reducibility was measured by means of TPR (Temperature Programmed Reduction). The experiments were performed on the calcined samples without any further pretreatment. All TPR-measurements involved heating at a rate of 10°C/min to 900°C with a feed gas consisting of 7% hydrogen diluted in argon. Quantitative calibration of the TPR-signal was carried out by a reference experiment on Ag_2O which is fully reducible under the conditions reported here.

Hydrogen chemisorption measurements were employed to investigate the cobalt dispersion. The H_2-isotherms were recorded using a Micromeritics ASAP 2010 volumetric glass apparatus. Prior to the measurements the catalysts were reduced in hydrogen at 350°C for 16 hours. Subsequently, the samples were outgassed in vacuum at 350°C for 30 minutes. The total adsorption isotherm was then measured at 100°C between 0.2 and 1 bar pressure.

Steady state isotopic transient kinetic (SSITKA) measurements were performed in order to determine the coverage with reactive intermediates and the intrinsic turnover frequency under steady state conditions [7-9]. The measurements were carried out in a differential quartz microreactor. Reduction was carried out *in situ* using the standard reduction conditions reported above. A mixture of 20 vol% synthesis gas (H_2:CO ratio 10:1) diluted in helium at 1.8 bar total pressure and 100 ml/min total feed rate was used as the reaction gas feed. The reaction temperature was 230°C. The CO-line was switched between an isotopically labelled (^{13}CO) and unlabelled supply (^{12}CO). The isotopic composition of the product stream was monitored with a

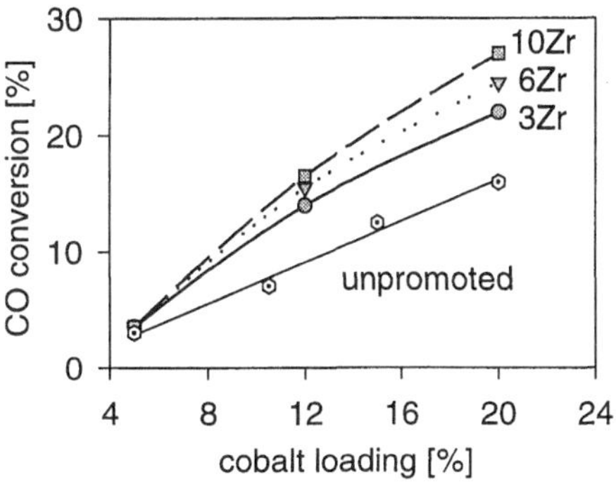

Figure 1. CO conversion as a function of cobalt loading for promoted and unpromoted catalysts.

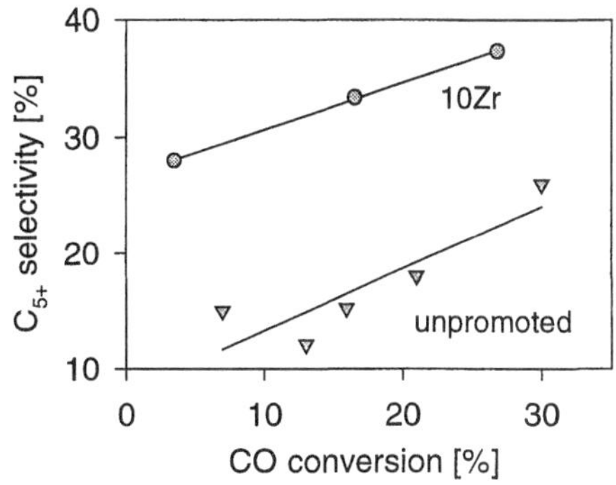

Figure 2. C_{5+} selectivity for 10% Zr-promoted and unpromoted catalysts as a function of CO conversion

Balzers QMG 420 masspectrometer. The assumptions and simplifications used in the calculations are described in [2].

3. Results and discussion

In Figure 1 the catalyst activity at 5 bar (in terms of CO conversion) is depicted as a function of the cobalt loading for three different zirconia loadings together with the unpromoted reference catalysts. A summary of all of the activity measurements at 5 bar is given in Table 1. For the promoted samples the activity increases with zirconia loading at constant cobalt amount. For constant zirconia loading the activity also increases with cobalt loading. However, the rise is not linear and levels out for high cobalt loadings. This is different for the unmodified system which shows a linear increase in activity even beyond the range of Co loadings presented here (25%).

Table 1
Results from characterisation and activity measurements at 5 bars

Catalyst	S_{BET} (m^2/g)	$r_{CO}\times10^{-5}$ (mol/g,s)	TOF (s^{-1})	S_{CH4} (%)	S_{C5+} (%)	Dispersion (%)
5Co3Zr	271	0.7	1.7	34	27	0.5
12Co3Zr	238	2.6	0.8	24	30	1.6
20Co3Zr	215	3.9	0.8	22	30	1.5
5Co6Zr	273	0.7	1.1	40	26	0.8
12Co6Zr	247	2.9	0.5	23	34	2.6
20Co6Zr	221	4.3	0.6	18	37	2.2
5Co10Zr	271	0.7	1.4	36	28	0.6
12Co10Zr	239	3.0	0.8	22	33	1.9
20Co10Zr	218	4.5	0.8	19	37	1.8
10.5Co	n.m.	1.4	0.3	37	15	2.5
15Co	n.m.	2.4	0.4	40	12	2.5
20Co	n.m.	2.9	0.3	37	15	2.7

To summarise these findings it can be concluded that zirconia modification is most effective with relatively high zirconia and cobalt loadings respectively, but proves to have no major effect on the activity when it comes to low cobalt loadings.

Promotion with zirconia also has profound effects on the product distribution. The yield of the C_{5+} fraction of the hydrocarbon product spectrum increases for all of the zirconia promoted samples (see Figure 2) compared to the unpromoted catalysts. However, the specific amount of Zr does not seem to lead to any major differences for the three different Zr loadings we investigated.

Total surface areas and hydrogen chemisorption data are reported in Table 1. Promotion with Zr does not lead to an increase in cobalt dispersion compared to the unmodified system. For most modified samples the observed chemisorption values are lower. For the three different Zr loadings the cobalt dispersion is highest for the 12% Zr samples, this sample gives a H:Co ratio similar to that of unmodified catalysts with similar cobalt loading. The total surface area varies with cobalt loading but there is little difference between the different Zr loadings.

The TPR-data demonstrates that zirconia modification has a profound effect on the reduction behaviour of the catalyst. In Figure 3 TPR-data for the samples 12Co10Zr, 12Co and for the modified support without cobalt are shown. Both Co catalysts show corresponding peaks but with different relative intensity. The low temperature peak around 550 K is absent on the modified support and only weak for the unmodified catalyst but proved to be of high intensity for all the Zr containing samples. The peak is very sensitive to calcination temperature and time and disappears after calcination around 400°C. It is assigned to nitrate that is still present on the catalyst after

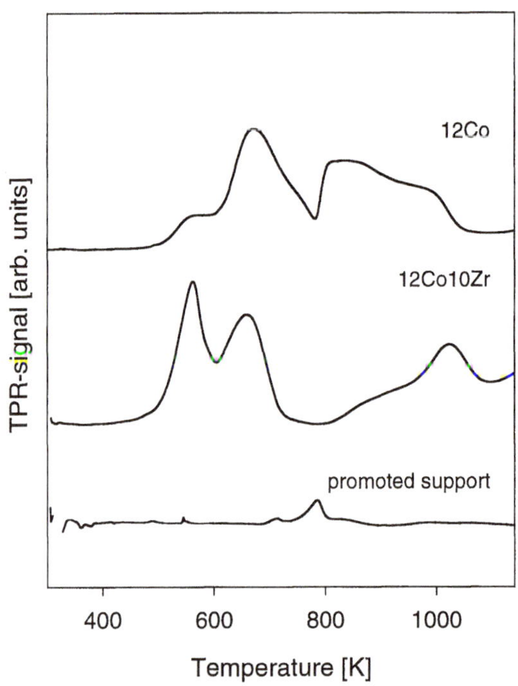

Figure 3. TPR-profiles for the catalysts 12Co and 12Co10Zr. TPR-data for the promoted alumina support without cobalt is also included.

Table 2
Turnover frequencies (TOF) obtained under different reaction conditions (s^{-1})

	SSITKA		Activity-testing
	Apparent TOF	Intrinsic TOF	Apparent TOF
20Co10Zr	0.018	0.13	0.75
12Co10Zr	0.020	0.13	0.79
12Co	0.016	0.13	0.31

calcination, as also proposed elsewhere [10,11]. However, the modified catalysts seem to have a higher affinity to nitrate reflected by the higher calcination temperatures necessary to remove the nitrate. The reason for this is not clear at the moment.

The broad peak around 660 K is attributed to bulk-like cobalt oxide (Co_3O_4). The two different oxidation states of the cobalt ions are not resolved. Zirconia promotion leaves this peak nearly unchanged. The broad feature between 750 and 1000 K is generally agreed to be a cobalt oxide surface species interacting strongly with the support. Zirconia modification leads to a substantial redistribution of the TPR-signal in this region. Basically the TPR-signal seems to shift to higher temperature for the promoted catalysts. This result is in line with the chemisorption data; the larger contribution of the high temperature peak indicates a large fraction of cobalt oxide that would not be reduced under standard reduction conditions.

On the basis of thermodynamic reasoning Zr should be present in oxidic form on the reduced catalyst. In order to check this we recorded TPR-data for the modified support alone. The TPR-profile indicates nearly zero reducibility (see Figure 3). However, we cannot fully exclude zirconia reduction on the cobalt containing sample where the metal could act as a reduction promoter.

A quantitative analysis of the TPR-data shows that the total TPR area and in turn the degree of reduction does not change with zirconia promotion. The degree of reduction under CO hydrogenation conditions may even be slightly lower for the promoted catalysts because of the accumulation of TPR-signal in the high temperature region. These cobalt species are less reducible under the standard reduction procedure at 350°C reported above. Therefore the increase in catalytic activity with zirconia promotion is clearly not caused by an increased cobalt reducibility.

In Table 1 the specific activities (turnover frequencies) are reported. The general trend is that the TOF is high for the 5% Co samples. These samples show very low activities and also low dispersions, indicating that very low cobalt loadings are not very suited for Fischer-Tropsch synthesis catalysts. For the samples with cobalt loadings above 10 wt% the results indicate that zirconia promotion gives improved activity without any improved dispersion, hence there is an apparent increase in the turnover frequency. In order to investigate the reason for this behaviour some samples were investigated further using the SSITKA technique. The experiments were performed at lower pressure (1.8 bar total pressure, thereof 20% syngas with a ratio hydrogen : CO = 10 : 1). Both steady state data and SSITKA-data (time constants for the surface sojourn) were recorded and are reported in Table 2. The measurements at the lower pressure do not reproduce the data recorded at the higher pressure very well for these catalysts. The catalyst activity increases slightly following the cobalt content, and there are only small differences in the turnover frequencies. Most significantly is the time constant (here expressed as $1/\tau$, the average intrinsic or true turnover frequency) equal for all the catalysts, indicating that any differences in specific activity must be linked with the surface coverage of intermediates. The zirconia-promoted samples show a slightly higher surface coverage, indicating an effect of zirconia on the adsorption properties of the cobalt. Similar results have been reported earlier for other promoters like Pt [2] and Re [12]. This indicates that these promoters when used in cobalt Fischer-Tropsch catalysts generally only affect the reducibility and the dispersion of the cobalt (metallic promoters) and also the ability to maintain a clean and hence active cobalt surface, observed as the surface coverage of active intermediates leading to products.

4. Conclusions

Active supported cobalt catalysts for the Fischer-Tropsch synthesis have been prepared by modifying the alumina support by impregnating zirconia on the surface. The catalysts show higher activities and improved selectivities to higher hydrocarbons in CO hydrogenation. The zirconia-promoted samples do not have higher dispersions or reducibilities compared to the unpromoted samples, and SSITKA-experiments at low pressures using a switch between ^{12}CO and ^{13}CO showed identical true turnover frequencies for samples with and without zirconia. The enhanced activity can then be attributed to differences in the surface coverage of reactive intermediates leading to hydrocarbon products, and the promoter effect of zirconia must be linked with this feature.

REFERENCES

1. E. Iglesia, S.L. Soled, R.A. Fiato and G.H. Via, J. Catal., 143 (1993) 345.
2. D. Schanke, S. Vada, E.A. Blekkan, A.M. Hilmen, A. Hoff and A. Holmen, J. Catal., 156 (1995) 85.
3. A. Kogelbauer, J.G. Goodwin and R. Oukaci, J. Catal., 160 (1996) 125.
4. A. Hoff, E.A. Blekkan, A. Holmen and D. Schanke, Stud. Surf. Sci. Catal., 75, (1992) 2067.
5. E. Iglesia, Appl. Catal., 161 (1997) 59.
6. S. Ali, B. Chen and J.G. Goodwin Jr., J. Catal., 157 (1995) 35.
7. J. Happel, I. Suzuki, P. Kokayeff and V. Fthenakis, J. Catal., 65 (1980) 59.
8. M. Rothaemel, K. Firing Hanssen, E.A. Blekkan, D. Schanke and A. Holmen, Catal. Today, 38 (1997) 79.
9. K.P. Peil, J.G. Goodwin Jr. and G. Marcelin, J. Catal., 132 (1991) 556.
10. A. Hoff, Ph.D.-thesis, Norwegian Institute of Technology, Trondheim, 1993.
11. P. Arnoldy and J.A. Moulijn, J. Catal., 93 (1985) 38.
12. S. Vada, A. Hoff, E. Ådnanes, D. Schanke and A. Holmen, Topics in Catalysis, 2 (1995) 155.

NATURAL GAS CONVERSION V
Studies in Surface Science and Catalysis, Vol. 119
A. Parmaliana et al. (Editors)

A precipitated iron Fischer-Tropsch catalyst for synthesis gas conversion to liquid fuels

Dragomir B. Bukur and Xiaosu Lang

Department of Chemical Engineering
Kinetics, Catalysis and Reaction Engineering Laboratory
Texas A & M University
College Station, TX 77843-3122, USA

Summary

A precipitated iron Fischer-Tropsch (F-T) catalyst with nominal composition 100 Fe/3 Cu/4 K/16 SiO_2 (in parts by weight) was tested in a stirred tank slurry reactor. The performance of this catalyst (activity and selectivity) under baseline activation (H_2 reduction at 240°C for 2 h) and process conditions (260°C, 1.48 MPa, 1.4 Nl/g-cat/h, H_2/CO = 2/3) was found to be comparable to that of the best state-of-the-art iron F-T catalysts [5]. In this study the repeatability of performance was demonstrated by performing multiple tests of the catalyst from the same preparation batch, whereas reproducibility of catalyst preparation procedure was demonstrated in tests of the catalyst from different batches.

Catalyst productivity (i.e. reactor space-time-yield) was increased by 40% (relative to baseline process conditions) by increasing reaction pressure from 1.48 MPa to 2.17 MPa, while simultaneously increasing gas space velocity in order to maintain a constant contact time in the reactor. The intrinsic activity of the catalyst was increased up to 75% through the use of different pretreatment procedures. The catalyst productivity in run SA-2186 at 260°C, 2.17 MPa, 3.4 Nl/g-cat/h and H_2/CO = 2/3 was 0.86 (g hydrocarbons/g-Fe/h) at syngas conversion of 79%, methane selectivity of 3% (weight percent of total hydrocarbons produced) and C_5+ hydrocarbon selectivity of 83 wt%. This represents a 75% increase in catalyst productivity relative to Rheinpreussen's slurry bubble column reactor [10].

The yield of liquid and wax hydrocarbons in different tests of our catalyst was 80-89%, and the Anderson-Schulz-Flory parameter was 0.92-0.94 (high alpha catalyst). This catalyst is ideally suited for production of high quality diesel fuels via hydrocracking of the F-T wax product.

1. INTRODUCTION

Fischer-Tropsch hydrocarbon synthesis from a coal derived synthesis gas has been practiced since mid 50's on commercial scale at SASOL plants in South Africa in fixed and fluidized bed reactors [1], and since 1993 in slurry bubble column reactors [2,3] utilizing potassium promoted iron catalysts. Two new commercial F-T plants, based on natural gas derived synthesis gas were constructed during this decade [4]. Mossgas plant in South Africa

This work was supported by U.S. DOE (Contract No. DE-AC22-94PC93069)

(using a promoted iron F-T catalyst and Sasol's circulating fluid bed reactor with capacity of ~7000 barrels/day) was commissioned in 1991, and Shell's plant at Bintulu, Malaysia was placed on stream in 1993 (tubular fixed bed reactors with cobalt catalyst processing about 100 MMSCFD of natural gas). Additional plants are expected to be built in a near future.

Further improvements in the catalyst performance (activity, stability and/or selectivity) would accelerate commercialization of this technology. Several catalysts ($Fe/Cu/K/SiO_2$) synthesized and tested at Texas A&M University (TAMU) in a stirred tank slurry reactor were found to be very active, while exhibiting high syngas conversion, good stability with time on stream and high selectivity to liquid and wax hydrocarbons [5]. Here, we report results from recent studies with one of these catalysts (100 Fe/3 Cu/4 K/16 SiO_2) illustrating: reproducibility of catalyst performance (multiple tests with the catalyst from the same preparation batch), repeatability of catalyst synthesis procedure (characterization and slurry reactor tests of catalysts from different preparation batches) and improvements in the catalyst performance through the use of either higher reaction pressure (increase in productivity) or different pretreatment procedures (increase in activity and improved catalyst stability).

2. EXPERIMENTAL

Experiments were conducted in a 1 dm^3 stirred tank slurry reactor (Autoclave Engineers). The feed gas flow rate was adjusted with a mass flow controller and passed through a series of oxygen removal, alumina and activated charcoal traps to remove trace impurities. After leaving the reactor, the exit gas passed through a series of high and low (ambient) pressure traps to condense liquid products. High molecular weight hydrocarbons (wax), withdrawn from a slurry reactor through a porous cylindrical sintered metal filter, and liquid products, collected in the high and low pressure traps, were analyzed by capillary gas chromatography. The reactants and noncondensible products leaving the ice traps were analyzed on an on-line GC (Carle AGC 400) with multiple columns and both flame ionization and thermal conductivity detectors [5-7].

Elemental analysis (by atomic absorption spectroscopy - in parts per weight) and textural properties (BET surface areas and pore volume) of catalysts from four different preparations are listed in Table 1. Details of catalyst preparation procedure were reported elsewhere [8].

Table 1
Elemental Analysis and Textural Properties of Synthesized Catalysts

Catalyst designation	S3416-1	S3416-2	S3416-3	S3416-4
Catalyst composition				
Fe	100	100	100	100
Cu	3.0	3.1	2.9	3.1
K	6.7	6.5	6.9	3.6
SiO_2	16	19	20	19
Surface area (m^2/g)	257	315	291	306
Pore volume (cm^3/g)	0.66	0.43	0.43	0.45

In all tests the catalyst was crushed and sieved to either 270/325 mesh particles (44-53 μm in diameter) or less than 270 mesh, prior to loading to a reactor. A purified n-octacosane was used as the initial liquid (slurry) medium in run SB-0261, whereas Ethylflo 164 oil (a hydrogenated 1-decene homopolymer, ~ C_{30} obtained from Ethyl Co.) was used in all other tests. The standard activation procedure was: H_2 at 240°C, 0.8 MPa, 7500 cm^3/min for 2 hours. After the pretreatment the catalysts were tested initially at: 260°C, 1.48 MPa, 1.4 Nl/g-cat/h (where, Nl/h, denotes volumetric gas flow rate at standard temperature and pressure) and $H_2/CO = 0.67\text{-}0.69$.

3. RESULTS AND DISCUSSION

3.1 Repeatability of Catalyst Performance - Multiple Tests with the S3416-1 Catalyst

During the first 100 hours of testing at the baseline conditions the syngas conversions were similar in all three tests (Fig. 1a). After 100 h, the syngas conversion started to decline in run SB-0261, and reached 76% at 150 h on stream. Between 160 and 240 h on stream the catalyst was tested at 263°C (results not shown), and upon returning to the baseline conditions the conversion was about 67%. After that, the catalyst became stable and the conversion did not change with time. On the other hand, in runs SB-0045 and SA-0705 the catalyst was stable up to 250 h, and the syngas conversion was about 81%. After that, in both tests, the reaction pressure and gas space velocity were increased proportionally to 2.17 MPa and 2.05 Nl/g-cat/h, respectively, in order to maintain constant gas residence time. In both cases, the conversion decreased slightly to about 79%, and in run SA-0705 the syngas conversion decreased from 79 to 75 % during the next 260 h of testing at higher reaction pressure. As a result of increase in gas space velocity and reaction pressure the reactor space-time-yield (STY) increased by about 40% (Table 2).

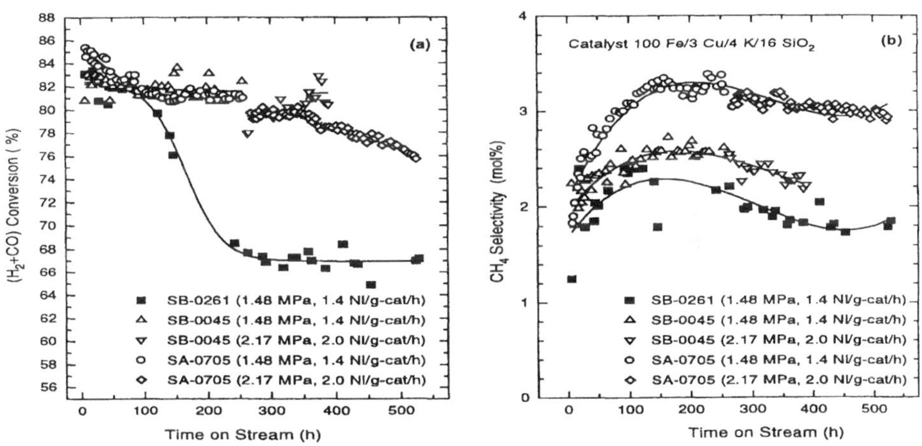

Figure 1. Changes in (H_2+CO) conversion (a) and methane selectivity (b) as a function of time.

Methane selectivity (mol % C basis) in runs SB-0261 and SB-0045, conducted in slurry reactor B, was between 1.9 and 2.4 mol%, whereas slightly higher selectivity (2-3 mol%) was obtained in run SA-0705 (Fig. 1b). Liquid and wax (C_5+ hydrocarbons) selectivities in all three tests were high: 83-89%.

In the original test of this catalyst (SB-0261) n-octacosane was used as the initial medium [5], whereas Ethylflo 164 oil was used as the start-up fluid in runs SA-0705 and SB-0045. However, the catalyst performance was similar in all three tests, except for a period of catalyst deactivation (100-240 h) observed in run SB-0261. The effect of reactor set-up (slurry A vs. slurry B reactor system) on the catalyst performance was within experimental errors (runs SB-0045 and SA-0705).

3.2 Reproducibility of Catalyst Preparation - Tests of Catalysts from Different Batches

Experimentaly determined copper and silica contents of all catalysts are in good agreement with each other, as well as with the nominal catalyst composition, whereas the potassium content of batch 4 (S3416-4) catalyst is lower than that of the other three batches (Table 1). Surface areas and pore volumes of catalysts from batches 2-4 are very similar to each other, whereas the catalyst from batch 1 (S3416-1) has a lower surface area and higher pore volume than the other three catalysts. In general, the differences in bulk composition and textural properties among the catalysts from different preparations are fairly small.

Results from slurry reactor tests of the 100 Fe/3 Cu/4 K/16 SiO_2 catalyst from four preparation batches are shown in Figure 2. Syngas conversion was similar in all seven tests (Fig. 2a), and it varied between 78 and 84% (i.e., 81 ± 3%). Methane selectivities (Fig. 2b) were also similar in all seven tests. For example, at about 100 h on stream, the mean value of methane selectivity from all seven tests was 2.6%, whereas the minimum value was 2.1% (SA-2715; batch 3 catalyst) and maximum 3.1% (runs SB-2145 and SA-0705). High syngas conversions and low gaseous hydrocarbon selectivities were maintained in runs SB-2145 and SA-1665 (batch 4 catalyst) which lasted 400-520 hours (not shown). Catalyst deactivation rate in these two tests was slightly higher than that observed in two recent runs with the catalyst from batch-1 (SB-0045 and SA-0075). On the basis of these results it is concluded that the reproducibility of catalyst preparation procedure has been successfully demonstrated.

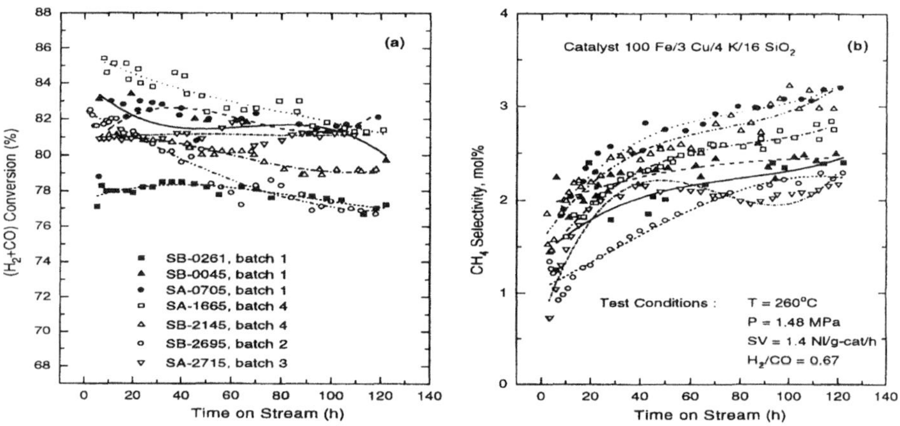

Figure 2. Synthesis gas conversion (a) and methane selectivity (b) as a function of time.

3.3 Pretreatment Effects on the Catalyst Performance

Significant improvements in the catalyst activity were obtained through the use of different pretreatment procedures. Results from slurry reactor tests after six pretreatment procedures have been reported elsewhere [11]. Here we present results illustrating performance of the catalyst (S3416-4) after the baseline hydrogen reduction (SA-1665), CO pretreatment at 280°C for 8 h (SA-0946) and a novel pretreatment procedure (TAMU pretreatment; run SA-2186).

Reaction temperature and gas feed composition were kept constant at 260°C and H_2/CO = 2/3, respectively, throughout the entire duration of all three tests, whereas changes in reaction pressure and/or gas space velocity were made during the tests as shown in Fig. 3a. During testing at the baseline process conditions the syngas conversion in all three tests was high

(76-80% up to 140 h on stream), reducing the need for expensive recycling of unused synthesis gas. Loss of conversion (catalyst deactivation) was minimal during this time period. After 250 h on stream, the catalyst in all three tests was tested at 2.17 MPa and gas space velocities of either 2 Nl/g-cat/h (SA-1665), 2.6 Nl/g-cat/h (SA-0946) or 3.4 Nl/g-cat/h (SA-2186) were employed. During testing at these conditions the syngas conversion decreased with time in run SA-1665 reaching 68% at 500 h, whereas in runs SA-0946 and SA-2186 it increased slightly with time (80-83%). Activities of the three catalysts are compared in Fig. 3b, in terms of values of an apparent first order reaction rate constant. The latter was calculated from experimental data assuming perfect mixing and negligible interphase and intraparticle transport resistances [7]. The catalyst used in run SA-2186 (TAMU pretreatment) was the most active, whereas the catalyst used in run SA-1665 (baseline procedure) was the least active. The use of TAMU pretreatment procedure resulted in 75% increase in activity relative to the baseline hydrogen activation, and 16.7% increase in activity relative to the use of CO activation. Also, in run SA-2186 the catalyst did not deactivate with time during 500 h of testing, and actually its activity increased with time during testing at 2.17 MPa.

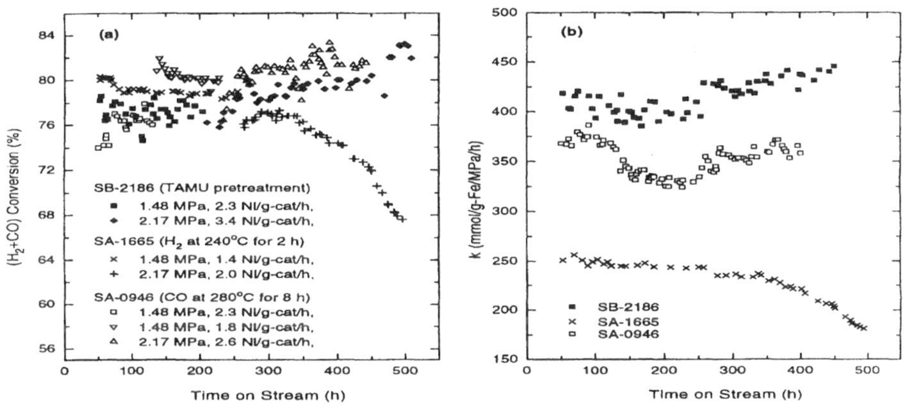

Figure 3. Variation in (H_2+CO) conversion (a) and apparent reaction rate constant (b) with time on stream.

Lumped hydrocarbon product distributions and product yields obtained in these three tests are shown in Table 2, together with results from Mobil's [9] and Rheinpreussen's [10] slurry bubble column reactor tests of precipitated Fe-Cu-K catalysts. Low methane and gaseous hydrocarbon selectivities (C_1+C_2 and C_2-C_4 hydrocarbons) were obtained in all three TAMU tests at reaction pressures of 1.48 and 2.17 MPa. Catalyst productivity (expressed as g hydrocarbons produced/g-Fe/h) at reaction pressure of 1.48 MPa in runs SA-0946 and SA-2186 was equal to or higher than that obtained in Mobil's and Rheinpreussen's bubble column reactor tests, and was markedly higher in both runs during testing at 2.17 MPa. The highest catalyst productivity, 0.86 (gHC/g-Fe/h), was achieved using TAMU's activation procedure, while maintaining the desired selectivity. This is the best performance to date, for catalysts developed for high wax production ("high alpha" catalysts).

The yield of liquid and wax hydrocarbons in TAMU tests was 80-87 wt%, and the Anderson-Schulz-Flory alpha parameter was 0.92-0.94 (high alpha catalyst). The latter is comparable to values for Shell's cobalt catalyst, and Sasol's precipitated iron catalyst [2], both of which are used commercially for production of high quality diesel fuels via hydrocracking of the F-T wax. It is clear that catalysts synthesized at TAMU are ideally suited for this purpose.

Table 2

Comparison of Catalyst Performance in Slurry Reactor Tests

Run ID	SA-1665		SA-0946		SA-2186		Mobil [9][a]	Rheinpreussen[a] [10]
Pretreatment	H_2, 240°C		CO, 280°C		TAMU pretr.		CT-256-13	
Test conditions[c]								
Pressure, MPa	1.48	2.17	1.48	2.17	1.48	2.17	1.48	1.20
Space velocity, Nl/g-Fe/h	2.3	3.4	3.0	4.4	3.9	5.8	2.3	3.1
Time on stream, h	220	361	216	419	145	314	475	
CO conversion, %	83.6	80.1	85.0	84.6	81.3	83.6	90	91
(H_2+CO) conversion, %	78.5	75.8	79.9	80.1	76.6	79.0	82	89
Hydrocarbon selectivites, wt%								
CH_4	3.2	3.0	3.6	2.6	3.9	3.0	2.7	3.2[b]
C_2-C_4	12.2	13.7	15.5	12.9	15.9	14.1	11.1	31.3
C_5-C_{11}	12.8	12.7	18.9	19.2	19.7	16.2	18.1	53.6
C_{12}+	71.9	70.5	62.0	65.3	60.6	66.6	68.1	11.9
C_1+C_2	6.6	6.8	8.1	6.2	8.3	7.0	5.6	6.8
Yields								
Nm^3/kg-Fe/h	1.8	2.6	2.4	3.5	3.0	4.5	1.9	2.8
g HC/g-Fe/h	0.38	0.53	0.49	0.71	0.58	0.86	0.39	0.49

[a]Slurry bubble column reactor test,[b]CH_4 + C_2H_6 ;[c]TAMU tests:260°C, H_2/CO=0.67-0.69, 3.4-4.7 wt% slurry; Mobil: 257°C, H_2/CO=0.73, 20 wt% slurry; Rheinpreussen: 268°C, H_2/CO=0.67, 35 wt% slurry.

Reactor space-time-yield can be substantially improved by operating at 3-4 MPa (STY is nearly proportional to reaction pressure), at lower syngas conversions (higher reaction rate) combined with recycle, and using a slurry bubble column reactor (less mixing relative to stirred tank slurry reactor). Excellent performance of our catalyst is due to a combination of several factors: catalyst preparation procedure, catalyst composition (amounts of promoters in the catalyst, i.e. potassium, copper, and silica) and catalyst activation procedure.

REFERENCES

1. M. E. Dry, in J. R. Anderson and M. Boudart (Editors), *Catalysis Science and Technology 1*, Springer, New York, 1981, p.159.
2. B. Jager and R. Espinoza, Catal. Today, 23 (1995) 17.
3. B. Jager, Stud. Surf. Sci. Catal., 107 (1997) 219.
4. J. Ansorge, Prep. Pap.-Am. Chem. Soc., Div. Fuel Chem., 43(2) (1997) 654.
5. D. B. Bukur, L. Nowicki and X. Lang, Chem. Eng. Sci. 49 (1994) 4615.
6. D. B. Bukur, S. A. Patel and X. Lang, Appl. Catal., 61 (1990) 329.
7. W. H. Zimmerman and D. B. Bukur, Can. J. Chem. Eng., 68 (1990) 292.
8. D. B. Bukur, X. Lang, D. Mukesh, W. H. Zimmerman, M. P. Rosynek and C. Li, Ind. Eng. Chem. Res., 29 (1990) 1588.
9. J. C. W., Kuo, Final Report on US. DOE Contract No. DE-AC22-83PC60019, Mobil Res. Dev. Corp., Paulsboro, New Jersey, 1985.
10. H. Kölbel, P. Ackerman and F. Engelhardt, 1955, Proc. Fourth World Petroleum Congress, Section IV/C, Carlo Colombo Publishers, Rome, 1955, p. 227.
11. D. B. Bukur, X. Lang and Y. Ding, Appl. Catal. (submitted).

NATURAL GAS CONVERSION V
Studies in Surface Science and Catalysis, Vol. 119
A. Parmaliana et al. (Editors)

Deposition of iron from iron-carbonyl onto a working Co-based Fischer Tropsch catalyst: The serendipitious discovery of a direct probe for diffusion limitation

K.P. de Jong[a], M.F.M. Post[b] and A. Knoester[b]

[a] Utrecht University, Department of Inorganic Chemistry and Catalysis, P.O. Box 80083, 3508TB Utrecht, The Netherlands

[b] Shell International Oil Products, Amsterdam, The Netherlands

During fixed-bed pilot plant runs to test silica-supported cobalt-based Fischer Tropsch catalysts deposition of iron from iron carbonyl was encountered due to some contamination of the carbon monoxide feed gas. From analysis of the used catalysts it turned out that the iron (1-2 %m/m) had been deposited preferentially at the center of the 1.4 mm-sized particles. Furthermore, from transmission electron microscopy it appeared that iron had reacted with silica to form iron hydro-silicates, thus not deteriorating the catalytic performance. The iron distribution over the individual particles is concluded to reflect the intraparticle CO profile originating from diffusion limitation during catalysis. Deposition of iron from $Fe(CO)_5$ during Fischer Tropsch catalysis can be considered as a probe for diffusion limitation and is a direct experimental evidence for the existence of concentration gradients in catalyst particles according to the classic Thiele-Wheeler concept.

1. INTRODUCTION

The conversion of synthesis gas to heavy alkanes, known as the Fischer Tropsch (FT) synthesis is a key step in the chemical liquefaction of natural gas. Over the last ten years the FT synthesis has received a great deal of attention using cobalt as the active metal for so-called low temperature FT leading to a heavy paraffinic product, often called wax. Since the wax is in the liquid form under reaction conditions the transport of reactant molecules through the pores will be relatively slow. The slow transport is related to both the low solubility of the reactant molecules H_2 and CO in the wax and the low diffusivity of the molecules in the liquid reaction medium. Accordingly, in fixed bed reactors with millimeter sized catalyst particles, diffusion limitation of the FT reaction has been noted by several workers [1-3]. Post et al. [2] have emphasized the occurrence of gradients of hydrogen over the particles, which affect overall activity. Under the conditions prevailing in the study of Post et al. the kinetics could be adequately described by first-order dependence on hydrogen whilst CO did not affect the reaction rate. Madon and Iglesia [3] have pointed to the concomitant occurrence of CO gradients, which will appear to be highly relevant for the present study.

Over the years we have tested many iron and cobalt based FT catalysts for extended periods of time in pilot plants. The catalysts that have been tested varied widely in terms of catalyst shape, size and composition and so did the process conditions [2]. After a particular test of an experimental cobalt-based FT catalyst, routine inspection revealed that a significant amount of iron had been deposited on the catalyst during testing. Careful inspection of fresh and spent catalyst as well as the testing conditions pointed to the decomposition of iron carbonyl as the cause of the unintentional iron deposition. A part of the careful inspection also involved characterization of the location and nature of the iron deposits in the spent catalyst particles. Surprisingly, we found that the iron was concentrated near the center of the particles.

This triggered what one might call a piece of 'forensic research' to deduce the details of iron deposition from iron carbonyl. The forensic nature of the research relates amongst others to the fact that the experimental details of the deposition were defined poorly. First of all the concentration of iron carbonyl was not known since its presence involved unintentional contamination of the feed gas. Second, during the pilot plant run many process conditions have been varied and only some average values can be used for the calculations.

In the subsequent parts of the paper we will take up the challenge to analyze what has gone on during our prolonged pilot plant run. This research will bring us the serendipitious discovery of a probe for diffusion limitation in FT. In fact the footprint of iron from iron carbonyl is a direct semi-quantitative evidence for the existence of CO gradients in catalyst particles according to the classic Thiele-Wheeler concept.

2. EXPERIMENTAL AND PHYSICAL PARAMETERS

For the specific pilot plant experiment to be analyzed, an experimental $Co/ZrO_2/SiO2$ catalyst was used with a spherical shape and an average diameter of 1.4 mm. The particle density of the catalyst amounted to 850 kg/m^3. The pilot plant run was continued for 2700 hours. For the average process conditions throughout the fixed bed reactor we will use the following data. The total pressure is 2.5 MPa; the ratio of partial pressures of H_2 and CO equals 2. The gas hourly space velocity (GHSV) is 1000 $Nm^3/(m^3.h)$ on the basis of STP conditions for the gas and on the bed volume for the catalyst. The temperature is 500 K.

The selectivity to higher alkanes was very high with an Anderson-Schulz-Flory (ASF) growth factor well above 0.90. The properties of the wax have been summarized in [2] and are tabulated in Table 1.

Table 1
Physical properties of the FT wax at reaction conditions (500 K)

Parameter	Symbol	Units	Value
Molecular weight	M	kg/kmol	300
Density	ρ	kg/m^3	650
Viscosity	μ	kg/(m.s)	6.10^{-4}

As has been mentioned in the introduction section both solubility and diffusivity are key in determining effective transport of species through the liquid filled pores. The transport of hydrogen and carbon monoxide is relevant for catalyst activity - and sometimes selectivity - while we need here to consider transport of iron carbonyl as well. To estimate the solubility of iron carbonyl we have

Table 2
Solubilities, liquid phase and effective diffusivities of species at 500K.

Species	Solubility mol/(Pa.m^3)	Liquid-phase diff. D_m, m^2/s	Eff. diffusivity D_e, m^2/s
H_2	$5*10^{-5}$	$1.4*10^{-8}$	$7*10^{-9}$
CO	$8*10^{-5}$	$1.2*10^{-8}$	$6*10^{-9}$
$Fe(CO)_5$	$8*10^{-4}$	$5*10^{-9}$	$2*10^{-9}$

extrapolated the vapor pressure data of Gilbert and Sulzmann [4] and applied Raoult's law. The solubility of hydrogen has been taken from [2] whilst more recent experimental data [5] have been used to estimate data for carbon monoxide. Please note that the solubility for CO has been obtained by multiplying the figure for H_2 by the experimental solubility ratio of CO and H_2 at 10 bar and 473 K of 1.6 [5]. The liquid-phase diffusivities have been obtained by using the correlation of Wilke and Chang as detailed in [2]. To obtain effective diffusivities we multiply the values of the liquid-phase diffusion coefficient, D_m, with the porosity (0.7) divided by the tortuosity (1.5) to give the effective diffusivity D_e of the species in question. The relevant data have been summarized in Table 2.

The spent catalysts have been extracted to remove most of the wax followed by careful oxidation in air. The bulk iron content of the samples has been determined by X-ray Fluorescence techniques. It was checked that the fresh catalysts contained no detectable amounts of iron. The distribution of iron over the catalyst particles has been obtained from Electron Microprobe Analysis (EMA) using a scanning electron microscope. EMA was performed after embedding the particles in a polymer matrix and making a cut through the center. Transmission electron microscopy has been done on samples following grinding and suspending in n-butanol. A fraction of the suspension was sprayed onto an ultra-thin carbon film. The TEM apparatus was a Philips EM-400T equipped with an EDAX system (9100/40) and operated at 100 keV.

3. RESULTS

Bulk analysis of the spent FT catalysts following 2700 h operation in the pilot plant have been carried out for both the top part of the reactor (at gas inlet) as well as the for the bottom part. The iron content was 1.6 %m/m for the sample from the top and 1.2 %m/m for the bottom sample. This implies that the decomposition of iron carbonyl present in the feed gas has occurred at a modest rate and conversion level. In case rapid decomposition had taken place a stronger gradient of iron contents over the reactor would have been apparent. For the sake of completeness we mention that extensive inspection of reactor walls after prolonged use for FT reaction has shown formation of iron carbonyl from corrosive attack of CO on the reactor wall to be unlikely. In other words we think that iron carbonyl has entered the reactor with the synthesis gas feedstock.

A representative line scan of iron over a spherical particle (ex-reactor bottom) obtained by EMA (Fig. 1) shows the iron deposition to have taken place mainly near the center of the particle. The line relates to a theoretical model developed in section 4.

Using TEM a detailed analysis of the structure of the iron deposits has been carried out. Since the main metal in the catalyst is cobalt care was taken to study also parts in the catalyst with limited amount of cobalt. An example of such an area is shown in Fig. 2. Clearly, platelike structures are observed next to small particles. From spot EDX measurements it follows that both structures contain iron rather than cobalt (or zirconium). The platelike structures probably consist of layered iron hydro silicates (HIS). From counting the number of particles (size ~2 nm) it appears that 10-20% of all the iron is present in particles of this size. We have not been able to quantify the amount of iron in IHS since the orientation of the platelike structure determines strongly the diffraction of the electron beam and thereby the 'visibility'. It may be that part of the iron is present in an amorphous phase or atomically dispersed in the support. In areas as those shown in Fig. 2 as well as in areas that seem to be free of IHS, EDX shows an iron content of 1.3 %m/m, which is very well in line with bulk analysis.

It is noted that the observed oxidic species of iron in the spent catalyst could originate from the fact that the spent catalysts - after extraction - have been oxidized. Still we think that already during FT catalysis the iron has been deposited as an oxide or as IHS. The reason is that during FT catalysis we have not observed deterioration of the selectivity to higher hydrocarbons. From experience with other supports, e.g. alumina, we know that deposition of iron from carbonyl leads to enhanced methane production related to the presence of metallic iron. The fact that we did not observe an increase in methane-make is an indication that the decomposition product of iron carbonyl is present in oxidized form on the support.

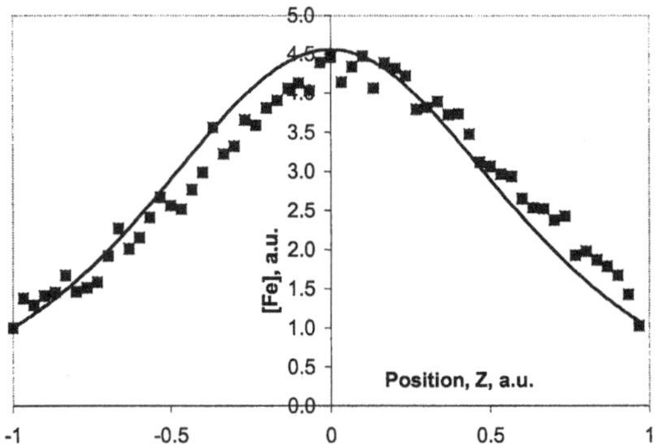

Figure 1: The iron concentration as a function along a line through a spent catalyst sphere going - from left to right - from edge to center to edge. Data points are experimental; the line is based on the model developed in section 4.

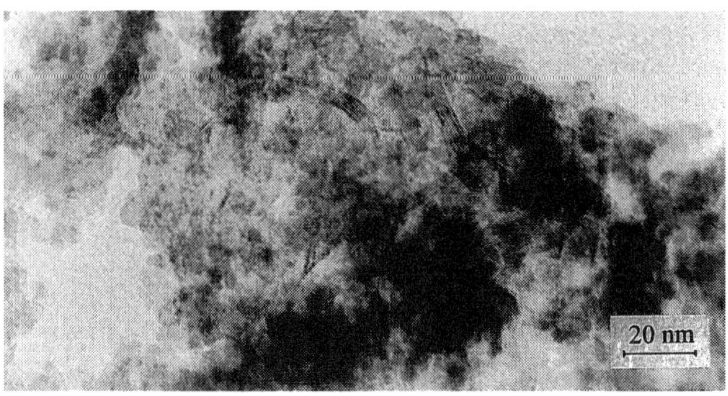

Figure 2: Transmission electron micrograph of an iron containing spent FT catalyst originating from the bottom bed of the reactor.

4. DISCUSSION

The decomposition of iron carbonyl to iron occurs according to the reaction:

$$Fe(CO)_5 \;\text{--->}\; Fe + 5\,CO$$

As has been argued, most likely the metallic iron thus formed reacts immediately with steam - abundantly available from the FT reaction - and the silica support to IHS. The first question that comes up is whether the decomposition has been equilibrium limited such that it can occur only at locations in the reactor with a low thermodynamic potential for CO, i.e. in the center of the catalyst particles.

We calculate first the gross rate of iron carbonyl deposition onto the (bottom bed) catalyst assuming (!) that the rate has been constant over the entire experiment. From the iron content on the catalyst of 1.2 %m/m accumulated over 2700 h with the given particle density we obtain an observed deposition rate of

$$r_{v,p} = 1.88*10^{-5} \ mol_{Fe}/(m^3{}_p.s)$$

Assuming complete decomposition of the carbonyl into deposits, at the given GHSV it follows that a minimum pressure of 2.3 Pa (0.91 ppmv) of iron carbonyl is required to arrive at this rate. In fact we do not deal with 'full conversion' of the iron carbonyl since we have only a modest gradient in iron content over the reactor. In fact one can estimate that the 'conversion' of iron carbonyl is limited to 25%. Therefrom we calculate the pressure of iron carbonyl at the reactor inlet amounts to 9.2 Pa and at the bottom to 6.9 Pa. From thermodynamic calculations it follows that at the prevailing temperature and CO pressure the equilibrium amount of iron carbonyl cannot exceed 0.01 Pa. Clearly, the pressure obtained from the mass balance for iron prove that the pressures have been 2 to 3 orders of magnitude above that expected for equilibrium. From this our first, and firm, conclusion is that the decomposition of $Fe(CO)_5$ is not dictated by thermodynamics but by kinetics.

From the pressure of $Fe(CO)_5$ at the reactor outlet we can now calculate the amount of dissolved iron carbonyl in the liquid phase, using the solubility figure of Table 2, to be C_i= $5.5*10^{-3}$ mol/m^3. The second question we now want to address whether or not it can be expected that the deposition of iron carbonyl has been affected by diffusion limitation for the carbonyl itself. To this end we calculate the parameter according to the Weisz criterion for spherical particles [6].

$$\Phi = \frac{R^2}{D_e} * r_{v,p} * \frac{1}{C_i}$$

with R being the radius of the particle and the other parameters as defined and calculated above. From the data given it follows that the value for Φ = 0.8 which implies that the decomposition of iron carbonyl has not been affected by the rate of intraparticle diffusion of the carbonyl itself.

The only parameters to affect the local rate of iron deposition from the decomposition of carbonyl are the concentration gradients of the reactants or products. These are hydrogen and carbon monoxide as reactants, and steam and hydrocarbons as products. It might be expected that steam accelerates the decomposition of $Fe(CO)_5$ since hydro silicates are formed to a significant extent (Fig. 2). For most of the experimental conditions over the FT run, however, a very significant axial steam gradient over the reactor has been apparent. This gradient should have caused an increase of the iron content going from top to bottom, the reverse of what has been observed. Hydrocarbons are not expected to have an influence either.

Clearly carbon monoxide is the most likely candidate to affect the local rate of carbonyl decomposition. Although the impact of CO is not via the thermodynamic equilibrium (vide supra), its local concentration may have an impact on the kinetics of the decomposition. We have not been successful in finding literature on the kinetics of the decomposition of $Fe(CO)_5$ under an appreciable partial pressure of CO. Work of Venter and Vannice [7], however, indicates that removal of the first ligand from $Fe(CO)_5$ is rate determining for its decomposition. From this we try a negative order (-1) in CO for the decomposition reaction. The dimensionless CO profile over the catalyst particles is deduced from the Thiele model for first-order kinetics and spherical particles [6]:

$$C(z) = \frac{\sinh(\phi.z)}{z.\sinh(\phi)}$$

C(z) is the concentration of CO at dimensionless position z in the particle and ϕ the Thiele modulus for the FT reaction. Assuming the local deposition rate of iron, D(z), to be inversely proportional to the CO partial pressure we obtain, in relative units

$$D(z) = \frac{1}{C(z)}$$

The theoretical line shown in Fig. 1 has been obtained for ϕ=3.45. Note that a value of ϕ=3 is quite realistic in view of the characteristics of the FT reaction as described before [2]. Previously, we have calculated the Thiele modulus on the basis of the diffusivity of H_2. In view of the H_2/CO feed ratio of 2, on the one hand, and the similar effective diffusivities and solubilities of H_2 and CO (Table 2), on the other hand, we assume the CO profile over the particles to be comparable to the H_2 profile.

We have not attempted to improve the fit of experiment and theory in view of the large uncertainties in the parameters used. The objective of showing the data in Fig. 1 is to convince the reader that we can model semi-quantitatively the 'egg yolk' type distribution of iron over the spent particles (Fig. 1) by involving the intraparticle concentration gradients of CO over the FT catalyst 'on duty'.

ACKNOWLEDGEMENT

This work has been carried under the auspices of NIOK, the Netherlands Institute for Catalysis Research, Report No. UU 98-3-003.

REFERENCES

1. R.B. Anderson, B. Seligman, J.F. Shultz, R. Kelly and M.A. Elliott, Ind. Eng. Chem. 44 (1951) 391.
2. M.F.M. Post, A.C. van 't Hoog, J.K. Minderhoud and S.T. Sie, AIChE Journal 35 (1989) 1107.
3. R.J. Madon and E. Iglesia, J. Catal. 149 (1994) 428.
4. A.G. Gilbert and K.G.P. Sulzmann, J. Electrochem. Soc.121 (1974) 832.
5. J.S. Chou and K.-C. Chao, Ind. Eng. Chem. Res. 31 (1992) 621.
6. C.N. Satterfield, "Mass Transfer in Heterogeneous Catalysis", Robert Krieger, Malabar (1981).
7. J.J. Venter and M.A. Vannice, J. Phys. Chem. 93 (1989) 4158.

NATURAL GAS CONVERSION V
Studies in Surface Science and Catalysis, Vol. 119
A. Parmaliana et al. (Editors)

IN SITU CHARACTERISATION OF COBALT BASED FISCHER-TROPSCH CATALYSTS : A NEW APPROACH TO THE ACTIVE PHASE

O. Ducreux, J. Lynch, B. Rebours, M. Roy, P. Chaumette

INSTITUT FRANÇAIS DU PETROLE, 1 et 4 avenue de Bois Préau, 92852 RUEIL MALMAISON Cedex , France

Introduction

Interest in the Fischer-Tropsch synthesis aimed at hydrocarbon production is increasing in the context of clean fuel production. Cobalt containing catalysts are known to be effective in Fischer-Tropsch synthesis, particularly for the production of higher molecular weight fractions.
The activity of cobalt catalysts is usually attributed to the active sites located on the surface of supported cobalt metal particles formed after reduction [1]. Various techniques have been employed and reported in the literature in order to characterise cobalt species. Information about size and concentration of oxide and metal crystallites supported on different oxides is plentiful (TiO_2, SiO_2, Al_2O_3 are the main ones) but difficult to correlate with the state of the surface phase during Fischer-Tropsch reaction.
In situ study of these catalysts is therefore particularly interesting. *In situ* X-ray diffraction allowes the changes in crystallographic phase of cobalt species during their treatment with hydrogen or under CO/H_2 mixture to be followed. Colley et al. observed transformation of f.c.c. to b.c.c. cobalt under 1:1 CO/H_2 mixture but showed no correlation with catalytic activity [2, 3]. In the case of iron, Jung and Thomson were able to correlate increase of methane formation with the formation of χ-$Fe_{2.5}C$ [4]. The purpose of this work is to perform real dynamic *in situ* physicochemical characterisation of model cobalt catalysts and correlate this data with catalytic performances.

Experimental methods

1. Catalysts

Different model catalysts have been studied : Co/SiO_2, Co/Al_2O_3, and $CoRu/TiO_2$ (table 1).
Co/SiO_2 and Co/Al_2O_3, catalysts were prepared via incipient wetness impregnation using aqueous solution of cobalt nitrate ($Co(NO_3)_2.6H_2O$). The samples were dried in air and calcined after the impregnation step.
The preparation procedure for $CoRu/TiO_2$ catalysts has been established following published data [5]. It is prepared by the stepwise impregnation with corresponding metal salt solutions of titania support (Degussa P25), composed by 80 ± 5% TiO_2 rutile and 20 ± 5% TiO_2 anatase. Appropriate chemical and thermal treatments of Co/TiO_2 described in ref. [5] were performed, before and after Ru addition.
Calcined samples were characterised by X-ray diffraction (XRD) and X-ray fluorescence. Co_3O_4 was the only crystalline cobalt phase observed after calcination.

Co-supported	wt% Co content	wt% Ru content	Particles size Co_3O_4 (Å) XRD	S_{BET} (m^2/g)
Co/SiO_2	13 ± 0.3	-	140 ± 25	460
$Co\text{-}Ru/TiO_2$	11.3 ± 0.3	0.18 ± 0.005	150 ± 25	22
Co/Al_2O_3	10.5 ± 0.3	-	200 ± 50	180

Table 1 : Supported cobalt catalysts

2. Characterisation measurements

In situ measurements were carried out using Cu(Kα) radiation on a Siemens D501 θ-2θ powder diffractometer equipped with an Anton Paar XRK reaction chamber. The gas handling system of the reaction chamber has been modified so as to be able to carry out analysis under realistic Fischer-Tropsch conditions. The powdered sample (0,2 g) is placed on a ceramic holder and pressed lightly. The inlet reactant flow passes through the sample. The reaction conditions have been chosen in order to avoid heavy wax formation : T = 503 K, P = 3 bars, $H_2/CO = 9$, $W/F \approx 8{,}4.10^{-2}$ g.h.l^{-1}.
An associated apparatus equipped with a microreactor and working under the same conditions has been constructed allowing the transfer of the catalyst after reaction to vacuum characterisation techniques such X-ray Photoelectron Spectroscopy (XPS: Kratos DS800 with Al Kα source) and Transmission Electronic Microscopy (TEM: Jeol 2010 with diffraction and analysis possibilities) without any contact with air. These methods are used to complete results obtained from XRD measurements. The results of comparative catalytic tests between microreactor and *in situ* cell shows similar conversion and selectivity.
Products are analysed at the exit by gas chromatography and the heaviest products are condensed after the cell. The activity is followed via CO consumption.

Results

1. Reduction under hydrogen - The activation step

Reduction is performed in the *in situ* chamber under pure hydrogen (T = 823 K, P = 1 bar, during two hours). In Fig.1, the XRD patterns of the Co/SiO_2 sample are shown. Silica is an amorphous phase and it is easier to see the cobalt metal pattern in this case.
The reduction of Co_3O_4 crystalline phase with hydrogen proceeds in two steps : first, reduction of Co_3O_4 to CoO (473-573 K) and then, consecutive reduction of the CoO phase to metal (523-773 K).

After the reduction step, the main part of the cobalt metal phase is the f.c.c. form, with some h.c.p. form detected. The intensity of the diffraction pattern is lower than expected considering cobalt content (about 13 wt%). Only 40% of the initial oxide cobalt signal is observed after the reduction process. Possible reasons for this low estimation of metallic cobalt content by XRD are :

- the formation of an amorphous cobalt phase during the reduction
- the formation of a highly disorganised system with structural faults

In situ diffraction studies by Srinivasan et al. also showed the presence of high stacking fault density in Co/SiO_2 catalysts leading to a mixture of f.c.c. and h.c.p. forms [6].

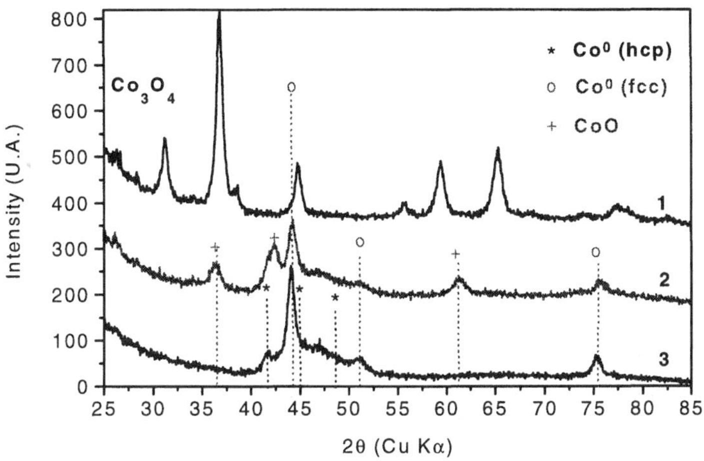

Fig. 1. XRD patterns of reduced Co/SiO_2 under hydrogen flow

1. before reduction - 2. T=573 K - 3 - T=773 K

Transfer from microreactor to TEM and XPS has been tested using these reduced samples. A first test transfer of the specimen in air confirmed that even short exposure to air degraded the surface by oxidation to CoO phase. Transfer under hydrocarbon from the microreactor to the TEM, and under inert gas from microreactor to the XPS apparatus showed no bulk or superficial oxidation.
After transfer without reoxidation, reduced samples prepared in the microreactor showed the presence of cobalt particles which produced a diffuse electron diffraction pattern (TEM). This is coherent with the low intensity of Co metal peaks observed in XRD and confirms the presence of a high defect density in the metal particles.

2. Continuous flow characterisation experiments

After reduction at 773 K, the sample was cooled to 503 K under hydrogen and the CO/H_2 mixture fed into the cell.
Under Fischer-Tropsch conditions, for the catalysts Co/Al_2O_3 and $Co\text{-}Ru/TiO_2$, XRD analysis (Fig.2) shows the formation of a new cobalt phase after only 70 hours on stream : a carbide phase Co_2C. At the same time, the intensity of Co metal phase signal decreases ; probably because of the transformation to carbide (Fig. 3).
On the contrary after 310 hours, studies on Co/SiO_2 and bulk Co_3O_4 revealed no particular structural transformation.
It has been proposed that carbidic carbon and/or CH_x are intermediates for the Fischer-Tropsch reaction [7]. These surface compounds could be identified using the chemical shift of the binding energy in XPS. Studies on Co/SiO_2 tested under flow, correlate the diffraction results. No visible spectral modification was found between the reduced catalyst and the catalysts after reaction, except for the surface carbon concentration (binding energy 284.4 eV), which increased. XPS

studies on Co-Ru/TiO_2, where Co_2C phase has been identified, show only a large carbon signal but without any chemical shift of the binding energy. The surface coverage by hydrocarbons is too high to allow detection of the carbide phase after test.

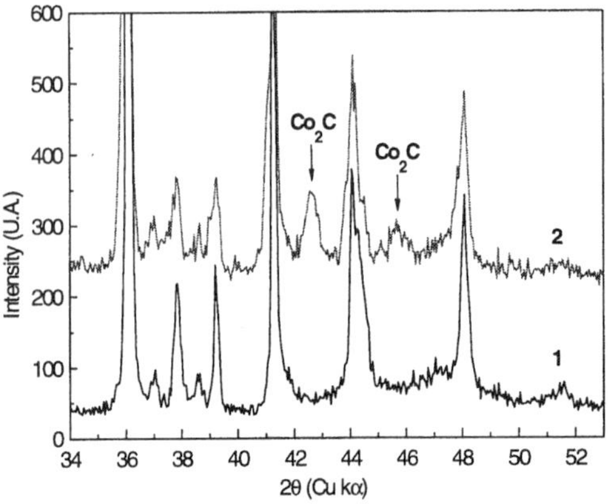

Fig. 2. XRD patterns of Co-Ru/TiO_2 1. after reduction (773 K) - 2. tested under CO/H_2 flow during 180h

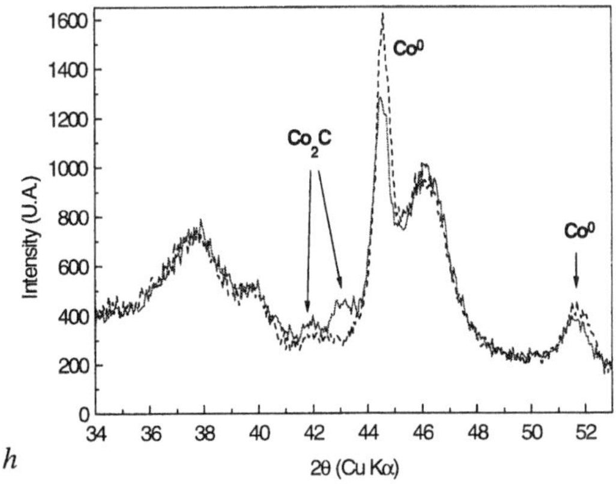

Fig. 3. XRD patterns of Co/Al_2O_3 after reduction (773 K) and tested under CO/H_2 flow during 245h

3. Evaluation of catalyst performance

An important possibility of this set-up is the correlation of structural modification with catalytic results.
Typical catalytic results obtained with our conditions are the following : conversion is about 20% (with CoRu/TiO_2), products are mainly linear alkanes from C_1 to C_{30}, with some alkenes and a small amount of primary alcohol (see table 2).

Hydrocarbon's selectivity	***(% wt)***
C_1	34.4
C_2-C_4	28.7
C_5-C_9	14.6
C_{10}-C_{15}	16.6
C_{15+}	5.7
Paraffins	79
Olefins	19
Alcohols	2
Average conversion(% CO)	18.5
Probability of chain growth	$\alpha = 0.75$

Table 2 : Catalytic results for Co-Ru/TiO_2 after 190h on stream ($T=503$ K, $P=3$ bars, $H_2/CO=9$, $W/F=8.4.10^{-2}$ $g.h.l^{-1}$)

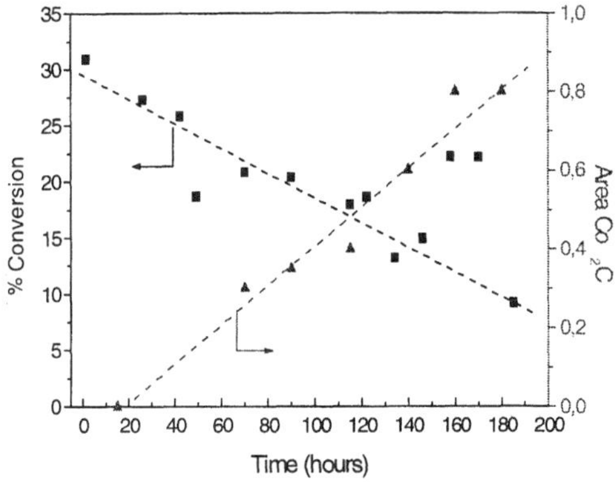

Fig. 3. Relationship between conversion (squares) and carbide XRD peak area (triangles) CoRu/TiO$_2$

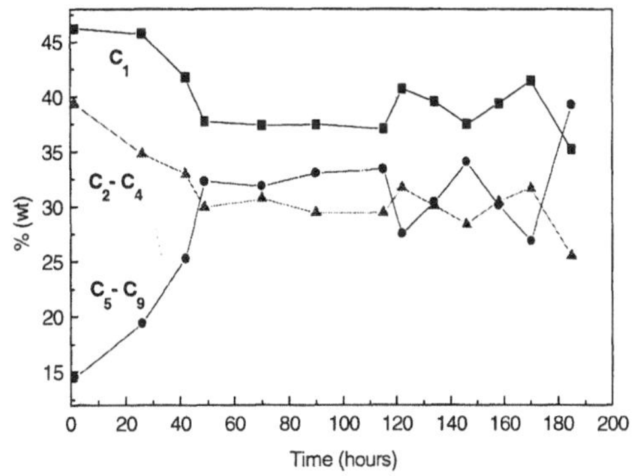

Fig. 4. Selectivity evolution during the test (gas phase) CoRu/TiO$_2$ (T=503 K, P=3 bars, H_2/CO=9, W/F=8.4.10^{-2} g.h.

During the test with CuRu/TiO$_2$, the main cobalt carbide pattern increases progressively, whilst the conversion, calculated from the fraction of CO in the gas exit phase, decreases (Fig.3). At the same time, the selectivity show an increase in the carbon chain growth towards C_5-C_9 products (Fig.4). For the Co/Al$_2$O$_3$ catalyst, both carbide formation and deactivation process are slower. Studies on Co/SiO$_2$ which showed no carbide formation revealed no major decrease in activity during the test (Fig 5). Selectivities also remained stable for this catalyst.

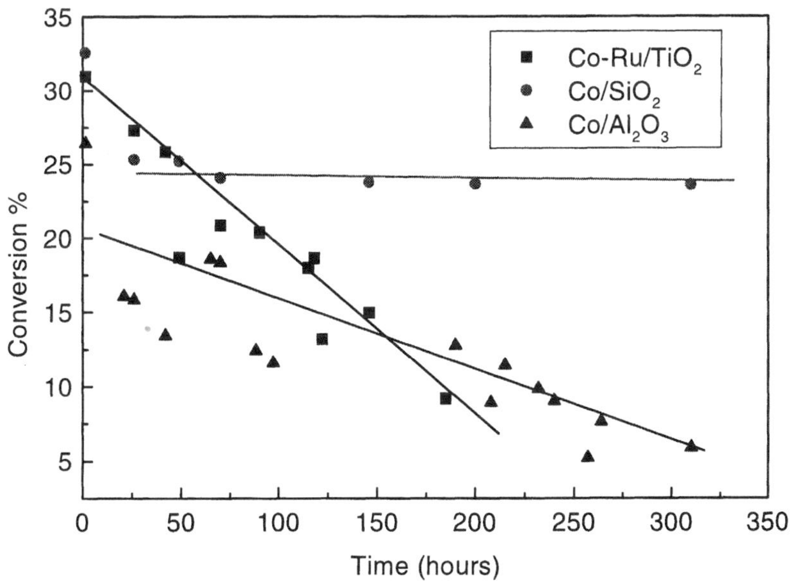

Fig. 5. Activity evolution during test for CoRu/TiO$_2$, Co/SiO$_2$ and Co/Al$_2$O$_3$ (T=503 K, P=3 bars, H_2/CO=9, W/F=8.4.10^{-2} g.h.l^{-1})

Discussion- Conclusion

Our results show a close correlation between structure modification and activity in the systems studied. Cobalt metal and CO can react to form a carbide Co_2C under conditions of temperature and pressure close to those of the Fischer-Tropsch reaction. Data from the literature indicate that in the case of iron and nickel, the formation of a carbide phase could be directly involved in the mechanism of paraffins production [4, 8]. But in our case, carbide formation seems to be more related with a deactivation process. Nevertheless the increase of C_{5+} selectivity (Fig.4) is also observed with carbide formation. When no carbide is formed there is no deactivation during the test (cobalt supported on silica : Fig.5). Further characterisation needs to be performed to correlate this carbide formation with the reduced cobalt phase structure and with the nature of the support. In view of our recent results, the f.c.c. cobalt metal form seems to be more predominant for alumina than silica. The formation of Co_2C rather than b.c.c. cobalt (on Co/Al_2O_3 or Co/MnO in the work of Colley et al.) may be due to the presence of stacking defects in our reduced catalysts. The presence of interstitial carbon formed by dissociation of CO may play an important role in the formation of the metastable Co_2C structure [9]. The interaction between the cobalt metal and the support is certainly also involved in the carbide formation.

The chemical and physical behaviour of this carbide, especially its formation and stability under carbon monoxide and hydrogen mixture, is of fundamental importance to the understanding of the mechanism of the Fischer-Tropsch reaction. Work in progress should provide further information on the working state of the catalyst and some indication of the deactivation mechanism.

Acknowledgement :
The authors acknowledge the financial support of the European Union (Contract JOF3-CT95-0016).

[1] Hindermann I.P., Hutchings G.J. , Kienneman A., *Catal. Rev. Sci. Eng.*, 35 (1), 1 (1993)
[2] Colley S.E., Copperthwaite R.G., Hutchings G.J., Terblanche S.P., Thackeray M.M., *Nature,* 339, 129 (1989)
[3] Colley S.E., Copperthwaite R.G., *Catal. Today* 9, 203 (1991)
[4] Jung H., Thomson W.J., *J. Catal.* 139, 375 (1993) and 134, 654 (1992)
[5] Iglesia E., Soled S.L., Fiato R.A.,Via G.H., *J. Catal.* 143, 345 (1993)
[6] Srinivasan R., De Angelis R. J., Reucroft P.J., Dhere A.G. and Bentley J., *J. Catal.* 116, 144 (1989)
[7] Nakamura J., Toyoshima I., *Surf. Sci.*, 201, 185 (1988)
[8] Barbier A., Brum Pereira E., Martin G.A., *Catal. Lett.* 45, 221 (1997)
[9] Hofer L.J.E., Peebles W.C., *J. Am. Chem. Soc.*, 69, 2497 (1947)

NATURAL GAS CONVERSION V
Studies in Surface Science and Catalysis, Vol. 119
A. Parmaliana et al. (Editors)
1998 Elsevier Science B.V.

Selective syn-gas conversion over a Fe-Ru pillared bentonite

R.Ganzerla[a], M.Lenarda[a], L.Storaro[a], R.Bertoncello[b].

[a]Dipartimento di Chimica, Università di Venezia "Ca' Foscari",
Calle Larga S.Marta, 2137, 30123, Venice, Italy
Fax: (+39)415298517, E-mail: ganzer@unive.it

[b]Dipartimento di Chimica Inorganica, Metallorganica ed Analitica, Università degli Studi di Padova, via Loredan 4, 35131-Padova (Italy)

SUMMARY

A micro-mesoporous catalyst was obtained by pillaring a natural bentonite with ruthenium doped iron oligomers. The material was characterized by XRD, XPS, TPR and nitrogen adsorption. Its catalytic activity for the syn-gas conversion was measured in 523 K-673 K temperature range. The catalyst showed, for reaction temperatures above 623 K, a good selectivity toward the production of light hydrocarbons with a high percentage of olefinic products.

1. INTRODUCTION

Pillared clays have been extensively used as catalysts for hydrocarbon cracking and other important reactions such as methanol or syngas conversion [1, 2]. The great interest arise essentially from the possibility of controlling the surface chemistry (acidity, redox properties) and the microtexture (surface area, micro and mesoporosity) of these catalysts. Thus silanized pillared clays impregnated with Ru, Fe and K [2], Cu nitrate impregnated Zr-pillared clay [3] and transition metal carbonyl clusters grafted on aluminum pillared clays resulted to be active catalysts for the conversion of CO/H_2 into hydrocarbons [4]. More recently Al-Fe pillared laponites showed good activity and selectivity in this reaction [5b]. Since iron oxide is known for its activity as Fischer-Tropsch catalyst and the doping with small amount of transition metals is known to be beneficial, in this work we report some results on the catalytic activity at relatively low temperature and at atmospheric pressure of a Fe-Ru pillared bentonite for the conversion of syn-gas to hydrocarbons.

2. EXPERIMENTAL

2.1 Materials

The clay used in this study was a natural calcium rich bentonite Detercal P1™ (montmorillonite 97%) of North African origin (Nador, Morocco), factory dried, ground and sieved, which was obtained from Industria Chimica Carlo Laviosa S.p.A. (Leghorn, Italy). (CEC=84 meq/100g).

The Fe-Ru pillared bentonite (FERUP) was prepared adding a basic aqueous solution containing $FeCl_3$+$RuCl_3$x3H_2O (Ru/Fe =0.05, OH/(Fe+Ru) = 2.0 to an acetone clay (30g) suspension (50% w/w) following a previously described procedure [6]. A thin film of the resulting cross-linked clay was air dried at 333 K, finely ground (> 40mesh) and calcined for a minimum of 18 h at 673 K in a ventilated oven.

The calcined pillared sample contained 0.70% w/w of Ru and 9.7% w/w of Fe. The interlayer spacing was d_{001}=2.24 nm and the specific surface area (S.A.) was 130m^2/g.

2.2 Characterization Methods

Elemental analyses were accomplished by Atomic Adsorption Spectroscopy with a Perkin-Elmer PE 3100 instrument.

Temperature programmed reduction (TPR) was performed in a conventional U-shaped quartz microreactor (bore = 6 mm, length = 200 mm) using a 5% H_2/Ar mixture flowing at 35 ml min^{-1} (STP). The temperature range was 293-873K and the heating rate 10 K min^{-1}. The reduction of CuO to metallic copper was used to calibrate the TPR apparatus for H_2 consumption.

Adsorption-desorption experiments using N_2 were carried out at 77 K on a Sorptomatic 1900 Carlo Erba porosimeter. Before each measurement the samples were outgassed at 423 K and $1.33 \cdot 10^{-3}$ Pa for 6 h. The N_2 isotherms were used to determine the specific surface areas (S.A.) using the BET equation.

X-Ray diffraction spectra were measured with a Philips diffractometer using the Cu-Kα radiation. The samples were disc shaped pressed powders and were previously treated at 673 K in a ventilated oven.

A Perkin Elmer PHI 5600ci spectrometer with monochromatized AlKα radiation (1486.6 eV) and non-monochromatized Mg Kα radiation (1253.6 eV) was used for the XPS analyses. The working pressure was less than 2x10^{-7} Pa. The standard deviation in the BE values of the XPS lines is 0.10 eV.

2.3 Catalytic syn-gas conversion

Catalytic experiments were performed in a tubular glass flow microreactor. Catalyst samples (250 mg) were pretreated for 10 h in hydrogen (15 ml min^{-1}) flow at 673 K. Experiments were performed at: CO/H_2 = 1; P = 0.1 MPa; T = 523÷673 K; Space Velocity = 500 h^{-1}. Reaction products were analysed using two GC (HP 5890) respectively equipped with TCD and a Poraplot Q (25 m, ϕ=0.53 mm) column or FID and a Al_2O_3/KCl column (50 m, ϕ=0.53 mm).

3. RESULTS

3.1 Catalysts characterization

The XPS spectrum of the FERUP clay calcined at 673 K (FERUP-673) showed a broad Ru 3p (3/2) peak at 462.3±0.2 eV attributable to a Ru(IV) species (the Ru 3p signal was used instead of the more frequently used 3d one because this last was partially obscured by the contamination carbon peak) and a Fe2p (3/2) peak at 711.8±0.2eV attributable to the α form

of Fe^{3+} hydrated oxide species. The spectrum of FERUP-673 after treatment at 673 K in H_2 (FERUP-673-red) showed a Ru 3p signal at 462.0±0.2 eV that we attribute to Ru(0). Nevertheless the presence of Ru(IV) species camnnot be excluded because the 3p signals are very close. In fact the Ru 3p signal of an authentic metallic ruthenium sample was found at 461.7±0.2 eV while that of a RuO_2 sample was found at 462.7±0.2 eV.
These findings were confirmed by the XRD spectrum of FERUP-673-red that showed the signals of metallic ruthenium, RuO_2, Fe_2O_3 (maghemite) and of Fe_3O_4 (magnetite). The ruthenium oxide presence after reduction can be explained by surface reoxidation during sample handling in air. The interlayer spacing d_{001}=2.24 nm was mantained after the H_2 pretreatment.

The H_2-TPR profile showed in the 373 - 573 K temperature range, a broad shoulder centered at 448 K attributable to the RuO_2 reduction to Ru metal and a broad band with the maximum at 510 K related to the reduction of Fe_2O_3 to Fe_3O_4 [7, 8].
It comes out from all these data that the pillars of FERUP-673 are most probably a mixture of Fe(III) oxides and Ru(IV) oxide. Hydrogen treatment at 673 K leads to the reduction of part of Fe_2O_3 to Fe_3O_4 and of RuO_2 to metallic ruthenium. The relatively low reduction temperature of the Fe(III) oxides calls for a ruthenium assisted process caused by the intimate contact between the species[8, 9].

3.2 Catalytic activity

The catalytic activity of FERUP-673-red in the syn-gas conversion was studied in the 523K -673 K temperature range.

The maximum activity was reached after a short induction period (3h) and afterwards remained constant for at least 20 h (Figure 1) as was found for similar catalytic systems [5].

Methane was the predominant reaction product at 523 K (Table 1, Figure 2) probably because iron species are less active in the FT synthesis at this temperature [9,10] and the catalyst behavior is dominated by the chemical properties of metallic ruthenium that is a well known very active methanation catalyst.

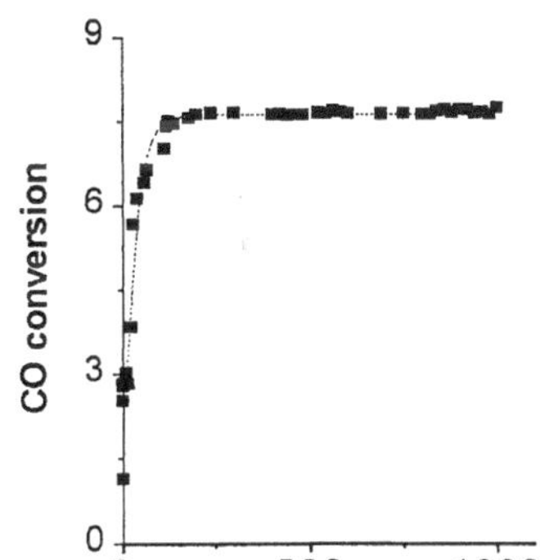

Fig 1. Syn gas conversion (moles of $CO \cdot g_{cat}^{-1}$ h^{-1} 10^3) over FERUP-673-red catalysts at 625 K.

Table 1: Syn-gas conversion on FERUP-673 red. [a]

T(K)	Conversion (mol g^{-1} h^{-1})10^3 CO	Selectivity (mole%) CH$_4$	C_2-C_3	C_4-C_6	CO_2	Olefinity (%) C_2-C_3
523	2.30	28.55	36.05	35.39	-	40.33
573	6.50	17.12	44.50	32.33	6.17	43.84
623	8.09	13.85	42.48	33.01	10.59	69.31
673	6.23	15.97	42.18	15.89	25.97	72.59

a) Reaction conditions: $CO/H_2 = 1$; P = 0.1 MPa; Space Velocity = 500 h^{-1}.

At higher temperatures short chain C_2-C_6 hydrocarbons are the dominant reaction products with a prevailing presence of C_2-C_3 species as was found for other iron pillared clays based catalysts [5b]. Olefinity (olefins/total hydrocarbons·%) increases (Table 1) with the increasing temperature and reaches a maximum at 673 K. No oxigenated products were found.
Carbon dioxide can be generated in these reaction conditions by Water Gas Shift Reaction (WGSR) or by the Boudouard equilibrium. Nevertheless this product resulted totally absent at 523 K, less than the 10% below 623 K and became important only at 673 K (Figure 2).
The hydrocarbons distribution in the C_1-C_6 range resulted not to follow the Anderson-Schulz-Flory statistics with a cutoff indicating an inhibition of the formation of > C_4 hydrocarbons, (the data of the experiments at 523 K and 623 K are reported in Figure 3 and 4).

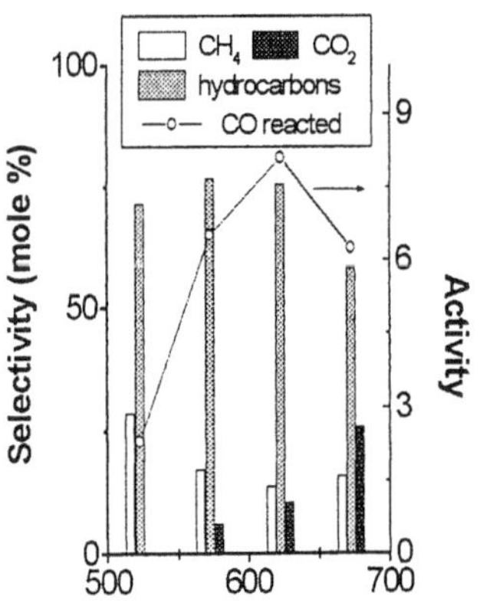

Figure 2: Hydrocarbon selectivity and CO conversion (moles of CO g$_{cat}^{-1}$ · h^{-1} · 10^3) red catalyst as function of temperature.

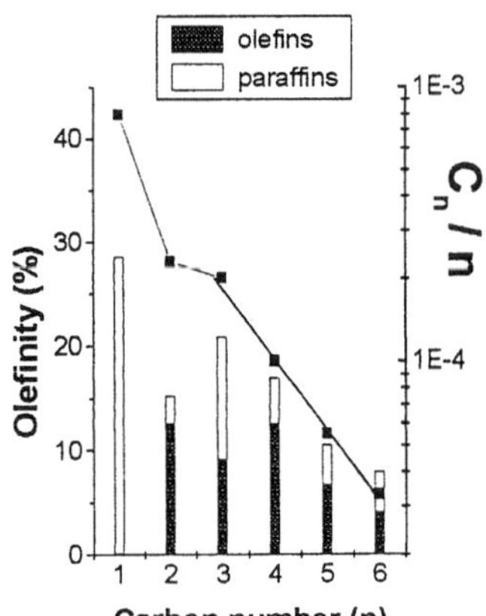

Figure 3. Olefinity and Anderson-Schulz-Flory plot for the hydrocarbon distribution of syn-gas conversion at 523 K.

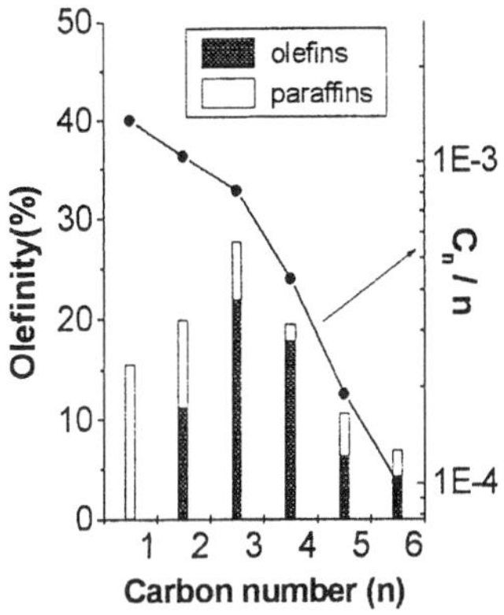

Figure 4. Olefinity and Anderson-Schulz-Flory plot for the syn-gas conversion at 623 K

The comparison between our data and those reported in the literature for analogous catalytic systems [5b, 9] is quite difficult because different operating systems were often used. A good selectivity for light alkenes and sufficient catalyst stability was found only for mixed metal Fe-Al pillared clays [5b] or zeolite encapsulated Fe and Co catalysts [10].

The behavior of FERUP-673-red clay confirmed this trend but, at 673 K, a good activity, a high selectivity for light alkenes, mainly C_3-C_4 olefins (fig.4), was found along with a very low conversion to methane and CO_2. These catalytic properties are probably attributable to the particular morphology of the interlayer spacing and to the presence of mixed Ru-Fe species on the pillars.

It was in fact reported also by other authors [11] that the addition of small quantity of Ru to FT cobalt based catalysts highly improved both the activity and the selectivity.

Temperatures above 673 K lead to FERUP-673-red clay structure collapse and the main products become CO_2 (60%) and CH_4 (15%).

Conclusions

Fisher-Tropsh synthesis of hydrocarbons was achieved over a Fe-Ru pillared bentonite in the 573-673 K temperature range. The catalyst showed a good activity and a high selectivity to light hydrocarbons in the whole examinated temperature range. The Fe-Ru pillared bentonite exhibited, in particular at 673K, a good activity, a high selectivity for light alkenes, mainly C_3-C_4 olefins, along with a very low conversion to methane and CO_2. The ruthenium presence, in very low amount, on the iron oxide pillars appeared to play a major role in determining the stability, activity and selectivity of the catalyst.

References:

1. R. Burch, C.I. Warburton, J. Catal.,97 (1986) 511.
2. M.P.Atkins, A.G. Ashton, European Patent Appl., EP 0.150.898 (1985).
3. E.P.Giannelis, E.C.Rightor, T.J.Pinnavaia, J.Am.Chem.Soc.110, (1988), 3880.
4. G.J.J.Bartley and R. Burch, Appl.Catal., 28 (1986) 209.

5. a)J. Barrault, C. Zivkov, F. Bergaya, L. Gatineau, N.Hassoun, H.Van Damme and D.Mari, J.Chem.Soc.Chem.Commun., (1988) 1403.
b) J.Barrault, L.Gatineau, N.Hassoun and F.Bergaya, Energy and Fuels, 6 (1992) 760
6. L.Storaro, M.Lenarda, R.Ganzerla and A.Rinaldi, Microporous Mater., 6 (1996) 55.
7. A.Jones, B.D.Mc Nicol, " Temperature programmed reduction for solid materials characterization", Marcell Dekker, New York, 1986.
8. D.J.Duvenhage, N.J.Coville, Appl.Catal A, 153(1997)43
9. E.G.Rightor, M.S. Tzou and T.J.Pinnavaia, J.Catal. 130 (1991) 29
10. L.F.Nazar,G.A.Ozin,F.Hughes, J.Godberg,D.Rancourt, Angew.Chem.Int.Ed.Engl. 22 (1983) 624.
11. E.Iglesia, Appl. Catal A 161 (1997) 59.

NATURAL GAS CONVERSION V
Studies in Surface Science and Catalysis, Vol. 119
A. Parmaliana et al. (Editors)
1998 Elsevier Science B.V.

Attrition-determining morphology changes on iron Fischer-Tropsch catalysts

Nancy B. Jackson and Lindsey Evans, Sandia National Laboratories, MS 0710, PO Box 5800, Albuquerque, NM 87185, United States

Abhaya Datye, University of New Mexico, Dept. of Chemical and Nuclear Engineering, Albuquerque, NM 87131, United States

Temperature programmed reduction and transmission electron microscopy were used to study the morphology changes of an iron Fischer-Tropsch catalyst during reaction, with an emphasis on potential attrition of the catalyst. In particular, the effect of potassium promotion was explored. Potassium appeared to minimize the formation of one type of carbide and, at low concentrations, limited graphitic carbon formation. At higher potassium levels (3%) graphitic carbon began to re-appear. Copper-promoted catalysts exposed to higher reaction temperatures (which is known to cause attrition) formed different carbides than those exposed to lower reaction temperatures. The potassium promoted catalysts investigated in this study did not produce the high temperature carbide present on catalysts that quickly attritted.

1. INTRODUCTION

The reactor of choice for the highly exothermic Fischer-Tropsch reaction is the three-phase bubble column reactor. By minimizing hot spots, this reactor allows for high selectivity towards wax production. However, to operate most efficiently, the catalyst must be small (about 70 μm) and uniform in size. Spray drying is the method most often used to synthesize catalysts for this type of reactor. If a catalyst attrits during reaction into small enough particles, particularly fines < 1 μm, the liquid phase of the bubble column becomes too viscous to operate properly and the catalyst cannot be separated from the product wax. Our studies have focused on the causes and potential prevention of attrition of iron-based Fischer-Tropsch catalysts, and in particular, the effect of promoters and reaction temperature on the attrition and morphology of the catalyst.

Attrition of iron catalysts in a synthesis gas environment is a chemically initiated process. In particular, morphology changes of the catalyst and carbon deposition on the catalyst during reaction are major contributors to this problem. We have used a combination of temperature programmed reduction (TPR) and high resolution transmission electron microscopy (TEM) to investigate this problem. We have confirmed that the active catalyst phase is an iron carbide and that the change from the starting material (iron oxide) to iron carbide requires a volumetric change that begins the process of attrition.[1,2] An iron oxide catalyst that is treated in synthesis gas goes through several morphological changes. First, the hematite catalyst, Fe_2O_3, is reduced to magnetite, Fe_3O_4, followed by a slow reaction to iron carbide. The iron carbide appears to form as "buds" off the main magnetite crystal. Carbonaceous material, CH_x, forms on the surface of the iron carbide particles but never on the oxide particles. See Figure 1. Eventually, as more and more iron carbide "buds" are formed, the catalyst attrits into submicron particles. At high temperatures, or under certain other conditions, graphitic carbon will also form.[1] Although HRTEM cannot differentiate among the different phases of iron carbide, TPR does quantify the different iron carbides, but does not directly identify the type of each carbide. Previous work has shown that TPR in H_2 of a reacted Fischer-Tropsch catalyst has three primary desorption regions. By halting reduction after each peak and looking at the sample in the HRTEM, we have identified the first peak at 270°C as the methane formed during the reduction of CH_x, the second group of (five) peaks is the methane formed from the reduction of the iron carbides (400-600°C), and the third peak (>650°C) is graphitic carbon. From our work in the HRTEM, the graphitic carbon has not always been pure graphite, but graphitic, ordered carbon structures, which is why it can be reduced at temperatures lower than typically expected for graphite.

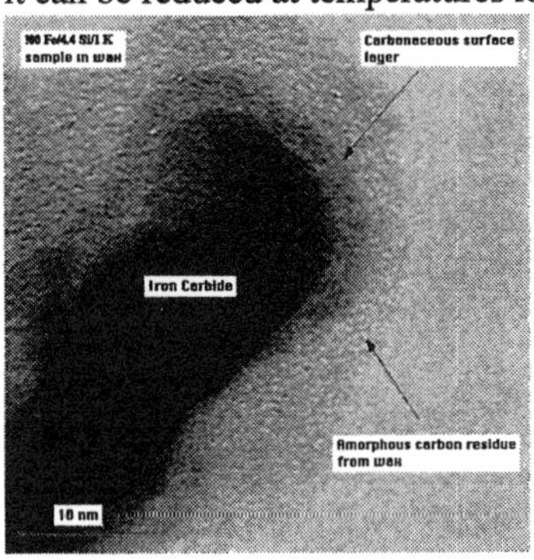

Figure 1. Typical TEM of a reacted iron-based catalyst. Iron carbide is predominant phase, carbonaceous material is present on the surface, and wax is difficult to completely remove. Identical morphology have been found on many iron catalyst following a wide range of conditions. [1,2,3]

2. EXPERIMENTAL

The iron catalysts used in the temperature programmed reduction experiments were synthesized using iron nitrate precipitated with a base. In the case of the copper/iron catalysts the base used was Na_2CO_3 and for the iron/potassium catalysts, NaOH was used to precipitate the oxide. The catalysts were repeatedly washed to remove the sodium present prior to calcination. No sodium was detected after the final wash using X-ray fluorescence spectroscopy. The potassium was added by incipient wetness using K_2CO_3 solution. The catalysts were calcined in air at 300°C and were characterized using N_2 adsorption with BET analysis. The temperature programmed reduction apparatus had a ¼ inch stainless steel tube catalyst bed. The catalyst was placed in the TPR reactor, and before TPR, the FT reaction was run in the same reactor. Therefore, the catalysts did not need to be passivated following reaction in synthesis gas and prior to TPR. The catalyst was loaded into the reactor and the effluent of the reaction was directed towards the vent with a cold trap in line to collect wax produced by the reaction. The catalyst was pretreated in H_2 for 2 h. Previous studies showed that this pretreatment leads to the formation of magnetite and does not reduce the catalyst to a metallic state.[1] The catalyst was then reacted in synthesis gas (H_2/CO=0.7) at 16 atm and 215°C for 24 h. Following reaction, the catalyst was cooled to 180-200°C and was flushed with He for anywhere from 4 h to 2 days until no more hydrocarbons could be detected in the helium stream by an FID detector. The reduction gas, 10% H_2/90% He, was introduced after the catalysts had been cooled to 100°C. The temperature was raised at 5°C/min until it reached 270°C where it remained for 1 h. Next the temperature was raised at 2°C/min until it reached 700°C. A TPR of an unpromoted iron catalyst is shown in Figure 2. In the case of the copper-containing catalysts this capability was limited to an upper temperature of 550°C. The catalyst was held at the high temperature for 30 minutes or until baseline was reached. An FID detector was used to measure the methane produced from hydrogen reacting with the catalyst carbon. The use of an FID detector, versus a TCD, allowed us to differentiate between desorption of water and a carbon species as well as realizing high sensitivity to the product.

TEM analysis was performed using a JEOL 2010 microscope operated at 200 keV. Other experimental conditions are identical to those described in Ref. 1. To prepare the sample for TEM analysis, the samples in the TPR reactor were flushed with helium at reaction temperature, cooled to room temperature and a mixture of 1% O_2 in He was slowly pulsed over the gas to make sure that were no noticeable exotherms.

3. RESULTS AND DISCUSSION

The addition of potassium to an iron Fischer-Tropsch catalyst has long been known to be a significant promoter of the reaction. Potassium will decrease the amount of methane and higher paraffins formed but will increase the number of olefins.[4] Carburization of iron in synthesis gas is more rapid on catalysts with potassium than without. Up to a maximum level, potassium increases the activity of a catalyst, eventually decreasing activity at higher levels. In contrast, selectivity improves (increasing longer chain selectivity) as potassium increases, even though overall activity may be decreasing.[5] Experiments performed using TPR have shown that even a small amount of potassium has a significant effect on the types of carbide formed during reaction. Figure 3 shows that for a series of potassium promoted catalysts between a 0.2% K catalyst and a 3% K, there is only a 75° range in the position or types of peaks forming from iron carbide reduction. However, the difference between a catalyst with no potassium promotion (Figure 2) and one with only 0.2% K is dramatic. The carbide most difficult to reduce (seen as a peak at

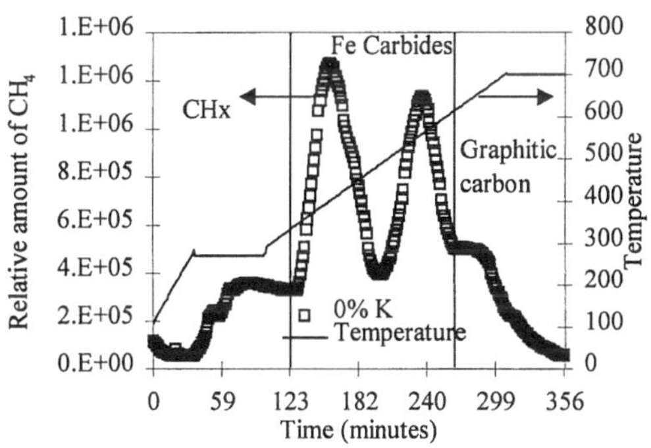

Figure 2. TPR of Fischer-Tropsch iron catalyst following reaction for 24 h at 215°C in H_2/CO = 0.7.

560°C in Figure 2) is not formed on potassium-promoted catalysts as seen in Figure 3. Since potassium-promoted catalysts are more active/selective (at least at low K loadings) than unpromoted iron, this would indicate that the more stable carbide is less active than the carbides that reduces in the 440°C ± 40 range.

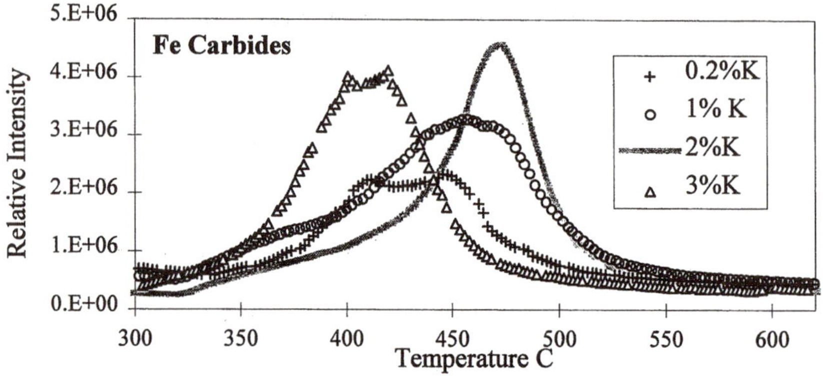

Figure 3. The carbide reduction peaks formed for all four potassium-promoted iron catalysts have peak maximums that fall between 400–475°C.

Another characteristic found in Figures 2 and 3 regarding potassium promoted catalysts is the formation of significant graphitic carbon on the unpromoted iron catalyst. The 0.2-2% potassium-promoted catalysts do not show substantial graphitic carbon formation in the TPR. Figure 4 shows the relative amount of graphitic carbon formed (unpromoted catalyst = 1.0) versus percent potassium following reaction for 24 h.

A minimum of graphite formation is found on the 1% K catalyst following 24 h of reaction. The activity of these particular catalysts has not been measured. However, many studies have been performed on the effect the amount of potassium promotion has on activity and selectivity.[6-8] The potassium loading which gives the maximum activity is dependent on many factors, including activation, silica content, and other promoters present. However, in many publications this maximum is found somewhere between 2 and 5 percent potassium. The formation of graphitic carbon above and below this range of potassium promoter may decrease activity by preventing a physical barrier to the catalyst. It appears that a certain amount of potassium is necessary to increase activity, but the potassium that increases activity also increases graphitic carbon formation, until eventually, the advantage of the activity provided by potassium is outweighed by the formation of graphitic carbon.

In order to examine whether there is a difference in the types and amounts of carbides formed under conditions known to cause attrition, several copper promoted catalysts were studied following reaction for 24 h at significantly different temperatures (215°C and 270°C) and reported in Figure 5. The temperature of reaction is known to have a profound effect on the attrition of an iron catalyst. The higher the temperature, the more quickly the catalyst attrits. Two catalysts were tested, 1% Cu/Fe and 2.5% Cu/Fe (atomic percent). By comparing each catalyst at different temperatures, we minimized the effects that non-chemical factors can have on the shift of TPR peaks such as surface area, porosity, and particle size.[9] Both catalysts showed nearly identical results. The higher reaction temperature eliminated the carbide that reduced at 350°C and added a carbide that reduced at a higher temperature (465°C). In summary, the carbides formed at a higher reaction temperature, which represent the composition of catalysts more susceptible to attrition, showed different carbide formation. Figure 4 shows the TPR of a 2.5% Cu/Fe catalyst that was reacted at both 215°C and 270°C and demonstrates the shift towards more stable carbides with higher reaction temperatures. Although copper is a valuable Fischer-Tropsch promoter, its enhancement on activity is not as significant as the addition of potassium. It is important to note that two types of carbides are formed on Cu-only promoted catalysts, whereas K-promotion appears to eliminate the higher-temperature-reducing, and less active carbide.

3. CONCLUSIONS

Three groups of carbides have been observed on the catalysts in this study. The ranges at which they reduce center around 400-475°C, 530°C and > 550°C. An unpromoted iron catalyst shows two carbide peaks with reduction temperatures at 430°C and >550°C. Addition of even a small amount of potassium, which is known to increase the activity of the catalyst, eliminates formation of the second carbide at the conditions used in this study. Copper promoted catalysts show two carbide reduction peaks, 350°C and 530°C. However, reaction at temperatures high enough to cause rapid attrition, 270°C, cause more-difficult-to-reduce carbides to form. The evidence

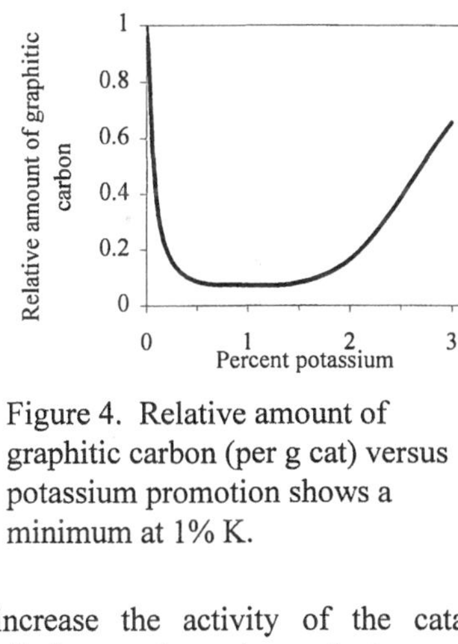

Figure 4. Relative amount of graphitic carbon (per g cat) versus potassium promotion shows a minimum at 1% K.

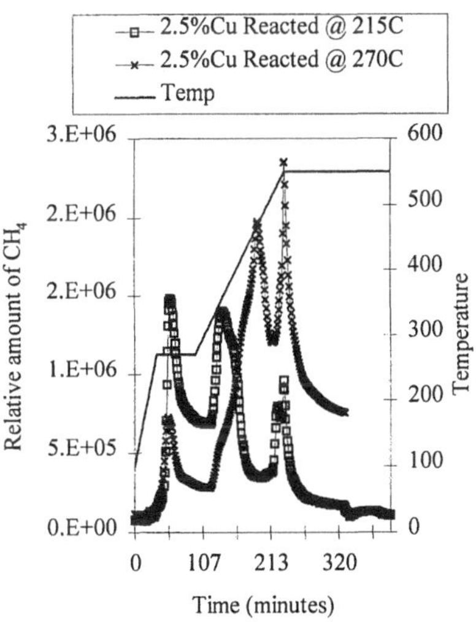

Figure 5. Copper-promoted iron catalysts reacted at 215°C and 270°C shows a shift in the reduction temperature of the carbides following reaction at higher temperatures.

from this study indicates that the most active carbides are the lowest reducing carbides, the ones that appear in the TPR below 500°C. It also appears that decreased formation of graphite on catalysts promoted with small amounts of potassium (versus no promotion or large amounts of potassium promotion) may play a role in the maximum activity coming from a catalyst with 2-5 % potassium promotion.

4. ACKNOWLEDGMENTS

This work was supported by the United States Department of Energy under Contract DE-AC04-94AL850000. Sandia is a multiprogram laboratory operated by Sandia Corporation, a Lockheed Martin Company, for the United States Department of Energy.

REFERENCES

[1] M. D. Shroff, D. S. Kalakkad, K.E. Coulter, S. D. Kohler, M. S. Harrington, N. B. Jackson, A. G. Sault, and A. K. Datye, *J. Catal.*, 156 (1995)185.
[2] D. S. Kalakkad, M. D. Shroff, S. D. Kohler, N. B. Jackson, and A. K. Datye, *Appl. Catal. A.*, 133 (1995) 335.
[3] N. B. Jackson, A. K. Datye, L. Mansker, R. J. O'Brien and B. H. Davis, in C. Bartholomew and Fuentes, (eds.), Catalyst Deactivation 1997, Elsevier, Amsterdam, 1997.
[4] R. A. Dictor and A. T. Bell, J. Catal., 97 (1986) 121.
[5] R. B. Anderson, The Fischer-Tropsch Synthesis, Academic Press, London, 1984, p. 144-149.
[6] M. E. Dry in J. R. Anderson and M. Boudart (eds.), Catalysis: Science and Technology, Vol. 1, Springer-Verlag, Berlin and New York, 1981, Ch. 4.
[7] D. G. Miller and M. Moskovits, J. Phys. Chem., 92 (1988) 6081.
[8] R. J. O'Brien, L. Xu, R. L Spicer, and B. H Davis, Division of Petroleum Chemistry Preprints, American Chemical Society, Washington, DC, 1996.
[9] R. J. Gorte, Catal. Today 28 (1996) 405.

NATURAL GAS CONVERSION V
Studies in Surface Science and Catalysis, Vol. 119
A. Parmaliana et al. (Editors)

Selective synthesis of C_2-C_4 olefins on Fe-Co based metal/oxide composite materials.

F. Tihay[a], G. Pourroy[b], A.C. Roger[a], and A. Kiennemann[a].

[a]LERCSI ECPM UMR CNRS 7515, 1, rue Blaise Pascal, 67008 Strasbourg, France.

[b]IPCMS UMR CNRS 046, 23, rue du Loess, 67037 Strasbourg, France.

Fe-Co based metal/oxide composite materials $(Co_\gamma Fe_{1-\gamma})_\alpha[Co_\beta Fe_{3-\beta}O_4]$ were synthesized with various values of γ, α and β via oxido-reduction processes between cobalt and iron ions leading to the simultaneous crystallization of two phases : a Fe-Co alloy, and a Fe-Co spinel. The materials were characterized by XRD, TGA-TDA and SEM. These catalysts are efficient in Fischer-Tropsch reaction to produce C_2-C_4 (50 wt %), CO_2 (<20 % molar selectivity), with high olefin/paraffin ratio (3.7).

1. INTRODUCTION

Fischer-Tropsch catalysts which produce hydrocarbons from syngas are highly diversified in their formulations depending on their efficiency in chain growing and/or on the olefin to paraffin ratio. The best catalysts given in the literature for the synthesis of light olefins are iron and cobalt metal on partially reduced oxide support [1,2]. The Fe-Mn and Co-Mn catalysts, with or without alkaline promoters, are the most studied. The high selectivity of Fe-Mn catalysts in C_2-C_4 has been correlated to the presence of a Mn-Fe spinel oxide. Unfortunately, carbide phases are formed during the catalytic tests and then catalysts have short lifetimes. Cobalt does not carburize as readily as iron under CO/H_2 atmosphere and numerous studies have confirmed its efficiency for obtaining light olefins [3]. The protection of the spinel phase under the reaction conditions and the presence of a metallic phase which does not carburize under test seem to be essential to produce C_2-C_4 olefins from CO/H_2.

The aim of this work is to combine the respective advantages of iron and cobalt based catalysts and synthesize mixed iron-cobalt catalysts able to improve the C_2-C_4 olefins selectivity with low CO_2 production. Indeed, an alloy between iron and cobalt could have an enhanced activity with respect to the individual metals [4]. Previous studies have shown that Co-Fe based metal/oxide composite catalysts are efficient to produce C_2-C_4 olefins from CO/H_2 when the cobalt ferrite is preserved under test[5]. The mastering of parameters of catalysts preparation leads to the control of the alloy composition and of the metal/spinel ratio and therefore allows an improvement of catalytic results.

2. EXPERIMENTAL

2.1. Preparation

The synthesis of Fe/Co based metal/oxide composites is based on the disproportionation of $Fe(OH)_2$ into Fe^0 and Fe_3O_4 in a concentrated KOH solution. When Co(II) is also involved, a part metallic iron reduces cobalt(II) into cobalt metal (E_0 (Fe^{2+}/Fe) = -0.409 V and E_0 (Co^{2+}/Co) = -0.280V) [6] which leads to the formation of a Co/Fe alloy. Unreduced Co(II) ions are involved in the spinel phase. Thus, aqueous solutions of Co(II) and Fe(II), $FeCl_2.4H_2O$ and $CoCl_2.6H_2O$, with $[Fe^{2+}]$ + $[Co^{2+}]$ = 3 M and Co/Fe ratios of 0.23 (sample1), 0.33 (samples 2a, 2b), 0.45 (sample 3), were prepared and heated at 60°C. An aqueous potassic solution with [KOH] =10 M was refluxed (115°C) in a stainless steel vessel. After addition of the metal chloride solution into the KOH solution, heating and stirring were maintained during one hour for maturation. The black precipitate formed was then filtered, washed with boiling and cold water, then ethanol and dried at 40°C under air. Sample 2b was prepared by annealing sample 2a at 435°C in argon for 15 hours.

2.2. Characterization

X-Ray diffraction data were collected at room temperature using of a D500 Siemens diffractometer equipped with a quartz monochromator (Co $K_{\alpha 1}$ = 1.78897 Å). The morphology of catalysts was observed by means of a JEOL scanning electron microscope. Co/Fe ratios were determined by EDXS analysis with an ultrathin-window spectrometer KEVEX. Thermogravimetric (TGA) and differential thermal (DTA) analyses were carried out in silica crucibles under air and under vacuum by using a Setaram 92 apparatus. The temperature was increased at a rate of 5°Cmin^{-1} from 25°C up to 980°C. The variation in weight was calculated after substracting the weight of the empty crucible. Specific surfaces areas were carried out by using the BET method based on the N_2 physisorption capacity at 77K on a Coulter SA 3100 apparatus. The specific surface areas are very low, below 10 $m^2.g^{-1}$.

2.3. Reactivity in CO hydrogenation

Catalytic tests were performed in a fixed bed reactor. The feed gas flow rate was adjusted by mass flow controllers. The pressure was regulated by coupling a pressure comparative and an electronic control valve. The catalyst (300 mg) was heated to 220°C with a rate of 0.2°Cmin^{-1} under nitrogen flow (2.3 Lh^{-1}). Nitrogen was then replaced by the CO/H_2 mixture (CO/H_2 = 1, total gas flow = 1.2 Lh^{-1}, GSHV = 3000 h^{-1}, P = 1 MPa) and the temperature increased to 230°C (0.2°Cmin^{-1}). The exit gases were analyzed by on-line gas chromatography. Two traps collect liquid products which were analyzed by gas chromatography. The reactivity of catalysts was studied at the steady state at 230, 240, 250 and 260°C. The catalysts were maintained 60 hours at each temperature.

3. RESULTS AND DISCUSSION

3.1. Characterization of catalysts before catalytic tests

The X-Ray diffraction pattern of sample 2a is presented in Figure 1a. It shows that the sample is highly crystallized and composed of two phases : a spinel phase with a lattice

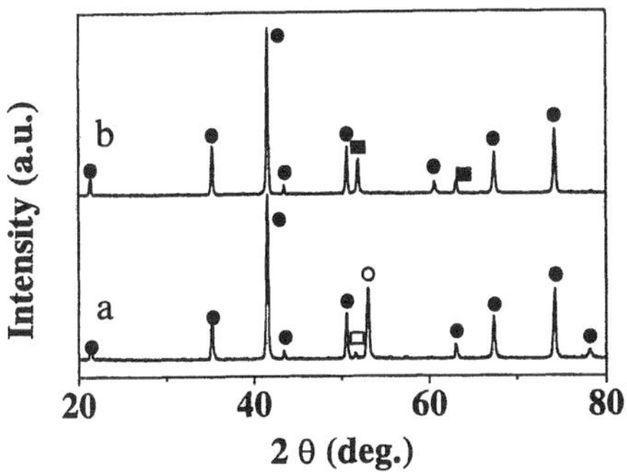

Figure 1. X-Ray diffraction patterns of catalysts a) catalyst 2a, b) catalyst 2b. ●, □, ○ and ■ represent respectively the spinel, Co-Fe alloy (α-Mn structure), Co-Fe alloy (b.c.c.), Co-Fe alloy (f.c.c.).

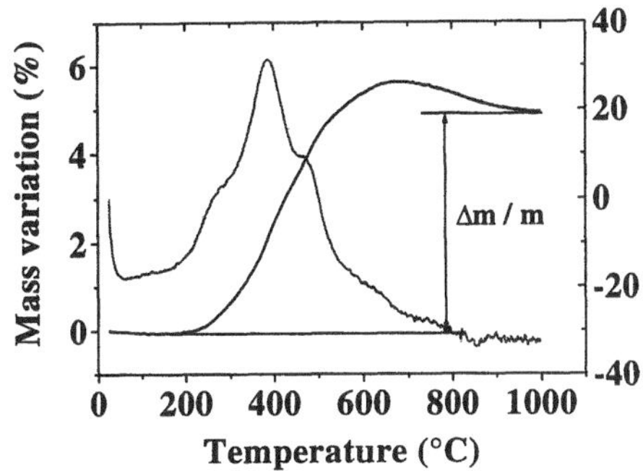

Figure 2. Thermal gravimetric analysis and differential thermal analysis of catalyst 2a.
(---) heat flow (mW).
(—) mass variation (%).

Table 1
Co/Fe ratio, metal lattice parameters, weight increase under air, total metal loading, and chemical formula of catalysts

Catalyst	Co/Fe	metallic phase $a_{b.c.c.}$ (Å)	metallic phase $a_{f.c.c.}$ (Å)	Δm/m in air (%)	M(%)	Formula
1	0.23	2.8490(3)	-	5.57	2.9	$(Co_{0.61}Fe_{0.39})^{0}_{0.12}[Co_{0.51}Fe_{2.49}O_4]$
2a	0.33	2.8420(4)	-	6.25	14.5	$(Co_{0.73}Fe_{0.27})^{0}_{0.68}[Co_{0.41}Fe_{2.59}O_4]$
2b	0.33	-	3.5493(8)	5.96	13.5	$(Co_{0.95}Fe_{0.05})^{0}_{0.62}[Co_{0.31}Fe_{2.69}O_4]$
3	0.45	2.8340(5)	-	4.95	5.8	$(Co_{0.80}Fe_{0.20})^{0}_{0.25}[Co_{0.86}Fe_{2.14}O_4]$

parameter always equal to 8.4025(21) Å and a metallic phase isomorphous with α-Fe, of b.c.c. structure. The lattice parameter (a) of the b.c.c. alloy varies owing to Co/Fe ratios [7]. For a = 2.8420(4) Å, the alloy corresponds to $(Co_{0.73}Fe_{0.27})^{0}$ (see Table 1). One peak, corresponding to the most intensive diffraction line of an iron-cobalt alloy isomorphous with α-Mn is also present but its low intensity shows that it is present in very low amount [8]. Samples 2b (treated at 435°C under Ar for 15 hours) is also made of a spinel phase and a metallic phase. While the lattice parameter of the spinel has not varied, the alloy of b.c.c. structure has been replaced by an iron-cobalt alloy of f.c.c. structure. Its lattice parameter a = 3.5493(7) Å shows that it is quasi-nearly cobalt. The thermal treatment of catalyst 2a into 2b results in a cobalt enrichment of the metallic alloy of the catalyst via a redox reaction.

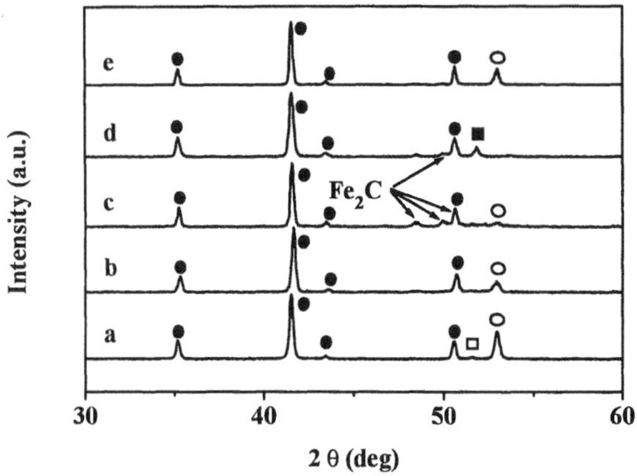

Figure 3. X-Ray diffraction patterns of catalysts : a) catalyst 2a before test, b) catalyst 1 after test, c) catalyst 2a after test, d) catalyst 2b after test, e) catalyst 3 after test. ●, □, ○ and ■ represent respectively the spinel, Co-Fe alloy (α-Mn structure), Co-Fe alloy (b.c.c.), Co-Fe alloy (f.c.c.).

The X-Ray diffraction diagrams of catalysts 1 and 3 are similar to that of catalyst 2a except for the alloy lattice parameter, i.e. for the alloy composition (see Table 1).

SEM observations of catalysts show that particles are monodisperse, of octahedral shape and of submicronic size. Unexpectedly, the analysis using backscattered electrons has not permitted to visualize pure metallic particles as it is the case for supported catalysts. That is the reason why we conclude that our catalysts consist in intimate mixture of the metallic and oxide phases : these are composite materials.

TG and TD analyses have been performed for each sample under air (Figure 2). The weight increase corresponds to the oxidation of the metallic phase and the Fe(II) ions of the spinel into Fe_2O_3 and CoO and a large exothermic peak in the DT curve. Therefore, taking into account the ratio r = Fe/Co determined by EDXS analysis, the cobalt ratio in the alloy determined by XRD analysis (γ) (assuming that the α–Mn has the same composition as the b.c.c. structure as previously shown [8]) and the weight increase under air t = Δm/m, the formula of these samples is written $(Co^0 Fe^0_{1-\gamma})_\alpha[Co_\beta Fe_{3-\beta}O_4]$ with $\alpha = \frac{4r+4}{p}$ and

$$\beta = 3 - \frac{4r}{p} - (1-\gamma) \text{ where } p = \frac{1+1.5r - 3.683t - 3.490rt}{1+t}$$

3.2. Catalytic tests

The catalytic results are summarized in Table 2. At 240°C, only catalysts 2a and 2b have a significant activity of 5.1%. At this temperature, catalyst 1 converts almost 1% of CO whereas the CO conversion for catalyst 3 is lower than 1%. This can be explained by the metal/spinel ratio of various composites : α is equal to 0.12 and 0.25 for catalysts 1 and 3 respectively whereas α is higher for catalysts 2a and 2b, 0.68 and 0.62 respectively.

Table 2
Result of catalytic tests of different catalysts at isoconversion of CO

Catalysts	1	2a	2b	3
Co/Fe ratio	0.23	0.33		0.45
Temperature (°C)	250	240	240	250
Total CO conversion (%)	5.9	5.1	5.1	5.2
Molar selectivity (%) into				
CO_2	18.4	23.6	25.1	15.2
Gaseous hydrocarbons	57.9	76.4	74.9	84.8
Liquid hydrocarbons	23.7	-	-	-
Hydrocarbons distributions (wt %) in the gaseous fraction C_1-C_9.				
C_1	21.5	21.1	24.3	25.0
$C_2 ; C_2^=$	6.0 ; 8.7	6.1 ; 6.3	7.4 ; 7.4	4.3 ; 9.0
$C_3 ; C_3^=$	2.4 ; 14.7	3.0 ; 12.9	3.4 ; 14.9	3.0 ; 15.1
$C_4 ; C_4^=$	1.7 ; 6.6	4.1 ; 11.2	4.6 ; 12.1	2.9 ; 13.5
C_5^+	38.5	35.3	26.0	27.2
Total C_2-C_4	40.1	43.6	49.7	47.8
Olefin to paraffin ratio in the C_2-C_4 fraction.	3.0	2.3	2.3	3.7

At a CO isoconversion of about 5 %, the reaction temperature is 250°C for the catalysts 1 and 3, and 240°C for the catalysts 2a and 2b. The CO/H_2 reaction produces only CO_2, water and hydrocarbons and no oxygenated products have been detected. The molar selectivity into CO_2 is always lower than 25 %, especially for the catalysts 1 and 3 for which the molar selectivity into CO_2 is only 18.4 and 15.2 respectively as reported in literature [9]. Th catalyst 1 is the only catalyst which produces liquid hydrocarbons (23.7 % molar). So, the molar selectivity into gaseous hydrocarbons is of 57.9 %, 76.4 % and 74.9 % for the catalysts 1, 2a, 2b respectively and reaches 84.8 % for the catalyst 3.

Within the C_1-C_9 gas fraction, the weight distribution into C_2-C_4 represents 40 % for the samples 1 and 2a, up to 50 % for the samples 2b and 3. In the C_2-C_4 fraction, the olefin to paraffin ratio (O/P) is largely in favor of olefins for catalyst 3 (O/P = 3.7) and remains satisfactory for the other systems (3.0 or 2.3).

To summarize the catalytic results, taking into account the molar selectivity into gaseous hydrocarbons, the C_2-C_4 weight selectivity in this fraction and the olefin to paraffin ratio, it can be concluded that the catalyst 3 appears to be the most efficient to produce C_2-C_4 olefins. The catalysts 2a and 2b, for which the only difference is the thermal treatment, present the same catalytic behaviors although the metallic phases have different compositions.

3.3. Characterization of catalysts after test

No morphological modifications have been observed by SEM, and the specific surface areas remain very low.

The X-Ray diffraction patterns of catalysts after test are presented in Figure 3, as well as that of catalyst 2a before test for a better understanding. In all cases, the spinel phase is preserved after test. The metallic phase remains in its initial b.c.c. structure for the catalysts 1 and 3 and no carbide phases have been detected for these two systems. However, for the catalysts 2a and 2b after test, characteristic diffraction lines of Fe_2C appear in the 2θ range 43° - 50°, especially for the catalyst 2a. This is consistent with the decrease of metallic peaks, more pronounced for the catalyst 2a (Figures 3c and 3d). This could be explained by the fact that the amount of metallic iron in the catalyst 2a is the highest one, 3.9 wt. % compared to about 1 wt. % for the catalyst 2b. However, the possibility of carburation of the spinel phase cannot be ruled out.

The similarity of catalytic behaviors of catalysts 2a and 2b tends to demonstrate that for the same metal loading (about 14 wt. %), the reactivity (activity, selectivity) is not affected by the alloy composition for $0.73 < \gamma < 0.95$ This could be due to the iron migration on the surface in both cases as it was observed by XPS analysis and reported in literature [10]. However, a higher metal loading results in the formation of iron carbides (catalysts 2a and 2b compared to catalysts 1 and 3).

However, for a given temperature, the catalyst 2a is more active than the catalyst 3 which has a lower metal loading, while both alloys have similar compositions. Furthermore, at isoconversion of CO, the latter is more efficient to produce the C_2-C_4 olefins and this could be due to the better stability of the metallic phase.

4. CONCLUSION

By this way of preparation and by the mastering of different parameters, very good Fe-Co catalysts with a stable metallic phase under operating conditions were prepared for the selective synthesis of C_2-C_4 olefins. The molar selectivity into hydrocarbons reaches 85 %, with 48 % in weight in the C_2-C_4, and an olefin to paraffin ratio equal to 3.7. Furthermore, a molar selectivity in CO_2 less than 20 % is obtained.

REFERENCES

1. D. Das, G. Ravichandranand and D.K. Chakrabarty, Appl. Catal. A, 107 (1993) 73.
2. D. Das, G. Ravichandranand and D.K. Chakrabarty, Appl. Catal. A, 131 (1995) 335.
3. M. van der Riet, G.J. Hutchings and R.G. Copperthwaite, J. Chem. Soc., Chem Commun., (1986) 798.
4. T. Ishihara, K. Eguchi and H. Arai, Appl. Catal., 30 (1987) 225.
5. C. Cabet, A.C. Roger, A. Kiennemann, S. Läkamp and G. Pourroy, J. Catal., 173 (1998) 64.
6. S. Läkamp and G. Pourroy, Eur. J. Solid State Inorg. Chem., 34 (1997) 295.
7. W.B. Pearson, “Handbook of Lattice Spacing and Structures of Metals”, Pergamon Press (1964).
8. G. Pourroy, S. Läkamp and S. Vilminot, J. Alloy Compounds, 244 (1996) 90.
9. T.A. Lin, L.H. Schwartz and J.B. Butt, J. Catal., 97 (1986) 177.
10. D.J. Duvenhage and N.J. Coville, Appl. Catal. A, 153 (1997) 43.

NATURAL GAS CONVERSION V
Studies in Surface Science and Catalysis, Vol. 119
A. Parmaliana et al. (Editors)

A Mathematical Model of Fisher-Tropsch Synthesis in a Slurry Reactor

V.A. Kirillov[a], V.M. Khanayev[a], V.D. Mescheryakov[a], S.I. Fadeev[b], and R.G. Lukyanova[b]

[a]Boreskov Institute of Catalysis, Novosibirsk 630090, Prosp. Acad. Lavrentieva 5, Russia
[b]Institute of Mathematics, Novosibirsk 630090, Prosp. Acad. Koptyuga 2, Russia

INTRODUCTION

Numerous experimental studies have persuasively shown the advantages of hydrocarbon fuel productions from synthesis gas via the Fisher-Tropsch synthesis (FTS) in slurry reactors [1,2]. The key advantages are: the effective temperature control and, as a result, almost temperature gradientless operation that permits to achieve high selectivity of C_{3+} hydrocarbons formation, a possibility to operate at low H_2/CO ratios without catalyst deactivation, lower operation expenses compare to tubular or fluidized bed catalyst reactors etc. To design large scale slurry reactor it is necessary to have reliable reactor model and knowledge about influence of various design and operation options on FTS.

The pioneering studies of the homogeneous hydrodynamic regimes in the slurry reactors were carried out in [3,4]. By contrast, we focus on the FTS that is carried out under heterogeneous hydrodynamic regime. This regime is of interest to FTS slurry reactors because it provides a rather weak dependence from hydrodynamic parameters of linear gas flow velocity. This is a rather important feature so far as in the course of synthesis the gas velocity is decreased along the reactor length by a factor of 2, therefore the hydrodynamic regime is also changed

1. A mathematical model

We assume that the following reactions proceed on the catalyst surface between the gas components dissolved in the liquid phase

$$\nu_{CO}CO + \nu_{H_2}H_2 \rightarrow \nu_{C_xH_y}C_xH_y + H_2O \qquad (1)$$

$$H_2O + CO \rightleftarrows CO_2 + H_2 \qquad (2)$$

The stoichiometric coefficients in eq. 1 depend on the probability of the chain growth α, and paraffin fraction γ in the reaction products [3]: $\Delta = 3+(1-\alpha^2)+\gamma\alpha(1-\alpha)$; $\nu_{CO}=-1/\Delta$, $\nu_{H_2O}=1/\Delta$, $\nu_{H_2}=[2+(1-\alpha)^2+\gamma\alpha(1-\alpha)/\Delta$, $\nu_{CH}=(1-\alpha)/\Delta$, $x=1/(1-\alpha)$, $y=2[x+(1-\alpha)+\gamma\alpha]$.
As for as the Fisher-Tropsch synthesis is carried out under isobar-isothermal conditions in slurry reactors, we shall not consider the equations for heat- and pulse transfer. The gas phase mass balance equation for the **i**-th component is written

$$\frac{d}{d\xi}\left(V_g C_{gi}\right) = \frac{d}{dz}\left[\phi D_g \frac{d}{dz}\left(C_{gi}\right)\right] - J_i. \qquad (3)$$

where $J_i = k_L a \cdot (C_{gi}/m_i - C_{li})$, m_i - solubility coefficient for component *i*.
The liquid phase mass balance for each component can be written

$$\frac{d}{d\xi}(V_l C_{li}) = \frac{d}{d\xi}\left[(1-\phi)D_l \frac{d}{d\xi}(C_{li})\right] + J_i + (1-\phi)C_S(\nu_i R_{FTS} + \vartheta_i R_{WGS}) \tag{4}$$

where R_{FTS} is the rate of synthesis (1) and R_{WGS} - rate of water-gas shift reaction (2).
Mass balance equation for the catalyst particle in the slurry phase is

$$\frac{d}{dz}\{[V_l - (1-\phi)V_S]C_S\} = \frac{d}{dz}\left[(1-\phi)D_S \frac{d}{dz}(C_S)\right] \tag{5}$$

Summing up the equations (3) for the all gas phase components, the equation for the gas velocity can be obtained as

$$\frac{d}{dz}\left(V_g \sum_1^N C_{gi}\right) = -\sum_1^N J_i \tag{6}$$

Substituting the gas phase state equation $\sum_1^N C_{gi} = \frac{P}{RT}$ into (6), the equation for gas velocity can be written as

$$\frac{P}{RT}\frac{dV_g}{dz} = -\sum_1^N J_i \tag{7}$$

The boundary conditions on equations (1-7) defined at $z = 0$:

$$\begin{aligned}
&V_g = V_g^0 \\
&V_g(C_{gi} - C_{g0}) = \phi D_g \frac{d}{dz}(C_{gi}) \\
&V_l(C_{li} - C_{li}^0) = (1-\phi)D_l \frac{d}{dz}(C_{li}) \\
&[V_l - (1-\phi)V_S]C_S = (1-\phi)D_S \frac{d}{dz}(C_S)
\end{aligned} \tag{8}$$

and at reactor outlet $z = L$:

$$\frac{d}{dz}(C_{li}) = \frac{d}{dz}(C_{gi}) = \frac{d}{dz}(C_s) = 0 \tag{9}$$

For closing system of equations it is need to define catalyst loading value G in the reactor

$$G = \int_0^V C_S dv \tag{10}$$

2. Parameter Estimation

To calculate gas hold up we use the hydrodynamic model obtained elsewhere [5]. The model assumes that the solid catalyst particles are uniformly distributed in the liquid. The gas phase is divided into the diluted phase (large-size bubbles) and dense phase (small-size bubbles). The total gas hold up in the slurry reactor is set equal to the sum of gas hold up in the dense phase (ϕ_{KR}) and diluted phase (ϕ_b):

$$\phi = \phi_{KR} + \phi_b,$$
$$\phi_{KR} = 2.16\exp\left(-13.1\rho_g^{-0.10}\mu_l^{0.16}\sigma^{0.11}\right)\exp(-5.8\phi_S),$$
$$\phi_B = 0.23D_K^{-0.166}\left(V_g - U_{KR}\right)^{0.56}, \quad U_{KR} = \phi_{KR}V_\mu, \tag{11}$$
$$V_\mu = 2.25\frac{\sigma}{\mu_l}\left(\frac{\sigma^3\rho_l}{g\mu_l^4}\right)^{-0.273}\left(\frac{\rho_l}{\rho_g}\right), \quad \phi_S = C_S/\rho_S.$$

The coefficient of axial dispersion in liquid phase is estimated by the correlation [3]

$$D_L = 0.68D_K^{1.4}V_g^{0.33} \tag{12}$$

The value of axial dispersion in gas phase is estimated by the correlation [6]

$$D_g = 5\times10^{-9}D_K^{1.5}\left(\frac{V_g}{\phi}\right)^3 \tag{13}$$

The coefficient of particle diffusion [7] is

$$D_S = 0.076D_K\sqrt{gD_K}\left[\frac{1+0.009Re_p\left(V_g/\sqrt{gD_K}\right)^{-0.8}}{1+8\left(V_g/gD_K\right)^{0.85}}\right]^{-1} \tag{14}$$

Here $Re_p = \dfrac{V_S d_p \rho_l}{\mu_l}$.

The coefficient of the gas-liquid mass transfer can be estimated by correlation [8]

$$K_La = AV_g^{0.67}\left(1-\frac{\phi_s}{0.58}\right) \tag{15}$$

According to ref. [8] for glass particles $d_p \le 0.1\,\mathrm{mm}$ in air-water system A = 0.39. Since the experimental conditions in the FTS slurry reactor are differed from [8], we have compared calculation results of this mathematical model with the experimental data of [10]. In calculations there were used relations of [6] for the physicochemical properties of the three-phase FT systems and coefficient A was estimated as A=0.78 (Figure 1).

3. Kinetic of the FTS

For the *Fe*-containing catalysts, at 220-260°C the better agreement with experimental data is gotten using a reaction synthesis rate equation [11]:

$$R_{FTS} = \frac{k_{FTS}C_{H_2}}{1+\dfrac{BC_{CO_2}}{C_{CO}}} \tag{16}$$

where

$$k_{FTS} = k^0_{FTS}\exp\left(-\frac{12407}{T}\right)\ m^3/kg_{cat}s\ ;\ B = 0.115;\ k^0_{FTS} = 1.4\times10^7\ m^3/kg_{cat}s$$

For *Co*-containing catalysts we used [12] $k_{FTS} = k^0_{FTS}\exp\left(-\frac{11677}{T}\right)$ $m^3/kg_{cat}s$; $k^0_{FTS} = (2.4 \div 5.2)\times10^6\ m^3/kg_{cat}s$; $B = 4.24 \times 10^{-6}$ mol/s.

For water-gas-shift reaction (2) we have used data of Ref. [13].

4. **Numerical analysis**

The most typical conditions of the FTS performance in a slurry reactor [5] are as follows: gas velocity under operation conditions $V_g = 0.08 - 0.14$ m/s, temperature in the reactor 210-270°C, pressure range 10-20 atm, reactor length 8-14 m, CO : H_2 ratio 0.4-1.5, liquid velocity at the reactor inlet 0-0.01 m/s, liquid saturation of synthesis gas at the reactor inlet 0-100%. *Co*-containing catalyst has particle sizes 50 ÷ 100 mkm, mass fraction of catalyst in slurry 0.1-0.4.

Using the mathematical model, we can determine the most optimal condition (with the viewpoint of maximum synthesis gas conversion) for carrying out of FTS in the slurry reactors. A key parameter, which determines the intensity of mass transfer processes in the reactor, is the gas velocity. Under heterogeneous hydrodynamic regime a rise of gas velocity leads to the increase of the gas-liquid mass transfer rate. At the same time it increases a gas flow axial dispersion and decreases a residence time of gas phase in the reactor so finally these factors provide a conversion decrease. Figure 2 shows the effect of this parameter. Other phenomens that markedly influence on the synthesis gas conversion are the catalyst loading, catalyst activity, liquid velocity and sizes of suspended catalyst particles. It was shown that the increasing of the catalyst concentration in slurry brings to an increase of the synthesis gas conversion. However with further catalyst concentration increase the effective viscosity of the suspension becomes higher. As a result, in accordance with (15), the gas-liquid mass transfer becomes worse, and consequently, conversion is decreased. For the conditions above, the best suited average catalyst concentration in suspension is 0.2. The catalyst activity influences similarly. The increase of CO conversion takes place when k^0_{FTS} is increased up to $5\cdot10^6$ $m^3/g_{cat}s$. With further increase of k^0_{FTS} the synthesis is controlled by mass transfer.

The CO/H_2 ratio affects markedly on the synthesis gas conversion. As the ratio is increased (that corresponds to the decrease in hydrogen concentration) X_{CO-H_2} is decreased. Preliminary saturation of liquid, introduced into reactor, by synthesis gas has weak influence and takes place only in bed parts that are close to the inlet of the reactor. The effect of liquid velocity on the conversion manifests mainly by changes in the distribution of catalyst concentration along the reactor length. If the liquid velocity is lower than the rate of particle precipitation, the maximum catalyst concentration is observed at the reactor part near of liquid velocity on the conversion manifests mainly by changes in the distribution of the inlet. If the liquid velocity is higher than the rate of precipitation, the maximum catalyst concentration can be easily obtained at the reactor outlet. Thus, we have numerically shown the existence of maximum with respect to conversions of synthesis gas as the function of the liquid velocity.

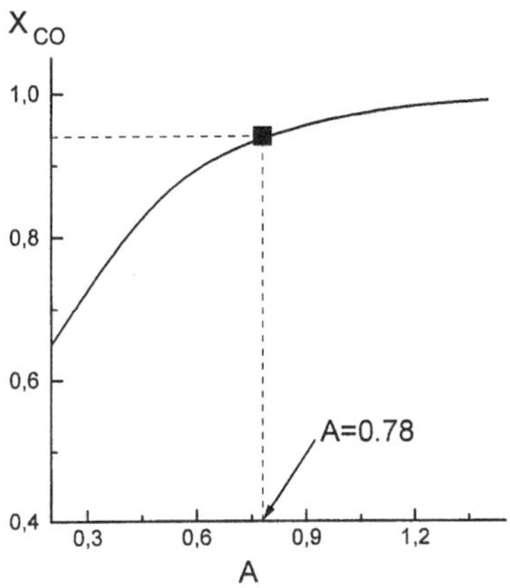

Figure 1. The dependence of CO conversion on coefficient A of Eq.15: $k^{0}_{FTS} = 1.4 \cdot 10^6\ m^3/kg_{cat}\ s$. Solid square is an experimental point of [10].

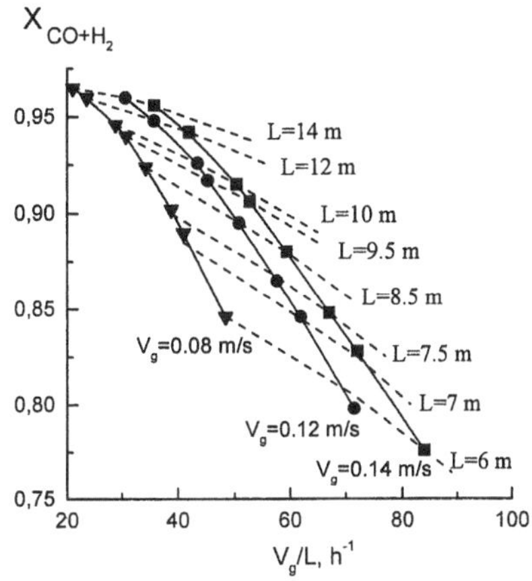

Figure 2. Effect of bed length (dash lines), gas flow velocity (solid lines) and gas residence time (L/V_g) on synthesis gas conversion in the FTS reactor: P=16 atm, CO/H_2=0.52, ρ_s=1200 kg/m^3, ρ_2= 650 kg/m^3, $k_{FTS} = 2.43 \cdot 10^6\ m^3/kg_{cat}\ s$.

CONCLUSION

The mathematical model of processes occurring in the FTS slurry reactors is proposed. We have estimated the coefficients of the mathematical model and shown that reaction of synthesis carried out in the transition region, when both the catalyst activity and gas-liquid mass transfer affect. The optimal conditions for the FTS were obtained numerically.

NOMENCLUATURE

C_{gi} component concentration in gas phase, mol/m^3
C_l component concentration in liquid phase, mol/m^3
C_s catalyst concentration, g/cm^3
d_p catalyst particle diameter, m
D_k reactor diameter, m
g free fall acceleration, m/s^2
G catalyst weight in reactor, kg
K_la gas-liquid mass transfer coefficient, s^{-1}
L length of catalyst bed, m

P pressure, atm
R universal gas constant
V_g gas superficial velocity, m/s
V_l liquid flow superficial velocity, m/s
V_s settling velocity of catalyst particles in swarm, m/s
z axial reactor coordinate, m
X conversion

Greek Letters

ϕ gas holdup
μ_l dynamic liquid viscosity, Pa S
ρ_g gas density, kg/m^3
ρ_s catalyst particle density, kg/m^3
σ surface tension coefficient, n/m
ρ_l liquid density, kg/m^2

REFERENCES

1. Dry M.E. Practical and theoretical aspects of catalytic Fischer-Tropsch process //Applied Catalysis A: General. 1996. V.138. N2,9. P.319-344.
2. Adesina A.A. Hydrocarbon synthesis via Fischer-Tropsch reaction: travails and triumphs // Applied Catalysis A: General. 1996. V.138. N2,9. P.319-344.
3. Stern D., Bell A.T., Heinemann H. Analysis of the Design of Bubble Column Reactors for Fischer-Tropsch Synthesis //IEC Proc.Des.Dev. 1995. V.24. P.1213-1219.
4. Stern D., Bell A.T., Heinemann H. Effects of Mass Transfer on the Performance of Slurry Reactors used for Fischer-Tropsch Synthesis //Chem.Eng.Sci. 1983. V.38. N4. P.597-605.
5. De Swart J.W.A. Scale-up of a Fischer-Tropsch Slurry Reactor /Ph. Dissertation. Amsterdam University. 1996.
6. Deckwer W.D., Louisi J., Zaidi A., Ralek M. Hydrodynamic Properties of the Fischer-Tropsch Slurry Process //IEC Proc.Des.Dev. 1980. V. 19. P.699-708.
7. Kato J., Nishiwaki A., Kago T., Fukuda T., Tonaka S. The Behaviour of Suspended Particles and Liquid Bubble Columns // J.Chem. Eng.Japan. 1972. V.5. P.112.
8. Nigam K.D.P., Schumpe A. Gas-Liquid Mass Transfer in a Bubble Column with Suspended Solids //AIChE Journal. 1987. V.33. N2. P. 328-330.
9. Kolbel H., Ralek M. The Fischer-Tropsch Synthesis in the Liquid Phase //Catal.Rev.-Sci.Eng. 1980. V.21. N2. P.225-274.
10. Storch G., Golambik N., Andersen R. Synthesis of Hydrocarbons from Carbon Oxide and Hydrogen, I.L., Moscow 1954.
11. Ledakowicz S., Nettelhoff H., Kokuun R., Deckwer W.D. Kinetics of the Fischer-Tropsch Synthesis in the Slurry Phase on a Potassium-Promoted Iron Catalyst //IEC Proc.Des.Dev. 1995. v.24. P.1043-1049.
12. Novel Fischer-Tropsch Slurry Catalysts and Process Concepts /DOE Report under Contract NO DE-AC22-84PC70030. 1986.
13. Newsome D.S. The Water-Gas-Sift Reaction//Catal.Rev.-Sci.Eng. !980, v. 21, N 2, P. 275-318

NATURAL GAS CONVERSION V
Studies in Surface Science and Catalysis, Vol. 119
A. Parmaliana et al. (Editors)

Mechanistic insights in the CO hydrogenation reaction over Ni/SiO_2

C. Márquez-Alvarez, G.A. Martin, and C. Mirodatos

Institut de Recherches sur la Catalyse, CNRS. 2 Av. Albert Einstein, F-69626 Villeurbanne Cédex, France.

Steady-state isotopic transient kinetic analysis (SSITKA) is used to determine the amount of adsorbed CO, intermediates CH_x and x value during the course of the methanation reaction catalysed by Ni/SiO_2. The results support a mechanism in which CO dissociative adsorption is followed by a stepwise hydrogenation of carbon, the hydrogenation of CH into CH_2 being the rate limiting step.

1. INTRODUCTION

In two recent publications [1,2], our group presented a thorough characterisation of a Ni/SiO_2 catalyst at different stages of the methanation reaction under a $CO+2H_2$ mixture, as well as a detailed kinetic study, performed by means of *in situ* transient techniques: SSITKA (steady-state isotopic transient kinetic analysis) and DRIFTS (diffuse reflectance infrared Fourier transform spectroscopy).

Characterisation by hydrogen chemisorption, magnetic measurements, electron microscopy, temperature programmed hydrogenation and DRIFTS browsed direct evidence that, in contact with the reacting mixture, after an initial period of carbon deposition, fast sintering and particle smoothing via nickel carbonyl transfer, leading to a preferential growth of Ni(111) planes, a surface nickel carbide monolayer was developed, with a stoichiometry $Ni_{2.5}C$. This carbide monolayer was shown to constitute a reservoir of reacting intermediates. A kinetic ensemble model [3] well fitted the kinetic data obtained under steady-state conditions at temperatures between 230 and 350°C, following the isotopic switch $^{13}CO+H_2/^{12}CO+H_2$. According to this model, and from DRIFTS and H_2/D_2 exchange results, it was concluded that the rate of hydrogenation is controlled by the probability for a hydrogen molecule to collide with an active site formed of one to two adjacent Ni atoms free from adsorbed CO. Several CO adspecies were identified by DRIFTS under reaction conditions, largely covering the metal surface, with an average stoichiometry of $CO/2Ni_s$, the concentration of active sites thus being statistically determined by the CO coverage. Carbon atoms belonging to the carbidic layer associated to the active site are hydrogenated by hydrogen activated on the free Ni atoms. The regeneration of the carbidic layer is in turn ensured by CO dissociation after methane desorption.

That study allowed us to experimentally determine important parameters which validate the proposed mechanism, such as the coverage of reaction intermediates, CH_x, as a function of temperature and the stoichiometry of adsorbed species, CO and CH_x. However, several aspects were not fully clarified concerning the exact rate limiting step and nature of the so called CH_x

species. In the present paper, the SSITKA study is extended and the H_2/D_2 kinetic isotope effect analysed in order to get a more advanced description of the reaction mechanism.

2. EXPERIMENTAL

The Ni/SiO_2 catalyst used in the present study was described in previous papers [1-3]. The precursor (20 wt.% Ni) was obtained by contacting silica (Aerosil Degussa, 200 m^2g^{-1}) with a solution of nickel nitrate hexamine. The solid was dried, then crushed to powder and reduced at 650°C for 15 h at 2°C min^{-1} heating rate in a hydrogen flow (5 l h^{-1}). The average Ni particle size was measured by hydrogen chemisorption, magnetic methods and electron microscopy and was found to be 4.2 nm after reduction with a good agreement between the three techniques [1].

The isotopic transient experiments under steady state conditions were carried out by changing rapidly the composition of the feeding mixture from $^{12}CO+2H_2$ to $^{12}CO+2D_2$ or from $^{13}CO+2D_2$ to $^{12}CO+2D_2$ and vice versa at the inlet of the microreactor. Before the series of switches was introduced, a pseudo steady-state was attained by contacting the catalyst with a $CO+2H_2$ mixture at 230°C for 4h. The conversion level was maintained below 10% at any temperature to ensure differential and isothermal conditions. The abrupt switches were obtained with an automatic four-way valve located just before the reactor. The signal distortion induced by the non-catalytic system (tubes, reactor...) was followed by the transient signal of a helium trace introduced in one of the feeds. The origin of the time scale was chosen as the time at which the inert tracer response starts. The gas composition at the reactor exit was continuously analysed with a VG-quadrupole mass spectrometer, and occasionally by gas chromatography using TCD and FID detectors. Details on the technique are reported in [4].

3. RESULTS

3.1. Inverse H/D isotope effect

Under the experimental conditions used in this study, i.e., a hydrogen to carbon monoxide ratio of 2 and temperature between 230 and 280°C, an inverse isotope effect is observed for the methanation reaction. The rate of methane formation with H_2 as a reactant (r_H) is smaller than that observed using D_2 (r_D) by a factor 0.6. This isotope effect is nearly constant in the temperature range 230-280°C (Figure 1). However, as expected, the effect is less pronounced at higher temperatures, the ratio r_H/r_D being c.a. 0.8 at 600°C [2]. Similar effects have already been reported for CO hydrogenation on different metals [5].

3.2. Steady-state isotopic transients $^{13}CO+2D_2/^{12}CO+2D_2$

The transient responses of ^{13}CO, $^{13}CD_4$ and He, following the stepwise switch from $^{13}CO+2D_2$(He) to $^{12}CO+2D_2$ have been obtained in the temperature range 250-280°C, and compared to the $^{13}CO+2H_2/^{12}CO+2H_2$ transient results reported elsewhere [2]. In both cases, small delays are observed in the CO decay with respect to the inert tracer (He), indicating the presence of a reservoir of adsorbed CO in equilibrium with CO in the gas phase during the course of the reaction [2,6]. The accumulation of active intermediates giving CD_4 (referred to as CD_x) is also revealed by significant delays in the $^{13}CD_4$ decay response. The experimental delays (τ_{CO}, τ_{CDx}) and the corresponding steady-state concentrations (N_{CO}, N_{CDx}) are shown in Table 1 for several temperatures.

Table 1
Time delay (τ), concentration (N) and coverage (θ) of CO and CD_x intermediates determined from SSITKA experiments following the switch $^{13}CO+2D_2/^{12}CO+2D_2$.

T (°C)	τ_{CO} (s)	N_{CO} (mmol g^{-1})	τ_{CDx} (s)	N_{CDx} (mmol g^{-1})	θ_{CO} (ML)	θ_{CDx} (ML)
250	1.03	0.182	38.4	0.166	0.70	0.29
270	0.88	0.154	22.6	0.233	0.59	0.40
280	0.85	0.146	15.1	0.244	0.56	0.42

From N_{CO} and N_{CDx} values, coverages are calculated by estimating the average stoichiometries of the adsorbed species. For this purpose, we make use of the rate equation (1) written according to the proposed ensemble model [2]:

$$r = K_o \exp(-E_o/RT)\, P_{H2}\, (1-\theta_{CO})^W \qquad (1)$$

This rate equation contains the term $(1-\theta_{CO})^W$ that represents the concentration of active sites, i.e., the fraction of the Ni surface occupied by reacting intermediates and, therefore, it equals θ_{CDx}. Then, if A and W are the stoichiometries Ni_s/CO and Ni_s/CD_x, respectively, it can be written:

$$N_{CDx} = (Ni_s/W)\,(1-A\,N_{CO}/Ni_s)^W \qquad (2)$$

where, Ni_s is close to 3.6 10^{20} surface Ni atoms per gram of catalyst [1].

Data in Table 1 well fit eq. (2), giving A=2.3 and W=1.0, in good agreement with stoichiometries calculated from the $^{13}CO+2H_2/^{12}CO+2H_2$ transients (Ni_s/CO=2.1 and Ni_s/CH_x=1.2, respectively [2]). The calculated coverages, in monolayer (ML), are shown in Table 1 and plotted in Figure 2 together with those of CO and CH_x (from $CO+H_2$ transients in ref. [2]).

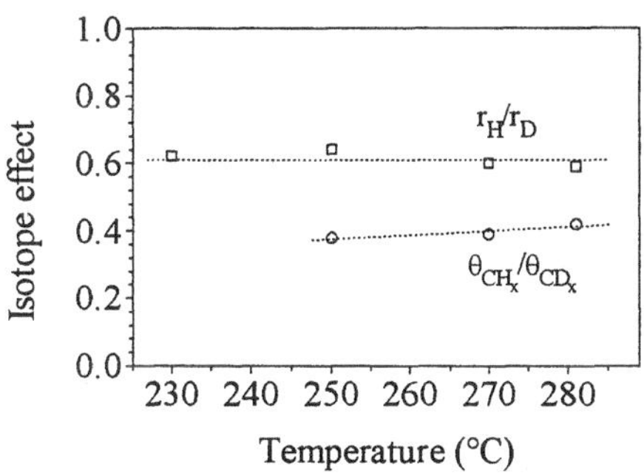

Figure 1. H/D isotope effect on reaction rate (r) and intermediate species coverage (θ).

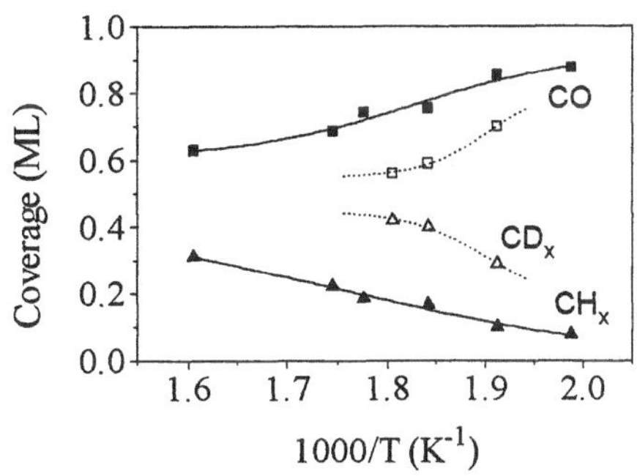

Figure 2. CO, CH_x and CD_x steady-state coverages under $CO+2H_2$ (full symbols) and $CO+2D_2$ (open symbols) reactions.

As shown in Figure 2, at any temperature the intermediate species coverage obtained in the presence of D_2 is higher and, correspondingly, the CO coverage smaller, than those obtained with H_2. This isotope effect on the steady-state concentration of intermediates slightly diminishes ($\theta_{CHx}/\theta_{CDx}$ ratio increases) at increasing temperatures (Figure 1). For the $CO+2D_2$ reaction, the whole carbide monolayer participates in the reaction at temperatures higher than 270°C, as at these temperatures θ_{CDx} stabilises at c.a. 0.4 ML, which corresponds to the measured C/Ni_s ratio under the reaction conditions [1]. For the $CO+2H_2$ reaction, that state is not reached but at temperatures higher than 350°C.

3.3. Steady-state isotopic transients $CO+2H_2/CO+2D_2$

The transient response of hydrogen and methane molecules has been followed when switching from H_2 to D_2 in the reaction mixture at 250°C. As it has been pointed out, a H/D isotope effect exists under this conditions, that provides different reaction rates (r_H=2.8, r_D=4.3 μmol s^{-1} g^{-1}) and coverages of intermediate species when changing between both gases. Therefore, this switch does not leave the steady-state strictly unchanged. For this reason, a sufficiently low temperature (250°C) was chosen in order to minimize the changes in the steady-state, while keeping the conversion level high enough for the products composition to be accurately analysed.

The normalized transient responses of methane molecules ($CH_{4-n}D_n$, with n=0,1,2,3,4), corresponding to fragments of m/e=16-20, are shown in Figure 3. The curves have been corrected for the contribution of H_2O, HDO and D_2O fragments to the signal intensity. The fragmentation patterns of all the methane molecules were also used to correct every main fragment signal intensity for heavier methane fragments contributions. The fragmentation patterns of CH_4 and CD_4 where obtained experimentally and those of CH_3D, CH_2D_2 and CHD_3 estimated by interpolation.

It can be seen in Figure 3 that a sequential transient from CH_4 to CD_4 is produced step by step, following the switch from H_2 to D_2, except for CH_3D and CD_4 species, that start to appear at the same time.

As for the hydrogen molecules, a transient production of HD is observed that reveals the fast dissociative adsorption/desorption equilibrium established between the gaseous hydrogen and the metallic surface [2].

4. DISCUSSION

The set of isotopic transients $^{13}CO+2H_2/^{12}CO+2H_2$ and $^{13}CO+2D_2/^{12}CO+2D_2$ have allowed us to determine the steady-state concentration of intermediates CH_x (or CD_x) during the methanation reaction on Ni/SiO_2. This study has verified experimentally that the concentration of those intermediates increases in the presence of D_2, as shown in Figure 2. This is in agreement with a higher D_2 adsorption coefficient on Ni with respect to H_2 [7], as the hydrogenation steps will be favoured in the presence of D_2 and the concentration of the corresponding reaction intermediate increased [8].

This increased concentration of reaction intermediates can explain the observed inverse isotope effect, according to Ozaki [9] who proposed that isotope effects could be due both to a kinetic effect on the rate-determining step and to a thermodynamic effect on the concentration of an intermediate. Our results confirm the calculations made by van Nisselrooij et al. [10] who estimated by statistical thermodynamics the rate and equilibrium constants isotope effect for a mechanism in which the dissociative adsorption of carbon monoxide and hydrogen is followed by the hydrogenation of adsorbed oxygen and carbon, and predicted a θ_{CH}/θ_{CD} ratio smaller than 1. They successfully reproduced their data on the kinetic isotope effect and proposed the hydrogenation of CH species as the rate limiting step.

The isotopic transient $CO+2H_2/CO+2D_2$ results have shown that, under methanation reaction conditions, a fast equilibrium of hydrogen dissociative adsorption takes place, as deduced from the transient formation of HD. Also, all the methane molecules, from CH_4 to CD_4, are produced when replacing H_2 by D_2. Both results support the mechanism of CO dissociation followed by a stepwise hydrogenation of surface carbon.

We have calculated the binomial distribution of $CH_{4-n}D_n$ molecules by taking the probability p for a H atom to be present in a certain methane molecule equal to the H/H+D ratio in the ensemble of $H_{2-i}D_i$ molecules at any time, and the probability for D atoms as 1-p. Such a distribution would be expected in the case that the hydrogenation steps (as well as the hydrogen dissociative adsorption equilibrium) were fast and the CO dissociation, the rate limiting step. The change of this H/H+D ratio with time (Figure 4, dotted line) has been calculated from the H_2, HD, and D_2 distribution obtained after H_2 was substituted by D_2 in the reaction mixture at 250°C (see reference [2]). The calculated $CH_{4-n}D_n$ binomial distribution (not shown for brevity) reveals a sequential deuteration from CH_4 to CD_4, all the $CH_{4-n}D_n$ transients being regularly spaced in time. With respect to these transients, the experimental data reported in Figure 3 show a delay in time and a quite lower CHD_3 intensity due to its simultaneous appearance with CD_4. This disagreement between the model and the experimental data rules out that CO hydrogenation is the rate determining step, and therefore reinforces the previous conclusion that the rate is controlled by the hydrogenation of carbon adspecies.

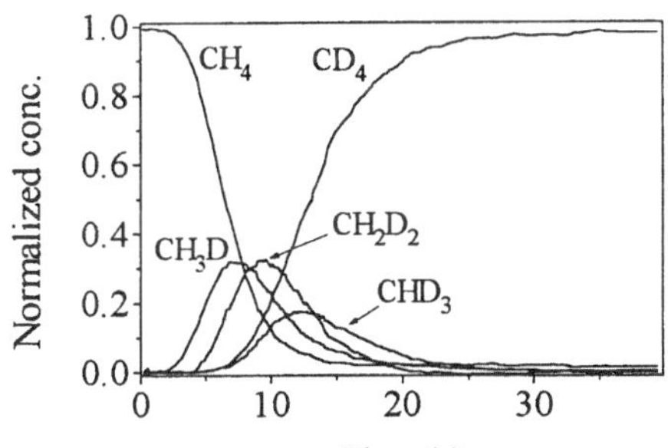

Figure 3. Normalized transient responses of methane molecules following the switch $CO+2H_2/CO+2D_2$ at 250°C.

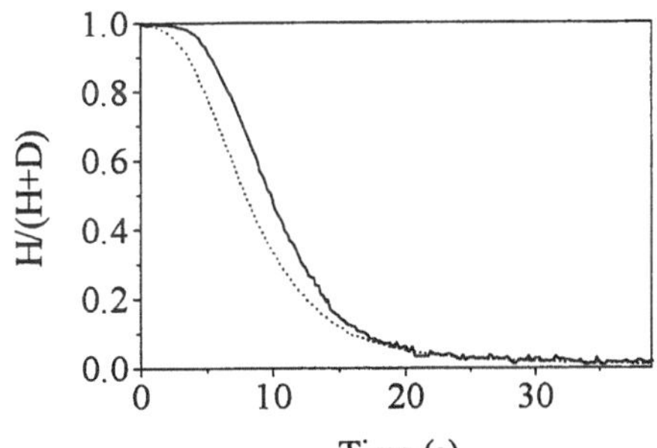

Figure 4. H/H+D atomic ratio in hydrogen (dotted line) and methane (full line) molecules for same switch as in Figure 3.

Further arguments in favour of the proposed mechanism can be found by comparing the H/H+D ratio in the ensemble of $CH_{4-n}D_n$ curves with that of $H_{2-i}D_i$ molecules (Figure 4, full and dotted lines, respectively). Methane being produced by hydrogenation of a pool of CH_x moieties, the delay observed in the H/H+D curve for methane species reveals the accumulation of H in that CH_x pool. The mean time for the substitution of those H atoms by D, given by the area between both curves in Figure 4, is $\tau_H = 2.3$ s. From this τ_H, the amount of H accumulated as CH_x species at 250°C is calculated as $N_H = 1.6\text{-}2.4\ 10^{19}$ H atoms per gram of catalyst (the uncertainty arising from the slight change in the methanation rate when replacing H_2 by D_2). At this temperature, by means of $^{13}CO+2H_2/^{12}CO+2H_2$ transients, the amount of CH_x species per gram of catalyst was determined [2] as $N_{CHx} = 3\ 10^{19}$. Therefore, the average value of x for the pool of CH_x species lies between 0.5 and 0.8 hydrogen atoms per carbon atom. This result shows that the CH is the most abundant surface intermediate, which in turn indicates that the hydrogenation of CH species is the rate limiting step, in agreement with the model developed by van Nisselrooij et al. [10].

5. CONCLUSION

The use of isotopic transient experiments has driven to a direct determination of the concentration of reaction intermediates and adsorbed (spectator) CO species under the steady-state methanation reaction catalysed by Ni/SiO_2. Moreover, it has been concluded that CH species are the most abundant surface intermediates. The results support a mechanism of CO dissociative adsorption, followed by a stepwise hydrogenation of carbon, in which the hydrogenation of CH into CH_2 is the rate limiting step.

Modelling of the transient substitution of H by D in the produced methane is at present being carried out in order to determine the kinetic constants for the elemental steps and further validate the proposed mechanism.

ACKNOWLEDGEMENTS

CMA acknowledges a postdoctoral fellowship granted by the Spanish Ministerio de Educación y Ciencia.

REFERENCES

1. M. Agnelli, M. Kolb and C. Mirodatos, J. Catal., 148 (1994) 9.
2. Agnelli, H.M. Swaan, C. Márquez-Alvarez, G.A. Martin and C. Mirodatos, J. Catal., in press.
3. J.A. Dalmon and G.A. Martin, J. Catal., 84 (1983) 45.
4. C. Mirodatos, J. Phys. Chem. 90 (1986) 481; Catal. Today, 9 (1991) 83; in Catalyst Characterization, B. Imelik and J.C. Vedrine (eds.), p. 651. Plenum Press, New York, 1994.
5. L. Luytens and J.C. Jungers, Bull. Soc. Chim. Belg., 54 (1945) 303; M.M. Sakharov and E.S. Dokukina, Kinet. Catal., 2 (1961) 639; C.S. Kellner and A.T. Bell, J. Catal., 67 (1981) 175; Y. Kobori, S. Naito, T. Onishi and K. Tamaru, J. Chem. Soc. Chem. Commun. (1981) 92; T. Mori, H. Masuda, H. Imai, A. Miyamoto, S. Baba and Y. Murakami, J. Phys. Chem., 86 (1982) 2753.
6. J.T. Gleaves, J.R. Ebner and T.C. Kuechler, Catal. Rev.-Sci. Eng., 30 (1988) 49.
7. T.P. Wilson, J. Catal., 60 (1979) 167
8. G. Henrici-Olivé and S. Olivé, The chemistry of the hydrogenation of carbon monoxide, Springer-Verlag, Berlin, 1984.
9. A. Ozaki, Isotopic studies of heterogeneous catalysis, Academic Press, New York, 1977.
10. P.F.M.T. van Nisselrooij, J.A.M. Luttikholt, R.Z.C. van Meerten, M.H.J.M. de Croon and J.W.E. Coenen, Appl. Catal., 6 (1983) 271.

NATURAL GAS CONVERSION V
Studies in Surface Science and Catalysis, Vol. 119
A. Parmaliana et al. (Editors)

Reactions of synthesis gas on CoIr/SiO_2 and CoRu/SiO_2

Dr. Marita Niemelä and Lic.Tech. Matti Reinikainen

VTT Chemical Technology, P.O. Box 1401, 02044 VTT, Finland

The work on carbonyl based silica supported Co, CoIr and CoRu catalysts provided evidence for the synergistic effect of Ir or Ru addition; the CO uptake and the tolerance to air-exposure increased, and the performance in CO hydrogenation was significantly altered.

1. INTRODUCTION

Cobalt catalysts being highly active in Fischer-Tropsch (FT) synthesis have gained a lot of attention, and many studies exist on the effect of precursor, method of preparation, pretreatment procedures, and promotion effects [1-4].

For example, Iglesia et al. [2] have recently obtained higher activities and C_{5+} selectivities by adding small amounts of ruthenium to cobalt catalysts. They report that the bimetallic interactions occurring upon the formation of mixed Co-Ru oxides during oxidation at high temperatures facilitated ruthenium promoted cleaning of the cobalt surface ensembles during Fischer-Tropsch synthesis thereby inhibiting deactivation [2]. The results of Matsuzaki et al. [4] have also indicated that iridium and ruthenium are indispensable promoters for a Co/SiO_2 prepared from acetate precursors, since they activate the catalyst by reducing the Co^{2+} species to Co^0 by hydrogen spill over. However, in case of Co/SiO_2 prepared from $Co_2(CO)_8$, they observed hardly any effect on the activity of the cobalt catalyst in connection with promotion [4].

Accordingly, the promotion of cobalt as well as different bimetallic catalyst compositions have also been an intriguing subject in our own research [5]. Our ruthenium-cobalt catalysts of cluster origin, however, failed to produce results as interesting as those of Iglesia et al. [2], Ichikawa et al. [6] or Xiao et al. [7]. However, the catalyst prepared from $Co_2(CO)_8$ and $Ru_3(CO)_{12}$ has showed some enhancement for the formation of oxygenated compounds [8] - a result in agreement with the findings of Ichikawa et al. [6] and Xiao et al. [7]. Much more interestingly, in regard to iridium promotion our results were different from those of Matsuzaki et al. [4], since $Ir_4(CO)_{12}$ appeared to be an effective promoter also for the Co/SiO_2 catalyst prepared from $Co_2(CO)_8$.

2. CATALYSTS

Co(5)/SiO_2, CoRu(5-2.7)/SiO_2, CoIr(5-5.2)/SiO_2, Ru(5)/SiO_2 and Ir(5)/SiO_2 catalysts, wherein the metal contents (wt%) are denoted in brackets were prepared as follows: The silica support was dry-impregnated with $Ru_3(CO)_{12}$ or $Ir_4(CO)_{12}$ in vacuum in oil bath, and reduced under hydrogen flow at 400°C. The CoRu/SiO_2 CoIr/SiO_2 and Co/SiO_2 were prepared by impregnating Ru/SiO_2, Ir/SiO_2 or silica, respectively, with $Co_2(CO)_8$ from a n-hexane solution under deoxygenated atmosphere. In the remainder of this paper the following abbreviations indicate the thus prepared catalysts: Co, CoRu, CoIr, Ru and Ir.

The catalysts were characterised in a pulse reactor system described in detail elsewhere [5]. They were reduced *in situ* (400°C), or air exposed (denoted by E) and reduced *in situ* (400°C). After reduction, the catalyst surfaces were saturated by CO at room temperature. The desorption of CO was determined by a Baltzers mass spectrometer as shown in Figure 1.

The results indicate that the amount of CO adsorbed on the promoted catalyst surface was significantly higher than could be expected based on the sum of the contribution of the respective monometallic catalysts. It should also be noted that this synergistic effect was much stronger for CoIr than for CoRu. Most likely, the presence of iridium and ruthenium enhanced the extent of reduction of cobalt [3,4], and as a result, the chemisorption capacity was increased. However, it is also possible that some bimetallic sites capable of adsorbing CO in high stoichiometry were formed at the perimeter of the cobalt and iridium entities or cobalt and ruthenium entities.

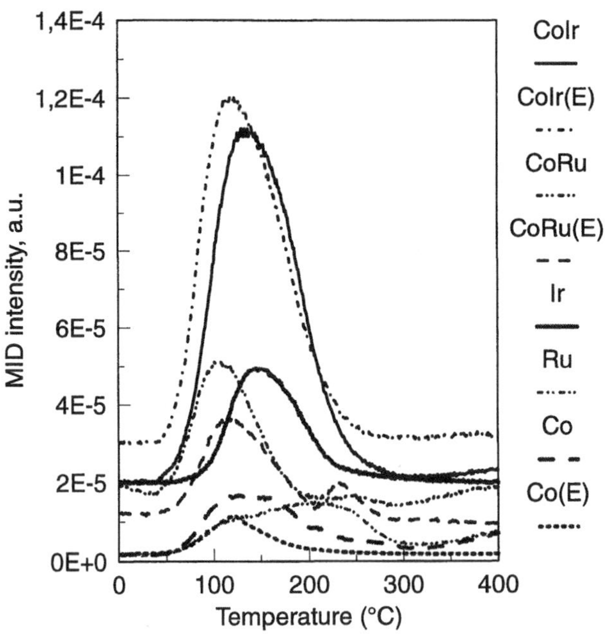

Figure 1. The desorption of CO from the fresh and air exposed catalyst surfaces.

The results also very interestingly indicate that particularly the iridium promotion appeared to protect the carbonyl based cobalt catalyst from the deleterious effect of air exposure, since similar amounts of CO were adsorbed on the directly reduced and air-exposed catalysts. In case of ruthenium promotion, this protection effect is less visible, see Figure 1. Nevertheless, the present results indicate that the air-sensitivity of the carbonyl catalysts can be reduced by suitable promotion thus making the handling of the catalyst much easier.

3. RESULTS IN CO HYDROGENATION

The activity of the catalysts, after *in situ* hydrogen reduction at 400°C and 2 h, was determined in CO hydrogenation. The results shown in Figure 2 demonstrate very clearly that the activity of the CoIr was high, being almost twice that of CoRu, and clearly higher than that of the sum of its individual components. Namely, the activity of Ir under similar operation conditions is almost negligible (X= 0.1% at 280°C). The higher activity of the iridium promoted catalyst was most likely due to the ability of iridium to reduce cobalt [4]. It is however, somewhat surprising that the activity of the CoRu was lower than that of Co or Ru, see also Figure 3. Namely, one would expect ruthenium to enhance the extent of reduction of cobalt [3,4], and thereby to increase the activity, since the extent of reduction for the unpromoted $Co_2(CO)_8$ derived Co/SiO_2 catalyst has been only 30% [M. Niemelä in 5].

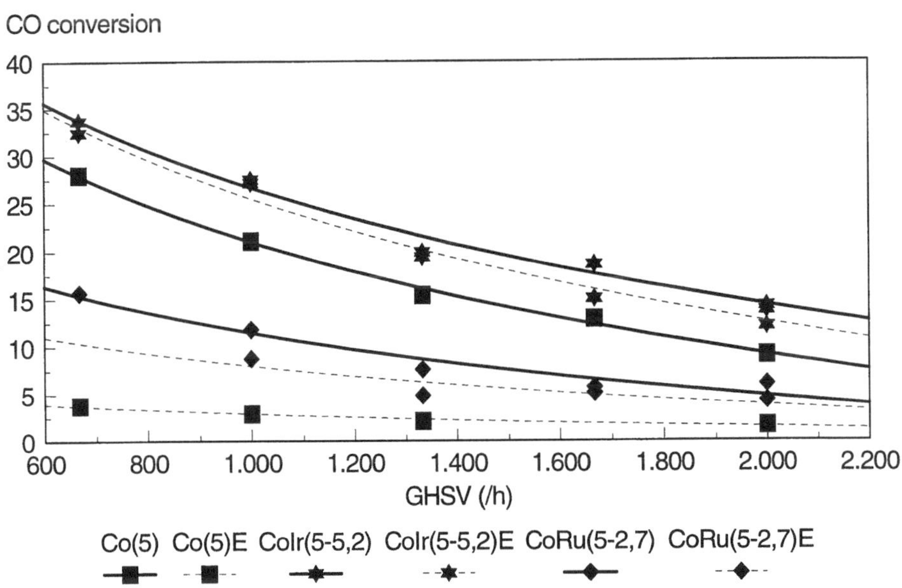

Figure 2. The activity of the catalysts at 233°C and 2 MPa using 1 g of catalyst and H_2:CO:Ar molar ratio of 6:3:1.

More importantly, the data shown in Figure 2 proves that the iridium promoted, and also the ruthenium promoted, catalyst could well withstand air exposure; the activities of the readily reduced and air exposed promoted samples were almost identical, whereas the activity of Co decreased tremendously after air exposure. Thus, the activity results for the readily reduced and air exposed sample are in good agreement with the respective CO chemisorption data, see also Figure 1.

The selectivities of the catalysts were determined at 233°C and 2MPa after *in situ* reduction at 400°C. The results depicted in Figure 3 indicate that the CoIr catalyst was different from its monometallic counterparts: it produced oxygenated compounds in higher yield. Instead, no significant improvement in performance was observed in connection with ruthenium promotion, although Ru was the most active with surprisingly high selectivity to oxygenates. Perhaps, the impurities of the support facilitated unintentional promotion [10].

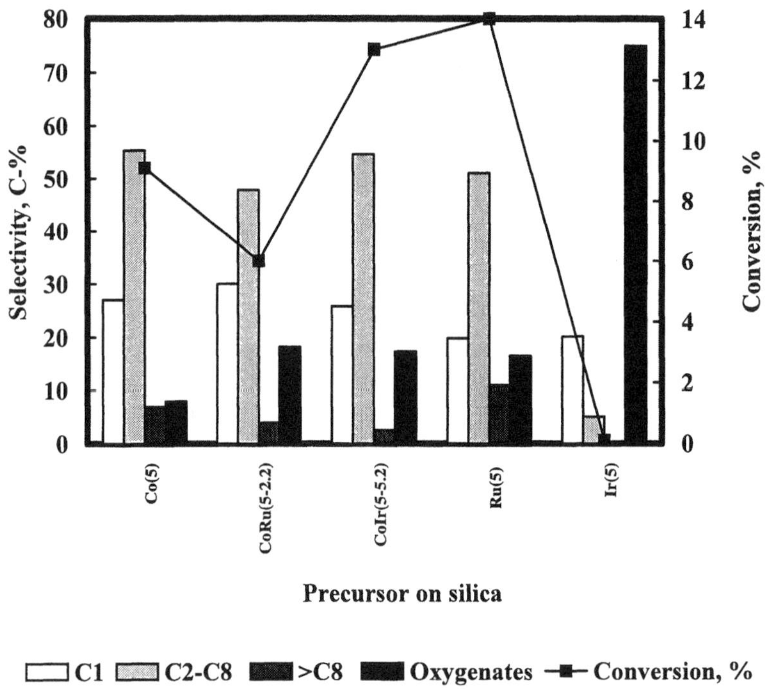

Figure 3. The selectivities of the catalysts at 233°C and 2MPa were determined using 1 g of catalyst and GHSV of 2000 h^{-1} with H_2:CO:Ar molar ratio of 6:3:1.

It is also interesting to pay closer attention to the oxygenated products; the more detailed composition of this product fraction is shown in Figure 4. Although the Ir catalyst is very selective towards the formation of methanol, no methanol was observed for CoIr. Instead, the CoIr catalyst produced a significantly higher amount of ethanol and methyl acetate than did Co - a result in agreement with Kintaichi et al. [9]. The results also indicate that although the total amount of oxygenates was quite similar for CoRu, Ru and Co, the type of products differ considerably. Thus, some bimetallic sites were present on both CoIr and CoRu, and these sites favoured a reaction route to ethanol and/or methyl acetate.

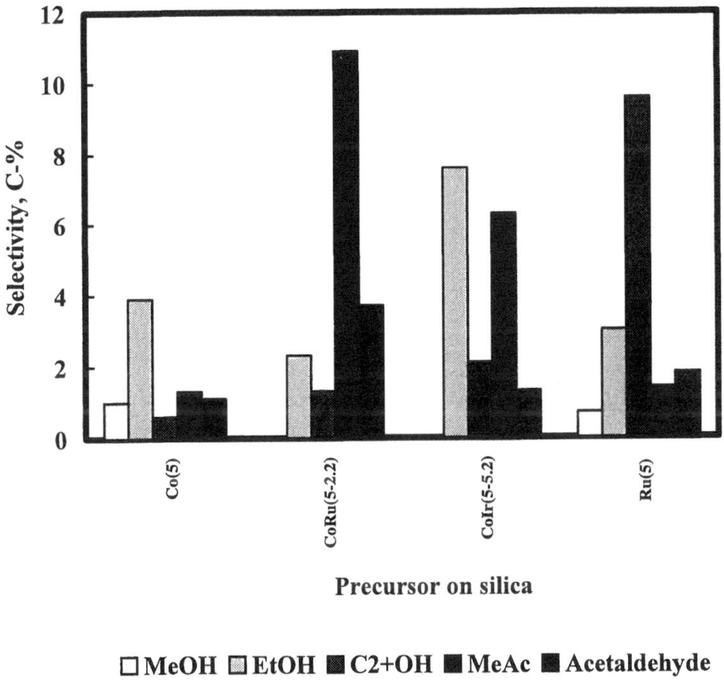

Figure 4. The brake down of the oxygenated products shown in Figure 3. In case of Ir only methanol was formed.

4. SUMMARY AND CONCLUSIONS

The studies on the promotion of $Co_2(CO)_8$ derived Co catalyst by iridium and ruthenium indicated that the amount of CO adsorbed on the catalyst surface was significantly higher for both of the promoted catalysts than could be expected based on the sum of the respective monometallic catalysts. This synergistic effect was much stronger for CoIr than for CoRu. In addition, particularly the iridium promotion appeared to protect the carbonyl based cobalt catalyst from the deleterious effect of air exposure, since similar amounts of CO were adsorbed on the directly reduced and air-exposed catalysts.

In agreement with the characterisations, the activity of the CoIr was high, being almost twice that of CoRu, and clearly higher than the sum of its components. Moreover, the CoIr catalyst produced oxygenated compounds in higher yield than the respective monometallic catalysts.

A detailed analysis of the oxygenated products revealed that CoIr produced no methanol, although Ir is selective to it, and a significantly higher amount of ethanol and methyl acetate than did Co. In addition, CoRu produced significantly more of methyl acetate than did Ru or Co.

In conclusion, some bimetallic sites were present on both CoIr and CoRu, and these sites were responsible for the increased CO chemisorption capacity, thus favouring the CO insertion reaction, which yields ethanol and/or methyl acetate.

REFERENCES

1. L. Guczi (ed), Stud. Surf. Sci. Catal. 64 (1991)
2. Kodansha, Progress in C1 Chemistry in Japan, Elsevier, 1989
3. E. Iglesia, S.C. Reyes, R.J. Madon, Adv. Catal. 39 (1993) 221-302, E. Iglesia, S.L. Soled, R.A. Fiato and G.H. Via, J. Catal. 143 (1993) 345 and E. Iglesia, S.L. Soled, R.A. Fiato and G.H. Via, Natural Gas Conversion II, Stud. Surf. Sci. Catal., 81 (1994), Elsevier, Amsterdam, 433
4. T. Matsuzaki, K. Takeuchi, T-a. Hanaoka, H. Arakawa and Y. Sugi, Appl. Catal. A:General, 105 (1993) 159-184 and T. Matsuzaki, T-A. Hanaoka, K. Takeuchi, H. Arakawa, Y. Sugi, K. Wei, T. Dong and M. Reinikainen, Catal. Today 36 (1997) 311-324
5. J. Kiviaho, PhD. Thesis, VTT Publications, 290, 1996, Espoo, Finland; M. Niemelä, Dr. Thesis, VTT Publications, 310, 1997, Espoo, Finland, M. Reinikainen, Dr. Thesis, to be published in 1998
6. M. Ichikawa, F-s. Xiao, C.G. Macpanty, A. Fukuoka, W. Henderson and D.F. Schriver, Stud. Surf. Sci. Catal. 61 (1991) 297
7. F-s. Xiao, A. Fukuoka and M. Ichikawa, J. Catal. 138 (1992) 206 and F-s. Xiao, J. Nat. Gas Chem. 3 (1994) 219
8. M. Niemelä, M. Reinikainen and J. Kiviaho, Accepted for publication in the Proceedings of the 7th International Symposium on Scientific Bases for the Preparation of Heterogeneous Catalysts, Belgium, September 1-4, 1998
9. Y. Kintaichi, Y. Kuwahara, H. Hamada, T. Ito and K. Wakabayashi, Chem. Lett. 1985, 1305
10. L.E.Y. Nonnemann, A.G.T.M. Bastein, V. Ponec and R. Burch, Appl. Catal. 62 (1990) L23 and L.E.Y. Nonnemann and V. Ponec, Catal. Lett. 7 (1990) 197

NATURAL GAS CONVERSION V
Studies in Surface Science and Catalysis, Vol. 119
A. Parmaliana et al. (Editors)

Comparison between Co and Co(Ru-promoted)-ETS-10 catalysts prepared in different ways for Fischer-Tropsch Synthesis

C.L. Bianchi, S. Vitali and V. Ragaini

Department of Physical Chemistry and Electrochemistry, University of Milan - Via Golgi, 19 - 20133 Milan (Italy)(tel. +39/2/26603266, fax: +39/2/70638129, e-mail: claudia@rs14.csrsrc.mi.cnr.it)

A particular molecular sieve containing titanium in the framework position, ETS-10 by Engelhard, was used as support for Fischer-Tropsch synthesis. It is discussed the different results in CO conversion and selectivity obtained if the active metal is introduced in the ETS-10 cages by ion-exchange or simply by impregnation. This comparison was made both on Co-based catalysts and on Ru-promoted samples.

1. Introduction

CO hydrogenation on classical Fischer-Tropsch (FT) catalysts leads to the formation of a wide range of products, according to Anderson-Schulz-Flory (ASF) chain-growth probability. Attempts to control the product selectivity have been made, involving molecular sieves as catalyst support for metals known to be active catalysts in FT synthesis [1 and references therein]. Geometric or diffusional constraints on the product molecules coupled with strong acidic functions greatly influenced the normal chain-growth process in FT reaction [2].

It was already shown by the authors [3-5] that very interesting results were obtained using ETS-10 (Engelhard [6]), a particular molecular sieve containing titanium in the framework position, as support for FT catalysts (commercial mixed sodium-potassium form, pore radius $r_p=4\times10^{-10}$m, BET area 325 m^2/g).

In a previous work [3] the authors showed the FT results performed on a sample prepared after Co ion-exchange and after promotion with different amounts of ruthenium, added in order to increase the Co reducibility and to improve the catalyst activity.

In the present paper it is discussed the different results in CO conversion and selectivity obtained if Co is introduced in the ETS-10 cages by ion-exchange or simply by impregnation. This comparison was made both on Co-based catalysts and on promoted ones.

2. Experimental

Two different procedures were followed for the exchanged or impregnated samples.

For the former samples, ETS-10 was previously calcined at 500°C for 4h. A suitable ion-exchange-form of ETS-10 was prepared by multiple ion-exchange with $Co(NO_3)_2.H_2O$ (Merck) solution (1.0 N, 20ml/g_{ETS} ≈ 5×4 meq/g_{ETS}) following the procedures proposed by Rabo et al. [7].

For the latter samples, they were prepared by a slurry impregnation of the supports with a solution of $Co(NO_3)_2 \times 6H_2O$ (Fluka) in distilled water. The support was stirred into the precursor solution at 40°C for 12h, then the excess water was evaporated under vacuum. The recovered powders were reduced in flowing H_2 (F_{H2} = 80 ml/min) at 350°C for 4h.

In both cases, a second impregnation was performed in slurry conditions using a solution of $Ru(NO)(NO_3)_3$ in pure ethanol, then evaporated again under vacuum. Then the samples were once more reduced in the same conditions reported before. The final content of active metals, measured by ICP-AES (Jobin Yvon JY24) was 5.8wt% for Co and 0.2wt% for Ru in the promoted samples.

All the samples were characterized by XPS (M-Probe, SSI), after *in situ* reduction performed at 350°C for 4 h with flowing hydrogen, and TPR (Thermo Quest Italia), performed using a reducing gas mixture (5 vol.% H_2/N_2) at a flow rate of 30 ml/min and increasing the temperature at a rate of 10°C/min. These two techniques were used to observe a possible difference in the active metal reduction due to the different position of the Co atoms in the zeolite framework.

Moreover TPD analyses (CE Instruments) were performed on the reduced catalysts to observe the different distribution of the active sites among the samples. They were performed saturating the samples with hydrogen as probe gas, introduced at 100 torr in a single dose at 100°C, and left for 1 h to reach the equilibrium conditions. The samples were then outgassed for 1h at the same temperature of adsorption to eliminate gas phase and weak hydrogen [8] and finally cooled at room temperature before starting the analysis. The TPD measurements were performed with a constant heating rate of 20°C/min.

Reaction tests were performed in a stainless steel tubular reactor, coated with copper, designed especially for Fischer-Tropsch synthesis of hydrocarbons (C_n: $n < 15$) and described elsewhere [9]. The reaction was carried out with a mixture of high purity CO and H_2 (SIAD); the H_2/CO molar ratio of the inlet mixture was 2. The catalysts (always 1g of fresh sample for each run) were tested at 548 K, 500 kPa and at a space velocity (S.V.) of 9.0×10^{-2} mmolCO/(mmolRu·s). The hydrocarbon products were analyzed on-line by gas-chromatography [9] and the C_4 fraction by means of GC-MSD (Hewlett Packard HP-5890 equipped with a mass selective detector HP-5971A, capillary column HP1) in order to quantify the amount of linear and branched hydrocarbons in the C_4 fraction [3].

3. Results and Discussion

Table 1 lists the ratio between oxides and metal species performed by XPS analysis after the *in situ* reduction step, as reported in the Experimental section. A complete reduction of Ru is always observed ; on the contrary, the reduction of cobalt is obtained only when the promoter is present.

Table 1
XPS data obtained hydrogenating the samples *in situ* at 350°C.

Sample	Ru (%wt.)	Co (% wt.)	Ru^{ox}/Ru^0	Co^{ox}/Co^0
Co-ETS-10 (exc.)	0	5.8	–	2.33
Co-ETS-10 (imp.)	0	5.8	–	0.71
Co(Ru)-ETS-10 (exc.)	0.2	5.8	0	0
Co(Ru)-ETS-10 (imp.)	0.2	5.8	0	0

This result is not new in literature. In a previous work [4] the authors of the present paper showed the influence of different amounts of Ru in Co ion-exchange catalysts to improve the catalyst activity. Moreover the change in Co reducibility due to the presence of small amounts of noble metals (Ru included) was also investigated by XPS [10].

The higher reducibility of Co due to the presence of Ru atoms was already shown by Goodwin et al. [11, 12] who correlate this effect to an increase in CO hydrogenation activity. They concluded that Ru seemed to inhibit the formation of hardly reducible Co species or to promote their easy reduction.

In the light of the difficulty for Co atoms to be reduced, it is not relevant to observe that in the absence of promotion, a complete Co reduction is never found, but it is really interesting to observe that if Co is introduced in the zeolite cages by ion exchange its reduction is more difficult than by simple impregnation. It is possible to think that in the first case Co atoms could go into the zeolite framework and occupy structural positions.

This different reducibility is also confirmed by TPR analysis (Fig.1). In the case of the impregnated catalyst a net hydrogen consumption can be observed at ca. 300°C (350°C is the temperature used to reduce the samples during the last step of the preparation). On the contrary, the exchanged sample shows a very broad peak with s lower intensity at a higher temperature (at 650° the zeolite collapses [6]).

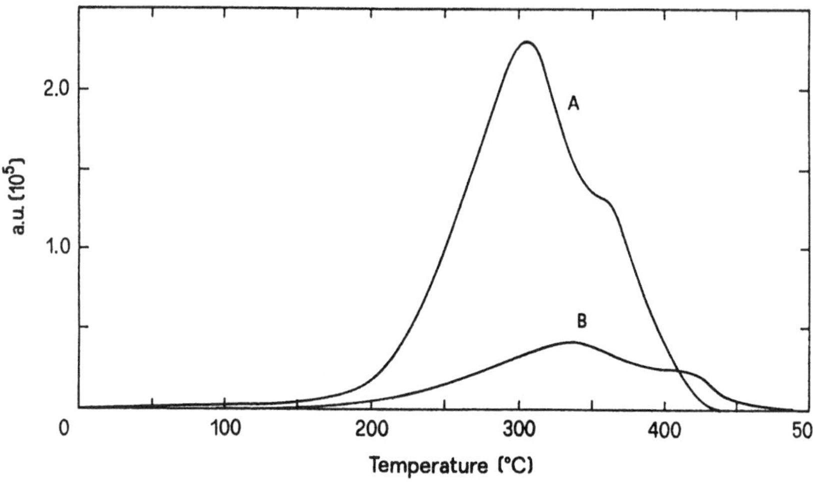

Figure1. TPR analysis of Co-ETS-10 (imp.) (A) and Co-ETS-10 (exc.) (B).

Tab. 2 lists the results of the CO hydrogenation obtained in this work

Tab.2
Fischer-Tropsch results at 235°C, 5 bar, $H_2/CO = 2$, s.v. 9×10^{-2} molCO/molRu×s

Samples	CO conversion %[a]	C_1%	C^+_2%[b]	$C_3^=/C_3$	$C_4^=/C_4$	iso C_4/C_4
Co-ETS-10 (exc.)	53.0	77.1	23.1	0.2	3.0	0.20
Co-ETS-10 (imp.)	71.2	32.1	67.9	0.3	2.9	0.15
Co(Ru)-ETS-10(exc.)	72.6	77.5	22.5	0.2	3.1	0.49
Co(Ru)-ETS-10(imp.)	98.9	58.4	41.6	∞	∞	0.42

[a] mean value (see Figure 3)
[b] C^+_2 stands for hydrocarbon with more than 2 carbon atoms

It is possible to observe that a higher CO conversion can be reached adding ruthenium as a promoter, both on exchanged and impregnated catalysts. This fact could be explained in the light of the XPS and TPR results as a different reduction degree of cobalt atoms in the samples.

The presence of Co atoms in the structure of the support, obtained by ion exchange, leads to the production of a very high C1 fraction.

An explanation for the large difference in both conversion and selectivity between impregnated and exchanged samples can be found in active site energy distribution as shown in Fig.2 which shows TPD analysis performed on Co-ETS-10 (imp.) (A) and Co-ETS-10 (exc.) (B).

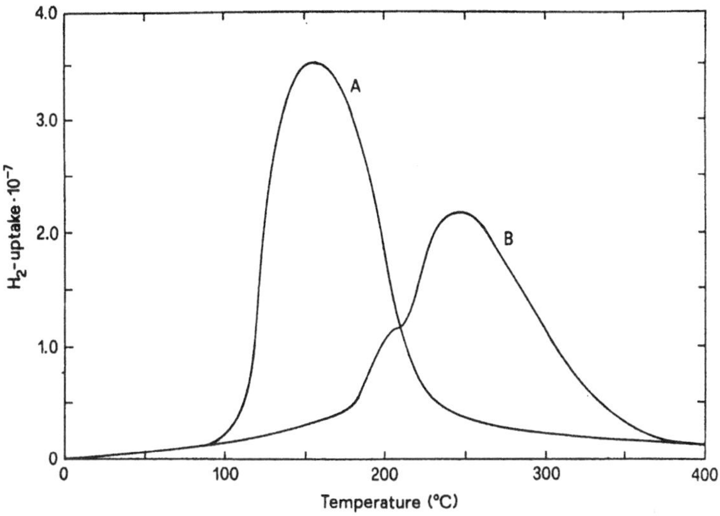

Figure 2. TPD analysis of Co-ETS-10 (imp.) (A) and Co-ETS-10 (exc.) (B).

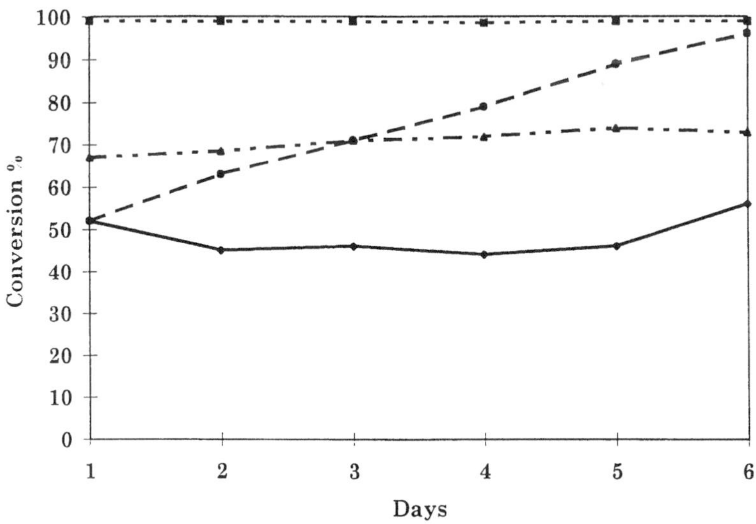

Figure 3. CO conversion (%) vs number of days of reaction : ---- Co(Ru)-ETS-10 (imp.) ;-- – -- – Co(Ru)-ETS-10 (exc.) ; – – – Co-ETS-10(imp.) ; —— Co-ETS-10(exc.).

The Co active sites of the impregnated catalyst can chemisorb (and thus desorb) dissociated hydrogen at lower temperature than the exchanged one ; in the light of this result the former sample should be a catalyst more active than the latter. Roughly estimating E_d for both the catalysts by means of a semiempirical correlation due to Knorr [13] [E_d (kJ/mol) $\approx$ 0.23 T_M (K), where T_M is the peak maximum temperature], it is clear that the exchanged sample has sites at higher energy : the hydrogen chemisorbed on these sites is less available to be used during the reaction.

Going back to the catalytic reaction, it is interesting to observe the change of CO conversion vs reaction time (Fig.3).

No deactivation occurs for all the samples. Moreover the conversion for Co-ETS-10(imp.) increases day after day probably due to a progressive reduction of Co oxides present on the sample (see Table 1), and therefore to an increase of Co active sites available for the reaction

Acknowledgment
The authors wish to thank Dr. M. Fadoni for TPR and Dr. L. Aina for TPD analyses. Moreover, Dr. C. Cavenaghi (Engelhard Italia) is gratefully acknowledged for fruitful discussions throughout this work.

References

[1] C.H. Bartolomew, in : New Trends in CO Activation, ed. L. Guczi, Studies in Surface Science and Catalysis, vol. 64 (Elesevier, Amsterdam), 1991
[2] E. Iglesia, S.C. Reyes, R.J. Madon and S.L. Soled, Adv. Catal., 39 (1993) 221
[3] C.L. Bianchi and V. Ragaini, J. Catal., 168 (1997) 70
[4] C.L. Bianchi, R. Carli, S. Merlotti and V. Ragaini, Catal. Lett., 41 (1996) 79
[5] R. Carli, C.L. Bianchi and V. Ragaini, Catal. Lett., 33 (1995) 49
[6] S.M. Kuznicki,(a) U.S. Patent 4,853,202 (1989), (b) U.S. Patent 4,938,939 (1990)
[7] J.A. Rabo, Zeolites : Sci. Technol., 80 (1984) 291
[8] R. Giannantonio, V. Ragaini and P. Magni, J. Catal., 1994, **146** 103
[9] V. Ragaini, R. Carli, C.L. Bianchi, D. Lorenzetti, G. Vergani, Appl. Catal. A : General 139 (1996) 17
[10] C.L. Bianchi, L. Aina, M. Fadoni and V. Ragaini, J. Catal., submitted
[11] A. Kogelbauer, J.C. Goodwin Jr. and JR. Oukaci, J. Catal., 160 (1996) 125
[12] A.R. Belambe, R. Oukaci and J.G. Goodwin Jr., J. Catal., 166 (1997) 8
[13] Z. Knorr, in Catalysis, Science and Technology, J.R. Anderson and M. Boudart (eds.), Springer–Verlag, Berlin, vol.3, p.231, 1982

NATURAL GAS CONVERSION V
Studies in Surface Science and Catalysis, Vol. 119
A. Parmaliana et al. (Editors)

Synthesis gas to branched hydrocarbons: a comparison between Ru-based catalysts supported on ETS-10 and on Al_2O_3 (doped with sulfated zirconia)

C.L. Bianchi, S. Ardizzone and V. Ragaini

Department of Physical Chemistry and Electrochemistry, University of Milan - Via Golgi, 19 - 20133 Milan (Italy) – tel. +39/2/26603266, fax: +39/2/70638129, e-mail: claudia@rs14.csrsrc.mi.cnr.it

The production of branched hydrocarbons directly from synthesis gas was performed by Fischer-Tropsch synthesis. A classical Ru/Al_2O_3 catalyst was compared to an hybrid catalyst composed by the classical catalyst and sulfated zirconia and the obtained results were themselves compared to a ruthenium catalyst supported on a titanium silicate (ETS-10).

1. Introduction

In the present paper the authors discuss the production of isomeric hydrocarbons from synthesis gas: Fischer-Tropsch synthesis (FTS) was combined with isomerization both over a hybrid catalyst composed of an FTS classical catalyst, such as Ru/Al_2O_3, and of sulfated zirconia [1], and over a ruthenium catalyst supported on a titanium silicate (ETS-10) tested both as Na + K, Na and H-form (all commercially available).

Sulfated zirconia is attracting much attention as a potential process catalyst because of increasing need for environmentally benign process. An example is the alkylation of isobutane with butenes to produce gasoline alkylates, usually obtained using HF or H_2SO_4 as both the catalyst and the reaction medium.

Sulfated zirconia catalysts also are being used for gasoline reformulation. The reduction in octane number resulting from the removal of aromatics could be mitigated by the addition of high-octane compounds such as highly branched hydrocarbons. The strong acidity and exceptional high activity of sulfated zirconia make it attractive as a catalyst in hydroisomerization or hydrocracking [2].

Interesting results for FTS were already obtained by the authors using ETS-10 as support [3, 4]. In the present work as well, the presence of ETS-10, both as Na+K, Na and H-form, causes significant change in C_3 and C_4 yields, olefin to paraffin ratios and even isocompounds formation in comparison both to the classical catalyst and to the hybrid one.

2. Experimental

The tested samples are reported in Table 1.

The hybrid catalyst was composed of an FTS component (Ru1% on γ-Al_2O_3 [Engelhard]) followed by an isomerization acid component (SO_4^{2-}/ZrO_2 prepared by treating a ZrO_2 hydrous precursor [5] with H_2SO_4 (1M) and subsequently calcining at 470°C for 5 h in flowing O_2 as described in [6]) ; the two materials are not mixed but are put in the reactor in two separate layer, thus the hydrocarbons generated on the FTS catalyst would isomerize on the acid catalyst before leaving the reactor.

The others were directly FTS catalysts prepared by slurry impregnation of the support (ETS-10 by Engelhard [7]) with a solution of $Ru(acac)_3$ in pure ethanol (all Fluka products). The support is stirred into the precursor solution at 40°C for 12h, then the excess ethanol was evaporated under vacuum. The recovered powders were reduced in flowing H_2 (F_{H2} = 80 ml/min) at 350°C for 4h.

The final content of ruthenium, measured by ICP-AES (Jobin Yvon JY24) was 1 wt% for all the samples.

Three particular commercial ETS-10 forms were used in this work characterized by a Si/Ti ratio of 5, a pores diameter of 8×10^{-10} m and a BET surface area of ca. 325 m^2/g . In particular, the Na+K form is characterized by : 8.8 wt.% of Na_2O and 4.6 wt.% of K_2O ; the H-form by 0.85 wt% of Na_2O and 3.7 wt% of K_2O ; the Na form by 8.3 wt.% of Na_2O and 0.72 wt.% K_2O.

The surface acidity features of the supports were determined by a modification of the Hammett-Bertolacini technique [8, 9]. This method relies on a selective adsorption of dye molecules, with different pK_a values, on the oxide surface. The operative techniques essentially consist in successive colorimetric tritation of the sample suspension in benzene under N_2 bubbling and stirring. The tritation has been performed starting from the indicator with the lowest (highest absolute) pK_a value, that adsorbs with its acid form on the sample surface and, successively, adding the titrant until the basic form of the indicator has been reached. The tritation route has been carried on by means of other dye molecules with less negative pK_a values on the same oxide suspension, until one of the indicators adsorbs in its basic form, in the absence of any titrant. The titrant was 0.1 N n-butylamine solution added dropwise by a microsyringe into the vial which was tightly protected from any contact with atmospheric moisture.

Reaction tests were performed in a stainless steel tubular reactor, coated with copper, designed especially for Fischer-Tropsch synthesis of hydrocarbons (C_n: $n < 15$) and described elsewhere [10]. The reaction was carried out with a mixture of high purity CO and H_2 (SIAD); the H_2/CO molar ratio of the inlet mixture was 2.

The catalysts (always 1g of fresh sample for each run) were tested at 548 K, 500 kPa and a space velocity (S.V.) of 9.0×10^{-2} mmolCO/(mmolRu·s) ; the hybrid catalyst was loaded in two separate layers in the reactor, 0.8g of Ru1%/Al_2O_3 and 0.2g of SO_4^{2-}/ZrO_2 [11, 12]. Blank tests using only SO_4^{2-}/ZrO_2 carried out under the same conditions gave no CO conversion, implying that only the FTS catalyst was responsible for CO hydrogenation. Each run was performed several times in order to check the reproducibility of both the CO conversion and the products selectivity.

The hydrocarbon products were analyzed on-line by gas-chromatography [7] and the C_4 fraction by means of GC-MSD (Hewlett Packard HP-5890 equipped with a mass selective detector HP-5971A, capillary column HP1) in order to quantify the amount of linear and branched hydrocarbons in the C_4 fraction [13]. Since CO is the only detectable reactant, the mass balance calculation is based on carbon, presuming that the amount entering the reactor is equal to the one leaving it. Therefore, the conversion is easily calculated by taking into account the total number of unreacted CO moles multiplied by the number of moles of carbon-containing species found at the exit.

3. Results and Discussion

Table 1 lists the CO conversion and the selectivity to light olefins and isocompounds obtained during the reaction.

Table 1
Fischer-Tropsch synthesis : activity and selectivity of the samples (T = 548 K, P = 500kPa, s.v. = 9.0 ×10^{-2} mmol CO/(mmol Ru×s), H_2/CO = 2, 1g of fresh sample).

N.	Sample	CO conversion (%)[a]	C_1 (%)	C^+_2[b] (%)	$C_3^=/C_3$	$C_4^=/C_4$	iso C_4/C_4
1	1%Ru/Al_2O_3	13.2	68.6	31.4	0.91	1.98	0.16
2	1%Ru/Al_2O_3+ SO_4^{2-}/ZrO_2	13.2	68.2	31.8	1.11	1.49	0.35
3	1%Ru/(Na+K)-ETS-10	11.3	46.9	53.1	3.24	0.73	0.61
4	1%Ru/Na-ETS-10	4.1	35.8	64.2	8.29	0.23	3.54
5	1%Ru/H-ETS-10	11.3	77.2	22.8	1.92	0.85	0.64

[a] steady state value

[b] hydrocarbons with more than 2 carbon atoms

It is interesting to compare the different behavior of the catalysts due on one side to the presence of acid sites on sulfated zirconia the catalyst coupled with Al_2O_3 or on the other side directly to the ETS-10 cages.

In the former case, no changes were found both in the CO conversion and in the C_1 and C_2^+ yield between 1% Ru/Al_2O_3 and 1% Ru/Al_2O_3+ SO_4^{2-}/ZrO_2. The second catalyst showed an higher production of C_4 isocompounds and this fact was expected because it is well known that acid catalysts enhance the isomerization reactions.

In the latter case, the presence of a strong acidity seems only to cause the breakage of the hydrocarbons chains due to cracking reactions and thus to lead to a higher production of C_1 fraction (ETS H-form).

The acid characterization of the powders is reported in Fig. 1.

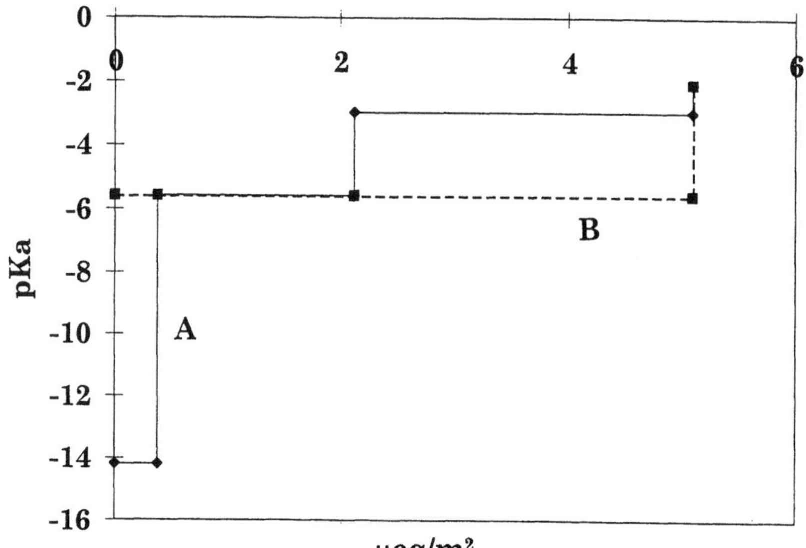

Figure 1. pK_a values versus density of sites of sulfated zirconia (A) and H-ETS-10 (B).

The 1M sulfated zirconia shows both strongly acid sites, with a broad energy distribution, and also a high density of sites [1]. On the contrary, the H-ETS-10 sample shows strong acid sites at a constant energy distribution. Na+K and Na-ETS-10 show no acidic function as already reported [4, 7].

In the light of this interpretation, the large production of isocompounds for the ETS-10 catalysts on which no acid sites are present can only be explained taking into account both the shape-selectivity due to the shape and size of the support cages and also to the contribution of electronic factors (e.g. electronpositivity, charge density, etc.) due to the presence in the zeolite framework of alkaline ions. Especially the presence of K seems to be interesting to explain both the decrease of CO conversion and increase of isocompounds formation [14].

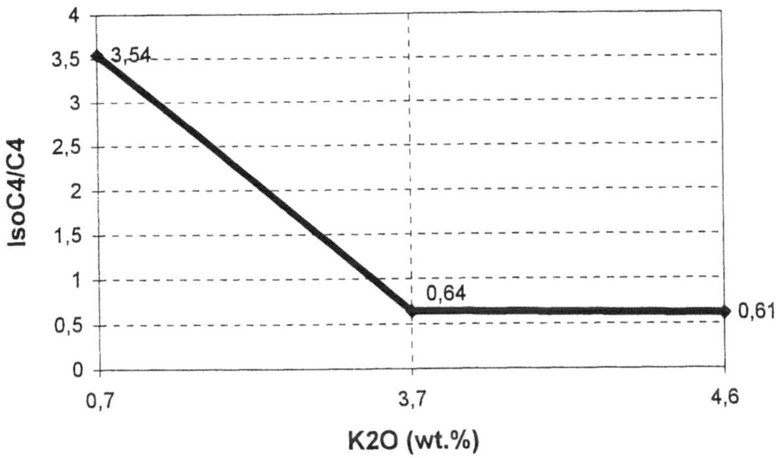

Figure 2. Correlation between concentration of K_2O in ETS-10 and isoC$_4$/C$_4$ ratio.

It is interesting to observe that in the case of the H-form, the concentration of potassium is quite the same as in the Na+K-form and that these two catalysts showed the same CO conversion and the same C_4 isoproduction. On the contrary in the Na-form, the K concentration is drastically lower and the catalyst showed a very low CO concentration, but coupled with the highest C_4 isoproduction, the highest ratio $C_3^{=}/C_3$ and the lowest production of C_1.

4. Conclusions

In the present paper, the comparison between different kind of catalysts for the isomerization of hydrocarbons directly from syngas is presented. Good results have been obtained by coupling a classical catalyst with sulfated zirconia which is nowadays considered an attractive compound for hydroisomerization and hydrocracking due to its intrinsic acidity.

But a better result was obtained using a molecular sieve as support for the FT metal (in this case ruthenium). With this kind of support, when a too strong acidity is present (H-ETS-10) the cracking reactions are promoted and a too high C_1 production is found. On the contrary, when no acid functions are present at all, the shape and size of the cages are able to promote isomerization reactions due to shape selectivity.

Further investigation will be performed to understand the large difference in CO conversion between the two non acid supports. In fact, at present no possible explanations have been found to justify such a low CO conversion for the Na-form, except the different amount of K in the framework of the two supports, element which is usually added in classical catalysts to modify the hydrocarbon selectivity [10].

Acknowledgment
The authors wish to thank Dr. C. Cavenaghi (Engelhard, Italy) for very fruitful discussions throughout this work.

References

[1] S. Ardizzone, C.L. Bianchi, W. Cattagni, V. Ragaini, Catal. Lett., in press
[2] B.H. Davis, R.A. Keogh, R. Srinivasan, Catal. Today, 20 (1994) 219
[3] C.L. Bianchi, R. Carli, S. Merlotti and V. Ragaini, Catal. Lett., 41 (1996) 79
[4] R. Carli, C.L. Bianchi and V. Ragaini, Catal. Lett., 33 (1995) 49
[5] S. Ardizzone, C.L. Bianchi, S. Carella, M.G. Cattania, Mater. Chem. Phys., 4 (1993) 154
[6] S. Ardizzone, C.L. Bianchi, E. Grassi, Colloids and Surf., in press
[7] S.M. Kuznicki,(a) U.S. Patent 4,853,202 (1989), (b) U.S. Patent 4,938,939 (1990)
[8] R.J. Bertolacini, Anal. Chem., 35 (1963) 599
[9] L. Forni, Catal. Rev., 8 (1974) 65
[10] V.Ragaini, R.Carli, C.L. Bianchi, D.Lorenzetti, G.Vergani, Appl. Catal. A: General 139 (1996) 17
[11] D.J.C. Yates, W.F. Taylor, J.H. Sinfelt, J. Am. Chem. Soc., 86 (1964) 2996
[12] X. Song, A. Sayari, Chemtech, Aug. (1995) 27
[13] C.L. Bianchi and V. Ragaini, J. Catal., 168 (1997) 70
[14] V.Ragaini, R.Carli, **C.L. Bianchi**, D.Lorenzetti, G.Vergani, Appl. Catal., 139 (1996) 17

NATURAL GAS CONVERSION V
Studies in Surface Science and Catalysis, Vol. 119
A. Parmaliana et al. (Editors)

α-Olefin Readsorption Product Distribution Model for the Gas-Solid Fischer-Tropsch Synthesis

G.P. van der Laan and A.A.C.M. Beenackers

Department of Chemical Engineering, University of Groningen,
Nijenborgh 4, 9747 AG Groningen, The Netherlands

1. INTRODUCTION

Increasing crude oil prices may cause a shift to coal and natural gas as the feed stock of the chemical industry and transportation fuels market. These can be converted into CO and H_2 by partial oxidation or steam reforming processes which subsequently can be converted to hydrocarbons in the Fischer-Tropsch (FT) process. The FT synthesis product spectrum consists of a complex multi-component mixture of linear and branched hydrocarbons and oxygenated products. Main products are linear paraffins and α-olefins.

The FT synthesis has been recognized as a polymerization reaction [1]. The reactants, CO and H_2, adsorb and dissociate at the surface of the catalyst and react to form chain initiator (CH_3), methylene monomer (CH_2), and H_2O. The most important growth mechanism for the hydrocarbon formation is the surface carbide mechanism by CH_2 insertion into adsorbed alkyl chains. Termination can take place by dehydrogenation to α-olefins or hydrogenation to paraffins [1,2].

The FT product yield decreases exponentially with increasing chain length. The so-called Anderson-Schulz-Flory (ASF) distribution is often used to describe the entire product range by a single parameter, α, the probability of the addition of a carbon intermediate to a chain [3]. However, significant deviations from the ASF distribution are reported in literature: (i) a relatively high yield of methane [4,5], (ii) a relatively low yield of ethene [4] relative to the ASF distribution, and (iii) an exponential decrease of the α-olefin to paraffin ratio and change in chain growth parameter, α_n, with increasing chain length. These deviations are caused by secondary reactions, readsorption and hydrogenation, of α-olefins [4,6]. However, secondary hydrogenation is strongly inhibited by CO and H_2O relative to readsorption [7]. Readsorption of α-olefins leads to chain initiation and results in a decrease of the olefin to paraffin ratio and an increase of the chain growth parameter with chain length.

A new product distribution model is presented to explain the deviations from the ASF distribution. This model combines a mechanistic model of olefin readsorption with kinetics of chain growth and termination on the same catalytic sites. In this study, the emphasis is on modeling the selectivity to linear olefins and paraffins.

2. THEORY

The α-Olefin Readsorption Product Distribution Model (ORPDM) accounts for secondary readsorption of α-olefins on FT growth sites on the precipitated iron catalyst (see Fig. 1). Here,

CO_G and CO_S denote the gas phase and the adsorbed CO, respectively. $C_{M,S}$ refers to adsorbed monomeric building units ($CH_{2,s}$), and $C_{n,S}$ is an adsorbed alkyl species with carbon number n. Conversion of CO to $C_{M,S}$ follows a sequence of elementary reaction steps, but is shown as a single step. Chain growth initiates by hydrogenation of $C_{M,S}$ to $CH_{3,S}$, while chain propagation proceeds via insertion of $C_{M,S}$ into adsorbed alkyl chains. Chain termination by dehydrogenation of adsorbed alkyl chains gives olefins, whereas paraffins are formed by hydrogenation of alkyl species. Based on the reaction network shown in Fig. 1, α-olefins may readsorb on growth sites and continue to grow via propagation with monomers or terminate as hydrocarbon product.

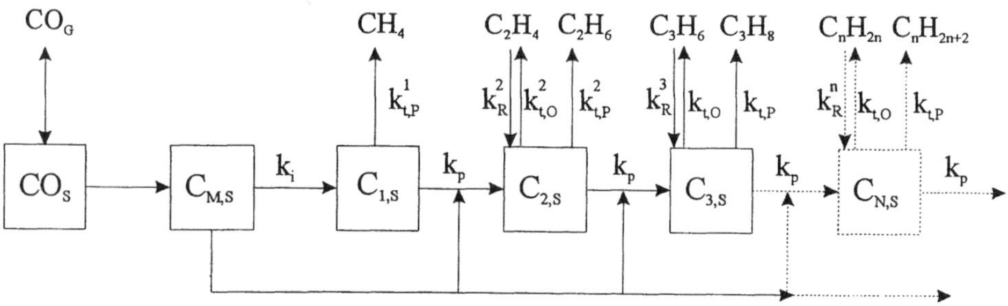

Figure 1. Reaction network α-Olefin Readsorption Product Distribution Model

Steady state mass balances for alkyl species with carbon number n, θ_n, can be derived to account for readsorption [4,7]:

$$k_p\theta_M\theta_{n-1} = (k_{t,O}\theta_v + k_{t,P}\theta_H + k_p\theta_M)\theta_n - k_R^* C_{C_nH_{2n}} \tag{1}$$

where θ_H is the surface coverage of adsorbed hydrogen and θ_v is the fraction of vacant catalytic sites. The actual concentration of the olefin at the catalyst surface, $C_{C_nH_{2n},S}$, can be related to the reaction rate:

$$R_{C_nH_{2n}} = k_{t,O}\theta_v\theta_n - k_R^* C_{C_nH_{2n},S} \tag{2}$$

The steady-state mass balance for α-olefins in an ideally mixed continuous reactor is:

$$R_{C_nH_{2n}} = \frac{\Phi_{v,R} P_{C_nH_{2n}}}{WRT_R} \tag{3}$$

where $P_{C_nH_{2n}}/RT_R$ is the gas-phase concentration of olefins in the reactor with carbon number n, $\Phi_{v,R}$ is the volumetric flow rate of the gas-phase at reactor conditions, and W is the weight of the catalyst.

The interfacial effects of reactive olefins at the gas-wax and wax-catalyst surface should be taken into account. Several authors stated that a greater solubility of larger hydrocarbons results in an increase of readsorption rates for larger olefins [5,6]. Vapor-liquid equilibria of Breman *et al.* [8] and Caldwell and Van Vuuren [9] show that the solubility of hydrocarbons increases exponentially with the chain length. Data on the adsorption of hydrocarbons on solids show that the enthalpy of adsorption increases linearly with carbon number [10]. Adsorption equilibria constants increase exponentially with chain length [4]. In multi-component mixtures (Fischer-Tropsch product spectrum) these effects result in a physisorbed layer with mainly long-chain hydrocarbons. Therefore we assume the olefin gas phase concentration to relate to the concentration at the catalyst surface as:

$$\frac{C_{C_nH_{2n},S}}{P_{C_nH_{2n}}/RT_R} \propto e^{cn} \tag{4}$$

where c is a positive constant depending on the exponential increase of the physisorption and solubility with chain length.

Rearranging and solving Eqs. 1 - 4 yields:

$$\frac{\theta_n}{\theta_{n-1}} = \frac{k_p\theta_M}{k_{t,O}\theta_v/(1+k_Re^{cn})+k_{t,P}\theta_H+k_p\theta_m} = \frac{p}{t_O/(1+k_Re^{cn})+1+p} = \alpha_n \tag{5}$$

where $k_R = k_R^* W P_R T_0/(\Phi_v P_0 T_R)$ (P_R= reactor pressure and Φ_v= flow rate at P_0= 0.1 MPa and T_0= 273 K). Propagation and termination to olefins are relative to the termination rate of paraffins: $p = k_p\theta_M/k_{t,P}\theta_H$, and $t_O = k_{t,O}\theta_v/k_{t,P}\theta_H$. The surface fractions of alkyl chains with carbon number n can be determined by successive calculation of the chain growth parameter with increasing carbon number:

$$\frac{\theta_n}{\theta_1} = \prod_{i=2}^{n}\alpha_i = \alpha_2\cdot\alpha_3\cdots\alpha_n \tag{6}$$

Solving Eqs. 2 and 3 gives the following molar selectivities:

$$m_{C_nH_{2n+2}} = \theta_1\prod_{i=2}^{n}\alpha_i \text{ and } m_{C_nH_{2n}} = \frac{t_O}{1+k_Re^{cn}}\theta_1\prod_{i=2}^{n}\alpha_i \tag{7}$$

The molar selectivity to product i is calculated from the experimental mole fraction, y_i, relative to the mole fractions of all products considered:

$$m_i = \frac{y_i}{\sum_i y_i} \tag{8}$$

Higher surface mobility or reactivity of C_1 and C_2 precursors and rapid readsorption of ethene give the most reasonable explanation for the deviations of the short-chain hydrocarbons from the ASF distribution. The selectivities of C_1 and C_2 products are calculated separately:

$$m_{CH_4} = t_P^1\theta_1\ ,\quad m_{C_2H_6} = t_P^2\theta_2 = t_P^2\alpha_2\theta_1\ ,\quad m_{C_2H_4} = \frac{t_O^2}{1+k_R^2}\theta_2 \tag{9}$$

with $t_P^1 = k_{t,P}^1/k_{t,P}$, $t_P^2 = k_{t,P}^2/k_{t,P}$, and $t_O^2 = t_P^2 t_O$ (see Fig. 1).

The model reduces to the ASF distribution when olefins can not readsorb, i.e. $k_R = 0$. Therefore, Eqs. 5-7 can also be used for the ASF distribution with substitution of $k_R = 0$.

The ORPDM accounts for the chain-length dependent readsorption of olefins on FT sites. The readsorption step depends on carbon number, resulting in a net decrease of the termination to olefins with increasing carbon number until no olefins are formed. At high carbon numbers, the chain growth parameter, α_n, approaches a maximum constant value of $\alpha_\infty = p/(1+p)$. The increased readsorption of long-chain olefins results in a decreasing olefin/paraffin ratio with chain length.

The accuracy of the fitted model relative to the experimental data was obtained from the $MARR$ (Mean Absolute Relative Residual) and RR (Relative Residual) functions:

$$MARR = \sum_i^n \left| \frac{m_i^{exp} - m_i^{mod}}{m_i^{exp}} \right| \frac{1}{n} x100 \quad RR = \frac{m_i^{exp} - m_i^{mod}}{m_i^{exp}} x100 \tag{10}$$

where n is the total number of optimized selectivities of all experiments together, m_i^{exp} is the experimental selectivity of the ith data point, and m_i^{mod} is the model prediction of the mole fraction.

3. EXPERIMENTAL

Fischer-Tropsch experiments were carried out with a gas-continuous Spinning Basket Reactor (SBR) with a reactor volume of 285 ml. The stainless steel reactor had the catalyst particles placed in four baskets mounted on the stirrer shaft. A detailed description of the experimental set up, analysis, and experimental procedures is given by Van der Laan [11]. The gaseous phase was analyzed with a Hewlett-Packard 5980A gas chromatograph. The gaseous components were linear paraffins C_1-C_{10}, α-olefins C_2-C_{10}, CO_2, H_2O, CO, and H_2.

The catalyst applied was a commercial precipitated iron catalyst (type LP 33/81) synthesized by Ruhrchemie AG (Oberhausen, Germany). The synthesis procedure was described by Frohning *et al.* [12]. The catalyst pellets were calcined in air at 573 K for 5 h and crushed and sieved to particle diameters between 0.125 and 0.160 mm. The baskets were loaded with 2.34 g of unreduced catalyst. The catalyst was pretreated with hydrogen with a flow rate of 0.83 10^{-3} Nm^3 kg_{cat}^{-1} s^{-1} according to Bukur *et al.* [13]. After reduction, synthesis gas was fed to the reactor at reference conditions of T= 523 K, P= 1.50 MPa, H_2/CO feed ratio= 2 and Φ_v/W= 1.51 10^{-3} Nm^3 kg_{cat}^{-1} s^{-1}. The experimental conditions were varied as follows: P= 0.8 - 4.0 MPa, H_2/CO feed ratio= 0.25 - 4.0, and Φ_v/W= 0.5 - 2.0 10^{-3} Nm^3 kg_{cat}^{-1} s^{-1} at a constant temperature of 523 K.

4. RESULTS AND DISCUSSION

16 kinetic experiments were carried out in the SBR with the Ruhrchemie precipitated iron catalyst at 523 K. Fig. 2 shows a typical distribution of the hydrocarbon products and the corresponding molar ratio of olefin to paraffins as a function of carbon number. We observed a decrease of the molar α-olefin/paraffin ratio with increasing carbon number and a curved line for the distribution of paraffins alone and paraffins and olefins combined.

The ASF model was optimized with two model parameters (p and t_O), within each experiment. The number of parameters in model ORPDM was equal to 7: $p, t_O, k_R, c, t_P^1, t_P^2$, and k_R^2 (see Eqs. 6-9). Four parameters were found to be independent of the experimental conditions (see table 1). Table 2 shows the accuracies of the optimized models expressed with the $MARR$ function for the paraffins and olefins, respectively as well as the total number of product mole fractions, n, for the complete set of experimental values at 523 K.

The ASF model results in large deviations between model and experiment. The curved paraffin distribution cannot be described with the ASF model. The model ORPDM describes n-dependent readsorption of olefins, resulting in a curved distribution of paraffins and a decreasing O/P ratio with carbon number. An example of a predicted product distribution with the optimized model values from model ORPDM is shown in Fig. 2a-b. The observed deviations from the ASF model are accurately described by our model, resulting in lower MARR values relative to the ASF model (see table 2).

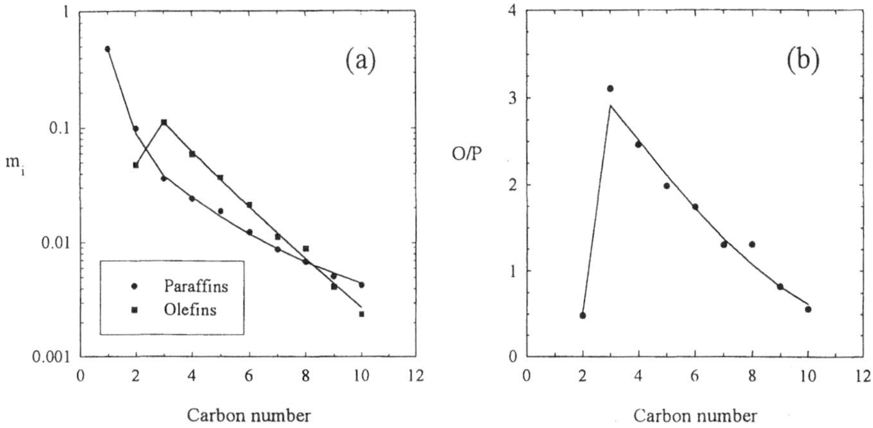

Figure 2. a. Product distribution as a function of carbon number, T= 523 K, P= 1.50 MPa, $(H_2/CO)_{feed}$=2, Φ_v/W= 1.5 10^{-3} Nm^3 kg_{cat}^{-1} s^{-1}, 1011 hours-on-stream. b. Corresponding molar O/P ratio. Lines are predictions of model ORPDM (p= 7.18, t_O= 9.18, k_R= 0.78). Symbols are experimental data points.

Table 1
Model parameters for ORPDM

Parameter	Value
t_P^1	(6.6 ± 1.9)
t_P^2	(1.6 ± 0.3)
k_R^2	$(12.6 \pm 3.5)\, k_R e^{2c}$
c	0.29 ± 0.07

Table 2
Accuracies of the kinetic models (Eq. 10)

model	MARR %		n
	paraffins	olefins	
ASF	49.9	42.5	443
ORPDM	10.1	9.1	443

Fig. 3 shows that the relative residuals between model and experiment, calculated with Eq. 10, are almost always within 25 % for all experiments. The residuals for methane, ethene, paraffins and olefins are shown separately to indicate that the model can accurately describe the well-known deviations of the ASF distribution. Fig. 4 shows the effect of carbon number on the chain growth parameter calculated with Eq. 5 according to model ORPDM for the experiment mentioned in Fig. 2. The calculated chain growth parameter is high at $n = 2$ due to rapid readsorption of ethene and increased termination to C_2 products, minimal for C_3 and increases to the asymptotical value of $\alpha_\infty = p/(1+p)$.

5. CONCLUSIONS

A product distribution model, which accounts for n-dependent olefin readsorption, proves to be able to describe accurately the deviations in the observed product distributions in both olefins and paraffins from ASF distributions: i.e. a relatively high yield of methane, a relatively low yield of ethene and an exponential decrease of the olefin to paraffin ratio and change of the chain growth parameter with chain length. For each experimental product distribution three parameters were

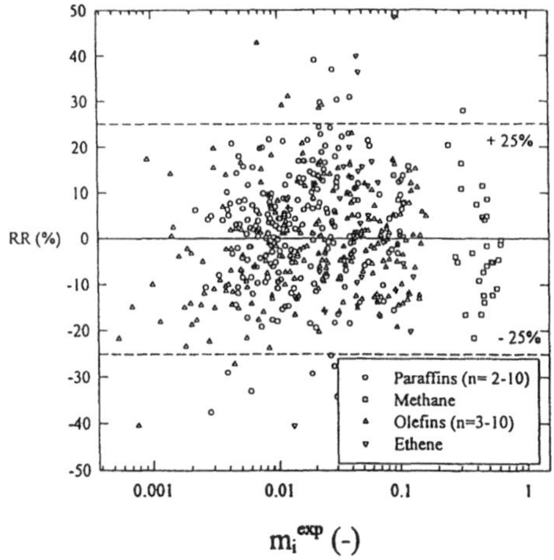

Figure 3. Relative residuals versus experimental values. Model ORPDM

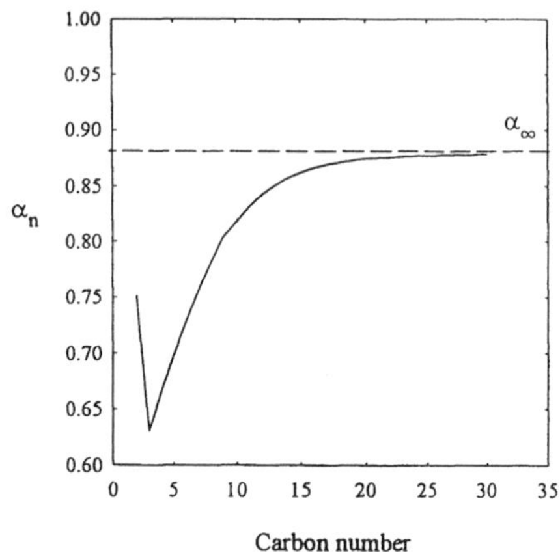

Figure 4. Calculated chain growth parameter as function of carbon number. Experimentalionditons, see Fig. 2. Model ORPDM

optimized, whereas four model parameters were optimized for the entire set of experiments. The superior accuracy of the olefin readsorption model in predicting experimentally observed product distributions is obtained from adding one extra parameter only, without the assumption of multiple catalytic chain growth sites.

REFERENCES

1. A.T. Bell, Catal. Rev.-Sci. Eng. 23 (1981) 203.
2. M.E. Dry and J.C. Hoogendoorn, Catal. Rev.-Sci. Eng. 23 (1981) 265.
3. L.S. Glebov and G.A. Kliger, Russ. Chem. Rev. 63 (1994) 185.
4. T. Komaya, and A.T. Bell, J. Catal. 146 (1994) 237.
5. E.W. Kuipers, C. Scheper, J.H. Wilson, and H. Oosterbeek, J. Catal. 158 (1996) 288.
6. E.W. Kuipers, I.H. Vinkenburg, and H. Oosterbeek, J. Catal. 152 (1995) 137.
7. E. Iglesia, S.C. Reyes, R.J. Madon, and S.L. Soled, Selectivity control and catalyst design in the Fischer-Tropsch synthesis: sites, pellets, and reactors. In E. Eley, H. Pines and P. Weisz (Eds.), Advances in Catalysis 39 (1993) 221.
8. B.B. Breman, A.A.C.M. Beenackers, E.W.J. Rietjens, and R.J.H. Stege, J. Chem. Eng. Data 39 (1994) 647.
9. L. Caldwell and D.S. van Vuuren, Chem. Eng. Sc. 41 (1986) 89.
10. D.M. Ruthven, *Principles of adsorption and adsorption processes*. New York, 1984.
11. G.P. van der Laan, PhD Thesis (in preparation), University of Groningen, The Netherlands.
12. C.D. Frohning, H. Kölbel, M. Ralek, W. Rottig, F. Schuur, and H. Schulz, Fischer-Tropsch-Synthese. In J. Falbe (Ed.), Chemierohstoffe aus Kohle, Chapter 8, Stuttgart, 1977.
13. D.B. Bukur, L. Nowicki, R.K. Manne, and X. Lang, J. Catal. 155 (1995) 366.

NATURAL GAS CONVERSION V
Studies in Surface Science and Catalysis, Vol. 119
A. Parmaliana et al. (Editors)

Surface Study of Pumice Supported Nickel Catalysts Used in the Hydrogenation of CO

A.M. Venezia*, A. Glisenti** and G. Deganello*[a]

*Istituto di Chimica e Tecnologia dei Prodotti Naturali (ICTPN-CNR), Via Ugo la Malfa 153, 90146 Palermo, Italy ;**Dipartimento di Chimica Inorganica, Metallorganica ed Analitica, Università di Padova, Via P. Loredan 4, 35131, Padova, Italy; [a]Dipartimento di Chimica Inorganica, Via Archirafi 26-28, 90126 Palermo, Italy

ABSTRACT

A series of pumice supported nickel catalysts used in the CO hydrogenation reaction were characterised by X-ray photoelectron spectroscopy. Qualitative and quantitative analysis of the XPS peaks have shown the effect of the calcination conditions on the chemical state of the nickel before hydrogenation and the particle size of the metal after reduction. Calcination at high temperature determined enrichment of sodium ions on the surface of the support and also on the metal particles. After exposure to the gas mixture CO/H_2, formation of nickel carbides and other carbon species was checked. The correlation found between the surface atomic ratio Na/Si and the activity and selectivity of the catalysts in the hydrogenation of CO substantiated the role of the alkali ions naturally present in the pumice support.

1. INTRODUCTION

Supported Nickel catalysts are widely used in CO hydrogenation, leading to formation of methane and higher hydrocarbons, very important for the production of synthetic fuel. The reaction has been considered in many studies as structure insensitive [1], however supports interacting with the metal have been shown to have an important role [2]. Addition of promoters such as alkali metal ions to nickel catalysts improves the selectivity toward alkenes and higher hydrocarbons [3]. The alkali promotion prevents metal sintering but increases the carbon deposits interacting with metallic nickel and forming surface and bulk carbides [3-4].

A recent study of the hydrogenation of CO on pumice supported Nickel catalysts has determined the effect of the support structure on the catalytic performance of the supported metal [5]. The ascertained donor properties of pumice [6], attributed to the presence of alkali ions in the carrier structure, seemed to determine a higher selectivity of these catalysts towards heavier hydrocarbons as compared to silica catalysts. Such effect would arise from a weakening of the C-O bond with the consequent increase of its dissociation rate. Moreover, a decrease of the CO conversion turn over frequency and of the C_{2+} selectivity with increasing catalyst dispersion has been also observed and attributed to a change of the metal ensemble.

In order to validate the above assumptions, the determination of the alkali ion location in the catalyst sample is necessary. In the present work, the surfaces of pumice supported nickel catalysts have been examined using X-ray photoelectron spectroscopy (XPS) with the goal of determining the chemical state of nickel [7] and the elemental distribution. Samples reduced at 773 K after calcination at different temperatures have been studied. Changes of the catalyst surfaces upon exposure to H_2 and to the CO/H_2 mixture have been also followed by XPS "in situ" experiments. In particular formation of nickel carbonyl or Nickel carbides was checked. Variations of the XPS intensity ratio of the Ni 2p / Si 2p and Na 1s / Si 2p were followed after different treatments and related to compositional changes.

EXPERIMENTAL

2.1. Catalysts Preparation

Pumice, obtained by Pumex Spa in Lipari, Italy, is characterised by a specific surface area of 5-7m^2g^{-1} [6]. Following a previous procedure, it was purified by treatment with diluted HNO_3 before being used as support [6]. Its bulk composition in terms of oxide percentages as determined by atomic absorption (AAA) and thermogravimetric analyses is the following: SiO_2= 85.5wt%, Al_2O_3= 6.8wt%; Na_2O=2.0wt%; K_2O= 3.2wt% and H_2O= 2.5wt%. The nickel catalyst was prepared by a homogeneous deposition-precipitation method developed by Van Dillen and co-workers for Ni on silica [8]; the method involved thermal decomposition of urea to CO_2 and ammonium hydroxide, at 363 K, and the subsequent precipitation of nickel hydroxide which slowly deposited on the hydroxylic sites of the pumice support [8]. The nickel weight loading as obtained from AAA, was 7.5 %. Precursor samples after being dried in oven at 373 K, were calcined in air, overnight at different temperatures. Subsequently they were reduced under a hydrogen flow of 30 cm^3/min at 773 K for 8 hours, with a heating rate of 10 K/min and then slowly cooled down to room temperature. The quite long period of reduction ensured, on the basis of TPR tests, a complete reduction of the samples. Moreover the X-ray diffractograms of the samples before and after reduction were typical of NiO in the first case and nickel metal in the second case. The nickel concentration was determined by atomic absorption spectroscopy with an accuracy of ± 10%.

2.2. XPS measurements

The XPS spectra were recorded using a Perkin Elmer PHI 5600ci spectrometer with a standard Al-Kα source (1486.6) working at 300W. The working pressure was around 1 x 10^{-6} Pa. Detailed spectra were recorded for the following regions: C 1s, O 1s, Si 2p, Ni 2p and Na 1s, at a pass energy of 11.75 eV and with an energy step of 0.05 eV. The precision on the binding energy (BE) values was ± 0.15 eV. After a Shirley type background subtraction, the raw spectra were fitted using non-linear least squares fitting program with Gaussian-Lorentzian peak shapes. As an internal reference for the absolute binding energies, the C 1s peak of hydrocarbon contamination set at 248.8 eV was used. The use of C 1s instead of the internal Si 2p gave more reliable and reproducible results. The atomic composition was evaluated using PHI sensitivity factors [9]. The experimental uncertainty on the reported atomic ratios is evaluated as ± 10%. The samples were analysed as pellets after being ground in a mortar.

A reaction chamber connected to the spectrometer allowed to expose the samples to H_2 or the gas mixture CO/H_2 (1:2 ratio) while heating up to 533 K. Exposure times of 30 min at pressure of 10^{-4} Torr were used. The extent of the reaction was ascertained by analysing the gas products with a quadrupole gas analyser directly connected to the reaction chamber.

3. RESULTS

In Table 1 the binding energies of the Ni 2p, Si 2p and O 1s of the samples after different treatments along with the Ni satellite separation (ΔE_{sat}) (eV) and Ni spin-orbit coupling (ΔE_{2p}) (eV) are reported. The Ni 2p spectra are characterised, as a function of the nickel chemical environment, by satellite peaks, so called "shake-up" satellites. The separation of such peaks from the primary photolines (ΔE_{sat}) decreases with increasing covalency of the metal-to-ligand bond. Moreover, due to Coulombic and exchange interactions between 3d unpaired electrons and 2p holes, multiplet splittings of the two components 2p $_{1/2}$ and 2p $_{3/2}$ occur [10]. The splitting is generally not resolved, but determines broad and asymmetric primary 2p holes. From these parameters, additional chemical information can be obtained. As shown in Figure 1 calcination of the precursor at different temperature modifies the Ni 2p spectrum.

All the nickel related parameters, in Table 1, indicate that, before calcination, nickel supported on pumice is present as $Ni(OH)_2$ [11]. Moreover, the related O 1s spectrum exhibits two peaks, the low energy one attributed to nickel hydroxide, the high energy one due to the support. Following the air treatment at 673 K the Ni 2p $_{3/2}$ splits into two components of binding energies typical of NiO. The corresponding O 1s peak appears as well. The calcination at high temperature, 1073 K, produces a broad Ni $2p_{3/2}$ peak. The fitting procedure yields curve at higher energy attributable to $NiAl_2O_4$ [7,11], and a curve at low energy probably due to unresolved NiO component. The variations in the Si 2p binding energy are within the experimental error. The binding energy of 1072.2 eV of the Na 1s is typical of ionic alkali metal and is not affected by the reduction treatment [12]. In table 2 the bulk and XPS derived atomic ratios of the nickel, sodium and oxygen to silicon are reported for the samples before and after different treatments. The comparison between the XPS and bulk related values, indicates surface enrichment of nickel, independently of the precursor treatments. Sodium migrates to the surface upon increasing the calcination temperature. The decrease of the XPS O/Si atomic ratio after calcination, is likely to be due to loss of the hydroxylic groups.

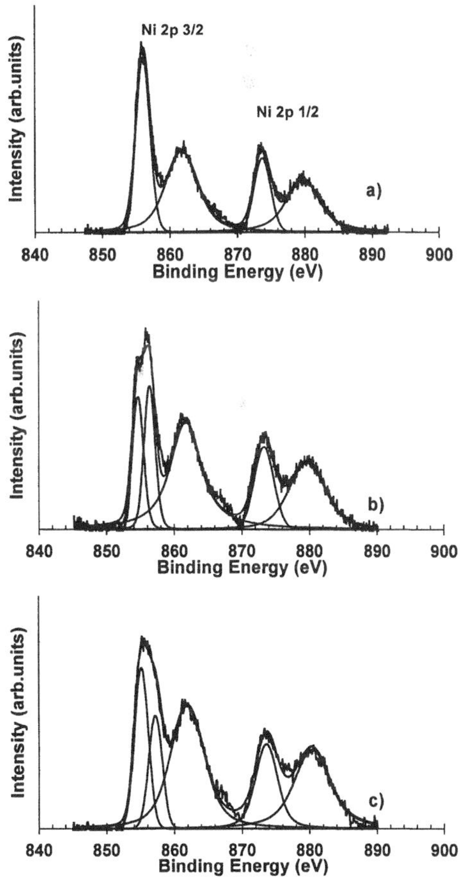

Figure 1. Ni 2p spectrum of the catalyst a) before calcination b) after calcination at 673 K and c) at 1073 K.

Table 1. Ni 2p, O 1s, and Si 2p binding energies (eV), Ni satellite separation (ΔE_{sat}) (eV),and Ni spin-orbit coupling (ΔE_{2p}) (eV) of the catalysts before the calcination and after calcination at 673 K and at 1073 K.

Sample	Ni $2p_{3/2}$	Ni (ΔE_{sat})	Ni (ΔE_{2p})	O 1s	Si 2p
7.5%Ni	856.0 (2.4)	17.7	5.7	531.3 (1.8)	102.4 (2.1)
				532.6 (1.8)	
7.5%NiC6	854.6 (1.9)	18.7	6.8	529.9 (1.5)	102.7 (1.9)
	856.4 (1.9)	16.9	5.1	531.5 (1.5)	
				532.6 (1.5)	
7.5%NiC10	855.1 (2.4)	18.5	6.8	530.4 (2.0)	102.8 (2.4)
	857.1 (2.4)	16.4	5.0	532.1 (2.0)	
				533.8 (2.0)	

Table 2. XPS derived atomic ratio of the catalysts before, after calcination (labeled with C) and after reduction at 773 K at low pressure (labeled with C and red)

Sample	Ni/Si_{bulk}	Ni/Si_{xps}	Na/Si_{bulk}	Na/Si_{xps}	O/Si_{bulk}	O/Si_{xps}
7.5%Ni	0.09	2.5	0.02	0.02	2.2	7.3
7.5%NiC6	0.09	2.9	0.02	0.09	2.2	5.4
7.5%NiC10	0.09	2.7	0.02	0.11	2.2	5.6
$7.5\%Ni^{red}$	0.09	2.4	0.02	0.07	2.2	8.9
$7.5\%NiC6^{red}$	0.09	2.6	0.02	0.10	2.2	4.7

The reducibility of the samples calcined at different temperature was checked by treating them at 773 K under a hydrogen pressure of 10^{-4} Torr for 2 h in a reaction chamber directly connected to the spectrometer. The spectra obtained after the reduction treatment of the uncalcined and calcined samples are shown in figure 2.

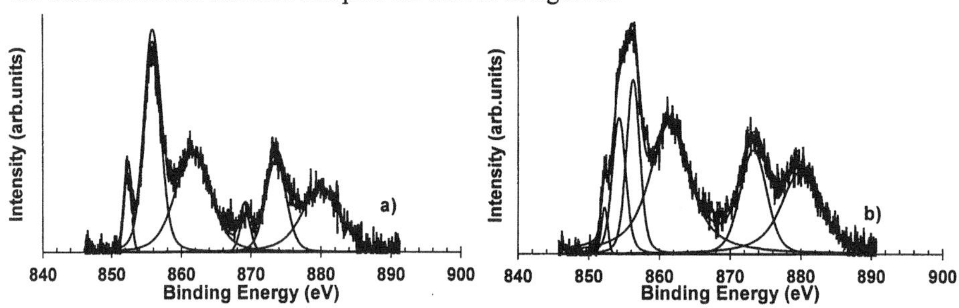

Figure 2. Ni 2p spectrum of sample reduced 'in situ 'a) without being calcined b) calcined at 673 K.

The Ni $2p_{3/2}$ binding energy, typical of metallic state, lays at 852.2 eV. The relative intensity of the metal to oxidised Ni components suggests that the catalyst is more easily reduced when has not been previously calcined.

Complete reduction of the samples was not performed in situ but was achieved under flowing hydrogen at atmosphere pressure at 773 K, as shown from TPR measurements [5]. The Ni $2p_{3/2}$ binding energy at 852.2-852.3 eV obtained for all samples was typical of the metallic state. The difference of such binding energies with the value of 852.6 eV reported for nickel metal [11], is observed for all samples, however, is within the experimental uncertainty, and in order to relate it to the interaction metal support, found for palladium supported on pumice [6], a wider set of data is needed.

In Table 3 the XPS derived Ni/Si_{xps} and Na/Si_{xps} atomic ratios of the samples calcined at various temperatures and reduced all at 773 K under atmospheric pressure of hydrogen, are also given. The variations of the XPS atomic ratios, Ni/Si, are consistent with the increase of the particle sizes. In order to locate the position of the sodium ions, the atomic ratios Na/Si of the various catalysts have been calculated taking into account the fact that supported nickel particles, assumed to be monodispersed, alter such ratio due to the different inelastic mean free path of the Na 1s and Si 2p electrons [13]. By calculation of the surface coverage, using appropriate nickel weight fraction, support specific surface area, and the nickel metal density [13], and taking the Na/Si intensity ratio of the corresponding calcined sample, the atomic ratios, $(Na/Si)_{calc}$ listed in Table 3 are obtained. The larger XPS experimental values with respect to the calculated ones suggest that sodium has migrated on top of the nickel particles.

Table 3. XPS derived atomic ratios, Ni/Si and Na/Si, of the sample calcined at various temperature and reduced at 773 K under atmospheric pressure of hydrogen and diameters d (Å) of the nickel particles as determined by the X-ray diffraction measurements [5].

Samples	Ni/Si_{xps}	Na/Si_{xps}	Na/Si_{calc}	d (Å)
7.5%NiC6R7	1.3	0.16	0.06	45
7.5%NiC8R7	1.3	0.15	0.07	56
7.5%NiC10R7	1.0	0.11	0.08	78
7.5%NiC12R7	0.9	0.22	0.11	250

During the hydrogenation of CO the activity and the selectivity of these catalysts decreased with the particle dispersion [5]. The reason for such behaviour was postulated to be a combination of both factors, electronic and geometric. Now the plots of the turnover frequency (TOF) for the CO conversion and of the yield ratio (C_{2+}/CH_4),versus the atomic ratio Na/Si_{xps} reported in Figure 3 and 4, confirm that the inverse dependence of the above quantity on the %Ni dispersion, observed before, is actually due to the presence of sodium ions.

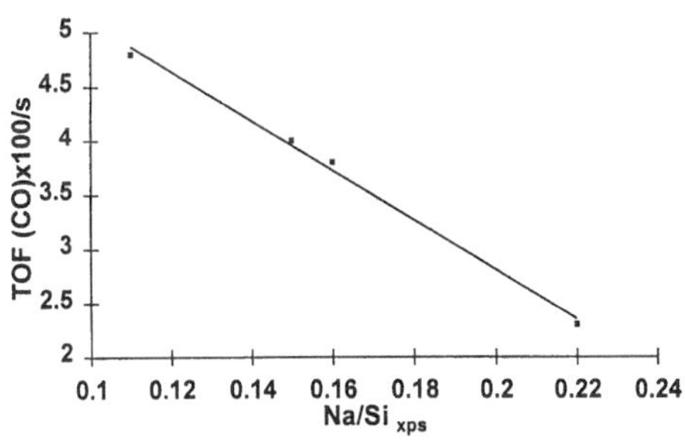

Figure 3. CO conversion turnover frequency (TOF) at 533 K versus the XPS derived atomic ratio Na/Si_{xps}.

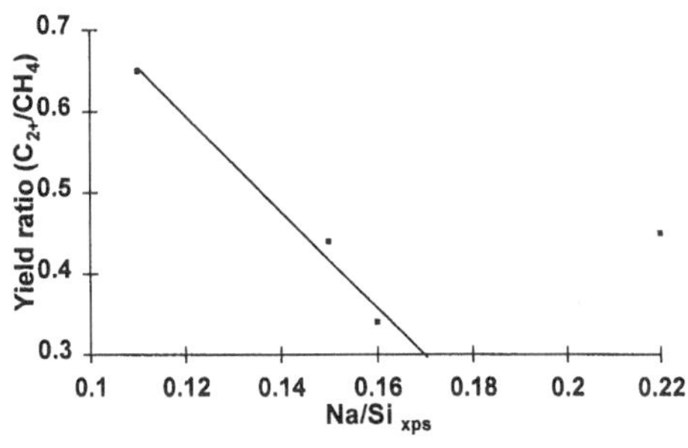

Figure 4. Ratio of the C_{2+} hydrocarbon yield and CH_4 yield versus the XPS derived atomic ratio Na/Si_{xps}

Following the exposure to the gas mixture in the 10^{-4} Torr range of pressure , the gas mass analysis of the products yielded essentially methane and C_{2+} hydrocarbon up to C_4 as obtained under real catalytic conditions [5]. The analysis of the carbon region did not reveal any additional C 1s component and no change in the carbon atomic concentration, signifying that the carbon species produced in the reaction are easily desorbed under high vacuum without leaving poisonous deposits. The XPS atomic ratios Ni/Si and Na/Si following gas exposure are listed in Table 4. Comparison of these ratios with those in Table 3 would rule out any syntering process.

Table 4. XPS derived atomic ratios, Ni/Si and Na/Si of the catalysts after exposure to the CO/H_2 gas mixture.

Samples	Ni/Si_{xps}	Na/Si_{xps}
7.5%NiC6R7	1.5	0.19
7.5%NiC8R7	1.9	0.16
7.5%NiC10R7	1.1	0.13
7.5%NiC12R7	1.2	0.25

Such result is in accord with the role played by the alkali ions which by hindering the migration of carbonyl species inhibit the particle agglomeration [14]. On the other hand, the increase of the surface sodium concentration would eventually result in an excess of alkali coverage leading to the catalyst deactivation.

4. CONCLUSION

The result of this study has confirmed the effect of the alkali ions, in particular sodium, naturally present in the support structure, on the catalytic properties of the pumice supported nickel catalyst used in the hydrogenation reaction of CO. Due to the catalyst pretreatment, sodium segregates on the support and on the metal particle surfaces favouring the dissociation of the CO and formation of the C_{2+} products. It is also confirmed that the metal particles do not sinter upon exposure to the reactant gases, probably due to the alkali ion presence.

REFERENCES

[1] D. W. Goodman, Accts. Chem. Res., 17 (1984) 194.
[2] A. Gil, A. Diaz, L. M. Gandia and M. Montes, Appl. Catal. A, 109 (1994) 167.
[3] C. Mirodatos, E. Brum Pereira, A. Gomez Cobo, J. A. Dalmon and G. A. Martin, Topics Catal., 2 (1995) 183.
[4] E. Brum Pereira and G. A. Martin, Appl. Catal. A, 103 (1993) 291.
[5] A. M. Venezia, A. Parmaliana, A. Mezzapica and G. Deganello, J. Catal., 172 (1997) 463.
[6] A. M. Venezia, A. Rossi, D. Duca, A. Martorana and G. Deganello, Appl. Catal., A, 125 (1995) 113.
[7] K. T. Ng and D. M. Hercules, J. Phys. Chem., 80 (1976) 2094.
[8] Van Dillen, J. A., Geus, J. W., Hermans, L. A. M., and Van der Mejden, J., in *Proc. 6th International Congress on Catalysis,* London, 1976, ed. by G.C. Bond, P. B. Wells and F. C. Tompkins, p. 677. The Chemical Society, London (1976).
[9] J. F. Moulder, W. F. Stickle, P. E. Sobol and K. D. Bomben in Handbok of X-Ray Photoelectron Spectroscopy. Physical Electronics, Ed. J. Chastain, Eden Prairie, Minnesota, 1992.
[10] J. C. Vedrine, G. Hollinger and T.M. Duc, J. Phys. Chem., 82 (1978) 1515.
[11] A. M. Venezia , R. Bertoncello and G. Deganello, Surf. Interface Anal., 23 (1995) 239.
[12] V. Pitchon, M. Guenin and H. Praliaud, Appl. Catal. 63 (1990) 333.
[13] A. M. Venezia, A. Rossi, L. F. Liotta, A. Martorana and G. Deganello, Appl. Catal. A, 147 (1996) 81.
[14] E. Brum Pereira and G. A. Martin, Appl. Catal. A, 103 (1993) 291.

NATURAL GAS CONVERSION V
Studies in Surface Science and Catalysis, Vol. 119
A. Parmaliana et al. (Editors)

Initial Episodes of Fischer-Tropsch Synthesis with Cobalt Catalysts

Hans Schulz, Zhiqin Nie and Michael Claeys

Engler-Bunte-Institute, University of Karlsruhe, Kaiserstraße 12, 76128 Karlsruhe, Germany

Knowledge about transient episodes in catalytic conversions can contribute to the understanding of their stationary state. In particular the initial transient episodes should be elucidative. Such investigations have been performed with cobalt catalysts. Time resolution of conversion and selectivity was obtained by momentaneous product recovery and the hereto adapted gas chromatography. A kinetic model was used for calculating normalized values of rates of elementary reaction steps. Several episodes of different kinetic regimes were observed which revealed how the catalytic Fischer-Tropsch regime is created by selective inhibition of individual steps of reactions.

1. INTRODUCTION

Several attempts are known in literature for defining the Fischer-Tropsch regime of CO-hydrogenation. Particularly a "primary oxymethylene complex" has been thought to be essential together with a condensation reaction between chemisorbed species under water elimination [1, 2]. These ideas have found merely support from work with modern tools of catalysis research.

In our previous work [3, 4] the following definition of the Fischer-Tropsch regime on the basis of its kinetic principles has been given: "Through **selective inhibition** several essential reaction steps, specifically those of product desorption, are slowed down and now the alternative reaction of combining species, the chain prolongation, becomes dominant, thus the growing of aliphatic chains being the governing phenomenon." Selective inhibition also concerns the step of methane formation (the associative desorption of methyl together with a H atom) and similarly the formation of paraffins. As a consequence α-olefins are the favored primary FT-products [5, 6]. By this definition the question about the nature of the "monomer" becomes less substantial as it appears well possible that e.g. CH_2, CO or ethylene are added to a growing chain and indeed all these reactions have been observed [7, 8, 9].

From the above definition of the FT-regime is follows that the actual selection of surface reactions among the many ones being imaginable, will be a dynamic process and should only develop with time from the beginning of the experiment. This has recently been investigated with an alkalized iron catalyst using H_2/CO and H_2/CO_2 synthesis gases. The deposition of carbon onto the iron catalyst was found to be an essential process [10, 11].

This investigation is addressed to the formation of the FT-regime with cobalt catalysts. From the beginning of the experiment conversion and selectivity were measured with high time resolution. Evaluation of results was performed on the basis of our kinetic model (non trivial surface polymerization [3, 6, 12]) and the multiplicity of product composition (hundreds of peaks in the chromatogram) converted into **probabilities of elemental reactions**. These probabilities of reactions are reported as a function of the carbon number of the involved species and each as a function of time. This now provides a picture of the establishing Fischer-Tropsch regime from its beginning up to the stationary state.

The attained higher level of insight could be of actual practical interest. Fischer-Tropsch-slurry phase conversion of syngas from natural gas to liquid fuels is predicted to become one of the important processes of fuel conversion in the near future [13].

2. EXPERIMENTAL

The FT-conversion was conducted in a fixed bed reactor with the powdered catalyst (d_p<0.1mm) covering larger fused silica particles (d_p=0.25-0.4mm) as an adhering layer. The weight ratio of catalyst to fused silica particles was 1:10. By this means an isothermal catalyst bed was provided, allowing for an uniform flow of the gas phase, minor pressure drop and no noticeable intraparticle mass transfer resistance.

The cobalt catalysts were prepared through quick precipitation from the hot solution of the nitrates of cobalt and promotors. The SiO_2-support was applied in the form of Aerosil 200 (Degussa, Hanau) being suspended in the solution. The precipitating agent was a 5-% aqueous NH_3-solution. The precipitate was washed and dried. In some cases a distinct amount of Pt (as reduction promotor) was added in the form of a $(NH_4)PtCl_4$-solution. The catalysts were calcined and reduced in situ. At zero time of a synthesis experiment the flow to the reactor was switched from argon to synthesis gas. To the product gas flow a stream of cyclopropane in nitrogen was added for the determination of conversion and yields directly from the chromatograms [4]. Momentaneous sampling from the hot gaseous product stream (523 K) was performed in small ampoules [14]. GC analysis of the ampoule samples covered the range of compounds C_1-C_{20}. Precolumn hydrogenation of olefins was used for the determination of chain branching of the product hydrocarbons.

3. RESULTS AND DISCUSSION

Most of the results presented and discussed below have been obtained with a catalyst of the composition 100Co:12Zr:100SiO_2:0.45Pt. The reaction conditions generally were 463 K, a total pressure of 0.5 MPa, a H_2/CO molar ratio of 2 and a GHSV of $300h^{-1}$. The experimental results represent the time dependence - from the beginning of the synthesis - of activity and selectivity. Selectivity is evaluated on the basis of elementary surface reactions with the help of the kinetic model [3, 6, 12].

The reaction products contained no CO_2, indicating the absence of any water gas shift reaction. CO-conversion (X_{CO}) was determined by flow measurement of the ingoing streams of CO and N_2 and the molar CO/N_2 ratio in the reactor exit stream. In the ampoule samples the yield of all volatile organic compounds (Y_{VOC}) was determined in relation to the organic reference compound. By substracting (on a carbon basis) the yield of volatile organic compounds from the CO-conversion, the difference $Y_{\Delta C}$ was obtained. This difference of the carbon mass balance is the yield of retardate which includes all carbon retained on the catalyst in any form and also the high molecular weight non-volatile product. The values of X_{CO}, Y_{VOC} and $Y_{\Delta C}$ are shown independence of reaction time in Figure 1. In the upper diagrams the time axis is extended in order to make the earliest fast changes visible and in the lower diagrams the time scale is compressed for showing the slower changes. In addition the breakthrough curve (F´(t)),

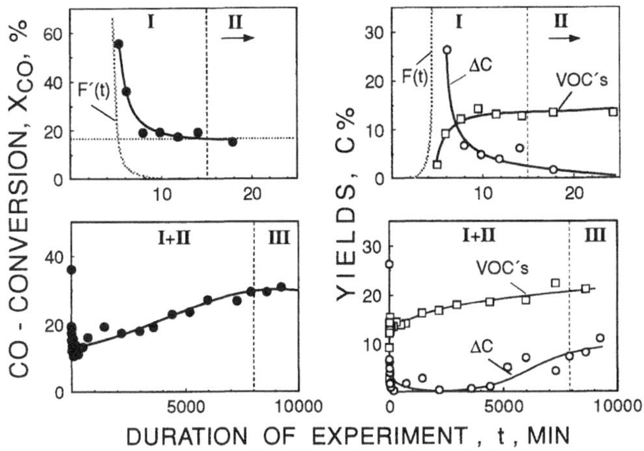

Figure1. CO-conversion (X_{CO}) and yields of volatile compounds (Y_{VOC}) and retardate ($Y_{\Delta C}$) as a function of time on stream. (Catalyst: 100Co:12Zr:100SiO$_2$:0.45Pt)

as representing the residence time distribution of the apparatus, is shown in the upper diagrams. Regarding the curves in Figure 1, three episodes of different behavior of the catalytic system can be distinguished. During the **episode I**, the first 10-15 minutes, there is a fast apparent CO-conversion which mainly reflects a high $Y_{\Delta C}$ as to be explained by the fast chemisorption of CO and hydrocarbon intermediates on the catalyst surface.

Then in **episode II** lasting from the 15th to about the 8000th minute (4-5 days), the CO-conversion (the activity) increases steadily from about 16 to about 36%. From the about 4000th minute onwards the yield $Y_{\Delta C}$ increases strongly as reflecting the now starting higher wax selectivity. It is concluded that this strong catalyst activation (because of its slow development, (4-5 days)) is associated with structural and even compositional catalyst changes, as reconstruction of the catalyst surface, particularly under the influence of the strongly chemisorbed CO. After about 5 days the stationary state of activity and selectivity is attained.

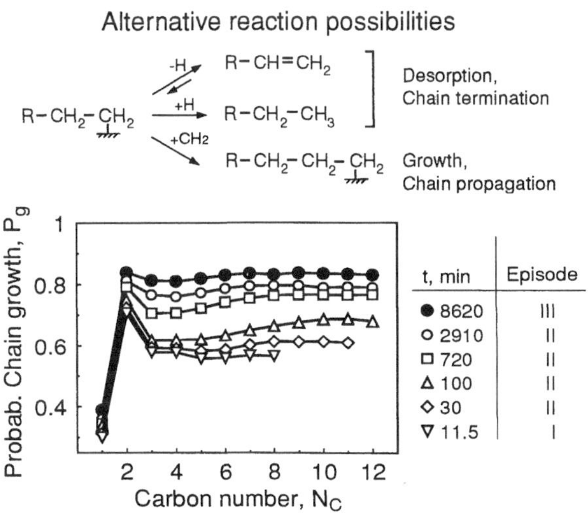

Figure 2. Chain growth probability in dependence of carbon number for different episodes of FT-synthesis catalyst: 100Co:12Zr:100SiO$_2$:0.45Pt

From the hydrocarbon chain-length distribution the **chain growth probability** has been calculated as a function of carbon number and is shown in Figure 2 for several values of time on-stream, together with the respective kinetic scheme of the alternative reactions of a surface species to desorb (irreversibly as a paraffin and reversibly as an olefin) or to grow to a species of next higher carbon number. It is to be noticed that e.g. in the range C_6-C_8 the prolongation probability increases from about 0.56 initially (at 11.5 minutes reaction time) to about 0.84 in the stationary state (after about 8000 minutes). This result

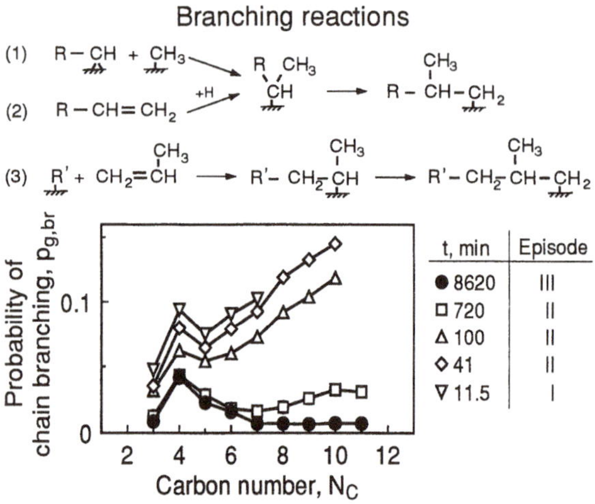

Figure 3. Chain branching probability in dependence of carbon number for different episodes of FT-synthesis catalyst: $100Co:12Zr:100SiO_2:0.45Pt$

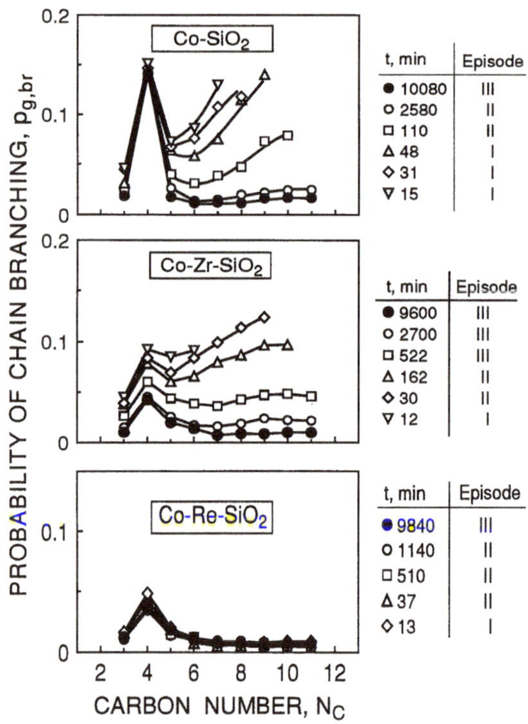

Figure 4. Influence of the promotors Zr and Re on chain branching in the different episodes of catalyst transformation during FT-synthesis

is important. It demonstrates the intrinsic feature of the Fischer-Tropsch regime: The chain growth - being possible through selective inhibition - develops only during hours and even several days under reaction conditions. Because the shape of the curves changes also characteristically with time, it can be concluded that the **polymerization nature of the conversion** (carbon number independent chaingrowth probability) is only approached in a relatively long lasting instationary kinetic episode.

The elementary reaction of **chain branching** is highly interesting for theoretical and practical reasons. Three individual branching reactions are pictured in the upper part of Figure 3. The curves in the diagram show the branching probability as a function of carbon number for several values of duration of the experiment. It is seen that the chain branching probability is low in the stationary state (only 0.01 at $N_C \geq C_7$), when the FT-regime with its selective inhibitions is already established. As any branching reaction has to be assumed as demanding in space, this fits with a densely populated surface of strongly chemisorbed species, many of them of appreciable chainlength. The first possible branching probability is commonly observed to be low [15], probably to be explained by a competitive reaction of the precursor at this carbon number.

It is remarkable how great is the branching probability in the beginning of the experiment. It is concluded that selective inhibition of branching is established with time on stream (whereas the Fischer-Tropsch activity increases). It is proposed that increasing spatial demands enforce this inhibition. The curve shape changes with time. The decline with carbon number in the stationary state is consistent with increasing the spatial demands due to longer chains of the chemisorbed species. In the early stage of the experiment surprisingly, the branching probability increases with increasing chainlength from C_5 onwards. It is proposed that now reaction two in Figure 3 is possible, which is the readsorption of α-olefins in position two. Chain growth of this species leads to branching. Readsorption of olefins in general (and thus also in two-position) is favored by increasing chain length. Readsorption of α-olefins in the two-position is then strongly inhibited in the fully developed FT- regime which favors the at-the-chain-end chemisorption of the hydrocarbon species.

Further results of **chain branching probability** with three different catalysts are shown in Figure 4. With the catalysts Co-SiO_2 and Co-Zr-SiO_2 an in principle similar behavior is observed (Figure 3). The catalyst Co-Re-SiO_2 exhibits a basically different time dependence with no preferred branching probability in the initial episode of the synthesis. Obviously the promoting action of the Rhenium is very intensive in spatial regards and this already from the beginning of the experiment.

The **olefinicity** of the product (Figure 5 left) increases with time on stream. It is concluded, that secondary olefin hydrogenation is increasingly inhibited. The same is observed and concluded for secondary olefin double bond shift (Figure 5 right).

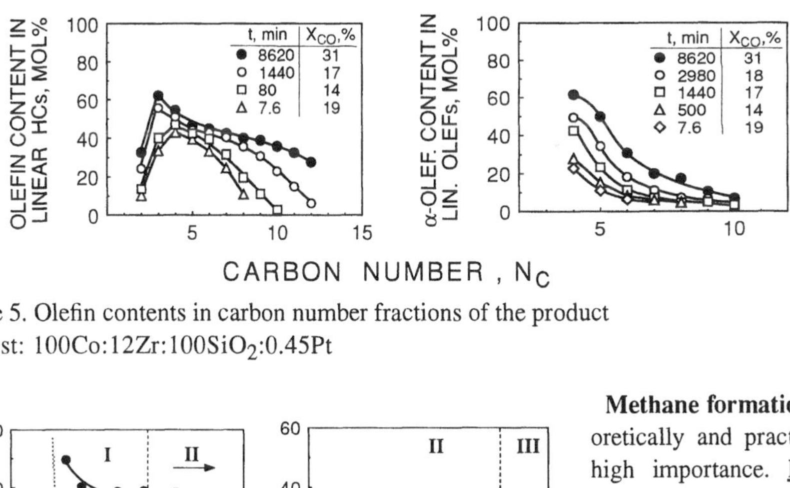

Figure 5. Olefin contents in carbon number fractions of the product catalyst: 100Co:12Zr:100SiO_2:0.45Pt

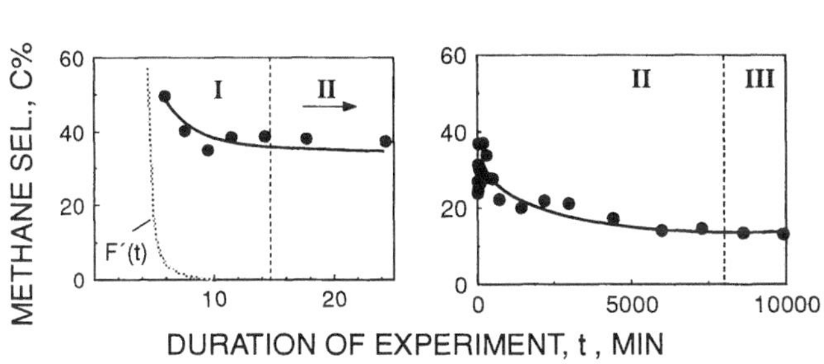

Figure 6. Methane selectivity as a function of time in episodes I to III (catalyst: 100Co:12Zr:100SiO_2:0.45Pt)

Methane formation is theoretically and practically of high importance. Figure 6 shows methane selectivity to decline drastically in both the instationary episodes I and II from about 40 to about 15%. This again is one of the important results of selective inhibition in the FT-regime. At the same time the CO-conversion even increases.

4. SUMMARY/CONCLUSIONS

Transient initial episodes of Fischer-Tropsch synthesis with cobalt catalysts have been investigated for changes in activity and selectivity. In a first episode of about 15 minutes the changes are fast and the overall conversion is not very "Fischer-Tropsch-like": The chain growth probability is low and the branching probability is high.

A second instationary episode lasts even 4 to 5 days during which the following changes haven been found:

— The catalyst activity (CO-conversion) increases by a factor of about two,
— The chain growth probability increases drastically,
— The chain branching probability decreases enormously and changes its nature,
— Olefinicity and α-olefin selectivity decline due to inhibition of secondary reactions,
— Methane selectivity declines strongly.

In the view of elementary reactions these changes correspond to an increasing reaction rate of chain growth whereas all other steps - chain termination, chain branching, secondary olefin reactions and methane formation - are slowed down.

As these changes of activity and selectivity in direction towards the ideal FT-system are remarkably slow in episode two. It is concluded that changes of the catalyst structure or/and of the composition of the catalyst are associated herewith.

REFERENCES

1. H. Storch., N. Golumbic, R. B. Anderson, "The Fischer-Tropsch and Related Syntheses", John Wiley & Sons Inc., New York, 1951
2. H. Pichler, Advances in Catalysis, Vol. IV, Eds. W. Frankenburg, E. Rideal, V. Komarewsky, (Academic Press Inc., New York, 1952) 271
3. H. Schulz, K. Beck and E. Erich, Stud. Surf. Sci. Catal. Vol. 36, Eds. D. Bibby, C. Chang, R. Howe und S. Yurchak; Elsevier, Amsterdam, 1988, p457
4. H. Schulz, E. van Steen and M. Claeys, Topics in Catalysis 2 (1995) 223
5. H. Schulz and H. Gökcebay, in "Catalysis of Organic Reactions", Eds. J. R. Kosak, M. Dekker, New York, 1984, 153
6. H. Schulz, K. Beck and E. Erich, Fuel Proc. Techn. 18 (1988b) 293
7. H. Schulz, B. R. Rao and M. Elstner, Erdöl und Kohle 22 (1970) 651
8. Y. T. Eidus, Russ. Chem. Rev., 36 (1967) 338
9. A. T. Bell, Catal. Rev.-Sci. Eng. 23, 1&2 (1981) 203
10. H. Schulz, E. Erich, H. Gorre and E. van Steen, Catal. Let. 7 (1990) 157
11. H. Schulz, M. Claeys, T. Riedel and S. Walter, "4th Int. Conf. on CO_2 Utilization", Kyoto, Japan (1997), in press
12. H. Schulz, K. Beck and E. Erich, Proc. "9th Int Congr. on Catalysis", Calgary, 1988, Vol. 2 Eds. M. Phillips, M. Ternan, The Chemical Institute of Canada, Ottawa, 1988c, p829
13. Oil and Gas J., June 23, (1997) 14
14. H. Schulz and S. Nehren, Erdöl und Kohle-Erdgas-Petrochemie, 39 (1984) 93
15. H. Schulz, E. van Steen and M. Claeys, Stud. Surf. Sci. Catal. Vol. 81, Eds. H. E. Curry-Hyde, R. F. Howe; Elsevier, Amsterdam, 1994, p 45

NATURAL GAS CONVERSION V
Studies in Surface Science and Catalysis, Vol. 119
A. Parmaliana et al. (Editors)

Scale Up of a Bubble Column Slurry Reactor for Fischer-Tropsch Synthesis

R. Krishna and C. Maretto[a]

Department of Chemical Engineering, University of Amsterdam, Nieuwe Achtergracht 166, 1018 WV Amsterdam, The Netherlands

[a] Advanced Engineering, EniTecnologie S.p.A, 20097 S. Donato Milanese (MI), Italy

1. INTRODUCTION

The conversion of natural gas to middle distillates involves three steps: (1) production of syngas by e.g. catalytic partial oxidation, (2) conversion of syngas to hydrocarbons, mostly paraffins in the range C_5 to C_{100+}, using Fischer Tropsch (FT) synthesis and (3) hydrocracking of the paraffinic hydrocarbons to middle distillates. The FT synthesis step is highly exothermic and the bubble column slurry reactor is the ideal reactor choice for this purpose because of the ability in this reactor type to achieve near isothermal conditions and also because of the relatively high heat transfer coefficients to cooling tubes [1,2]. However the scale up of the bubble column slurry reactor requires a carefully planned strategy in which the scale dependent hydrodynamic parameters are determined on a relatively large scale by use of cold-flow experimental studies. This is then combined with catalysis and kinetic studies that are scale independent. One objective of this paper is to emphasise the scale up issues for the FT slurry reactor. In order to develop the scale up information extensive experimental work was carried out at the University of Amsterdam [3-7] in columns of 0.05, 0.1, 0.174, 0.19, 0.38 and 0.63 m diameter with a variety of liquids (water, paraffin oil, tetradecane) and slurries of varying concentration of silica particles (skeleton density = 2100 kg m^{-3}; pore volume = 1.05 mL g^{-1}; particle size distribution, d_p: 10% < 27 μm; 50% < 38 μm; 90% < 47 μm) in paraffin oil (density = 790 kg m^{-3}; viscosity = 0.0029 Pa s; surface tension = 0.028 N m^{-1}). The results of these experimental studies are used to develop a model for carrying out the slurry reactor design and optimisation.

2. BUBBLE COLUMN SLURRY REACTOR HYDRODYNAMICS

The churn-turbulent regime of operation is the most optimal one for FT synthesis [1]; in this regime there is a wide distribution of bubble sizes. Small bubbles in the size range 1 - 10 mm co-exist along with fast-rising large bubbles that are in the size range of 20 to 70 mm [7]. The hydrodynamics is pictured in Fig. 1. From a design and scale up point of view one of the

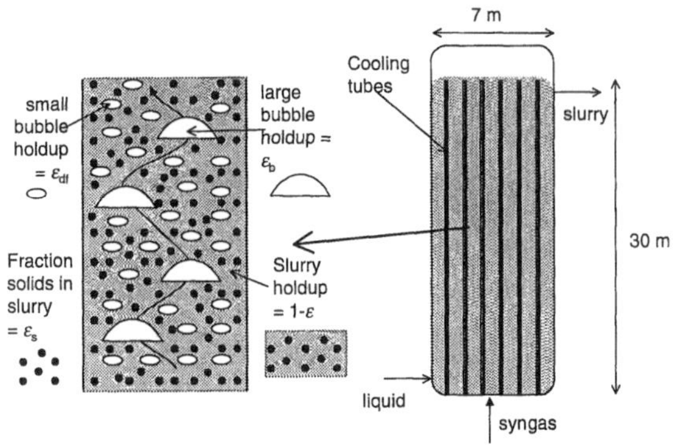

Fig. 1. Hydrodynamic picture of bubble column slurry reactor.

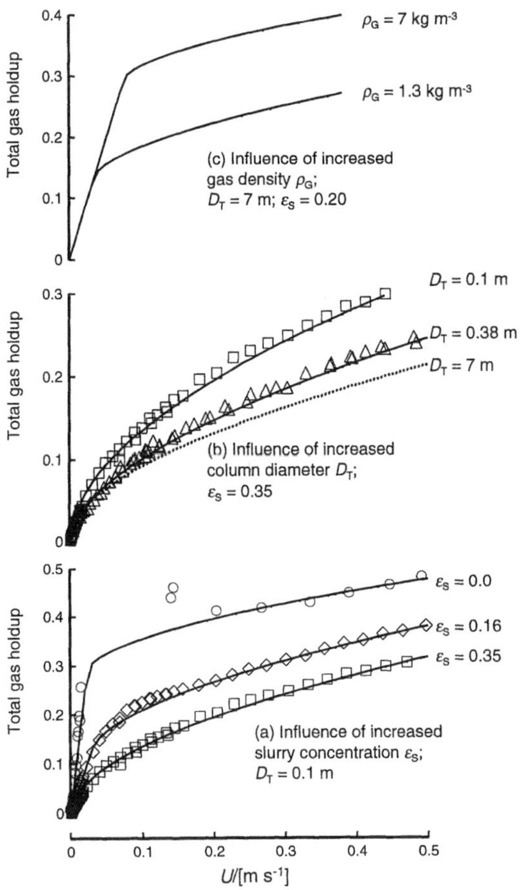

Fig. 2. Influence of (a) slurry concentration, (b) column diameter and (c) increased gas density on gas holdup.

most important parameters which needs to be estimated is the holdup of the gas bubbles in the reactor. The slurry phase and the cooling tubes occupy the remainder of the reactor. The gas holdup in the reactor is affected very significantly by the catalyst concentration, ε_s, expressed here as volume fraction in the slurry (gas-free basis). In calculating the value of ε_s it is assumed that the pores of the catalyst are filled with liquid and that the catalyst filled with this liquid forms the catalyst phase.

Figure 2 (a) shows the strong influence of increased catalyst concentration on the gas holdup [5]. The addition of fine particles, in this case of mean diameter of 38 μm, enhances the coalescence of small bubbles and the holdup of this population decreases. Due to enhanced coalescence of small bubbles the rise velocity of the small bubbles will increase with increasing solids concentration $V_{small}=V_{smal,ref}\left(1+0.8\varepsilon_S/V_{small,ref}\right)$ where the value of $V_{small,ref}$ = 0.095 m/s is taken from experimental studies [5]. The holdup of the large bubble population is virtually constant. At a slurry concentration of 38 vol %, the small bubble population ε_{df} is almost completely destroyed. The decrease in the small bubble holdup with increasing catalyst concentration is given by $\varepsilon_{df}=\varepsilon_{df,ref}\left(1-0.7\varepsilon_S/\varepsilon_{df,ref}\right)$ where the value of the small bubble holdup in "pure" liquid, $\varepsilon_{df,ref}$ has to be determined experimentally. For paraffin oil $\varepsilon_{df,ref}$ = 0.27 from the cold-flow experimental studies. This value is also expected to hold for conditions prevailing in the FT synthesis reactor.

The average size of the large bubbles is in the size range of 20 - 70 mm, increasing with increasing superficial gas velocity U [7]. The rise velocity of these bubbles is strongly influenced by the column diameter. Figure 2 (b) compares the total gas holdups in the 0.1 m and 0.38 m diameter columns with 35 vol% slurry. The strong column diameter influence is evident. For slurry concentrations $\varepsilon_s > 0.16$, the large bubble holdup can be estimated from $\varepsilon_b{=}0.3\left(U-U_{df}\right)^{0.58}/D_T^{0.18}$. In small diameter columns, say smaller than 0.1 m diameter, slug flow is attained at $U > 0.3$ m/s. In larger diameter columns, $D_T > 1$ m slug flow is not possible. The total gas holdup is calculated from $\varepsilon{=}\varepsilon_b{+}\varepsilon_{df}\left(1-\varepsilon_b\right)$, following the model of Krishna and Ellenberger [4].

Also shown in Fig. 2 (b) are the estimations of the large bubble holdup for a commercial scale reactor of 7 m using the model of Krishna *et al.* [5]. This model assumes that the column diameter dependence of the large bubble holdup, $\varepsilon_b{=}0.3\left(U-U_{df}\right)^{0.58}/D_T^{0.18}$, persists only up to 1 m. For larger column diameters, we set $D_T = 1$ in the correlation. From Fig. 2 (b) it should be evident that scaling up from say a 0.1 m diameter pilot plant to a commercial scale reactor of 7 m diameter is not straightforward. To obtain representative conversion levels in the pilot scale and commercial scale reactors the smallest scale pilot plant should be 1 m in diameter. This explains perhaps why the Exxon pilot plant for FT synthesis is 1.3 m in diameter [8]; the results from this pilot plant can be used directly for commercial plant design. Another approach [3] is to study scale effects in cold-flow units and use a smaller diameter pilot plant, of 0.1 - 0.2 m diameter say, just to obtain process information. The scale up to commercial reactor sizes is then carried out on the basis of a mathematical model that includes the influence of column diameter of the hydrodynamics and mass transfer.

The FT synthesis is carried out at a pressure in the range of 30 - 40 bar. The syngas density at 30 bar is 7 kg/m^3. Most cold-flow hydrodynamic studies are carried out at atmospheric pressure with air or nitrogen, with a gas density of 1.3 kg/m^3. Increased gas density tends to have two effects (1) it delays transition from homogeneous to heterogeneous regime and (2) the gas holdup at the regime transition point is increased [6]. The correlation of Reilly *et al.* [9] is recommended for the estimation of the regime transition point, if no experimental data are available. This correlation predicts that the holdup of the small bubbles at the regime transition point increases with $\rho_G^{0.48}$. The effect of increased gas density is therefore incorporated in the following manner $\varepsilon_{df}{=}\varepsilon_{df,ref}\left(\rho_G/\rho_{G,ref}\right)^{0.48}\left(1-0.7\varepsilon_S/\varepsilon_{df,ref}\right)$. The strong influence of gas density on the total gas holdup is demonstrated in Fig. 2 (c) for the case of a commercial scale reactor operating with slurry concentration $\varepsilon_s = 0.20$.

In estimating the mass transfer from large bubbles it is necessary to appreciate that the large bubbles suffer frequent coalescence and break-up [7]. The frequency with which the coalescence and break-up occurs varies between 2 and 16 Hz, increasing with increasing bubble size. This has important consequences for interphase mass transfer because the really large sized bubbles in the population have only a momentary existence and their mass transfer characteristics are not poor despite their large size. We recommend that the k_La for the large bubble population be estimated from the following relationship [10]:

$\left(\left(k_La\right)_{large}/\varepsilon_b\right)_{ref} = 0.5\ \mathrm{s}^{-1}$; $\left(\left(k_La\right)_{large}/\varepsilon_b\right) = \left(\left(k_La\right)_{large}/\varepsilon_b\right)_{ref} \times \sqrt{Đ_L/Đ_{ref}}$ where $Đ_{ref} = 2\times10^{-9}$ m^2/s is the value of the diffusivity at reference conditions, ε_b is the large bubble holdup and

$Đ_L$ is the diffusivity of the transferring component at the reaction temperature. The diffusivities of CO and H_2 will be different and so will the corresponding mass transfer coefficients. At the reaction temperature of 240 °C the value of $Đ_L$ for diffusivity of CO is 17.2×10^{-9} and for H_2 is 45.5×10^{-9} m^2/s. For mass transfer from small bubbles we assume a similar relation but take $\left((k_L a)_{small}/\varepsilon_{df}\right)_{ref} = 1\,s^{-1}$.

3. REACTOR MODELLING AND OPTIMIZATION STUDIES

We carried out a simulation of a commercial scale reactor with diameter D_T = 7 m, dispersion height H = 30 m, pressure p = 30 bar and temperature T = 240 °C. For the chosen reactor dimensions, the liquid phase can be considered to be well mixed and the conditions will be practically isothermal. For the purposes of property estimation the liquid phase is taken to be $C_{16}H_{34}$. Other properties estimated for the liquid phase are thermal conductivity: 0.113 W/m/K, heat capacity: 1500 J/kg/K, viscosity: 0.00029 Pa s, surface tension: 0.01 N/m. The catalyst chosen is Co/MgO supported on silica (average particle size of 50 μm) which has a nominal composition: 21.4 wt% Co and 3.9 wt % Mg. Other properties of the catalyst in the simulations are: skeleton density 2030, particle density 647 kg/m^3, thermal conductivity 1.7 W/m/K, heat capacity 992 J/kg/K, pore volume 0.00105 m^3/kg. The catalyst is expected to be well dispersed in the liquid and there will be no solids gradient along the reactor height. Such gradients can be expected only in tall narrow pilot plant reactors with small diameters of say 0.1 to 0.2 m. Syngas with a molar ratio H_2/CO = 2 enters the reactor. The distribution coefficients, defined by $c_G^* = m c_L$, for H_2 and CO are respectively estimated to be 2.96 and 2.48. The appropriate reactor model which emerges from the hydrodynamic studies at Amsterdam [2-7] is shown in Fig. 3 (the stirrers shown are only conceptual !). The large bubbles are assumed to traverse the column in plug flow with a superficial gas velocity $U - U_{df}$ where U_{df} is the superficial gas velocity through the small bubbles. We take U_{df} to be equal to the gas velocity at the regime transition point. The transition velocity is calculated from $U_{df} = V_{small}\varepsilon_{df}$ where the small bubble holdup and rise velocity, ε_{df} and V_{small}, are estimated from the model of Krishna *et al.* [5], after introducing a correction factor to account for the increase in the holdup at the regime transition point due to increased gas density as described above. The properties of the slurry were determined using the recommendations of Deckwer [11].

The Fischer Tropsch synthesis can be described by the simple reaction scheme $CO + 2\,H_2 \rightarrow \text{-}(CH_2)\text{-} + H_2O$. The Yates and Satterfield kinetics [12] for the reaction rate of CO is used: $(-R_{CO}) = a\, p_H p_{CO}/(1+b\, p_{CO})^2$ where R_{CO} is the consumption rate of CO expressed in mol CO per kg of supported Co/MgO catalyst per second, a = 0.0088533 exp[4494.41× (1/493.15-1/T)] mol/(s kg_{cat} bar^2) and b = 2.226 exp[-8236×(1/493.15-1/T)] bar^{-1}. It is must be underlined that the Yates-Satterfield kinetics were determined for a narrow temperature range 220-240 °C and hydrocarbon selectivity was not included in their model. To describe the catalyst selectivity the Anderson-Schulz-Flory for the carbon number distribution was chosen. Considering that most of the hydrocarbon products are paraffins, the mole fraction of each species C_nH_{2n+2} is obtained as follows x_n = (1-α_{ASF}) α_{ASF} where α_{ASF} is the probability factor of hydrocarbon chain growth. The higher the α_{ASF}.factor the higher is the fraction of heavy paraffins. A value of α_{ASF}=0.9 is chosen which is a typical value for Co catalyst. The consumption ratio of CO and H_2 is 2. As the feed ratio of CO and H_2 was set equal to the

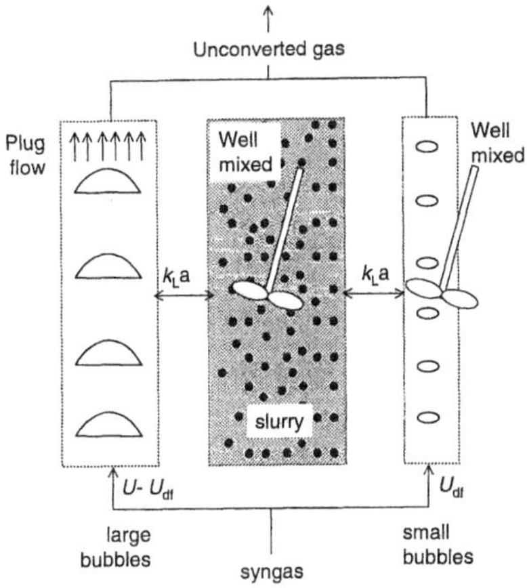

Fig. 3. FT slurry reactor model

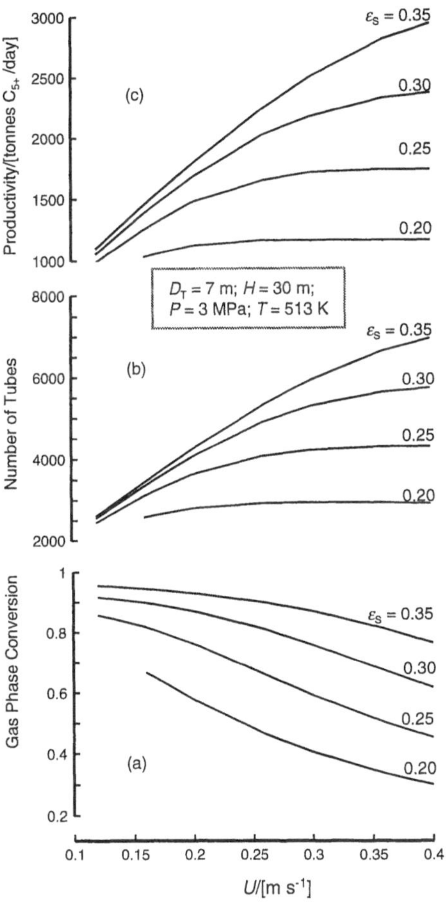

Fig. 4. FT reactor simulation results.

consumption ratio, the conversion of CO and H_2 are both equal to one another, χ_{CO+H}. The amount of inerts in the entering gas phase was taken to be 5% and the gas contraction factor (for 100% syngas conversion) can be calculated as $\alpha = -0.648$. The superficial gas velocity varies with conversion as $U(1+\alpha\chi_{CO+H})$.

For removal of reaction enthalpy ΔH = -0.172 MJ/(mol CO), vertical cooling tubes of 50 mm diameter are installed with a constant coolant (steam) temperature of 230 °C. The heat transfer coefficient from slurry to the coolant was estimated using the correlation of Deckwer [11]. The pitch for the vertical cooling tubes will depend on the number of tubes to be installed. In the calculations the pitch varied from 0.12 to 0.19. This pitch size is considered to be large enough not to influence the bubble size, bubble holdup or slurry phase backmixing.

The main results of the simulations are presented in Fig. 4 for a range of inlet superficial gas velocity U = 0.12 to 0.4 m/s and catalyst concentrations in the range ε_s = 0.2 – 0.35. Increasing U causes a decrease in syngas conversion. The reactor productivity, expressed as tonnes per day of C_{5+} hydrocarbons produced, increases with increasing U, as does the required number of cooling tubes. It is evident that for the highest reactor productivity only a moderate conversion level is reached, and the non-reacted syngas should be recycled to the reactor. In practice it is desirable to operate at conversion levels of about 90%. This means that the maximum superficial gas velocity has to be restricted to below 0.3 m/s; see Fig. 4.

Increasing the slurry concentration ε_s increases the conversion and the reactor capacity, as well as the number of required cooling tubes to be installed in the reactor. The influence of ε_s is not only on the kinetic term, which is proportional to the catalyst loading, but also on the total gas holdup.

Increasing ε_s reduces the total gas holdup, making more room available for the catalyst. Therefore increasing ε_s has more than a proportional influence on the reactor conversion and productivity. From the reactor performance point of view it is advisable to use the highest catalyst concentrations consistent with ease of handleability. From the experience gained at the University of Amsterdam, we consider $\varepsilon_s = 0.40$ the maximum slurry concentration which can be used in commercial practice.

An economically viable FT complex would need to have a high production capacity, of the order of 5000 t/day of middle distillates, which can be considered to be C_{10+} hydrocarbon products. For the assumed Anderson-Schultz-Flory distribution with the probability chain growth factor $\alpha_{ASF} = 0.9$ we estimate that 80% of the C_{5+} products will be in the middle distillates range. From the results presented in Fig. 4 (c), we find that operation at a superficial gas velocity at the inlet of 0.3 m/s and a slurry concentration of 35 vol% would require a total of three reactors in parallel in order to produce 5000 t/day of middle distillates. Three reactors allow a good degree of flexibility on operating conditions. In each of these three reactors we would need to install about 6000 vertical cooling tubes at a pitch of about 0.15 m.

Increasing, or decreasing the interphase mass transfer coefficient from the base case value has a negligible effect on reactor performance. Increasing the Yates-Satterfield kinetic parameter *a* by a factor 2 results in a 60% increase in reactor productivity. It can be concluded that the FT reactor is kinetically controlled. If the catalyst activity is twice as high as given by Yates-Satterfield, then the number of reactors in parallel required for a 5000 t/day middle distillates complex will be two instead of three. The importance of improved catalyst formulations in developing a viable FT reactor technology is evident.

REFERENCES

1. J.W.A. de Swart, R. Krishna and S.T. Sie, Proceedings of the 4th International Natural Gas Conversion Symposium, Kruger Park, South Africa, November 19-23, 1995
2. Jager, B., Proceedings of the 4th International Natural Gas Conversion Symposium, Kruger Park, South Africa, November 19-23, 1995
3. R. Krishna, J. Ellenberger and S.T. Sie, *Chem. Eng. Sci.*, **51** (1996) 2041
4. R. Krishna and J. Ellenberger, *A.I.Ch.E.J.*, **42** (1996) 2637.
5. R. Krishna, J.W.A. de Swart, J. Ellenberger, G.B. Martina and C. Maretto, *A.I.Ch.E.J.*, **43** (1997) 311
6. H. M. Letzel, J.C. Schouten, C. M. van den Bleek and R. Krishna, *Chem.Eng. Sci.*, **52** (1997) 3733
7. J.W.A. de Swart, R.E. van Vliet and R. Krishna, *Chem. Eng. Sci.*, **51** (1996) 4619.
8. B. Eisenberg, L.L. Ansell, R.A. Fiato and R.F. Bauman, Advanced gas conversion technology for remote natural gas utilization, Paper presented at the 73rd Annual GPA convention, New Orleans, Louisiana, March 7-9, 1994.
9. I.G. Reilly, D.S. Scott, T.J.W. De Bruijn and D. MacIntyre, *Canad. J. Chem. Engng.*, **72** (1994) 3
10. D.J. Vermeer and R. Krishna, *Ind. Eng. Chem. Process Des. Dev.*, **20** (1981) 475.
11. W.D. Deckwer, Y. Serpemen, M. Ralek and B. Schmidt. *Ind. Eng. Chem. Process Des. Dev.*, **21** (1982) 231
12. I.C. Yates and C.N. Satterfield, *Energy & Fuels*, **5** (1991) 168.

NATURAL GAS CONVERSION V
Studies in Surface Science and Catalysis, Vol. 119
A. Parmaliana et al. (Editors)

DISPERSION AND REDUCIBILITY OF Co/SiO_2 AND Co/TiO_2

Roberto Riva[a], Hans Miessner[a#], Gastone Del Piero[a], Bernadette Rebours[b], Magalie Roy[b]

[a]Eniricerche, via Maritano 26, I-20097 San Donato Milanese (MI), Italy

[b]Institute Francais du Petrole, B.P.311, 92852 Rueil-Malmaison Cedex, France

[#]present address: Gesellschaft zur Förderung der naturwissenschaftlich-technischen Forschung in Berlin-Adlershof, Germany

INTRODUCTION

The interaction of cobalt with various supports has been widely studied, as cobalt has important catalytic properties both in hydrodesulfurization reactions and in the Fischer Tropsch synthesis (1-8). Much effort has been devoted to understanding the relationship between the dispersion of cobalt and the activity of the catalyst in the Fischer Tropsch synthesis (9,10). The formation of surface compounds between cobalt and the support has been reported to decrease the activity of the catalyst (2,5,11). Moreover, strong metal-support interaction has been found to affect the dispersion of supported metals (12). According to the literature, the interaction of cobalt with titania is much stronger than with silica.

The present study deals with the interaction between cobalt and the support, either silica or titania. It aims to understand how the interaction with the support affects both the reducibility and the dispersion of cobalt. The response of cobalt to reduction is studied with TPR experiments, in which the temperature is raised at a steady rate, and with XPS after reduction treatments at constant temperature. The dispersion of cobalt is studied with XPS.

EXPERIMENTAL

Preparation of the samples

Johnson&Matthey Co_3O_4 was used as a reference compound for XPS spectra. The quality of the sample was checked by X-ray diffraction (XRD) before XPS analysis.
Silica supported samples with various degrees of cobalt loading (from 2wt% to 27wt%) were prepared following the incipient wetness impregnation method. After impregnation the samples were calcined at 400°C in air for 4 hours. The surface area of the Merck silica was 430 m^2/g, its particle size being in the range 15-45 mm with an average pore radius of 35Å.

Titania supported samples containing 12wt% Co were prepared with the same procedure, using Degussa P25 titanium dioxide. After this treatment the surface area of the support was found to be ca. 40 m^2/g with an average particle size around 0.1 mm.

X-Ray Diffraction (XRD)

The XRD data were collected at ambient conditions using a Philips diffractometer with monochromatic Cu Ka radiation (l=1.5418Å). Qualitative phase analysis was carried out using the Siemens Diffrac AT package run on a IBM PC330 P-75. For titania supported samples, the quantitative phase analysis was carried out by using the Rietveld profile fitting method (13) with the procedure proposed by Hill and Howard (14). Structural data were taken from Wyckoff (15). For silica supported samples, the conventional method reported by Klug and Alexander was used (16). Crystal size was calculated from line broadening applying the Scherrer equation (16).

Temperature Programmed Reduction (TPR)

TPR experiments were performed in a U-shaped tubular quartz reactor. After loading the sample, the reactor was flushed with He at 150°C for 1 hour, then cooled down to 50°C in flowing He. The gas flow (2%H_2-He) was adjusted for each sample in such a way as to maintain a roughly constant ratio between the amount of cobalt contained in the sample and the H_2 available. The temperature was then raised at the constant rate of 10°C/min from 50°C to the desired temperature (700-900°C). The content of H_2 and H_2O in the outflowing gas was monitored with a VG-Fisons quadrupole mass spectrometer.

X-Ray Photoelectron Spectroscopy (XPS)

The XPS spectra were collected with a VG Escalab MKII spectrometer. A non-monochromatic Al X-ray source was used. The binding energy values given in the literature for the following peaks were used as a reference: Si 2p 103.3 eV for silica supported samples, Ti 2p 458.7 eV for titania supported samples, O 1s 530 eV for unsupported Co_3O_4 (17,18). A reaction chamber connected to the vacuum system of the spectrometer allowed the samples to be transferred into the measurement chamber without exposure to air after reducing and oxidizing treatments. The reducing treatments were carried out in 3%H_2-Ar at various temperatures and for various lengths of time. The oxidizing treatments were done in synthetic air at 400°C for 5 hours at least. The Co 2p and the Si 2s or Ti 2p peaks were used for the quantitative analysis, by assuming the composition of the sample to be uniform throughout the volume probed by XPS (18,19,20). The dispersion of cobalt over the two supports was studied by analysis of Co/Si and Co/Ti atomic ratios respectively.

RESULTS AND DISCUSSION

Unsupported cobalt

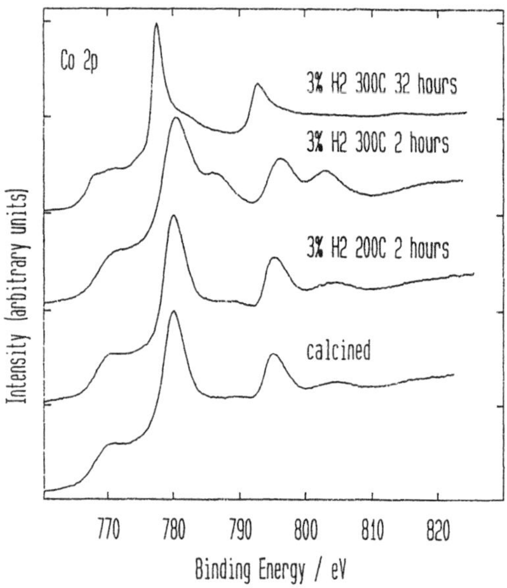

Figure 1: XPS Co 2p peak of unsupported cobalt.

Figure 1 shows how the Co 2p photoelectron peak is affected by reducing treatments. Co_3O_4 is stable up to 200°C and is reduced completely at 300°C. The reduction occurs in two steps: first Co_3O_4 is reduced to CoO (third curve from the bottom), then CoO is reduced to metallic Co (curve at the top). Metallic cobalt is easily distinguished from oxidized cobalt because of the large difference in binding energy. The difference in lineshape make it possible to distinguish between Co_3O_4 and CoO. In fact a satellite peak appears on the high binding energy side of both Co $2p_{3/2}$ and Co $2p_{1/2}$, due to multiplet splitting. These assignments are in agreement with literature data (1,21). Reportedly, the lineshape of CoO applies to Co^{2+} in general, even when cobalt forms silicate or titanate through reaction with the support.

Silica supported cobalt

Only cubic Co_3O_4 is detected by XRD in silica supported calcined samples. The amount of this phase increases with the increase in cobalt loading, the quantity being always close to that calculated from chemical analysis data. The size of the Co_3O_4 crystallites, evaluated by XRD, tends to increase with increasing cobalt loading (from 120 to about 160 Å), even though the values are rather scattered . The same measurements have been made on reduced and passivated (1%O_2-N_2 at room temperature for 2 hours before exposure to air) samples. At high temperature (900°C) cobalt crystallizes as cubic metal, while at lower temperature (400°C) a fraction of Co crystallizes also in the hexagonal form and some residual CoO is present, probably due to the passivation process. Crystal size tends to increase with cobalt loading, as found for calcined samples, with a strong dependence on the reduction temperature.

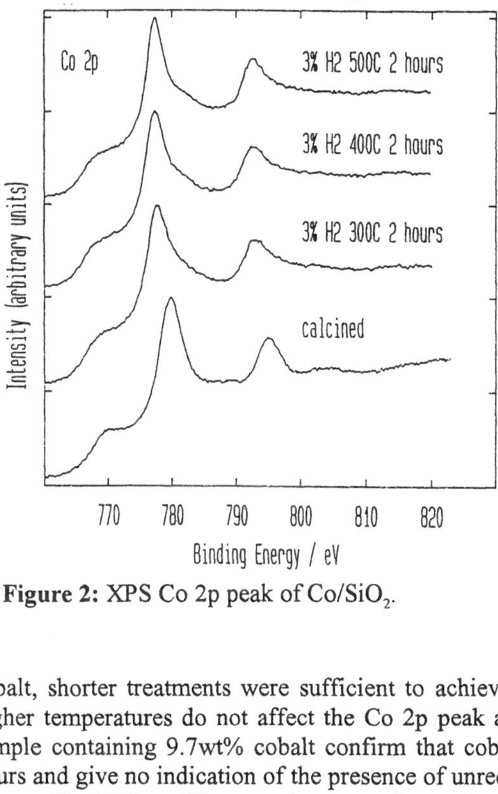

Figure 2: XPS Co 2p peak of Co/SiO_2.

Reducing treatments of the 18wt%Co-SiO_2 sample, studied with XPS, show that the surface cobalt oxide is completely reduced at 300°C, just like unsupported cobalt (fig. 2). Compared to unsupported cobalt, shorter treatments were sufficient to achieve the complete reduction. Treatments at higher temperatures do not affect the Co 2p peak any more. Reduction experiments on the sample containing 9.7wt% cobalt confirm that cobalt is completely reduced at 300°C in 2 hours and give no indication of the presence of unreducible cobalt.

The TPR profiles contain two major peaks at 340°C and 430°C, and a broad peak at higher temperatures (not shown). The two major peaks are similar to those obtained with pure Co_3O_4 . The ratio between the H_2 consumed at 430°C and that consumed at 340°C is 3:1. A similar behaviour was observed by other authors (8,22,24). It is generally agreed that this represents the reduction of Co_3O_4 particles to metallic cobalt through the CoO step, as already pointed out for unsupported Co_3O_4.

The response of the dispersion of cobalt to reducing and reoxidizing treatments has also been studied. The samples were exposed to air after 700°C-TPR, reduced again for 2 hours at 400°C in the reaction chamber connected to the XPS spectrometer. The samples were then reoxidized and their spectra were collected again. The measured Co/Si ratios are listed in table 1 and deserve some comment. The Co/Si ratios of the samples containing 5.1 and 9.7 Co wt% are not significantly affected by reduction and reoxidation. On the other hand the Co/Si ratios of the samples containing 18.4 and 22.8 Co wt% decrease appreciably after reduction and reoxidation. This leads to the conclusion that sintering of the cobalt particles occurred in the two samples with the highest content of cobalt. The tendency of the supported particles towards sintering proves that the interaction between cobalt and silica is not strong, since sintering causes the area of the interface between the two phases to decrease. Table 1 also shows that the Co/Si ratio increases strongly when the content of cobalt changes from 9.7% to 18.4%. Then the Co/Si ratio levels off at a constant value. This behaviour is attributed to the progressive development of a Co rich outer shell on the surface of the SiO_2 particles, followed by the onset of the growth of the cobalt particles. SEM, TEM and XPS data, not shown in this paper, support this conclusion.

TABLE 1: Silica supported samples: XPS Co/Si atomic ratios after different treatments.

Co wt%	calcined	reduced & reoxidized
5.1	0.15	0.14
9.7	0.20	0.19
18.4	0.72	0.56
22.8	0.70	0.52

Titania supported cobalt

The XRD spectra of several titania supported samples (with ca 12% Co and Rutile/Anatase ratio ranging from 76/24 to 85/15) indicate that all the cobalt contained in the calcined samples is in the form of crystalline Co_3O_4 . After reduction and passivation most of the cobalt is amorphous and only a small fraction crystallizes as cubic Co (table 2), whereas neither the rutile to anatase ratio nor the morphology of the support changes. Therefore, the reduction treatment affects the phase composition of cobalt quite strongly, turning the oxidized crystalline phase into a mainly amorphous phase after reduction.

TABLE 2: Titania supported samples (12wt%Co-TiO_2): cobalt phases composition (XRD).

sample	Co_3O_4 (%)	cubic Co(%)	amorphousCo(%)	crystal size(Å)
85% rutile - calc.	100%	-	0%	300(Co_3O_4)
red. & passivated	-	17%	83%	220(Co cubic)
76% rutile - calc.	100%	-	0%	400(Co_3O_4)
red. & passivated	-	17%	83%	190(Co cubic)

The response of cobalt to reducing treatments has been studied with XPS. The results are shown in figure 3. Co_3O_4 is readily reduced to Co^{2+} with a 2 hour treatment at 300°C, but the complete reduction of Co^{2+} to metallic cobalt is not accomplished even after 66 hours at 300°C. In fact, a high binding energy shoulder indicates that a fraction of the cobalt is not reduced and is probably in the Co^{2+} oxidation state. This behaviour is markedly different from that of unsupported Co_3O_4 and is probably due to the partial formation of cobalt titanate, which is less reducible than Co_3O_4 , according to the literature. Treatments at higher temperatures increase the degree of reduction. This behaviour is confirmed by tests on samples that were prepared in different batches and can be regarded as XPS evidence of the well known metal-support interaction.

Table 3 gives the atomic ratio and the binding energy values obtained after two consecutive reducing and oxidizing treatments. The increase in the dispersion of cobalt is very strong after the first reduction-reoxidation step, since the Co/Ti ratio increases from 0.53 to 0.94. The Co/Ti ratio does not vary appreciably after the second redox treatement (final Co/Ti ratio 0.92), which means that the second redox step does not affect the dispersion of cobalt any more. It must be remarked that the reduction step is necessary in order to obtain an increase in the dispersion of cobalt. In fact, treating the calcined samples in air at 400°C for 10 hours does not affect the Co/Ti atomic ratio. This behaviour is consistent with the model proposed by Horseley, which depicts the metal-support interaction as an electron exchange between a partially reduced support and the metal (23).

TABLE 3: Response of 12wt%Co-TiO_2 to various treatments.

treatment	atomic Co/Ti
calcined	0.53
reduced & reoxidized	0.94
twice reduced & reoxidized	0.92

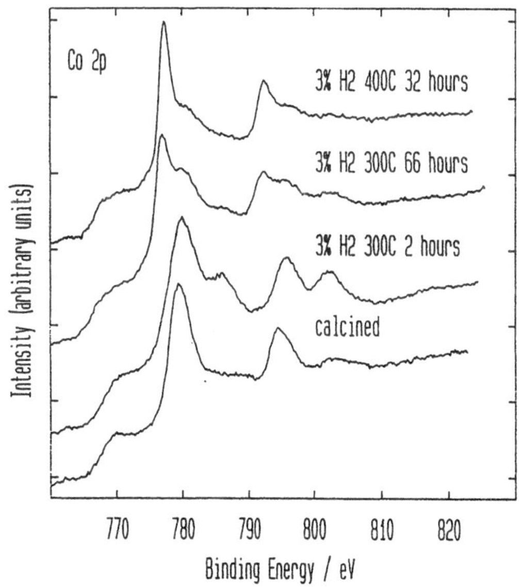

Figure 3: XPS Co 2p peak of Co/TiO_2.

The TPR profiles of titania supported samples are quite different from those of silica supported samples: only two peaks are detected and their maxima occur at higher temperatures, 380-400°C and 500-600°C respectively, the latter being very broad.

The conclusion that a reaction occurs between the cobalt particles and titania during the reduction treatment is supported by the following arguments:

- XRD data indicate that cobalt is prevailingly amorphous in the reduced and passivated samples, while it had completely crystallized as Co_3O_4 in the calcined samples.
- Contrary to both unsupported Co_3O_4 and silica supported Co_3O_4, XPS reduction tests show that titania supported cobalt is not completely reducible at 300°C in 3%H_2 . Moreover, the dispersion of cobalt (Co/Ti atomic ratio) increases appreciably after reduction and reoxidation, compared to the starting calcined samples.
- The TPR peaks fall at higher temperatures for titania supported samples than for silica supported samples.

CONCLUSIONS

This study has addressed the interaction of cobalt with two different kinds of support: silica and titania. The formation of a surface compound between cobalt and titania that is more resistant to reduction than Co_3O_4 shows that the interaction is much stronger in the case of titania. On the contrary, the behaviour of silica supported samples is very similar to that of unsupported Co_3O_4 under reducing treatments. The different reactivity of cobalt with silica and titania explains why reducing and reoxidizing treatments have opposite effects on the dispersion of cobalt depending on whether it is supported on SiO_2 or TiO_2 . The low reactivity of cobalt with silica favours sintering effects after reduction and reoxidation treatments. In contrast, the level of dispersion of titania supported cobalt tends to increase after the same treatments owing to the high reactivity of cobalt with titania.

REFERENCES

1. Okamoto,Y.; Hajime,N.; Imanaka,T; Teranishi,S. Bull. Chem. Soc. Jpn. 48(4) (1975) 1163
2. Zowtiak,J.M.; Bartholomew,C.H. J. Catal. 83 (1983)107
3. Reuel,C.R.; Bartholomew,C.H. J. Catal. 85 (1984) 78
4. Paryjczak,T.; Rynkowski,J.; Karski,S. J. Chromatog. 188 (1980) 254
5. Chin,R.L.; Hercules,D.M. J. Phys. Chem. 86 (1982) 360
6. Castner,D.G.; Santilli,D.S. ACS Symposium Series 248; American Chemical Society, Washington, D.C. (1984) chapter 3
7. Ming,H.; Baker,B.G. Appl. Catal. 123 (1995) 23
8. Okamoto,Y.; Nagata,K.; Adachi,T.; Imanaka,T.; Inamura,K.; Takyu,T. J. Phys. Chem. 95 (1991) 310
9. Iglesia,E.; Soled,S.L.; Fiato,R.A. J. Catal. 137 (1992) 212
10. Iglesia,E.; Soled,S.L.; Baumgartner,E.J.; Reyes,S.C. J. Catal. 153 (1995) 108
11. Sato,K.; Inoue,Y.; Kojima,I.; Miyazaki,E.; Yasumori,I. J. Chem. Soc., Faraday Trans. 1 80 (1984) 841
12. Stevenson,S.A.; Dumesic,J.A.; Baker,R.T.K.; Ruckenstein,E. editors Metal-Support Interactions in Catalysis, Sintering and Redispersion; Van Nostrand Reinhold Company: New York (1987)
13. Young,R.A. The Rietveld Method; Oxford University Press: Oxford (1993)
14. Hill,R.J.; Howard,C.J. J. Appl. Cryst. 20 (1987) 467
15. Wyckoff,R.W.G. Crystal Structures; Interscience Publishers: New York (1963)
16. Klug,H.P.; Alexander,L.E. X-ray Diffraction Procedures; John Wiley & Sons: New York (1974)
17. Moulder,T.F.; Stickle,W.F.; Sobol,P.E.; Bomben,K.D. Handbook of X-ray Photoelectron Spectroscopy; published by Perkin Elmer Corporation: Eden Prairie (1992)
18. Briggs,D.; Seah,M.P. Practical Surface Analysis, 2nd ed.; John Wiley & Sons: Chichester (1990)
19. Ertl,G.; Kuppers,J. Low Energy Electrons and Surface Chemistry; VCH Verlagsgesellschaft: Weinheim (1985)
20. Niemantsverdriet,J.W. Spectroscopy in Catalysis; VCH Verlagsgesellschaft: Weinheim (1995)
21. Chuang,T.J.; Brundle,C.R.; Rice,D.W. Surf. Sci. 59 (1976) 413
22. Castner,D.G.; Watson,P.R.; Chan,I.Y. J. Phys. Chem. , 94 (1990) 819
23. Tauster,S.J.; Fung,S.C.; Baker,R.T.K.; Horseley,J.A. Science 211 (1981) 1121
24. Sexton,B.A.; Hughes,A.E.; Turney,T.W. J. Catal. 97 (1986) 309

NATURAL GAS CONVERSION V
Studies in Surface Science and Catalysis, Vol. 119
A. Parmaliana et al. (Editors)

Characterization of Bubble Column Slurry Phase Iron Fischer-Tropsch Catalysts

Yaming Jin and Abhaya K. Datye

Center for Microengineered Materials and Department of Chemical & Nuclear Engineering, University of New Mexico, Albuquerque, NM 87131, USA

Abstract

The cross-sectional transmission electron microscopy (XTEM) method and x-ray diffraction (XRD) were used for phase analysis of bubble column slurry phase iron Fischer-Tropsch catalysts. For the deactivated LGX-171, the carbide phase shows mono-dispersion characteristic. The carbide particles have well-defined shape, spherical or rectangular, and give a distinctive χ-carbide XRD pattern. The average particle size, 39.4 nm, by XRD was in good agreement with statistical value of 37.5 nm by TEM. On the other hand, two carbide phases are found to coexist in the active catalyst: big twisted particles (20-40 nm) and highly dispersed carbide particles (less than 10 nm). High resolution TEM work shows that the big distorted carbide particles belong to ε'-carbide, while the highly dispersed carbide phase is most probably a mixture of ε'-carbide and χ-carbide. For this particular set of iron catalysts, big faceted magnetite single crystals were found to be present in the catalyst both at the active and deactivated states. From these, we conclude that the ε'-carbide must represent the active phase in Fe F-T catalysts.

1. Introduction

Fischer-Tropsch Synthesis is recognized as a viable route for conversion of syngas to liquid fuels (**1**). This study is directed at understanding mechanisms of catalyst deactivation in Fe catalysts used in a slurry phase reactor. We report here analyses of end-of-run catalysts from two F-T runs. The catalyst reactivity has been reported elsewhere (**2**), but in the previous work, we were unable to conclusively identify the causes of catalyst deactivation.

It is recognized that loss of surface area of the catalytically active phase and deposition of unreactive carbonaceous deposits must constitute possible mechanisms for catalyst deactivation (**2,3**). However, there is as yet no consensus on the nature of the active phase in Fe F-T catalysts. Previous work has tried to correlate bulk phase information with catalytic reactivity based on results of analytical methods such as Mössbauer spectroscopy (MS) and X-ray diffraction (XRD) techniques. Besides the intrinsic limitations of these bulk techniques for the quantification of a highly dispersed iron phase (with particles less than 10 nm), there are several other experimental difficulties in determining an accurate phase composition of slurry bubble column F-T catalysts. The catalyst removed from the reactor is dispersed in product wax at a loading of 5 wt%, hence a wax removal step is usually performed prior to analysis. We have recently (**4**) shown that Soxhlet extraction, the commonly used wax removal procedure, can cause oxidation of the reduced iron phases. Furthermore, the wax is often crystalline and interferes with the diffraction peaks from the iron carbide phases of interest. We have also found that there are enormous differences is scattering factors for the various iron phases making quantitative analysis based on peak heights completely unreliable (**4**). We believe that some of these experimental difficulties have resulted in the generally accepted conclusion that there is no clear relationship between catalyst activity and the phase composition of the working catalyst.

In order to get accurate phase information of working FTS catalysts, and to minimize problems with surface oxidation of the reduced iron phases, we feel it is necessary to characterize working iron catalysts protected by the hydrocarbon wax (**4**). Quantitative Rietveld structure refinement analysis allowed us to obtain useful phase information of the wax-containing iron catalyst sample. Nevertheless, the peak overlaps between the crystalline wax and iron phases of interest, and the low diffraction peak intensities suggested that the XRD analysis be corroborated with other methods. In this paper, we report a cross-section transmission electron microscope (XTEM) study of the phases present in working Fe FTS catalysts. The TEM results along with XRD analysis helps to provide a more comprehensive picture of these catalysts.

2. Experimental

The catalyst sample studied in this paper has a chemical composition of 100Fe/4.4Si/1.0K and was provided to us by Dr. Burtron Davis at the Center of Applied Energy Research (CAER), University of Kentucky, after use in FT synthesis runs. In run LGX-171, the precipitated oxide precursor was pretreated with syngas at 1 atm and 270°C for 24 hours, then underwent FTS at 270°C, 175 psig. The wax-mixed catalyst sample was removed from the slurry reactor after time-on-stream (TOS) 3164 hours. The catalyst activity was high for the first 2800 hours of this run, but over the last few hundred hours there was a rapid deactivation and the catalyst was removed at the end of run where CO conversion was 20%. On the other hand, in run LGX-175, the catalyst precursor was pretreated in CO at 1 atm and 270°C for 24 hours and underwent FTS at the same conditions as LGX-171. The catalyst sample was removed after TOS 1160 hours while its CO conversion was 79%. Detailed reactivity data were reported elsewhere (**2,5**).

XRD data were obtained on a Scintag PAD-V powder diffractometer using Cu-Kα radiation (λ=1.5406 Å). Scans were taken from 10° to 90° in step-scan mode, 0.04° per step, 0.4° per minute. For cross section TEM, the wax-mixed iron catalysts were first embedded in Spürr's low viscosity epoxy. After curing the epoxy, sections with thickness about 40-60 nm were prepared. The microtomed sample sections were mounted on TEM grids with holey carbon film and examined in a JEOL 2010 HRTEM microscope operated at 200 KeV. These thin sections allow us to get high resolution images that permit detailed phase identification, and also to get particle size distributions for each iron phase.

3. Results

The XRD spectra of LGX-171 and LGX-175 with wax partly removed are shown in figure 1. Two phases, magnetite and χ-carbide, can be clearly identified in the XRD spectrum of LGX-171 (the low activity catalyst). The average particle size estimated by Scherrer's equation is 27.2 nm and 39.4 nm for the two phases respectively. The catalyst from run LGX-175 (high activity) also shows pronounced magnetite peaks with an estimated average particle diameter of 30.3 nm. However, the nature of the carbide peaks is very different from those in LGX-171. From the intensity of the (002) diffraction peak of χ-carbide at $2\theta=41.45^{\circ}$ which does not interfere with the other peaks, we can see

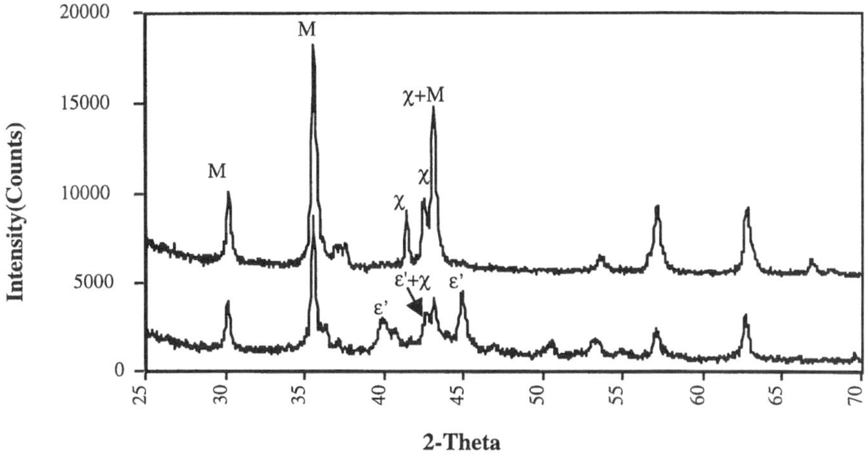

Fig.1 XRD spectra of LGX-171 (upper) deactivated catalyst and LGX-175 (lower), active catalyst after partial removal of hydrocarbon wax.

that very little χ-carbide is present in LGX-175. On the other hand, the broad diffraction peaks at $2\theta=44.96\ ^{\circ}$ and $40.00\ ^{\circ}$ match the (121) and (120) diffraction lines of ε'-carbide. By using Scherrer's equation, we can estimate the average particle size of ε'-carbide to be 19.2 nm.

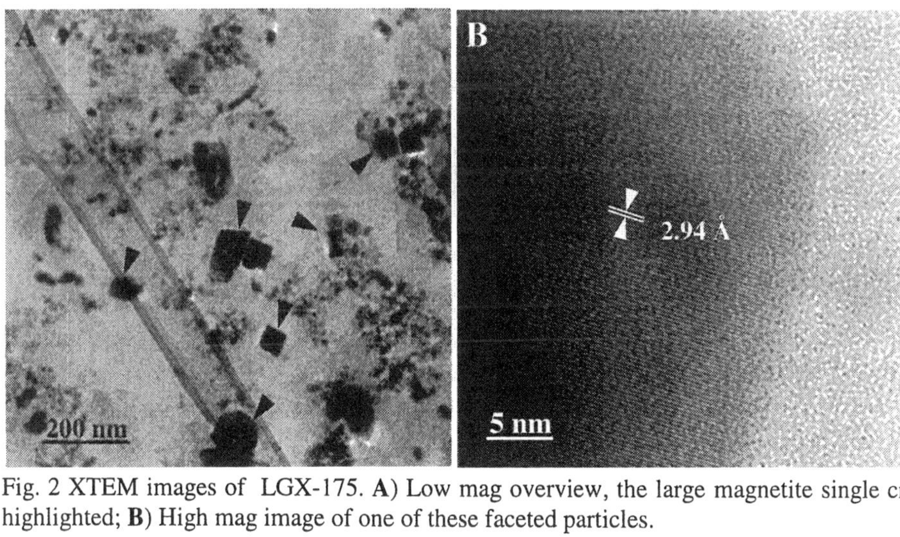

Fig. 2 XTEM images of LGX-175. **A**) Low mag overview, the large magnetite single crystals are highlighted; **B**) High mag image of one of these faceted particles.

Figure 2 shows XTEM images of the sample from LGX-175 (figure 2-A). Large (around 50 nm) faceted particles are seen to be sparsely dispersed in this image. By analyzing the high resolution images, we can conclusively identify these particles as magnetite single crystals, as shown in figure 2- B (2.94 Å corresponds to the (220) plane of magnetite). The rest of the catalyst is a complex mixture of iron phases. A typical view is shown in figure 3-A. Closer examination shows three different types of iron phases to be present in this particular sample. There are big distorted particles (20 to 40 nm) always with a surface layer or smaller particles sticking to them (figure 3-A). High resolution images of these iron phases have lattice spacings of 2.26 Å and 2.02 Å correspond to d-values of (120) and (121) planes of ε'-carbide. Therefore, we attribute this phase to be ε'-carbide. The rest of the catalyst contains highly dispersed particles (usually less than 10 nm) as shown in figure 3-A. Those without a surface layer are identified as fine magnetite particles, shown in figure 3-B (the 2.53 Å spacing corresponds to the d-value of the (311) plane of magnetite). The fact that magnetite lattice fringes are also seen on the surface of the carbide suggests that the highly dispersed magnetite may actually be a passivating oxide formed due to air exposure. The particles with a surface layer marked in figure 3-A are identified as highly dispersed iron carbide. High resolution imaging has revealed that these particles exhibit lattice fringes of ε'-carbide as well as χ-carbide. These carbide particles do not necessarily coexist with the fine magnetite particles as seen in figure 3-A. In fact this phase was found to be well dispersed everywhere in the sample.

In contrast to sample LGX-175 (the active catalyst), sample LGX-171 (the deactivated catalyst) only showed the presence of χ-carbide in addition to the magnetite, both by XRD and TEM. These

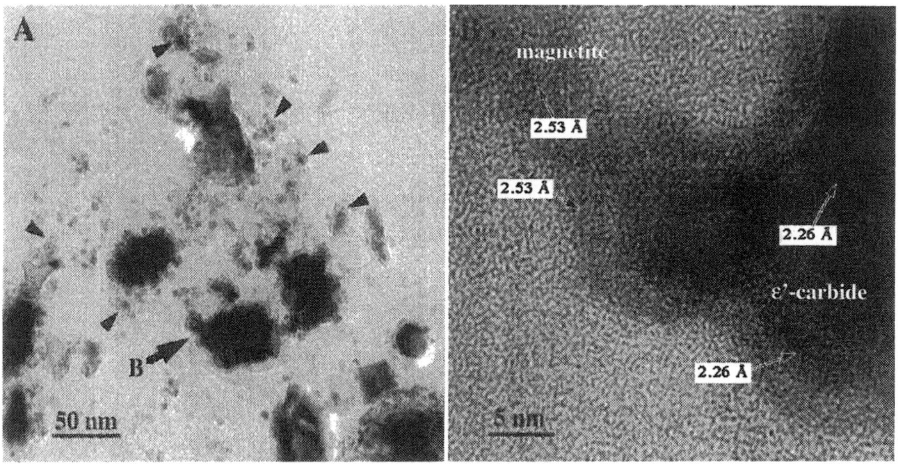

Fig. 3 TEM images showing the complex microstructure of sample LGX-175. **A**) Low mag view of 3 distinct phases: large distorted ε'-carbide particles, fine carbide particles (some are highlighted), and highly dispersed magnetite particles. **B**) High mag view showing the coexistence of these 3 phases.

χ-carbide particles had smooth surfaces, and well defined shapes - rectangular or spherical. All of these particles have a uniform surface layer having a thickness of about 3-4 nm (fig. 4-A). The average particle diameter, 39.4 nm by XRD, is consistent with the estimated average diameter of 37.5 nm by TEM. There are also large magnetite single crystals dispersed in this sample similar to those seen in LGX-175. In addition, there is a polycrystalline magnetite phase distributed throughout this catalyst sample. However, unlike the LGX-175 (active) sample, the size of these magnetite particles is larger (about 10 to 20 nm), and those particle always form agglomerates as shown in Figure 4-A.

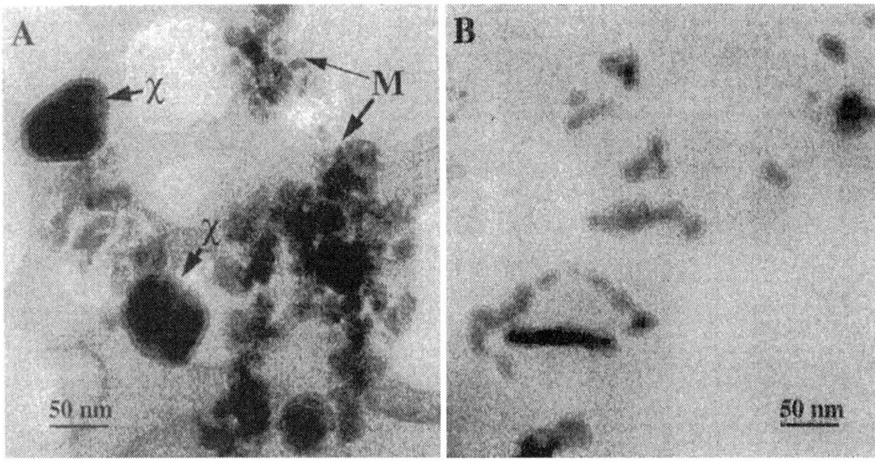

Fig.4. Comparison of carbide phases in LGX-171 (deactivated) and LGX-175 (active).
A) Regular shaped χ-carbide particles in LGX-171. B) Highly dispersed carbide particles seen in LGX-175 (active catalyst) are missing in sample LGX-171.

4. Discussion

In this work, we have analyzed two used FT synthesis catalyst samples by XRD and TEM. The XRD shows ε'-carbide to be present in LGX-175 but only χ-carbide to be present in LGX-171. It also appears that the ε'-carbide peak seen in XRD comes mainly from the distorted, large ε'-carbide particles seen in LGX-175 and the highly dispersed carbide particles appear to be missed completely by XRD. However, much of the surface area must come from the highly dispersed carbide phase that is seen in the active catalyst but totally absent from the deactivated catalyst. Hence, analysis of the active phase must consider this highly dispersed phase. Considering the particle size of this carbide phase, we feel that this phase has most probably been missed by bulk analytical methods such as XRD and Mössbauer (particularly at room temperature where it would be superparamagnetic). This highly dispersed phase is actually a mixture of ε'-carbide and χ-carbide (as determined from lattice fringe spacings) and it is difficult to quantify the ratio of these two kinds of carbide by TEM alone (which only looks at a small region of the sample). It would be necessary to perform Mössbauer spectroscopy of these samples at liquid He temperatures and correlate with the TEM images to get at the true distribution of these phases. Without this data, we can state that the high activity of catalyst LGX-175 must be related, at least in part, to the presence of the highly dispersed carbide phase. Furthermore, since ε'-carbide is the major bulk carbide phase seen in the active catalyst (by XRD), and it is completely absent in the deactivated catalyst, we feel comfortable in stating that the bulk ε'-carbide must be more active than the bulk χ-carbide.

In the above analysis, we have ignored the presence of the magnetite: the large single crystals as well as the highly dispersed magnetite. With H_2O and CO_2 being by-products of F-T synthesis (their ratio being determined by the extent of water gas shift), we can expect some magnetite to be present as a result of oxidation caused by these molecules. However, in a previous study (**4**) where the hot wax had been carefully removed under an inert blanket, we found neither the large magnetite crystals nor the dispersed phase. Other recent work (**6**) involving samples that were not exposed to air before analysis also confirms the absence of magnetite in the catalyst used for F-T synthesis. Hence, we suggest that the magnetite seen here may have two different origins. Since these two samples were both removed from the reactor without an inert protective environment while the slurry was still hot, we think part of the fine magnetite phase could arise due to oxidation of the iron carbide during slurry removal. In the case of LGX-175, we have also seen a magnetite layer on the surface of the iron carbide., analogous to a passivating layer of oxide seen in our previous work on air exposed samples (**7**). The large faceted magnetite single crystals are present in both the active catalyst (LGX-175) and inactive catalyst (LGX-171). They could not arise from an oxidation of the iron phase during slurry removal. We conclude that these magnetite particles represent unreduced magnetite from the catalyst precursor. Our previous work (**8**) has shown that magnetite that had a Swiss-cheese (porous) morphology could be readily reduced to iron carbide. The more perfect, faceted magnetite particles must be harder to reduce and hence they persist even after 3000 hours of reaction in syngas, a possibility we will investigate in our future work.

To summarize, we can conclude that the active F-T catalysts consists of a mixture of highly dispersed carbide phase and bulk ε'-carbide. In the deactivated catalyst, both ε'-carbide and the dispersed carbide were missing. This leaves open the question of how and why the catalyst in run LGX-171 deactivated, since it appeared to run at high conversion for over 2800 hours. Since, no samples are available of the catalyst in its active state, we can only speculate that a process upset may have caused transformation of the active dispersed carbide phase into a mixture of magnetite and χ-carbide. It is clear that in its deactivated form the catalyst contains plenty of χ-carbide, but show very low activity. Therefore, it appears that the ε'-carbide must represent the active phase in Fe F-T catalysts. Future work will attempt to correlate these observations with Mössbauer spectroscopy which should permit quantification of the phase composition of these working F-T catalysts. We conclude that XTEM provides valuable insight for understanding the microstructure of Fe catalysts used for F-T synthesis.

Financial support for this work from the US Department of Energy, FETC university coal research grant DE-FG-22-95PC95210 is gratefully acknowledged. We thank Dr. Burtron Davis for providing the catalyst samples whose analysis is reported here.

References

1. Xu, L., Bao, S., O'Brien, R.J., Raje, A. and Davis, B.H., ChemTech 47, 1998.
2. Jackson, N.B., Datye, A.K., Mansker, L.D., O'Brien, R.J. and Davis, B.H., in Catalyst Deactivation, (Bartholomew, C.H. and Fuentes, G.A. eds.), Stud. Surf. Catal., 111, 501 (1997).
3. Eliason, S.A. and Bartholomew, C.H., ibid, 517 (1997).
4. Mansker, L. D., Jin, Y., Bukur, D. B. and Datye, A. K., submitted to Appl. Catal., 1998.
5. O'Brien, R.J., Xu, L., Spicer, R.L. and Davis, B.H., Energy and Fuels, 10(4), 921 (1996).
6. Mahajan, D., et al., Energy and Fuels, in press
7. Shroff, M.D. and Datye, A.K., Catal. Lett., 37, 101 (1996)
8. Shroff, M.D., Kalakkad, D.S., Coulter, K.E., Kohler, S.D., Harrington, M.S., Jackson, N.B., Sault, A.G., and Datye, A.K., J. Catal., 156(2), 185 (1995).

NATURAL GAS CONVERSION V
Studies in Surface Science and Catalysis, Vol. 119
A. Parmaliana et al. (Editors)

Effect of silica on iron-based Fischer-Tropsch catalysts

K. Jothimurugesan,[a] James J. Spivey,[b] Santosh K. Gangwal,[b] and James G. Goodwin, Jr.[c]

[a]Department of Chemical Engineering, Hampton University, Hampton, VA 23668, USA

[b]Research Triangle Institute, Research Triangle Park, NC 27709, USA

[c]Department of Chemical and Petroleum Engineering, University of Pittsburgh, Pittsburgh, PA 15260, USA

ABSTRACT

The effect of silica addition via coprecipitation and as a binder to a doubly promoted Fischer-Tropsch (F-T) synthesis iron catalyst (100 Fe/5 Cu/4.2 K) was studied. The catalysts (nominally 50 µm particles) were prepared by coprecipitation, followed by binder addition and spray drying at 250 °C in a 1 m diameter, 2 m tall spray dryer. The binder silica content was varied from 0 to 20 wt%. A catalyst with 12 wt% binder silica was found to have the highest attrition resistance. F-T reaction studies over 100 hours in a fixed-bed reactor showed that this catalyst maintained around 95% CO conversion with a methane selectivity of less than 7 wt% and a C_5^+ selectivity of greater than 73 wt%. The effect of adding precipitated silica from 0 to 20 parts by weight (pbw) to this catalyst (containing 12 wt% binder silica) was also studied. Addition of precipitated silica was found to be detrimental to attrition resistance and resulted in increased methane and reduced wax formation.

1. INTRODUCTION

The conversion of natural gas into liquid fuels is an increasingly important means of meeting the energy needs of developing regions and transporting energy from remote gas fields to the existing refining and distribution infrastructure [1]. These plants rely on F-T synthesis to convert synthesis gas into liquid fuels. Although F-T plants based on natural gas-derived syngas (e.g., Shell's Bitulu plant) use, or plan to use, cobalt-based catalysts because of their generally higher yield of linear, saturated paraffins, iron-based catalysts have also been used for F-T synthesis of fuels from gas-derived syngas due to their lower cost and lower methane selectivity [2-4].

The development of attrition-resistant iron catalysts for commercial slurry bubble column reactors (SBCRs) is a critical challenge. A comparison of silica with other supports for iron-based F-T catalysts showed that silica-supported materials have generally higher activity and wax selectivity [5]. Dry [5] and Yang and Goodwin [6] report improved mechanical stability with silica supports, but no systematic variation of the silica content has been undertaken.

The literature on the effect of silica on these catalysts is limited, but increasing levels of precipitated silica have been shown to decrease the F-T activity of 100 Fe/5 Cu/4.2 K catalysts, although the stability increased [7]. Water-gas shift activity decreased with silica content in this study, and product selectivity also changed—less branched hydrocarbons and total olefins (but more internal olefins) were produced, consistent with the results of Egiebor and Cooper [8]. These effects were ascribed to a lower degree of reduction of the iron and a decrease in the surface basicity.

Studies on spray-dried iron F-T catalysts are also limited. Shroff et al. [9] describe in detail the effect of activation procedures on spray-dried Fe/Cu/K/Si catalysts, but they did not study the effect of silica content on attrition resistance or activity.

The objective of the work reported here is to develop 50 to 90 μm attrition-resistant iron-based F-T catalysts for SBCRs and to compare the effects of precipitated silica and binder silica.

2. EXPERIMENTAL

2.1. Catalyst preparation

Two types of silica-containing iron catalysts were prepared. The first series of catalysts contained binder silica but no precipitated silica and had a composition of 100 Fe/5 Cu/4.2 K (plus binder silica). These catalysts were prepared by coprecipitation using an aqueous solution containing $Fe(NO_3)_3 \cdot 9H_2O$ and $Cu(NO_3)_2 \cdot 2.5 H_2O$ in the desired Fe/Cu atomic ratio, which was precipitated by adding ammonium hydroxide. The resulting precipitate was then filtered and washed three times with deionized water. The potassium promoter was added as aqueous $KHCO_3$ solution to the undried, reslurried Fe/Cu coprecipitate. To this catalyst, five different levels of binder silica were added: 4, 8, 12, 16, and 20 wt%. These catalysts were then spray dried at 250 °C using a large bench-scale spray drier and calcined at 300 °C for 5 h in a muffle furnace. These catalysts are designated Fe-bSi(x), denoting that they contain x wt% binder silica. A standard Ruhrchemie catalyst (identified as Batch 52119) was obtained from the United States Department of Energy (DOE), Pittsburgh, PA, as a benchmark catalyst.

The second series of catalysts contained both precipitated and binder silica. Four such catalysts were prepared containing 5, 10, 15, and 20 pbw precipitated silica (yielding catalysts of the composition 100 Fe/5 Cu/4.2 K/y SiO_2, where y is 5, 10, 15, or 20). The precipitated silica was added as a 1.0 M solution of $Si(OC_2H_5)_4$ to the nitrate solution described above. To each of these catalysts, 12 wt% binder silica was added. These catalysts are designated Fe-pSi(y). These catalysts were then spray dried and calcined in the same way as those above.

2.2. Catalyst characterization

The BET surface areas of the catalysts were determined by N_2 physisorption using a Micromeritics Gemini 2360 system. The hydrogen reduction behavior was measured in a Micromeritics 2705 system. X-ray powder diffraction patterns were obtained using a Phillips PW1800 X-ray unit using a CuKα radiation. The attrition of the iron-based catalysts was tested using method ASTM-D-5757-95 in a 3-hole attrition tester. [The ASTM test is more severe than what would be experienced in an SBCR.] The SEM micrographs were taken using a Cambridge Stereoscan 100.

2.3. Activity testing

The F-T activity of the catalysts was measured in a fixed-bed microreactor system at 250 °C, 1.48 MPa, and 2.0 NL/g cat·h, at an H_2/CO ratio of 0.67. The catalyst charge was 1.3 g, and the catalyst was reduced in-situ in CO at 0.1 MPa, 280 °C, 35 cc/min for 16 h. After reduction, the reactor temperature was lowered to 50 °C. The system was then pressurized to 1.48 MPa, the carbon monoxide flow was cut off, and synthesis gas (H_2/CO=0.67) was introduced at a gas space velocity of 2 NL/(g cat·h). The reactor temperature was then increased gradually to 250 °C. The catalysts were then tested over a period of 100 h.

3. RESULTS

3.1. Characterization

Table 1 shows the BET surface areas of the fresh and reduced catalysts, the hydrogen uptake, and the attrition resistance of all the catalysts synthesized. The BET surface area decreased as the catalyst was reduced, but decreased less after reduction as the silica content (either precipitated or binder) increased, as expected. This trend is consistent with the results of Bukur et al. [7], as are the absolute values of the surface areas of comparable catalysts reduced in CO.

The hydrogen uptake generally decreased with silica content, though the effect of the precipitated silica is much less than the effect of the binder silica. Figure 1 shows the hydrogen uptake pattern of one of the catalysts. There were slight variations among the catalysts, with all showing peaks at 320 and 750 °C. The peak at 320 °C corresponds to the reduction of $Fe_2O_3 \rightarrow Fe_3O_4$, and the peak at 750 °C corresponds to the reduction of Fe_3O_4 to metallic iron [10]. The small shoulder peak at roughly 250 °C is due to the reduction of $CuO \rightarrow Cu$ [10]. The higher H_2 consumption by the Fe-bSi catalysts compared to the Fe-pSi catalysts indicates a greater extent of reduction for catalysts containing binder silica. SEM micrographs of the Fe-bSi(4) catalyst show that the catalyst particles are roughly spherical and approximately 50 to 90 μm in diameter (Figure 2).

The attrition resistance is perhaps the most interesting result. There is a minimum in the weight loss as a function of binder silica content, with the most attrition-resistant material being the Fe-bSi(12) catalyst, containing 12 wt% binder silica. For this reason, this material was used as the basis for preparing the Fe-

Table 1
Chemical and physical properties of iron-silica catalysts

Catalyst designation	Binder silica (wt%)	Precipitated silica (pbw)	BET surface area (m^2/g) Fresh	BET surface area (m^2/g) Reduced[a]	H_2 TPR (mmol H_2/g cat)	Attrition loss[b] (wt%) 1 h	Attrition loss[b] (wt%) 5 h
Fe-bSi(4)	4	-0-	80.3	35.6	24.3	24.4	32.6
Fe-bSi(8)	8	-0-	95.7	50.8	23.0	25.7	35.4
Fe-bSi(12)	12	-0-	121	68.7	20.6	12.8	22.7
Fe-bSi(16)	16	-0-	151	103	19.0	22.0	30.1
Fe-bSi(20)	20	-0-	172	98.9	18.4	34.9	35.0
Fe-pSi(5)	12	5	163	116	18.8	24.2	37.3
Fe-pSi(10)	12	10	168	144	17.9	31.0	39.6
Fe-pSi(15)	12	15	189	163	17.7	42.1	c
Fe-pSi(20)	12	20	218	181	17.6	39.1	c

[a] Reduced in CO, 27 cc/g cat·min, 0.1 MPa, 280 °C, 16 h.
[b] Measured using ASTM-D-5757-95. The higher the weight loss, the poorer the attrition resistance. For comparison, a commercial FCC catalyst was tested; wt% loss was <1% (1 h) and 2% (5 h).
[c] Attrition was severe enough to plug the tester; wt% loss was >40 wt%.

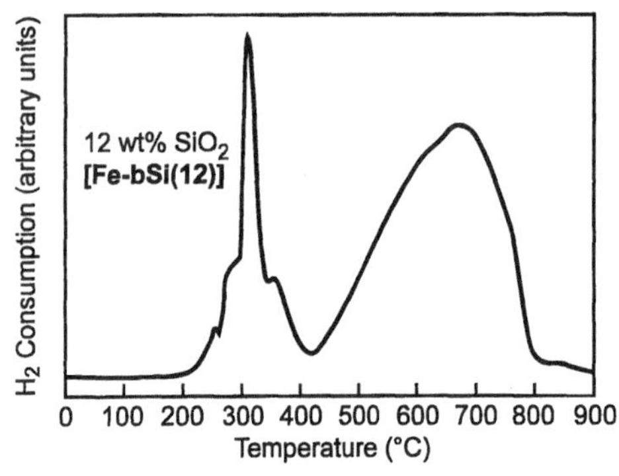

Figure 1. Hydrogen TPR of catalyst containing 12 wt% binder silica but no precipitated silica.

pSi(y) series of catalysts with various levels of precipitated silica. The catalysts containing precipitated silica were less attrition-resistant than those containing only binder silica. None of the catalysts, as expected, were as attrition-resistant as a commercial FCC material, however.

XRD of all the fresh catalysts showed only Fe_2O_3, while all the reduced catalysts showed only the $\chi Fe_{2.5}C$ and Fe_3O_4 phases. This result is similar to that of Bukur et al. [10], who also observed $\chi Fe_{2.5}C$ after a similar reduction procedure. Shroff et al. [9] also show the formation of an unspecified carbide phase after CO activation at 270 °C, but their short (2 h) activation left a substantial proportion of unreduced Fe_3O_4.

3.2. Fischer-Tropsch activity tests

Table 2 shows the CO conversion and hydrocarbon selectivity for the various catalysts, along with data obtained on a Ruhrchemie catalyst for comparison. [Data

Figure 2. SEM micrograph of reduced Fe-bSi(4), containing 4 wt% binder silica.

taken from 0 to 90 h showed little variation in CO conversion after an initial induction period, during which the CO conversion reached a steady value. Hydrocarbon selectivities were not measured during this period. There was no significant change with time in the CO conversions or hydrocarbon selectivities reported in Table 2 over the 100 h test duration for either the Fe-bSi or Fe-pSi catalysts.] All catalysts tested were more active than the Ruhrchemie catalyst. Except for the Fe-pSi(16) and Fe-pSi(20) catalysts, the CO conversion was essentially the same, 94%. However, the selectivity varied with the silica content. There was a beneficial effect of binder silica up to 8 to 12 wt% on selectivity (reduced methane, nearly constant C_{12}^{+}). However, as the binder silica content increased above 12 wt%, the C_1, C_2-C_4, and C_5-C_{11} selectivities

Table 2
Catalyst activity and selectivity

Catalyst	Ruhrchemie	Catalysts with binder silica Fe-bSi (4)	(8)	(12)	(16)	(20)	Catalysts with precipitated silica[a] Fe-pSi (4)	(8)	(12)	(16)
Time-on-stream (h)	42.7	52	42	65	90	67	68	66	66	66
CO conversion (%)[b]	86	94.3	94.1	94.3	95.5	94.5	95.5	94.4	90.1	88.2
Hydrocarbon selectivity (wt%)										
C_1	8.3	7.4	6.8	6.8	9.9	9.6	8.8	10.2	10.2	9.5
C_2-C_4	21.3	18.1	17.6	19.6	25.0	23.5	23.2	23.5	22.4	20.1
C_5-C_{11}	14.3	12.7	13.0	12.8	17.3	17.6	22.0	26.5	30.5	32.8
C_{12}^{+}	56.1	61.8	62.5	60.8	47.8	49.3	46.0	39.8	36.9	37.7

[a] These catalysts also contain 12 wt% binder silica.
[b] Measured at 250 °C, 1.48 MPa, 2 NL/g cat·h, $H_2/CO=0.67$.

increased, while the C_{12}^+ selectivity decreased. As the precipitated silica content increased (at 12 wt% binder silica), the selectivities of all the C_1-C_{11} products increased. However, the C_5-C_{11} selectivity for the catalysts containing precipitated silica was generally higher than the selectivity for those catalysts containing only binder silica. The chain growth parameter (α, from C_{30}-C_{40} yields) was around 0.9 for all catalysts. The relatively low methane selectivity of the most attrition-resistant catalyst, Fe-bSi(12), at 94% CO conversion and the high selectivity of C_{12}^+ suggest the addition of a binder silica that not only improves the attrition resistance but also results in a catalyst with higher F-T activity and C_5^+ selectivity than the Ruhrchemie material.

4. CONCLUSION

The addition of binder silica to a precipitated 100 Fe/5 Cu/4.2 K F-T catalyst, followed by spray drying, increases the attrition resistance significantly. Within the range of catalysts tested here, the optimum binder silica content is approximately 12 wt%. The F-T activity of this catalyst is higher than a Ruhrchemie catalyst at 250 °C, 1.48 MPa. The addition of precipitated silica to a catalyst containing 12 wt% binder silica decreases its attrition resistance and increases its methane selectivity.

ACKNOWLEDGMENT

The authors are grateful for U.S. DOE/FETC support of this research under grant DE-FG22-96PC96217 and for the technical guidance of Project Officer Dr. Richard E. Tischer.

REFERENCES

1. Eilers et al., *Catal. Lett.*, 7, 253-270, 1990; *Natural Gas Week,* 12 (42), 8, 1996; *Petr. Intell. Week.,* 35 (42), 1, 1996.
2. Schulz et al., *Nat. Gas. Conv. II, Elsevier, Surf. Sci. Ser.,* v. 81, 455, 1994.
3. Ravavarapu et al., op. cit., 413.
4. Hedden et al., Erdol ErErdgas Kohle, 7/8, 318-321, 1994.
5. Dry, M.E., in Catalysis Science and Technology, J.R. Anderson and M. Boudart, eds., v. 1., Springer-Verlag, New York, 159-256, 1981.
6. Yang, C.H., and J.G. Goodwin, *Can. J. Chem. Eng.*, **61**, 213-217, 1983.
7. Bukur, D.B., X. Lang, D. Mukesh, W.H. Zimmerman, M.P. Rosynek, and C. Li, *Ind. Eng. Chem. Res.*, **29**, 1588-1599, 1990.
8. Egiebor, N.O., and W.C. Cooper, *Can. J. Chem. Eng.*, **63**, 81-85, 1985.
9. Shroff, M.D., D.S. Kalakkad, K.E. Coulter, S.D. Kohler, M.S. Harrington, N.R. Jackson, A.G. Sault, and A.K. Datye, *J. Catal.*, **156**, 185-207, 1995.
10. Bukur, D.B., K. Okabe, M.P. Rosynek, C. Li, D. Way, K.R.P.M. Rao, and G.P. Huffman, *J. Catal.*, **155**, 353-365, 1995.

NATURAL GAS CONVERSION V
Studies in Surface Science and Catalysis, Vol. 119
A. Parmaliana et al. (Editors)

CO_2 Hydrogenation for the Production of Light Alkenes over K-Fe-Mn/silicalite-2 catalyst

Xu Longya*, Wang Qingxia, Liang Dongbai, Wang Xing, Lin Liwu, Cui Wei and Xu Yide

State Key Laboratory of Catalysis,
Dalian Institute of Chemical Physics, Chinese Academy of Sciences,
P. O. Box 110, Dalian 116023, P. R. China

MnO and K_2O are two kinds of promoters to Fe/Silicalite-2 (denoted as Fe/Si-2) catalyst for light olefin formation from CO_2 hydrogenation. It has been revealed that CO_2 adsorption capacity was enhanced by the addition of MnO, and that K_2O could further increase both the adsorption capacity and bonding strength of the Fe-Mn/Si-2 catalyst to CO_2 and CO species. Based on a two step reaction mechanism in which a reversible water gas shift reaction and a Fischer-Tropsch reaction are involved, the promotional effects of these two kinds of promoters for the formation of C_2H_4 and C_3H_6 can be explained.

1. INTRODUCTION

Study on CO_2 hydrogenation for the formation of hydrocarbons, especially light alkenes, has attracted the interest of many researchers[1-3]. This reaction can provide an alternative route to produce basic chemical feed stocks such as ethylene and propylene from non-petroleum sources. Promising progress has been made in selective CO_2 hydrogenation for the production of light olefins[2-4]. In our previous papers[1,2], development of an iron catalyst supported on a silicalite-2 type zeolite (denoted as Fe/Si-2) and promoted by alkali and MnO additives for the production of light olefins by CO_2 hydrogenation has been reported. It is well-known that MnO and K_2O promoters possess novel effects for improving the catalytic activity and olefin selectivity of iron catalyst for CO hydrogenation[5]. In our previous research publications[4,6,7], the promotions of MnO and K_2O to Fe/Si-2 catalyst has been studied particularly for the selective production from CO hydrogenation. In additional, the Chemical-Physics state of Fe and Mn, including the reduction degree, dispersion and value-state have been investigated clearly related to the performance of K-Fe-Mn/Si-2 catalyst for CO hydrogenation. In the present paper, CO_2-TPD, CO_2/H_2-TPSR and CO/H_2-TPSR techniques were employed to study the promotion effects of MnO and K_2O on the behavior of the Fe/Si-2 catalyst for CO_2 hydrogenation.

*To whom correspondence should be addressed

2. EXPERIMENTAL

2.1. Preparation of the catalysts

Supported K-Fe-Mn/Si-2 catalysts were prepared by impregnating a solution with a given amount of $Fe(NO_3)_3$ and $KMnO_4$ or KNO_3 onto a Silicalite-2 zeolite, with addition of 60% of Al_2O_3 as a binder, under vacuum condition. After drying at 400 K for 8 h and calcined at 800 K for 10 h, the catalysts were cooled to room temperature, crushed and screened. Catalyst samples with a size of Φ1.2 mm x 2.5 mm were used for the CO_2 hydrogenation reaction tests.

2.2 CO_2 hydrogenation reactions

The CO_2 hydrogenation performance of catalyst, with charge of 6 ml was evaluated in a small fixed-bed flow reactor made of stainless steel. After being reduced in situ with flowing H_2 at 700 K, 0.5 MPa and GHSV 2000 h-1 for several hours, the catalyst proceeded CO_2 hydrogenation under the following reaction conditions: reaction temperature of 620 K, pressure of 2.0 MPa, GHSV of 1000 h^{-1}, and CO_2/H_2=1/2-1/4(mole ratio). The products of CO_2 hydrogenation were analyzed on line by gas chromatography with a TCD detector.

2.3 Temperature programmed desorption and surface reaction

CO_2-TPD and CO_2/H_2-TPSR were carried out in a flow reactor system equipped with an ANELVA TE-150 multichannel mass spectrometer and an IBM PC computer. Each desorption species was monitored and recorded by one channel of the TE-150 mass spectrometer at a given sensitivity which remained the same for all experiments.

CO_2-TPD: 0.3 g of catalyst, after reduction in situ under H_2 at 770 K, was switched to a He stream containing CO_2 for 10 minutes at room temperature. CO_2-TPD was performed from room temperature to 900 K at a heating rate of 16 K/min.

CO_2/H_2-TPSR: 0.3g of catalyst, after reduction in situ, was exposed to CO_2 for 10 minutes at room temperature. The gas flow was then switched to H_2. CO_2/H_2-TPSR was carried out at a heating rate of 16 K/min.

3 RESULTS and DISCUSSION

3.1 CO_2 hydrogenation over the K-Fe-Mn/Si-2 catalysts

From Tables 1 and 2, it can be seen that both the CO_2 conversion and the light olefin selectivity increased with the additions of MnO, and K_2O further, onto the Fe/Si-2 catalyst. Meanwhile the formations of CH_4 and CO decreased with MnO adding, declined much more with K_2O introduced additionally. The same trends have been observed for CO hydrogenation over the same Fe/Si-2 catalyst[2,4]. It can be visualized according to the two-step mechanism that the promotion effects of the MnO and K_2O to the Fe/Si-2 catalyst should be essentially the same for both the CO hydrogenation and the CO_2 hydrogenation. Meanwhile, it diminishes the formation of CH_4 and CO during CO_2 hydrogenation. The results of our study is in accordance with this visualization, thus provided evidence for the validity of the two-step mechanism of CO_2 hydrogenation.

Table 1
Effects of MnO and K_2O on Fe/Si-2 catalyst for CO_2 hydrogenation*.

Catalyst	Catalyst composition (wt.%)			CO_2 conv. (%)	Distribution of products (mole%)	
	Fe	Mn	K		CO	HC
9Fe/Si-2	9.0	0.0	0.0	40.1	28.3	71.7
9Fe2Mn/Si-2	9.0	2.0	0.0	38.9	27.1	72.9
9Fe9Mn/Si-2	9.0	9.0	0.0	44.1	24.2	75.8
3K9Fe9Mn/Si-2	9.0	9.0	3.0	47.0	21.6	78.4
8K9Fe9Mn/Si-2	9.0	9.0	8.0	52.3	18.4	81.6

* Reaction conditions: 620 K, 2.0 MPa, 1200 h^{-1} and CO_2/H_2=1/3.

Table 2
Effects of MnO and K_2O on olefin selectivity of CO_2/H_2 reaction over Fe/Si-2 catalyst*.

Catalyst	Distribution of hydrocarbon products(wt%)							
	CH_4	C_2H_4	C_2H_6	C_3H_6	C_3H_8	C_4H_8	C_4H_{10}	$C_2^=$-$C_4^=$
9Fe/Si-2	39.2	16.5	14.6	13.9	6.9	6.3	2.6	36.7
9Fe2Mn/Si-2	38.4	18.9	10.1	17.8	3.9	8.9	2.0	45.6
9Fe9Mn/Si-2	37.8	22.3	4.3	19.5	1.5	13.0	1.6	54.8
3K9Fe9Mn/Si-2	29.8	24.6	3.5	22.4	1.7	16.3	1.7	63.3
8K9Fe9Mn/Si-2	25.0	25.8	3.5	24.0	1.9	18.1	1.7	67.9

* Reaction conditions: 620 K, 2.0 MPa, 1200 h^{-1} and CO_2/H_2=1/3.

3.2 Effects of MnO and K_2O promoters on CO_2-TPD of the Fe/Si-2 catalyst

In our previous studies, it has been found that MnO and K_2O promoters were essential for the selective production of light olefins from CO hydrogenation over the Fe/Si-2 catalyst. This was envisaged as mainly resulted from an increase in the adsorption ability and adsorption capacity of the Fe/Si-2 catalyst for CO with the addition of the MnO and K_2O promoters[8]. It would be of interest to probe also the effects of the addition of MnO and K_2O on the adsorption behaviors of the Fe/Si-2 catalyst with respect to CO_2. Figure 1 shows the CO_2-TPD profiles from Fe/Si-2 catalysts with the addition of MnO and MnO plus K_2O. It can be seen that two groups of desorption peaks of CO_2, CO and O_2 appeared on these CO_2-TPD profiles, one at lower temperatures corresponding to weak CO_2 adsorption, while the other at higher temperatures due to strong CO_2 adsorption. It is interesting to note that all of the high temperature peaks of CO_2, CO and O_2 shifted to higher temperatures with the addition of MnO or MnO plus K_2O. Furthermore, the shifts of the high temperature peaks of the CO species were more pronounced than that of the CO_2 or O_2 species. Especially, the areas of the higher temperature peaks for CO and O_2 species increased greatly with the addition of the K_2O promoter. There is evidence[9] that the C_{ad} species on catalyst surface formed from the strong CO adsorption and its subsequent hydrogenation are responsible for the formation of methane and ethylene, while more favorable for the formation of ethylene with high Cad concentration on the catalyst surface. So, the functions of MnO and K_2O additive on the Fe/Si-2 catalyst are to enhance the capacity for and the strength of CO_2 or CO adsorption and to increase the reactivity of CO_2 hydrogenation.

Thus, we can conclude that the addition of both MnO and K_2O onto the Fe/Si-2 catalyst can result in a remarkable increase both in the adsorption capacity as well as in the adsorption ability of the Fe/Si-2 catalyst for both CO_2 and CO, and these are very important factors for the formation of light olefins from CO_2 or CO hydrogenation.

3.3 Effects of MnO and K_2O addition on CO_2/H_2-TPSR over the Fe/Si-2 catalyst

CO_2/H_2-TPSR studies were carried out to investigate the effects of the MnO and K_2O promoters on the surface reaction behaviors in CO_2 hydrogenation over the Fe/Si-2 catalyst. Figure 2 shows the CO_2/H_2-TPSR profiles over Fe/Si-2 catalysts with the addition of MnO and K_2O. Only CO, O_2 and H_2O were produced during the CO_2/H_2-TPSR, and no hydrocarbon formation was detected. These products can be considered as to be formed from the decomposition of CO_2:

$$2CO_2 \text{------>} 2CO + O_2$$

as well as from the reversible water gas shift reaction. Evidently, no CO hydrogenation occurred in the surface reactions.

Some interesting phenomena have been observed from the CO_2/H_2-TPSR on the addition of MnO and K_2O over the Fe/Si-2 catalyst. First, the CO desorption peak shifted to a higher temperature when MnO was added. With the further addition of K_2O an additional shift of CO desorption peak was observed. Second, H_2O start to form at 400 K, while CO_2 and O_2 species vanished at about 600 K. Third, CO_2 and O_2 disappeared quickly with the addition of MnO, but more quickly with the further adding of K_2O. These results again support the conclusion that the addition of MnO and K_2O enhanced the CO and CO_2 adsorption ability and capacity of the Fe/Si-2 catalyst, thus promoted the conversion of CO_2 in a H_2 gas stream, and caused the reversible water-gas shift reaction to move to the right hand direction, resulting in the formation of CO and H_2O. Since stronger CO adsorption is favorable for the selectivity of light olefins formation as well as for the prohibiting of CH_4 formation during CO hydrogenation[9,10], therefore, the addition of MnO and K_2O are favorable for the enhancement of the activity and selectivity of the Fe/Si-2 catalyst.

3.4 Effects of MnO and K_2O on the CO hydrogenation over the Fe/Si-2 catalyst

Table 3
Effects of MnO and K_2O on olefin selectivity of CO/H_2 reaction over Fe/Si-2 catalyst*.

Catalyst	CO	Selectivity of hydrocarbon (wt%)		
	conv.(%)	CH_4	$C_2^=$-$C_4^=$	C_2^o-C_4^o
9Fe/Si-2	76.3	48.9	14.9	36.2
9Fe2Mn/Si-2	77.4	48.1	29.4	22.5
9Fe9Mn/Si-2	77.8	45.0	39.0	16.0
3K9Fe9Mn/Si-2	82.3	38.6	47.8	13.6
8K9Fe9Mn/Si-2	90.4	22.4	69.6	8.0

* Reaction conditions: 620 K, 2.0 MPa, 1000 h^{-1} and CO/H_2=1/2.

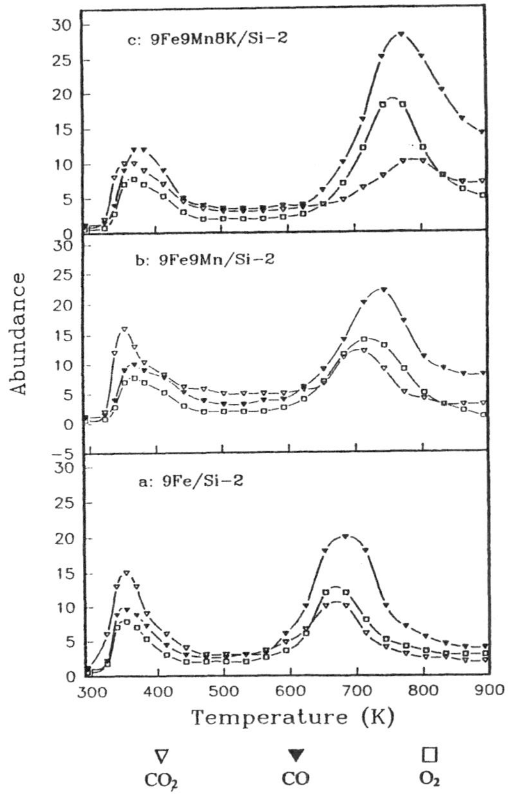

Figure 1. Profiles of CO_2-TPD over the K-Fe-Mn/Si-2 catalyst

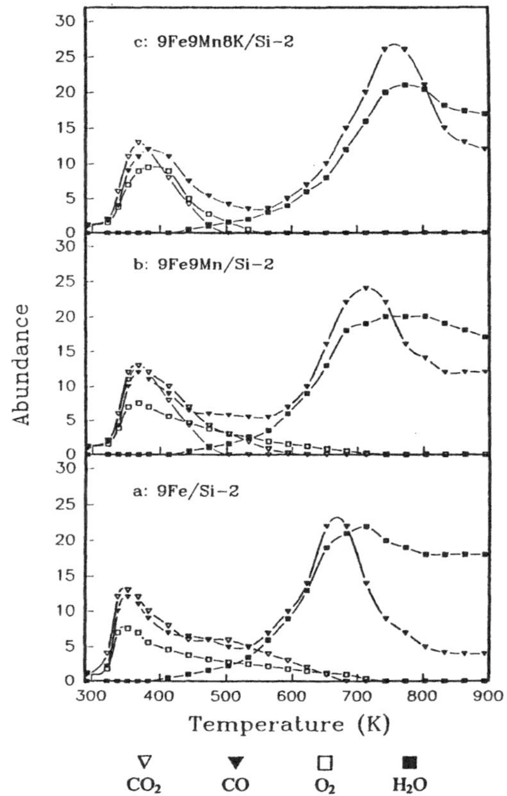

Figure 2. Profiles of CO_2/H_2-TPSR over the K-Fe-Mn/Si-2 catalyst

Table 3 lists the catalytic performances of K-Fe-Mn/Si-2 catalysts for the production of light olefins from CO hydrogenation. Over Fe/Si-2 catalyst, the selectivity to $C_2^{=}$-$C_4^{=}$ alkene is only 14.9% at the CO conversion of 76.3%. The addition of MnO promoter onto the Fe/Si-2 catalyst results in a remarkable improvement in the selectivity to light alkenes, The selectivity to light alkenes is about 39% over 9Fe9Mn/Si-2 catalyst. Meanwhile, the formation of CH_4 and the CO conversion are hardly affected by incorporating MnO additive. The addition of K_2O as a second promoter further enhances the activity and selectivity of Fe-Mn/Si-2 catalysts as shown in Table 3. With increasing the K_2O amount, the CO conversion increases from 77.8 to 90.4%, while the formation of CH_4 decreases from 45 to 22.4%. Selectivity for $C_2^{=}$-$C_4^{=}$ alkene as high as 69.6% at the CO conversion of 90.4% can reached over the 8K9Fe9Mn/Si-2 catalyst. Therefore, both MnO and K_2O promoters are crucial and of significance for enhancing the reactivity and selectivity of the Fe/Si-2 catalyst for the production of light olefins from CO hydrogenation and from CO_2 hydrogenation.

Generally, according to the following two-step reaction mechanism involving a reversible water gas shift reaction and a Fischer-Tropsch reaction:

$$CO_2 + H_2 \rightleftharpoons CO + H_2O \quad (1)$$

$$CO + (m/2n + 1)H_2 \rightleftharpoons 1/nC_nH_m + H_2O \quad (2)$$

The MnO and K_2O promoters may have the same promoter effects for light olefin formation from CO_2 hydrogenation as from CO hydrogenation over Fe-Mn/Si-2 catalyst, which can prove, in turn, the two-step reaction mechanism.

REFERENCES

[1] Z.J. Lu, P.Z. Lin, D.B. Liang and L.W. Lin, Chem. Eng. Nat. Gas, 18(1993), 23.
[2] L.Y. Xu, Q.X. Wang, D.B. Liang and Z.J. Tian, J. Nat. Gas Chem. 4(1995), 167.
[3] M.S.Iyengar, 8th Annual Pittsburgh Coal Conf., 1991, P508.
[4] L.Y. Xu, Q.X. Wang, Y.D. Xu and J.S. Huang, Catal. Lett., 24(1994), 177.
[5] H.Kolbel, Deutsche Offen., 507, 647(1976).
[6] L.Y. Xu, Q.X. Wang, Y.D. Xu and J.S. Huang, Catal. Lett., 31(1995), 253.
[7] L.Y. Xu, G.Q.Chen and G.Y. Cai, J. Catal. (in Chinese), 11(1990), 442.
[8] L.Y. Xu, Q.X. Wang, Y.D. Xu and J.S. Huang, Studies in Surface Science and Catalysis, Natueal Gas Conversion II, Netherlands 81(1994), 473.
[9] 10, H. Ahlafi, D. Bianchi and C.O. Bennett, Appl. Catal., 66(1990),99.
[10] L.Y. Xu, Q.X. Wang, Y.D. Xu and J.S. Huang, Catal. Lett., 31(1995), 253.

Figure 1 Profiles of CO2-TPD on the Fe/Si-2 catalyst with MnO and K2O promoters.

Figure 2 Profiles of CO2/H2-TPSR on the Fe/Si-2 catalyst with MnO and K2O promoters.

NATURAL GAS CONVERSION V
Studies in Surface Science and Catalysis, Vol. 119
A. Parmaliana et al. (Editors)

Steady-State production of olefins and aromatics in high yields from methane using an integrated recycle reaction system

J. H. Lunsford, E. M. Cordi, P. Qiu and M. P. Rosynek
Department of Chemistry, Texas A&M University
College Station, Texas 77843, U.S.A.

ABSTRACT

Methane can be converted to olefins (principally ethylene) in high yields using an integrated recycle system that includes a catalyst for the oxidative coupling reaction. A membrane contactor has been used to facilitate the complexation of olefins with aqueous Ag^+ ions. The olefins are released from the complexes at elevated temperatures and are recovered in nearly pure form. Alternatively, the olefins may be converted to aromatics by replacing the membrane contactor with a Ga/H-ZSM-5 zeolite catalyst. With both variations of the system, product yields greater than 70% have been obtained during continuous operation.

INTRODUCTION

Although the oxidative coupling of methane (OCM) is a promising method for the direct conversion of natural gas to more useful chemicals and fuels, the ethylene yields are generally <10% with a single-pass flow reactor operating under steady-state conditions. After more than a decade of research on OCM catalysts and reactor designs, the empirical evidence suggests that this yield may be a realistic upper limit. The ethylene yield, however, may be significantly improved by using a recycle system in which the desired product is continuously removed.

Vayenas and co-workers [1] utilized a closed-loop recirculation system with continuous removal of C_{2+} olefins. The olefins were adsorbed in a 5A molecular sieve at 30 °C and subsequently desorbed in an inert gas steam by heating the sieve to 400 °C. Using this two-step technique, they obtained an ethylene yield of 85%. Subsequently, the system was modified so that CH_4 and O_2 could be continuously added, and an ethylene yield of 50% was attained at a CH_4 conversion of 76% [2]. Hall and Meyers [3] employed a cryogenic trap to remove both C_2H_4 and C_2H_6 and, in a separate experiment, a polyvinyl alcohol membrane, impregnated with $AgNO_3$ to remove C_2H_4. Oxygen, but not methane, was continuously added to their system. With the cryogenic separation method, the total C_2 yield was 59% and with the membrane separation technique, the C_2H_4 yield was 47%. Ethylene was produced at an overall rate of 0.2 ml/min, More recently, Mashocki [4] has described a recycle-type reactor that contained an adsorbent for continuous removal of olefin products, which were subsequently recovered by thermal treatment. Using this system, he obtained a C_2H_4 yield of 60%.

Selective removal of olefins from a recycle stream clearly is an important factor in the development of a process which increases olefin yield above that of single-pass operation [5]. In this paper we describe results obtained by the continuous separation and removal of the desired products by two methods: (1) complexation of olefins with aqueous Ag^+ ions in a membrane contactor, and (2) selective conversion of olefins to condensible aromatic hydrocarbons [6,7]. In both cases, it has been possible to achieve product yields >70% with continuous addition of CH_4 and O_2. The olefins or aromatics are delivered in nearly pure form at 1 atm.

2. EXPERIMENTAL

In both recycle systems, the OCM catalyst was Mn(2 wt%)/Na_2WO_4(5 wt%)/SiO_2. This catalyst, which has been described separately, was chosen because of its superior performance in the single-pass mode (80% C_{2+} selectivity at 20% CH_4 conversion) [8,9]. Approximately 0.5 g of the catalyst was placed in a 6 mm i.d. alumina tube and heated to 800 °C. During the exothermic OCM reaction, a small volume at the entrance of the catalyst bed is at a considerably higher temperature [10].

As shown in Figures 1 and 2, the catalytic reactor is followed by an open 1.5 cm i.d. fused-quartz tube that was heated to 800 °C. The purpose of this reactor was to effect the further dehydrogenation of C_2H_6 to C_2H_4 and H_2. Trace amounts of C_2H_2 also were formed. The CH_4/O_2 ratio was such that all of the O_2 was consumed in the OCM reactor. It was observed that the CH_4 inhibits the thermal dehydrogenation of C_2H_6; therefore, the advantage of the second reactor was not as large as one might expect. A third reactor, containing 0.5% Pd/α-Al_2O_3 at 80 °C, was used to remove the small amount of C_2H_2 that was produced during C_2H_6 dehydrogenation. This is important since acetylene reacts with silver to form an irreversible silver acetylide complex that is explosive in the dry form. Carbon dioxide and water vapor were removed with KOH and cold traps, respectively.

The membrane contactor (Figure 1), manufactured by Hoechst Celanese, consisted of Celguard X-30 microporous polypropylene hollow fibers which were arranged in a shell-and-tube configuration. The gases passed through the tube side and a concentrated solution containing $AgNO_3$ (4 M) and $NaNO_3$ (6.2 M) passed through the shell side at *ca.* 30 °C. The presence of $NaNO_3$ decreased the solubility of CH_4 in the solution. The olefins formed a complex with the Ag^+ ions and were transported across the membrane to the aqueous liquid phase. The solution was recirculated through a regeneration column where it was boiled to release the complexed olefins into the gas phase.

To achieve the conversion of olefins to aromatics, the membrane contactor was replaced by a fourth reactor that contained 1.0 g of 5 wt % Ga/H-ZSM-5 at 520 °C. An in-line trap, located downstream from this reactor and cooled to -78 °C, continuously removed condensible products generated by the Ga/H-ZSM-5 catalyst.

For both systems, the O_2 flow rate was fixed at a desired level, and the CH_4 was adjusted to achieve a constant pressure that was slightly greater than 1 atm. After passing through the

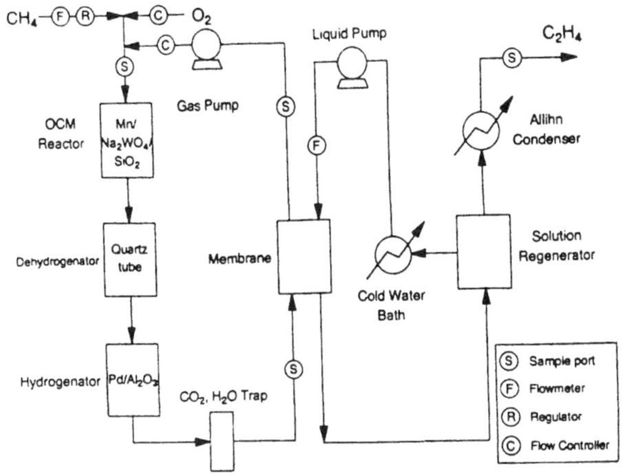

Figure 1. Schematic diagram of integrated reaction/separation system for conversion of methane to ethylene and propene (Ref. 6).

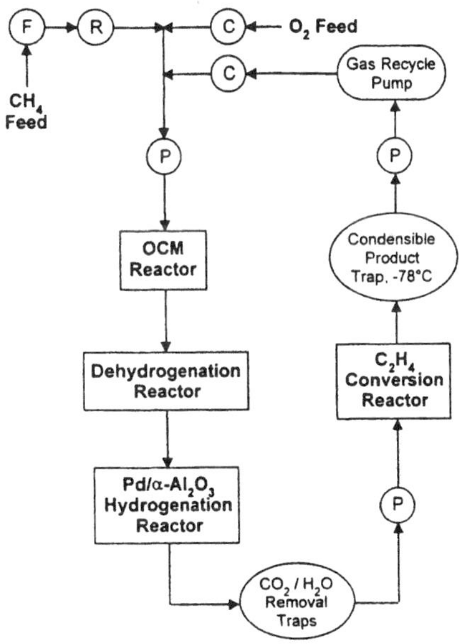

Figure 2. Schematic diagram of integrated recycle system for conversion of methane to aromatics. C=mass flow controller; F=flowmeter; P=gas sampling port; R=pressure regulator (Ref. 7).

series of reactors and traps, the unreacted CH_4, as well as C_2H_6 and a small amount of H_2, was recirculated over the OCM catalyst.

3. RESULTS AND DISCUSSION

3.1. Olefin yields with the membrane contactor

Ion-facilitated transport membranes have been widely used, at least on a laboratory scale, to remove olefins from gas streams [11]. In this study, C_2H_4 separation efficiencies of 94% have been achieved at a flow rate of 120 ml/min. Thus, only a very small amount of the C_2H_4 is lost by recirculation over the catalyst.

The results obtained with the continuous removal of C_2H_4 in the membrane contactor are shown in Figure 3. With increasing O_2 flow rate, the olefin yield decreased from 75% to 63% in a linear manner. Concomitantly, the rate of olefin production increased from 2.6 to 4.6 ml/min. Typically, the final product stream contained 79% C_2H_4, 6% C_3H_6, 13% CH_4 and 2% CO_x at a pressure of *ca.* 760 Torr. The CH_4 conversion, based on the amount of CH_4 added to the system, was *ca.* 97%; the remaining portion was transported through the membrane and dissolved in the aqueous solution.

In evaluating a recycle system the recycle ratio is an important factor. As shown in Figure 4, the olefin yield and productivity reached a maximum at a recycle flow rate of about 125 ml/min. The maximum occurs because the olefin separation efficiency begins to decrease at larger flow rates. Olefins recycled to the OCM reactor become oxidized to CO_x, thereby decreasing olefin yield. At the maximum olefin yield the recycle ratio was *ca.* 12.

Whereas the olefin yields obtained in the recycle system with the contact membrane are impressive, several limitations become apparent. The most significant of these is the low throughput, which is often the case with membrane separations. The recycle ratio also is rather large, although one could decrease this value with only a modest sacrifice in yield. Operation of the various components in the system at different temperatures may be a problem; however, it should be noted that the required heat can be derived from the OCM reaction. In addition, the dehydrogenator and the subsequent hydrogenator (Figure 1) can be omitted from the system, with a loss of only about 5% in olefin yield.

3.2. Aromatic yields with the Ga/H-ZSM-5 catalyst

In an effort to improve product throughput and perhaps to reduce the recycle ratio, while maintaining high yields, the membrane contactor and the liquid recycle loop was replaced by a reactor containing a Ga/H-ZSM-5 zeolite catalyst. Over such a catalyst, olefins (C_2H_4) are efficiently converted to aromatic products, as shown in Table 1. In these experiments a simulated mixture of 3% C_2H_4 in CH_4 was used as the reagent. The catalyst slowly deactivates due to coke formation, but it can be reactivated by burning off the coke.

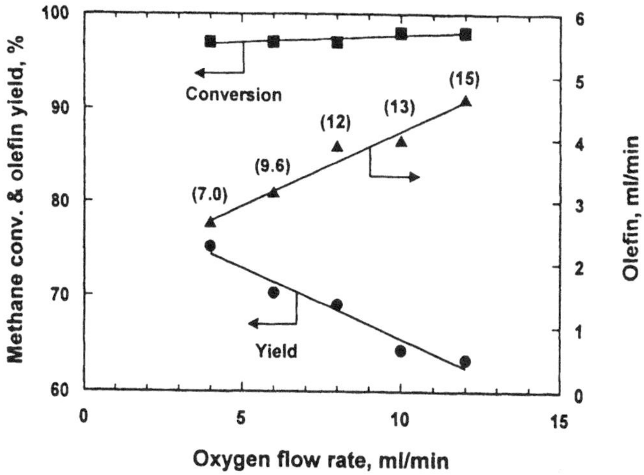

Figure 3. Effect of O_2 flow rate on olefin rate, olefin yield, and CH_4 conversion at a recycle flow rate of 120 ml/min; liquid recycle=49 ml/min. Methane inlet flow rates are shown in parentheses (Ref. 6).

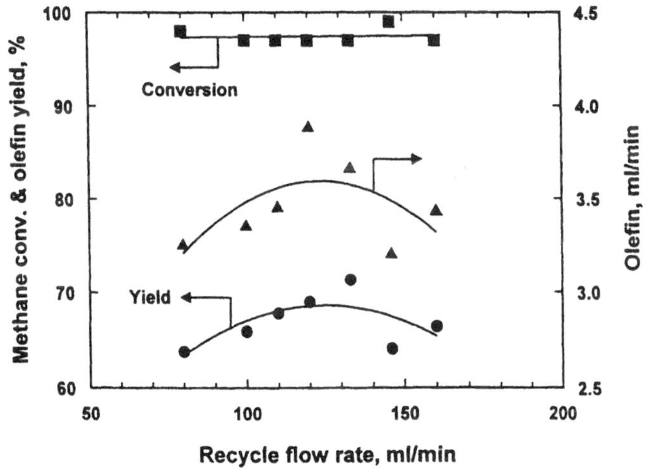

Figure 4. Effect of recycle flow rate on olefin rate, olefin yield, and CH_4 conversion at an O_2 flow rate of 8 ml/min; liquid recycle=49 ml/min (Ref. 6).

Table 1
Ethylene conversion and product selectivity over Ga-containing H-ZSM-5[a]

Ga (wt%)	C_2H_4 conv. %	Selectivity (%) non-aromatics	aromatics[b] Bz	Tol	C_8	Aromatics select. %	Aromatics yield %
0.0	40	80	6	9	0	15	6
0.1	65	28	32	28	10	69	45
0.5	91	9	38	31	7	79	72
2.0	92	7	35	31	7	82	75
5.0	93	6	34	31	8	80	74
10.0	92	12	34	31	7	80	73

[a] Si/Al= 25, 1.0 g of catalyst; reaction temp.=520°C; CH_4=100 ml/min, C_2H_4=3 ml/min; data taken after 70 min. on stream.
[b] Bz=benzene, Tol=toluene; small amounts of other aromatics also were present.

Subsequent work has shown that this is a bifunctional catalyst [12]. Initial activation of ethylene occurs at Brönsted acid sites inside the zeolite channels, leading to oligomerization reactions that generate C_4, C_6 and possibly higher hydrocarbons. The initial oligomerization products undergo a complex sequence of acid-catalyzed isomerization, cracking, re-oligomerization and cyclization reactions that result in an array of C_1-C_6 non-aromatic products. The role of the gallium is to catalyze dehydrogenation of the various acid catalyzed intermediates and to promote cyclization reactions.

With Ga/H-ZSM-5 in the recycle system, the results depicted in Figure 5 were obtained. The aromatic product yield decreased linearly from 80% at an O_2 flow rate of 5 ml/min to 50% at an O_2 flow rate of 37 ml/min. During these experiments, the recycle flow rate was constant at 120 ml/min. As the amount of O_2 increased, the total rate of CH_4 conversion also increased, but above ~ 20 ml/min (CH_4/O_2 = 6) the rate of CH_4 conversion to aromatic products remained essentially constant. This behavior is due to the fact that, with decreasing CH_4/O_2 reactant ratio, an increasing fraction of the recycled CH_4 (and unreacted C_2H_6) was converted into CO_x in the OCM reactor. Thus, the O_2 flow rate needed to optimize aromatics production under these operating conditions was ~ 12 to 16 ml/min, for which the aromatic product yield exceeded 70%, and the rate of CH_4 conversion to aromatic products was >10 ml/min. The CH_4 recycle ratio under these conditions was ~ 8-10. Of course, 100% of the added CH_4 was converted.

The effect of CH_4 recycle flow rate on system performance was less pronounced than that of O_2 flow rate, as shown in Figure 6. At a fixed O_2 flow rate of 12 ml/min, the overall rate of CH_4 conversion was virtually unaffected by the recycle flow rate. Increasing the CH_4 recycle rate,

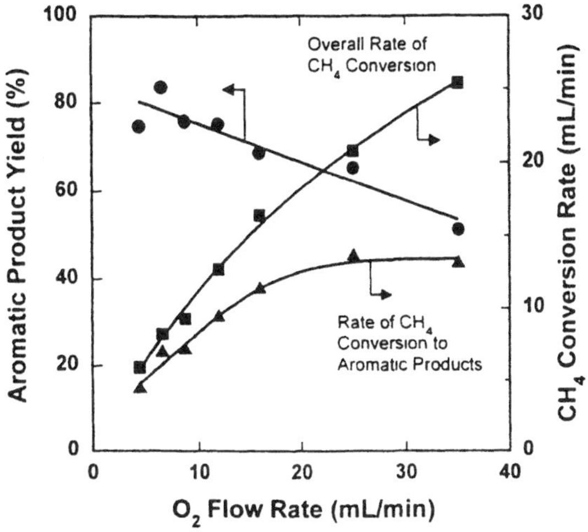

Figure 5. Effect of O_2 flow rate on CH_4 conversion rates and aromatic product yield at a CH_4 recycle flow rate of 120 ml/min (Ref. 7).

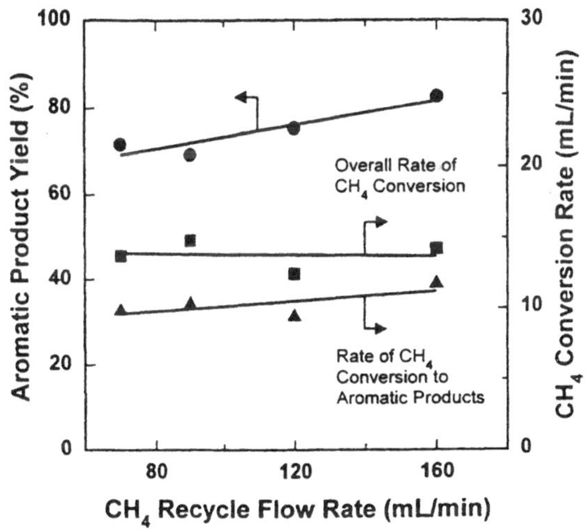

Figure 6. Effect of CH_4 recycle flow rate on CH_4 conversion rates and aromatic product yield at an O_2 flow rate of 12 ml/min (Ref. 7).

however, corresponds to an increase in CH_4/O_2 reactant ratio, resulting in improved C_2 selectivity in the OCM reactor and a corresponding increase in the rate of CH_4 conversion to aromatics and the yield of aromatic products. At a methane recycle flow rate of 70 ml/min, which corresponds to a recycle ratio of 7, the aromatic product yield was 72% and the rate of conversion of CH_4 to aromatics was 10 ml/min.

4. CONCLUSIONS

Product yields in excess of 70% can be achieved in continuous-feed mode by using a recycle reactor with removal of ethylene by complexation with silver ions or by converting the ethylene to condensible aromatics. The results demonstrate that recycling ethane over the OCM catalyst does not seriously affect the selectivity and, therefore, the product yield. The advantage of the membrane contactor system is that higher value products (olefins) can be produced in nearly pure form. The advantage of the catalytic system for converting olefins to aromatics is that more economically transportable liquid products are formed, making it more suitable for a remote natural gas-producing site. It appears that comparable product yields can be obtained at a smaller recycle ratio by reacting the olefins to aromatics. In addition, scale up may be easier with the Ga/H-ZSM-5 catalyst than with the membrane contactor.

ACKNOWLEDGMENT

The authors gratefully acknowledge financial support of this research by the Energy Research Laboratories Division of the Canada Centre for Mineral and Energy Technology.

REFERENCES

1. Y. Jiang, I. V. Yentekakis and C. G. Vayenas, Science, 264 (1994) 1563.
2. I. V. Yentekakis, M. Makri, Y. Jiang and C. G. Vayenas, ACS Div. Petr. Chem. Prepr., 41 (1996) 119.
3. R. B. Hall and G. R. Myers, in: Methane and Alkane Conversion Chemistry, M. M. Bhasin and D. W. Slocum (eds.), Plenum, New York, 1995, pp. 123-130.
4. A. Mashocki, Appl. Catal. A, 146 (1996) 391.
5. S. C. Reyes, R. W. Borry and E. Iglesia, 14th North American Meeting of the Catalysis Society, Snowbird, UT, 1995, T-284.
6. E. M. Cordi, S. Pak, M. P. Rosynek and J. H. Lunsford, Appl. Catal. A, 155 (1997) L1.
7. P. Qiu, J. H. Lunsford and M. P. Rosynek, Catal. Lett., 48 (1997) 11.
8. X. Fang, S. Li, J. Lin, J. Gu and D. Yang, J. Mol. Catal. (China), 6 (1992) 427.
9. D. Wang, M. P. Rosynek and J. H. Lunsford, J. Catal., 155 (1995) 390.
10. S. Pak and J. H. Lunsford, Appl. Catal. A, in press.
11. R. D. Hughes, J. A. Mahoney and E. F. Steigelmann, Recent Developments in Separation Science, vol. 9, CRC Press, Cleveland, 1986, p. 173.
12. P. Qiu, J. H. Lunsford and M. P. Rosynek, Catal. Lett., in press.

NATURAL GAS CONVERSION V
Studies in Surface Science and Catalysis, Vol. 119
A. Parmaliana et al. (Editors)

Methane transformation into aromatic hydrocarbons by activation with ethane over Zn-ZSM-11 zeolite

Liliana B. Pierella(*), Griselda A. Eimer(+) and Oscar A. Anunziata(*)

CITeQ (Centro de Investigación y Tecnología Química) Facultad Córdoba, Universidad Tecnológica Nacional. CC 36 -SUC 16, (5016) Córdoba, ARGENTINA.
e-mail: liliana@sa.frc.utn.edu.ar - anunziat@sa.frc.utn.edu.ar - FAX:054-51-690585

ABSTRACT
Very high levels of methane (C1) conversion to aromatic hydrocarbons were obtained by interaction with ethane (C2) (molar fraction in the feed: C1/C1+C2 = 0.4-0.8) over Zn-ZSM-11 (molar fraction Zn/Zn+H=0.86) at 550°C and total pressure of 1 atm. The yield in aromatic hydrocarbons was about 10-40 mol% C.

1.INTRODUCTION
Direct conversion of methane (C1) to more valuable compounds, such as liquid hydrocarbons is not only a promising approach for the utilization of natural gas (NG) resource but also a challenging technical project. C1 (the main compound of NG) conversion, under non-oxidizing conditions is a great task to catalysis science. Walsh et al. reported [1] the formation of aromatic-rich hydrocarbons from DPO of C1 with O_2 over ZSM5 in the presence of small amounts (0.2-0.4 mol%) of light hydrocarbons additive, such as propane (C3) or propylene (C3=) in the feed. The reaction sequence involves DPO of C1 to methanol (C1OH) followed by the C1OH-gasoline reaction and the hydrocarbons are comes from, alkenes or alkanes precursors present in the feed to initiate the MTG reaction. Han et al. [2,3] reported similar results showing product selectivity to COx >80% and to liquid hydrocarbons products >13%. Wada [4] reported that unpromoted rare earth oxides were active for the oxidative methylation of C2 with C1, indicating that Eu_2O_3-Ox catalyst gave the maximum yield of C3 products under selected reaction conditions. Thus the maximum C3 compound was 8%. Recently Wang [5] and Pierella [6] reported the aromatization of C1 in the presence of small amount of light hydrocarbons under non-oxidizing conditions over Mo-Zeolite at low pressure (1-2atm). Commercial NG can contain up to 10 % of ethane (C2). Solymosi and Szoke [7] reported high ethane conversion and benzene selectivity using MoC/ZSM-5 at 700°C. In this work, the activation of methane with ethane and the transformation of pure methane and pure ethane, using HZSM-11 and Zn loaded ZSM-11 zeolite, have been studied. The final objective is the transformation of NG into petroleum-chemicals products.

2.EXPERIMENTAL
Catalytic reactions were carried out in a continuous flow quartz reactor with an inner diameter of 10 mm at atmospheric pressure. Products were withdrawn periodically from the

---

(*) CONICET-Researcher; (+) CONICET Fellowship. Research Grants: PID-CONICET N° *6963/96* and CONICOR N° *3663/96*.

outlet of the reactor and analyzed by on-line gas chromatography equipped with a FID detector. The following feeds were used in this study: high purity methane (>99.97%) ethane (>99.997%) supplied by AGA. Commercial Natural Gas (analysis: C1= 82.8; C2= 16.1; C3= 0.8; C4= 0.5 mol%) was supplied by ECO-GAS (Argentine). The studies with methane were carried out at GHSV= 2224, 820 and 590 ml/gh. For ethane GHSV= 2224 and 820ml/gh were employed and molar fraction x(C2) = 1 and x(C2)= 0.36 using N_2 as diluyent. Natural gas was evaluated at 748ml/gh (620ml/gh for C1 and 120 ml/gh for C2). The reaction products were analyzed using a 2 m Porapak Q column. Conversion and product distribution were expressed on a carbon-atom basis. H-ZSM-11 and Zn-ZSM-11 catalyst (Si/Al=17) with Zn molar fraction = 0.86, was synthesized and characterized in our laboratory [8,9].

3.RESULTS AND DISCUSSION

3.1 Natural Gas Studies

The results of NG conversion and products distribution at 550°C and total pressure of 1atm, over Zn-ZSM-11 zeolite are summarized in table 1. As we can see C1 was not converted in the reaction conditions analyzed, meanwhile C2 was transformed on 40%.

Table 1: Natural Gas conversion and products distribution using Zn-ZSM-11 catalyst at 550°C and Total pressure, Tp= 1atm

	Molar Fraction in the feed	GHSV ml/gh	Conversion mol % (C)	Products distribution mol % (C)
C1	0.828	620	0	78.09
C2	0.161	120	46.53	14.54
C3	0.008	5.96	78.78	0.43
C4	0.003	2.23	100	0
C2=				2.77
AH				4.17

3.2 Methane and Ethane studies

Table 2 gives catalytic data and results conversion and the reaction products distribution using two different feeds: a)C1 and b) C2/C2+N2 a t 550 °C over Zn-ZSM-11 as a function of space velocity and the molar fraction of C2/C2+N2. C2 conversion and aromatics yield increase as the molar fraction of C2 in the feed increases and decreasing the space velocity [10]. Methane was not converted under the same reaction conditions and not even at lower space velocity (590ml/g.h.)

3.3 Methane + Ethane additive studies

In table 3 we can be seen the results obtained using C1+C2 as feed over H-ZSM11 and Zn-ZSM11 zeolites at different molar fraction of C2 and space velocities. Zn-ZSM11 zeolite appeared as a good material for C1 activation with C2 at C2/C2+C1= 0.6 and 2240 ml/g.h. The aromatics yield reaches a maximum about 40% at 810mol/g.h. H-ZSM11 zeolite activates methane only about 2% at the better reaction condition. Taking into account above results we choice Zn-ZSM11 zeolite to study the effect of reaction condition over C1 transformation and product distributions.

Table 2: C1 and C2 conversion and products distribution at different space velocity and molar fraction of C2 (C2/C2+N2)

GHSV (ml/gh)	2240		810	
C1 Conversion(*)	0		0	
C2 Molar fraction(**)	0.36	1	0.36	1
C2 (Conversion)	6.5	12	10.75	18
C2= (Mol% C)	4.8	5.8	5.5	6.2
AH(Mol% C)	1.25	3.5	4	7.75
Other (Mol% C)	0.45	2.7	1.25	4.05

(*) Feed: C1, Tp =1atm; (**) Feed: C2/C2+N2, Tp=1atm

Table 3: H-ZSM-11 and Zn-ZSM-11 catalytic activity using C1+C2 as feed.

GHSV (ml/hg)	2240				810			
Catalyst	Zn-ZSM-11		H-ZSM-11		Zn-ZSM-11		H-ZSM-11	
C2/C1+C2 in the feed (*)	0.6	0.26	0.63	0.27	0.6	0.26	0.63	0.27
C2 Conversion, mol % C	35.47	24.16	1.91	1.18	52.25	37.05	2.06	1.25
C1 Conversion, mol % C	21.85	5.02	1.99	0	39.55	10.9	3.34	0.08
Products distribution mol% (C)								
C1	18.84	56.16	22.41	57.83	15.95	51.57	21.57	57.63
C2	48.92	30.98	75.66	41.82	35.15	26.52	76.08	41.79
C2=	5.63	10.77	1.93	0.36	3.95	5.83	2.35	0.58
AH	22.51	1.52	0	0	38.88	15.7	0	0
Others	4.08	0.57	0	0	6.06	0.39	0	0

(*)Temperature: 550°C; Total pressure 1 atm.

C1 and C2 conversion increased as the molar fraction of C1 diminished. C2 was converted more efficiently in presence of C1 than in presence of N2 at the same molar fraction (Table 2). Furthermore, C1 was activated raising to excellent conversion levels (40%), at molar fraction 0.4 and GHSV= 810ml/g.h. We suggest that C1 could be activated by C2, initiating in this way its transformation [3]. The progress of the catalytic reactions of a gas mixture (C1+C2) at 550°C can be observed in Fig. 1 and 2. These figures show C1 and C2 conversion at different molar fraction of C1 and GHSV=2240 and 810. According to the data showed in figure 1 and 2, C1 conversion decreases as its molar fraction in the feed increases, reaching a value of C1 conversion = 0 at x(C1)>0.82. Upon this molar fraction C1 was not converted. This has been supported by the data showed in table 1, where the molar fraction of C1 in NG was 0.828 and C1 was not activated even at lower GHSV than the space velocity used for plots 1 and 2. In Figure 3 and 4 show propane (C3), ethylene (C2=), butane (C4) and aromatic hydrocarbons (AH) yields at different GHSV and as a function of C1 molar fraction. C3 and AH yields were increased by decreasing molar fraction of C1.

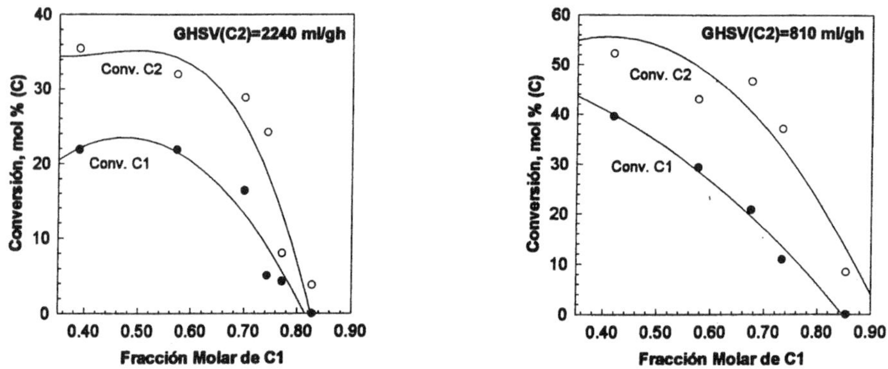

Figures 1-2: C1 and C2 conversion against the molar fraction of C1 in the feed, at Tp= 1 atm and 550°C, over Zn-ZSM-11 zeolite.

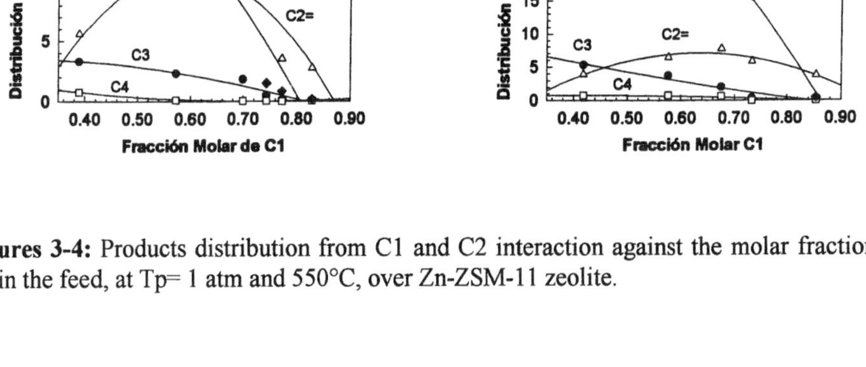

Figures 3-4: Products distribution from C1 and C2 interaction against the molar fraction of C1 in the feed, at Tp= 1 atm and 550°C, over Zn-ZSM-11 zeolite.

Thus, methane activation could occur through the interaction with ethane (or C2+ carbenium ions) toward aromatization steps. Aromatic hydrocarbons are the main products obtained at C1/C1+C2 = 0.4<x<0.74 reaching values of 15% (BTX=7.5-5-2) and 25%(BTX=12-9-3) at GHSV=2240 and 810ml/gh respectively.

The product distribution vs C1 or C2 conversion levels [11] obtained by varying the molar fraction of C1/C1+C2 in the feed at GHSV=810ml/gh and 550°C over Zn-ZSM-11 zeolite are showed in figures 5 and 6. According to those results, C2= would be a primary and unstable reaction product and aromatics would be secondary and stables ones. The effect of C2 added to C1 is significantly illustrated by data showed in figure 5 where C3 appeared as primary reaction products from C1 conversion. C3 seems to be the primary product when C1 is activated with C2 owing to the interaction with C2= or ethyl-carbenium ion. The Lewis acid sites present in Zn-Zeolite [8,9], as EDA(electron-donor-acceptor) complex would allow the interaction of reactive intermediates with C1 [10,12].

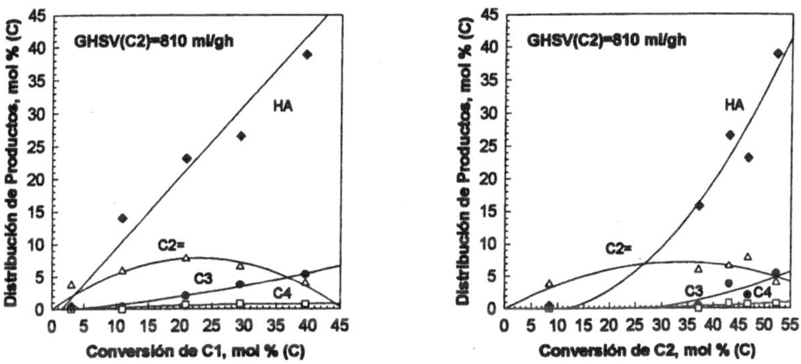

Figures 5-6: Products distribution from C1 and C2 interaction against the conversion levels of C1 and C2 at Tp= 1 atm and 550°C, over Zn-ZSM-11 zeolite.

Figures 7-9 show the effect of time on stream over C1 and C2 conversion and product

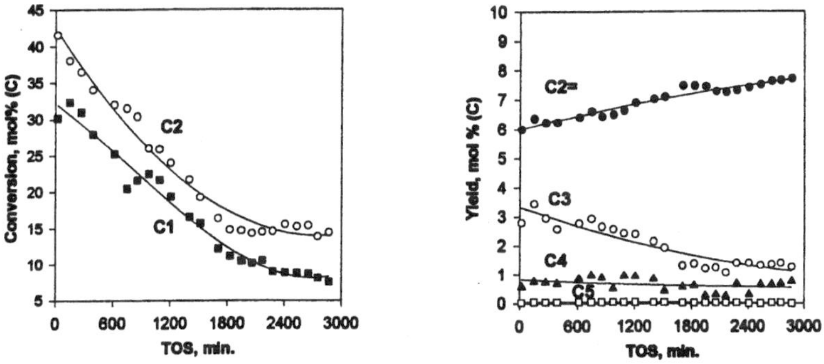

Figures 7-8: Change in C1 and C2 conversion and gaseous products distribution as a function of TOS at Tp= 1 atm and 550°C, over Zn-ZSM-11 zeolite.

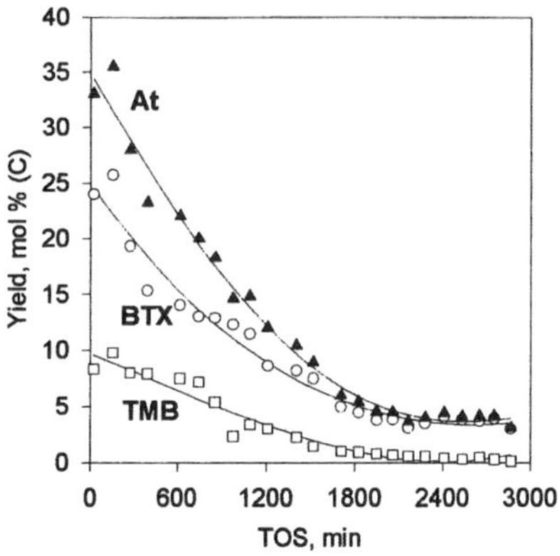

Figure 9: Aromatics distribution as a function of TOS at Tp= 1 atm and 550°C, over Zn-ZSM-11 zeolite from C1+C2 interaction.

distribution. C1 and C2 conversion decrease quickly as well as aromatics yield with TOS, whereas C2= and C4 increase slowly. C3 yield decrease very slowly. These results are in accordance with the fact that C3 is the probable primary product of C1 activation with C2, C2= is a primary product of C2 dehydrogenation, C4 would be the first oligommer from C2= and others are secondary.

4.CONCLUSIONS

Catalytic transformation of C2 is improved by increasing its molar fraction in the feed, using N2 as diluyent. Methane was not converted under the same reaction conditions. The ethane is able to activate methane at low molar fraction of C1 in the feed. C1 conversion was very high, (40 %) at 550°C, GHSV(C2)=810ml/gh and total pressure 1 atm, over Zn-ZSM-11. The formation of aromatic-rich hydrocarbons has been allowed at molar fraction of C1 C1/C1+C2 = $0.4<x<0.74$.

REFERENCES

1. D.E.Walsh, S.Han, R.E. Palermo, J.Chem.Soc.Chem.Commun. 18(1991)1259.
2. S.Han, R.E.Palermo, J.A.Pearson, D.E.Walsh, Catal. Lett., 16(1992)263-268
3. S.Han, D.J.Martenak, R.E.Palermo, J.A.Pearson, D.E.Walsh, J.Catal. 136(1992)578.
4. K.Wada, Y.Watanabe, F.Saitoh, T.Suzuki, Appl.Catal., 88(1992)23-28
5. L. Wang, L. Tao, M. Xie, G. Xu, Catal. Lett. 21 (1993)35.
6. L.B. Pierella, L. Wang, O.A. Anunziata, React.Kinet.Catal.Lett., 60 (1997) 101.
7. F. Solymosi and A. Szoke, Appl.Catal., 166(1)(1998)225-235.
8. O.A. Anunziata, L.B .Pierella, Catal. Lett. 19 (1993) 143.
9. O.A. Anunziata, L.B. Pierella, R.G. Marino, Appl.Catal., 165(1997)35-49
10. L.B. Pierella, G.A. Eimer, O.A. Anunziata, React.Kinet.Catal.Lett.,63(1998)271-278.
11. J. Abbot, A. Corma, B. Woiciechowski, J. Catal. 92 (1985) 398.
12. T. Inui, J. Og The Japan Petrol. Inst.- Sekiyu Gakkaishi 33(4)(1990) 198.

NATURAL GAS CONVERSION V
Studies in Surface Science and Catalysis, Vol. 119
A. Parmaliana et al. (Editors)

Catalytic Dehydroaromatization of Methane with CO/CO_2 towards Benzene and Naphthalene on Bimetallic Mo/Zeolite Catalysts: Bifunctional Catalysis and Dynamic Mechanism

Shetian Liu[a], Linsheng Wang[b], Qun Dong[c], Ryuichiro Ohnishi and Masaru Ichikawa*

Catalysis Research Center, Hakkaido University, Kita-Ku, N-11, W-10, Sapporo 060, Japan

1. INTRODUCTION

In recent years a great challenge and intriguing problem in heterogeneous catalysis is the catalytic dehydrocondensation of natural gas into useful petrochemical feed stocks such as ethylene by the oxidative coupling of methane over Li/MgO and Sm_2O_3 catalysts[1] and the lower hydrocarbons by the two-step conversion of methane on Co and Ru catalysts[2], while they have not been feasible yet due to the lower yields and selectivities. Since 1993, the non-oxidative conversion of methane to benzene has been reported[3-8] on Mo/HZSM-5 catalysts in conjunction with their reactivities, reaction mechanism and catalyst characterization. Recently, we have preliminary reported that the Fe/Co modified Mo/HZSM-5 catalysts exhibit the higher conversion of methane to aromatics such as benzene and naphthalene with a large yield of hydrogen[9],and the CO/CO_2 addition to methane effectively improve their catalyst stability[10]. This paper will present the promotion role of CO/CO_2 addition to methane feed to effectively reduce the coke formation and improve their catalyst stability on Mo/HSM-5 and Fe/Co modified Mo/HZSM-5. The remarkably higher rates and selectivities of benzene formation in the catalytic dehydrocondensation of methane were obtained by increasing the pressure (1-5 atm) and flow rates of methane at 973K on Mo/HZSM-5 catalysts. In addition, we will discuss on the bifunctional catalysis of Mo carbide and HZSM-5 support, which have been characterized by XAFS, TG/DTA/Mass and IR spectroscopy.

2.1 Materials and catalysts preparation

3-6wt%Mo loading Mo/HZSM-5 and Fe/Co modified Mo/HZSM-5 catalysts were prepared by impregnation(incipient wetness) of NH_4ZSM-5(SiO_2/Al_2O_3=20~1900; surface area, 780-925 m^2/g; Toso Co. and CRI Zeolyst, Inc.) with $(NH_4)_6Mo_7O_{24}4H_2O$, $Co(NO_3)_2$ or $Fe(NO_3)_3$ aqueous solutions. The resulting materials are dried at 393K and calcined at 773K for 6 h, similarly as reported in the previous literature[9,10]. Powdered Mo carbide(Mo_2C; 100-500 mm size) was purchased from Japan New Metal Co. and used as it. The hybrid catalysts of 3wt% Mo/SiO_2+HZSM-5 and Mo_2C+HZSM-5 were prepared by mechanically grinding both components and by the co-impregnation from the slurry solution.

2.2. Catalytic measurement and kinetic analysis

The catalytic tests were carried out under 1-5 atmospheric pressure of methane in a continuous flow system with a quartz reactor of 8mm i.d. which is charged with 0.15-0.30 g of catalyst pellets of 20-42 mesh. The feed gas mixture of 98%CH_4(99.9% purity) and 2%Ar as internal standard for analysis was introduced into the reactor at 7.5-20 ml/min[space velocity =

a) On leave from Hebei University of Science and Technology, Hebei, China;b) Dalian Institute of Chemical Physics, Dalian, China;c) Taqie Institute of Petrolium Chemistry, Taqie, China.

1000-20000 ml(g-cat.h)$^{-1}$] through a mass flow controller(Brooks 5850E) after flashing with He at 973K for 40min. Hydrocarbon products including C_2-C_4 alkanes(and/or alkenes) and condensable C_6-C_{12} aromatics such as benzene, toluene, xylene and naphthalene derivatives were analyzed by the on-line FID gas chromatography using a six-way sampling valve heating at 533K on a Porapak-P column and the one-line TCD gc for H_2, Ar, CO, CH_4 and CO_2 on an activated carbon column. The condensable materials such as benzene, toluene, xylene, naphthalene, methylnaphthalene and anthracene which were identified by using the on-line GC-MS(Perkin-Elmer, Auto System GC with 910 Q-Mass). Using an internal standard analyzing method, methane conversion, selectivities of hydrocarbon products and coke formation formed on the catalysts were evaluated according to the mass balance for carbon and hydrogen[9,10].

3. RESULT AND DISCUSSION

3.1 Methane aromatization on Mo/HZSM-5 and Fe/Co modified Mo/HZSM-5

The non-oxidative conversion of methane was conducted by flowing methane(1-5 atm) on the catalysts in the fixed bed at 973K. Using 2%Ar as the internal standard for the reaction analysis, the yields and selectivities of benzene and coke formation were carefully evaluated according to the previous work[9,10]. C_2-hydrocarbons and aromatic products such as benzene and naphthalene were continuously obtained with a considerable evolution of hydrogen (H_2/ benzene=9-18 mol/mol) on the Mo/HZSM-5, Fe/Co modified Mo/HZSM-5 catalysts. The conversion of methane(ca. 12 %) usually decreases rapidly at the initial stage and moderately at the later stage of the reaction down to 4% in 30 h possibly due to the coke formation. The major hydrocarbons produced on Mo/HZSM-5 consist of benzene(50-68%), naphthalene(45-12%) and toluene (2-5 %) as the aromatics, and C_2 products(4-12%) such as ethylene and ethane. The other carbon-containing products such as CO, propene, butene, and methylnaphthalene, phenanthrene, anthracene and tetracene as poly-condenced aromatics were produced in a small amount (totally less than several % selectivity). Further bulky substituted aromatic compounds are minorly obtained on the catalysts at 973K because of the higher activity for hydrogenolysis and molecular shape selectivity due to the limited size of micropores of HZSM-5(5-6±) accessible only to benzene and toluene. Our obtained results on the Mo/HZSM-5 catalyst show that the benzene selectivities based on the consumed methane did not exceed 42% at the maximum value owing to the formation of coke(32-42% sel) and polycondensed by-products.

From Fig.1a and 1b it is clear that the addition of Fe as for Co, to Mo/HZSM-5 catalysts

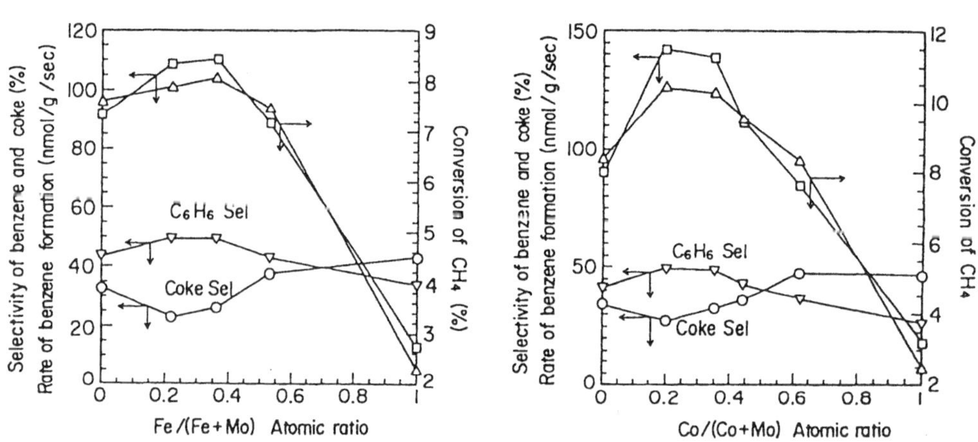

Fig. 1. Catalytic performances of (a) Co-Mo//HZSM-5 and (b) Fe-Mo/HZSM-5 for methane dehydroaromatization at 973K and 1 atm as the function of the Co/Mo and Fe/Mo composition.

yields a considerable enhancement of aromatic products such as benzene and naphthalene in the reaction at 973K. A maximum rates(120-150 nmol/h/g-cat) and selectivity(52-54% sel) of benzene formation and minimizing coke formation(less than 20%) was attained by the optimum addition of Co(and/or Fe) to Mo/HZSM-5. Further addition of Fe or Co to Mo/HZSM-5 results in a decrease of the aromatic products owing to substantial increase of coke formation in the methane conversion. The characterization by XRD, TG/DTA/Mass and EXAFS suggest that the binary FeMo and CoMo carbides are highly dispersed in the micro-channels of HZSM-5 zeolite in the methane conversion at 973K, implying that Fe and Co act as anchoring elements to promote the dispersion of Mo carbide as the active species for the reaction.

3.2. Effects of pressure and flow rate

Fig.2 shows the dependency of rates of products and coke formation in the methane reaction on 3%Mo/HZSM-5 at 973K by varying the methane pressure of 1-5 atm, respectively. It is interesting to find that the formation rates of benzene and C_2 hydrocarbons increase moderately by increasing the methane pressure. The maximum yields and selectivity (65% sel) of benzene were attained at 3-4 atm of methane by the constant flow rate of 2500 (ml/h/g-cat) at 973K, although the methane conversion is thermodynamically unfavorable for the pressurized condition($6CH_4= C_6H_6+9H_2$). Moreover, it is of worthy to note that the coke formation is substantially suppressed and the catalyst stability was relatively improved by increasing the methane pressure above 3 atm on the catalyst, as shown Fig.2. This may be related with the sufficient supply of hydrogen from methane and the concentration of the surface carbon species CH_x for the formation of aromatic products such as benzene and naphthalene on the catalysts.

On the other hand, by increasing the flow rates of methane refereed to as the specific velocity (ml/h/g-cat) from 1000 to 20000 the rate of benzene formation(nmol/s/g-cat) under the 3 atmosphere of methane at 973K on 3%Mo/ HZSM-5 is remarkably enhanced by ca. 10 times as shown in Fig. 3. The methane conversion at the 5 h time-on stream moderately decreased from 8 to 6 % by changing the specific velocity from 1000 to 20000(ml/h/g-cat). These results suggest that benzene is produced under the diffusion control on the catalysts and the methane conversion to benzene is close to the thermodynamical equilibrium(benzene conversion for H_2/

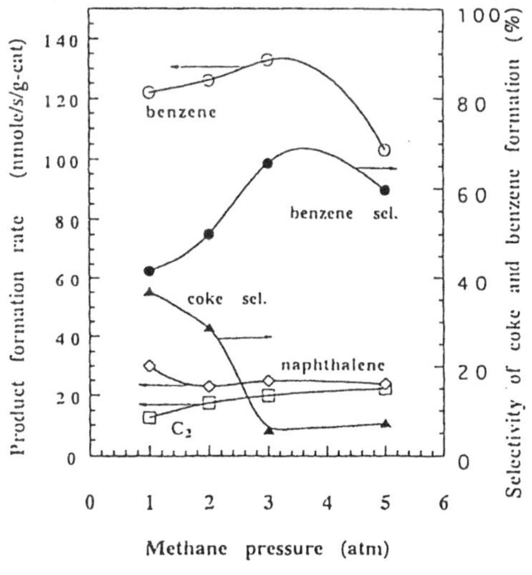

Fig. 2 Catalytic performances of 3%wt Mo/ HZSM-5 for methane dehydroaromatization varying the methane pressure from 1 to 5 atm. at 973K and SV=1440 ml/h/g-cat.

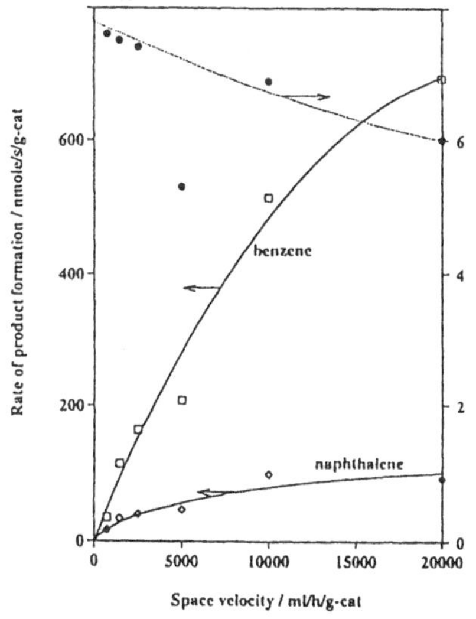

Fig.3. Product formation rates and methane conversion on 3%wtMo/HZSM-5 by varying specific velocity of methane(ml/h/g-cat) at 3 atm and 973K

C_6H_6=12 at 973K is calculated as 8.2% and 1 atm of methane) under the reaction conditions. The effective transportation of the aromatic products results in the higher conversion of methane and suppressing the coke formation through the micro-channels of HZSM-5.

3.3. Selective dehydroaromatization of methane by CO/CO_2 addition

The addition of CO to methane feed gas exerts a significant effect on the catalytic performances of Mo/HZSM-5 and Fe/Co modified Mo/HZSM-5 for the aromatization reaction to promote the benzene formation and catalyst stability. Using pure CH_4 as the feed gas, methane conversion and benzene formation rate were decreased greatly during 24 hours of reaction time on stream, as shown in Fig. 4. While with the addition of CO in the feed gas, the methane conversion drop was substantially moderated regardless of varying the partial pressure of CO (from 1.7% to 12.3%). After attaining a increase of benzene formation in the early stage of the methane conversion with CO, benzene formation rate kept almost constant during the prolonged reaction up to more than 24 hours. It is worthy to note that in the steady reaction stage, the amount of CO is almost kept constant before and after the reaction in all the case of CO addition. The results imply that CO is regenerated possibly due to the reaction of the surface carbon with the active oxygen species derived from CO dissociation as shown in Scheme 1.

By addition of CO to a 1 atm methane flow at 973K the Co and Fe modified Mo/HZSM-5 catalysts at 973K and 1 atm of methane exhibit remarkable catalytic activities for the benzene formation in high selectivities (75-85% sel. at 10-13% CH_4 conv.), as shown in Fig. 4. The rate of benzene formation decreases moderately but keep the higher rates and lower coke selectivity(less than 10%) for 100 h. Similarly, the stability of the catalyst was also improved by adding a few per cent of CO_2 into the feed gas. The addition of 1.6% CO_2 yielded a higher methane conversion and hydrogen formation rate. Nevertheless, the benzene formation was greatly suppressed by the exceeding addition of CO_2(more than 12%), but it yields the improvement of the catalyst stability in the methane aromatization during the prolonged time. It is suggested that CO_2 is converted with CH_4 to produce double amount of CO and H_2 under the reaction condition by the reforming process ($CO_2+CH_4=2CO+2H_2$). The formed CO may exert a similar promotion of products formation and the catalytic stability. The excess CO_2 may greatly consume the surface active carbon species such as CH_x and C_2 species for the reaction.

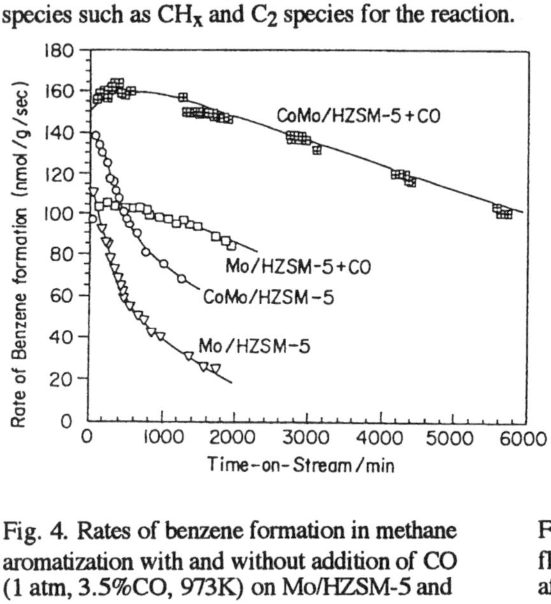

Fig. 4. Rates of benzene formation in methane aromatization with and without addition of CO (1 atm, 3.5%CO, 973K) on Mo/HZSM-5 and Co-Mo/HZSM-5 catalysts.

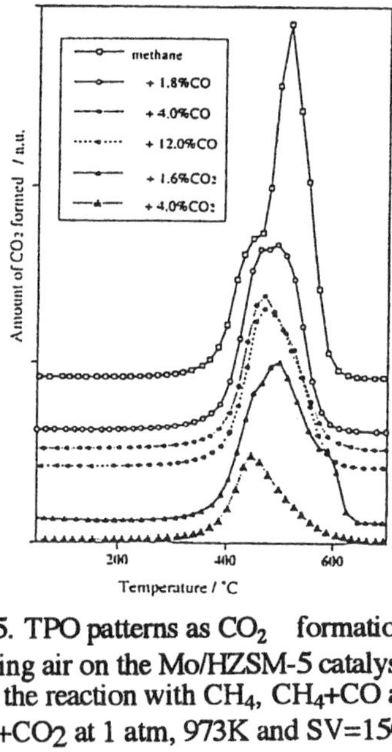

Fig.5. TPO patterns as CO_2 formation by flowing air on the Mo/HZSM-5 catalysts after the reaction with CH_4, CH_4+CO and CH_4+CO_2 at 1 atm, 973K and SV=1500

As shown in Fig.5, it was demonstrated by the temperature-programmed oxidation(TPO) experiments(ramping temperature rate:2K/min, air flow of 20 ml/min) that the amount of coke deposition on the catalyst surface was greatly reduced by adding various amount of CO or CO_2 to the methane feed gas. The increase of CO concentration from 1.7% to 12.0% in methane flow resulted in a marked suppression of the total coke formation on the catalyst surface, particularly the irreversible or inert coke which was oxidized to CO_2 at the higher temperature above 773K in TPO experiment. The increase of CO_2 partial pressure(1.4 to 4.0%) in methane feed gas decreased effectively the coke formation on the catalyst to a much lower level, compared with that of CO. This may be related to its higher activity of the dissociated oxygen in the reaction with surface carbon to form CO(CO_2 + C ---> 2CO) as shown in scheme 1.

To understand the role of CO in methane aromatization reaction, the $^{13}CO+CH_4$ reaction was conducted in a closed circulating reaction system. The starting composition of the mixture reactant is 8%^{13}CO+92%CH_4 with a total pressure around 220 torr. Even after 5min of the closed circulating reaction at 973K, ^{13}C atoms involved in methane ($^{13}CH_4$) reached the value of 8%. The isotopic abundance of ^{13}C incorporated in benzene molecule were detected by GC-Mass analysis as the random distribution ($^{13}CC_5H_6$-47.6%, $^{13}C_2C_4H_6$-36.7%, $^{13}C_3C_3H_6$-13.0%, $^{13}C_4C_2H_6$-2.4%, $^{13}C_5CH_6$-0.3%) regardless of the reaction time of methane on Mo/HZSM-5. These results suggest that the carbon derived from CO dissociation is efficiently incorporated into the benzene formation and the isotopic scrambling reaction between ^{13}CO and $^{12}CH_4$ proceeds rapidly under the reaction condition at 973K. Based on the above results we suggest the mechanism of an unique role of CO in methane aromatization reaction to stabilize the catalyst performances and promote the benzene formation as shown in scheme 1. Firstly, CO dissociates on the Mo sites to form the active carbon species CH_x and C_2 species through reaction (2) and (3); followed with their oligomerization to form higher hydrocarbons such as benzene and naphthalene in reaction (1) on the catalyst;The dissociated oxygen species [O] from CO may react with the surface inert carbon species (coke) to regenerate CO, resulting in the suppression of the coke formation on the catalyst through reaction (4).

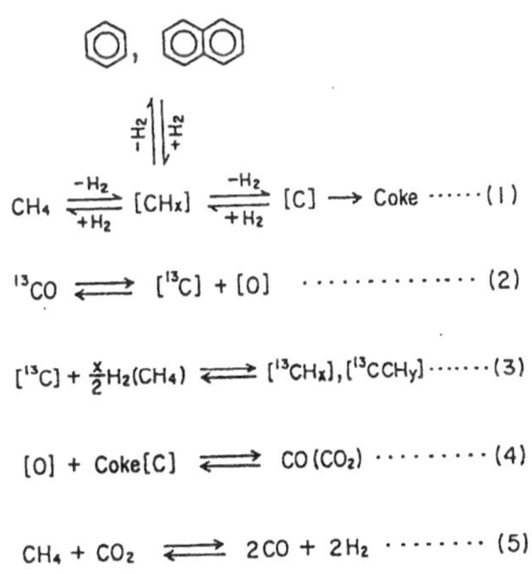

Scheme 1

Table 1. Catalytic performances of the hybrid catalysts of MoC_2(or MoO_3/SiO_2)+HZSM-5 in the methane aromatization at 1 atm and 973K

catalyst	CH_4conv(%)	V(Benzene)^a
HZSM-5	1.2	0.6
Mo_2C	0.8	3
MoC_2+HZSM-5	5.7	150
MoO_3/SiO_2	5.3	412
MoO_3/SiO_2+HZSM-5	6.2	2019
3%Mo/HZSM-5	8.5	3162

a: Specific rates of benzene formation(nmol/Mo/sec)

3.4. Bifunctional catalysis by Mo carbide and ZSM-5 zeolite support

EXAFS(Mo K-edge) and TG/DTA/Mass studies have been performed in methane conversion(1 atm, 300-973K) on 6wt% Mo/HZSM-5 using the SOR facility(KEK-10 line) and Mac Science system(TG-DTA2020S), respectively. It was demonstrated that Mo oxide on HZSM-5 is converted with methane at 923K by the evolution of H_2 and CO to Mo carbide cluster(Mo_2C; Mo-Mo; C.N.= 2.1-3.5, R-2.98±), where the methane aromatization reaction initiates to form C_6H_6, $C_{10}H_8$ and C_2H_4 in evolving hydrogen above 923K. The results imply that Mo carbide on HZSM-5 support is an active species for the catalytic methane aromatization similarly as it has been proposed previously by Solymosi and Lunsford[5,6,8].

To understand the bifunctional role of Mo carbide and HZSM-5 support the following experiments were conducted. As indicated in Table 1, it was of interest to find that the specific rates of benzene formation(C_6H_6/Mo/sec) on a hybrid catalysts of MoC_2(or MoO_3/SiO_2) +HZSM-5 at 973K were greatly enhanced(50-100 times) by the mechanical mixture of each component, which shows negligible or low activity. These imply the bifunctional catalysis for methane aromatization at the interface of Mo carbide and ZSM-5 support. The results by in-situ FTIR and EXAFS studies suggest that methane is activated on Mo carbide sites to form CH_x and C_2-species(e.g. CH_3-C;2967, 2878, 1468 and 1382 cm^{-1}) which may migrate at the interface of ZSM-5 acidic support, followed by the dehydroaromatization of C_2-intermediates to benzene and naphthalene. As it is discussed in the session 3.3., the common carbon species CH_x by the dissociation of CO and methane on MoC_2 are converted to C_2-hydrocarbons, benzene and naphthalene on the acidic sites of the ZSM-5 support. The influence of SiO_2/Al_2O_3 ratio(20-1900) of the HZSM-5 zeolite on benzene production and coke formation were also investigated. It was found that there exists a sharp optimum SiO_2/Al_2O_3 ratio of 40 yielding highest rates of benzene formation and lowest coke formation. The HZSM-5 supports having too smaller or larger Si/Al ratios beyond 40 produce the Mo/HZSM-5 catalysts of poor activity for methane aromatization. The proper Bronsted acidity of HZSM-5 promotes the conversion of C_2-intermediates to benzene and naphthalene and suppress the coke formation. This bifunctional catalysis of Mo_2C/HZSM-5 for methane aromatization reaction is discussed by analogy with the synergetic promotion on Pt/Al_2O_3 catalyst for the dehydroaromatization of alkanes.

4. Conclusion

(1) Catalytic dehydroaromatization of methane with CO and/or CO_2 proceeds at 973K under the higher pressure of methane(1-5 atm) on Mo/HZSM-5 and Fe/Co modified Mo/HZSM-5 towards aromatic products such as benzene and naphthalene with higher yields and stable selectivities (above 85 %) and minimizing the coke formation in the prolonged time of 100 h.
(2) The catalyst stability by adding CO and/or CO_2 to the methane feed is associated with effectively suppressing the coke formation due to CO dissociation on Mo/HZSM-5 catalysts.
(3) A remarkable synergetic promotion for benzene formation is obtained on the hybrid catalysts consisting of Mo/SiO_2+HZSM-5 and Mo_2C+HZSM-5, whereas each component shows a marginal activity in the reaction.
(4) Bifunctional catalysis of Mo/HZSM-5 is discussed by the mechanism that methane is activated on Mo sites to form CH_x and C_2-species, which may migrate at interface of ZSM-5 having the optimum acidity for the dehydrocyclization to benzene and naphthalene.

REFERENCES

[1] J. Lunsford, Catal. Today, 6(1990) 235.
[2] L. Guczi; R.A. van Santen,and K.V. Sayma, Catal. Rev.-Sci. Eng., 38(1996) 249.
[3] L. Wang, L. Tao, M. Xie, G. Xu, J. Huang, Y. Xu, Catal. Lett., 21(1993) 35.
[4] Y. Xu, S. Liu, L. Wang, M. Xie, X. Guo, Catal. Lett., 30(1995)135.
[5] F. Solymosi, A. Szöke, J. Cserényi, Catal. Lett., 39(1996) 157.
[6] F. Solymosi, J. Cserényi , A. Szöke, T. Bánsagi, A. Oszkó, J. Catal., 165 (1997) 156.
[7] S.-T. Wong, Y. Xu, L. Wang, S. Liu, G. Li, M. Xie, X. Guo, Catal. Lett., 38 (1996) 39.
[6] Y. Xu, Y. Shu, S. Liu, J. Huang, X. Guo, Catal. Lett., 35 (1995) 233.
[8] D. Wang, J. H. Lunsford, M. P. Rosynek, J. Catal., 169 (1997) 347.
[9] S. Liu, Q. Dong, R. Ohnishi, M. Ichikawa, J.C.S. Chem. Commun.,(1997)1455.
[10] S.Liu, Q. Dong, R. Ohnishi and M. Ichikawa, J.C.S.,Chem. Commun.(1998) in press.

NATURAL GAS CONVERSION V
Studies in Surface Science and Catalysis, Vol. 119
A. Parmaliana et al. (Editors)

Study of the hydrogenation step in the non-oxidative oligomerization of methane on Pt/SiO_2 (EUROPt-1)

E. Marceau [a], J. M. Tatibouët [a] *, M. Che [a +] and J. Saint-Just [b]

[a] Laboratoire de Réactivité de Surface - UMR CNRS 7609 - Université Pierre et Marie Curie, Paris VI - 4 place Jussieu - 75252 Paris Cedex 05 - FRANCE

[b] Gaz de France - Direction de la Recherche - CERSTA - BP33 - 93211 Saint-Denis la Plaine Cedex - FRANCE

Abstract

Hydrocarbons can be produced from methane through a two-step procedure on Pt/SiO_2 catalyst EUROPt-1 : (1) adsorption of methane and dehydrogenation in flow conditions ; (2) subsequent hydrogenation of the carbonaceous deposit. Methane adsorption was performed at 300°C ; it was shown by TPO experiments that only 55% of the carbonaceous species deposited were hydrogenated at that temperature after 15 minutes in hydrogen. Hydrogen reacted with one half of the reactive carbonaceous species to give back methane and with the other half to form higher alkanes. The first two minutes of hydrogenation mainly lead to ethane and n-pentane. Between 2 and 30 minutes the rate of production in C_2-C_5 alkanes appeared to be steady with time and their instantaneous selectivity decreased when the number of carbon atoms in the chain increased. Both steps of the non-oxidative conversion of methane contribute to the creation of C-C bonds between carbonaceous species, with distinct slates in alkanes.

1. INTRODUCTION

Methane oligomerization under non-oxidative conditions at moderate temperature was first described seven years ago simultaneously by the groups of Amariglio in France and van Santen in the Netherlands [1, 2]. Both reported that light hydrocarbons could be formed by hydrogenation of carbonaceous species previously deposited on a metallic surface (Co, Ru, Pt) via methane adsorption. A two-step sequence involving a solid reactant was used in both cases to circumvent the unfavorable thermodynamics of the homologation reaction

$$n\,CH_4 \longrightarrow C_nH_{2n+2} + (n-1)\,H_2\,.$$

---

* Current address : Laboratoire de Catalyse en Chimie Organique - UMR CNRS 6503 - Université de Poitiers - 40 avenue du recteur Pineau - 86022 Poitiers Cedex - FRANCE
+ Institut Universitaire de France.

Step 1 : methane adsorption

$$CH_{4(g)} + \text{catalyst} <===> \text{carbonaceous species}_{ads} + H_{2(g)}$$

Step 2 : hydrogenation of the carbonaceous deposit

$$\text{carbonaceous species}_{ads} + H_{2(g)} <====> C_nH_{2n+2(g)} + \text{catalyst.}$$

Van Santen et al. performed hydrogenation *at a lower temperature* than methane adsorption in order to favour the thermodynamics of each step. The C-C bonds were supposed to be created between CH_x units during the hydrogenation step [3]. Amariglio et al. performed both steps *at the same temperature*. The formation of C-C bonds between the hydrogen-deficient CH_x fragments was supposed to be initiated as soon as methane adsorbed, in contrast with van Santen's group procedures and hypotheses [4]. In order to shed some light on the nature of the limiting step for C-C bond formation, we carried out the non-oxidative homologation of methane to higher hydrocarbons in isothermal conditions on Pt/SiO_2 catalyst EUROPt-1. The duration of exposure of the metal to methane was kept constant and the duration of hydrogenation of the carbonaceous deposit formed in these conditions was varied to check its possible influence on the production of hydrocarbons.

2. EXPERIMENTAL

200 mg of EuroPt-1 catalyst (Pt (6.3 wt%) / SiO_2, dispersion = 60% [5]) were used in a tubular glass reactor. Step 1 (methane adsorption) consisted in flowing a CH_4/He mixture (CH_4/He = 1 / 2 vol% ; total flow rate = 15 $mL.min^{-1}$) during 5 min through the catalyst bed maintained at 300°C. A two-minute purge was performed with helium (30 $mL.min^{-1}$) before hydrogenation. During step 2 (hydrogenation at 300°C, variable time) (H_2/He = 1 / 2 vol% ; total flow rate = 45 $mL.min^{-1}$), formed hydrocarbons which were heavier than methane were trapped on activated carbon cartridges Carbotrap™ 200 (Supelco) and released afterwards upon heating. They were subsequently separated and their amount was determined by gas chromatography (GC) (chromatograph Delsi DI 700, column Chrompack PLOT 25 x 0.53 m, coated with alumina deactivated by KCl, detection by FID).

Carbonaceous species that did not react with hydrogen and remained on the catalyst surface were titrated by total oxidation at 400°C using pure O_2 (flow rate = 10 $mL.min^{-1}$) or temperature - programmed oxidation (TPO) experiments with a O_2/Ar mixture (O_2/Ar = 1 / 10 vol % ; total flow rate = 5.5 $mL.min^{-1}$). In TPOs, the reactor was quenched from 300 to 40°C in inert gas and the temperature ramp in dilute oxygen from 40 to 400°C was 20°$C.min^{-1}$. A mass spectrometer Delsi Nermag Anagaz 200 was used to monitor and integrate the CO_2^+ peak (m/e = 44), after calibration with pure CO_2 pulses. The catalyst was regenerated by a one-hour treatment in dilute hydrogen at 400°C. No deactivation of the catalyst was noticed. All gases were supplied by Air Liquide and methane (N45) was checked not to contain more than 0.05 ppmv. of heavier hydrocarbons.

3. RESULTS AND DISCUSSION

Titrations with pure oxygen of the residual carbonaceous species remaining on the catalyst surface after methane adsorption and hydrogenation showed that hydrogenation was not a fast process (Fig. 1). 45% of the total carbonaceous species (counted in μmol of C atoms) had not yet reacted after 15 min of hydrogenation ; 55% had thus reacted with hydrogen to yield methane and higher hydrocarbons. Two groups of species could be identified by TPO experiments (Fig. 2) : one could be oxidized at 40°C, giving an irregular oxidation peak, and another one was oxidized at higher temperatures (100-250°C). TPO experiments performed after 15 minutes of hydrogenation showed that both groups had been consumed in equal proportions (55%) through their most reactive components toward oxygen.

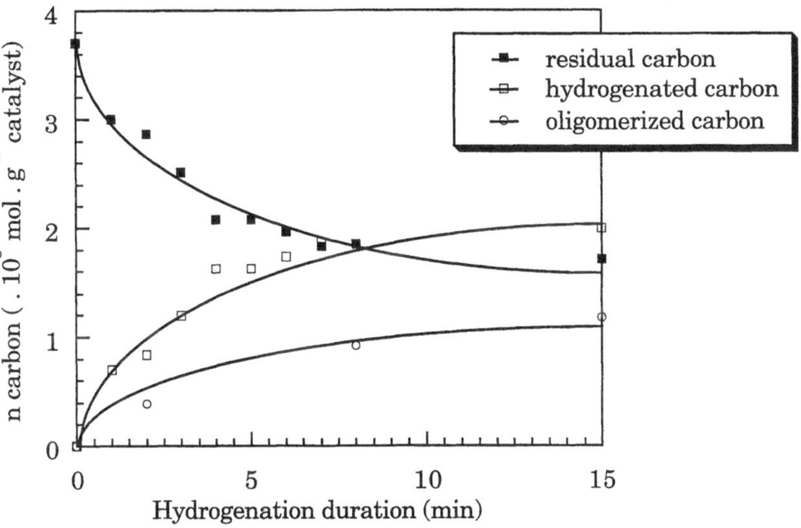

Figure 1. Effect of the hydrogenation step duration on the consumption and oligomerization of carbonaceous species deposited on the catalyst (results expressed in μmol C atoms . g^{-1} catalyst).

Only one half of hydrogenated carbonaceous species oligomerized and lead to a production of C_2-C_8 linear and branched alkanes dosed by GC (Fig. 1), the main product obtained by hydrogenation being thus CH_4. Linear alkanes were always produced in higher amounts than branched ones. The production slate in alkanes was strongly dependent on the hydrogenation duration. During the first two minutes of hydrogenation, the main products given by coupling of carbonaceous species and hydrogenation were ethane and n-pentane, whether in terms of μmol of C atoms oligomerized (Table 1) or in terms of μmol of C_n alkanes released (Fig. 3).

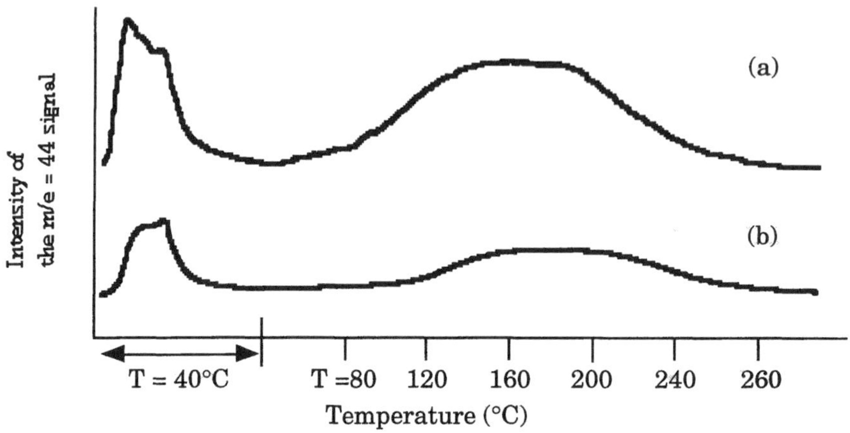

Figure 2. TPO profiles obtained : (a) after methane adsorption and cooling in inert gas ; (b) after 15 minutes of hydrogenation and cooling in inert gas (signal measured by mass spectrometry : m/e = 44, CO_2^{+}).

Table 1
Cumulative amounts (and selectivities) of methane oligomerization products (alkanes) during hydrogenation of carbonaceous species deposited on EUROPt-1 at 300°C (results expressed in µmol C atoms . g^{-1} catalyst).

tH_2 (min)	0-2	0-8	0-15	0-30
C_2	0.108 (27.5%)	0.326 (35.5%)	0.444 (37.5%)	0.584 (36%)
C_3	0.030 (7.5%)	0.129 (14%)	0.180 (15%)	0.270 (17%)
branched C_4	0.004 (1%)	0.024 (2.5%)	0.032 (3%)	0.060 (4%)
linear C_4	0.012 (3%)	0.060 (6.5%)	0.088 (7.5%)	0.148 (9%)
Σ branched C_5	0.030 (7.5%)	0.070 (7.5%)	0.090 (7.5%)	0.125 (8%)
linear C_5	0.190 (48.5%)	0.200 (22%)	0.215 (18%)	0.217 (13.5%)
ΣC_6	0.018 (5%)	0.072 (8%)	0.078 (6.5%)	0.120 (7.5%)
ΣC_{7-8}	0 (0%)	0.037 (4%)	0.052 (5%)	0.082 (5%)

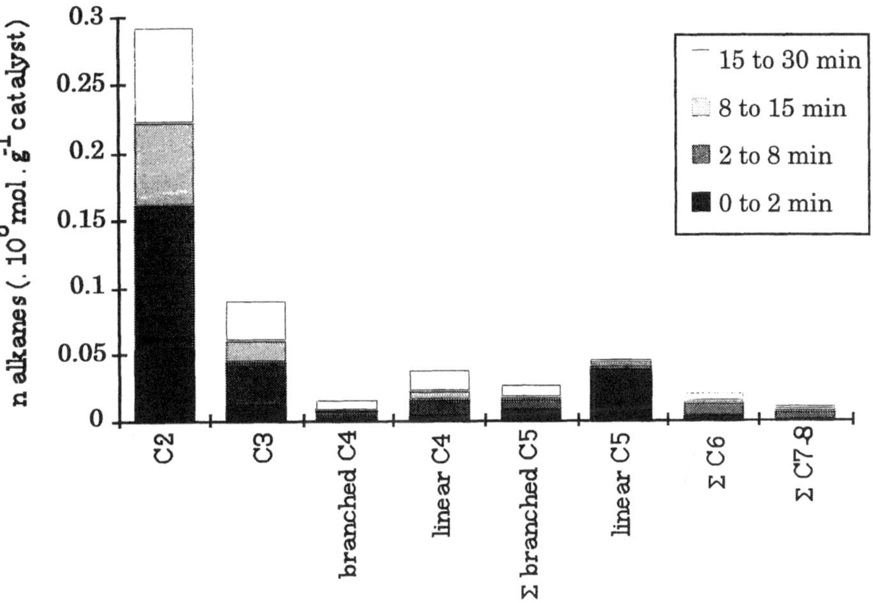

Figure 3. Production in alkanes as a function of hydrogenation duration (results expressed in μmol alkanes . g^{-1} catalyst).

Ethane and n-pentane were also observed to desorb during the purge in helium that was carried out immediately after methane adsorption, proving that C-C bonds actually formed during step 1, as stated by Belgued et al. [1, 4].

Between 2 and 30 minutes of hydrogenation, C_n alkanes ($2 \leq n \leq 5$) were produced steadily. Lighter hydrocarbons were produced in higher yields ($C_2 > C_3 > \Sigma C_4 > \Sigma C_5$) and the selectivity was constant with time (Fig. 3).

C_6-C_8 alkanes were produced mainly after long exposures to hydrogen. These heavier hydrocarbons were produced in less reproducible yields than lighter hydrocarbons.

These results show that C-C bonds are created in both steps of the non-oxidative oligomerization of methane on supported platinum. The remarkable selectivity in n-pentane has already been mentioned in the literature [4, 6-7], but it has not been linked so far to the specific coupling of carbonaceous species during step 1. The use of a sintered Pt/SiO_2 catalyst with larger platinum particles leads to a lower n-pentane selectivity, suggesting that the C-C bond formation during step 1 is structure-sensitive [8].

After the active sites on platinum particles have been freed, i. e. after the first minutes of hydrogenation and the release of n-pentanes in the gas phase, a dynamic coupling of carbonaceous species can occur steadily in hydrogen flow (C-C bonding during step 2). The slate of products (the lighter the alkane, the higher the yield) could be characteristic of a statistical mechanism of coupling, similar to the Fischer-Tropsch mechanism, as assumed by Koerts et al. for low coverages of carbon on supported metals [9].

Finally, it is likely that the poor reproducibility of the production of the heaviest hydrocarbons originates from the slow hydrogenolysis of heavy carbonaceous species with low reactivity. These species may be part of the 50% of hardly hydrogenable carbon. As no deactivation of the catalyst has been noticed even after some thirty cycles of reaction, it must be assumed that these high molecular weight compounds are not located on the sites producing lighter alkanes.

4. CONCLUSION

The hydrogenation of carbonaceous species deposited on supported platinum (EUROPt-1) via methane adsorption is a slow process. It is confirmed that three mechanisms exist for the production of higher alkanes during the hydrogenation step :

- the hydrogenation of hydrogen-deficient C_nH_y species formed during methane adsorption (giving mainly ethane and n-pentane)
- the coupling of CH_x species within the course of hydrogenation
- the hydrogenolysis of "graphitic" carbonaceous species.

These results should be extended to carbonaceous deposits obtained at temperatures other than 300°C. The different slates in higher alkanes given by these deposits upon hydrogenation could be explained in terms of their different reactivity to hydrogenation according to the three above-mentioned mechanisms.

REFERENCES

1. M. Belgued, H. Amariglio, P. Paréja and A. Amariglio, Nature 352 (1991) 789 ; H. Amariglio and J. Saint-Just, French Patent application 9,009,340 (July 20th 1990) ; H. Amariglio and J. Saint-Just, US Patent application 5,414,176 (May 9th 1995).
2. T. Koerts and R. A. van Santen, J. Chem. Soc., Chem. Comm. (1991) 1281 ; T. Koerts and R. A. van Santen, UK Patent application 2,253,858A (March 21st 1991).
3. T. Koerts and R. A. van Santen, J. Mol. Catal. 74 (1992) 185.
4. M. Belgued, A. Amariglio, P. Paréja and H. Amariglio, H., J. Catal. 159 (1996) 441 and 449.
5. G.C. Bond and Z. Paál, Appl. Catal. A: General 86 (1992) 1.
6. E. Mielczarski, S. Monteverdi, A. Amariglio and H. Amariglio, Appl. Catal. A: General 104 (1993) 215.
7. H. Amariglio, M. Belgued, P. Paréja and A. Amariglio, Catal. Lett. 31 (1995) 19.
8. E. Marceau, J. M. Tatibouët, M. Che and J. Saint-Just, to be submitted for publication.
9. T. Koerts, M. J. A. Deelen and R. A. van Santen, J. Catal. 138 (1992) 101.

NATURAL GAS CONVERSION V
Studies in Surface Science and Catalysis, Vol. 119
A. Parmaliana et al. (Editors)

Preparation of Fluidized Catalysts by Spray-Dry Method and their Catalytic Performance for the Oxidative Coupling of Methane

T.Wakatsuki[1], H.Okado[1], K.Chaki[1], S.Okada[1], K.Inaba[1], M.Yamamura[1], T.Takai[2] and T.Yoshinari[3]

[1] JAPEX Research Center, Japan Petroleum Exploration Co. Ltd.
1-2-1 Hamada Mihama-ku, Chiba 261-0025, Japan
[2] Technology Research Center, Japan National Oil Corporation
1-2-2 Hamada Mihama-ku, Chiba 261-0025, Japan
[3] Research & Development Center, Cosmo Research Institute
1134-2 Gongendo Satte, Saitama 340-0112, Japan

1. Introduction

A number of the engineering approaches by the use of a bubbling fluidized-bed reactor for the oxidative coupling of methane (OCM) have been reported [1, 2]. In general, it is necessary that a catalyst for a bubbling fluidized-bed reactor has specific particle properties (e.g. a spherical shape, smooth surface, attrition resistance, particle size, bulk density, etc.). In this point of view, the preparation of fluidized catalysts are very significant. However, it seems that anyone has never investigated the preparation of OCM catalysts for a fluidized-bed reactor.

We have reported the catalytic performance of natural calcium compounds (NCC) catalysts [3], using a fluidized-bed reactor [4]. We study the preparation of the fluidized catalysts by spray-dry method on the basis of the NCC catalysts. Then, in this paper, we describe the preparation of fluidized catalysts by spray-dry method, and their catalytic performance.

2. Experimental

2.1. Preparation of catalysts

The fluidized NCC (Fluid-NCC) catalysts were prepared as follows : seashells as NCC, were washed and then dried at 393 K for 24 h, followed by calcining at 1173K for 10 h in air. The calcined seashells were ground to fine powder (under 1 μ m), and mixed with some binders and water. The obtained materials were spray-dried, and then were calcined again at sintering temperature.

2.2. The OCM reaction in a fixed-bed reactor

The catalysts were tested in a conventional fixed-bed alumina tube reactor of 11 mm i.d. at atmospheric pressure. The Fluid-NCC catalysts was plugged in the center of the reactor. A reaction gas mixture of methane and oxygen (CH_4/O_2 mole ratio = 9) without a diluent was introduced to the reactor (flow-rate = 100 Nml/min) which was heated to a reaction temperature of 873 - 1173 K. After about 30 min, a part of the effluent gas was introduced into an on-line gas chromatograph for analysis. The products with carbon number of more than two (ethane, ethylene, C_3 hydrocarbons and higher) are hereafter defined as C_2+ hydrocarbons.

2.3. The OCM reaction in a fluidized-bed reactor

The Fluid-NCC catalysts were also tested in a bubbling fluidized-bed quartz tube reactor of 22 mm i.d. at atmospheric pressure. A gas distributor was made of quartz frits with an average pore size of about 100 μ m. The minimum fluidization velocity (Umf) of the Fluid-NCC catalysts was 0.3 - 1.0 cm/sec. A reaction gas mixture of methane and oxygen (CH_4/O_2 mole ratio = 4 - 10) without a diluent was introduced to the reactor which was heated to a reaction temperature of 923 - 1123 K. Gas velocity of the reactor was in the range from 2.3 to 31.4 cm/sec at reaction temperature.

3. Results and Discussion

3.1. Catalytic performance of the Fluid-NCC in the fixed-bed reactor

The Fluid-NCC catalysts were prepared by spray-dry method with a binder such as SiO_2, Al_2O_3, SiO_2-Al_2O_3, ZrO_2, clay, etc. and a flux such as alkali metal elements, boron, lead glass, etc. All Fluid-NCC catalysts were spherical and rigid. Fig. 1 shows SEM micrograph of a typical Fluid-NCC catalyst (JC-F). Table 1 shows physical properties of the Fluid-NCC catalysts and catalytic performance in the fixed-bed reactor. The Fluid-NCC catalyst prepared with only binders (JC-A), high content of boron (JC-E) and lead glass (JC-G) had poor catalytic performance. The poor catalytic performance of JC-E is due to calcium borate which is detected by XRD and is inert for the OCM, and that of JC-G is due to complete oxidation of hydrocarbons by lead. While, catalytic performance of the Fluid-NCC catalysts with both of binders and flux (JC-B, C, D, F) were just a little bit low compared with that of the original NCC catalyst. The Fluid-NCC catalyst with lithium (JC-B) had excellent catalytic performance, although its catalytic performance decreased with time on stream. This decreased catalytic performance is attributed to decrease surface area by melting of lithium carbonate.

JC-C, D and F have good specific particle properties. And, JC-F has the greatest particle density of 0.94 g/cm^3 which is one of dominant properties in the fluidized OCM reaction. The particle density of the catalyst, which is affected by calcination temperature, has an influence on diffusion of reactants and products (bubble phase - cloud phase and cloud phase - emulsion phase) in the catalyst bed. Then, the physical properties and the catalytic performance of JC-F were studied by changing the binders content, the flux content and the calcination temperature, as summarized in table 2. The surface area and the OCM performance decreased with increasing the binders content and the flux content, however, the particle density and a average particle size

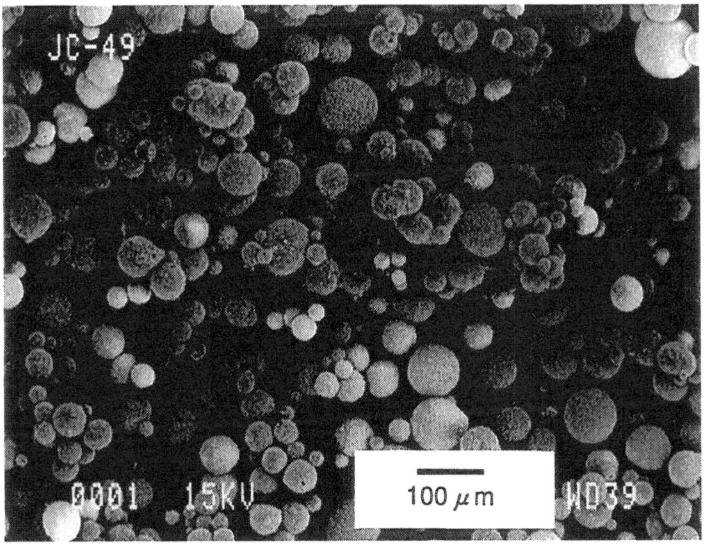

Fig. 1 SEM micrograph of typical Fluid-NCC (JC-F).

increased. From observation of SEM, JC-F1 is agglomerated by calcination, because of the excess binders and flux. For JC-F2 - F4, the surface area decreased with increasing calcination temperature, whereas the particle density increased. The OCM performance showed a maximum at calcination temperature of 1473 K. It can be seen that the OCM performance is dependent on the surface area, which is good agreement with the results of Iwamatsu and Aika[5]. Then, a maximum C_2+ yield of 13.2 %

Table 1 Physical properties of the fluid-NCC catalysts and their catalytic performance in the fixed-bed reactor.

	components (mol%)								recalc.	surface	particle	CH_4*	C_2+*
catalyst	NCC	binder	flux						temp.	area	density	conv.	selec.
			Li	Na	K	B	Pb	others	(K)	(m^2/g)	(g/cm^3)	(%)	(%)
JC-A	94.6	5.4							1273	3.5	0.27	11.9	58.6
JC-B	76.5	4.7	18.8						873	0.4	0.68	15.7	78.8
JC-C	84.7	5.2		10.1					1173	1.8	0.54	13.0	68.9
JC-D	87.8	5.3			6.9				1173	2.0	0.56	15.8	74.3
JC-E	73.5				8.9	17.6			1373	0.3	0.55	11.5	62.6
JC-F	69.3	16.1		0.5	6.5	4.8		2.8	1473	0.5	0.94	15.2	69.0
JC-G	74.6	14.3		0.8	7.9		2.4		1223	0.6	0.56	11.4	61.4
NCC	100								1123	1.1	0.85	16.6	77.3

* : CH_4/O_2 = 9 (without a diluent), flow-rate = 100 Nml/min, GHSV = 4000 h^{-1}.

Table 2 Physical property of the fluid-NCC catalysts and their catalytic performance in the fixed-bed reactor.

catalyst	components (mol%) NCC	binder	flux	recalc. temp. (K)	particle size (μm)	surface area (m^2/g)	particle density (g/cm^3)	CH_4* conv. (%)	C_2+* selec. (%)
JC-F1	69.3	16.1	14.6	1473	112.3	0.5	0.94	15.2	69.0
JC-F2	79.4	10.3	10.3	1273	88.3	3.9	0.55	16.6	72.8
JC-F3	79.4	10.3	10.3	1373	76.1	2.4	0.64	16.0	76.5
JC-F4	79.4	10.3	10.3	1473	78.7	1.3	0.65	17.1	77.3
JC-F5	79.4	10.3	10.3	1573	92.0	0.5	0.97	14.9	71.1

* : $CH_4/O_2 = 9$ (without a diluent), flow-rate = 100 Nml/min, GHSV = 4000 h^{-1}.

(methane conversion = 17.1 %, C_2+ selectivity = 77.3 %) was obtained by JC-F4. The catalytic performance of the JC-F4 was nearly equal to that of the original NCC catalyst.

3.2. Catalytic performance of the Fluid-NCC in the fluidized-bed reactor

The OCM in the bubbling fluidized-bed reactor was carried out over JC-F4. Good fluidization for JC-F4 at flow-rate over an Umf of 0.37 cm/sec was obtained without particle agglomeration. Fig. 2 shows the relationship between contact time (GHSV)

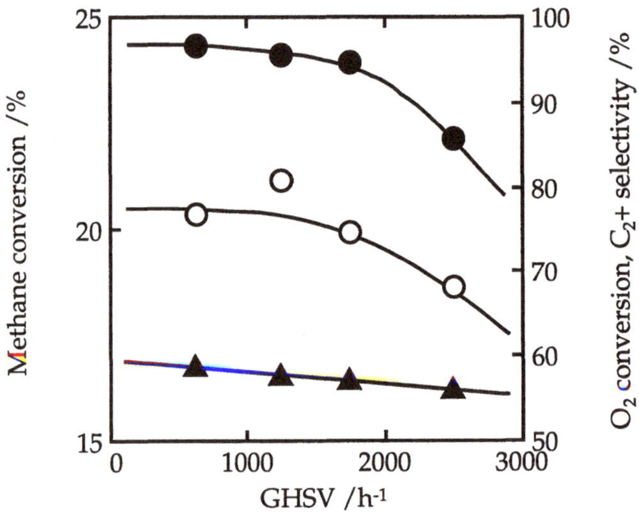

Fig. 2. Dependence of methane and oxygen conversion and C_2+ selectivity on contact time over JC-F4. ○: methane conversion, ●: oxygen conversion, ▲: C_2+ selectivity. Conditions : reaction temperature = 1073 K, CH_4/O_2 mole ratio = 5.

and the catalytic performance of JC-F4. At a CH_4/O_2 mole ratio of 5 and reaction temperature of 1073 K, the effect of GHSV on catalytic performance was investigated by changing the flow-rate of reaction gas. The conversions of methane and oxygen decreased with increasing the GHSV. In the fixed-bed reactor, oxygen was almost spent at GHSV of 4000 h^{-1}, however, oxygen was almost consumed at GHSV of 1800 h^{-1} in the fluidized-bed reactor. The fluidized OCM reaction requires longer contact time for the consumption of oxygen than in the fixed-bed reaction. From this results, it can be seen that the diffusion of reactants and products between the bubble phase and the emulsion phase influences the OCM performance in the fluidized-bed reaction. On the other hand, the C_2+ selectivity decreased gradually with GHSV. Residual oxygen in off-gas deceased the C_2+ selectivity. It is suggested that a back-mixing of residual oxygen and/or the non-selective gas-phase reactions of C_2+ hydrocarbons in a dilute phase influence the C_2+ selectivity.

Fig. 3 shows the OCM performance of JC-F4 as a function of CH_4/O_2 mole ratio at GHSV of 1250 h^{-1}, reaction temperature of 1073 K. The methane conversion decreased with increasing CH_4/O_2 mole ratio, whereas C_2+ selectivity increased with the increase in CH_4/O_2 mole ratio. A maximum C_2+ yield in the fluidizied-bed reactor of 12.2 % (methane conversion =22.9 %, C_2+ selectivity = 53.4 %) was obtained at CH_4/O_2 mole ratio =5. This result shows that C_2+ hydrocarbons are oxidized to CO_x consecutively at the high oxygen concentration.

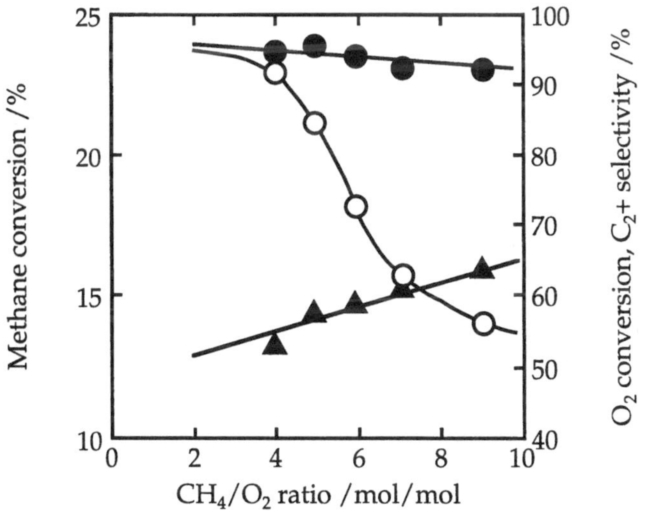

Fig. 3. Dependence of methane and oxygen conversion and C_2+ selectivity on CH_4/O_2 mole ratio over JC-F4. ○: methane conversion, ●: oxygen conversion, ▲: C_2+ selectivity. Conditions : reaction temperature = 1073 K, flow-rate = 25 NL/h, GHSV = 1250 h^{-1}.

From these results, it can be concluded that the diffusion of reactants and products and the non-selective gas-phase reactions of hydrocarbons in bubbles, as they are very important factors in the fluidized OCM reaction, are likely to be related to the decrease in C_2+ hydrocarbon selectivities at the high oxygen concentration.

4. Summary

The following conclusions are drawn from the results of the preparation of the Fluid-NCC catalysts by spray-dry method and their catalytic performance for the OCM reaction in the bubbling fluidized-bed reactor.

(1) The Fluid-NCC catalysts have excellent catalytic performance and good particle properties for the OCM reaction.

(2) The maximum C_2+ yield in the fixed-bed reactor was 13.2 % (methane conversion = 17.1 %, C_2+ selectivity = 77.3 %), under the following conditions ; reaction temperature = 1073 K, CH_4/O_2 mole ratio = 9, flow-rate = 100 Nml/min, GHSV = 4000 h^{-1}.

(3) The maximum C_2+ yield in the fluidized-bed reactor was 12.2 % (methane conversion = 22.9 %, C_2+ selectivity = 53.4 %), under the following conditions ; reaction temperature = 1073 K, CH_4/O_2 mole ratio = 5, flow-rate = 25 NL/h, GHSV = 1250 h^{-1}.

(4) The catalytic performance in the fluidized OCM reaction is related to the diffusion of reactants and products and the non-selective gas-phase reactions of hydrocarbons in bubbles.

This work is a part of the project named "Direct Conversion of Natural Gas to Liquid Fuels", a special program of Japan National Oil Corporation.

We wish to thank Mr. S. Fujii, Mr. K. Oohama and their colleagues(Catalysts & Chemicals Industries Co., Ltd.) for their helpful discussions on the preparation of fluidized catalysts by the spray-dry method.

References

[1] J. H. Edwards, R. J. Tyler and S. D. White, Energy & Fuels, 4 (1990) 85.
[2] R. Andorf, L. Mleczko, D. Shcweer and M. Baerns, Can. J. Chem. Eng., 69 (1991) 891.
[3] M.Yamamura, H.Okado, N.Tsuzuki, K.Chaki, T.Wakatsuki, K.Inaba, S.Suzuki and S.Kitada, Stud. Surf. Sci. Catal., 81 (1994) 253.
[4] T. Wakatsuki, M. Yamamura, H. Okado, K. Chaki, S.Okada, K. Inaba, S. Suzuki and T.Yoshinari, Stud. Surf. Sci. Catal., 107 (1997) 319.
[5] E. Iwamatsu and K. Aika, J. Catal., 117 (1989) 416.

NATURAL GAS CONVERSION V
Studies in Surface Science and Catalysis, Vol. 119
A. Parmaliana et al. (Editors)

Mechanism of " Chloro-Pyrolysis " of Methane

Paul-Marie MARQUAIRE* and Marwan AL KAZZAZ

Département de Chimie Physique des Réactions
CNRS UMR 7630, ENSIC - INPL Université

address: DCPR-ENSIC - BP 451 - 1 rue Grandville - 54001 NANCY Cedex - FRANCE
fax: 33- 383 37 81 20

A detailed radical mechanism of "chloro-pyrolysis" of methane is proposed, it explains the formation of the vinyl chloride at high temperature (around 1000°C) in agreement with our experimental results and other kinetic studies. The analysis of mechanism allows to find the best operating conditions for increasing the vinyl chloride yield.

1. INTRODUCTION

Due to the interest in natural gas upgrading, numerous studies have been carried out on the methane conversion into higher value, transportable materials. Vinyl Chloride is today manufactured from petroleum via ethylene, but the Natural Gas could be an alternative feedstock by the new *Methane to Vinyl Chloride (MTVC) process.* It is a two step process in which the first step involves the chlorination of methane (or methanol) using any well known methods of the art. The second step converts the methyl chloride to vinyl chloride by a new "chloro-pyrolysis" reaction, it is a CH_3Cl / Cl_2 *gas phase reaction* at high temperature, under no flame condition ... In specific conditions [1], the reaction produces mostly vinyl chloride, acetylene and HCl :

$$CH_3Cl + x\,Cl_2 \rightarrow C_2H_3Cl,\ C_2H_2,\ HCl$$

The "chloro-pyrolysis" produces C_2 hydrocarbons even though it is well known that the thermal reaction between CH_3Cl and Cl_2 is a chlorination reaction which produces chloromethanes.

A first experimental study of CH_3Cl / Cl_2 thermal reaction [2] has shown that the temperature effect on the selectivities is very strong. At 950°C, a space time of 40ms and 5% of chlorine, the CH_3Cl conversion is about 15% and the chlorine conversion is total; we obtain a vinyl chloride selectivity of 30%, and 30% for the C_2 hydrocarbons (C_2H_2+ C_2H_4). Other products are CH_2Cl_2 , CH_4 , C_4H_4 and HCl. For these operating conditions, the detail of selectivities is given Table 1.

Table I : Selectivities (%) of CH_3Cl chloro-pyrolysis at 950°C, 5% Cl_2 , 40ms.

C_2H_3Cl	C_2H_2	C_2H_4	C_4H_4	CH_4	CH_2Cl_2
32	18	9	6	15	20

In this paper, a detailed radical mechanism is proposed, it explains *qualitatively* our experimental results of "chloro-pyrolysis", and it is in agrement with other kinetic studies [3-8]. The analysis of mechanism will allow to find the best operating conditions for increasing the vinyl chloride yield.

As our "Chloro-pyrolysis of Methyl Chloride", other gas phase processes use chlorine for methane activation : - the Benson process [9]: methane/chlorine flame - the Gorin process [10]: pyrolysis of CH_3Cl - the "CCOP" Senkan process [11]: oxy-pyrolysis of CH_3Cl . These "chlorine catalysed" processes use the chlorine to produce C_2 non-chlorinated hydrocarbons, even when our reaction produces a chlorinated hydrocarbon, the vinyl chloride.

2. MECHANISM

The *primary mechanism* allows to understand the reaction, in particular the very strong influence of temperature and the formation of the vinyl chloride at high temperature (around 1000°C).

At low temperature (500°C) the reaction is a long chain reaction of chlorination, the mechanism proposed by Kurtz [3] is:

initiationsteps:

$$Cl_2 + M \rightarrow 2\ Cl\cdot + M \quad (1)$$

$$CH_3Cl \rightarrow CH_3\cdot + Cl\cdot \quad (1')$$

propagation steps

$$Cl\cdot + CH_3Cl \rightarrow HCl + CH_2Cl\cdot \qquad (2)$$

$$CH_2Cl\cdot + Cl_2 \rightarrow CH_2Cl_2 + Cl\cdot \qquad (3)$$

termination step

$$CH_2Cl\cdot + CH_2Cl\cdot \rightarrow C_2H_4Cl_2 \qquad (4)$$

With a negative "activation energy", the primary chain length decreases when the temperature increases.

At high temperature, it is a short chain reaction with radical coupling reactions, and the radicals concentrations are:

$$CH_2Cl\cdot > CH_3\cdot >> Cl\cdot$$

according to these recombination reactions (4) and (5):

$$CH_2Cl\cdot + CH_3\cdot \rightarrow C_2H_5Cl \qquad (5)$$

The methane formation comes from:

$$CH_3\cdot + CH_3Cl \rightarrow CH_4 + CH_2Cl\cdot \qquad (6)$$

$$CH_3\cdot + HCl \rightarrow CH_4 + Cl\cdot \qquad (7)$$

The *secondary* reactions explain the formation of others products. The decomposition of $C_2H_4Cl_2$ and C_2H_5Cl leads to vinyl chloride and ethylene very rapidly:

$$C_2H_4Cl_2 \rightarrow C_2H_3Cl + HCl \qquad (8)$$

$$C_2H_5Cl \rightarrow C_2H_4 + HCl \qquad (9)$$

The major path of consumption of these products are the molecular decomposition by HCl elimination. Reactions with radicals such as $CH_2Cl\cdot$ and $Cl\cdot$ also contribute to the destruction process, but only to a minor extent at high temperature.

The *next* reactions are the decomposition of vinyl chloride lead to acetylene by the same dehydrochlorination:

$$C_2H_3Cl \rightarrow C_2H_2 + HCl \qquad (10)$$

and the formation of vinylacetylene C_4H_4. Two pathways are possible, the addition of $C_2H_3\cdot$ or $C_2H\cdot$ radicals to acetylene [12-13], or the polymerization pathway involving $CH_2Cl\cdot$ [14]. As noted above, $CH_2Cl\cdot$ is the principal radical in our experimental conditions. The addition of $CH_2Cl\cdot$ to C_2H_2 forms a chloropropenyl radical (11) that can isomerize to an allylic radical by a 1-3 H atom (or Cl atom) shift reaction (12). The recombination (13) is followed by further dehydrochlorinations to produce the vinylacetylene:

$$CH_2Cl\cdot + C_2H_2 \rightarrow \cdot CH=CH\text{-}CH_2Cl \quad (11)$$

$$\cdot CH=CH\text{-}CH_2Cl \rightarrow C_3H_4Cl\cdot \quad (12)$$

$$CH_2Cl\cdot + C_3H_4Cl\cdot \rightarrow C_4H_6Cl_2 \quad (13)$$

$$C_4H_6Cl_2 \rightarrow C_4H_4 + 2\,HCl \quad (14)$$

The major reaction channels of the "chloro-pyrolysis" of methane have been identified, and they are presented in this figure:

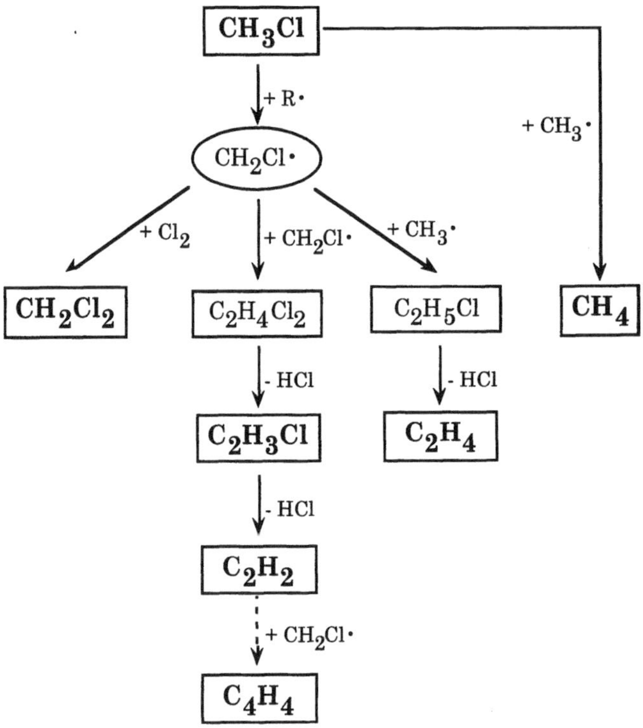

The formation of methane and ethylene indicates that the methyl radicals are important too in our system. Consequently, the polymerization pathway involving $CH_3\cdot$ can produce also the vinylacetylene.

According to the proposed mechanism, the vinyl chloride is the main precursor of acetylene, but another possibility is the ethylene.

It is particularly important to note that $CH_2Cl\cdot$ radical concentration controls the C_2 selectivity by the competition between two elementary reactions:

$$CH_2Cl\cdot + CH_2Cl\cdot \rightarrow C_2H_4Cl_2 \quad (4)$$

and

$$CH_2Cl\cdot + Cl_2 \rightarrow CH_2Cl_2 + Cl\cdot \quad (3)$$

3. DISCUSSION

In this paper, we only dicuss in a qualitative way the most important elementary processes, which are thought to explain the formation of the products and the variations of selectivities with the temperature. In a work in progress, a detailed quantitative modeling is developed; this model would be useful for an improvement of the reactor configuration and of the operating conditions in order to increase the vinyl chloride yield.

Our experimental results on CH_3Cl / Cl_2 reaction can be interpreted by a competition between two major pathways :

- the "chloro-pyrolysis" (short chain reaction) :

$$CH_3Cl + Cl_2 \rightarrow C_2H_4Cl_2 \rightarrow \boxed{C_2H_3Cl} \rightarrow \boxed{C_2H_2} \dashrightarrow \boxed{C_4H_4}$$

- the "classical" chlorination (long chain reaction) :

$$CH_3Cl + Cl_2 \rightarrow \boxed{CH_2Cl_2}$$

At 950°C and 5% of chlorine, the chlorine conversion is total for a space time of 40 ms, and the formation of acetylene is important; for increasing the vinyl chloride yield, there are two ways:

- to decrease the space time, with keeping a total conversion of chlorine,
- to increase the efficiency of the quench in order to reduce the decomposition of vinyl chloride to acetylene.

4. CONCLUSION

A detailed radical mechanism of "chloro-pyrolysis" of methane is proposed, it explains qualitatively the formation of the vinyl chloride at high temperature (around 1000°C) in agrement with our experimental results and other kinetic studies. The analysis of mechanism allows to find the best operating conditions for increasing the vinyl chloride yield.

At high temperature the reaction CH_3Cl / Cl_2 is not a chlorination reaction but a "chloro-pyrolysis" similar to the pyrolysis. The presence of chlorine induces a pyrolysis at lower temperature and/or lower reaction time. In these conditions, the addition of chlorine has a beneficial effect on the pyrolysis of CH_3Cl : the reaction can produce C_2H_3Cl with a limited decomposition to acetylene, but the chlorine gives two by-products : CH_2Cl_2 and HCl.

ACKNOWLEDGEMENT
This work has been funded by the CNRS and Gaz de France (GDF).

REFERENCES

1. P.M. Marquaire, Y. Muller and M. Al Kazzaz, Fr Patent No 2 711 649 (1995).
2. P.M. Marquaire, M. Al Kazzaz, Y. Muller and J. Saint Just, Studies in Surface Science and Catalysis,107 (1997) 269.
3. B.E. Kurtz,Ind. Eng. Chem. Process Des. Develop., 11 (1972) 332.
4. M. Weissman and S.W. Benson, Int. J. Chem. Kin., 16 (1984) 307.
5. S.B. Karra and Senkan S.M., I&EC. Res., 27 (1988) 1163.
6. R. Yildirim and S.M. Senkan , I&EC. Res., 34 (1995) 1842.
7. E. Ranzi, M Dente M. Rovaglio, T.Faravelli and S.B. Karra , Chem. Eng. Comm., 117 (1992) 17.
8. J.F. Roesler, R.A. Yetter and F.L Dryer , Combust. Sci. and Tech., 101 (1994) 199.
9. S.W. Benson, US Patent No. 4 199 533 (1980).
10. E. Gorin, US Patent No. 2 320 274 (1943).
11. S.M. Senkan, US Patent No. 4 714 796 (1987).
12. M. Frenklach, D.W. Clary, T. Yvan, W.C. Jr. Gardiner and S.E. Stein, Combust. Sci. and Tech., 50 (1986) 79.
13. M. Frenklach, J.P. Hsu, D.L. Miller and R.A. Matula, Combustion and Flame, 64 (1986) 141.
14. M. Weissman and S.W. Benson, Prog. Energy Combust. Sci., 15 (1989) 273.

NATURAL GAS CONVERSION V
Studies in Surface Science and Catalysis, Vol. 119
A. Parmaliana et al. (Editors)

Mechanistic Study of Benzene Formation in CH_4-CO Reaction over Rh/SiO_2

Shuichi Naito*, Tadahiko Karaki, Toshiaki Iritani and Masaru Kumano

Department of Applied Chemistry, Faculty of Engineering, Kanagawa University, 3-27-1, Rokkakubashi, Kanagawa-ku, Yokohama, 221, Japan.

Benzene was formed selectively among hydrocarbons in CH_4-CO reaction over silica supported Rh catalysts, at 573-723K under atmospheric pressures. Accumulation of surface carbons, which come from both CO and methane, is important for the formation of benzene. C_1 building blocks for benzene formation (CH) are different from those for other hydrocarbons (CH_2), whose concentration depends strongly on the amount of surface carbon and hydrogen.

1. INTRODUCTION

Many attempts have been made to activate methane under non-oxidative conditions and to convert it into higher hydrocarbons. They are divided into two main groups depending on the catalysts employed. One approach is a two step sequence of decomposition and hydrogenation of methane over group VIII transition metals [1-3]. Its first step involves dissociative adsorption and decomposition of methane to leave various kinds of carbonaceous species on the reduced metal surface, which are hydrogenated to higher hydrocarbons in the second step. This latter step seems to be analogous to the carbon-carbon bond formation process in Fischer-Tropsch synthesis. Accordingly, the formed hydrocarbons are not selective, obeying the Schulz-Flory equation. The other approach is to achieve selective formation of benzene directly from methane at higher temperatures over ZSM-5, silica or alumina supported MoO_3 catalysts, where molybdenum carbide is considered to be responsible for the production of ethylene, and the formed ethylene would be trimerized to benzene on the acidic sites of the ZSM-5 or alumina support [4-7].

Recently, we have found that benzene is formed selectively among hydrocarbons in CH_4-CO reaction over silica supported Rh, Ru and Pd catalysts, under atmospheric pressures, although more than 90% of the products was CO_2 [8]. When CH_4 alone was introduced onto the freshly reduced catalysts, ethane and ethylene were the only products, and when only CO was introduced, CO_2 was the only product. No benzene was detected in either case. These results indicate that both CH_4 and CO are required for benzene formation.

These reactions are not particularly promising when considering practical use, because the selectivity for benzene was at most 10% over any of the catalysts investigated. But from a mechanistic point of view, it seems to be very interesting, because the catalytic behavior looks more like the case of molybudenum/ZSM-5 catalysts than group VIII transition metals, although we are not using acidic supports like ZSM-5. Accordingly, we focused our attention on this unique catalytic behavior, and tried to elucidate the mechanism of benzene formation in connection with that of CO hydrogenation over silica supported Rh metal catalysts.

2. EXPERIMENTAL

The catalysts were prepared by a conventional impregnation method, employing metal chloride salts as catalyst precursors. After drying, they were reduced by hydrogen and 5 wt % and 20 wt % supported catalysts were prepared. For catalyst characterization, transmission electron microscopy was employed to estimate the metal particle sizes, and hydrogen adsorption at room temperature was measured to determine metal dispersions. Table 1 summarizes the particle sizes of Rh metals, determined by TEM photograph and metal dispersion, from which turnover frequencies (TOF) were estimated.

The reaction was carried out in a closed gas circulation system, using a liquid N_2 cold trap to gather the primary reaction products and shift the reaction equilibrium. The composition of the gas phase as well as the trapped product was analyzed by three different columns of gas chromatography. The formation of benzene was confirmed by mass spectroscopy as well as infrared spectra.

Table 1. Dispersion and particle size of Rh/SiO_2.

Loading	Particle size(A)		Dispersion(%)	
	TEM	H_2 ads.	TEM	H_2 ads.
5wt%	20.2	22.4	54	49
20wt%	--	47.8	--	23

In the case of infrared spectroscopic experiments, the catalyst was pressed into a 20 mm diameter disk and put into the infrared cell, which was connected to a closed gas circulation system. Infrared spectra were recorded with a JEOL Diamond 20 Fourier-transform IR spectrometer, with a liquid nitrogen cooled HgCdTe detector. Spectral resolution was 2 cm^{-1} in the region of 4400-400 cm^{-1}.

3. RESULTS AND DISCUSSION

Figure 1(a) shows the time courses of the CH_4-CO reaction over freshly reduced 5 wt % Rh/SiO_2 catalyst at 623K. At the initial stage of the reaction, a considerable amount of CO_2 was formed accompanied with the decrease of gaseous CO. The amount of accumulated carbon on the catalyst surface was estimated from the mass-balance of the gas phase. The benzene formation

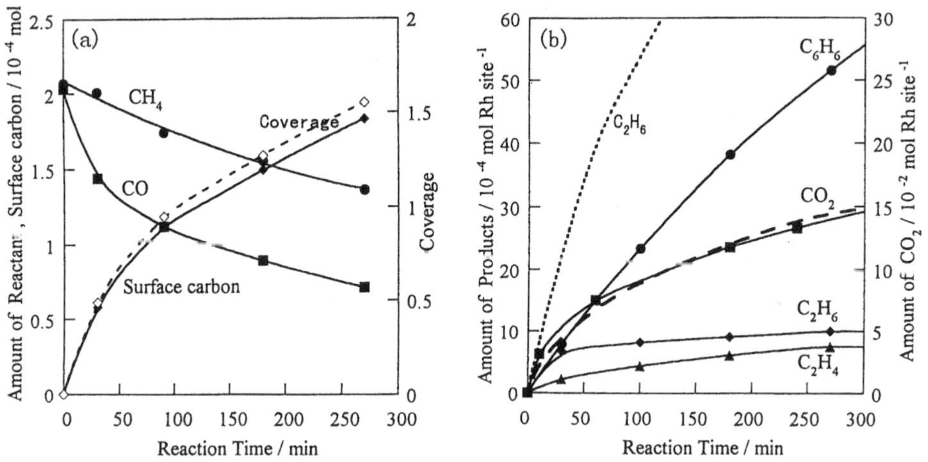

Figure 1(a) and 1(b). CH_4-CO reaction over 5 wt% Rh/SiO_2 at 623K.

rate was two orders of magnitude slower than CO_2 formation at the initial stage, but increased linearly although the rates of other products decreased drastically after a few hours.

Figure 1(b) enlarges the time courses of the product formation in Figure 1(a). The rates of CO_2, ethane and ethylene formation decreased considerably after two hours, but the rate of benzene formation stayed almost constant . We therefore defined this surface as a steady state surface of CH_4-CO reaction, and all the following kinetical data were taken on this steady state surface by reintroducing the reaction gas after 2 hours. The broken line in the figure represents the time courses of CO_2 formation when only CO was introduced onto the freshly reduced Rh catalysts, which coincides well with the CO_2 formation in CH_4-CO reaction. These results indicate that at the initial stage surface carbon is mainly accumulated by the disproportionation of CO. The dotted line in the figure represents the time courses of ethane formation when only CH_4 was introduced onto the freshly reduced surface, which increased almost linearly with time and is different from the ethane in CH_4-CO reaction.

Table 2 summarizes turnover frequencies of CH_4-CO reaction as well as the disproportionation of CO and the coupling of methane over 5 and 20 wt % Rh/SiO2 at 623K. As summarized in Table 1, the particle sizes of 5 and 20 wt % catalysts are about 20 and 50 A respectively, and affect the TOF of disproportionation of CO and coupling of methane in different ways. Dissociation of CO seems to be faster on larger particles of Rh, whereas dissociation of methane is easier on smaller particles of Rh. The TOF of CO_2 formation was six times larger, but that of benzene formation was considerably smaller over larger particles of Rh metals. Accordingly, the selectivity for benzene formation was much better over 5 wt % catalysts and went up to about 10 % (more than 80 % in hydrocarbons) at the later stage of the reaction.

Pressure dependence of the initial rate of benzene formation upon the partial pressure of CH_4 and CO was investigated at 623K over a steady state surface of 5 wt % Rh/SiO_2, as shown in Figure 2. The empirical reaction orders for CH_4 and CO were 0.97 and -0.99 respectively, indicating that strongly adsorbed CO may inhibit the activation of methane. It is interesting to note that the addition of a small amount of hydrogen during CH_4-CO reaction accelerated the formation of benzene several times, which suggests that the supply of hydrogen from methane is the rate determining step in this reaction.

Table 2. TOFs of various reactions over 5 and 20 wt% Rh/SiO_2 at 623K.

Catalysts	Reactions	TOF ($x10^{-8}$ sec^{-1})				Selectivity (%)			
		C_2H_6	C_2H_4	C_6H_6	CO_2	C_2H_6	C_2H_4	C_6H_6	CO_2
5 wt%	CH_4-CO	0.24	0.28	3.33	28.2	0.8	0.9	10.4	88.0
	CH_4 coupling	13.3	0.02	0	0	99.8	0.2	0	0
	CO disprop.	0	0	0	300	0	0	0	100
20 wt%	CH_4-CO	0.37	0.06	0.56	194	0.2	0.03	0.3	99.5
	CH_4 coupling	1.94	0	0	0	100	0	0	0
	CO disprop.	0	0	0	444	0	0	0	100

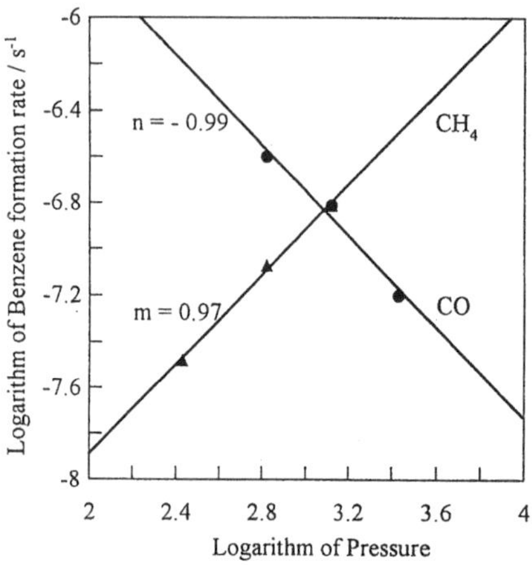

Figure 2. Pressure dependence of CH_4-CO reaction over 5wt% Rh/SiO_2 at 623K.

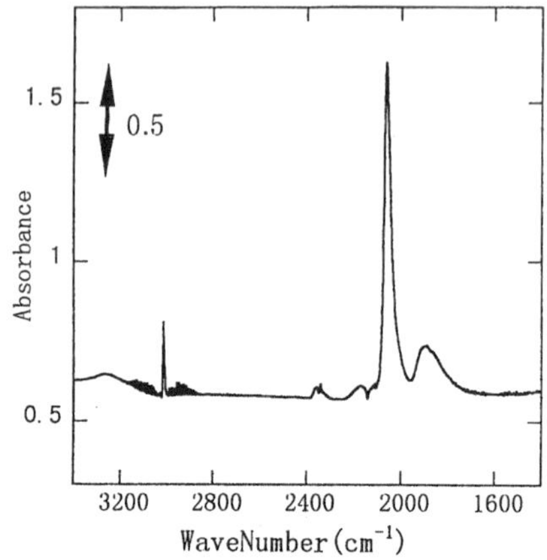

Figure 3. Infrared spectra of adsorbed species during CH_4-CO reactions over 5 wt%/SiO_2.

Figure 3 represents the infrared spectra of CH_4-CO reaction over 5 wt% Rh/SiO_2. CO was the only adsorbed species during the reaction, whose intensity decreased with time because of the accumulation of carbon on the metal surface. Accordingly, we utilized the decrease in intensity of adsorbed CO to estimate the amount of carbon accumulated on the metal surface and compared it with CO-H_2 reaction. The spectra in Figure 4 show the changes in intensity of the adsorbed CO during CH_4-CO and CO-H_2 reactions over Rh/SiO_2 catalysts at 623K. The spectra (A)-1 and (B)-1 were taken when the reaction gas was introduced at room temperature. (A)-2,-3,-4 and (B)-2,-3,-4 represent the spectra after 30, 60 and 120 min. of the CH_4-CO and CO-H_2 reactions, respectively. Each spectrum was taken after lowering the sample

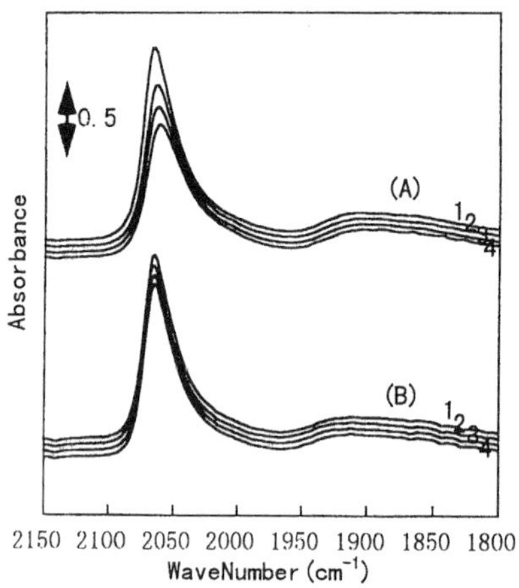

Figure 4. Spectral change of adsorbed CO during(A) CH_4-CO, and (B) CO-H_2 reactions over 5wt% Rh/SiO_2.

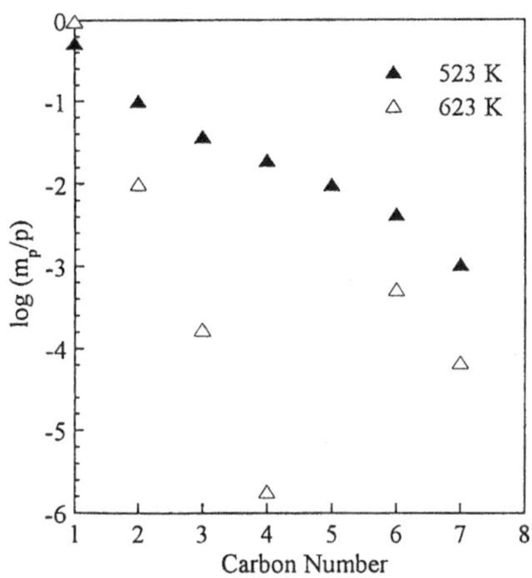

Figure 5. Schultz-Flory plots of CO-H_2 reaction over steady state surface of CH_4-CO reaction over 5 wt% Rh/SiO_2.

temperature to room temperature. In the case of CO-H_2 reaction, the amount of adsorbed CO did not change much, but for CH_4-CO reaction the intensity was reduced to about 60% after 120 min. of the reaction, indicating the accumulation of carbon on the Rh surface. Since the total amount of accumulated carbon could be estimated to be at 80-100 % of the surface Rh metal, some of the carbon must have been accumulated on the support, as well. No C-H bands at around 3000 cm^{-1} were observed during the reaction, which indicates that accumulated surface carbon species does not contain any hydrogen in it.

To clarify the mechanistic difference between these two reactions, CO-H_2 reaction was carried out over steady state CH_4-CO surfaces of 5 wt% Rh/SiO_2, and the product distribution was analyzed by the Schulz-Flory equation as shown in Figure 5. At 523K, the product distribution obeyed the Schulz-Flory plots up to C_7 hydrocarbons, indicating that ordinary CO-H_2 reaction would proceed even on a CH_4-CO steady state surface. When the reaction temperature was raised to 623K, the slope of the plots became much steeper, which indicates a reduction of the chain growth probability by a decrease in the concentration of chain carrier on the surface. At the same time a considerable amount of benzene was formed which did not obey the Schulz-Flory plots. These results indicate that benzene is produced through the different chain carriers from other hydrocarbons.

Table 3 summarizes the turnover frequency as well as the selectivity of CO-H_2 reaction over steady state CH_4-CO surfaces of 5 wt % Rh/SiO_2 catalyst, and compares them with those of CH_4-CO reaction. The TOF of benzene formation for CO-H_2 reaction was 1.5 times larger than that for CH_4-CO reaction, but its selectivity in hydrocarbons (except methane) was only 1.5 %. The TOFs of various products when only methane or CO was introduced onto the steady state CH_4-CO surface were also listed in Table 3. In the case of CO, a small amount of benzene was detected by the replacement between adsorbed species, but when methane was introduced onto the steady state surface, comparable amount of benzene with CH_4-CO reaction was detected at the initial stage, which is different from the case of freshly reduced catalysts. This result strongly suggests a that certain amount of building blocks for benzene formation is accumulated on the Rh surface during CH_4-CO reaction, which is hydrogenated to benzene by the supply of hydrogen from methane dissociation.

Table 3. TOFs of various transient response reactions over steady state CH_4-CO surface of 5 wt% Rh/SiO_2 at 623K.

Reactions	TOF (x10 $^{-7}$ sec^{-1})			selectivity(%)		
	C_2H_6	C_2H_4	C_6H_6	C_2H_6	C_2H_4	C_6H_6
CO-H_2	152	165	4.72	47.2	51.3	1.5
CH_4-CO	0.24	0.28	3.33	6.2	7.3	86.5
CH_4	17.4	1.43	3.43	78.2	6.4	15.0
CO	0	0	0.7	0	0	100

To investigate the reaction pathways for benzene formation more clearly, $^{12}CH_4$-^{13}CO reaction (^{13}C purity =90%) was carried out at 623K over 5 wt% Rh/SiO_2 catalyst, and isotopic distribution in the reactant and products were followed by mass spectroscopy. After 450 min, isotopic distribution in formed benzene was as follows: $^{13}C_6H_6$=20%, $^{13}C_5{}^{12}CH_6$=55%, and $^{13}C_4{}^{12}C_2H_6$=25%, which indicates that carbon atoms in benzene mainly come from CO, but some from methane as well. On the other hand, 24% of formed CO_2 contains ^{12}C atoms, while 100% of the ethane comes from methane carbon. Isotopic mixing of methane and CO carbons also took place during $^{12}CH_4$-^{13}CO reactions as shown in Figure 6. This carbon mixing may proceed through the various possible reactions between methane, CO and CO_2 as follows: $2CO \longrightarrow C(a) + CO_2$, $CO + 6H(a) \longrightarrow CH_4 + H_2O$, $2CH_4 + 7O(a) \longrightarrow CO + CO_2 + 4H_2O$, $CO_2 + 2H(a) \longrightarrow CO + H_2O$. The isotopic distribution of accumulated surface carbon was estimated by the hydrogenation of the deposited carbon. The formed methane contained approximately 35 % of ^{12}C and 65 % of ^{13}C. The isotopic distribution of formed benzene may be explained statistically from this isotopic distribution of accumulated active surface carbon, which indicates that C_2 hydrocarbons are not the building blocks for benzene formation.

In the case of CO-H_2 reaction, dissociatively adsorbed carbon is hydrogenated in sequence, forming CH(a), CH_2(a), and CH_3(a) species. And carbon-carbon bond formation takes place by the insertion of methylene chain carriers to the alkyl species. This is why produced hydrocarbons are not selective and obey the Schulz-Flory equation. On the other hand, in the case of CH_4-CO reaction, isotopic tracer experiments indicate the mixing of CO carbon and methane carbon in formed benzene and CO_2 as well as reactant CO and methane. Accordingly it is reasonable to suppose a certain carbon island on the metal surface, which mainly consists of C(a) and CH(a) species. Benzene may be formed in this carbon island.

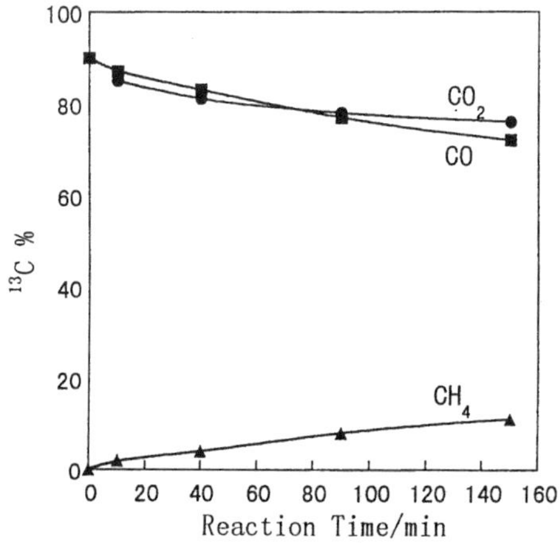

Figure 6. Time courses of the isotopic distribution of CO, CO_2 and CH_4 during $^{12}CH_4$-^{13}CO reaction.

4. CONCLUSION

(1) Accumulation of a certain surface carbon island, which originates from both CO and methane, is important for the formation of benzene in CH_4-CO reaction over Rh/SiO_2.
(2) Building blocks for benzene formation are not the C_2 hydrocarbons but probably C_1 species (CH), which are different from those for other hydrocarbons (CH_2). Their concentration strongly depends on the amount of surface carbon and hydrogen.
(3) The rate determining step for benzene formation may be the supply of surface hydrogen by the dissociation of methane.

REFERENCES

1. M. Belgued, P. Pareja, A. Amariglio and H. Amariglio, Nature, **352**, 789 (1991).
2. T. Koerts, J.A.G. Deelen, and R. A. van Santen, J. Catal., **138**, 101 (1992).
3. L. Wang, L. Tao, M. Xie, and G. Xu, Catal., Lett., **21**, 35 (1993).
4. L. Wang, Y. Xu, S-T. Wong, W. Cui, and X. Guo, Appl. Catal., **152**, 173 (1997).
5. D. Wang, J.H. Lunsford, and M.P. Rosynek, J. Catal., **165**, 150 (1997).
6. F. Solymosi, A. Erdoheyi, and A. Szoke, Catal. Lett., 32, 43 (1995).
7. F. Solymosi, J. Cserenyi, A. Szoke, and A. Oszko, J. Catal., 165, 150 (1997).
8. S. Naito, T. Karaki, and T. Iritani, Chem. Letts., 877 (1997).

NATURAL GAS CONVERSION V
Studies in Surface Science and Catalysis, Vol. 119
A. Parmaliana et al. (Editors)
1998 Elsevier Science B.V.

Simulation of the Non-oxidative Methane Conversion with a Catalytically Active Carbonaceous Overlayer

M. Wolf, O. Deutschmann, F. Behrendt, and J. Warnatz

Universität Heidelberg,
Interdisziplinäres Zentrum für Wissenschaftliches Rechnen,
Im Neuenheimer Feld 368, D-69120 Heidelberg, Germany

Abstract:

The oxygen-free conversion of methane on transition-metal catalysts could be an interesting alternative to the oxygen-containing conversion (OCM) due to its higher selectivity towards higher hydrocarbons (C_{2+}). So far, the main obstacle has been the low conversion rate compared to the OCM process. This is mostly ascribed to the fact that oxygen-free CH_4-conversion is accompanied by a quick deposition of a carbonlayer on the metal surface under atmospheric pressure. This carbonlayer consists of several different species and has been held responsible for catalyst poisoning by the catalyst models discussed in literature. Kinetic simulations with a new catalyst model assuming one of the species forming the carbonlayer, a carbidic-like species C(s), being catalytically active towards CH_4-conversion yields qualitatively good accordance with experimental data.

1. Introduction

The investigation of conversion of methane to higher hydrocarbons (commonly described as C_{2+}) has focused mainly on oxidative conversion. Here, yields are much higher than for the non-oxidative or de-hydrogenative coupling of methane which is accomplished by using a two-step process. The motivation for research on non-oxidative conversion is its much higher selectivitiy with respect to C_{2+} than for oxidative conversion where the main products are CO and CO_2. In order to optimise the relatively low yield of non-oxidative conversion, one has first to understand the detailed surface mechanism. A kinetic model has been developed for this process, with carbidic carbon being catalytically active in promoting CH_4 adsorption. This model describes qualitatively well the experimental data of Belgued et al. [1].

2. Model and Simulation

The simulation is based on a flow reactor where the gas flow can be described as stagnation-point flow. The corresponding governing equations for the gas phase can be used in their one-dimensional form keeping the computational requirements low. The transport to the surface as well as in the gas phase is described by a detailed transport model. Details on the governing equations and boundary conditions can be found in Deutschmann et al. [2,3]. The chemical reactions in the gas phase and at the surface are modelled using elementary steps. The surface mechanism used consists of approximately 85 elementary reactions (forward and reverse reactions) containing C_1 and C_2 species. The thermodynamical data of a surface species (standard enthalpy of formation $\Delta H_{f(s)}^0(298\ K)$) are calculated from the standard enthalpy of formation of that species in the gas phase, considering its adsorption enthalpy on the metal surface (e.g., platinum). Apart from necessary adaptations for generating a consistent data set or due to lacking data, the adsorption enthalpies for all species used in the C_1 and C_2 part of the surface mechanism are taken from Shustorovich et al. [4].

The thermodynamical data calculated for standard states (T = 298 K) are used within a temperature range of 300-800 K without a temperature correction. This procedure should not cause any trouble because only the reaction enthalpies (the difference between the standard formationenthalpies of surface species involved in a reaction) are relevant and ΔC_p (difference of heat capacities) is assumed to be small due to compensation effect, see Benson [5]. The most important reactions within this surface mechanism are given in Table 1. Reactions R3 and R4 are accelerated with increasing carbon coverage due to increasing lateral repulsion between adsorbed carbon atoms, see Tontegode et al. [6]. In the simulation the C(s) coverage dependent repulsion is represented by an increase of the standard enthalpy of formation for C(s) of up to 30 kJ/mol of carbon atoms involved in R3 and R4. In R2 no C(s)-coverage dependence has been regarded due to the fact that one of the product species, ethylidyne CCH_3(s), is bound similarly to the Pt surface as C(s). Hence, we assumed that CCH_3(s) experiences the same lateral repulsion by an increasing C(s) coverage. Therefore, taking into account a carbon coverage dependent activation enthalpy for forward and reverse reaction will have the same impact on the results as taking into account none.

Table 1 Surface reaction mechanism for the adsorption of methane on platinum (units: preexponential factor A [mol, cm, s], activation enthalpy E_a [kJ/mol], sticking coefficient S^0 [-]). The index (s) describes a surface species; Pt(s) denotes bare surface sites; $\Theta_{C(s)}$ describes the dependence of the activation enthalpy on the C(s) coverage. CCH_3 is the ethylidene species, $CHCH_3$ is a carbene-like C_2 species.

	reaction number			A	E_a	S^0
R1	CH_4 + 2 Pt(s)	-->	CH_3(s) + H(s)		72.2	$5.0\cdot10^{-4}$
R2	CH_4 + C(s) + Pt(s)	-->	CCH_3(s) + H(s)		71.5	$1.6\cdot10^{-3}$
R3	CH_4 + C(s)	-->	$CHCH_3$(s)		50.2	$2.0\cdot10^{-7}$
				$\Theta_{C(s)}$	-30.0	
R4	CH_4 + 2 C(s)	-->	CCH_3(s) + CH(s)		159.4	$4.0\cdot10^{-2}$
				$\Theta_{C(s)}$	-60.0	
R5	CCH_3(s) + Pt(s)	-->	CH_3(s) + C(s)	$1.37\cdot10^{22}$	46.0	
R6	CH_3(s) + C(s)	-->	CCH_3(s) + Pt(s)	$1.37\cdot10^{22}$	46.9	

Comparison of the present simulation with the results of Belgued et al. is possible because in the chosen experimental temperature range and reaction conditions the surface reactions dominate and transport effects are less important.

3. Results and Discussion

In Fig. 1 the measured and simulated rate of formation of C_2H_6 are shown together. The results compare qualitatively well. While the experimental rate of formation of C_2H_6 at T = 593 K decreases with time, as it would be expected from the work of Somorjai [7] (under atmospheric pressure the catalyst surface is covered by a carbon layer within minutes diminishing catalytic activity), the experimental curve at T = 523 K exhibits a maximum. This does not fit with a model describing a deactivation by carbon. Belgued at al. [1] do not give any explanation of this maximum. The model used here gives an explanation for the development of the temperature-dependent formation rate with time, assuming a catalytically-active carbidic carbon C(s) at the surface formed by CH_3(s) decomposition. Due to its radical-like character, C(s) enhances the sticking probability for CH_4 on the Pt surface (the CH_4 sticking coefficient on bare Pt lies between 10^{-10} and 10^{-12} [8]). The model discussed here gives an auto-catalytic explanation for the maximum of the C_2H_6 formation rate at T = 523 K. Due to the stochiometry of R2 the rate reaches its maximum when the C(s) and Pt(s) coverage is 0.5. Therefore, the formation rate of C_2H_6 is strongly correlated with the C(s) coverage, see Fig. 2. At higher temperatures, the C(s) deposition accelerates and the carbon coverage increases much faster compared with lower temperatures. Consequently, the rate maximum is shifted to very short times. Hence, it can be assumed that also the experimental data for T = 593 K exhibit a maximum which is reached so early in time that it could not be resolved by Belgued et al.

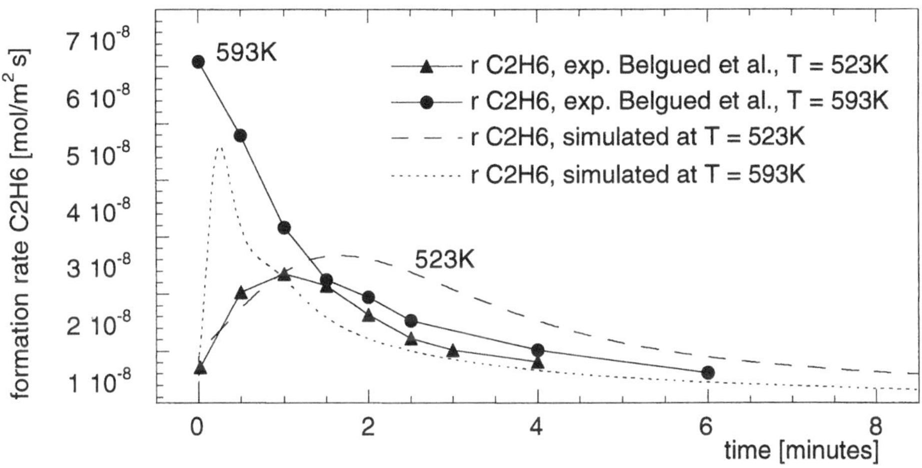

Fig. 1: Comparison of experimental (lines with symbols, Belgued et al. [1]) and calculated (lines) rates of C_2H_6 formation (p = 1 bar, flow rate = 400 $cm^3 min^{-1}$).

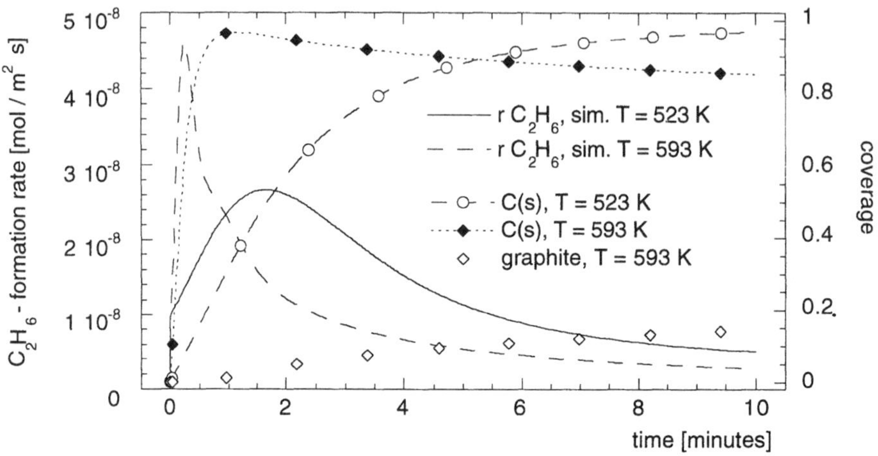

Fig. 2: Calculated formation rates of C_2H_6 at T = 523 K and T = 593 K (lines without symbols), calculated carbon coverage at both temperatures and calculated graphite at T = 593 K (symbols with and without lines) (p = 1 bar, flow rate = 400 $cm^3 min^{-1}$).

To demonstrate that the present model of a catalytically active C(s) species also fits quite well with other experimental results than formation rates of C_2H_6, in Fig. 3 C(s) and CH_x coverages are shown. The difference in designation (carbon C(s), CH_x) can be understood by comparing the different models of carbonlayer established by different groups [9-11]. All of these models are similar.

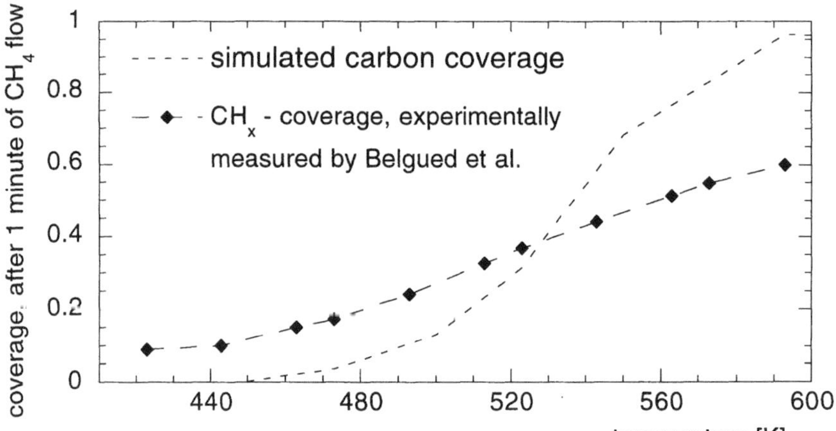

Fig. 3: Comparison of experimental (line with symbols, Belgued et al. [1]) and calculated (dotted line) coverage of carbon or CH_x after 1 minute of CH_4 flow at different temperatures (p = 1 bar, flow rate = 400 $cm^3 min^{-1}$).

The flow analysis made for the simulation at T = 523 K for different times-on-stream (TOS) shows interesting details about the mechanism. Within this mechanism a CH_3 (s) accumulation is possible only for low temperatures (200K). Above T = 250 K only C(s) and H(s) are present on the platinum surface due to a fast CH_3(s) decomposition [10]. The rate-determining step (RDS) within the CH_4 conversion is the adsorption of CH_4 (low sticking probability). Our model adds three possible adsorption reactions to the conventional one on platinum, see Table 1. In the beginning of the reaction, after 2 s TOS, the dominating adsorption channel for CH_4 is R1 (96.2 %) whereas R2 consumes only 3.7 % of CH_4 due to the minimal C(s) coverage in the beginning. As it is visible in Fig. 4 the influence of R2 increases with an increasing C(s) coverage up to 49 % after 90 s TOS, whereas the importance of R1 decreases. With the increasing influence of R2 the formation rate of C_2H_6 passes through a maximum (autocatalytic effect). At the same time (90 s TOS), the RDS changes from R1 to R2. After 300 s TOS, R3 seems to become the dominating reaction but R3 runs quickly into a partial equilibrium (after 60 s TOS) with the result that nearly all of the $CHCH_3$(s) formed is transformed back into CH_4 and C(s). Hence R2 and its product CCH_3(s) (ethylidyne) keeps dominating the overall process. Being the most sensitive reaction, it is not only dominating the C_2H_6 selectivity of the process (most important reaction channel) but also the C_2H_6 yield. The ethylidyne species formed in R2 is part of another partial equilibrium (R5 and R6) between ethylidyne and its dissociation products C(s) and CH_3(s). This partial equilibrium provides the process with a small, but continous, amount of CH_3(s) recombining to C_2H_6. Therefore, it is responsible for the maintenance of catalytic activity over a long period.

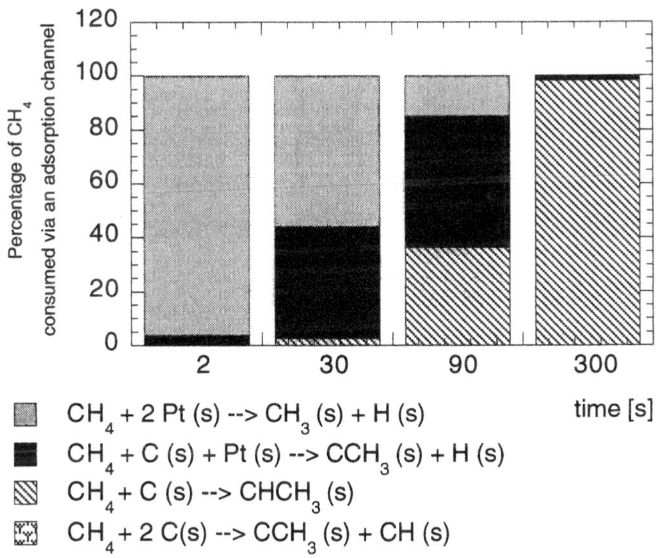

Fig. 4: Reaction flow analysis for the CH_4 consumption at different times-on-stream (p = 1 bar, flow rate = 400 $cm^3\,min^{-1}$, T = 523 K).

4. Conclusion

The model discussed assumes a carbidic surface species C(s) (named as α carbon in Koerts et al. [9]) to be a catalytically active part in the chemisorption of methane. The autocatalytical behaviour of C(s) which is deposited on platinum by chemisorption of CH_4 and followed by quick decomposition of CH_3(s) enhances the sticking probability of CH_4 on the surface compared to bare platinum. The model is qualitatively quite successful in reproducing the temporal behaviour of the formation rates of ethane whereas the conventional model can only explain the decline of the formation rate but not the presence of a maximum. Methane adsorption on free Pt sites is still of importance but only during the initial phase of the process, then the main adsorption channel changes. For higher temperatures this initial phase is shortened due to a quicker carbon deposition, therefore, the rate maximum is shifted to very short times complicating experimental resolution.

The maintenance of catalytic activity for hours can be ascribed to the quickly established partial equilibrium of reactions R5 and R6. This equilibrium always supplies the process with a constant surface concentration of CH_3(s) with the ability of forming ethane. The rate maximum is a direct consequence of the stochiometry of the overall dominant reaction R2. The maximum appears when the platinum surface is half-covered by carbon with decreasing rates for higher coverages. Therefore, the decline of the C_2H_6 formation rate with TOS is not directly caused by carbon deactivating the Pt surface but more indirect due to stochiometry of the most sensitive reaction in the process (R2) after 90 s TOS.

Acknowledgement

This work was supported by the Deutsche Forschungsgemeinschaft (DFG) within the Sonderforschungsbereich 359 „Reaktive Strömung, Diffusion und Transport".

References

[1] M. Belgued, A. Amariglio, P. Paréja, H. Amariglio, J. Catal. 159 (1996) 441-448.

[2] O. Deutschmann, F. Behrendt, J. Warnatz, J. Catal.Today 21 (1994) 461.

[3] O. Deutschmann, F. Behrendt, U. Maas, J. Warnatz, JVST A 13 (1995) 1373.

[4] E. Shustorovich, Metal surface reaction energetics, VCH, Weinheim, 1991, pp 191-223.

[5] S. W. Benson, Thermochemical Kinetics (2nd ed), Wiley-Interscience, NY, 1976, p 22.

[6] A.Ya.Tontegode, Progress in Surface Science 38 (1991) 201-429.

[7] G. A. Somorjai, Introduction to Surface Science, Chap. 7, Wiley-Interscience (1993).

[8] F. Zaera, Chem. Rev. 95 (1995) 2651-2693 .

[9] T. Koerts, R. A. van Santen, Proceedings of the 10th International Congress on Catalysis, 1992, p 1065.

[10] H. P. Bonzel, H. J. Krebs, Surf. Sci. 91 (1980) 499-513.

[11] S. M. Davis, F. Zaera, G. A. Somorjai, J. Catal. 77 (1982) 439-459.

NATURAL GAS CONVERSION V
Studies in Surface Science and Catalysis, Vol. 119
A. Parmaliana et al. (Editors)

Direct Conversion of Methane to Methanol with Micro Wire Initiation (MWI)

Y. Sekine* and K. Fujimoto*

*Department of Applied Chemistry, Faculty of Engineering, The University of Tokyo, Hongo, Tokyo, Japan 113-8656

Methane is oxidized by oxygen in the absence of catalyst with micro wire initiation (MWI; the concept is shown in Fig. 1) method, which is composed of electrically heated small wire and low temperature reaction zone. With the existence of initiation reaction, methanol and CO was formed even under 400 K gas phase temperature. Main products of the reaction were methanol, hydrogen, CO, CO_2 and water. We claimed a new model for the oxidation of methane with MWI.

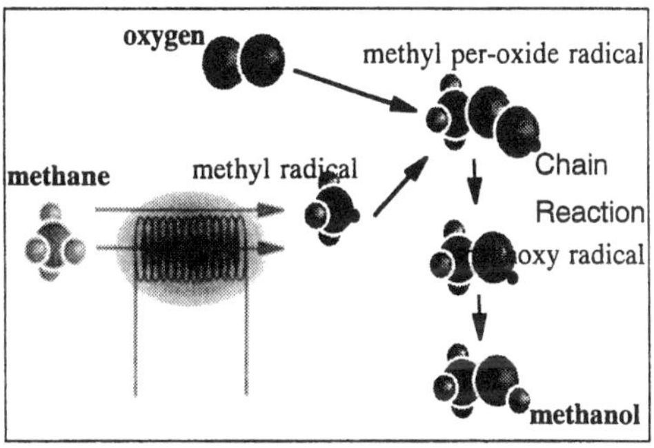

Fig. 1 Concept of thermal dissociation of methane on and around high temperature micro filament wire and following chain reaction.

1. INTRODUCTION

Methane is a main component of natural gas and one of the most abundant carbon resources. Although the chemical utilization has been highly desired, it mostly pass through the reforming of methane to synthesis gas, because of its high stability. Therefore the development of direct conversion process of methane to methanol or other oxygen compound has been expected for a long time. Up to now, so many trials have been made for the completion of the process [1]~[2]. Some papers about the catalyzed system have been reported but its conversion level is still low. Ga_2O_3/MoO_3 hybrid catalyst system shows higher activity than Ga_2O_3 or MoO_3 and MeOH selectivity is 22 % while methane conversion is 3 % [3]. But some studies have reported that there are no advantage in using catalysts in the gas phase oxidation of methane [4]~[5]. It is because that non catalytic gas phase radical reactions are playing strong role in the oxidation of methane.

In the non catalytic system, there are so many reports in homogeneous oxidation of methane and some of them show high yield but the reproducibility of the results are not enough. Recently, in homogeneous gas phase methane oxidation, main topics are; process cost efficiency, the role of reactor surface, partial pressure effect, total pressure effect and the decomposition of formed methanol and formaldehyde. Surface/volume ratio is very sensitive to methanol selectivity [6]. Total pressure affects on either methanol selectivity or methane conversion [7]. Higher pressure leads to high methanol selectivity and high methane conversion. It depends on the stability of CH_3OO radical. The decomposition of formed methanol is a serious problem. No decomposition of methanol occurred in helium flow [8] but if oxygen molecules exist then the decomposition of methanol occurred quickly. High oxygen to methane ratio leads methanol molecules to successive oxidation to carbon oxide [9].

In case of the non catalyzed oxidation process, the serious problem is that it requires high temperature to activate methane which also promote side reactions to make carbon dioxide. Present authors demonstrated that the micro wire initiation (MWI) was quite effective for the low temperature partial oxidation of methane with oxygen [10]. The MWI reaction of methane was operated with a batch type reactor, which gave a very selective (over 80 %) formation of methanol at the first stage of the reaction (Fig. 2), but with the process time, methanol was decomposed quickly to CO and CO_2. From the correlation of oxygen conversion and methanol selectivity, high methanol selectivity can be obtained only at low oxygen conversion. We regard that methanol formation is very quick reaction but the consecutive oxidation by the remaining dioxygen should decompose of methanol.

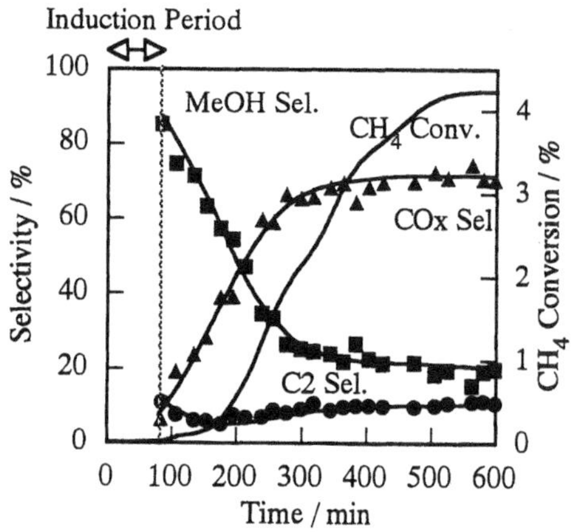

Fig. 2 Products and reaction time on batch reactor with MWI.
Reaction Condition: Gas Phase Tempertature = 450 K, CH_4/Air = 4/1 MPa, MWI-Temperature = 1023 K.

In the present study, we try to separate the location of the initial activation of methane at high temperature (>973 K) and of the successive chain reaction at low temperature (<473 K). The newly designed reactor is shown in Fig. 3, where methane and oxygen are introduced into the MWI zone and should be activated on and around the micro filament wire which is heated at high temperature by electricity. The activation of methane should proceed either on the filament surface and in the gas phase, which involve the dissociation of methane to methyl radical and the oxidative decomposition of methane to methyl radical. The formed free radical species will smoothly move to the chain reaction zone within <0.1 s. The chain reaction zone is separated from the initiation zone with a pin hole (ϕp = 0.1 mm). In the chain reaction chamber, mild oxidation will occur at temperature as low as <473 K.

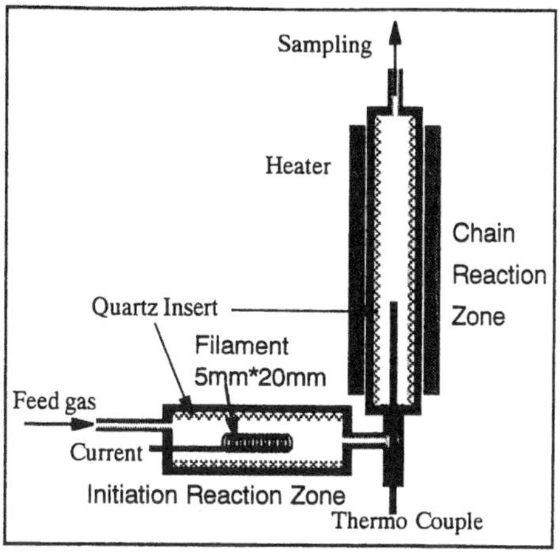

Fig. 3 Initiation-Chain Reaction Separated Flow MWI Reactor;

1st. stage reactor volume = 5 cm^3,
2nd. stage reactor volume = 10 cm^3,
made of stainless steel 316.

2. EXPERIMENTAL

For the flow type MWI experiments, a new design reactor was made (Fig. 3). An initiation zone and a chain reaction zone are connected in series. All the materials are stainless steel 316 and inner walls are covered with quartz insert. In the initiation reaction zone, a home-made micro filament made of NiCr wire (ϕ=0.5 mm) is installed. The filament wire is bound up by original method with a machine and its bound size of micro wire are; 5 mm diameter; 20 mm length; resistance is about 11 ohm. The chain reaction zone is made of a stainless steel pipe; 60 cm length; 4 mm inner diameter. At the bottom of the chain reaction zone, a thermo couple is inserted to measure the reaction temperature. Gas phase temperature is controlled by an electric furnace installed outside of the reactor. Products flowed out through a back pressure valve were analyzed by online gas chromatography. The material balances are calculated for each experiments and the values are ; C, H: within 97 %, O: within 90 %.

3. RESULTS AND DISCUSSION

Standard reaction conditions are; gas phase temperature 473 K, feed gas ratio is CH_4/Air = 280/70 cm^3*min^{-1}(NTP), pressure is 4 MPa, surface temperature of the micro filament wire is about 1043 K. At the initial stage of the reaction under the standard conditions, CO and methanol were formed while in the steady state methane conversion was about 3.7 %, oxygen conversion is about 98 % and methanol selectivity was about 28 %.

3.1. Pressure Effect.

In the homogeneous gas phase methane oxidation, high pressure is favorable for the methanol formation [1]. Figure 4 shows the pressure effect on the products formation with MWI at 473 K gas phase temperature and 1043 K micro wire filament surface temperature. This figure demonstrates that while either methane conversion or oxygen conversion shows complicated behavior against the operation pressure. Under the atmospheric pressure condition, methane conversion was about 1 % and hydrogen yield also 1 %, and main product was CO. With the increasing pressure, CO and hydrogen yield decreased and the selectivity of methanol rose up. At 6 MPa, methane conversion was 1.4 % and hydrogen yield; 0.2 %, the selectivity of methanol; 27 %. From these results, we estimated as follows; methoxy radical (CH_3O) is a key species for products distribution. In case of the atmospheric pressure, CH_3O immediately decomposes thermally and turns into CH_2O, CHO then finally CO forms (1).

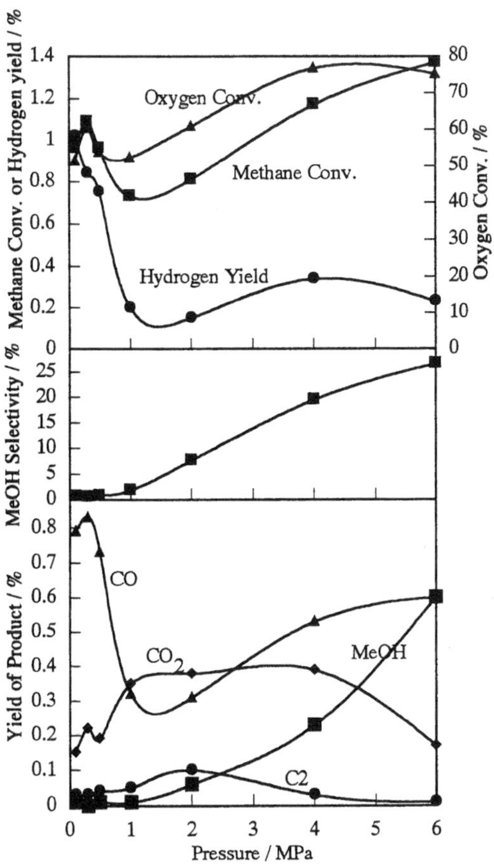

Fig. 4 Pressure effect with MWI reactor, Reaction condition:
Micro Wire Initiation, 37.5 W,
Gas Phase Temp. 473 K,
CH_4/Air = 280/35 $cm^3 * min^{-1}$(NTP).

$$CH_3O \rightarrow CH_2O \rightarrow CHO \rightarrow CO \qquad (1)$$

On the other hand, under higher pressurized conditions, CH_3O collides with CH_4 or other hydrogen containing species, and turns to CH_3OH because of higher collision frequency (2).

$CH_3O + HX (CH_4 \text{ etc.}) \text{->} CH_3OH + X (CH_3 \text{ etc.})$ (2)

3.2. Effect of the Gas Phase Temperature in the Chain Reaction Zone

MWI activates the methane molecule and the activated species are smoothly introduced into the chain reaction zone which is connected downstream to the initiation zone. Table 1 shows the effect of temperature of the chain reaction under the same initiation reaction conditions. With the MWI system, methanol was formed at the temperature as low as 393 K. Below 653 K where no reaction proceed without MWI, there are no obvious differences in either methane conversion or product distribution for different reaction temperatures. Over 653 K, methane conversion increased with increasing reaction temperatures. It means that while chain reaction zone is operated at only 473 K wall temperature, apparently it is effective for promoting chain reactions to produce oxygen containing species. The elementary reactions as shown in (3)~(6) are very quick reaction even in low temperature gas phase as 473 K.

Table 1 Effect of Gas Phase Temperature in the Chain Reaction Zone

Gas Phase Temp. / K	CH_4 Conv. / %	O_2 Conv. / %	MeOH Yld. / %	C2 Yld. / %	CO Yld. / %	CO_2 Yld. / %
393	1.3	81.0	0.34	0.05	0.76	0.15
473	1.4	83.6	0.36	0.05	0.79	0.14
683	2.0	100.0	0.97	0.02	0.74	0.26

With Micro Filament Wire Initiation Method, 35 W, 0.4 mm d.m. 100 cm length, NiCr wire, CH_4/ Air = 280/35 cm^3*min^{-1}(NTP), Pressure 4 MPa.

$CH_3OO + CH_4 \text{->} CH_3OOH + CH_3$ (3)
$CH_3OO + CH_3 \text{->} 2CH_3O$ (4)
$CH_3O + CH_4 \text{->} CH_3OH + CH_3$ (5)
$CH_3OOH \text{->} CH_3O + OH$ (6)

It is apparent that if MWI activation exists, successive chain reaction never require the high temperature and therefore only 393 K (maintains not to condense the water in the reaction chamber) is good enough for methane oxidation with MWI.

3.3. Estimated Oxidation Scheme of Methane with MWI

From these results, we postulate a simplified reaction mechanism for MWI system as follows; Methane conversion mainly depends on the filament temperature, the gas composition, the existence of the chain reaction zone chamber. Methane activation occurs on and around the hot filament in the activation zone and the methanol formation proceeds mainly in the chain reaction chamber which is maintained at low temperature. Elementary reactions which are involved in methane activation, is similar to those involved in high temperature gas phase oxidation of methane. Major activated species should be methyl radical, methyl peroxy radical, methoxy radical, methyl hydroperoxide, formate radical, hydroperoxy radical, hydroxy radical. In the initiation reaction zone the formed methyl radical will meet smoothly with oxygen and make methyl peroxide radical. Some of the methyl radicals will combine to each other and make C2 hydrocarbon.

4. CONCLUSIONS

The MWI method for methane oxidation could convert methane into methanol and CO where methane and small amount of oxygen are reacted at ~1073 K in an initiation reaction zone, and the intermediates are introduced to a chain reaction chamber whose temperature are maintained at 393 K~473 K. It could be well explained that some activated species such as methyl radical, methyl-peroxy radical or hydroperoxy radical are introduced from the initiation chamber to the chain reaction chamber to proceed the methane oxidation to methanol and CO via conventional gas phase chain reaction mechanism. MWI process could obtain same conversion and yield with much lower energy consumption than conventional method. The problems are that the conversion level is still low and the separation of products and recycle after the reaction are required.

REFERENCES

1. Krylov, O. V. Catalysis today 1993, 18 (3), 209-302.
2. Gesser. H. D.; Hunter, N. R.; Prakash, C. B. Chemical Reviews 1985, 85 (4), 235-244.
3. Hargreaves, J. S.; Hutchings, G. J.; Joyner, R. W.; Taylor, S. H. J. Soc. Chem. Commun. 1996, 523-524.
4. Arutyunov, V. S.; Basevich, V. Y.; Vedeneev, V. I. Ind. Eng. Chem. Res. 1995, 34, 4238-4243.
5. Walsh, D. E.; Martenak, D. J.; Han, S.; Palermo, R. E. Ind. Eng. Chem. Res. 1992, 31, 1259-1262.
6. Thomas, D. J.; Willi, R.; Baiker, A. Ind. Eng. Chem. Res. 1992, 31, 2272-2278.
7. Rytz, D. W.; Baiker, A. Ind. Eng. Chem. Res. 1991, 30, 2287-2292.
8. Chun, J. W.; Anthony, R. G. Ind. Eng. Chem. Res. 1993, 32 (5), 788-795.
9. Chun, J. W.; Anthony, R. G. Ind. Eng. Chem. Res. 1993, 32 (5), 796-799.
10. Sekine, Y.; Fujimoto, K. Energy & Fuels 1996, 10 (6), 1278-1279.

NATURAL GAS CONVERSION V
Studies in Surface Science and Catalysis, Vol. 119
A. Parmaliana et al. (Editors)

Active site generation by water for the activation of methane over non-reducible oxide catalysts: A study of MgO system

Takashi Karasuda, Katsutoshi Nagaoka, and Ken-ichi Aika*

Department of Environmental Chemistry and Engineering, Interdisciplinary Graduate School of Science and Engineering, Tokyo Institute of Technology
4259 Nagatsuta, Midori-ku, Yokohama 226, Japan
Fax No.: 81-45-924-5441 E-Mail: kenaika@chemenv.titech.ac.jp

MgO was proposed to form an active structure [vacancy + O^-] above 973K. The temperature dependency of H_2 desorption, XPS, and $^{18}O_2$ isotopic exchange results supported the model and on such an active site, the oxidative coupling of methane (OCM) can be initiated. It was proposed that adsorbed (and absorbed) water was responsible for the active site generation. The addition of water promoted C_2 production during the OCM and the water isotope effect on C_2 production could also be observed.

1. Introduction

So far, much research has been carried out on the oxidative coupling of methane (OCM). However, very few studies have pointed out the significant role of water in the reaction [1-4]. A model O^- species produced on the uv-irradiated MgO [5] is destroyed above 523 K [6], and no active oxygen species can be detected by ESR on the MgO surface activated above 973 K where the OCM reaction occurs. At these temperatures hydrogen was observed to desorb leaving an O^--like oxygen anion with a defect [1-3]. The source of the defect has been proposed to be a water molecule absorbed in MgO [1]. In this paper it is shown that the OCM activity is remarkably increased when water is added to MgO. Moreover, the hydrogen isotope effect (i.e.) of adding water could be observed, casting doubt on the conclusion of past works reporting that the C-H rupture is the main cause of the i.e. Here, a pure MgO system was examined by various methods.

2. Experimental

MgO and Li/MgO were evacuated at 1273 K for 24 hours, then cooled to ambient temperature, and again the samples were heated to 1173 K at a rate of 10 degrees per minute. The gas evolved from the samples were analyzed (TPD). The XP spectra was recorded for the samples evacuated at various temperatures. Oxygen isotopic exchange between $^{18}O_2$ and MgO was also analyzed at various temperatures. A water-He mixture (pulse form) was injected on the samples, and CH_4/O_2/He gas (CH_4/ Air/He = 16/20/20, total pulse volume: 0.5 cm^3, CH_4: 5.8 μmol) was repeatedly added to observe the OCM activity.

3. Results and Discussion

3.1. Defect formation proved by TPD

Hydrogen was generated above 973 K accompanied with H_2O desorption from MgO (Fig.1). Oxygen was not observed up to 1273K, but it is known to be evolved at higher temperatures. The adsorbed (and absorbed) water (or OH species) in the MgO lattice is believed to cause the evolution of H_2 leaving a (Mg^{2+}) defect and a neutral oxygen atom, which then accepts one electron from the lattice oxygen ion. The $O^-□O^-$ site is considered to be formed as shown in Eq. 1-2. Table.1 shows two kinds of symbols.

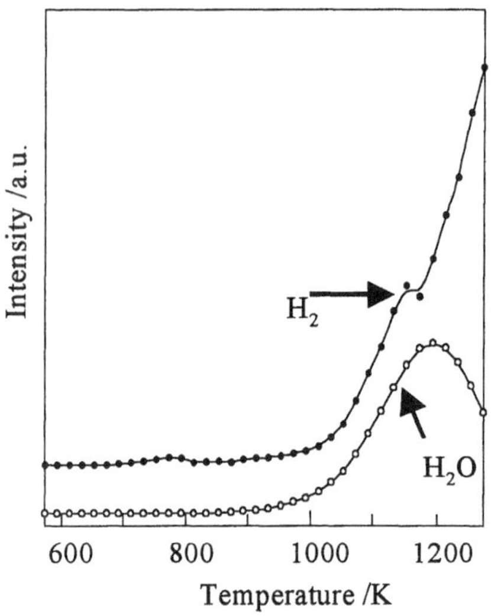

Fig.1 The bulk dissolved H_2O and H_2 TPD spectra from MgO

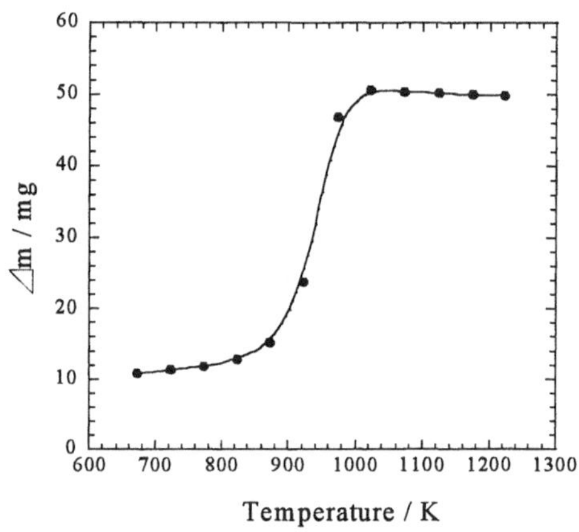

Fig.2 Apparent dielectric constant of MgO sample as a function of the temperature expressed by the field forces (Δm). The data was taken about 3 min after reaching the indicated temperature. The surface is positively charged against the bulk [1].

Defect formation with H_2O

$$Mg_{Mg}{}^{x} + O_o{}^{x} + H_2O_i{}^{x} \rightarrow Mg_{Mg}{}^{x} + 2(OH)o^{\cdot} + V_{Mg}{}^{''} \qquad (1)$$

Hole or O^- formation ($Oo^x + h^{\cdot} = Oo^{\cdot}$)

$$Mg_{Mg}{}^{x} + 2(OH)o^{\cdot} + V_{Mg}{}^{''} \rightarrow Mg_{Mg}{}^{x} + 2\,Oo^{\cdot} + V_{Mg}{}^{''} + H_2 \qquad (2)$$

Table 1. Symbols

Kroger-Vink terminology	$Vo^{\cdot\cdot}$	Oo^x	$(OH)^{\cdot}$	$Oo^{\cdot}$	$Mg_{Mg}{}^{x}$	$V_{Mg}{}^{''}$
Chemical symbols	O defect	O^{2-}	OH^-	O^-	Mg^{2+}	Mg defect

3.2. Dielectric constant measurements

Freund et al [1] measured the dielectric constants of MgO and found that the surface became positively charged when it was heated above 973K, as is shown in Fig.2. This means the surface is rich in O^- ($O_0^{\cdot}$) and poor in Mg vacancy ($V_{Mg}{}^{''}$). This infers the defect structure [vacancy + O^-] has been generated at 973K.

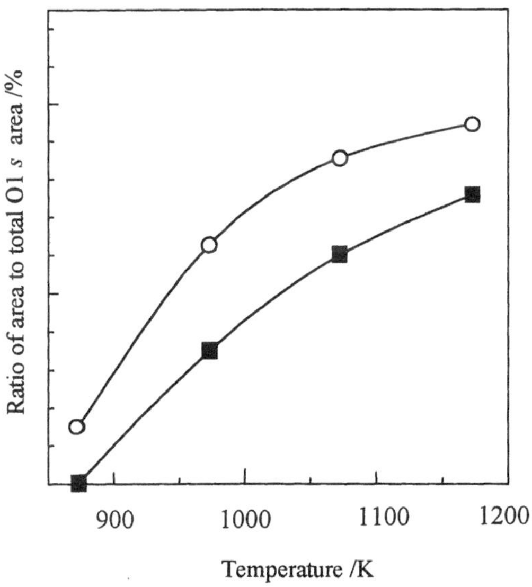

Fig.3 Ratio of the O1s shoulder peak(○) as a function of the evacuation temperature. Actually the peak area is divided into the O^{2-} and shoulder (O^- (■) and $CO_3{}^{2-}$).

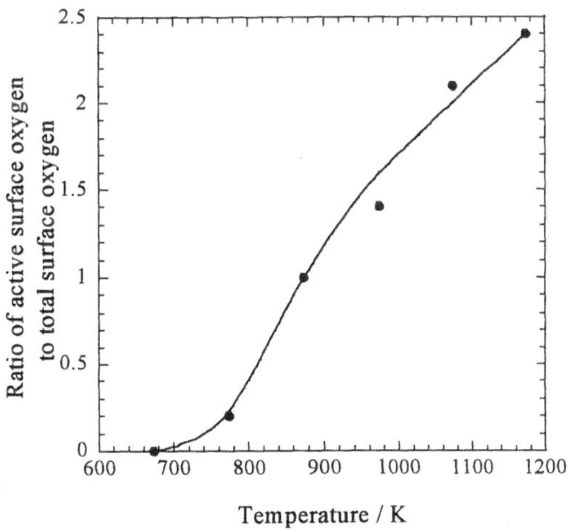

Fig.4 Temperature dependence of active surface oxygen ($N_{MgO(a)}$) per surface oxygen ($N_{MgO(s)}$) calculated by a two stage exchange model [4].

3.3. XPS and ESR

The O^- produced in this way was not identified by ESR, but the XPS binding energy of oxygen had a shoulder at a higher B.E. side, showing the formation of O^--like species. The lattice oxygen (O^{2-}) of MgO shows XPS B.E. at 531. 3 eV , while the shoulder peak appears at 533.6 eV when it is evacuated above 973K. This shoulder includes two kinds of O1s (CO_3^{2-} and O^-). The amount of CO_3^{2-} was calculated from the C1s data of CO_3^{2-}. Fig.3 shows the relative amounts of O1s shoulder peak (○) and O^- (■) as a function of the temperatures.

3.4. Isotope exchange

The active surface oxygen can exchange with gaseous $^{18}O_2$ above 973 K. The amounts of the exchangeable oxygen of MgO increase with the increase in temperature, as is shown in Fig. 4 [4].

3.5. OCM Reaction

The oxidative coupling reaction was carried out on 0.2g of MgO. The results are shown in Fig. 5. It was found that C_2 compounds ($C_2H_6 + C_2H_4$) are formed above 973K.

These measurements (3.1 to 3.5) show that all phenomena occur only when MgO is heated above 973K. However, they occur not from the temperature effect of the reaction (activation energy), but from the generation of the active site. The active site is believed to be the [defect + O^-] structure. Interestingly these O^- cannot be identified by ESR since the O^- probably exists in high concentration or have exchanging electrons around them.

Even if these active species are considered to be generated properly at higher temperatures than 973K, it does not react with methane after it is quenched at room temperature. However, the isolated O^- anions produced by uv-irradiation and N_2O contact react easily with methane to form methyl radicals and methoxides on the surface at room temperature [9]. The nature of the two kinds of active sites are different.

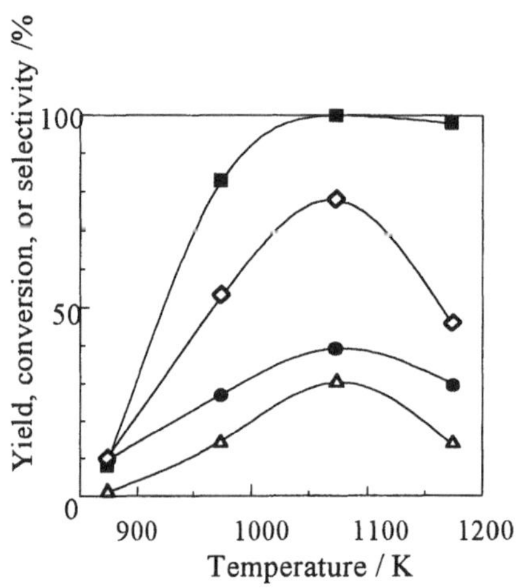

Fig.5 Temperature dependence of MgO activity for OCM reaction; reactant gas (CH_4 / Air / He = 16/20/20 cm^3 min^{-1}), MgO = 0.2g, CH_4 conversion(●), O_2 conversion(■), C_2 selectivity(◇), C_2yield(△).

3.6. Water effect and its isotope effect on the OCM

If the cause of active center generation is actually water, the OCM activity should be increased when water is added to MgO. Table 2 shows that the added water increases C_2 (C_2H_6 and C_2H_4) production. An i.e. of 1.5±0.2 was also observed (H_2O vs. D_2O). The active site is formed by the O-H rupture of hydroxyls in and on MgO, which must be the cause of the i.e.. O-D rupture is harder than O-H.

Table 2. Water treatment effect on O_2 conversion (%) and C_2 hydrocarbon formation (mol) using 0.2g MgO with pulsed gas (CH_4/Air/He=16/20/20ml min^{-1}) of 0.5 mL sample size

	nontreatment		H_2O treatment		D_2O treatment	
Temperature /K	O_2 conv.	C_2 hydrocarbon	O_2 conv.	C_2 hydrocarbon	O_2 conv.	C_2 hydrocarbon
873	-	-	-	-	-	-
973	100	1.19×10^{-8}	100	2.32×10^{-8}	100	2.10×10^{-8}
1073	100	7.54×10^{-8}	100	1.82×10^{-7}	100	1.53×10^{-7}
1173	100	8.98×10^{-8}	100	2.45×10^{-7}	100	1.77×10^{-7}

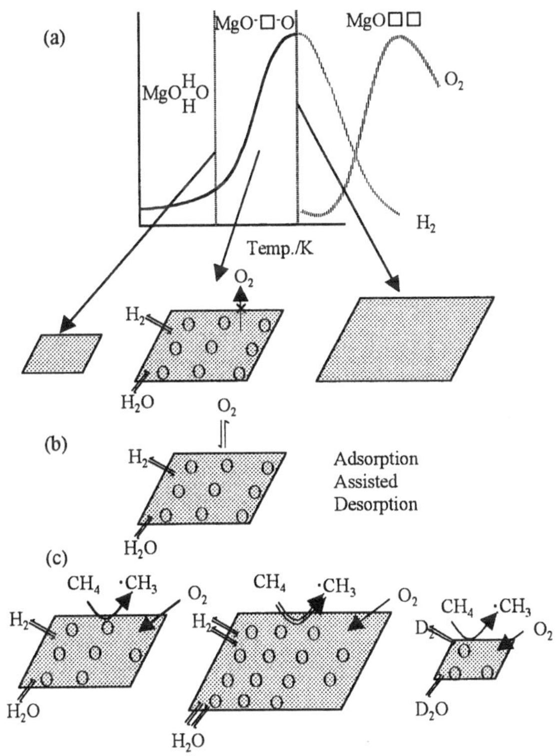

FIG.6. Active site generation[a], oxygen activation[b], and OCM reaction model[c].

3.7. A new explanation for the CH_4/CD_4 isotope effect

The reaction rate of CH_4 is known to be faster than CD_4 in the OCM reaction. An i.e. of about 1.5 reported by Cant *et al.* [10] seemed too high for the i.e. by C-H (or C-D) rupture at 1023 K. We can propose a new explanation for this phenomenon. The produced water in the OCM reaction continuously generates the active site for the OCM reaction. The observed i.e. in the CH_4/CD_4 experiment must be caused partly by the C-H rupture of methane (kinetic i.e.) but mainly by the O-H rupture in the regeneration of the active sites (equilibrium i.e.). Such a model is also shown in Fig.6(c). Once the active site number is decided by the temperature and H_2O pressure (the equilibrium conditions), the active site oxygen can easily exchange with gaseous O_2 (recovering the consumed surface oxygen through O_2 during the OCM reaction) as in eqs. 3 and 4. The detailed analysis will be disclosed elsewhere using a kinetics proposed by us [8].

The methyl radicals have two chances: one is to produce C_2 hydrocarbons (eq.5) and one is to be converted to CO_2 (eq.6).

$$O_2 + 2[e^-] \rightarrow 2O^- \quad (3)$$

$$O^- + CH_4 \rightarrow \cdot CH_3 + OH^- \quad (4)$$

$$2\cdot CH_3 \rightarrow C_2H_6 \quad (5)$$

$$\cdot CH_3 + xO_2 \rightarrow CO_2 + H_2O \quad (6)$$

References

1. M. M. Freund, F. Freund, F. Batllo, Phys. Rev. Lett., 63 (1989) 2096.
2. I. Balint and K. Aika, Natural Gas Conversion II, H. E. Curry-Hyde and R. F. Home Eds., Elsevier, Amsterdam, (1994) pp.177-186.
3. I. Balint and K. Aika, J. Chem. Soc., Faraday Trans., 91(1995) 1805.
4. T. Karasuda and K. Aika, J. Catal., 171, (1997) 439.
5. K. Aika and J. H. Lunsford, J. Phys. Chem., 81 (1977) 1393.
6. M. Iwamoto and J. H. Lunsford, J. Phys. Chem., 84 (1980) 3079.
7. A. Goto and K. Aika, Bull. Chem. Soc. Jpn., 71 (1998) 95.
8. E. Iwamatsu and K. Aika, J. Catal., 117 (1989) 416.
9. K. Aika and T. Karasuda, in "Catalysis in Petroleum Refining and Petrochemical Industries 1995" M. Absi-Halabi et. al. Eds., Elsevier, Amsterdam, (1996) pp.397-406.
10. N. W. Cant, C.A. Lukey, P. F. Nelson, R. J. Tyler, J. Chem. Soc., Chem. Commun., (1988) 766.

NATURAL GAS CONVERSION V
Studies in Surface Science and Catalysis, Vol. 119
A. Parmaliana et al. (Editors)

Oxidative Coupling of Methane over a Sm/C and Mg/C Catalysts Using N_2O as Oxidant

M. Bajus[a] and M.H.Back[b]

[a]Slovak Technical University, Faculty of Chemical Technology, Radlinského 9, 812 37 Bratislava, Slovakia.
[b]University of Ottawa, Department of Chemistry, Ottawa, Ontario K1N 6N5, Canada

ABSTRACT

This paper reports on the oxidative coupling of methane to C_2 hydrocarbons over magnesium and samarium - promoted carbon catalysts in the absence and presence of nitrous oxide. The kinetics of the methane conversion were studied in a flow system at the temperature 850°C. Conversion of methane in the presence of Sm and Mg on active carbon without nitrous oxide was very low. In the presence of nitrous oxide, the conversion is greatly increased, and selectivity to C_2 hydrocarbons is also increased. With increased residence time and increasing amounts of nitrous oxide, ethane was dehydrogenated, forming ethylene.

1.INTRODUCTION

The conversion of natural gas to higher valued or more easily transportable substances is a goal of considerable scientific and practical interest. One of the promising routes is the oxidative coupling of methane to C_2^+ hydrocarbons in the presence of a catalyst and at temperatures from 600 to 900°C. Much of this effort has been related to the search for better catalyst, motivated by the hope of finding catalyst of sufficient selectivity and activity to enhance the commercial prospects of the reaction [1-5]. One such catalyst is carbon [6]. Carbon is a well established, commercially available catalyst support. Activated carbon is essential as a support material for precious metal catalysts, which are widely used in the synthesis of high - value - added chemical products. A recent study from Japan [7] described the effect of carbon fibers on the reactivity of methane, where a substantial accelerating effect was observed.

In addition to the NO decomposition, the decomposition of nitrous oxide over various metal and oxide catalysts has received much attention for its unique behavior as an oxidizing agent compared to conventional oxidizing agent, such as molecular oxygen. N_2O and O_2 as oxygen donors for oxidative coupling of methane has been compared [8].

In previous studies on the OCM reaction we have reported that Li/carbon catalyst effectively catalyse the oxidative coupling of methane [9]. Nitrous oxide had a strong promoting effect on the homogenous coupling of methane a less notice able effect in the presence of the carbon catalyst. Nitrous oxide improved the selectivity to C_2 hydrocarbons. The purpose of the present study was first to explore the effect of the incorporation of samarium and magnesium as a metal oxides into the carbon support. Secondly, the effect of nitrous oxide on the coupling of methane was investigated.

2. EXPERIMENTAL

2.1. Apparatus

The arrangement of the pyrolysis apparatus was a typical flow system. It had provision for the introduction of three reactants, a tubular quartz reactor, diameter 10 mm, maintained at constant temperature in a Lindberg three-zone furnace model # 54957, a condenser, traps and a sample collector for analysis of products. The temperature in the pyrolysis oven was regulated by a control console, Lindberg model # 597744A. The movable NiCr-Ni thermocouple was placed in thermotubes by the side of the reaction vessel. The flow-rates of the reactants were controlled by flow controllers, measured on rotameters, and calibrated by a soap bubble flowmeter. The reactant gases were pre-mixed before entering the reactor. Nitrogen was used as the diluent. Equimolar mixtures of methane and nitrogen were fed to the reactor with a total pressure of 101.2 kPa. Flow-rates varied 4.7 to 255 $cm^3 min^{-1}$.

2.2. Analysis

Products were analyzed by gas chromatography a using Hewlett-Packard instrument 5710A with FID detector and 5750A with TC detector. The gaseous products from the pyrolysis of methane consisted of ethane, ethylene, propane, propene and acetylene. Separation was achieved on a column of n-octane on Porasil C, 5.5 m of 1/4 in. + 2 m of 3/8 in., maintained at 60°C. An FID detector was used. The products H_2, CO, CO_2 and N_2O were analyzed using a column of carbosieve 511, 10 ft. X 1/8 in. maintained at 100°C and using a thermal conductivity detector.

2.3. Materials

Instrument grade methane (99.7%) and nitrous oxide (U.H.P. 99.99%) were obtained from Matheson Gas Products Canada and used without further purification. Other reactants and their sources were as follows: activated carbon, Darco, 20-40mesh, granular, surface area 1500m^2 g^{-1}, pore volume 1.5 cm^3 g^{-1}, Aldrich Chemical Company, Inc. [7440-44-0]; Magnesium sulfate ($MgSO_4.7H_2O$), crystals, AC-5568, Anachema, Chemicals Ltd., Montreal, Toronto; Samarium (III) chloride hexahydrate ($SmCl_3$. $6H_2O$); 99% + irritant; 24, 880-0 [13465-55-9], Aldrich Chemical Company, Inc.

2.4. Preparation and treatment of the carbon catalyst

The carbon catalyst was heated in a flow of nitrogen at 900°C for several h before experiments were commenced. The Mg-promoted carbon catalyst was prepared by adding activated carbon and $MgSO_4$. $7H_2O$ to deionized water and evaporating the water, while stirring, until only a thick paste remained. The paste was dried at 140°C for more than 5 h. The $MgSO_4.7H_2O$/carbon thus obtained was then converted to the magnesium-promoted carbon by heating in the reactor at 465°C for 1 h under an oxygen flow of 0.83 cm^3 s^{-1}. Procedure of the preparation Sm/carbon catalyst was the some. The catalysts comprised 7% by weight magnesium or samarium.

3. RESULTS AND DISCUSSION

Methane was pyrolyzed over Sm/carbon and Mg/carbon catalysts at the temperature 850°C alone and in the precence of nitrous oxide. Some representative results of the reaction under a variety of conditions as well in the absence of catalysts are summarized in Table 1. The main products analyzed were ethane, ethylene, hydrogen, carbon monoxide, carbon dioxide and acetylene. Yields of propane and propene are not included since they were usually negligible and never more than 0.05 mol - %. The yields of carbon was calculated in case the balance of hydrogen excess. Hydrogen was not originated only from dimerization of methane to hydrocarbons C_2 but also by decomposition of methane to carbon and hydrogen. In spite of that the activity of the carbon catalysts had not changed during of use, the carbon catalyst or deposited carbon may act as a reactant and become incorporated into the products. Nitrous oxide is known to oxidize carbon at much lower temperatures than we used. The carbon reacts with N_2O to form products, presumably CO and CO_2. Conversion of methane alone over the Sm/carbon catalyst at 850°C was low (~ 1.5%). The main products were ethane and hydrogen but also carbon monoxide appeared even when nitrous oxide was absent (Table 1). The effect of residence time on the selectivity of the main products, ethane, hydrogen and carbon monoxide is shown in Figure 1. The selectivity of ethane was 60% at a reacion time of 0.1 - 0.2 s, but it decreased with increasing reaction time. Over the some time range the selectivity of hydrogen passed through a maximum and decreased. The addition of nitrous oxide to methane increased the conversion substantially. In the presence of nitrous oxide (4.5 to 9.5%) the effect of residence time on the conversion, selectivity and yields of C_2 hydrocarbons is shown in Figure 3. When conversion is increased selectivity to C_2 hydrocarbons decreased. The formation of ethylene is secondary with respect to ethane. Ethylene is cleary formed from ethane. Under the same conditions the yields of carbon monoxide were 6.7% and carbon dioxide 0.4 - 1.3%.

Table 1
Yields of products (mol-%) in the pyrolysis of methane at 850°C

Conditions of experiments							
Catalyst	-	Sm/C			Mg/C		
N_2O (vol-%)	5.13	-	4.53	9.53	-	4.27	8.31
Residence time (s)	2.04	1.91	0.10	1.70	1.48	0.10	1.58
Conversion (%)	7.16	1.49	10.9	21.2	2.53	8.30	21.3
Ethane	1.90	0.78	5.96	1.25	0.92	4.80	1.40
Ethylene	2.14	0.04	0.14	2.04	0.0	0.44	6.89
Acetylene	0.05	0.02	0.02	0.09	0.01	0.0	0.70
Hydrogen	0.45	0.57	2.07	11.1	1.17	1.09	8.20
Carbon monoxide	0.0	0.46	1.76	6.73	0.45	0.84	3.19
Carbon dioxide	0.0	0.0	1.33	0.37	0.0	0.86	0.89
Carbon	0.0	0.0	0.0	2.72	0.11	0.0	0.0
N_2O decomposed (%)	46	-	97.5	100	-	100	100
N_2O converted to CO_x (%)	0	-	4.4	7.4		2.6	5.0

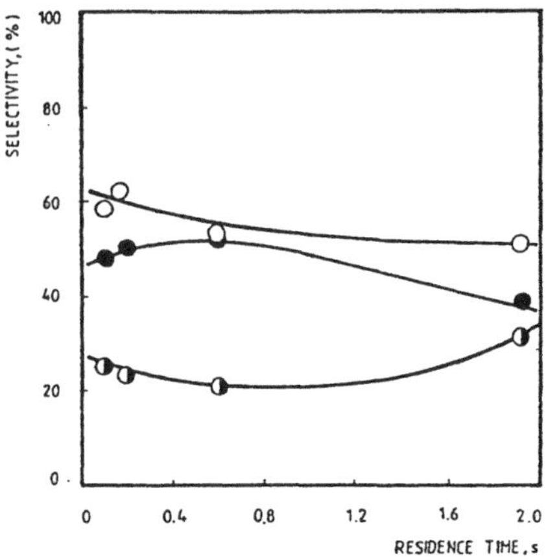

Fig.1. Effect of residence time on the selectivity of ❍ ethane,❍ hydrogen and ❍ carbon monoxide at 850°C over Sm/ /carbon catalyst.

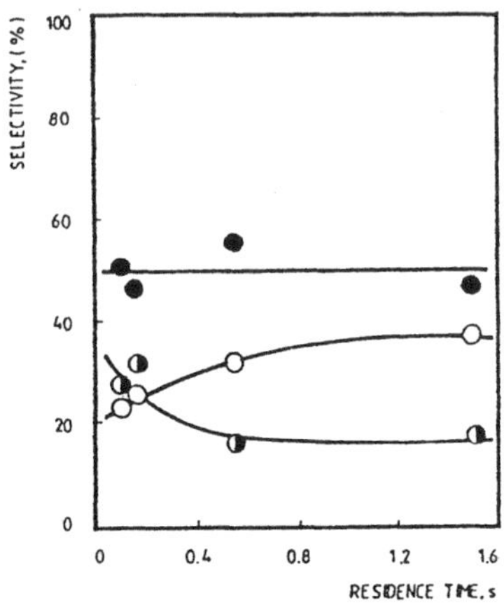

Fig. 2. Effect of residence time on the selectivity of ❍ ethane, ❍ hydrogen and ❍ carbon monoxide at 850 °C over Mg/carbon catalyst.

Conversion of methane alone during pyrolysis over Mg/carbon catalyst at 850°C was 2.5%, slight greater than conversion over the Sm/carbon catalyst under similar conditions. Again the main products were ethane, hydrogen and carbon monoxide. Their selectivity as a function of residence time is shown in Figure 2. The effect of residence time on the yields of products in the presence of 4.3 - 8.3 % nitrous oxide is shown in Figure 4. In the presence of nitrous oxide, conversion and selectivity of C_2 hydrocarbons is shown as a function of residence time.

The most important results of the present study is the demonstration of the effectiveness of carbon as a catalyst for the decomposition of methane. Conversion of methane in the presence Mg and Sm on active carbon without oxidant (nitrons oxide) was very low. In the presence of nitrous oxide, however, the conversion is greatly increased, and the selectivity to ethane is also increased. With increased residence time and increasing amounts of nitrous oxide, ethane was dehydrogenated, forming ethylene. Nitrous oxide was an important agent for the coupling of methane. Increasing content of nitrous oxide in the reactant led a higher percentage conversion of methane, but the selectivity of the C_2 hydrocarbons fell and carbon monoxide became an important products. With Mg/carbon nitrous oxide had a significant promoting effect on the conversion and at short reaction times the selectivity to C_2 hydrocarbons remained high. As the proportion of nitrous oxide increased, ethylene became the major C_2 hydrocarbon product and yield of ethane was reduced. The maximum selectivity for C_2 hydrocarbons was achieved before the maximum conversion of methane, as shown in Figure 4. Although carbon monoxide was also formed, it remained a minor product.

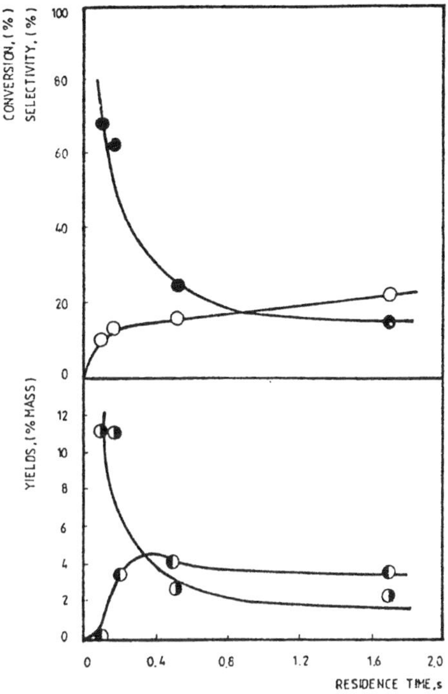

Fig.3. Influence of residence time on the ❍ conversion,● selectivity of C^{+}_{2} hydrocarbons and yields of ◐ ethylene, ◑ ethane at 850°C over Sm/carbon catalyst in the presence of 4.5 - 9.5% (vol.) nitrous oxide.

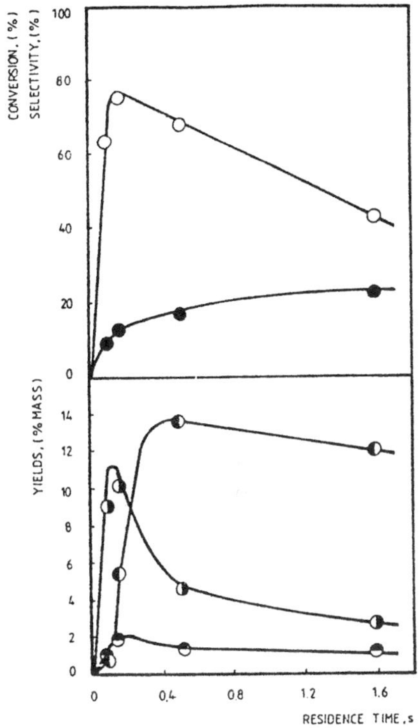

Fig.4. Influence of residence time on the ● conversion, ❍ selectivity of C^{+}_{2} hydrocarbons and yields of ◑ ethane, ◐ ethylene at 850°C over Mg/carbon catalyst in the presence of 4.5 - 8.3 (vol.) nitrous oxide.

REFERENCES

1. Studies in Surface Science and Catalysis 100 (Catalyst in Petroleum Refinery and Petrochemical Industries), Elsevier, Amsterdam, 1996.
2. H.E. Curry - Hyde and R.F. Howe (Eds.), Studies in Surface Science and Catalysis 81.(Natural Gas Conversion II), Elsevier, Amsterdam, 1994.
3. D.Wang, M.P. Rosynek and J.H. Lunsford, J.Catal.**155** (2) (1995) 390.
4. Y.Zeng and Y.S. Lin, Ind.Eng.Chem.Res. **36**(1997) 277.
5. L.M. Ramachandra and D.Moser, J.Membr. Sci **116** (2) (1996) 25.
6. H.Marsh, E.A. Heintz and F.Rodriguez - Reinoso (eds.), Introduction to Carbon Technologies, University of Alicante, Alicante, 1997.
7. I.Mochida, Y.Aoyagi, S.Yatsunami and H.Fujitsu, J.Anal.Appl.Pyrol.,**21** (1991), 95.
8. H.Yamamoto, Y.Ch.Hon, X.Mingting, S.Chunlei and J.H.Lunsford, J.Catal.**142** (1993) 325.
9. M.Bajus and M.H.Back, Applied Catalysis A.: General **128** (1995), 61.

NATURAL GAS CONVERSION V
Studies in Surface Science and Catalysis, Vol. 119
A. Parmaliana et al. (Editors)

"ONE-STEP" METHANE CONVERSION UNDER NON OXIDATIVE CONDITION OVER Pt-Co/NaY CATALYSTS AT LOW TEMPERATURE

L. Guczi[1]*, L. Borkó[1], Zs. Koppány[1] and I. Kiricsi[2]

[1]Department of Surface Chemistry and Catalysis, Institute of Isotope and Surface Chemistry, Chemical Research Center, Hungarian Academy of Sciences, P. O. Box 77, H-1525, Budapest, Hungary, *
[2]Department of Applied Chemistry, József Attila University, Rerrich B. tér 1, H-6720, Szeged, Hungary

ABSTRACT

Here is the first report on the non oxidative conversion of methane to larger hydrocarbons *in a "one-step" process* over Pt-Co/NaY bimetallic samples at low temperature. When the methane is pulsed in a H_2/helium mixture at 250°C, the maximum activity being at 1.3 vol. % hydrogen content, higher hydrocarbons are produced during the *hydrogen assisted methane chemisorption*. The "one-step" process is primarily assigned to the metallic cobalt particles whose reduction is facilitated by the presence of platinum. The effect of hydrogen on the "one-step" methane conversion is discussed in terms of the CH_x surface species in which $2 \geq x > 0$ controlled by the relative hydrogen coverage.

1. INTRODUCTON

In previous works the non-oxidative conversion of methane in "two-step" has been investigated when in the first step the methane was chemisorbed on the metal sites followed by the second step in which the surface CH_x species were hydrogenated into various hydrocarbons [1-8]. In mechanistic studies a correlation was found between the hydrogen content of the surface CH_x species (the optimum value for x being around 2) and the chain length of the hydrocarbons produced in the second step. Pt-Co/NaY and Ru-Co/NaY proved to be the best catalysts on which the C_{2+} selectivity was found to be in the range of 80-90 %. The major concerns in these studies was that the hydrogen atoms formed in the dissociation of methane, were removed from the surface during the first step, thus, the methane lost most of its hydrogen atoms and the carbon was largely irreversibly chemisorbed. Despite this problem the amount of methane converted into surface CH_x species could be determined and the yield was calculated on the basis of the amount of CH_x converted into higher hydrocarbons.

To overcome these difficulties a so called "one-step" methane conversion was suggested in which methane pulses were flushed over Pd-Co/SiO_2 catalysts under

* corresponding author, phone/fax: (36)-1-395-9001; e-mail: guczi@alpha0.iki.kfki.hu

non oxidative conditions at low temperature [9]. When some amounts of hydrogen (1.3 vol. %) was admixed to helium and methane pulses were introduced, the conversion based upon the amount of methane converted into higher hydrocarbons, increased compared to that measured in "two-step" process.

The present paper deals with a mechanistic study for the "one-step" methane conversion over Pt-Co/NaY samples. The samples were characterized by TPR and chemisorption and their structure was compared with those prepared earlier [10-11].

2. EXPERIMENTAL

2. 1. Catalysts

The catalysts were prepared by successive ion exchange method [10]. Three catalyst samples were employed. First, Co^{2+} ions using $Co(NO_3)_2$ precursor was ion exchanged into NaY zeolite at 60°C for 24 h followed by thorough washing. The cobalt containing sample was soaked in a solution of $Pt(NH_3)_4(NO_3)_2$ ions and the water was evaporated. The sample is denoted as **(I)**. In the second process Pt^{2+} ions using $Pt(NH_3)_4(NO_3)_2$ were exchanged first into NaY zeolite, then the sample was treated with cobalt nitrate solution. It is denoted as sample **(II)**. Preparation of sample **(III)** was similar to catalyst **(II)**, except the platinum complex exchanged was decomposed by heat treatment in air with a temperature ramp rate of 3°C min^{-1}, then cobalt was introduced into the sample. In this way the amount of cobalt taken up by the platinum containing zeolite was larger than in **(II)**. The metal loading was occasionally checked by X-ray fluorescence spectroscopy. In Table 1 the metal loading of the samples are presented. The basic difference between $Pt_{77}Co_{23}$/NaY denoted by **(S)** [10, 11] and samples **(I)**-**(III)** was that in the former case exchange was performed in a solution with pH = 6.5, whereas the latter samples had higher pH value at the preparation.

2. 2. Catalyst characterization

The samples were characterized by temperature programmed reduction (TPR) using 1 vol. % hydrogen/argon mixture with 10°C min^{-1} ramp rate. An apparatus SORBSTAR equipped with QMS type Hiden HAL 02/100, was employed for the TPR measurements. Generally, two peaks were observed in the TPR experiments, the first one measured around 100°C characteristic of pure platinum reduction, and the second one at 350°C which is assigned to the reduction of bimetallic samples (Table 1).

2. 3. Catalytic reaction

The reaction for methane conversion was performed by means of temperature programmed reaction and in a flow system detailed elsewhere [9]. 100 mg catalyst was first reduced in hydrogen at 400°C for 1 h, then one 0.5 cm^3 (22.3 μmole) methane pulse (in some cases 10 pulses) was introduced into the system at various temperatures in a mixture of H_2/He (1.3 vol. % to 80 vol. % hydrogen) with a total flow rate of 100 cm^3 min^{-1}, unless otherwise indicated. The products were collected in a cold trap and after warming up they were analyzed by means of a gas chromatograph type CHROMPACK CP 9002 using a 50 m long plot fused silica column (0.53 mm I.D) with a stationary phase of CP-Al_2O_3/KCl with a temperature programmed mode. The reaction was characterized by the amount of C_{2+} products in μmoles (in methane equivalents) or the rate was calculated in μmole s^{-1} g_{cat}^{-1} by calculating the contact time from the flow rate and the volume of the methane pulse. Selectivity was calculated by (C_i/C_{2+}) x 100 from i = 1 - 8.

Table 1
Metal loadings and hydrogen uptake in TPR

Sample	Co, wt %	Pt, at. %	Co, at. %	H_2 uptake at 100°C	H_2 uptake at 350°C
(S) [10]	1. 0	77	23	-	-
(I)	6. 5	9	91	124	73
(II)	5. 1	14	86	55	131
(III)	5. 9	12	88	110	39

3. RESULTS AND DISCUSSION

3. 1. Temperature programmed reaction (TPRE)

TPRE [9] was applied to compare the effect of hydrogen on the production of ethane when the methane was pulsed into He or 1 vol. % H_2/Ar mixture over sample (I). 200°C was chosen as by TPRE this temperature appeared optimum for the reaction. As illustrated in Fig. 1. (a), five methane pulses (4 μmole each) and sequentially five hydrogen pulses were repeatedly introduced in a 1 vol. % H_2/Ar mixture and measured by TCD signals. Simultaneously CH_4 at m/e=16 and C_2H_6 at m/e=30 were recorded by QMS as indicated in part (b) of Fig. 1. The TCD signal for CH_4 was higher when blank run (without catalyst) was carried out. Similar experiment was pursued with methane pulses in He (TCD results is not plotted here)and the QMS results are presented in Fig. 1. (c). The major difference shown by (b) and (c) in Fig. 1, is that the amount of the ethane produced during the hydrogen pulses is significantly higher when methane is deposited in 1 vol. % H_2/Ar mixture. The most plausible suggestion is that in the presence of H_2/Ar the surface CH_x species contains more hydrogen, consequently the surface is carbonized to a lesser extent than in the He used as carrier gas. (in He experiments the amount of CH_4 increases within the 5 pulses shown in (c) of Fig. 1.).

3. 2. One-step methane conversion to larger hydrocarbons

In preliminary experiments carried out over $Pt_{77}Co_{23}$/NaY sample [4] methane (22.3 μmole each) was pulsed into a stream of H_2/He mixture. In the first set of experiments at 250°C, the H_2 content was reduced in the sequence of 80, 4.8 and 0 vol. % and the amount of C_{2+} hydrocarbons was diminished in the series of 0.47,

0.07 and 0.01 μmol/100mg catalyst, respectively, and ethane selectivity increased from 59 to 90 %.

At zero hydrogen content only ethane was formed and with increasing hydrogen content the amount of C_{3+} increased. At the hydrogen content of 4.8 and 80 vol. % the C_{3+} values was found to be 26 and 41 %, respectively.

In the second set of experiments the temperature dependence of the methane conversion was studied at 50, 150 and 250°C in a 80 vol. % H_2/He mixture. The high selectivity in the C_6 and C_7 formation at 150°C is supposed to be due to an optimum surface coverage of the hydrogen atoms, which controls dissociation of the CH_4 to surface CH_x species with not too high hydrogen deficiency. On the other hand, at low coverage of surface hydrogen the chain growth of the CH_x species is facilitated at the expense of the hydrogenation and a dedicated balance is set between chain growth leading to C_{2+} hydrocarbons and the desorption of small hydrocarbons.

The "one-step" and "two-step" processes are compared in Table 2.

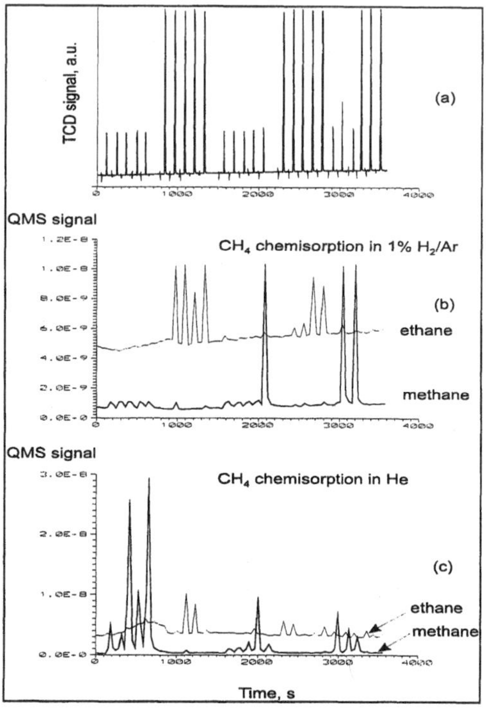

Fig. 1. TPRE on CH_4 conversion at 200°C. (a) in 1 % H_2/Ar; (b) and (c): QMS signals for C_2 species

Table 2
Comparison of one- and two-step methane conversion performed at 250°C over 100 mg $Pt_{77}Co_{23}$/NaY (sample S) [10] using 4.8 vol. % H_2/He mixture

Method	Products in μmole	Selectivity C_2, %	Selectivity C_{3+}, %	Comment
"one-step"	1.16	59.1	40.9	10 CH_4 pulses in H_2/He
"two-step" 1st step: CH_4 ads.	0.11	90.0	10.0	10 CH_4 pulses in pure He
2nd step: H_2	0.04	40.5	59.5	H_2 for 10 min at 250°C

As was indicated [9], during 10 successive methane pulses the catalyst was slowly deactivated, therefore, in the further experiments single methane pulse was applied. The result for $Pt_{12}Co_{88}$/NaY (sample (**III**)) is presented in Table 3.

From these results the we can conclude that the "one-step" process performed over NaY supported bimetallic catalysts, is superior to those measured on the same catalysts in a "two-step" process. In the former case the conversion of methane is about 5 % referred to one gram of catalyst, whereas in the two-step process it is about 8 times less. In the case of the $Pt_{12}Co_{88}$/NaY catalyst the C_2 selectivity is low, but the contribution of the higher hydrocarbons is significantly higher.

Table 3
Comparison of one- and two-step methane conversion performed at 250°C over sample (III) (100 mg) using 1.3 vol. % H_2/He mixture

Method	Products C_{2+} in μmole	Selectivity C_2, %	Selectivity C_{3+}, %	Comment
"one-step"	0.08	24.4	75.6	1 CH_4 pulse in H_2/He
"two-step" 1st step: CH_4 ads.	0.007	24.8	75.2	1 CH_4 pulses in pure He
2nd step: H_2	0.003	51.4	48.6	H_2 for 10 min at 250°C

Finally, we wish to show a comparison among the samples (I) - (III) in the "one-step" reaction with regard to the selectivity, activity and the effect of additional hydrogenation at 400°C. The data are presented in Table 4.

Table 4
Comparison of samples (I) to (III) in "one-step" methane conversion performed at 250°C over 100 mg catalysts using 1.3 vol. % H_2/He mixture

Sample	Products C_{2+} in μmole	Rate μmole s^{-1} g^{-1}	C_2-C_4 in %	C_{5+} in %	Condition
(I)	0.013	0.43	25,2	74.8	
(II)	0.043	1.44	65.9	34.1	
(III)	0.082	2.74	24.8	75.2	
(I)	0.037	-	71.7	28.3	H_2 at 400°C
(II)	0.025	-	22.5	77.5	H_2 at 400°C
(III)	0.062	-	11.9	88.1	H_2 at 400°C

The results in Table 4 is explained in the following way. In sample (I) the cobalt is reduced to a small extent as, due to the preparation technique, platinum is located at the outer surface of zeolite [11] similar to Ru-Co/NaY system [12]. The rate is low due to the limited number of surface cobalt sites. Once the cobalt ions are exchanged after the platinum ions having been inserted (samples (II) and (**III**)), they are reduced to a larger extent, thus, the rate of CH_4 conversion increases. Due to the intimate contact between Pt and Co in sample (**II**), the C_2-C_4 selectivity is highest because hydrogen activation takes place easily over Pt sites surrounding cobalt atoms. When this type of contact between Pt and Co does not exist, the effect of cobalt is amplified indicated by the enhanced C_{5+} selectivity (sample (**I**) and (**III**)). The estimated rate of methane conversion is calculated by assuming a square shape methane pulse. Additional hydrogenation at 400°C for 1 h results in further removal of mainly long chain hydrocarbons from the surface. The difference between the data shown Tables 2 and 3 are presumably attributed to the difference between the preparation techniques, i. e. the mixing of the two ions inside the zeolite appears to be more complete when the exchange is carried out at pH = 6.5 (sample (**S**)).

This is the first evidence that “one-step” reaction in the non oxidative methane conversion at low temperature is a feasible process and the achievable conversion and selectivity of the C_{2+} hydrocarbons formed, are superior to that obtained in the “two-step” process. According to the proposed mechanism it is suggested that small amount of H_2 in He during methane chemisorption maintains certain hydrogen coverage on the metal surface, consequently the total hydrogen removal from the surface CH_x species are prevented. When the x value is $2 \geq x > 0$, the surface chain growth already starts during methane chemisorption. When hydrogen coverage is too high, $x \geq 3$, consequently dissociation of CH_4 is hampered. When hydrogen coverage is too low, significantly larger fraction of the chemisorbed methane is transferred into surface carbon whose conversion and subsequent hydrogenation is hardly impossible.

4. ACKNOWLEDGMENTS

The authors are indebted to the National Science and Research Fund (grant # T-022117) and to the COST Program (grant # D5/001/93). The help of Miss A. Tamási in preparation of the bimetallic samples is greatly acknowledged.

5. REFERENCES

1 T. Koerts, M. J. A. Deelen, and R. A. Van Santen, *J. Catal.*, **138** (1992) 101
2 A. Erdőhelyi, J. Cserényi, and F. Solymosi, *J. Catal.*, **141** (1993) 287
3 L. Guczi, R. A. van Santen and K. V. Sarma, *Catal. Rev. Eng. Sci.*, **38** (1996) 249
4 L. Guczi K. V. Sarma, and L. Borkó, *Catal. Lett.*, **39** (1996) 43
5 G. Boskovic, J. S. M. Zadeh and K. J. Smith, *Catal. Lett.*, **39** (1996) 163
6 L. Guczi, Zs. Koppány, K. V. Sarma, L. Borkó and I. Kiricsi, Progress in Zeolite and Microporous Materials (Eds.: H. Chon, S. -K. Ihm and Y. S. Uh), *Stud. Surf. Sci. Catal.*, **Vol. 105**, pp. 861, Elsevier Sci. Publ. Co., Amsterdam, 1997
7 M. Belgued, A. Amariglio, P. Pareja and H. Amariglio, *J. Catal.*, **159** (1996) 449
8 L. Guczi, K. V. Sarma and L. Borkó, J. *Catal.*, **167** (1997) 495
9 L. Guczi, L. Borkó, Zs. Koppány and F. Mizukami, *Catal. Lett.*, submitted
10 G. Lu, T. Hoffer and L. Guczi, *Catal. Lett.*, **14**, 207 (1992)
11 Z. Zsoldos, G. Vass, G. Lu and L. Guczi, *Appl. Surf. Sci.*, **78**, 467 (1994)
12 L. Guczi, R. Sundararajan, Zs. Koppány, Z. Zsoldos, Z. Schay, F. Mizukami and S. Niwa, *J. Catal.*, **167**, 482 (1997)

NATURAL GAS CONVERSION V
Studies in Surface Science and Catalysis, Vol. 119
A. Parmaliana et al. (Editors)

Heteropolyacid-Catalyzed Partial Oxidation of Methane in Trifluoroacetic Acid

Tsugio Kitamura, Dung-guo Piao, Yuki Taniguchi, and Yuzo Fujiwara
Department of Chemistry and Biochemistry, Graduate School of Engineering,
Kyushu University, 6-10-1 Hakozaki, Fukuoka 812-8581, Japan

The new catalytic system has been examined for the partial oxidation of methane in liquid phase. It is found that the vanadium containing heteropolyacids/$K_2S_2O_8$/$(CF_3CO)_2O$/CF_3COOH catalyst system converts methane to methyl trifluoroacetate in 95% yield based on methane.

1. INTRODUCTION

The lower alkanes such as methane are the most abundant of the hydrocarbons but the least reactive. Thus, the partial oxidation of methane is of great practical interest which is appealing for the liquefaction of natural gas and the chemical conversion to more useful chemical products. In continuing studies on C-H bond activations [1,2], we have found that methane in trifluoroacetic acid (TFA) can be also converted to methyl trifluoroacetate (**1**) along with a small amount of methyl acetate (**2**) in the presence of a catalytic amount of $H_5PV_2Mo_{10}O_{40}$ as a catalyst and $K_2S_2O_8$ as an oxidizing agent, and $(CF_3CO)_2O$ (TFAA) under mild conditions (Eq 1).

$$CH_4 \xrightarrow[\text{TFA}]{\text{V cat., } K_2S_2O_8\text{, TFAA}} \underset{\mathbf{1}}{CF_3COOCH_3} + \underset{\mathbf{2}}{CH_3COOCH_3} \qquad (1)$$

2. EXPERIMENTAL

In a 100-mL stainless steel autoclave fitted with an 85-mL glass tube and a magnetic stirring bar, catalyst, $K_2S_2O_8$, TFA and TFAA were added, successively. The autoclave was closed and then pressurized to 20 atm with CH_4. The mixture was heated with stirring at 80℃ for 20 h. After cooling the autoclave was opened and the mixture was analyzed by GLC. Methyl trifluoroacetate (**1**) was obtained along with a small amount of methyl acetate (**2**) as a by-product by this reaction.

3. RESULTS AND DISCUSSION

At first, we examined the activity of various catalyst systems for the partial oxidation of methane in TFA solution. Consequently, some vanadium-substituted Keggin type heteropoly-acids such as $H_5PV_2Mo_{10}O_{40}$ gave good results as shown in Table 1. As is apparent from the table, the oxidation of methane with $K_2S_2O_8$ proceeds to afford a small amount of esters **1** and **2** even in the absence of the catalyst (entry 1). By the addition of the catalyst, especially vanadium containing heteropolyacids, the yields of the products (**1** and **2**) increased. Among

Table 1
Heteropolyacid Catalyzed Partial Oxidation of Methane[a]

Entry	Catalyst	TON	Yield/%[b]	Ratio/**1**:**2**
1	none	-	0.4	79:21
2	$H_3PMo_{12}O_{40}$	5	0.1	76:24
3	$H_4PVMo_{11}O_{40}$	80	2.4	83:17
4	$H_5PV_2Mo_{10}O_{40}$	161	4.7	88:12
5	$H_6PV_3Mo_9O_{40}$	92	2.8	86:14
6	$H_7PV_4Mo_8O_{40}$	76	2.1	73:27
7	$H_8PV_5Mo_7O_{40}$	62	1.8	67:33
8	$H_3PW_6Mo_6O_{40}$	17	0.4	73:27
9	$H_3PW_{12}O_{40}$	22	0.5	76:24
10	$H_4SiW_{12}O_{40}$	15	0.4	72:28
11	$H_4SiW_4Mo_8O_{40}$	6	0.1	80:20
12	$H_5SiVW_{11}O_{40}$	161	3.5	73:27
13	$H_4SiMo_{12}O_{40}$	19	0.5	74:26
14	$H_4PVW_{11}O_{40}$	50	1.0	76:24
15	$H_5PV_2W_{10}O_{40}$	241	6.4	73:27
16	$H_6PV_3W_9O_{40}$	139	3.3	55:45
17	V_2O_5	57	4.0	83:17
18	$NaVO_3$	17	1.5	54:46
19	MoO_3	5	0.3	78:22
20	Na_2WO_4	16	1.0	96: 4

a) Reaction conditions: CH_4 (20 atm), catalyst (50 mg), $K_2S_2O_8$ (5.00 mmol), TFAA (10.0 mmol), TFA (5.0 mL), 80°C, 20 h.

b) GLC yield based on CH_4.

them, $H_5PV_2W_{10}O_{40}$ gave the highest turnover number (TON) and the highest yields of esters (entry 15). It is noted that the yield and selectivity of the products decrease with the increasing numbers of vanadium atom in the heteropoly anion (entries 4-7 and 15-16). Vanadium(V) oxide is also more effective than MoO_3 and Na_2WO_4 (entries 17-20), suggesting that an oxo-vanadium moiety of the heteropolyanion acts as the active site. On the basis of the yield and the selectivity of the product (entry 4), we chose $H_5PV_2Mo_{10}O_{40}$ as the catalyst.

The present partial oxidation of methane required a highly acidic solvent, trifluoroacetic acid, that gave the best result. The presence of TFAA accelerated the reaction rate. The role of TFAA seems to be the activation of the catalyst by removal of water from the hydrated catalyst since all the heteropolyacids usually exist as the 30 hydrates.

Figure 1 shows the effect of the amount of $H_5PV_2Mo_{10}O_{40}$ catalyst. The yield of products increased with increasing amount of the catalyst. The highest yield of products (**1** and **2**) was obtained by using 0.013 mmol (TON=490) of the catalyst. Excess use of the catalyst resulted in lower yields because of the further oxidation.

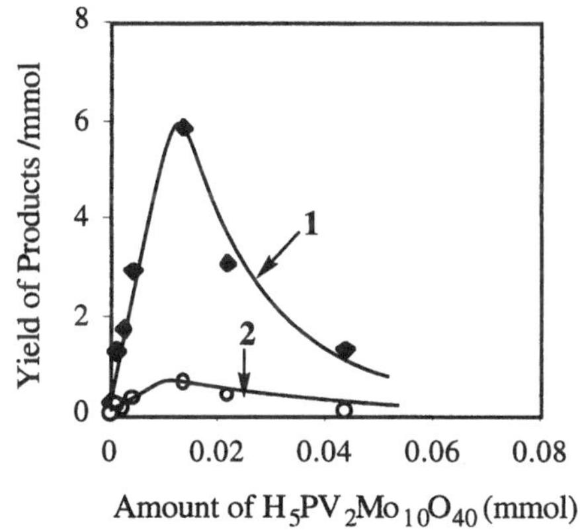

Figure 1. Effect of the Amount of Catalyst

Conditions: CH_4(20 atm), $K_2S_2O_8$ (5.00 mmol), TFA (5.0 mL), TFAA (10.0 mmol), 80°C, 20 h.

Several oxidants were examined in the reaction using a $H_5PV_2Mo_{10}O_{40}$ catalyst in TFA. $K_2S_2O_8$ gave the best result in the partial oxidation of methane. Uses of other oxidants such as $Na_2S_2O_8$, $(NH_4)_2S_2O_8$, MnO_2, $KMnO_4$, and H_2O_2 in lieu of $K_2S_2O_8$ resulted in the inferior results. Figure 2 shows the effect of the amount of $K_2S_2O_8$. The yield of products increased with increasing amount of $K_2S_2O_8$. The best result was obtained by using 5 or 6 mmol of $K_2S_2O_8$. Excess use of the oxidant resulted in lower yields because of lower efficiency of stirring.

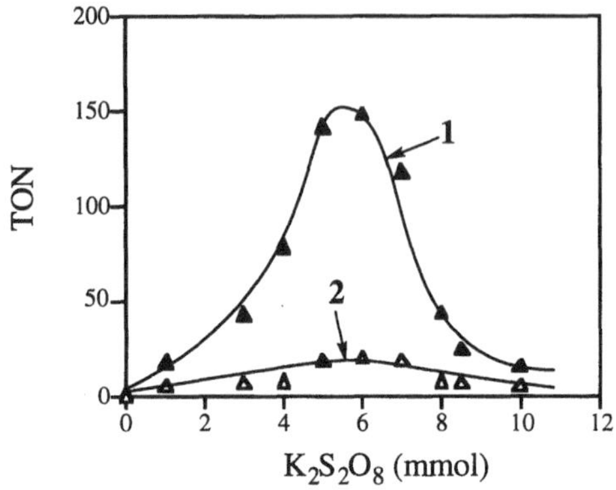

Figure 2. Effect of the Amount of $K_2S_2O_8$

Reaction conditions: CH_4 (20 atm), $H_5PV_2Mo_{10}O_{40}$ (0.022 mmol), TFA (5.0 mL), TFAA (10 mmol), 80°C, 20 h.

Time course of the reaction at several temperatures under the same conditions is shown in Figure 3. The initial reaction rate (within 10 h) increases as temperature increases. The TON of the catalyst at 100℃ increases up to 10 h and then decreases rapidly. The best TON was obtained in the reaction at 80℃ for 20 h.

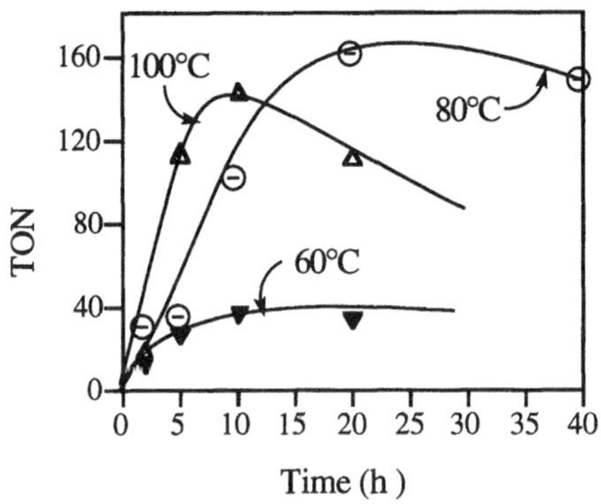

Figure 3. Time Course of the Reaction

Conditions: CH_4 (20 atm), $H_5PV_2Mo_{10}O_{40}$ (0.022 mmol), $K_2S_2O_8$(5.00mmol), TFA (5.0 mL), TFAA (10.0 mmol).

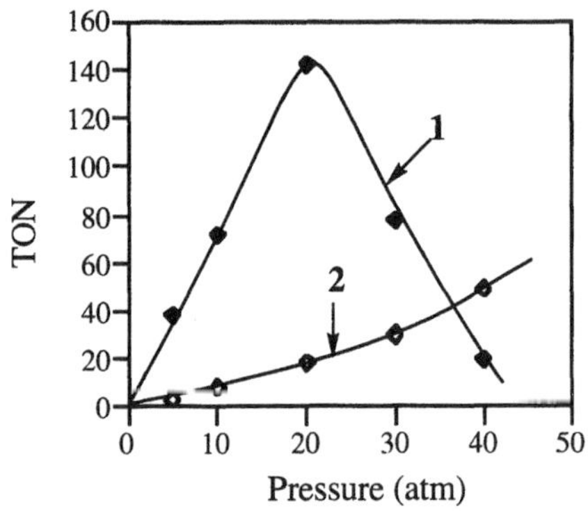

Figure 4. Effect of Pressure of CH_4

Conditions: $H_5PV_2Mo_{10}O_{40}$ (0.022 mmol), $K_2S_2O_8$ (5.00 mmol), TFA (5.0 mL), TFAA (10 mmol), 80°C, 20 h.

Next, we investigated the effect of the pressure of CH_4. The representative results are summarized in Figure 4. The TON of the catalyst increased with increasing pressure of methane. The best result was obtained at 20 atm of CH_4. The reaction under the pressure higher than 20 atm caused a decrease in the yield of the ester **1**. Interestingly, the yield of ester **2** increased with increasing pressure of CH_4.

In order to improve the yield based on methane, the reaction using a 25-mL autoclave was examined. The representative results are listed in Table 2. The yield based on methane increased with increasing amount of the solvent (entries 1-3). The reaction at 6 atm of methane

Table 2
Quantitative Conversion of Methane to Methyl Trifluoroacetate[a]

Entry	TFA (mL)	TON (**1**+**2**)	Product (**1**+**2**)/mmol	Yield[b]/%
1	5	36	0.79	13.9 (93 : 7)
2	7.5	118	2.57	54.3 (93 : 7)
3	9	128	2.72	67.7 (89 : 11)
4	9[c]	98	2.11	87.6 (89 : 11)
5	9[d]	81	1.80	95.0 (82 : 18)

a) Reaction conditions: 25-mL autoclave, $H_5PV_2Mo_{10}O_{40}$ (0.022 mmol), CH_4 (10 atm), $K_2S_2O_8$ (5.00 mmol), TFAA (10.0 mmol), 80°C, 20 h.
b) GC yield based on CH_4. Numbers in parentheses are the ratio of **1** and **2**.
c) CH_4 (6 atm).
d) N_2 (5 atm) added in the reaction mixture, CH_4 (5 atm).

afforded the product in 87.6% yield (entry 4). Furthermore, we succeeded in the quantitative conversion of methane to methyl trifluoroacetate by pressurizing with inert nitrogen (5 atm) (entry 5).

The detailed results on partial oxidation of methane by vanadium catalysts are presented and the mechanistic implication is discussed.

REFERENCE

[1] Y. Fujiwara, K. Takaki, and Y. Taniguchi, *Synlett*, 591 (1996) and references cited therein.
[2] A. Sen, *Platinum Metals Rev.*, **35**, 126 (1991).

NATURAL GAS CONVERSION V
Studies in Surface Science and Catalysis, Vol. 119
A. Parmaliana et al. (Editors)

Performance of Na_2WO_4-Mn/SiO_2 catalyst for conversion of CH_4 with CO_2 into C_2 hydrocarbons and its mechanism

Yu Liu Ruiling Hou Xuxia Liu Jinzhen Xue and Shuben Li

State Key Laboratory for Oxo Synthesis and Selective Oxidation,
Lanzhou Institute of Chemical Physics, Chinese Academy of Sciences,
Lanzhou 730000,P.R.China

Na_2WO_4-Mn/SiO_2, a better catalyst for oxidative coupling of methane, was used to investigate its performance for conversion CH_4 with CO_2. About 5% CH_4 and 95% selectivity to C_2 were obtained at 820 ℃. The reaction temperature favored CH_4 conversion, but unfavored the selectivity to C_2. This result is due to that the surface lattice oxygen(desorbed at 800 ℃) is responsible for selective oxidation of methane to C_2, whereas, the bulk lattice oxygen(desorbed at 850 ℃) is responsible for deep oxidation. In addition, from the results of O_2-TPD and CH_4 and CO_2 pulses reaction, a possible mechanism for CO_2 activation and CH_4 reaction with CO_2 was suggested.

1. INTRODUCTION

Direct conversion of methane to C_2H_4 and C_2H_6 by oxidative coupling of methane has been studied over a wide variety of oxide catalysts. It has been established that the heterogeneous and homogeneous reaction co-existent in oxidative coupling of methane. During the homogeneous reaction, methyl radical is inevitable to react with gaseous oxygen to form CO_2. Obviously, this side reaction will decrease C_2 selectivity. In order to reduce gas phase reaction, besides membrane and other reactors[1,2], CO_2 has been used as oxidant instead of O_2, since CO_2 does not react with methyl radical in the gas phase.

The earlier work of Aika et al[3] had succeeded in utilizing CO_2 as an oxidant over PbO/MgO and alkaline earth metal-doped CaO catalysts. Recently, Asami et al[4] systematically performed this reaction over seventeen metal oxides. They reported that C_2 hydrocarbons could be formed over many oxides and rare earth oxides with high selectivity. However, no detailed work has been published to elucidate the mechanism over related catalysts. On the other hand, the above catalysts possess some strong basic sites which would easily react with CO_2 Thus, it would probably influence their activity for this reaction. In this paper, Na_2WO_4-Mn/SiO_2 catalyst, developed by Li's group[5], as the most promising catalyst for oxidative coupling of methane, is selected to investigate its performance for conversion of CH_4 with CO_2 as well as the possible mechanism. It seems to us, this work is of great significance, because this catalyst does not react with CO_2 at elevated temperature[6].

2. EXPERIMENTAL

The preparation method of Na_2WO_4-Mn/SiO_2 catalyst and the testing system were described previously[5]. The data processing method has also been described elsewhere[4]. CO_2 pulse reaction was carried out in a 45 ml/min He flow at 1093 K after the surface and bulk lattice oxygen on the catalyst was desorbed at 1153 K for half an hour. CO_2 and CH_4 alternative pulse reaction were also carried in a 45 ml/min He flow at 1093 K. O_2-TPD was performed in a 45 ml/min He flow with a heating rate of 20 K/min before the desorption of oxygen on the catalyst and after the treatment of the catalyst with CO_2. Pulse size was 0.3 ml and pulse interval was 1.5 min. The effluents were analyzed using an on-line Finnigan Mat 700 ion trap mass detector (ITD)[7,8].

3. RESULTS AND DISCUSSION

The results of the effect of temperature on CH_4 conversion with CO_2 over Na_2WO_4-Mn/SiO_2 catalyst were listed in Table 1. As may be seen, the conversions of CH_4 and CO_2 as well as the selectivity to C_2 hydrocarbons increase with reaction temperature. In addition, the ratio of C_2H_4 to C_2H_6 also increases with reaction temperature, which indicated that some C_2H_4 was formed from the thermal dehydrogenation of C_2H_6. 4.73% conversion of CH_4 and 94.5% C_2 selectivity were obtained at 820 ℃. It should be point out that no reaction of CH_4 with CO_2 was observed in the absence of catalyst at 850 ℃(Table1: blank). The results indicated that temperature favored the reaction over Na_2WO_4-Mn/SiO_2 catalyst. On the other hand, the results implied that temperature also favored the activation of CO_2 on the surface of Na_2WO_4-Mn/SiO_2 catalyst.

Table 1

Effect of temperature on CH_4, CO_2 conversion and C_2 selectivity.

Temperature (℃)	Conversion(%) CH_4	Conversion(%) CO_2	Selectivity to C_2(%)	C_2H_4/C_2H_6
650	0.11	0.03	-	-
700	0.31	0.12	67.1	0.35
750	0.35	0.14	65.2	0.32
800	4.37	2.17	93.8	0.69
820	4.73	2.32	94.5	0.72
850	4.88	2.35	89.2	0.80
850 (blank)	0	0	-	-

Catalyst charge: 0.15g, total pressure: 0.1MPa, CO_2/CH_4=2;
total flow rate: 20ml/min. The results were obtained after 5 h reaction.

Clearly, when reaction temperature increased up to 850 ℃, the C_2 hydrocarbon selectivity decreased from 94.5% (820 ℃) to 89.2%, though the conversion of CH_4 and CO_2 increased. In order to explain the above phenomena and deduce the possible mechanism over Na_2WO_4-Mn/SiO_2 catalyst, O_2-TPD, CO_2 and CH_4 pulse reaction were carried out.

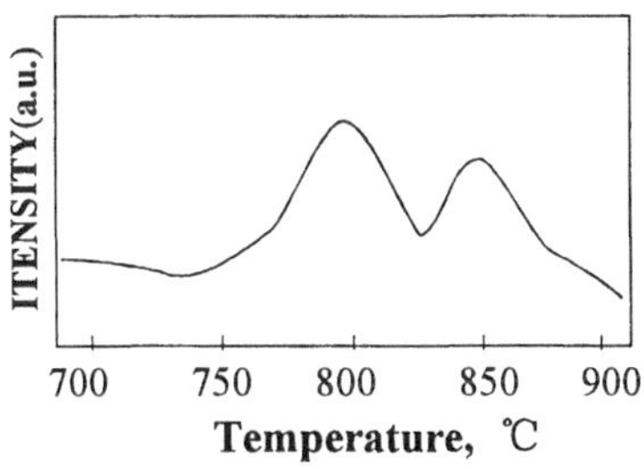

Figure 1 O_2-TPD profile of Na_2WO_4-Mn/SiO_2 catalyst.

The O_2-TPD profile (Fig 1) showed that there were two oxygen desorption peak, one at 800 ℃, another at 850 ℃. According to Li et al[5], the former peak is attributed to surface lattice oxygen and the latter one is attributed to bulk lattice oxygen. The results of CH_4 pulse reaction showed that the production of C_2H_4 and C_2H_6 at 800 ℃ were higher than that at 850 ℃. On the contrary, only trace amount of CO could be detected at 800 ℃ and large amount of CO is formed at 850 ℃. It could be concluded that the surface lattice oxygen(800 ℃) is responsible for methane activation, whereas the bulk lattice oxygen is partially responsible for both methane activation and oxidation. This result revealed why C_2 selectivity decreased at 850 ℃(Table 1).(Another reason may be attributed to the effect of temperature on the reaction rates, which is being confirmed in our lab now.)

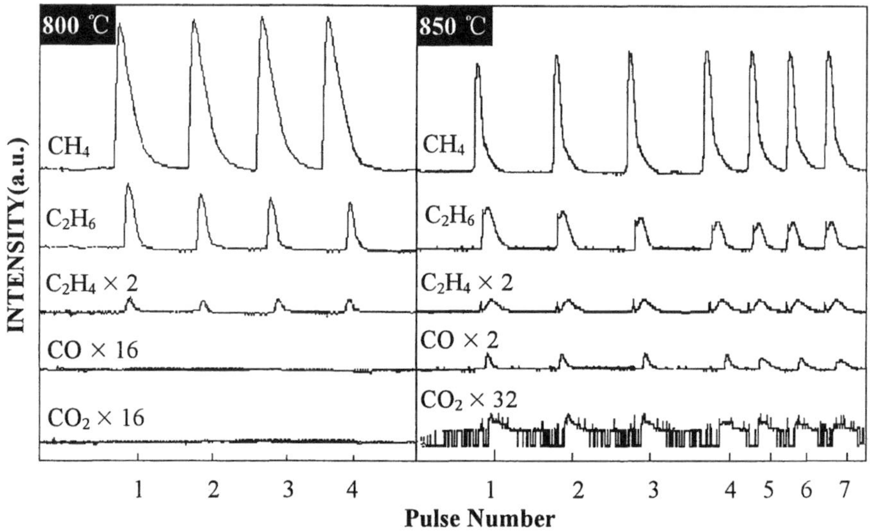

Figure 2 CH_4 pulse over Na_2WO_4–Mn/SiO_2 catalyst at 800 ℃ and 850 ℃.

The results of CO_2 pulse reaction over the oxygen desorbed catalyst at 820 ℃ showed that CO was formed on the surface of the catalyst. This means that CO_2 could be decomposed to CO and active oxygen on the catalyst, $CO_2 \rightarrow CO + O^*$. The O_2-TPD profile of the catalyst which underwent oxygen desorption and then treated with CO_2 showed only one very small peak at 800 ℃. Obviously, the mount of surface lattice oxygen dominates CH_4 and CO_2 conversion. This is why methane conversion is lower when CO_2 is used as oxidant than O_2.

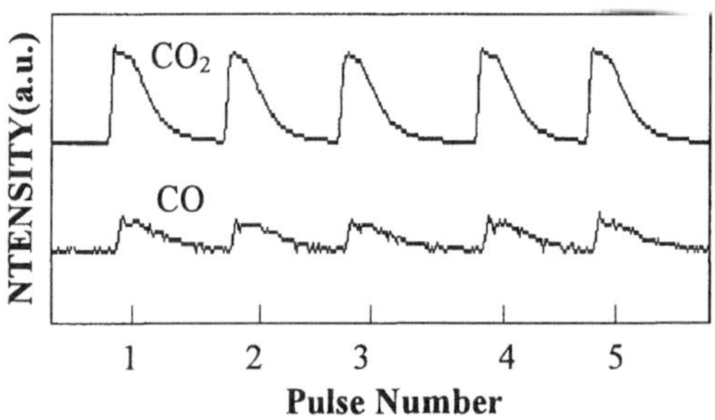

Figure 3 CO_2 pulse over Na_2WO_4–Mn/SiO_2 catalyst at 820 ℃.

Based on the above results, the possible main mechanism of conversion of CH_4 with CO_2 on Na_2WO_4-Mn/SiO_2 catalyst is as follows:
$CO_2 \rightarrow CO + O^*$, $O^* + CH_4 \rightarrow CH_3\cdot + HO$, $2CH_3\cdot \rightarrow C_2H_6$, $C_2H_6 + O^* \rightarrow C_2H_4 + HO$, $C_2H_6 \rightarrow C_2H_4 + H_2$

REFERENCES

1. K.Omata, S.Hashimoto, H.Tominaga, K.Fujimoto, Appl.Catal.,52(1989)L1.
2. C.A.Jones, J.J.Leonard, J.A.Sofranko, Energy Fuels, 1(1987)12
3. T.Nishiyama, Ken-Ichi AiKa, J.Catal.,122(1990)346.
4. K.Asami, T.Fujita, K.Kusakabe et al, Appl.Catal.,126(1995)245.
5. X.P.Fang, S.B.Li, J.Z.Lin, et al, J.Mol.Catal.(China),6(1992)255.
6. J.H.Lunsford, Angew.Chem.Int.Ed.Engl.,34(1995)970.
7. Yu Liu and Shikong Shen, Appl.Catal,121(11995)57.
8. Yu Liu, C-C Yu, X-X Liu, J-Z, Xue, B. Zhang and S-K Shen, Chem.Lett.,(1996) 1127.

NATURAL GAS CONVERSION V
Studies in Surface Science and Catalysis, Vol. 119
A. Parmaliana et al. (Editors)

Oxidative coupling of methane to ethylene in a reaction system with products separation and gas recirculation

A. Machocki and A. Denis

Department of Chemical Technology, University of Maria Curie-Sklodowska,
3 Maria Curie-Sklodowska Square, 20-031 Lublin, Poland

The paper presents the effect of the reaction temperature and catalyst contact time on product output distribution of the OCM reaction carried out in a reaction system which involves adsorptive separation of the products and gas recirculation. With methane conversion reaching 94%, the yield of C_{2+} hydrocarbons was up to 75%, while that of ethylene up to 60%. The selectivity of C_{2+} hydrocarbons achieved the level of 80% with ethylene constituting 90-95%.

1. INTRODUCTION

The application of selective adsorption of the products of oxidative coupling of methane (OCM) and recirculation of unreacted methane and nonadsorbed ethane makes it possible to obtain high methane conversion, high yield of C_{2+} hydrocarbons, and high ethylene content in produced hydrocarbons [e.g. 1-7]. The large amount of formed carbon dioxide may raise some objections, however one may expect that the selectivity of the whole process may be improved by optimising the reaction conditions. A decrease in oxygen concentration in the methane-oxygen mixture — which with a single passage of the reagents through the reactor usually improved overall selectivity — has brought few advantages in a recirculation process with a continuous supply of raw materials [7]. Oxidative dehydrogenation of ethane to ethylene, occurring together with methane conversion in the recirculation process in the same reactor and with the same catalyst, produces additional amounts of carbon dioxide which are the higher, the greater amount of ethane is fed back again into the reactor (as it occurs in the case of oxygen-poor reaction mixtures). In consequence, the overall selectivity to C_{2+} hydrocarbons, especially to ethylene, is not greatly dependent on the composition of the methane-oxygen mixture. Additional negative consequences of employing a low oxygen concentration concern a decreased ratio of the amount of ethylene to that of ethane in the obtained hydrocarbons and, primarily, a very low productivity of the reaction system.

The aim of the present paper is to demonstrate the effects that can be achieved by optimising parameters of the recirculation OCM process, i.e., the temperature and the rate of circulation changing the time of contact of the reactants with the catalyst. The paper also shows some effects resulting from the better efficiency of the adsorptive separation of the products after improving the properties of the adsorbent by replacing calcium ions with silver ions in the molecular sieve 5A.

2. EXPERIMENTAL

The OCM process was carried out at 700-800°C in the reactor-adsorber system with gas recirculation into which methane and oxygen were continuously supplied (Figure 1).

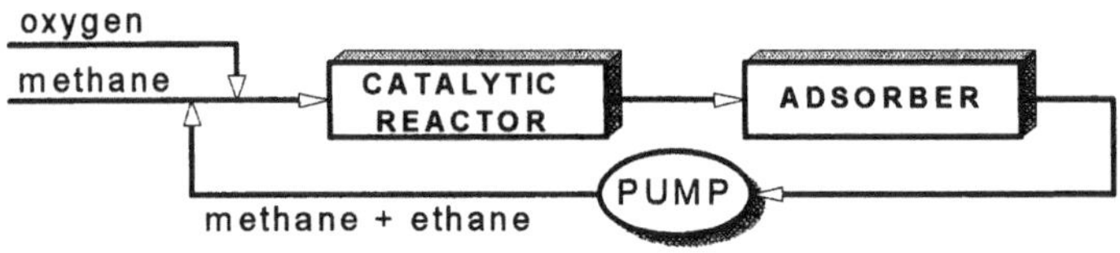

Figure 1. The idea of the OCM process in the reactor-adsorber system with gas recirculation

The oxygen concentration in the reaction mixture entering the reactor was 4.76 vol.%, which corresponds to the methane-to-oxygen ratio equal to 20:1. The contact time W/F (W - catalyst weight, 0.5 g; F - rate of the gas circulation, 37.5-300 cm^3/min) was changed from 0.1 to 0.8 s·g/cm^3. The catalyst of the reaction was Na^+/CaO (1.7 wt.% of Na^+), obtained by calcium carbonate impregnation with a solution of sodium carbonate, and for products separation 10 g of the molecular sieve 5A (Fluka) was employed. Recirculation experiments were performed after catalyst stabilisation for 4-5 hours under reaction conditions, the reactor operated in the single pass mode.

3. RESULTS

The adsorption of the products of the OCM process was continued until the full utilisation of the sorption capacity of the molecular sieve, i.e., till the moment of its breakthrough by ethylene, C_{3+} hydrocarbons and CO_2. It occurred much later than the adsorbent breakthrough by ethane [4]. This adsorption time defined as above as well as the time of the whole experiment depends crucially on the OCM reaction conditions and in particular on the W/F contact time (Figure 2). It was longer when the OCM was performed with longer contact times, i.e., with lower circulation rates and, hence, lower rate of the flow of post-reaction gases through the adsorbent.

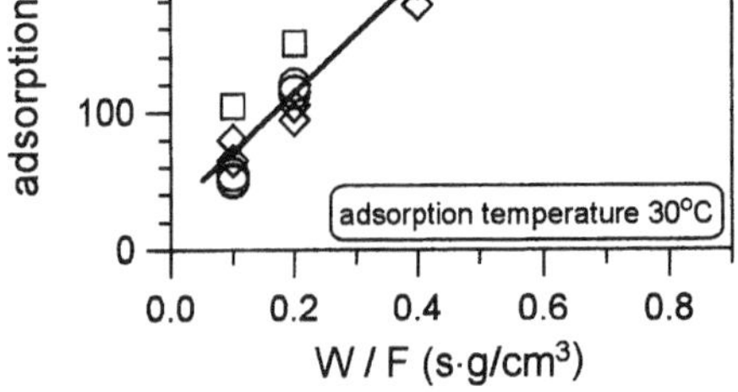

Figure 2. The time of products adsorption vs. conditions of the OCM process.

Methane conversion increases slightly with reaction temperature and contact time (Figure 3). In view of the mode of carrying out the process it depends on the amount of methane retained by the adsorbent. The higher amounts of the strongly adsorbed

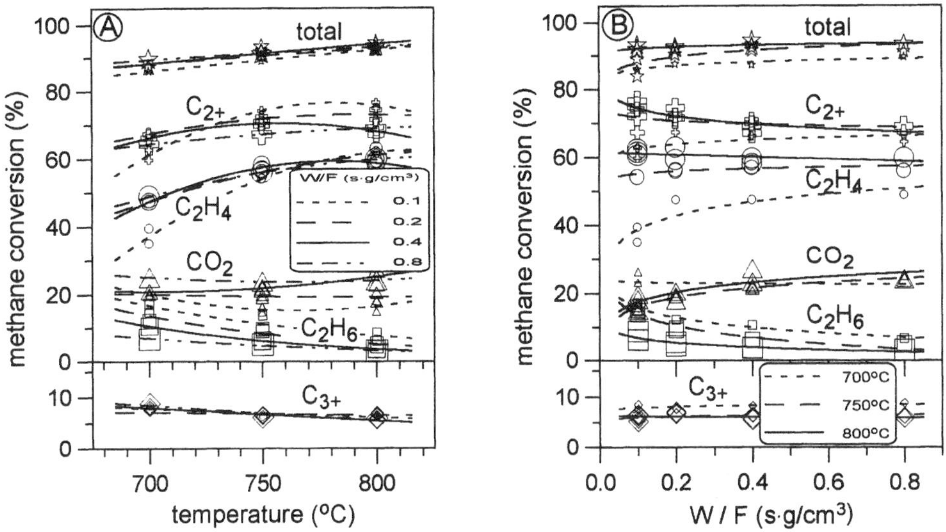

Figure 3. Methane conversion dependence on OCM reaction temperature (A) and contact time (B) of the gas recycle system with product separation

process products (ethylene, carbon dioxide, water vapour) in post-reaction gases formed at higher temperatures and at longer W/F facilitates the displacement of adsorbed methane from the adsorbent and thus, a greater portion of it undergoes the reaction.

The yield of C_{2+} hydrocarbons reaches about 75% (Figure 3). The highest values are observed at temperatures 750-800°C and at shorter contact times. At the same temperature range the yield of the main product of the process, ethylene, changes in a similar way; its maximum values reached about 60%. Evidently the smallest portion of methane was converted to ethylene at the temperature of 700°C. Ethane yield decreases with increasing temperature and contact time and it is opposite to the case of carbon dioxide.

The selectivity of methane conversion (Figure 4) to C_{2+} hydrocarbons reaches the highest values, i.e., about 80%, at short contact times, with a slight maximum at the temperature of 750°C, which implies that these may be the most advantageous conditions for the OCM process. In comparison with our previous results obtained at 800°C [3,4], by optimising the reaction conditions without changing the kind of the catalyst, it has been possible to improve the overall selectivity of the process by several percent. The selectivity of the process towards ethylene remains very similar at temperatures 750-800°C and longer contact times. At shorter contact times the ethylene selectivity is the highest at 800°C, approaching the value of about 67%. The temperature of 700°C is definitely much too low for the OCM process oriented towards ethylene production. The results indicate that for a high ethylene selectivity in the recirculation process it should be high in a single passage of the methane-oxygen mixture through the catalyst. The results also suggest that the slackening of the reaction conditions — lowering of the temperature in the range in which the selectivity to the sum of C_{2+}

hydrocarbons in a single pass mode of operation is similar (here 750-800°C) — and the shortening of the contact time prevent full oxidation of some part of circulating ethane.

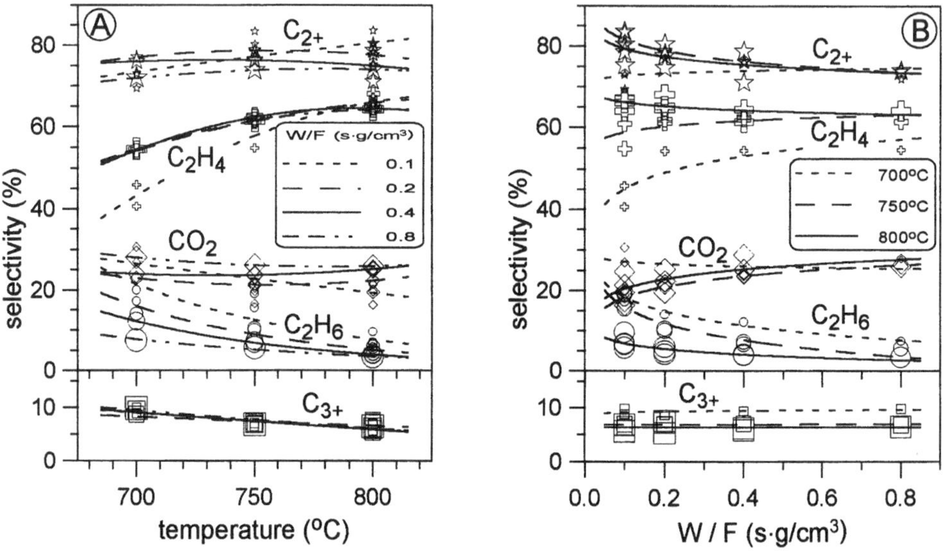

Figure 4. Selectivity dependence on OCM reaction temperature (A) and contact time (B) of the gas recycle system with product separation.

The temperature of the process and the contact time also change the composition of the obtained C_{2+} hydrocarbons (Figure 5). The lowering of the temperature and the shortening of the contact time decrease the content of ethylene from 85-90 % to 68-80 % and increase the content of ethane from 5-10 % to 10-25 %, which is reflected in an appropriate change in the ethylene/ethane ratio. Further improvement of the composition of the obtained C_{2+} hydrocarbons may be achieved by modifying the properties of the adsorbent by the replacement of calcium ions with silver ions in the molecular sieve 5A. The content of ethylene among C_{2+} hydrocarbons produced at the temperatures 750-800°C and with contact times of 0.2-0.4 s·g/cm^3 is then increased to 95%. The increase in the ethylene capacity of the adsorbent modified with silver brings about the increase in the ethylene/ethane ratio from 6-14 to 40-50. The exchange of ions in the molecular sieve 5 A also extends (1.3 -1.7 times) the time after which ethylene breaks through it. In all conditions the contents of C_{3+} hydrocarbons is lower than 10 % while the ratios of propylene and propane are similar to those in a single passage of the reagents through the catalyst.

The above results indicate that, when ethylene is the desired final product of the process, the conditions which are more favourable for the improvement of overall selectivity and yield of the process are less advantageous for the composition of the obtained hydrocarbons. Nevertheless, it seems that for many reasons the suppression of carbon dioxide formation is a

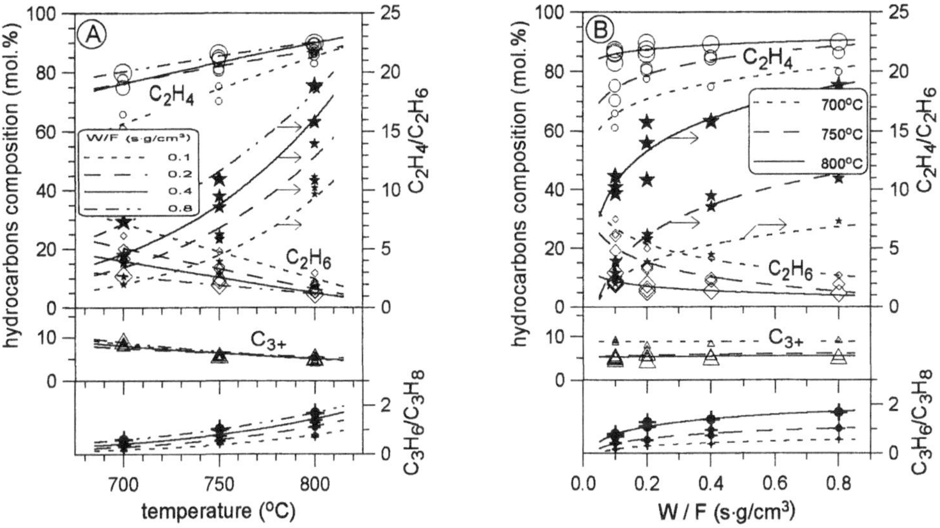

Figure 5. Dependence of the composition of formed hydrocarbons on OCM reaction temperature (A) and contact time (B) of the gas recycle system with product separation.

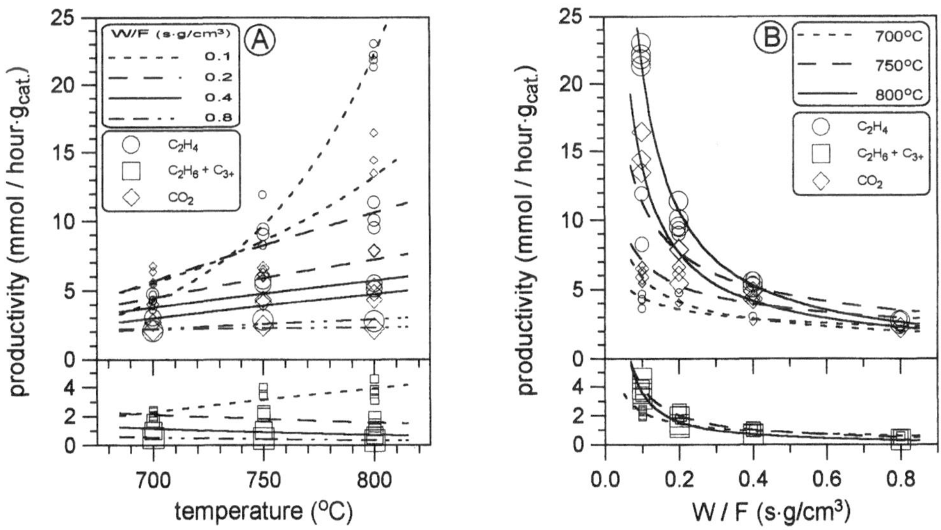

Figure 6. Influence of OCM reaction temperature (A) and contact time (B) on the productivity of the gas recycle system with product separation.

superior aim and its realisation may facilitate the commercialisation of the process of direct conversion of methane into ethylene. For that reason it should be assumed that the optimum conditions of the process are those in which the selectivity to the sum of hydrocarbons is the highest and then one should look for other ways of improving the hydrocarbons composition.

The final choice of the conditions of the process will also depend on the rate of the ethylene production as well as the other hydrocarbons and carbon dioxide. The results presented in Figure 6 illustrate the advantages of the application of temperatures higher than 750°C and contact times shorter than 0.4 $s \cdot g/cm^3$ in the OCM process with gas recycle and product separation.

4. CONCLUSIONS

The choice of conditions of the recirculation process oriented towards ethylene production must take into account a number of factors, the most important of which would certainly be the most advantageous economical index of the process. For the Na^+/CaO catalyst it seems that the best results would be obtained at the temperatures of 750-800°C and the contact time of the order of 0.1-0.4 $s \cdot g/cm^3$.

The improvement of the effects of the recirculation OCM process with a continuous supply of raw materials may be expected, among others, from an improvement of the effectiveness of the adsorptive separation of the products. A decrease in the amount of adsorbed methane should raise the degree of methane conversion, while the lowering of ethane adsorption improve the composition of the obtained hydrocarbons, increasing the amount of ethylene in it.

ACKNOWLEDGEMENTS

Financial support by the State Committee for Scientific Research (Grant No 3 T09B 043 11) is acknowledged.

REFERENCES

1. Y.Jiang, I.V.Yentekakis, C.G.Vayenas, Science, 264 (1994) 1563.
2. M.Makri, Y.Jiang, I.V.Yentekakis, C.G.Vayenas, 11th International Congress on Catalysis, Baltimore 1996, J.W.Hightower, W.N.Delgass, E.Iglesia, A.T.Bell (Eds.), Stud. Surf. Sci. Catal. v.101, Elsevier, Amsterdam, 1996, p.387.
3. A.Machocki, A.Denis, 11th International Congress on Catalysis, Baltimore 1996, Book of Abstracts, Po-295.
4. A.Machocki, Appl.Catal.A:General, 146 (1996) 391.
5. I.V.Yentekakis, Y.Jiang, M.Makri, C.G.Vayenas, Natural Gas Conversion IV, South Africa 1995, M.de Pontes, R.L.Espinoza, C.P.Nicolaides, J.H.Scholtz, M.S.Scurrell (Eds.), Stud. Surf. Sci. Catal. v.107, Elsevier, Amsterdam, 1997, p.307.
6. Y.Jiang, C.Yu, W.Li, J.Yan, Y.Ji, Natural Gas Conversion IV, South Africa 1995, M.de Pontes, R.L.Espinoza, C.P.Nicolaides, J.H.Scholtz, M.S.Scurrell (Eds.), Stud. Surf. Sci. Catal. v.107, Elsevier, Amsterdam, 1997, p.339.
7. A.Machocki, A.Denis, Advances and Challenges in the Catalytic Activation and Functionalisation of Light Alkanes - NATO Advanced Study Institute, Vilamoura, Portugal, 1997, E.G.Derouane, J.Haber, M.Guisnet, F.Ramoa Ribeiro, F.Lemos (Eds.), p.C51.

NATURAL GAS CONVERSION V
Studies in Surface Science and Catalysis, Vol. 119
A. Parmaliana et al. (Editors)
1998 Elsevier Science B.V.

The Oxidative Coupling of Methane and the Aromatization of Methane Without Using Oxidants

Xie Mao-song*, Chen Wen-heng, Wang Xue-lin, Xu Gui-fen, Yang Xu, Tao Long-xiang and Guo Xie-xian

Dalian Institute of Chemical Physics, The Chinese Academy of Sciences, P.O.BOX: 110 Dalian Liaoning 116023

Keywords: methane, oxidative coupling, aromatization, benzene, HZSM-5

It is found that the water in co-feed gas had three kinds of effects on the oxidative coupling of methane: 1) decrease in the temperature of catalyst bed and increase in the selectivity of coupling products 2) a rise in the temperature of catalyst bed 3) oscillating in the temperature of catalyst bed. Those mainly depend on the properties of the catalysts and operation conditions. We have developed the radial flow fixed bed reactor for the oxidative coupling of methane, which may maintain the good performance of catalysts on the larger test scale. It is proposed that the methane aromatization on the Mo/ZSM-5 zeolite catalyst without using oxidants is a structure sensitive reaction and the deactivation of the catalyst is initially related to the reduction of the Mo species at higher valence and loss of surface oxygen. The equation of the coke forming is listed.

1. Introduction

The utilization of methane is always an attractive subject. The work of keller and Bhasin[1] described the oxidative conversion of methane to higher hydrocarbons and we reported the highly selective (benzene: above 98%) and highly active (CH_4 conversion: above 7%) aromatic conversion of methane by non-oxidative activation over Mo/HZSM-5 zeolite catalysts[2][3].

One of the problems suffered in the oxidative coupling of methane (OCM) is uncontrolled temperature rise in the catalyst bed when the space velocity, the ratio of O_2 to CH_4 and the thickness of the catalyst bed increase. Especially, a fact that the good performance of a catalyst can not be maintained on the larger test scale is notable owing to a big temperature rise of catalyst bed. For non-oxidative activation of methane the shorter life of catalysts and coke formation on catalysts affect the

This project is supported by NNSF China.

*To whom correspondence should be addressed.

application of the process.

The developing of a new radial flow fixed bed reactor for OCM and the study of aromatic mechanism and coke formation over Mo/ZSM-5 catalysts for non-oxidative activation of methane make progress in the utilization of methane.

2. Experimental

A Li-Sm/MgO catalyst was prepared by method of co-deposition and immersion [4]. M/HZSM-5(M=Mo, Zn and so on) catalysts were prepared by impregnating NH_4ZSM-5 with ammonium molybdate or an aqueous solution of corresponding metal nitrates [3][5][6]. Catalytic tests were described elsewhere [4][5][6].

3. Results and discussion

3.1 The effect of water in cofeed gas on OCM

Most catalysts used in OCM are nonconductive and the heat accumulated in the process of the oxidative coupling reaction of methane would give a rise in the temperature of the catalyst bed. The relationship between the temperature rise of the catalyst bed and operation conditions is experimentally set [4].

When gas mixture of 50l/h O_2 and 200l/h CH_4 enters the reactor in which charged 20ml Li-Sm/MgO catalyst at 780 ℃, the center temperature of catalyst bed rapidly rise to 1050℃ and then maintains at 980℃. If adding amount of 108ml/h water to above gas mixture, the center temperature of catalyst bed may keep a lower level at 950℃ and selectivity of C_2 hydrocarbons increases from 56.0% to 64.6%.

For one of modified Li-Sm/MgO catalysts the phenomenon that the temperature of catalyst bed increase from 920℃ to 990℃ after adding water to gas mixture of CH_4 and O_2 has been observed.

Third effect of water in cofeed gas on OCM (fig.1) was found that on the employed

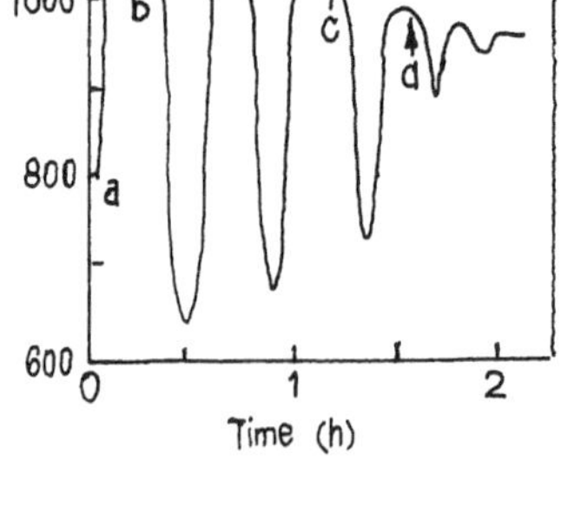

Fig. 1 temperature osillation of The catalyst bed.
CH_4: 3.7×10^5ml/h

a. O_2: 6.4×10^4ml/h, H_2O: 108ml/h

b. O_2: 5.5×10^4ml/h, H_2O: 108ml/h

c. O_2: 4.8×10^4ml/h, H_2O: 44ml/h

d. O_2: 4.8×10^4ml/h, No water

Li-Sm/Mgo catalyst the temperature of the catalyst bed would oscillate due to adding water to the feed gases and the amplitude of temperature increased with the increase of water. The phenomenon of temperature oscillation disappeared when the adding of water was stopped.

3.2 Radial flow fixed bed reactor for OCM

Using the commercial fixed bed reactors, the thickness of the catalyst bed always increased with increasing the catalyst amount charged resulting in the temperature rise of catalyst bed and destruction of good performance of the catalyst.

It can be seen from Table 1 that decreasing the thickness of the catalyst bed to a rational degree for a certain amount of catalyst may be a way to lower the temperature rise.

Table 1. The effect of the thickness of the catalyst bed on the bed temperature

Diameter of reactor(mm)	Thickness(mm)	T(℃)	C_2 Selectivity(%)
Φ51×3	10	780	42.9
Φ40×3	22	890	47.3
Φ30×2.5	40	900	42.1
Φ20×1.5	115	1200	---

(Space velocity of CH_4: 7000h^{-1}, CH_4/O_2=3.3, 20ml catalyst)

We have developed the radial flow fixed bed reactor for OCM [7], namely, the reactants pass catalyst bed by means of radial flow.

Using the radial flow catalytic reactor, the axial length of the reactor may meet the needs for a large amount of catalyst charged. Thin catalyst bed may be kept and the center temperature of the catalyst bed would not be too high.

The radial flow catalytic reactor with amount of 200ml La-BaO catalyst , 5000/hr space velocity of CH_4 and 780℃ oven temperature showed obvious advantages: although the catalyst amount charged is big, the performance of the catalyst is still better and the conversion of CH_4 , the selectivity and the yield of C_2 hydrocarbons achieved to 25.2%, 67.4% and 17.0% respectively.

3.3 Acid centers of the catalyst don't take an active part in the methane aromatization without using oxidants.

Comparing the result of catalytic performance (Table 2) with that of NH_3-TPD patterns (Fig.2) of the 2% Mo/HZSM-5 (Si/Al=50), it is obvious that the catalyst is still active after 2 hours reaction, but acidity of the catalyst is dramatically lost after

Table 2. Catalytic performance of 2% Mo/HZSM-5 catalysts at 973

Catalyst	yield of benzene (%)		
	1*	2*	3*
Mo/HZSM-5(Si/Al=50)	12.3	8.8	7.3

CH_4 space velocity: 1300/hr , catalyst charge: 1ml

*Reaction time(hr)

half hour reaction. It means that the idea of the acid sites actively taking part in the methane aromatization without using oxidants is doubtable.

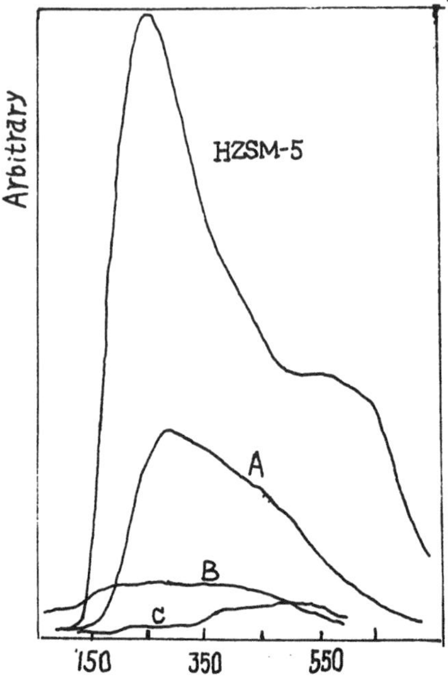

Fig.2. NH_3-TPD patterns of the 2%Mo/HZSM-5 catalysts after different reaction time
A. Fresh catalyst
B. After 10 min. reaction.
C. After half an hour reaction

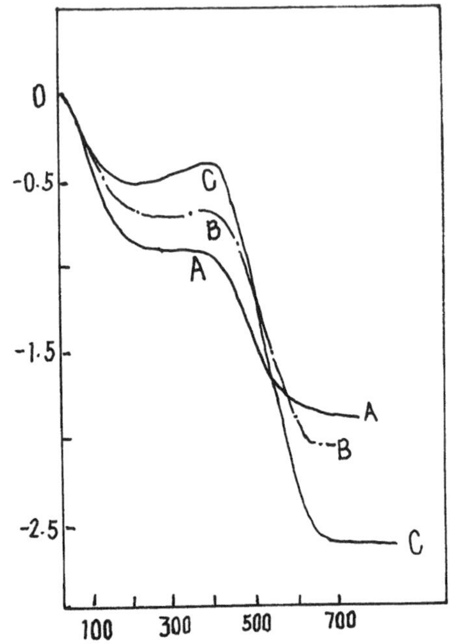

Fig. 3. TPO profiles of the 2%Mo/HZSM-5 catalysts after different reaction time
A. Reaction for 0.5 h
B. Reaction for 1.5 h
C. Reaction for 3.5 h

3.4 Methane aromazation on Mo/ZSM-5 catalyst is a structure-sensitive reaction

XRD patterns showed that the MoO_3 crystallite pattern can not be detected if Mo loading is less than 10% and the crystallinity of HZSM-5 zeolite decreases with increasing Mo loading [6]. This gives evidence that there is some kind of interaction between Mo species and the HZSM-5 zeolite, which leads to the destruction of the zeolite skeleton in some extent and to the introduction of Mo species into the zeolite channels in the process preparing Mo/HZSM-5 catalyst. The Mo species as catalytic sites are highly dispersed on the HZSM-5 zeolite surface and it's channels.

By X-ray power diffraction, Jia-ching Lin et al [8] indicated that the Cs^+ cations locations are directly related to locations of Bronsted acid sites and the aluminum atom is located in the four-membered ring near the two-channel intersections.

Liu Zhen-Yi et al [9] described the position of $Ni^{2+}(2)$ cations in dehydrated ZSM-5, which is in the sinusoidal channel at the edge of the intersection between the straight and sinusoidal channels.

K.D.Huddersman[10]considered that the Ti^{+} cations are located in partially occupied sites in the intersection between the straight and sinusoidal channels.

The result of NH_3-TPD patterns (Fig.2) shows that the Bronsted acid site density of 2%Mo/HZSM-5(Si/Al=50) catalyst is lower than that of HZSM-5(Si/Al=50). It means that Mo species would cover or replace the Bronsted acid sites, which are near the two-channel intersection [8] after impregnating HZSM-5 with ammonium moliybdate.

According references [8] [9] [10]and our NH_3-TPD result we proposed that Mo (VI)cations as active sites of Mo/ZSM-5 catalyst are located in the intersection between the sinusoidal and straight channels and construct an active sites group.

The site number of active group may be severely restricted because a good performance of the Mo/HZSM-5(Si/Al=50) catalyst only was maintained in the rather small range of Mo loading (1.7%-2.5%) that would affect the site number of active group. The products distribution of methane aromatization (Table 3) showed that the ethene is initially formed from methane and can not be easily transformed into benzene on the HZSM-5 zeolite at 973K, but the benzene is exclusively formed from methane at 973K over the 2%Mo/ZSM-5 catalyst. This is obvious that benzene directly form from methane on the Mo active site group in this case.

Table 3. The products distribution of methane aromatization at 973K on 2% Mo/ZSM-5 and HZSM-5

Reaction time (hr)	products distribution					
	ethene %		ethane %		benzene %	
	2%Mo/ZSM-5	HZSM-5	2%Mo/ZSM-5	HZSM-5	2%Mo/ZSM-5	HZSM-5
1	0.96	1.40	NO	1.65	8.01	1.02
2	1.40	1.00	1.17	1.71	5.26	NO
3	1.42	0.67	1.18	1.66	4.37	--
4	1.67	0.70	1.24	1.78	3.57	--

The result that the selectivity of ethene and ethane was increasing and the selectivity of benzene on 2%Mo/ZSM-5 catalyst was decreasing with reaction time also observed from Table3. This may be that the big active site group of Mo species, which is needed for benzene forming from CH_4, is getting destruction with reaction time, but ethen and ethane can be formed on a smaller active site group.

We proposed that the methane aromatization on the Mo/ZSM-5 zeolite catalyst without using oxidants is a structure-sensitive reaction. Rationally, in this case the aromazation of methane does not mainly follow in the train of the formation of ethene and ethane.

3.5 The deactivation of the catalyst is initially related to the reduction of the Mo(VI) species .

Temperature-programmed oxidation (TPO) profiles of coking catalysts during reaction at different time intervals are shown in Fig 3.

Curves A, B, C show the TPO profiles of the same batch of catalyst samples coked during methane aromatization reaction at 973K for 0.5h, 1.5h and 3.5h respectively.

Comparing curves A, B and C, it was observed that the amount of coke increases with the reaction time.

For the catalysts reacted for 1.5 and 3.5 hours, their TPO profiles in the region360-440 ℃ show an increase in weight (especially for 3.5h). We suggest that this increasing is due to the oxidation of Mo species reduced during the methane aromatization. The catalyst after reacting for 3.5h loss greater part of its activity.

After the deactivated catalyst is low temperature oxidized at 340 ℃ (without burning off the coke) for 0.5h, the catalytic performance of the catalyst was so restored that the conversion of methane is improved from 2.5% (in the deactivated catalyst) to 9.2% (in the regenerated catalyst).

This might indicate that the deactivation of the catalyst is initially related to the reduction of the Mo active sites to lower oxidation states.

Polycyclic aromatic hydrocarbons, which was detected by GC-MS combined system, form graphite coke at high temperature owing to loss of their hydrogen, that was verified by NMR.

Through a dynamic analysis of the coke during its depositing on the catalysts which were activated at the different temperature or possessed the different Si/Al ratio, the equation of the coke forming is got as follows: $W= a \times T^n$.

W: the amount of coke (mg)/the weight of catalyst(g), T: reaction time(min),
a: constant, n: the power of T is related to the reaction temperature and the Si/Al ratio. The experimental result showed that LgW is proportional to LgT.

For 2%Mo/HZSM-5(Si/Al=50) catalyst reacted at 700℃, n is equal to 0.441.

References

1. G. E. Keller and M. M. Bhasin, J. Catal., 73 (1982) 9
2.L. Wang, L. Tao, M.Xie, G.Xu, J. Huang and Y. Xu, Catal. Lett., 21 (1993) 35.
3.M.-S. Xie, L. Wang and L. Tao et al. Chinese Patent, 93115889.3
4.M.-S. Xie, X.Wang and G.Xu et al. Journal of Natural Gas Chemistry, Vol4, No.2, 212 (1995).
5.M.-S. Xie, Xu Yang, W.-H.Chen, L.-X. Tao, X. -L. Wang and G.-F. Xu et al. Journal of Natural Gas Chemistry, Vol.5, No.3, 210 (1996)
6.M.-S. Xie, X.Yang, W.-H. Chen, et al. Studies in Surface Science and Catalysis, Vol.105, Part B, 869, (1997)
7.M.-S. Xie, X.Wang and G. Xu et al. Chinese Patent, 93115959.8
8.J.-C. Lin and K.J. Chao Zeolites, Vol.11, 376, (1991)
9.Z.-Y. Liu, W.-J. Wan and C.-H. Jiang et al. Chinese Journal of Catalysis, Vol.7, No.1,46 (1986)
10.K.D.Huddersman and L.V.C.Rees, Zeolites, Vol.11, 270 (1991)

NATURAL GAS CONVERSION V
Studies in Surface Science and Catalysis, Vol. 119
A. Parmaliana et al. (Editors)

The Effect of Compositional Changes on Methane Oxidation Processes with Structurally Invariant Catalysts

H. Hayashi[a], S. Sugiyama[a], and J.B. Moffat[b]

[a]Department of Chemical Science and Technology, Faculty of Engineering, The University of Tokushima, Minamijosanjima, Tokushima 770, Japan

[b]Department of Chemistry and the Guelph-Waterloo Centre for Graduate Work in Chemistry, University of Waterloo, Department of Chemistry, Waterloo, ON N2L 3G1, Canada

ABSTRACT

The crystallographic structure of hydroxyapatites is retained during variations in composition, while catalytic properties are altered. In the oxidation of methane, the conversion, the products formed and their selectivities are strongly influenced by changes in the stoichiometry of the hydroxyapatites, the nature of the cation and the introduction of a second cation.

1. INTRODUCTION

Solids whose crystallographic structures remain invariant under changes in composition are of value in understanding the nature and source of the properties of a heterogeneous catalyst.

Hydroxyapatites are solids which fulfill such expectations but additionally display catalytic functionalities which are composition dependent (1). The apatite $[M_{10}(XO_4)_6Z_2]$, where M is a metal, X is, for example, P, As, Si, S, and Z is an hydroxyl or halide is evidently a solid of considerable versatility (2). With M as calcium, X as phosphorus and Z as the hydroxyl group, calcium hydroxyapatite $[Ca_{10}(PO_4)_6(OH)_2$, abbreviated as CaHAp], a solid which does not readily lose hydroxyl groups and whose lattice is believed to be stable up to 1000°C is formed (3) (Fig. 1).

Hydroxyapatite has a hexagonal structure constructed from columns of Ca and O atoms which are parallel to the hexagonal axis. The hexagonal unit cell of hydroxyapatite contains 10 cations located on two sets of nonequivalent sites, 4 on site I and 6 on site II. The calcium ions on site I are aligned in columns, while those on site II are in equilateral triangles centered on the screw axes.

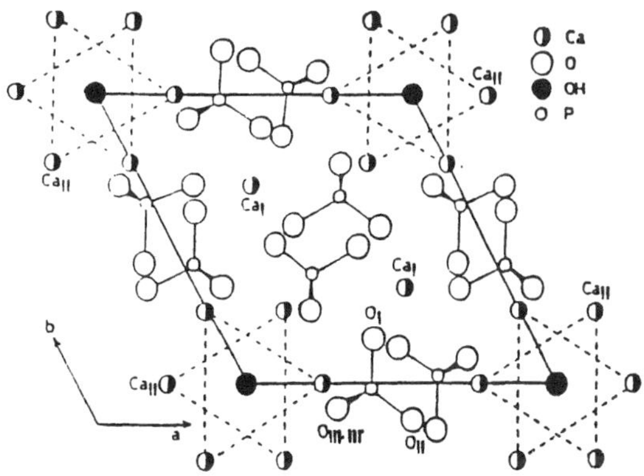

Fig. 1 Structure of stoichiometric calcium hydroxyapatite structure projected on a, b plane (Suzuki).

The present work is concerned with an examination of the changes in the properties, particularly those related to the conversion of methane, with the stoichiometry, nature of the cations and ion-exchange of hydroxyapatite.

2. EXPERIMENTAL

Stoichiometric and nonstoichiometric calcium hydroxyapatites were prepared from $Ca(NO_3)_2{\cdot}4H_2O$ (BDH AnalaR) and $(NH_4)_2HPO_4$ (BDH AnalaR) (4). Strontium hydroxyapatites (SrHAp) were similarly prepared from the corresponding strontium compound. Lead-calcium hydroxyapatites (PbHAp) were prepared from $Na_2HPO_4{\cdot}12H_2O$, $Ca(CH_3COO)_2{\cdot}H_2O$ and $Pb(CH_3COO)_2{\cdot}3H_2O$ (5). Compositions of the prepared solids were found from ion chromatographic analysis (Dionex 4500 i) of the solutions remaining from the synthesis and/or atomic absorption spectrometry and/or inductively coupled plasma spectrometry (SPS-1700, SEIKO).

^1H and ^{31}P MAS NMR was obtained with a Bruker AMX-500, with an external reference of benzene or $(NH_4)_2HPO_4$, at room temperature and a spinning rate of 6.5-8 kHz. IR spectra were recorded on a Bomen MB-100 FTIR spectrometer. Surface analyses (XPS) employed a Shimadzu ESCA-1000AX spectrometer under conditions described previously (1). Powder X-ray diffraction (XRD) patterns were recorded with a Rigaku RINT 2500X using monochromatized MgKα radiation. Surface areas were calculated by application of the infinite layer BET equation to adsorption isotherms measured on a conventional BET nitrogen adsorption apparatus (Shibata P-700). Reactions were performed in a conventional fixed-bed continuous flow reactor operated under atmospheric pressure. Details of the method of operation and analytical procedures have been described previously (1).

3. RESULTS AND DISCUSSION

The Ca/P ratio in calcium hydroxyapatite may be varied from stoichiometric to nonstoichiometric values by appropriate adjustment of the relative quantities of the preparative reagents. Thus, values of Ca/P ranging from 1.53 to 1.72 are attainable while the apatitic XRD pattern is maintained (Fig. 2). This has been variously attributed to [1] undetected phases, [2] surface adsorption, [3] lattice substitution, [4] intercrystalline mixtures of CaHAp and octacalcium phosphate (1,2). Although the surface areas of CaHAp vary relatively little with changes in the Ca/P ratio nevertheless the values appear to pass through a minimum at the stoichiometric composition (Table 1).

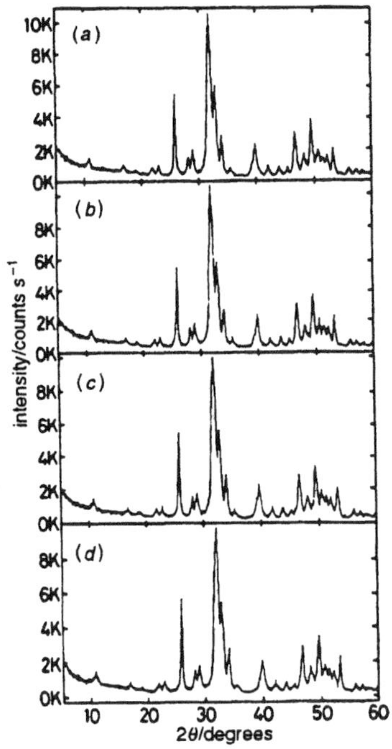

Fig. 2. XRD patterns of the fresh calcium hydroxyapaptites: (a) $CaHAp_{1.72}$, (b) $CaHAp_{1.68}$, (c) $CaHAp_{1.64}$. (d) $CaHAp_{1.53}$.

Although the conversion of methane at 973K on CaHAp remains at 10 ± 2% regardless of the Ca/P ratio the selectivities show significant changes with the Ca/P ratio

Table 1
Surface Areas[a] of Stoichiometric and Nonstoichiometric CaHAp and SrHAp

X[b]/P	CaHAp	SrHAp
1.73		65.5
1.72	72.9	
1.70		67.9
1.68	65.3	
1.67		60.3
1.64	73.8	
1.61		72.4
1.53	82.6	

[a] m^2/g
[b] X = Ca or Sr

(Fig. 3). For Ca/P values of 1.64 and below the selectivity to CO is approximately 70% while for higher Ca/P the latter selectivity decreases while that to CO_2 increases until at the stoichiometric composition the two selectivities are each approximately 40%. The C_{2+} selectivity remains at 20 ± 2% for all stoichiometries of CaHAp.

As with CaHAp the XRD patterns of SrHAp with Sr/P ratios ranging from 1.61 to 1.73 are essentially indistinguishable and match that for $Sr_{10}(PO_4)_6(OH)_2$ (JCPDS33-1348) (not shown). The surface areas for the various SrHAp compositions, as with those for CaHAp, appear to reach a minimum for the stoichiometric composition. For a reaction temperature of 973K and similar conditions employed to generate the data in Fig. 3 relatively little differences in the conversions and selectivities are observed as the Sr/P ratio is altered. However C_2 hydrocarbons are observed and the selectivities to these, while relatively low, reach a maximum for the stoichiometric composition (not shown). In contrast, however, if the reaction temperature is decreased to 873K selectivities to CO greater than 90% for conversions of methane greater than 10% are observed with stoichiometric strontium hydroxyapatite.

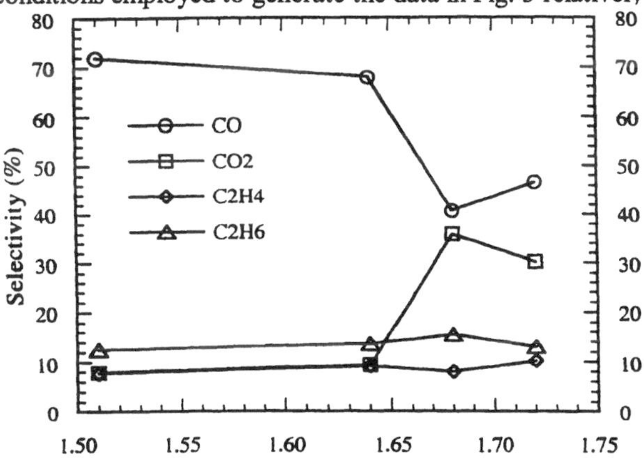

Fig 3. Selectivities to products from the oxidation of methane on CaHAp of various stoichiometries. Reaction temperature, 973 K; W = 0.5 g; F = 30 ml min^{-1}; CH_4/O_2 = 7/1; time-on-stream = 30 min.

Ion-exchange of the cations in CaHAp may be affected with little or no structural alteration. Thus the replacement of a portion of the calcium cations with lead cations yields solids whose XRD patterns are essentially identical and can be attributed to the $Ca_{10}(PO_4)_6(OH)_2$ structure (not shown). The shift of the strongest peak in each catalyst to lower diffraction angles with increasing lead substitution is associated with an enlargement of the unit cell, consistent with the difference in the ionic radii (Ca^{2+}, 0.99 Å; Pb^{2+}, 1.20 Å) (Table 2). Up to approximately 45%, lead fills almost exclusively site II while beginning to fill site I at higher concentrations (see Fig. 1) (5).

Table 2
Surface Areas and XRD 2θ values for CaHAp and CaPbHAp

Composition[a]	A_s [b]	2θ [c]
0.0	72.8	31.772
0.0085	69.1	21.700
0.057	74.7	31.690
0.12	65.6	31.660
0.20	51.0	31.460
0.35	51.5	30.920
1.0	–	30.114

[a] Pb/[Pb + Ca]; [b] Surface area (m^2/g); [c] (211) plane.

Somewhat surprisingly, ^{31}P MAS NMR spectra show only one peak for the CaPbHAp samples, indicating that the three crystallographically nonequivalent phosphate ions in the unit cell have very similar environments, either with or without the substitution of lead. The ^{31}P chemical shift increases continuously as lead is added to CaHAp, at least up to a substitution of 35% (Fig. 4).

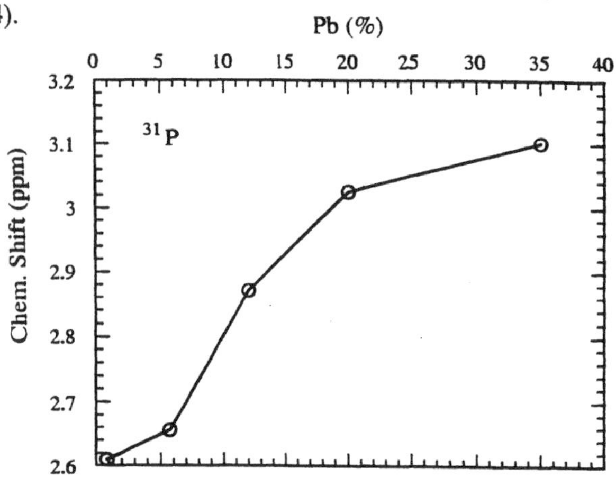

Fig. 4. Chemical shift from ^{31}P MAS NMR for CaHAp with ion-exchanged Pb.

While little or no structural change occurs on ion-exchange of Ca^{2+} by Pb^{2+}, the catalytic properties for the conversion of methane are considerably altered (Fig. 5). The introduction of as little as 1% lead increases the production of C_{2+} hydrocarbons while decreasing that of CO. Further increase of the lead content to approximately 10% generates a selectivity to C_{2+} hydrocarbons of 80%, which remains relatively unchanged for further increases in the concentration of lead.

Lead can also be ion-exchanged into strontium hydroxyapatite with results similar to those observed in the substitution of lead for calcium. The effects on the catalytic properties in the conversion of methane are semiquantitatively similar to those found on the partial replacement of calcium by lead. Where very high selectivities to CO and little or no C_{2+} hydrocarbons were found with SrHAp, the addition of small quantities of lead produces substantial selectivities to C_{2+} hydrocarbons (up to 50%) and the virtual elimination of CO from the product stream in favor of CO_2 with concomitant doubling of the conversion from 6 to 12%. As with CaPbHAp only minor changes in the conversion and selectivities with further increases in lead content are observed (not shown).

Evidence from the present and previous work suggests that at temperatures of 600°C CaHAp may partially dehydrate as

$Ca_{10}(PO_4)_6(OH)_2 \rightarrow Ca_{10}(PO_4)_6(OH)_{2-2x}O_x\Box_x + xH_2O$
($\Box$ = vacancy, x<1)

to produce an oxygen vacancy, $\Box$, the presence of which may be necessary, but not sufficient, for the oxidative catalysis (1). Oxygen was found to be strongly adsorbed on stoichiometric CaHAp at room temperature in contrast to the reversible adsorption on the nonstoichiometric composition (1), suggesting that the aforementioned oxygen vacancies are fewer in number on the latter in comparison with the former material. This appears to be substantiated by the results from XPS measurements of the surface compositions for stoichiometric and nonstoichiometric CaHAp (Fig. 6). Fresh samples of CaHAp show a minimum in the surface O/P concentration ratio at approximately stoichiometric Ca/P compositions but a maximum with samples previously used for the oxidation of methane, suggesting that the stoichiometric material has a larger number of oxygen vacancies which tend to be occupied by oxygen during the oxidation process.

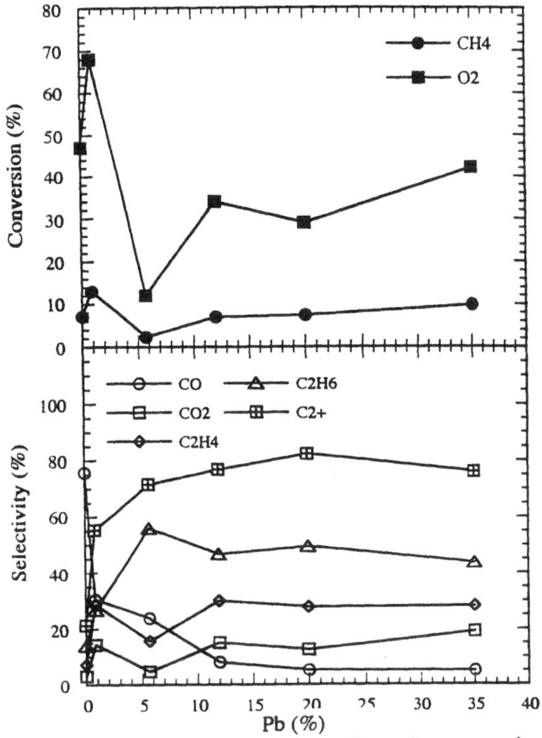

Fig. 5. Selectivities to products from the conversion of methane on CaHAp with ion exchanged Pb. Reaction temperature = 700°C; W=0.30 g; CH_4/O_2 = 7/1; F=15 ml min^{-1}; time on stream = 3 hr.

The introduction of lead is believed to provide sites for the stabilization of methyl radicals. A minimum number of these sites are required in order to locate the stabilized methyl radicals in sufficient proximity to each other so that ethane molecules may be formed.

Although the results obtained with SrHAp are semiquantitatively similar to those obtained with CaHAp, nevertheless significant disparities can be observed. Stoichiometric SrHAp provides selectivities to CO of 90% or greater while CaHAp generates approximately equal selectivities (40%) of each of CO and CO_2. The introduction of lead to the calcium compound produces selectivities to C_2 hydrocarbons in excess of 80% but approximately half of this with the strontium hydroxyapatite. Since the structures of the two hydroxyapatites are similar, either with or without added lead it is tempting to attribute the differences in their catalytic properties to the dissimilarities in the sizes of the cations (Ca^{2+}, 0.99 Å, Sr^{2+}, 112 Å). Since Ca^{2+} has the larger charge density it can be expected to have the stronger influence on the catalytic sites, that is, the oxygen vacancies. It is probable, however, that such a rationalization is unduly simplistic and other factors, such as the relative stabilities, undoubtedly also must play a role in any explanations of the differences between CaHAp and SrHAp.

4. CONCLUSIONS

1. The crystallographic structure ofhydroxyapatite is invariant under changes in composition and stoichiometry; **2.** With calcium deficient CaHAp the conversion of methane produces a selectivity to CO as high as 70% whereas the stoichiometric analogue produces equal quantitites of CO and CO_2; **3.** With stoichiometric SrHAp selectivities to CO greater than 90% can be attained; **4.** Ion-exchange of the cations in CaHAp can be performed without structural alterations; **5.** With the replacement of small fractions of Ca by Pb the selectivities to higher hydrocarbons from methane increase to greater than 80%; **6.** The catalytic properties of the hydroxyapatites may result at least in part, from oxygen vacancies which are generated at 600°C or higher.

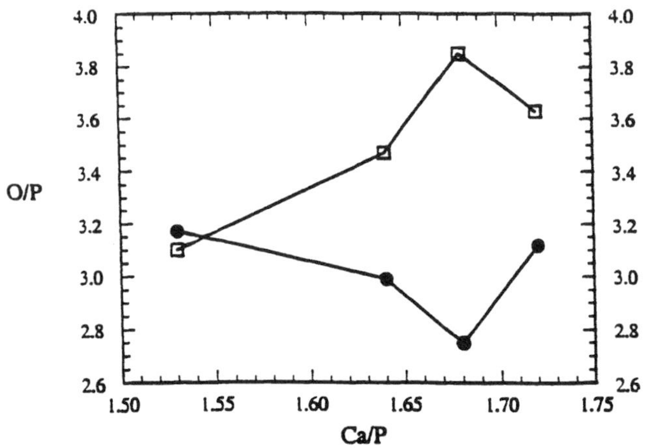

Fig. 6. Surface (XPS) composition of stoichiometric and nonstoichiometric CaHAp, fresh (●) and previously employed (□) for the oxidation of methane.

5. ACKNOWLEDGEMENTS

The financial support of the Natural Sciences and Engineering Research Council of Canada is gratefully acknowledged.

6. REFERENCES

1. J.B. Moffat, S. Sugiyama and H. Hayashi, Catal. Today 37 (1997) 15 and references therein.
2. J.C. Elliott, Structure and Chemistry of the Apatites and Other Calcium Orthophosphates, Elsevier, Amsterdam, 1994.
3. D.E.C. Corbridge, Studies in Inorganic Chemistry 10, Phosphorus, An Outline of its Chemistry, Biochemistry and Technology, 4th Ed, Elsevier, Amsterdam, 1990.
4. E. Hayek and H. Newesely, Inorg. Synth. 7 (1963) 63.
5. A. Bigi, M. Gandolfi, M. Gazzano, A. Ripamonti, N. Roveri and S.A. Thomas, J. Chem. Soc., Dalton Trans. (1991) 2883.

NATURAL GAS CONVERSION V
Studies in Surface Science and Catalysis, Vol. 119
A. Parmaliana et al. (Editors)

New Stage of Oxidative Coupling Reaction of Methane: Development of Novel Catalysts by Modification of Solid-Super Acid

Kazuhisa Murata*, Takashi Hayakawa, Satoshi Hamakawa, and Kunio Suzuki

National Institute of Materials and Chemical Research,
1-1, Higashi, Tsukuba, Ibaraki 305-8565, Japan

Tungsten oxide/ zirconia (W/Z)-based catalysts were prepared and their performances were tested for the oxidative coupling of methane (OCM) reaction. Two types of new catalysts were found; One is prepared by impregnation of both Eu or Ce and Li_2CO_3 over W/Z, and the other is by impregnation of both Ce or Mn and NaCl. The former catalysts showed long lifetime over 100h, but modest C2 yield(ca.18%), while the latter showed excellent C2 yield over 30%, but rather short lifetime (ca.30h).

1. INTRODUCTION

In this decade, much attention has been paid to the conversion of methane into more valuable products under oxidative [1] or non-oxidative [2] conditions. Recently we have reported novel catalyst systems prepared from lithium-doping of sulfated-zirconia for the oxidative coupling of methane (OCM). The aim of this study is to improve not only C_2 yield, but also lifetime of catalysts. Thus, tungsten oxide-modified zirconia (abbreviated as W/Z) were chosen as starting solid super-acid [3]; The W/Z could be expected to be more thermally stable than sulfated zirconia. Then, the W/Z was modified by transition metals (Ce, Eu, Mn, etc.) and/or alkali metals (Li_2CO_3, NaCl, etc.). The effectivenesses of these catalyst systems on the OCM reaction were investigated.

2. EXPERIMENTAL

The WO-doped catalysts, which were prepared by impregnation of $(NH_4)_6(H_2W_{12}O_{40})$ to amorphous zirconia (AZr) [4], followed by calcination in air for 3 h at 1103K, were further impregnated with an aqueous solution of alkali and/or transition metal compounds such as Li_2CO_3, NaCl, $Mn(NO_3)_2$, $Ce(NO_3)_3$. This was then evaporated to dryness, drying overnight at 373K, calcination in air at 1123K for 16h, to form alkali and/or transition metal-doped WZ. In the Cat.B (Table 1), both $(NH_4)_6(H_2W_{12}O_{40})$ and NaCl were first impregnated to AZr and, then, after calcination at 1103K, $Ce(NO_3)_3$ and NaCl were second impregnated to the NaCl/W/Z compounds and calcinated at 1123K.

The catalytic runs were carried out under atmospheric pressure and in a fixed bed vertical flow reactor constructed from an alumina tube (i.d. 6mm) packed with catalyst and mounted inside a tube furnace. The catalyst was pretreated in nitrogen at 873 K for 1h. And, then, the

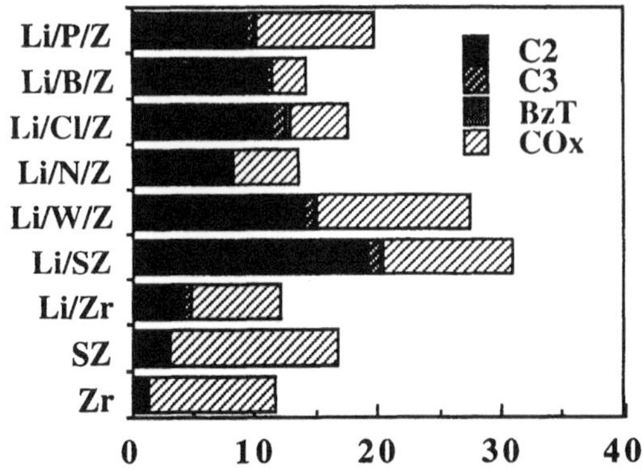

Fig.1 The OCM reaction over various ZrO_2 catalysts.
Reaction temp.: 1073K
Gas Comp.: $CH_4/O_2/N_2$=15/5/80

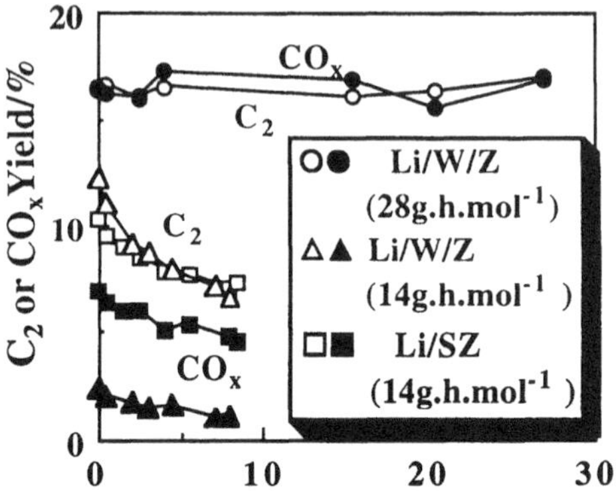

Fig.2 Catalytic Activity of OCM as a Function of Time.
Temp.:1023K

reactant gas mixture was introduced at W/F=28 g-cat.h.mol^{-1} (1023K). The products were analyzed on-line by gaschromatography (GC). XRD and TPD measurements were carried out by procedures described previously [5].

3. RESULTS AND DISCUSSION

3.1 Preliminary Study on Catalytic Activity

First, various ZrO_2-based super-acids were tested as supports. The reaction was carried out at 1073K using Li_2CO_3 /acid-promoted ZrO2 catalysts, which were prepared by impregnation of acid precursors (6 wt%) to AZr, followed by impregnation of Li_2CO_3. The results are presented in Fig.1. The C_2 yield was in the order: $(NH_4)_2SO_4$ (Li/SZ,19.0)> 1N H_2SO_4 (19.0)> $(NH_4)_6(H_2W_{12}O_{40})$(Li/WZ, 14.1)> NH_4Cl (Li/Cl/Z, 11.33)> trimethylene borate (Li/B/Z, 10.5)> $(NH_4)_3PO_4$ (Li/P/Z, 9.12)> NH_4NO_3 (Li/N/Z, 7.80)> none (Li/Zr, 4.13). The performances of the Li/SZ and Li/W/Z catalysts were briefly tested at 1023K as a function of reaction time. At W/F=14 g.h.mol^{-1}, the C_2 yields decreased as the reaction proceeded (Fig.2), while the yield remained constant when the Li/W/Z catalyst was used at W/F=28 g.h.mol^{-1}. The nature of the Li/W/Z catalyst were similar to that of the Li/SZ catalyst[5], as ascertained by XRD patterns and NH_3 and CO_2 TPD analyses: In the XRD patterns, mixed oxide phases such as Li_2ZrO_3 and Li_2WO_4 were observed, in addition to monoclinic ZrO_2 phase with a very small amount of WO_3 phase. And the TPD analyses indicated that the Li/W/Z surface could not be of a super-acid character, due to, at least, partially neutralization by impregnation of Li_2CO_3. The results in the use of W/Z as support will be hereinafter described.

3.2 Li_2CO_3/W/Z-Based Catalysts

3.2.1 Effect of transition metal additives.

A correlation between CH_4 conversion and C_2 selectivity is shown in Fig.3, where these data were taken after 1-h reaction at 1073K. By additions of these metals, CH_4 conversion was increased, whereas C_2 selectivity was decreased. Of these metals employed, Ce, Ti, and Eu seem to be favobable for an increase in C2 yield. Unfortunately, the increase in Ce content of 10wt% to 20wt% resulted in the decrease in both C_2 selectivity and CH_4 conversion.

3.2.2 Effect of Reaction Temperature.

Figure 4 shows effect of reaction temperature on the OCM reaction, where the Eu/Li/W/Z (□,■) and Mn/NaCl/W/Z (○,●) catalysts were used. In Eu/Li/W/Z catalysts, the CH_4 conversion increased with the increase in reaction temperature, whereas, in the Mn/NaCl/W/Z catalysts, the conversion abruptly increased with the increase in reaction temperature:The conversion at 973K was as high as >40%. And then it leveled off at 1023K and finally decreased with the temperature increase. The C_2 yields increased with the increase in the temperature and leveled off at 1073K (Mn/NaCl/W/Z cat.). Thus, catalyst performances were found to be gretly improved, in particular, in the temperature below 1000K, by use of NaCl-containing catalysts. The findings in the temperature below 1000K could be explained by the view that the activation energy of C_2 formation is of second order for CH_3 radicals, whereas that of CO_x formation is of first order reactions [1].

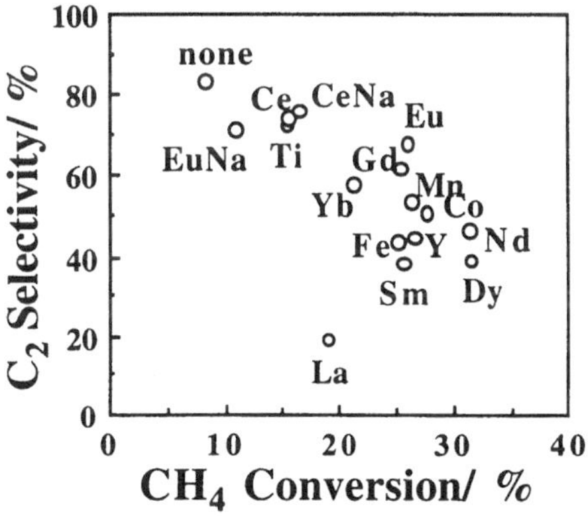

Fig.3 Effect of Transition Metals for the OCM over Li_2CO_3/W/Z catalysts.
Catalyst: 10wt%M/5wt%Li/13wt%W/Z
Reaction temp.: 1073K
Gas Comp.: $CH_4/O_2/N_2$=10/5/85
Na: Na_2CO_3 instead of Li_2CO_3.
Co,Mn,Fe: 5wt%

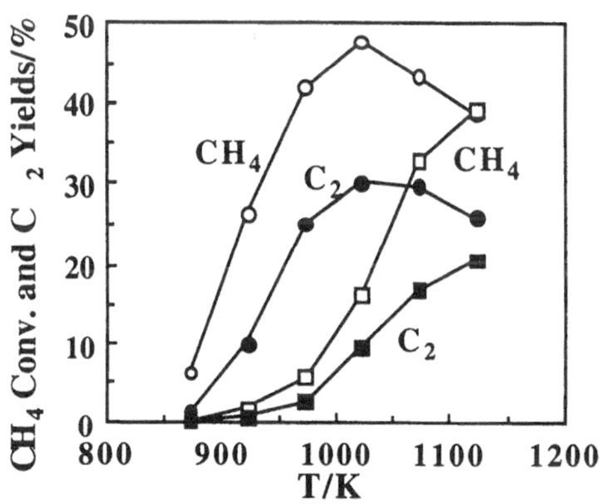

Fig.4 Temperature dependence on the OCM Reaction.
Cat.: Eu/Li/W/Z (□,■)
Mn/NaCl/W/Z(○,●)
Gas: CH_4/ O_2/ N_2 = 10/5/85

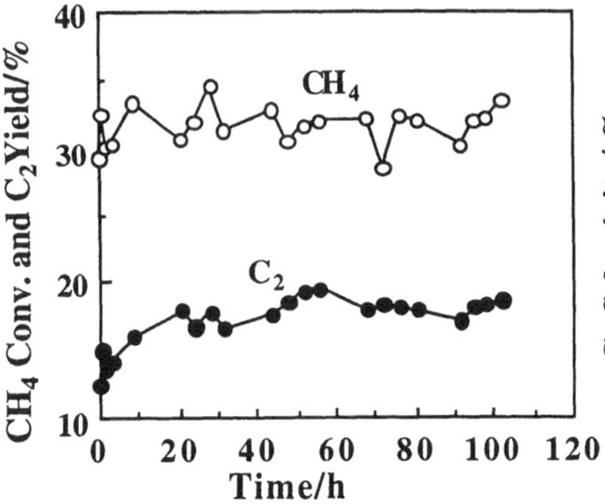

Fig.5 Catalytic Activity of OCM Reaction as a Function of Time. Temp.:1023K, Cat.: Eu/Li/W/Z

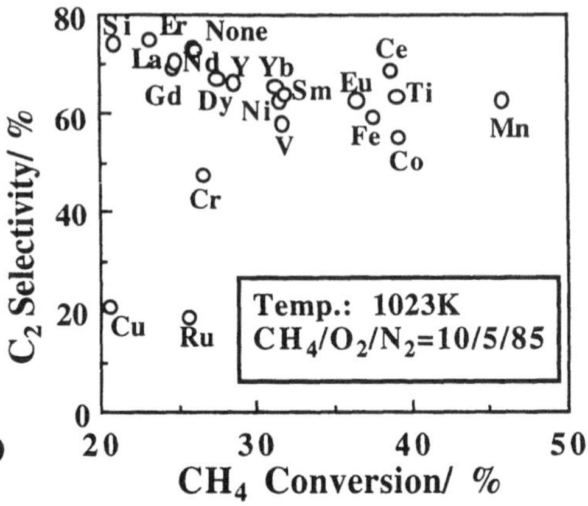

Fig.7 Effect of Metal Additives.
Cat.:NaCl/M/W/Z(Na/Zr=0.33)
2wt%:Si,Cr,Cu,Ru,V,Ni,Fe,Co,Ti,Mn
10wt%:Er,La,Nd,Gd,Dy,Y,Yb,Sm,Eu,Ce

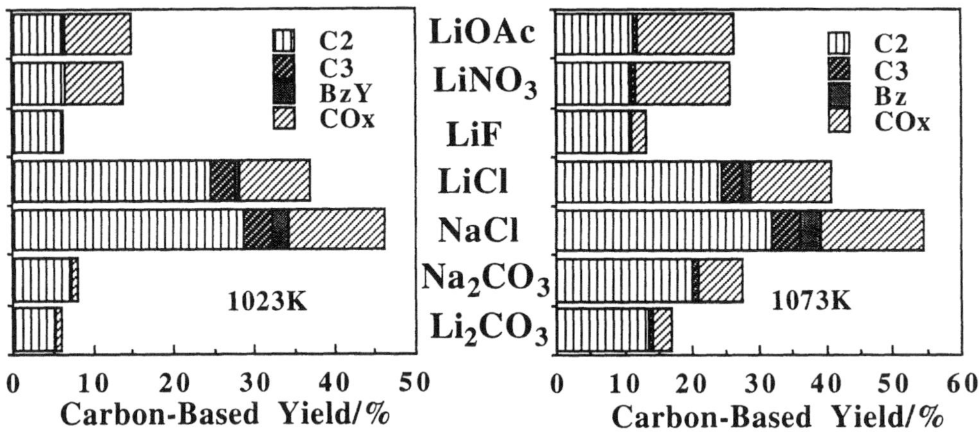

Fig.6 Effect of Alkali Metals.
Cat.: Alkali/Ce/W/Z Gas: $CH_4/ O_2/ N_2$ = 10/5/85

3.2.3 Catalyst Performances as a Function of time.

The Eu/Li/W/Z catalysts was prepared by impregnation with an aqueous solution of $Eu(NO_3)_3$(Eu:10wt%) and Li_2CO_3(Li:5wt%) over W/Z support, followed by calcination at 1123K for 100h. The reaction was carried out at 1073K under a mixed gas flow of $CH_4/O_2/N_2$ (=10/5/85). It was found that both the CH_4 conversion and C_2 yield remained unchanged over the range of 100 h (Fig 5(O,□)). At this point, the x-ray diffraction of the used Eu/Li/W/Z catalyst exhibited a pattern similar to the fresh catalyst.

3.2.4 Effect of other alkali compounds instead of Li_2CO_3.

The alkali-doped Ce/W/Z catalysts, which were prepared by impregnation of Ce(10wt%) and alkali (A/Zr=0.33 (molar)), were examined for the reaction. In the W/Z-based catalyst systems, alkali chlorides (NaCl and LiCl) were also found to be effective (Fig.6); In particular, in the NaCl/Ce/W/Z system, the C_2 yields were 28.6% (1023K) and 31.7% (1073K), being comparable with those of the most excellent performances previously reported [5]. On the contrary, LiF was not so effective for our catalyst systems [6]. Thus, in our next stage, the effectivenesses of the NaCl/W/Z systems were examined.

3.3 NaCl/W/Z-Based Catalysts

3.3.1 Effect of Transition Metal additives.

As shown in Fig.7, where a correlation between CH_4 conversion and C_2 selectivity was plotted and each data was taken after 1-h reaction at 1023K, the addition of some metals was found to be effective for the raise in CH_4 conversion, although their C_2 selectivities were slightly decreased: In particular,.the additions of Mn and Ce seem to be quite favorable.

3.3.2 Effects of Catalyst Components.

Using Mn/NaCl/W/Z catalysts, the effects of catalysts components on the reaction were investigated. The C_2 yields were greatly dependent on these components: The maximum yield (>30%) was achieved at W content of 13wt%, Mn content of 2wt%, Na/Mn molar ratio of 10. In case of Ce/NaCl/W/Z catalysts, similar trends resulted, except for Ce content of 6-10wt%. The catalysts performances of Mn/W/Z systems were also affected by alkali metal chlorides employed. Sodium chloride seemed to be the most favorable, as shown in Fig.8, where the performances were plotted against alkali ion radius.

3.3.3 Catalyst Performances as a Function of time.

Three catalysts, which were prepared by different impregnation methods of NaCl (Table 1), were used for the reaction, which was carried out under a mixed gas flow of $CH_4/O_2/N_2$ (=10/5/80). In Cat.A, the C_2 yield was decreased from 35.5% to 22.5% (after 26h)) (O). on the contrary, in Cat.B, the C_2 yield decreased slowly from 285.% to 25.7 (after 28h)(■). In reaction at 973K using Cat.C, the yield decreased only from 25% to 20% even after 100h. Thus, two-step impregnation of NaCl was found to be effective for an improvement of catalyst

Table 1 Zr-based catalyst preparation methods and reaction conditions for life tests.

	Impregnation 1	Calcination 1	Impregnation 2	Calcination 2	Reaction Temp.	W/F g.mol.h^{-1}
Cat.A	13wt%W	1103K, 3h	6wt%Ce/ NaCl (Na/Ce=10(molar))	1123K, 16h	1023K	28
Cat.B	13wt%W/NaCl (Na/W=11.5)	1103K, 3h	6wt%Ce/ NaCl (Na/Ce=20(molar))	1123K, 16h	1023K	56
Cat.C	13wt%W/ NaCl (Na/W=5.73)	1103K, 3h	6wt%Ce/ NaCl (Na/Ce=10(molar))	1123K, 16h	973K	56

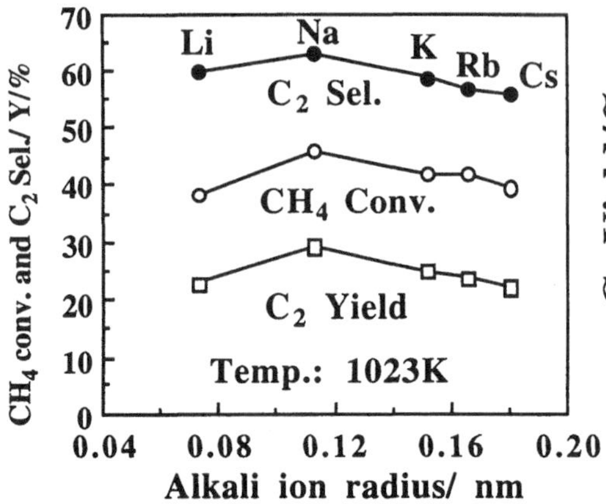

Fig.8 Effect of Alkali metal chlorides.
Catalyst: Zr/13wt%W/2wt%Mn
Gas Comp.: $CH_4/O_2/N_2$=10/5/85

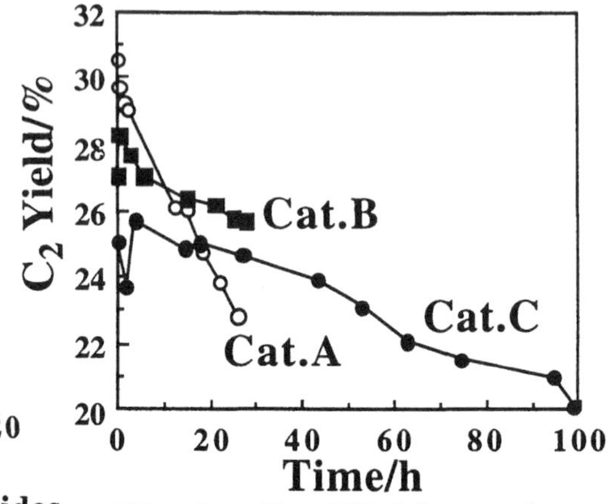

Fig.9 C_2 Yield vs.time.
Cat.A,B,C: See Text.

life. These findings would be consistent with the view that a surface NaCl could play an important role in CH_4 activation at reaction temperature as low as 973K and a deactivation could be associated with, at least in part, a loss of NaCl. Further study is currently underway to clarify the role of NaCl in NaCl/W/Z-based catalysts.

3.4 Summary

In summary, tungsten oxide/ zirconia (W/Z)-based catalysts were prepared and their performances were tested for the OCM reaction. Two types of catalysts were found; One is Li_2CO_3-containing, and the other is NaCl-containing. The former catalysts such as europium or cerium/Li_2CO_3/W/Z showed long lifetime over 100h, but modest C2 yield(ca.18%). And the latter such as cerium or manganese/NaCl/W/Z showed excellent C2 yield over 30%, but rather short lifetime (ca.30h), although the C_2 yields decreased from 35.5% to 22.5% (26h) with increase in reaction time. The two-step impregnation of NaCl for Ce/W/Z systems was found to be effective for an improvement of catalyst life. Further work is in progress to improve the catalyst performances.

REFERENCES

[1] J. H. Lunsford, Angew. Chem. Int. Ed. Engl., 34(1995) 970.
[2] K. Murata and H. Ushijima, J. Chem. Soc., Chem. Commun., (1994) 1157.
[3] H. Hino, K. Arata, J. Chem. Soc., Chem. Commun., (1988)1259.; M. Hino, K. Arata, J. Chem. Soc., Chem. Commun., (1979) 1148.
[4] K. Murata, T. Hayakawa and K. Fujita, J. Chem. Soc., Chem. Commun.,(1997) 221.
[5] K. Otsuka and T. Komatsu, J. Chem. Soc., Chem. Commun., (1987) 388.
[6] M. Fukai, K. Noro, K. Nomura, Y. Nakamura, and K. Otsuka, Sekiyugakkaishi, 40(2), 65-70 (1997).
[7] M. Fukai, K. Noro, K. Nomura, Y. Nakamura, and K. Otsuka, Sekiyugakkaishi, 40(2), 65-70 (1997).

NATURAL GAS CONVERSION V
Studies in Surface Science and Catalysis, Vol. 119
A. Parmaliana et al. (Editors)

Effect of the Na addition to PrO_x/MgO on the Reactivity and Selectivity in the Oxidative Coupling of Methane.

Graciela T. Baronetti, Cristina L. Padró, Alberto A. Castro and Osvaldo A. Scelza.

Instituto de Investigaciones en Catálisis y Petroquímica (INCAPE), Facultad de Ingeniería Química (Universidad Nacional del Litoral)-CONICET, Santiago del Estero 2654, (3000) Santa Fe, ARGENTINA.

ABSTRACT

The Na addition to PrO_x (10 wt%)/MgO produces a very important modification in the activity and selectivity to ethane + ethylene in the oxidative coupling of methane (OCM), which was carried out by injecting pulses of pure methane on the samples at 1023 K. The specific activity decreases as the Na content increases, but the selectivity to C_2 dramatically increases. According to XRD, SEM-EDX and XPS results, the behaviour in OCM can be related to the different surface species. For low Na contents, reduced surface PrO_x species were found, while a Pr-Na compound was detected by XRD for high Na concentrations in the samples. The high initial selectivity to C_2 hydrocarbons for samples with high Na/Pr ratio could be related to the presence of the above mentioned Pr-Na species.

1. INTRODUCTION

The oxidative coupling of methane (OCM) is an important route for the direct transformation of methane (from natural gas) to more valuable hydrocarbons. Different metallic oxides catalysts were used for the above reaction. In particular, the rare-earth oxides showed a good performance. Thus, it was reported in a previous paper [1] that alkaline-promoted bulk PrO_x can be used as a redox agent (in the absence of oxygen) or as a catalyst for the methane oligomerization (CH_4+O_2 feed). When PrO_x was used as a redox agent, it showed a higher activity in OCM than other lanthanide oxides due to the rapid interconversion of the oxidation states, rapid diffusion of the oxygen in the bulk and the high oxidation potential. The addition of alkaline compounds to bulk PrO_x was found to enhance the yield to higher hydrocarbons in the OCM [1]. Besides, the rare-earth oxides have a high thermal stability, adequate basicity and low surface area, conditions which are necessary for this process [2].

Several authors have reported the use of the promoted and non-promoted reducible bulk oxides (such as PrO_x) for this reaction, but the literature is not abundant for supported oxides [3]. Hence, a study on the effect of the addition of increasing amounts of Na to PrO_x supported on MgO is analyzed in this paper by using XRD, SEM-EDX and XPS measurements. Moreover, the reactivity and selectivity of the different samples in the oxidative coupling of methane was determined by injecting pulses of pure methane in the absence of oxygen.

2. EXPERIMENTAL

Na-doped PrO_x (10 wt% Pr)/MgO samples were prepared by adding $Pr(NO_3)_3.6 H_2O$ and NaOH to MgO suspended in deionized water. The NaOH concentration in the solution was such as to obtain the desired Na loading in the samples (Na/Pr molar ratio: 1, 2, 4 and 10). The selection of the Na-precursor was made taking into account that, according to the literature [4], NaOH showed a similar behaviour than Na_2CO_3 in OCM. The slurry was stirred at room temperature and then the remaining water was eliminated by heating, under stirring, until a thick paste was obtained. Samples were dried at 393 K and calcined in an air stream in two steps: first at 738 K (to decompose $Mg(OH)_2$) for 1 h, and second at 1023 K (reaction temperature) for 3 h.

Samples were tested in the oxidative coupling of methane at 1023 K and at P = 1 atm. The OCM was carried out in a quartz fixed-bed reactor by injecting pulses (0.5 ml STP) of pure CH_4 (without O_2) in a He stream. Samples were kept in a pure He stream (without O_2) between pulses. Before reaction, calcined samples (0.15 g) were heated from 298 K up to the reaction temperature (1023 K) in flowing pure He (99.99%) and maintained at this temperature during 2 h. Reaction products were analyzed by GC. The carbon mass balance was carried out after each CH_4 pulse was injected into the reactor. By considering the basic properties of the samples, a fraction of the carbon amount fed into the reactor as CH_4 could be retained by the solid, probably as CO_2 (reaction product) strongly adsorbed on the surface. Hence, the selectivity to C_2 hydrocarbons (C_2H_4 and C_2H_6) was calculated taking into account both the CO and CO_2 amounts in the gas phase and the quantity of CO_2 retained by the sample (obtained from the carbon mass balance).

After He-treatment (or at the beginning of reaction), samples were immediately quenched from 1023 K to room temperature in order to keep the oxidation state of the PrO_x [5]. Then, samples were characterized by XRD, SEM-EDX and XPS. XRD patterns were taken in a diffractometer Rich Seifert ISO-Debyeflex 2002. The characterization of samples by SEM-EDX was carried out in a Jeol JSM 35-C Scanning Electron Microscope equipped with an EDAX analyzer. X-ray photoelectron spectra were obtained with an ESCA 750 Shimadzu instrument, using Mg-Kα radiation. Samples were previously evacuated in a chamber directly connected to the analysis system. Surface charging was observed for all the samples. Binding energy (BE) for the Pr $3d_{5/2}$ line was referred to Mg 2p level at 50.8 eV [6]. Multiple peaks were fitted by the van Citter's method using Gaussian and Lorentzian components.

3. RESULTS AND DISCUSSION

In all OCM experiments the reaction products were: C_2H_6, C_2H_4, CO, CO_2, and H_2O. Table 1 shows the values of the initial specific activity (A: conversion of CH_4 /mg Pr), selectivity to C_2 (C_2H_6+C_2H_4) (S_{C_2}) and the yield to C_2 (Y_{C_2}) corresponding to the first CH_4 pulse for PrO_x/MgO samples doped with different Na contents. Table 1 also includes the values obtained for the undoped sample [7]. It can be seen that the A

Table 1: Initial (first pulse) specific activity (A), selectivity (S_{C_2}) and yield to C_2 hydrocarbons (Y_{C_2}) of PrO_x-Na/MgO (10 wt % Pr) samples with different Na contents.

Sample $(Na/Pr)_{molar}$	Specific activity A: X_{CH4}/mg Pr	S_{C_2}, %	Y_{C_2}, %
0	2.44	3.1	1.1
1	5.52	29.2	21.9
2	2.97	47.2	19.2
4	2.08	49.5	14.0
10	1.48	80.9	16.2

value increases for samples with low Na concentrations with respect to the PrO_x/MgO sample, but it decreases for Na/Pr molar ratio > 4. Moreover, it can be observed a dramatic increase of the selectivity to C_2 after Na addition. In fact, the S_{C_2} increases about 9 folds for the sample with Na/Pr =1 and 27 folds for the sample with Na /Pr= 10, with respect to the undoped one. Furthermore, it can be observed that the addition of a small quantity of Na to PrO_x/MgO sharply enhances the yield to C_2 hydrocarbons. It must be indicated that the activity of samples decreased along the successive pulses, while the S_{C_2} increased with the successive injection of CH_4 pulses. Moreover, an additional co-feed experiment (CH_4/O_2 molar ratio= 2) carried out on the sample with Na/Pr molar ratio=1 showed a lower yield to C_2 (8.0 %) than the initial one for the pulse experiments with pure methane.

XRD patterns of samples (after the He-treatment) showed that different crystalline Pr oxide phases are present when the Na content is modified. MgO and $Mg(OH)_2$ characteristics lines were observed in all samples irrespective of the Na content. For the undoped sample only a $PrO_{1.83}$ phase was found [7]. When a low amount of Na was added to PrO_x/MgO, a $PrO_{1.75}$ phase was detected, as well as the $PrO_{1.83}$ species. These results would indicate that the promotion of PrO_x/MgO samples with small Na amounts favours the development of more oxygen-deficient praseodymium oxide structures. For the highest Na content, weak characteristics lines of PrO_2Na compounds were also detected, as well as the $PrO_{1.83}$ and $PrO_{1.75}$ lines.

SEM-EDX results showed morphological changes when increasing Na amounts were added to PrO_x/MgO. Thus, the sample with a Na/Pr molar ratio=1 (Figure 1a)

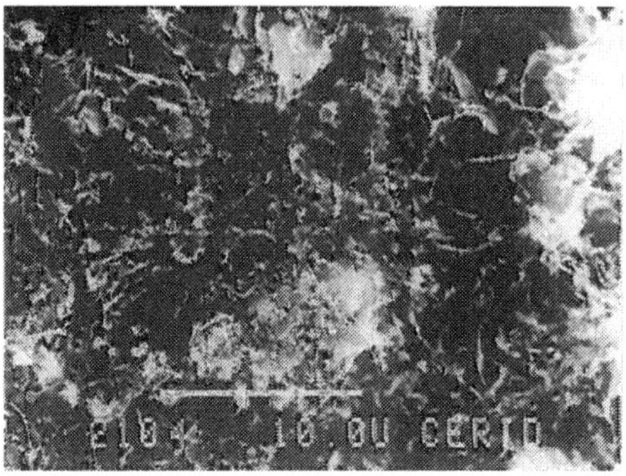

(a)

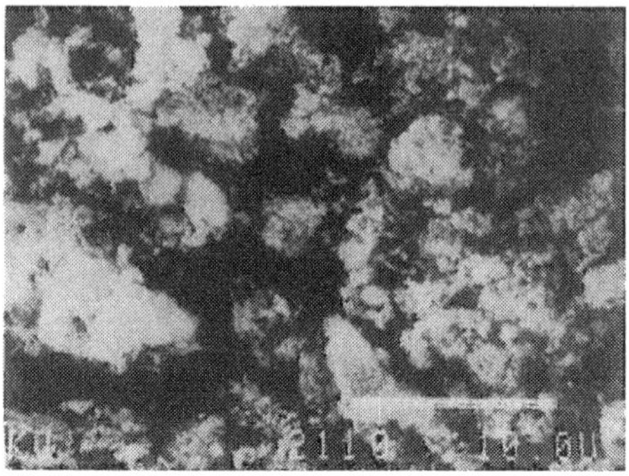

(b)

(c)

Figure 1. SEM photographs of different Na-PrO_x/MgO samples. (a) : Na/Pr = 1 , (b) : Na/Pr = 4 and (c): Na/Pr = 10.

shows the characteristic "flakes" structure similar to that of pure MgO [6]. Besides, a small quantity of agglomerates was observed (Figure 1a). EDX analysis of this sample showed the presence of Pr and Mg in both zones ("flakes" and agglomerates). The presence of the "flakes" structure and the results of reactivity in OCM for this sample could be related to a high Pr dispersion on the MgO matrix. In fact, taking into account the results of Table 1, it can be observed that the sample with Na/Pr molar ratio = 1 showed a specific activity higher than that of the undoped one. It was reported in a previous paper that the undoped PrO_x/MgO with a high PrO_x dispersion showed a specific activity higher than that of a PrO_x/MgO with a low PrO_x dispersion [7]. Hence, the addition of low amount of Na to PrO_x/MgO appears to increase the PrO_x dispersion on the MgO with respect to the PrO_x/MgO without Na, and consequently the activity is also increased. No "flakes" structure was observed for samples with Na/Pr molar ratio ≥ 4. In these cases the SEM microphotographs (Figures 1b-c) revealed two zones, one of them shows a picture of small particles (particle size< 0.1μ), and the other one corresponds to large particles (6-8μ). The EDX analysis of the sample with Na/Pr = 10 revealed that the small particles are composed by Pr and Mg, and the big ones showed an important concentration of Pr and Na.

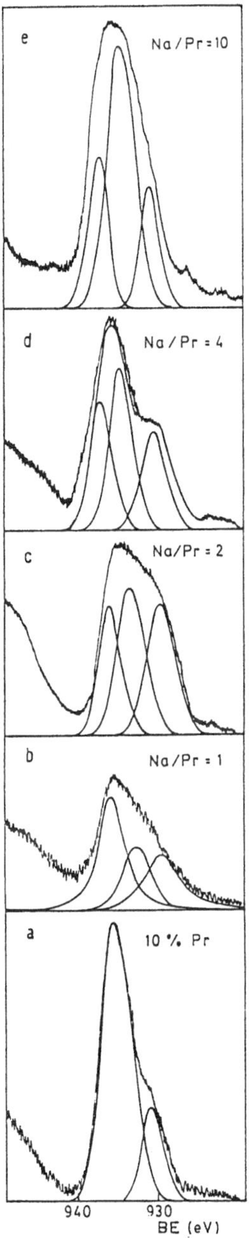

Figure 2. XPS profiles of Na-PrO_x/MgO samples. (a): PrO_x/MgO

XPS profiles (Figure 2) showed the presence of surface praseodymium oxide species with different oxidation states. The XPS profile of the Pr $3d_{5/2}$ line depends on the Na content. In a previous paper we reported that the Pr $3d_{5/2}$ lines for PrO_2 is positioned at 936.3 eV for PrO_2 in samples with high PrO_x dispersion on the MgO matrix and at 935.2 eV for samples with low PrO_x dispersion. Peaks positioned at low binding energies (< 931.8 eV) corresponded to a more reduced PrO_x (x< 1.83) species [7]. Figure 2a includes the profile corresponding to the undopped sample. In this profile we observed one peak at 935.5 eV corresponding to PrO_2 with low dispersion. The another one at 931.3 eV can be assigned to more reduced species. For samples with low Na contents (Na/Pr molar ratios =1 and 2), three peaks were observed (Figures 2b and 2c). The peaks positioned at 933.3-933.0 eV and 929.8 eV can be assigned to more reduced PrO_x species (x<1.83). The peak at the highest binding energy (936.0 eV) would correspond to a high dispersed PrO_2 species. The ratio between oxidized species and the more reduced one decreases from 4.2 to 0.56 when the Na/Pr molar ratio increases from 0 to 2. In the case of the samples with high Na/Pr molar ratio (4 and 10), XPS profiles showed three peaks. The peaks at the lowest and the highest BE would correspond to PrO_2 and PrO_x (x<1.83) species, respectively. The peak at 934.5 markedly increases for Na/Pr=10 sample. This peak could be assigned to a praseodymium species with an oxidation state lower than that corresponding to Pr^{4+} species. In order to assign this peak, it must be noted that the sample with Na/Pr=10 showed a very high selectivity to C_2 hydrocarbons, in contrast with the samples with lower Na contents. Taking into account this result and that the XRD pattern shows the presence of PrO_2Na species, and considering that the bulk compounds between Pr and Na display a very high selectivity to C_2 hydrocarbons in the OCM reaction [1], the peak at 934.5 eV in XPS profile for the sample with Na/Pr= 10 could be due to a surface Pr-Na species.

From the above results it can be concluded that the effect of Na addition to PrO_x/MgO can be associated with a modification of the rare-earth oxide phase by Na, and this effect changes with the Na loading in the sample. The results of the reactivity in OCM show that the specific activity decreases when the Na content increases, but the selectivity to C_2 markedly increases with the Na content. The specific activity can be related to the PrO_x dispersion. In fact, it was found a high PrO_x dispersion at low Na content on the matrix, but the Pr species appears to be agglomerated as large particles for higher Na contents. The increase of the selectivity to C_2 hydrocarbons when Na/Pr ratio increases can be related to the nature of the surface species. For low Na/Pr ratio, the Pr-Na compound was not detected. In these cases, the ratio between more oxidized and more reduced PrO_x species decreases with respect to the undoped sample. The addition of low Na amounts produces a decrease in the concentration of PrO_x species with high oxidation states. According to the literature, the more oxidized PrO_x species displays a low selectivity to C_2 hydrocarbons [8,9] and the addition of alkali metal (such as lithium) to bulk PrO_x suppressed the formation of Pr^{4+} species, which can be associated with labile lattice oxygen responsible for the surface oxidation of methane [8,10]. For samples with high Na/Pr molar ratio, besides the presence of reduced PrO_x species, the formation of a Pr-Na species appears to have a very important contribution to the high selectivity to C_2 hydrocarbons.

4. ACKNOWLEDGMENTS

This work was carried out with the financial support of the National University of Litoral (Program CAI+D). Authors thank the Japan International Cooperation Agency for the donation of the ESCA spectrometer.

5. REFERENCES

1. A.M. Gaffney, C.A. Jones, J.J. Leonard, and J.A. Sofranko, J. Catal. 114 (1988) 422.
2. O. Forlani and S. Rossini, Mat. Chem. Phys., 31 (1992) 155.
3. M.Y. Sinev, Y.P.Tyulenin, B.V Rozentuller, Kinetika i Kataliz 32 (1991) 896.
4. J.A.S.P. Carreiro and M. Baerns, J. Catal. 117 (1989) 396
5. K. Otsuka, M. Kunitomi, and T. Saito, Inorganica Chimica Acta, 115 (1986) L31.
6. C.L. Padró, W.E. Grosso, G.T. Baronetti, A.A. Castro and O.A. Scelza, Studies in Surface Science and Catalysis 82 (1994) 411.
7. G.T. Baronetti, W.E. Grosso,S.P. Maina, C.L. Padró, A.A. Castro, O.A. Scelza and J.M. Palacios Latasa, J. Chem.Technol. Biotechnol. 70 (1997) 141.
8. J.M. De Boy and R.F. Hicks, Ind. Eng. Chem. Res. 27 (1988) 1577.
9. A. Ekstrom and J.A. Lapszewicz, J. Phys Chem. 93 (1989) 5230.
10. M.G. Poirier, R.Breault, S. Kaliaguine, and A. Adnot, Appl. Catal. 71 (1991)103.

NATURAL GAS CONVERSION V
Studies in Surface Science and Catalysis, Vol. 119
A. Parmaliana et al. (Editors)

Study of the catalytic performance, surface properties and active oxygen species of the fluoride-containing rare earth-alkaline earth oxide based catalysts for the oxidative coupling of methane †

Wei Zheng Weng, Ruiqian Long, Mingshu Chen, Xiaoping Zhou, Zisheng Chao and Hui Lin Wan*

State Key Laboratory for Physical Chemistry of Solid Surface, Department of Chemistry and Institute of Physical Chemistry, Xiamen University, Xiamen, 361005, CHINA

Addition of SrF_2 and BaF_2 to Ln_2O_3 and LaOF improved their catalytic performances for the OCM reaction. The promotion effect of the fluorides in the catalysts may be principally related to the extent of phase-phase interaction between fluoride and oxide. Comparing with the corresponding alkaline earth oxide promoted Ln_2O_3 catalyst system, an alkaline earth fluoride promoted Ln_2O_3 catalyst system is less basic and will therefore be favorable to prevent the CO_2 inhibition. However, there is no simple correlation between the acidity/basicity of an OCM catalyst and its catalytic performance. The results of *in situ* FTIR characterization show that $\underline{O}_2^-$ is the active oxygen species or one of the active species for OCM reaction over LaOF and BaF_2/LaOF catalysts.

1. INTRODUCTION

The oxidative coupling of methane (OCM) to make ethylene has been a subject of intensive study since Keller and Bhasin reported their early work on the reaction in 1982[1]. After more than 15 years of study, many catalyst systems have been developed. Among these catalyst systems, the alkaline earth oxide and/or rare earth oxide based irreducible metal oxide (and carbonate) catalyst systems, mostly of the host-dopant type, have been intensively investigated[2,3], and it was found that the OCM performance of many catalysts could be significantly improved by chlorine presented either in the form of a chloride component built into the catalyst such as Cl^- promoted Li^+/MgO or as a volatile chlorinated compound (organic or inorganic) in the reactant feed[3-5].

Drawing inspiration from the promoting effect of Cl^-, we considered that it would also be interesting and of fundamental significance to have a detail investigation on the influence of other halides, especially fluorides, since the alkaline earth or rare earth fluorides are usually more stable than the corresponding chlorides under the OCM reaction conditions. In this paper, the catalytic performance of a series of SrF_2 and BaF_2 promoted rare earth sesquioxide and lanthanum oxyfluoride catalysts, the nature of the promotion effects of fluorides and the relationship between acidity/basicity of the fluoride containing catalysts and their catalytic behavior will be reported and discussed. The results of *in situ* FTIR characterization of the

---

† Supported by the National Natural Science Foundation of China, Doctoral Foundation of Education Committee of China, the Fujian Provincial Natural Science Foundation of China and a grant from SINOPEC.
* Correspondence author.

active oxygen species for the OCM reaction over the LaOF and BaF_2/LaOF catalysts are also presented.

2. EXPERIMENTAL

The details of catalyst preparation and catalytic performance evaluation were the same as reported previously[6-7]. The XRD experiments were performed on a Rigaku Rotaflex D/Max-C using CuK_α radiation with λ=1.5406 Å. The CO_2-TPD experiments were carried out on a Perkin-Elmer Auto System XL gas chromatograph equipped with TCD. In each experiment, 100.0 mg of catalyst sample was placed in a quartz microreactor (4.0 mm ID and c.a. 35 cm in length), the rest part of the reactor was filled with quartz chip (40-60 mesh). The sample was first treated in a flow of He (30 ml/min) at 800 °C for 2 h followed by cooling to 700 °C under He and treating with a flow of CO_2 (30 ml/min) at 700 °C for 30 min. The sample was then cooled to 50 °C under CO_2 atmosphere and purging with He at 50 °C until the base line of GC was flat, and finally heated in a flow of He (30 ml/min) at the heating rate of 15 °C/min from 50 to 800 or 1000 °C. The IR experiments were performed on a Perkin-Elmer Spectrum 2000 system equipped with a liquid nitrogen cooled MCT detector and a home-built transmittance IR cell with BaF_2 windows designed to treat the sample *in situ*. The catalyst was pressed into a self-supported disk and treated in O_2 at 750 °C for 30 minute followed by evacuation (10^{-2} ~10^{-3} Torr) at same temperature for at least 4 h before the introduction of the reactants to the catalysts. For the experiments of CO_2 adsorption, the treated sample was exposed to CO_2 (50 Torr, 50 min, for La_2O_3 and BaF_2/La_2O_3; 1 atm, 5 min, for LaOF and BaF_2/LaOF) at 750 °C followed by brief evacuation at same temperature. For the experiment of O_2 adsorption, the treated sample was exposed to 1 atm of O_2 at 750 °C for 10 min followed by brief evacuation at same temperature. The O_2 pre-adsorbed sample was then exposed to definite amount of CH_4 in a close IR cell at 750 °C to study the reaction of oxygen adspecies with CH_4. The IR spectrum was taken with 36 scans at the resolution of 4 cm^{-1} over 0.5-1 minute. The reported absorbances were obtained from spectra referenced to the background spectra which were taken under vacuum at 750 °C prior to the introduction of the reactants to the catalysts.

3. RESULTS AND DISCUSSION

3.1. Promotion effect of fluorides in rare earth-alkaline earth oxide and oxyfluoride based catalysts

The OCM performances of a series of SrF_2 and BaF_2 promoted Ln_2O_3 and LaOF catalysts are shown in Table 1. Under the reaction conditions as indicated in the Table 1, relatively low C_2 yield was observed over pure Ln_2O_3. With the addition of certain amount of SrF_2, which showed poor OCM activity, to Ln_2O_3, the OCM performance of the catalysts was found to be improved. For La_2O_3, Nd_2O_3, Sm_2O_3, Eu_2O_3 and Gd_2O_3, the promotion effect of SrF_2 is remarkable, whereas for Dy_2O_3, Ho_2O_3, Er_2O_3, Tm_2O_3 and Yb_2O_3, the improvement in the C_2 yield is less significant. In General, the C_2 yield of the SrF_2/Ln_2O_3 catalysts decreased with the increasing in atomic number of lanthanide, i.e. La, Nd > Sm > Gd, Eu > Dy > Ho > Er > Tm > Yb, which is consistent with the successive decreasing of the conductivity of Ln_2O_3[8]. At the temperature of OCM reaction, the cations and/or anions in the oxides may become substantially mobile and the ionic conductivity becomes an important component for the conductivity[9]. This implies that the morbility of lattice oxygen for the rare earth susqioxides may also decrease in the same sequence. So the difference in the promotion effect of SrF_2 on Ln_2O_3 may be explained by that the interaction or ionic exchange between the SrF_2 and Ln_2O_3 phases is favored for the Ln_2O_3 with higher lattice oxygen mobility.

Table 1
The OCM performance of SrF_2 and BaF_2 promoted Ln_2O_3 and LaOF

Catalyst (mol%)	Temp. (°C)	Conv. (%) CH₄	O₂	Selectivity (%) CO	CO₂	C₂H₄	C₂H₆	C₂	Yield C₂ (%)
La_2O_3 [a]	700	29.4	100	10.5	52.3	21.5	15.7	37.2	10.9
La_2O_3 [b]	800	27.7	86.4	36.2	40.1	15.5	8.2	23.7	6.6
20%SrO/La_2O_3 [a]	700	27.0	/	3.9	50.6	25.4	20.1	45.5	12.3
20%SrF_2/La_2O_3 [a]	700	34.2	98.1	6.4	36.3	36.1	21.2	57.3	19.6
Nd_2O_3 [a]	750	27.2	99.1	7.2	53.5	20.1	19.2	39.3	10.7
50%SrF_2/Nd_2O_3 [a]	750	34.3	98.9	4.0	38.9	33.1	24.0	57.1	19.6
Sm_2O_3 [a]	800	26.3	99.2	8.9	52.3	21.6	17.2	38.8	10.2
50%SrF_2/Sm_2O_3 [a]	800	34.0	99.5	3.9	40.3	33.1	22.7	55.8	19.0
Eu_2O_3 [a]	750	26.3	98.4	13.8	53.1	21.8	11.3	33.1	8.7
50%SrF_2/Eu_2O_3 [a]	750	33.1	99.0	6.7	40.4	31.9	21.0	52.9	17.5
Gd_2O_3	750	29.5	99.7	17.5	49.0	20.1	13.4	33.5	9.9
50%SrF_2/Gd_2O_3 [a]	750	34.4	99.5	5.9	39.5	32.0	22.6	54.6	18.8
Dy_2O_3	750	31.3	99.5	12.4	45.0	25.0	17.6	42.6	13.3
50%SrF_2/Dy_2O_3 [a]	750	32.6	96.3	8.5	40.0	32.0	19.5	51.5	16.8
Ho_2O_3 [a]	750	28.8	99.3	18.2	48.8	19.9	13.1	33.0	9.5
50%SrF_2/Ho_2O_3 [a]	750	30.8	97.3	8.6	45.5	26.7	19.1	45.8	14.1
Tm_2O_3 [a]	750	26.4	99.4	18.2	53.8	16.4	11.6	28.0	7.4
50%SrF_2/Tm_2O_3 [a]	750	28.6	99.5	8.5	51.2	21.2	19.1	40.3	11.5
Yb_2O_3 [a]	750	28.8	99.4	20.3	48.5	19.3	11.9	31.2	9.0
50%SrF_2/Yb_2O_3 [a]	750	28.7	99.4	12.7	51.6	19.9	15.8	35.7	10.3
30%BaO/La_2O_3 [b]	800	25.3	97.2	3.5	43.5	29.3	23.7	47.5	12.0
30%BaF_2/La_2O_3 [b]	800	26.4	95.0	8.0	38.2	34.3	19.5	53.8	14.2
LaOF [b]	800	24.9	97.5	14.5	46.3	23.6	15.6	39.2	9.8
15%BaF_2/LaOF [b]	800	31.0	91.7	4.9	34.8	37.3	23.0	60.2	18.7

a) Feed: CH_4:O_2=3:1, GHSV=20000h^{-1}. b) Feed: CH_4/O_2=4:1, GHSV=15000h^{-1},
The data were obtained after 30 min on stream.

As the result of ionic exchange, new phases such as oxyfluoride (LnOF) and lattice defects such as anionic vacancies (e.g. in the cases of substituting O^{2-} for F^- and/or the cation with lower valence for the cation with higher valence) may be formed. For the oxide or oxyfluoride catalysts with stable cation valence, the presence of anionic vacancies in the catalyst may create suitable sites available for oxygen adsorption, while a catalyst with higher concentration of anionic vacancy and high anionic mobility will also be favorable to the migration of OH^- group generated in the reaction to close surface sites, and then eliminating as H_2O to regenerate the anionic vacancy. The XRD experiments have proved that LnOF phases are formed in the fresh or reacted SrF_2/Nd_2O_3, SrF_2/Sm_2O_3, BaF_2/Eu_2O_3 and SrF_2/Gd_2O_3 catalysts, while for the other catalysts, except SrF_2/La_2O_3, XRD only detects the alkaline earth fluoride and Ln_2O_3 phases. For the SrF_2/La_2O_3 catalyst, since the diffraction lines of tetragonal LaOF and SrF_2 overlap each other, it is difficult to judge from the XRD experiment that if the LaOF phase is formed. But the XRD have detected the formation of $SrCO_3$ phase in the fresh SrF_2/La_2O_3 catalyst, indicating that partially anionic exchange between the SrF_2 and La_2O_3 phases takes place in the catalyst leading to the formation of SrO. These results suggest that the promotion effects of the fluorides in the catalysts may be principally related to the extent of phase-phase interaction between the fluoride and oxide phases.

Comparing with the corresponding alkaline earth oxide (SrO, BaO) modified rare earth oxide catalyst systems, e.g. SrO/La_2O_3 or BaO/La_2O_3, a fluoride promoted catalyst such as BaF_2/La_2O_3 or SrF_2/La_2O_3 also showed better OCM performance. Similar promotion effects were observed in the BaF_2 modified tetragonal LaOF (an ionic conductor with superstructure of fluorite containing intrinsic anionic Frenkel defects and anionic vacancies). This catalyst demonstrated excellent catalytic performances for both OCM[7] and ODE[10] reactions.

3.2. Surface acidity/basicity of the fluoride-containing catalyst and their relationship with catalytic property

The modification effect of fluoride on the acid/base properties of the catalyst was studied by CO_2-TPD and FTIR spectrum of CO_2 adsorption. Figure 1 shows the CO_2-TPD profiles of La_2O_3, LaOF, 30mol% BaO/La_2O_3, 30mol% BaF_2/La_2O_3 and 15mol% BaF_2/LaOF. Based on the desorption temperature and peak intensity of the CO_2-TPD profiles, it is clear that the basicity of the catalysts decreases in the sequence $BaO/La_2O_3 > La_2O_3 > BaF_2/La_2O_3 > BaF_2/LaOF > LaOF$. Similar conclusion can be made from the IR spectra of CO_2 adsorption performed at 750 °C over the La_2O_3, LaOF, BaF_2/La_2O_3 and BaF_2/LaOF catalysts (Figure 2), in which the IR peak intensities of surface lanthanum oxycarbonate ($La_2O_2CO_3$) also decrease in the same sequence. These results indicate that, comparing with the alkaline earth oxide promoted rare earth oxide (e.g. BaO/La_2O_3), the surface of alkaline earth fluoride promoted rare earth oxide (e.g. BaF_2/La_2O_3) is less basic.

From the list of the early OCM catalysts, the strongly basic oxides of alkali metals, alkaline earth metals and some rare earth metals constitute the best of the catalytic materials[2-3]. It was found that the OCM performance including C_2 selectivity for the alkaline earth oxides increased with the increase of basicity[3]. Therefore, a parallel relationship between the catalyst basicity and its OCM performance has been suggested[3]. For an oxide or complex oxide with stable cation valence, its Lewis basicity, or p-type conductivity, or electron donating ability of lattice oxygen is parallel to its ability of adsorbing and activating oxygen, and to its catalytic activity. However, during the OCM process, CO_2 generated by the side reaction will instantly react with surface Lewis base sits (even with O_2^{2-} species) forming surface carbonate, especially for the catalysts promoted with strongly basic oxides such as BaO, SrO and Li_2O. As a result, the catalyst's ability to adsorb oxygen decreases and catalytic activity will also be affected, while the selectivity improves. In order to maximize the OCM performance of the catalyst, higher reaction temperature is required. When the alkaline earth fluoride was added to the rare earth oxide such as La_2O_3, besides the formation of anionic vacancies due to the ionic exchange as discussed above, the basicity of the catalyst is also modified, so that the catalytic activity usually is not or not very seriously affected by the CO_2. Similar function of Cl^- on the basicity and carbonate formation of the Li^+/MgO system was found by Lunsford et al.[4-5]. The improvement of selectivity of a fluoride-promoted catalyst may be attributed to the isolation of surface oxygen adspecies by fluoride and the formation of active oxygen adspecies in the form of less rich in electron such as O_2^-, which is much less active than O^- and O_2^{2-}.

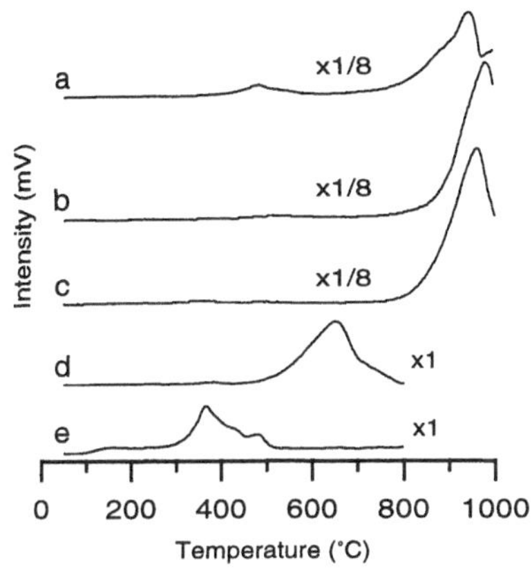

Figure 1. CO_2-TPD profiles of the catalysts.
a) 30mol%BaF_2/La_2O_3,
b) 30mol%BaO/La_2O_3,
c) La_2O_3, d) 15mol%BaF_2/LaOF, e) LaOF.

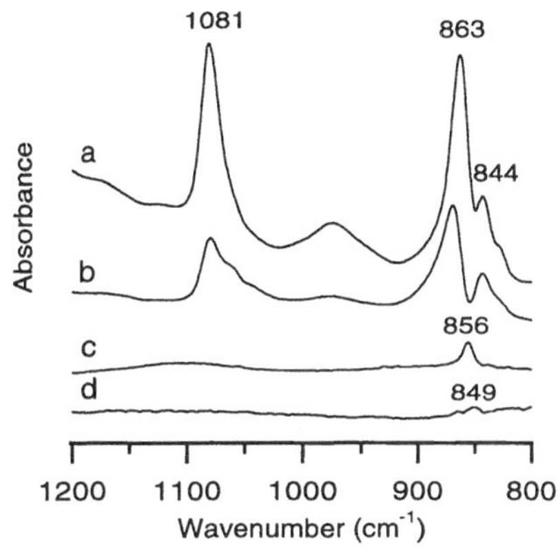

Figure 2. IR spectra of CO_2 adsorption over the catalysts at 750 °C.
a) La_2O_3, b) 30mol%BaF_2/La_2O_3,
c) 15mol%BaF_2/LaOF, d) LaOF.

From the data of catalytic performances listed in Table 1 and the results of CO_2 adsorption shown in Figure 1 and 2, it can be found that there is no simple correlation between the acidity/basicity of Ln_2O_3, alkaline earth fluoride promoted Ln_2O_3, LnOF and alkaline earth fluoride promoted LnOF catalysts and their OCM performances. For the pure La_2O_3, addition of BaF_2 decreased the basicity, whereas for the LaOF, addition of BaF_2 increased the basicity, but results of catalytic performance evaluation reveal that 30mol% BaF_2/La_2O_3 and 15mol% BaF_2/LaOF demonstrate better OCM performance than the corresponding La_2O_3 and LaOF. This result again indicates that a good OCM catalyst needs not be strongly basic, and that the relationship between the acidity/basicity of an OCM catalyst and its catalytic properties is complicated

3.3. in situ FTIR characterization of active oxygen species for OCM

After a LaOF sample was exposed to 1 atm O_2 at 750 °C followed brief evacuation at same temperature, two broad IR bands with maximum at 1090 cm^{-1} and a shoulder at 1180 cm^{-1} were observed (Figure 3 b). By comparing with the IR spectra of CO_2 adsorbed LaOF recorded at 750 °C (Figure 2 d), the possibility of attributing these bands to the surface oxycarbonate species can be excluded. Since the positions of the bands are consistent with those of superoxide ligands in the matrixes[11], complex[12] and oxide surface[13], they were assigned to the adsorbed superoxide species ($\underline{O_2}^-$) located in two different microchemical environments.

When a LaOF sample, which has been pre-adsorbed with O_2 followed by brief evaluation to removed the gas phase O_2, was exposed to a definite amount CH_4 at 750 °C in a close IR cell, the IR absorbance at c.a. 1090 cm^{-1} gradually decreased while the IR band of gas phase C_2H_4 at 950 cm^{-1} increased (Figure 3), and a parallelism between the decrease of the IR band of $\underline{O_2}^-$ (at 1090 cm^{-1}) and the increase of the IR band of gas phase C_2H_4 (at 950 cm^{-1}) was found (Figure 4). Similar results were also observed over the 15mol% BaF_2/LaOF. These results suggest that $\underline{O_2}^-$ is the active oxygen species or one of the active oxygen for the OCM reaction over the corresponding catalysts.

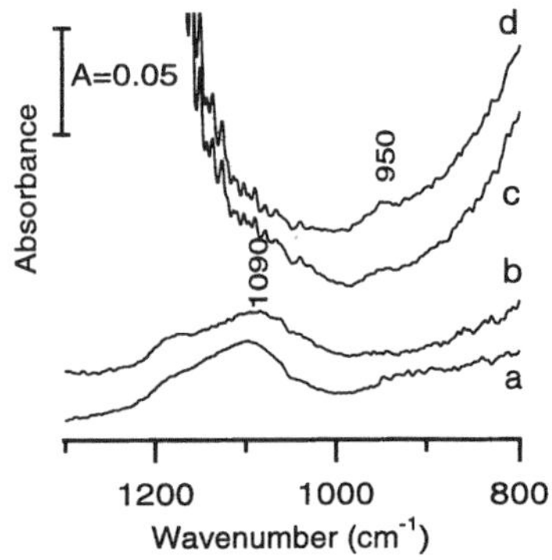

Figure 3. IR spectra of O_2 pre-adsorbed a) 15mol% BaF_2/LaOF and b) LaOF, and the reaction of b) with CH_4 at 750°C for c) 1 min and d) 25 min.

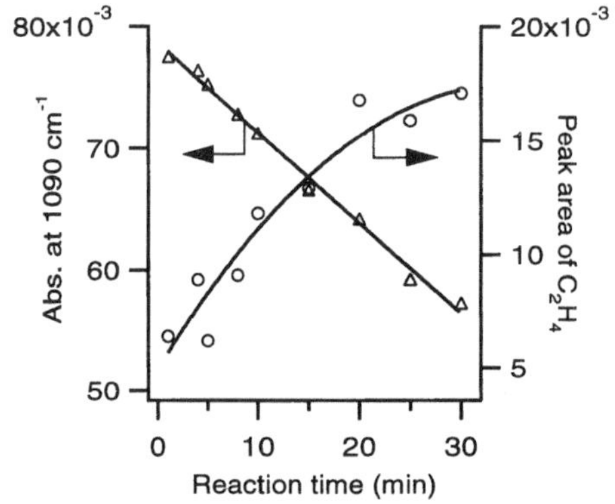

Figure 4. Plots of the IR absorbance of $\underline{O_2}^-$ at 1090 cm^{-1} and the peak area of C_2H_4 at 950 cm^{-1} vs. the reaction time for the reaction of O_2 pre-adsorbed LaOF with CH_4 at 750 °C.

REFERENCES

1. G. E. Keller and M. M. Bhasin, J. Catal., 73 (1982) 9.
2. Y. Amenomiya, V. I. Birss, M. Goledzinowski, J. Galuszka and A. R. Sanger, Catal. Rev. Sci. Eng., 32 (1990) 163.
3. A. M. Maitra, Appl. Catal., A 104 (1993) 11.
4 J. H. Lunsford, P. G. Hinson, M. P. Rosynek, C. Shi, M. Xu and X. Yang, J. Catal., 147 (1994) 301.
5. J. H. Lunsford, Angew. Chem. Int. Ed. Engl., 34 (1995) 970.
6. R. Q. Long, S. Q. Zhou, Y P. Huang, W. Z. Weng, H. L. Wan and K. R. Tsai, Appl. Catal., A: General, 133 (1995) 269.
7. Z. S. Chao, X. P. Zhou, H. L. Wan and K. R. Tsai, Appl. Catal., A:General, 130 (1995) 127.
8. K. A. Gschneidner, jr and L. Eyring (eds.), Handbook on the Physics and Chemistry of Rare Earth, Vol. 3, North-Holland, 1979, p. 385.
9. Z. Zhang, X. E. Verykios and M. Baerns, Catal. Rev. Sci. Eng., 36 (1994) 507.
10. X. P. Zhou, Z. S. Chao, J. Z. Luo, H. L. Wan and K. R. Tsai, Appl. Catal., A: General, 133 (1995) 263.
11. L. Andrew, J. Chem. Phys., 54 (1971) 4935.
12. L. Vaska, Acc. Chem. Res., 9 (1979) 175.
13 M. Che and A. J. Tench, Adv. Catal., Vol. 32, p.1.

NATURAL GAS CONVERSION V
Studies in Surface Science and Catalysis, Vol. 119
A. Parmaliana et al. (Editors)

Transition metal catalyzed acetic acid synthesis from methane and carbon monoxide

Yuzo Fujiwara[a], Tsugio Kitamura[a], Yuki Taniguchi[a], Taizo Hayashida[a], and Tetsuro Jintoku[b]
[a]Department of Chemistry and Biochemistry, Graduate School of Engineering, Kyushu University, Fukuoka 812-8581, Japan
[b]Idemitsu Petrochemical Co., Tokuyama 745-0843, Japan

The new catalytic systems other than the Pd/Cu system have been examined for the acetic acid synthesis from methane and CO. It is found that the $VO(acac)_2/K_2S_2O_8/CF_3COOH$ catalyst system converts methane and CO to acetic acid in 93% yield based on methane.

1. INTRODUCTION

Activation and functionalization of C-H bonds of saturated hydrocarbons have attracted a great attention recently in connection with resources for energy and chemicals production. Particularly methane, a major component of natural gas, is the most abundant organic molecule, so its selective and high yield conversion into useful liquid or solid materials is strongly desired in industry. However, it is difficult to achieve a functionalization of alkanes since they are unreactive to most conventional synthetic methods, and if the expected products are formed, they are reactive than alkanes in general, which causes selectivity problems.

In continuing studies on C-H bond activations [1], we have found that methane, ethane, and propane give rise to the corresponding acetic, propionic, and butyric acids, respectively when allowed to react with CO using the $Pd(OAc)_2/Cu(OAc)_2/K_2S_2O_8/$ CF_3COOH catalyst system [2]. The yields of propionic (27% yield on C_3H_8) and butyric (42%) acids are higher than that of acetic acid (ca. 1%). We have found that oxygen can be used as an oxidant in lieu of $K_2S_2O_8$ in the Pd/Cu system [2e].

The present work had two primary aims. Firstly, to extend the Pd/Cu based catalyst system to other transition metal catalyst systems and secondly, to improve the yield of acetic acid from methane and CO.

2. RESULTS AND DISCUSSION

In order to improve the acetic acid yield, we tested the effect of several transition metal catalysts including heteropolyacids on the reaction of methane and CO to form acetic acid, and found that vanadium compounds gave good results as shown in Table 1. Among heteropolyacids, $H_7PV_4Mo_8O_{40}$•$30H_2O$ and $H_5SiVW_{11}O_{40}$•$29H_2O$ gave high turnover numbers

Table 1
Acetic acid synthesis from methane and CO[a]

$$CH_4 + CO \xrightarrow[CF_3COOH]{catalyst} CH_3COOH$$

catalyst	TON[b]	
none	-	(0.00%)[b]
$H_3PMo_{12}O_{40}$•$30H_2O$	0.2	
$H_4PVMo_{11}O_{40}$•$30H_2O$	22.2	
$H_5PV_2Mo_{10}O_{40}$•$30H_2O$	24.5	
$H_6PV_3Mo_9O_{40}$•$30H_2O$	24.2	
$H_7PV_4Mo_8O_{40}$•$30H_2O$	29.2	
$H_8PV_5Mo_7O_{40}$ •$30H_2O$	24.2	
$H_3PW_6Mo_6O_{40}$•$30H_2O$	0.4	
$H_3PW_{12}O_{40}$•$30H_2O$	2.1	
$H_4SiW_{12}O_{40}$•$30H_2O$	0.1	
$H_5SiVW_{11}O_{40}$•$29H_2O$	31.1	
$H_4SiW_4Mo_8O_{40}$•$29H_2O$	0.8	
$H_4SiMo_{12}O_{40}$•$28H_2O$	0.6	
$VO(acac)_2$	27.5	
$VOSO_4$•$3H_2O$	39.9	(1.03%)[b]
V_2O_5	25.5	
$NaVO_3$	28.2	
Pr_6O_{11}	1.6	
Gd_2O_3	3.8	

a) CH_4 (40 atm), CO (20 atm), catalyst (0.05 mmol), $K_2S_2O_8$ (5.00 mmol), CF_3COOH (5 ml), 80°C, 20 h in a 120-ml autoclave.

b) Based on CH_4.

(TON), 29.2 and 31.1, respectively. Furthermore, simple vanadium compounds such as $VO(acac)_2$, $VOSO_4 \cdot 3H_2O$, and $NaVO_3$ were found to have high catalytic activity. We selected $VO(acac)_2$ as catalyst for the reaction of methane and CO to convert to acetic acid because of its stability, solubility and simplicity of the structure.

Firstly, the solvent effect on the reaction was studied using $VO(acac)_2$ as catalyst and $K_2S_2O_8$ as oxidant at 80℃ for 20 h. The results are shown in Table 2. The date in the table

Table 2

Effect of solvents on the reaction of methane and CO to acetic acid[a]

Solvent	TON
CF_3COOH	27.5
H_2O	5.9
2N HCl	2.2
2N CF_3COOH	3.5
2N H_2SO_4	2.4
2N NaOH	0.1

a) CH_4 (40 atm), CO (20 atm), $VO(acac)_2$ (0.05 mmol), $K_2S_2O_8$ (5.0 mmol), solvent (5.0 ml), 80°C, 20 h in a 120-ml autoclave.

show that CF_3COOH (TFA) is the best solvent for the reaction.

Secondly, the effect of oxidants was studied using a $VO(acac)_2$ catalyst and TFA as solvent (Table 3). From the data in the table, one can see that $K_2S_2O_8$ is the best oxidant for

Table 3

Effect of oxidants on the reaction of methane and CO to acetic acid[a]

Oxidant	TON
none	0.1
$K_2S_2O_8$	33.3
KIO_3	2.3
$Me_3N(O)$	0.4
H_2O_2 (35%)	2.5
tert-BuOOH(70%)	9.4
NaClO(10%)	6.9

a) CH_4 (40 atm), CO (20 atm), oxidant (5.0 mmol), $VO(acac)_2$ (0.05 mmol), TFA (5.0 ml), 80°C, 20 h in a 120-ml autoclave.

the $VO(acac)_2$ catalyst.

Then, the effect of the amount of $K_2S_2O_8$ was studied. Figure 1 shows the plot of the $K_2S_2O_8$ amount vs. TON of the catalyst for the formation of acetic acid from methane and CO. Finally, the effect of methane pressure was studied (Figure 2). Figure 2 shows the plot

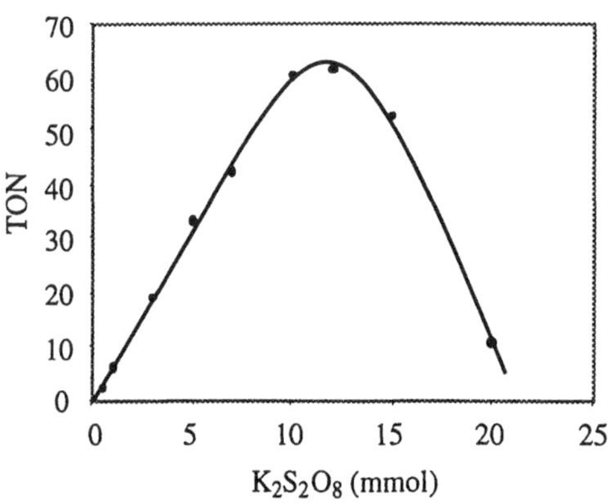

Figure 1. Effect of the amount of $K_2S_2O_8$.[a]

a) CH_4 (40 atm), CO (20 atm), $VO(acac)_2$ (0.05 mmol), TFA (5.0 mL), 80°C, 20 h in a 120-mL autoclave.

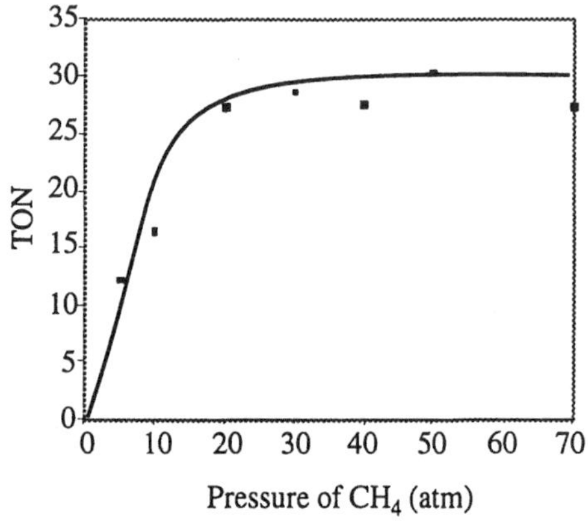

Figure 2. Effect of CH_4 pressure.[a]

a) CO (20 atm), $VO(acac)_2$ (0.05 mmol), $K_2S_2O_8$ (5.0 mmol), TFA (5.0 mL), 80°C, 20 h in a 120-mL autoclave.

of CH_4 pressure vs. TON of the catalyst of the reaction of methane with CO to form acetic acid. From the figure, one can see that TON increases as the increasing methane pressure until 20 atm of methane pressure.

From these results, a $VO(acac)_2/K_2S_2O_8/CF_3COOH$ catalyst system was found to be one of the best catalyst systems which converts methane and CO to acetic acid.

Using this catalyst system in a 25-ml autoclave, very high yield of acetic acid was achieved as shown in eq. 1. This is the highest yield of acetic acid from methane and CO [3]. However, $K_2S_2O_8$ should be substituted with cheaper O_2 for industrial processes.

$$\underset{\substack{5\ \text{atm} \\ (0.945\ \text{mmol})}}{CH_4} + \underset{\substack{20\ \text{atm} \\ (3.78\ \text{mmol})}}{CO} \xrightarrow[\substack{\text{TFA (20 ml)} \\ 80^{\circ}\text{C, 20 h}}]{\substack{VO(acac)_2\ (0.05\ \text{mmol}) \\ K_2S_2O_8\ (10.0\ \text{mmol})}} \underset{\substack{0.87\ \text{mmol} \\ \text{TON: 18} \\ 93\%\ \text{Yield on } CH_4}}{CH_3COOH} \qquad (1)$$

Although the mechanism of the reaction is not yet clear at this stage, the high oxidation state oxo-vanadium species V(V)=O would abstract H• from CH_4 to form a methyl radical (CH_3•) which reacts with CO to give an acetyl radical (CH_3CO•). The CH_3CO• radical would give acetic acid via oxidation by V(V)=O to CH_3CO^+ [1b].

The present reaction has advantage that it employs cheap methane and CO for the acetic acid synthesis, and has a potential to be industrialized in near future.

3. EXPERIMENTAL

The reaction was carried out in the following way; in a 25-ml stainless steel autoclave were placed $VO(acac)_2$ (0.05 mmol), $K_2S_2O_8$ (10 mmol), TFA (20 ml), and a Teflon-coated magnetic stirring bar. The autoclave was closed, flushed with three times with CO, and pressurized to 20 atm with CO (3.80 mmol), and then 5 atm with CH_4 (0.95 mmol). The mixture was heated at 80℃ with stirring for 20 h. After cooling and venting of residual gases, the autoclave was opened and the mixture was analyzed directly by GLC (Unisole 10T + H_3PO_4) without aqueous work-up using butyric acid as an internal standard to give acetic acid (0.87 mmol) in 93% yield based on CH_4.

ACKNOWLEDGEMENT

This work was supported in part by Grant-in Aid for Scientific Research on Priority Area No. 283 "Innovative Synthetic Reactions" and Scientific Research (A) No. 09355031 from Monbusho.

REFERENCES

1. a) Y. Fujiwara, T. Jintoku, and K. Takaki, *Chemtech*, 636 (1990). b) Y. Fujiwara, K. Takaki, and Y. Taniguchi, *Synlett*, 591 (1996).
2. a) T. Nishiguchi, K. Nakata, K. Takaki, and Y. Fujiwara, *Chem. Lett.*, 1141 (1992). b) K. Nakata, T. Miyata, T. Jintoku, A. Kitani, Y. Taniguchi, K. Takaki, and Y. Fujiwara, *Bull. Chem. Soc. Jpn.*, **66**, (1993) 3755. c) K. Nakata, Y. Yamaoka, T. Miyata, Y. Taniguchi, K. Takaki, and Y. Fujiwara, *J. Organomet. Chem.*, **473**, (1994) 329. d) K. Nakata, T. Miyata, Y. Taniguchi, K. Takaki, and Y. Fujiwara, *J. Organomet. Chem.*, **489,** (1995) 71. e) M. Kurioka, K. Nakata, T. Jintoku, Y. Taniguchi, K. Takaki, and Y. Fujiwara, *Chem. Lett.*, 224 (1995).
3. a) M. Lin and A. Sen, *Nature*, **368,** (1994) 613. b) A. Sen and M. Lin, *J. Chem. Soc., Chem. Commun.*, 508 (1992); 892 (1992).

NATURAL GAS CONVERSION V
Studies in Surface Science and Catalysis, Vol. 119
A. Parmaliana et al. (Editors)

Study of the reactions of ethylene on supported Mo_2C/ZSM-5 catalyst in relation to the aromatization of methane

F. Solymosi and A. Szőke

Institute of Solid State and Radiochemistry, University of Szeged and Reaction Kinetics Research Group of the Hungarian Academy of Sciences, P. O. Box 168, H-6701, Szeged, Hungary

The reactions of ethylene are investigated on the catalysts found to be active in the aromatization of methane at high temperature. In contrast to the ZSM-5, pure Mo_2C with low surface area exhibited a little activity for the cracking and transformation of ethylene into other hydrocarbons. Preparation of Mo_2C in highly dispersed state on ZSM-5 enhanced the conversion of ethylene, and only slightly decreased the selectivity to aromatics measured for pure ZSM-5 under similar experimental conditions. A more significant effect of Mo_2C was observed when it was deposited on silica rather inert for the activation and reactions of ethylene.

1. INTRODUCTION

Aromatization of saturated and unsaturated hydrocarbons has been the subject of extensive research in heterogeneous catalysis. This reaction proceeds rather easily from C_2 compounds, but much more restrictedly from CH_4. Recently, however, it was found that MoO_3 deposited on ZSM-5 is an effective catalyst for the conversion of CH_4 into benzene with high selectivity (70-85%) at about 10% of conversion [1-7]. More detailed studies showed that not the MoO_3, but Mo_2C, formed in the reaction between CH_4 and MoO_3, is the key compound which activates the CH_4 molecules to produce hydrocarbon fragments, CH_3, CH_2, which dimerize into ethane and/or ethylene [4-7].These compounds may undergo aromatization on the acidic sites of ZSM-5. The conversion of ethylene into aromatics on a pure H-ZSM-5 is an important and well studied reaction [8-11].

In a previous work, we examined the aromatization of ethane on Mo_2C/ZSM-5 catalyst [12]. In the present paper we report the influence of Mo_2C on the reactions of ethylene on ZSM-5 and on SiO_2.

2. EXPERIMENTAL

Catalytic reactions were carried out at 1 atm of C_2H_4 + Ar gas mixture containing 10% of ethylene in a fixed-bed, continuous flow reactor consisting of a quartz tube (17 mm i.d.) connected to a capillary tube [5].The flow rate was 12 ml/min. Generally 0.5 g of loosely compressed catalyst sample was used. Reaction products were analyzed gas chromatographically using a Hewlett-Packard 5890 gas chromatograph and a Porapak QS column. In some cases the initial interaction between the catalysts and the reacting gases was followed by mass spectrometry using a Balzers quadrupole mass spectrometer (type OMS 200).The ethylene and methane conversion was calculated from the hydrogen balance. The selectivity values of product formation represent the fraction of ethylene or methane that has been converted into specific products taking into account the number of carbon atoms in the molecules. A pulse reactor was also employed (8 mm o.d. quartz tube), which was incorporated between the sample inlet and the column of the gas chromatograph.

XPS measurements were performed in a Kratos XSAM 800 instrument at a base pressure of 10^{-9} Torr using Mg K_α primary radiation (14 kV, 15 mA). To compensate for possible charging effect, binding energies (BE) were normalized with respect to the position of the Si(2p) in SiO_2 (BE = 103.4 eV) for supported samples and to Fermi-level for the Mo_2C. The pass energy was set at 40 eV; and an energy step width of 50 meV and dwell time of 300 ms were used. Typically 10 scans were accumulated for each spectrum. Fitting and deconvolution of the spectra were made using the VISION software (Kratos). Infrared spectroscopic measurements were made in a vacuum IR cell using self-supporting wafers of catalyst powders. Spectra were recorded with a Biorad (Digilab.Div.) Fourier transform IR spectrometer FTS 155).

The gases used were of commercial purity (Linde). Ar (99.996 %) and H_2 (99.999 %) were deoxygenated with an oxytrap. The other impurities were adsorbed by a 5A molecular sieve at the temperature of liquid nitrogen. The H-ZSM-5 support was obtained by five times repeated ion exchange of Na-ZSM-5 ((Si/Al = 55.0) with an aqueous solution of ammonium nitrate (1N), and calcined in air at 863 K for 5h. Before catalytic measurements each sample was oxidized in an O_2 stream at 973 K in situ and then flushed with Ar for 15 min. Hexagonal Mo_2C was prepared by the method of Lee et al. [13]. Briefly, about 0.5 g of MoO_3 was heated in 1:4 methane-H_2 mixture flowing at 300 ml (STP)/min in a quartz cell with two stopcocks. Preparation temperature was increased rapidly to 773 K and at 30 K/h between 773 and 1023 K, and maintained at 1023 K for 3 h. Following the suggestion of Lee et al [13], the sample was deactivated at 300 K with air, or used in situ for catalytic studies. The BET surface area of this sample is 4.6 m^2/g. A detailed description of the properties of Mo_2C samples can be found in the reviews of Oyama and Haller [14,15].

Supported Mo_2C was produced by the carburation of supported MoO_3 in the catalytic reactor, in a similar way as described above for the preparation of bulk Mo_2C. MoO_3/ZSM-5 was prepared by impregnating the supports with a basic solution of ammonium paramolybdate to yield a nominal 2wt% of MoO_3. The suspension was dried at 373 K and calcined at 873 K for 5 h. As Mo_2C always contains excess carbon, the catalyst was treated with H_2 at 873 K before the catalytic measurements to remove this carbon species. Some characteristic data for the catalysts are collected in Table 1.

3. RESULTS AND DISCUSSION

3.1. XPS characterization of Mo_2C catalysts

The binding energy of Mo($3d_{5/2}$) for pure Mo_2C was found to be 227.7 eV, which agrees well with the values reported before [4,5]. The C(1s) peak was rather broad indicating the coexistence of several forms of carbon. For the main peak we obtained a value of 283.8 eV.

Practically the same binding energies were measured when Mo_2C was prepared on ZSM-5 or on SiO_2.

Table 1
Some chracteristic data for the catalysts

Sample	Surface area, m^2/g	Micropore area m^2/g	Micropore volume ml/g
Mo_2C	4.6	0.32	0.0001
ZSM-5	311.1	147.2	0.08
Mo_2C/ZSM-5	276.0	117.4	0.06

3.2. Reactions on ZSM-5 and on SiO_2

The reaction of ethylene has been first investigated on pure ZSM-5. Flowing ethylene on ZSM-5 the formation of CH_4, C_2H_6, C_3H_8, C_4H_{10}, C_5H_{12}, and unsaturated C_3-C_4 has been observed at rather low temperature, around 573 K. Benzene formed only in trace amounts.

At 773 K, the conversion of ethene was about 80% which did not alter in the time of stream. The rates of the formation of various products were also constant. Based on the selectivity, the main product was propane followed by toluene, butane, benzene, ethane and methane. Raising the temperature to 873 K, the rate and selectivity of benzene formation increased. At 973 K, where the aromatization of methane occurred readily, the main products were benzene and toluene, the selectivity of their formation attained a value of 60% and 20%, resp.. In Figure 1 we plotted the conversion of ethylene and the selectivities of the formation of various products. In the section A, only the C-containing gaseous products were taken into account, whereas in the section B, the amount of hydrogen was also considered. For ZSM-5 catalyst, the selectivity values are almost the same in both cases indicating that only a limited amount of hydrogen is produced.

Some measurements have been performed on pure silica. A very little reaction (~0.05 % of conversion) of ethylene was observed 773 K. At 973 K, however, the formation of all products observed for ZSM-5 occurred. The highest selectivity, ~30%, was measured for butane, followed by ethane (~23%) and benzene (~17%) (Fig. 1).

The results obtained correspond well to those reported before showing that ZSM-5 an effective catalyst for the ethylene conversion. At lower temperature, ethylene undergoes rapid isomerization and oligomerization. At higher temperature the formation of aromatics comes into prominence. On silica with no Bronsted acidity the aromatization of ethylene is much more restricted.

3.3. Reactions on Mo_2C

On pure Mo_2C, an ethylene conversion of 0.3% was attained at 773 K, which increased to 5% at 973 K. The conversion was rather constant up to 120 min. All the products detected on pure ZSM-5 appeared in this case, too. The highest selectivity value (30% each) was obtained for ethane and butane. The selectivity to benzene was only 10%. Calculation based on all gaseous products gave somewhat lower values (Fig. 1). Accordingly, pure Mo_2C exhibits a relatively little activity towards the cracking and transformation of ethylene into other compounds, which is mainly due to its low surface area.

3.4. Reactions on Mo_2C/ZSM-5

Deposition of Mo_2C on ZSM-5 appreciably affected the reaction pathways of ethylene observed on previous samples. At 773 K, it somewhat decreased the conversion obtained for pure ZSM-5. Toluene formed with the highest selectivity (~ 33%), followed by propane (~ 25%) and benzene (~ 15%). A complete conversion of ethylene into other products was experienced at 973 K. The selectivity to benzene production was practically the same as for Mo_2C-free ZSM-5. Toluene and propane formed with lower selectivities, while the methane production was significantly higher. Some data are displayed in Fig. 1. On the basis of these features we can conclude that Mo_2C deposited in highly dispersed state on ZSM-5 influences only slightly the aromatization of ethylene on the acidic sites of ZSM-5. The decrease in the selectivity of aromatics may be attributed to the reduction of the number of acidic sites, or to blocking some of the active area of the ZSM-5. Bearing these features in mind, we are safe to conclude that the primary role of Mo_2C in the aromatization of methane is its mild activation to produce CH_3 and CH_2 radicals which are coupled into ethylene.

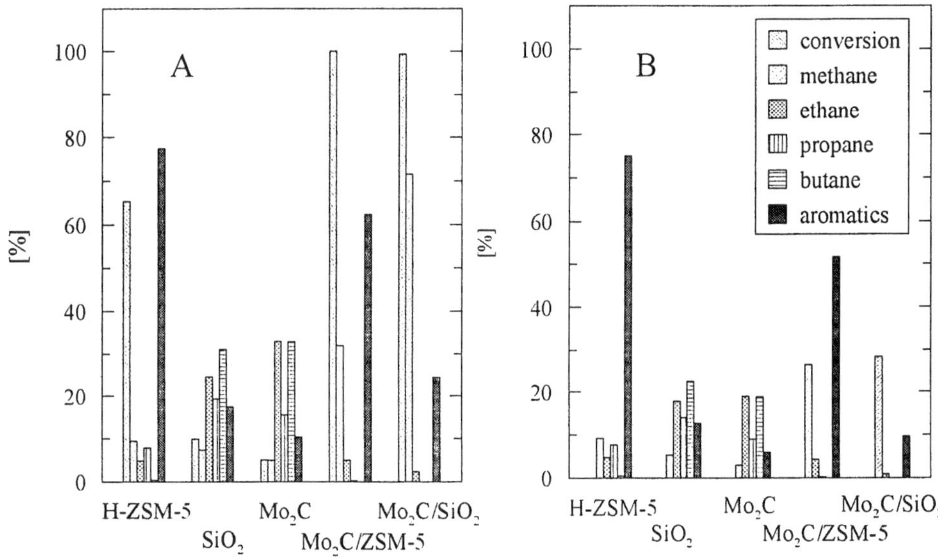

Figure 1. Conversion of ethylene and selectivities of the formation of various products on different catalysts at 973 K. (A) without H_2, (B) with H_2.

3.5. Reactions on Mo_2C/SiO_2

Deposition of Mo_2C on silica exerted a dramatic influence on the reaction of ethylene. Whereas on pure silica we observed no reaction at 773 K, on Mo_2C/ZSM-5 we detected all the products reported before. After an initial fast decay, the conversion of ethylene was about 10%. The main gaseous products were H_2, C_2H_6 ($S \approx 40\%$), propane ($S \approx 23\%$) and benzene ($S \approx 10\%$). Raising the temperature to 973 K caused the complete decomposition of ethylene. Besides H_2, CH_4 ($S \approx 70\%$) and C_6H_6 ($S \approx 20\%$) were identified. Changes in the conversion and selectivities in time of stream are presented in Fig. 2 A and B.The results suggest that

Mo_2C in highly dispersed state can activate the ethylene molecule on the rather inactive silica support. The enhanced formation of benzene over Mo_2C/SiO_2 may be the result of the creation of acidic sites on the silica surface.

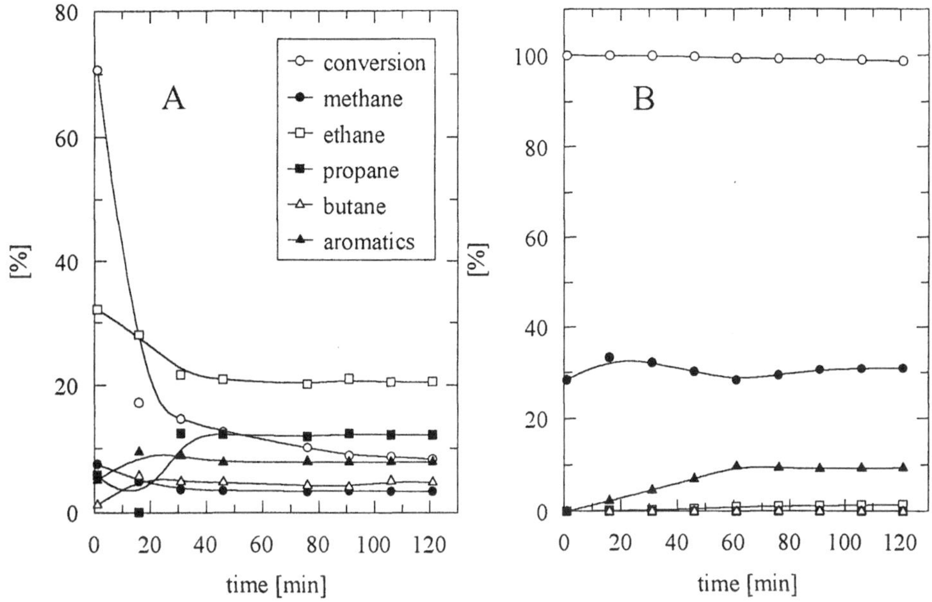

Figure 2. Conversion of ethylene and the selectivities of the formation of various products on Mo_2C/SiO_2 at (A) 773 K and (B) 973 K.

3.6. Hydrogenation of surface carbon formed

The evolution of hydrogen indicated that a fraction of ethylene has been decomposed to carbon. The reactivity of this carbon has been investigated toward hydrogen. On ZSM-5 the hydrogenation of this carbon, after the reaction at 773 K, produced methane with Tp = 730 and 1080 K. Ethane in traces has been also detected above 730 K. On Mo_2C/ZSM-5, following the cracking of ethylene at 773 K, we obtained only one CH_4 peak at 900 K. Evolution of ethane occurred in the same way as for ZSM-5. After the reaction of ethylene at 973 K, the peak temperature for CH_4 formation moved to 1025 K.

3.7. Reactions of CH_2 and CH_3 species

As the primary step in the aromatization of methane is its activation and the production of CH_3, and then CH_2 species, we performed detailed measurements concerning the chemistry of these species on the components of the catalysts. The methods used were FTIR spectroscopy combined with mass spectrometry and pulse technique. The hydrocarbon fragments have been generated by the thermal decomposition of the corresponding halogenated compounds. The main results are as follows: CH_3 and CH_2 are coupled rapidly into higher hydrocarbons on ZSM-5 above 573 K Methane formation was minimal. On pure Mo_2C, their decomposition to surface carbon and self-hydrogenation into methane are the main processes. When Mo_2C is deposited on ZSM-5, the dominant reaction pathways observed for pure ZSM-5 were only slightly affected by Mo_2C. These results suggest that during the aromatization of methane the hydrocarbon fragments formed in the primary steps of methane activation may migrate from Mo_2C onto ZSM-5, where they are recombined without their significant decomposition.

4. REFERENCES

1. L. Wang, L. Tao, M. Xie, and G. Xu, Catal. Lett. 21 (1993) 35.
2. Y. Xu, S. Liu, L.Wang, M. Xie and X. Guo, Catal. Lett. 30 (1995) 135.
3. F. Solymosi, A. Erdõhelyi and A. Szõke, Catal. Letts 32 (1995) 43.
4. F. Solymosi and A. Szõke, Catal. Letts. 39 (1996) 157.
5. F. Solymosi, J. Cserényi, A. Szõke, T. Bánsági and A. Oszkó, J. Catal. 165 (1997) 150.
6. D. Wang, J.H. Lunsford, M.P. Rosynek, Topycs in Catal. 3 (4) (1996) 299.
7. D. Wang, J.H. Lunsford and M.P. Rosynek, J. Catal. 169 (1997) 347.
8. N.Y. Chen and T.Y. Yan, Ind. Eng. Chem. Process Des. Dev. 25 (1986) 151.
9. N.Y. Chen and W.O. Haag, in: Hydrogen Effects in Catalysis: Fundamentals and Practical Applications, Z. Paal and P.G. Menon, eds. (Dekker, New York, 1988) p. 695.
10. M. Guissent, N.S. Gnep and F. Alario, Appl. Catal. A 89 (1992) 1; M. Guissent, N.S. Gnep,D. Aittaleb and Y.T. Doyemet, Appl. Catal. A 87 (1992)255.
11. L.A. Dufresne and R. Le Van Mao, Catal. Lett. 25 (1994) 371.
12. F. Solymosi and A. Szõke, Appl. Catal. A. 166 (1998) 225.
13. J.S. Lee, S.T. Oyama and M. Boudart, J. Catal. 106 (1987) 125.
14. S.T. Oyama and G.L. Faller, "Catalysis, Specialist Periodical Reports". G. C. Bond and G. Webb, eds. Vol. 5 p. 333, The Chemical Society. London, 1981.
15. S.T. Oyama, Catal. Today 15 (1992) 179.

NATURAL GAS CONVERSION V
Studies in Surface Science and Catalysis, Vol. 119
A. Parmaliana et al. (Editors)

Experimental Investigations on the Interaction between Plasmas and Catalyst for Plasma Catalytic Methane Conversion (PCMC) over Zeolites

Chang-jun Liu, Lance L. Lobban and Richard G. Mallinson

Institute for Gas Utilization Technologies and School of Chemical Engineering and Materials Science, The University of Oklahoma, 100 E. Boyd Street, Room T335, Norman, OK 73019, USA

ABSTRACT

In this study, temperature programmed CO_2 desorption, a two-step plasma catalytic methane conversion (PCMC), and temperature programmed oxidation of carbonaceous species were carried out to investigate the interactions between the plasma and heterogeneous catalysts. The experiments demonstrated that NaOH treated Y, NaY and NaX zeolites, which have significant basicity, stabilize sustained streamer corona discharges at low temperatures leading to better and longer lived PCMC. The experiments also confirmed that the basicity, polarity and reactivity of zeolites may be increased by the plasma leading to improved behavior for PCMC. TPO of carbonaceous deposits show that NaX, NaY, NaA and 5A zeolites generate less coke and it is oxidized below about 650 K, while NaZSM-5 generates significantly more coke which oxidizes at higher temperature. This more refractory coke appears to destabilize the plasma leading to arcing and poorer PCMC performance. A two-step PCMC in which methane is adsorbed without plasma and then the plasma is generated with co-reactants but without methane in the gas phase showed that active plasma species interacting with the catalyst surface are necessary for the selective production of higher hydrocarbons.

1. INTRODUCTION

Plasma chemical processing is a promising route for the synthesis of chemicals, for the cleanup of waste streams, and for the modification of material surfaces. Many kinds of reactive particles, electrons, free radicals, ions, metastable species and photons, are produced in a plasma chemical processing system. Among all of the plasma generated species, free radicals are believed to be most important for chemical reactions. Control and manipulation of free radical reactions are therefore essential for the successful application of plasmas for organic synthesis [1]. However, the electrons within plasmas serve principally to excite and decompose the gas molecules at a high rate but in a non-selective fashion. Multi-pathway chemical reaction mechanisms may therefore not be avoidable and lead to production of complex product distributions.

One way to surmount this difficulty is to introduce a heterogeneous catalyst into the plasma [2-16]. Reactions combining a low temperature, non-equilibrium plasma and catalyst can be carried out at very low gas temperatures which inhibit the formation of by-products which occur at high temperatures. Such a combination has led to products that are very difficult or very expensive to achieve by regular thermal-catalytic technologies [4].

PCMC with zeolites achieving yields up to 32 percent over zeolites has recently been reported [17]. The structure of the zeolite was shown to have a significant impact on the activity and longevity of PCMC, but the reasons for these variations were not evident. In this paper, results of an experimental investigation of the interactions between plasma and catalyst for PCMC to higher hydrocarbons over zeolites is reported.

2. EXPERIMENTAL

2.1. Reactor System

The activity for plasma catalytic methane conversion over all of these zeolites has been previously reported and a detailed description of the plasma catalytic micro-reactor system can be found elsewhere [5,8,17]. The reactor is a quartz tube with an inside diameter of 7.0 mm. A lower plate electrode is mounted perpendicular to the tube axis with holes in it to allow gas flow. Catalysts are supported on this electrode. The upper electrode is a wire suspended above the catalyst bed and the distance between the electrodes is 10 mm. The discharge is created by a Bertan Associates Model 210-50R high voltage dc power supply with the lower electrode as the ground. An oven surrounds the reactor tube to provide controlled temperatures when needed. Gas flows are established by Porter Instruments mass flow controllers and calibrated with a bubble meter.

2.2. Catalysts

The zeolite catalysts used in this work include NaX, Na-ZSM-5, NaY, NaA, Linde type 5A zeolite and NaOH treated NH_4Y zeolite. The NaY, NaA, NaX and NH_4Y were obtained from Aldrich and the first three zeolites were used as received. The Na-ZSM-5 was obtained from Chemie Uetikon, and Linde type 5A from Matheson Coleman and Bell. Both of these two zeolites were also used as received. The NaOH treatment of NH_4Y zeolite has been discussed elsewhere [8,17].

2.3. CO_2 Temperature Programmed Desorption (TPD)

Temperature Programmed Desorption (TPD) was performed with a tubular temperature-programmable plasma catalytic reactor. The reactor system is the same as the plasma catalytic micro-reactor system mentioned above. Helium at 200 ml/min was supplied to the reactor at 200°C for two hours to desorb any contaminants from the surface of the catalyst before TPD characterization. The catalyst was then cooled to 50°C or heated up to 500°C for CO_2 chemisorption in 8% CO_2/helium at a total flowrate of 25 ml/min. A gas chromatograph (HP5890) with a thermal conductivity detector (TCD) was used to detect the CO_2 in the effluent. The CO_2 chemisorption was not terminated until a constant CO_2 level was observed in the effluent. The system was then cooled down quickly to room temperature by flowing compressed air between the furnace and the outer quartz reactor. Helium at 200 ml/min was supplied to the reactor again to purge out any gaseous CO_2 residue before the TPD measurement was started. During the TPD characterizations, the heating rate was 7.5 °C/min and helium flow rate was 20 ml/min for desorption. The maximum temperature was typically 750°C. The GC was also used to measure the desorbing CO_2.

2.4. Temperature Programmed Oxidation (TPO)

PCMC can cause coke formation on the catalyst surface. Information about the carbonaceous species which have formed during PCMC is helpful towards understanding the mechanism of the plasma-heterogeneous reactions. It has been found experimentally that some carbonaceous species have little effect on plasma reactions, while some have a significant negative effect. TPO can give information about the nature of the carbonaceous species on the catalysts. The setup for TPO is the same as for TPD described above. The procedure for TPO is:

a. 200 ml/min helium is used to purge the zeolite for two hours at 200°C;
b. The system is cooled down to the desired temperature for plasma catalytic reactions;
c. Plasma reactions are conducted for two hours or more at 50°C;
d. The plasma is stopped and the reactor is quickly cooled down to room temperature;
e. 200 ml/min helium is used to purge residual gases from the system;

f. The TPO experiment is begun at room temperature with 33.3% oxygen in helium with a total flow of 20 ml/min. The temperature increase rate is also 7.5°C per minute.

2.5. Two-step PCMC for the Investigation of the plasma surface reactions

To investigate the plasma surface reactions of chemisorbed methane, a two-step plasma catalytic methane conversion was conducted. In the first step, methane chemisorption without plasma was performed at 250°C with 5% CH_4/helium with a total flow of 100 ml/min over NaY zeolite. The amount of chemisorbed methane on the surface was determined by conducting step one and then oxidizing the carbon fragments at 523 K for 10 minutes [18] in a separate experiment. To accomplish this, helium (200 ml/min) was flowed first to purge any non-chemisorbed methane. During the oxidation, the carbon dioxide which was formed was adsorbed in a molecular sieve column. The amount of carbon dioxide collected on the molecular sieve was then determined by running a TPD measurement, as described above. In the second step, a co-reactant gas, hydrogen, oxygen or carbon dioxide, but no methane, was passed through the catalyst bed together with helium at a total flow of 20 ml/min. The corona discharge was generated to initiate plasma catalytic reactions. The products evolved from the plasma catalytic reactions were detected by the HP5890 GC with TCD detector.

3. RESULTS AND DISCUSSION

Previous results for PCMC have demonstrated different plasma catalytic activities for methane conversion to higher hydrocarbons with different zeolites [17]. The highest methane conversion has been achieved over NaOH treated Y zeolite, while the best selectivity for C_2 hydrocarbon products was obtained with NaX zeolite. A difference between these two zeolites is their base properties [19,20,21]. Both hydroxyl groups and oxygen ions of $(AlO_4)^-$ tetrahedra contribute to the total basicities of zeolites, whereas the latter exhibit Lewis basicity alone [19]. In this work, CO_2 TPD was used to characterize the base properties and plasma effects on these properties because of the difficulty in using conventional characterization techniques in the presence of a plasma. CO_2 chemisorption has been used for the characterization of the acid-base nature and coverage of the support materials by the active phase of a catalyst [22] and for the neutralization of basic sites in zeolites [23]. The CO_2 TPD conducted here show that the base strength of the zeolites decreases from NaX, NaA, Linde type 5A, NaOH treated Y, NaY to Na-ZSM-5, while the density of basic sites (under the influence of corona discharges) decreases from NaOH treated Y, NaX, Na-ZSM-5, NaY, NaA to Linde Type 5A, as shown in Table 1 (the amount of CO_2 adsorbed is referred to as the density of base sites). There is a significant plasma enhancement of zeolite basicities with NaOH treated Y, NaY and NaX. Both basicity and density of base sites were enhanced. However, a significant reduction in the base site density was observed with NaA and Linde type 5A. This could be the reason that somewhat lower PCMC activities were obtained with NaA and Linde type 5A. Na-ZSM-5 possesses neither acidic nor basic hydroxyls. No significant change is observed in the base sites of Na-ZSM-5. These results suggest that a significant plasma effect occurs on zeolites containing acidic or basic hydroxyls. This also supports a mechanism in which the role of OH groups is important for PCMC [8]. The order of base site density with zeolites under the influence of plasmas are almost in the same order of plasma catalytic activity from high to low, as reported previously [17]. Previous results [8] have suggested that the interaction between plasmas and catalysts induces a significant change both in the plasma and in the catalyst surface reactions. Secondary electron emissions are thought to be responsible for part of the unusual PCMC performance of zeolites. Two paths could lead to significant secondary electron emissions. Itoh *et al.* [24] have correlated the overall acid-base property of the zeolite with the arithmetic mean of the electrostatic potentials of the cations. Ward [20,21,25] also analyzed the variation of the hydroxyl content and Bronsted acidity of the various alkaline earth forms and considered [25] that the small cations with their associated high

Table 1. CO_2 adsorbed on zeolites (applied voltage: 6 kV; applied power: 7 w for plasma CO_2 chemisorption; the reactive conditions for methane conversion see ref.[17])

zeolite	CO_2 chemisorption condition	gas temperature for CO_2 chemisorption	amount of CO_2 adsorbed (μmol/g)	plasma effect on CO_2 adsorbed amount	CH_4 [18] conversion at 50°C(%)	C_2 [18] selectivity (%)
NaOH treated Y	no discharge	50°C	21.8		0.0	0.0
NaOH treated Y	with gas discharge	50°C	25.3	+16.1%	50.0	25.0
NaOH treated Y	no discharge	500°C	18.2		0.0	0.0
NaOH treated Y	with gas discharge	500°C	22.5	+23.7%	50.0	25.0
NaX	no discharge	50°C	18.4		0.0	0.0
NaX	with gas discharge	50°C	19.4	+5.4%	48.1	33.7
NaZSM-5	no discharge	50°C	17.2		0.0	0.0
NaZSM-5	with gas discharge	50°C	17.0	-1.1%	44.6	31.2
NaY	no discharge	50°C	12.4		0.0	0.0
NaY	with gas discharge	50°C	16.7	+34.3%	45.4	30.1
NaA	no discharge	50°C	16.6		0.0	0.0
NaA	with gas discharge	50°C	15.9	-4.2%	41.2	24.8
Linde type 5A	no discharge	50°C	11.2		0.0	0.0
Linde type 5A	with gas discharge	50°C	10.8	-3.6%	39.9	27.6

electrostatic field and polarizing power would result in the equilibrium moving towards dissociation, while larger cations would be expected to produce less dissociation. When the electrostatic potential is sufficiently high, an intense electric field will be formed within the zeolitic framework and electron emissions are induced.

The electron emissions from the zeolite appear to significantly enhance the PCMC reactions in the plasma phase, while the zeolite also provides a large surface area for heterogeneous reactions, either by the generation of radical cations, or by reactions occurring on polarized or non-polarized acidic or basic sites.

Different carbonaceous species have also been found to have been formed on different zeolites during PCMC. The carbonaceous deposits formed during PCMC over NaZSM-5 have a negative effect on PCMC and at the same time the stable streamer discharges shift to arc-like discharges. The active plasma region is thereby reduced significantly and PCMC activity is observed to decrease, as has been discussed previously [8,17]. The carbonaceous species formed on NaX, NaY, NaA or Linde type 5A are much less in amount and do not negatively affect the PCMC activity, compared to NaZSM-5. The carbonaceous deposits on NaZSM-5 may act as electron scavengers and decrease its capability for electron emissions. TPO characterization has been conducted and the TPO spectra show more carbonaceous deposits formed on NaZSM-5 compared to the other two zeolites, NaX and NaOH treated Y zeolite. NaZSM-5 shows a specific TPO peak between 600K and 840K with a small low-temperature shoulder. The TPO spectrum for NaX does not contain this peak, while the spectrum for NaOH treated Y zeolite shows a small peak falling into this temperature scope. This large peak may be responsible for the shift of streamer discharges to the arc-like discharges and for the reduction in PCMC activity. Both NaX and NaOH treated Y zeolite show a wide smaller peak between 400K and 750K. These carbonaceous species evidently do not cause reduction in the PCMC activity during 4-hour PCMC experiments.

It was mentioned above that PCMC includes heterogeneous and homogeneous (plasma phase) reactions. The TPO experiments show that different carbonaceous species exist on the catalyst surface. These carbonaceous species may be involved in plasma reactions. Also, the plasma regeneration of used zeolites from PCMC, using 5% O_2 in helium with a total flowrate of 50 ml/min at 50 – 200°C, produced not only carbon dioxide, but also hydrogen, carbon monoxide, acetylene, ethylene and ethane. In order to better understand the reactions which occur on the catalyst surface under the influence of plasmas, a two-step PCMC over NaY was conducted to investigate plasma heterogeneous reactions of chemisorbed methane. The amount of chemisorbed methane over NaY zeolite during the first step was measured using the procedure described above and is 33.23 μmol/g. In the second step, a corona discharge was generated over the NaY zeolite with chemisorbed methane to start the plasma promoted catalytic reactions. Only methane was detected (without any detectable C_2 hydrocarbons) using a pure helium plasma, but essentially 100 percent methane conversion was achieved when other co-reactants were used. Table 2 summarizes the reaction conditions and selectivities of products of the two-step PCMC with different co-reactants. The selectivities of products in Table 2 were evaluated directly from the GC peak areas. It is shown that the hydrogen containing plasma induces the most selective production of C_2 hydrocarbons. CO_2 and O_2 containing plasmas also produce some C_2 products. This result suggests that active plasma species, like O^-, H, $O(^1D)$ and so on, play an important role in the formation of higher hydrocarbons during the plasma heterogeneous methane conversion.

Table 2. Measured Selectivities for Two-step PCMC over NaY

feed for the 2nd step	feed ratio	feed rate (ml/min)	gas temp. for 2nd plasma reaction step (°C)	selectivities (%)			
				C_2H_6	C_2H_4	CH_4	CO
CO_2/He	1/39.2	20	250	5.1	2.2	84.0	8.7
O_2/He	1/49.0	20	150°	3.1	0.0	79.3	17.6
He	–	20	150	0.0	0.0	100.0	0.0
H_2	–	20	150	10.0	4.1	85.9	0.0

4. CONCLUSIONS

The presence of certain zeolites with significant basicity, such as NaX and Y zeolites tested here, can stabilize sustained streamer corona discharges at low temperatures for methane conversion. The PCMC reactions form carbonaceous deposits on zeolites. The formation of such carbonaceous species depends upon the zeolite used. The carbonaceous deposits on NaZSM-5 have a negative influence on PCMC, while the "low temperature" coke formed on NaX and NaOH treated Y zeolite do not have a negative effect during a four hour reaction period. The basicity, polarity and thereby the reactivity of zeolites are modified by the plasma. This modification can be explained as a polarization effect on the electrostatic potential of zeolites and from the irradiation of the zeolite during plasma reactions. The OH groups in the zeolites significantly promote modification. The active species in the plasma phase are very helpful for the formation of higher hydrocarbons during plasma heterogeneous chemical processing, while a hydrogen containing plasma leads to the most selective production of higher hydrocarbons.

ACKNOWLEDGEMENT

Support from the US Department of Energy (under Contract No. DE-FG21-94MC31170) is greatly appreciated. The assistance from graduate students, Bobby Joe Hill, Chris Gordon, Phil Howard, Terence Caldwell and David Larkin, is also appreciated.

REFERENCES

1. J. Huang, M.V. Badani, S.L. Suib, J.B. Harrison and M. Kablauoi, J. Phys. Chem., 98(1994)206.
2. M.B. Kizling and S.G. J„r†s, Appl. Catal. A, 147(1996)1.
3. B. Eliasson, U. Kogelschatz, B. Xue and L.M. Zhou, Application of dielectric-barrier discharges to the decomposition and utilization of greenhouse gases, 13th International Symposium on Plasma Chemistry (ISPC-13), Beijing, August ,1997.
4. M. Venugopalan and S. Veprek S., in F.L. Boschke (Editor), Topics in Current Chemistry, Springer-Verlag, New York, 1993, p.3.
5. A. Marafee, C.-J. Liu, G.-H. Xu, R. Mallinson and L. Lobban, Ind Eng. Chem. Res., 36(1997)632.
6. D.E. Rapakoulias, S. Cavadias and D. Mataras, High Temp. Chem. Processes, 2(1993)231.
7. J.A. Cairns, J.P. Coad, E.W.T. Richards and L.A. Stenhouse, Nature, 288(1980)686.
8. C.-J. Liu, A. Marafee A., B.J. Hill, R. Mallinson and L. Lobban, Appl. Catal. A, 164(1997)21.
9. S.L. Suib and R.P. Zerger, J. of Catal., 139(1993)383.
10. V.F. Kiselev and O.V. Krylov, Adsorption processes on semiconductor and dielectric surfaces I, Springer-Verlag, New York, page 214(1985).
11. K. Jogan, A. Mizuno, T. Yamamoto and J.-S. Chang, IEEE Trans. on Ind. Appl., 29(1993)876.
12. T. Yamamoto, K. Mizuno, I. Tamori, A. Ogata, M. Nifuku, M. Michalska and G. Prieto, IEEE Trans. on Ind. Appl., 32(1996)100.
13. T. Yamamoto, K. Ramanathan, P.A. Lawless, D.S. Ensor, J.R. Newsome, N. Plaks and G.H. Ramsey, IEEE Trans. on Ind. Appl., 28(1992)528.
14. R.-H. Zhang, T. Yamamoto and D.S. Bundy, IEEE Trans. on Ind. Appl., 32(1996)113.
15. B. Eliasson and U. Kogelschatz, IEEE Trans.on Plasma Sci., 19(1991)1063.
16. J.G. Birmingham and R.R. Moore, US Patent 4,954,320.
17. C.-J. Liu, R. Mallinson and L. Lobban, Under revision for J. Catalysis, June 1998.
18. L. Guczi, K.V. Sarma and L. Bork¢, Catal. Lett., 39(1996)43.
19. W. Przystajko, R. Fiedorow and I.G. Dalla Lana, Zeolites, 7(1987)477.
20. J.W. Ward, J. of Catal., 22(1971)237.
21. J.W. Ward, J. of Catal., 14(1969)365.
22. F.M. Mulcahy, K.D. Kozminski, J.M. Slike, F. Ciccone, S.J. Scierka, M.A. Eberhardt, M. Houalla and D.M. Hercules, J. of Catal., 139(1993)689.
23. C. Mirodatos, P. Pichat and D. Barthomeuf, J. of Phys. Chem., 80(1996)1335.
24. H. Itoh, T. Hattori, K. Suzuki and Y. Murakami, J. of Catal., 79(1983)21.
25. J.W. Ward, J.of Catal., 10(1968)34.
26. C. Benndorf, P. Joeris and R. Kr"ger, Pure & Appl. Chem., 66(1994)1195.
27. A. Grill, Cold Plasma in Materials Fabrication: from Fundamentals to Applications, IEEE Press, New York, 1994.
28. D.W. Werst, E.E. Tartakovsky, E.A. Piocos and A.D. Trifunac, J. Phys. Chem., 98(1994)10249.
29. D.W. Werst, E.A. Piocos, E.E. Tartakovsky and A.D. Trifunac, Chem. Phys. Lett., 229(1994)421.
30. D.W. Werst and P. Han, Catal. Lett., 45(1997)253.
31. X.-S. Liu, G.-H. Zhang and J.K. Thomas, J. Phys. Chem. B., 101(1997)2182.
32. X.-S. Liu, G.-H. Zhang and J.K. Thomas, J. Phys. Chem. B., 99(1997)10024.
33. J.K. Thomas, Chem. Rev., 93(1993)301.
34. C.-J. Liu, A. Marafee, B.J. Hill, G.-H. Xu, R. Mallinson and L. Lobban, Ind. Eng. Chem. Res., 35(1996)3295.

NATURAL GAS CONVERSION V
Studies in Surface Science and Catalysis, Vol. 119
A. Parmaliana et al. (Editors)

Oxidative Methylation of Acetonitrile to Acrylonitrile with CH_4

Wenmin Zhang and Panagiotis G. Smirniotis*
Chemical Engineering Department, University of Cincinnati,
Cincinnati, OHIO 45221-0171, USA

ABSTRACT

Several basic catalysts have been tested for the oxidative methylation of acetonitrile to acrylonitrile. We found that Li/MgO catalysts with nominal lithium content in the vicinity of 25 wt % are effective for this reaction. Such optimum performance is associated with the maximum concentration of the [Li^+O^-] surface centers of Li/MgO. The coupling reaction takes place via radical mechanism between methane and the α-carbon of acetonitrile and, leads to propionitrile, which is further transformed via oxidative dehydrogenation to acrylonitrile. Experimental evidences indicate that the reaction is Langmuir-Hinselwood. It was concluded that the above active centers are primarily responsible for the advantageous behavior observed and not the basicity of the catalysts. Other supports and catalysts which were found to be very active by other researchers in generating methyl radicals from methane, proved to be ineffective for the transformation of acetonitrile to acrylonitrile. For example mono- and bimetallic combinations of alkali metals on Sm_2O_3, La_2O_3, CaO, and Bi_2O_3 catalyzed primarily the oxidation of acetonitrile to CO_x.

1. INTRODUCTION

Acetonitrile is the main byproduct during the production of acrylonitrile via the ammoxidation of propene and/or propane. The selective transformation of acetonitrile to acrylonitrile is a promising method for the utilization of the former nitrile since the market's demand is significantly lower than the availability. A possible route for this reaction can take place over basic catalysts via the addition of a carbon atom at the α-position of the acetonitrile. Several investigators in the past used molecules such as formaldehyde, methanol, and methane as methylating agents for this reaction. Alkali hydroxides [2] and alkali phosphates [3,4] and were successful catalysts for the production of acrylonitrile from the reaction of formaldehyde with acetonitrile. Methanol was used as methylation agent for the selective transformation of acetonitrile to acrylonitrile over magnesia-based catalysts [5,6,7]. Others [8,9] described the synthesis of acrylonitrile via the oxidative methylation of acetonitrile with methane over K-, Ba-, and Mg-halides supported on quartz.

2. EXPERIMENTAL

The catalysts employed in the present study were synthesized from high purity oxides and the corresponding precursors (M_xCl_y, where M=Na, K, Rb, Cs, Y, Sm, and La). The appropriate amount of the precursor(s), distilled water, and the support (on dry basis) were mixed vigorously and heated at 95 °C until all the water evaporated. Then, the catalysts were ground, sieved, and calcined in dry air for a period of 14 hours at 700 °C to facilitate the diffu-

---

* author to whom correspondence should be addressed

sion of the dopants into the lattice of the support. A final step of grinding and sieving followed. The catalysts were stored in a vacuum drier because some of them were hydroscopic and/or adsorb carbon dioxide from the atmosphere. The basicity of the catalysts was characterized with FT-IR, CO_2-TPD, as well as titration. The chemical composition and the BET surface area of the catalysts prior and after the reaction was determined with ICP spectroscopy (after the catalysts were dissolved in HNO_3) and a BET apparatus, respectively. More details on the synthesis and characterization of the catalysts (Table 1) can be found elsewhere [10].

The catalytic experiments were performed in a differential plug flow reactor. The catalyst was supported between two quartz wool plugs in the middle of a ceramic alumina tube. The tube was placed horizontally in a controllable furnace. Methane, oxygen and helium (balance) were supplied to the reactor with mass flow controllers at the appropriate concentrations. Acetonitrile or the other nitriles used in the kinetic studies were introduced into the reactor with a computerized liquid infusion pump through a heated line. The effluent of the reactor was analyzed *on-line* with a gas chromatograph (Hewlett-Packard, 5890 Series II) equipped with a mass spectrometer (Hewlett-Packard, 5792). Three columns were used in series and allowed the separation of all the products.

Table 1. Properties of oxides and oxide-based catalysts studied for the methylation of acetonitrile to acrylonitrile with methane.

Catalyst	Nominal Loading of metal(s),wt %		BET surface area, m^2/g	Concentration of metal(s) after calcination, wt %	
	1st metal	2nd metal		1st metal	2nd metal
Sm_2O_3	-	-	2.4	-	-
La_2O_3	-	-	0.3	-	-
MgO	-	-	7.7	-	-
Li/MgO	20	-	2.1	8.0	-
Li/Sm_2O_3	10.0	-	0.3	0.2	-
Li/La_2O_3	20	-	0.7	0.1	-
CaO	-	-	0.3	-	-
Na/CaO	5.4	-	0.4	3.7	-
Cs/Na/CaO	5.5	5.5	2.9	0.5	2.1
Cs/K/CaO	5.3	5.3	3.5	0.08	0.8

Note: The calcination step was carried out at 700 °C for 14 hours in air

3. RESULTS AND DISCUSSION

An effective catalyst for the selective transformation of acetonitrile with methane to acrylonitrile and propionitrile must activate the α-carbon of acetonitrile. Acetonitrile possesses an electron-withdrawing group (-CN) and thus basic catalysts can potentially activate selectively its α-carbon. Simultaneously, the catalyst must activate methane. Taking the above into consideration, one should consider oxidative coupling of methane (OCM) catalysts as potential systems for the present reaction. The open literature describes numerous oxide-based catalysts which can effectively generate methyl radicals from methane at elevated temperatures [11,12,13] and further lead to C_{2+} hydrocarbons by gas phase coupling of individual radicals.

Several catalysts considered in the present study for the oxidative transformation of acetonitrile to acrylonitrile in the presence of methane (Table 1). Several catalysts made of monometallic and bimetallic combinations of alkali metals on oxide supports such as Sm_2O_3, La_2O_3, CaO, and BiO_2 were tested during this study.

Sm_2O_3 and La_2O_3 are effective catalysts for the formation of methyl radicals for OCM reactions [14,15,16] at elevated temperatures. Specifically, lanthana was the most active oxide among all the rare earth metal oxides in producing methyl radicals as detected by EPR spectroscopy [17]. However, those catalysts were active in oxidizing the methyl radicals to the undesired CO and CO_2. Both oxides were tested for the transformation of acetonitrile in the presence of methane in the range of 550 to 740 °C. It was found that the conversion of acetonitrile rises abruptly above 660 °C (Figure 1). The acrylonitrile produced over both oxides acquired very low concentrations. Whereas methane was not oxidized even at the highest temperatures, the acetonitrile was transformed to carbon monoxide and carbon dioxide. HCN was not observed as a final product of the oxidation while the nitrogen of the converted acetonitrile was transformed to molecular nitrogen. It was found that the relative concentration of CO and CO_2 is controlled by the water-gas shift equilibrium. Moreover, it was concluded that when oxygen is the limiting reactant, lattice oxygen of the support is used for the oxidation of the nitrile at temperatures above about 660 °C. Lanthana shifts the temperature window to lower temperatures in comparison with samaria.

Li loaded on MgO was found to be an effective catalyst in activating methane. It was proposed that the active sites for the formation of methyl radicals via hydrogen atom abstraction are $[Li^+O^-]$ surface centers generated by the substitution of Mg^{2+} of the support with Li^+ [1,18]. This substitution is feasible because Mg^{2+} (0.66 Å) is comparable in size with Li^+ (0.68Å). We found that Li/MgO with nominal composition of the lithium is in the vicinity of about 20 wt % can effectively utilize acetonitrile and methane towards acrylonitrile in the temperature range of 660 to 720 °C (Figure 1). For higher reaction temperatures the acetonitrile is transformed completely to CO and CO_2. Small amounts of propionitrile were detected as well. Analysis of the reactor effluent with high resolution chromatography and FT-IR (for the condensed products) indicates that other nitriles were not generated. MgO without lithium did not produce any acrylonitrile but oxidized acetonitrile to CO and CO_2. A study of the lithium content on MgO indicates that this advantageous behavior occurs for Li contents (nominal) in the vicinity of 25 wt%. Catalysts with high loadings of lithium were not effective for the transformation of acetonitrile and methane to acrylonitrile. This observation is additional evidence that the $[Li^+O^-]$ surface centers are responsible for the simultaneous activation of methane and acetonitrile. Lunsford and coworkers [1] found that a maximum of the $[Li^+O^-]$ sites exists for Li/MgO catalysts with 13 wt % lithium content. The maximum rate of generation of methyl radicals observed during OCM reactions was strongly associated with these sites. On the other hand if lithium is loaded on samaria or lanthana, the rate of the reactant (acetonitrile) oxidation significantly decreases. However, over Li/Sm_2O_3 and Li/La_2O_3 acrylonitrile is not generated. This observation indicates that an additional role of lithium is to decrease the rate of the nitrile oxidation by the support. From a carbon mass balance we concluded that under the present operating conditions (oxygen was limited reactant) over all the above catalysts the oxidation of methane does not occur. Hence, the generated CO and CO_2 comes from the oxidation of acetonitrile and the produced nitriles.

Additional experiments were performed to validate the formation of radicals from acetonitrile and its coupling with methyl radicals over Li/MgO leading to acrylonitrile [19]. It was concluded that indeed this catalyst favors the abstraction of a hydrogen atom from the α-carbon of acetonitrile which will be further react with methyl radicals to form propionitrile (reactions 1.1, 1.2). The latter nitrile will be transformed to acrylonitrile via its oxidative dehydrogenation (reaction 1.3). Our study showed that oxygen was absolutely necessary for the reaction. However, the oxygen concentration should be kept below the stoichiometric (based on acetonitrile) in order to suppress the undesired oxidation. It is worth noting, that in addition to propionitrile, CO was found to be a primary product of oxidation reactions of acetonitrile and the produced nitriles.

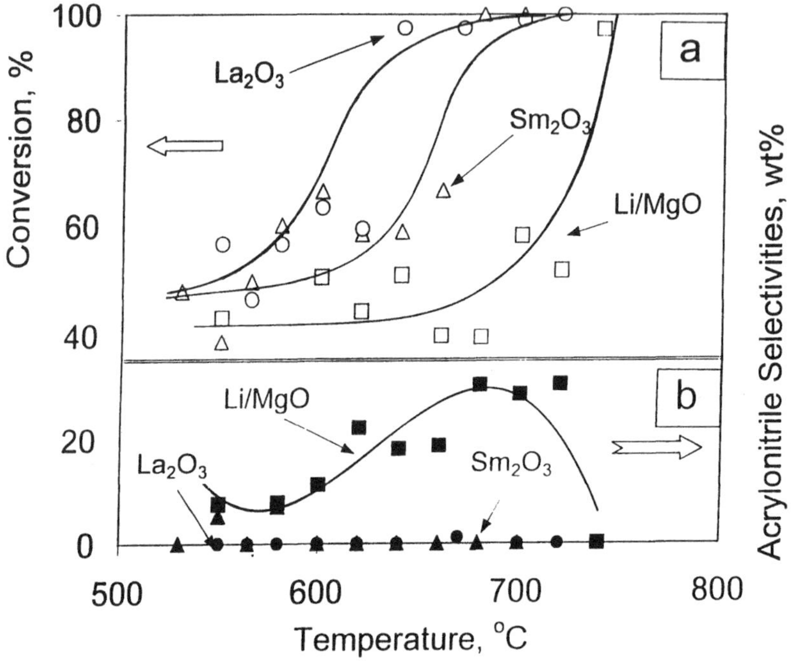

Figure 1. Acetonitrile conversion (top) and acrylonitrile selectivity (bottom) versus reaction temperature over Li/MgO (20 wt% Li), S_2O_3, and La_2O_3 (TOS= 1 hour, WHSV= 0.92 h^{-1} CH_4/O_2/Acetonitrile=5/1/2).

$$CH_4 \longrightarrow HC_3^{\bullet} + H^{\bullet} \tag{1.1}$$

$$CH_3C{\equiv}N \longrightarrow {}^{\bullet}CH_2C{\equiv}N + H^{\bullet} \quad \text{(hydrogen abstraction)} \tag{1.2}$$

$$HC_3^{\bullet} + {}^{\bullet}CH_2C{\equiv}N \longrightarrow CH_3CH_2C{\equiv}N \xrightarrow{O_2} CH_2{=}CHC{\equiv}N + 2\,{}^{\bullet}OH \tag{1.3}$$

MgO loaded with other alkali metals was also considered for this reaction. It was found that lithium resulted in the most efficient catalyst. At comparable levels of conversion the acrylonitrile selectivity decreased monotonically with the increase of the ionic radius of the alkali metal. This behavior is probably a result of the lack of the effective substitution of Mg^{2+} (0.66Å) by the other alkali metal used due to the significant difference in size. As described above this was not the case for lithium. The incorporation of a second metal on Li/MgO, namely Sm, La, Pr, and Cs, was also attempted. It was found that the addition of the second metal decreases the efficiency of the catalyst. FT-IR and TPD studies of the catalysts after they were exposed to CO_2 indicated that the basicity of the Li/MgO was significantly lower than that of other oxides or oxide catalysts (Figure 2). This supports the fact that the [Li^+O^-] surface species of Li/MgO are primarily responsible for the transformation of acetonitrile to acrylonitrile. By studying the basicity properties of all the catalysts involved in the present study we concluded that their capability to transform acetonitrile and methane to acrylonitrile is inversely proportional to the number and strength of basic sites.

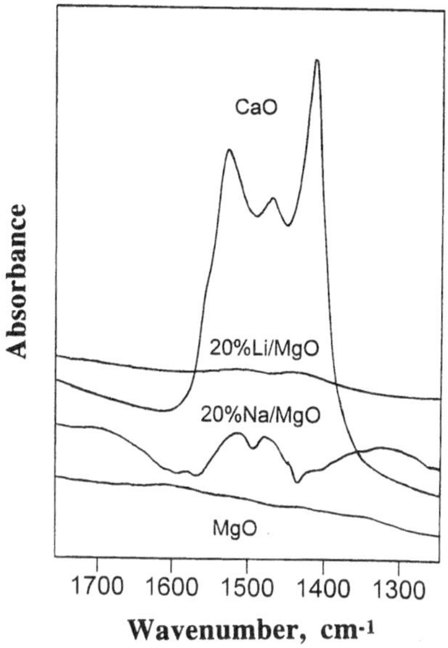

Figure 2. FT-IR spectra of CO_2 on selected oxide supports and oxides doped with alkali metals. The bands in the range of 1400-1600 cm^{-1} correspond to different types of surface carbonate species

CaO promoted with selected alkali metals can be a good catalyst for OCM reactions [20]. This was attributed to the fact that K and Na can effectively substitute Ca in the lattice. In our studies we also considered CaO and its mono- and bimetallic combinations with alkali metals. Surprisingly, this category of catalysts demonstrated deleterious behavior in transforming methane and acetonitrile to acrylonitrile and/or propionitrile. CaO oxidized partially acetonitrile to CO and CO_2 without producing any acrylonitrile. Under identical conditions when mono- and bimetallic combinations of alkali metals (Na, K, and Cs) were added to the support, the catalyst oxidized completely acetonitrile. C_{2+} hydrocarbons which could be generated by the oxidative coupling of methane, were not detected. These observations indicate that when methane and acetonitrile coexist, the active sites of the catalyst are utilized to oxidize acetonitrile (less stable) rather than activate methane. Bi_2O_3 and Y_2O_3/Bi_2O_3 were considered as well but both catalysts completely oxidized acetonitrile.

From the above discussion, it is evident that many catalysts which are efficient for the oxidative coupling of methane to higher hydrocarbons cannot catalyze the transformation of acetonitrile and methane to acrylonitrile and/or propionitrile. Li/MgO catalyzes effectively the activation and coupling of acetonitrile and methane. The time on stream behavior of Li/MgO (20 wt% lithium) was investigated as well. If the feed stream does not contain any water it was found that the acetonitrile conversion increases slightly with time. On the other hand, the selectivity of acrylonitrile decreases monotonically and approaches the performance of individual MgO. This behavior was attributed to the sublimation of lithium. At the end of the run the concentration of lithium on the catalyst was at low levels and the catalyst performance was very similar to that of MgO. When steam is cofed into the reactor, the acrylonitrile selectivity acquires lower values and decreases much faster with time on stream. The chemical composition of the spent catalysts indicates that this behavior is associated with the rapid loss of lithium from the catalyst. It is worth nothing, that among all the supports considered in the present study, MgO retained the largest amount of metals/promoters added.

Our kinetic investigations [19] showed that acetonitrile and methane are transformed to the corresponding radicals by hydrogen abstraction. Both radicals remain on the surface and react for the formation of a larger nitrile which finally desorbs from the surface.

4. CONCLUSIONS

Several oxides doped with mono- and bimetallic combinations of metals (primarily alkali metals) were tested for the oxidative transformation of acetonitrile to acrylonitrile with methane. It was found that lithium-doped MgO favors the formation of acrylonitrile. The reaction proceeds via the coupling of the radical formed at the α-carbon of acetonitrile with methyl radicals formed from methane and leads as a first step to propionitrile. The latter nitrile is transformed instantaneously to acrylonitrile. This advantageous behavior of Li/MgO was associated with the effective abstraction of hydrogen atoms from the reactants catalyzed by its [Li^+O^-] surface sites and the decreased rates of the reactant acetonitrile oxidation. The addition of a second alkali metal did not improve the catalyst's performance. Other supports, namely CaO, Sm_2O_3, La_2O_3, and Bi_2O_3 promoted with metals, were not nearly as effective as Li/MgO. One can conclude that the basicity of the catalyst is not a crucial factor on its performance for this reaction.

REFERENCES

1) D. J. Driscoll, W. Martir, J.-X. Wang, and J. H. Lunsford, J. Am. Chem. Soc., 107, (1985) 58.
2) Y. Yamazaki, and T., Kawai, Sekiyu Gakkai Shi, 12 (1969) 693.
3) K. E. Khcheyan, O. M. Revenko, N. E. Mak, and M. P. Tikhonova, Khim. Proc. (Moscow), 47(10) (1971) 730.
4) B. D. Bader, M. Cerghitescu, P. Bader, and M. Toader, Revista de Chimie, 24(4) (1973) 247.
5) W. Ueda, T. Yokoyama, Y. Moro-oka, and T. Ikawa, Ind. Eng. Chem. Prod. Res. Dev., 24 (1985) 340.
6) H. Kurokawa, T. Kato, W. Ueda, Y. Morikawa, Y. Moro-oka, and T. Ikawa, J. Catal., 126 (1990a) 199.
7) H. Kurokawa, T. Kato, T. Kuwabara, W. Ueda, Y. Morikawa, Y. Moro-oka, and T. Ikawa, J. Catal., 126 (1990b) 208.
8) K. E. Khcheyan, O. M. Revenko, A.N. Shatalova, and E. G. Gel'perina, Neftekhimiya, 17 (1977) 594.
9) K. E. Khcheyan, A.N. Shatalova, O. M. Revenko, and L. I. Agrinskaya, Neftekhimiya, 20 (1980) 876.
10) W. Zhang, and P. G. Smirniotis, (submitted to J. Catalysis), 1998
11) J. S. Lee, S. T. Oyama, Catal. Rev.- Sci. Eng., 30(2) (1988) 249.
12) Y. Amenomiya, V. I. Birss, M. Goledzinowski, J. Galuszka, and A. R. Sanger, Catal. Rev.- Sci. Eng., 32(3) (1990) 163.
13) E. N. Voskresenskaya, V. G. Roguleva, and A. G. Anshits, Catal. Rev.-Sci. Eng. 37(1) (1995) 101.
14) Lin, C.-H., Campbell, K. D., Wang, J.-X., and Lunsford, J. H., J. Phys. Chem , 90 (1986) 534.
15) K. Otsuka, K. Jinno, and A. Morikawa, Chemistry Letters, pp. 499, (1985).
16) K. Otsuka, Q. Liu, M. Hatano, and A. Morikawa, Chemistry Letters, pp. 467 (1986).
17) K. D. Campbell, H. Zhang, and J. H. Lunsford, 10th North American meeting of the Catalysis Society, Paper C-5, San Diego, (1987).
18) T. Ito, J.-X. Wang, C.-H. Lin., and J. H. Lunsford, J. Am. Chem. Soc., 107 (1985) 5062.
19) P. G. Smirniotis and W. Zhang, Applied Catalysis A, (accepted for publication), 1998
20) J. A. S. P., Carreiro, and M. Baerns, React. Kinet. Catal. Lett., 35 (1987) 349.

NATURAL GAS CONVERSION V
Studies in Surface Science and Catalysis, Vol. 119
A. Parmaliana et al. (Editors)

New Directions for COS Hydrolysis: Low Temperature Alumina Catalysts

John West[1], B. Peter Williams[2], Nicola C. Young[3], Colin Rhodes[1] and Graham J. Hutchings[1]

[1]Department of Chemistry, University of Wales Cardiff, PO Box 912, Cardiff, CF1 3TB, UK.
[2]ICI Katalco, Clitheroe, Lancashire, BB7 4QB, UK.
[3]ICI Katalco, PO Box 1, Billingham, Cleveland, TS23 1LB, UK.

Abstract

Carbonyl sulphide hydrolysis using two alumina catalysts (surface areas 150 m^2g^{-1} and 300m^2g^{-1}) has been studied. Previous studies have focused upon concentration ranges for the reactant ([COS] typically > 1000 ppm) and temperatures (typically > 100 °C) that are much higher than those encountered commercially. In the study presented here, the concentration of carbonyl sulphide is 150 ppm, reactor pressure (125 – 187 kPa) and temperatures in the range of 30 – 250 °C. At the higher temperature (250 °C) the data collected are in agreement with previous studies and the reaction follows Langmuir Hinshelwood kinetics with the surface hydrolysis of a thiocarbonate being the rate determining step. Data obtained at lower temperature (30 – 60 °C) indicates that the rate of COS hydrolysis decreases monotonically with increasing [H_2O]. These data are consistent with Langmuir Hinshelwood kinetics where the products are not adsorbed and either (a) the adsorption of COS is rate determining or (b) the surface reaction of adsorbed COS and an intermediate derived from H_2O is rate determining. The current experimental data cannot differentiate between these possibilities. Using present test conditions it is apparent that the surface area of the alumina catalyst is an important design parameter, a feature not observed in previous high temperature studies.

Introduction

In the last two decades it has been recognised that emissions of sulphur compounds, including COS, into the atmosphere have been unacceptably high. Similarly, it has become apparent within industry itself that the effects of sulphur have far-reaching consequences environmentally and economically. The detrimental effect that sulphur has on industrial heterogeneous catalysts is well known [1] and levels of sulphur as low as 1ppm have been shown to effectively poison the activity of a modern bi-metallic reforming catalyst. As little as 4 mg of sulphur per gram of catalyst on the surface of the Fe-Cu-K catalyst used in the Fischer Tropsch Synthesis process decreases measured activity by 50% [2]. In addition, it is known that the presence of feedstock sulphur leads to increased corrosion of the reactors used in refinery processes [3]. Consequently, well established procedures have been in place for many years to remove sulphur compounds by hydrodesulphurisation [4]. Raw materials that have undergone desulphurisation are significantly less hazardous, less corrosive and can be used for the manufacture of odour free products.

With the introduction of stringent legislation to reduce sulphur emissions, fresh impetus is being given to modifying and improving existing desulphurisation technology. However, hydrodesulphurisation does not remove or significantly hydrogenate one sulphur containing compound, namely carbonyl sulphide. Typically an alternative technology has been used for the removal of carbonyl sulphide and this is based on the formation of hydrogen sulphide by hydrolysis:

$$COS + H_2O \longleftrightarrow H_2S + CO_2 \quad (1)$$

Numerous studies have investigated this reaction [5-16] and to date alumina and/or titania materials have been identified as potential catalysts. Developments include the investigation and optimisation of the catalyst morphology and new formulations are increasingly being introduced to maximise the active site density or to reduce the likelihood of site blockage. Recent improvements have included increasing the average pore diameters of the alumina catalysts from 600 - 750 nm to diameters in excess of 1μm to reduce the possibility of site blockage by sulphur adsorption [17].

While a large number of studies have been made of the hydrolysis reaction it is observed that virtually all utilise reaction conditions that are significantly removed from the conditions typically found in commercial process streams. In particular, the requirement is to treat low concentrations of carbonyl sulphide (typically ≤ 150 ppm) at near ambient temperature (30 - 60 °C).

The conditions used in previous studies are shown in Figure 1 (150 ppm = 0.0225 kPa partial pressure at 1.5 bara) and it is apparent that few of the previous studies investigate these conditions but tend to use much higher carbonyl sulphide concentrations (typically *ca.* 5000 ppm) and higher temperatures (typically > 100 °C).

In this paper we present a study of alumina catalysts used for the hydrolysis of carbonyl sulphide for which the concentration of carbonyl sulphide investigated was 150 ppm together with temperatures in the range 30 – 250 °C and pressures in the range of 125 – 187 kPa. The aim was to investigate the reaction mechanism under industrially relevant conditions.

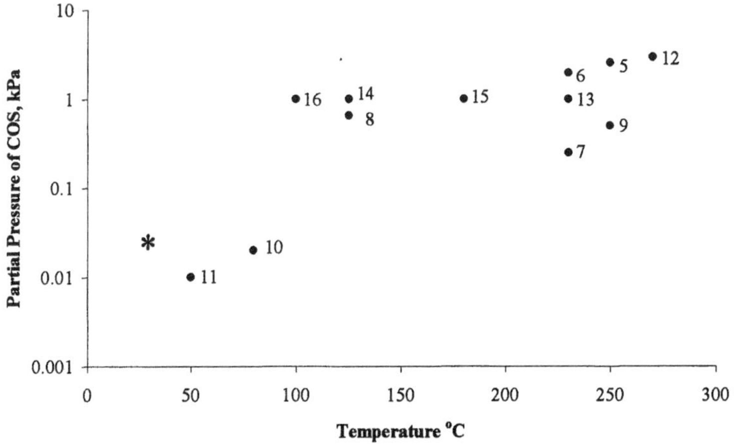

Figure 1 Reaction conditions for previous studies examining COS hydrolysis at atmospheric pressure, the numeric labels correspond to quoted reference, with * indicating the partial pressure of COS being used in this study.

Experimental

Two γ-alumina samples were selected for study. The first was supplied by Brockmann and had a surface area of 150 m^2g^{-1} and the second was an ICI alumina with a surface area of 300 m^2g^{-1}; both were used without further treatment. The catalysts were tested in a fixed bed microreactor (Figure 2) at gas hourly space velocities of 100000 – 500000 h^{-1}. Carbonyl sulphide (BOC, 1% in N_2) and nitrogen were fed to a saturator via calibrated mass flow controllers.

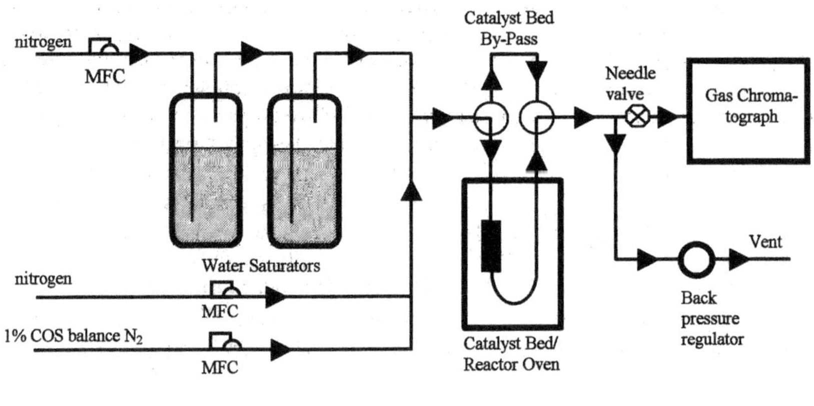

Figure 2 Schematic diagram of microreactor apparatus.

The water saturator comprised two bubblers, one being maintained at 20°C and the second at a controlled temperature and in this way the concentration of water in the diluted carbonyl sulphide gas stream was maintained at the saturated vapour pressure of water at the temperature set in the second saturator. This was found to be the most effective method for delivering low, controlled concentrations of water to the reactor. The exit gases from the reactor were analysed using a Varian 3400 gas chromatograph fitted with a Porapak T and Porapak Q column and a pulsed flame photometric detector.

Results and Discussion

The initial experiments were carried out to determine the lifetime of alumina catalysts and the stabilisation period required prior to the acquisition of kinetic data. It was found that the catalysts operated at steady state following a period of *ca.* 5 h prior to data collection. Hence, the catalysts were allowed to stabilise for 5 h prior to data collection. Representative data for the effect of time on stream on COS conversion are shown in Figure 3.

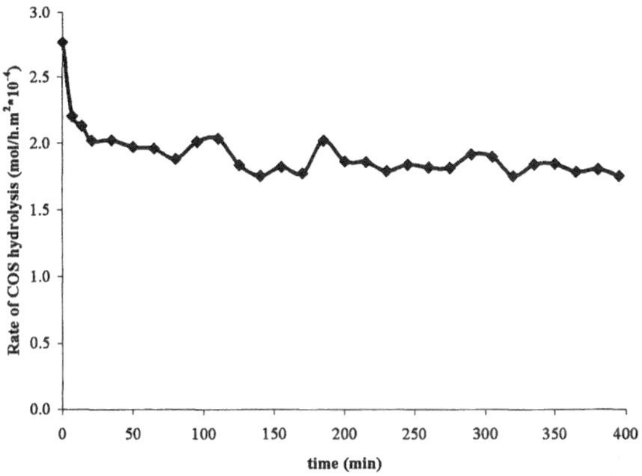

Figure 3 COS conversion as a function of time on stream. Reaction conditions: alumina (0.5g, 300 m^2g^{-1}), 30°C, 1.6 bara, GHSV 100000 h^{-1}, 250 ppm COS, 1200 ppm H_2O.

Experiments were then carried out with the lower surface area alumina (150 m^2g^{-1}, Brockmann). The hydrolysis of COS was investigated with a fixed concentration of COS (150 ppm) but with varying water concentration at a fixed total gas hourly space velocity. The results for two temperatures are shown in Figure 4. At the higher temperature the results are consistent with the results of the previous studies [1-3,5,8,9]. Clearly the alumina is very active under these conditions and the data are consistent with the mechanism proposed by Hoggan *et al.* [18] in which the reaction proceeds via the formation of a surface thiocarbonate upon adsorption of COS and the rate determining step is the reaction of this surface species with a surface hydroxyl group. The data are consistent with Langmuir Hinshelwood kinetics where the products of reaction are not strongly adsorbed, equation 2,

$$\text{rate} = \frac{k_1[COS][H_2O]}{(1 + K_2[COS] + K_3[H_2O])^2} \quad (2)$$

where k_1 is the rate constant ($k_1 = k.K_{COS}.K_{H_2O}.XS^2$ where XS is the total concentration of available active sites) and K_2 and K_3 are equilibrium constants for the adsorption of COS and H_2O respectively. At 250°C the alumina is a very active catalyst with rates of COS hydrolysis in the range 4 – 6 × 10^2 mol $m^{-2}h^{-1}$ being observed. As expected, at the lower temperature of 30 °C the rate is significantly lower. However, the rate of COS hydrolysis is observed to decrease significantly as the concentration of water is increased. This indicates that the rate determining step at the lower temperature is different than that determined for the high temperature study. It is apparent that water is acting as an inhibitor and the data can be described using Langmuir Hinshelwood kinetics with the adsorption of COS being the rate determining step or the surface reaction of adsorbed COS and an intermediate derived from

H_2O being rate determining, equation 3, k_1 is the rate constant (k_1 = k.XS where XS is the total concentration of available active sites). The data in Figures 5 and 6 show good agreement between experimental and calculated results for both of the models presented (equations 2 and 3).

$$\text{rate} = \frac{k_1[COS]}{(1 + K_3[H_2O])} \qquad (3)$$

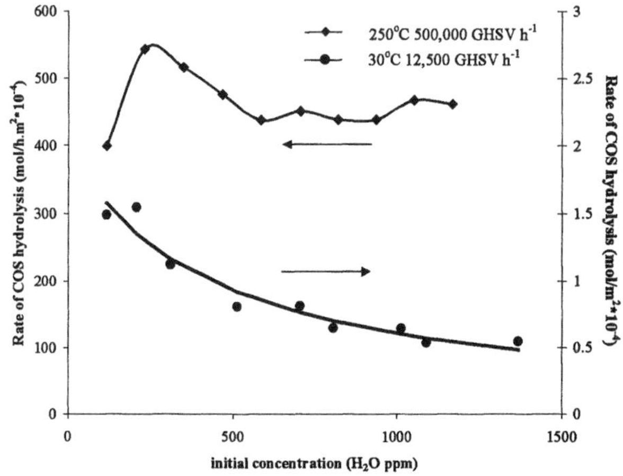

Figure 4 Rate of COS hydrolysis as a function of $[H_2O]$ at constant [COS] = 150 ppm, alumina 150 m^2g^{-1}.
Key: ♦ = 250 °C, 1.6 bara, 500000 h^{-1}; • = 30 °C, 1.6 bara 12500 h^{-1}

The higher surface area alumina was also investigated at 60 °C and the same relationship between the rate of COS hydrolysis and increasing water concentration at constant space velocity was observed (Figure 5) although the rate was significantly higher. The rate of COS hydrolysis was examined at constant $[H_2O]$ with increasing [COS] at constant space velocity and temperature and the results are shown in Figure 6.

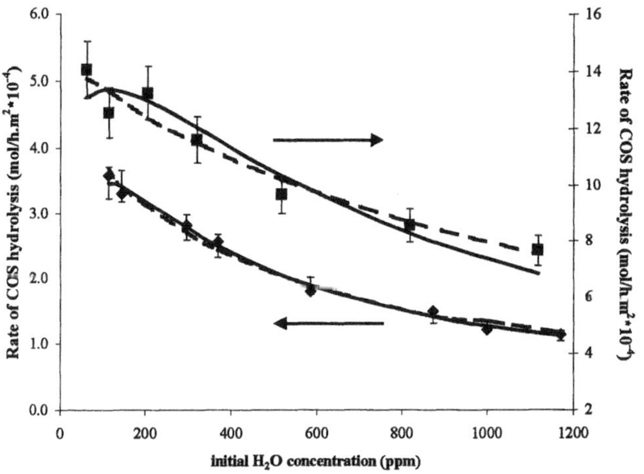

Figure 5 Rate of COS hydrolysis as a function of $[H_2O]$ at constant [COS] = 150 ppm, 1.6 bara. Key: experimental data ♦ 30 °C, GHSV 100000 h^{-1}, ■ 60 °C, GHSV 500000 h^{-1}, calculated data for COS adsorption limited model (equation 2) -----, calculated data for surface reaction type model (equation 3) ——.

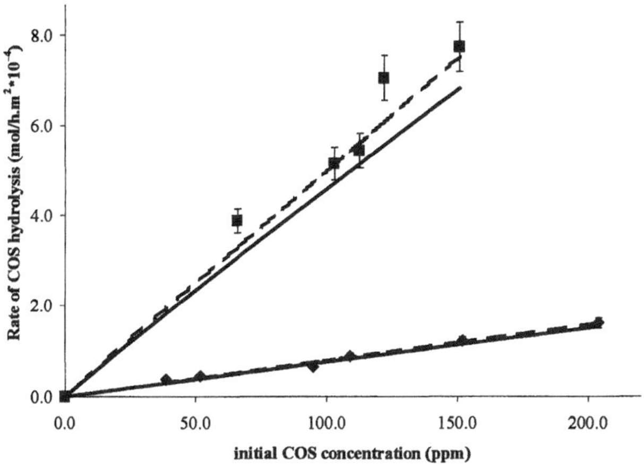

Figure 6 Rate of COS hydrolysis as a function of [COS] at constant $[H_2O]$ = 1200 ppm, 1.6 bara. Key: experimental data ♦ 30 °C, GHSV 100000 h^{-1}, ■ 60 °C, GHSV 500000 h^{-1}, calculated data for COS adsorption limited model (equation 2) -----, calculated data for surface reaction type model (equation 3) ——.

The relationship between the rate of COS hydrolysis and increasing reactor pressure was investigated using the higher surface area alumina at 30 °C with 150 ppm COS and 1200 ppm H_2O. Although the experimental error is such that no firm conclusion can be drawn, the model suggests a slight positive effect of increasing reactor pressure upon the rate of COS hydrolysis (Figure 7). These data was collected for relatively high partial pressure of water which we have shown previously (Figure 5) to inhibit the rate by extensive coverage of the active sites.

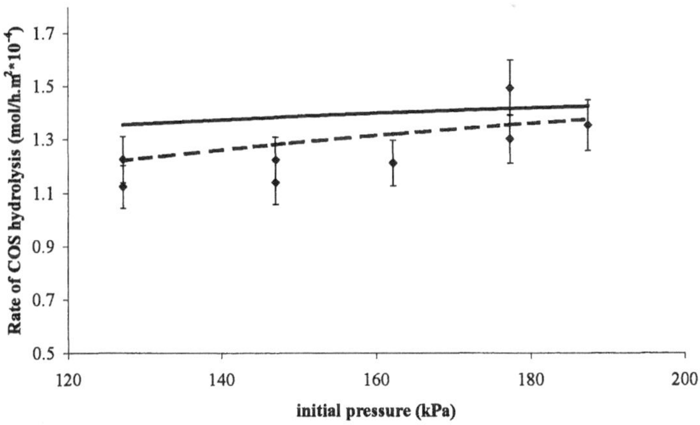

Figure 7 Rate of COS hydrolysis as a function of pressure at constant [COS] = 150 ppm, $[H_2O]$ = 1000 ppm, 30 °C, 1.6 bara, GHSV = 100000 h^{-1}. Key: experimental data ♦ , calculated data for COS adsorption limited model (equation 2) -----, calculated data for surface reaction type model (equation 3) ——.

It is therefore clear that the rate determining step of COS hydrolysis on alumina catalysts is dependent on the reaction temperature. At high temperatures, which are of limited industrial interest, the rate determining step is the surface reaction and the results we have obtained are in agreement with previous studies [18]. However, at lower temperatures the competitive adsorption between the reactants dominates with either the rate of COS adsorption being rate limiting or the surface reaction of adsorbed COS and an intermediate derived from H_2O being involved in the rate determining step.

Under these circumstances, it is clear that an alumina with a high surface area will give the highest activity. The design of active catalyst for low temperature COS hydrolysis will therefore depend mainly on methodology that can yield high surface aluminas that maintain high surface area when exposed to relatively high but variable concentrations of water vapour. In addition, this study shows that the previous investigations have concentrated on inappropriate reaction conditions and that further studies should concentrate on the study of lower COS concentrations and temperature <100 °C.

Conclusions

The hydrolysis of carbonyl sulphide over two alumina catalysts (surface areas 150 m^2g^{-1} and 300 m^2g^{-1}) has been studied. At the higher temperature (250 °C) the data obtained are in agreement with previous studies and the reaction follows Langmuir Hinshelwood kinetics with the surface hydrolysis of a thiocarbonate being the rate determining step [18]. However, at lower temperature (30°C) this mechanism is not operative and the reaction is observed to follow Langmuir Hinshelwood kinetics with the adsorption of carbonyl sulphide and/or the surface reaction of adsorbed COS and an intermediate derived from H_2O being the rate determining step. The current experiments cannot differentiate between these possibilities at present and future work examining the nature of the species upon the catalyst surface at low temperature is currently underway. Under these conditions it is the surface area of the alumina catalyst that is the important design parameter, a feature not reported in previous high temperature studies.

References

1. C.N. Satterfield, "Heterogeneous Catalysis in Industrial Practice" McGraw Hill, 1991, p378.
2. Z-T Liu, J-L Zhou and B-J Zhang, J.Mol.Cat., 94 (1994) 255.
3. C.N. Satterfield, "Heterogeneous Catalysis in Industrial Practice" McGraw Hill, 1991, p379.
4. R. Raheel, K. Holder and G.J. Hutchings, Catal. Today, in press.
5. Z.M. George, J. Catal., 32 (1974) 261.
6. Z.M. George, J. Catal., 35 (1974) 218.
7. R. Fiedorow, R. Leute and I.G. Dalla Lana, J. Cat., 85 (1984) 339.
8. R.S. Coward and W.M. Skaret, Oil & Gas J., Apr. 8 (1985) 86.
9. J. Nougayrede, Oil & Gas J., Aug 10 (1987) 65.
10. S. Tan, C. Li, S. Liang and H. Guo, Catal. Lett. 8 (1991) 155.
11. K. Miura, K. Mae, T. Inoue, T. Yoshimi, H. Nakagawa and K. Hashimoto, Ind. Eng. Chem. Res., 31 (1992) 415.
12. S. Tong, I.G. Dalla Lana and K.T. Chuang, Can. J. Chem. Eng., 71 (1993) 392.
13. J. Bachelier, A. Aboulayt, J.C. Lavalley, O. Legendre and F. Luck, Catal. Today, 17 (1993) 55.
14. H.M. Huisman, P. Van der Berg, R. Mos, A.J. van Dillen and J.W. Geus, ACS Symp.Series, 552 (1994) 393.
15. C. Morterra and G. Magnacca. Catal. Today, 27 (1996) 497.
16. A. Aboulayt, F. Mauge, P.E. Hoggan and J.C. Lavalley, Catal. Lett. 39 (1996) 213.
17. A. Maglio and P.F. Schubert, Oil & Gas J., Sept 12, (1988) 85.
18. P.E. Hoggan, A. Aboulayt, A. Pieplu, P. Nortier and J.C. Lavalley, J. Catal., 149 (1994) 300.

NATURAL GAS CONVERSION V
Studies in Surface Science and Catalysis, Vol. 119
A. Parmaliana et al. (Editors)

Kinetic nature of limited yield of principal products at heterogeneous-homogeneous oxidation of methane.

V.S.Arutyunov, V.Ya.Basevich, O.V.Krylov, and V.I.Vedeneev.

Semenov Institute of Chemical Physics, Russian Academy of Science,
117334, Kosygina 4, Moscow, Russia

Kinetic analysis of the high temperature oxidative catalytic conversion of methane shows the close link between gas phase and surface processes and lets to suppose that, at any rate in the first approximation, the surface processes are predominantly provided the necessary rate of radicals generation. For an effective catalysis of high temperature oxidative conversion of methane in the most important cases the catalyst should provide the rate of generation of methyl radicals that exceeds the rate of their thermal gas phase generation at least on 5-6 orders of magnitude. But never the less there exists the maximal yield of principal products which can not be overcome by the further increase of the rate of generation of radicals.

INTRODUCTION

Apart of the reactions of complete conversion of methane into CO_2 and H_2O, and steam, oxygen or dry reforming of methane into syngas, a significant attention is paid to catalytic processes of direct partial oxidative conversion of methane in other useful products. The most important are the oxidation of methane into methanol, oxidation of methane into formaldehyde and oxidative coupling of methane into ethane and ethene. Because of the high temperatures that are needed for the catalytic reactions of such type, they may be considered as high temperature catalytic oxidative processes. It was shown by many authors that kinetic behaviour of these very different catalytic reactions has similar features [1]. In this paper we discuss some important examples and analyze possible roots of these features on the ground of kinetic modeling of gas phase oxidation of methane with an additional generation of methyl radicals.

MAIN FEATURES OF THE PARTIAL CATALYTIC OXIDATION OF METHANE

The most important common features of different reactions of partial catalytic oxidation of methane are:

i. High and very similar values of activation energy (~200 kJ/mol);
ii. Proportionality of the rate of the reaction to the first power of methane partial pressure;
iii. Exhibition of the kinetic isotope effect (KIE) (in the most cases $k_{CH4}/k_{CD4} > 1$);
iv. Limited yield of final products (excepting complete oxidation and conversion to syngas).

Some important examples will be considered below.

At steam reforming of methane as well as at oxygen and dry reforming the kinetic equation of the reaction for very different catalysts looks like [2-4]

$$\mathbf{r = k}P_{CH4} \tag{1}$$

The activation energy of the steam reforming of methane on NiO.MgO catalyst attains 240 kJ/mol. Kinetic isotope effect k_{CH4}/k_{CD4} on Ni/Al_2O_3, Ni/La_2O_3 [5] and other catalysts may achieve value 2 at oxygen and dry reforming. Usual temperature of oxygen reforming 700-800°C, of steam reforming - 800-900°C, and that of dry reforming - 900-1000°C, but in [6] it was shown that on Rh/MgO all of these three reactions proceed to the end with 100% yield at the same temperature 580°C.

Catalytic oxidation of methane into methanol proceeds with low yields. At atmospheric pressure usual yields of methanol are 0.5-1.5%, and at enhanced pressure 5 MPa, which is close to the pressure of the gas phase partial oxidation of methane into methanol, the maximal yield on the best catalysts (Fe-Na-sodalite and silicoferrate) is equal to 4% at 410-430°C [7]. At catalytic oxidation of methane into formaldehyde maximal yields are somewhat higher - 2-4% [8]. The best catalyst is presumably V_2O_5/SiO_2 which lets to obtain such yields at 550-600°C [8]. The activation energy of this process on MoO_3/SiO_2 catalyst is 189 kJ/mol [9]. Several reports about significantly higher yields in these two processes were not confirmed in following studies. In the vast majority of studies the kinetics of these reactions followed to the equation (1)

Literature data on catalysts of oxidative coupling of methane were observed in [10], where it was shown that in all cases the yield of C_2-hydrocarbons did not exceed 25%. As in previous case, some reports about higher yields up to ~30% and even higher were not confirmed. Kinetic data on this reaction are more numerous than on other catalytic reactions of methane oxidation. It was shown that this reaction proceeds via oxidative-reductive mechanism including the interaction of methane with O^- or O_2^{2-} on surface with formation of CH_3 radicals and OH_s surface groups and following reoxidation of surface by oxygen. Kinetic isotope effect $k_{CH4}/k_{CD4} > 1$ and the validity of the equation (1) were also observed for this reaction. Experimental values of the activation energy for catalytic oxidative coupling of methane are situated in the range of 170-290 kJ/mol.

These examples let to suppose that the first and limiting step of the most catalytic processes of methane oxidation is the interaction of CH_4 molecule with surface with formation of CH_3 radical:

$$\mathbf{CH_4 + O_s \rightarrow CH_3 + OH_s} \tag{2}$$

Than the surface reoxidizes in catalytic cycle, probably, by the interaction of surface OH groups with O_2.

$$\mathbf{4OH_s + O_2 \rightarrow 2H_2O + 4O_s} \tag{3}$$

But in the most cases the reoxidation does not limit the total rate of the process. The exhibition of the kinetic isotope effect $k_{CH4}/k_{CD4} > 1$ confirms that the rupture of C-H bond really may be the rate controlling step of reaction.

KINETIC MODELING THE PARTIAL OXIDATION OF METHANE WITH AN ADDITIONAL GENERATION OF RADICALS.

Our previous kinetic simulation of gas phase oxidative coupling of methane [11] has shown (fig.1) that at the typical experimental conditions the increase of the rate of generation of CH_3 radicals increases the yield of C_2-hydrocarbons up to the maximal value 22%, which practically coincides with the best results of catalytic coupling on the most active catalysts [1,10]. This maximum attains at the optimal rate of generation of CH_3 radicals that exceeds the rate of their thermal generation by approximately 6 orders of magnitude.

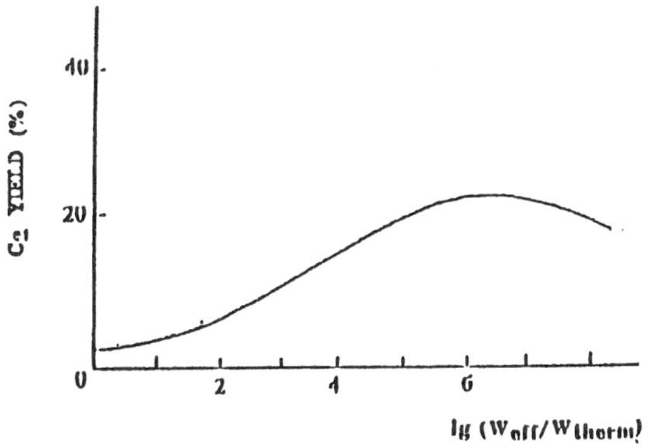

Figure 1. Simulation of C_2-hydrocarbon yield vs. $\lg(W_{eff}/W_{therm})$, where W_{therm} is the rate of homogeneous thermal generation of CH_3 radicals and W_{eff} is an effective rate of generation of radicals at simulation. T=1073 K, P=1 atm, CH_4:O_2:N_2 = 10:2:12.5.

It should be noted that the most active catalyst of methane oxidative coupling Sm_2O_3 at 700°C provides the rate of CH_3 radicals generation $\approx 1.4{\times}10^{22}$ molecule/m^2s what is very close to the optimal theoretical value mentioned above. So in this case the optimal rate of generation of CH_3 radicals is provided by heterogeneous process, and the maximal C_2-hydrocarbons yield is controlled by subsequent gas phase kinetics.

Analogues investigations both theoretical and experimental [12] show that the maximal methanol yield ~2% at heterogeneous-homogeneous partial oxidation of methane to methanol is also determined by gas phase reaction, while catalyst provides the necessary rate of CH_3 radicals generation. At high pressures exceeding certain critical value P_{cr}, which depends on temperature and some other conditions, this reaction turn to another quasistationary regime by means of a very sharp change (fig.2). In this new regime the rate of oxidation exceeds the usual rate in 10^3-10^4 times [13], because the reaction becomes self-accelerating, and the generation of new chains does not determined now by slow gas phase reaction

$$\mathbf{CH_4 + O_2 \rightarrow CH_3 + HO_2} \qquad (4)$$

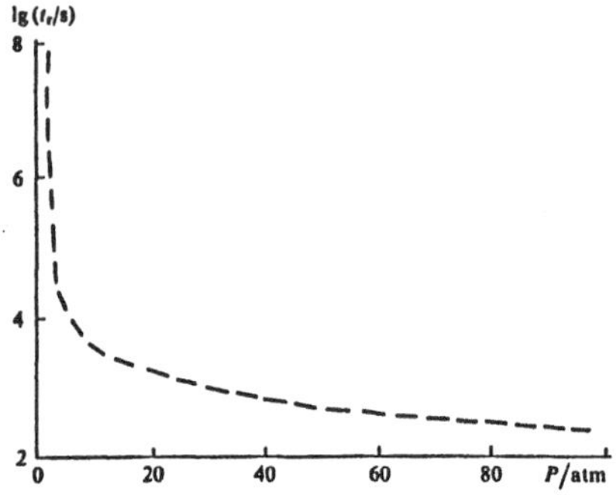

Fig. 2. Simulation of the pressure dependence of reaction time. T=650 K, CH_4:O_2 = 9:1.

or heterogeneous-catalytic reactions (2)-(3), but by a chain branched process with a rate of radicals generation exceeding that of thermal generation (4) in 10^3-10^4 times. At high pressures steady state chain branched gas phase reaction itself provides the rate of CH_3 radicals generation that in the most cases is higher than the catalytic generation of radicals [13]. That makes the role of catalyst in this case minor. The catalyst is able to influence significantly on such self-accelerating process only if the rate of catalytic generation of radicals exceeds the rate of thermal generation (4) no less than in 10^4 times. This was confirmed by our kinetic calculations of the dependence of the rate of the process and the yield of methanol on the rate of additional generation of methyl radicals by the same way as it was done in [11] (fig. 3).

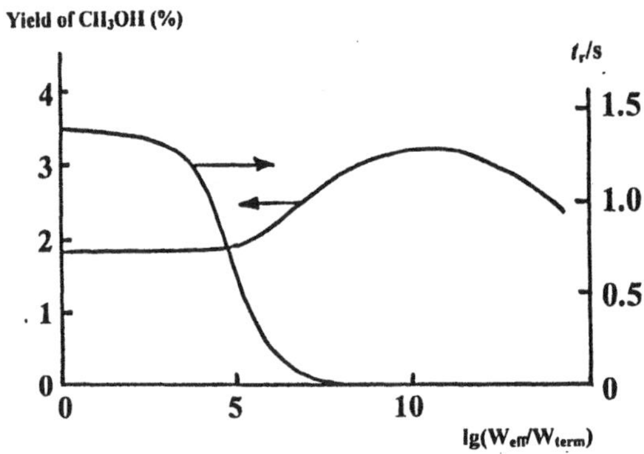

Fig. 3. Simulation of CH_3OH yield and reaction time t_r vs. lg(W_{eff}/W_{therm}). T=683 K, P=100 atm, CH_4:O_2 = 19:1.

Very similar behavior was observed at kinetic simulation of methane oxidation to formaldehyde (fig.4).

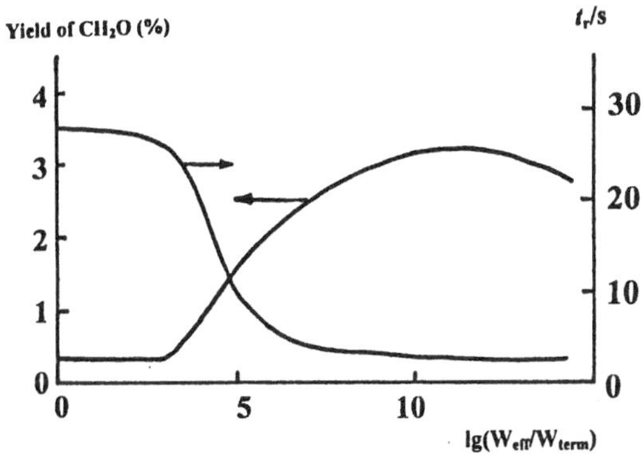

Fig. 4. Simulation of CH_2O yield and reaction time t_r vs. lg(W_{eff}/W_{therm}). T=753 K, P=1 atm, CH_4:O_2 = 9:1.

Although product yield in gas phase process is low enough (≈0.4%), the increase of CH_3 radicals generation enhances this yield up to 3-4% which is very close to the best experimental results [8,14]. And as well as at oxidative coupling, here we also observe limited yield of the product which can not be overcome by more intensive generation of radicals.

These examples indicate the close link between gas phase and surface processes of methane oxidation, the last is predominantly provided the necessary rate of radicals generation. There were some attempts to develop a common scheme of homogeneous-heterogeneous oxidation of methane, e.g. a scheme of methane oxidative coupling, taking into account elementary reactions of free radicals with the surface [15,16]. But these attempts are based on different assumptions and correlations and can not be regarded as reliable enough. Analysis of diffusion processes of atoms and free radicals in porous catalyst [17] have shown that CH_3 radicals are the only active particles outgoing into the gas phase and initiating all subsequent conversion processes at catalytic oxidation of methane. All other atoms and free radicals are decay at their interaction with surface. So this fact along with the results of our study lets to make some assumption on the role and necessary characteristics of catalyst at high temperature oxidative conversion of methane.

CONCLUSIONS

In the majority of high temperature catalytic processes of oxidative conversion of methane the role of catalyst, at any rate in the first approximation, is restricted by its ability to generate (and decay) radicals.

For an effective catalysis of the processes of oxidative conversion of methane catalyst should provide the rate of generation of methyl radicals exceeding the rate of their thermal gas phase generation at least on 5-6 orders of magnitude.

There exists the maximal yield of desirable products which can not be overcome by the further increase of the rate of initiation of the process.

REFERENCES

1. O.V.Krylov, Catal.Today, 18 (1993) 209.
2. M.I.Temkin, Adv.in Catalysis. 28 (1979) 175.
3. A.M.De Groot, G.F.Froment, Appl. Catalysis, A138 (1996) 245.
4. L.Ma, D.L.Trimm, C.Jiang, Appl. Catalysis, A138 (1996) 275.
5. Z.Zhang, X.E.Verykios, Natural Gas Conversion IV, Studies in Surface Science and Catalysis, Vol. 107, M.de Pontes, R.I.Espinoza, C.P.Nicolaides, J.H.Scholz, and M.S.Scurrell (eds.), Elsevier, Amsterdam, 1997, P.511.
6. Y.N.Wang, R.O.Hermann, K.Klier, Surface Sci., 279 (1992) 33.
7. V.A.Durante, D.W.Walker, W.Seitzer, J.E.Lyons, Proc.Symp. on Methane Activation (Intern. Chem. Congress on Pacific Basin), Honolulu, 1989, P.20.
8. A.Parmaliana, F.Arena, F.Frustery, Catal.Today, 24 (1995) 231.
9. A.Parmaliana, N.Giordana, F.Arena, J.Catalysis, 167 (1997) 57.
10. J.G.McCarty, A.B.McEwen, and M.A.Quinlan, in: New Developments in Selective Oxidation, G.Centi, F.Trifiro (eds.), Elsevier, Amsterdam, 1990, P.393.
11. V.I.Vedeneev, O.V.Krylov, V.S.Arutyunov, V.Ya.Basevich, M.Ya.Gol'denberg, M.A.Teitel'boim. Appl. Catalysis, A127 (1995) 51.
12. V.S.Arutyunov, V.Ya.Basevich, V.I.Vedeneev. Russian Chemical Reviews. 65 (1996) 197.
13. V.I.Vedeneev, V.S.Arutyunov, V.Ya.Basevich, M.Ya.Gol'denberg, M.A.Teitel'boim, N.Yu.Krymov. Catal.Today, 21 (1994) 527.
14. S.Irusta, E.A.Lombardo, and E.E.Miro. Catal.Letters, 29 (1994) 389.
15. M.Yu.Sinev, Catal.Today, 24 (1995) 231.
16. C.A.Mins, R.Mauti, A.M.Dean, K.D.Rose, J.Pys.Chem., 38 (1994) 13357.
17. P.M.Couwenberg, Qi Chen, G.B.Marin, Ind.Eng.Chem.Res., 21 (1996) 415.

NATURAL GAS CONVERSION V
Studies in Surface Science and Catalysis, Vol. 119
A. Parmaliana et al. (Editors)

Pb-substituted Hydroxyapatite Catalysts Prepared by Coprecipitation Method for Oxidative Coupling of Methane

Kwan-Young Lee,[a] Yoon-Chull Han,[b] Dong Jin Suh,[b] and Tae-Jin Park[b]

[a] Department of Chemical Engineering, Korea University, Anam-dong, Sungbuk-ku, Seoul 136-701, Korea

[b] Division of Chemical Engineering, Korea Institute of Science and Technology, P.O.Box 131, Cheongryang, Seoul 136-791, Korea

Pb-substituted hydroxyapatites ($Ca_{(10-x)}Pb_x(PO_4)_6(OH)_2$, x=0-10) prepared by coprecipitation method were studied as catalysts for oxidative coupling of methane using gas flow reactor in the temperature range of 675 to 800°C. They showed much higher activity for the production of C_2 compounds than those of unsubstituted hydroxyapatites. Especially, $Ca_{9.5}Pb_{0.5}(PO_4)_6(OH)_2$ showed the highest activity among them, which revealed maximum C_2 yield of about 22%. As compared to the catalysts prepared by ion-exchange method, these catalysts had much higher activity and better thermal stability. Such high activity and stability of the catalysts could be confirmed by XPS, XRD, ICP, etc. to be resulted from the optimum amount of Pb substitution, uniformity of structure, and the maintenance of their own structure during the reaction.

1. INTRODUCTION

As a kind of effective utilization of natural gas, oxidative coupling of methane has been a primary target to many researchers since 1980's [1]. Various catalysts have been reported to show quite good performance even though yield of C_2-compounds still remains below 25% [2]. Among the effective catalysts, supported PbO has been much interested in since the research by Baerns et al. showed C_2 selectivity higher than 80% [3,4]. However, Pb component tends to be lost during the reaction and the catalyst becomes deactivated [5]. Recently, several attempts to overcome the problem have been reported using nonvolatile salts or mixed oxides containing Pb. Among them, lead phosphate, lead sulfate [4], and Pb ion-exchanged hydroxyapatite [6] have been reported to show some activity but the maximum C_2 yield was as low as 13%.

Based on the author's previous experiment for substituted hydroxyapatites [7], the uniformity and stability of Pb ion-exchanged hydroxyapatite were questionable. Thus, coprecipitation method instead of ion-exchange was used for the preparation of the substituted hydroxyapatite catalysts in this study. As a result of using coprecipitation method for the preparation of Pb-substituted hydroxyapatite, higher catalytic activity and better thermal stability were achieved [8]. In this study, reactivity and stability of Pb-substituted hydroxyapatites ($Ca_{(10-x)}Pb_x(PO_4)_6(OH)_2$, x=0-10, which will be abbreviated hereafter by Ca(10-x)Pbx) prepared by coprecipitation has been studied more systematically and the characterization to elucidate the reason for the substantial effect of preparation method on them has been carried out.

2. EXPERIMENTAL

Pb-substituted hydroxyapatites were prepared by coprecipitation of $Ca(NO_3)_2$, $Pb(NO_3)_2$, and $NH_4H_2PO_4$ aqueous solutions as reported in the literature [9]. In N_2 atmosphere, the $Ca(NO_3)_2$ and $Pb(NO_3)_2$ solutions (0.2 M, each) were titrated simultaneously into the $NH_4H_2PO_4$ solution. During the titration, pH was adjusted to 10.3-10.8 by addition of aqueous ammonia. After reflux at 90°C overnight, the precipitates were filtered and washed three times repeatedly, and then dried at 110°C for 3 h. The dried powder catalysts were heated in O_2 atmosphere with increase of temperature at the rate of 5°C/min up to 300°C, and calcined at 800°C for 2 h subsequently.

For the comparison, $Ca_{7.6}Pb_{1.46}(PO_4)_6(OH)_2$ (which will be abbreviated by Pb25Ap1.51 as in ref. 6), the best catalyst in the report by Matsumura et al. was selected and prepared by ion-exchange according to Matsumura's method [6].

The reaction was performed using a continuous flow reactor made of quartz. Methane, oxygen, and He as a carrier gas were introduced at the feed rate of 8, 4, and 25 ml/min, respectively. 0.36 g of catalyst (W/F = 0.582 g s cm^{-3}) was used and its reactivity was investigated in the temperature range of 675 to 800°C using on-line GC. XRD, XPS, ICP, AA, BET, and TG/DTA were performed for the characterization of the catalysts.

3. RESULTS AND DISCUSSION

Reactivity of hydroxyapatite catalyst and the effect of Pb substitution are shown in Fig. 1. Calcium hydroxyapatite ($Ca_{10}(PO_4)_6(OH)_2$) without substitution of Pb showed quite high conversion near 30% but very low C_2 selectivity at 725°C. On the other hand, Pb-substituted hydroxyapatite catalysts showed much higher selectivity than the unsubstituted one. Especially, Ca9.5Pb0.5 showed the highest activity among them to reach the C_2 yield of 20% at 725°C as shown in Fig. 1. It is not clear now why this catalyst showed the best result. However, considering that the hydroxyapatite structure can be changed more easily with the increase of the amount of cation-substitution[7], the incorporation of Pb with maintaining the hydroxyapatite structure seems to be responsible for the high activity. Temperature dependency of conversion, selectivity, and yield for the catalyst is shown in Fig. 2. Conversion of methane, yield of C_2 compounds, and selectivity to ethylene increased with temperature up to 725°C, while selectivity to ethane decreased. It seems likely that the ethylene is formed at high temperature as a secondary product from the ethane by oxidative dehydrogenation as Ross[5], Hinsen[2], and Ito[10] have discussed previously. No further changes were observed above 725°C. As stated previously, there seems to be a certain limit to the maximum yield of C_2 products according to McCarty et al. as shown in Fig. 3 [2,11]. The results of this study were added at the figure for comparison, indicating that Ca9.5Pb0.5 catalyst has comparably high catalytic activity for oxidative coupling of methane.

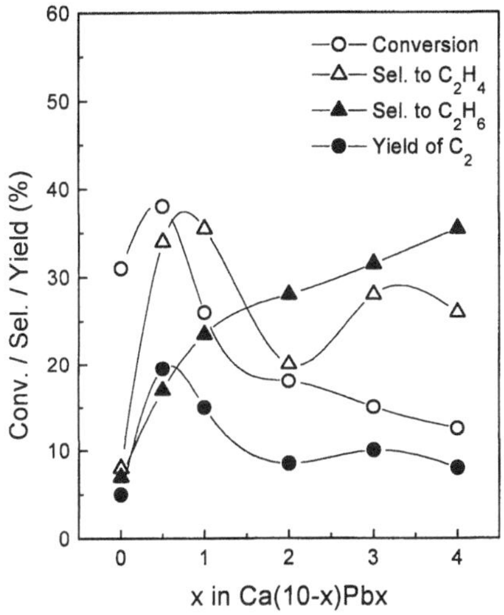

Figure 1. Effect of Pb content on reactivity of Pb-substituted hydroxyapatite catalysts at 725°C.

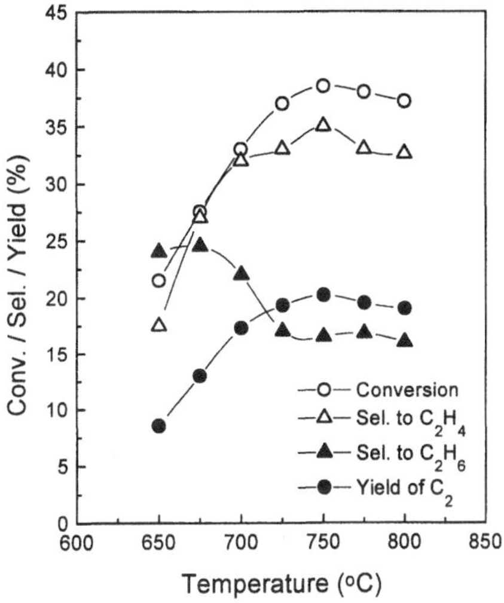

Figure 2. Temperature dependency of conversion, selectivity, and yield for Ca9.5Pb0.5 catalyst.

The initial motive to select Pb-substituted hydroxyapatite as catalyst was to maintain the reactivity by keeping Pb-contents during the reaction. Under this consideration, life test of this catalyst was performed and the results are shown in Fig. 4. In this figure, the results of Pb25Ap1.51 catalyst were plotted together for comparison. At 750°C, Ca9.5Pb0.5 showed 38.8% conversion and 21.1% C_2 yield at the beginning of the reaction and the reactivity was maintained for over 40 h, while Pb25Ap1.51 exhibited 25.8% conversion and 16.3% C_2 yield at the beginning and was deactivated quickly to show only 18% conversion and 10.9% C_2 yield after 40 h. In the case of Ca9.5Pb0.5, the similar results were obtained at 800°C as well, at which 40% conversion and 22% C_2 yield did not noticeably change even after the reaction for 40 h. Even though further studies, for example, at elevated oxygen partial pressure, are necessary to show the feasibility in practice, these results indicate that Ca9.5Pb0.5 catalyst is a promising catalyst with high activity and thermal resistance for the oxidative coupling of methane, for which the preparation method is very important.

By using several analytical tools, the differences between two catalysts prepared differently were investigated. Table 1 shows the changes of BET surface area, and Ca/P and Pb/P molar ratios during the reaction at 800°C for 40 h. Decrease of surface area of Ca9.5Pb0.5 was relatively small comparing to that of Pb25Ap1.51. Bulk structure and composition of Ca9.5Pb0.5 measured by ICP and AA were also more stable. However, the differences of changes of surface area and composition between two catalysts were not enough to explain the substantial differences of activity and stability in Fig. 4. Fig. 5 shows XPS spectra for the fresh catalysts and used ones at 800°C, which indicates that both catalysts hardly lost Pb contents at surface after reaction. TG/DTA also showed that no significant loss of Pb occurred in the both cases. Fig. 6 shows the XRD patterns of fresh and used catalysts. Not only fresh Ca9.5Pb0.5 but also used one had the peaks of hydroxyapatite only. Peak at 31.7° in 2θ was assigned as a major peak of hydroxyapatite due to (211) and (121) planes (JCPDS file, No. 9-432). On the other hand, fresh Pb25Ap1.51 showed much different pattern, which could be assigned to the mixed phases of tricalcium phosphate, $Ca_3(PO_4)_2$ (JCPDS files, No. 29-359 and No. 9-169) and lead hydroxyapatite, $Pb_5(PO_4)_3(OH)$ (JCPDS file, No. 8-259). This catalyst changed to a similar structure to hydroxyapatite after reaction. Bigi et al. have reported the main peak of hydroxyapatite shifted to low 2θ value with incorporation of Pb [12]. Considering their results and higher Pb contents for Pb25Ap1.51 than Ca9.5Pb0.5, the peak at 31.4° in 2θ can be assigned to hydroxyapatite. However, as shown in Fig. 6(A), there also appeared many peaks which could not be assigned to hydroxyapatite in the used Pb25Ap1.51, indicating that mixed phases were formed. Therefore, it can be concluded that such high activity and stability of Ca9.5Pb0.5 resulted from the optimum amount of Pb substitution, uniformity of structure, and the maintenance of its own structure during the reaction.

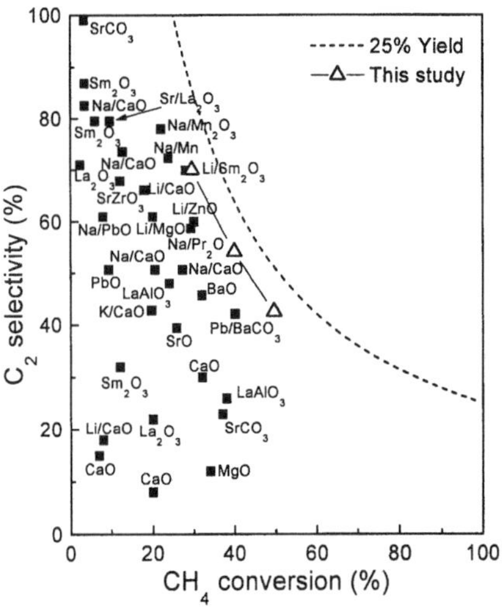

Figure 3. Comparison of catalytic activity of Ca9.5Pb0.5 to the other published results.

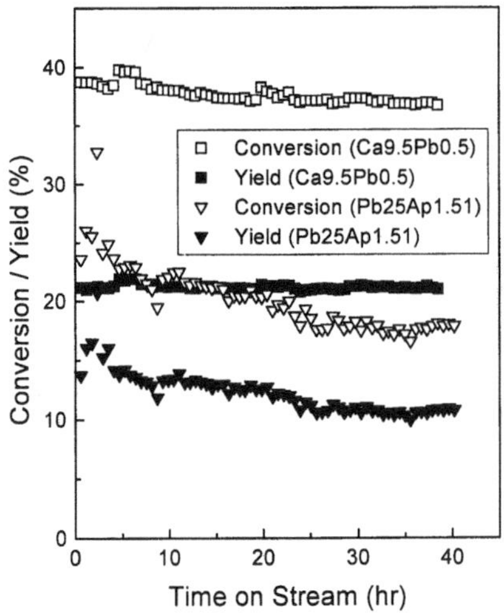

Figure 4. Comparison of stability of Ca9.5Pb0.5 and Pb25Ap1.51 at 750°C.

4. CONCLUSION

Pb-substituted hydroxyapatite was very efficient catalyst for oxidative coupling of methane. The reactivity depended strongly on Pb contents and preparation method. Among Pb-substituted hydroxyapatites, Ca9.5Pb0.5 prepared by coprecipitation showed the best catalytic activity. With this catalyst, C_2 yield over 20% was achieved at relatively low temperature, 725°C. As compared to Pb25Ap1.51 that was the best one among the catalysts prepared by ion-exchange method in the previous report, Ca9.5Pb0.5 showed much higher activity as well as better stability. The difference between two catalysts was confirmed not due to the loss of Pb. Such high activity and stability of Ca9.5Pb0.5 catalyst came from the optimum amount of Pb substitution, uniformity of structure, and the maintenance of its own structure during the reaction realized by proper preparation method.

ACKNOWLEDGEMENT

This study has been supported by the Research-Aid Fund of the Ministry of Science and Technology of Korea.

Table 1
BET surface area and bulk composition of fresh and used catalysts

Catalysts	BET surface area / $m^2 g^{-1}$		Changes of molar ratio[a]					
			Ca/P			Pb/P		
	fresh	used	intended[b]	fresh	used	intended[b]	fresh	used
Ca9.5Pb0.5	28.9	18.1	1.58	1.52	1.52	0.083	0.085	0.086
Pb25Ap1.51	54.1	16.1	1.27	1.46	1.39	0.24	0.36	0.34

[a] Ca, Pb were measured by ICP, while P by AA.
[b] stoichiometry intended as preparation.

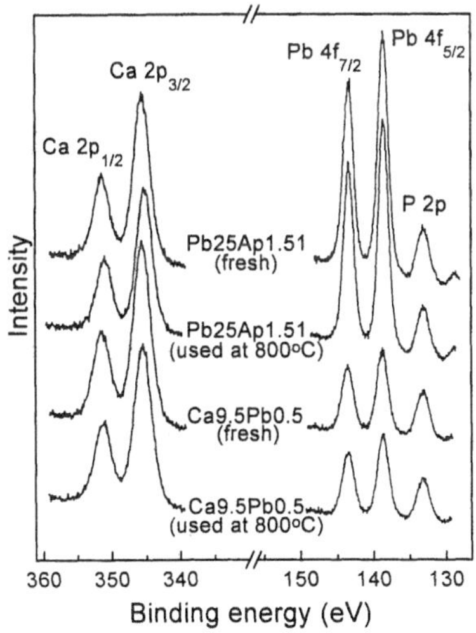

Figure 5. XPS spectra for the fresh catalysts and used ones at 800°C.

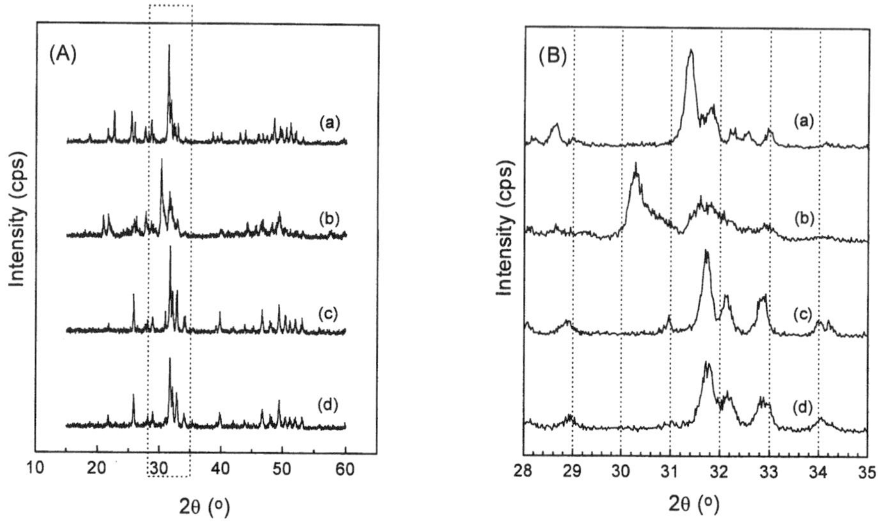

Figure 6. XRD patterns of fresh and used catalysts in 2θ of 15-60° (A) and 28-35° (B). (a) Pb25Ap1.51 used in the reaction at 800°C for 40 h, (b) Pb25Ap1.51 calcined at 500°C for 1 h (fresh), (c) Ca9.5Pb0.5 used in the reaction at 800°C for 40 h, (d) Ca9.5Pb0.5 calcined at 800°C for 2 h (fresh).

REFERENCES

1. G. E. Keller and M. M. Bashin, *J. Catal.*, **73**, 9 (1982).
2. O. V. Krylov, *Catal. Today*, **18**, 209 (1993).
3. W. Hinsen, W. Bytyn, M. Baerns, *Proc. 8th ICC*, **3**, 581 (1984).
4. J. A. S. P. Carreiro and M. Baerns, *Reac. Kinet. Catal. Lett.*, **35**, 349 (1987).
5. J. A. Roos, A. G. Bakker, H. Bosch, J. G. van Ommen, and J. R. H. Ross, *Catal. Today*, **1**, 133 (1987).
6. Y. Matsumura and J. B. Moffat, *Catal. Today*, **17**, 197 (1993).
7. K.-Y. Lee, M. Houalla, D. M. Hercules, and W. K. Hall, *J. Catal.*, **145**, 223 (1994).
8. K.-Y. Lee, D. J. Suh, and T.-J. Park, *Abstract of 3rd World Congress on Catalysis*, E-3 (1997).
9. K. Yamashita, H. Owada, H. Nakagawa, T. Umegaki, and T. Kanazawa, *J. Am. Ceram. Soc.*, **69**, 590 (1986).
10. T. Ito, J-X. Wang, C-H. Lin, and J. H. Lunsford, *J. Am. Chem. Soc.*, **107**, 5062 (1985).
11. J. G. McCarty, A. B. McEven, M. A. Quinan, *New Developments in Selective Oxidation, Proc. Intern. Congress (Rimini, Italy, 1989)*, Amsterdam, Elsevier, 1990, p.393.
12. A. Bigi, A. Ripamonti, S. Brückner, M. Gazzano, N. Roveri, and S. A. Thomas, *Acta Cryst.*, **B45**, 247 (1989).

NATURAL GAS CONVERSION V
Studies in Surface Science and Catalysis, Vol. 119
A. Parmaliana et al. (Editors)

The production of hydrogen through methane conversion over reagent catalysts. An evaluation of the feasibility of catalytic cracking unit utilization for methane conversion

M.I.Levinbuk[a] and N.Y.Usachev[b]

[a]Gubkin State Academy of Oil and Gas, 65 Leninsky prospect, 117 917 Moscow, Russia.
[b]Insitute of Organic Chemistry, 47 Leninsky prospect, 117 913 Moscow, Russia.

1. INTRODUCTION

Contemporary requirements imposed on the quality of petroleum products stipulated an increase in hydrogen consumption for hydrofining process from 0.3 to 1.8 wt. % of the quantity of crude oil consumed by refinery plants [1]. However, the reformulated gasoline program introduced in the USA in the 90s demands for a decline in reforming capacities, which considerably reduces the potential for hydrogen production at refinery plants [2]. The steam conversion of natural gas is a very power-intensive process affording a mixture of CO and H_2, which requires a subsequent separation of components for hydrogen to be utilized in hydroprocesses. The introduction of variable valence metals (nickel oxides) into cracking catalysts containing pentasil zeolites instead of Y-zeolites initiates methane partial oxidation reactions, the continuity of which is ensured by the catalytic cracking technology [3].

The catalyst used in this process circulates between the reactor and the regenerator thus partially oxidizing the feed in the first vessel while further oxidation of the reduced metal oxide occurs in the second one. We call it as a "reagent catalyst" since it acts as the carrier of the reagent (oxygen in this case) too. It is necessary to calculate the material and heat balances of the reactor-regenerator vessels of old Russia's TCC (Thermofor Catalytic Cracking) units to determine whether they can be utilized for methane conversion into hydrogen.

2. EXPERIMENTAL

Nickel incorporating pentasil (ZSM-5) zeolites (3÷12 wt. % of NiO) were prepared by impregnating zeolite samples having various SiO_2/Al_2O_3 molar ratios (29÷210) with a $NiNO_3$ solution. Methane catalytic cracking into carbon and hydrogen was examined over nickel-containing pentasil zeolites in a vacuum circulation laboratory unit [4] that lets catalytic, adsorption and acidic properties of these samples be investigated in one reactor (with a 0.2 g sample loaded). Catalytic methane conversion was studied at catalyst/feed mass ratios varying in the range between 5 and 12 and at reaction temperatures of 475-570°C. Adsorption, acidic and catalytic properties of nickel-containing pentasil zeolites were investigated upon high-temperature (570÷600°C) pretreatment by hydrogen and oxygen. Oxygen, hydrogen, nitrogen and carbon oxide adsorption over pentasil zeolite samples was studied at the liquid nitrogen temperature (-196°C). Acidic properties of pentasil zeolite samples were investigated by high-temperature desorption of ammonia adsorbed at 20°C.

3. RESULTS AND DISCUSSION

The purpose of the present is only to demonstrate the most important experimental results, so there are described catalytic and adsorption properties of pentasil zeolites with SiO_2/Al_2O_3 molar ratios of 29÷210 and a NiO content of 8.0 wt. %. Methane conversion was not observed over pentasil zeolites with a NiO content of 3.0 wt. % at a reaction temperature of 550°C. Principal rules of catalytic methane cracking into carbon and hydrogen are depicted in Fig.1 and 2.

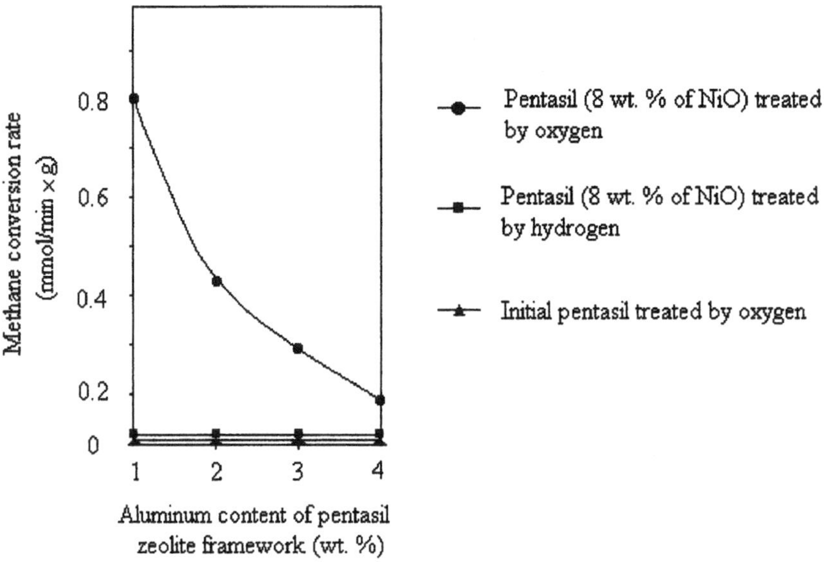

Figure 1. Methane conversion rate vs. the aluminium content of zeolite framework and kind of its preliminary treatment.

Fig.1 shows the dependence of methane cracking rate on the aluminum content of zeolite pentasil framework at 550°C and a catalyst/feed mass ratio of 5.0. Fig.1 implies that *the decline in the aluminum content (i.e. the growth of SiO_2/Al_2O_3 molar ratio) causes a fourfold increase in the rate of methane cracking into carbon and hydrogen* over investigated samples of nickel-containing pentasil zeolites. Nickel-containing pentasil zeolites pretreated by hydrogen were not active in methane conversion.

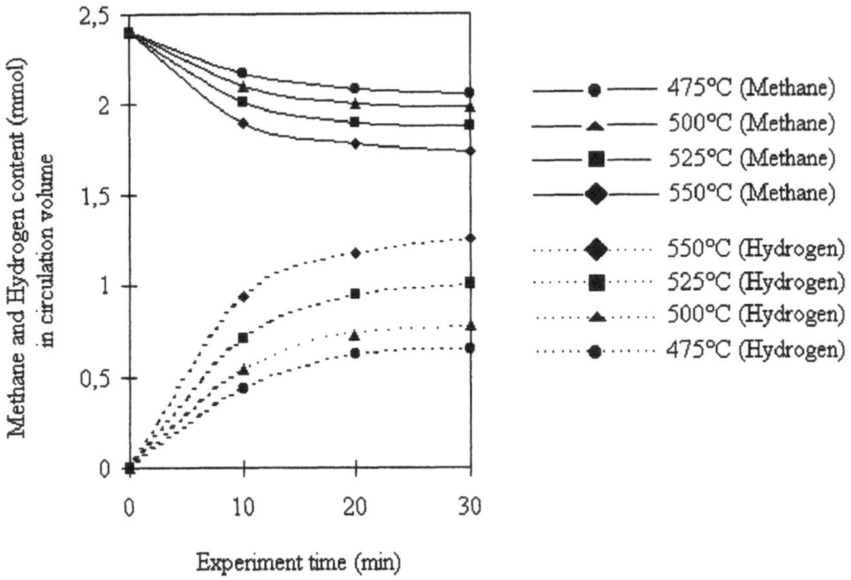

Figure 2. Kinetic characteristics of methane conversion and hydrogen yield at different temperatures of methane decomposition reaction on pentasil zeolite (SiO_2/Al_2O_3 = 80) with 8 wt% of NiO.

Fig.2 represents kinetic curves for methane conversion and the hydrogen yield over an oxidized sample of nickel-containing pentasil zeolite (a SiO_2/Al_2O_3 molar ratio of 80) at 475 ÷550°C reaction temperatures and a catalyst/feed mass ratio of 5.0. It follows from Fig.2 that methane converts into carbon and hydrogen *with no oxidation products present in the gas phase*. 2 moles of H_2 were obtained per mole of converted CH_4 in all the experiments carried out. Hence, probably, carbon either accumulates on the catalyst surface or form compounds with the nickel component.

The investigation of adsorption properties of nickel-containing pentasil zeolites (with a SiO_2/Al_2O_3 molar ratio of 80) revealed that the quantities of adsorbed oxygen and hydrogen change upon high-temperature pretreatment of samples by oxygen and hydrogen within the investigated range of equilibrium adsorbate pressures (Fig.3). The difference between oxygen adsorption over oxidized and reduced forms of the Ni-pentasil zeolite samples is 0.32 mmol/g, which is equal to 30 m^2/g when converted to nickel oxide surface. So the nickel oxide surface actually active in methane cracking is probably 30 m^2/g. The values of carbon oxide and nitrogen adsorption on Ni-containing pentasil zeolites do not change after pretreatment by oxygen and hydrogen.

The results obtained testify to the feasibility of using the reactor-regenerator vessels of catalytic cracking units in order to obtain pure hydrogen by natural gas conversion. On the

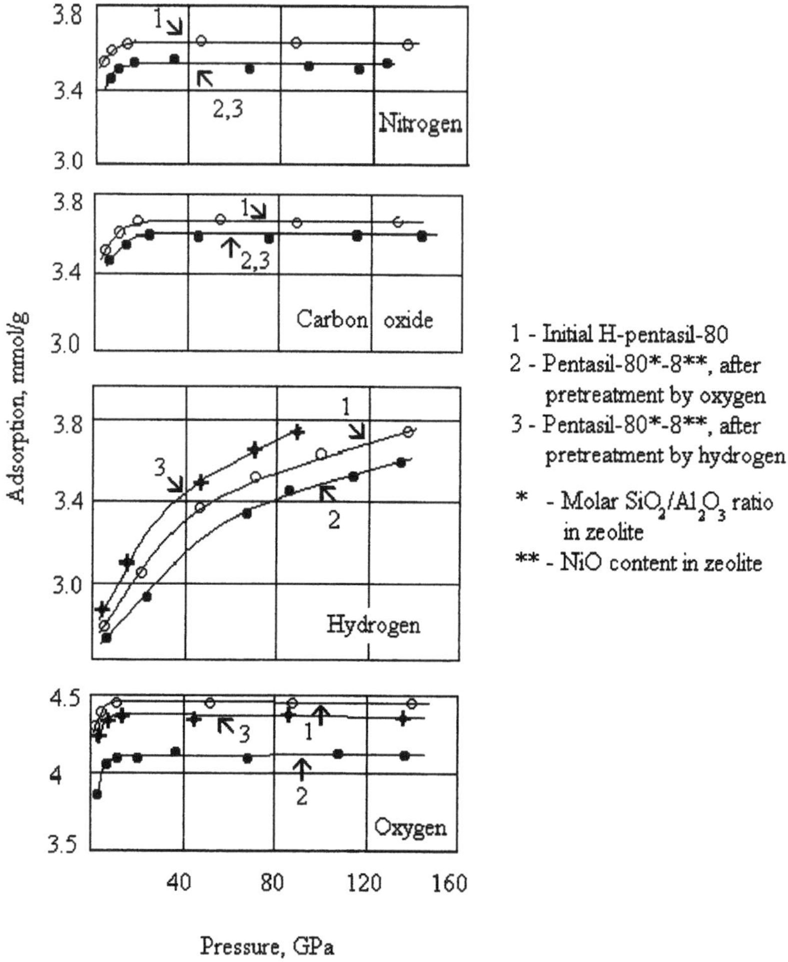

Figure 3. Adsorption isotherms of various substances at -196°C over nickel-containing pentasil zeolites.

basis of the experiments carried out, an optimum nickel-containing pentasil zeolite (a SiO_2/Al_2O_3 molar ratio of 210, a 10 wt. % NiO content in zeolite) was selected to minimize the modernization of industrial catalytic cracking units required to adapt them to this process. Introduction of 20 wt. % of the mentioned zeolite into the cracking catalyst matrix ensures a 70 % methane conversion at 580°C and a Ni-pentasil/methane mass ratio of 6.0.

The results of calculations of material and heat balance of a TCC unit and those for a unit minimally modified to adapt its reactor-regenerator vessels for the conversion of methane into hydrogen are represented in Table 1.

Table 1
The results of calculations of material and heat balance of the operation of the reactor-regenerator vessels of a TCC unit for vacuum gas oil and methane cracking respectively (the operation conditions of the regenerator vessel are not changed)

#	Parameters	Conditions of cracking on reactor-regenerator vessels of TCC units	
		Vacuum gas oil feed	Methane feed
1	Coke burning-out capacity of the regenerator, kg/h	1400	1400
2	Feed rate, ton/h	35.0	2.7
3	Catalyst circulation rate, ton/h	60	120
4	Catalyst/feed mass ratio	1.4	30
5	Feed cracking heat, kcal/kg	50	3100
6	Feed conversion, wt. %	60	70
7	Cracking temperature, °C	460	580
8	Gas volume rate at the exit from the reactor, m^3/sec	5.9	5.3
9	Purpose product yield, ton/h	12.2 (gasoline)	0.46 (hydrogen)
10	Temperature of feed preheating, °C	480	600

It is follows form the Table 1 that the major problem in the way of utilization of the reactor-regenerator vessels of TCC units for natural gas conversion is the heating of the reactor, because the specific heats of vacuum gas oil cracking and methane cracking differ considerably.

Consequently, the reactor-regenerator vessels of TCC units have to be equipped with new compressors in the pneumatic transport system so as to increase the rate of catalyst circulation and the tube furnace has to be modified to preheat the feed up to 600°C rather than 480°C. A mixture of hydrogen and unconverted CH_4 leaves the reactor. The mixture of CH_4 and H_2 can be separated easier than that of H_2 and CO; besides, small CO admixtures are less desirable than those of CH_4 when hydrogen is used in hydroprocesses.

In the case of FCC (Fluid Catalytic Cracking) units only the feed preheating tube furnace has to be modernized since the catalyst circulation rate in these units reaches 1000 ton/h. Hence the reactor-regenerator vessels of a TCC unit (with a vacuum gas oil capacity of 300000 ton/year) can produce up to 4000 ton of pure hydrogen per year. For comparison, 4500 ton of hydrogen per year were produced at a reforming unit with a 300000 ton/year capacity.

4. CONCLUSION

The introduction of Ni-containing pentasil zeolites into TCC and FCC catalyst matrices makes it possible to generate pure hydrogen from natural gas by using the existing technology of the production of such catalysts and the reactor-regenerator vessels of cracking units, with only minor modifications required.

REFERENCES

1. J.Bousquet, M.Valas., Large Chemical Plants, 9th Int. Symp., Belgium, Antwerpen, (1995) 241
2. R.H.Gilman, Oil and Gas J., Sept. 3 (1990) 44
3. M.I.Levinbuk, N.Y.Usachev et al., 3rd World Congr.on Oxid. Catalysis, Studies in Suface Science and Catalysis, San-Diego, Elsevier, v.110 (1997) 731
4. U.V.Shumovski et al., React. Kinet. Catal. Lett., 21 (1983) 3

NATURAL GAS CONVERSION V
Studies in Surface Science and Catalysis, Vol. 119
A. Parmaliana et al. (Editors)

Catalytic Partial Oxidation of Methane at Extremely Short Contact Times: Production of Acetylene

L. D. Schmidt[a], K. L. Hohn, and M. B. Davis

[a]Department of Chemical Engineering and Materials Science, University of Minnesota, Minneapolis, Minnesota 55455

The adiabatic and autothermal oxidative coupling of methane over Pt on α-Al_2O_3 foam monoliths has been studied at space velocities above 10^5 hr^{-1} and residence times of < 5 milliseconds. While the selectivity to C_2's is very low on Rh, increasing from <0.5% to 5% as space velocity is doubled, C_2 selectivities as high as 20% have been observed on Pt. Acetylene is the favored C_2 on Pt under most conditions. High space velocities, low fuel to oxygen ratios, and low N_2 dilution lead to coupling selectivities near 20%. The driving force for C_2 production appears to be the temperature profile which appears at high space velocities. As the space velocities is increased, the front temperature of the monolith decreases to less than 200°C while the back temperature increases to nearly 1500°C. It is suggested that the high temperature at the back of the monolith may be high enough that homogeneous reactions compete with surface reactions. A combustion model is used to predict C_2 selectivities from homogeneous reactions alone, and model predictions are similar to experimental data.

1. INTRODUCTION

The conversion of natural gas into value-added chemicals has received great attention in the last twenty years. There have been two distinct approaches to methane conversion. One approach is to convert methane into more valuable products, such as higher hydrocarbons, in a single step. Methane can be converted to higher hydrocarbons through oxidative coupling, where methane and oxygen are fed over a metal or metal oxide catalyst [1] or through methane homologation, where methane is fed to produce a carbon layer and then hydrogen is fed to desorb higher hydrocarbons [2,3]. Both of these processes produce high selectivities to higher hydrocarbons with low methane conversions

The other method of methane conversion has been to produce synthesis gas, from which a wide range of products can subsequently be produced. Synthesis gas can be produced with high selectivity over Rh monoliths at millisecond contact times [4]:

$$CH_4 + 1/2\ O_2 \rightarrow CO + 2\ H_2 \quad (1)$$

Less than 0.5% C_2 hydrocarbons are produced in this process at space velocities around 10^5 hr^{-1}. However, it has recently been found that as the space velocity is increased, significant amounts of C_2 hydrocarbons are produced [5]. C_2 selectivities as high as 10% were found. We find that using Pt instead of Rh means at least twice the coupling selectivity.

The formation of any higher hydrocarbons in a syngas process would be undesirable because of coke formation problems. However, a high selectivity to acetylene could be desirable, because this simple process could be used to generate acetylene catalytically.

2. EXPERIMENTAL

Catalysts were prepared as detailed previously [6]. Ceramic foam α-Al_2O_3 monoliths (18 mm diameter, 10 mm long, 45 pores per linear inch) were saturated with an aqueous solution of H_2PtCl_6 and dried overnight. The catalysts were then calcined in air at 600°C and reduced in H_2 at the same temperature. Metal loadings were 2 to 3 % Pt by weight.

Experiments were carried out in a tubular quartz reactor 40 cm long and 18 mm in diameter. To better approximate adiabatic operation, insulation was wrapped around the exterior of the reactor and uncoated ceramic monoliths were placed before and after the catalyst to reduce axial radiative heat losses. The temperature of the catalyst was measured with a Pt/13%Rh thermocouple placed at the exit and a chromel/alumel thermocouple placed at the entrance of the catalyst.

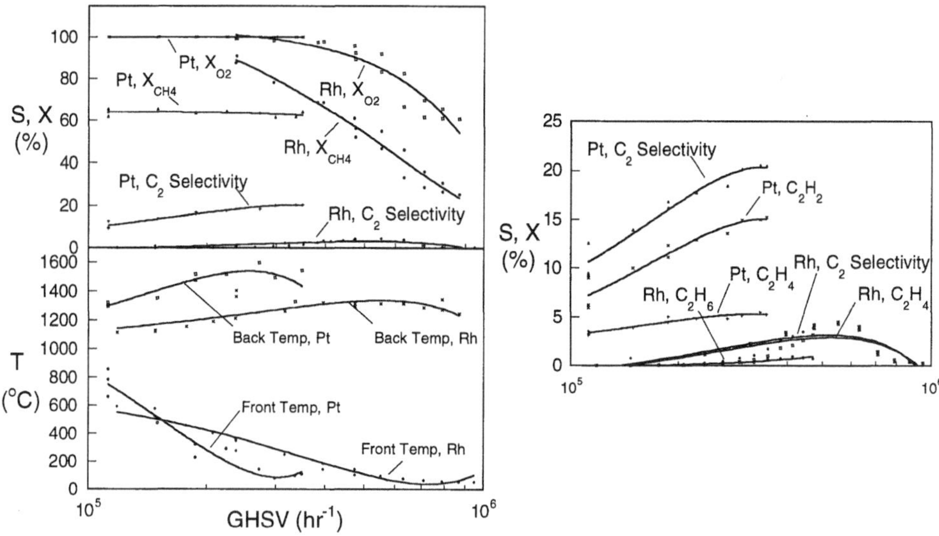

Figure 1. Effect of gas hourly space velocity (GHSV) on methane coupling. Panel a) shows CH_4 and O_2 conversions and C_2 selectivity vs. GHSV on Rh and Pt. Panel b) shows front and back temperatures vs. GHSV on Rh and Pt. Panel c) shows selectivities for different C_2 hydrocarbons vs. GHSV on Rh and Pt.

Reactant gases were controlled by mass flow controllers with an accuracy of ± 0.01 standard liter per minute, slpm. The CH_4/O_2 ratio was 1.7 and N_2 dilution was 20% for all experiments unless otherwise stated. Product gas compositions were measured with an HP 5890 gas chromatograph equipped with thermal conductivity and flame ionization detectors and integrated by an on-line computer. Methane conversions and product selectivities are determined from molar flow rates and all mass balances closed within ± 5%.

3. RESULTS AND DISCUSSION

3.1. Effect of Flow Rate

The conversions of O_2 and CH_4 and the total coupling selectivity are shown in Figure 1 panel a) as a function of GHSV for Rh and Pt at CH_4/O_2=1.7. On Rh, both O_2 and CH_4 conversions fall as GHSV is increased above 4 x 10^5 hr^{-1}. CH_4 conversion drops from above 80% to around 30% at GHSV of 9 x 10^5 hr^{-1}. O_2 conversion drops from 100% to 60% at GHSV of 9 x 10^5 hr^{-1}. Coupling selectivity increases from less than 0.1% to around 5% before dropping back to near zero at the highest space velocities. On Pt, the O_2 and CH_4 conversions remain nearly constant at 100% and 62%. Coupling selectivity rises from 10% at GHSV of around 10^5 hr^{-1} to 20% at space velocities near 2.5 x 10^5 hr^{-1}. Panel b) shows front and back temperatures as a function of GHSV on Rh and Pt. As space velocity is increased, the exit temperature increases while the entrance temperature decreases for both metals. The exit temperature reaches nearly 1500°C while the entrance temperature decreases to under 200°C for Pt. On Rh, the exit temperature reaches 1300°C while the entrance temperature decreases to almost room temperature.

The panel to the right in Figure 1 shows the selectivities of C_2H_2 and C_2H_4 on Pt and Rh. Unlike Rh, where C_2H_4 is the favored C_2, Pt favors C_2H_2 production. C_2H_2 selectivities increase from 7% at 10^5 hr^{-1} to 15% at 3.5 x 10^5 hr^{-1} on Pt. C_2H_4 selectivities remain relatively constant at 4-5% for all space velocities. C_2H_6 selectivities were less than 1%, and C_3 and C_4 hydrocarbons were not observed.

3.2 Effect of CH_4/O_2 Ratio

The CH_4/O_2 ratio was varied between 1.4 and 2.4 with a constant GHSV of 1.8 x 10^5 hr^{-1}. Maximum coupling selectivities were observed at the lowest CH_4/O_2 ratios. The coupling selectivity increased from 15% at CH_4/O_2 of 1.7 to 19% at CH_4/O_2 of 1.4. At ratios higher than 1.7, coupling selectivity steadily decreases, going to only 1% at a CH_4/O_2 ratio of 2.4. As the ratio changes, the relative selectivities to C_2H_2, C_2H_4, and C_2H_6 change. C_2H_2 selectivity is at a maximum of 15% at a CH_4/O_2 ratio of 1.4. C_2H_4 selectivity is at a maximum of 6% at CH_4/O_2 ratios around 1.7, but tails off to a selectivity of 3% at a ratio of 1.4, and 1% at a ratio of 2.4. The selectivity to C_2H_6 is under 1% at a CH_4/O_2 ratio of 1.4, but steadily increases as the CH_4/O_2 ratio increases. It reaches a maximum selectivity of 2.5% at a CH_4/O_2 ratio of 2.4.

3.3 Effect of Nitrogen Dilution

The nitrogen dilution was varied between 10% and 40% at a CH_4/O_2 ratio of 1.7 and a constant GHSV of 1.8 x 10^5 hr^{-1}. Both CH_4 and O_2 conversions fall with increasing dilution. CH_4 conversion goes from 64% at a N_2 dilution of 10% to just over 30% at 40% dilution. Oxygen conversion falls from 100% at 10% dilution to less than 80% at 40% dilution. As dilution increases the entrance temperature remains relatively constant at ~ 400°C while the exit temperature decreases from 1500°C to under 1400°C. Total coupling selectivity decreases from 16% at 15% N_2 dilution to 4% at 40% dilution. As was the case for CH_4/O_2 ratio, changing the amount of diluent changes the relative selectivities of the C_2's. The selectivity to C_2H_2 reaches a maximum of 11 % at a dilution of 15%, but decreases to almost zero at N_2 dilutions above 30%. C_2H_4 selectivity is a maximum of 7% at 30% dilution. That selectivity drops to around 5% at 10% dilution, and only 1% at 40% dilution. C_2H_6 selectivity increases with increasing dilution, increasing from around 0% at 15% dilution to 3% at 40% dilution.

3.3 Homogeneous Reactions

Homogeneous reactions may be crucial in interpreting these experiments, as we will discuss. During experiments where the flow rate was varied, we observed a flame downstream of the catalyst at space velocities greater than 2.5×10^5 hr^{-1} at a CH_4/O_2 ratio of 1.7. These bright yellow to white flames were several millimeters in diameter and extended several inches down the center of the tube. It did not appear that the catalyst was acting like a flame holder, as we never observed flat flames just behind the catalyst. At space velocities less than 2.5×10^5 hr^{-1}, flames were never observed, no matter what the CH_4/O_2 ratio or nitrogen dilution was.

Since the maximum coupling selectivities of 20% were obtained at conditions just below and including the flow rates at which the flame appeared, it seemed possible that the flame, and not the catalyst, was responsible for producing the C_2's. In order to examine this possibility, 12 uncoated monoliths were placed downstream of the back heat shield to suppress conventional flames. With this setup, no flames were observed at any flow rate, and at least 20% C_2 products were obtained at space velocities around 2.5×10^5 hr^{-1}. At space velocities above 2.5×10^5 hr^{-1}, the conversions of methane and oxygen as well as the coupling selectivity dropped. These results suggest that the reported results at lower space velocities are all occurring within the monolith or the back heat shield. However, at the highest space velocities, the flame is responsible for much of the observed chemistry. Significant amounts of oxygen and methane which are not converted in the monolith are being consumed by the flame to form mainly acetylene.

4. MECHANISM OF OXIDATIVE METHANE COUPLING

We suggest that the temperature profile that develops as the space velocity is increased is responsible for the production of acetylene. At low space velocities, the temperature throughout the monolith is relatively constant at ~1000°C. At this temperature, the dominant reaction is the direct partial oxidation of methane to syngas, and less than 0.5% coupling products are observed. The temperature everywhere in the catalyst is too low for gas phase reactions (simulations suggests < 3%) to be significant and too high for carbon to exist on the surface.

For the temperature profile at high space velocities, two possible mechanisms could explain the formation of C_2's from methane. In the first possible mechanism, a layer of carbon or hydrocarbon first absorbs on the colder front region of the monolith. This carbon layer can then desorb through interaction with gaseous species to form the observed coupling products, $C_s + CH_4 \rightarrow C_2H_2 + H_2$. This mechanism is similar to the methane homologation reaction reported on Pt at 500-800°C [2,3]. The biggest difference is that there is oxygen in our system, which will compete with any coupling reactions on the surface, but which provides the heat to drive the process.

The second possible mechanism at these high space velocities is that coupling products are formed through gas phase reactions in the back of the monolith where the temperature is in excess of 1400°C. The colder front region of the monolith catalyzes complete oxidation of CH_4 to CO_2 and H_2O rather than to CO and H_2 because of the lower temperatures. This highly exothermic reaction releases enough energy to heat up the back of the monolith to temperatures which are high enough to allow gas phase reactions to proceed on similar time scales as surface reactions. Surface reactions lead to partial and complete oxidation, while gas phase reactions produce the observed C_2's.

4.1 Model of Homogeneous Reactions

To better understand which mechanism may be responsible for C_2 production over Pt monoliths, our system was modeled using a kinetic homogeneous oxidation mechanism developed by Mimms and Dean [7] to model gas phase reactions. This model includes 447 reactions and 120 species. Isothermal plug flow is assumed, with no surface reactions.

Simulations were run at 1400°C with a CH_4/O_2 ratio of 1.7 and the predictions were compared to experimental data. Predicted selectivities of C_2H_2, C_2H_4, and C_2H_6 are shown versus time in Figure 2, as are CH_4 and O_2 conversions. At times of less than 0.5 milliseconds, the selectivity to C_2H_6 rises sharply to nearly 40%. C_2H_6 selectivity then begins to decrease as it dehydrogenates to C_2H_4. C_2H_4 selectivity reaches a maximum of nearly 40% at around 0.5 milliseconds. It then begins to dehydrogenate to C_2H_2, and its selectivity decreases. The selectivity to C_2H_2 steadily increases until it is around 20%, at which time it still increases, but at a much slower rate. The total coupling selectivity follows these trends, as it peaks at nearly 60% at 0.3 milliseconds when the methane and oxygen conversions are low and C_2H_6 and C_2H_4 selectivities are high. It then decreases until it reaches roughly 20%,

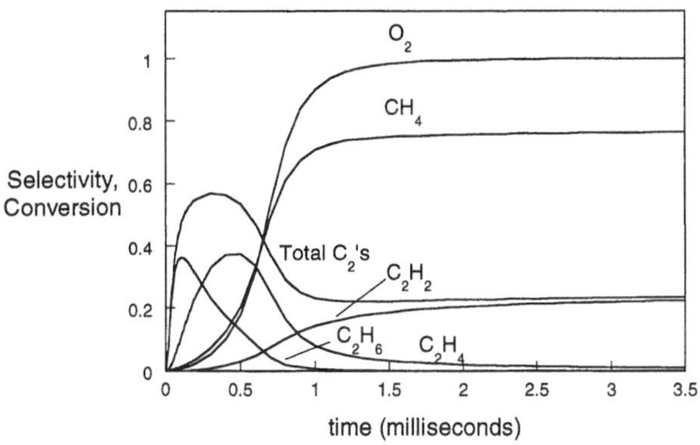

Figure 2. Simulated conversions and selectivities at 1400°C with CH_4/O_2=1.7.

almost all of which is acetylene. The simulated coupling selectivity of just over 20% after several milliseconds is similar to the experimental value of 20%, suggesting that homogeneous reactions alone can explain the observed results. It is interesting to note that the model predicts coupling selectivities as high as 60% if the process could be frozen at ~0.4 milliseconds. Note that these results are transient. Equilibrium calculations at these conditions predict CO and H_2 as the major products.

Simulations run to predict the effect of CH_4/O_2 ratio did not agree with the experimental results. The model predicts higher coupling selectivity with higher CH_4/O_2 ratio while experimentally the opposite is observed. This suggests that homogeneous chemistry alone can not explain the observed trend. It is possible that the observed increase in coupling selectivity with decreasing CH_4/O_2 ratio may be due to thermal effects. CH_4/O_2 ratios closer to stoichiometric will lead to higher temperatures which could lead to increased production of C_2's.

5. CONCLUSIONS

Coupling selectivities as high as 20% have been observed on Pt coated α-Al_2O_3 monoliths at high space velocities, low CH_4/O_2 ratios and low N_2 dilution levels. As the space velocity is increased, the exit of the monolith reaches a temperature of almost 1500°C, which is high enough that the rates of homogeneous reactions may be as fast or faster than heterogeneous reactions. A kinetic homogeneous oxidation model was run isothermally at temperatures similar to the experimental exit temperatures to determine whether homogeneous reactions alone could explain the observed coupling selectivities. Simulated coupling selectivities were similar to experimental results, suggesting that homogeneous reactions may be the primary mechanism for acetylene formation over Pt monoliths at extremely short contact times.

REFERENCES

1. Amenomiya, Y., Birss, V. I., Goledzinowski, M., Galuszka, J., and Sanger, A. R., *Catalysis Review - Science and Engineering* **32**, 163-227 (1990).
2. Amariglio, H., Saint-Just, J., and Amariglio, A., *Fuel Processing Technology* **42**, 291-323 (1995).
3. Koerts, T., and van Santen, R. A., *Journal of the Chemical Society, Chemical Communications* 1281-1283 (1991).
4. Hickman, D. A., Haupfear, E. A., and Schmidt, L. D., *Catalysis Letters* **17**, 223-237 (1993).
5. Witt, P. M., and Schmidt, L. D., *Journal of Catalysis* **163**, 465-475 (1996).
6. Hickman, D. A., and Schmidt, L. D., *Journal of Catalysis* **136**, 300-308 (1992).
7. Mimms, C. A., Mauti, R., Dean, A. M., and Rose, K. D. , *J. Phys. Chem.* **98**, 13357 (1994).

NATURAL GAS CONVERSION V
Studies in Surface Science and Catalysis, Vol. 119
A. Parmaliana et al. (Editors)

Non-oxidative catalytic conversion of methane with continuous hydrogen removal

Richard W. Borry III, Eric C. Lu, Young-Ho Kim*, and Enrique Iglesia

Materials Sciences Division, E.O. Lawrence Berkeley National Laboratory and
Department of Chemical Engineering, University of California, Berkeley CA 94720, USA

Simulations predict maximum C_2-C_{10} yields of 14% for both homogeneous and surface-initiated CH_4 pyrolysis at 1038 K. Yields are limited by thermodynamics, kinetic inhibition by H_2, and carbon formation. Continuous H_2 removal from the system can overcome these constraints and increase maximum predicted yields to 88%. CH_4 pyrolysis experiments at 950 K on 4 wt% Mo/H-ZSM5 achieves near-equilibrium conversions (10-12%) with >90% C_2-C_{10} selectivity by catalyzing both C_2H_4 formation from CH_4 (on MoO_xC_y) and C_2H_4 aromatization (on H^+), while restricting chain growth to benzene and naphthalene. An H-transport membrane reactor of dense $SrZr_{0.95}Y_{0.05}O_3$ thin (10-100 μm) films can be used to overcome these thermodynamic constraints. Self-supporting, thick (1000 μm) disks were prepared via combustion synthesis methods, which form denser membranes than powders formed via co-precipitation. Membrane reactor experiments using thick $SrZr_{0.95}Y_{0.05}O_3$ disks had H-transport rates insufficient to affect CH_4 pyrolysis reactions.

1. INTRODUCTION

The direct conversion of methane to fuels and petrochemicals remains a formidable challenge. Unrestricted chain growth during endothermic pyrolysis leads to undesired carbon and polynuclear aromatics. [1] Oxidative methane coupling avoids thermodynamic constraints by kinetically coupling C-H bond activation with removal of hydrogen via oxidation with O_2. These reactions, however, are unselective and lead to high CO and CO_2 yields. [2, 3] C-H bond activation and hydrogen removal steps can also be coupled by transporting hydrogen atoms across a dense ceramic membrane. This approach preserves the stoichiometry of oxidative coupling without direct contact between CH_4 and O_2 and allows the use of air as the oxidant. Electrochemical attempts to implement this approach have not led to practical yields, because of electrical and oxygen anion conductivity in $SrCe_{0.95}Yb_{0.05}O_3$ conductors and because of carbon deposition at electrodes. [4, 5] In the non-electrochemical approach proposed here (Figure 1), catalytic methane pyrolysis on Mo/H-ZSM5 forms C_{2+} hydrocarbons and H_2 on one side, H-atoms are transported across a dense oxide film, and reacted with air on the other side. Active Mo/H-ZSM5 catalysts allow this system to operate below 973 K, where homogeneous carbon formation is minimal. Here, we describe a rigorous analysis of reactor behavior using detailed kinetic-transport models, the characterization of the structural requirements and pathways for methane pyrolysis on Mo/H-ZSM5, and the synthesis and evaluation of proton conductors of $SrZr_{0.95}Y_{0.05}O_3$ composition.

* permanent address: Kunsan National University, Kunsan, Chonbuk 573-701, Korea

2. METHODS

2.1. Simulation of membrane reactors

A gas-phase kinetic model with 65 elementary steps involving 29 species describes accurately pyrolysis rates and selectivity below 1100 K, including the formation of C_{10+} hydrocarbons. [6] Surface reactions include CH_4 conversion to methyl radicals, recombination of H-atoms to give H_2, and transport of H atoms across the membrane. Methyl radical formation rates were estimated using linear free energy relations between C-H bond activation rates and O-H bond energies and O-H bond strengths in MoO_x-H. Reported diffusivities for $SrZr_{0.95}Y_{0.05}O_3$ were used to calculate hydrogen transport rates. [7] Molecules larger than naphthalene were assumed to form solid carbon.

Figure 1. CH_4 aromatization with hydrogen removal

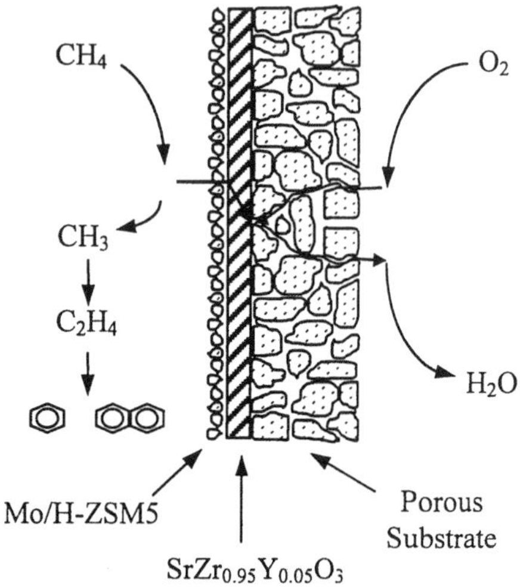

2.2. Mo/H-ZSM5 catalysts for methane pyrolysis

Mo/H-ZSM5 was prepared from physical mixtures of MoO_3 (Johnson Matthey, 99.5% purity) and H-ZSM5 (Zeochem, Si:Al=14.5). The effects of oxidation pre-treatment on catalyst structure and performance were examined using MoO_3/H-ZSM5 mixtures (0.3 g, 0-8 wt% Mo) dried at 623 K for 24 h in 20% O_2/Ar (100 cm^3/min). Samples were heated at 10 K/min to 973 K and H_2O evolution was measured by mass spectrometry (Leybold Inficon, model THP-TS200), using Ar as an internal standard. Samples were held at 973 K for 0.5 h, then cooled to 300 K. The number of exchangeable H atoms (mainly Bronsted acid sites) in treated MoO_3/H-ZSM5 samples was obtained by heating these samples from 300 K to 973 K (10 K/min) in 5% D_2/Ar (100 cm^3/min) and measuring the evolution of HD and H_2 by mass spectrometry. [8]

Catalytic CH_4 reactions were carried out at 950 K in a tubular reactor with plug-flow hydrodynamics (1.0 g, 25 cm^3/min, 1:1 CH_4/Ar, 1.08 bar). Product streams were analyzed on-line using heated transfer lines (400 K) and gas chromatography {HP6890 GC; Carboxen 1000 packed column (3.2 mm x 2 m, Supelco) with thermal conductivity detector and HP-1 capillary column (0.32 mm x 50 m, Hewlett-Packard) with flame ionization detector}. Catalysts were treated in 20% O_2/He (100 cm^3/min) at 950 K for 2 h before catalytic reactions. Selectivities are reported on a carbon basis, as the percentage of the converted CH_4 appearing as a given product, using Ar as an internal standard in order to ensure accurate mass balances. The carbon missing within the measured products (1-10%) is treated as solid carbon in reporting yields.

2.3. Membrane materials and synthesis methods

Dense $SrZr_{0.95}Y_{0.05}O_3$ membranes can transport hydrogen with perfect selectivity at 600-1000 K. [9] $SrZr_{0.95}Y_{0.05}O_3$ powders were prepared by co-precipitation of metal hydroxides (Y, Zr) or carbonate (Sr) at a pH of 9 from an aqueous solution of the metal nitrates using NH_4OH and $(NH_4)_2CO_3$. The perovskite structure was detected by X-ray diffraction after air treatment at 1223 K. Membrane precursor powders were also prepared using glycine-nitrate [10] and glycolate [11] combustion methods, which result in smaller and easier-to-sinter particles with perovskite structure after oxidation at 1223 K. Powders were pressed into disks (25 mm dia. x 1.5 mm) and densified at 1823 K for 4 h in flowing air. This procedure led to smaller disks (16 mm dia. x 1 mm) with densities of 85-100% of the skeletal $SrZrO_3$ density. Disk densities were obtained by weighing and measuring with calipers. Powder surface areas were measured by N_2 physisorption at 77 K using the BET method (Quantachrome Autosorb-6).

3. RESULTS AND DISCUSSION

3.1. Simulation of membrane reactors

The activation of a C-H bond in CH_4 to form CH_3 and H radicals limits homogeneous pyrolysis rates and initiates a sequence of chain growth reactions. [12] The major stable molecular products are ethylene, benzene, and naphthalene at 823-1073 K. For this discussion, the reaction pathways can be simplified without loss of accuracy as:

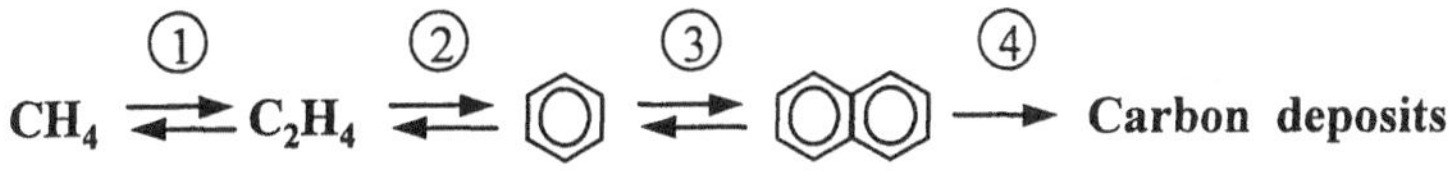

Reverse rates are important even at low CH_4 conversion (<5%), because of thermodynamic limitations (23% equilibrium conversion to C_2-C_{10} at 1038 K) and of the large amount of H_2 produced in sequential dehydrogenation steps.

Detailed simulations of CH_4 pyrolysis at 1038 K lead to the results in Figure 2. The maximum C_2-C_{10} yield is 14% for homogeneous reactions and does not change when a faster heterogeneous methyl radical generation function is added. This reflects the sequential nature of the reaction scheme above and of the full reaction mechanism. An increase in the rate of methyl radical formation leads to a faster equilibration of step 1 (limited to 8% conversion at

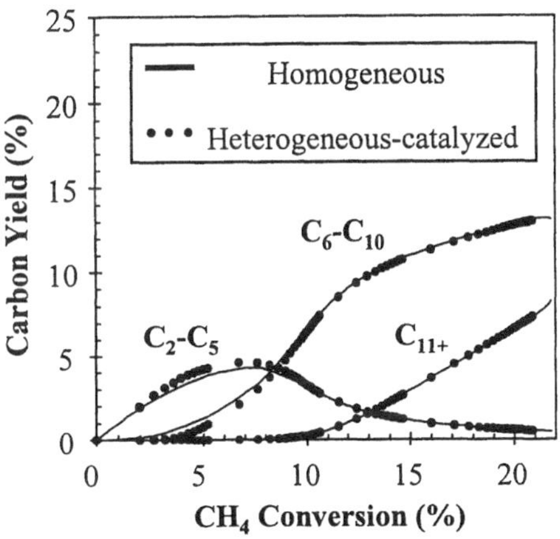

Figure 2. Simulation of CH_4 pyrolysis without H-removal (1038 K, 0.59 bar)

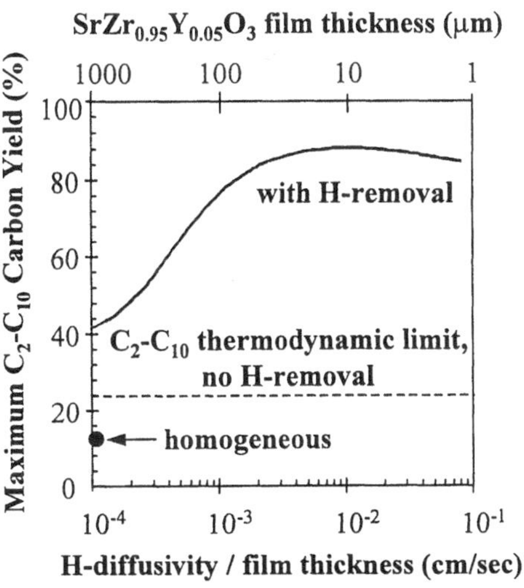

Figure 3. Effect of H-removal rate on CH_4 pyrolysis simulation (1038 K, 0.59 bar)

1038 K) and ethylene conversion to aromatics (step 2) becomes rate-limiting. Therefore, surface-initiated CH_4 reactions do not influence product selectivity above 10% CH_4 conversion, because homogeneous chain-growth (steps 2-4) determines the product distribution in the simulations.

Figure 3 shows the simulated effect of hydrogen removal on CH_4 pyrolysis yields. Removal of hydrogen at rates expected for a $SrZr_{0.95}Y_{0.05}O_3$ membrane of 1.0 mm thickness leads to maximum C_2-C_{10} yields of about 42% (at 85% CH_4 conversion). Higher conversions achieved by increasing residence times lead to lower C_2-C_{10} yields (<25%), because intermediate C_2-C_{10} products convert to coke precursors via slower chain-growth reactions (step 4). Thinner membrane films (5-50 μm) increase the rate of H removal without affecting the rate of carbon formation via chain-growth reactions and lead to maximum C_2-C_{10} yields near 90%.

Simulations also show that gas-phase CH_4 pyrolysis with continuous hydrogen removal leads to high C_2-C_{10} yields only at low temperature (<1000 K), above which unselective homogeneous pathways lead preferentially to carbon. Below 1000 K, achieving near-equilibrium CH_4 conversion within practical residence times requires a catalyst; this catalyst must restrict chain growth in order for yields to exceed those in homogeneous reactions and must also be stable at the severe reducing/carburizing conditions of CH_4 pyrolysis. H_2 transport rates must approach those of CH_4 reactions in order to maintain low H_2 concentrations, because H_2 inhibits CH_4 conversion. In addition, membrane materials must not reduce or carburize during operation.

3.2. Chain-limiting catalytic pyrolysis of methane on Mo/H-ZSM5

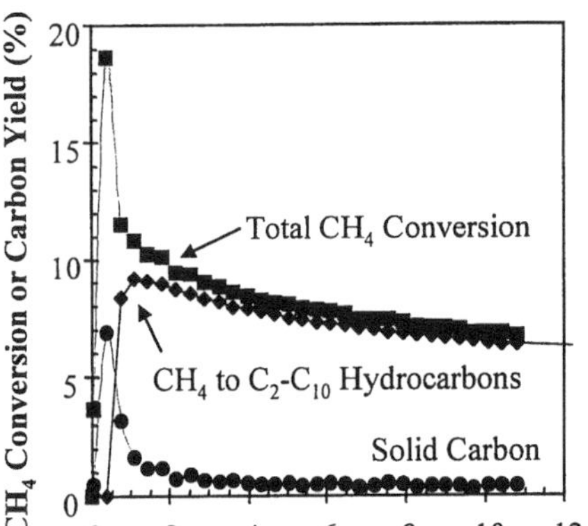

Figure 4. CH_4 pyrolysis on 4 wt% Mo/H-ZSM5 at 950 K

Recent studies have shown that Mo/H-ZSM5 restricts chain growth, increases reaction rates, and leads to CH_4 pyrolysis below 1000 K with low selectivity to carbon. [13, 14] Thermodynamic constraints, however, limit benzene yields to about 12% at 973 K. Our catalytic data (Figure 4) confirm these results using catalysts prepared via simple exchange of Mo^{+6} from MoO_3 onto H-ZSM5 by surface and gas-phase transport.

Mo/H-ZSM5 (4 wt% Mo) forms CO_2, CO, H_2O and carbon during initial contact with CH_4 at 950 K, as Mo^{+6} cations are converted to oxycarbide species (MoO_xC_y) that activate CH_4. Steady-state CH_4 conversions (9-11% at 750 cm^3 CH_4/g cat-h) and benzene selectivities (75%) are reached after 1 h. Deactivation decreases CH_4 pyrolysis rates (to 6% conversion after 40 h), but treatment in 20% H_2/He at 950 K restores initial rates and selectivities, without the activation period observed on fresh catalysts. Regeneration by temperature-programmed oxidation (400→950 K, 5 K/min) restores both initial induction periods and steady-state reaction rates.

Mo/H-ZSM5 appears to satisfy the requirements for low-temperature CH_4 activation catalysts suggested by our simulations of membrane reactors. Unfavorable thermodynamics for endothermic methane pyrolysis, however, preclude rate increases by further catalyst modifications. Conversions greater than ~10% will require higher temperatures, lower methane pressure, or the continuous removal of one of the products during reaction. (Table 1)

Table 1. Thermodynamic equilibrium calculations for CH_4 conversion to C_2-C_{10} (950 K)

CH_4 pressure (bar)	1.0	0.05	1.0	1.0	1.0
H removal (%)	0	0	25	50	75
CH_4 conversion (%)	10.4	31.3	38.1	66.4	94.7

3.3. Structure and function of Mo cations in Mo/H-ZSM5

Pyrolysis rates on Mo/H-ZSM5 prepared by solid-state reaction of MoO_3/H-ZSM5 mixtures are similar to those on samples prepared via impregnation of H-ZSM5 with ammonium heptamolybdate. [13] During air oxidation of either sample, we expect isolated MoO_x species to move into the ZSM5 pore structure and exchange with H atoms located at framework acid sites. The rate of surface migration of MoO_x species into H-ZSM5 channels becomes significant above the Tammann temperature (534 K) [15] and gas-phase transport occurs above 623 K [16].

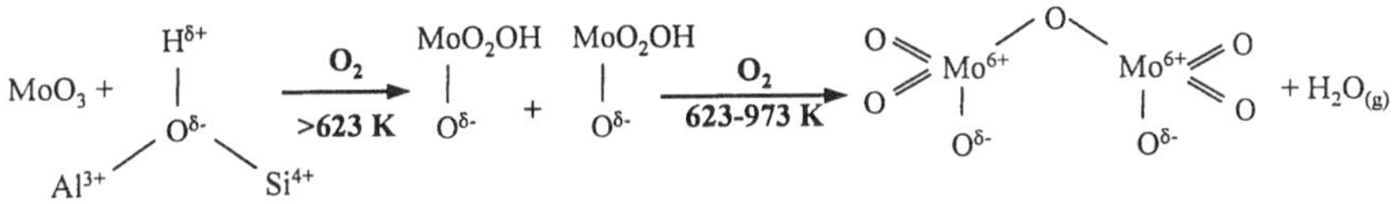

The number of H_2O molecules desorbed during heating in air corresponds to the number of protons exchanged by migration of MoO_x to exchange sites in H-ZSM5. (Figure 5, slope = 1.07 H/Mo) Also, the number of remaining H acid sites obtained by isotopic exchange of D_2 with surface OH groups in the zeolite decreases linearly as Mo concentration increases. (Figure 6a, slope = -1.02 H/Mo) Thus, one H^+ is lost from H-ZSM5 for each Mo exchanged, up to a Mo concentration of 5.1 wt% Mo, beyond which surface O-H groups disappear. Higher Mo concentrations lead to sublimation of the excess MoO_x. The stoichiometries shown by the data in Figures 5 and 6a, charge balance requirements, and preliminary X-ray absorption and NMR results [17] are consistent with isolated $(Mo_2O_5)^{2+}$ dimers interacting with two cation exchange

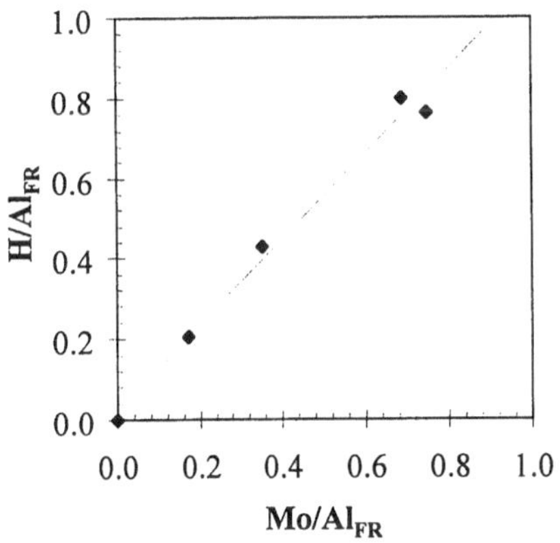

Figure 5. Number of H per framework Al (Al_{FR}) desorbed as H_2O during oxidation of MoO_3/H-ZSM5 mixtures

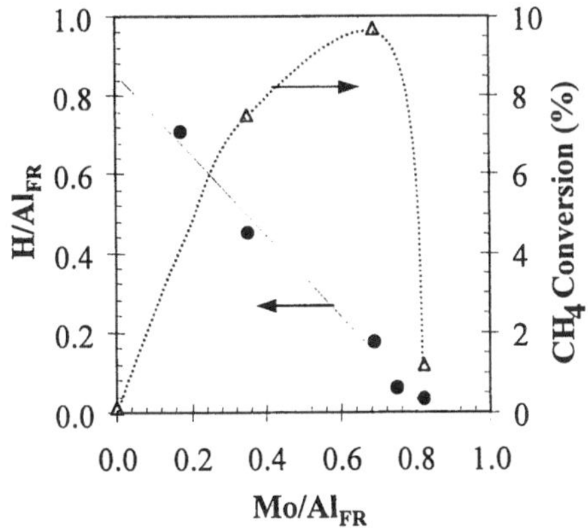

Figure 6. a) Number of exchangeable H per Al remaining on oxidized MoO_3/H-ZSM5 samples b) CH_4 conversion at 950 K vs. Mo loading

sites, which convert to MoO_xC_y during CH_4 reactions. CH_4 conversion rates reach a maximum value at intermediate Mo concentrations (Figure 6b), suggesting a requirement for both Mo species and H^+ in CH_4 conversion pathways.

3.4. Methane activation and acid-catalyzed oligomerization reactions

Figure 7 shows product selectivities on 4 wt% Mo/H-ZSM5 at 950 K as conversion changes by varying space velocity (150-1500 cm^3/g cat-h), catalyst loading (0.5-1.0 g), or the extent of deactivation (1-96 h on stream). The coincidence of deactivation and residence time results show that deactivation occurs by loss of active sites without modification of intrinsic site chemistry. These selectivity/conversion curves demonstrate that CH_4 reacts sequentially to form C_2H_4 as the initial reactive product, C_6H_6 as a secondary product, and naphthalene as the kinetic end point (except for its slow conversion to carbon). Earlier studies [13, 14] also suggested sequential pathways involving CH_4 activation on Mo sites to produce CH_3 radicals, C_2H_6, and then C_2H_4 via homogeneous pathways, followed by acid-catalyzed chain growth reactions of C_2H_4 within shape-selective ZSM5 channels (~5.5 Å diameter), which limit the size of polymeric products. C_2H_4 reacts via oligomerization and cyclization reactions on residual protons in Mo/H-ZSM5 to form single-ring aromatics and naphthalene. The instantaneous removal of C_2H_4 (via aromatization) removes thermodynamic and kinetic constraints that limit CH_4 to C_2H_4 reactions to low conversions (3.8 %) at 950 K.

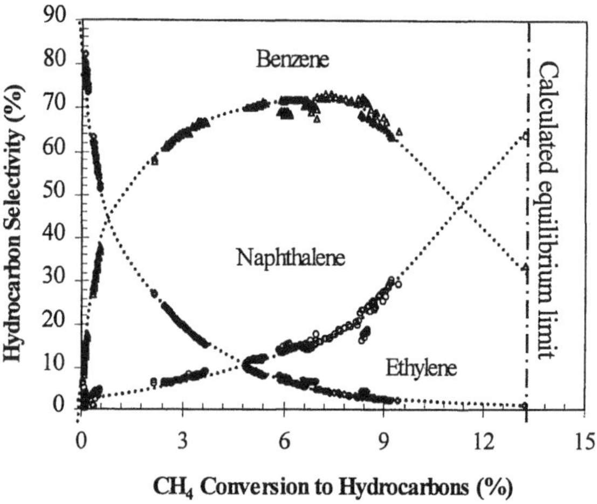

Figure 7. CH_4 conversion vs. product selectivity at different space velocities and levels of deactivation (950 K, 0.5-1.0 g 4% Mo/H-ZSM5, 5-50 cm^3/min 1:1 CH_4/Ar)

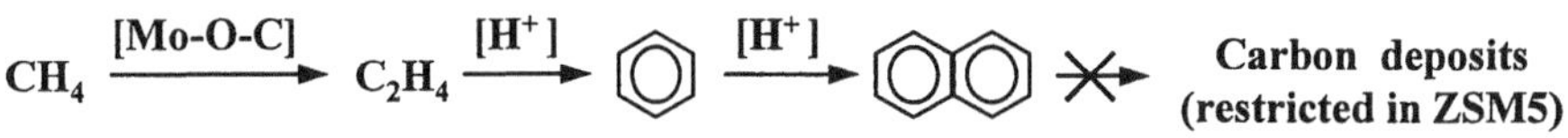

The effect of the relative numbers of $Mo^{\delta+}$ and H^+ sites on CH_4 conversion (Figure 6b) confirms the bifunctional nature of reaction pathways. CH_4 conversion rates increase as Mo concentration increases (0-4 wt% Mo), because the initial formation of C_2H_4 limits overall rates as long as H^+ sites are available to convert C_2H_4 into more stable aromatics. When H^+ sites disappear at Mo concentrations above 4 wt%, CH_4 conversion decreases sharply, because it is limited by the unfavorable equilibrium of C_2H_4 formation.

3.5. Synthesis and characterization of membrane materials

Metal membranes (Pd, Pd/Ag) cannot be used in our proposed scheme (Figure 1), because they cause rapid carbon formation during CH_4 pyrolysis. [18] $SrCe_{0.95}Yb_{0.05}O_3$ proton-transport membranes exhibit oxide ion mobility at 1023 K [5] and lose oxygen in reducing environments.

$SrZr_{0.95}Y_{0.05}O_3$ has H-diffusivities similar to those in $SrCe_{0.95}Yb_{0.05}O_3$ [7], but remains stable in reducing environments. [19] Our simulation results (section 3.1) show that practical membranes must transport hydrogen at rates comparable to reaction rates. These constraints translate into a requirement for the synthesis of thin (10-100 μm) films of $SrZr_{0.95}Y_{0.05}O_3$ perovskite.

$SrZr_{0.95}Y_{0.05}O_3$ is difficult to form into dense membranes. [20] The synthesis of dense ceramics depends critically on the size and uniformity of the powders used to form the compressed porous structure ("green" body) that must then be sintered into disks or films with densities above 95% of the skeletal perovskite density. Small crystallites of uniform size favor desired densification processes, which form gas-tight structures, over sintering processes that form large pores and weak structures with low bulk density. [21] Table 2 shows some representative properties of $SrZr_{0.95}Y_{0.05}O_3$ powders prepared by three different methods. Combustion synthesis methods lead to loose powders with low density and smaller crystallites with higher surface area, because the rapid exothermic decomposition of metal nitrate/organic precursors cause the formation of a large number of isolated nuclei during expansion and quenching of the reacting mixture. Individual nuclei grow by consuming reactants within a surrounding diffusion radius without significant agglomeration of crystallites.

Table 2
Properties of $SrZr_{0.95}Y_{0.05}O_3$ powders before densification into H-transport membranes

Method	"Loose" powder density (g/cm^3)	BET surface area (m^2/g)	"Green" disk density (% of $SrZrO_3$ theoretical)
Co-precipitation	1.34	4.79	44
Glycine-nitrate	0.016	19.1	48
Ammonium glycolate	0.070	12.3	51

The packed density of "green" disks after isostatic compression at 138 MPa is similar for the three synthesis methods. (44-51%, Table 2) After sintering at 1673-1923 K for 4 h, however, materials prepared by combustion methods lead to disks with much higher densities (Figure 8). Disks from co-precipitated powders did not densify at all at these conditions. Even after mechanical grinding and hydrostatic compression at 1240 MPa, their bulk densities were much lower than those of powders prepared by glycolate and glycine-nitrate combustion methods.

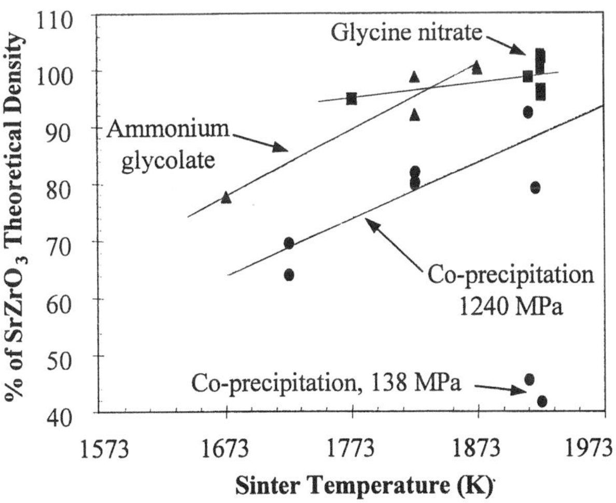

Figure 8. Effect of sintering temperature on final membrane density for three $SrZr_{0.95}Y_{0.05}O_3$ powders

$SrZr_{0.95}Y_{0.05}O_3$ disks (0.9-1.2 mm thick) with >95% theoretical density were sealed to the end of a 19 mm O.D. alumina tube using a ceramic paste (Aremco Ceramabond 571). Mo/H-ZSM5 (0.5 g, 4 wt% Mo) was placed on the methane side of the disk, while the opposite side was left exposed to air. This membrane reactor was heated at 0.5 K/min to 950 K, leak-tested using He, and exposed to CH_4 (150 cm^3 CH_4/g cat-h). CH_4 conversion rates and product selectivity were similar to those measured in conventional tests (without hydrogen removal). The estimated H_2 transport rate at 950 K for a $SrZr_{0.95}Y_{0.05}O_3$ disk with 1.5 cm^2 area is 44 μmol/h, which is significantly lower than the measured rate of H_2 production from CH_4 (930 μmol/h). Thus, less than 5% of the H_2 formed was removed and reaction rate enhancements were not detectable. These initial experiments and the simulation results (section 3.1) show that H_2 removal rates must increase by about a factor of 20, either by increasing the surface area or decreasing the thickness of the membrane. The latter approach can be implemented by sequential spin coating of viscous $SrZr_{0.95}Y_{0.05}O_3$ slurries onto porous Al_2O_3 or ZrO_2 supports. [22] The use of sequential coating/oxidation steps with powders formed via combustion methods has led to the synthesis of thin films with high densities, which are being tested in our membrane reactor.

4. ACKNOWLEDGEMENTS

The authors thank Drs. Anthony Dean and Sebastian Reyes of the Corporate Research Labs at Exxon Research and Engineering for some of the FORTRAN subroutines and kinetic and thermodynamic data used in the kinetic simulations. Richard Borry was supported by a National Science Foundation Fellowship. Dr. Young-Ho Kim was supported by the Korean Science and Engineering Foundation (KOSEF) during his sabbatical leave. Project funding was provided by the Federal Energy Technology Center (U.S. Department of Energy, contract DE-AC03-76SF00098) under the technical supervision of Dr. Daniel Driscoll.

REFERENCES

1. Chen, C.-J., M.H. Back, R.A. Back, *Can J Chem* **54** 3175 (1976)
2. Labinger, J.A., *Catal Lett* **1** 371 (1988)
3. Reyes, S.C., E. Iglesia, C.P. Kelkar, *Chem Engr Sci* **48** 2643 (1993)
4. Hamakawa, S., T. Hibino, H. Iwahara, *J Electrochem Soc* **140** 459 (1993)
5. Langguth, J., R. Dittmeyer, H. Hofmann, G. Tomandl, *Appl Catal A* **158** 287 (1997)
6. Borry, R.W., E. Iglesia, *Chem. Eng. Sci. (submitted for publication)* (1998)
7. Schober, T., J. Friedrich, J.B. Condon, *Solid State Ionics* **77** 175 (1995)
8. Biscardi, J.A., G.D. Meitzner, E. Iglesia, *J Catal (submitted for publication)* (1998)
9. Shin, S., Huang, H.H., Ishigame, M., Iwahara, H., *Solid State Ionics* **40/41** 910 (1990)
10. Chick, L.A., I.R. Pederson, G.D. Maupin, J.L. Bates, L.E. Thomas, G.J. Exarhos, *Mater Letters* **10** 6 (1990)
11. Soled, S.L., E. Iglesia, S. Miseo, B.A. DeRites, R.A. Fiato, *Top Catal* **2** 193 (1995)
12. Dean, A.M., *J Phys Chem* **94** 1432 (1990)
13. Wang, D., J.H. Lunsford, M.P. Rosynek, *Topics in Cat* **3** 289 (1996)
14. Solymosi, F., J. Cserenyi, A. Szoke, T. Bansagi, A. Oszko, *J Cat* **165** 150 (1997)
15. Satterfield, C.N., Heterogeneous Catalysis in Industrial Practice, (New York:McGraw-Hill, Inc.) 1991
16. Brewer, L., Molybdenum: Physico-chemical properties of its compounds and alloys, (Vienna:International Atomic Energy Agency) 1980
17. Borry, R.W., A. Huffsmith, G. Meitzner, J. Reimer, E. Iglesia, *(in preparation)* (1998)
18. Andersen, A., I.M. Dahl, K-J. Jens, E. Rytter, A. Slagtern, A. Solbakken, *Cat Today* **4** 389 (1989)
19. Yajima, T., H. Suzuki, T. Yogo, H. Iwahara, *Solid State Ionics* **51** 101 (1992)
20. Iwahara, H., T. Yajima, T. Hibino, K. Ozaki, H. Suzuki, *Solid State Ionics* **61** 65 (1993)
21. Rahaman, M.N., Ceramic Processing and Sintering, (New York:Marcel Dekker, Inc.) 1995
22. de Souza, S., S.J. Visco, L.C. De Jonghe, *Solid State Ionics* **98** 57 (1997)

NATURAL GAS CONVERSION V
Studies in Surface Science and Catalysis, Vol. 119
A. Parmaliana et al. (Editors)

Synthesis gas formation by catalytic partial oxidation of methane in a heat-integrated wall reactor

A. Piga[a], T. Ioannides[b] and X. E. Verykios[a]

[a] Department of Chemical Engineering, University of Patras, GR-26500 Patras, Hellas

[b] Institute of Chemical Engineering and High Temperature Chemical Processes (ICE/HT-FORTH), P.O. Box 1414, GR-265 00 Patras, Hellas

Partial oxidation of methane to synthesis gas has been carried out in a new type of reactor, designated as heat-integrated wall reactor (HIWAR). The reactor comprises a ceramic tube in the inner and outer surface of which, a metal catalyst film is deposited. The proposed reactor offers the possibility of reducing drastically the magnitude of hot spots created during partial oxidation of methane. The performance of the HIWAR reactor is compared to that of a conventional wall reactor (CWR), which contains the catalyst film on the inner surface of the reactor tube only.

1. INTRODUCTION

In recent years, catalytic partial oxidation of methane to synthesis gas has been studied as an alternative route for methane conversion to synthesis gas. Pena et al. [1], have recently reviewed work in this field. Metals of group VIII, such as Rh, Ni, Ru, Pt, Ir and Pd, are active catalysts for this reaction.

Methane conversion to synthesis gas may take place either via the indirect or via the direct reaction scheme. According to the indirect reaction scheme, part of methane is combusted by supplied oxygen producing CO_2 and H_2O. The reforming reactions take place consecutively. In the direct reaction scheme , methane is directly oxidized to synthesis gas.

Direct oxidation of methane to synthesis gas has been suggested in the case of Ru/TiO_2 catalyst [2,3], as well as in the case of Rh- or Pt- loaded monolith catalysts at short residence times [4,5]. In most cases, the partial oxidation of methane to synthesis gas, follows the indirect reaction scheme, in which the strongly exothermic combustion of methane occurs initially. As a result of this, hot spots of considerable magnitude may develop in the reactor. The applicability of several types of reactors, such as fluidized-bed reactors, [6,7], dual-bed or mixed-catalyst bed reactors [8], the Hot-Spot reactor [9], hydrogen-selective membrane reactors [10,11] and monolithic reactors [4,5,13] has been investigated.

In an earlier publication [14], preliminary results of the application of a new type of reactor, designated as *heat-integrated wall reactor (HIWAR)*, in the reaction of the partial oxidation of methane were presented. This reactor offers the possibility of controlling the temperature in the combustion zone and, in its experimental mode, is shown schematically in Figure 1.

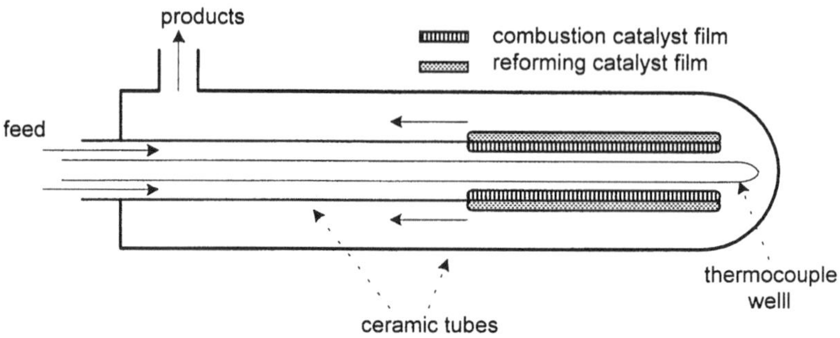

Figure 1. Shematic diagram of the Heat-Integrated Wall Reactor (HIWAR)

The reactor comprises a non-porous ceramic tube of high thermal conductivity and a catalyst film is deposited on its inner and outer surface. The design of the reactor offers the possibility to deposit different catalysts on the inner and outer surface of the tube. The ceramic tube is enclosed in a larger ceramic (or quartz) tube. The reactants enter the inner tube where the exothermic methane combustion takes place. A large amount of the heat generated is transported through the tube wall towards the outer surface of the tube, where the endothermic reforming reactions take place. As a result of this, the temperature in the combustion zone can be controlled and the magnitude of hot spots can be reduced. The feed is preheated as the products of the reaction are carried away from the reactor,.

In this work, new results of experiments carried out in the HIWAR reactor are presented. Particular attention has been paid to the investigation of the performance of conventional wall reactors (CWR), which contain the catalyst film on the inner surface of the tube alone, as they are the reference basis for the evaluation of the performance of the HIWR reactors.

2. EXPERIMENTAL

A detailed description of the laboratory-scale heat-integrated wall reactor can be found elsewhere [14]. A Rh/Al_2O_3 catalyst film was used in the reactor and it was deposited via dip-coating with a Rh/Al_2O_3 sol. Two different Rh/Al_2O_3 sols were prepared, with Rh loadings of 4.6 wt.% and 10wt. %, respectively. The procedure of preparation of the Rh/Al_2O_3 sols can be found in Ref [14]. The Rh/Al_2O_3 powder derived from the first sol has a surface area of 250 m^2/g after calcination at 500°C and a Rh dispersion of 0.5 (as determined by equilibrium hydrogen chemisorption). Following calcination at 1000°C for 2 hours, the surface area drops to 55 m^2/g.

Two different procedures were followed for the deposition of the catalyst film on the reactor tube. In the first procedure, which was used for the preparation of reactor CWR-A, the reactor tube was dipped in the 4.6% Rh/Al_2O_3 sol and then was gradually elevated from the surface of the sol at a rate of 1 cm/min. In the second procedure, which was used for the preparation of reactors CWR-B and HIWAR, the reactor tube was dipped in the 10% Rh/Al_2O_3 sol and then was immediately removed from the sol. After each dip-coating, the reactor tube was dried at 100°C and subsequently calcined at 500°C for 2 h.

In the case of CWR-A, seven catalyst layers were deposited on the inner surface of the tube over a length of 10cm at one end of the alumina tube. The reactor was tested under reaction conditions after deposition of each catalyst layer. Regarding the CWR-B and HIWAR reactors, the catalyst was deposited over a length of 11cm at one end of the alumina tube. Two catalyst layers were deposited on the inner surface of the tube in the case of the CWR-B. After testing under reaction conditions, one catalyst layer was deposited on the outer surface of CWR-B. This reactor configuration is called HIWAR-1. Additional amounts of catalyst were deposited on the outer surface of the same reactor tube by consecutive dip-coatings (HIWAR-2, HIWAR-3). The characteristics of all the reactors which were tested are shown in Table 1.

The reaction feed consisted of CH_4 and O_2 with a 2:1 ratio and the feed flow rate was in the range of 60-1950 cc(STP)/min. Flow rates of reactants were measured and controlled by thermal mass flow controllers (MKS Instruments). Experiments were conducted starting from the lowest flow rate and then increasing it stepwise until the hot spot in the reactor became of the order of 1000°C. In each flow rate, one hour was allowed for the reactor to reach steady-state operation. Experiments were carried out at furnace temperatures of 600°C and 700°C. Temperature profiles were measured along the reactor tube by a thermocouple as shown in Figure 1. At a furnace temperature of 700°C, the temperature at the middle of the catalyst film was 700°C under nitrogen flow while at the beginning as well as at the end of the catalyst film was 60°C lower. Before testing, the catalyst film was reduced under hydrogen flow at 500°C.

Table 1
Characteristics of reactors

Reactor	No. of layers	Dip-coating	Rh loading	Length	Position	Catalyst loading (mg) inner	Catalyst loading (mg) outer
CWR-A	1-7	Withdrawal at 1cm/min	4.6%	10cm	Inner surface	1.5-6.5	--
CWR-B	2	Immediate withdrawal	10%	11cm	Inner surface	2	--
HIWAR-1	2+1	Immediate withdrawal	10%	11cm	Inner and outer surface	2	1
HIWAR-2	2+2	Immediate withdrawal	10%	11cm	Inner and outer surface	2	3
HIWAR-3	2+3	Immediate withdrawal	10%	11cm	Inner and outer surface	2	6

3. RESULTS AND DISCUSSION

3.1. Conventional reactor CWR-A

The temperature profiles of the conventional wall reactor CWR-A with varying number of deposited layers are presented in Figure 2. The experiments were conducted at a furnace temperature of 700°C and feed flow rate of 450cc/min. The residence time for this flow rate is 73 ms, while the true contact time is estimated to be of the order of 15 ms. The point at z=11 cm denotes the position of the outer tube enclosing the reactor tube. Generally, a hot spot is observed due to the occurrence of the exothermic combustion of methane. The position of the hot spot is near the end of the catalyst film for the reactor with one catalyst layer and it shifts gradually towards the left as additional catalyst layers are deposited on the inner surface of the reactor tube. For an intermediate number of layers (four layers) the temperature profile is rather flat and the hot spot temperature is lower compared to all other cases. At high catalyst loadings, the hot spot is located at the beginning of the catalyst film. Experiments were also carried out at a furnace temperature of 600°C. In this case, the hot spot was always observed at the end of the catalyst film. It is believed that the observed trends in temperature profiles are related to the dip-coating procedure used for the deposition of the catalyst film in CWR-A. The slow withdrawal of this tube during dip-coating has probably led to non-uniform deposition of catalyst along the tube, with more catalyst having been deposited towards the end of the tube. It was noted that the tube had a more intense black color close to its end.

Methane conversion obtained in the CWR-A, at furnace temperatures of 700°C and 600°C and at a flow rate of 450cc/min, is presented in Figure 3 as a function of the number of catalyst layers deposited on the inner surface of the tube. It is obvious that methane conversion does not change appreciably as additional catalyst layers are deposited on the reactor tube and is of the order of 60% and 55% at furnace temperatures of 700°C and 600°C, respectively. Oxygen conversion was 100% in all cases. The minimum in methane conversion observed for the reactor with four catalyst layers correlates well with the corresponding temperature profile (Fig. 2), which exhibits the smallest hot spot. This observation implies that the reaction essentially takes place in the region of the hot spot and the contribution of the cooler sections of the catalyst film is small.

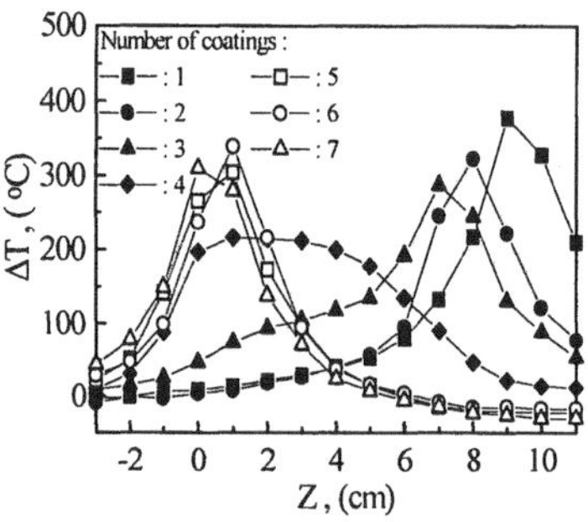

Figure 2. Temperature profiles along CWR-A under reaction conditions as a function of the number of catalyst layers (furnace temperature: 700°C)

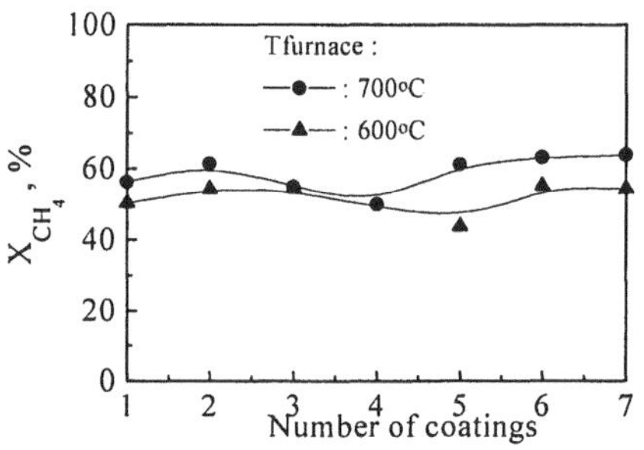

Figure 3. Methane conversion obtained in the CWR-A under reaction conditions as a function of the number of catalyst layers (feed flow rate: 450cc/min)

3.2. Reactors CWR-B and HIWAR

Experiments were also carried out employing the CWR-B and the HIWAR reactors. Temperature profiles of the CWR-B reactor and the HIWAR reactors are presented in Figure 4, at a furnace temperature of 700°C and feed flow rate of 450cc/min. The observed profiles show that deposition of catalyst on the outer surface of the reactor tube, reduces the magnitude of hot spots compared to those observed in the CWR-B. This is due to the endothermic reforming reactions that take place on the outer surface of the reactor. Due to the larger amount of catalyst in the outer surface of HIWAR-3 the hot spot temperature is significantly lower compared to that of CWR-B and HIWAR-1 and HIWAR-2. Another interesting observation is that two temperature maxima develop in the case of the HIWAR-3, one located at the end of the catalyst film. This probably implies that methane combustion is completed at the end of the catalyst film. The presence of two maxima might be attributed to different rates of heat production and consumption along the reactor tube. The changes induced in the temperature profiles upon deposition of catalyst layers on the outer surface of the reactor tube can be explained as follows: The occurrence of endothermic reforming reactions on the outer surface leads to the decrease of the wall temperature. As additional catalyst layers are deposited on the outer surface the wall temperature decreases further. This leads to a decrease of the rate of combustion, so that combustion cannot any more be completed at the first part of the inner catalyst film and the hot spot moves gradually towards the end of the catalyst film.

Methane conversion obtained over the CWR-B and the HIWAR reactors is presented in Figure 5 as a function of feed flow rate at a furnace temperature of 700°C. The maximum feed flow rate employed for each reactor was determined by the observed hot spot temperature. Thus, in the case of CWR-B the hot spot was close to 1000°C at a flow rate of 450 cc/min, while in the case of HIWAR-1 and HIWAR-2 a hot spot of ~1000°C was obtained at flow rates of 1050 and 600 cc/min, respectively. In the case of HIWAR-3, the hot spot was only 925°C at a flow rate of 1950 cc/min. Higher flow rates were not used in this case due to limitations of the mass flow controllers. It can be seen that for the feed flow rates examined, methane conversion in the heat integrated wall reactors is higher than methane conversion in CWR-B. Comparing the performance of the HIWAR-2 and HIWAR-3, it is obvious that in the case of HIWAR-2, conversions are higher than those of HIWAR-3. This can be correlated to the higher temperatures prevailing in the first part of HIWAR-2 compared to HIWAR-3. It is also very interesting to note the essentially constant methane conversion of 80% obtained in HIWAR-3 irrespective of the feed flow rate. This can be explained by taking into account that, as the feed flow rate increases the reactor temperature also increases due to the higher net rate of heat production.

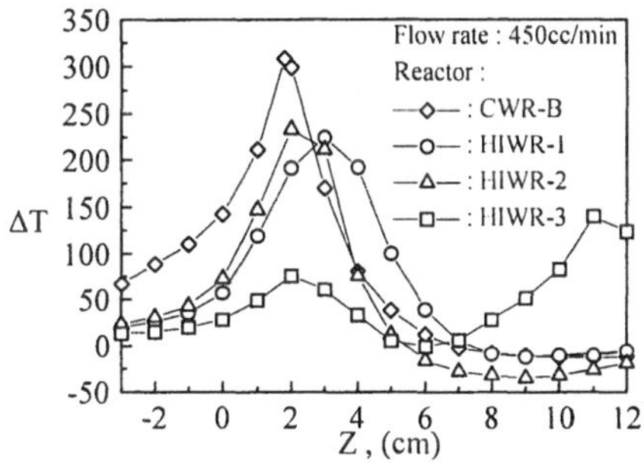

Figure 4. Temperature profiles along The CWR-B and the HIWAR reactors under reaction conditions at a furnace temperature of 700°C.

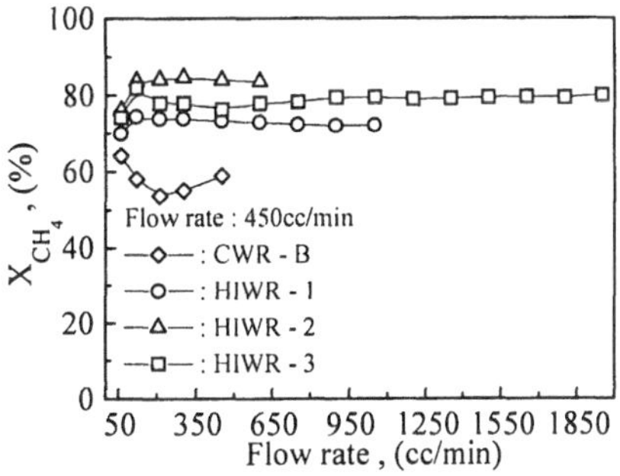

Figure 5. Methane conversions obtained in the CWR-B and the HIWAR reactors at a furnace temperature of 700°C as a function of feed flow rate.

As mentioned previously, tests were carried out in the CWR-A and CWR-B reactors following deposition of each catalyst layer. The temperature profiles along these two reactors at a furnace temperature of 700°C and flow rate of 450cc/min, were presented in Figures 2 and 4. If the performance of CWR-A and CWR-B containing two catalyst layers on the inner surface of each rector tube is compared, it can be seen that in the case of CWR-A the hot spot is observed at the end of the catalyst film. In the case of CWR-B, on the other hand, the hot spot is observed at the first part of the catalyst film. This difference is probably due to the different procedures which were followed for the deposition of the catalyst film in each reactor. Another contributing factor is the higher Rh loading (10% vs. 4.6%) of the sol employed in the case of CWR-B.

4. CONCLUSIONS

The following conclusions can be derived from the results of the present study:

- The heat-integrated wall reactor offers the possibility of reducing significantly the magnitude of hot spots which are created during the partial oxidation of methane.
- Employing 8 mg of a 10% Rh/Al_2O_3 catalyst in the form of a film in the heat-integrated wall reactor, methane conversions of the order of 80% could be obtained at total feed flow rates of ~ 2 liters/min and near atmospheric pressure. The observed reactor performance was accompanied by low hot spot temperatures, not exceeding 925°C in the gas phase, when the reactor was operated inside a furnace maintained at 700°C. By comparison, methane conversions obtained in a conventional wall reactor were of the order of 60% at feed flow rates of 450 cc/min and the hot spot in this case was ~1000°C.
- The heat-integrated wall reactor can be further optimized by fine-tuning of the relative quantities of catalyst deposited on the inner and outer surface of the reactor tube.

REFERENCES

1. M. A. Penã, J. P. Gomez and J. L. G. Fierro. Appl. Catal. A 144 (1996) 7.
2. Y. Boucouvalas, Z. L. Zhang and X. E. Verykios, Catal. Lett. 40 (1996) 189.
3. Y. Boucouvalas, Z. L. Zhang, A. M. Efstathiou and X. E. Verykios in *Studies in Surface Science and Catalysis,* Vol. 101, eds. J. W. Hightower, W. N. Delgass, E. Iglesia and A. T. Bell (Elsevier Amsterdam, 1996) p. 443.
4. D. A. Hickman, E. A. Howpfear and L. D. Schmidt, Catal. Lett., 17 (1993) 223.
5. P. M. Torniainen, X, Chu, and L. D. Schmidt, J. Catal., 146 (1994) 1.
6. S. S Bharadwaj and L. D. Schmidt, L. Catal.,146 (1994) 11.
7. U. Olsbye, E. Tangstad and I. M. Dahl, in Natural Gas Conversion II, eds. H. E. Curry-Hyde and R. F. Howe (Elsevier, Amsterdam, 1994) p. 303.
8. L. Ma and D. L. Trimm, Appl. Catal. A, 138 (1996) 265.
9. J. W. Jenkins and E. Shutt, Platinum Metals Rev.,33 (3) (1989) 118.
10. U. Balachandran, J. T. Duesk, R. L. Mieville, R. B. Poeppel, M. S. Kleefisch, S. Pei, T. P. Kobylinski, C. A. Udovich and A. C. bose, Appl. Catal. A,133 (1995) 19.
11. Santos, J. Coronas, M. Menendez and J. Santamaria, Catal. Lett., 30 (1995) 189.
12. Ioannides and X. E. Verykios, Catal. Lett., 36 (1996) 165.
13. M. Witt and L. D. Schmidt, J. Catal., 163 (1996) 465.
14. Ioannides and X. E. Verykios, Catal. Lett., 47 (1997) 183.

NATURAL GAS CONVERSION V
Studies in Surface Science and Catalysis, Vol. 119
A. Parmaliana et al. (Editors)

Ceramic Membrane Reactors for the Conversion of Natural Gas to Syngas

Carl A. Udovich

Amoco Research Center, 150 W. Warrenville Road, Naperville, Illinois 60566 (USA)

Introduction

Traditionally, natural gas conversion has been applied primarily for the manufacture of commodity chemicals such as methanol and ammonia. However, there is an increasing shift to apply natural gas conversion as a tool to develop otherwise unmarketable remote gas and associated gas. Competitor analysis indicates that natural gas conversion to added-value products is changing from strictly chemical products to a combination of chemicals, fuels and fuel related materials that provide not only salable products but solutions to gas production and business related issues as illustrated in Figure 1. This change has been brought about in part by technological improvements and the implementation of new technology that has reduced manufacturing costs, allowing gas conversion to move from its traditional role within the highly fragmented chemical industry to a new role as a tool for large-scale centralized fuel product manufacture. However, the economic competitiveness remains dependent on a combination of relatively low gas cost, crude oil prices near \$20/bbl, and the availability and marketability of an incremental and sustainable added-value product. In order to enhance and sustain economic competitiveness of gas-to-liquid, GTL, products such as Fischer-Tropsch, a significant improvement in the economics is needed. Such an improvement is possible through the technology we are developing.

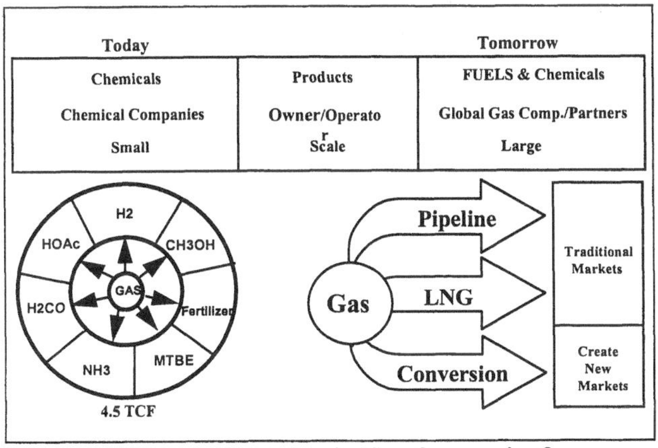

Figure 1: What is Changing in Gas Conversion?

New Approach to Syngas Production

The manufacture of products from natural gas consists of three steps. Of these three steps, the synthesis gas (syngas) preparation step is the single most costly amounting to approximately 60% of the total capital costs as shown in Figure 2. Advances in reducing the cost of this step will lead to the greatest reduction in the final cost of any marketable natural gas conversion product.

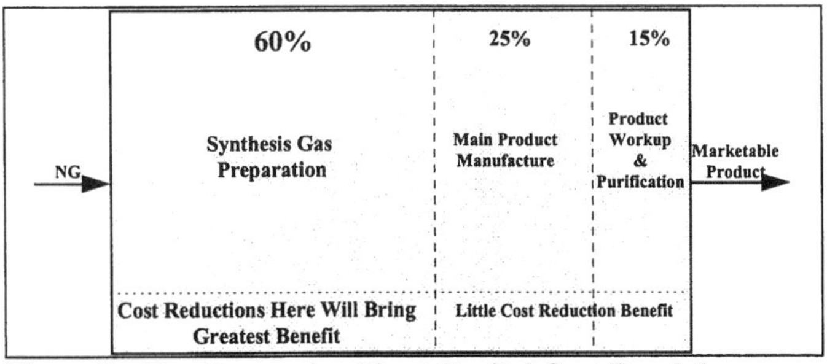

Figure 2: Gas Conversion Plant Capital Cost Breakdown

Existing commercial processes for syngas generation involve reaction with either steam or oxygen (or a combination thereof) to partially oxidize natural gas to a mixture of hydrogen and carbon monoxide The reaction with oxygen alone is called partial oxidation and requires a costly air separation plant. The ratio of hydrogen to carbon monoxide that results is suitable for further conversion of the syngas into synthetic fuels. The process that uses steam (steam reforming) instead of oxygen requires high temperature heat addition and the ratio of hydrogen to carbon monoxide that results is too high for synthetic fuels but preferred for hydrogen and ammonia production.

Five international companies including Amoco, BP Chemicals, Praxair, Sasol and Statoil have formed an alliance to unite their resources and expertise to develop and commercialize a unique technology that addresses overall cost reduction in the generation of syngas. The combined efforts of the Alliance are highly focused and competitive, thereby increasing the probability of success beyond that achievable by a single company alone. The Alliance is developing an innovative technology that can convert natural gas into syngas in high yield and with exceptionally high selectivity. The novel technology, known as OTM Syngas (Oxygen-Transport-Membrane Synthesis Gas) integrates the separation of oxygen from air, steam reforming and natural gas oxidation into one step thereby eliminating the need for a separate and costly oxygen plant as shown in Figure 3. OTM Syngas has the potential to significantly reduce the energy and capital costs of syngas production and thus enable the major cost reduction and energy efficiency improvement required to make GTL conversion sustainably competitive and economically attractive.

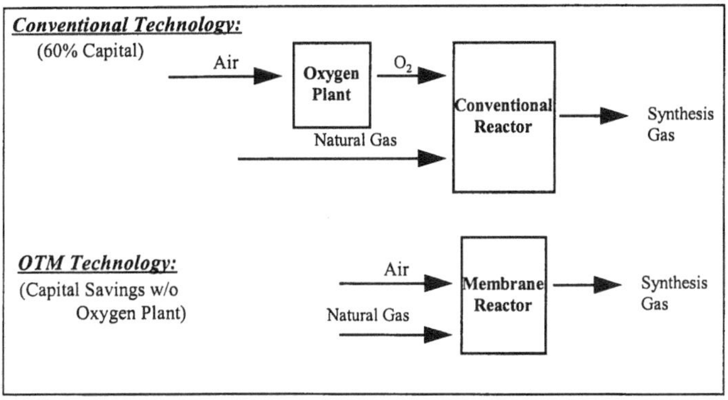

Figure 3: Synthesis Gas Generation ("Syngas", CO/H_2)

Oxygen permeable ceramic membranes are gaining considerable attention for the separation of oxygen from air. Dense phase, mixed conductivity ceramic oxides have been developed which separate oxygen from air with high flux in the same temperature region as required for typical reforming of natural gas. These novel ceramic materials can be fabricated into reactor components such as tubes in which both the separation of oxygen from air and the oxidation of natural gas can be combined. The membrane materials are non-porous and are composed of mixed metal oxides that conduct both oxide ions and electrons through their oxygen deficient lattice structure at elevated temperatures. When driven by an imposed oxygen partial pressure gradient, oxygen ions are conducted through the dense non-porous membrane at high flux and high selectivity. The oxygen flux is inversely proportional to the thickness of the membrane and directly proportional to the pressure gradient as illustrated in Figure 4.

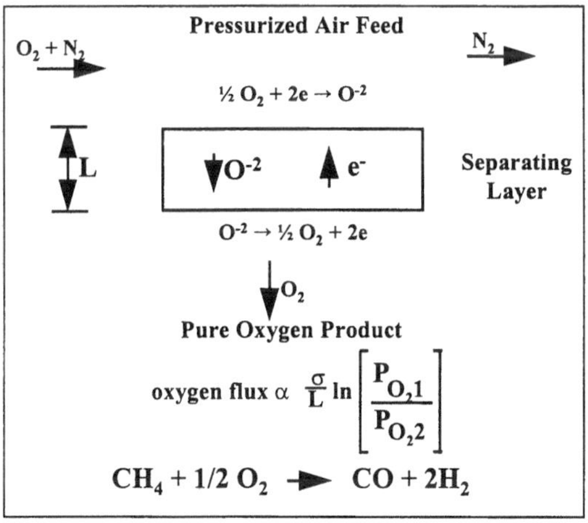

Figure 4: Oxygen Separation by Ion Transport

The process is illustrated in Figure 5 where air is introduced on one side of the membrane and natural gas on the other. Oxygen from air is adsorbed on the surface of the membrane, dissociated and ionized. Oxygen ions diffuse through the OTM material and react with natural gas in the presence of a reforming catalyst on the fuel side to give partial oxidation products. Electrons are released during the oxidation process and are simultaneously conducted in the reverse direction to maintain electrical neutrality. The consumption of oxygen by natural gas creates the oxygen partial pressure gradient across the membrane which maintains the flux. The separation and reaction of oxygen can be controlled to give high selectivity and yield because the flux can be controlled as a function of membrane thickness, ionic conductivity, oxygen partial pressures and membrane temperature. Since oxygen is transported by an ionic mechanism, the product syngas will be nitrogen free.

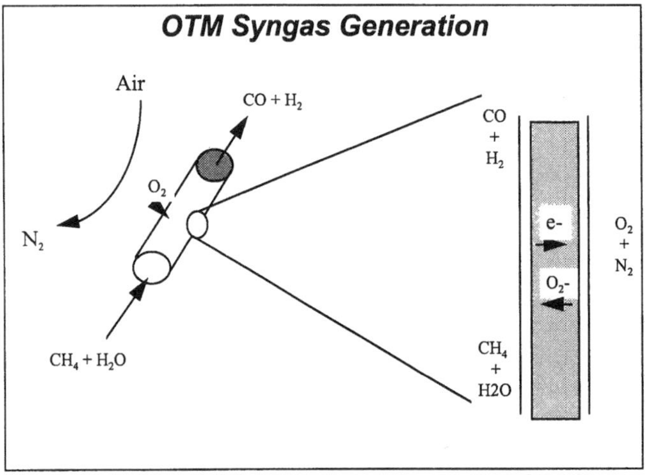

Figure 5: Integrated Oxygen Separation and Reforming

The dense ceramic materials are related to inorganic perovskites which have structures wherein the oxide ions and the large cations form a ccp array with the smaller cation occupying the octahedral holes formed exclusively by oxide ions. This type of structure is often slightly distorted and is adopted by a great many ABO_3 oxides in which one cation is comparable in size to the oxygen anion and the other much smaller with the cation charges being variable. Substitution of the highly charged cations with other cations of lower valencies causes a charge imbalance which results in higher oxygen vacancies to maintain electrical neutrality. By appropriate selection of cations, oxygen ion site deficiency can be affected influencing the overall conductivity and flux of oxygen.

A number of early OTM material candidates were initially developed separately by BP in the late eighties and nineties and by Amoco through a DOE supported cooperative effort with Argonne National Labs in the early-to-mid nineties. To demonstrate the behavior of these materials, an early candidate with a composition and stoichiometry of Sr Fe Co_5 O_x and referred to as SFC can be used as an example. This particular material does not have a perovskite structure but demonstrates some very favorable characteristics of OTMs. SFC material was fabricated into tubes by a plastic extrusion process, sintered under proprietary conditions and evaluated in a laboratory device. The oxygen transport performance characteristics are demonstrated in Figure 6.

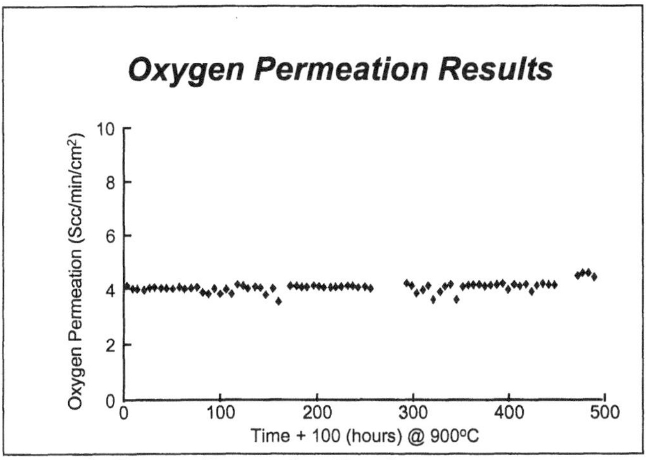

Figure 6: Oxygen Permeation Results

The oxygen permeation rate of SFC remains fairly constant during the 500 hour run at 4.0 scc/min/cm^2 wherein the oxygen permeation rate is defined as the flow of oxygen transporting through the membrane divided by the exposed area. The overall conversion of methane was maintained at a high level and the relative hydrogen to carbon monoxide ratio under operating conditions is approximately two. The SFC material has a high ionic transference number of 0.4 over a large temperature range of 650-950 degrees Celsius. The ionic conductivity of this material is nearly equal to the electronic conductivity illustrating another unique property.

Since the reactor tubes of these ceramic materials are composed of mixed metal oxides , they can experience problems which are not typical of conventional metallurgical reactor tubes. Metal oxides can be reduced in a syngas environment resulting in catastrophic failure of the reactor tube. The membrane material of an OTM reactor is exposed to a strongly reducing gas on one side and to air on the other side at high temperatures. For maximum economic benefit, the process also needs to be conducted at high pressures. Long-term stability under such severe conditions requires a very robust material which must not undergo reduction on the natural gas side. In addition, the chemical composition gradient across the wall of the reactor cannot vary significantly as to cause such a chemical mismatch wherein the reactor wall could fail from chemical expansion-contraction differences.

Major technical challenges abound in the areas of material performance, fabrication processing-reliability, ceramic reactor design-seals, process integration, engineering-scale-up and cost competitiveness.

The Alliance development program is focused on establishing element reliability and process operability issues as shown in Figure 7.

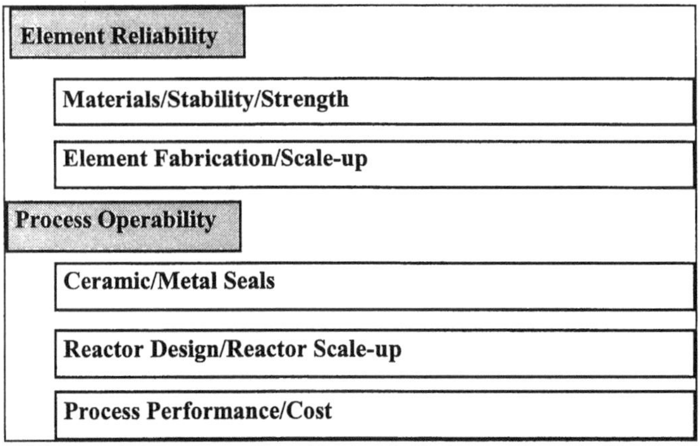

Figure 7: Fast Track Development Program in Place

Advances have been made wherein oxygen flux is no longer an issue and materials have been identified with enhanced stability characteristics. Progress is being made in other technically challenging areas. Although the risk remains high, the Alliance participants believe that we are in a strong position to resolve the technical challenges and move the program towards commercialization over the next several years.

NATURAL GAS CONVERSION V
Studies in Surface Science and Catalysis, Vol. 119
A. Parmaliana et al. (Editors)

Oxidehydrogenation of propane to propylene in catalytic membrane reactor: a model for the interpretation of experimental data.

G. Capannelli[a], A. Bottino[a], D. Romano[a], O. Monticelli[a], A. Servida[a], F. Cavani[b], A. Bartolini[c] and S. Rossini[c]

[a]Dipartimento di Chimica e Chimica Industriale, Università di Genova,
Via Dodecaneso 31, 16146 Genova, Italy

[b]Dipartimento di Chimica Industriale e dei Materiali, Università di Bologna,
Viale Risorgimento 4, 40136 Bologna, Italy

[c]Snamprogetti SpA,
Via Maritano 26, 20097 San Donato Milanese (MI), Italy

This paper describes the oxidehydrogenation reaction in catalytic membrane reactors using V_2O_5 as catalyst active phase. A simplified model of the catalytic membrane reactor is outlined and used to rationalize the kinetic data. Despite its simplicity, the model is capable to describe the effect of the relevant process variables: temperature, residence time, and oxygen-to-hydrocarbon molar ratio. Further work is in progress to better evaluate the model under wider experimental conditions.

1. INTRODUCTION

Catalytic partial oxidation plays an important role in the chemical process industry. Light paraffins are currently used as feedstocks in direct dehydrogenation processes that exhibit the drawback of being thermodynamically limited. A way of overcoming this problem is to use the oxidative dehydrogenation process that also avoids frequent catalyst regeneration. In oxidative dehydrogenation, the rates of the undesired reactions are faster than that of the desired reaction, and there is a strong thermodynamic driving force towards the complete oxidation. This calls for innovative reactor designs to improve the selectivity and productivity. Our work is along this direction and aims at evaluating the performance of catalytic membrane reactors (CMR) in oxidehydrogenation of light paraffins. In a previous work [1] the preliminary results for propane were discussed and showed the potential of CMR's that appeared to better perform than conventional packed bed reactors in terms of selectivity and catalytic reactivity. The prime goal of the paper is to outline and discuss a CMR model for the rationalization of the experimental data obtained for the oxidehydrogenation of propane.

2. EXPERIMENTAL

The pilot plant used for the kinetic runs has been fully described in the previous work [1]. Here, it is enough to say that the CMR was a tubular multilayer membrane (SCT-France), 15 cm long and with an inner diameter of 0.67 cm, constituted of three α-Al_2O_3 layers and one γ-Al_2O_3 layer deposited onto the internal surface. The catalyst phase, V_2O_5, was deposited on the γ-Al_2O_3 as reported in [1]. The CMR was characterized in terms of morphological, structural, and mass transport properties. In Figure 1 a schematic of the multilayer tubular membrane is shown along with its peculiar characteristics.

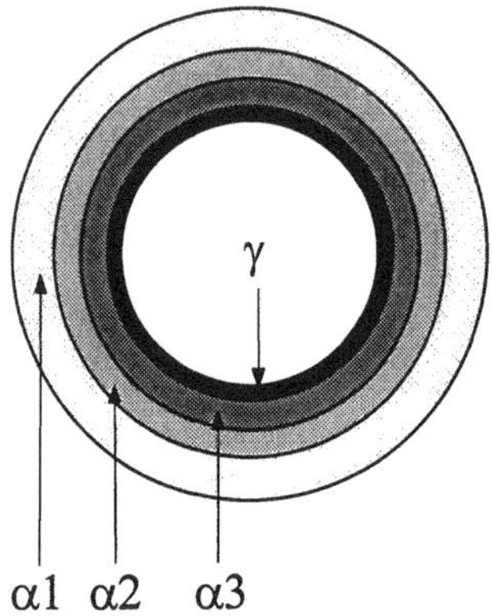

layer	Thickness [μm]	d_{pores}[nm]	ε/τ
α1	2000	10000	0.08
α2	40	640	0.08
α3	20	160	0.08
γ*	2	5	0.05

*Catalyst loading: 1.5 wt.%

Figure 1. Schematic of the multilayer membrane

All the experimental kinetic runs were obtained by running the CMR in the monolith-like configuration as depicted in Figure 2. The performance of the CMR was evaluated by investigating the effect of the three relevant process variables: residence time, τ_R, temperature, and hydrocarbon-to-oxygen molar ratio, HC/O.

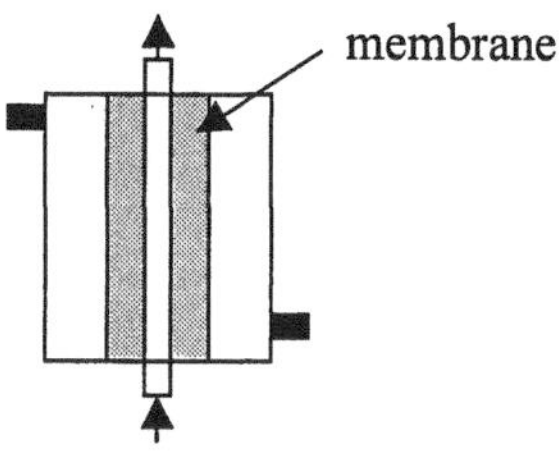

Figure 2. Catalytic membrane reactor in the monolith-like configuration.

3. CATALYTIC MEMBRANE REACTOR MODEL

In this section we outline the catalytic membrane reactor model developed to identify the role of chemical and mass transport phenomena in oxidehydrogenation of alkanes. In particular, the model has been used to rationalize the experimental data previously reported [1] for the monolith-like configuration. Here, we simply summarize the basic assumptions adopted: i) the CMR is viewed as a plug-flow reactor with reactive and permeable walls; ii) the reaction takes place only within the γ-Al_2O_3 layer; iii) the external mass transfer resistance is neglected.

The model equations are constituted of the continuity equations, written for the main external gaseous stream, and of the diffusion-reaction equations, written for the fluid within the porous matrix. The two sets of mass balances are coupled through the diffusional fluxes evaluated at the fluid/γ-Al_2O_3 layer interface. Furthermore, since the thickness of the γ-Al_2O_3 layer is much more smaller than the inner membrane radius, the diffusion-reaction equations have been written in plane co-ordinates. With the outlined assumptions the model equations are as follows.

Continuity Equation for the external fluid:

$$A\tilde{u}\frac{dc_i}{dz} = J_i\, 2\pi r_\gamma \qquad c_i = c_i^0, \quad z = 0 \tag{1}$$

where the fluxes J_i are computed accordingly to:

$$J_i = -D_i^e \left.\frac{dc_i}{dy}\right|_0 \tag{2}$$

Diffusion-Reaction equation for the fluid within the γ-Al_2O_3 layer:

$$D_i^e \frac{d^2\tilde{c}_i}{dy^2} + \sum_{j=1}^{r} \nu_{i,j} r_j = 0 \qquad \tilde{c}_i = c_i, \quad y = 0 \qquad \frac{d\tilde{c}_i}{dy} = 0, \quad z = \delta \tag{3}$$

The mass transport across the membrane is described in an effective way through the Bonsaquet formula:

$$D_i^e = \frac{\varepsilon}{\tau}\left\{ 1/\left(1/D_i^m + 1/D_i^k \right)\right\} \tag{4}$$

where D_i^k and D_i^m represent the Kundsen and the molecular diffusivity of the ith species, respectively, D_i^e is the effective diffusivity, τ is the tortuosity factor, and ε the porosity.

As first attempt, the oxidehydrogenation of propane has been described through the reaction scheme recently proposed by Stern and Grasselli [2] in which the direct oxidation to CO_x was neglected on the basis of the experimental observations. Since no acrolein was detected at the operating conditions of our experiments, we assumed that all the acrolein, eventually formed, is immediately oxidized to CO_x. The simplified kinetic scheme is depicted in Figure 3 along with the adopted lumped reaction rate laws.

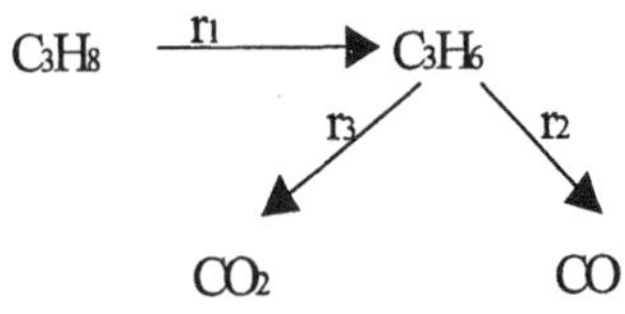

$$r_1 = k_1 c_{C3H8}^{a} c_{O2}^{b}$$
$$r_2 = k_2 c_{C3H6} c_{O2}^{c}$$
$$r_3 = k_3 c_{C3H6} c_{O2}^{d}$$

Figure 3. Adopted simplified reaction scheme.

It is worth pointing out that while the transport parameters (diffusivity coefficients and ε/τ) have been either estimated *a priori* or evaluated from permeability experiments, the kinetic parameters (reaction rate constants and apparent reaction orders) have been computed as fitting parameters.

4. RESULT AND DISCUSSIONS

The preliminary experimental kinetic runs were carried out over an operating window spanning a rather wide range of the three relevant process variables: residence time (3-8.4 s), hydrocarbon-to-oxygen molar ratio (0.1-5.84), and temperature (400-480°C). However, the kinetic parameters of the model have been computed by fitting the experimental data obtained for fixed conditions: residence time of 8.4 s and temperature of 400°C. The fitting strategy consisted of trying to recover the experimental selectivities and conversion as function of the ratio HC/O. The process variable HC/O was selected as independent one in order to test the capability of the model to describe the oxidehydrogenation over a wide range of operating conditions varying from lean (low HC/O values) to rich conditions (high HC/O values).

In Table 1 the transport parameters adopted in the simulations are reported, while the computed kinetic parameters are shown in Table 2. The reaction rate constants are reported in terms of the non-dimensional Thiele moduli, Φ_i, computed on the basis of the propane effective diffusivity.

Table 1
Transport parameters [$m^2/sec\ 10^9$] adopted in the simulations

D^e_{C3H8}	D^e_{C3H6}	D^e_{CO}	D^e_{CO2}	D^e_{H2O}	D^e_{He}	D^e_{O2}
0.613	0.628	0.786	0.615	0.972	2.07	0.729

Table 2
Computed kinetic parameters

a	b	c	d	Φ_1	Φ_2/Φ_1	Φ_3/Φ_1
0.8	0.5	0.5	1.0	0.046	6.9	37.4

The fitting results are summarized in Figures 4 and 5 where the comparison between model predictions (lines) and experimental data (symbols) is reported in terms of conversion and product selectivities. We have also tried to account for the direct oxidation of propane to CO_x without observing any significant improvement, and thus, we have neglected it for this preliminary analysis.

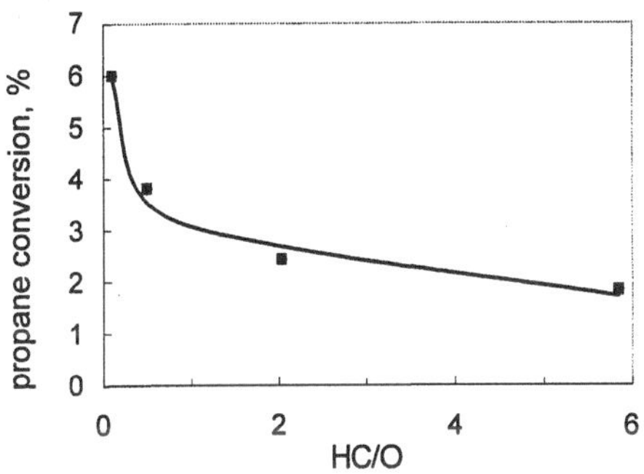

Figure 4. Propane conversion vs. HC/O

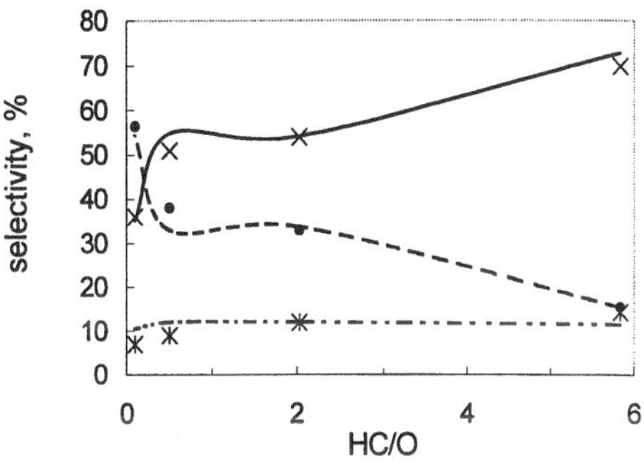

Figure 5. Selectivity vs. HC/O
(×) C_3H_6, (*) CO, (•) CO_2

The results show that the developed model is capable to capture the main physico-chemical processes taking place in the CMR. In particular, the simplified kinetic scheme appears suitable to describe the chemistry of the system over a wide range of HC/O. It is worth pointing out that the predicted selectivities well compare with the experimental ones. The predictive character of the CMR model has been assessed by comparing the model predictions with experimental data not used for the parameters fitting. The results are summarized in Figures 6 and 7 where propane conversion and propylene selectivity are shown as function of the residence time for the three values of the HC/O ratio investigated.

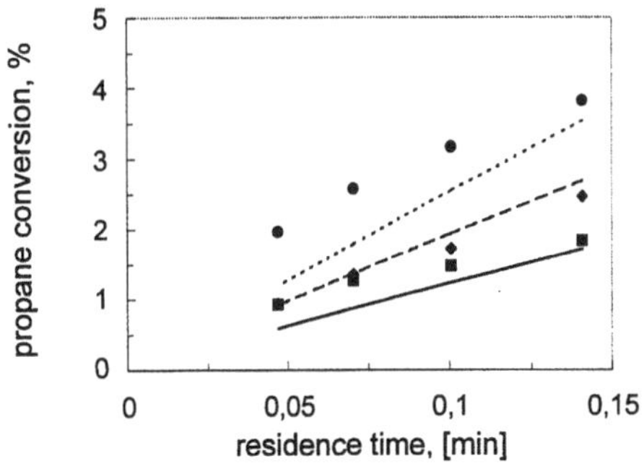

Figure 6. Propane conversion vs.residence time
(■)5.84, (◆) 2.03, (•) 0.5.

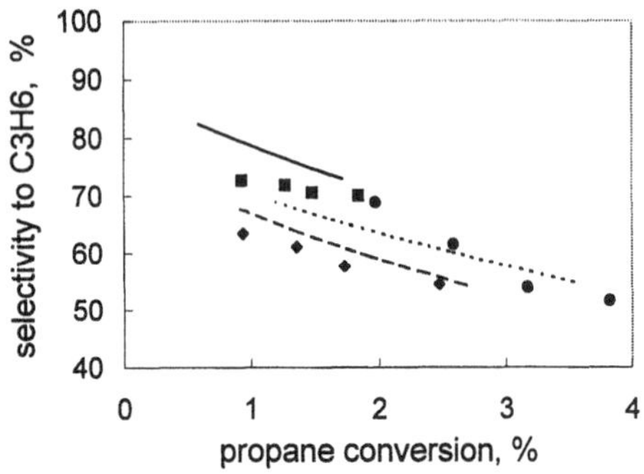

Figure 7. Propylene selectivity vs. propane conversion
experimental data: as in Fig. 6.

The results show that the catalytic reactor model, developed to describe the oxidehydrogenation of propane, performs reasonably well in predicting the CMR behavior over a relatively wide range of residence times and HC/O ratios. In particular, the model seems to be unable to fully capture the fact that in large excess of hydrocarbon, namely for high values of the ratio HC/O, the selectivity to propylene seems to slightly depend on propane conversion.

5. CONCLUSIONS

A simplified model of catalytic membrane reactors has been proposed for the interpretation of the oxidehydrogenation of propane to propylene. Even though a simplified kinetic scheme has been adopted, the developed model is capable to capture the essential characteristics of the CMR and to well describe the interactions between mass transport and chemical processes. The results show that within the layer of γ-Al_2O_3 the reactions take place mainly in a kinetic regime and that the short diffusional path offered by the thin γ-Al_2O_3 layer limits the formation of the complete oxidized species. This can be advocated as explanation for the better performance of CMR's with respect to conventional packed bed reactors as previously reported [1]. Work is in progress for better characterizing the performance of CMR in oxidehydrogenation reactions. In particular, new experimental runs are planned over a wider range of operating variables to better understanding the system chemistry.

REFERENCES

1. G. Capannelli, E. Carosini, F. Cavani, O. Monticelli and F. Trifirò, Chem. Eng. Sci., 51, (1996) 1817.
2. D. L. Stern, R. K. Grasselli, J. Catal. 167, (1997) 560.

NATURAL GAS CONVERSION V
Studies in Surface Science and Catalysis, Vol. 119
A. Parmaliana et al. (Editors)

Partial Oxidation of Ethane in a Three-Phase Electro-Fenton System

E.R.Savinova[1], A.O.Kuzmin[1], F.Frusteri[2], A.Parmaliana[2] and V.N.Parmon[1]

[1]Boreskov Institute of Catalysis, Novosibirsk 630090, Russian Federation, Fax: +7(3832) 35 57 56, E-mail: Elensav@catalysis.nsk.su *

[2]Institute CNR-TAE, Via S.Lucia, 39, Messina, Italy, Fax: 39 90 624247, E-mail: Parmaliana@itae.me.cnr.it

Partial oxidation of ethane to acetaldehyde and ethanol in a three-phase electrocatalytic system under mild conditions (1 bar, 360 K) is demonstrated. Factors controlling the rate and selectivity of the process have been evaluated and the radical-chain oxidation mechanism has been proposed.

1. INTRODUCTION

Creation of new efficient methods for production of oxygenates from light paraffins under mild conditions is one of the most challenging chemical problems. Recently, an electrochemical approach based on reductive cathodic activation of dioxygen has been suggested for the oxidation of various organic substrates [1-3]. To avoid consecutive deep oxidation of hydrocarbons, a three-phase catalytic membrane reactor has been proposed, allowing an efficient separation of the products from the oxidants through a partially hydrophobic carbon membrane [4, 5]. The present work demonstrates that the advantages of the electrochemical and membrane systems can be combined to achieve selective oxidation of ethane with dioxygen under mild conditions.

2. EXPERIMENTAL

All solutions were prepared from the "chemical pure" grade chemicals and "Milli-Q RG" (Millipore) water. Carbon membranes were prepared according to procedure described elsewhere [4] and consisted from hydrophobic carbon paper with the partially hydrophobic composite (Carbon Ketjenblack, 950 m^2/g + Teflon) catalytic layer on one side. Nitrogen adsorption and Hg-porosimetry have shown that carbon membranes have 197 m^2/g specific surface area and developed pore structure with macropores ca. 30 μm in diameter. Nafion-H (1100 EW product, Dupont, Wilmington, D.E.) was deposited on carbon

* Financial support from INTAS (grant N. 93-2402) is gratefully acknowledged.

membrane from the solution in isopropyl alcohol and then dried at 180°C. Cu and Ag were deposited on the electrode surface via the impulse galvanostatic reduction of $CuSO_4$ and $Ag[NH_3]_4^+$, respectively.

Partial ethane oxidation was carried out in a three-compartment electrochemical cell (Fig. 1) in galvanostatic regime. The liquid compartments were separated with a Nafion membrane. The mixture of ethane and dioxygen was introduced into the cathode compartment of the cell. The reaction products were collected in the cold (2°C) trap. The current density was referred to the geometric area of the electrode.

Liquid oxygenates were analysed with HPLC and gas chromatography using a column from graphitised soot impregnated with 2 wt.% of diglycerine. Gaseous products were analysed using Porapak S and NaX columns.

Concentration of H_2O_2 in the electrolyte was followed by periodic titrations with permanganate solution (0.01 eq/dm^3).

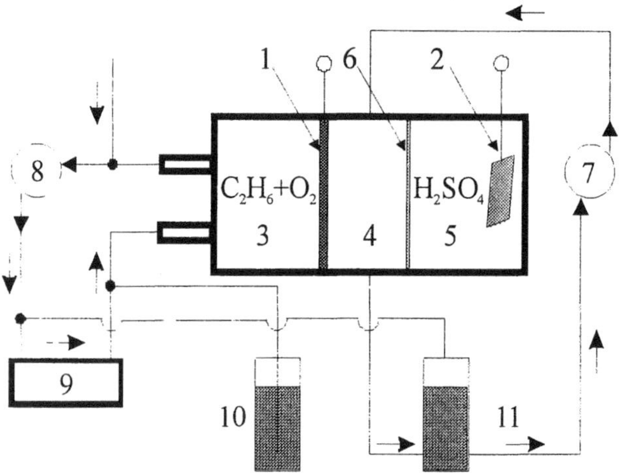

Figure 1. Schematic diagram of the experimental set-up.
1 - carbon cathode,
2 - Pt anode,
3 - gas compartment,
4 - liquid cathode compartment,
5 - liquid anode compartment,
6 - Nafion membrane,
7 - liquid recycling pump,
8 - gas recycling pump,
9 - cold trap,
10 - gas escape,
11 - liquid vessel.

3. RESULTS AND DISCUSSION

Ethane oxidation with dioxygen has been found to start upon cathodic polarisation of the carbon electrode [6]. Partial oxidation products, namely acetaldehyde and ethanol, were formed and collected in the cold trap. An examination of the condensate and electrolyte solution with HPLC and gas chromatography proved an absence of any other oxidation products. Selectivity of the partial oxidation was 95÷98%; the amount of CO_2 formed due to the deep C_2H_6 oxidation not exceeding 5%. Current efficiency of the ethane oxidation ranged from 1 to 3 %, depending on the experimental conditions. The reaction did not occur under the open circuit.

1.1 Dioxygen reduction at carbon gas-diffusion cathode

Formation of H_2O_2 has been detected in the electrolyte upon cathodic polarisation of the working electrode. The dependence of H_2O_2 production rate on the current density was characterised by a volcano-shaped curve (Fig.2). Maximum at 13.3 mA/cm^2 can be explained by the existence of two consecutive electrode reactions: dioxygen reduction to hydrogen peroxide and subsequent reduction of the latter to water.

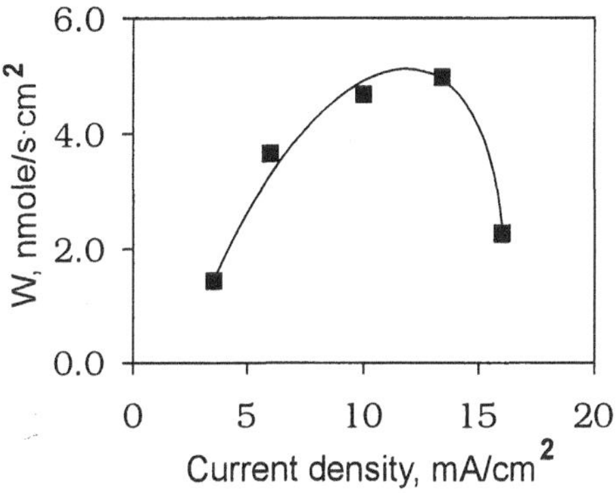

Figure 2. Effect of the current density on H_2O_2 formation rate (W) over carbon membrane (20 wt.% Teflon, 20 wt.% Nafion) at 353 K, p_{O_2} = 1.2 bar, $2 \cdot 10^{-3}$ M H_2SO_4.

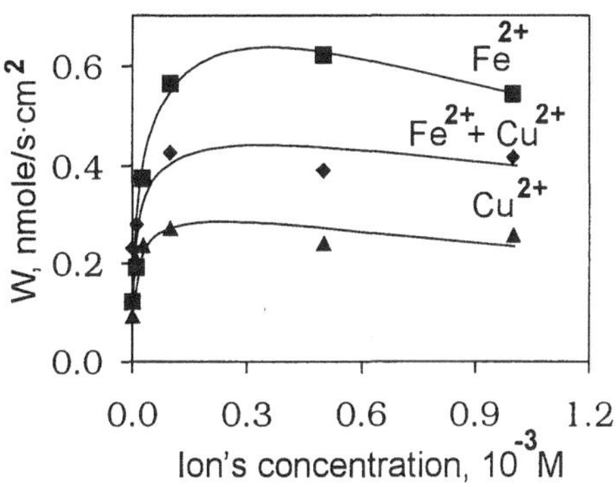

Figure 3. Effect of $Fe^{2+}$and Cu^{2+}on C_2H_6 partial oxidation rate (W) over carbon membrane (20 wt.% Teflon) at 363 K, 15 mA/cm^2. p_{O_2} = 0.3 bar, $p_{C_2H_6}$ = 0.7 bar, H_2SO_4 0.2 M.

1.2. Factors controlling the rate of partial oxidation

The presence of *transition metal ions (Fe^{2+} and Cu^{2+})* appeared to be of crucial importance for the partial ethane oxidation. The oxidation rate on a bare carbon electrode in the absence of catalytic additives was rather low (Fig. 3, Table 1). An addition of 10^{-4} M $FeSO_4$ to the electrolyte solution increased the reaction rate by a factor of 5, while the selectivity to acetaldehyde did not change drastically. Cu^{2+} or a mixture of Cu^{2+} and Fe^{2+} also increased the oxidation rate (Fig. 3). However, the maximum rate was lower in the latter cases.

The radical inhibitors (traps) of organic (2-naphthol) and inorganic (Cl^-) nature have been found to inhibit the reaction of the ethane oxidation (Fig. 4). However, a considerable quenching of the reaction was observed only in the presence of high concentration of 2-naphthol ($2 \cdot 10^{-3}$ M) or Cl^- (0.1 M). Note that the rate constant of Cl^- interaction with OH-radical is ca.$1 \cdot 10^{10}$ vs. $3.7 \cdot 10^8$ for ethane [10]. The induction period, which is a direct indication of the radical-chain mechanism, was not observed on the kinetic curves. It is remarkable that the rate of the ethane oxidation in the presence of $2 \cdot 10^{-4}$ M 2-naphthol was rather high.

The data obtained indicate that the reaction of the ethane oxidation proceeds in the pores of the gas-diffusion electrode presumably on the three-phase boundary between the electrode, electrolyte and the gas phase. Since the diffusion coefficient of ethane in the gas phase is much higher than that of 2-naphthol in the electrolyte, the reaction is inhibited only in the presence of high concentration of the latter. The results on the effect of the radical inhibitors favour the radical-chain mechanism of the paraffin oxidation but do not prove it unambiguously.

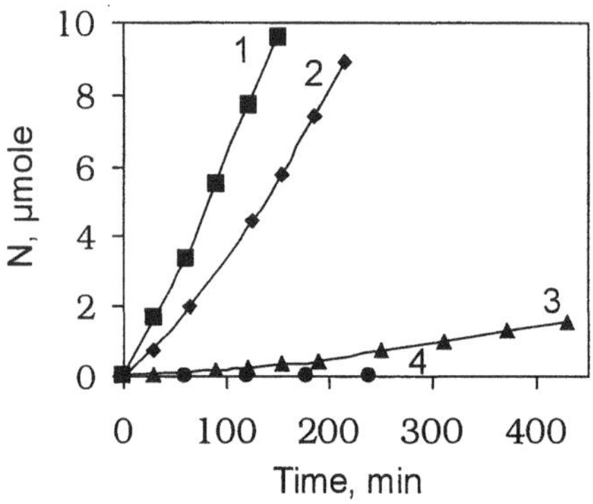

Figure 4. Kinetics of C_2H_6 oxidation over carbon membrane (20 wt.% Teflon) at 363 K, 15 mA/cm^2, p_{O_2}=0.3bar, $p_{C_2H_6}$=0.7bar, H_2SO_4 0.2 M: (1) – without radical trap, (2) - $2 \cdot 10^{-4}$ M 2-naphthol, (3) - $2 \cdot 10^{-3}$ M 2-naphthol, (4) - 0.1 M Cl^-.

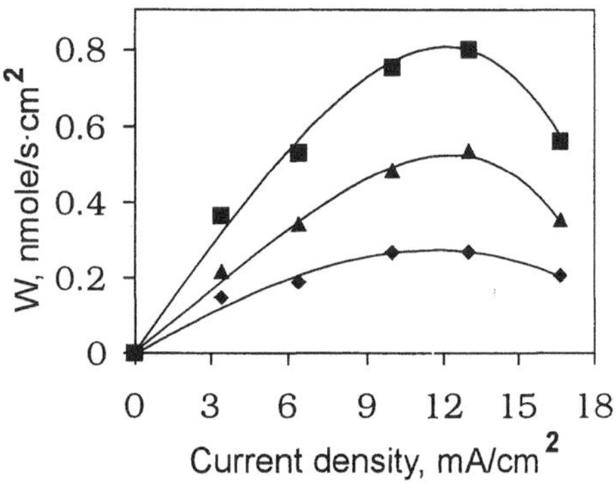

Figure 5. Effect of the current density on C_2H_6 partial oxidation rate (W) over carbon membrane (20 wt.% Teflon, 20 wt.% Nafion) at 353 K, p_{O_2} = 0.3 bar, $p_{C_2H_6}$=0.7 bar, $2 \cdot 10^{-3}$ M H_2SO_4.

Promotion of the carbon membrane with the *Nafion-H* film has been found to result in an increase of the partial oxidation rate (Table 1). This effect was most pronounced when Fe^{2+} was not added to the electrolyte. Meanwhile, carbon membranes have been found to contain ca. 1.3 μg/cm^2 of iron as an impurity. Since Nafion-H is a cation-exchange resin, it may act as a support for the positively charged Fe^{2+} ions, protecting the leakage of iron into the electrolyte and providing its high concentration in the reaction zone.

Current density (i) has been found to be of key importance for the oxidation process. The dependence of the partial oxidation rate on i was characterised by a volcano-shaped curve (Fig. 5). Selectivity to acetaldehyde increased with an increase of the overall oxidation rate. Comparison of the influence of i on C_2H_6 oxidation and H_2O_2 production rates (Figs. 2 and 5) proves that H_2O_2 is a direct precursor of the active oxygen species participating in the process of C_2H_6 partial oxidation.

The effect of *C_2H_6 partial pressure* on the oxidation rate has been studied at constant oxygen pressure (0.3 bar). The reaction order to C_2H_6 varied from 1 at low (< 0.4 bar) to 0 at high (> 0.6 bar) C_2H_6 partial pressure.

Substrate selectivity was investigated via comparison of C_2H_6 and C_3H_8 partial oxidation reactions. Propane oxidation gave rise to the formation of a variety of the oxidation products: ethanol (4%), acetaldehyde (14%), isopropanol (23%), n-propanol (7%), acetone (35%) and propionic aldehyde (17%). The ratio between the rates of C_2H_6 and C_3H_8 oxidation was 1 : 1.4. The ratio between the rates of oxidation of the primary and secondary carbon atoms was 1 : 2. Low substrate selectivity observed in the electrochemical system under study is typical for the radical mechanism of the paraffin oxidation [10].

Table 1

Effect of Nafion-H on the partial oxidation of ethane*.

Nafion-H, wt.%	$FeSO_4$ in the liquid phase, M	Reaction rate, nmole/s·cm²	Selectivity (%)	
			CH_3CHO	CH_3CH_2OH
0	0	0.12	~75	~25
20	0	0.36	50	50
0	10^{-4}	0.56	67	33
20	10^{-4}	0.83	65	35
0	10^{-3}	0.54	60	40
20	10^{-3}	0.53	73	27

*20 wt.% Teflon, 363 K, p_{O_2} = 0.3 bar, $p_{C_2H_6}$ = 0.7 bar, 15 mA/cm², 0.2 M H_2SO_4.

It has been found that besides transition metal ions dispersed metals (Cu and Ag) also catalyse partial ethane oxidation in the electrochemical system. The maximum reaction rate for Cu-modified electrode (3 mg/cm² Cu) was higher than that observed for the bare carbon membrane in the presence of $FeSO_4$. It is of interest that the reaction rate was not affected by the addition of $FeSO_4$ to the electrolyte.

1.3. Partial ethane oxidation in the fuel cell mode

Ethane has been found to be oxidised selectively in H_2/O_2 fuel cell [11] with co-generation of electricity and valuable chemicals (ethanol, acetaldehyde). The reaction was carried out in the gas electrochemical cell with a thin layer electrolyte (84% H_3PO_4 + 1 mM Fe^{2+} + 1 mM Cu^{2+}) at 343 K by flowing C_2H_6 and O_2 mixture through the cathode compartment and H_2 through the anode compartment. Current density equalled ca.5 mA/cm² and was stable during the time of the experiment (6 hours). The rate of the ethanol and acetaldehyde production was 0.1 and 0.14 nmole/s·cm², respectively.

1.4. Discussion of the reaction mechanism

The first step of the paraffin oxidation in the three-phase electrochemical system is dioxygen reduction to hydrogen peroxide. The latter is catalytically decomposed in the presence of Fe^{2+} and Cu^{2+} giving rise to the formation of OH and HO_2-radicals [7, 8]. OH-radicals are able to oxidise various organic substrates including paraffins [8, 10]. There are a number of the experimental results on the kinetics of C_2H_6 and C_3H_8 oxidation that favour the radical-chain mechanism of the reaction. These are: (a)- a considerable increase of the reaction rate in the presence of Fe^{2+} and Cu^{2+} ions; (b)- quenching of the reaction in the presence of the radical traps; (c)- low substrate selectivity; (d)- formation of a variety of C_2 - C_3 oxygenates from C_3H_8; (e)- dependence of the reaction order towards ethane on its partial pressure.

We thus infer that the oxidation mechanism involving OH-radicals is very likely and the system under study may be recognised as an "electro-Fenton". High selectivity of the partial oxidation which is not typical for radical-chain oxidation processes is provided by the efficient separation of easily oxidised products (ethanol, acetaldehyde) and strong oxidants (OH-radicals) in the three-phase membrane reactor. It should be noted however that due to the complex structural organisation of the present system we were not able to obtain direct evidence of the radical-chain mechanism of the partial oxidation.

Further studies are needed to clarify the mechanism of the ethane oxidation on Ag- and Cu-modified electrodes.

REFERENCES

1. R.L. Cook, A.F. Sammels, J. Electrochem. Soc. 137 (1990) 2007.
2. K.W. Freese, Jr, Langmuir 7 (1991) 13.
3. K. Otsuka, Y. Shimizu, I. Yamanaka, J. Electrochem. Soc., 137 (1990) 2076.
4. A. Parmaliana, F. Frusteri, F. Arena, N. Giordano, Cat. Let. 12 (1992) 353.
5. A. Parmaliana, in: New Developments in selective oxidation, Vol. 55, eds. G. Genti and F. Trifiro, Elsevier Science Publishers B.V., Amsterdam, 1990.
6. F.Frusteri, E.N.Savinov, A.Parmaliana, E.R.Savinova, V.N.Parmon and N.Giordano, Catal.Lett. 27 (1994) 355.
7. Advances in Catalysis and Related Subjects, v. IV , Eds. W.G.Frankenburg, V.I.Komarewsky and E.K.Rideal, 1952.
8. C.Walling, Accounts Chem.Res., 8, (1975) 125.
9. A.K.Pikaev, S.A.Kabakchi, Rate constants of the primary products of water radiolysis, Moscow, 1982 (in Russian).
10. E.S.Rudakov, Reactions of the alkanes with oxidants, metal complexes and radicals in solutions. Naukova Dumka, 1985 (in Russian).
11. A.O. Kuzmin, E.R. Savinova and V.N. Parmon, React. Kinet. Catal. Lett., 1998.

NATURAL GAS CONVERSION V
Studies in Surface Science and Catalysis, Vol. 119
A. Parmaliana et al. (Editors)

Hydrocarbons catalytic combustion in membrane reactors

A. Bottino[a], G. Capannelli[a], A. Comite[a], F. Ferrari[a], O. Monticelli[a], D. Romano[a], A. Servida[a], F. Cavani[b] and V. Chiappa[c]

[a]Dipartimento di Chimica e Chimica Industriale, Università di Genova, Via Dodecaneso 31, 16146 Genova, Italy

[b]Dipartimento di Chimica Industriale e dei Materiali, Università di Bologna, Viale Risorgimento 4, 40136 Bologna, Italy

[c]IPLOM SpA, Via C. Navone 3/b, 16017 Busalla (Genova), Italy

This paper reports preliminary results concerning the emission control of volatile organic compounds (VOC) by combustion in catalytic ceramic membrane tubes. Toluene was used as model species of VOC. The performance of the innovative catalytic combustor system was assessed as a function of relevant operating variables such as, temperature, residence time, and oxygen-to-hydrocarbon ratio.

1. INTRODUCTION

Volatile Organic Compounds (VOC) represent a very important class of pollutants that are present in various kind of industrial streams. VOC include a wide variety of chemical compounds such as for example alcohols, aldehydes, ketones, aliphatic and aromatic hydrocarbons, etc. The most common technologies for reducing and controlling VOC are reviewed in Ref. [1] and include: thermal and catalytic oxidation, adsorption, absorption, condensation, flaring, boiler/process heaters, biofiltration, membrane processes, and UV oxidation.

Recently, an innovative reactor, based on the forced permeation of VOC streams through a Pt/γ-Al_2O_3 catalytic membrane, has been proposed [2,3] for the control of VOC emissions. The results show that complete VOC (toluene and methyl ethyl ketone) combustion is achieved at temperatures much lower than those required in conventional monolith reactors. However, the catalytic system exhibits the disadvantage of a noticeable pressure drop due to the membrane permeation resistance.

In this paper we report preliminary results on the VOC emission control through a catalytic membrane combustor operating in a monolith-like flow configuration. The performance of the catalytic combustor was evaluated with respect to the following process variables: temperature, residence time, and oxygen-to-hydrocarbon ratio.

2. EXPERIMENTAL

Catalytic membranes were obtained by depositing the catalyst (Pt) onto the internal active layer of Al_2O_3 multilayered porous tubes supplied by SCT (France) and Schumacher (Germany). Both tubes had a length of 150 mm, an outer diameter of 10 mm, and a thickness of 1.5 mm. The tubes were sealed by a vitrification process for an extent of 25 mm at both ends. The active layer of SCT tubes was made of γ-Al_2O_3 (average pore diameter 5 nm) while that of Schumacher tubes was made of TiO_2 (average pore diameter 5 nm). A scanning electron microscope (SEM, Leica Stereoscan 440) was used for the morphological characterization of the membrane tubes.

The catalyst (Pt) was deposited on the active layers by the impregnation method described elsewhere [4]. The catalyst loading was measured by atomic adsorption after the chemical attack of the sample with aqua regia. The specific surface area of the active layer was measured by N_2 adsorption/desorption measurements (Micromeritics ASAP 2000).

The catalytic activity of the membranes was evaluated in the 100-350 °C temperature range with a toluene concentration between 800-5000 ppm. Toluene was selected as model species of VOC. A schematic of the experimental set-up used for the catalytic combustion tests is shown in Fig. 1. The carrier gas (N_2 or air) was saturated with toluene by using two saturators (A) arranged in series that were maintained at constant temperature. The saturated gas stream was mixed with O_2 and/or air in order to reach the desired O_2/toluene ratio. The pre-mixed reactants stream was fed to the reactor (B) located within the oven (C) in order to operate at isothermal conditions. The gas flowrates were controlled through the mass flow meters (D). The feed flowed tangentially to the membrane surface (i.e. without permeating through the membrane) and the combustion products were analyzed by using an on line gas chromatograph (E).

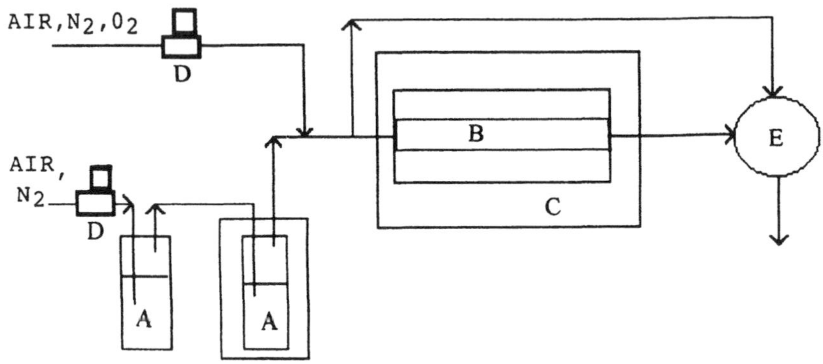

Figure 1. Schematic diagram of the experimental set-up used to perform catalytic combustion tests.

3. RESULTS AND DISCUSSION

As an example of SEM characterizations, Fig. 2 shows a micrograph of the upper part of the cross-section of a Schumacher membrane. The active layer appears to be crack-free and very thin.

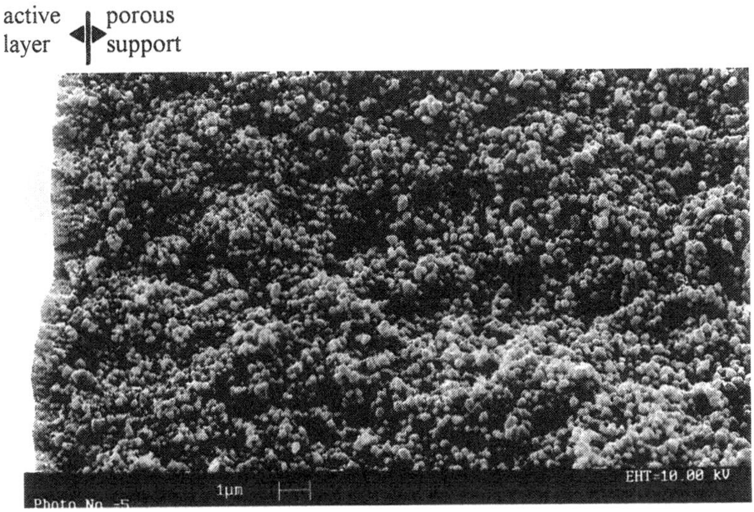

Figure 2. Scanning electron micrograph of the upper part of the cross-section of the Schumacher membrane.

The average thickness of the active layer, evaluated from this and other cross-section micrographs, is reported in Table 1 along with the BET surface area value. For comparison purposes, the characteristics of the SCT membrane are also reported. The results show that the two membranes exhibit comparable BET surface area but different active layer thickness.

Table 1
BET surface area (m^2/g) and thickness (µm) of the active layers of SCT and Schumacher tubular membranes

Membrane	Active layer composition	BET surface area	Thickness
SCT	γ-Al_2O_3	244	2
Schumacher	TiO_2	254	0.7

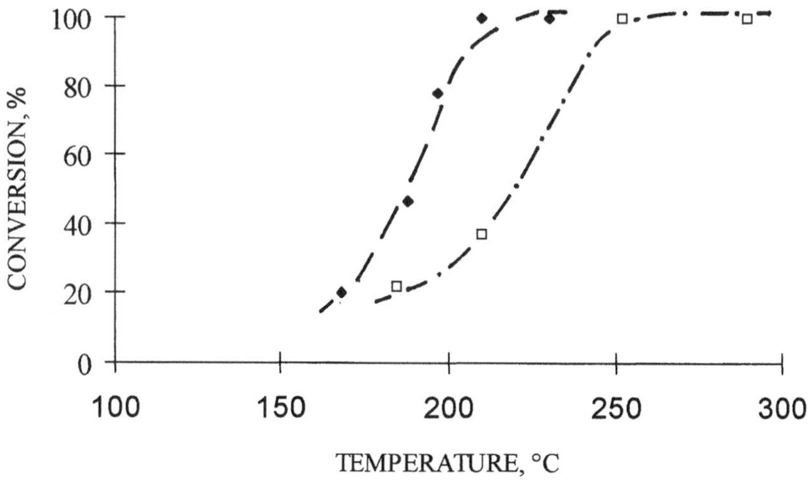

Figure 3. Toluene conversion vs. temperature for the SCT (□) and the Schumacher (◆) membranes.
Toluene concentration: 4760 ppmv. Residence time: 10 s. O_2/Toluene molar ratio: 29.

A first series of combustion experiments was carried to compare the performance of Schumacher and SCT membranes. The results are shown in Fig. 3 where the toluene conversion is reported as a function of temperature. For both the membranes by increasing the operating temperature the conversion increases first slowly, then more rapidly, and finally levels off. The results show that in the Schumacher membrane, complete VOC combustion is achieved at temperatures lower than those required for the SCT one. Indeed, the Schumacher membrane exhibits a light-off temperature, i.e. the temperature measured at a conversion of 50 %, about 30°C lower than that of SCT membrane. On the basis of these preliminary tests the Schumacher membrane was selected for further kinetic studies.

The effect of the inlet toluene concentration on the reactor performance is shown in Fig. 4. The results refer to catalytic tests carried out at a residence time of 2.5 s and using air as carrier gas. They show that an increase in toluene inlet concentration decreases the reactor performance by shifting the conversion curve towards higher temperatures. This is in agreement with the results obtained by Pina et al. [2,3] concerning the toluene combustion in catalytic membrane reactors operating in the flow-through mode.

The effect of the oxygen-to-toluene molar ratio on the reactor performance is shown in Fig. 5 where the toluene conversion is reported for two different temperatures (169 and 203 °C). The results show that at high temperature the O_2-to-toluene ratio does not have any significant effect on the toluene conversion, while at low temperature it improves the reactor performance. In particular, as Fig. 6 clearly indicates, operating at 203°C, the catalytic membrane reactor exhibits such a high catalytic activity that complete toluene conversion is attained even at low contact times (less than 1 s).

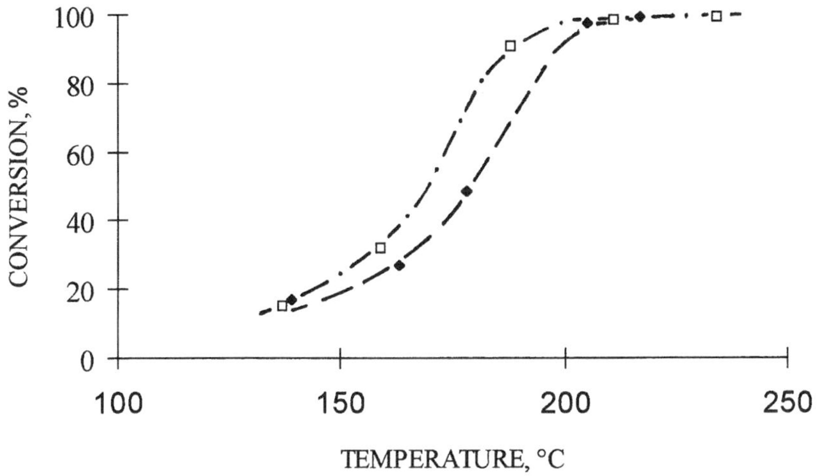

Figure 4. Toluene conversion vs. temperature for different toluene concentration in the feed. Toluene concentration: (□) 890 ppmv, (◆) 1610 ppmv. Residence time: 2.5 s.

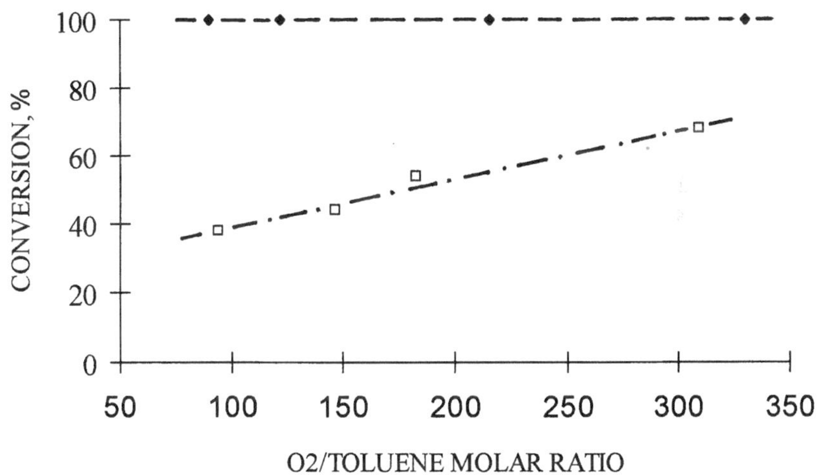

Figure 5. Toluene conversion vs. O_2/toluene molar ratio.
Operating temperature: (□)169 °C, (◆) 203 °C. Toluene concentration: 890 ppm. Residence time: 2.5 s

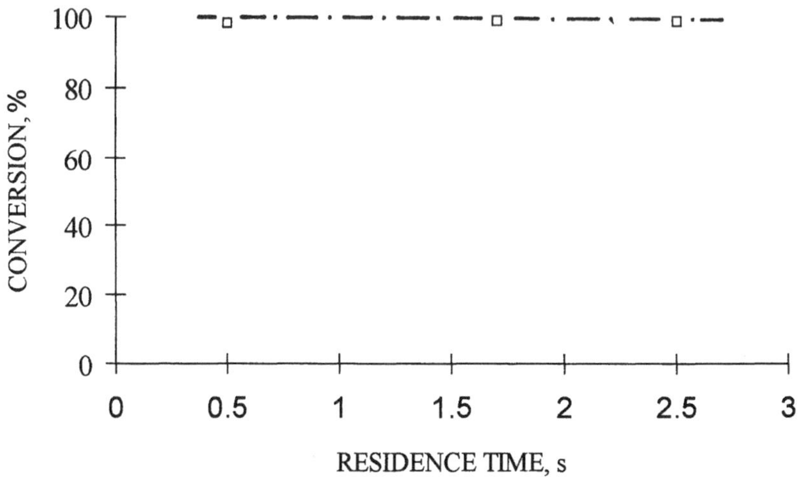

Figure 6. Toluene conversion vs. residence time at 203 °C operating temperature. Toluene concentration: 890 ppmv.

4. CONCLUSIONS

Catalytic combustion of toluene has been investigated by using a catalytic membrane reactor operating in a monolith-like flow configuration. The results indicate that in catalytic membrane reactors complete combustion of toluene can be achieved at temperatures lower than those usually found in conventional monolith reactors. This may be due to the surface area of catalytic membrane that is about one order of magnitude higher than that of conventional monoliths. The preliminary results indicate that operating the membrane combustor in the tangential flow configuration, complete combustion is achieved at temperatures slightly higher than those required for the all-through configuration [2, 3]. Work is in progress to identify optimal membrane characteristics and operating conditions in order to improve the performance of the catalytic membrane combustor.

REFERENCES

1. E.C. Moretti and N. Mukhopadhyay, Chem. Eng. Prog., 89 (1993) 20.
2. M.P. Pina, M. Menéndez and J. Santamaria, Appl. Catal.B, 11 (1996) 19.
3. M.P. Pina, S. Irusta, J. Santamaria, R. Hughes and N. Boag, Ind. Eng. Chem. Res., 36 (1997) 4557.
4. G. Capannelli, A. Bottino, G. Gao, A. Grosso, A. Servida, G. Vitulli, A. Mastrantuono, L. Lazzaroni and P. Salvadori, Catal. Lett., 20 (1993) 287.

NATURAL GAS CONVERSION V
Studies in Surface Science and Catalysis, Vol. 119
A. Parmaliana et al. (Editors)

Syngas Formation by Partial Oxidation of Methane in Palladium Membrane Reactor

E. Kikuchi and Y. Chen

Department of Applied Chemistry, School of Science & Engineering,
Waseda University, 3-4-1 Okubo, Shinjuku-ku, Tokyo 169-8555, JAPAN

The partial oxidation of methane (POM) occurred at a low temperature of 500℃ by reaction of an oxygen-deficient CH_4/O_2 mixture over supported precious metal catalysts. The catalytic activities decreased in the order of Rh, Pt, Pd, Ir, and Rh/Al_2O_3 also showed a high catalytic selectivity for the oxidative conversion of CH_4 to CO and H_2. On application of a hydrogen-permeable membrane reactor, CH_4 conversion and production of CO and H_2 were promoted by removing H_2 from the reaction system. In the membrane reactor, it was also found that deposition of coke began being exactly at the consumption of H_2O. Addition of steam to the reactant flow could effectively depress the coke deposition and improve the yield of H_2 via steam reforming and water-gas shift reaction. It was shown that the reaction path over these catalysts involved a sequence of following reactions: the initial complete oxidation of part of the methane to CO_2 and H_2O, followed by the highly endothermic steam reforming (SRM) and CO_2 reforming of unconverted methane, and water-gas shift reaction (WSR), establishing an equilibrium. The equilibrium of air POM in the membrane reactor gives a product mixture of H_2, CO_2, and N_2. This can provide an internal heating process to produce pure H_2 without formation of CO. H_2 thus produced is separated from CO_2 and N_2, followed by the reverse water gas shift reaction (RWSR) to adjust the H_2/CO ratio in a conventional reactor. This provides a more economical POM process as the separation of CO_2 from N_2 in the unpermeated gas from the membrane reactor should be less costly than air separation.

1. INTRODUCTION

Partial methane oxidation, leading to valuable oxygen-containing compounds, such as methanol, formaldehyde, synthesis gas, and the oxidative coupling of methanol into ethane, ethylene and other hydrocarbons, is one of the most rapidly developing and practically attractive field of catalysis. Especially studies of the catalytic POM to syngas (1), which is a versatile feedstock for Fischer-Tropsch synthesis, as well as methanol and ammonia syntheses, has been developing quickly in the last years [2-12].

$$CH_4 + 1/2O_2 = CO + 2H_2 \qquad \Delta H_{298} = -8.5 \text{ kcal / mol} \qquad (1)$$

Direct conversion of methane to syngas is a slightly exothermic reaction different from SRM. The endothermicity of SRM requires energy input, while a reactor based on the exothermic direct POM would be more energy efficient. It is of great practical importance, that the stoichiometry of POM gives a H_2/CO ratio of 2, that is a desirable feedstock for the production of methanol or Fischer-Tropsch synthesis.

It is generally accepted [1-4] that POM results from an initial reaction of complete combustion of a part of methane (2),

$$CH_4 + 2O_2 = CO_2 + 2H_2O \qquad \Delta H_{298} = -191.7 \text{ kcal / mol} \qquad (2)$$

which consumes all the oxygen. The produced water vapour and carbon dioxide are reduced by the residual methane or hydrogen (RWSR):

$$CH_4 + H_2O = CO + 3H_2 \qquad \Delta H_{298} = 49.3 \text{ kcal/mol} \qquad (3)$$
$$CH_4 + CO_2 = 2CO + 2H_2 \qquad \Delta H_{298} = 59.1 \text{ kcal/mol} \qquad (4)$$
$$CO_2 + H_2 = CO + H_2O \qquad \Delta H_{298} = 9.9 \text{ kcal/mol} \qquad (5)$$

Therefore, high reaction temperatures above 750℃ are normally required for complete conversion of methane.

In contrast, Hickman and Schmidt [5-8] reported the direct POM for the contact time of 10^{-2} to 10^{-4}10 sec on Pt and Pt-Rh monoliths catalysts with high selectivities. Choudhary and co-workers [9, 10] also reported that high selectivity to CO and H_2 could be achieved at temperatures in a wide range (300 - 800℃) on a variety of catalysts with much greater gas space velocities (GHSV: 10^5 - 10^6 h^{-1}). Above 700℃, the observed CH_4 conversion and CO selectivity were close to the equilibrium ones. However, below 700℃, these values were much higher than that obtained at the reaction equilibrium. These findings, however, are in some dispute as claimed by Dissanayake et al. [11] or Chang and Heinemann [12].

In this paper, the effects of applying a hydrogen-permeable membrane reactor to POM with air are discussed. In our early works, experimental and theoretical studies using palladium membranes, which selectively permeate hydrogen, for SRM [13], WSR [14], aromatization of propane [15], and CO_2 reforming of methane [16], demonstrated that the performance of palladium membranes could be used as a separation medium and the reaction conversion was improved even in lower temperatures for the equilibrium-limited endothermic reversible reactions.

Commercial POM unavoidably employs pure O_2 to produce syngas, because the separation of N_2 from the syngas is not more economical. An alternative route to produce syngas without air separation will be formulated by use of a membrane reactor. We also determined the effect of adding steam to the feedstock and operating molar ratio of H_2O/CH_4 favored in this reaction system to gain high selectivity in pure hydrogen formation.

2. EXPERIMENTAL

2.1 Catalysts

A series of catalysts were prepared by impregnating an alumina support with an aqueous solution of noble metals to yield a nominal 1wt% metal. The following salts of metals were used: $H_2PtCl_6X_6H_2O$, $PdCl_2$, $RhCl_3$ $3H_2O$ and $[IrCl(NH_3)_5]Cl_2$. The impregnated powders were dried at 60℃ and oxidized at 500℃ for 2 hr. The fragments of catalyst pellets to 32-65 mesh were reduced at 500℃ for 1 hr. After reduction, the sample was flushed with argon and settled to the reaction temperature.

2.2 Apparatus

The palladium membrane reactor, a double tubular type in a continuous flow system, used in this study was the same as previously described [13]. Catalyst particles were uniformly packed outside the membrane (the reaction side) and hydrogen permeation through membrane was restricted only in the part of catalyst bed. Permeated hydrogen can be evacuated by use of a rotary vacuum pump.

2.3 Methods

Reactions in the membrane reactor was carried out using 6 g catalyst and under the space velocity of 5000-25000 g-catalyst min CH_4-mol^{-1}. The molar ratio of CH_4 to O_2 in the reacting gas mixture was 1 : 0.5 and additional steam was supplied from a saturator. The exit gases were analyzed with a TCD gas chromatograph. The system was operated at a total pressure of 1 atm. Reactions were also done in a conventional flow reactor and at the temperature of 500°C to compare the catalytic activities.

3. RESULTS AND DISCUSSION

3.1 Membrane reactor effects

Firstly, the reaction was investigated in a conventional reactor with some noble metal catalysts supported on alumina, where was found catalytic activities are, as presented in Fig.1. in the following order:

$Rh/Al_2O_3 > Pt/Al_2O_3 > Pd/Al_2O_3 > Ir/Al_2O_3$.

Conversion of CH_4 on Rh catalyst in the membrane reactor was remarkably higher than that in conventional reactor, as shown in Fig.2. An increase in CH_4 conversion approaching to 100% was observed with increasing contact time, since selective removal of H_2 from the reaction system shifted the thermodynamic equilibrium toward the product side. In conventional rector under the same reaction conditions, the approach to the equilibrium of CH_4 conversion was fast, leaving excess CH_4 in the product, while the available oxygen was consumed in this phase of reaction.

Equilibrium yields are attributed to the reaction sequence in which the feed CH_4 is converted to H_2O and CO_2, and the remaining CH_4 is converted to CO and H_2 via reforming reactions. In the membrane reactor, CH_4 conversion increased leading to a obvious promotion of equilibrium yields, because the H_2 produced was selectively removed from the reaction system continuously. The results are illustrated in Fig.3, showing the relation of CH_4 conversion to product selectivities in both conventional and

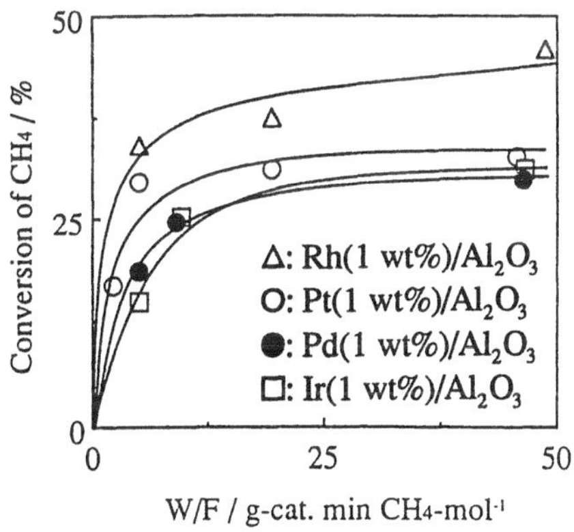

Figure 1. Conversion of CH_4 on various catalysts in conventional reactor as a function of time factor W/F.
Reaction temperature, 773 K.
Molar Air/CH_4, 2.5.

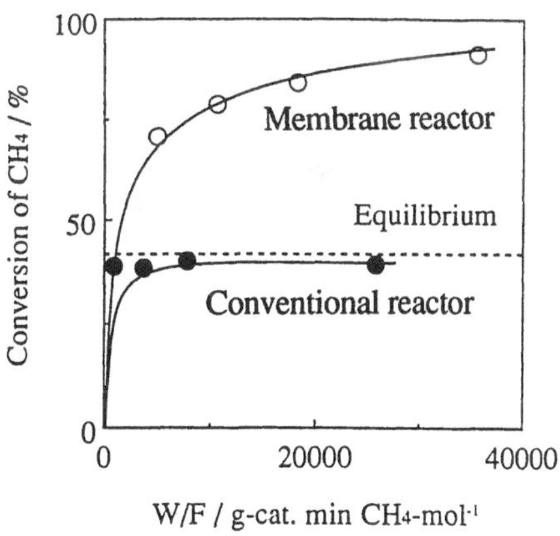

Figure 2. Conversion of CH_4 as a function of time factor W/F in conventional reactor and that in membrane reactor.
Reaction temperature, 773 K.
Molar Air/CH_4, 2.5.
Catalyst: Rh(1 wt%)/Al_2O_3.

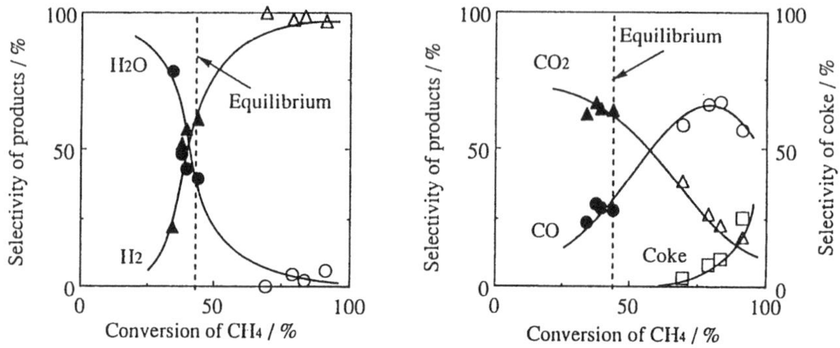

Figure 3. Selectivities of H_2, H_2O, CO, and CO_2 dependence on conversion of CH_4. Reaction temperature, 773 K; molar Air/CH_4, 2.5. Catalyst: Rh(1 wt%)/Al_2O_3. ●, ▲: Conventional reactor; O, Δ, □: Membrane reactor.

membrane reactors. It was also found at high CH_4 conversion that, using a membrane reactor, carbon began to deposit. The carbon deposition became more pronounced with higher CH_4 conversion, accompanied with consumption of H_2O in the reaction system which brought about the decrease in the ratio of CO_2/CO in the reaction zone leading to the carbon forming condition of the reaction (6) favorable thermodynamically.

$$2CO = C + CO_2 \tag{6}$$

3.2 Effect of adding steam

To attain higher selectivity of pure hydrogen formation still further, steam can be added into the reaction system to convert CO to CO_2 and to product H_2 by WSR.

Figure 4 shows the changes of reaction heats versus the molar H_2O/CH_4 ratio in POM. With the increase in the molar H_2O/CH_4 ratio, the evolution of reaction enthalpy will increase to a positive value. It is also entirely possible to keep the reaction enthalpy value in negative by adjusting the partial pressure of H_2 remained in reaction side.

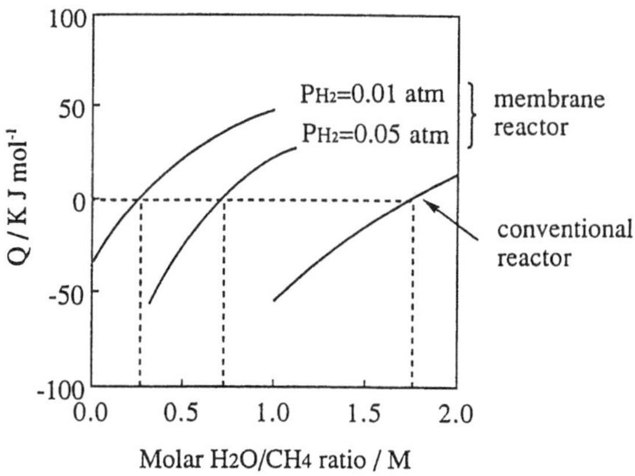

Figure 4. Relation between Q at equilibrium and molar H_2O/CH_4 ratio in partial oxidation reaction.
Reaction temperature, 773 K.
Total pressure, 1 atm.
Molar $(H_2O+N_2)/O_2/CH_4$, 2/0.5/1.

The permeation of hydrogen from the reaction system brings the increase of partial pressure of CO in reaction zone, so that the Boudouard reaction (6) would be suggested to promote coke formation easily in the catalyst bed. Figure 5 describes the relation between molar H_2O/CH_4 ratio and P_{CO_2}/P^2_{CO} at equilibrium in POM in the membrane reactor. The equilibrium constant for the reaction (6) at 500℃ was calculated as 232. It is obviously to see adding steam to the reaction system, coke formation can be depressed thermodynamically.

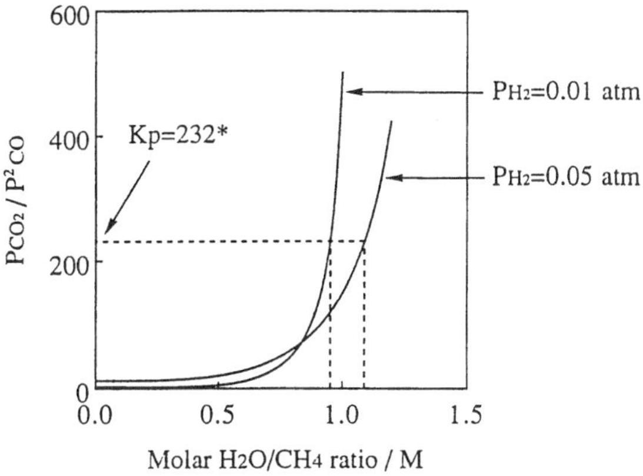

Figure 5. Relation between molar H_2O/CH_4 ratio and P_{CO_2}/P^2_{CO} in partial oxidation reaction in membrane reactor at equilibrium.
Reaction temperature, 773 K.
Total pressure, 1 atm.
Molar $(H_2O+N_2)/O_2/CH_4$, 2/0.5/1.
*Kp=232 in $2CO \rightleftarrows C+CO_2$

As demonstrated in Fig.6, adding H_2O to the feed in membrane reactor, not only the coke deposition was depressed but also the CH_4 conversion was enhanced significantly. With increase in the molar H_2O/CH_4 ratio above 0.5, the coke deposition is depressed completely and the increase in the H_2O/CH_4 ratio further to 1 enables the CH_4 conversion to reach 100% and to gain still the most high yield of H_2.

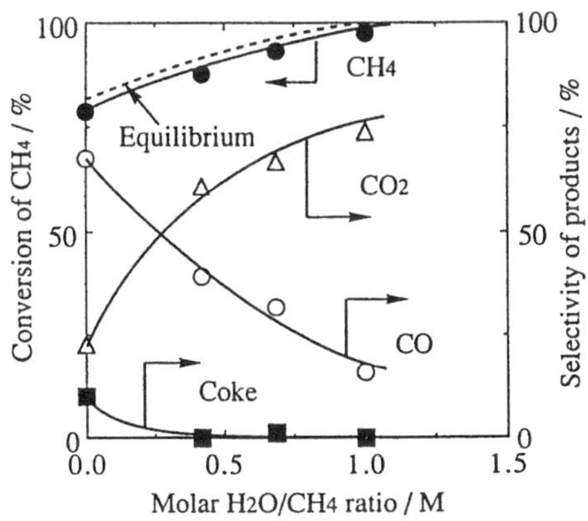

Figure 6. Conversion of CH_4 and selectivities of products in membrane reactor vs. molar ratio H_2O/CH_4.
Reaction temperature, 773 K; molar ratio $CH_4/O_2/(Ar+H_2O)$, 2/1/4; W/F, 10000 g-cat. min/mol-CH_4; total pressure, 1 atm. Catalyst, Rh(1 wt%)/Al_2O_3.

3.3 Application of membrane POM to syngas production

These results show that the reaction path involves a sequence of initial complete oxidation of a part of CH_4 to CO_2 and H_2O, followed by highly endothermic steam and CO_2 reforming of unconverted CH_4 to CO and H_2 and WSR, establishing the equilibrium. Therefore, it is possible to produce H_2 and CO_2 by air POM. This can provide an internal heating process to produce pure H_2 without formation of CO, a favorable reformer applicable to the fuel cell system. Addition of steam to the reactant flow could also be effective to improve the yield of H_2 via SRM and WGS, and also to depress carbon deposition.

Utilization of air instead of pure O_2 is extremely beneficial, since air separation is unnecessary. Commercial partial oxidation process unavoidably employs pure O_2 to produce syngas, because the separation of N_2 from the syngas is not more economical. An alternative route to produce syngas without air separation can be effected by use of a membrane reactor: H_2 and CO_2 are produced in the membrane reactor which separates H_2 from a mixture of CO_2 and N_2, followed by RWSR to adjust the H_2/CO ratio in a conventional reactor. Separation of CO_2 from N_2 in the unpermeated gas from membrane reactor should be less costly than air separation.

REFERENCES

[1] M. Prettre, CH. Eichner and M. perrin, Trans., Faraday Soc. 43(1946) 355.

[2] A.T. Ashcroft, A.K. Cheetham, J.S. Foord, M.L.H. Green, C.P.Grey, A.J. Murrell and P.D.F. Vernon, Nature 344(1990) 319.

[3] P.D.F. Vernon, M.L.H. Green, A.K. Cheetham and A.T. Ashcroft, Catal. Today 13(1992) 417.

[4] D. Dissanayake, M.P. Rosynek, K.C.C. Kharas and J.H. Lunsford, J. Catal. 132(1991) 117.

[5] D.A. Hickman and L.D. Schmidt, J. Catal. 138(1992) 267.

[6] D.A. Hickman, E.A. Haupfear and L.D. Schmidt, Catal. Lett. 17(1993) 233.

[7] D.A. Hickman and L.D. Schmidt, AIChE Journal 39(1993) 1164.

[8] P.M. Torniainen, X. Chu and L.D. Schmidt, J. Catal. 146(1994) 1.

[9] V.R. Choudhary, A.M. Rajput and V.H. Rane, J. Phys. Chem. 96(1992) 8686.

[10] V.R. Choudhary, A.M. Rajput and B. Prabhakar, J. Catal. 139(1993) 326.

[11] D. Dissanayake, M.P. Rosynek and J.H. Lunsford, J. Phys. Chem. 97(1993) 3644.

[12] Y.-F. Chang and H. Heinemann, Catal. Lett. 21(1993) 215.

[13] S. Uemiya, N. Sato, H. Ando, T. Matsuda and E. Kikuchi, Appl. Catal. 67(1991) 223.

[14] S. Uemiya, N. Sato, H. Ando and E. Kikuchi, Ind. Eng. Chem. Res. 30(1991) 589.

[15] S. Uemiya, I. Koike and E. Kikuchi, Appl. Catal. 76(1991) 171

[16] E. Kikuchi and Y. Chen, Stud. in Surf. Sci. Catal. 107(1997) 547

NATURAL GAS CONVERSION V
Studies in Surface Science and Catalysis, Vol. 119
A. Parmaliana et al. (Editors)
1998 Elsevier Science B.V.

PARTIAL OXIDATION OF LIGHT PARAFFINS ON SUPPORTED SUPERACID CATALYTIC MEMBRANES

F. Frusteri [1]*, F. Arena[2], C. Espro[2], N. Mondello[1] and A. Parmaliana [2]

[1] Istituto CNR-TAE, Via Salita S. Lucia 39, I-98126 – S. Lucia, Messina-Italy
[2] Dipartimento di Chimica Industriale, Università degli Studi di Messina
Salita Sperone c.p. 29, I-98166, S. Agata – Messina, Italy

ABSTRACT

Superacid supported catalytic membranes were found to be active and very selective in the partial oxidation of light paraffins (C_1-C_2) to the corresponding alcohols and aldehydes with H_2O_2 under mild conditions (T_R: 80-110 °C; P_R: 1,4 bar) in a three phase catalytic membrane reactor (3PCMR). Among different catalytic membranes investigated, Nafion based ones showed the best performance in terms of both activity and selectivity. Addition of Fe^{2+} ions in the liquid phase enhances the reaction rate. A reaction pathway based on the electrophilic hydroxylation of the C-H bond on superacid sites and subsequent reaction of the activated paraffin with OH radicals has been proposed.

1. INTRODUCTION

Light paraffins are very unreactive molecules as demonstrated by their low acidity , low basicity and high C-H bond strength . As a consequence they can be made to react only with very reactive species like radicals that being obtained under drastic reaction conditions, generally do not allow the attainment of high selectivity to oxygenates. Therefore, better results can be expected from activation under mild conditions. During the last decade several groups have claimed the electrophilic activation of CH_4 on noble metal catalysts in different reaction media /1-3/:the easy re-oxidation of the metal limits the practical suitability of these catalytic systems. The activation of the C-H bond of methane by Pt black in aqueous solution of ferric sulphate /4/ has been also reported, however the yield to partially oxidised products is limited by their subsequent oxidation /1-4/. Recently, Periana et al /5/ proposed a homogeneous system for the selective oxidation of methane to methanol with sufuric acid catalysed by Hg^{2+} ions. The net reaction is the oxidation of CH_4 by concentrated sulfuric acid to produce methyl bisulfate, water and SO_2. Separate hydrolysis of methyl bisulfate and reoxidation of the sulfur dioxide with air provides a potentially practical scheme for the oxidation of methane to methanol with molecular oxygen. A molar yield to CH_3OH of 43% has been obtained. This is the best result till now obtained in the selective oxidation of CH_4 to CH_3OH, however technological problems linked with the employment of concentrated sulphuric acid and the high cost of the oxidant are two drawbacks while hinder the development of this system.

The aim of this work is to explore the potential of the three phase catalytic membrane reactor (3PCMR) /6/ in the selective partial oxidation of light paraffins (C_1-C_2).

1. EXPERIMENTAL

Catalysts. Catalytic membranes were prepared by :
a) impregnation of the bare membranes with a solution of the active species; and b) deposition of a paste containing active species and teflon on a carbon paper acting as a physical support. Bare membranes were obtained by deposition of the carbon-teflon paste on carbon paper and subsequent activation at 300°C in N_2 atmosphere. Ketjenblack carbon (EC 600 Akzo Chemie, BET S.A., 950 m^2/g) was used as support.
Superacid based membranes were obtained by incipient wetness impregnation of the bare membrane with an isopropanol solution of Nafion-H (1100 EW product, Dupont Wilmington, DE), phosphomolybdic acid (HPMo) and phosphotungstic acid (HPW). The Cs (CsM) based membrane was prepared by deposition of $H_{0.5}Cs_{2.5}PW_{12}O_{40}$–teflon- ethanol past (60/10/30, wt/wt/wt) on carbon paper and subsequent activation at 300 °C in N_2.
Apparatus and procedure. Catalytic membranes have been tested in the 3PCMR elsewhere described /6/. The membrane was sandwiched between two teflon plates with the catalyst side turned towards the liquid phase. Catalytic measurements have been performed in the range 80-110 °C with methane-nitrogen or ethane-nitrogen mixture ($P_{Hydr.}/P_{N2}$ = 3.7) at 140kPa absolute pressure operating in batch mode with separate recirculation of both the gas and the liquid phases. The liquid phase was constituted by a H_2O_2 solution ($7.25 \cdot 10^{-2} \div 1.17$mol/l) containing Fe^{2+} ions ($[Fe^{2+}] = 1 \cdot 10^{-6} \div 2.7\ 10^{-4}$ mol/l). Oxygenates formed during reaction were trapped at 2°C down-stream of the reactor and analysed by a GC equipped with a FID detector using a Carbopack B 3%SP1500 column (l = 2.5 m; i.d. = 2 mm) operating at 50 °C. H_2O_2 concentration change was followed by periodical titration with permanganate solution (0.1 mol/l). The acidic properties of the membranes were comparatively evaluated by ammonia chemisorption tests carried out in a Micromeritics 2900 TPR/TPD unit equipped with TCD, while the morphological features and the surface chemical composition of the catalytic layer, were analysed by Scanning Electron Microscopy by using a Philips XL20 equipped with EDS micro-probe which allowed to accomplish the surface chemical map.

3. RESULTS and DISCUSSION

3.1. *NH_3 Adsorption*

In Table 1 are listed the values of NH_3 uptake of several catalytic membranes. By comparing the acidic properties of various membranes with the same surface loading of the active species, it comes out that Nafion based sample (A-3 sample) is much less acidic than HPMo and HPW ones. For HPMo series, the NH_3 uptake increases with loading going from 20 to 40 wt% (C-2 and C-4 samples) and it levels off for 60 wt% loaded membrane (C-4 sample). HPW membrane (D-2 sample) shows a similar NH_3 uptake of the same loaded HPMo membrane (C-2 sample), while the amount of NH_3 adsorbed on CsM membrane (referred to the unity weight of active species), in spite of its high surface loading, is of the same order of that of Nafion based membrane.

3.2. *Scanning Electron Microscopy-EDX*

The surface chemical composition of Nafion based membranes has been evaluated by EDX

Table 1. NH_3 uptake data of various superacid membranes.

Sample	Active Species loading (mg/cm^2)	NH_3 uptake (μmol/g $_{membrane}$)
A-3	1.02	6.01
C-2	1.02	36.2
C-4	2.0	126.7
C-6	3.05	133.6
D-2	1.02	20.7
CsM	8.0	78.4

measurements. The results reported in Table 2, in terms of atomic percentage of the various elements, indicate a progressive decrease of C and a corresponding increase in O, F and S with Nafion loading. The increase of S with Nafion loading reflects the increase in the concentration of sulfonic groups ($-SO_3H$) which likely controls the acidic properties of the Nafion membranes.

Table 2. SEM-EDX analysis of carbon supported Nafion catalytic membranes.

Sample	C	O	F	S	Cl	K
	(mol %)					
Bare-membrane	99.196	0.178	0.597	0.017	0.008	0.001
A-1	97.491	0.296	2.048	0.122	0.010	0.017
A-2	95.593	0.503	3.571	0.251	0.015	0.045
A-4	89.381	1.246	8.599	0.633	0.020	0.100

3.3 Ethane Partial Oxidation (EPO)

The selective partial oxidation of ethane with H_2O_2 on Nafion membranes in presence of Fe^{2+}, in the T range 80-110 °C, leads to the formation of acetaldehyde (C_2H_4O) and ethanol (C_2H_6O). No CO_2 has been observed.

The result of a typical run at 100 °C (membrane sample, A-3; [Fe^{2+}] = 5.6 x 10^{-5} mol/l and [H_2O_2] = 0.29 mol/l), expressed in terms of cumulative amount of oxygenated products (C_2H_4O and C_2H_6O) formed in both liquid and gas phases and H_2O_2 concentration versus the reaction time, is reported in Fig. 1. The selectivity values to C_2H_4O and C_2H_6O were 88 and 12% respectively. It can be observed that at the beginning of reaction the oxygenated products distribute in both gas and liquid phases until their concentration in liquid phase reaches the saturation equilibrium value, afterwards all products formed are continuously collected in gas phase. The rate of formation of oxygenated products remains constant with the reaction time. This finding indicates that Nafion membrane operates without any deactivation.

The time of the attainment of saturation equilibrium concentration of oxygenated products in liquid phase (t_c) depends upon the reaction temperature (T_R). Higher is T_R lower is t_c, namely t_c at 80 and 110 °C is 150 and 90 min respectively. This finding could be explained considering that the membranes are not perm-selective, therefore the permeation of reaction products from the liquid phase containing the oxidant is controlled by their boiling points. As a consequence, in the gas phase, relevant amount of H_2O (b.p., 100 °C) and negligible amount of H_2O_2 (b.p. , 152,4°C) are collected together with C_2H_4O (b.p., 20.2 °C) and C_2H_6O (b.p., 78.4 °C).

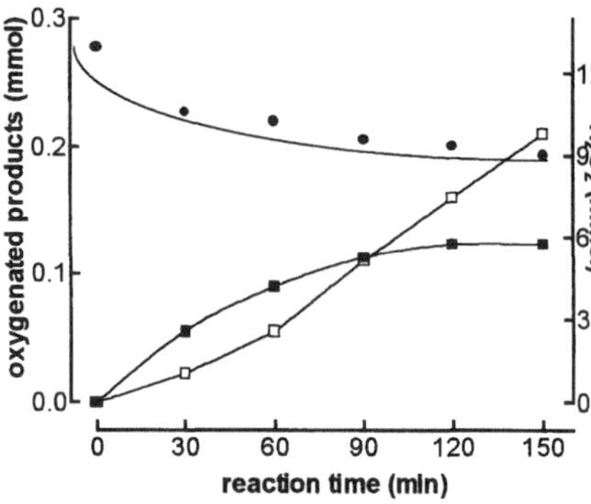

Fig.1. EPO on 5% Nafion/C catalytic membrane at 100°C. Product distribution in the gas and liquid phases vs. reaction time: ●) H_2O_2; ■) Liquid phase; □) Gas phase

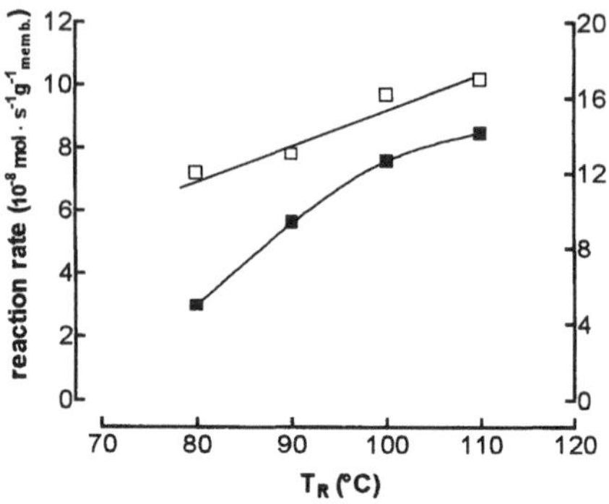

Fig.2. EPO on 5% Nafion/C catalytic membrane. Influence of reaction temperature on reaction rate and H_2O_2 yield.
■) Reaction rate; □) H_2O_2 yield

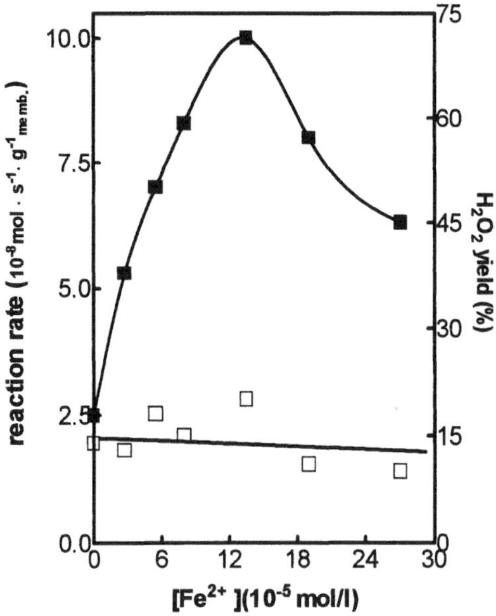

Fig.3. EPO on 5% Nafion/C catalytic membrane. Influence of Fe^{2+} concentration on reaction rate and H_2O_2 yield.
□) H_2O_2 yield; ■) Reaction rate

The effect of reaction temperature on reaction rate and H_2O_2 yield is shown in Fig 2. The reaction rate monotonically rises with T_R. The evolution of the reaction is governed by several factors such as [Fe^{2+}], [H_2O_2] and C_2H_6 concentration in liquid phase. By keeping constant value of [Fe^{2+}] and [H_2O_2] it has been ascertained that the concentration of C_2H_6 in liquid phase is the controlling factor of the reaction rate. Thus, since the solubility of C_2H_6 in the T range 80-110 °C decreases with T going from $1.1 \cdot 10^{-3}$ mol/l at 80 °C to $0.8 \cdot 10^{-3}$ mol/l at 110°C, it results that such physical factor could explain the deviation of the kinetics of the system from the formal kinetic laws accounting for the quasi linear relationship between T_R and r. The influence of Fe^{2+} concentration on reaction rate and H_2O_2 yield has been evaluated at 80°C performing a series of experiments with the A-3 sample at different [Fe^{2+}].

The results shown in Fig. 3 indicate that [Fe^{2+}] enhances the reaction rate according to a volcano-shape relationship whose maximum corresponds to a [Fe^{2+}] of $13.5 \cdot 10^{-5}$ mol/l. The H_2O_2 yield is slightly affected by [Fe^{2+}]. On the basis of our previous findings /7/ it can be stated that the promoting role of Fe^{2+} is associated with the generation of OH radicals according to the reaction: $Fe^{+2} + H_2O_2 \rightarrow OH^{\bullet} + OH^{-} + Fe^{+3}$

Therefore, it can be inferred that the reaction proceeds according to a radical mechanism which involves the activation of ethane on superacid sites (S*) and the subsequent reaction of activated paraffin with OH radicals:

$$C_2H_6 + S^* \rightarrow C_2H_6^* \quad (a)$$

$$C_2H_6^* + OH^{\bullet} \rightarrow S^* + CH_3CH_2OH + CH_3CHO \quad (b)$$

In order to explain the volcano-shape relationship between reaction rate and Fe^{2+} concentration the reaction mechanism of Fe^{+2}-H_2O_2 system must be considered /8/:

$$Fe^{+2} + H_2O_2 \rightarrow OH^{\bullet} + OH^{-} + Fe^{3+} \quad (1)$$

$$OH^{\bullet} + Fe^{2+} \rightarrow OH^{-} + Fe^{+3} \quad (2)$$

$$OH^{\bullet} + H_2O_2 \rightarrow HO_2^{\bullet} + H_2O \quad (3)$$

$$HO_2^{\bullet} + Fe^{+2} \rightarrow HO_2^{-} + Fe^{3+} \quad (4)$$

$$HO_2^{\bullet} + Fe^{3+} \rightarrow O_2 + H^{+} + Fe^{2+} \quad (5)$$

From this reaction network it emerges that OH radicals generated in the reaction (1) can further react with Fe^{+2} to form OH^{-} and Fe^{+3} (reaction 2) and with H_2O_2 to form $HO_2^{\bullet}$ and H_2O (reaction (3)). Therefore, in principle it can be assumed that in our reaction system activated paraffin ($C_2H_6^*$), Fe^{2+} and H_2O_2 compete for OH radicals.

3.3 Methane Partial Oxidation(MPO)

The activity of the various superacid catalytic membranes has been comparatively determined in the MPO with H_2O_2 in presence of Fe^{2+} ions at 80 °C. The values of reaction rate, shown in Table 3, indicate that the activity of the Nafion based membrane (sample A-3, $r_2 = 2.6 \cdot 10^{-9}$ mol $\cdot$ $g^{-1}_{memb.}$ s^{-1}) is one order of magnitude higher than that of HPMo, HPW and CsM membrane samples ($r_2 = 0.1$-$0.15\ 10^{-9}$ mol $g^{-1}_{membr.}$ s^{-1}). Such different catalytic behaviour cannot be rationalised in terms of acidic properties of the various system as probed by NH_3 adsorption. In fact, among the investigated membranes (Table 1), A-3 sample is characterised by the lowest NH_3 uptake whilst it reveals the highest catalytic activity. Furthermore, neither the acidic strength of the various active species, expressed in terms of Hammett function value (Ho), can be invoked to rationalise the catalytic behaviour of such catalytic membranes. Indeed, on the basis of the Ho values, heteropolyacid based system (HPMo, HPW and CsM membranes) should be stronger acids than Nafion /9/. Therefore, the reason for the different catalytic behaviour of our system could lye in their different stability in H_2O_2 media. Nafion is a very stable catalyst in the presence of H_2O_2 , while heteropolyacid system could react with H_2O_2 to form peroxophosphates ($PW_xO_y^{2-}$ $PMo_xO_y^{2-}$) which do not enable any catalytic action /9/.

A linear relationship between the reaction rate, referred to the weight unity of the membrane, and the concentration of acidic sites (-SO_3-H) estimated by EDX analysis is found (Fig. 4).

This linear trend is diagnostic of the fact that the reaction occurs at the exposed surface (liquid side surface) of the membrane and therefore the fraction of active species distributed into the pore structure of the catalytic layer, being inaccessible to the gas reactant, is not catalytically effective.

Table 3. Partial oxidation of CH_4 to methanol on different carbon supported catalytic membranes. $[H_2O_2]$=0.29 mol/l; T_R=80°C.

Sample	Active phase	Rate (10^{-9} mol/s×g $_{membr.}$)
A-3	Nafion/C	2.60
C-4	HPMo/C	0.2
D-2	HPW/C	0.1
CsM	$H_{0.5}\ Cs_{2.5}\ PW_{12}\ O_{40}$	1.2

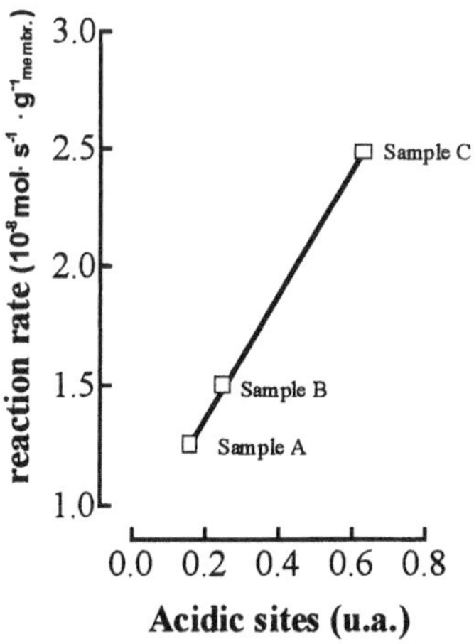

Fig. 4. MPO on Nafion/C catalytic membranes Reaction rate vs. concentration of acidic sites: Sample A (5%Nafion/C); Sample B (10%Nafion/C); Sample C (30%Nafion/C).

4. CONCLUSIONS

Superacid based membranes catalyse the selective partial oxidation of paraffins (C_1-C_2) to oxygenates with H_2O_2 in the presence of Fe^{2+} ions under mild conditions (T_R = 80-110°C; P = 1.4 bar). Nafion based membrane results to be the most active system probably due to its excellent chemical stability in the presence of H_2O_2. Fe^{2+} ions enhance the reaction rate according to a volcano-shape relationship. Such trend was interpreted on the basis of Fe^{2+}-H_2O_2 reaction mechanism which involves the reaction of Fe^{2+} with OH radicals (side reaction) that results in a lowering of the reaction rate of paraffin oxidation. Reaction proceeds according to a radical mechanism which entails the activation of paraffin on superacid sites and the subsequent reaction of activated paraffin with OH radicals.

REFERENCES

1. Yu.V.Geletii and A.E.Shilov, Kinet.Katal. 24 (1983) 486
2. A.Sen , JACS, 109 (1987) 8109
3. Moiseev, JCS, Chem. Comm. Chem.Commun. 1049 (1990)
4. P.J. Sienberg and L.B.Kool, Symposium on structure of Ject Fuels III, San Francisco Meeting, April 5-10 1992
5. R.A.Periana, D.J. Taube, E.R.Evitt, D.G.Löffler, P.R.Wentrcek, G.Voss and T.Masuda, Science, 259 (1993) 341
6. A.Parmaliana, F.Frusteri, F.Arena and N.Giordano, Cat.Lett. 12 (1992) 353
7. F. Frusteri, E.N. Savinov, A. Parmaliana, E.R. Savinova, J.N. Parmon and N. Giordano, Catal. Lett., 27 (1994) 355
8. J.H. Baxendale, Advances in Catalysis, Edited by W.G. Frankenburg, V.I. Komarewsky, E. K. Rideal, Vol IV(1952)31
9. I.V. Kozhevnikov, Catal. Rev.- Sci. Eng., 37(2)(1995)311

NATURAL GAS CONVERSION V
Studies in Surface Science and Catalysis, Vol. 119
A. Parmaliana et al. (Editors)
1998 Elsevier Science B.V.

An experimental study of the partial oxidation of methane in a membrane reactor

A. Basile[a], S. Fasson[b], G. Vitulli[c], and E. Drioli[a,b]

[a] Research Institute on Membranes and Modelling of Chemical Reactors, CNR-IRMERC
I-87030 Arcavacata di Rende (CS), Italy

[b]Department of Chemical Engineering and Materials, University of Calabria
I-87030 Arcavacata di Rende (CS), Italy

[c] CNR-CSMSOA, I-50126 Pisa, Italy

In this work, both a Membrane Reactor using a composite palladium membrane operating in the Knudsen regime and a Traditional Reactor have been used for studying the effect of the reaction temperature on the Catalytic Partial Oxidation of Methane to produce syngas.

1. INTRODUCTION

The most common method for converting natural gas to produce syngas currently used in industry is the process of steam reforming of methane.

Another approach to produce syngas is the Catalytic Partial Oxidation of Methane (CPOM), that involves an exothermic overall reaction. Unlike the reforming process, the heat required is not transferred from external sources across the reactor wall but is generated internally by combustion through the consumption of only a relatively small part of the feed.

Recently, various researchers [1-4] suggested that the route for the syngas generation from CPOM would be a viable alternative to steam reforming.

Recently, by some researchers [5, 6], in order to overcome some thermodynamic constraint, a completely different approach has been considered: the Membrane Reactor (MR). A MR combines the separation properties of membranes with the characteristics typical of catalytic reactions. In these systems, by selectively removing one of the products from the reaction mixture, the conversion of CH_4 should increase.

A MR for CPOM to syngas has been recently used by Santamaria *et al.* [5]. In order to elucidate the potential of a MR for the CPOM to syngas, Mleczko *et al.* [6] have developed a reaction model in which the membrane properties play an important role in the reaction process.

In this work experimental results of the performance of both Traditional Reactor (TR) and MR using Ni-based catalyst are compared. In particular, the effect of temperature on the activity of the two reactors is investigated.

2. DESCRIPTION OF THE PROCESS

2.1 Reactor concept and composite membrane preparation

In this work, two reactors, TR and MR, both having the same geometrical dimensions (length = 25 cm; *i.d.* = 0.67 cm) have been used. The starting material and the introduction of the metal phase on the inner surface of a commercial tubular membrane have been described elsewhere [8]. In this work a new palladium membrane has been utilized where a third Pd layer has been formed on the previous one using the same co-condensation technique. The resulting thickness of the three palladium layers is of the order of 0.2 - 0.3 micron.

2.3 Experimental apparatus

Fig. 1 represents the schematic flow diagram of the experimental reaction system.

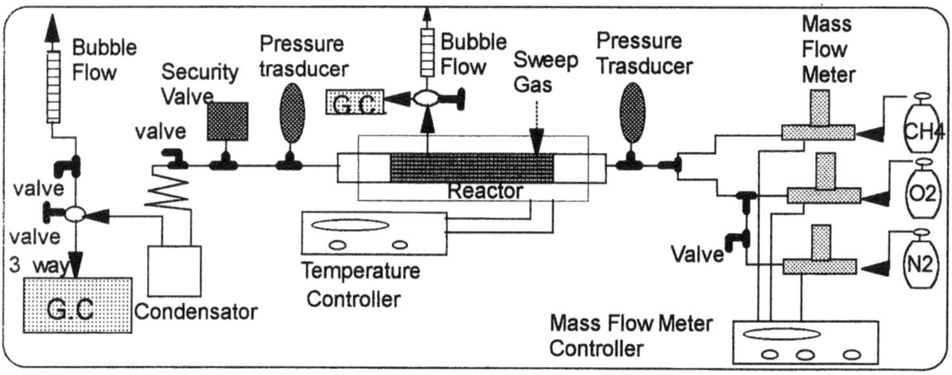

Fig.1 Schematic flow diagram of the experimental reaction system

The lumen of the reactor was packed both with: a) catalyst particles Ni-5256 E 3/64" (4 g) containing very highly dispersed nickel on a silica support; and b) glass particles (4 g) 2 mm diameter. The catalyst was pretreated using N_2 for 6 hours at 480°C using a flow rate of $1.4 \cdot 10^{-2}$ mol/min. Reactions were performed using two different feed gas mixtures:

- $CH_4/O_2/N_2$=2/1/14.0, CH_4/O_2=2, CH_4=$1.92 \cdot 10^{-3}$ mol/min, O_2=$9.92 \cdot 10^{-4}$ mol/min.
- $CH_4/O_2/N_2$=3/1/18.7, CH_4/O_2=3, CH_4=$2.30 \cdot 10^{-3}$ mol/min, O_2=$7.50 \cdot 10^{-4}$ mol/min.

Both feed gas mixtures have the following gas flow rate:

N_2= $1.4 \cdot 10^{-2}$ mol/min. CH_4+O_2=$3.0 \cdot 10^{-3}$ mol/min, $CH_4+O_2+N_2$=$1.7 \cdot 10^{-2}$mol/min

The permeabilities of pure gases (hydrogen, nitrogen, oxygen and argon) and of a CH_4/O_2 mixture have been performed with the same experimental apparatus.

The permeate stream pressure (shell side) was held continually at 1 atm. The feed gas pressure ranged from 1.59 bar to 1.63 bar. During reaction, the maximum temperature difference on the module length was 2°C between the feed gas and permeate gas; the maximum pressure difference was 0.02 bar between the feed gas and retentate gas. The reaction temperature range was 300-580°C. N_2 was used as an equi-current sweep gas: $7 \cdot 10^{-3}$ mol/min. The CH_4 and O_2 conversions were calculated considering for the MR both permeate and retentate flows. All gases were used >99.99%. Separation of both permeate gas and retentate gas from H_2O occurred on a glass column containing H_2O vapor adsorbent (drierite). The mass balance closures were within +/- 5% in all experiments reported in this work.

3. RESULTS AND DISCUSSION

3.1 Traditional reactor (TR)

The effect of a cycle of sequential changes in reaction temperature, T, on conversions of CH_4 and O2 for CH_4/O_2=3 is presented on Fig. 2. In the range T=490-547°C, X_{CH4} is only 10%, while X_{O2} is in the range 33.3-38.4%. As the T is increased to 580°C, X_{CH4} increases to 76% while X_{O2} reaches 100%. Decreasing T from 580°C to 504°C, X_{CH4} decreases without retracing the pathways observed during the initial increase of T; vice versa, X_{O2} remained 100%. The effect of a cycle of sequential changes in T on CO selectivity during X_{CH4} is shown on Fig. 3. CO selectivity is 0.0% in the range T = 490-547°C, and reaches 54.1% at 580°C, and decreases when T is decreased from 580°C to 540°C. In both cases, the same trend was observed by Boucouvalas *et al.* [2], Chu Y. *et al.* [3] and by Dissanayake *et al.* [7] studying the same reaction. For example, at 500°C, Dissanayake *et al.* [7] obtained a CO selectivity of 30%; in our case it is 31%. They also obtained the same trend of X_{CH4} and X_{O2} as a function of T. A quantitative determination of surface carbon contents considering a mass balance on carbon has been done:

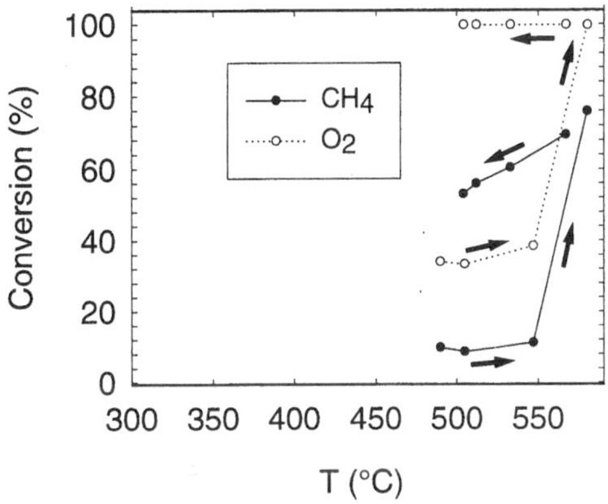

Fig. 2. Effect of reaction temperature on X_{CH4} and X_{O2}. TR. CH_4/O_2=3.

C_{graf}= C_{feed} -$C_{permeate}$-$C_{retentate}$. Results of this calculation for each T are shown on Fig. 4.

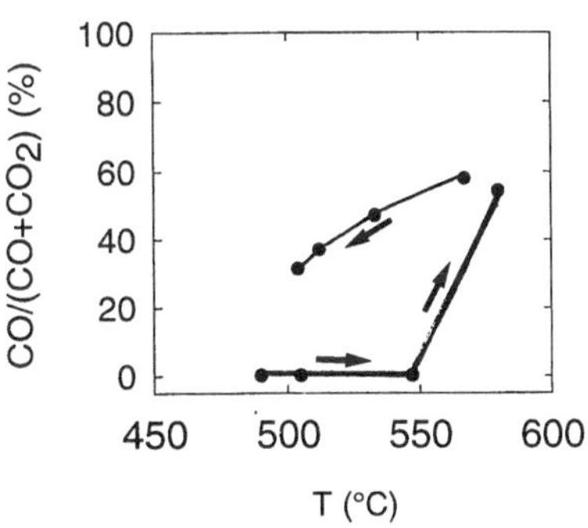

Fig 3. Effect of reaction temperature on CO selectivity.TR. CH_4/O_2=3.

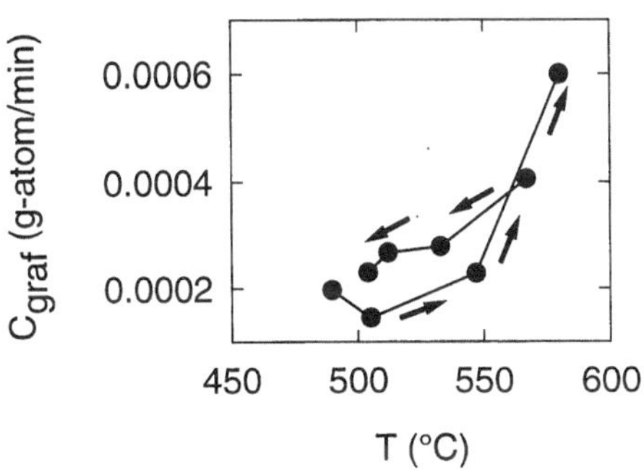

Fig. 4. Surface carbon contents versus reaction temperature. TR. CH_4/O_2=3.

3.2 Membrane Reactor (MR)

It has been verified at four different temperatures (593 K, 613 K, 623 K, 633 K) that the permeation of pure H_2 through the composed three-Pd-layered membrane results in a rate of H_2 permeation that follows the Sievert law. The selectivity H_2/N_2 results to be infinite for all temperature tested. The temperature dependence of the hydrogen removed from the reactor through the composite palladium membrane is an Arrhenius type: $P_{H2}= 5.57 \cdot 10^{-3}\exp(-14.4/RT)$ [mol/s m^2 $kPa^{0.5}$]. Nevertheless, during the reaction experiments, the gas permeabilities have changed. In particular, both CH_4/O_2 for a mixture and O_2/Ar for pure gas selectivities are close to Knudsen regime. The effect of a cycle of sequential changes in reaction temperature on X_{CH4} and X_{O2} for CH_4/O_2=3 is shown on Fig. 5. At the reaction temperature range of 300-543°C, X_{O2} is always 100%, while X_{CH4} has a maximum in the range 500-540°C: 80.24%. ison between MR and TR gives: a) MR is able to work at less temperature than traditional reactor; in particular, at 540°C: MR works better than TR: 61% versus 80%; b) the hysteresis area is very small. Both the effects are due to the separation properties of the membrane.

The effect of a cycle of sequential changes in reaction temperature on CO selectivity during conversion of CH_4 is shown on Fig. 6. CO selectivity =20.8% at T=450°C, reaches 75.4% at 540°C. CO selectivity decreases when temperature is decreased from 540°C to 300°C. The maximum CO selectivity was 87.7%. For each temperature the CO selectivity for MR is higher than for TR, while the trend is similar to the TR one.

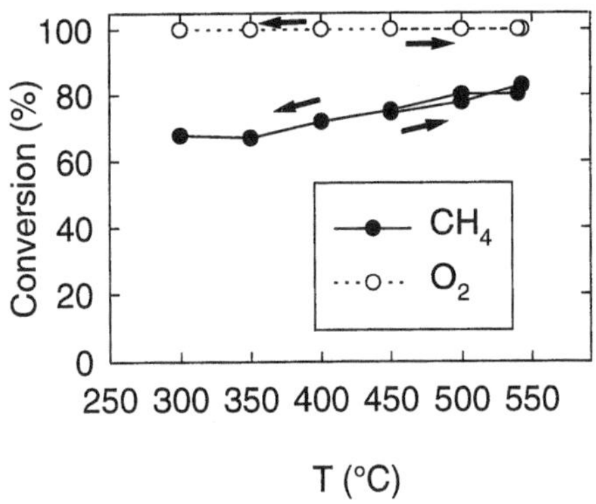

Fig. 5. Effect of reaction temperature on X_{CH4} and X_{O2}. MR. CH_4/O_2=3.

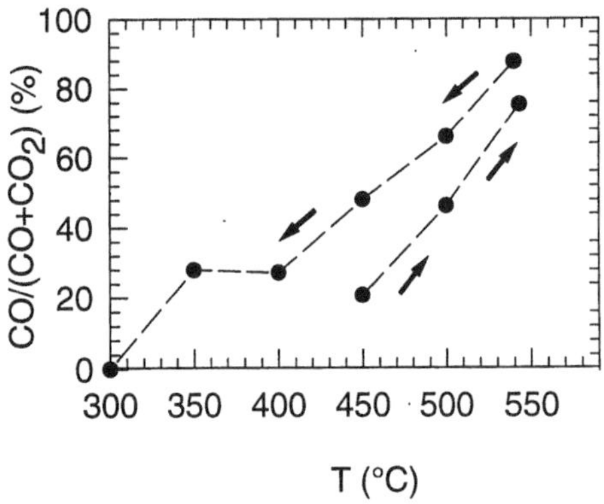

Fig. 6. Effect of reaction temperature on CO selectivity. MR. CH_4/O_2=3.

On Fig. 7 is shown the effect of reaction temperature on X_{CH4} and X_{O2} for a CH_4/O_2=2 molar feed ratio. At 540°C the maximum X_{CH4} (88.58%) is obtained, while X_{O2} is always 100%. Using both feed gas mixture with CH_4/O_2=2 and CH_4/O_2=3, MR shows a 100% of X_{O2}, while X_{CH4} is higher for CH_4/O_2=2. In particular, at 500°C, X_{CH4}= 88.58% for CH_4/O_2= 2, while X_{CH4}=77.74% for CH_4/O_2=3, as we expected.

On Fig. 8 the effect of reaction temperature on CO selectivity for CH_4/O_2=2 is shown. Using both feed gas mixture with CH_4/O_2=2 and CH_4/O_2=3, MR shows a maximum CO selectivity =87.7% at 540°C for CH_4/O_2=3, while for CH_4/O_2=2 the selectivity of CO is only 54%.

3.2 Membrane Reactor (MR)

It has been verified at four different temperatures (593 K, 613 K, 623 K, 633 K) that the permeation of pure H_2 through the composed three-Pd-layered membrane results in a rate of H_2 permeation that follows the Sievert law. The selectivity H_2/N_2 results to be infinite for all temperature tested. The temperature dependence of the hydrogen removed from the reactor through the composite palladium membrane is an Arrhenius type: $P_{H2}= 5.57 \cdot 10^{-3} \exp(-14.4/RT)$ [mol/s m^2 kPa$^{0.5}$]. Nevertheless, during the reaction experiments, the gas permeabilities have changed. In particular, both CH_4/O_2 for a mixture and O_2/Ar for pure gas selectivities are close to Knudsen regime. The effect of a cycle of sequential changes in reaction temperature on X_{CH4} and X_{O2} for CH_4/O_2=3 is shown on Fig. 5. At the reaction temperature range of 300-543°C, X_{O2} is always 100%, while X_{CH4} has a maximum in the range 500-540°C: 80.24%. ison between MR and TR gives: a) MR is able to work at less temperature than traditional reactor; in particular, at 540°C: MR works better than TR: 61% versus 80%; b) the hysteresis area is very small. Both the effects are due to the separation properties of the membrane.

The effect of a cycle of sequential changes in reaction temperature on CO selectivity during conversion of CH_4 is shown on Fig. 6. CO selectivity =20.8% at T=450°C, reaches 75.4% at 540°C. CO selectivity decreases when temperature is decreased from 540°C to 300°C. The maximum CO selectivity was 87.7%. For each temperature the CO selectivity for MR is higher than for TR, while the trend is similar to the TR one.

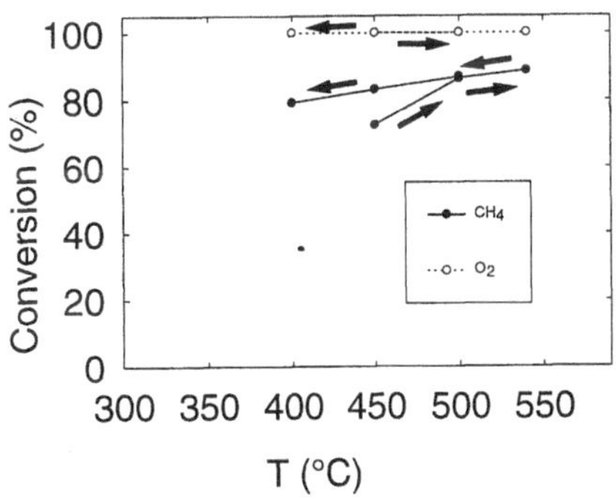

Fig. 7. Effect of reaction temperature on X_{CH4} and X_{O2}. MR. CH_4/O_2=2.

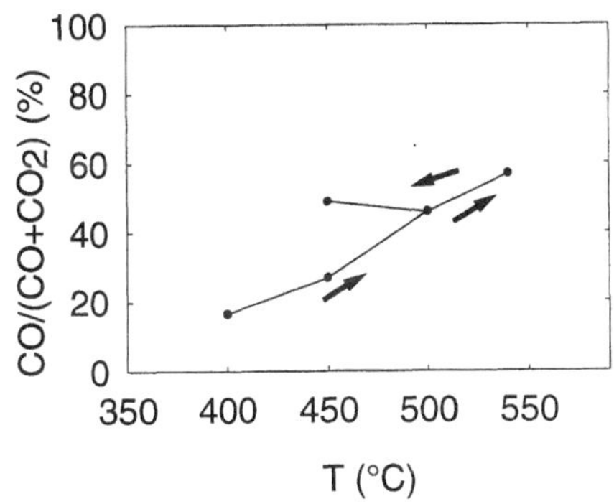

Fig. 8. Effect of reaction temperature on CO selectivity. MR. CH_4/O_2 = 2.

On Fig. 7 is shown the effect of reaction temperature on X_{CH4} and X_{O2} for a CH_4/O_2=2 molar feed ratio. At 540°C the maximum X_{CH4} (88.58%) is obtained, while X_{O2} is always 100%. Using both feed gas mixture with CH_4/O_2=2 and CH_4/O_2=3, MR shows a 100% of X_{O2}, while X_{CH4} is higher for CH_4/O_2=2. In particular, at 500°C, X_{CH4}= 88.58% for CH_4/O_2= 2, while X_{CH4}=77.74% for CH_4/O_2=3, as we expected.

On Fig. 8 the effect of reaction temperature on CO selectivity for CH_4/O_2=2 is shown. Using both feed gas mixture with CH_4/O_2=2 and CH_4/O_2=3, MR shows a maximum CO selectivity=87.7% at 540°C for CH_4/O_2=3, while for CH_4/O_2=2 the selectivity of CO is 54%. On the following table a comparison of X_{CH4} is presented.

Table 1
Methane conversion at various temperature, time factor (W/F) and feed gas composition.

X_{CH4}%	T (°C)	W/F (g-cat·min/cm^3)	CH_4/O_2/gas-carrier	Reference
53.0	800	$5.56 \cdot 10^{-5}$	2/1/48	[2] - TR
96.4	800	$1.40 \cdot 10^{-1}$	2/1/1	[5] - MR
75.0	500	$2.80 \cdot 10^{-2}$	1.78/1/25	[7] - TR
83.1	543	0.48	2/1/14	This work - MR
11.2	547	0.48	3/1/19	This work - TR
88.6	540	0.36	3/1/19	This work - MR

From Table 1, it appears still quite difficult to try any conclusion due to the different experimental conditions of the various works. However, from experimental data on MR and TR of this work, it is evident that X_{CH4} is higher using a MR than a TR.

4. CONCLUSIONS

On the basis of the results reached in this work, a Membrane Reactor using a composite palladium membrane gives higher CO selectivity and higher methane conversion than a Traditional Reactor having the same geometrical dimensions and operating at the same experimental conditions.

Research is in progress for analyzing and optimizing the Membrane Reactor behavior in the various possible experimental conditions.

REFERENCES

1. De Groote A.M., Froment G.F., 138 (1996) 245-264.
2. Boucouvalas Y., Zhang Z., Verykios X.E., Catalysis Letters, 40 (1996) 189-195.
3. Chu Y., Li S., Lin J., Gu J., Yang Y., Applied Catalysis A: General, 134 (1996) 67-80.
4. Basile F., Basini L., Fornasari G., Matteuzzi D., Trifirò F., Vaccari A., XV Simposio Iberoamericano De Catalisis, Universidad Nacional De Cordoba, Cordoba - Argentina, Ed. Herrero E., Anunziata O., Perez C., Vol. 3, pp. 1843-1848.
5. Santos S., Coronas J., Menéndez M., Santamaria J., Catalysis Letters, 30 (1995) 189-199.
6. Mleczo L., Ostrowski T., Wurzel T., Chemical Eng. Sci., 51(11) (1996) 3187-3192.
7. Dissanayake D., Rosynek M.P., Kharas K.C.C., Lunsford J.H., J. Catalysis, 132 (1991) 117-127.
8. Basile A., Drioli E., Santella F., Violante V., Capannelli G., Vitulli G., Gas Sep. Purif., 10(1) (1966) 53-61.

NATURAL GAS CONVERSION V
Studies in Surface Science and Catalysis, Vol. 119
A. Parmaliana et al. (Editors)
1998 Elsevier Science B.V.

Progresses on the partial oxidation of methane to syngas using a membrane reactor

A. Basile[a], S. Fasson[b]

[a] Research Institute on Membranes and Modelling of Chemical Reactors, CNR-IRMERC
I-87030 Arcavacata di Rende (CS), Italy

[b]Department of Chemical Engineering and Materials, University of Calabria
I-87030 Arcavacata di Rende (CS), Italy

In a previous work [1], a comparative study of the Catalytic Partial Oxidation of Methane to synthesis gas in a traditional and a membrane reactors have been done.

In this work we continue to study the effect of the reaction temperature on methane and oxygen conversions, for the same reaction carried out using other two different membrane reactors.

1. INTRODUCTION

One the difficulties in making today the Catalytic Partial Oxidation of Methane (CPOM) commercially viable is given by its thermodynamics: *e.g.* with increasing pressure, the equilibrium conversion of methane is dramatically reduced.

The chemistry of producing synthesis gas from methane by CPOM using a traditional reactor is well known [2-6]: synthesis gas reactors operate very close to thermodynamic equilibrium.

On the other hand, recently studies devoted to improve the performance of traditional reactors in terms of methane conversion using a Membrane Reactor (MR) have been initiated.

The performance of a MR is dramatically influenced by the permselective characteristics of the system gases-membrane, allowing one or more of the reaction product to permeate out of the reactor, thus shifting equilibrium towards higher conversions.

In particular, for the CPOM reaction to produce syngas, Basile *et al.* [1] used a composite palladium membrane; Santamaria and his co-workers [7] proposed a modified commercial ceramic membrane. Ostrowski *et al.* have performed a comparative study based on simulation and experimental work in fixed-bed and fluidized-bed Membrane Reactors [8].

In this work, firstly, a MR using a composite palladium membrane, prepared by using an electroless technique, operating in a better regime selectivity than the Knudsen one have been used for studying the effect of the reaction temperature on the partial oxidation of methane to produce syngas. Secondly, a commercial tubular composite (Pd-Ag/ceramic) for the same reaction has been used. The performance of both the reactors with literature data on both traditional and membrane reactors studied reported in [1] are compared.

2. DESCRIPTION OF THE PROCESS

2.1 Reactor concept

The membrane reactor configuration developed in this work is shown on Fig. 1:. the composite palladium membrane was housed in a stainless steel module and four thick graphite rings for gas tightness were used.

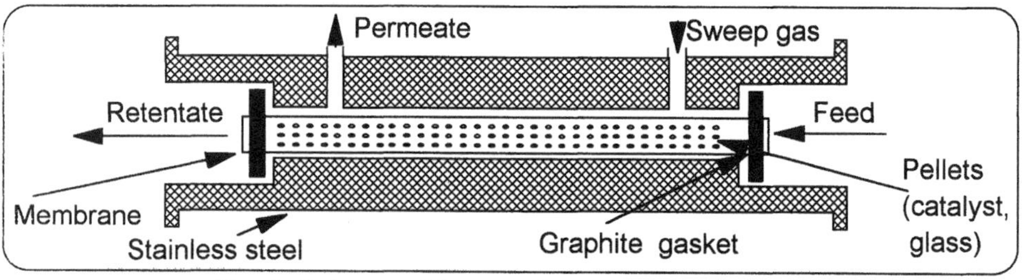

Fig. 1. Scheme of the membrane reactor module

2.2 Composite membrane preparation (MR1)

The support material is a commercial tubular ceramic membrane (length=25 cm; *i.d.*=0.67 cm; *o.d.* 1.02 cm, inside nominal average pore diameter = 5 nm). At first the support was cleaned using different solutions in order to remove different impurities. The activation was obtained using Sn^{2+} and Pd^{2+} solutions. The electroless solution consists of: source of Pd^{2+} ions; reducing agent; complexing agent; pH buffer (NH_3 aqueous solution, 28 wt%).

During the deposition, the palladium density and film thickness depend on the deposition time. The membrane used in this work results in a $5 \cdot 10^{-6}$ m palladium thickness. Hereinafter, this composite membrane will be indicated as MR1.

2.3 Commercial composite membrane (MR2)

The commercial tubular composite membrane was furnished by Johnson-Matthey: a stainless tube in which a tubular membrane is allocated (L_{tot}=19 cm; L_{eff}= 17,3 cm, ceramic tube *o.d.*=2.5 cm). The separation layer consists in a Pd-Ag (23% wt%) having thickness 7.5 micron. Hereinafter, this composite membrane will be indicated as MR2.

Catalyst weight: for both the reactor it was the same (4 g), but the reactors were filled with different weight of glass particle: 4 g for MR1 and 80 g for MR2.

2.4 Experimental apparatus

The experimental apparatus and the experimental details have been described elsewhere [1]. Reactions were performed using two different feed gas mixtures of $CH_4/O_2/N_2$, as reported on [1].

3. RESULTS AND DISCUSSION

3.1 Membrane Reactor 1 (MR1)

In Fig. 2 the dependence of O_2 and N_2 pure on temperature for MR1 is shown. On the same figure, the ideal separation factor $\alpha_{O2/N2}$ is also indicated and compared with Knudsen

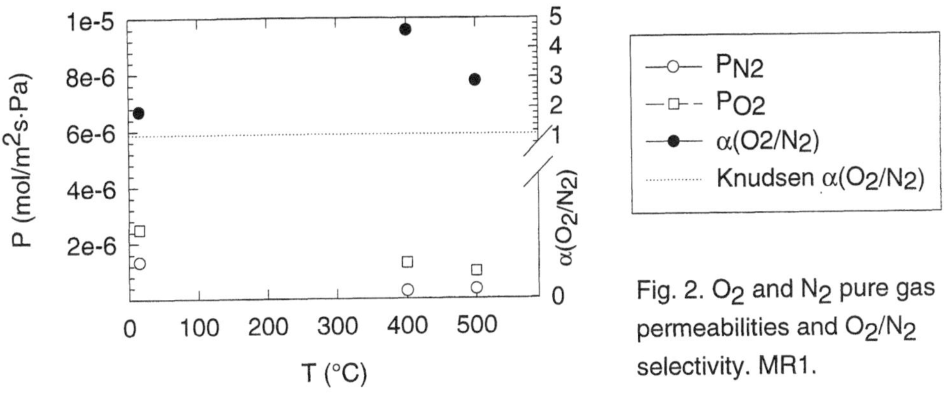

Fig. 2. O_2 and N_2 pure gas permeabilities and O_2/N_2 selectivity. MR1.

ideal selectivity. The experiments have been carried out before reaction. In this figure: with increasing temperature, the permeation rate of both O_2 and N_2 decreased, suggesting a close-to-Knudsen diffusion regime through MR1. After reaction both the permeability and ideal separation factor remained almost the same.

The effect of a cycle of sequential changes in reaction temperature on conversions of methane and oxygen for $CH_4/O_2=3$, and for $CH_4/O_2=2$ is presented on Fig. 3. To confirm some experimental results, two different series of experiments for $CH_4/O_2=3$ were carried out. At the reaction temperature of 350°C, conversion of CH_4 was 65.5% and conversion of O_2 was 90%. As the temperature was increased to 500°C, the conversion of methane increased to 93% while oxygen reached 100% at 400°C. Decreasing the temperature from 500°C to 300°C, the conversion of CH_4 decreased from 92.7% to 78.3%, without retracing the pathways observed during the initial temperature increase; vice versa conversion of O_2 remained always 100%. The maximum methane conversion for $CH_4/O_2=2$ is 93%, while $CH_4/O_2=3$ is 87%, at the same temperature of 500°C.

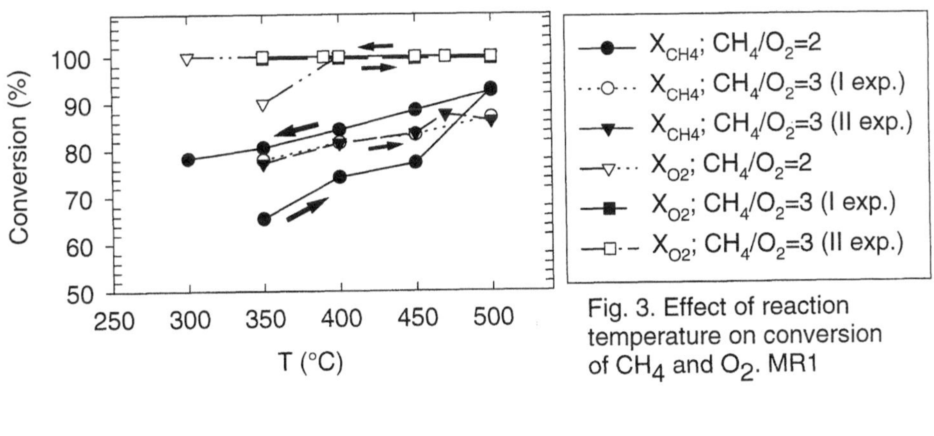

Fig. 3. Effect of reaction temperature on conversion of CH_4 and O_2. MR1

The effect of a cycle of sequential changes in reaction temperature on CO selectivity during conversion of CH_4 on Fig. 4 is shown. CO selectivity =0.0% in the range T=300-400°C, reaches 67% at 500°, and decreases from 75.5% to 0.0% when temperature is decreased from 500°C to 300°C.

Considering both figures (Fig. 3 and Fig. 4), the same trend was observed by Boucouvalas *et al.* [3], Chu Y. *et al.* [4] and by Dissanayake *et al.* [6] studying the same reaction. For example, at 500°C, Dissanayake *et al.* [6] obtained a CO selectivity of 30%; in our case using a membrane reactor, CO selectivity is 75.5%. They also obtained the same trend of conversions of methane and oxygen as a function of reaction temperature.

3.2 Membrane Reactor 2 (MR2)

The H_2 permeation through the MR2 is an activated process. In fact, the temperature dependence of the H_2 permeation rate for this composite membrane before reaction was [9]:

$$P_{H2} = Po \exp(-Ep/RT) \quad (1)$$

where: $Po = 3.41 \cdot 10^{-10}$ $m^3 \cdot m/m^2 \cdot s \cdot Pa^{0.5}$, Ep = 10.3 kJ/mol., $\alpha_{H2/N2}$ = infinite.

After reaction, the hydrogen permeability increased significantly, while the ideal selectivity $\alpha_{H2/N2}$ was reduced from infinite to 30. Probably the Pd/Ag film was damaged during experimental operations.

The effect of the reaction temperature on conversions of methane and oxygen for CH_4/O_2=3 is shown on Fig. 5. At the reaction temperature range of 300-450°C, conversion of oxygen is increasing from 85% to 100%, while methane conversion increased from 85% to 90%. The experiments were carried out only by increasing the temperature. Due to the reduced selectivity $\alpha_{H2/N2}$ from infinite to 30 we decided to stop reaction experiments on this membrane reactor.

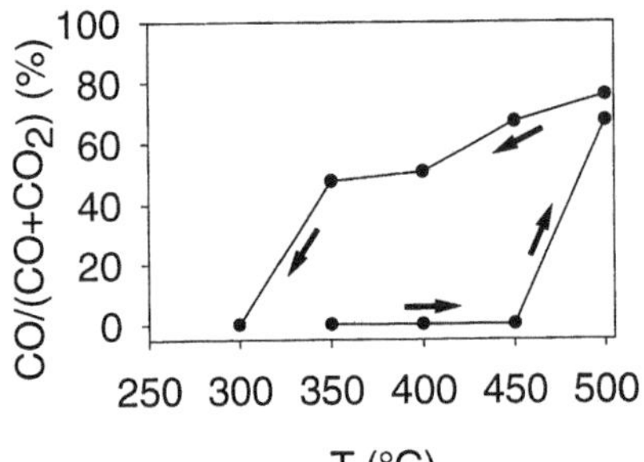

Fig. 4. Effect of reaction temperature on CO selectivity. MR1. CH_4/O_2=2.

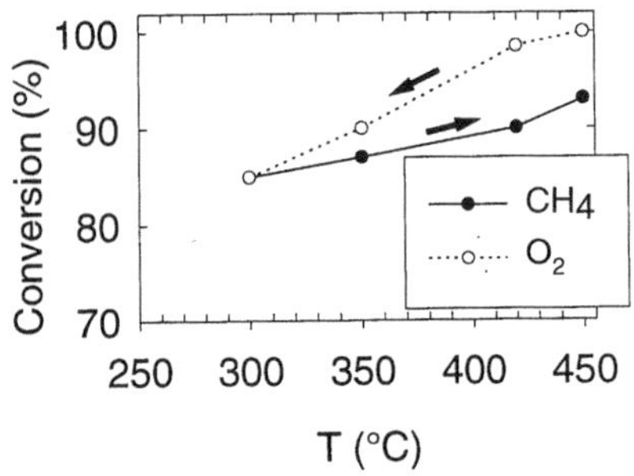

Fig. 5. Effect of reaction temperature on conversion of CH_4 and O_2. MR2.

The effect of an higher methane conversion, compared with MR1 and other reactors presented in a previous study [1] are due to the very high separation properties of the membrane MR2. It is important, however, to observe that in the case of MR2, the CO selectivity was always 0.0%. No syngas was produced, but only CO_2 and H_2O. This means that among the three possible reaction:

- Combustion: $CH_4 + 2\ O_2 \rightarrow CO2 + 2\ H_2O$
- Steam reforming: $CH_4 + H_2O = CO + 3\ H_2$
- Carbon dioxide reforming: $CH_4 + C\ O_2 = 2\ CO + 2\ H_2$

only the first one takes place.

A Membrane Reactor seems to produce a greater carbon formation compared to a Traditional Reactor. Besides an increased methane conversion at lower temperature, the removal of reaction products from the reaction gas influences the potential of carbon formation via:

$CH_4 = C(s) + 2H_2$ $(-\Delta H^\circ_{298} = -75$ kJ/mol), and/or

$2CO = C(s) + CO_2$ $(-\Delta H^\circ_{298} = -173$ kJ/mol)

Quantitative determinations of surface carbon contents were made in both the reactors MR1 and MR2 considering for each temperature a mass balance on carbon: $C_{graf} = C_{feed} - C_{permeate} - C_{retentate}$.
Results of this calculation for each temperature are shown on Fig. 6 and compared with a TR presented by Basile et al. [1]. The surface carbon deposition was almost constant varying temperature for MR2: the higher surface carbon deposition was $2.1 \cdot 10^{-3}$ g-atom/min at 450°C. MR1 presented a range from $1.24 \cdot 10^{-3}$ to $1.52 \cdot 10^{-3}$ g-atom/min for $CH_4/O_2=3$ and $CH_4/O_2=2$ in the temperature range of 300-500°C. TR presented a less content of surface carbon deposition, varying from $0.2 \cdot 10^{-3}$ g-atom/min at 490°C to $0.6 \cdot 10^{-3}$ g-atom/min at 580°C. Vice versa Dissanayake *et al.* [6] did not found surface carbon under 700°C.

On the following table a comparison of X_{CH4} is presented.

Table 1
Methane conversion at various temperature, time factor (W/F) and feed gas composition.

X_{CH4}%	T (°C)	W/F (g-cat·min/cm^3)	CH_4/O_2/gas-carrier	Reference
53.0	800	$5.56 \cdot 10^{-5}$	2/1/48	[3] – TR
96.4	800	$1.40 \cdot 10^{-1}$	2/1/1	[7] – MR
75.0	500	$2.80 \cdot 10^{-2}$	1.78/1/25	[6] – TR
83.1	543	0.48	2/1/14	[1] – MR
11.2	547	0.36	3/1/19	[1] – TR
88.6	540	0.36	3/1/19	[1] – MR
93.0	500	0.48	2/1/14	This work - MR1
87.0	500	0.36	3/1/19	This work - MR1
90.0	500	0.36	3/1/19	This work – MR2

From Table 1, some partial conclusion could be tried. In fact, from experimental data on MR and TR of [1] and MR1 and MR2 of this work, it is evident that X_{CH4} is higher using a MR than a TR. Considering the same CH_4/O_2/gas-carrier, in the case of MR, the best methane conversion was obtained by the composite palladium membrane reactor.

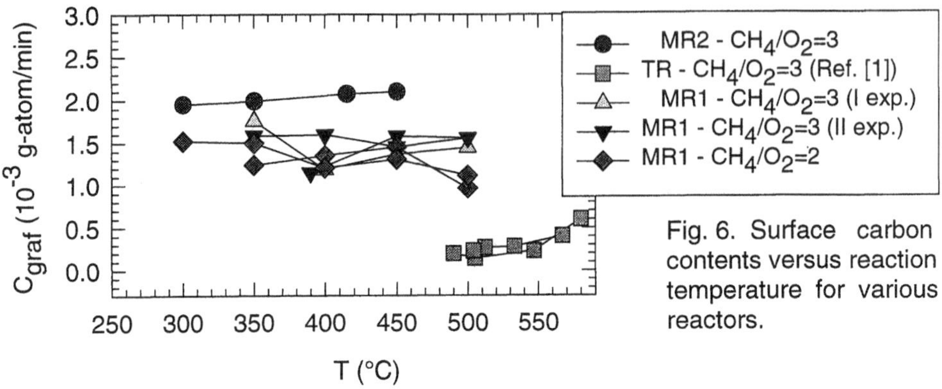

Fig. 6. Surface carbon contents versus reaction temperature for various reactors.

4. CONCLUSIONS

On the basis of the results presented in this work, a Membrane Reactor is able to reach higher methane conversion than a traditional one operating at the same experimental conditions. 100% conversion might be obtained using a membrane reactor having a stable permselective layer.

On the other hand, in general a Membrane Reactor seems to produce a greater carbon formation compared to a Traditional Reactor one.

Attention must be paid with the partial oxidation of methane to produce syngas using a membrane reactor to the stability of the Pd or Pd-Ag film which can be easily destroyed due to high temperature that locally can be reached.

ACKNOWLEDGMENT

We wish to express our thanks to Dr. A. Gordano for preparing the active layer of the composite membrane, and Dr. A. Bruno in taking care of the experimental apparatus.

REFERENCES

1. Basile A., Fasson S., Vitulli G., Capannelli G., Drioli E., Paper presented at the 5th Natural Gas Conversion Symp., Taormina (Italy), Sept. 20-25, 1998.
2. Blanks R.F., Witting T.S., Peterson D.A., Chemical Eng. Science, 45(8) (1990) 2407-2413.
3. Boucouvalas Y., Zhang Z., Verykios X.E., Catalysis Letters, 40 (1996) 189-195.
4. Chu Y., Li S., Lin J., Gu J., Yang Y., Applied Catalysis A: General, 134 (1996) 67-80.
5. Basile F., Basini L., Fornasari G., Matteuzzi D., Trifirò F., Vaccari A., XV Simposio Iberoamericano De Catalisis, Universidad Nacional De Cordoba, Cordoba - Argentina, Ed. Herrero E., Anunziata O., Perez C., Vol. 3, pp. 1843-1848.
6. Dissanayake D., Rosynek M.P., Kharas K.C.C., Lunsford J.H., J. Catalysis, 132 (1991) 117-127.
7. Santos S., Coronas J., Menéndez M., Santamaria J., Catalysis Letters, 30 (1995) 189-199.
8. T. Ostrowski, A. Girior-Fendler, C. Mirodatos, L. Mleczko, Proc. of the II Int. Conf. on Catalysis in Membrane Reactors, Sept. 24-26 1996, Moscow (Russia), p. 24.
9. G.S. Madia, Steam Reforming of Methane, Dissert. Thesis, Univ. of Calabria, 1997, p.140.

NATURAL GAS CONVERSION V
Studies in Surface Science and Catalysis, Vol. 119
A. Parmaliana et al. (Editors)

Isobutanol Synthesis from Syngas

W. Falter, C.-H. Finkeldei, B. Jaeger, W. Keim and K.A.N. Verkerk

Institut für Technische Chemie und Petrolchemie der Rheinisch-Westfälischen Technischen Hochschule Aachen, Worringerweg 1, 52074 Aachen, Germany

Abstract
The increased demand for MTBE (methyl-t-butyl ether) as gasoline additive has attracted attention to alternative pathways for their production. Within this respect the synthesis of higher alcohols by CO hydrogenation has gained new interest, since the selective production of isobutanol-methanol mixtures could offer a possible route. Using a Zr/Zn/Mn/K/Pd catalyst at 400-450 °C, 250 bar pressure, 20.000 GHSV/h^{-1} space time yields of 700-750 $g \cdot l^{-1} \cdot h^{-1}$ of isobutanol could be obtained. A comparison was made using a fixed bed (tubular) and a CSTR reactor system.

Introduction

The hydrogenation of carbon monoxide to hydrocarbons or oxygenates is a field of substantial academic and industrial interest. Especially intriguing from the standpoint of resources is the straight forward availability of synthesis gas from various sources such as: natural gas, oil and oil residues, coal, tar sands, oil shale, bio-mass and many more. The advent of combined power plants based on coal could even make CO/H_2 available in nearly unlimited amounts. In addition, the abundant supply of natural gas has attracted natural gas derived syngas as feedstock for chemicals and fuels. For instance, the synthesis of hydrocarbons via Fischer-Tropsch is practiced in Shell's SMD process. Figure 1 summarizes the direct and indirect conversion pathways.

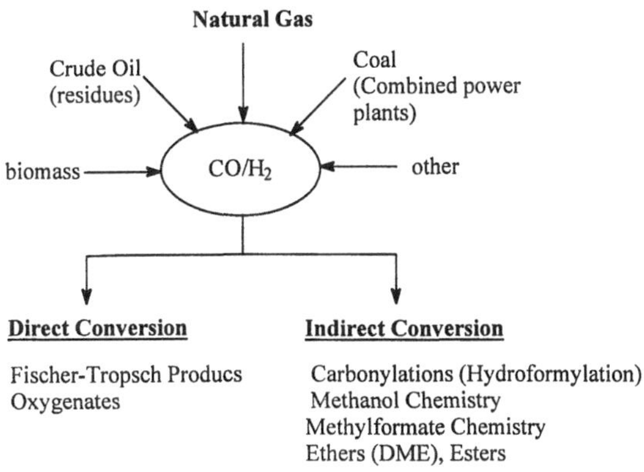

Figure 1. Conversion pathways of synthesis gas

While methanol synthesis from CO/H_2 is a well established process, work on the synthesis of higher oxygenates (e.g. alcohols) has been less successful.
Since many years we are interested in the synthesis of oxygenates [1], especially isobutanol, which easily can be converted to methyl-tert-butylether (MTBE), an important automotive fuel(additive).

$$CH_4 \longrightarrow CO/H_2 \longrightarrow \text{isobutanol} \xrightarrow{CH_3OH} \text{MTBE}$$

The „Isobutyloel Synthese“ was practiced by BASF up to 1953 and at Leuna (former DDR) up to 1990. Table 1 contains typical data based on various catalysts.

Table 1
Isobutanol synthesis

Process	Catalyst	Reaction Conditions			Space Time Yield	
		T/°C	P/bar	GHSV/h^{-1}	i-BuOH$^{a)}$	total Alcohols$^{a)}$
BASF	Zn/Cr/K	420	325	15000	125	735
Klier	Cu/Zn/Cr/Cs	325	76	5330	139	179
Lurgi-Octamix	Cu/Zn/Cr Promoters	300	90	3000	47	460
Keim/Falter	Zr/Zn/Mn/K/Pd	445	250	20000	740	1160
Stiles	Cu/Mn/Zn/Co/Cr/K/Cs	410	175	40000	358	1773
Sofianos	Cu/Zn/Cr/Zr/Mn	350	100	8000	81	240
Snamprogretti	Zn/Cr				70 % CH_3OH	
Topsoe-Anic					30 % higher alcohols from which 15 % i-butanol	

a) $g \cdot l^{-1} \cdot h^{-1}$

Research at Aachen
Our Work at Aachen embraced various objectives:

- catalyst development
- fixed bed technique
- slurry reactor technique.

1. CATALYST DEVELOPMENT

Three methods shown in Figure 2 were used to prepare various catalysts:

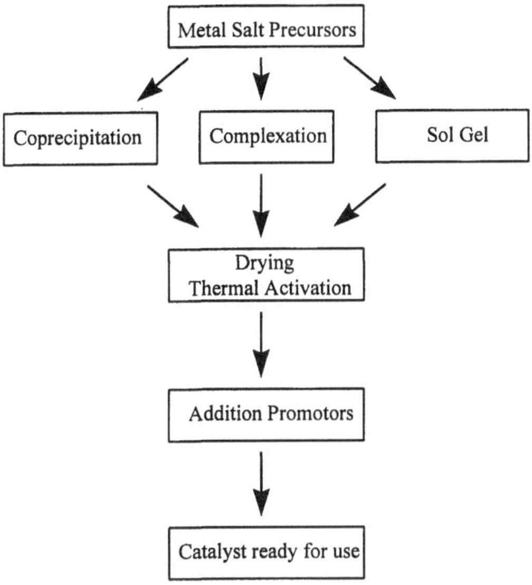

Figure 2. Catalyst synthesis methods

Metal salt precursors were chosen among: Zr, Zn, Mn, Pd, Cr, Cu, Mo, K, Na, Cs. The best catalyst proved to be the potassium promoted $ZrO_2/ZnO/MnO$.
A typical catalyst was prepared by coprecipitation of the metal nitrates with potassium hydroxide at 80°C, keeping the pH constant at 11 ± 0.2. The precipitate was washed, pelletized and dried at 130°C. After calcination at 450°C for 3 hours (heating rate: 4°/min) the resulting catalyst was powdered, pressed and sieved to sizes desired. Potassium loading could be influenced by pH variations. Pd was introduced by impregnating. BET surfaces ranged from 100-200 m^2 g and a maximum in pore radius distribution was 3-4nm.
Besides coprecipation also complexation and sol gel technique have led to good catalysts [1g].

2. CATALYST TESTING

For catalyst testing a fixed bed tubular reactor and a slurry bed reactor have been used.
The application of slurry reactors for exothermic reactions has gained considerable interest in the industrial and academic world. For synthesis gas conversions the LPMeOH process by Air Products is a well known example [2]. This prompted us to apply this technology to isobutanol synthesis which runs under more extreme reaction conditions.

Because of much easier and faster handling and the necessity to compare results from slurry reactions with those obtained in fixed bed reactions cited in the literature, fixed bed reactor systems are preferred for catalyst development and optimization [3]. The main differences e.g. temperature gradients, backmixing and particle sizes will obviously have major influences on a reaction which mechanistically consists of a network from parallel and consecutive steps.
The fixed bed turbular reactor continuous unit was set up to be operated at temperatures up to 500°C and pressures up to 40 Mpa. The unit was constructed to switch directly from fixed bed to slurry reactor operation. A process management system and on line GC analysis with gas partitioner as well as an automatic product sampler for off-line analysis allowed continuous operation.
Catalyst activations and reactions have been carried out in a fixed bed reactor from stainless steel with an inner diameter of 9 mm. Catalyst particles have been mixed with an equal amount of copper particles of the same size. Glass particles were added on top as a preheating zone.
The catalyst was activated in situ by pressurizing with hydrogen (30 Nl·h^{-1}, 3 Mpa) and heated to 225°C with a rate of 4°/min keeping this temperature constant for 120 min. Subsequently the reactor was pressurized with H_2/CO (1/:1) to 25 Mpa while the temperature was increased up to reaction conditions with 4°/min.
All reactions were conducted at 25 MPa varying temperature the from 370°C to 430°C in steps of 30°. Afterwards the measurements at 430°C were repeated to test reproducibility. At each temperature linear gas velocity was changed in six steps from 24 to 227 Nl·h^{-1}. All setpoints were allowed to come to steady state for 75 min. Then the first on-line measurement was started followed by collecting one off-line sample. A second on-line measurement, 75 minutes later, ended each analysis.
The slurry lab scale reactor consisted out of a 300 ml stirred reactor (speed up to 3500 rpm) with reflux. The catalyst was powdered (< 160 μm). Decalin was used as inert solvent. The reflux condenser is important because the reaction was carried out near the critical temperature of decalin at which a significant vaporisation of decalin occured.
Many catalytic runs were carried out and the parameters temperature, pressure, GHSV and particle size were investigated in more detail.
The **temperature** has the greatest impact. Temperatures between 400-450°C are optimal. Temperatures below 400° lead to a rapid decrease in i-butanol yield (STY).
Also the **pressure** applied is of great influence. To obain reasonable i-butanol yields pressure around 250 bars must be used.
Best **gas-hourly-space velocities** (GHSV/h^{-1}) range between 20.000 - 100.000.
The **particle size** of the catalyst is not specifically limited, although an increasing particle size causes an increase in mass transfer limitations. Preferably the particle size is within the range of 0.1 to 5.0 mm, more preferably within the range of 0.25 to 2.0 mm and most preferably within the range of 0.25 to 0.50 mm. Typically the catalyst particles show a pore radius distribution having a maximum of between 0.5 to 5 nm, preferably at about 3 nm.
Great emphasis was placed on a comparison of a fixed bed tubular reactor with a continuously stirred tank reactor (CSTR). Typical results are shown in Table 2.

Table 2
Comparison of tubular and CSTR reactors, Pressure 25 MPA, Temp. 400 °C

Reactor GHSV [h^{-1}]	Tubular 75650	CSTR 75650
C_{CO} [%]	21	34
S_{CO2} [%]	39	63
STY [g/l·h)]		
Methanol	2196	922
Ethanol	30	199
n-Propanol	41	132
i-Butanol	751	219
2-Methylbutanol-1	98	43

Obviously the tubular fixed bed reactor is better suited for the production of i-butanol.
To maximise the yield of i-butanol impregnation with palladium is necessary. Table 3 lists a comparison of two catalysts A and B.

A ZrO_2/ZnO/MnO/K
B ZrO_2/ZnO/MnO/K/**Pd**

Table 3
Influence of Pd-impregnation on product composition in a Tubular Reactor

	A (without Pd)	B (with Pd)
GHSV [h^{-1}]	45250	49500
T°C	430	430
P (bar)	250	250
STY [g/l·h^{-1}]		
CH_4	354	209
CH_3OH	815	1169
Ethanol	46	29
n-Propanol	51	45
iso-Butanol	221	600
2-Methyl-butanol-1	32	76
C_{CO} [%]	33	32
S_{CO2} [%]	41	40
MeOH/i-BuOH	3.7	1.9

Addition of Pd increases the activity to methanol and isobutanol significantly, but does not affect the activity of ethanol and n-propanol.
Mechanistic proposals for the formation of isobutanol in the literature are contrary.
The most accepted reaction network for higher alcohol synthesis given by Klier at al. [3] for cesium promoted Cu/Zn-oxide catalysts describes the mechanistic differences between the reaction paths to methanol and isobutanol. The first step in higher alcohol synthesis is hydrogenation of CO to a surface intermediate, which is very similar to methanol. Linear

primary alcohols are built by linear chain growth including CO insertion steps. Isobutanol and 2-methylbutanol-1 origin from β-addition including aldolic condensation. Formation of 1-propanol can be reached via both pathways as shown in Figure 3.
Isobutanol and 2-methylbutanol-1 do not undergo consecutive reactions following this network. They cannot react further in aldolic condensations and the probability of linear chain growth is low. On the other hand the linear alcohols are able to undergo linear chain growth as well as β-addition.

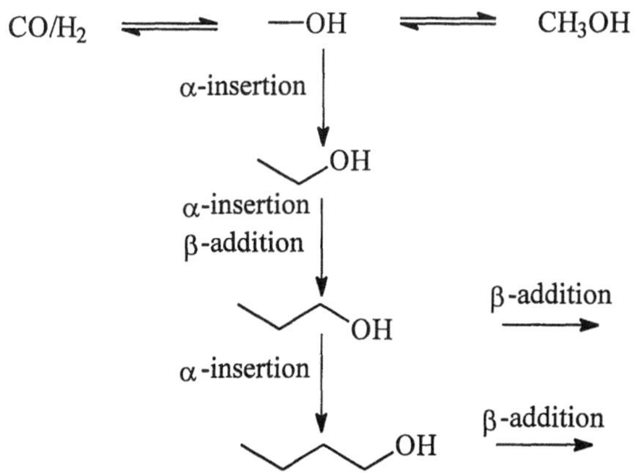

Figure 3. Reaction network for alcohol synthesis from CO/H_2

ACKNOWLEDGMENT

We gratefully thank the US Department of Energy and Air Products and Chemicals, Inc. For the support of this work.

REFERENCES

1. a. W. Keim and W. Falter, Catal. Lett. 3 (1989) 59-64.
   b. W. Keim and W. Falter, DE 3 810 421 (1989), DE 3 524 317 (1989).
   c. J. Seibring, Dissertation, RWTH Aachen (1985).
   d. G. Kolle-Görgen, Dissertation, RWTH Aachen (1985).
   e. B. Jaeger, Dissertation, RWTH Aachen (1997).
   f. C.H. Finkeldei, Dissertation, RWTH Aachen (1996).
   g. K. Verkerk, Dissertation, RWTH Aachen (1997).
   h. W. Keim, B. Jaeger, C.H. Finkeldei and K. Verkerk, Preprints of papers, presented at the national meeting of the American Chemical Society Division of fuel chemistry, Bd. 41 (3) (1996) 875-879.

2. G.W. Roberts, D.M. Brown, T.H.Huisung and J.J. Lewnard, Chem. Eng. Sci. 45(8) (1990) 2713-2720.
3. K.J. Smith, C.-W. Young, R.G. Hermann and K. Klier, Ind. Eng. Chem. Res. 30 (1991) 61-71.

NATURAL GAS CONVERSION V
Studies in Surface Science and Catalysis, Vol. 119
A. Parmaliana et al. (Editors)

Synthesis of Higher Alcohols. Enhancement by the Addition of Methanol or Ethanol to the Syngas

M. Lachowska and J. Skrzypek

Institute of Chemical Engineering, Polish Academy of Sciences, PL 44-100 Gliwice, ul.Ba³tycka 5, Poland

The addition of methanol or ethanol to the syngas strongly increases yields of propanols, butanols, pentanols, hexanols. These yields are even more than ten times higher in comparison with the yields of alcohols in the synthesis only from syngas. The addition of ethanol is more efficient than that of methanol.

1. INTRODUCTION

Higher aliphatic alcohols C_1-C_6 are of current interest as blending stocks for motor gasoline. It is a clean fuel and it can be an example of sustainable technology in the nearest future.

The reactions occur as follows:

$$CO + 2H_2 \Leftrightarrow CH_3OH \quad (1)$$

$$2CO + 4H_2 \Leftrightarrow C_2H_5OH + H_2O \quad (2)$$

The formation of higher alcohols proceeds according to the general formula :

$$nCO + 2nH_2 \Leftrightarrow C_nH_{2n+1}OH + (n-1)H_2O \quad (3)$$

The injections of methanol and ethanol into synthesis gas cause that these alcohols incorporated into the synthesis form higher alcohols [1-4].

$$CH_3OH + CO + 2H_2 \Leftrightarrow C_2H_5OH + H_2O \quad (4)$$

$$CH_3OH + nCO + 2nH_2 \Leftrightarrow C_{n+1}H_{2n+3}OH + nH_2O \quad (5)$$

$$C_2H_5OH + CO + 2H_2 \Leftrightarrow C_3H_7OH + H_2O \quad (6)$$

$$C_2H_5OH + nCO + 2nH_2 \Leftrightarrow C_{n+2}H_{2n+5}OH + nH_2O \quad (7)$$

The water gas shift reaction is always present in this process.

$$CO + H_2O \Leftrightarrow CO_2 + H_2 \quad (8)$$

2. EXPERIMENTAL

A catalyst for higher alcohol synthesis from syngas was developed in our laboratory. The catalyst consists of CuO (50-60),ZnO (25-39), ZrO_2 (7-14), Fe_2O_3 (1-4), MoO_3 (7-15), ThO_2 (1-3) and Cs_2O (0.5-1.5), wt%. The optimum conditions of the process over the catalyst investigated were: P~10 MPa, T~600 K, H_2/CO~1, GHSV~8000h^{-1}. The yield obtained was about 130 g/kg$_{cat}$/h of liquid product that contained about 40 wt% of methanol and 25-25 wt% of higher alcohols. By-products especially hydrocarbons were practically absent but traces of methane were detected. The catalyst exhibits a remarkable stability during one-year experiments and high selectivity toward alcohols.

The surface area of the catalyst was measured by BET method using argon. The surface area of the oxidised catalyst was 20.3 m^2/g and increased to 30.4 m^2/g after reduction by hydrogen. The mean pore radius was 160•10^{-10}m. The Cu metal surface area of reduced catalyst was measured using the pulse N_2O decomposition technique. Assuming a Cu atom density of $1.46 \cdot 10^{19}$ atoms/m^2, the Cu surface area was 6.2 m^2/g. The reversible adsorption of carbon oxide fitted very closely the values of parameters of Langmuir isotherm listed in the table 1.

Table 1. Parameters of Langmuir isotherm

Adsorption temperature [K]	$a_\infty \cdot 10^6$ [mol/m$_{cat}^2$]	b [Pa^{-1}]
423	0.60	41.3
473	0.70	28.4
523	0.71	4.05

The reactions were carried out with the addition of methanol or ethanol separately to the CO/H_2 mixture. Experiments were conducted in tubular high-pressure fixed bed reactor at temperature 600 K and the pressure of 4-7 MPa and GHSV 20 000 h^{-1} and CO/H_2=0.22.

3. DISCUSSION

It was observed that the addition of methanol and ethanol to the syngas strongly increases yields of all C_{3+} alcohols that is shown in the table 2 and figs 1-6.

Table 2.
Enrichment factors for products - calculated as the ratio of the product's yield with methanol or ethanol injection to its yield without injection.
* data after J.G.Nunan et al.[2].

Product	Ethanol injection 1750 g/kg$_{cat}$/h	Ethanol injection 460 g/kg$_{cat}$/h	Ethanol injection 690 g/kg$_{cat}$/h	Ethanol injection* 193 g/kg$_{cat}$/h	Ethanol injection* 193 g/kg$_{cat}$/h	Ethanol injection* 193 g/kg$_{cat}$/h	Methanol injection 304 g/kg$_{cat}$/h	Methanol injection 624 g/kg$_{cat}$/h
	7MPa, 600K	4MPa, 600K	4MPa, 600K	7.6MPa ,533K	7.6MPa, 553K	7.6MPa, 573K	7MPa, 600K	4MPa, 600K
Methyl acetate	-	-	-	46.8	8.2	0.7	-	-
Propanol	4.8	3.7	4.1	21.73	7.0	2.32	1.7	2.7
Butanol	15.3	12.3	17.2	-	10.6	2.58	1.6	2.8
Pentanol	4.3	5.0	5.4	-	-	-	1.6	2.9
Hexanol	4.5	7.7	9.5	-	-	-	1.6	3.4
Heptanol	2.3	6.1	4.2	-	-	-	1.5	4.7

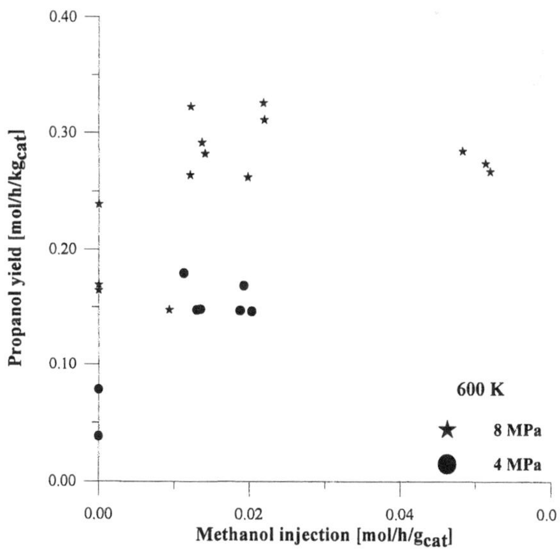

Figure 1.
The propanol yield as a function of methanol injection for the pressure 7 and 4 MPa, 600 K, 20g of catalyst and the syngas composition: CO-19 H_2-74, CO_2-0, N_2-7 %mol.

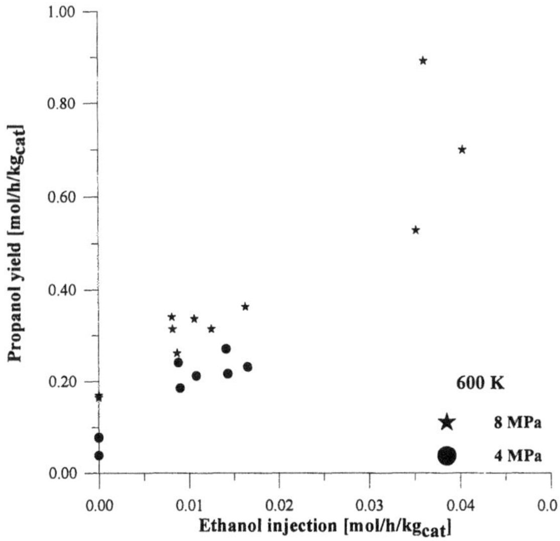

Figure 2.
The propanol yield as a function of ethanol injection for the pressure 7 and 4 MPa, 600 K, 20g of catalyst and the syngas composition: CO-19 H_2-74, CO_2-0, N_2-7 % mol.

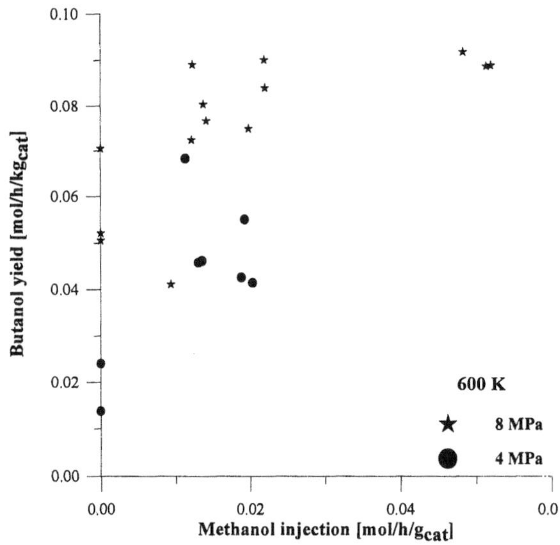

Figure 3.
The butanol yield as a function of methanol injection for the pressure 7 and 4 MPa, 600 K, 20g of catalyst and the syngas composition: CO-19 H_2-74, CO_2-0, N_2-7 % mol.

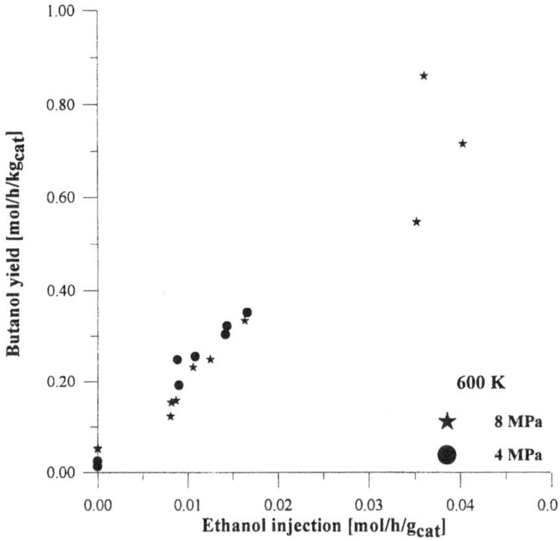

Figure 4.
The butanol yield as a function of ethanol injection for the pressure 7 and 4 MPa, 600 K, 20g of catalyst and the syngas composition: CO-19 H_2-74, CO_2-0, N_2-7 % mol.

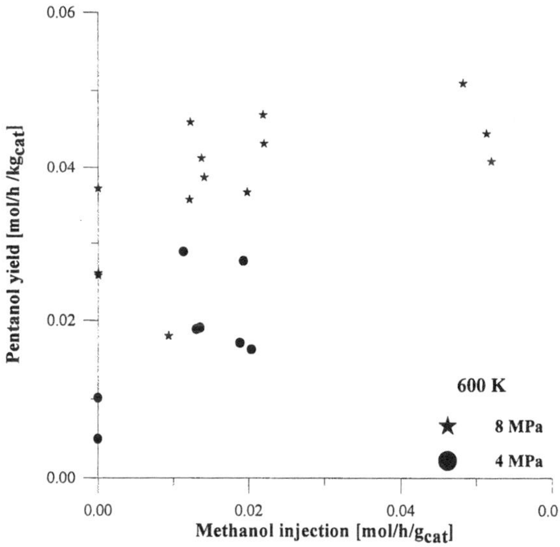

Figure 5.
The pentanol yield as a function of methanol injection for the pressure 7 and 4MPa, 600 K, 20g of catalyst and the syngas composition: CO-19 H_2-74, CO_2-0, N_2-7 % mol.

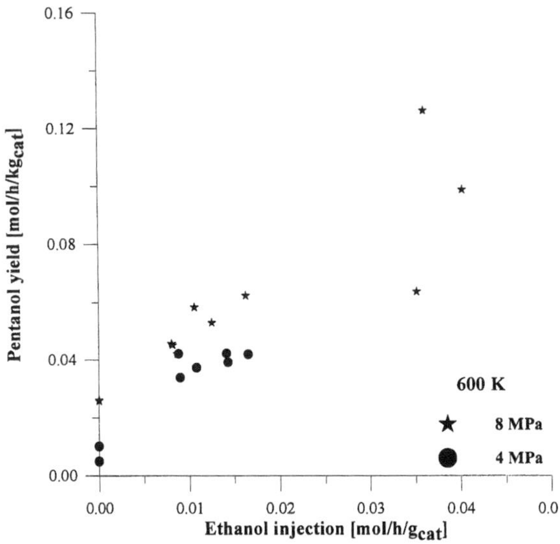

Figure 6.
The pentanol yield as a function of ethanol injection for the pressure 7 and 4 MPa, 600 K, 20g of catalyst and the syngas composition: CO-19 H_2-74, CO_2-0, N_2-7 % mol.

This enrichment of the C_{3+} alcohols yields is even more than ten times in comparison with synthesis of alcohols only from syngas and it is quite similar like in the J.G.Nunan's [2] paper, but there are still no by-products (only the traces of methane) in our product. The addition of methanol is less efficient than that of ethanol. Our results were not consistent with those reported by Majocchi et al. [5], where the addition of C_1 to feed stream did not produce any significant change in the formation of higher alcohols, only the injection of C_2 was accompanied by a significant promotion of propanol production.

The results indicate that the higher alcohols are obtained by carbon chain growth, probably with COH→CCOH (α-addition) as the slow initial step of chain growth and the rate determining step. This problem is of considerable practical interest that could find application in industry.

REFERENCES

1 A.Kienneman, H.Idriss, R.Kieffer, P.Chaumette, D.Durand, Ind. Eng.Chem.Res., 30 (1991) 1130.
2 J.G.Nunan, Ch.E.Bogdan, K.Klier, K.J.Smith, Ch-W.Young, R.G.Herman, J.Catal. 116 (1989) 195.
3 J.G.Nunan, Ch.E.Bogdan, .G.Herman, K.Klier, Cat.Lett. 2 (1989) 49.
4 D.Z. Wang, Chin. J. Fuel Chem. Techn., 22 (1) (1994) 63.
5 L.Majocchi, L.Lietti, A.Beretta, E.Micheli, P.Forzatti, EUROPACAT-III, Kraków 1997.

NATURAL GAS CONVERSION V
Studies in Surface Science and Catalysis, Vol. 119
A. Parmaliana et al. (Editors)

ROLE OF CR IN FE BASED HIGH TEMPERATURE SHIFT CATALYSTS

J. Koy[a], J. Ladebeck[a] and J.-R. Hill[b]
[a]SÜD-CHEMIE AG Waldheimerstr. 13 83052 Bruckmühl-Heufeld
[b]Molecular Simulations Inc. 9685 Scranton Road San Diego, CA 92121-3752

1. INTRODUCTION

The water-gas shift (WGS) reaction is the conversion of carbon monoxide and steam to form carbon dioxide and hydrogen (eq. 1), it is a reversible, exothermic reaction and usually assisted by a catalyst. The reaction enthalpy amounts to -40.6 kJ mol^{-1}.

$$CO + H_2O \leftrightarrow CO_2 + H_2 \quad (1)$$

The water-gas shift reaction is an important step in many industrial processes, for example ammonia, hydrogen and synthesis gas production. Due to its industrial importance, WGS and involved catalysts were objects of various investigations [1]. In industrial HTS converters Fe based catalysts are preferably applied due to their high stability.
For iron based catalysts a regenerative mechanism was proposed. This mechanism can be described as follows [2, 3]:

$$H_2O + * \leftrightarrow H_2 + O* \quad (2)$$
$$CO + O* \leftrightarrow CO_2 + * \quad (3)$$

where * is the active site with oxygen vacant and O* a surface containing oxygen. Surface cations which can change their oxidation state are required for the regeneration mechanism.

1.1. HTS CATALYSTS BASED ON FE

The precursors of the iron based HTS catalysts are α-FeO(OH) or α-Fe_2O_3 or a mixture of α- and γ-Fe_2O_3 depending on the preparation method [4]. A subsequent thermal treatment will transform the precursors completely into anti-ferromagnetic, most stable hexagonal close-packed hematite α-Fe_2O_3. During activation hematite is reduced to magnetite. Magnetite, with its less stable cubic inverse spinel structure, is not thermoresistant enough and will re-crystallise quite rapidly at temperatures applied in HTS reaction. Therefore, it cannot be used as an industrial HTS catalyst without a structure stabiliser.

1.2. CR AND ITS FUNCTION

According to XRD analyses, HTS catalyst precursors as supplied are solid solutions of α-Fe_2O_3, in which Fe is substituted by Cr in the lattice. However, above 14% of Cr_2O_3 in Fe_2O_3, Cr_2O_3 is forming a separate phase [1]. The active phase of the catalyst is stated as a Fe_3O_4 structured material. Pure Fe_3O_4 is not alone viable due to sintering or over-reduction. A structure stabiliser must be added to make the catalyst effective. Cr is effective and contributes to performance improvement in several ways. It has been found that Cr_2O_3 prevents iron oxide from high temperature sintering and loss of surface area. The stabilising effect of Cr is based on the substitution of Fe^{3+} in the lattice by Cr^{3+}. The influence of the Cr substitution on the iron oxide lattice was described in the literature on the basis of practical investigations and knowledge.

This contribution will present the results of molecular modelling experiments. The Cr-substitution and the influence on the bulk and surface properties were calculated systematically. The most likely chromium sites were then studied by electron structure calculations to develop a better understanding of the electronic properties which govern the chromium incorporation in iron oxides.

The thermostability of a catalyst is also a property of great interest. It is rather difficult to find microscopic properties which correlate with the thermostability. The diffusion of the cations in the iron oxide lattice and possible segregation of the dopant on the surface are effects which could be related to thermostability. Diffusion can be studied using molecular dynamics while segregation effects can be investigated using surface energy calculations.

Based on the modelling work it should be possible to find an explanation for the role of Cr substitution in iron oxides and to prove that modelling can be used as a tool in solid state chemistry and catalysis research in comparison to practical experience.

2. Calculations / Simulations

2.1. Modelling of Bulk Properties

A systematic study of the function of Cr in the hematite and magnetite lattice required the calculation of conformational energies for a large number of structures. A shell-model potential [5] was used for these calculations with parameters from [6]. The ions making up the system are thought of consisting of a core and a shell which are differently charged and connected by a harmonic spring. Electrostatic interactions between core and shell of the same ion are excluded. Since core and shell can be at different positions in the space it is possible to describe polarisation effects with this potential, which are essential for oxides. The positions of the ions in the model are optimised by searching for a minimum of the energy (geometry optimisation). The unit cell was optimised as well by that procedure. The absolute values of the energies calculated this way do not have any physical meaning. It is therefore only meaningful to compare relative energies for different configurations of the same system.

The electronic structure of the systems was determined using ASW calculations (Augmented Spherical Wave) [7]. ASW is a density functional method [8] designed to determine the electronic structure of close-packed solids. ASW calculations allow the computation of isomeric shifts which are observable with Mößbauer spectroscopy, since the isomeric shift, δ, is related to the electron density at the nucleus.

Molecular dynamics calculations [9] were used to study the diffusion of cations. Molecular dynamics calculations study the evolution of a molecular system with time. The starting positions were the positions obtained in the geometry optimisation. The starting velocities were randomly drawn from a Maxwell-Boltzmann distribution for a given temperature.

2.2. MODELLING OF SURFACE PROPERTIES

A systematic study of the effect of Cr on the stability of surfaces required the calculation of surface energies for a large number of structures. These calculations required also larger systems than simulations of the bulk where usually a single unit cell is appropriate. The surface calculations used a two-dimensional slab which was, due to limited computational resources, around 45Å thick. Most of the atoms in this slab were held fixed at the positions optimised for the bulk model. Only the layers of atoms closest to the surface up to a depth of around 10Å were allowed to move. All surface calculations were performed by using a rigid ion potential in two steps. First, the energy of the unrelaxed surface was calculated for a certain Cr distribution. Structures which had the same surface energies for their unrelaxed surfaces were eliminated assuming that these surfaces would also yield the same energy after relaxation. Secondly, all surfaces with different surface energies were fully relaxed. All surface calculations were performed using the program MARVIN (Minimization and Relaxation of Vacancies and Interstitials Near Surfaces Program) [10] which is able to handle full two-dimensional periodicity.

Only the surfaces of pure and doped magnetite were investigated, since magnetite is believed to be the active phase in the catalytic process, which runs at 350°C.

To study ion mobility on the surface, molecular dynamics calculations have been performed. A three-dimensional stack of slabs was constructed and used for this type of calculations with GULP (General Utility Lattice Program) [11]. The stack was built so that there was approximately 10Å vacuum between each surface. Molecular dynamics calculations were run for 10ps of equilibration followed by 100ps of data collection. The time step used was 1fs.

3. RESULTS AND DISCUSSION

3.1. BULK PROPERTIES

The calculations were started with the well known structure of hematite, α-Fe_2O_3, (one unit cell, $Fe_{12}O_{18}$) and two iron atoms were replaced by chromium. Two Cr atoms per unit cell correspond to 10.9 weight-% Cr. All possible sites, Cr could occupy in the hematite lattice, were investigated.

Since Cr is supposed to occupy the octahedral sites in the magnetite structure only, four chromium atoms (= 11.3 weight % Cr) were placed on octahedral positions in the structure (one unit cell, $Fe_{24}O_{32}$) and all possible sites Cr could occupy were systematically studied. The pure hematite and magnetite as well as the most stable Cr substituted structures were also used in ASW calculations to obtain knowledge of the electronic structure of these systems.

3.1.1. STRUCTURE STABILITY

a) Hematite

The most stable structures for the Cr substituted hematite have the dopants as close together as possible. The dopants form "pairs". The pairs are ordered in layers in the crystal. As the energy differences between the most stable and the least stable structure are very small (5.6 kJ/mol), one can conclude that there is no preferred substitution site for Cr. The density functional calculations performed on hematite allow a comparison of the stability of the pure and the Cr substituted hematite. The binding energy is the energy obtained when the atoms are moved from infinite separation to their positions in the crystal. Table 1 lists the calculated binding energies, obtained from density functional calculations.

Table 1
Calculated binding energies for pure and substituted hematite

Compound	$E_{binding}$ [kJ/mol]
hematite	-2321.8
Cr-hematite	-2333.7

As table 1 shows the Cr substituted hematite structure is slightly more stable than the pure hematite. The energy difference between the most and least stable Cr substituted hematite is extremely small (0.04 kJ/mol).

b) Magnetite
If magnetite is substituted by Cr the most stable configuration was found to be the one where all four Cr atoms are placed in the same cube thus forming "pairs" as in hematite. The energy differences obtained between the most and the least stable structures are generally much higher in magnetite (122.8 kJ/mol) than in hematite.

3.1.2. ELECTRONIC STRUCTURE

Mößbauer isomeric shifts have been obtained from ESOCS calculations. Quadrupolar splitting was not considered in these calculations. Table 2 summarises the results of the calculations.

Table 2
Mößbauer isomeric shifts, δ, calculated from ESOCS results with respect to ^{57}Fe

Compound	Substitution		δ [mm/s]
hematite	pure		0.85
	Cr, most stable		0.85 - 0.91
	Cr, least stable		0.84 - 0.94
magnetite	pure	Fe(II)	1.11
		Fe(III)	1.18
	Cr, most stable	Fe(II)	0.95 - 0.96
		Fe(III)	1.56 - 1.60
	Cr, least stable	Fe(II)	0.95 - 0.98
		Fe(III)	1.58 - 1.61

Substitution of Fe atoms by Cr has an impact on the electronic structure of the system. Experimentally determined Mößbauer shifts have been reported e. g. [12]. Cr substitution in hematite leads to a higher isomeric shift on Fe atoms (lower electron density). The highest isomeric shift occurs for the Fe atoms closest to Cr while Fe atoms in the next co-ordination sphere already have a shift similar to the pure hematite.
Fe(II) and Fe(III) ions behave differently on Cr substitution in magnetite. The isomeric shift of Fe(II) decreases while it increases for Fe(III) compared to pure magnetite. As a result Fe(II) and Fe(III) become electronically more distinguished if Cr is introduced. There is not much difference on the isomeric shift in dependence of the distance from Cr atoms. The isomeric shifts do not vary much between the most and least stable substitution patterns.

3.1.3. THERMOSTABILITY

As a first step in studying the thermostability of a catalyst system, molecular dynamics simulations were performed for different temperatures (350°C and 450°C) for the ideal solid (i.e. without considering lattice defects). A common way to analyse the results of a molecular dynamics simulation is to calculate the mean square displacement of the atoms as a function of time. In case of diffusion of an atom the slope of this function can be used to determine the diffusion coefficient. All the simulations carried out show no diffusion of the atoms but only vibration.

3.2. SURFACE PROPERTIES OF THE MAGNETITE STRUCTURE

3.2.1. SEGREGATION ENERGIES

To analyse the effect of a dopant on the surface, calculations were performed which first placed a single dopant ion in the bulk and than as close as possible to the surface. The energy difference for having the dopant as close as possible to the surface and in the bulk provides information about the likelihood to find the dopant on the surface. Two opposing trends will determine whether an ion is more stable on the surface or in the bulk. The first are the lattice distortions caused by this ion which will make the whole system more stable if the ion is moved to the surface. The second is the lack of co-ordination or bond partners on the surface which will make the whole system more stable if the ion is in the bulk. For the Cr ion the second trend is more important. The results ($E_{difference\ (111)\ surface}$ = 77.7 kJ/mol; $E_{difference\ (110)\ surface}$ = 4.3 kJ/mol) show that Cr on the surface destabilises the system.

3.2.2. THE (111) SURFACE

In the most stable Cr substituted surfaces the Cr atoms cluster together as in the bulk and as a result "bands" of Cr are formed as part of the surface. A number of Cr atoms is exposed on the surface. Cr substitution on the (111) surface gives a broad distribution of surface energies for the unrelaxed surface. After relaxation the distribution narrows significantly.

3.2.3. THE (110) SURFACE

Cr forms a similar "band" structure as on the (111) surface, but none of the Cr atoms is exposed on the surface. The distribution of surface energies is rather narrow for the unrelaxed surface. That distribution widens a little bit on relaxation, but both distributions are much more similar than for the (111) surface. It is interesting to compare the stabilities of the different surfaces between pure and doped magnetite. While in pure magnetite the (110) surface is slightly more stable than the (111) surface, Cr substitution destabilises the (110) surface. On average both surfaces become nearly equally stable.

3.2.4. THERMOSTABILITY / SURFACE ION MOBILITY

In a final step the surface ion mobility on the surface was studied with molecular dynamics calculations. The simulations were performed at 350°C. In the case of the least stable Cr substituted (110) surface the mean square displacement is on average constant. This is similar to the results obtained for the bulk where only vibrations could be observed. However in pure magnetite the curves are not parallel to the time-axis which means that there diffusion plays a role. The slope of the mean square displacement was used to determine diffusion coefficients for Fe ($1.05 * 10^{-11} m^2/s$) and O ($2.02\ 10^{-11} m^2/s$). Millot et al. determined the oxygen diffusion coefficients for magnetite under various atmospheres [13]. Their results vary between 3.22 *

10^{-11} m^2/s and 21.5 * 10^{-11} m^2/s depending on the partial pressure of oxygen. Considering that these diffusion coefficients have been obtained at higher temperature, the presented simulation result appears to be the right order of magnitude.

4. Summary and Conclusions

The goal of our investigations was to study the influence of Cr incorporation on the bulk and surface properties of iron oxide based HTS catalyst with modelling techniques and compare the results with the findings based on practical experience. It has been shown that Cr fits very well into the hematite and also magnetite lattice. The substitution pattern for the most stable configuration of Cr in hematite and magnetite shows a framework of its own. The dopants in the most stable configuration are as close together as possible, forming pairs. The pairs are ordered in layers in the crystal. The bands of Cr just beneath the surface could be responsible for the higher thermostability of the Cr doped catalyst. Cr is not distributed randomly , but forms a superstructure in itself. We could not observe any cation diffusion in the Cr doped magnetite. The calculated Mößbauer isomeric shifts show that the introduction of Cr makes the Fe(II) and Fe(III) more distinguished.

Cr substitution has also an effect on the stabilisation of different surfaces. Cr destabilises the (110) surface compared to pure magnetite and does not affect the (111) surface.

It can be concluded that the calculations / simulations showed that there are two main effects of Cr in iron oxide based HTS catalysts:

- The first one is the superstructure of the dopant itself in the iron oxide lattice and the resulting thermostability
- The second one is the electronic influence on the Fe cations in the lattice and the resulting catalytic performance.

With these theoretical investigations it was possible to obtain a better understanding of the role of Cr in Fe based HTS catalysts and the achieved knowledge gives an additional explanation of the practical experiences.

References

[1] David S. Newsome, Catal. Rev.-Sci., **21(2)**, 275 (1980)
[2] D.G. Rethwisch and J.A. Dumesic, Appl. Catal. **21**, 97 (1986)
[3] D.G. Rethwisch and J.A. Dumesic, J. Catal. **101**, 35 (1986)
[4] G.C. Maiti and S.K. Gosh, Indian Journal of Technology, **19**, 35 (1981)
[5] B.G. Dick and A.W. Overhauser, Phys. Rev., **112**, 90 (1958)
[6] G.V. Lewis, C.R.A. Catlow, J. Phys. C, Solid State Phys., **18**, 1149 (1985)
[7] J.Kübler and V.Egert, Electronic and Magnetic Properties of Metals and Ceramics, Part 1, VCH, Weinheim, 1992
[8] J. Labanowski and J. Andzelm Eds., Density Functional Methods in Chemistry, Springer, New York, 1991
[9] M.P. Allen, D.J. Tildesly, Computer Simulation of Liquids, Clarendon Press, Oxford, 1987
[10] D.L. Gay and A.L. Rohl, J. Chem. Soc., Faraday Trans., **91**, 925 (1995)
[11] J.D. Gale, J. Chem. Soc., Faraday Trans., **101**, 1248, (1997)
[12] G. Doppler, A.X. Trautwein, H.M. Ziethen, E. Ambach, R. Lehnert and M.J. Sprague, Appl. Catal., **40**, 119 (1988)
[13] F. Millot, J.C. Lorin, B. Klossa, Y. Niu and J.R. Torento, Ber. Bunsenges. Phys. Chem., **101**, 1351 (1997)

NATURAL GAS CONVERSION V
Studies in Surface Science and Catalysis, Vol. 119
A. Parmaliana et al. (Editors)

Isoalcohol synthesis from CO/H_2 feedstocks

C. R. Apesteguía[a], S. Miseo[b], B. De Rites[b] and S. Soled[b]

[a]INCAPE, UNL-CONICET, Santiago del Estero 2654, (3000) Santa Fe, Argentina

[b]Exxon Research and Engineering Company, Corporate Research Science Laboratories, Annandale, New Jersey 08801, USA.

The synthesis of methanol and isobutanol from synthesis gas over copper-containing $Mg_yCe(Y)O_x$ catalysts was studied. The influence of the catalyst composition, temperature and contact time on isoalcohol productivity was established. The reaction pathway for isoalcohol synthesis was studied by adding methanol, ethanol or propanol to the reactant feed.

1. INTRODUCTION

The selective production of isobutanol-methanol mixtures from synthesis gas is a potentially attractive technology for the synthesis of methyl-tert-butil ether (MTBE). Modified methanol synthesis catalysts, such as $Mn(Zn)O/Cr_2O_3$/alkali operated at high temperatures and CuO/ZnO/alkali operated at low temperatures, exhibit promising catalytic performances for a one step synthesis of isobutanol [1-4]. Catalysts consisting of palladium supported on a coprecipitated manganese, zinc, zirconium, lithium oxide support show high isobutanol synthesis productivity at high temperatures (>400°C) and pressures (>100 atm) [5]. However, even at the high reactor temperatures and pressures required, current catalytic approaches for isoalcohol synthesis produce relatively low yields and selectivities. Hence, practical application of such catalytic process still requires significant improvements, and one approach involves developing novel catalytic materials capable of more selectively producing methanol and isobutanol mixtures at moderate reactor temperature and pressure. Recently [6,7], we have found that catalysts based on coprecipitated mixtures or solid solutions of alkaline earth oxides and rare earth oxides (also including yttrium oxide), which may contain copper oxide and an alkali dopant, are active and selective for isoalcohol synthesis at less severe conditions than those required by prior art catalysts. In this paper we intentionally introduce reaction products (methanol, ethanol, 1-propanol) into the CO/H_2 feed streams and study their effect on isoalcohol selectivity. The catalytic tests were performed on copper-containing $Mg_yCe(Y)O_x$ mixed oxides.

2. EXPERIMENTAL

Catalysts were prepared by coprecipitation of rare earth oxides and alkaline earth oxides under controlled pH conditions as described elsewhere [7]. Copper oxide was either coprecipitated or added by impregnation. The precursor was then decomposed in air at 400-600°C for 4 h. The supported catalysts are designated by a slash, e.g. $Cu/M_yN_zO_x$ designates Cu supported on a $M_yN_zO_x$ mixed oxide with Cu loading given in wt % and y, z, and x represent g-atom quantities. Coprecipitated catalysts are designated without a slash, e.g. $Cu_pM_yN_zO_x$, indicates a coprecitated oxide where p, y, z, and x represent g-atom quantities of the respective elements.

Powder X-ray diffraction patterns (XRD) were collected on a Rigaku diffractometer using monochromatic CuK_α radiation. The chemical composition of the samples was measured by an Inductively Coupled Plasma (ICP) spectrometer (Jarrel Ash). BET surface areas (S_g) were measured by N_2 adsorption at 77 K in a Omnisorp sorptometer.

Samples were tested in a plug-flow packed-bed reactor using a 1:1 CO/H_2 feed. Both on-line and off-line samples were analyzed by GC-MS for CO_2, methanol, ethanol, linear C_3-C_6 primary alcohols, branched C_4-C_6 primary alcohols, secondary alcohols (isopropanol, 2-butanol), methane, ethane, linear C_3-C_{12} aliphatic hydrocarbons, branched C_4-C_6 hydrocarbons, ethylene, linear C_3-C_5 olefinic hydrocarbons, dimethylether (DME) and esters (methyl acetate, methyl propanoate, methyl i-butanoate).

3. RESULTS AND DISCUSION

Table 1 shows the chemical composition and BET surface areas of impregnated and coprecipitated catalysts studied here. All the catalysts contained less than 10 wt% of copper and had surface areas between 75 and 125 m^2/g.

Table 1
Catalysts: Chemical composition and BET surface areas

Catalyst	Elemental Analysis (wt%)[a]					S_g
	Cu	Mg	Ce	Y	K	m^2/g
Cu/CeO_2	6.83	-	72.15	-	-	75
$Cu_{0.5}Mg_5CeO_X$	7.71	28.40	32.82	-	-	102
$K/Cu_{0.5}Mg_5CeO_X$	7.71	28.40	32.80	-	0.86	90
Cu/Mg_5YO_x	9.62	35.03	-	23.81	-	125
$Cu_{0.5}MgYCeO_x$	7.81	6.63	37.12	25.84	-	88

The XRD patterns, which are not shown here, indicate that sample $Cu_{0.5}MgYCeO_x$ was a solid solution which contains the CuO and MgO substituted into a crystalline CeO_2 structure. In samples with a high magnesium loading ($Cu_{0.5}Mg_5CeO_x$ and Cu/Mg_5YO_x catalysts), the solubility of the MgO in the rare earth oxide host was exceeded, so we observed both crystalline MgO and doped cerium (yttrium) oxide. Copper oxide was not detected in any sample, thereby suggesting that copper was finely dispersed into the mixed oxide matrix.

Catalytic tests were carried out at 50 atm. During the 250 hour catalytic runs, the temperature was varied between 260 and 360°C and the space velocity (GHSV) between 460 and 1850 cm^3(STP)/g_{cat}h. In Table 2 we present the catalytic results obtained on Cu/Mg_5YO_x, $Cu_{0.5}Mg_5CeO_X$, and $Cu_{0.5}MgYCeO_x$ catalysts after 120 h on stream, at 320°C and 50 atm. Methanol and isobutanol were the major products in the oxygenate fraction. Neither the alcohols nor the hydrocarbons followed a Schulz-Flory distribution; branched alcohols were readily formed as indicated by the high branched alcohol/linear alcohol ratio. Table 2 shows

Table 2
Alcohol synthesis (productivities, expressed in $g/kg_{cat}h$)
320°C, GHSV = 1832 $cc/g_{cat}\cdot h$, $CO:H_2 = 1$, P = 50 atm. Data measured after 120 on stream

Product	Cu/Mg_5YO_x	$Cu_{0.5}Mg_5CeO_x$	$Cu_{0.5}MgYCeO_x$
Methanol	60.8	64.0	58.8
Ethanol	2.90	0.77	1.87
1-propanol	2.32	1.02	1.44
1-butanol	0.32	0.07	0.25
1-pentanol	0.17	0.33	0.26
1-hexanol	0.15	0.18	0.09
Isobutanol	4.35	7.07	5.76
2-m-1-butanol	0.54	0.42	0.58
2-m-1-pentanol	0.26	0.22	0.23
DME	1.34	2.82	0.33
Methane	10.0	9.98	10.3
X_{CO}(%)[a]	19.9	20.8	18.7
Alc/Hyd[b] (%C)	1.9	2.3	1.6
Bra/Lin[c]	1.2	3.7	2.0

[a] CO conversion
[b] Total alcohols/Total hydrocarbons ratio
[c] Branched C_4-C_6 alcohols/Linear C_2-C_6 alcohols

that, at a similar CO conversion level, the $Cu_{0.5}Mg_5CeO_X$ catalyst yields the highest productivity to isobutanol and the Alc/Hyd and Bra/Lin ratios. Thus, additional studies were performed using catalysts containing copper, ceria, magnesia and potassium. Fig. 1 presents the catalytic selectivities obtained on Cu/CeO_2, $Cu_{0.5}Mg_5CeO_X$ and $K/Cu_{0.5}Mg_5CeO_X$ catalysts. The Cu/CeO_2 catalyst promoted selectively the formation of isobutanol but the production of short-chain linear alcohols was significant, leading to a relatively low Bran/Lin ratio of 0.9. When copper oxide was coprecipitated with magnesia and ceria, the catalytic activity as well as selectivity and productivity to isobutanol clearly increased. Compared with Cu/CeO_2, CO conversion increased from 16.4 to 20.8% and isobutanol productivity from 5.1 to 7.1 $g/kg_{cat}\cdot h$. As shown in Fig. 1, the selectivity to isobutanol also increased and, on a methanol-free basis, isobutanol represented about 75% of the alcohol fraction. The Bran/Linear ratio increased from 0.9 (with the Cu/CeO_2 catalyst) to 3.7. The addition of 0.86% potassium to $Cu_{0.5}Mg_5CeO_x$ catalyst blocks surface acid sites and selectively decreases the formation rate of methane and DME. Fig. 1 shows that the selectivity to methane on $K/Cu_{0.5}Mg_5CeO_x$ was substantially lower as than on undoped $Cu_{0.5}Mg_5CeO_x$. The decrease in methane formation was accompanied by a simultaneous increase of the Alc/Hyd ratio from 2.8 to 4.1. The K-doped $Cu_{0.5}Mg_5CeO_x$ catalyst had similar isobutanol and methanol productivities as the undoped catalyst.

Over all the catalysts, increasing the temperature from 260 to 340°C increased the isoalcohol formation rate while methanol yield decreased due to thermodynamic constraints. Therefore, the C_{2+} linear alcohol selectivity passed through a maximum. As a consequence, the oxygenate fraction became depleted in intermediates (ethanol, propanol) and enriched in

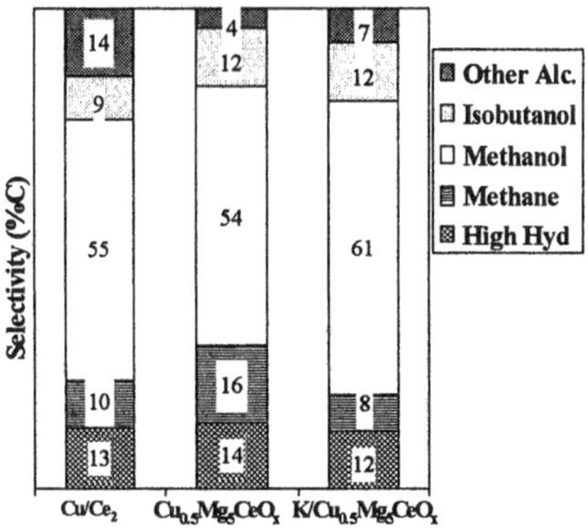

Figure 1. Alcohol synthesis on the Cu/Ce/Mg/K system. T = 320°C, P = 50 atm

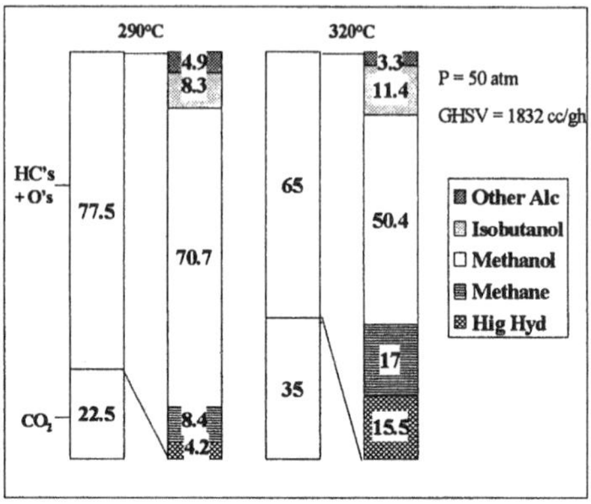

Figure 2. Effect of temperature on catalyst selectivity. $Cu_{0.5}Mg_5CeO_x$; selectivity in %C

branched alcohols. However, the production of CO_2 and hydrocarbons also increased at higher temperatures. In Fig. 2 we have presented the selectivities to carbon dioxide, total alcohols, total hydrocarbons, methane, methanol, and isobutanol using the $Cu_{0.5}Mg_5CeO_x$ catalyst at 290 and 320°C. The selectivity to CO_2 at 290°C (22.5%) increased to 35% at 320°C. On a CO_2-free basis, selectivities to methane and higher hydrocarbons increased from 8.4 and 4.2 respectively, at 290°C, to 17 and 15.5 at 320°C. On the other hand, increasing contact time favored isobutanol and branched alcohol formation while decreasing methanol slightly. Thus, the isobutanol/methanol ratio can be somewhat controlled by the temperature and space velocity of the reaction.

Table 3 shows how adding methanol, ethanol or propanol to the reactant feed influences alcohol productivity. Addition of propanol selectively increases the formation of isobutanol; the production of linear and branched C_{3+} alcohols is also promoted. The productivity to 1-butanol drastically increases when ethanol is added. Finally, the addition of methanol

Table 3: Effect of the alcohol addition to CO/H_2 feedstocks on isoalcohol synthesis
Productivities in g/kg cat/h. Alcohol added: 1-propanol, 0.774 mmol C/h; methanol, 1.29 mmol C/h; ethanol, 1.68 mmol C/h. $Cu_{0.5}Mg_5CeO_x$, $CO:H_2 = 1$, P = 50 atm

Product	Alcohol added					
	None	Propanol	None	Methanol	None	Ethanol
	290°C		300°C		310°C	
Methanol	153.0	143.4	110.4	-	69.4	51.5
Isobutanol	4.94	15.9	3.80	6.14	2.39	2.40
2-me-1-butanol	0.59	0.71	0.51	0.70	0.48	0.90
2-me-1-pentanol	0.21	1.00	0.21	0.40	0.26	0.86
Ethanol	1.17	1.10	0.78	1.10	0.80	-
1-propanol	1.24	-	0.91	1.19	0.86	5.51
1-butanol	0.11	0.28	0.08	0.11	0.11	26.21
1-pentanol	0.08	0.44	0.07	0.11	0.10	1.30
1-hexanol	0.04	0.08	0.05	0.05	0.05	1.85

promotes linear and branched chain growth pathways, and thereby increases the productivities to C_{2+} alcohols. Alcohol mixtures of methanol/1-propanol and methanol/ethanol where also added to the $CO:H_2$ feed; results are given in Table 4. It is shown that the addition of

Table 4
Effect of the addition of alcohol mixtures to CO/H_2 feedstocks on isoalcohol synthesis
Productivities in g/kg cat/h. Alcohol mixture added: (methanol + 1-propanol), 3.970 mmol C/h; (methanol + ethanol), 2.064 mmol C/h
$Cu_{0.5}Mg_5CeO_x$, $CO:H_2 = 1$, P = 50 atm, GHSV = 1832 cc/g/h

Product	Alcohol mixture added			
	After 48 h in run		After 230 h in run	
	None	Methanol + 1-propanol	None	Methanol + ethanol
Methanol	157.2	-	142.9	-
Isobutanol	8.08	45.8	5.76	5.32
2-me-1-butanol	0.92	1.02	0.56	1.78
2-me-1-pentanol	0.35	3.32	0.34	0.39
Ethanol	1.72	1.80	1.39	-
1-propanol	1.87	-	1.55	6.23
1-butanol	0.14	0.54	0.12	16.6
1-pentanol	0.16	3.22	0.23	0.53
1-hexanol	0.06	0.64	0.06	0.39

methanol/1-propanol selectively promotes the formation of isobutanol while the mixture methanol/ethanol mainly increases the productivity to 1-butanol.

The formation of higher alcohols from CO/H_2 occurs through a sequential mechanism involving four main steps [7,8]: i) synthesis of methanol and formation of a C_1 intermediate species, ii) formation of the primary carbon-carbon bond, probably via the coupling of two C_1 intermediates, followed by a linear chain-building process (**L**) dominated by aldol coupling and C_1 insertion pathways which produce a C_{n+1} alcohol from a C_n alcohol, iii) aldol-type addition of the C_1 intermediate to the β carbon of a linear C_n alcohol to produce a 2-methyl branched C_{n+1} alcohol (**A-C$_1$**), iv) addition of C_2 (**A-C$_2$**) and C_3 (**A-C$_3$**) intermediates via self-condensation and cross-coupling reactions to produce linear C_{2+n} and C_{3+n} alcohols from a C_n alcohol. Fig. 3 presents a simplified reaction network of synthesis of methanol and higher alcohols from synthesis gas.

In Fig. 3, isobutanol is formed by the aldol-type C_1 addition to C_3 intermediate species. The addition of 1-propanol to the reactants increases the concentration of surface C_3 species and, as a consequence, increases isobutanol productivity. The predominant formation of isobutanol via the direct reaction between C_1 and C_3 species is confirmed by the results in Table 4 which show that the addition of a methanol/1-propanol mixture selectively increases the production of isobutanol. The addition of propanol also increases the formation of higher 2-methyl alcohols and linear C_{4+} alcohols. The increase in 2-me-1-butanol and 2-me-1-pentanol therefore results from a higher 1-butanol and 1-pentanol selectivity and by **A-C$_2$** propanol/ethanol addition or self-condensation of 1-propanol (**A-C$_3$** addition).

As expected, the addition of methanol increases both linear and aldol condensation chain growth rates by C_1. Nevertheless, the ratio of increase in isobutanol productivity is significantly higher than the increase in 1-butanol. This shows that the aldol **A-C$_1$** reactions are the predominant chain growth pathways on $Cu_{0.5}Mg_5CeO_x$. However, Table 3 shows that

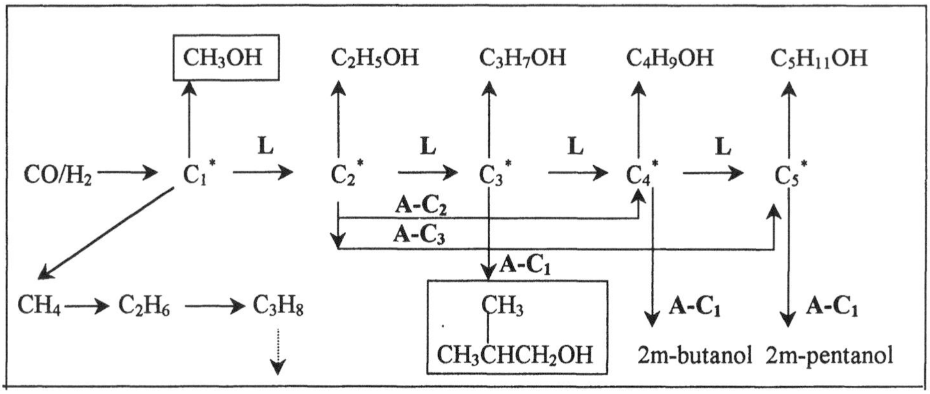

Figure 3. Simplified reaction network of synthesis of methanol and higher alcohols
L: Chain growth
A-C$_1$, A-C$_2$, A-C$_3$: 1-, 2-, and 3- carbon addition, respectively

the addition of ethanol selectively increases the formation of 1-butanol via an **A-C$_2$** condensation mechanism. Similar qualitative result was obtained when a methanol/ethanol mixture was added to the reactants (Table 4). The self-condensation of ethanol is a bimolecular reaction between adjacent adsorbed species and requires a high density of basic sites. The selective formation of 1-butanol therefore reflects a high concentration of surface C_2 species derived from the addition of ethanol. The high coverage in C_2 intermediates blocks the C_1/C_3 aldol condensation reactions and, as a result, the isobutanol productivity is not significantly changed by adding ethanol in relatively high concentrations.

Results show that our copper-containing $Mg_yCe(Y)O_x$ catalysts selectively catalyze the low-temperature isoalcohol synthesis from CO/H_2. The catalyst formulation combines the hydrogenation function required to form methanol with the basic/aldol condensation function needed to promote branching. Magnesium oxide provides the basic sites needed for the formation of 2-methyl branched alcohols via aldol condensation reactions. Copper promotes methanol formation and hydrogenation-dehydrogenation reactions. $Ce(Y)O_x$ is a high surface area matrix which contains finely dispersed metallic copper.

REFERENCES

1. R. di Pietro and A. Paggini, Fr. Patent 2,490,215 (1982).
2. P. Forzatti and E. Tronconi, Catal. Rev. Sci. Eng., 33 (1991) 109.
3. M. Schneider, K. Kochloefl and O. Bock, Eur. Pat. Appl. 152, 809 (1985).
4. J.G. Nunan, R.G. Herman and K. Klier, J. Catal., 116 (1989) 222.
5. C.R. Apesteguia, S.L. Soled and S. Miseo, U.S. Patent 5,387,570 (1995); US Patent 5,508,246, (1996), Eur. Pat. Appl. 94303184.9 (1994).
6. C.R. Apesteguia, B. De Rites, S. Miseo, and S.L. Soled, Catal Lett., 44 (1997) 1.
7. T.J. Mazanec, J. Catal., 98 (1986) 115.
8. J.G. Nunan, C.E. Bogdan, K. Klier, K.J. Smith, C.-W. Young and R. Herman, J. Catal., 113 (1988) 410.

NATURAL GAS CONVERSION V
Studies in Surface Science and Catalysis, Vol. 119
A. Parmaliana et al. (Editors)

Alcohols carbonylation to alkyl formates catalyzed by strongly basic resins

C.Carlini[1], M. Di Girolamo[2], M. Marchionna[2], A.M. Raspolli Galletti[1], G. Sbrana[1]

[1] Dipartimento di Chimica e Chimica Industriale,
via Risorgimento 35, 56126 Pisa

[2] SNAMPROGETTI S.p.A., Research Laboratories,
via F. Maritano 26, 20097 San Donato Milanese (MI)

Summary

Heterogeneous basic polymeric resins were checked as catalysts for the carbonylation of methanol and higher alcohols to alkyl formates and their activity compared with that of the conventional homogeneous systems. The data were interpreted in terms of different morphology, basicity and swelling degree of the strongly basic resins.

1. Introduction

The direct synthesis of methanol from CO and H_2 is a well established industrially applied reaction. Although thermodinamically favoured at low temperatures, it suffers for kinetic limitations, high temperatures and pressures being necessary for this process. In order to develop a more economical technology for the conversion of natural gas to methanol, new synthetic processes, *via* the intermediate production of syn-gas, have been investigated [1].

In fact, methanol may be also obtained from syn-gas by a two step process, consisting of the carbonylation of an alcohol into the corresponding formic ester followed by hydrogenolysis with production of methanol [2]:

$$ROH + CO \rightleftharpoons HCOOR$$
$$HCOOR + H_2 \longrightarrow ROH + MeOH$$

$$CO + 2\,H_2 \longrightarrow MeOH$$

The above reactions are generally performed using MeOH as ROH to avoid transesterification reactions and subsequent separation problems.

With the aim to improve this two steps process, our attention has been focused on the first step, *i.e.* the carbonylation reaction. Indeed, methanol carbonylation to methyl formate is an equilibrium hexothermic reaction industrially carried out in the homogeneous liquid phase using sodium methoxide as catalyst at a temperature of 80°C. About 95 % of CO but only 30 % of methanol are converted under plant conditions, but nearly quantitative conversion of methanol can be achieved by recycling the unreacted alcohol [3].

The most significant drawback of the current industrial process is represented by the progressive deactivation of the methoxide, due to the presence of traces of water and CO_2 in the feed, thus causing its transformation, respectively, into the weaker bases sodium formate and sodium methoxycarbonate, both insoluble in the reaction medium.

Finally, methyl formate represents an interesting product itself and a valuable intermediate for the production of other chemicals such as formic acid, formamide and N,N-dimethylformamide.

Therefore, with the aim to get more active catalysts, in order to work at lower temperatures where the equilibrium is favoured, and simplify the separation steps, the performances of heterogeneous strongly basic resins in the carbonylation of methanol and higher alcohols have been investigated.

2. Results and discussion

2.1 Methanol carbonylation

In a preliminary study we reported that some basic ion exchange resins could be used as catalysts for methanol carbonylation [4]. In this context, the catalytic performances of several basic resins with different characteristics, in terms of crosslinking, morphology and nature of functional groups (Table 1) have been tested. Each resin was carefully activated to assure the complete exchange of the Cl^- anions with the OH^- and subsequently with the CH_3O^- groups.

Table 1 - Characteristics of basic ion exchange resins

Resin	Morphology	Functional group [a)]	Moisture (%)	Declared exchange capacity (meq/g)	Max. operating T (°C)
Amberlyst A26	macroporous	I	61	4.4	60
Amberlyst A27	macroporous	I	45	2.6	60
IRA 900	macroporous	I	59	3.7	60
IRA 400	gel	I	46	4.4	60
IRA 416	gel	II	50	3.8	35

[a)] Functional group: I = $-CH_2N(CH_3)_3^+Cl^-$, II = $-CH_2N(CH_3)_2(CH_2CH_2OH)^+Cl^-$

The experiments were initially carried out batchwise. The catalytic activity of each tested resin was compared with that of the homogeneous systems, by adopting a high alcohol/base ratio (350) for a better discrimination of the activity of the different systems (kinetic control).

As reported in Table 2, macroporous resins displayed a very different behaviour from each other. Indeed, Amberlyst A26 (run 3) and Amberlite IRA 900 (run 5) resulted more active with respect to both the homogeneous systems, the industrial catalyst sodium methoxide (run 1) and the homogeneous counterpart tetrabutylammonium methoxide (run 2).

On the other hand, the macroporous Amberlyst A27 resulted slightly less active than the homogeneous sodium methoxide (compare run 4 with run 1) and after the first catalytic cycle it was recovered as a fine powder deposited on the walls of the mechanically stirred reactor. This behaviour may be addressed to the high porosity of Amberlyst A27, double as compared with Amberlyst A26, and hence to its fragility.

Table 2 - Methanol carbonylation with different basic catalysts [a)]

Run	Catalyst	Conv. (%)	TOF [b)] (h^{-1})
1	MeONa	32	22.4
2	$MeONBu_4$	15	10.5
3	Amberlyst A26	40	28.0
4	Amberlyst A27	29	20.3
5	IRA 900	34	23.8
6	IRA 400	40	28.0
7	IRA 416	19	13.3

[a)] Reaction conditions: MeOH 50 ml (1.25 mol); P_{CO}: 5 MPa; base: 3.57 meq; T: 60 °C; time: 5 h. [b)] TOF = turnover frequency: MF mol/eq cat x h

Finally two gel-type resins, Amberlite IRA 400 and IRA 416 were tested. The first one, functionalized with the quaternary ammonium groups, showed a catalytic activity (run 6) higher than that of the homogeneous systems and similar to that of the most active macroporous matrices. This behaviour seems to suggest that the catalytic activity does not depend on the polymer morphology, all types of polymeric matrices in the presence of methanol as reaction medium reaching a similar swelling degree. On the contrary, the gel resin Amberlite IRA 416, functionalized with ethanolamine groups (run 7), displayed a lower catalytic activity, which may be related to the lower basic strenght of its functional groups.

In all the batch catalytic runs carried out with heterogeneous catalysts only traces of dimethylether (DME) were formed. This by-product derives from the reaction of the anchored methoxide anion with methyl formate according to eq. 1:

$$\text{®-}CH_2N(CH_3)_3^+MeO^- + HCOOMe \longrightarrow \text{®-}CH_2N(CH_3)_3^+HCOO^- + DME \quad \text{(eq. 1)}$$

where ®- is the resin matrix. The above reaction is responsible for catalyst deactivation, the formate co-product being catalytically inactive.

In order to achieve a deeper insight on the stability of the basic resins towards temperature and poisons (mainly CO_2 and H_2O), the methanol carbonylation was also accomplished in experiments where methanol was continuously fed to the reactor and the reaction mixture removed at the same rate, thus maintaining a nearly constant liquid phase volume in the reactor.

As shown in Figure, for the reaction catalyzed by Amberlyst A26, the concentration of methyl formate initially increases because the reaction rate is faster than the feeding rate; however, at longer reaction times, methyl formate concentration decreases due to deactivation effects. The resin activity can be completely restored by its regeneration with caustic washing.

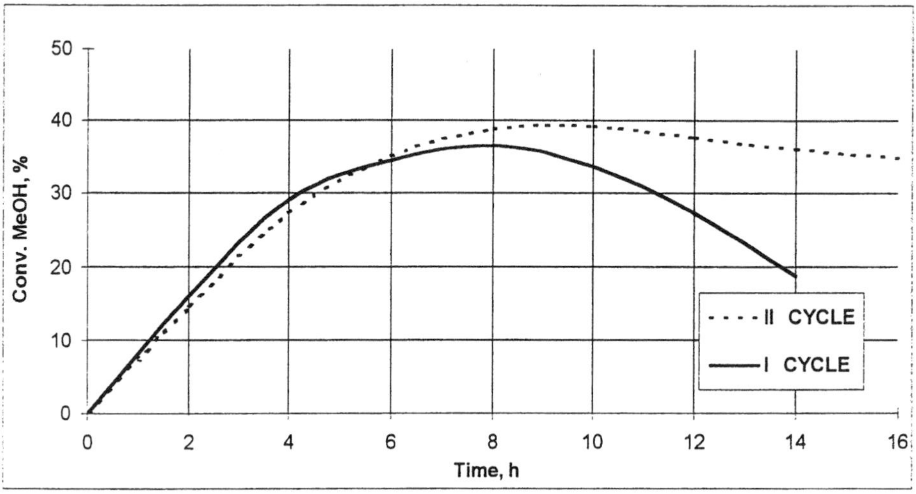

Figure. *R*eversible deactivation in continuous runs. Reactions conditions: LHSV : 2 h^{-1}; P_{CO} : 5 MPa; T : 48 °C;

In order to check irreversible thermal deactivation processes, a few runs were carried out with the same sample of Amberlyst A26, regenerated after each cycle, at 67°C. Also under these adopted conditions, only a completely reversible deactivation was observed although this type of resin, in the OH^- form, is claimed to be unstable at temperatures higher than 60 °C, due to loss of ionic structure, as a consequence of the Hofmann degradation.

2.2 Higher alcohols carbonylation

The activity of basic resins was also tested in the carbonylation of higher alcohols to verify if, analogously to homogeneous sodium alkoxides [5], this reaction would be favoured by increasing the alcohol chain length.

As reported in Table 3, the results for ethanol carbonylation confirm the superiority of the strongly basic resins (Amberlyst A26, IRA 900, IRA 400) with respect to the homogeneous ethoxide, independently of the matrix morphology.

An increase of the rate of ethanol carbonylation was observed with respect to methanol, analogously to what found with homogeneous catalysts.

Table 3 - Ethanol carbonylation with basic resins[a)]

Run	Catalyst	Conv. %		TOF [b)]
		3h	5h	(h^{-1})
8	EtONa	27	56	31.5
9	Amberlyst A26	40	75	46.6
10	Amberlyst A27	18	28	21.0
11	IRA 900	53	82	61.8
12	IRA 400	49	84	57,2
13	IRA 416	13	23	15.2

[a)] Reactions conditions: EtOH : 60 ml (1mol); P_{CO} : 5 MPa; T : 60 °C; catalyst : 2.85 meq. [b)] Turnover after 3h

The reaction rate resulted further enhanched when *n*-hexanol was used as substrate (Table 4), and this alcohol also allowed to discriminate the catalytic behaviour of the resins with different morphology.

Table 4 - *n*-Hexanol carbonylation with basic resins [a)]

Run	Catalyst	Conv. %		TOF [b)]
		3h	5h	(h^{-1})
14	Amberlyst A26	68	85	79.3
15	Amberlyst A27	16	27	18.6
16	IRA 900	47	61	54.8
17	IRA 400	84	89	98.0
18	IRA 416	12	20	14.0

[a)] Reactions conditions: *n*-HexOH : 1 mol; P_{CO} : 5 MPa; T : 60 °C; catalyst : 2.85 meq. [b)] Turnover after 3h

In fact, when *n*-hexanol is the reaction medium, the polymeric matrices, depending on their morphology, undergo to a different extent both the shrinkage and swelling effects, due to the contemporary presence in the alcohol of the lyophobic hydroxy moiety and the lyophilic long alkyl chain. Moreover, in the case of porous resins, shape selectivity may play also a significant role.

The positive role played by *n*-hexanol on the reaction rate was furtherly evidenced by carbonylation experiments carried out on methanol/*n*-hexanol mixtures (Table 5) in the presence of the macroporous Amberlyst A26 resin, at a constant alcohols/base molar ratio. Indeed, when a molar ratio methanol/*n*-hexanol = 0.5 was adopted (run 22) the carbonylation rate of *n*-hexanol resulted substantially the same as that obtained in run 14, where pure *n*-hexanol was used. Moreover methanol conversion was more than doubled with respect to the run 3, carried out on pure methanol, in all cases the same overall alcohol/base ratio being adopted.

Table 5- MeOH/n-Hexanol mixtures carbonylation in the presence of Amberlyst A26 [a)]

Run	Substrate	Molar ratio	Conv. MeOH %		Conv. *n*-HexOH %	
			3h	5h	3 h	5h
3	MeOH	-	20	40	-	-
19	MeOH/*n*-HexOH	2	18	36	22	34
20	MeOH/*n*-HexOH	1	24	54	29	52
21	MeOH/*n*-HexOH	0.67	38	68	41	72
22	MeOH/*n*-HexOH	0.5	63	90	65	88
23	MeOH/*n*-HexOH	0.33	75	90	67	88
14	*n*- HexOH	-	-	-	68	85

[a)] Reaction conditions: T: 60 °C; P_{CO} : 5 MPa; Σ mol alcohols/ eq. base : 350.

3. Conclusions

On the basis of the obtained results, it may be concluded that the use of heterogeneous basic resins appears as a valuable tool for significantly improving the current industrial process of methyl formate production, thus favouring the economical balance of the two steps processes for the synthesis of methanol. Continuous runs have evidenced that resin activity can be completely restored by caustic washing, no loss of ionic structure being observed at high temperature.

References

1. M. Marchionna, M. Lami, A.M. Raspolli Galletti, Chemtech, april 1997, p. 27.
2. J.A. Christiansen, J.C. Gjaldbaek, K. Dan. Vidensk. Selsk., Mat.-Fys. Medd., 20 (1942), 1
3. W. Reutemann, M. Kieczka, Ulmann Encyclop. of Ind. Chem., 5th Ed., Vol. A12 (1989) p. 13.
4. M. Di Girolamo, M. Lami, M. Marchionna, D. Sanfilippo, M. Andreoni, A.M. Raspolli Galletti, G. Sbrana, Catal. Lett. 38 (1996) 127.
5. S.P. Tonner, D.L. Trimm, M.S. Wainwright, N.W. Cant, J. Mol. Catal. 18 (1983) 215.

NATURAL GAS CONVERSION V
Studies in Surface Science and Catalysis, Vol. 119
A. Parmaliana et al. (Editors)

Kinetics of Higher Alcohol Synthesis over low and high temperature catalysts and simulation of a double-bed reactor

L. Majocchi[a], A. Beretta[a], L. Lietti[a], E. Tronconi[a], P. Forzatti[a], E. Micheli[b], L. Tagliabue[b]

[a]Dipartimento di Chimica Industriale e Ingegneria Chimica " G. Natta", Politecnico di Milano, piazza L. da Vinci 32, 20133 Milano, Italy

[b]SNAMPROGETTI SpA via Maritano 26, 20097 S. Donato Milanese (MI), Italy

Simplified kinetic models of higher alcohol synthesis over a Cu-based and a Cu-free catalyst were developed, based on previous data. Low-T kinetics and high -T kinetics were combined to simulate a double bed reactor. It was verified that this configuration can significantly improve isobutanol production with respect to standard single-stage reactors.

1. INTRODUCTION

The catalytic hydrogenation of carbon monoxide to methanol and higher alcohols involves a huge variety of intermediates and products, which differ in chemical nature, carbon atom number and molecular structure. The product distribution and the reaction network are specific of the catalytic system used. CO-insertion steps and formation of linear alcohols prevail over the Fisher-Tropsh catalysts [1]. Aldol-type $C_n + C_1$, or $C_n + C_m$ condensations are active over both low temperature Cu-based catalysts and high temperature zinc chromite-based catalysts; these routes give rise to the formation of both linear and branched alcohols [2-4]. Presently, methanol and isobutanol mixtures produced *via* HAS are believed to be potential reactants for the synthesis of MTBE (methyl-tert-butil ether). Beretta et al. [5] have shown experimentally that the production of isobutanol can be significantly increased by coupling the catalytic properties of a Cs-doped Cu-containing catalyst and a Cs-doped Cu-free Zn/Cr/O catalyst in a double bed configuration. This arrangement exploits the high activity of the Cu-based catalyst in producing short-chain alcohols, by using it as first low-temperature bed wherein the chain-growth process is initiated. The zinc-chromite catalyst, operating as second high temperature bed, terminates HAS by converting the ethanol and propanol enriched feed stream into isobutanol.

In previous works, the authors performed HAS experiments over both a Cs/Cu/Zn/Cr/O catalyst and a Cs/Zn/Cr/O catalyst [6, 7]. Operating conditions were searched which maximized the production of short-chain alcohols and branched alcohols over the former and latter catalyst, respectively. In this work, simplified kinetic models were derived and fitted to the data. Also, low-T kinetics and high -T kinetics were combined to simulate a double stage reactor and the attainable improvement of isobutanol production was quantified in comparison with standard single-stage reactors.

2. KINETICS OF HAS OVER A Cs/Cu/Zn/Cr/O CATALYST

The HAS reacting system over the Cs-doped Cu/Zn/Cr/O catalyst was schematized as a mixture of CO, H_2, CO_2, H_2O, methanol, ethanol, isobutanol, C_{4+}-alcohols, methane, ethane, methyl formate and methyl acetate. The following reactions were assumed to account for their formation:

(r1) $CO + 2\,H_2 \leftrightarrow CH_3OH$ (r2) $CO + H_2O \leftrightarrow CO_2 + H_2$
(r3) $2\,CH_3OH \rightarrow C_2H_5OH + H_2O$ (r4) $CH_3OH + C_2H_5OH \rightarrow C_3H_7OH + H_2O$
(r5) $CH_3OH + C_3H_7OH \rightarrow i\text{-}C_4H_9OH + H_2O$
(r6) $(N_{HA}\text{-}2)\,CH_3OH + C_2H_5OH \rightarrow HA_{C4+} + (N_{HA}\text{-}2)\,H_2O$
(r7) $CO + 3\,H_2 \rightarrow CH_4 + H_2O$ (r8) $C_2H_5OH + H_2 \rightarrow C_2H_6 + H_2O$
(r9) $CO + CH_3OH \leftrightarrow HCOOCH_3$ (r10) $CO + C_2H_5OH \leftrightarrow CH_3COOCH_3$

The chain growth was assumed to proceed *via* C_1-addition steps, with C_1=methanol [4]. Based on the results of chemical enrichment experiments [6], ethanol was treated as intermediate in the formation of C_{4+} oxygenates and ethane. The syntheses of methyl formate and methyl acetate were also included in the model as these species were very abundant in the product mixture over the Cu-based catalyst.

Reversible kinetic expressions were used for methanol synthesis and the water gas shift reaction.

$$r_1 = k_{MeOH}\left(P_{CO}P_{H2}^2 - P_{MeOH} / K_{EQ,MeOH}\right) \qquad r_2 = k_{SHIFT}\left(P_{CO}P_{H2O} - P_{CO2}P_{H2} / K_{EQ,SHIFT}\right)$$

Second order kinetics were adopted for the $C_1 \rightarrow C_2$, $C_2 \rightarrow C_3$ and $C_3 \rightarrow iC_4$ steps of the chain growth process, whereas first order kinetics were defined for C_{4+}-higher alcohol synthesis.

$$r_3 = k_{C1-C2}P^2_{MeOH} / P_{H2} \qquad r_4 = k_{C2-C3}P_{MeOH}P_{EtOH} / P_{H2}$$

$$r_5 = k_{C3-iC4}P_{MeOH}P_{PrOH} / P_{H2} \qquad r_6 = k_{C2-HAC4+}P_{EtOH}$$

First order dependencies on the partial pressure of hydrogen and ethanol were used to explain methane and ethane formation, respectively.

$$r_7 = k_{CH4}P_{H2} \qquad r_8 = k_{C2H6}P_{EtOH}$$

Reversible kinetics expressions were defined for methyl formate and methyl acetate synthesis.

$$r_9 = k_{MF}\left(P_{MeOH}P_{CO} - P_{MF} / K_{EQMF}\right) \qquad r_{10} = k_{MAc}\left(P_{EtOH}P_{CO} - P_{MF} / K_{EQMF}\right)$$

The parameters were estimated by fitting the model to the results of HAS runs previously performed over a Cs/Cu/Zn/Cr/O catalyst [6] and accounting for the effect of temperature, GHSV and H_2/CO feed ratio (catalyst preparation and experimental apparatus have been described in [6]). Figure 1 shows the experimental and calculated effect of reaction temperature on the concentration of methanol, ethanol, propanol and isobutanol. Methanol mole fraction decreased with increasing temperature; its synthesis reaction was governed by thermodynamic equilibrium in the whole T-range investigated. The concentration of ethanol, propanol and isobutanol showed distinct maxima which were shifted towards higher temperature with increasing C-chain length. The production of C_1-C_3 alcohols was really high, whereas the concentration of isobutanol and the other branched alcohols was only minor. While a precise simulation of the experimental bell-shaped trends presented difficulties likely associated to the simplifying character of the kinetic model, however the final fit was especially adequate in the range T<300°C, that is the range of potential industrial interest both for amount of ethanol produced and for catalyst stability.

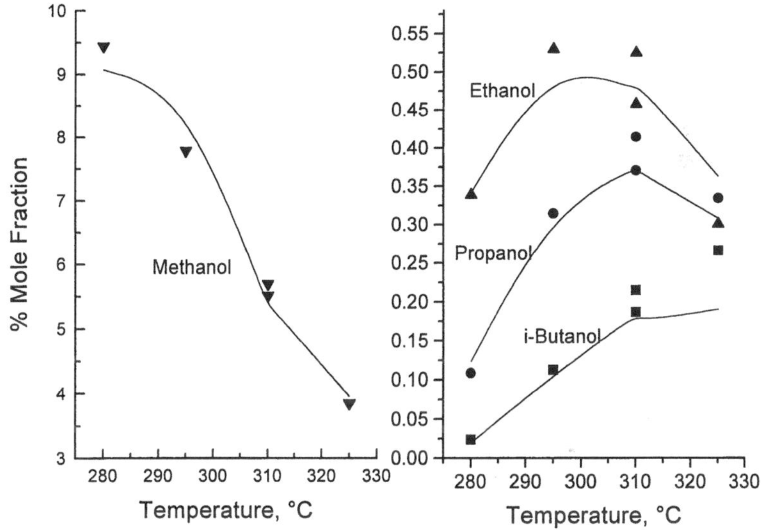

Figure 1 - Experimental (symbols) and calculated (solid lines) effect of reaction temperature on the distribution of methanol, ethanol, propanol and isobutanol. GHSV=10000 Ncc/h/g_{cat}, H_2/CO=0.75, P=7.5 MPa. The following form was used: $k_i=k_i°\exp[-E_{att,i}/R(1/T - 1/568)]$. $k°_{MeOH}$=2.87E-3, $k°_{C1-C2}$=4.25, $k°_{C2-C3}$=33.8, $k°_{C3-iC4}$=20.9, ($E_{att,MeOH}$/R)=10.47, ($E_{att,C1-C2}$/R)=14, ($E_{att,C2-C3}$/R)=10, ($E_{att,C3-iC4}$/R)=12. r_i (moli/g_{cat}/h).

3. KINETICS OF HAS OVER A Cs/Zn/Cr/O CATALYST

In the case of the high-temperature Cs-doped Zn/Cr/O catalyst, the following reactions were introduced to schematize the reaction network:

(r1) $CO + 2\,H_2 \leftrightarrow CH_3OH$ (r2) $CO + H_2O \leftrightarrow CO_2 + H_2$
(r3) $2\,CH_3OH \rightarrow C_2H_5OH + H_2O$ (r4) $CH_3OH + C_2H_5OH \rightarrow C_3H_7OH + H_2O$
(r5) $CH_3OH + C_3H_7OH \rightarrow i\text{-}C_4H_9OH + H_2O$

(r6) $(N_{HA}\text{-}3)\ CH_3OH + C_3H_7OH \rightarrow HA_{C4+} + (N_{HA}\text{-}3)\ H_2O$
(r7) $CO + 3\ H_2 \rightarrow CH_4 + H_2O$ (r8) $N_{HYD}CO + 2N_{HYD}\ H_2 \rightarrow HYD_{C2+} + N_{HYD}H_2O$

Contrary to the low-T kinetics for the Cu-based catalyst, the formation of C_{4+} higher oxygenates was attributed to the evolution of propanol, which is the most abundant short-chain intermediate over the high-T catalyst [8]. Also, it was observed that the rate of production of C_{2+} hydrocarbons was independent of the concentration of oxygenates and the general stoichiometry (r8) was thus introduced.

In analogy with the low-T kinetic model, reversible kinetic expressions were adopted for methanol synthesis and the water gas shift reaction.

$$r_1 = k_{MeOH}\left(P_{CO}P_{H2}^2 - P_{MeOH} / K_{EQ,MeOH}\right) \qquad r_2 = k_{SHIFT}\left(P_{CO}P_{H2O} - P_{CO2}P_{H2} / K_{EQ,SHIFT}\right)$$

First order kinetic expressions were then defined for the single steps of the chain growth process. They were found more adequate to describe the experimental product distribution than second order kinetics as those used for modeling HAS over the Cu-based catalyst.

$$r_3 = k_{C1-C2}\frac{P_{MeOH} / P_{H2}^{0.5}}{\left(1 + K_{H2O}P_{H2O} + K_{HA}P_{HA}\right)} \qquad r_4 = k_{C2-C3}\frac{P_{EtOH}}{\left(1 + K_{HA}P_{HA}\right)}$$

$$r_5 = k_{C3-iC4}\frac{P_{PrOH}}{\left(1 + K_{HA}P_{HA}\right)} \qquad r_6 = k_{C3-HA}\frac{P_{PrOH}}{\left(1 + K_{HA}P_{HA}\right)}$$

Finally, kinetics for the formation of methane and hydrocarbons were assumed to be dependent on the single partial pressure of hydrogen, in line with the results of a previous mechanistic model of HAS over the Cs/Zn/Cr/O catalyst [5].

$$r_8 = k_{CH4}P_{H2} \qquad r_9 = k_{HYD}P_{H2}$$

The kinetic parameters were fitted to 15 HAS runs over a Cs/Zn/Cr/O catalyst. Catalyst preparation and description of the testing apparatus have been reported elsewhere [7]. Figure 2 reports the experimental and calculated effect of reaction temperature on the distribution of methanol, ethanol, propanol and isobutanol. The formation of higher alcohols over the zinc-chromate catalyst was observed at temperatures higher than 350°C; in particular, the production of isobutanol was appreciable over 380°C. By comparison with the data and the calculations reported in Figure 1, it can be observed that over the high-T catalyst the outlet concentration of short chain alcohols was always much lower than over the Cu-based catalyst, while the production of isobutanol and other branched oxygenates was significantly higher. Also, the formation of isobutanol became more and more selective, with increasing temperature. However, too high temperatures have no practical interest due to an undesired consistent formation of hydrocarbons.

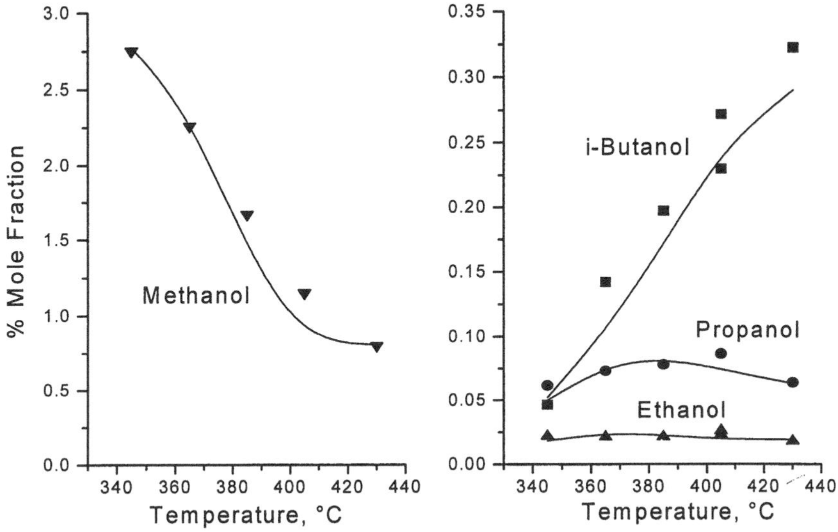

Figure 2 - Experimental (symbols) and calculated (solid lines) effect of reaction temperature on the distribution of methanol, ethanol, propanol and isobutanol. GHSV=5000 $Ncc/h/g_{cat}$, $H_2/CO=1.0$, P=8.0 MPa. The following form was used: $k_i=k_i°exp[-E_{att,i}/R(1/T - 1/678)]$. $k°_{MeOH}=2.7E-4$, $k°_{C1-C2}=3.0E-2$, $k°_{C2-C3}=0.18$, $k°_{C3-iC4}=1.8E-2$, $(E_{att,MeOH}/R)=9.7$, $(E_{att, C1-C2}/R)=14.5$, $(E_{att, C2-C3}/R)=5.8$, $(E_{att, C3-iC4}/R)=4.1$.

4. SIMULATION OF A DOUBLE BED REACTOR

The kinetic models derived for the single Cu-based and Cu-free catalysts were combined for the simulation of a double bed reactor, consisting of two reactors in series: the first one operating with the Cs/Cu/Zn/Cr/O catalyst in the temperature window which was found above to be optimal for the production of intermediate species; the second one operating with the Cs/Zn/Cr/O catalyst at the higher temperature which were found to secure a high rate in the synthesis of isobutanol from ethanol and propanol. The single reactors were assumed as isothermal with plug flow; the feed stream entering the second reactor was assumed to have the composition of the product mixture from the first low-T reactor. A partial separation of methanol downstream from the first reactor was also simulated, given its very high production over the Cu-based catalyst.

Figure 3 shows the simulation of a double-bed reactor wherein a value of GHSV=10000 $Nl/kg_{cat}/h$ was assumed for each bed. It shows a progressive formation of methanol, ethanol and propanol along the first bed operating at 295°C. Once entered the second bed, operating at 420°C, methanol partly decomposes and the (lower) equilibrium concentration corresponding to the higher temperature is reached. Thus methanol partial split upstream from the high-T bed serves the purpose of minimizing the amount of C_1-alcohol decomposed. In the second reactor, ethanol is also quickly consumed to propanol which in turn promotes the formation of isobutanol. The final product mixture obtained from the double-bed reactor contains almost twice the amount of iso-butanol that would be guaranteed by the single zinc chromite catalyst at a total GHSV = 5000 $Nl/kg_{cat}/h$

5. CONCLUSIONS

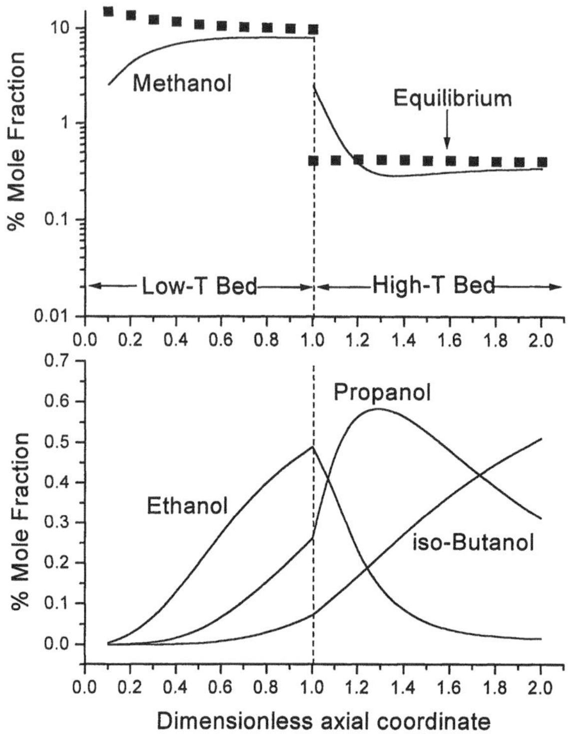

Figure 3 - Simulation of a double bed reactor with an inlet H_2/CO feed ratio of 0.75.

The present work gives a quantitative interpretation of the experimental data reported by Beretta et al [5]. Kinetic models were developed which reproduce the ability of Cu-based catalyst in producing short-chain alcohols and the high activity of the zinc-chromite catalyst towards isobutanol. It was shown that a synergetic effect can be obtained by using the two catalyst in series. Simulations, valuable for process design applications, confirmed the potential advantage offered by the double-bed configuration for improving isobutanol productivity.

REFERENCES

1. J.P. Hindermann, G.J. Hutchins, A. Kiennemann, Catal. Rev. Sci. Eng., 35 (1993) 1.
2. C.R. Apesteguia, B.De Rites, S. Miseo, S. Soled, Catal. Lett., 44 (1997) 1.
3. P. Forzatti, E. Tronconi, I. Pasquon, Catal. Rev. Sci. Eng., 33 (1991) 109.
4. J.G. Nunan, C. E. Bogdan, K. Klier, K.J. Smith, C.W. Young, R.G. Herman, J. Catal., 113 (1988) 410.
5. A. Beretta, Q. Sun, R.G. Herman, K. Klier, Ind. Eng. Chem. Res., 35 (1996) 1534.
6. L. Majocchi, L. Lietti, A. Beretta, P. Forzatti, E. Micheli, L. Tagliabue, Appl. Catal. A: General, 166 (1998) 393-405.
7. E. Tronconi, L. Lietti, G.Groppi, P. Forzatti, P. Pasquon, J. Catal. 135 (1992) 99.
8. A. Beretta, E. Tronconi, P. Forzatti, I. Pasquon, E. Micheli, L. Tagliabue, G.B. Antonelli, Ind. Chem. Eng. Res., 35 (1996) 2144.

NATURAL GAS CONVERSION V
Studies in Surface Science and Catalysis, Vol. 119
A. Parmaliana et al. (Editors)

Developing highly active iridium catalysts for methanol synthesis

S. Marengo[a], R. De Castro[a], R. Psaro[b], C. Dossi[b], R. Della Pergola[b], L. Sordelli[b] and L. Stievano[c]

[a]Stazione Sperimentale per i Combustibili, 20097 S. Donato Milanese, Italy

[b]CNR Center "CSSCMTBSO" and Dipartimento di Chimica Inorganica, Metallorganica e Analitica, 20133 Milano, Italy

[c]Dipartimento di Chimica Fisica, 30123 Venezia, Italy

1. INTRODUCTION

The increasing demand for methanol from both the chemical industry and the fuel market, continues to stimulate research efforts aimed at developing innovative processes for the efficient conversion of syngas [1,2].

It has been proposed that uniformly dispersed alloys having a predetermined composition can be obtained by controlled thermal decomposition of supported mixed metal carbonyl clusters.

We present here the most significant results of an investigation on iron-promoted iridium catalysts for methanol synthesis. Mixed-metal clusters of the type $[Et_4N]_2[Fe_2Ir_4(CO)_{16}]$ or $[Et_4N]_2[Fe_2Ir_2(CO)_{12}]$ were utilised as precursors of the supported metal species in the catalysts; bimetallic systems prepared by sequential impregnation of carbonyl clusters of iridium and iron ($Ir_6(CO)_{16}$, $Fe_3(CO)_{12}$) were also studied with the aim of comparing different precursors and of evaluating catalysts with Fe/Ir ratio above 1.

2. EXPERIMENTAL

2.1. Catalyst preparation

$[Et_4N][FeIr_3(CO)_{12}]$, $[Et_4N]_2[Fe_2Ir_4(CO)_{16}]$, $[Et_4N]_2[Fe_2Ir_2(CO)_{12}]$ were prepared according to known procedures [3]. The support (MgO or SiO_2, both with surface area of 200 m^2/g) was evacuated at 10^{-2} mbar and impregnated under N_2 at 298 K for 1 h with acetone solution of the bimetallic cluster; after drying, the materials were stored in N_2. The samples prepared from mixed-metal clusters are designated in the text as $FeIr_3$, Fe_2Ir_4 and Fe_2Ir_2.

The catalysts prepared by sequential impregnation of $Ir_6(CO)_{16}$ and $Fe_3(CO)_{12}$ are designated as FeIr in the text, followed by the value of the Fe/Ir ratio (1:1 or

2:1). For a comparison, bimetallic samples obtained by impregnation of aqueous solutions of the inorganic salts $IrCl_3.3H_2O$ and $Fe(NO_3)_3.9H_2O$ were also tested.

Samples were prepared at two different total metal (Fe+Ir) loadings of 1 and 4 wt %; metal contents were then determined by AAS (Fe) and ICP-OES(Ir) upon chemical dissolution.

2.2. Apparatus and methods for catalytic measurements

The CO hydrogenation was studied in an automated microreactor, which allows remote control and programming of the process parameters [4].

The freshly adsorbed cluster (0.4-0.6 g) was transferred into a stainless steel tubular reactor under inert atmosphere, and reduced in flowing H_2 at 623 K for 2 h prior to catalytic experiments.

2.3. Catalyst characterization

H_2 chemisorption was determined by a static volumetric method with an automated apparatus. The sample was reduced in flowing hydrogen at 623 K for 2 h and evacuated at 633 K for 2 h.

EXAFS measurements were carried out at the DCI storage ring at LURE (Orsay, France). An in-situ cell that allowed heating from room temperature to 873 K under controlled atmosphere has been used.

The radioactive ^{193}Os feeding the ^{193}Ir Mössbauer transition was prepared by neutron activation of isotopically enriched ^{192}Os metal. The Mössbauer measurements with ^{193}Ir were carried out at 4.2K in a liquid helium bath cryostat, while ^{57}Fe Mössbauer measurements, at temperatures from 80K to room temperature, were carried out in a gas-flow cryostat. Due to the different sample thickness required by the two Mössbauer isotopes, the samples could not be measured in an in-situ cell. In order to prevent decomposition and to simulate in-situ conditions, the samples, kept under controlled atmosphere, were quenched in liquid nitrogen before handling and subsequently handled inside of a liquid nitrogen bath.

3. RESULTS AND DISCUSSION

3.1. Perfomance of the cluster-derived catalysts under reaction conditions

The supported clusters were inactive when exposed to a CO-H_2 mixture up to 523 K, so that a proper activation procedure had to be adopted. Direct reduction in H_2 at 623 K proved more efficient than oxidation at either 423 or 623 K, followed by reduction. In the case of salt-derived samples, the two procedures gave comparable results.

The cluster-derived catalysts exhibited the highest activity right after pretreatment in H_2. A partial deactivation was observed during the first 10 h on stream, mainly due to surface fouling, as indicated by the fact that initial activity could be restored by treatment in H_2 at 623 K.

An opposite behavior was exhibited by the salt-derived FeIr/MgO samples: the initial activity after activation was very low and rose remarkably during the initial 100 h. Similar results are reported in the literature for silica supported Fe-Ir catalysts [5].

3.2. Effect of composition on the catalytic properties

The silica-supported catalysts showed quite lower activity and selectivity compared to their MgO-supported homologous (Table 1). This difference is more evident with the Fe_2Ir_2 precursor. In the class of MgO-supported clusters, an increase in the Fe/Ir ratio enhances remarkably the methanol productivity. Upon raising metal loading from 1 to 4 wt %, high specific activity is maintained.

The sample obtained by sequential impregnation of monometallic clusters (FeIr 2:1) exhibited the highest selectivity. The catalyst prepared from salt precursors showed relatively good selectivity, but quite lower activity compared to the corresponding cluster-derived samples.

Table 1
Catalyst functionality under steady-state conditions[a]

Catalyst (Fe+Ir=1wt%).	Rate[b]	Selectivity (% C efficiency)	
	CO	MeOH	HC
$FeIr_3/SiO_2$	5.4	47.5	44.2
Fe_2Ir_2/SiO_2	4.3	56.5	41.2
$FeIr_3/MgO$	6	76.9	17.6
Fe_2Ir_4/MgO	11.3	79.6	20.4
Fe_2Ir_2/MgO	24.8	62.8	32.2
Fe_2Ir_2/MgO [c]	25.8	63.8	29.4
FeIr/MgO 2:1 [c]	19.7	95.8	2.0
FeIr/MgO 1:1 [c] salt-derived	9.9	85.1	11.3

[a] T=523 K, P=1.1 MPa, GHSV=1200 h^{-1}, $H_2/CO=2/1$

[b] Rate=$mol_{(converted\ CO)}/mol_{Ir}.h$; [c] Fe+Ir=4%

3.3. Optimization of reaction conditions

In order to optimize reaction conditions, the rate dependences on CO and H_2 partial pressures were estimated by fitting a power-law expression of the form:

$$rate = ae^{-Ei/RT}\, P_{H_2}{}^{x}\, P_{CO}{}^{y}$$

The results show that the presence of iron lowers the hydrogen dependence from 1.6 to 0.5; this suggests an easier activation on the catalyst surface, as confirmed by the values of the CO dependence, which turn from negative (-0.7) to slightly positive for methanol, indicating the lifting of inhibition effects. These data suggest that high partial pressure of H_2 and CO favors activity and selectivity to methanol in the iron-promoted samples.

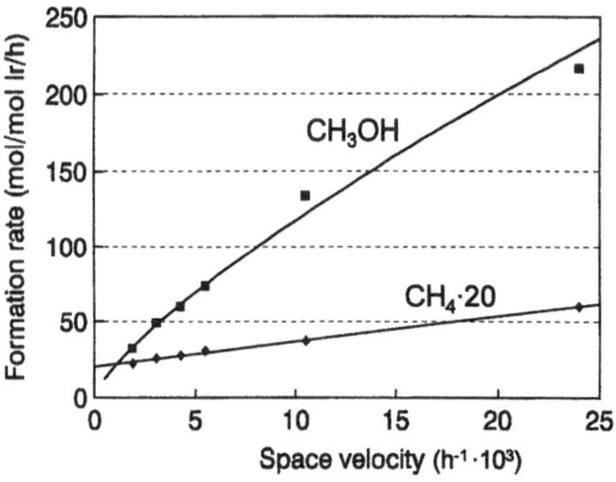

Figure 1. Effect of space velocity on product formation rate with FeIr/MgO 1:1. T = 563 K, P = 6.1 MPa

Experiments of parameter programmed reaction in which space velocity was automatically changed stepwise while monitoring the effluent composition [4], showed that contact time plays a key role in the catalyst performance (Figure 1). In fact, methanol productivity exhibits a five-fold increase on raising GHSV from 2000 to 20000 h^{-1}. However, values of space velocity above 25000 could not be achieved due to excessive pressure drop in the catalyst bed.

Under optimized reaction conditions (T = 563 K; P = 6.1 MPa; GHSV = 24000 h^{-1}; H_2/CO = 4/1), 98 % selectivity to methanol was obtained, with specific activity much higher than that obtained under the "standard" conditions of Table 1. The methanol productivity was remarkable for all of the cluster-derived samples, as shown in Table 2; the catalysts prepared by sequential impregnation of monometallic complexes (FeIr), though, exhibited better properties than the sample derived from mixed-metal clusters (Fe_2Ir_2).

Table 2
Performance under optimized conditions[a]

Catalyst	CO conv.	Selectivity [%]		Rate [$mol_{CO\ conv}$ /mol Ir/h]		
(Fe+Ir = 4%)	[%]	OXY	MeOH	MeOH	CH_4	EtOH
Fe_2Ir_2	5.77	94	92	197	11.2	1.7
FeIr 1:1	4.88	99	98	217	3.0	0.8
FeIr 2:1 [b]	4.25	99	98	213	2.0	0.8

[a] T = 563 K; P = 6.1 MPa; GHSV = 24000 h^{-1}; H_2/CO = 4/1
[b] T = 543 K

A comparison with the best results reported in the literature shows that the catalysts of this study (prepared from either bimetallic or monometallic clusters) reveal much higher activity and selectivity than SiO_2 -supported materials with similar composition; also for other classes of methanol catalysts, the reported specific activities are far lower (Table 3). The commercial copper-based catalyst gives the highest methanol yield per liter or Kg of catalyst, but the activity per mole of active metal is over one order of magnitude lower compared to FeIr/MgO.

Table 3
Activity of the cluster-derived FeIr/MgO catalyst compared with the best results reported for various methanol catalysts

Catalyst	Conditions		MeOH selectivity	mol MeOH/ mol_{Me} h	kgMeOH/ kg_{cat} h	Ref.
	P (bar)	T(K)				
FeIr/MgO [a]	61	563	98	217	1.1	this study
$FeIr/SiO_2$	40	542	84	20.8	0.14	[5]
$Cu/ZnO/Al_2O_3$	50	523	99+	6.1	1.2 [b]	[6]
promoted Pd	20	553	98.5	121	n.a.	[7]
Mo sulfide	55	600	83 [c]	n.a.	0.22 [c]	[8]

[a] Fe+Ir 4 %, Fe/Ir 1:1; prepared by impregnation of $Fe_3(CO)_{12}$ and $Ir_6(CO)_{16}$
[b] (kg/lcat h); [c] C_1-C_4 oxygenates

3.4. Investigation on the surface structure of the active catalysts

Extensive physicochemical characterization was performed in order to clarify the nature of the surface sites responsible for the remarkable catalytic activity.

Chemisorption measurements showed that iron enhances the dispersion of iridium; the most active catalysts with 4 wt% metal loading, exhibited H/Ir ratios ranging from 0.6 for Fe_2Ir_2 to 1 for FeIr 1:1.

In the first step of the structural characterization, we focused on the freshly reduced Fe_2Ir_2/MgO. HRTEM confirmed the high metal dispersion; in fact 70 % of the metal particles had diameter below 20 Å. *In-situ* EXAFS measurements on the Ir edge give coordination numbers (CNs) of 7.0 for Ir-Ir and 1.4 for Ir-Fe indicating the presence of highly dispersed bimetallic particles. However, the difference between the CNs suggests that only a small fraction of the iridium atoms is interacting with the iron. Data on the Fe edge reveal that the major part is present on the surface as oxidic phases. These findings, together with the absence of Ir-O contributions, are consistent with the model of small iridium particles stabilized on the surface of MgO by interfacial iron.

The ^{193}Ir Mössbauer spectrum of the *in-situ* reduced Fe_2Ir_2/MgO consists of a broad single line with a full width at half maximum of about 2.7(1) mm/s. The best fit yields a mean isomer shift (IS) of 0.17(2) mm/s, which indicates that the metallic iridium is alloyed with the iron. The alloy has the fcc structure of the iridium γ-phase and an approximate average composition of about 40% of iron [9]. The broad linewidth could have different origins: it may be due to an inhomogeneous composition of the alloy forming the crystallites, or it may as well be caused by surface effects or by direct interaction of the crystallites with the support.

The ^{57}Fe Mössbauer spectra of the reduced Fe_2Ir_2/MgO consist of two quadrupole doublets, attributable to Fe(II) in a (Fe^{2+}, Mg^{2+})O phase [10] and to Fe(III) most probably forming oxidic phases, and of a single broad line with a relative resonance area of about 50% attributable to the iron-iridium γ-phase. The unknown Lamb-Mössbauer factors of the different components allow only a rough estimate of the amount of iron in the different phases from the relative resonance areas.

A comparison of the Mössbauer and of the EXAFS results suggests that about half of the iron is oxidized and it is most probably present at the surface of the support, where it is stabilizing the iridium crystallites. The remaining iron is alloyed in the iridium crystallites. As long as the EXAFS measurements point to the presence of pure iridium metallic particles, it is likely that the iron is present only at the surface of the metallic iridium particles, forming an interfacial alloy.

The low activity and selectivity observed for silica-supported catalysts is probably due to the quick fragmentation on silica surface of the bimetallic clusters, before the bimetallic particles are formed. The role of the basic MgO support is to prevent the disruption of the Fe-Ir metal bonds via oxidative addition of surface hydroxyls with segregation of iron, during the hydrogen pretreatment.

Work is in progress to elucidate the surface features responsible for the better properties of the FeIr/MgO catalysts obtained by sequential impregnation of monometallic clusters, compared to the samples derived from bimetallic clusters. The higher density of active sites revealed by hydrogen chemisorption could be one of the factors determining this performance.

REFERENCES

1. M. Marchionna, M. Lami, A. M. Raspolli Galletti, CHEMTECH, (1997) 27.
2. I. Puskas, CHEMTECH, (1995) 43.
3. R. Della Pergola, A. Ceriotti, L. Garlaschelli, F. Demartin, M. Manassero, N. Masciocchi and M. Sansoni, Inorg. Chem., 32 (1993) 3277; b) R. Della Pergola, L. Garlaschelli, F. Demartin, M. Manassero, N Masciocchi and M. Sansoni, J. Chem. Soc. Dalton Trans., (1990) 127; c) R. Della Pergola, L. Garlaschelli, S. Martinengo, F. Demartin, M. Manassero and N. Masciocchi, J. Chem. Soc. Dalton Trans., (1988) 2037.
4. S. Marengo, S. Martinengo and L. Zanderighi, Chem. Eng. Sci., 47 (1992) 2793.
5. C. Koningsberger, C. P. J. H. Borgmans, A. M. J. van Elderen, B. J. Kip and J. W. Niemantsverdriet, J. Chem. Soc., Chem. Commun., (1987) 892.
6. C. Chinchen, P. J. Denny, J. R. Jennings, M. S. Spencer and K. C. Waugh, Appl. Catal., 36 (1988) 1.
7. S. Hahm and W. Y. Lee, Appl. Catal., 65 (1990) 1.
8. Z. Liu, X. Li, M. R. Close, E. L. Kugler, J. L. Petersen, D. B. Dadyburjor, Ind. Eng. Chem. Res., 36 (1997) 3085.
9. R. L. Mössbauer, M. Lengsfeld, W. von Lieres, W. Potzel, P. Teschner, F .E. Wagner and G. Kaindl, Z. Naturforsch, 26a (1971) 343.
10. M. Boudart, A. Debouille, J. A. Dumesic, S. Khammouma and H. Topsoe, J. Catal, 37 (1975) 486.

NATURAL GAS CONVERSION V
Studies in Surface Science and Catalysis, Vol. 119
A. Parmaliana et al. (Editors)

Isobutanol and Methanol Synthesis on Copper Supported on Alkali-Modified MgO and ZnO Supports

Marcelo J.L. Gines, Hyung-Seuk Oh, Mingting Xu, Anne-Mette Hilmen, and Enrique Iglesia

Dept. of Chem. Eng., University of California at Berkeley, Berkeley, CA 94710, USA

K-$Cu_yMg_5CeO_x$ and Cs-Cu/ZnO/Al_2O_3 catalyze isobutanol and methanol synthesis from H_2-CO mixtures, but synthesis rates are inhibited by CO_2 and H_2O formed during reaction. The initial C-C bond in ethanol forms directly from CO on K-$Cu_{0.5}Mg_5CeO_x$ catalysts, but via CH_3OH coupling reactions on Cs-Cu/ZnO/Al_2O_3 catalysts. Reactions of pure ethanol on these catalysts show that chain growth occurs via bifunctional aldol-type condensation pathways requiring Cu metal crystallites and basic sites on alkali-modified MgO and ZnO. On both types of catalysts, cross-coupling reactions of aldehydes and ^{13}C-labeled methanol suggest that chain growth occurs by condensation of C_2^+ alcohols or aldehydes with a reactive intermediate derived from methanol.

1. INTRODUCTION

Isobutanol synthesis provides a potential route to methyl-tert-butyl-ether (MTBE) via isobutanol dehydration to isobutene and reactions of isobutene with methanol. Li-Pd/ZrO_2/MnO_2/ZnO catalysts produce significant isobutanol yields, but only at very high temperatures (750-800 K) and pressures (10-30 MPa) [1]. Alkali-promoted Cu/MgO/CeO_x and Cu/ZnO/Al_2O_3 catalyze the higher alcohol synthesis at lower temperatures (~573 K) and pressures (~5 MPa), but isobutanol yields are significantly lower [2-4]. Our work includes CO_2 addition studies, isotopic tracer probes of the initial C-C bond formation pathways, and kinetic and isotopic analyses of alcohol dehydrogenation and condensation steps leading to chain lengthening of C_2^+ alcohols. In this study, we focus on the comparison of the behavior of copper supported on alkali-modified MgO and ZnO supports on the synthesis of branched alcohols from H_2-CO mixtures.

2. EXPERIMENTAL

K-CuMgCeO_x and Cs-Cu/ZnO/Al_2O_3 samples were prepared by pH-controlled coprecipitation and incipient wetness impregnation methods [3-5]. Copper surface areas were measured by N_2O decomposition on pre-reduced samples. The density and strength of basic sites were determined using a new method based on the isotopic exchange of $^{13}CO_2/^{12}CO_2$ mixtures [4,6]. Characterization data are shown in Table 1 for all catalysts.

High-pressure alcohol synthesis from H_2-CO mixtures (4.5 MPa, 583 K, CO/H_2=1) was performed in a packed-bed reactor. Isotopic tracer studies of alcohol synthesis pathways were carried out on K-$Cu_yMg_5CeO_x$ and Cs-Cu/ZnO/Al_2O_3 catalysts using $^{13}CO/H_2/^{12}CH_3OH$

reactant mixtures. Ethanol reactions were carried out in a gradientless recirculating batch reactor at 573 K and atmospheric pressure. Reduced catalyst samples were exposed to the reactant mixture $C_2H_5OH/CH_4/He$ {4.0/2.7/94.6 kPa}. Cross-coupling reactions of $^{13}CH_3OH$ (^{13}C: 99 %, ^{18}O: < 1 %) with acetaldehyde and propionaldehyde were carried out using similar procedures and $^{13}CH_3OH/^{12}C_2H_4O(^{12}C_3H_6O)/CH_4/He$ {7.3/4.0(4.0)/2.7/87.3 kPa} reactant mixtures. Mass spectrometric analysis after chromatographic separation was used to determine ^{13}C content, position, and distribution in reactants and products.

Table 1
Composition, Surface Area, and Basic Site Density of Mixed Metal Oxides

Sample	Cu [wt %]	K (Cs)[a] [wt %]	BET area [m^2 / g]	Cu[b] Dispersion	Exchangeable CO_2 (at 573 K) [μmol / m^2]
$Cu_{0.5}Mg_5CeO_x$	7.7	1.0	147	0.14	2.3
$Cu_{7.5}Mg_5CeO_x$	49	1.2	92	0.05	3.3
$Cu/ZnO/Al_2O_3$	44	(2.9)	62	0.05	1.1

[a] Bulk composition measured by atomic absorption. Values in parenthesis are for catalysts promoted with Cs instead of K. [b] Cu dispersion was determined by N_2O decomposition at 363 K.

3. RESULTS AND DISCUSSION

3.1. Isobutanol synthesis from CO/H_2 mixtures

Methanol, isobutanol, and CO_2 are the most abundant alcohol products of H_2-CO mixtures on K-$Cu_yMg_5CeO_x$ and Cs-$Cu/ZnO/Al_2O_3$ catalysts. Ethanol, 1-propanol, 1-butanol, methyl formate, methyl acetate dimethyl ether, and hydrocarbons were also formed [4,7]. Oxygen rejection during higher alcohol synthesis occurred predominantly by CO_2 formation [8]. The inhibition effects of CO_2 were examined by adding CO_2 to H_2-CO on K-$Cu_yMg_5CeO_x$ and Cs-$Cu/ZnO/Al_2O_3$. The results are shown in Table 2.

Table 2
Effect of CO_2 on Methanol and Isobutanol Synthesis

Catalyst	Methanol Productivities [g/kg-cat.h]		Isobutanol Productivities [g/kg-cat.h]	
	No CO_2 added	CO_2 added	No CO_2 added	CO_2 added
K-$Cu_{0.5}Mg_5CeO_x$	81.1 (0.03)	45.1 (0.10)	4.0 (0.03)	1.2 (0.10)
K-$Cu_{7.5}Mg_5CeO_x$	129.4 (0.06)	133.3 (0.11)	6.4 (0.06)	3.7 (0.11)
Cs-$Cu/ZnO/Al_2O_3$	164.8 (0.06)	158.5 (0.14)	14.9 (0.06)	10.6 (0.14)

[583 K, 4.5 MPa, $H_2/CO=1$, 3000 $cm^3/g/h$]. Values in parenthesis are the average CO_2 pressure in MPa.

Methanol and isobutanol synthesis steps are inhibited by CO_2 (Table 2). On catalysts with high density of Cu surface atoms (K-$Cu_{7.5}Mg_5CeO_x$ and Cs-$Cu/ZnO/Al_2O_3$), methanol synthesis rates decreased only slightly when CO_2 was added. In contrast, on catalysts with lower density of Cu surface sites (K-$Cu_{0.5}Mg_5CeO_x$), methanol synthesis rates were strongly inhibited by CO_2. This reflects the closer approach to thermodynamic equilibrium on catalysts with higher Cu surface densities, which render them less sensitive to a loss of active sites by CO_2 inhibition. These catalysts maintain equilibrium methanol concentration even after a large fraction of the Cu surface is covered by oxygen adatoms during steady-state alcohol synthesis.

As a result, detectable CO_2 inhibition effects require higher CO_2 concentrations on catalysts with high Cu loading. CO_2 inhibits isobutanol synthesis steps on catalyst with low Cu atom surface densities more strongly than on those with higher Cu site densities. This reflects the reversible titration of Cu and basic sites, both of which are required for isobutanol synthesis. Aldol-type coupling reactions of alcohols require both Cu and basic sites [9]; thus, blockage of Cu surface atoms by adsorbed oxygen can decrease isobutanol synthesis rates, even when the rate of quasi-equilibrated methanol synthesis steps is unaffected by this reversible titration of active Cu surface atoms. On catalysts with low Cu content, chain growth reactions are limited by the availability of Cu sites, required for hydrogen removal of H-atoms formed in C-H activation steps on basic sites; consequently, reversible oxidation of surface Cu atoms by CO_2 decreases chain growth rates.

3.2. Isotopic tracer studies of alcohol synthesis pathways

The synthesis of higher alcohols from CO/H_2 mixtures on alkali-modified methanol synthesis catalysts has been widely studied, but the pathways required for the formation of initial carbon-carbon bonds remain controversial. Several mechanisms for the formation of the initial ethanol C-C bond on (Cs)-Cu/Zn/(Al)O catalysts have been proposed [10,11].

Isotopic tracer studies of chain growth pathways on 1.0 wt % $K\text{-}Cu_{0.5}Mg_5CeO_x$ catalysts using $^{13}CO/H_2/^{12}CH_3OH$ have shown that the ^{13}C content in ethanol (extrapolated to zero residence time) is 94% (Figure 1a); reflecting the predominant formation of ethanol from ^{13}CO, without significant contribution from $^{12}CH_3OH$ [12]. Also, at longer residence times, ethanol forms via reverse aldol reactions of higher alcohols, which contain lower ^{13}C contents because $^{12}CH_3OH$ participates in their formation. This reaction leads to the observed decrease in the ^{13}C content of ethanol as residence time increases [12]. The ^{13}C content in 1-propanol calculated by assuming it forms only via methanol aldol-coupling with ethanol (dashed line in Figure 1a) is lower than experimental values. This shows that some of the 1-propanol molecules are produced by carbonylation of C_2H_5OH with ^{13}CO. This is consistent with slow linear chain growth pathways that also lead to the formation of 1-butanol during CO hydrogenation on $K\text{-}Cu_{0.5}Mg_5CeO_x$ [7].

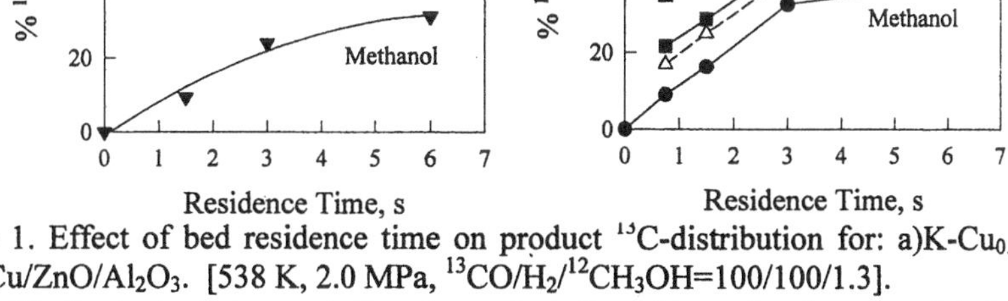

Figure 1. Effect of bed residence time on product ^{13}C-distribution for: a)$K\text{-}Cu_{0.5}Mg_5CeO_x$; b)$Cs\text{-}Cu/ZnO/Al_2O_3$. [538 K, 2.0 MPa, $^{13}CO/H_2/^{12}CH_3OH$=100/100/1.3].

Similar isotopic tracer studies shown that ethanol is predominantly unlabeled (14% ^{13}C content) at short bed residence times on $Cs\text{-}Cu/ZnO/Al_2O_3$ catalysts (Figure 1b), suggesting that it forms by direct coupling of methanol-derived C_1 species, as reported previously by others [11]. The small amounts of labeled ethanol observed even after extrapolation to zero residence time could reflect minority pathways involving ^{13}CO and reaction intermediates similar to those detected on K-$Cu_{0.5}Mg_5CeO_x$ catalysts. This is significantly different from the predominant pathway involving the direct reactions of CO on K-$Cu_{0.5}Mg_5CeO_x$.

This feature might be attributed to different properties of alkali promoters. Cs^+-O^{2-} cation-anion pair is able to break the C-H bond of formaldehyde to form surface $Cs^+(CHO^-)$ species [11], which then react with formaldehyde species to form the initial C-C bond. K^+-O^{2-} species, however, may not cleave C-H bonds in formaldehyde and may be unable to stabilize CHO- intermediates because K^+ is less acidic than Cs^+ and the oxygen anion is less basic than in Cs_2O. The other possibility is that the larger Cu surface area of Cs-$CuZnAlO_x$ compared to K-$Cu_{0.5}Mg_5CeO_x$ enhances methanol dehydrogenation rate and therefore increases the local concentration of very reactive formaldehyde intermediates, which may favor direct coupling of methanol-derived C_1 species.

3.3. Condensation reactions of alcohols

The role of Cu and basic sites on dehydrogenation and condensation reactions was probed by measuring ethanol reaction rates on K-$Cu_yMg_5CeO_x$ and Cs-Cu/ZnO/Al_2O_3 catalysts. Initial rates of dehydrogenation and condensation reactions are given in Table 3.

Table 3
Effects of Cu- and basic sites on Ethanol Conversion and Product Formation

	Initial Ethanol Dehydrogenation [a)]	Initial Condensation Rate [b)]
K-$Cu_{0.5}Mg_5CeO_x$	0.22	0.5
K-$Cu_{7.5}Mg_5CeO_x$	0.24	4.5
Cs-Cu/ZnO/Al_2O_3	0.14	4.2

[(a)] Turnover rates per Cu surface atom, in s^{-1}; [(b)] Product formation rates, in x 10^{-8} mol / m^2-support • s

Aldol condensation rates increase with increasing Cu content in the catalysts, even though ethanol dehydrogenation turnover rates (per Cu surface atom) are very similar on Cs-Cu/ZnO/Al_2O_3, K-$Cu_{0.5}Mg_5CeO_x$, and K-$Cu_{7.5}Mg_5CeO_x$ catalysts. This suggests that Cu sites catalyze ethanol dehydrogenation steps leading to acetaldehyde, and also they are involved in rate-determining steps required for ethanol condensation reactions. Cu sites provide ethanol dehydrogenation sites and also dispose of H-atoms abstracted by oxygen anions during direct C-H and O-H activation reactions of ethanol [4,9]. Reactions of C_2H_5OH/D_2 mixtures [9,13] show that Cu sites increase the rate of hydrogen adsorption-desorption, a step that is effectively irreversible (rate-determining) in ethanol reactions on Cu-free K-Mg_5CeO_x catalysts.

Ethanol reactions lead mainly to acetone and n-butyraldehyde, which form only 2-propanol and 1-butanol after subsequent hydrogenation during CO/H_2 reactions. 2-Propanol and 1-butanol cannot, however, form isobutanol precursors, such as isobutyraldehyde and propionaldehyde. ^{13}C-tracer studies of methanol-acetaldehyde and methanol-propionaldehyde cross-coupling reactions were carried out in order to examine reaction pathways that can actually lead to the C_3 and C_4 oxygenate precursors required for isobutanol formation. Table 4 shows ^{13}C-distribution in products of $^{13}CH_3OH$-C_2H_4O reactions on K-$Cu_{7.5}Mg_5CeO_x$ and Cs-Cu/ZnO/Al_2O_3 catalysts. Although reaction pathways for ethanol formation from

$^{13}CO/H_2/^{12}CH_3OH$ differ on these two catalysts, reaction pathways for C_2^+ formation from $^{13}CH_3OH/^{12}C_2H_4O$ are similar on both catalysts.

Table 4
^{13}C-distribution in products of $^{13}CH_3OH$-C_2H_4O reactions on K-$Cu_yMg_5CeO_x$ and Cs-Cu/ZnO/Al_2O_3

Catalysts	K-$Cu_{7.5}Mg_5CeO_x$						Cs-Cu/ZnO/Al_2O_3					
Number of ^{13}C	0	1	2	3	4	%$^{13}C^{a)}$	0	1	2	3	4	%^{13}C
Methanol	0.0	100	--	--	--	100(100)	0.0	100	--	--	--	100
Acetaldehyde	99.1	0.9	0.0	--	--	0.5(0.2)	99.3	0.7	0.0	--	--	0.4
CO_2	35.6	64.4	--	--	--	64(86)	31.7	68.3	--	--	--	68
Ethanol	98.4	0.6	1.0	--	--	1.3(0.8)	98.7	0.4	0.9	--	--	1.1
Propionaldehyde	16.8	73.4	9.8	0.0	--	31(33)	6.7	82.6	10.7	0.0	--	35
i-Butyraldehyde	0.0	18.9	60.9	20.2	0.0	50(53)	0.0	3.2	66.3	30.5	0.0	57
n-Butyraldehyde	83.5	16.5	0.0	0.0	0.0	4.1(1.5)	82.0	18.0	0.0	0.0	0.0	4.5

a) Values in parenthesis are %^{13}C in the products of $^{13}CH_3OH$-C_2H_4O reactions on K-$Cu_{0.5}Mg_5CeO_x$; [573 K, 101.3 kPa total pressure, 8.0 kPa methanol, 4.0 kPa acetaldehyde, balance He, 21 % methanol conversion].

On both catalysts, propionaldehyde molecules using $^{13}CH_3OH$-C_2H_4O mixtures contain predominantly one ^{13}C atom, suggesting that they form via condensation of acetaldehyde with methanol-derived C_1 species. Isobutyraldehyde contains predominantly two ^{13}C atoms, suggesting that it forms by condensation of propionaldehyde (with one ^{13}C) with a C_1 species derived from methanol. Isotopic analyses of the reaction products of propionaldehyde-$^{13}CH_3OH$ mixtures confirm these condensation pathways [9]. These studies also show that aldol formation and dehydration-decarbonylation pathways are reversible and that reverse aldol condensation steps after aldol-keto interconversion provide a pathway for acetone formation and for isotopic scrambling. The rate of reverse aldol condensation steps increased with increasing Cu content on these catalysts. This indicates that the rate of hydrogen transfer increases with Cu content. The increase in the ^{13}C content of propionaldehyde as a contact time increase is consistent with the scheme shown in Figure 2, which includes reverse aldol reactions after aldol-keto isomerization.

4. CONCLUSIONS

Methanol and isobutanol are the most abundant alcohol products of CO hydrogenation on K-$Cu_yMg_5CeO_x$ and Cs-Cu/ZnO/Al_2O_3 catalysts at relatively low temperature (583 K) and pressure (4.5 MPa). CO_2, formed in CO hydrogenation and water-gas shift reactions, strongly inhibits methanol and isobutanol synthesis. The formation of the initial C-C bond in ethanol occurs directly from ^{13}CO, without significant involvement of methanol on K-$Cu_{0.5}Mg_5CeO_x$ catalysts. In contrast, Cs-Cu/ZnO/Al_2O_3 catalysts lead to ethanol synthesis via condensation reactions of the initial methanol product of CO hydrogenation. Ethanol dehydrogenation and coupling reactions at atmospheric pressures show that Cu sites are involved, through a bifunctional mechanism, in aldol condensation steps. Reactions of acetaldehyde with $^{13}CH_3OH$ lead to singly-labeled propionaldehyde, consistent with chain growth via addition of a methanol-derived C_1 species to acetaldehyde on both catalysts. Similarly, isobutyraldehyde molecules formed from propionaldehyde-$^{13}CH_3OH$ mixtures contain mostly one ^{13}C, showing that chain growth proceeds mainly via by aldol-type addition of a methanol-derived C_1 species to ethanol and higher alcohols or to the corresponding aldehydes.

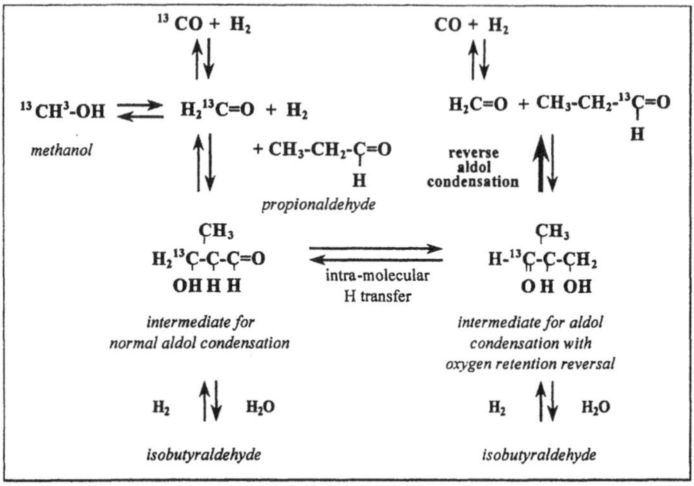

Figure 2. Reaction scheme of the labeled propionaldehyde production.

Acknowledgments

This work was supported by the Division of Fossil Energy of the United States Department of Energy under Contract Number DE-AC22-94PC94066. H.-S. Oh acknowledges a post-doctoral fellowship from the Korea Research Foundation. M.J.L. Gines acknowledges a post-doctoral fellowship from the Universidad Nacional del Litoral, Santa Fe, Argentina.

REFERENCES

1. W. Keim, and W. Falter, Catal. Lett. 3 (1989) 59.
2. J.G. Nunan, R.G. Herman and K. Klier, J. Catal. 116 (1989) 222.
3. C.R. Apesteguia, S.L. Soled, and S. Miseo, US Patent 5,387,570 (1995).
4. M. Xu, M. J.L. Gines, B.L. Stephens, A.-M. Hilmen and E. Iglesia, J. Catal. 171 (1997) 130.
5. C.R. Apesteguia, B. DeRites, S. Miseo and S.L. Soled, Catal. Lett. 44 (1997) 1.
6. M. Xu, S.C. Reyes and E. Iglesia, to be submitted to Journal of Physical Chemistry.
7. A.-M. Hilmen, M. Xu, M.J.L. Gines and E. Iglesia, Appl. Catal., in press (1998).
8. P. Forzatti, E. Tronconi and I. Pasquon, Catal. Rev. Sci. Eng. 33 (1991) 109.
9. M.J.L. Gines and E. Iglesia, J. Catal., in press (1998).
10. D.J. Elliot and F.J. Pennella, J. Catal. 114 (1988) 90.
11. J.G. Nunan, C.E. Bogdan, K. Klier, K.J. Smith, C.-W. Young and R.G. Herman, J. Catal. 113 (1988) 410.
12. M. Xu and E. Iglesia, Catal. Lett., in press (1998).
13. E. Iglesia, D.G. Barton, J.A. Biscardi, M.J.L. Gines, and S.L. Soled, Catal. Today 38 (1997) 339.

NATURAL GAS CONVERSION V
Studies in Surface Science and Catalysis, Vol. 119
A. Parmaliana et al. (Editors)

Direct Synthesis of Dimethyl Ether form Synthesis Gas

T. Shikada[a], Y. Ohno[a], T. Ogawa[a], M. Ono[a], M. Mizuguchi[a], K. Tomura[a], and K. Fujimoto[b]

[a]NKK Corporation, 1 Kokan-cho, Fukuyama 7210931, Japan

[b]The University of Tokyo, Hongo, Bunkyou-ku, Tokyo 1130033, Japan

Optimum synthesis gas conversion and high selectivity to DME were attained at the initial H_2/CO ratio of 1 in feed gas using hybrid catalyst in a slurry bubble column reactor. The study of process economics led to the estimation that the production cost per a heating value in the DME synthesis was reduced by 20% compared with a conventional methanol synthesis.

1. INTRODUCTION

Dimethyl ether (DME) has been more attracting industrial attention because it is used as an aerosol propellant instead of ozone-destroying chlorofluorocarbons [1], a raw material for synthesis of methyl acetate or acetic anhydride [2], and an intermediate in the production of gasoline by MTG process [3]. Recently, DME has been noticed as potential fuels for transportation, domestic use, and power generation because of its physical properties and combustion characteristics [4-8]. DME, which is a colorless gas with a boiling point of -25° C, is chemically stable and easily liquefied under pressure. Since the properties of DME are similar to those of propane and butane, which are main components of LPG, DME can be handled and stored like LPG.

In the present work the effects of operational factors, especially initial H_2/CO ratio in feed gas, on the conversion and the selectivity to DME were investigated using hybrid catalyst in a slurry bubble column reactor. The process for large scale DME production from natural gas and the process economics were also studied in comparison with methanol synthesis process.

2. DME SYNTHESIS FROM SYNTHESIS GAS

2.1. Thermodynamics

Direct DME syntheses from synthesis gas follow mainly the two overall reactions: one is Eq.(1) and the other is Eq.(2). The Eq.(1) consists of three steps that are a methanol synthesis reaction (3), a dehydration of methanol (4), and a water-gas shift reaction (5). Without the shift reaction, the reaction can be carried out following Eq.(2).

$$3CO + 3H_2 \rightarrow CH_3OCH_3 + CO_2 \quad , +245.7 \text{ kJ/mol} \qquad (1)$$

$$2CO + 4H_2 \rightarrow CH_3OCH_3 + H_2O \quad , +204.8 \text{ kJ/mol} \quad (2)$$
$$CO + 2H_2 \rightarrow CH_3OH \quad , +90.7 \text{ kJ/mol} \quad (3)$$
$$2CH_3OH \rightarrow CH_3OCH_3 + H_2O \quad , +23.4 \text{ kJ/mol} \quad (4)$$
$$CO + H_2O \rightarrow CO_2 + H_2 \quad , +40.9 \text{ kJ/mol} \quad (5)$$

Figure 1 shows equilibrium conversions of synthesis gas (CO conversion plus H_2 conversion) for the two DME synthesis reactions, Eq.(1) and Eq.(2), and for the methanol synthesis reaction, Eq.(3), as a function of initial H_2/CO ratio. In each reaction, the equilibrium conversion has its maximum peak where the H_2/CO ratio in the feed gas corresponds to the stoichiometric value, that is, with H_2/CO ratio of 1.0 for Eq.(1) and 2.0 for both Eq.(2) and (3). The maximum equilibrium conversion for Eq.(1) is much higher than those for Eq.(2) and (3).

2.2. Experimental Results of DME Synthesis form Synthesis Gas

The DME synthesis reaction was carried out using a slurry bubble column reactor with an inner diameter of 90 mm and a height of 2 m. The catalyst employed was a physical mixture of $CuO\text{-}ZnO\text{-}Al_2O_3$ and γ-alumina-supported copper. The slurry was prepared by mixing the fine powders (<120 mesh) of the catalyst with n-hexadecane.

Figure 2 shows the effect of H_2/CO ratio on synthesis gas conversion and product selectivities. Optimum synthesis gas conversion was obtained at the H_2/CO ratio of 1.0 as expected from the thermodynamic study. The selectivity to DME decreased slightly with an increase in H_2/CO ratio while that of methanol, major by-product, increased. Methane was also formed but its selectivity was less than 1%. The formation amount of CO_2 was approximately equal in moles to that of DME implying that DME was mainly synthesized following Eq.(1).

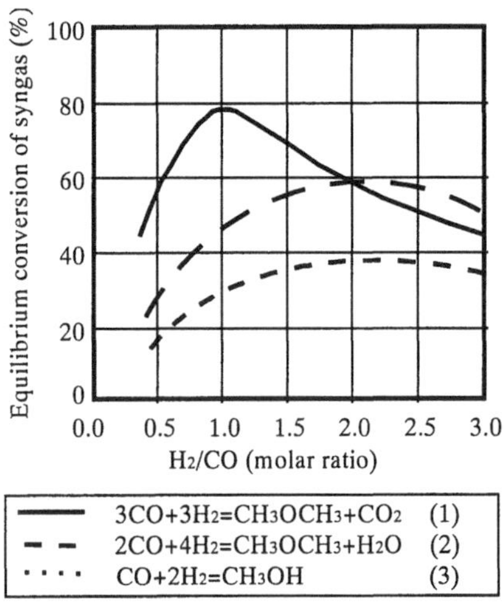

Figure 1. Equilibrium conversion of synthesis gas at 280°C, 50atm

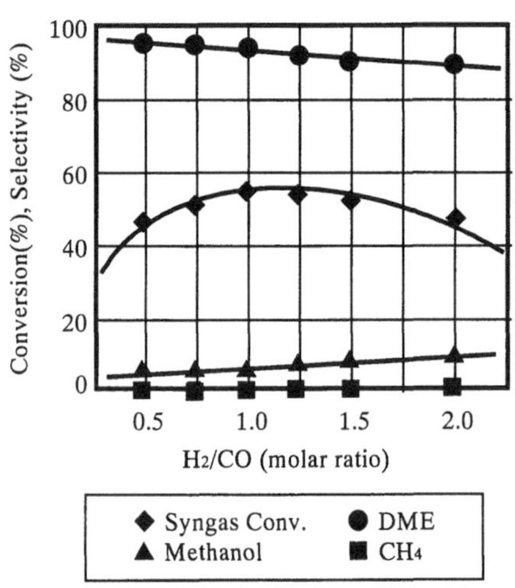

Figure 2. Conversion and selectivity as a function of H_2/CO ratio at 260°C, 50atm

3. DME SYNTHESIS FROM NATURAL GAS

In large scale production of DME as well as methanol the plant consists of three major process units and a utility section. Process flow for methanol and DME plants is quite similar although the operating conditions of separation and purification are slightly different because of differences in their boiling points. In this section detailed comparison of processes and economics is discussed for the methanol, the combined methanol/DME and the DME syntheses.

3.1. Synthesis Gas Production form Natural Gas

The reactions concerning the production of synthesis gas from methane and their reaction heats are shown below:

$$CH_4 + H_2O \rightarrow CO + 3H_2 \quad , -206.1\ kJ/mol \quad (6)$$

$$CH_4 + CO_2 \rightarrow 2CO + 2H_2 \quad , -247.0\ kJ/mol \quad (7)$$

$$CH_4 + 1/2O_2 \rightarrow CO + 2H_2 \quad , +36.0\ kJ/mol \quad (8)$$

Since the steam and the carbon dioxide reforming reactions are highly endothermic, it is important to supply the reaction heat. Steam reforming has some disadvantages that the size of multi-tubular reactor is quite limited and the gases obtained are always rich in hydrogen. In autothermal reforming it is easy to adjust the H_2/CO ratio but it is difficult to lower the oxygen plant cost. Reforming with CO_2 is suitable for DME synthesis adopted in the present study because synthesis gas with H_2/CO ratio of 1 is obtained.

In the present reaction system, equimolar amount of CO_2 is formed stoichiometrically accompanied with the formation of DME, if the water-gas shift reaction is accomplished. By-product CO_2 can be utilized for the reforming of natural gas. As shown in Eq.(9), however, one-third mole of CO_2 is insufficient per 1 mole of methane.

$$CH_4 + 1/3CO_2 \rightarrow 2/3CH_3OCH_3 \quad , -124.6\ kJ/mol \quad (9)$$

It is favorable to use the autothermal reforming combined with CO_2 reforming in viewpoints of compensating the short amount of CO_2 and supplying the reaction heat. Natural gas with a high concentration of CO_2 can be also used for this purpose.

3.2. Process Parameters at Synthesis Section

Table 1 shows the process parameters at the product synthesis sections for the three processes. In the methanol synthesis process as a base case and the combined methanol/DME synthesis process, the parameters were estimated form the operational data of

Table 1 Process parameters in syntheses of methanol, combined methanol/DME, and DME

	Methanol	Methanol/DME	DME
Recycle Ratio	5	5	2
Purge Gas Ratio (%)	23	15	5
Per-pass Conversion (%)	14	18	50
Total Conversion (%)	77	85	95

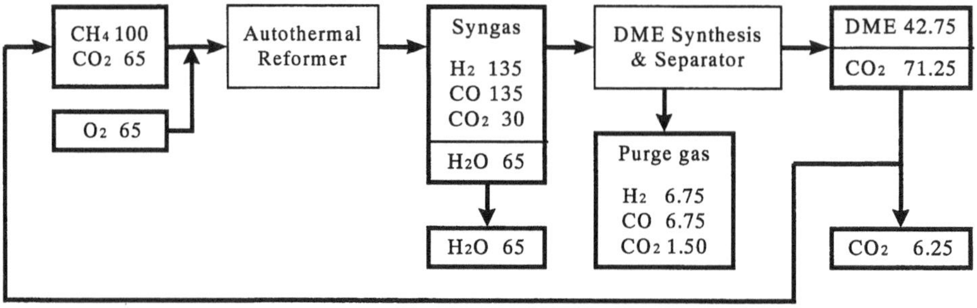

Figure 3. Stream flow of DME synthesis

ICI's quench converter [9] and the stream data of example described in Topsoe's patent [10], respectively. In above two processes per-pass conversions are quite low compared with their equilibrium conversions. It is assumed that the low conversions are due to the high recycle ratios to stabilize reaction temperatures. In the case of present DME synthesis a slurry reactor is used to be easy to maintain reaction temperature homogeneously and hence high conversion per pass can be achieved. Therefore, overall conversion of around 95% can be obtained and size of equipment can be reduced.

3.3. Material Balance

Figure 3 shows the stream flow form natural gas to product DME for the DME synthesis process. An equipment of CO_2 removal is usually placed after the reformer. However, CO_2 recovered at ordinary pressure from the equipment should be pressurized again to be utilized for reforming. It is desirable for synthesis gas to contain CO_2 of 5 to 10%. These concentration levels of CO_2 are required for inhibiting an excessive reduction of catalyst. Therefore, the equipment of CO_2 removal can be favorably eliminated. In the synthesis section, the amount of purge gas is relatively small because of high conversion (95%).

Apparent heat efficiencies and the units of CO_2 emission per calorific value for three processes are given in Table 2. Higher apparent heat efficiency and lower CO_2 emission are estimated in the DME synthesis process. The amount of CO_2 emission in DME production and its combustion (88.4 mg-C/kcal) is comparable to that in combustion of diesel fuel (86.8 mg-C/kcal).

3.4. Estimation of Production Cost

It has been reported that the relative investment and natural gas consumption for the

Table 2 Apparent heat efficiencies and CO_2 emission

	Methanol	Methanol/DME	DME
Apparent Heat Efficiency (%)	55.0	63.2	70.7
CO_2 Emission (mg-C/kcal)			
Production	34.9	22.7	12.7
Utilization as Fuel	78.8	75.7	75.7
Total	113.7	98.4	88.4

combined methanol/DME and the DME synthesis plants are lower than that for the conventional methanol synthesis plant [5]. A typical investment breakdown by sections are as follows: 53% synthesis gas production, 25% utilities, 14% distillation section, and 8% synthesis section [5].

In this paragraph more detailed process economics are discussed. On the assumption that the relative investment is proportional to the two-third power of gas flow rate and the natural gas consumption is proportional to the reciprocal of overall conversion, the following equations are used in calculation.

$$A=Ao(Xo/X)^{2/3} \quad (10)$$
$$B=Bo(Xo/X \cdot (1+R)/(1+Ro))^{2/3} \quad (11)$$
$$C=Co \quad (12)$$
$$D=Do(A+B+C)/(Ao+Bo+Co) \quad (13)$$
$$G=Go(Xo/X) \quad (14)$$

where A, B, C, and D, respectively, denote the relative investment of syngas production, synthesis, purification, and utilities. X represents the overall conversion and G denotes the relative natural gas consumption. The subscript 0 indicates the value for the base-case methanol synthesis.

The results of calculation are shown in Table 3. The investment in the combined methanol/DME and the DME syntheses are reduced by 5 and 14% compared with the methanol synthesis, respectively. The natural gas consumption is also reduced by 9 and 19% which are in fair agreement with the values described in the literature [5].

Table 3 Relative investment and relative natural gas consumption

	Methanol	Methanol/DME	DME
Relative Investment	100	94.7	85.9
Synthesis Gas Preparation /A	53	49.6	46.0
Synthesis /B	8	7.5	4.4
Separation and Purification /C	14	14	14
Utilities /D	25	23.7	21.5
Relative Natural Gas consumption /G	1.00	0.906	0.811

Table 4 Estimation of production cost

	Methanol	Methanol/DME	DME
Capital Cost (US$/t-methanol eq.)	63.0	59.7	54.1
Natural Gas (US$/t-methanol eq.)	46.2	41.9	37.5
Total (US$/t-methanol eq.)	109.2	101.6	91.6
Capacity (t/d)	2500	1984	1797
Production Cost (US$/t)	109.2	128.0	127.4
Production Cost (US ¢ /Mcal)	2.29	2.07	1.85

The estimation of production cost was carried out supposing 260 US$ million of the capital investment for the production of 2500 tons methanol per day and 1.5 US$/million Btu of natural gas price. As the amount of methanol equivalent, 659 tons methanol and 1325 tons DME per day are produced in the combined synthesis and 1797 tons DME per day in the DME synthesis. The production costs estimated are shown in Table 4. It can be seen that the production cost per a heating value in the combined methanol/DME syntheses is reduced by 10% and that in the DME synthesis by 20% in comparison with the methanol synthesis.

4. CONCLUSIONS

The hybrid catalyst containing $CuO\text{-}ZnO\text{-}Al_2O_3$ and copper-modified γ-alumina was found to show a high activity for the synthesis of DME in slurry phase. Optimum synthesis gas conversion (56%) and high selectivity to DME (94%) were achieved at the initial H_2/CO ratio of 1.

Based on the study on the comparison of processes and economics, the following conclusions were obtained.

1) The DME synthesis was more favorable in the economical viewpoint than the methanol and the combined methanol/DME syntheses.

2) The capital investment in the DME synthesis plant is reduced to about 86% of a conventional methanol synthesis plant.

3) The natural gas consumption in the DME synthesis is 19% lower than that in the methanol synthesis.

4) The production cost per a heating value in the DME syntheses is reduced by 20% in comparison with the methanol synthesis.

REFERENCES

1. F.M.H. Debets, Aerosol Rep., 29 (1990) 16.
2. T. Shikada, K. Fujimoto, M. Miyauchi, H. Tominaga, Appl. Catal., 7 (1983) 361.
3. C.D. Chang, Cat. Rev.-Sci. Eng., 25 (1) (1983) 1.
4. S.C. Sorenson and S.E. Mikkelson, SAE Paper 950064 (1995).
5. Ib Dybkjaer, J.B. Hansen Proc. 4th International Natural Gas Conversion Symposium, Kruger National Park, South Africa, p.99 (1995).
6. Y. Ohno, T. Shikada, T. Ogawa, M. Ono, M. Mizuguchi, K. Fujimoto, Preprints, 213th ACS Annual Meeting, San Francisco, USA, p.705 (1997).
7. T. Ogawa, M. Ono, M. Mizuguchi, K. Tomura, T. Shikada, Y. Ohno, K. Fujimoto, Proc. 14th Annual Internatioanl Pittsburg Coal Conference & Workshop, Shanxi, People's Republic of China, 30-3004 (1997).
8. Y. Adachi, M. Komoto, I. Watanabe, Y. Ohno, Proc. ASCOPE '97 Conference, Jakarta, Indonesia (1997).
9. R.P. Dry, Ind. Eng. Chem. Research, 27 (1988) 616.
10. B. Voss, F. Joensen, J.B. Hansen, International Patent WO 96/23755.

NATURAL GAS CONVERSION V
Studies in Surface Science and Catalysis, Vol. 119
A. Parmaliana et al. (Editors)

Dimethyl ether conversion to light olefins over SAPO-34: Deactivation due to coke deposition

De Chen[1,2], Hans Petter Rebo[1], Kjell Moljord[2], and Anders Holmen[1]

[1]Department of Industrial Chemistry, Norwegian University of Science and Technology (NTNU), N-7034 Trondheim. [2]SINTEF Applied Chemistry, N-7034 Trondheim, Norway.

The effect of coke deposition on dimethylether (DME) conversion to light olefins (DTO) was studied over SAPO-34 with crystal sizes of 0.25 and 2.5 μm using a Tapered Element Oscillating Microbalance (TEOM) reactor. DME diffusivity was found to decrease exponentially with the coke content, while the intrinsic rate constant went through a maximum. The reaction rate on the 2.5 μm crystals was significantly influenced by diffusion. A kinetic model taking into account the effects of diffusion and the nature of coke was developed. The coke selectivity and catalyst capacity of SAPO-34 were also studied.

1. INTRODUCTION

The conversion of gas to olefins (GTO) is a promising way of producing high value chemicals from natural gas via synthesis gas and methanol. Methanol is converted to light olefins over SAPO-34 [1], with dimethyl ether as a key reaction intermediate. DME can be produced directly from syngas over a bifunctional catalyst [2,3]. It has been shown that the methanol synthesis and the subsequent conversion to gasoline (MTG) steps can be integrated into a single synthesis loop, which offers lower investments and lower energy consumption [4]. A conceptual design [2] has shown that DME conversion to gasoline (DTG) offers advantages over the Mobil MTG process in several areas, including heat duty, hydrocarbon yield and selectivity, syngas conversion and overall efficiency. DME is therefore an attractive alternative source also for olefin production [3].

In a similar way as in the MTO reaction, coke deposition during DTO over SAPO-34 is a very rapid process. An oscillating microbalance (TEOM) reactor has been shown to be a suitable tool for studying the reaction and the coke deposition simultaneously for such a rapidly deactivating process [5,6]. The present work deals with a kinetic study of coke deposition and catalyst deactivation during the DTO reaction over SAPO-34 with crystal sizes of 0.25 and 2.5 μm using the TEOM reactor.

2. EXPERIMENTAL

The reactions were carried out in a TEOM reactor where changes in the reaction rate and the catalyst weight can be studied simultaneously. The set-up is similar to what has been described previously [5]. The MTO and DTO reactions were studied at 425°C, WHSV from 79 to 564 h^{-1}

and a methanol or DME partial pressure of 8 kPa, with helium as diluent. The experimental procedures are similar to that described previously [6]. SAPO-34 with crystal sizes of 0.25 μm and 2.5 μm but with identical properties were obtained from SINTEF-Oslo. A detailed description of the preparation and characterization of the catalyst can be found elsewhere [7]. The unit cell composition of SAPO-34 is $(Si_{2.86}Al_{17.49}P_{15.66})O_{72}$. The conversions were calculated on a CH_2 basis and the selectivities were obtained from the normalised hydrocarbon mole distribution.

3. RESULTS AND DISCUSSION

3.1 Diffusion and reaction in SAPO-34

Olefins (C_2-C_6) were the major products and in addition small amount of methanol was formed during the DTO reaction. Fig. 1 shows the conversion of DME and the coke deposition over 0.25 μm and 2.5 μm crystals. The DME conversion increased with decreasing crystal size. This clearly indicates that the conversion of DME was influenced by diffusion.

Olefin formation during the DTO reaction is assumed to be a first order reaction. The first order reaction rate constant was calculated by Eq. (1) on a carbon basis:

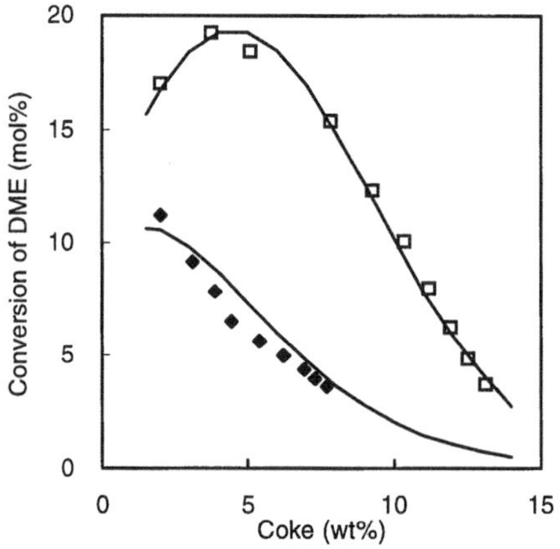

Fig. 1 DME conversion versus coke contents over 0.25(□) and 2.5 (◆) μm crystals at WHSV: 395 h^{-1}, 425 °C and P_{DME}: 8 kPa. Lines: predicted by model (Eq. 1 and 4).

$$k_{app} = -\ln(1-X)\frac{F_{A0}\rho_C}{C_{A0}W} \qquad (1)$$

where X is the conversion of oxygenates, C_{A0} ($kmol/m^3$) is the initial reactant concentration calculated from the ideal gas law, ρ_c is the catalyst density (800 kg/m^3), W (kg) is the catalyst weight and F_{Ao} (kmol/s) is the molar flow rate of the reactant. The apparent rate constant k_{app} (s^{-1}) is smaller than the intrinsic rate constant k, if diffusion effects cause the effectiveness factor η to be less than unity. Assuming the SAPO-34 crystals to be spherical, the effectiveness factor is given as:

$$\eta = \frac{3}{\Phi}\left(\frac{1}{\tanh\Phi} - \frac{1}{\Phi}\right) \qquad (2)$$

where Thiele modulus $\Phi = R(k/D)^{1/2}$, R is the crystal radius and D is the diffusivity of reactant.

When the observed rate constants differ for two crystals with radius R_1 and R_2, the parameters η_1, η_2 and Φ_1, Φ_2 can be uniquely determined by the method of triangulation in the ln η and ln Φ plot [8]. The intrinsic rate constant k and the diffusivity D can thus be calculated, assuming that k and D are identical for the different crystals. This method was applied to calculate k and D on SAPO-34 containing coke, based on the assumption that the coke was randomly deposited inside the crystals for all the samples. As shown in Fig. 2, the diffusivities decreased more rapidly than the intrinsic reaction rate constants due to coking, which resulted in an increasing Thiele modulus with increasing coke content.

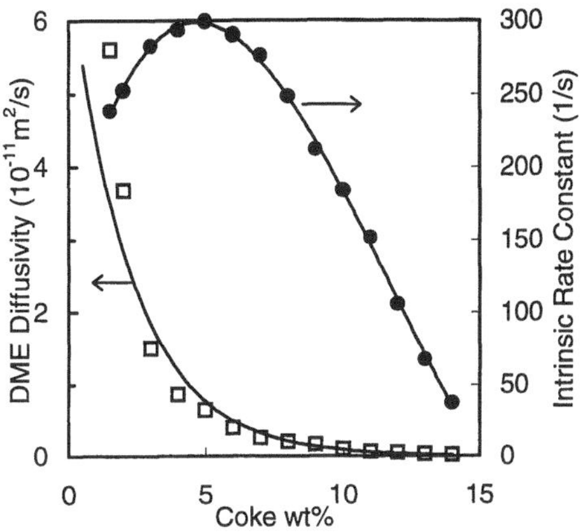

Fig. 2 Effective DME diffusivity and intrinsic rate constant of DTO versus coke content at 425 °C.

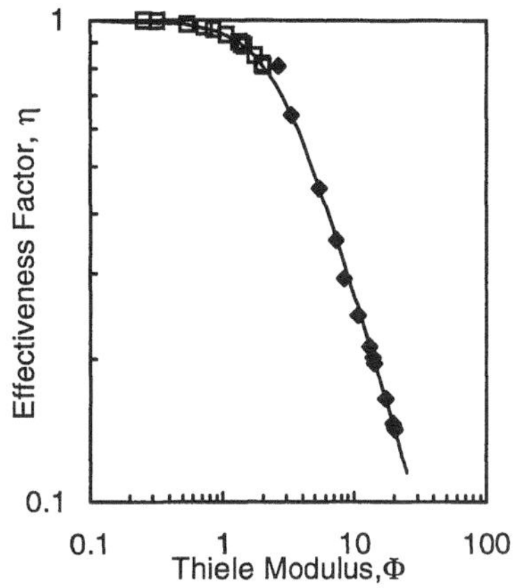

Fig. 3 Effectiveness factor versus Thiele modulus for different coke contents at 425 °C. □:0.25 μm,◆: 2.5 μm. Line: calculated by Eq. 2.

The Thiele modulus and the effectiveness factor (Fig. 3) were calculated based on the intrinsic rate constants and effective diffusivities at different coke contents. The effectiveness factor for the smallest crystals was larger than 0.95 for coke content less than 4 wt%, and decreased to 0.8 for about 14-15 wt% coke (Fig. 3). This corresponds with a situation where the reaction changed from a kinetically controlled regime to a diffusion influenced regime, due to the intracrystalline coke deposition. All the effectiveness factors were less than 0.8 on the 2.5 μm crystals, indicating that DTO was influenced by diffusion on these crystals. As more coke was deposited, the effectiveness factor decreased much faster on the larger crystals than on the smaller crystals, which resulted in the faster deactivation on the 2.5 μm crystals (Fig. 1). It can therefore be concluded that due to strong diffusion limitation large particles should be avoided in order to reduce the deactivation rate.

3.2 A kinetic model for deactivation of DTO over SAPO-34

Fig. 2 shows that the intrinsic rate constant increased initially with the coke content, and went through a maximum at a coke content of about 5 wt%. This result indicates that some of the species accumulated inside the pores has a promoting effect on the DME conversion, and these species are defined as active coke [6]. All the mass increase on the catalysts is referred to as coke in the present work. The nature of coke is normally very complex, and it depends on the catalysts, the nature of reactants and products, and on the operation conditions. Coke could be heavy hydrocarbons, aromatics, or simply some of the strongly adsorbed reactants and products. It was observed in our experiments at relatively low temperatures that the adsorption of methanol and DME on SAPO-34 is quite strong and partly irreversible. Part of the irreversible adsorbed species might remain on the surface as coke at reaction conditions. However, the nature of the active coke is still not very clear, and is a subject for detailed studies.

In order to describe the changes in activity with coke content for different sized crystals, a proper kinetic model should take into account the promoting effects of coke as well as the diffusion effects. The promoting effects are accounted for in the kinetic model by a term for the interaction between active coke and DME, in addition to the term for the main reaction. The total apparent reaction rate of DME can then be written as:

$$r_{DME} = (k_1^0 \cdot \phi + k_2^0 \cdot C_{AC}) \cdot \eta \cdot C_{DME} = k_{app} \cdot C_{DME} \quad (3)$$

where C_{DME} is the concentration of DME, C_{AC} (wt%) is the amount of active coke and index 0 refers to initial conditions. At the initial stage, i.e., without any coke present, the initial intrinsic rate constant equals k_1^0. When more coke is formed, the conversion of DME will be influenced by site coverage and pore blocking caused by contributions from the active and the inactive coke. The rate constant k_1 decreases with coke deposition, which could be accounted for by a deactivation function (ϕ). As pointed out by Froment and Bischoff [9], the catalyst deactivation should be related to the true cause of the deactivation such as the coke content, and not to the time on stream. In the present work deactivation has therefore been expressed as a function of the total coke content, i.e. the weight increase of the catalyst bed. However, the rate constant k_2 for the reaction between DME and active coke is not assumed to be affected by coke, but the rate depends on the amount of active coke C_{AC}, which is expected to change with the total coke content.

Different types of deactivation functions ϕ [9], and active coke functions C_{AC} have been proposed and tested and the best fit was obtained by using the parameters in Eq. 4.

$$r_{DME} = (171.17\,(1 - 0.0684C) + 55.96\,C\exp(-0.0153C^2))\eta \cdot C_{DME} \quad (4)$$

where C is the total coke content (wt%) and η can be calculated by Eq. 2. The changes in effective DME diffusivities with coke as presented in Fig. 2 can be described by Eq. 5:

$$D = 7.2512 \cdot 10^{-11} \cdot \exp(-0.4381C) \quad (5)$$

Variations in DME conversion with coke for the two SAPO-34 crystals can thereby be predicted as shown in Fig. 1, and the result is very satisfactory. The apparent first order rate constant for DME conversion over the 0.25 μm crystals at different space velocities and different coke contents were calculated by Eq. 1, and plotted against the accumulated coke in Fig. 4. The apparent rate constants at different space velocities are almost constant, supporting the assumption of a first order reaction. The consistency between experimental and predicted apparent reaction constants is also good as shown in Fig. 4.

The coking rate depends on the crystal size, and a higher coking rate was found on 0.25 μm crystals. In the present work, however, the kinetic study focuses only on the coke deposition over the 0.25 μm crystals, where the rate is almost free of diffusion limitation. In Fig. 5 the amount of coke over the 0.25 μm SAPO-34 crystals at different space velocities is plotted against the amount of DME fed to the catalyst bed. The coking rate clearly depends on the DME conversion. A lower space velocity resulted in a higher DME conversion, and thus a higher coking rate. A kinetic model which has been developed previously [10] for MTO was used to describe the coke deposition during DTO. The kinetic model (Eq. 6) takes into account the effect of conversion (X), the ratio of amount of DME to catalyst (RDC) and the deactivation of the coking reactions (ϕ_C):

$$dC / dt = k_C^0 \cdot RDC \cdot X \cdot \phi_C \quad (6)$$

where k_C^0 is the initial coking rate constant. Different types of deactivation functions [9] for coking have been suggested and tested. $\phi_C = \exp(-\alpha_C \cdot C)$ gives the best representation of the experimental results, and differs from the deactivation function for the main reaction. The integrated formula is shown in Eq. 7:

$$C = \frac{1}{\alpha_C} \ln[1 + \alpha_C \cdot k_C^0 \cdot AF_{DME}(1 - \exp(-\frac{k^0 \cdot C_{A0} \cdot W}{\rho_C \cdot F_{A0}}))] \quad (7)$$

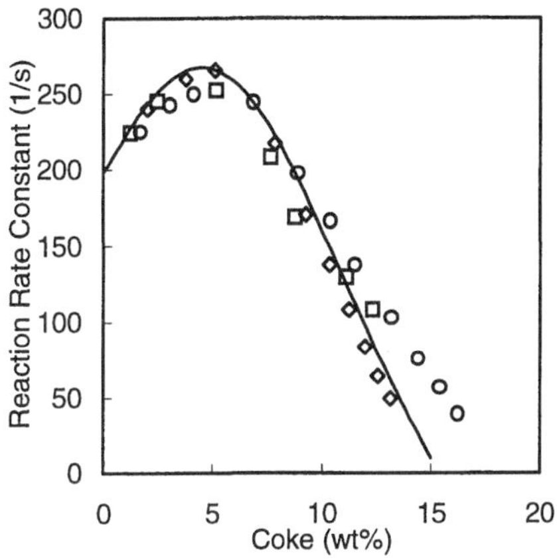

Fig. 4 Apparent first order rate constant versus coke content over 0.25 μm crystals at WHSV: 564 h^{-1}(□), 393 h^{-1} (◇), 158 h^{-1} (○). Line: predicted by model (Eq. 4).

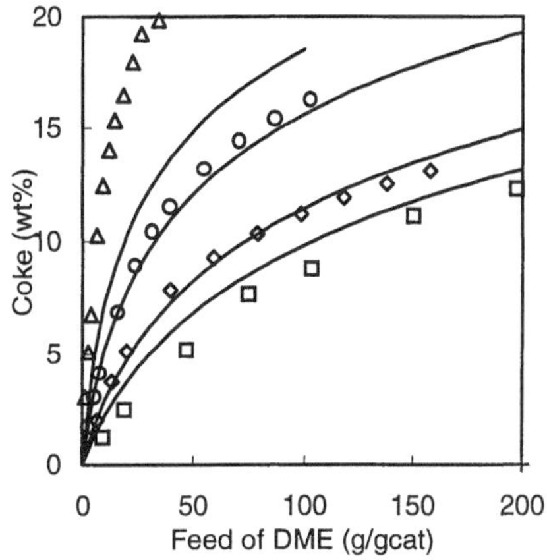

Fig. 5 Coke deposition during DTO over 0.25 μm crystals at WHSV: 564 h^{-1} (□) , 393 h^{-1} (◇), 158 h^{-1} (○), 79 h^{-1} (Δ). Line: predicted by model (Eq. 7).

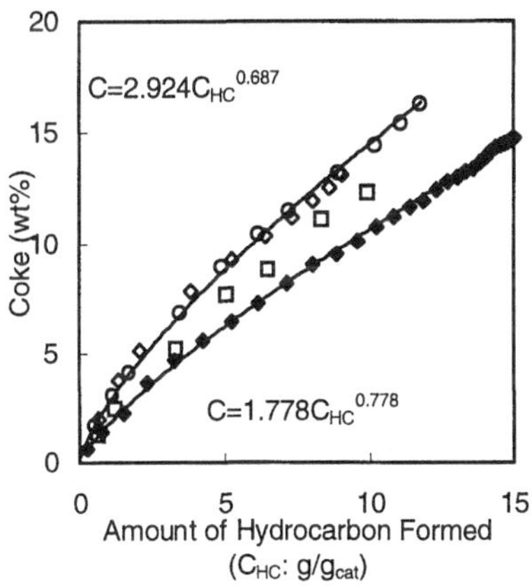

Fig.6 Coke contents versus C_{HC} (g/g_{cat}) on SAPO-34 (0.25 μm) at 425 °C during DTO: 564 h^{-1} (□) , 393 h^{-1} (◇), 158 h^{-1} (○),MTO: 385 h^{-1}(◆).

AF_{DME} (g/g) is the cumulative amount of DME fed per catalyst mass. The parameters in Eq. 11 (α_C = 0.1795, k_C^0 = 3.21 and k^0 =171.17) were estimated by curve fitting. The model fitted the experimental coke deposition rather well as shown in Fig. 5, except at a space velocity of 79 h^{-1}. The higher experimental coking rate than predicted at this space velocity is probably caused by a temperature rise in the catalyst bed, due to the high conversion and the highly exothermic nature of the reaction.

3.3 Comparison between MTO and DTO

The olefin distribution during DTO is independent of the crystal size, which is in good agreement with the results from the MTO reaction [11]. A very similar distribution of olefins was observed for DTO and MTO, but the selectivity to coke was different.

The relationship between the coke content and the amount of hydrocarbons formed (C_{HC})

during DTO was almost identical, regardless of the space velocity (Fig. 6). The time averaged coke selectivities and the catalyst capacities can be directly obtained from such relationships. If a coke content of 15 wt% is defined as the final state, the catalyst capacity for olefin formation is 10.8 and 15.5 g/g_{cat} and the total coke selectivity is 1.39 and 0.95 wt% for DTO and MTO, respectively. The catalyst capacity for olefin formation is 1.44 times higher for MTO than for DTO at 425 °C. The difference in coke selectivity between MTO and DTO might be explained by the concentration of water inside the crystals. According to the stoichiometry of the reaction, the molar ratio of hydrocarbon to water for DTO (CH_2/H_2O = 2) is twice that of MTO (CH_2/H_2O = 1). Therefore, it can be assumed that cofeeding of water and DME can reduce the coke selectivity and increase the catalyst capacity, and an equilibrium mixture of methanol, DME and water directly from the one-step syngas conversion over a bifunctional catalyst might be an alternative source for olefin production.

4. CONCLUSIONS

The DTO reaction was influenced by diffusion for the 2.5 μm crystals of SAPO-34 , while the reaction was free of diffusion limitation for the 0.25 μm crystals at low coke contents. The reaction deactivated faster for the large crystals. The effective DME diffusivity decreased exponentially with the coke content, while the intrinsic rate constant went through a maximum, which could be caused by the participation of a type of active coke in the reaction. A kinetic model was developed to account for the effect of diffusion and deactivation. Very good consistency was found between the kinetic model and the experimental data. The coke selectivity was higher and the catalyst capacity for olefin formation during DTO was lower compared to MTO over SAPO-34.

ACKNOWLEDGEMENTS

The authors thank the Norwegian Research Council and Norsk Hydro ASA for supporting this work, and Terje Fuglerud, Norsk Hydro for fruitful discussions.

REFERENCES:

1. B. V. Vora, T. L. Marker, P. T. Barger, H. E. Fullerton, H. R. Nilsen, S. Kvisle, T. Fuglerud, Stud. Surf. Sci. Catal., 107 (1997) 87.
2. S. Lee, M. Gogate, C. J. Kulik, Fuel Sci. Tech. Int'l, 13 (1995) 1039.
3. G. Cai, Z. Liu, R. Shi, C. He, L. Ynag, C. Sun, Y. Chang, Appl. Catal., 125 (1995) 29.
4. J. Topp-Jorgensen, J. R. Rostrup-Nielsen, Oil & Gas J., May 19 (1986) 68.
5. D. Chen, A. Grønvold, H. P. Rebo, K. Moljord, A. Holmen, Appl. Catal., 137 (1996) L1.
6. D. Chen, H. P. Rebo, K. Moljord, A. Holmen, Stud. Surf. Sci. Catal., 111 (1997) 159.
7. R. Wendelbo, D. Akporiaye, A. Anderson, I. M. Dahl, H. B. Mostad, T. Fuglerud, submitted to Microporous Materials.
8. W. O. Haag, R. M. Lago, P. B. Weisz, Chem. Soc. Faraday Dis., 72 (1981) 317.
9. G. B. Froment, K. B. Bischoff, Chemical Reactor Analysis and Design, Jonh Wiley & Sons, New York, 1979.
10. D. Chen, Ph. D thesis, Department of Industrial Chemistry, NTNU, 1998.
11. D. Chen, K. Moljord, T. Fuglerud, A. Holmen, submitted to Microporous Materials.

NATURAL GAS CONVERSION V
Studies in Surface Science and Catalysis, Vol. 119
A. Parmaliana et al. (Editors)

Chain Growth Reactions of Methanol on SAPO-34 and H-ZSM5

Enrique Iglesia, Tom Wang, and Sara Y. Yu

Department of Chemical Engineering, University of California at Berkeley, Berkeley CA 94720

ABSTRACT

Reactions of $C_3H_6/^{13}CH_3OH$ mixtures show that chains grow via methylation of adsorbed intermediates with CH_3OH on H-ZSM5 and SAPO-34 and alkenes desorb via β-scission steps. Turnover rates are faster on H-ZSM5 than on SAPO-34 and they increase with residence time and with the addition of alkenes on H-ZSM5, because β-scission rates are higher for the larger growing chains favored at such conditions. Rates do not increase with alkene concentration on SAPO-34 because transport processes control the rate at which alkenes enter the gas phase. Ethene is a minor product on H-ZSM5 because β-scission pathways favor the formation of C_3+ products. The lower turnover rates, faster deactivation, and higher ethene selectivities on SAPO-34 reflect diffusional constraints imposed by the small connections between intracrystalline cavities. Light alkenes are selectively extracted from equilibrated alkene mixtures formed within elliptical cavities in SAPO-34. The reactivity of alkenes of varying length and structure during $^{13}CH_3OH$ reactions confirms these conclusions. Intermediate transport restrictions lead to maximum ethene selectivities and appear to require small SAPO-34 crystals with few external acid sites.

1. INTRODUCTION

Medium-pore pentasil zeolites (H-ZSM5) with low Al content [1] and small crystallites of SAPO-34, a silicoaluminophosphate with chabazite structure [2], are effective in methanol conversion to light alkenes [3,4]. SAPO-34 shows higher selectivity to ethylene, but it deactivates rapidly and requires frequent regeneration [5]. Several reactive intermediates (carbenes, oxonium ylides, radicals, oxoniums) have been proposed for initial C-C bond formation steps on H-ZSM5 [6]. Some reports conclude that ethylene is the first alkene product [7-9], but others favor the initial desorption of larger alkenes [10-12]. On small-pore SAPO-34, the slow diffusion of products can lead to selective sieving of smaller ethene products from a complex mixture of alkenes [13]. Extracrystal acid sites and transport restrictions can obscure mechanistic details on SAPO-34, but recent C_2H_5OH-$^{13}CH_3OH$ co-feed studies suggest that light alkenes can form via β-scission of larger alkenes [14,15].

In this study, reaction pathways required for methanol conversion to light alkenes on H-ZSM5 and SAPO-34 are probed using kinetic and isotopic tracer methods at conditions leading to high C_2-C_5 alkene selectivity (> 70%). Our data show that methylation and β-scission of large carbocations leads to the formation of alkenes. Ethene formation is not favored by these β-scission steps; ethene selectivity increases as methylation/β-scission steps approach equilibrium at long bed or intraparticle residence times.

2. METHODS

H-ZSM-5 was obtained by NH_4NO_3 exchange of Na-ZSM-5 (Si/Al=175; 0.5, 1, 2, and 4 μ average crystal diameter) [16] and treatment in dry air at 773 K. The Al content is 0.096 mmol/g (2 μ sample); it agrees well with the number of NH_3 desorbed during decomposition of exchanged NH_4^+ (0.11 mmol/g) and corresponds to a H^+ density of 0.55 H^+/u.c. (0.11 H^+/nm^3). A sample with Si/Al ratio of 14.5 (Zeochem) was exchanged using the same procedure [17].

SAPO-34 was prepared by the procedure of reference [2]. The Si content was 1.15 mmol/g and the NH_3 uptake was 1.03 mmol/g (H^+ density: 2.5 H^+/u.c., 1.0 H^+/nm^3). Scanning electron microscopy showed cubic crystallites of 0.3 to 0.7 μ diameter. X-ray diffraction and electron micrographs showed that H-ZSM5 and SAPO-34 samples had excellent crystallinity.

Methanol reactions were carried out in a gradientless batch reactor [17] on H-ZSM5 (673 K, 11-12 mg) and SAPO-34 (653 K, 5-6 mg) at conversions per pass below 2-3%. Samples were treated in air at 823 K for 1 h before reaction. He (85 kPa, >99.95%) and H_2O (5 kPa) were used to dilute CH_3OH (10 kPa, Fisher Certified). C_2H_4, C_3H_6, 1-C_4H_8, and iso-C_4H_8 (1 kPa, C.P., >99.5%) were mixed with $^{13}CH_3OH$ (10 kPa, 99% ^{13}C, Cambridge Isotopes) in co-feed studies. Chemical and isotopic contents were measured by chromatography using flame ionization and mass selective detection. Isotopomer distributions were obtained from mass fragmentation data [18].

3. RESULTS AND DISCUSSION

Selectivities and methanol conversion rates per acid site are shown in the Table on H-ZSM5 and SAPO-34 at similar reactant (CH_3OH + CH_3OCH_3) conversion. SAPO-34 gives lower turnover rates and C_6+ selectivity than H-ZSM5 and much higher ethene/propene ratios. Alkene/alkane ratios are very high (>10) on both samples at all conversions. Turnover rates are initially low on H-ZSM5, but reach the values in the Table as conversion increases with increasing contact time (Figure 1). No initial induction was observed on SAPO-34 (Figure1).

Table. Methanol conversion to light alkenes on H-ZSM5 (Si/Al=175) and SAPO-34 [673K, 10kPa CH_3OH, 5 kPa H_2O, 38.5-39%methanol/DME conversion]

	H-ZSM5	SAPO-34
Site Density (mmol/g)	0.096	1.15
Turnover Rate (s^{-1})	0.64	0.15
Selectivity (C%)		
C_1	0.44	0.70
C_2	1.4	29.1
C_3	34.5	48.1
C_4	22.6	16.5
C_5 (iso/normal ratio)	13 (0.74)	4.2 (0.09)
C_6+	26.8	1.1
Ethene/Ethane	75	238
Propene/Propane	370	13

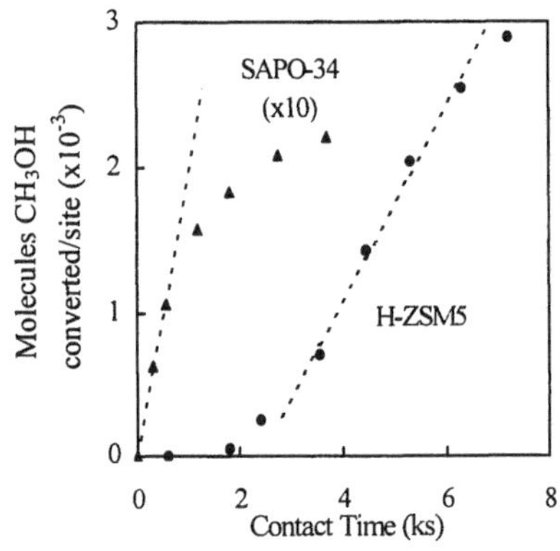

Figure 1. CH_3OH conversion turnovers on H-ZSM5 and SAPO-34 (x10) in gradientless batch reactor [673 K, 10 kPa CH_3OH, 5 kPa H_2O]

The evolution of products with contact time resembles that reported by others [1]. CH_3OH-CH_3OCH_3 equilibrium is fast, alkene selectivities reach a maximum at intermediate conversions, and aromatics and larger alkanes become more abundant as contact time increases. These trends are similar on SAPO-34 and H-ZSM5, but ethene and C_6+ selectivities differ on the two catalysts. The effect of conversion on C_2H_4/C_3H_6 ratios is shown in Figure 2. On SAPO-34, ethene selectivity increases slightly as conversion increases; ethene becomes a favored product only as alkenes approach equilibrium at long contact times. On H-ZSM5 (2 μ, Si/Al=175), C_2H_4/C_3H_6 ratios are much smaller and increase slightly with contact time, except at very low conversions, where this ratio is very high and decreases sharply with increasing contact time. These high initial ethene selectivities have been misinterpreted as evidence for ethene as the initial alkene formed in methanol reactions.

Ethene/propene ratios on H-ZSM5 did not depend on crystal size for samples with low Al content, but reached higher values on Al-rich H-ZSM5 (Figure 3). The x-axis in Figure 3 [the product of L^2 (L, crystal diameter) and acid sites per unit volume] consists of a Thiele parameter that reflects the severity of intracrystal transport restrictions. The higher C_2H_4/C_3H_6 ratios obtained at high values of this parameter reflect the intracrystalline equilibration of alkene mixtures as intracrystal residence time increases with increasing severity of transport restrictions. Ethene is abundant in equilibrated alkene mixtures, but β-scission kinetics favor the desorption of C_3+ alkenes. High ethene selectivities on SAPO-34 reflect diffusional restrictions that become more severe for larger alkenes and lead to equilibrated alkene mixtures within SAPO-34 cavities.

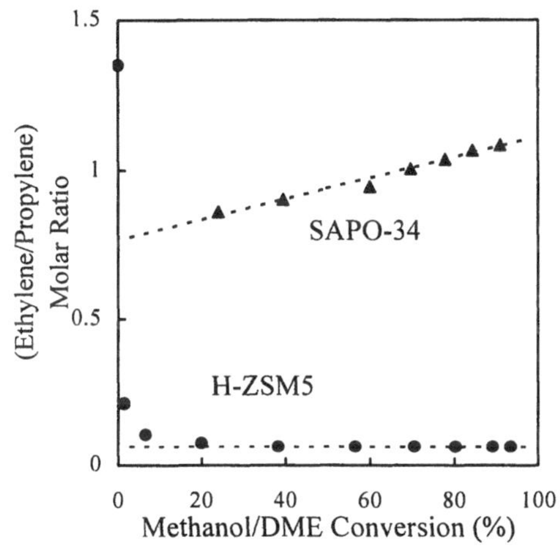

Figure 2. Ethene/Propene ratio in reaction products [673 K, 10kPa CH_3OH, 5 kPa H_2O]

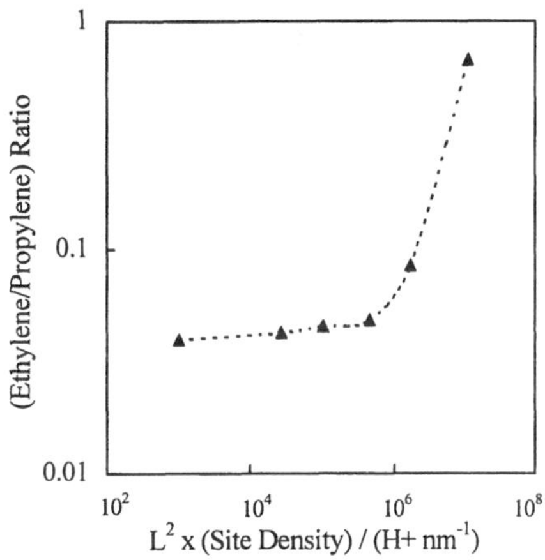

Figure 3. Effect of transport restrictions {L^2 x (site density)} on ethene/propene ratio [673K, 10 kPa CH_3OH, 5 kPa H_2O, 45-55% methanol/dimethyl-ether conversion]

Chain growth pathways were probed using $C_3H_6/^{13}CH_3OH$ reactant mixtures. The Scheme shows two chain growth paths and the isotopomer distributions expected for butenes formed via each path. Chains growing by methylation of carbocations can terminate as alkenes by deprotonation or by β-scission chain transfer steps. The latter path preserves a smaller carbocation and avoids the need to re-form chains from methanol after each turnover. β-Scission from intermediates that undergo fast isomerization and intramolecular scrambling of carbons (from C_3H_6 or $^{13}CH_3OH$) would lead to binomial isotopomer distributions. Deprotonation removes an intact chain, which can only form from a specific number of C_3H_6 or $^{13}CH_3OH$ molecules. Thus, butenes can only contain either four ^{13}C atoms (from $^{13}CH_3OH$) or one ^{13}C (from one $^{13}CH_3OH$ and one C_3H_6).

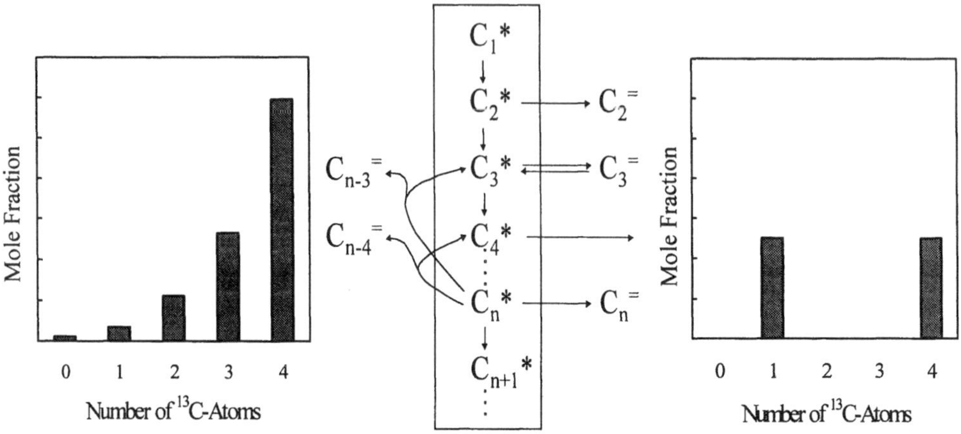

Scheme. Chain growth pathways and expected isotopomer distributions in butene formed from $C_3H_6/^{13}CH_3OH$ mixtures

$C_3H_6/^{13}CH_3OH$ reactants lead to binomial isotopomer distributions in all C_{4+} olefins (Figure 4) on H-ZSM5 (Si/Al=175), except at the low conversions within initial induction periods, suggesting that products form by sequential methylation and β-scission of large carbocations. Turnover rates, selectivities, and isotopomer distributions were not affected by crystal size on H-ZSM5 (Si/Al=175) or by contact time (after the initial induction period), confirming that these data reflect primary chain growth pathways, uncorrupted by intracrystalline transport restrictions and secondary reactions within channels. The hexene fraction contains only molecules with three or more ^{13}C atoms; thus, alkene oligomerization does not occur during methanol conversion on kinetic-limited H-ZSM5 catalysts.

These methylation-cracking pathways avoid the need for the chain initiation from CH_3OH after each alkene formation turnover, but they lead to low ethene selectivity, because of the low ethene selectivity of β-scission pathways, and to an increase in reaction rate as the average growing chain becomes larger with increasing alkene concentration. These "living" intermediates form as small chains hydrogen transfer steps that also form methane and ethane, the most abundant products along with ethene during the initial induction period. At low conversions, chains are small because alkenes are unavailable to readsorb and maintain long chains at steady-state. Thus, termination occurs predominantly by deprotonation at low conversions and reaction rates are slow because chains must be initiated using CH_3OH-CH_3OCH_3 equilibrated mixtures after each turnover. The isotopomer distribution in 1-butene formed from $C_3H_6/^{13}CH_3OH$ mixtures at low conversions (Figure 4) contains a larger than statistical concentration of butenes with one ^{13}C, which can only form by the intact desorption of butyl cations formed from one C_3H_6 and one $^{13}CH_3OH$.

Binomial isotopomer distributions were also obtained from alkene-$^{13}CH_3OH$ reactants on SAPO-34, but their assignment to methylation-cracking pathways is difficult, because transport restrictions and secondary reactions are not negligible on these small-pore catalysts. Added alkenes are less reactive on SAPO-34 than on H-ZSM5, because they diffuse through intercavity ports with much greater difficulty that methanol. Products form preferentially from the faster diffusing $^{13}CH_3OH$ molecules in the reactant mixture. $C_3H_6/^{13}CH_3OH$ gives a higher than statistical fraction of singly labeled 1-C_4H_8. This isotopomer becomes more abundant on less crystalline SAPO-34 samples, suggesting that it forms via parallel pathways on extracrystalline weak acid sites. These sites are less reactive but more accessible to added alkenes than intracrystalline acid sites.

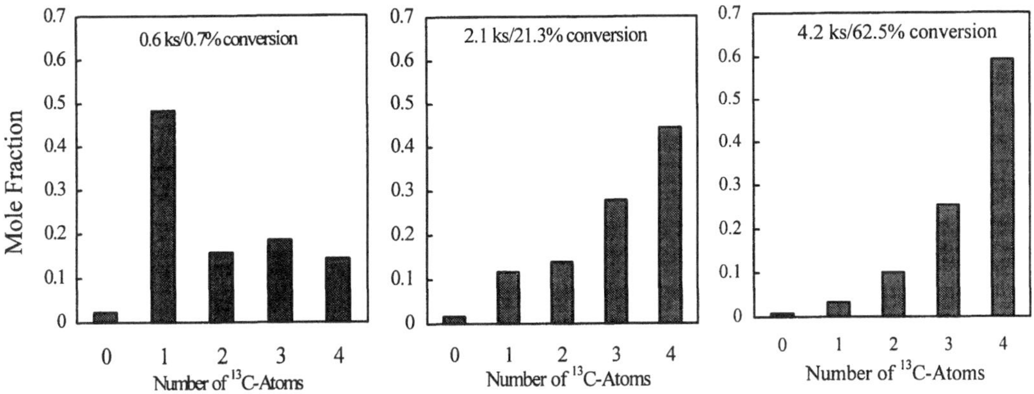

Figure 4. Isotopomer distribution in the butene products of $C_3H_6/^{13}CH_3OH$ reactant mixtures (legend: contact time/methanol+DME conversion) [H-ZSM5, Si/Al=175, 2 μ, 673 K, 10 kPa CH_3OH, 5 kPa H_2O]

The ^{12}C-content in C_{n+1} alkenes formed from $C_nH_{2n}/^{13}CH_3OH$ reactants reflects the relative reactivity of methanol and C_n alkenes. The ^{12}C content in products is much higher on H-ZSM5 than on SAPO-34, even though reactions occur via carbocation pathways on both catalysts, because transport restrictions limit the availability of added alkenes for chain growth. The reactivity of alkenes increases with chain length on H-ZSM5 (Fig. 5), as expected in acid-catalyzed pathways. The opposite trend, however, is observed on SAPO-34 (Fig. 5). Reactivity differences between H-ZSM5 and SAPO-34 become greater for larger alkenes. These differences reflect the effectiveness factor for alkene reactions on diffusion-limited SAPO-34. Isobutene is less reactive than n-butene on SAPO-34, even though the kinetics of alkene reactions on acid sites favor branched alkenes (as observed on H-ZSM5). Alkenes formed from CH_3OH within SAPO-34 cavities must also overcome these diffusional constraints as they exit the crystals. Lower alkenes are consequently sieved by SAPO-34 crystals, but larger or branched alkenes remain and approach methylation/β-scission and isomerization equilibrium, in agreement with in-situ NMR studies [13]. Isopentane to n-pentane ratios in products of CH_3OH reactions are near equilibrium on H-ZSM5 (0.74, Table), but very low (0.09) on SAPO-34, because of the selective sieving of linear chains over branched products by the small apertures in SAPO-34. The sieving of alkenes from equilibrated mixtures leads to high selectivity to ethene, a minority product of β-scission kinetics, and to low selectivity to branched alkenes on SAPO-34 but not on H-ZSM5. Such transport restrictions may also lead to the lower methanol turnover rates measured on SAPO-34 (Table). Low initial rates are not observed on SAPO-34 (Figure 1), because reaction rates are not limited by the kinetics of β-scission, but by transport rates.

Sieving from equilibrated methanol-alkene mixtures leads to product selectivities that depend on the relative diffusivity of alkenes through intercavity apertures in SAPO-34 (0.43 nm). The average chain length within this equilibrated mixture, however, depends on the relative diffusivities of methanol reactants and alkenes because the effective pressure of alkenes within SAPO-34 cavities (0.67 x 1.01 nm) increases with increasing diffusion pathlength (crystal size) and methanol reaction rates (acid site density). Severe diffusional restrictions lead to sieving of light alkenes, but from mixtures of larger alkenes and with significantly greater propensity for unreactive polymeric residues. Kinetically-limited small SAPO-34 crystals (or H-ZSM5 crystals with channels larger than SAPO-34) lead to poor sieving and to the formation of the C_3+ alkenes favored by primary β-scission pathways. Optimum ethylene selectivities appear to require intermediate levels of transport restrictions, while higher stability requires unencumbered transport. An optimum compromise appears to be reached on relative small (<0.5 μ) SAPO-34 crystals [3]. H-ZSM5 (0.51-0.56 channel diameter), while more stable, does not provide any sieving selectivity for alkenes during methanol conversion, because the kinetic diameters of linear C_2-C_4 hydrocarbons (0.4-0.5 nm) are smaller than the channel dimensions in H-ZSM5, but similar in size to the intercavity ports in SAPO-34.

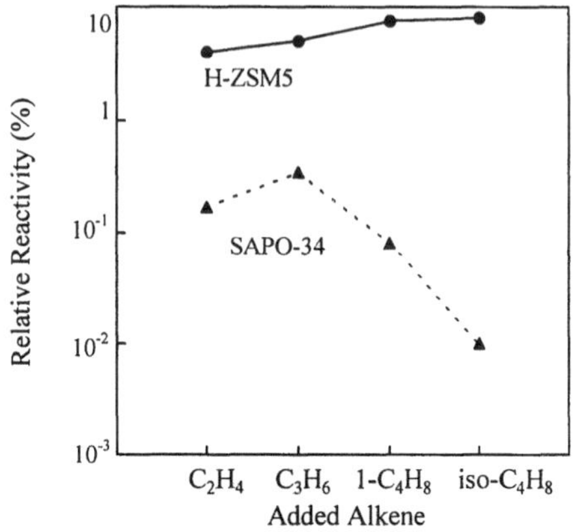

Figure 5. Relative alkene/methanol reactivity (mole basis) obtained from the isotopic content of the C_{n+1} alkene products of $^{12}C_nH_{2n}/^{13}CH_3OH$ mixtures [673 K, 10kPa $^{13}CH_3OH$, 0.96 kPa $^{12}C_nH_{2n}$, 5 kPa H_2O]

REFERENCES

[1] C.D. Chang, Catal. Rev. Sci. Eng. 26 (1984) 233.
[2] B.M. Lok, C.A. Messina, R.L. Patton, R.T. Gajek, T.R. Cannan, E.M. Flannigen, J. Am. Chem. Soc. 106 (1984) 6092; U.S. Patent 4 440 871.
[3] S.W. Kaiser, Arab. J. Sci. Eng. 10 (1985) 361; P.T. Barger, S.T. Wilson, J.S. Holmgren, U.S. Patent 5 126 308 (1982).
[4] A.J. Marchi, G.F. Froment, Appl. Catal. 71 (1991) 139.
[5] B.V. Bora, T.L. Marker, P.T. Barger, H.R. Nilsen, S. Kvisle, Stud. Surface Sci. Catal. 107 (1997) 87.
[6] G.J. Hutchings, R. Hunter, Catal. Today 6 (1990) 279.
[7] W.W. Kaeding, S.A. Butter, J. Catal. 61 (1980) 155.
[8] W.O. Haag, R.M. Lago, P.G. Rodenwald, J. Mol. Catal. 17 (1982) 161.
[9] M.M. Wu, W.W. Kaeding, J. Catal. 88 (1984) 478.
[10] R.M. Dessau, R.B. LaPierre, J. Catal. 78 (1982) 136.
[11] R.M. Dessau, J. Catal. 99 (1986) 111.
[12] B. Sulikowski, J. Klinowski, Appl. Catal. 89 (1992) 69.
[13] M.W. Anderson, B. Sulikowski, P.J. Barrie, J. Klinowski, J. Phys. Chem. 94 (1990) 2730.
[14] I.M. Dahl, S. Kolboe, Catal. Lett. 20 (1993) 329.
[15] I.M. Dahl, S. Kolboe, J. Catal. 149 (1994) 458.
[16] J.P. Verduijn, Intern. Pat. WO 93/08124 (1993).
[17] J.A. Biscardi, E. Iglesia, Catal. Today 31 (1996) 207.
[18] G.L. Price and E. Iglesia, Ind. Eng. Chem. Res. 28 (1989) 839.

NATURAL GAS CONVERSION V
Studies in Surface Science and Catalysis, Vol. 119
A. Parmaliana et al. (Editors)

The investigation of the processes of organic products synthesis from natural gas via syngas

V.M. Mysov[a], K.G. Ione[a] and A.V. Toktarev[b]

[a]Technology Institute of Catalytic and Adsorption Processes "Zeosit", Pr. Lavrentieva, 5, Novosibirsk, 630090, Russia

[b]Boreskov Institute of Catalysis, Pr. Lavrentieva, 5, Novosibirsk, 630090, Russia

The selectivity of syngas conversion for the synthesis of organic products such as aromatic hydrocarbons, dimethyl ether (DME) and high-octane gasoline on bifunctional zeolite-containing catalysts in two-step processes was investigated. The catalysts were prepared using H-form zeolites ZSM-5, ZSM-12 and β zeolite (modified by Si, Mg-compounds or nonmodified), CuO-ZnO- or ZnO-Cr_2O_3-catalysts and Al_2O_3 as a binder. The influence of temperature (280-440^{0}C), pressure (40-80 atm) and syngas composition on the selectivity of hydrocarbon synthesis in a fixed bed flow-type reactor was studied. The routes of highly selective aromatics syntheses from mixture syngas + toluene and syngas + 2-methylnaphthalene (2-MN) were shown. It is established that the selectivity of process depends on the type of zeolites. The selectivity to p-xylene – 70-91% and the one to p-ethyltoluenes – 76-97% via alkylation of toluene with syngas were achieved. It was shown that dimethylnaphthalenes (DMN) were the main products in 2-MN alkylation with syngas. In all cases the selectivity depends upon molecular sieve effects of the zeolite component of the catalysts.

1. INTRODUCTION

In conventional technologies the predominant rout of chemical transformation of natural gas is a production of synthesis gas with its further conversion to other chemical products and, first of all, to methanol. Methanol, in turn, is a raw material for production of many other organic substances such as formaldehyde, acetic acid and can also be used in new processes: high octane gasoline production, DME synthesis, and synthesis of alkyl aromathics [1-3]. It is known that synthesis of methanol in industry is carried out in thermodynamically unfavorable conditions due to low value of the equilibrium constant of the reaction of methanol formation from carbon oxides and hydrogen. That is why the search for new thermodynamically favorable paths of organic substances syntheses directly from syngas is a strategic area for many investigations [4-7].

The selectivity of syngas conversion for the synthesis of organic products such as aromatic hydrocarbons, DME and high-octane gasoline on bifunctional zeolite-containing catalysts in two-step processes shown in Figure 1 in dependence on catalysts composition and reaction conditions was studied.

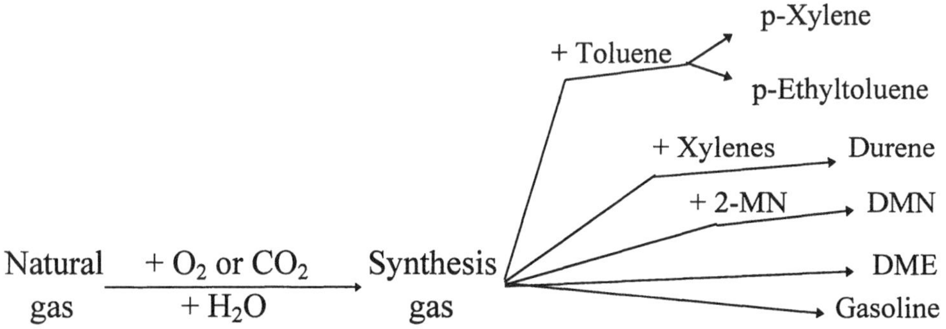

Figure 1. Scheme for synthesis of organic products from natural gas via syngas.

2. EXPERIMENTAL

ZSM-5 type zeolites with $SiO_2/Al_2O_3 = 27, 75, 100$ as well as ZSM-12 ($SiO_2/Al_2O_3 = 200$) and β zeolite ($SiO_2/Al_2O_3=75$) were prepared by hydrothermal syntheses. Mg,Si- and Si-zeolites were prepared by impregnating of H-zeolites with a solutions of Si, Mg-compounds. For preparation of bifunctional catalysts powders of zeolite component and CuO-ZnO- or ZnO-Cr_2O_3 components in desired proportion were homogenized in a mortar followed by further pressing, crushing and sieving. The particles fraction of 0.25-0.5 mm was used in the reaction.

The catalytic experiments were carried out in a fixed bed flow-type reactor in the gas phase at 280-440^{0}C under 40-80 atm. The reaction products were analyzed by three gas chromatographs equipped with integrators. H_2, CO, CO_2, N_2 and CH_4 were analyzed by the thermal conductivity detector (TCD) with an activated carbon column. Methanol, DME, water were analyzed with a TCD using column filled with chromosorb-102. The gaseous hydrocarbons were analyzed with a TCD using column packed with Al_2O_3. The liquid hydrocarbons were analyzed by the TCD using a 3-m-long column with Benton-34/SP-1200 on 100/120 Supelcoport or by the flame ionization detector (FID) with a 50-m-long capillary column. The identification of products was conducted by GC-MS.

3. RESULTS AND DISCUSSION

3.1. Hydrocarbons synthesis from carbon oxides and hydrogen mixtures

Whatever the method of preparation, bifunctional catalysts based on methanol synthesis component and synthetic zeolite may be considered as a system with non-additive, synergistic properties [8, 9]. It is known that bifunctional catalyst activity and selectivity in the process of syngas conversion depend on N_{BAC} value which characterizes a relative concentration of acidic centers in the bifunctional catalyst. As N_{BAC} value of bifunctional catalyst grows, the acidic function of the catalyst also increases with corresponding acceleration of reaction of methanol conversion to DME and hydrocarbons.

Recently we have shown the influence of N_{BAC} value on selectivity of synthesis gas conversion at low (280°C) and medium (380°C) temperatures [10]. The main organic product of synthesis gas conversion in the wide range of N_{BAC} values at 280°C was DME. Its content in total products went through the maximum with the increase of acidity of CuO-ZnO/ZSM-5 catalyst. As a result of catalyst composition optimization, as well as optimization of process parameters (pressure, temperature and space velocity of syngas), selectivity towards DME over 96% was achieved.

When the process is carried out on $ZnO-Cr_2O_3$/ZSM-5 catalyst at temperatures over 350°C, synthesis gas is converted mainly to hydrocarbons with selectivity to C_{5+} from 55 to 76% [4, 11]. We have studied the influence of pressure, temperature and syngas composition on productivity of $ZnO-Cr_2O_3$/ZSM-5 catalyst towards C_{5+} hydrocarbons (Table 1) and their distribution (Figure 2).

Table 1
Dependence of C_{5+} productivity (relative units) upon pressure and temperature

Temperature, ^0C / Pressure, atm	40	60	80
320	not measured	not measured	1,0
340	1.0	1.2	1.6
360	1.1	2.2	3.7
380	1.3	3.3	4.2
400	2.2	5.0	5.9

One can see from Table 1 that productivity of $ZnO-Cr_2O_3$/ZSM-5 catalyst steadily grows with the increase of pressure from 40 to 80 atm and temperature from 320 to 400°C. Figure 2 shows that the increase of hydrogen content in the gas mixture results in lowering of aromatic hydrocarbons content in gasoline fraction. On the base of results of the present work it may be concluded that the process of conversion of synthesis gas to hydrocarbons on studied catalysts goes according to the scheme of non-trivial bifunctional catalysis over the stages of methanol and DME formation. It is possible to obtain DME, aromatics or iso-paraffins with high selectivity depending on catalyst's acidity, process temperature and syngas composition.

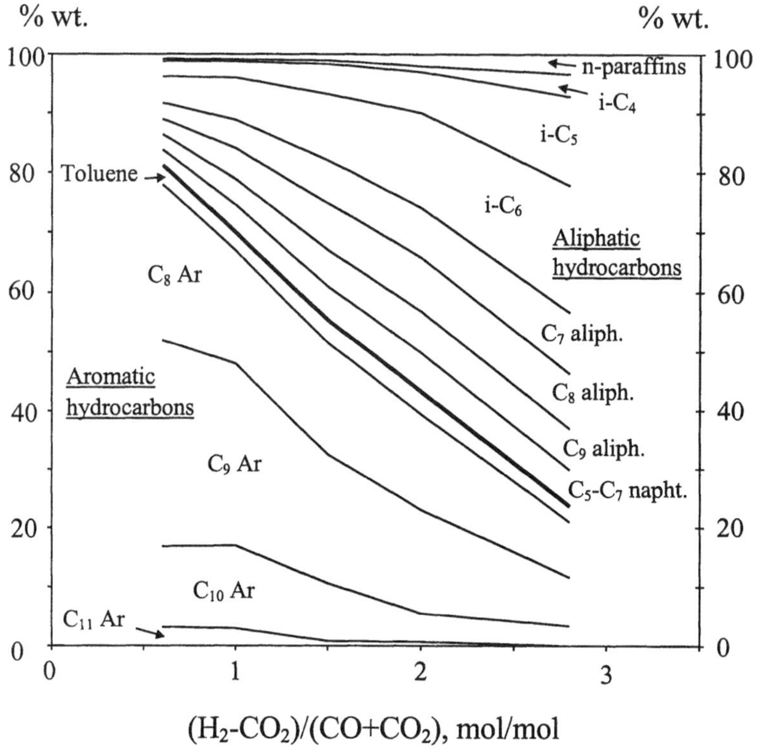

Figure 2. Influence of syngas composition on the C_{5+} hydrocarbon distribution.

3.2. Alkylation of aromatics with synthesis gas

Alkylation of toluene with methanol or ethene on modified by different elements ZSM-5 type zeolites to produce p-xylene or p-ethyltoluene is studied in detail previously [12-15]. In many papers it was shown that synthesis gas may be used as an alkylation agent instead of methanol. It was admitted [16, 17] that the efficiency of alkylation by methanol being formed in the catalyst bed continuously, was significantly higher compared with that in the case of the conventional methylation with methanol. But, Namba *et al.* showed that although alkylation of toluene with methanol on modified with boron ZSM-5 zeolites proceeded with as high as 94% selectivity to p-xylene [16], the use of the same modified zeolites bound with Zn-Cr component in the bifunctional catalyst gave only 43% selectivity with toluene conversion of 6.5% only in alkylation of toluene with synthesis gas.

We showed that the selectivity of the process to p-xylene was 52% with toluene conversion more than 40% when bifunctional catalysts were not modified [10]. Moreover, the composite catalysts containing ZSM-5 zeolite modified with Mg and Si oxides exhibited selectivity to p-xylene and p-ethyltoluene nearly 70-91% and 76-97%, respectively. Toluene conversion was 20-30% but syngas conversion was over 30%.

We studied the bifunctional catalysts containing ZSM-5, ZSM-12 and β zeolites for reaction of 2-MN methylation with synthesis gas. The main alkylation products were dimethylnaphthalenes (65-93% in total naphthalene products), but naphthalene, trimethylnaphthalenes and syngas conversion products were formed too. It was shown that selectivity of 2-MN alkylation process depends on temperature, contact time, catalyst composition, dimensions of zeolite channels and acidity of zeolite. The selectivity to (2,6+2,7)-DMN for ZSM-5 and the one to 1,2-DMN for β zeolites changed with process time. Effect of zeolite channels dimensions on the process selectivity is shown in Table 2.

Table 2
Alkylation of 2-methylnaphthalene with synthesis gas

Type of zeolite	ZSM-5		ZSM-12		β zeolite	
Dimensions of the channels, Å	5.3×5.6		5.5×6.2		6.4×7.6	
Hours on stream	2	12	2	12	2	12
DMN distribution, wt %						
2,6 + 2,7	48.2	91.5	18.1	16.6	25.6	4.1
1,7	26.8	1.1	15.7	15.8	8.0	15.7
1,3 + 1,6	18.5	3.4	21.1	21.4	35.5	14.7
2,3 + 1,4	2.6	0.8	9.9	10.1	16.8	7.7
1,5	trace	-	trace	-	trace	trace
1,2	3.9	3.2	35.2	36.1	14.1	57.8

Reaction temperature - 440°C, pressure - 80 atm.

From Table 2 one can see that increased selectivity to (2,6+2,7)-DMN is achieved on ZSM-5 and increases more with process time. On wide pore zeolites the distribution of formed DMN is in accordance either with thermodynamic equilibrium (β zeolite, first 2 hours), or with an increased amount of 1,2-isomer.

CONCLUSION

It was shown conversion of syngas on bifunctional zeolite-containing catalysts in two-step processes may result in synthesis of p-xylene (70-91% selectivity), p-ethyltoluene (76-97% selectivity), DMN (65-93% selectivity), DME (over 96% selectivity), aromatics and iso-paraffins. These data may be applied for synthesis of these compounds from natural gas via its conversion to synthesis gas.

REFERENCES

1. C.D. Chang and A.J. Silvestri, J. Catal., 47 (1977) 249.
2. T. Hibino, M. Niwa and Y. Murakami, J. Catal., 128 (1991) 551.
3. T. Komatsu, Y. Araki, S. Namba and T. Yashima, in "Zeolites and related microporous materials: State of the art 1994" (J. Weitkamp et al. eds), Stud. Surf. Sci. Catal., Elsevier, Amsterdam, 84 (1994), 1821.
4. F. Simard, U.A. Sedran, J. Sepulveda, N.S. Figoli and H.I. de Lasa, Appl. Catal., 125 (1995) 81.
5. R.A. Comelli and N.S. Figoli, Ind. Eng. Chem. Res., 32 (1993) 2474.
6. M.F.M. Post and S.T. Sie, Process For Carrying Out Catalytic Conversions, US Patent No. 4 403 044 (1983).
7. K. Fujimoto, T. Shikada, Y. Yamaoka and T. Sumigama, Method Of Producing Dimethyl Ether, US Patent No. 5 466 720 (1995).
8. C.D. Chang, W.H. Lang and A.J. Silvestri, J. Catal., 56 (1979) 268.
9. K. Fujimoto, Y. Kudo and H. Tominaga, Pan-Pacific Synfuels Conference, Tokyo, 1982, V.1, 123.
10. K.G. Ione and V.M. Mysov, in "Catalysis in coal conversions" (B.N. Kuznetsov and V.V. Lunin eds), Proceedings of the Third International Symposium, Russia, Novosibirsk, 1997.
11. F. Simard, A. Mahay, A., A. Ravella, G. Jean and H.I. de Lasa, Ind. Eng. Chem. Res., 30 (1991) 1448.
12. W.W. Kaeding, C. Chu, L.B. Young, B. Weinstein and S.A. Butter, J. Catal., 67 (1981) 159.
13. W.W. Kaeding, L.B. Young and C.C. Chu, J. Catal., 89 (1984) 267.
14. T. Hibino, M. Niwa and Y. Murakami, J. Catal., 128 (1991) 551.
15. V.N. Romannikov and K.G. Ione, React. Kinet. Catal. Lett., Vol. 51, № 1 (1993) 151.
16. Namba S., Yamagishi T. and Yashima T., Nippon Kagaku Kaishi, № 3 (1989) 595.
17. Yashima T., Yamagishi T. and Namba S., Proc. 9th Int. Congr. Catal., Vol.1, Ottawa, 1988, 453.

NATURAL GAS CONVERSION V
Studies in Surface Science and Catalysis, Vol. 119
A. Parmaliana et al. (Editors)

A review of low temperature methanol synthesis

M. Marchionna[a], M.Di Girolamo[a], L. Tagliabue[a], M.J. Spangler[b] and T.H. Fleisch[c]

[a] SNAMPROGETTI, Research Laboratories,
Via Maritano 26, 20097 San Donato Milanese, Italy

[b] AMOCO Research Center,
150 West Warrenville Road, Naperville, IL 60563-8460, USA

[c] AMOCO Production Company,
PO Box 3092, Houston, Texas 77253-3092, USA

It has been more than ten years since the discovery of highly active catalysts which allow low temperature (< 150°C), single pass conversion of syngas to methanol. Detailed studies including catalyst refinements, reactor engineering, and process development have been conducted at the Snamprogetti and Amoco laboratories. The presence of electron-rich metal hydrides is the key to the unusual activity of these catalytic systems. Catalyst deactivation phenomena have till now hampered the development of a commercially viable process.

1. INTRODUCTION

For many decades methanol has been regarded as "liquid C_1" that is the vector for natural gas, particularly from remote areas, to markets [1-3]: feedstock for chemical production (methanol based chemistry), fuel for transportation or power stations, octane booster blending component (alone or with solubilizers) in gasoline, neat (M100 or M85) alternative fuel for spark or diesel engines, and intermediate for the production of fuel additives for reformulated gasoline such as MTBE (methyl t-butyl-ether). Currently the latter is used on a very large scale.

Recent interest in methanol for energy applications has stimulated research toward cheaper production processes. In existing technologies the major costs (> 60%) are associated with syngas production. A significant reduction in these costs would be achieved using partial oxidation with air, but because of the nitrogen dilution, a much higher per pass conversion would be needed than is currently obtained in processes based on heterogeneous catalysts. As conversion to methanol is adversely affected by increasing temperature, several research groups have addressed their efforts toward development of active catalysts able to operate at low temperatures (60-150°C) where the equilibrium is more favorable [4].

Investigations of these catalytic systems have shown that the synthesis consists of two steps - methanol carbonylation to methyl formate followed by hydrogenolysis of the methyl formate (or strict derivatives) to two methanol molecules. The reactions can be carried out either concurrently in a single reactor (single stage process) or in two separate reactors (two-step process) [4].

Research reviewed here is only on the most promising single stage process. The catalytic systems are composed of a metallic component (nickel, copper, cobalt, palladium, or combinations of these metals) in the presence of an excess of a strong base, typically alkaline alkoxides. Nickel homogeneously-catalyzed systems have been studied by several companies and institutions (Amoco, Snamprogetti, Shell, Mitsui Petrochem., NKK) [4-6]. Copper systems (both homogeneous and heterogeneous) have attracted the attention of several companies (Snamprogetti, Mitsui Petrochem., Shell, Union Carbide, NKK) and institutions (in Norway, in USA, in Japan and in China) [4-6]. Other metal systems have generated interest; among them, cobalt [5,6], palladium [6], molybdenum [7] and chromium [8]. Mixtures of all these metals have shown catalytic activity although it is worth noting that the most active catalysts are based on nickel and copper.

Reported here will be some of the main results independently achieved by Amoco (in collaboration with Brookhaven National Lab, BNL) on the nickel system and by Snamprogetti (in collaboration with the University of Pisa) on the copper system.

2. EXPERIMENTAL

Experimental procedures using copper systems were reported elsewhere [9]. For the nickel systems, *in-situ* infrared spectroscopy experiments to investigate the intermediate catalytic species were reported in ref. [10]. To verify catalytic activity, a dedicated laboratory scale pilot plant reactor unit was constructed by Amoco for the evaluation of the former BNL technology. The unit consisted of a 50 ml CSTR Autoclave reactor with once-through continuous gas and catalyst supply; operations were unattended with automatic sampling for liquids and gases. Extraordinary safety precautions were adopted for containment and monitoring of CO and especially $Ni(CO)_4$.

3. MECHANISM AND CATALYTIC SPECIES INVOLVED

For nickel systems, *in-situ* infrared spectroscopy monitoring of the $Ni(CO)_4$/MeONa system was performed under the same experimental conditions of the catalytic runs [9]; it illustrated the involvement of the $[HNi(CO)_3]^-$ species in the catalysis. The overall results led to a hypothetical catalytic cycle, in which methanol is carbonylated to methyl formate with MeONa as the catalyst, whereas $[HNi(CO)_3]^-$ catalyzes the hydrogenolysis of methyl formate to methanol.

For copper systems, composed of copper (I) salts and alkaline methoxides, the catalyst is not completely homogeneous, with the alkaline alkoxide in solution and the copper present as a dispersed solid. The solid catalyst, generated under reaction conditions by the interaction of CuCl with excess MeONa, contains Cu in the +1 oxidation state as observed by ^{65}Cu NMR. When lipophilic alkoxides such as $CH_3OCH_2CH_2ONa$ are used, the resulting copper system is completely homogeneous. Mechanistic explanations of the results concerning the characterization and reactivity of the catalytic species are as follows: in the case of copper, a complex copper-sodium cluster species, $\{HNa_xCu(OMe)_x\}_y$, is likely involved. Also catalyst activation has been rationalized according to the following scheme (Figure 1).

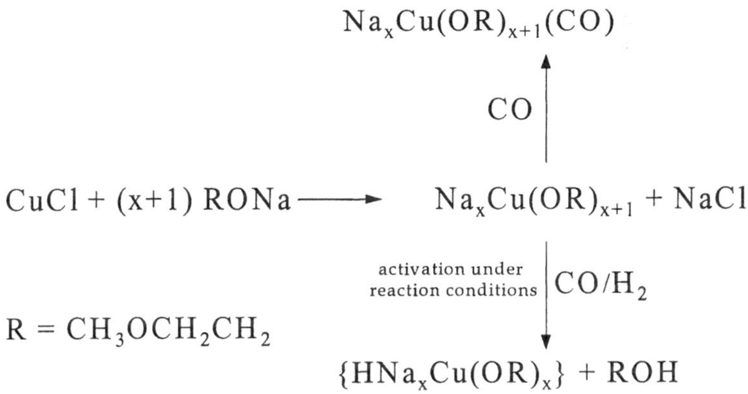

Figure 1. Mechanism of the $CH_3OCH_2CH_2ONa/CuCl$ catalytic system.

The extremely mild reaction conditions under which this methanol synthesis proceeds mainly result from the easier reduction of the activated C-O bond in methyl formate compared to that of carbon monoxide. The presence of electron-rich metal hydrides is the key to the high hydrogenating activity of these catalytic systems.

4. CATALYTIC ACTIVITY AND DEACTIVATION PHENOMENA

To demonstrate this process, a dedicated laboratory scale reactor unit was constructed by Amoco for evaluation of the nickel system. Different parameters were investigated: catalyst composition, MeOH/co-solvent ratio, temperature, pressure, N_2 dilution of syngas, space velocity, and poisons (CO_2, H_2O) effect.

Under conditions of continuous single-pass operation, high syngas conversions (> 90%) with high selectivities to methanol (95-99%) were achieved.

It was observed that, for a given nitrogen content, methanol productivity increased almost linearly with increasing syngas pressure. However, when the nitrogen content was increased (Table 1), and despite the fact that thermodynamic equilibrium depends only on CO and H_2 partial pressure, the productivity diminished because of kinetic effects.

Table 1
Nickel catalytic system performances.

T (°C)	MPa	% N_2	Syngas Conversion (%)	Productivity (mol MeOH/h/l)
110	2	5	96	1.7
130	2	5	92	3.2
145	2	5	91	4.7
130	4	5	93	8.1
130	2	40	86	1.3
130	4	40	87	3.1

$H_2/CO=2$; GHSV:100-600 sccm; Catalyst:$Ni(CO)_4$/KOMe; Solvent: MeOH/triglyme

Similar results were achieved by Snamprogetti on copper systems. In order to increase the reaction rate at low pressures or with high nitrogen content, an increase of catalyst concentration was necessary; also an increase of the H_2/CO molar ratio was beneficial.

In both the systems, the rate of methanol synthesis increased with increasing metal (Ni or Cu) concentration, thus suggesting that the hydrogenolysis step may be the rate limiting one. The systems are highly active (initial Turnover Frequencies (TF) of 100-600 h^{-1}; initial MeOH productivities of 1-9 M/h) under very mild conditions (T = 60-150°C; P = 0.5-4 MPa).

Unfortunately, the strong base component, the use of which is unavoidable, tends to deactivate with time because of the effect of acid poisons such as CO_2 and H_2O or of side-reactions. Actually, methoxide is transformed to weaker bases such as methyl carbonate and formate through different reactions:

$$MeONa + CO_2 \longrightarrow MeOCOONa \quad (1)$$

$$MeONa + H_2O \longrightarrow MeOH + NaOH \quad (2)$$

$$NaOH + CO \longrightarrow HCOONa \quad (3)$$

$$MeONa + HCOOMe \longrightarrow HCOONa + MeOMe \quad (4)$$

In the copper systems deactivation may proceed also due to a metal-catalyzed mechanism: in fact, model stoichiometric reactions with copper hydrides (HCuL; L = PPh_3, pyridine or absent) have pointed out the following reactivity:

$$HCuL + H_2O + HCOOMe + L \longrightarrow (HCOO)CuL_2 + MeOH + H_2 \quad (5)$$

The poisons would need to be limited in the feed streams to part per million levels, thus requiring special syngas generation reaction conditions and additional equipment for acid gas removal.

5. ECONOMIC EVALUATIONS

Furthermore, in our preliminary economic evaluations it was discovered that methanol synthesis, in order to be economic, must be carried out under very low pressure (< 2 MPa) because, even though air is a very cheap oxygen source, the costs for its compression from atmospheric to synthesis reactor pressure are very high (much higher than those for the CO_2 removal unit). Using these conditions and the most advanced syngas preparation processes, i.e. autothermal reforming or rhodium catalyzed partial oxidation, the production cost of methanol in this process is the same or slightly less than in a conventional adiabatic multistage process, but the difference is too small to compensate for the technical and commercial risks needed to develop such a novel process (Figure 2).

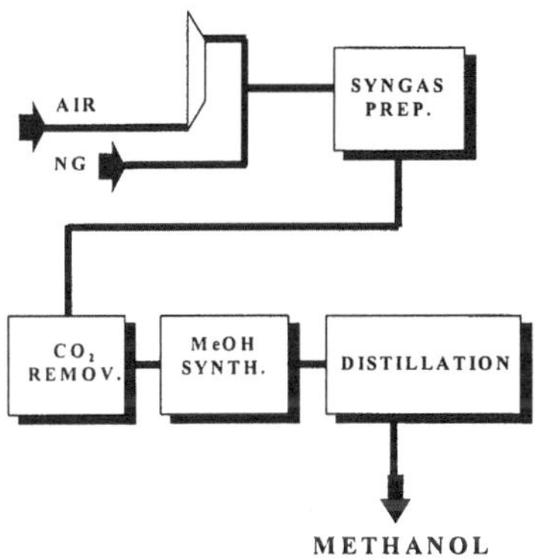

Relative investment costs

	CAM	ULT
Syngas prep.	73.1	57.5
MeOH Synth.	19.0	19.4
Separation	6.3	4.7
Auxiliary	1.6	7.1
total	100	88.7

CAM = conventional adiabatic methanol

Consumption (Gcal/t)

process	7.8	8.2
fuel	0.6	1.6
total	8.2	9.8

Snamprogetti, Internal analysis

Figure 2. Low Temperature Methanol Synthesis flow diagram and economics.

Economic evaluations done by Amoco came to similar conclusions. In addition to the concerns addressed above, their evaluation cited disadvantages from inadequate utilization of low-quality heat (because of the lower reaction temperature), and additional capital costs for $Ni(CO)_4$ containment and the larger reactors needed because of lower methanol synthesis rates than in the conventional process.

6. FURTHER DIRECTIONS

Economic evaluations clearly display that improvement could be derived from further savings in syngas preparation and by reducing auxiliary costs, mainly related to a CO_2 removal unit. Present research is focused toward these two directions - more strict integration of the syngas preparation and methanol synthesis steps and improvement in catalyst tolerance to impurities through either *in-situ* or *ex-situ* regeneration of the base components.

With regard to the first aspect, it is worth noting that there is a general effort toward improving the syngas preparation step by means of new production methods, i.e. catalytic partial oxidation, the use of ceramic membranes which could simplify the N_2/O_2 separation, or by other means [11]. It is likely that, in the near future, low temperature methanol synthesis could benefit from advances in this area.

With regard to the second aspect, it is mandatory to overcome the problem of the disposal of the deactivated weak base, which tends to accumulate. Two solutions are possible: either the base is auto-repaired *in-situ* or is regenerated *ex-situ*.

Some examples of *in-situ* regeneration of the deactivated base have been recently reported, both in the direct synthesis of methanol from syngas [12,13] or in the related reaction of methanol carbonylation to methyl formate [13]. The main problem to overcome in these cases is the low reaction rate under low pressures. *Ex-situ* regeneration has the advantage of decoupling the regeneration step from the reaction one thus maintaining, on one side, the high activity of the methanol synthesis catalyst and obtaining, on the other side, higher regeneration yields. Some attractive results have been recently achieved in this area [14].

ACKNOWLEDGEMENTS

Besides all the coworkers involved in the different projects, the authors wish to thank Dr. Y. Ohno (of NKK) for information on recent Japanese research.

REFERENCES

1. Europ. Chem. News, 12-18 January, 1998, 35.
2. J.-P. Lange, Ind. Eng. Chem. Res., 36 (1997) 4282.
3. M. J. Gradassi, N. W. Green, Fuel Processing Technology, 42 (1995) 65.
4. M. Marchionna, M. Lami, A.M. Raspolli Galletti, Chemtech, April 1997, 27 and references therein contained.
5. M. Marchionna; Braca, G. (ed.), Oxygenates by Homologation or CO Hydrogenation with Metal Complexes, Kluwer, Dordrecht, **1994**, pp. 57-61.
6. S. Ohyama, Shokubai, 38 (1996) 326.
7. R.S. Sapienza, W.A. Slegeir, T.E. O'Hare, D. Mahajan, US 4 619 946 (1986).
8. K. Fujiwara, H. Matsuda, T. Daimatsuzawa, H. Sugawara, S. Akamatsu JP 09 249 593, (1997).
9. M. Marchionna, M. Lami, F. Ancillotti, EP 375 071, (1990).
10. M. Marchionna, L. Basini, A. Aragno, M. Lami, F. Ancillotti, J. Mol. Catal., 75 (1992) 147.
11. Europ. Chem. News, Chemscope, September 1997, 24.
12. J.W. Tierney, I. Wender, W.M. Palekar, USP 5,384,335 and 5,385,949 (1995).
13. D. Mahajan, N. Gupta, AIChE Spring Nat.Meet., Houston (1993), Paper 61-a.
14. A.M. Raspolli Galletti, M. Di Girolamo, M. Marchionna, unpublished results.

NATURAL GAS CONVERSION V
Studies in Surface Science and Catalysis, Vol. 119
A. Parmaliana et al. (Editors)

CHARACTERIZATION AND SELECTIVITY OF Ru/MoO_3 CATALYSTS FOR THE FORMATION OF OXYGENATES FROM CO + H_2. INFLUENCE OF THE TEMPERATURE OF REDUCTION.

M. Dufour [a], F. Villars [a], L. Leclercq [a] ,G. Leclercq [a], M. J. Pérez-Zurita [b], M. L. Cubeiro [b] , M. R. Goldwasser [b] and G. C. Bond [c]

[a] Université des Sciences et Technologies de Lille, Laboratoire de Catalyse Hétérogène et Homogène, URA CNRS n°402, Bâtiment C3, 59655 Villeneuve d'Ascq, France (e-mail : catalyse@univ-lille1.fr).

[b] Universidad Central de Venezuela, Facultad de Ciencias, Escuela de Quimica, Apartado Postal 47102, Caracas, Venezuela.

[c] Brunel University, Chemistry Department, Uxbridge, Middlesex, UB8 3PH, U.K.

Ru/MoO_3 catalysts containing 0.5, 1 and 2 wt% Ru have been prepared using two kinds of MoO_3, and their reduction by H_2 has been followed by XRD and TGA. Ru catalyses the reduction of the MoO_3; a hydrogen Mo bronze is first formed, and reduction to Mo metal is complete at 773 K. Catalysts reduced at 623 K or above give ~ 25-30% of C_1-C_3 alcohols from syngas; the proportion of C_2 or C_3 alcohols decreases with reduction temperature, and when this is 723 K only methanol is formed. Reduced MoO_x species promote alcohol synthesis either through an SMSI effect or through a dual-site mechanism.

INTRODUCTION

Synthesis of higher alcohols, particularly C_2 and C_3 alcohols, from syngas ($CO+H_2$) in acceptable yields is of potential economical interest and several types of catalyst have been studied for this purpose. The approach adopted in this work was to select ruthenium (Ru) as the catalytically active metal, because of its capability to construct C-C bonds and to produce mainly long chain hydrocarbons. The challenge is to modify this intrinsic property of Ru in order to produce alcohols by partly impeding the breaking of the C-O bond in CO and favouring CO insertion. From the available literature, much of which concerns other metals such as rhodium [1], it appeared that the most promising modifiers or promoters would be ions or oxo-species of elements such as molybdenum, tungsten, titanium, vanadium, etc. The formation of oxygenates by modified Ru catalysts has also been stated [2-6] to give high yield of straight-chain alcohols. The pre-treatment of the catalyst seems to be a crucial point. Of a number of catalysts tested, Ru/MoO_3 gave the best alcohol selectivity, and we found that the more reduced the support was, the better alcohol production we got. In this study we report on the influence of the specific surface area of MoO_3 and of the extent of its reduction on its catalytic performance in syngas conversion.

1. EXPERIMENTAL

Preparation and characterisation of catalysts : two kinds of supports were used in this study : commercial MoO_3 (2 m^2/g) supplied by Merck and MoO_3 (13 m^2/g) prepared in our laboratory according to Fransen's procedure [7]. Both were impregnated with an aqueous $RuCl_3$ (Fluka) solution to give nominal Ru contents of 0.5, 1 or 2 wt %. Catalyst samples were characterised by chemical analysis, surface area (measured by a single point BET apparatus), thermogravimetry in flowing H_2 (TGA), temperature programmed X-ray diffraction in flowing H_2, and by H_2 and CO chemisorption.

Thermogravimetry: thermogravimetric analyses were performed with a Sartorius vacuum microbalance. All treatments were carried out with a temperature gradient of 150 K/h and ended with a 3 h isothermal period at 773 K with the same gas flow rate of 3 l/h. Two treatments were applied: direct reduction in flowing H_2 or three step-procedure consisting first in drying in He, followed by cooling to room temperature under He and finally reduction in H_2 (773 K, 3l/h).

X-ray diffraction: the temperature programmed XRD experiments in flowing H_2 were run in a Siemens D5000 diffractometer. A diffractogram was recorded at room temperature, then at every 50 K (120 K/h) up to 873 K, and again after reduction at room temperature.

H_2 and CO chemisorptions : the chemisorptions were performed at room temperature in a conventional glass volumetric system pumped by a turbomolecular pump. The samples were first reduced at 523 K for 3 h, then outgassed at 523 K overnight at 10^{-6} Torr. For further calculations, the amounts of H_2 and CO chemisorbed were obtained by extrapolation of the isotherms to zero pressure.

Catalytic Activity Measurement : catalysts (1 g) were tested in CO hydrogenation at high pressure (5 MPa), in a stainless steel fixed-bed flow reactor, with a space velocity (VVH) of 6000 h^{-1} and feed gas composition of H_2 :CO = 2 :1. After reduction in flowing H_2 at one of three temperatures in the range 523-723 K, and at atmospheric pressure, the catalysts were cooled to the reaction temperature (473 K). Products were sampled into an on-line gas chromatograph equipped with TCD and FID detectors respectively with a CTR-1 and a Tenax column. The activities and selectivities were determined in the range of 453-493 K at a total pressure of 5 MPa. The selectivity to each product is defined on the basis of CO consumed.

2. RESULTS AND DISCUSSION

Chemical analysis and surface area : chemical analyses and total surface areas are reported in Table 1.

No significant difference was found between the specific surface areas of Ru catalysts supported on the commercial MoO_3 and that of the support alone, but there was a decrease by a factor of two with the home-made MoO_3.

TGA measurements : a study of the reduction of MoO_3 (Merck) and of the Ru/MoO_3 (Merck) catalysts has been performed by TGA. The results obtained after direct reduction, and after drying in He flow at 773 K, followed by reduction, differed in that the former showed clear evidence for the desorption of water at low temperature; otherwise they were similar, but we will discuss the results obtained by reducing the dried samples, since they were simpler. In Table 2, stoichiometries of the reduced products from support and catalysts versus time and temperature are reported. Total weight loss of dried samples was about the same for MoO_3 and for the catalysts, viz. 23-24 %.

Table 1
Composition and specific surface areas

sample	Mo /wt %	Ru /wt %	Cl /wt %	Fe /ppm	surface area /m^2g^{-1}
MoO_3 (Merck)	69.8	-	-	120	1.6
0.5wt%Ru/MoO_3 (Merck)	70.4	0.9	0.7	140	1.7
1wt%Ru/MoO_3 (Merck)	72.0	0.7	0.6	210	1.4
2wt%/MoO_3 (Merck)	69.8	1.4	1.4	140	1.4
MoO_3	65.4	-	-	-	13
1wt%Ru/MoO_3	66.3	1.0	0.5	-	7

With each catalyst, a slight weight increase was first observed, beginning at about 443 K. Such an increase was never found with the support alone, and should be ascribed to the formation of a hydrogen molybdenum bronze, H_xMoO_3 [9-10]. The weight increase did not seem to be much influenced by Ru content, the approximate formula being close to $H_{0.5}MoO_3$. Then about 473 K a sharp weight decrease started; this is primarily due to dehydration of the bronze phase. Reduction of the Ru salt must occur before (perhaps only just before) the bronze is formed, as the metal is needed to provide H atoms. The weight loss due to Ru salt reduction is probably superimposed on the weight gain due to bronze formation. Finally, from about 493 K, a slower weight loss occurred, similar to that obtained with MoO_3 alone, but at a significantly lower temperature. Figure 1, which compares the reduction of the support alone with the 2%Ru/MoO_3 catalyst shows the strong catalytic effect of Ru on the reduction of MoO_3. After 3 h of reduction at 773 K, MoO_3 was not completely reduced to Mo metal, its global formula corresponding approximately to $MoO_{0.6}$. A constant final degree of reduction to Mo, whatever the Ru content, was observed and agreed with previous results of our laboratory on Pt-Mo/SiO_2 [8].

X-ray diffraction : Before reduction the two supports MoO_3 (Merck) and 2wt%Ru/MoO_3 (Merck) exhibited the same crystallographic phase, i.e. molybdite MoO_3. Diffraction peaks due to $RuCl_3$ are indistinguishable from those of the support. For MoO_3 alone, no change occurred below 673 K, but at 723 K, monoclinic MoO_3 appeared ; Mo metal was observed only at 873K. Concerning 2wt%Ru/MoO_3, some authors [9] have shown that below 473 K, there is the formation of hydrogen molybdenum bronzes (obtained in particular conditions and confirmed by TGA), H_xMoO_3, with $0.3<x<0.9$. At 473 K the approximate formula of the bronze on 0.5wt%Ru/MoO_3 should be the same than on 2wt%Ru/MoO_3, since the diffraction patterns look alike. The addition of the precious metal enhanced the reduction of materials containing Ru, compared to MoO_3 alone, since reduction step started at the lower temperature of 473 K. These results confirmed those obtained by thermogravimetric analyses.

Table 2
TGA results for reduction of dried catalysts and support (Merck MoO_3)

samples	Temperature /K	time /min	weight loss (%)	oxide formulae
MoO_3	713-773	170-250	12	$MoO_{1.6}$
	773	250-280	1.5	$MoO_{1.5}$
	773	370	9	$MoO_{0.6}$
0.5wt%Ru/MoO_3	493-523	85-95	6.5	$MoO_{1.8}$
	523-773	95-205	13	$MoO_{0.7}$
	773	370	4	$MoO_{0.4}$
1wt%Ru/MoO_3	483-521	75-90	6	$MoO_{1.7}$
	521-773	90-125	16	$MoO_{0.4}$
	773	370	5.5	Mo
2wt%Ru/MoO_3	486-523	75-90	5.5	$MoO_{1.9}$
	523-773	90-205	13	$MoO_{0.8}$
	773	370	3	Mo

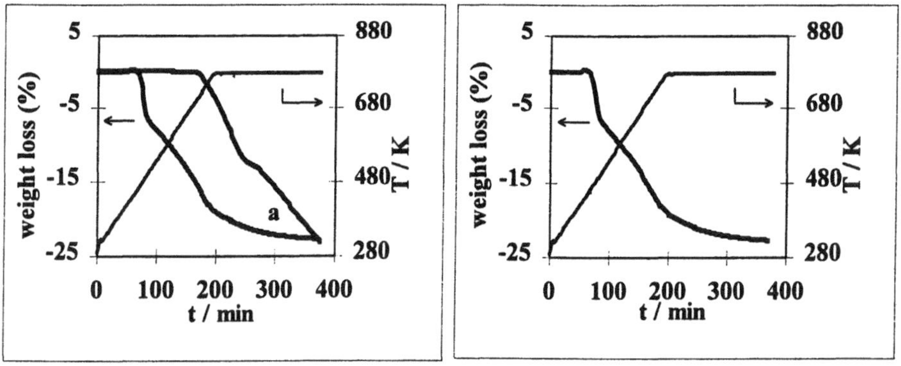

Figure 1. Thermoreduction of MoO_3 (Merck) **a** and 2wt%Ru/MoO_3 (Merck) **b** after drying in He

Chemisorption : the amounts of H_2 and CO chemisorbed on the three Ru/MoO_3 (Merck) catalysts are plotted in Figure 2 and are quite similar whatever the Ru content. The chemisorption capacity increased smoothy with temperature of reduction. The amount of the chemisorbed CO was always higher than that of H_2, which showed that obviously CO did not chemisorb only on Ru metal with a stoichiometry of 1 CO per Ru atom at the surface. When using a fresh sample of catalyst and performing a direct reduction at 673 K, the same values of H_2 and CO uptake were obtained.

Catalytic test : TGA and XRD experiments showed that MoO_3 reduction continuously increased with time and temperature. Thus, we decided to study the influence of the temperature of reduction on the activity and selectivity of Ru/MoO_3 catalysts with the same Ru content (1 wt % Ru) based on commercial MoO_3 and MoO_3 prepared in the laboratory.

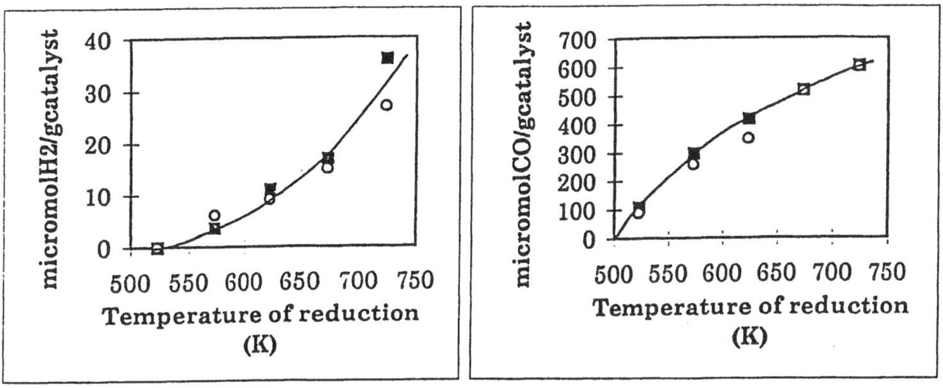

Figure 2. Development of chemisorption capacity versus temperature of reduction
■ 2wt%Ru/MoO_3 (Merck) O 0.5wt%Ru/MoO_3 (Merck)

An increase of the reduction temperature from 523 K to 623 K increased the selectivity to alcohols, but a further increase to 723 K did not bring any change to the total alcohol selectivity. Methanol was the only alcohol formed at 723 K (Table 3).

Table 3
Influence of the reduction temperature on the activity and the selectivity of 1%Ru/MoO_3 catalysts after 20h on stream at 473 K.

catalyst		Ru/MoO_3			Ru/MoO_3 (Merck)		
Temperature of reduction / K		523	623	723	523	623	723
Conversion (%)		0.6	0.3	0.8	0.5	0.3	0.3
Alcohol Selectivity (%)		17	23.8	24.6	23.6	29.9	29.1
Atom ratios	$\frac{C_1\,OH}{\Sigma\,C_n\,OH}$	71.6	73.3	100	69.9	91.1	100
	$\frac{C_2\,OH}{\Sigma\,C_n\,OH}$	28.4	26.7	0	20.1	8.9	0
	$\frac{C_3\,OH}{\Sigma\,C_n\,OH}$	0	0	0	10	0	0

A temperature of reduction of about 623 K was optimal for obtaining higher alcohols. A direct reduction at 623 K followed by a reaction test performed under the same conditions (473 K) gave an identical distribution of alcohols. In each case the specific surface area of the catalysts was higher after they had been used (increases respectively from 7.3 to 7.6 $m^2.g^{-1}$ and from 1.4 to 2.7 $m^2.g^{-1}$), but apparently, the area did not influence the catalytic performances, since the two sets of results were quite similar.

CONCLUSION

In conclusion it appears that formation of methanol and CO insertion to give higher alcohols are sensitive in different ways to the reduction state of the catalysts. The CO insertion needs a more oxidised state than the methanol formation. The reduction state of the support is a very important parameter to optimise the promoter effect of Mo suboxide and to control the selectivity toward the formation of longer chain alcohols than methanol. These results can be explained by a promotional effect which may follow either from a cooperation between metal particles and promoter sites at the periphery of the metal or in the manner proposed to account for the so-called "strong metal support interaction" (SMSI): promoter ions or oxo-species may migrate from the support into the metal [11-12]. As suggested earlier [13], the best selectivity may occur when the surface of the Ru particles is partially covered by oxo-Mo species in a way which maximises the concentration of Ru-Mo pair-sites.

ACKNOWLEDGEMENT

We are grateful to the European Union Council for the provision of funds which made our collaboration possible (grant n°CI1*-CT92-0093).

REFERENCES

1. C.S. Kellner and A.T. Bell, J. Catal. 71 (1981) 288.
2. M. Inoue, T Miyake, Y Takegami and T. Inui, Appl. Catal. 29 (1987) 285 ; J. Molec. Catal. 45 (1987) 285; J. Chem. Soc., Chem Comm. (1983) 70.
3. K.J. Smith and R.C. Everson, J. Catal. 99 (1986) 319.
4. M. Inoue, A. Kurusu, H. Wakamatsu, K. Nakajima and T. Inui, Appl. Catal. 29 (1987) 361.
5. Yu-Wen Chen, J. Chin. Inst. Chem. Eng. 17 (1986) 13 ;.Canad. J Chem. Eng. 64 (1986) 875.
6. H.C. Foley and A.J. Hong, Appl. Catal. 61 (1990) 351.
7. T. Fransen, P.C. Van Berge and P. Mars, React. Kinet. Catal. Letter, 5, 4 (1976) 445.
8. A. Elgharbi, PhD Lille (1991).
9. G.C. Bond and J.B.P. Tripathi, J. Less-Common Met. 36 (1974) 31 ; G.C. Bond and P.A. Sermon, J. Chem Soc., Faraday Trans. I 72 (1976) 730 ; G.C. Bond and J.B.P. Tripathi « Chemical Uses of Molybdenum, Proc. 1st Conf., 1973 » p.17, (Ed. P.C.H. Mitchell], Climax Molybdenum Co., London (1974).
10. R. Benaly, C. Hoang-Van and P. Vergnon, Bull. Soc. Chim. Fran. 3 (1985).
11. V. Ponec, Catal. Today, 12 (1992) 227.
12. M. A. Vannice, Catal. Today, 12 (1992) 255.
13. M.J. Perez Zurita, I.S. Henriquez, M.R. Goldwasser, M.L. Cubeiro and G.C. Bond, J. Molec. Catal. 88 (1994) 213.

NATURAL GAS CONVERSION V
Studies in Surface Science and Catalysis, Vol. 119
A. Parmaliana et al. (Editors)

High Yields in the Catalytic Partial Oxidation of Natural Gas to Formaldehyde: Catalyst Development and Reactor Configuration

A. Parmaliana[1], F. Arena[1], F. Frusteri[2] and A. Mezzapica[2]

[1] Dipartimento di Chimica Industriale, Università di Messina - Italia
[2] Istituto CNR-TAE, Messina - Italia

The research strategy for the development of silica based oxide catalysts suitable for the industrial exploitation of the Partial Oxidation of Natural Gas to Formaldehyde (MPO) is presented. A precipitated commercial SiO_2 sample, selected as the more suitable catalyst system has allowed to achieve very high HCHO yields (15-25 mol %) at 650-690 °C in a mini-plant reactor device operated in continuous flow recycle mode. Several engineering and technological aspects related to the process configuration, reactor material and geometry, HCHO absorption system and catalyst form have been addressed.

1. INTRODUCTION

Formaldehyde is currently produced by two commercial processes: a) oxidation-dehydrogenation of CH_3OH with air on a Ag catalyst and b) oxidation of CH_3OH with air on a metal oxide catalyst (Formox process), about a third of the total demand of methanol (25.5 MMT for the 1996) being the feedstock of such processes . Generally, it is stated that the technology for manufacturing HCHO consists of a multi-step process: $CH_4 \rightarrow CO/H_2 \rightarrow CH_3OH \rightarrow HCHO$. This is a highly costly and energy-intensive sequence process for converting natural gas into a commodity chemical. Among the various catalytic routes proposed during the last decades for the conversion of the natural gas to higher hydrocarbons, fuels or oxygenates, the direct partial oxidation to formaldehyde (MPO), being a potentially technological break through has attracted a great research interest. All the studies have been mainly addressed to discover active and/or selective catalysts as well to ascertain the nature of the reaction pathway /1/, while no or limited attention has been devoted to devise the reactor and process configuration as well to optimize the operating conditions, although these issues are crucial for the attainment of high yields to HCHO.

A great variety of transition metal oxide and multicomponent oxide catalysts, in bulk or supported form, has been claimed to be effective /1/. However, a significant progress in this search has been made after 1986 focusing the studies on MoO_3 and V_2O_5 containing catalysts. Since the catalytic activity of such oxide systems at a certain T depends upon the contact time and the CH_4/oxidant ratio, we have proposed the adoption of the STY (space-time-yield) as the key parameter for evaluating the performance of the MPO catalysts /2/. On this account, the superior performance of silica based oxide catalysts has been documented /2-4/. We have largely contributed to this issue highlighting how critical is the own reactivity of the bare SiO_2 support in controlling the reactivity of silica supported MoO_3 and V_2O_5 catalysts /4-5/. In particular, comparing the activity of various commercial SiO_2 samples, it has been disclosed that both the preparation method and the extent of impurities control the reactivity of the SiO_2 surface /4/. The working mechanism of bare and low loaded (< 5wt%) MoO_3 and V_2O_5 promoted silica catalysts in the MPO involves the activation of gas-phase oxygen on surface reduced sites stabilized under reaction conditions, whilst highly loaded (>5wt%) MoO_3 and V_2O_5 silica catalysts acting according to a redox mechanism result in less effective MPO systems / 6 /.

Although the excellent STY $_{HCHO}$'s of SiO_2 and V_2O_5/SiO_2 catalysts, the corresponding values of HCHO yield per pass (mol%) so far reported do not exceed 3-4 % being therefore quite far from the target for industrial exploitation. In order to attain higher HCHO yields, several original catalytic and reaction engineering approaches have been recently reported /7-10/. Sojka et al /7/ claimed the use of redox couples, constituted by a centre for activating oxygen ($Cu^{1+/2+}$) and another center ($Fe^{3+/2+}$) for generating methyl radicals from methane, as MPO catalysts (i.e., CuO-FeO supported on ZnO). Sun et al /8/ proposed the use of a double catalyst bed, where the first catalyst consisted of a 1% SrO/La_2O_3 while the second bed was a 2% MoO_3/SiO_2 system, for enhancing the productivity of CH_2O. This approach has permitted to triple the STY_{HCHO} at 630°C /8/. Herman et al /9/ have pointed out that the addition of steam in the methane/air reacting mixture improves the STY $_{HCHO}$ and the HCHO selectivity of V_2O_5/SiO_2 catalysts. Recently, Yu et al /10/ have claimed single pass HCHO yield values as high as 20.5% in a catalysis-separation reaction system where CH_4 is oxidised to HCHO with air or O_2 by a sort of homogeneous-heterogeneous catalytic system constituted by $Na_2B_4O_7$ deposited on the wall of the quartz reactor and NO. The H_2O vapour, used as diluent of the reaction mixture, tends to adsorb rapidly the HCHO produced in a cooled zone of the reactor ensuring then the achievement of very high HCHO yields. The practical suitability of this approach is limited by the low concentration (0.5-1wt%) of the HCHO solution condensed downstream of the reactor.

Therefore, as the limited HCHO yields or the existence of practical and technological drawbacks, none of the above approaches appears still mature for industrial exploitation. Then, it is evident that further research efforts should be aimed to face engineering and technological aspects of the MPO process since catalysts with adequate specific activity are today available.

This paper is aimed to overview the main goals achieved during the last decade in the development of silica based oxide catalysts for MPO reaction as well as to address the crucial role of the reactor configuration and operation mode on the HCHO yield.

2. EXPERIMENTAL

2.1 Catalyst

Bare, acid-washed, Fe doped and heat-treated commercial SiO_2 samples, prepared by pyrolysis, precipitation or sol-gel methods, along with several series of differently loaded (1-50 wt%) silica supported MoO_3 and V_2O_5 catalysts have been studied. All the details related to the physico-chemical characteristics of the SiO_2 samples, the catalyst preparation methods and the notations used to identify the catalyst samples have been reported elsewhere /3,5,6,11/.

2.2 Catalyst Characterisation

H_2-TPR, CH_4-TPR, O_2 chemisorption (Low Temperature O_2 chemisorption (LTOC) and High Temperature O_2 chemisorption (HTOC), NH_3-TPD and Diffuse Reflectance UV-Vis techniques have been used for evaluating the reducibility, the redox properties, the oxide dispersion, the density of reduced sites and the structure of the surface oxide species of MoO_3/SiO_2 and V_2O_5/SiO_2 catalysts. The procedures adopted for characterizing the oxide catalysts and the related results are presented elsewhere /4-6,11/.

2.3 Catalytic Measurements

The activity of silica based oxide catalysts in the partial oxidation of methane or natural gas with O_2 or air has been evaluated in a wide range of experimental conditions (P_R: 1 bar, T_R: 500-750°C; CH_4/O_2 : 2/1 – 9/1; GHSV : 1,000 – 200,000 h^{-1}) by a RFBR (recirculation flow batch reactor), a TPR (temperature programmed reactor) and several Plug Flow Reactors (characterised by different reactor volume ;V_R = 1-50 cm^3) operating in continuous flow (CFR) and continuous flow with recycle (CFRR) modes. The apparatus and the experimental procedures are described in our previous papers /3-6, 11/.

RESULTS and DISCUSSION

a – Development of Silica Based Oxide Catalysts

The research strategy for the development of MPO catalysts has been focused to the identification and/or design of oxide systems which entail high selectivity along with high productivity to HCHO. Apart from few exceptions, transition metal oxides or multicomponent oxides, sometime already used in other selective oxidation reactions, have been claimed as effective catalysts /1/. Table 1 summarizes the best activity data so far reported for various catalytic systems in terms of STY_{HCHO}. The highest HCHO productivity values are associated with silica based oxide catalysts and notably the more significant progress in the achievement of high HCHO productivity has been made during the last 5 years. Low-medium loaded (2-5 wt%) V_2O_5/SiO_2 systems are the more active catalysts, while the high STY_{HCHO} 's, recently claimed, are the result of testing operations carried out at very high GHSV's which entail low HCHO molar yields (< 3-4 %) not adequate for industrial exploitation.

Our first contribution to the development of MPO catalysts lies in the comparative study of the catalytic performance of several commercial SiO_2 samples /3-4/. It has been disclosed the following reactivity scale for the various silicas on the basis of the preparation method: precipitation> sol-gel > pyrolysis /3-4/.

The influence of (i) the extent of impurities, (ii) the annealing treatment and (iii) the doping with transition metal ions on the catalytic behaviour of the studied SiO_2 samples has been also ascertained /4-6/. Since the superior performance /3-5/, the precipitated Si4-5P SiO_2 has been selected as the more suitable SiO_2 support and further investigated as MPO catalyst /6,11/. The promoting role of V_2O_5 and MoO_3 on the reactivity of the precipitated silica has been evaluated /11/. In particular the influence of the oxide loading and the catalytic action of the promoters along with the identification of the surface active oxide species and the reaction pathway have been assessed /11/. It has been argued that V_2O_5 at any loading exerts a promoting action while MoO_3 depresses the activity of the SiO_2 surface at low T becoming a promoter at rather high T(>650°C). On this account, the activity of differently loaded MoO_3 and V_2O_5 catalysts as a function of the reaction temperature is outlined in Fig. 1 in terms of normalised specific surface activity ($NSSA = SSA_i/SSA_{PS}$ where SSA_i and SSA_{SiO2} are

Table.1. Formaldehyde Productivity in the Partial Oxidation of Methane on Oxide Catalysts (Selected Data from the Literature)

Catalyst	Reaction Mode	T_R (°C)	STY_{HCHO} (g Kg_{cat}^{-1} h^{-1})	Year	Reference
5.0% MoO_3/SiO_2	FBR[a]	220	7.2	1990	/26/ in Ref /3/
9.0% BeO-9.0% B_2O_3/SiO_2	PFR	600	24	1987	/4/ in Ref /3/
2.3% V2O5/SiO2	PFR	600	77	1989	/9/ in Ref /3/
No catalyst	PFR	625	15[b]	1991	/22/ in Ref /3/
2.7% MoO_3/SiO_2	PFR	650	102	1988	/8/ in Ref /3/
Mo-Sn/SiO_2	PFR	700	80	1989	/10/ in Ref /3/
MgO	PFR	750	270	1990	/12/ in Ref /3/
Cu/Fe/ZnO	PFR[c]	750	75	1991	/7/
SiO_2	PFR	730	267	1992	/12/
Sr/La_2O_3/MoO_3/SiO_2	DBC[d]	630	187	1992	/8/
$Na_2B_4O_7$-NO	PFR[e]	750	23.0	1998	/10/
2%V_2O_5/SiO_2	PFR[f]	600	1179	1997	/9/
SiO_2	TPR[g]	750	880	1995	/5/
4%MoO_3/SiO_2	TPR	800	1050	1995	/5/

Note

a) FBR, Fluidized Bed Reactor under UV irradiation - b) Yield quoted is gHCHO l^{-1} h^{-1} - c) PFR, Plug Flow Reactor, double redox catalyst - d) PFR, Plug Flow Reactor; DBC, double bed catalyst- e) PFR , Plug Flow Reactor provided with an internal cold coil for HCHO separation - f) PFR, Plug Flow Reactor; addition of steam to the reaction mixture - g) TPR, Temperature Programmed Reactor

the specific surface activity of the catalyst and bare SiO_2 support, respectively). These data clearly account for the superior activity of the 5% V_2O_5/SiO_2 catalyst at any T /11/.

All silica based catalysts did not present any deactivation problem during long-term operations (2,500-3,000 h) in the T range 600-700 °C.

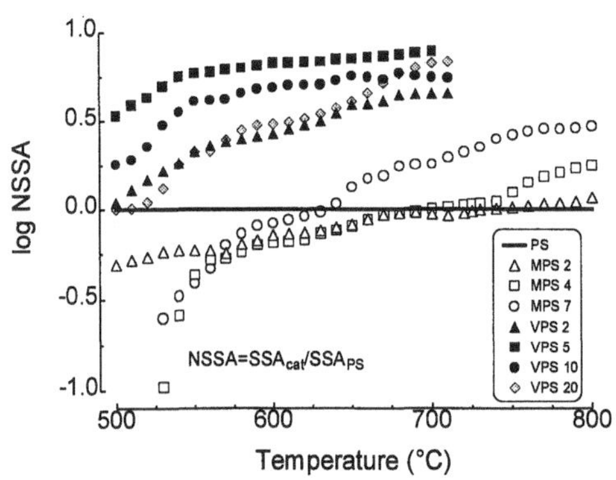

Fig. 1. Partial Oxidation of Methane to Formaldehyde on various silica supported MoO3 and V2O5 catalysts.
Logarithm of the normalised specific surface activity (NSSA) vs T.

On the whole, it can be stated that both unpromoted precipitated Si4-5P SiO_2 and the 5% V_2O_5/SiO_2 catalysts ensure the attainment of rather high STY $_{HCHO}$ (0.2-1.0 Kg $_{HCHO}$ Kg_{cat}^{-1} h^{-1}) /5/. However, in spite of its high activity and HCHO productivity, at a given level of CH_4 conversion the 5% V_2O_5 promoted SiO_2 is less selective than the bare SiO_2 support. Such a different catalytic behaviour reflects the lower oxidizing power of the SiO_2 surface which hinders the further oxidation of HCHO. This is a critical feature when MPO catalytic systems are used in continuous flow mode at moderate space velocities. For this reason, precipitated Si4-5P SiO_2 has been considered the more suitable system for further investigation aimed to highlight the feasibility of the MPO process.

Table 2. Partial Oxidation of Natural Gas to Formaldehyde on SiO2 $_{Si4-5P}$ Catalyst. HCHO yield and Productivity in Mini-Plant Operations.

Operating Conditions

Run	Reactor Material$^{\alpha}$	Reactor *Φ(mm)*	Reaction Mixture $CH_4/O_2/N_2$	Flow Rate Nm^3/min	Recycle Ratio R	Wcat$^{\beta}$ (g)
a	Q	10	33.3/14.9/51.8	75	53	1.5
b	Q	15	33.3/14.9/51.8	75	53	3
c	Q	15	20/17.6/62.4	62.5	64	3
d	SS	18	20/17.6/62.4	62.5	64	3
e	Q	10	20/17.6/62.4	62.5	64	1.5
f	Q	15	22.7/16.8/60.5	65.6	61	3
g	Q	15	22.7/16.8/60.5	65.6	61	3
h	Q	15	22.7/16.8/60.5	65.6	61	3

*α: Q, quartz; SS,Stainless Steel-β: for all the runs the SiO2 catalyst was in granular form (20- 40 mesh) with the exception of the run **g** for which pellets (Φ, 3mm) have been used.*

Activity Data

Run	TR (°C)	CH4conv (mol%)	O2conv. (mol%)	HCHO selectivity (%)	YHCHO (mol %)	STY HCHO (g Kg_{cat}^{-1} h^{-1})	HCHO Productivity g/h$^{\alpha}$	HCHO conc. (wt%)$^{\beta}$
a	680	35.0	70	56	19.6	240	0.92	37.0
b	670	40.0	95	49.0	19.6	120	1.03	35.1
c	670	59.0	88	40.5	23.9	73.3	0.74	29.7
d	663	48.0	90	4.0	1.9	5.89	0.54	3.3
e	690	50.0	72	49	24.5	150	0.64	35.1
f	645	47.0	72	47	22.1	80.7	0.72	33.9
	660	52.0	87	42	21.8	79.7	0.78	30.7
	678	54.0	90	40	21.6	79.0	0.81	29.4
g	643	47.7	74	34.3	16.4	59.7	0.70	25.9
	655	51.5	85	31.6	16.3	59.3	0.75	23.8
	669	53.3	90	30.2	16.1	58.7	0.77	22.9
h	670	52.0	84	47.4	24.6	90.1	0.79	34.1

α – hourly production of HCHO solution
β – concentration (wt%) of the HCHO solution

B – Mini-Plant Operations: HCHO Productivity in Simulated Industrial Conditions

The industrial exploitation of a new catalytic process is generally pursued by fixing several targets for the catalyst performance and for the efficiency of the various units (Reactor, Separation and /or Purification Systems, etc..) that in front of the capital and variable costs ensure a return of investment more rewarding than that of the current technology. Several additional factors such as the disposability and low cost of the feedstock, and the lower environmental impact play a deciding role in promoting the development of a novel technology. Since the low yields (3-5%) so far attained for the MPO reaction, no techno-economical analyses of the process feasibility have been made and therefore no striking targets for the molar yield and HCHO selectivity have been posed. However, considering the current exploitation of other NG conversion routes, it comes out that HCHO yield higher than 20% with HCHO selectivity higher than 40% are the minimal targets.

On the basis of the excellent intrinsic reactivity of the precipitated Si4-5P SiO_2 /4/, we have designed, realised and run a mini-plant system operating in continuous (CFR) and continuous flow recycle (CFRR) modes. The activity of the SiO_2 catalyst has been evaluated by this reactor system in simulated industrial conditions in the T_R range 600-720 °C. The influences of the recycle ratio ($R = r_f/r_r$ where r_f and r_r refer to feed flow rate and recycle flow rate, respectively) , T_R, GHSV, feed composition, reactor material and geometry, catalyst form and absorption system on the HCHO yield (Y_{HCHO} molar yield (%)) and HCHO productivity (hourly production of HCHO solution and concentration of the HCHO solution) have been assessed.

An accurate design of the operating conditions has allowed to achieve very high HCHO yields (15-25 mol%) and to attain high HCHO productivity in terms of hourly production of HCHO solution along with high HCHO concentration (Table 2). The efficiency of the HCHO absorption system has been higher than 80%. At 690 °C a HCHO yield of 24.6% (CH_4 conv, 50%, HCHO sel, 49 %), corresponding to a STY_{HCHO} of 150 g_{HCHO}/Kg_{cat} h, has been attained. This is the best result so far reported for the MPO reaction.

The high HCHO yields above reported, the use of air as oxidant and the simple process scheme are the milestones which will drive the development of a research strategy aimed to assess the techno-economical feasibility of the MPO process .

References

1. O.V. Krylov, Catal. Today 18(1993)209
2. A. Parmaliana, F. Frusteri, A. Mezzapica, M.S. Scurrell and N. Giordano, J. Chem. Soc., Chem. Commun., 751(1993)
3. A. Parmaliana, F. Frusteri, A. Mezzapica, D. Miceli, M.S. Scurrell and. N. Giordano, J. Catal. 143(1993)262
4. A. Parmaliana, V. Sokolovskii, D. Miceli, F. Arena and N. Giordano, J. Catal. 148(1994)514
5. F.Arena, F. Frusteri, A. Parmaliana and N. Giordano, Appl. Catal., A: General 125(1995)39
6. A. Parmaliana, F. Arena, F. Frusteri, G. Martra, S. Coluccia and V. Sokolovskii, 3rd World Congress on Oxidation Catalysis , Stud. Surf. Sci. Catal. 110(1997)347
7. Z. Sojka, R.G. Herman, K. Klier, J. Chem. Soc., Chem. Commun. 185(1991)
8. Q. Sun, J.I. Di Cosimo, R.G. Herman, K. Klier and M.M. Bhasin, Catal. Lett. 15(1992)371
9. R.G. Herman, Q. Sun, C. Shi, K. Klier, C.B. Wang, H. Hu, I.E. Wachs and M.M. Bhasin, Catal. Today, 37(1997)1
10. L. Yu, S. Yuan, Z. Wu, J. Wan, M. Gang, G. Pan, Y. Chen, Appl. Catal., A: General, 171(1998)L171
11. F. Arena, N. Giordano and A. Parmaliana, J. Catal. 167(1997)66
12. Q. Sun, R.G. Herman and K.Klier, Catal. Lett. 16(1992)251

NATURAL GAS CONVERSION V
Studies in Surface Science and Catalysis, Vol. 119
A. Parmaliana et al. (Editors)

Concurrent synthesis of methanol and methyl formate catalyzed by copper-based catalysts in a slurry phase

Xing-Quan Liu Yu-Tang Wu Wen-Kai Chen Zuo-Long Yu*

Chengdu Institute of Organic Chemistry, The Chinese Academy of Sciences
Chengdu, Sichuan 610041 P.R.China

The concurrent synthesis of methanol and methyl formate from syngas at low temperature and pressure in a slurry reactor have been investigated. The results showed that the catalysts prepared by different technique have different activities and selectivities of MF, and the catalyst prepared by complex precipitation method has higher activity of methanol and MF synthesis. Temperature and pressure have a significant influence on the activity and MF selectivity of the catalysts. The activity of Cu-Cr catalyst increases with the augmentation of temperature and pressure. The MF selectivity of Cu-Cr catalyst decreases with the increase of temperature. Nevertheless, the selectivity of MF increases with the increase of pressure.

1. INTRODUCTION

Methanol (CH_3OH) is a principal chemical product which is widely applied in most fields, and methyl formate ($HCOOCH_3$, MF) is honoured with a new building block in C_1 chemistry[1]. The traditional production method is a two-step process which comprises technique of carbonylation of methanol first and hydrogenation of methyl formate then. At present, the industrial method used for synthesizing methanol from syngas ($CO+H_2$) is a one-step gas-phase process[2] such as ICI technique which is shown as follows:

$$CO + 2H_2 \longrightarrow CH_3OH \qquad (1)$$

Currently, the most economical method of producing methyl formate is carbonylation of methanol catalyzed by alkali methoxide such as CH_3ONa and CH_3OK and so on[3]. But the concurrently direct synthesis of methyl formate from syngas, which is based on the following reaction equation:

$$2CO + 2H_2 \longrightarrow HCOOCH_3 \qquad (2)$$

has not been reported in the literature so far.

This report has published our recent investigation on concurrent synthesis of methanol and methyl formate from syngas catalyzed by copper-based complex catalysts in a slurry phase.

* To whom correspondence should be addressed

2. EXPERIMENTAL

2.1. Preparation of catalysts

CuCl catalyst was prepared by directly using CuCl chemical reagent, the further treatments such as dehydrated drying and removal of $CuCl_2$ were carried out before use.

Cu-Cr oxide catalysts were prepared in terms of reference[4]. Cu-Cr-Mn oxide catalysts were prepared by coprecipitation of the mixed aqueous solution of copper nitrate, chromium nitrate and manganese nitrate using sodium carbonate or ammonium carbonate. Cu-Mn oxide catalysts were prepared as Cu-Cr-Mn oxide catalysts with corresponding nitrate. All the catalysts were further treated with filtration. drying, pulverization and thermal decomposition etc.

2.2. Evaluation of activity and MF selectivity of catalysts

The activity assessment and MF selectivity investigation were carried out in a slurry reactor with mechanically magnetic agitator. The Cu-based catalyst charge was 6.0 g and the volume of the reactant mixture was 200 ml. Xylene, dioxane, tetrahydrofuran and decalin were used as solvents, respectively. 170 ml of solvents was added each time. The reaction was performed at 90~150°C for 2 h twice. The resultant mixture was analyzed by SC-3A and GC-103 gas chromatograph with micro-processor.

3. RESULTS AND DISCUSSION

3.1. Comparison of relative activity of catalyst between different technique

The relative activity of catalyst prepared with different technique was shown in Fig.1. We can see from Fig. 1 that the catalytic activity of catalyst prepared by our technique is about 5 times than that of catalyst prepared by ICI technique and about 4 times than that of catalyst prepared by recent US patent technique[5]. The results obtained with the same conditions demonstrate that our technique of catalyst preparation is superior to recently reported ones. Because the catalysts prepared by ICI and USP 5,221,652 technique almost have not activities at low temperature.

3.2. Catalytic activity and selectivity for MF of different catalysts

The catalytic activity and the selectivity of MF of different catalysts were shown in Fig.2. From Fig.2, we can find that different catalyst has different activity and selectivity. Of them, Cu-Cr catalyst has the highest activity, however, its selectivity of MF is the lowest and the selectivity of MF of Cu-Mn catalyst is the highest. By comparison, Cu-Cr-Mn and Cu-Mn catalysts have higher activity and selectivity than CuCl and Cu-Cr catalysts.

3.3. Effect of temperature

The effect of temperature was shown in Fig.3. It can be found from Fig.3 that the activity of CuCl catalyst decreases with the increase of temperature (Curve 1), but the activity of Cu-Cr catalyst increases with the increase of temperature (Curve 3). In the meantime, the selectivity of CuCl catalyst decreases from 363~383 K and then increases after 383 K (Curve 2), but the selectivity of Cu-Cr catalyst decreases regularly with the increase of temperature (Curve 4). Therefore, we can deduce that the mechanism of Cu-Cr catalyst is different from that of CuCl catalyst.

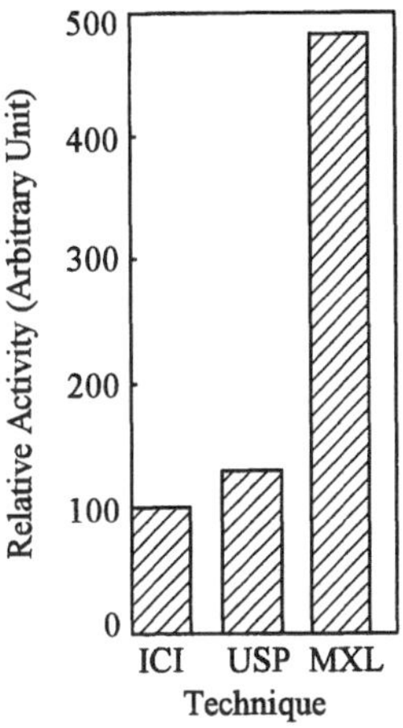

Fig.1 Comparison of relative activity between different technique

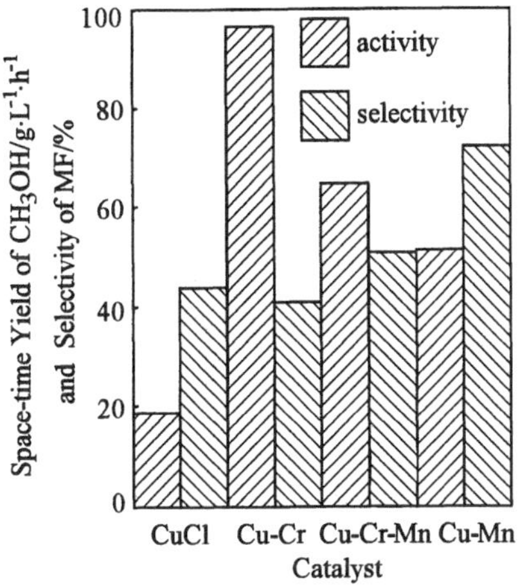

Fig.2 Catalytic activity and MF selectivity of different catalysts

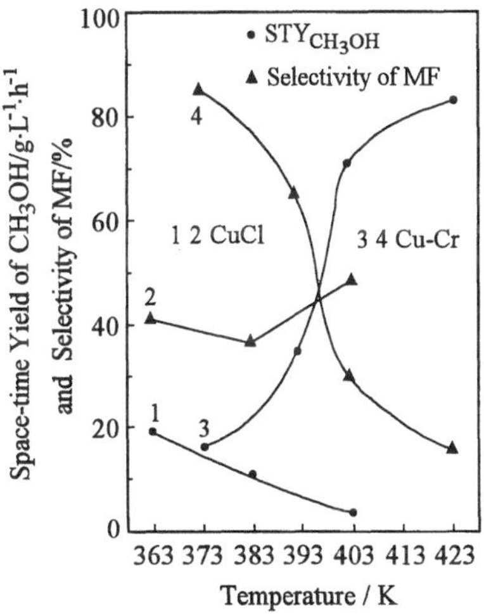

Fig.3 Effect of temperature on catalyst activity and selectivity
[Catalyst]=20 g/l, P=4.0 MPa
Solvent: THF, H_2/CO=1.96

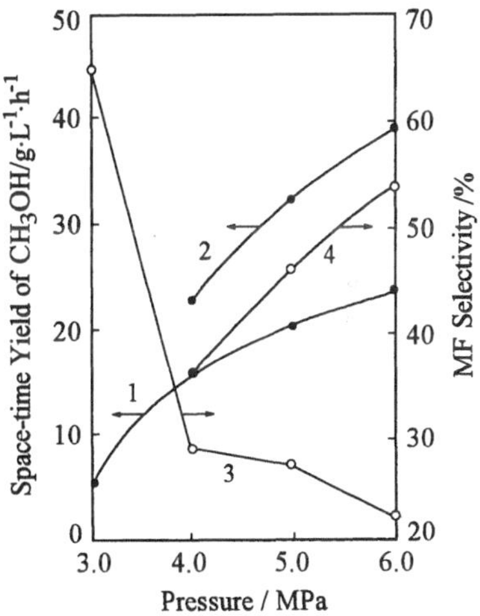

Fig.4 Effect of pressure on the activity and selectivity of catalyst
[Catalyst]=20 g/l, T=383 K
Solvent: Xylene, H_2/CO=1.96

3.4. Effect of pressure

Fig.4 shows the effect of pressure. It demonstrates that the activity of CuCl catalyst (Curve 1) and Cu-Cr catalyst (Curve 2) increases regularly with the increase of pressure. However, the selectivity tendency of CuCl catalyst and Cu-Cr catalyst was very different. We can see from Fig.4 that the selectivity of CuCl catalyst decreases sharply when the pressure is less than 4.0 MPa (Curve 3), but the selectivity of Cu-Cr catalyst increases regularly with the increase of pressure in the range of the experimental pressure (Curve 4). These results indicate that the catalyst mechanism of CuCl are probably very different from that of Cu-Cr catalyst. Practically, the reaction catalyzed by Cu-Cr catalyst take place in the form of concurrent mechanism, which is in accordance with the theoretical results.

4. CONCLUSIONS

From above described results, the following conclusions can be drawn:

1. The catalyst prepared by copper-ammonia complex technique has higher activity of methanol and methyl formate synthesis from syngas and is better than the ones prepared by other techniques.
2. Temperature has remarkable effects on the activity and selectivity of MF of catalysts. As to Cu-Cr catalyst, the activity increases with the increase of temperature and the selectivity of MF decreases with the increase of temperature. The result of CuCl catalyst is opposed to that of Cu-Cr catalyst.
3. Pressure has significant effect on the activity and selectivity of MF of catalysts. The activity of Cu-Cr and CuCl catalysts increases with the increase of pressure. The selectivity of CuCl catalyst decreases with the increase of pressure, but the selectivity of Cu-Cr catalyst increases with the increase of pressure.

ACKNOWLEDGMENTS

The authors are grateful to the State Science and Technology Commission of China and the Chinese Academy of sciences for the financial support.

REFERENCES

1. J.S.Lee, J.C.Kim, and Y.G.Kim, *Appl.Catal.*, **57** (1990) 1
2. G.C.Cinchen, P.J.Denny, J.R.Jenings, M.S.Spencer, and K.C.Waugh, *Appl.Catal.*, **36** (1988) 1
3. L.J.Kaplan, *Chem.Eng.*, **89** (1982) 71
4. Y.T.Wu, W.K.Chen, X.Q.Liu, S.Z.Luo, Z.X.Jia, and Z.L.Yu, *Chinese Patent CN* 1136979A (1996)
5. J.W.Tierney, *USP* 5, 221, 652 (1993); *USP* 5, 384, 335 (1995)

NATURAL GAS CONVERSION V
Studies in Surface Science and Catalysis, Vol. 119
A. Parmaliana et al. (Editors)

Paraffins as raw materials for the petrochemical industry

F. Cavani and F. Trifirò

Dipartimento di Chimica Industriale e dei Materiali, Viale Risorgimento 4, 40136 Bologna

The reactions of oxidative transformation of paraffins to valuable chemicals for the petrochemical industry are reviewed. The oxidation of different alkanes is compared in terms of selectivity to the desired product.

1.The oxidation of alkanes catalyzed by heterogeneous systems

Most intermediates and monomers in the petrochemical industry are currently produced starting from alkenes and aromatics as building blocks, but research is now being aimed at the replacement of some of these raw materials with alkanes, which are more economical than the corresponding olefins, readily available and with low toxicity as compared to aromatics (1-8). Moreover, several processes which are no longer sustainable from the environmental point of view will be replaced by processes possibly starting from paraffins.

Table 1. Processes under study or development for the oxidative transformation of light paraffins (C_1-C_6) with molecular oxygen

Raw material	Product
Methane	Methanol
Methane	Syngas
Methane	Ethylene
Ethane	1,2 Dichloroethane, Vinyl chloride
Ethane	Acetaldehyde
Ethane	Acetic acid
Ethane	Ethylene
Propane	Acrolein, Acrylic acid
Propane	Acrylonitrile
Propane	Propylene
Isobutane	Methacrylic acid
Isobutane	Isobutene
n-Pentane	Phthalic anhydride
Cyclohexane	Cyclohexanone

1.1 O/N/Cl insertion on paraffinic substrates

Amongst the several possible applications (Table 1), the oxidation of **n-butane** currently represents the most important example of exploitation of a natural gas component for the production of an intermediate for the petrochemical industry. For many years it has been replacing the process starting from benzene, owing to the lower toxicity and to the lower cost of the paraffin with respect to the aromatic compound (9). The successful development of this process has been made possible by the discovery of a catalytic system able to activate the paraffin and to insert selectively oxygen yielding maleic anhydride in a single step. This system is made of a V/P mixed oxide, $(VO)_2P_2O_7$. The fundamental property of this structure is its ability to catalyze a variety of reactions, necessary to perform the complex transformation of the paraffin to a highly functionalized molecule. One key factor is the surface acidity, both Broensted- and Lewis-type, useful to activate the C-H bond in the secondary carbon atom (the first step of alkane transformation) and to favour the desorption of maleic anhydride.

Another successful example, recently scaled up to the level of a demonstrative plant, is the ammoxidation of **propane** to acrylonitrile (10-11). The process, developed by BP Americas, makes use of a fluid-bed reactor. The catalyst is a V/Al/Sb mixed oxide; the main component is $VSbO_4$, an oxide with the rutile structure. Vanadium in the $VSbO_4$ is responsible for the activation of the paraffin and for the formation of propylene, while excess antimony oxide, present as a dispersed microcrystalline oxide over the rutile surface, plays its role in the allylic ammoxidation with formation of the nitrile, either via O-insertion and formation of intermediate acrolein, or by direct N-insertion. Several other rutile-based mixed oxides have been claimed in recent years (Ni/Sb/O, Fe/Sb/O, Sn/Sb/O, Ti/Sb/O) as catalysts for the ammoxidation of propane (12).

Other reactions of interest, which have not yet reached the commercialization step, are listed in Table 1. The challenge of **methane** transformation to liquid fuels has directed research first towards i) the oxidative coupling to ethane and ethylene (with also an interest in pyrolytic dimerization of methyl chloride, the latter being produced by methane chlorination or oxyhydrochlorination) and ii) the oxidation to methanol, while more recently the oxidation to syngas appears as one of the most promising routes for valorization of methane (13-16). Particular reactor configurations include monolithic and short-bed reactor types. When noble-metal based catalysts and oxygen-poor atmospheres (the methane-to-oxygen-ratio is 2, corresponding to the stoichiometric one) are used in these reactor types high selectivities to CO/H_2 are obtained, due to a combination of favourable factors which include i) very short residence times (of the order of milliseconds), that allow the products to be quickly carried out of the reaction zone, and ii) considerable temperature difference between the catalyst surface and the gas phase, which contributes to give a very high reaction rate and minimize the contribution of undesired homogeneous reactions (17). These peculiarities are also of utility in the oxidehydrogenation of paraffins to olefins. Indeed relevant results have also been reported in monolith-type reactor configurations in oxidehydrogenation of C_2-C_5 paraffins.

The oxidation of **ethane** aimed at the production of acetic acid in one step usually yields ethylene with better selectivity than acetic acid, since the olefin is very stable (18). For the same reason, one possible commercial exploitation of ethane consists in a reaction where the olefin produced becomes the raw material for the transformation to a valuable chemical, either in the same reactor (it is necessary that the following reaction can be carried out under conditions which are close to those necessary for ethane activation), or in an *integrated process*, where the ethylene-containing stream exiting from the first reactor constitutes the feedstock (after necessary make-up) for the second reactor. An example is the single-step or two-step (via intermediate oxidehydrogenation of ethane) oxychlorination of ethane to 1,2-dichloroethane and to vinyl chloride, where the by-product ethylene may easily be transformed to the chlorinated products. This process was studied in the 1970's, but did not reach commercial application. Interest is now growing again, as documented by recent patents issued by companies which are PVC producers (19). The concept of process integration can apply to other olefins as well, made by oxidative dehydrogenation from alkanes.

Currently, studies are being published which describe attempts to transform in a single step **isobutane** into methacrolein and methacrylic acid (a process that might replace the highly environmentally unfriendly aceton-cyanohydrin technology), and **propane** to acrylic acid (as an alternative to the two-step oxidation of propylene) (20-25). Interest from industrial companies for these reactions indeed goes back to the beginning of the 1980's (26). In these reactions the main problem concerns the high reactivity of the desired products, which under conditions necessary for alkane activation are not stable and undergo unselective oxidative transformations. Moreover, a problem concerning isobutane is its current scarce availability as a pure component. Its separation from the C_4 fraction is expensive, and it is preferably carried out by making a selective transformation into products that are easily separated from the other components, i.e. by dehydrogenation followed by condensation with methanol to yield MTBE (the demand for which is great, since it is used in reformulated gasolines) in very large plants and thus with very favourable economics. Indeed, polymer-grade isobutene can be made with favourable economics by decomposition of MTBE. Nevertheless, the reaction of isobutane selective oxidation is of considerable interest, and becomes economically competitive with other processes of low environmental impact (i.e., the two-step oxidation of isobutene) provided conversions of isobutane at least higher than 15% and selectivity to methacrylic acid + methacrolein around 65-70% are reached, under fuel-rich conditions (oxygen as the limiting reactant). The best performing catalysts are P/Mo Keggin-type heteropolycompounds, containing vanadium, alkali metals and copper as dopants. The best results have been obtained for heteropolycompounds also containing arsenic, which however obviously represents an impracticable solution (27).

The availability of C_5 paraffins, as a consequence of environmental regulations which dictated the lowering of the vapour pressure of gasolines, and the discovery that **n-pentane** can be transformed with unexpected selectivity to valuable compounds, i.e. maleic anhydride and phthalic anhydride, opened new prospects in the exploitation of saturated raw materials (28-32). In this case the key factor for industrial commercialization is the achievement of a high selectivity to phthalic anhydride, possibly close to that obtained from o-xylene.

The formation of phthalic anhydride from n-pentane in relatively large amounts (selectivity as high as 40% at 50-60% pentane conversion) is a fascinating demonstration of how catalyst polyfunctionality allows very complex transformations (which would formally need several synthetic steps) to be carried out. Indeed, low amounts of phthalic anhydride are also obtained in the selective oxidation of n-butane. Different reaction mechanisms have been proposed for the formation of phthalic anhydride, and one possible route involves the formation of intermediate pentadiene, followed by its cyclodimerization (likely catalyzed by Lewis acid centres), aromatization and oxidation (with also decarboxylation) to the final product (33).

1.2 Oxidative dehydrogenation of paraffins

The huge number of papers devoted to the reaction of paraffin **oxidative dehydrogenation** is a demonstration of the current scientific and industrial interest for alternatives to catalytic and thermal dehydrogenation/cracking reactions which suffer from several drawbacks: i) thermodynamic limitations on paraffin conversion; ii) side reactions such as thermal cracking; iii) strongly endothermic reactions to which large amounts of heat must be supplied at temperatures above the reaction temperature; iv) formation of coke on the catalyst which requires frequent regeneration (1,6,34-36). With the goal of overcoming these limitations, research is proceeding along some directions, of which the following are the most likely to be implemented: i) optimization of current dehydrogenation technologies, to obtain more selective, stable and environmentally safe catalysts and lower investments and utility costs; ii) dehydrogenation coupled with hydrogen oxidation, to supply the heat of reaction inside the catalytic bed while avoiding overheating and to shift the equilibrium toward the desired products; iii) oxidative dehydrogenation, to overcome thermodynamic limitations, operate at low temperature with an exothermic reaction, and avoid frequent catalyst regeneration, and iv) membrane-assisted dehydrogenation, to obtain high conversion at low temperatures and to conduct the reactions and separations in the same equipment. In oxidehydrogenation the advantage gained from the use of a cheaper raw material and of an exothermic process has to deal with drawbacks such as i) the loss of valuable hydrogen (coproduced in dehydrogenation and in steam-cracking), ii) the difficulty in separation of CO from the paraffin (in the case of ethane oxidehydrogenation), and iii) the formation of traces of corrosive by-products. Moreover, while in the case of ethane the problem mainly concerns the low reactivity of the molecule (the selectivity to the olefin which can be achieved is high, due to the low ethylene reactivity and to the nature of the mechanism involved), in the case of oxidehydrogenation of propane and of n-butane to the corresponding olefins the selectivity problem is of main concern.

The catalysts utilized for oxidative dehydrogenation are not simply modifications of dehydrogenation catalysts, achieved by addition of oxidant components; and neither, is the polyfunctional nature of dehydrogenation catalysts exploited, by carrying on the reaction in the presence of oxygen. On the contrary, brand new classes of catalysts have been developed in recent years for the dehydrogenation of paraffins in the presence of molecular oxygen (1,6,34-36). Different classes of catalysts have been developed, namely:

1) Catalysts based on oxides of Group IA and IIA metals, which are also active for methane oxidative coupling. These are used mainly in ethane oxidehydrogenation, which requires very high reaction temperatures (higher than 600°C), and the mechanism involves the generation of an ethyl radical on the catalyst surface which is then transferred to the gas phase, where it further reacts. The catalyst is thus only involved in the scission of the C-H bond. The most successful system is the one based on Li/Mg/O. High selectivity to ethylene can be obtained when chlorine-containing compounds are also fed to the reactor, or when the catalyst is doped with halides. Chlorine radicals are thought to favour the homogeneous decomposition of ethyl radicals to ethylene.

2) Catalysts based on transition metal oxides, including rare earth oxides. While rare earth oxides are claimed mainly for ethane oxidehydrogenation, vanadium oxide is the main component in most mixed oxides particularly active and selective in propane and butane oxidehydrogenation. In the latter case the catalyst becomes reduced by interaction with the hydrocarbon, through a classical redox cycle. Some of the most studied systems include alumina- or silica-supported vanadium oxide, V/Mo/Nb/O, V/Mg/O, V-zeolites, Ni/Mo/O, Mg/Mo/O.

Particular technical solutions include those where the hydrocarbon is put in contact with the reducible catalyst (based on a transition metal oxide) in the reactor, and the reduced catalyst is reoxidized in a separate fluidized or mobile bed. This circulating bed configuration has been proposed by Atlantic Richfield Co and by Phillips Petroleum Co. for ethane oxidehydrogenation, with catalysts based on SiO_2-supported Na/P/Mn/O systems (37).

2. Reasons for low selectivity in paraffin oxidation

2.1 The control of consecutive reactions on the desired products

The main by-products obtained in the reaction of selective oxidation of paraffins are carbon oxides, which are usually formed both by parallel reactions upon the reactant, and by consecutive reactions upon the desired product. The latter, in fact, is usually much more reactive than the reactant itself under reaction conditions which are necessary to activate the paraffin. Exceptions to this rule are those reactions where the product exhibits particular stability, such as:

1) the oxidehydrogenation of ethane to ethylene, where the C-H bond is stronger in ethylene than in ethane;

2) the ammoxidation of propane to acrylonitrile; the nitrile is relatively stable if compared to propane, and is much more stable than the corresponding unsaturated oxygenated compounds, i.e. acrolein and acrylic acid;
3) the oxidation of n-butane to maleic anhydride; the anhydride is relatively stable towards decarboxylation. The same occurs in the case of n-pentane oxidation to maleic and to phthalic anhydrides.

When the reaction is aimed at the synthesis of unsaturated carboxylic acids, i.e. acrylic acid from propane or methacrylic acid from isobutane, respectively, selectivities not higher than 40-50% are usually obtained, and moreover it is necessary to operate either at low hydrocarbon conversion, or at high hydrocarbon-to-oxygen ratio, thus in conditions at which oxygen is the limiting reactant. It has been argued (38) that when the dissociation enthalpy of the weakest bond in the product is no more than 30-40 kJ/mole weaker than the bond dissociation enthalpy of the weakest bond in the reactant, then the product is obtained with high selectivity; otherwise the selectivity falls drastically with increasing reactant conversion.

2.2 The susceptibility of reactants and intermediates towards nonselective oxidative attacks

A second factor which remarkably affects the selectivity to the product of partial oxidation is the nature of the reactant itself, and its susceptibility to nonselective oxidative attacks. This problem mainly concerns alkanes containing tertiary carbon atoms, such as isobutane. In this case the high reactivity of the C-H bond makes it susceptible to the formation of an hydroperoxide, which may be the precursor for the formation of either other oxygenated products (precursors of carbon oxides) or isobutene, depending on reaction conditions (i.e., temperature). The presence of two adjacent secondary carbon atoms in n-butane, instead, facilitates the reaction of H-abstraction with formation of intermediate 2-butene (first intermediate in the mechanism leading to maleic anhydride), with respect to other nonselective routes.

A third factor which determines the selectivity derives from the reaction mechanism, i.e. at a molecular level in the adsorbed state, especially when the reaction involves a number of transformations, such as in the oxidation of alkanes to O- or N-containing compounds. In this case, the reaction pathway leading to the formation of the desired product consists of several highly reactive intermediates, which further evolve leading to different kinds of compounds, depending on the nature of the neighbouring active site (i.e., either a Lewis-type acid site, or a proton-donor site, or an O-insertion site, or an H-abstraction site, etc.). This is the case of the reaction of n-butane transformation to maleic anhydride, where intermediate 2-butene can undergo different transformations: i) allylic H-abstraction, with formation of butadiene; ii) allylic O-insertion, with formation of unsaturated aldehydes and acids, iii) double bond protonation, followed by β-scission, with formation of lower fragments, iv) electrophilic O-insertion, with formation of CO_x precursors, or v) desorption into the gas phase.
The relative ratio between the rates for the different transformations is not only a function of the surface properties of the catalyst (multifunctionality), but also of the surface availability of O^{2-} insertion sites, which in turn depends on i) the hydrocarbon-to-oxygen ratio in the feed (either fuel-lean or fuel-rich conditions), ii) the extent of interaction between the hydrocarbon and the Lewis-type acid centres on the catalyst surface, and iii) the rate of O_2^- (nonselective oxygen species) evolution towards nucleophilic, selective O^{2-} insertion sites, with respect to the rate of O^{2-} depletion from the catalyst surface as a consequence of the interaction with the hydrocarbon. The oxidation state of the catalyst under reaction conditions is one fundamental factor in affecting catalytic performance, and determines the choice of the operating parameters (temperature, feed composition).

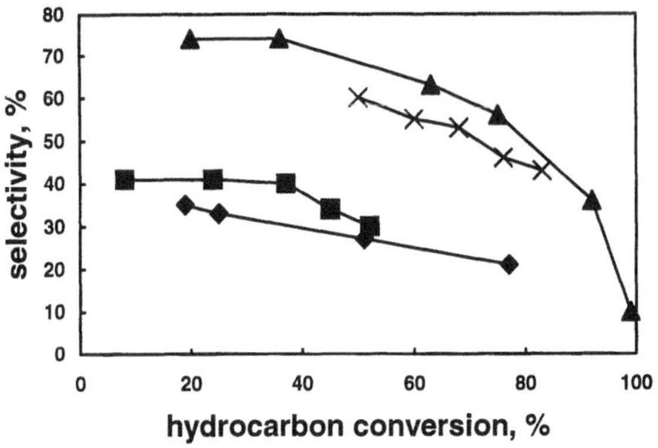

Figure 1. Selectivity to acrylonitrile in propane ammoxidation (✕), to phthalic anhydride in n-pentane oxidation (■), to methacrylic acid in isobutane oxidation (◆) and to maleic anhydride in n-butane oxidation (▲). In all cases results were collected by varying the residence time. See text for catalysts and experimental conditions.

Trends relative to the selectivity to the product of partial oxidation as functions of the conversion of the limiting reactant in the oxidation of several alkanes are reported in Figure 1, for some of the most widely investigated reactions. The following examples are reported:

1) The ammoxidation of propane to acrylonitrile over V/Sn/Sb/O 0.2/1/1 (atomic ratios) catalyst (12) at the following conditions: temperature 450°C, feedstock composition 8% propane, 20% oxygen, 8% ammonia, remainder inert (ammonia is the limiting reactant). The selectivity extrapolated at low conversion is relatively high, suggesting that propane undergoes parallel combustion to a fair extent. The product undergoes successive combustion reactions.

2) The oxidation of n-butane to maleic anhydride over a nonequilibrated $(VO)_2P_2O_7$ catalyst at the following conditions: temperature 307°C, feed composition 0.52% n-butane, 12% oxygen, remainder helium (n-butane is the limiting reactant) (39). The initial selectivity is relatively high, and is stable up to approximately 60% conversion, pointing out that the parallel reactions of combustion upon n-butane (or upon adsorbed intermediates) have negligible rates. This occurs only under conditions at which the number of O^{2-}-insertion sites available at the surface are enough to favour the quick transformation of the adsorbed olefin. The selectivity then rapidly decreases with further increasing n-butane conversion, and this makes it necessary to operate at hydrocarbon conversion lower than 60%. At high n-butane conversion (at the end of the catalytic bed) the atmosphere is strongly oxidizing, and the catalyst becomes nonselective. Conditions poorer in oxygen are considered more selective due to lower carbon oxides formation. Indeed the consecutive reaction is kinetically unfavoured, since the catalyst works in a more reduced state. However, under these conditions other by-products form (crotonaldehyde, methylvinylketone), starting from intermediate unsaturated compounds. On the contrary, when hydrocarbon-lean conditions are used, the only products are maleic anhydride, carbon oxides and low amounts of phthalic anhydride. This facilitates the purification of maleic anhydride, which can be obtained with high degrees of purity for polymer applications.

3) The oxidation of n-pentane to phthalic anhydride over an equilibrated $(VO)_2P_2O_7$ at the following conditions: temperature 340°C, feed composition 1% n-pentane in air (n-pentane is the limiting reactant) (31). In this case the initial selectivity to phthalic anhydride is low, due to the presence of the parallel reactions of maleic anhydride and carbon oxides formation. This reaction is analogous to the oxidation of n-butane, except for the additional step which involves the cyclodimerization of intermediate pentadiene to yield a C_{10}-

alkylaromatic, precursor of phthalic anhydride, which requires surface acidity (possibly Lewis-type). The surface availability of O^{2-}-insertion sites and of acid sites thus may determine the maleic-to-phthalic anhydride selectivity ratio. A second important factor is the temperature, since at low temperature the reversible exothermic reaction of pentadiene cyclodimerization is thermodynamically more favoured than at temperatures higher than 340°C, and therefore at low temperature the formation of phthalic anhydride is preferred with respect to that of maleic anhydride. Hydrocarbon-rich conditions can not be used, because many by-products are formed.

4) The oxidation of isobutane to methacrylic acid over an Fe-doped potassium/ammonium salt of P/Mo Keggin-type heteropolycompounds, at the following conditions: temperature 350°C, feed composition: 26% isobutane, 13% oxygen, 12% water, remainder helium (oxygen is the limiting reactant) (21). Methacrylic acid is very reactive, and in order to obtain a relatively high selectivity to this product it is necessary to operate under hydrocarbon-rich conditions. Under hydrocarbon-lean conditions the selectivity to methacrylic acid is very poor, because the heteropolycompound works in a fully oxidized state. On the contrary, it is claimed that a partial reduction of molybdenum is necessary in order to obtain acceptable selectivity to methacrylic acid and form low amounts of carbon oxides (25). In contrast to that occurring from n-butane, the nature of the by-products obtained is the same under both isobutane-rich and isobutane-lean conditions.
The initial selectivity is low, and this indicates that the contribution of undesired parallel reactions of combustion upon isobutane is very important. This probably arises from the presence of the very reactive tertiary carbon atom in the reactant.

3. Conclusions

Very different reaction conditions are necessary to obtain the best selectivity in the selective oxidation of various alkanes. These conditions are a function of i) the reactant nature, ii) the product nature, iii) the kind of mechanism involved, that is the susceptibility of intermediates to undergo different oxidative attacks or to desorb (then adsorbing on nonselective sites), iv) the polyfunctional nature of the catalyst and v) the working state of the catalyst under reaction conditions. Improvements in catalyst compositions and optimization of reaction conditions can be reasonably expected to lead to further improvements, but it is likely that considerable improvements can only be obtained by developing brand new catalytic systems, and possibly new reactor configurations where the control of selectivity is achieved through control of fluodynamic and diffusional characteristics.

References

1) F. Cavani and F. Trifirò, in "Catalysis Vol. 11", Royal Society of Chemistry, (1994), p. 246
2) S. Albonetti, F. Cavani and F. Trifirò, Catal. Rev.-Sci. Eng., 38 (1996) 413
3) B. Delmon, P. Ruiz, S.R.G. Carrazán, S. Korili, M.A. Vicente Rodriguez and Z. Sobalik, in "Catalysts in Petroleum Refining and Petrochemical Industries 1995", M. Absi-Halabi et al., (Eds.), Elsevier Science, Amsterdam, 1996, p. 1
4) L.D. Schmidt, M. Huff and S.S. Bharadwaj, Chem. Eng. Sci., 49 (24A) (1994) 3981
5) H.H. Kung, P. Michalakos, L. Owens, M. Kung, P. Andersen, O. Owen and I. Jahan, in "Catalytic Selective Oxidation", S.T. Oyama and J.W. Hightower (Eds.), ACS Symp. Series 523, 1993, p. 387
6) E.A. Mamedov and V. Cortes Corberan, Appl. Catal., A: General, 127 (1995) 1
7) F. Cavani, F. Trifirò, in "3rd World Congress on Oxidation Catalysis", (R.K. Grasselli, S.T. Oyama, A.M. Gaffney, J.E. Lyons, Eds.), Elsevier Science, Amsterdam, 1997, p. 19

8) J. Haber, in "Heterogeneous Hydrocarbon Oxidation", B.K. Warren and S.T. Oyama, (Eds.), ACS Symp. Series 638, 1996, p. 20
9) F. Cavani and F. Trifirò, Chemtech, 24 (1994) 18
10) Y. Moro-oka and W. Ueda, in "Catalysis Vol. 11", Royal Soc. of Chemistry, 1994, p. 223
11) V.D. Sokolovskii, A.A. Davydov and O. Yu. Ovsitser, Catal. Rev.-Sci. Eng., 37 (1995) 425
12) S. Albonetti, G. Blanchard, P. Burattin, F. Cavani and F. Trifirò, EP 691,306 A1 (1995); EP 420,025 A1 (1996); France Patent Appl. 95 11680 (1995), all assigned to Rhone Poulenc Chimie
13) J.H. Lunsford, in "Natural Gas Conversion II", H.E. Curry-Hyde and R.F. Howe (Eds.), Elsevier Science, Amsterdam, (1994), p. 1
14) Z. Zhang, X.E. Verykios and M. Baerns, Catal. Rev.-Sci. Eng., 36 (1994) 507
15) S.S. Bharadwaj and L.D. Schmidt, Fuel Process. Techn., 42 (1995) 109
16) M.A. Peña, J.P. Gómez and J.L.G. Fierro, Appl. Catal., A: General, 144 (1996) 7
17) D.A. Hickman and L.D. Schmidt, J. Catal., 138 (1992) 267
18) M. Roy, M. Gubelmann-Bonneau, H. Ponceblanc and J.C. Volta, Catal. Lett., 42 (1996) 93
19) GB Patent Appl. 9318507.2 (1993), assigned to EVC
20) F. Cavani, E. Etienne, M. Favaro, A. Galli, F. Trifirò and G. Hecquet, Catal. Lett., 32 (1995) 215
21) G. Busca, F. Cavani, E. Etienne, E. Finocchio, A. Galli, G. Selleri and F. Trifirò, J. Molec. Catal., 114 (1996) 343-359
22) N. Mizuno, M. Tateishi and M. Iwamoto, Appl. Catal., A: General, 118 (1994) L1, ibidem 128 (1995) L165
23) N. Mizuno, D.-J. Suh, W. Han and T. Kudo, J. Molec. Catal., A: Chem., 114 (1996) 309
24) L. Jalowiecki-Duhamel, A. Monnier, Y. Barbaux and G. Hecquet, Catal. Today, 32 (1996) 237
25) W. Ueda, Y. Suzuki, W. Lee and S. Imaoka, in "11th International Congress on Catalysis-40th Anniversary", J.W. Hightower et al. (Eds.), (1996), p. 1065
26) H. Krieger and L.S. Kirch, US Patent 4,260,822 (1981), assigned to Rohm and Haas Co.
27) K. Nagai, Y. Nagaoka, H. Sato and M. Ohsu, EP 418,657 (1990), assigned to Sumitomo
28) G. Centi, J. Lopez Nieto, D. Pinelli and F. Trifirò, Ind. Eng. Chem., Res., 28 (1989) 400
29) G. Calestani, F. Cavani, A. Duran, G. Mazzoni, G. Stefani, F. Trifirò and P. Venturoli, in "Science and Technology in Catalysis, 1994", (Y. Izumi, H. Arai, M. Iwamoto, Eds.), Kodansha Ltd., Tokyo, Elsevier, Amsterdam, 1995, p. 179
30) S. Albonetti, F. Cavani, F. Trifirò, P. Venturoli, G. Calestani, M. Lopez Granados and J.L.G. Fierro, J. Catal., 160 (1996) 52
31) F. Cavani, A. Colombo, F. Giuntoli, E. Gobbi, F. Trifirò and P. Vazquez, Catal. Today, 32 (1996) 125
32) U.S. Ozkan, T.A. Harris and B.T. Schilf, Catal. Today, 33 (1997) 57
33) F. Cavani and F. Trifirò, Appl. Catal. A: General, 157 (1997) 195-221
34) H.H. Kung, Adv. Catal., 40 (1994) 1
35) F. Cavani and F. Trifirò, Catal. Today, 24 (1995) 307
36) T. Blasco and J.M. Lopez Nieto, Appl. Catal., A: General, 157 (1997) 117
37) R.G. Gastinger, A.C. Jones and J.A. Sofranko, Eur. Patent 253,552 (1987), assigned to Atlantic Richfield Co.
38) F.E. Cassidy and B.K. Hodnett, in DGMK Conference "C4 Chemistry- Manufacture and Use of C4 Hydrocarbons", DGMK, (1997), p. 149
39) F. Cavani, G. Centi and F. Trifirò, Appl. Catal., 15 (1985) 151

NATURAL GAS CONVERSION V
Studies in Surface Science and Catalysis, Vol. 119
A. Parmaliana et al. (Editors)

Acetonitrile by Catalytic Ammoxidation of Ethane and Propane: A New Reaction of Alkane Functionalization

G. Centi,* and S. Perathoner

Dip. Chimica Industriale, Univ. of Messina, Salita Sperone 31, 98166 Messina, Italy.

Acetonitrile can be selectively synthetized by ammoxidation of ethane or propane over Nb-Sb mixed oxides supported on alumina, with selectivities to acetonitrile of about 50-55% up alkane conversions of about 30%. In both cases, CO forms in about a 1:1 molar ratio with acetonitrile, due to a parallel reaction from a common intermediate. Feeding n-butane the selectivity to acetonitrile halfes. The differences in the behavior of this catalyst with that of analogous V-Sb-oxide catalysts, but selective to acrylonitrile from propane, and some aspects of the possible reaction mechanism are also briefly discussed. In particular, the absence of allylic type reactivity is evidenced.

Keywords: acetonitrile, ethane, propane, niobium, antimony.

1. Introduction

Acetonitrile is currently produced by catalytic dehydration of acetamide or dehydrogenation of ethylamine, and as a side product of the ammoxidation of propylene to acrylonitrile, but the development of an effective and economic process for its synthesis from a low cost hydrocarbon such as ethane can considerably expand its use. In fact, acetonitrile can be selectively hydrated over solid Brϕnsted acid catalysts to give the corresponding amide easily further transformed to acetic acid, and thus acetonitrile can be a first step for a new process of acetic acid synthesis. Acetonitrile can be also selective oxidized to glycolonitrile or glycolamide or can be converted to acrylonitrile by reaction with methanol [1,2]. The latter can be thus a new process for acrylonitrile synthesis starting from ethane, alternative to that commercial from propene or that under development from propane [3].

It has recently been shown that propane can be selectively ammoxidized to acrylonitrile using V-Sb-oxide based catalysts [3], but this catalyst is poor active and selective in the ammoxidation of ethane due to the different characteristics of C2 and C3 alkanes and their intermediates (for example, only primary H atoms are present in ethane and in the ethylene intermediate are absent allylic H atoms such as in propene). A different catalyst and reaction mechanism is thus necessary to ammoxidize ethane to acetonitrile than to convert propane.

In a preliminary communication was shown that acetonitrile can be synthetized from ethane on alumina-supported Nb-Sb-oxides [4]. Recently, Armor et al. [5] observed that also cobalt-exchanged MFI zeolites are active and selective in this reaction with a comparable behavior,

* corresponding author: fax: +39-90-391518, e-mail: centi@scirocco.unime.it

but Nb-Sb-oxides allow a more direct comparison with V-Sb-oxide based catalysts which shown a selective behavior in propane ammoxidation to acrylonitrile [3]. In the latter catalysts, it was demonstrated that the selective behavior in propane ammoxidation to acrylonitrile is correlable to the presence of a defective vanadium-antimonate phase having a rutile-like structure and in which antimony is present as Sb^{5+} only and vanadium both as V^{3+} and V^{4+}, plus V^{5+} supported over the vanadium-antimonate crystals [6]. Niobium belongs to the same row of vanadium, but due to the large ionic radius and absence of available *d* orbitals, cannot react with antimony oxide to form a structure analogous to that of vanadium antimonate. However, Nb can react with antimony to form an antimony-niobate structure in which, however, antimony is present as Sb^{3+}. It should be also observed that in vanadium-antimonate both vanadium and antimony play a catalytic role, the latter especially in the mechanism of selective insertion of oxygen (or nitrogen) in the organic substrate [3]. Furthermore, niobium based catalysts are known for their behavior in alkane oxidative dehydrogenation [7]. Therefore, using Nb instead of V in Me-Sb-oxide based catalysts may be expected to convert selectively alkanes, but through a different reaction mechanism.

In this work the catalytic behavior of Nb-Sb-oxides supported on alumina in the ammoxidation of C2-C4 alkanes as well as of propene is reported, in order to outline its peculiar reactivity characteristics and use for a process of acetonitrile synthesis from propane, as well as the differences with respect to V-Sb-oxide based catalysts.

2. Experimental

The Al_2O_3-supported Nb-Sb-oxide catalysts was prepared by a slurry method, which involves refluxing Sb hydroxide, Nb_2O_5 and Al_2O_3 in an ammoniacal solution for 24 h. The Nb:Sb atomic ratio in the solution was 1:5 and the amount of Al_2O_3 70% w.t. The solid was then separated by filtration, washed, dried at 150°C for 24 h, and calcined in air at 620°C for 6 h. The final surface area was 233 $m^2\ g^{-1}$. The method of preparation is analogous to that already reported for V-Sb-oxide catalysts supported on alumina [8]. Hereinafter, the sample will be called $NbSbO/Al_2O_3$.

Catalytic tests were made at atmospheric pressure in a tubular fixed-bed quartz flow reactor, using 3 g of catalyst (particle dimensions 0.2-0.4 mm) diluted with quartz powder. The feed composition was: 3.6% ethane, 5.9% oxygen, 5.2% ammonia and the remaining He. Analogous feed compositions were used for the other hydrocarbons. The space velocity was $457 \cdot g \cdot h\ (\text{mole of ethane})^{-1}$. Catalyst showed a stable activity, at least in a time scale of about 24 h. The products of reactions were detected by gas chromatographic on line analyses, but ammonia and HCN were determined by titration after adsorption. Further details were reported previously [8]. Selectivity and conversion data are reported on a carbon basis.

3. Results and Discussion

3.1 Characterization of the Catalyst

The catalyst after calcination is amorphous by X-ray diffraction (XRD) analysis and also infrared spectra evidence the presence of only the broad bands associated to alumina lattice vibrations which mask the detection of possible Nb and Sb species. By calcination at higher temperature (800°C), together with diffraction lines of α-Sb_2O_4 weak lines indicating the formation of $AlSbO_4$ rutile phase are detected, but not lines of niobium-oxide or of its possible

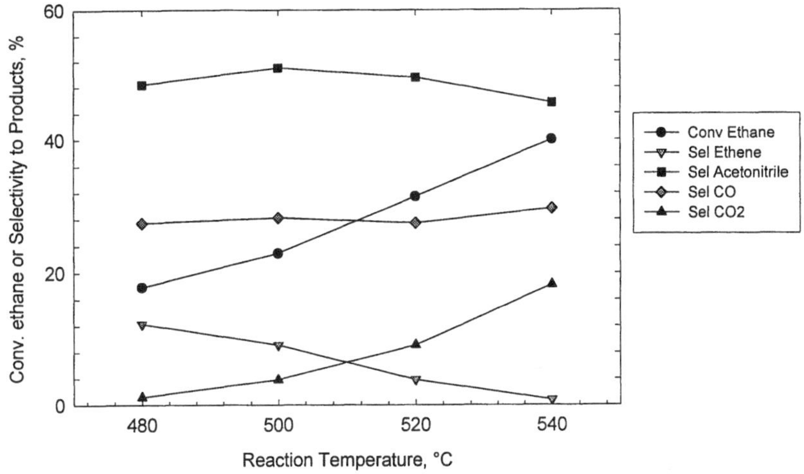

Fig. 1 Catalytic behavior of $NbSbO/Al_2O_3$ in ethane ammoxidation. Reaction conditions as indicated in the experimental part.

compounds. This indicates that niobium is rather disperse and strongly interacting with alumina surface, whereas Sb-oxide could react with alumina itself to form a rutile phase.

In order to check the possible formation of compounds between niobium and antimony, samples having a Nb:Sb ratio of 1:1 and 1:5 were prepared by calcination at 650°C (12 h) of mechanical mixtures of Nb_2O_5 and Sb_2O_3. Whereas for an analogous treatment, V_2O_5/Sb_2O_3 mechanical mixtures form an XRD monophasic $VSbO_4$ rutile phase or a biphasic $VSbO_4$ + α-Sb_2O4 system depending on the V:Sb ratio [3], in Nb_2O_5/Sb_2O_3 catalyst the solid state reaction at 650°C also leads to a disappearence of reflections of Nb_2O_5 (partial in 1:1 sample and complete in 1:5 sample), but without the corresponding formation of an antimony-niobate phase. The latter phase could be evidenced to be present in small amounts by XRD analysis, but not in the amount corresponding to that expected from the disappearence of Nb-oxide reflections. This indicates that the large part of niobium is probably present supported over the antimony-oxide crystallites. From these data, it may be tentatively concluded that $NbSbO/Al_2O_3$ catalyst is formed mainly by microcrystalline (XRD amorphous) α-Sb_2O_4 or $AlSbO_4$ phases on the alumina surface which act as a support for a well spread niobium oxide.

3.2 Behavior of $NbSbO/Al_2O_3$ in ethane ammoxidation

Reported in Figure 1 is the catalytic behavior of $NbSbO/Al_2O_3$ in ethane ammoxidation. Ethane is selectively converted to acetonitrile with selectivities of about 50% up to an ethane conversion of about 30-35%. In addition to acetonitrile, CO is also formed and the trend of the selectivity to CO against conversion parallels that of acetonitrile (for $T \leq 520°C$). CO forms in about equimolar amounts of acetonitrile and thus with approximately half selectivity on a carbon basis. Ethene is also detected especially at the lower reaction temperatures, whereas CO_2 forms above 500°C. Besides to these products, HCN is also detected with a selectivity of around 5% or less.

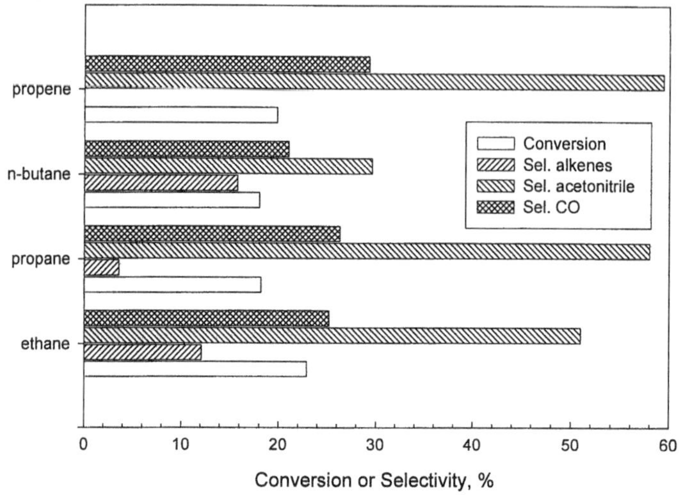

Fig. 2 Comparison of the catalytic behavior of $NbSbO/Al_2O_3$ in C2-C4 alkane and in propene ammoxidation at 500°C. Experimental conditions as indicated in Fig. 1.

Differently from the case of CO_2, CO does not form by consecutive oxidation of acetonitrile ($T \leq 520°C$), but derives probably from a competitive pathway on a reaction intermediate common with the pathway to acetonitrile, as indicated from the fact that the CO/acetonitrile ratio remains nearly constant in the studied range of conversion and temperature (Fig. 1). This is confirmed also from tests changing the space-velocity and conversion at a constant reaction temperature. It may be also note that the CO/CO_2 ratio at temperatures equal or below 500°C is ten or more, whereas in the selective ammoxidation of alkanes the ratio is usually much below one [3]. This indicates the difference in the mechanism of alkane conversion of $NbSbO/Al_2O_3$ catalyst with respect V-Sb-oxide catalysts, for example. It should be also evidenced that the latter are very few active in ethane ammoxidation and forming mainly carbon oxides instead of acetonitrile.

When a catalyst of only Sb-oxide or Nb-oxide supported on alumina is tested in the same conditions of Sb-Nb-mixed oxides on alumina, acetonitrile is found is the first case and propene in the second case as main products, but catalyst activity in considerably lower. It is thus possible to suggest that Nb plays a primary role in the selective oxidative dehydrogenation of propane and Sb in its consecutive (amm)oxidation, but a synergetic effect of the two components is present.

3.3 Comparison of the behavior of $NbSbO/Al_2O_3$ in C_2-C_4 hydrocarbon ammoxidation

The catalytic behavior of $NbSbO/Al_2O_3$ catalysts in the ammoxidation of different light alkanes is compared in Figure 2. For sake of conciseness, the behavior is compared only at 500°C which corresponds to the temperature of maximum selectivity in acetonitrile from ethane (Fig. 1), but the comparison at different reaction temperatures do not leads to significatively different conclusions.

$NbSbO/Al_2O_3$ catalyst is also active in propane ammoxidation to form acetonitrile with slightly higher selectivity than from ethane. The formation of CO parallels again that of acetonitrile: the molar amounts of acetonitrile and CO are nearly the same and thus selectivity ratio is 2/3 on a carbon basis. Acrylonitrile also forms in propane ammoxidation on $NbSbO/Al_2O_3$ cataalyst, but with maximum selectivity lower than about 3-4%, whereas selectivities up to 60% are obtained on V-Sb-oxide based catalysts [3]. While on V-Sb-oxide based catalysts the rate of ethane depletion is significantly lower than that of propane, on $NbSbO/Al_2O_3$ the rate of propane conversion is similar or slightly lower than that of ethane conversion (Fig. 2), evidencing again the difference between the two type of catalysts.

Propene is also detected in propane ammoxidation over $NbSbO/Al_2O_3$ catalyst. In order to check the possible role of alkene intermediate, propene was also used as reactant. The results are summarized in Figure 2. The behavior feeding propene is very similar to that feeding propane, both in term of product conversion and rate of hydrocarbon depletion. This indicates that the alkene is a possible intermediate in the mechanism of transformation, but also that the presence of reactive hydrogen on the hydrocarbon substrate (allylic hydrogens in propene) does not significantly alters the reactivity. This suggests that the rate determining step in the reaction could be the attack of electrophilic oxygen to the π bond of alkene, and not the abstraction of a hydrogen followed by nucleophilic type oxygen addition to hydrocarbon.

Using n-butane as the reactant for ammoxidation lower selectivities to acetonitrile are instead found (Fig. 2). It is worth noting, however, that in this case butenes and propene (in about a 3:1 ratio) are detected as products of oxidative dehydrogenation, whereas butadiene formation was not observed. This confirms the absence of allylic type oxidation activity (1-butene to butadiene) of this catalyst and the activity in C-C bond breaking. However, propionitrile forms with selectivities in the 5-10% range.

3.4 Aspects of the reaction network and mechanism

Although present data do not allow detailed indication on the reaction mechanism, they indicate some peculiar characteristics of the reactivity of $NbSbO/Al_2O_3$ with respect to analogous V-Sb-oxide based catalysts [3 and references therein]. Three are the most striking evidences: (i) Nb-Sb-oxide based catalyst, differently from V-Sb-oxide based catalysts, give oxidation with carbon chain rupture and nearly equimolar formation of CO and acetonitrile from a probable common intermediate, (ii) the order of reactivity, both regards the increase of the carbon chain series and the difference between alkane and alkene reactivity (the case of propane and propene) is different from that observed for V-Sb-oxides and (iii) allylic type reactivity (H abstraction and O insertion) in absent in $NbSbO/Al_2O_3$, differently from V-Sb-oxides.

The detection of alkene from alkanes and the decrease of their selectivity with increasing conversion and reaction temperature (Fig. 1) suggests that the alkane oxidative dehydrogenation to corresponding alkenes is the first step of the reaction over $NbSbO/Al_2O_3$ catalysts, similarly to V-Sb-oxide based catalysts [3]. This is reasonable also taking into account that niobium-based catalysts are known for their behavior in alkane oxidative dehydrogenation [7]. However, differently from V-Sb-oxide based catalysts, in $NbSbO/Al_2O_3$ catalyst are absent sites for allylic type oxidation. This is reasonably connected to the fact that antimony must be in the Sb^{5+} state to be active in allylic type oxidation and when reduced to

Sb^{3+} his reoxidation is very slow, if not catalyzed from redox elements such as iron or vanadium [9]. Niobium does not catalyze this reaction, although is in a good contact with antimony oxide as indicated from the fact that antimony maintains it in a highly dispersed state over alumina.

On V-Sb-oxides it was shown [10] that two competitive routes are present in propane ammoxidation: the propene intermediate can either (i) react with Brϕnsted acid sites to give an isopropylate intermediate which easily degradate to acetate and formiate species responsible of acetonitrile and CO formation, respectively, and (ii) give allyl alcoholate type species further transformed to acrolein and acrylonitrile. If the latter type of active sites is absent in $NbSbO/Al_2O_3$, the first pathway of transformation dominates the reactivity with formation of acetonitrile and CO from propane as main products of reaction and in a nearly equimolar amount.

In ethane ammoxidation, after the first step of ethene formation, the alkoxy intermediate may give an analogous reactivity path, but being the C-O bond formed at a methylic groups instead that a methylenic position as in isopropylate, a second oxygen attack at the same carbon atom to form an acetate species or at the nearlying carbon atom to form two formate species is possible. Both these intermediates further then fast transform to acetonitrile and CO, respectively. The relative rates of transformation along these two pathways is about 2:1 according to the selectivity ratio. In n-butane ammoxidation, instead, a more complex situation is present, because the alkoxide intermediate may either evolve to C3 + C1 fragments (the first responsible to the C3 type products detected) or to C2 products. For this reason, the selectivity to acetonitrile from n-butane is lower than in the case of ethane and propane.

In conclusion, Nb-Sb-oxide based catalysts are a promising system to develop a process of ethane ammoxidation to acetonitrile which can be also used to convert in the same conditions propane to acetonitrile, allowing to use directly a LPG stream. The catalytic behavior is significantly different from that of other catalysts for alkane ammoxidation such as V-Sb-oxide based catalyst, and can be related to the properties of niobium oxide maintained in a highly dispersed state by the combined action of antimony oxide and alumina support and absence of allylic type reactivity.

References

1. W. Ueda, T. Yokoyama, Y. Moro-oka, T. Ikawa, Ind. Eng. Chem. Prod. Res. Dev., 24 (1985) 340.
2. J.F. Brazdil, R.G. Teller, W.A. Marritt, J.C. Gleaser, A.M. Ebner, J. Catal., 100 (1986) 516.
3. G. Centi, S. Perathoner, F. Trifirò, Appl. Catal. A: General, 157 (1997) 143.
4. R. Catani, G. Centi, J. Chem. Soc. Chem. Comm., (1991) 1081.
5. Y. Li, J.N. Armor, J. Chem. Soc. Chem. Comm., (1997) 2013.
6. G. Centi, P. Mazzoli, S. Pperathoner, Appl. Catal. A: General, 165 (1997) 273.
7. R.H. Smits, K. Seshan, J.R. Ross, in New Developments in Selective Oxidation by Heterogeneous Catalysts, P. Ruiz and B. Delmon Eds., Studies Surface Science and Catal. Vol, 72, Elsevier Science: Amsterdam 1992, p. 221.
8. R. Catani, G. Centi, F. Trifirò, R.K. Grasselli, Ind. Eng. Chem. Research, 31 (1992) 107.
9. G.Centi, F.Trifirò, Catal. Rev.-Sci. Eng., 28 (1986) 165.
10. G. Centi, S. Perathoner, CHEMTECH, 2 (1998) 13.

NATURAL GAS CONVERSION V
Studies in Surface Science and Catalysis, Vol. 119
A. Parmaliana et al. (Editors)

Partial Oxidation of Hydrocarbons: an Experimental and Kinetic Modeling Study

T. Faravelli[a], A. Goldaniga[a], E. Ranzi[a], A. Dietz[b], M. Davis[b] and L. D. Schmidt[b]

[a]CIIC Department, Politecnico, Milan, 20132 Italy

[b]University of Minnesota, Minneapolis, MN 55455 USA

Partial oxidation offers an attractive alternative route to form olefins from alkanes in simple and low cost reactors. Using Pt catalyst on α-alumina support in a reactor with a contact time of a few milliseconds, olefin selectivities of up to 70% have been achieved at large alkane and O_2 conversion without catalyst fouling and deactivation. Under suitable conditions, cyclohexane oxidative dehydrogenation to form cyclohexene prevails over the decomposition reactions toward butadiene and smaller olefins.
A general and extensive pyrolysis and oxidation reaction scheme, involving about 150 species and 3000 reactions, has been used to verify the importance of homogeneous reaction paths in explaining the experimental results of oxidative pyrolysis of alkanes from propane up to hexanes. For higher alkanes, relevant cracking occurs even at short contact times and considerable ethylene and propylene yields are obtained.
Comparisons between model predictions and experiments at different temperatures (1000-1300 K), residence times (1-5 ms), and alkane to oxygen ratios (0.5-1.2) suggest that surface reaction steps play a significant role in this process especially for the smaller alkanes and at the lower temperatures (higher alkane to oxygen ratios). In these conditions, catalyst activity promotes the reactivity of the system and improves CO_2 formation as well as oxidative dehydrogenation. Few simplified catalytic steps coupled with the homogeneous reaction scheme allow us to explain the experimental data. This model is a useful tool for investigating and optimizing operating conditions and reactor configurations.

1. INTRODUCTION

Ethylene and olefins are usually produced by the steam cracking process with an indirect heating of the hydrocarbon feed in pyrolysis furnaces. Several research attempts have been and are continuously spent to find alternatives to this conventional process.
Oxygen addition or oxidative pyrolysis of alkanes seems of industrial relevance in several practical situations. The BASF process and its various modifications are a typical example. A 1.7/1 methane/oxygen premixed flame preheated at about 900 K produces C_2H_2 with an overall yield of about 25 %. The two stage HOECHST High Temperature Pyrolysis process allows productions of acetylene and ethylene with hydrocarbon feeds heavier than methane. A fuel gas is burned with oxygen in near stoichiometric conditions in a combustion chamber and the hot combustion gases, mixed with hydrocarbon feed and dilution steam, enter the reaction cracking zone with residence times of about 1 ms.

There are several process alternatives and their common feature is the possibility to compensate for the endothermic cracking reactions with exothermic oxidation reactions.

- Advanced cracking reactors - (Union Carbide, Kureha & Chiyoda)

A multi-port burner produces a heat carrier gas of about 2000 °C by the combustion of fuel and oxygen in the presence of preheated steam. The initial temperature is about 1600 °C and the final temperature at the exit of the reactor is 700-900 °C after a residence time of 10-30 ms. The cracked gas is then quenched by oil in a quench cooler. High temperatures and low contact times allow very selective cracking conditions.

- Thermal regenerative cracking - (Gulf, Stone & Webster)

The heat needed for feedstock cracking is transferred directly from hot micron-size solid particles in a moving stream. The residence times are of the same order of coil cracking furnaces and the steam to hydrocarbon ratio is 0.2 to 0.4. Coke and fouling on hot particles can be burnt in a regenerative section, analogous to a riser FCC loop [1].

- Oxygen addition to the feed of industrial crackers

Partial oxidation of a small fraction of the feedstock in the cracking coils can generate a significant amount of heat which can be directly used by the endothermic pyrolysis reactions. The heat-transfer limitations in the industrial crackers are partially removed. The oxygen addition increases the alkane conversion, while the olefin yields decrease only marginally. The oxygen addition produces several advantages in the industrial cracker sections, like a decrease of the residence time or an increase of the possible throughput, higher severity or reduced tubeskin temperatures, and, finally, longer coil lifetime [2]

The aim of this work is to analyze the catalytic oxidative pyrolysis of alkanes, to verify the effect of oxygen addition to the feed, to clarify the extent of the homogeneous reactions, and the role of the catalyst.

2. EXPERIMENTAL DATA

Experimental data on several hydrocarbons up to n-pentane, n-hexane and cyclo-hexane have been obtained in a cylindrical quartz reactor containing a 10 mm long Pt coated α-alumina foam monolith (45 pores per inch) 10 mm long [3]. Experimental data, completely reported and deeply discussed elsewhere [4], always refer to the air oxidation of hydrocarbons at a contact time of about 5 ms. Reactor temperature, strongly dependent on the alkane to oxygen ratios, was measured using a Pt-Pt/13%Rh thermocouple axially located between the catalytic monolith and a rear radiation shield. The calculated adiabatic reaction temperature was within 50 K of the measured reaction temperature suggesting that the reactor is nearly adiabatic.

3. HOMOGENEOUS REACTION SCHEME

The detailed reaction scheme used for the simulation of pyrolysis and partial oxidation of the different paraffins has been deeply discussed elsewhere [5]. This kinetic model, able to describe both the oxidation and combustion of commercial gasolines in internal combustion engines and the soot precursor formation in fuel rich flames, is constituted by more than 150 species involved in more than 3000 elementary and lumped reactions. As a first modeling exercise we analyze the relative reactivity of the different paraffins in pyrolysis and oxidative pyrolysis (alkane/O_2 = 1) conditions. It seems worthwhile to mention that the effect of oxygen addition to the different hydrocarbons strongly decreases with an increase of the number of C atoms. As a matter of fact, while the ratio of contact times to reach 60% ethane conversion in pyrolysis and partial oxidation is more than 15, the same ratio becomes about 3 for n- and iso-butane and less than 2 for n-hexane.

The apparent decomposition rate in these operating conditions can be roughly approximated, for n-paraffin heavier than n-pentane, by the following general expression:

$$r = 3x10^{12}\left(n_c - 3\right)^{0.5}\left(n_c - 1\right)\exp\left(-\frac{50500}{RT}\right)c_{hyd}^{1.5}\left[1 + 1.2c_{O_2}^{0.33}\exp\left(\frac{6500}{RT}\right)\right] \qquad (\text{kmole/m}^3\text{s})$$

where n_c is the number of carbon atoms of the hydrocarbon feed, c_{hyd} and c_{O2} are the hydrocarbon and O_2 concentrations. This rate expression disregards the presence of an induction time and fairly reproduce the predictions of the detailed scheme at about 60% of hydrocarbon conversion. The difference in activation energy mainly reflects the role of the H-abstraction reactions of OH• radicals. In fact, while metathesis reactions of alkyl as well as H• radicals require about 10000 cal/mole , OH• radicals can abstract H-atoms with less than 3500 cal/mole.

4. RESULTS AND DISCUSSION

Figure 1 shows a simple comparison between model predictions and experimental data on n-pentane conversion at different hydrocarbon to O_2 ratios. The experimental data refer to the oxidation of n-pentane in air over a Pt coated foam monolith (45ppi) while the model predictions refer to an ideal isothermal plug flow reactor at a contact time of 5 ms. The reactor temperature spans from about 1000 to 1200 K. The quite good agreement between experimental and calculated selectivity to olefins (ethylene, propylene and 1-butene) observed in figure 2 confirms that the process can be mainly explained by the homogeneous reaction scheme. At high temperatures (higher than 1100 K), where the CO selectivity is higher than that of CO_2, the process is mostly homogeneous even though the radical pool can also be sustained by surface reactions.

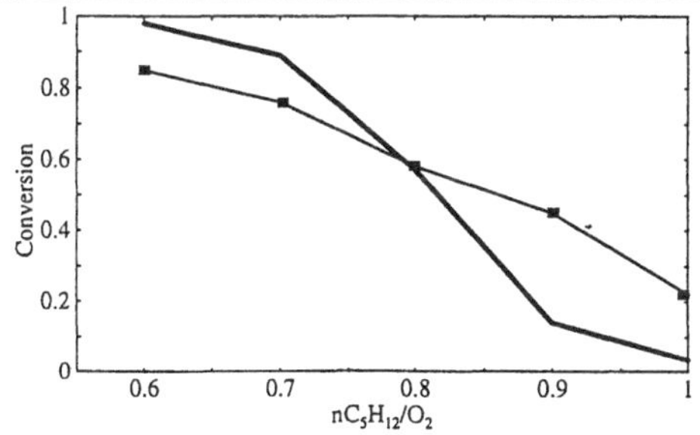

Figure 1: *n*-Pentane oxidation. Comparison between experimental (points) and calculated conversion.

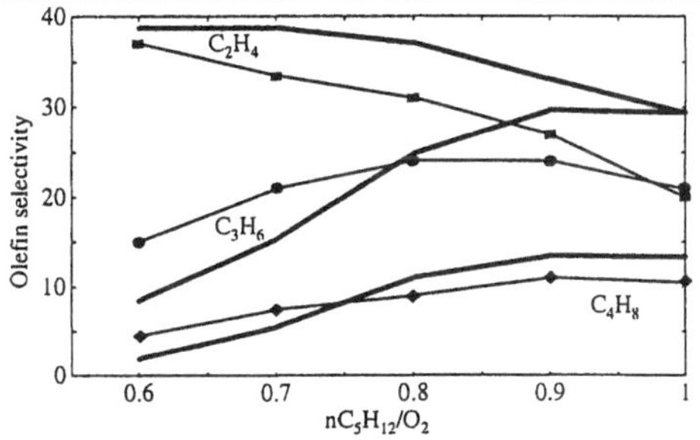

Figure 2: *n*-Pentane oxidation. Comparison between experimental (points) and calculated selectivity.

At low temperatures the conversion levels predicted by the homogeneous model indicate that only catalytic reaction steps can justify the overall conversion and the large CO_2 production. These and similar comparisons between model predictions and experiments at different temperatures, residence times and alkane to oxygen ratios suggest that the catalyst or better the surface reaction steps play a dominant role in this process especially for the smaller alkancs and at the lower temperatures (high alkane to oxygen ratios).

The experimental selectivities to olefins (moles of products/ 100 moles of converted hydrocarbon) are close to those predicted by the homogeneous model. At low hydrocarbon to oxygen ratios, that is at high temperatures, the conversion is overpredicted by the homogeneous model. This excess of reactivity could be attributed both to the assumption of ideal conditions in model simulations and to the lack of possible radical recombination reactions promoted by the catalyst surface. At low temperatures (or high hydrocarbon to oxygen ratios) the experimental conversions can only be explained by the catalyst effect. The CO_2 production, which cannot be justified by the homogeneous model, indicates the heterogeneous mechanism also at higher temperatures.

Predictions of the homogeneous kinetic model are compared in Fig. 3 with experimental data. These data refer to the catalytic partial oxidation of *n*-hexane at 1100-1200 K. It is evident from the analysis of figure 3 that olefin selectivities are well predicted by the homogeneous model up to *n*-hexane to oxygen ratios of about .9 (temperatures higher than 1150 K). As this ratio increases, conversion and reactor temperature decrease and catalyst effect in terms of hexenes and CO_2 formation becomes relevant and reduces the selectivities of smaller olefins.

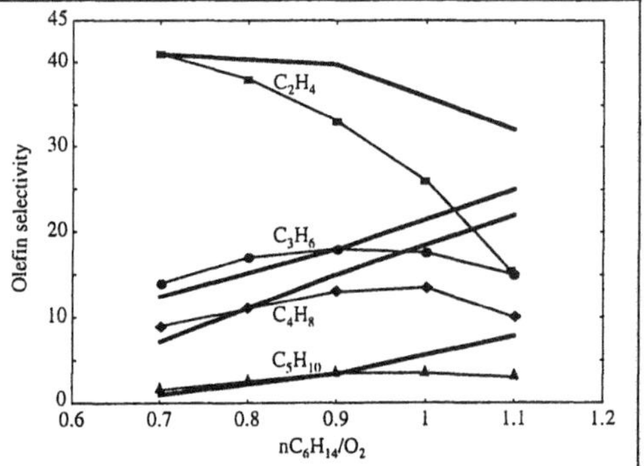

Figure 3. *n*-hexane oxidative pyrolysis. Comparison between experimental (points) and calculated selectivities.

Similar comparisons and discrepancies have been observed for different paraffins such as propane, butanes and cyclohexane. Catalyst effect is more evident in these cases, particularly in cyclo-hexane dehydrogenation.

4.1- Cyclohexane pyrolysis and oxidation

Decomposition of cyclohexane mainly goes through an initial isomerization reaction to form 1-hexene, which decomposes to form allyl and n-propyl radicals.

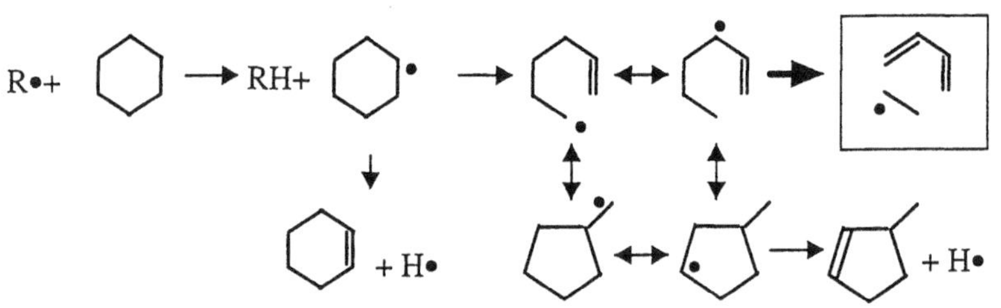

Figure 4. H-abstraction reactions on cyclo-hexane

Once these radicals are formed they abstract H atoms from cyclohexane and the resulting cyclohexyl radicals can isomerize and decompose following the reaction paths described in figure 4. The formation of 1,3-butadiene and ethyl radicals is the prevailing decomposition path while the dehydrogenation reactions to form cyclohexene or methyl cyclopentene are less favored due to their higher activation energy. Similarly, decomposition reactions of cyclohexyl radicals to form allyl radicals and propylene or methyl radicals and pentadiene are of minor importance. The complete reaction path of figure 4 can be lumped into the following equivalent metathesis reaction:

$$R\bullet + CyC_6H_{12} \rightarrow RH + \alpha(C_4H_6+C_2H_5\bullet) + \beta(CyC_6H_{10}+H\bullet) + \gamma(C_5H_8+CH_3\bullet) + \delta(C_3H_6+C_3H_5\bullet)$$

where R• stands for all the possible H-abstracting radicals and the kinetic parameters are easily predicted on the basis of the 12 secondary H atoms of cyclohexane [6].
The stoichiometric coefficients of this equivalent reaction can be evaluated from the relative importance of the competitive rate constants. In the temperature ranges of interest, they can be easily estimated: $\alpha = 0.75$, $\beta = 0.1$, $\gamma = 0.08$, and $\delta = 0.07$
In order to extend the previous detailed kinetic scheme to cover also cyclohexane it is then necessary to include also the following elementary reactions:

		A[#]	E[#]
CyC_6H_{12}	$= 1C_6H_{12}$	$1.50\ 10^{16}$	82500
CyC_6H_{12}	$= CyC_6H_{10}+H_2$	$0.60\ 10^{14}$	70000
CyC_6H_{10}	$= C_2H_4+ C_4H_6$	$1.00\ 10^{15}$	67000
CyC_6H_{10}	$= C_6H_6+2\ H_2$	$1.00\ 10^{14}$	69000
$H\bullet+ CyC_6H_{10}$	$= C_3H_6+ aC_3H_5\bullet$	$0.36\ 10^{11}$	2000
$OH\bullet+ CyC_6H_{10}$	$=$ Acrolein$+ 1C_3H_7\bullet$	$0.50\ 10^{10}$	0

[#]Units are cal, mole, l, s

Cyclohexene (CyC_6H_{10}) forms benzene through H-abstraction reactions, via cyclohexadiene:

$$R\bullet + CyC_6H_{10} = RH + C_6H_6 + H_2 + H\bullet$$

Frequency factors and activation energies of these elementary reactions are assumed on the basis of well tested analogy and similarity rules [5].
The experimental data obtained at 950-1100 K show a relevant effect of heterogeneous reactions. The catalyst surface promotes the reactivity of the system with a selective dehydrogenation towards cyclohexene as well as CO_2 formation. It is possible to reproduce these data quite well, by simply adding three empirical reaction steps related to the adsorption of oxygen on the catalyst surface and its successive interactions with cyclohexane (to form cyclohexene and CO_2) and with butadiene:

		A[#]	E[#]
$O_2 + *$	$\rightarrow O_2^*$	$2.0\ 10^3$	5000
$O_2^* + H_2$	$\rightarrow OH + OH + *$	$3.0\ 10^6$	4000
$O_2^*+ CyC_6H_{12}$	$\rightarrow CO_2 + 0.35CyC_6H_{10}$ + products + *	$1.5\ 10^7$	4000

[#]Units are cal, mole, l, s

where * and O_2^* stand respectively for the active site and the adsorbed oxygen.
Figure 5 shows the overall comparison between experimental data and model predictions. These catalytic reactions coupled with the detailed homogeneous model allows to explain the complete set of experimental data. This kinetic model is a useful tool to investigate and suggest new operating conditions and reactor configurations which may lead to optimum olefin yields.

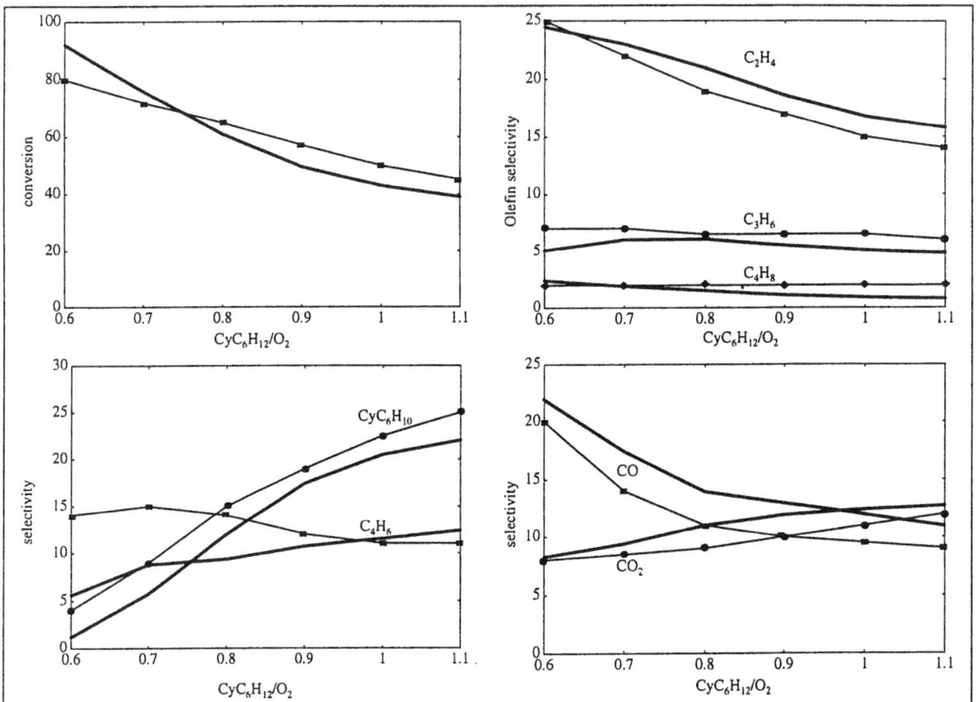

Figure 5. Cyclohexane oxidative pyrolysis. Comparison between experimental data (points) and model predictions

REFERENCES

1. Y. C. Hu, Hydrocarbon processing, Nov., (1982) 109-117.
2. Q. Chen, E. J. A. Schweitzer, P. F. Van Den Oosterkamp, R. J. Berger, C. R. De Smet and G. B. Marin, Ind. Eng. Chem. Res., 36 (1997) 3248-3251,.
3. D. A. Hickman and L. D. Schmidt, J. Catal. 138 (1992) 267-282.
4. A. G. Dietz III, A. F. Carlsson and L. D. Schmidt, J. Catal. in press, (1998).
5. E. Ranzi, T. Faravelli, P. Gaffuri, A. Sogaro, A. D'Anna and A. Ciajolo, Combust. Flame 108 (1997) 24-43.
6. E. Ranzi, M. Dente, T. Faravelli, and G. Pennati, Combust. Sci. and Tech. 95 (1994) 1-50.

NATURAL GAS CONVERSION V
Studies in Surface Science and Catalysis, Vol. 119
A. Parmaliana et al. (Editors)

Supercritical-phase oxidation of isobutane to t-butanol by air

Li Fan, Takashi Watanabe and Kaoru Fujimoto

Dept. of Applied Chemistry, School of Engineering, The University of Tokyo,
Hongo 7-3-1, Bunkyo-ku, Tokyo, 113-8656, Japan

Direct introduction of air to supercritical-phase isobutane exhibited high activity for selective synthesis of tertiary butyl alcohol, especially on SiO_2-TiO_2 and Pd/carbon catalysts. A kind of critical phenomenon, the remarkable enhancement of activity around the critical point, was observed if the reaction phase changed to supercritical state.

1. INTRODUCTION

A lot of effort has been recently made to achieve successful applications of supercritical fluid in catalytic reactions including both homogeneous and heterogeneous systems, such as Fischer-Tropsch reaction, alkylation, and carbon dioxide conversion [1-4]. But few organic oxidation synthesis reactions in supercritical phase medium have been studied until now, especially where air is used directly as oxidant. It is of theoretic interest as well to investigate the reaction behavior of oxygen species such as peroxide in supercritical fluids. It is expected that high solubility, high diffusion capacity, cluster structures around solute molecules, and special ionization effects of the supercritical fluid can exert strong influence on reaction mechanism itself.

As initial material in methyl tertiary butyl ether (MTBE) production and methyl methacrylate (MMA) synthesis where MTBE is a high-octane-number gasoline additive and MMA is a resin material, tertiary butyl alcohol (TBA) is of increasing importance recently [5,6]. On the other hand, TBA can be simply converted to isobutene via dehydration reactions. Commercially, production of isobutene from isobutane through direct dehydrogenation reaction needs high reaction temperature such as 773K-873K, in which the catalyst is deactivated quickly. Here we report a new synthesis method of TBA where air is used directly as oxidant to convert isobutane to TBA. This synthesis reaction can be conducted efficiently if isobutane is at supercritical phase, on selected catalysts.

2. EXPERIMENT

A typical flow-type fixed-bed reactor was employed where isobutane was fed by a high-pressure pump. Isobutane has a critical temperature of 408.13 K and a critical pressure of 36.48 bar. The amorphous SiO_2-TiO_2 catalyst is commercially available (Fuji Silysia Chemical Co.). Pd/CeO_2 (4wt%) and Pd/C (2.5wt%) catalysts were prepared by impregnating $PdCl_2$ acidic solution on CeO_2 (Soekawa Chemicals, 99.99%) or active carbon (Shirasagi, Takeda Pharmacy Co.) and were dechlorinated in flowing hydrogen at 400°C for 3 h.

All products was confirmed on GC-MS (Shimadzu GCMS 1600) and analyzed by gas chromatography (Shimadu GC-14B or Hitachi 163). A capillary column (TC-1) with flame ionization detector was used for the analysis of oxygenates and heavy hydrocarbons. Molecular sieve column and active carbon column, with thermal conductivity detector, were utilized for inorganic gas analysis. Lighter hydrocarbons were determined by active alumina column with flame ionization detector. The standard reaction conditions were as follows: mole ratio of iC_4 / air = 3/1; W/F (total) = 10 g.h/mol; catalyst weight = 0.5 g; reaction temperature = 426 K. The total pressure for gas phase reaction was 44 bar and that for supercritical phase was 54 bar. Correspondingly, the partial pressure of isobutane was 33 bar and 41 bar, respectively.

3. RESULTS AND DISCUSSION

Table 1 compares the reaction performances of 3 catalysts and of the non-catalytic case at supercritical and gas phases. In either the non-catalytic reaction or the catalyzed reaction, changing of the existing state of isobutane from gas phase (44 bar) to supercritical phase (54 bar) resulted in the remarkably enhanced conversions of isobutane and oxygen. Generally, selectivity of the target products (TBA and isobutene) increased slightly while

Table 1. Reaction Performance of Catalytic Oxidation of iso-Butane by Air in Supercritical Phase or Gas Phase

Catalyst	Total Pressure (bar)	i-C4H10 Conv. (%)	O2 Conv. (%)	TBA Sel.(%)	i-C4H8 Sel.(%)	TBA and i-C4H8 Yield (%)
non	44	0.3	2.5	55.0	7.0	0.2
non	54	1.2	9.9	58.1	8.1	0.8
SiO2-TiO2	44	2.9	24.0	59.0	5.2	1.9
SiO2-TiO2	54	4.9	40.6	61.2	6.3	3.3
SiO2-TiO2	12	0.0	0.0	0.0	0.0	0.0
SiO2-TiO2*	54*	0.1	1.1	55.5	7.7	0.1
Pd/C	44	0.5	4.2	61.2	2.1	0.3
Pd/C	54	3.1	25.6	64.8	20.1	2.7
Pd/CeO2	44	1.8	17.1	36.5	17.4	1.0
Pd/CeO2	54	3.2	28.4	56.5	3.7	1.9

i-C4/air = 3/1, W/F = 10 g.h/mol, catalyst weight: 0.5 g, temperature: 426 K

*Liquid-phase reaction where the reaction temperature was 403 K

one changed from gas phase to supercritical phase. Exceptionally, Pd/C catalyst gave tremendous variance in both conversions and selectivities of the reaction if a phase change happened. When the total pressure was increased from 44 bar to 54 bar, the total yield of TBA and isobutene on Pd/C catalyst was enhanced from 0.31% to 2.70%. For the high isobutene selectivity on Pd/C catalyst in the supercritical phase, we inferred that the TBA formed was dehydrated rapidly on this catalyst, leading to high selectivity of isobutene. But the reason was not clear at the present stage.

All the catalyzed reactions showed remarkably higher activities than non-catalytic reactions, which proved the promotional role of the catalysts in the reaction.

As exhibited in Table 1, if the total pressure of the gas phase reaction was as low as 12 bar, the reaction could not proceed, even if most active SiO_2-TiO_2 catalyst was used. Similarly, for the liquid phase reaction on SiO_2-TiO_2, the reaction rate was very low.

Generally, the main byproduct of the reaction was acetone. Methanol was also formed. The combustion products, CO_2, only formed in very small amounts. CO was not detected. Lighter hydrocarbons (C1-C3) formed but amounted very tiny; over 98% of them were methane. The total amount of all C1 species (methanol, CO_2, C1-C3) was equal to that of acetone, which indicates the decomposition of C4 to C3 species (acetone) and C1 fractions. No formic acid or formaldehyde was detected.

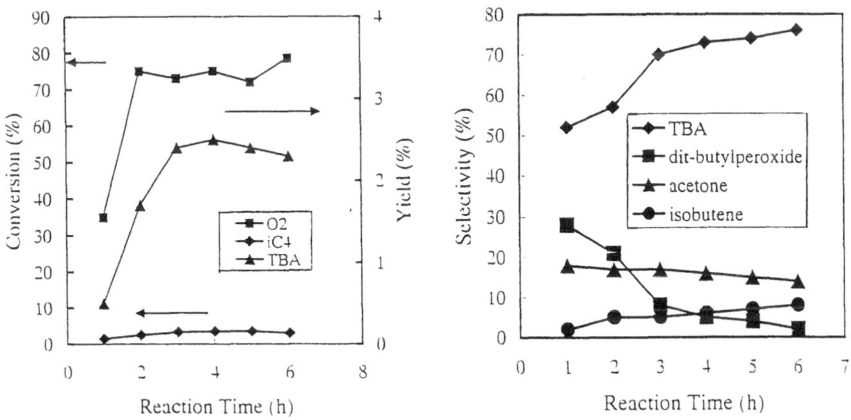

Fig. 1 Reaction performance change with time on stream

SiO_2-TiO_2 catalyst, 54bar, 426 K, W/F = 10 g.h/mol, catalyst weight: 0.5 g, i-C_4/air = 3/1

In Fig. 1, it is exhibited the time-on-stream change of reaction performances. O_2 conversion and isobutane conversion reached steady state at 2h, which indicates the existence of reduction period at the initial stage of the oxidation reaction. More interestingly, di-t-butyl peroxide (ROOR) formed with the selectivity of 30% at the initial stage. While the reaction proceeded, di-t-butyl peroxide selectivity decreased and correspondingly TBA selectivity increåsed. It is known that di-t-butyl peroxide should be produced from coupling of t-butyl peroxide radical or from quenching of t-butoxy radical.

$$2ROO\cdot = ROOR + O_2$$

$$2RO\cdot = ROOR$$

Formation of ROOR provides an evidence of the peroxide radical mechanism of this reaction. ROH was considered to form through the decomposition of ROOR and the subsequent hydrogen transfer from isobutane to radicals.

As clearly indicated in Table 1, this reaction depended greatly on temperature and pressure, which implies some effect of the reaction phase.

Figure 2 showed the jump of the activities on SiO_2-TiO_2 catalysts around the critical point when the reaction system changed from gas phase to supercritical phase. The total yield of TBA and isobutene increased sharply from 2.2% to 3.6%, while the total pressure was slightly emhanced from 47 bar to 49 bar, indicating the obvious critical phenomenon around the critical point (48 bar). Correspondingly, the conversions of O_2 and isobutane were enhanced to a great extent with this slightly-enhanced pressure. More interestingly, further increase of the total pressure in the supercritical phase was not favorable to the reaction on SiO_2-TiO_2 catalyst. It is clear that the activity at 54 bar was lower than that at 49 bar. It seems that higher pressure in the supercritical phase reaction suppressed the decomposition of ROOR and led to the lower TBA yield.

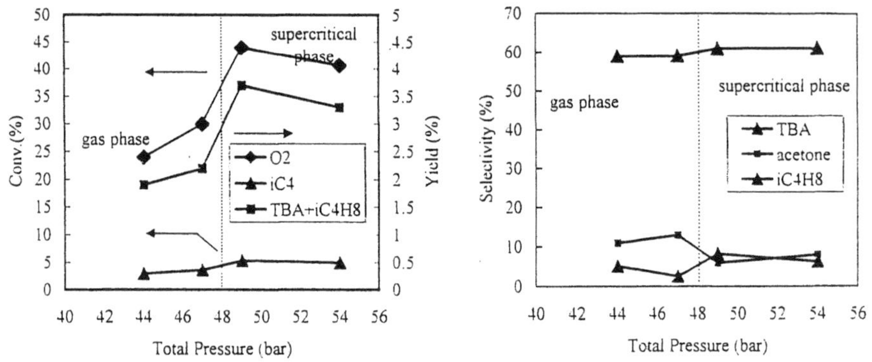

Fig.2 Comparison of the reaction performances in supercritical phase and gas phase. SiO_2-TiO_2 catalyst, 426 K, W/F = 10 g.h/mol, catalyst weight: 0.5 g, i-C_4/air = 3/1

Fig. 3 compared different reaction performances where partial pressure of air was changed. Isobutane/air ratio was change as 2/1 or 3/1. It is clearly demonstrated that the elevated air pressure increased the selectivity of TBA and isobutene, while the selectivity of acetone was suppressed. This trend was more obvious in the area of supercritical phase. As a result, TBA yield was increased if the air partial pressure was enhanced.

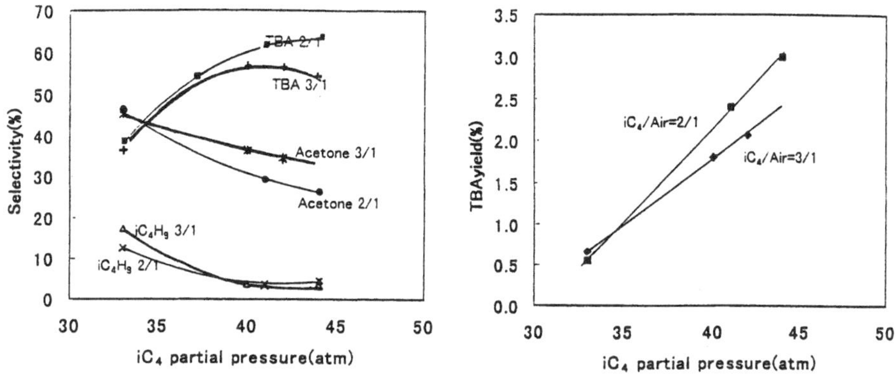

Fig. 3 Effect of air partial pressure on the product distribution in supercritical phase and gas phase

Pd/CeO_2 catalyst, 426 K, W/F = 10 g.h/mol, catalyst weight: 0.5 g, i-C_4/air = 3/1;2/1

For the oxidation mechanism in supercritical fluid, we consider that dioxygen can attack the most active hydrogen of isobutane to form tertiary butyl hydroperoxide (TBHP, (ROOH). TBHP is known to be used as an oxygen donor in epoxidation of olefins [7]. It is inferred that TBHP can form in the supercritical phase isobutane coexisting with dioxygen. This auto-oxidation step can proceed without catalyst [8], which easily happened in the inductive period at the initial stage of the reaction. ROOR was an intermediate in this radical reaction process. TBHP can decompose homolytically, resulting in tertiary butoxy radical and hydroxide radical [9, 10]. Tertiary butoxy radical combined with a third hydrogen from another isobutane to form TBA. TBA can be dehydrated on acidic sites, leading to the increase of isobutene selectivity. For the main byproduct acetone, it should be derived from decomposition of tertiary butoxy radical, accompanied by C1 compounds such as CH_3OH, CO_2 or CH_4.

The role of the catalyst is inferred to be that the catalyst could efficiently improve oxidation of isobutane with TBHP, resulting in the obviously enhanced conversion.

Catalyst deactivation was not observed during the supercritical phase reaction (426K) even if it reacted continuously for 30 h.

4. SUMMARY

SiO_2-TiO_2, Pd/carbon catalysts showed remarkably higher activity for tertiary butyl alcohol synthesis from supercritical-phase isobutane and air than from gas-phase or liquid-phase isobutane. Remarkable enhancement of activity around the critical point was observed if the reaction phase changed to supercritical state. Accompanied by supercritical fluid, catalyst deactivation was not observed in this reaction where reaction temperature was lower (426K).

REFERENCES

[1] L. Fan, K. Yokota and K. Fujimoto, AIChE J., 38(1992)1639.
[2] P. G. Jessop, T. Ikariya and R. Noyori, Nature, 368(1994)231.
[3] P. E. Savage, S. Gopalan, T. Mizan, C. Martino and E. Brock, AIChE J., 41(1995)1723.
[4] L. Fan, I. Nakamura, S. Ishida and K. Fujimoto, Ind. Eng. Chem. Res., 36(1997)1458.
[5] European Chemical News, 1993, Feb. 15.
[6] Chemical Week, 1993, Feb. 17.
[7] N. Emanuel, E. Denisov and Z. Maizus, Liquid Phase Oxidation of Hydrocarbons, Plenum Press, New York, 1967.
[8] A. Bielanski and J. Haber, Oxygen in Catalysis, Marcel Dekker, New York, 1991, p. 452.
[9] L. Fan, T. Watanabe and K. Fujimoto, Appl. Catal., 158(1997)L41
[10] L. Fan, Y. Nakayama and K. Fujimoto, Chem. Comm., (1997)1179

NATURAL GAS CONVERSION V
Studies in Surface Science and Catalysis, Vol. 119
A. Parmaliana et al. (Editors)

Isobutane Dehydrogenation and Pt L_{III}-edge XAFS Studies on γ-Al_2O_3 Supported Pt-containing Catalysts*

Jifei Jia[a], Liwu Lin[a], Yuan Kou[b], Zhusheng Xu[a], Tao Zhang[a], Jianzhong Niu[b], and Longya Xu[a]

[a]State Key Laboratory of Catalysis, Dalian Institute of Chemical Physics, Chinese Academy of Sciences, P.O.Box 110, Dalian 116023, China

[b]State Key Laboratory for Oxo Synthesis and Selective Oxidation, Lanzhou Institute of Chemical Physics, Chinese Academy of Sciences, Lanzhou, 730000, China

γ-Al_2O_3 supported Pt-containing catalysts have been investigated by Isobutane dehydrogenation and XAFS analyses. The results indicated that Pt species is highly dispersed on the γ-Al_2O_3 support for the γ-Al_2O_3 supported Pt-containing catalysts after reduction. Pt is in amorphous metallic state, and no alloy of Pt-Sn or Pt-Fe was formed after reduction for the Pt-Sn/γ-Al_2O_3 and Pt-Fe/γ-Al_2O_3 catalysts. Pt anchored γ-Al_2O_3 support via promoter-SnO or FeO, which resulted in an enhanced dehydrogenation performance for the both catalysts.

1. INTRODUCTION

Supported dual-component catalysts have been widely used in petroleum industry for more than 30 years[1,2]. Based on many characterization results, Researchers have found that many supported dual-component catalysts are composed of a complex system which are a metal with reduced state, a metal oxide (promoter) and support in the real industrial conditions[2-7]. The local structure of supported dual-component catalysts are proposed that the reduced metal is anchored to the support via metal oxide for Pt-Sn/γ-Al_2O_3, Ru-Fe/TiO_2, Co-Mn/SiO_2, Co-Mn/Al_2O_3 et al.[3-7]. It has been reported that γ-Al_2O_3 supported Pt-Sn and Pt-Fe dual-component catalysts are all good dehydrogenation catalysts[2,8]. So we employed Isobutane dehydrogenation and Pt L_{III}-edge XAFS to study the nature of γ-Al_2O_3 supported Pt, Pt-Fe and Pt-Sn catalysts.

2. EXPERIMENTAL DETAILS

The preparation, evaluation and XAFS analyses of the catalysts had been given previously[2,8-10]. In all platinum-containing catalysts, the Pt content was 0.375 wt%. The Sn(Fe)/Pt of 5:1 atomic ratio for γ-Al_2O_3 supported Pt-Sn and Pt-Fe catalysts.

---

*The project supported by FORD and NSFC NO.09412302. We are very grateful to the Photon Factory in Tsukuba, Japan, for use of the BL-7c facilities. We thank Dr. T. Tanaka Tanaka for experimental assistance.

3. RESULTS AND DISCUSSION

Fig.1 showed the Isobutane dehydrogenation of γ-Al_2O_3 supported Pt, Pt-Fe and Pt-Sn catalysts. The results indicated that the conversion of Isobutane and selectivity of Isobutene (by-products are methane, ethane, ethylene, propane and propylene) for γ-Al_2O_3 supported Pt-Fe and Pt-Sn dual-component catalysts are higher than mono Pt catalyst during the reaction time.

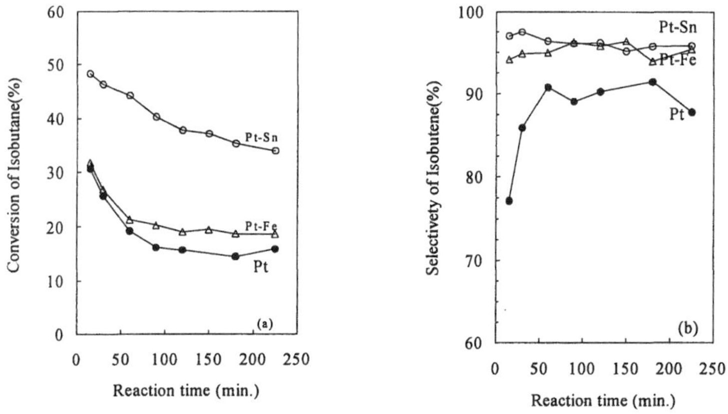

Figure 1. The Isobutane dehydrogenation on γ-Al_2O_3 supported Pt, Pt-Fe and Pt-Sn catalysts (a) conversion of Isobutane (b) selectivity of Isobutene

Fig. 2 summarized the Pt L_{III}-edge XANES spectra of Pt and PtO_2, as well as Pt/γ-Al_2O_3, Pt-Sn/γ-Al_2O_3 and Pt-Fe/γ-Al_2O_3 catalysts before reduction. It can be seen from Fig.2 that there are no difference in energies of ca. 1.5 eV for the main peak of the samples. There is mainly one peak for the PtO_2 sample, but a main peak with two satellite peaks(the satellite peaks are in higher energies) for the Pt sample. The XANES spectra of the Pt/γ-Al_2O_3, Pt-Sn/γ-Al_2O_3 and Pt-Fe/γ-Al_2O_3 catalysts before reduction are all similar to that of the PtO_2 sample either in energy position or in intensity. This implied that before reduction the Pt species on these catalysts were all in the oxidized states.

Fig. 3 summarized the Pt L_{III}-edge XANES spectra of Pt and PtO_2, as well as Pt/γ-Al_2O_3, Pt-Sn/γ-Al_2O_3 and Pt-Fe/γ-Al_2O_3 catalysts after reduction. It is clear from Fig.3 that the XANES spectra of the Pt/γ-Al_2O_3, Pt-Sn/γ-Al_2O_3 and Pt-Fe/γ-Al_2O_3 catalysts after reduction are all similar to that of the Pt foil. However, their satellite peaks are much smaller than that of Pt. These results indicate that the Pt species in the Pt-containing catalysts after reduction all existed in the metallic states, and no more in the oxidized states. J.H.Sinfelt[11] had reported similar results and concluded that the metal dispersion could affect the shapes of the XANES spectra. However, his comparison of the Pt/γ-Al_2O_3 catalyst with large Pt crystallites was near the limits of uncertainty in the determination. The much smaller satellite peaks of our γ-Al_2O_3

supported Pt-containing catalysts implied that the dispersion of the Pt species on our catalysts was higher than that investigated by J.H.Sinfelt.

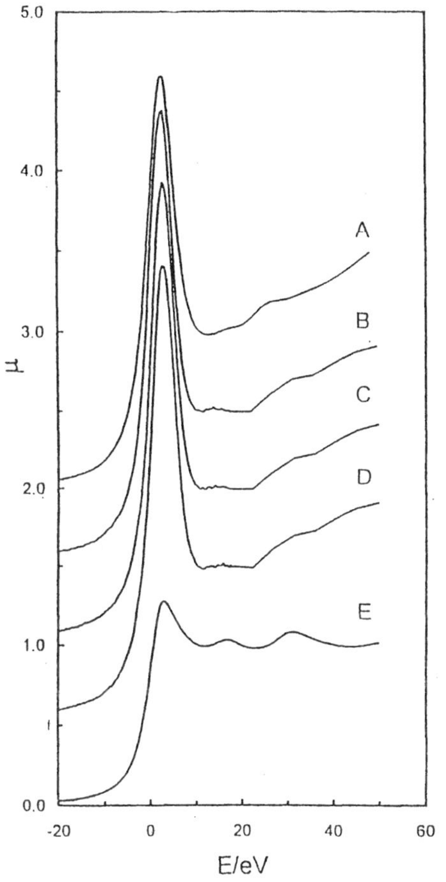

Figure 2. Pt L_{III}-edge XANES spectra of the catalysts before reduction compared with those of Pt and PtO_2 (A) PtO_2 (B) Pt/γ-Al_2O_3 (C) Pt-Sn/γ-Al_2O_3 (D) Pt-Fe/γ-Al_2O_3 (E) Pt

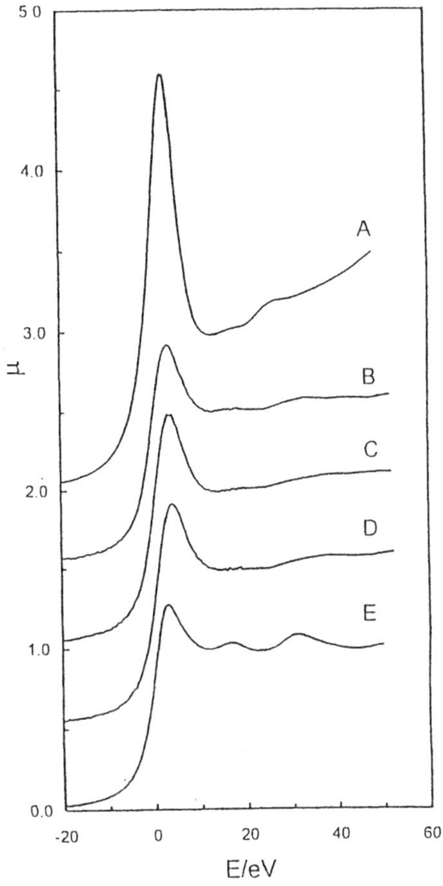

Figure 3. Pt L_{III}-edge XANES spectra of the catalysts after reduction compared with those of Pt and PtO_2 (A) PtO_2 (B) Pt/γ-Al_2O_3 (C) Pt-Sn/γ-Al_2O_3 (D) Pt-Fe/γ-Al_2O_3 (E) Pt

The Fourier transforms of the Pt L_{III}-edge EXAFS of Pt-Sn/γ-Al_2O_3 catalyst before and after reduction, Pt-Fe/γ-Al_2O_3 catalyst before and after reduction, Pt/γ-Al_2O_3 catalyst before and after reduction, and their comparison with PtO_2 and Pt foil, are shown in Fig. 4. Peaks of highest intensity centered around 1.85? (not corrected for phase shift) were contributed by the nearest oxygen neighbors, while the remaining ranged from 2 to 3? were mainly caused by the nearest platinum neighbors of the Pt-Sn/γ-Al_2O_3, Pt-Fe/γ-Al_2O_3, Pt-Fe/γ-Al_2O_3 catalyst before and after reduction and PtO_2. In the case of the Pt foil, the most intense peak centered at 2.38? was given rise by the nearest platinum neighbors. The intensity of the peak for the nearest neighbor oxygen in the Pt/γ-Al_2O_3, Pt-Fe/γ-$Al_2O_3$and Pt-Sn/γ-Al_2O_3 catalysts after reduction were lower than the catalysts before reduction and PtO_2 and this was due to the fact that the Pt species had lost more oxygen neighbors and converted to the metallic state after reduction. Consequently, the intensity of nearest neighbor Platinum of the above catalysts after reduction were lower than that of the Pt foil. It is in good agreement with Pt L_{III}-edge XANES results in this aspect.

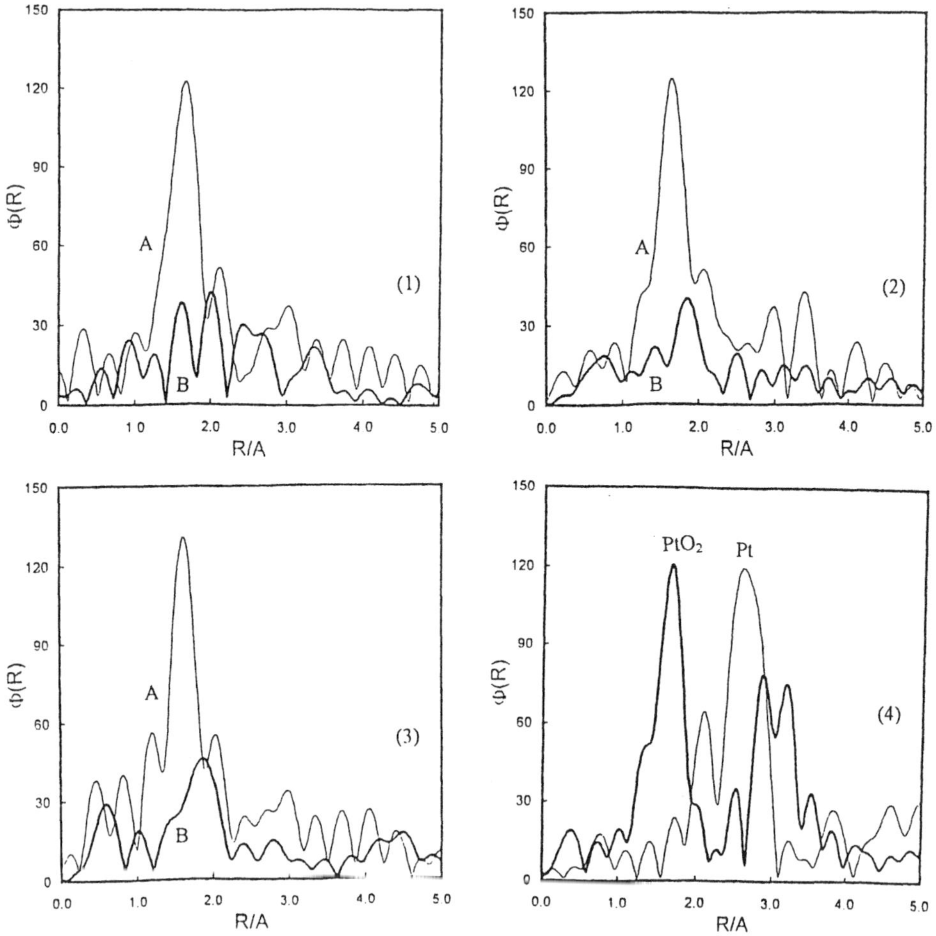

Figure 4. Fourier transforms of Pt L_{III}-edge EXAFS of the γ-Al_2O_3 supported Pt-containing catalysts before and after reduction compared with PtO_2 and Pt Foil. (1)Pt/γ-Al_2O_3, (2) Pt-Sn/γ-Al_2O_3, (3) Pt-Fe/γ-Al_2O_3, (4) Pt and PtO_2, (A) before reduction, (B) after reduction.

The best fits to the Fourier-filtered Pt L_{III}-edge EXAFS results of Pt/γ-Al_2O_3, Pt-Sn/γ-Al_2O_3, and Pt-Fe/γ-Al_2O_3 catalysts after reduction compared with Pt foil are summarized in Fig. 5 and Table 1. The best fits indicated that the Pt species are highly dispersed on the surface of γ-Al_2O_3 support for the catalysts after reduction. No Pt-Fe or Pt-Sn alloys are formed for Pt-Fe/γ-Al_2O_3, and Pt-Sn/γ-Al_2O_3 catalysts after reduction. The results also near the EXAFS results of Pt-Sn/γ-Al_2O_3 catalyst observed by J.H Sinfelt et al.[12] and the dispersion of Pt species on our catalysts are higher than that investigated by J.H.Sinfelt et al..

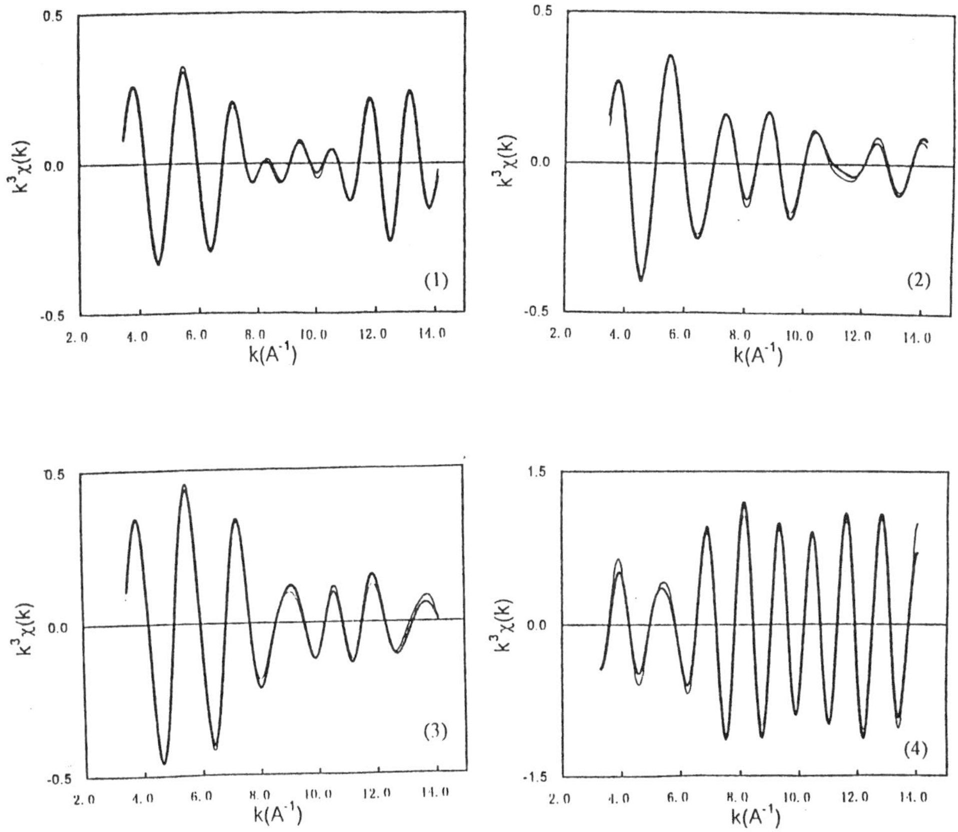

Figure 5. Best fits to Fourier-filtered Pt L_{III}-edge EXAFS of the γ-Al_2O_3 supported Pt-containing catalysts after reduction compared with Pt Foil. (1) Pt/γ-Al_2O_3, (2) Pt-Sn/γ-Al_2O_3, (3) Pt-Fe/γ-Al_2O_3, (4) Pt foil (in the range of 1.00-3.07 Å) with the weight of κ^3.

The EXAFS results also implied that the Pt species was in an amorphous state after reduction for the catalysts. In this regard, it is similar to the Pt-Re/γ-Al_2O_3 catalyst observed by B.C. Gates et. al.[13]. Combined with our formal work, we can infer that Fe and Sn species are in oxidized state[14], Pt is in an amorphous metallic state, and no alloy of Pt-Fe or Pt-Sn was formed after reduction on Pt-Fe/γ-Al_2O_3 and Pt-Sn/γ-Al_2O_3 catalysts. It has been reported that there are interactions between Pt, FeO(SnO) and γ-Al_2O_3 Support, which resulted in an enhanced dehydrogenation performance for the Pt-Fe/γ-Al_2O_3 and Pt-Sn/γ-Al_2O_3 catalysts, which is better than that of the Pt/γ-Al_2O_3 catalysts[8]. So we conclude this interaction is that Pt anchored Support γ-Al_2O_3 via FeO(SnO)-promoter. It is called "sandwich" geometric structure by Prof. Liwu Lin[2]. That is the promoter elements such as Sn and Fe site on the surface of alumina support as an oxidation state (SnO and FeO). And the active platinum element highly disperses on the promoter layer (SnO or FeO) to form new active centers Pt-SnO(FeO) which may enhance the activity and selectivity of paraffin dehydrogenation. These new active centers may also reduce the adsorption ability of unsaturation hydrocarbons which can cause the carbon deposition on the catalyst surface.

Table 1
Pt L_{III}-edge EXAFS Derived Coordination Number (CN), Shell Radius (R), and Debey-Waller Factor (DW) for Pt/γ-Al_2O_3, Pt-Fe/γ-Al_2O_3, and Pt-Sn/γ-Al_2O_3 catalysts after reduction compared with Pt foil

catalyst	shell	CN	R(?)	DW(?$^{-1}$)	R factor
Pt/γ-Al_2O_3	Pt-O	0.8	1.97	0.0000	0.062
	Pt-O	1.0	2.22	0.0000	
	Pt-Pt	0.6	2.70	0.0010	
Pt-Sn/	Pt-O	0.6	1.81	0.0031	0.100
γ-Al_2O_3	Pt-O	2.0	2.15	0.0033	
	Pt-Pt	0.9	2.54	0.0062	
Pt-Fe/	Pt-O	0.8	1.97	0.0000	0.089
γ-Al_2O_3	Pt-O	1.8	2.17	0.0000	
	Pt-Pt	0.6	2.75	0.0044	
Pt Foil	Pt-Pt	12.2	2.76	0.047	0.133

REFERENCES

1. J.H. Sinfelt, Bimetallic Catalysts: *Discoveries, Concepts, and Applications*; Wiley: New York, (1983) 130.
2. L. Lin, HUAXUTONGBAO (in Chinese), 9(1994) 14.
3. W. Yang, L. Lin, Y. Fan and J. Zhang, Calal. lett., 12(1992) 267.
4. O. A. Bariås, A. Holmen, and E. A. Blekkan, J.Catal., 158(1996) 1.
5. S.R. Adkins, and B.H. Davis, J.Catal., 89(1984) 371.
6. F. J. Berry, C. Xu, S. Jobson, and R. J.Strange, Chem. Soc. Faraday Trans., 85(1989) 3891.
7. B. J. Tan, K. J. Klabunde, and P. M. Sherwood, A. J. Am. Chem. Soc., 113(1991) 855.
8. J. Jia, Z. Xu, T. Zhang, and L. Lin, Chinese J. Catal., 18(1997), 97.
9. Y. Kou, H. Wang, J. Niu, and W. Ji, J.Phys. Chem., 100(1996) 2330
10. J. Jia, J. Y. Kou, L. Lin, Z. Xu, and T. Zhang, React. Kenet Catal. Lett., 1998, in press.
11. J.H. Sinfelt, and G.D. Meitzner, Acc. Chem. Res., Vol.26, No. 1 (1993), 1.
12. G. Meitzer, G. H. Via, F. W. Lytle, J. H. Sinfelt, J. Chem. Phys., 87(1987) 1.
13. S. Rezgui, R. Jentoft, and B. C. Gates, J. Catal., 163(1996) 496.
14. J. Jia, Y. kou, Z. Xu, T. Zhang, B. Zhang, and L. Lin, Chinese J. Catal., 18(1997) 263.

NATURAL GAS CONVERSION V
Studies in Surface Science and Catalysis, Vol. 119
A. Parmaliana et al. (Editors)

Oxidative dehydrogenation of ethane over Na_2WO_4-Mn/SiO_2 catalyst using oxygen and carbon dioxide as oxidants

Yu Liu Jinzhen Xue Xuxia Liu Ruiling Hou and Shuben Li*

State Key Laboratory for Oxo Synthesis and Selective Oxidation,
Lanzhou Institute of Chemical Physics, Chinese Academy of Sciences,
Lanzhou 730000,P.R.China

The performance of Na_2WO_4-Wn/SiO_2 catalyst for oxidative dehydrogenation of ethane was studied by using oxygen and carbon dioxide as oxidants. The >70% ethane conversion and selectivity to C_2H_4 could be achieved at 700 ℃, with the space velocity higher than 30000h^{-1}. The stability test of the catalyst for ODE showed that ethane conversion and selectivity remained constant during 100 h. CO_2 could be used as an oxidant instead of O_2 for oxidative dehydrogenation of ethane. 53.3% ethane conversion and 97% selectivity to ethylene could obtained at 800 ℃. The selective decreased when temperature was higher than 800 ℃. This is attributed to that the surface lattice oxygen is responsible for selective oxidative dehydrogenation of ethane, whereas the bulk lattice oxygen is responsible for deep oxidation of ethane.

1. INTRODUCTION

The oxidative dehydrogenation of ethane(ODE) to ethylene over Li promoted catalysts, such as Li^+/MgO, has been studied at low space velocities (200~2000h^{-1})[1~3]. Recently, Li et al [4,5] reported a good catalyst — Na_2WO_4-Mn/SiO_2 for oxidative coupling of methane, which had been repeated and studied by Lunsford et al[6]. From the results of Li [4;5], it can be concluded that this catalyst has the advantages of high space velocity, high selectivity to ethylene and high ratio of ethylene to ethane. The results implied that Na_2WO_4-Mn/SiO_2 might be an effective catalyst for the oxidative dehydrogenation of ethane, either. The present work was undertaken to explore its performance for the oxidative dehydrogenation of ethane at high space velocity. Furthermore, since Mn-containing catalysts were found to provide high conversions both of C_2H_6 and CO_2[7], in order to reduce the possible gas phase reaction of $C_2H_5\cdot$with O_2, dehydrogenation of C_2H_6 with CO_2 was also investigated over this catalyst.

* Corresponding author

2. EXPERIMENTAL

The preparation method of Na_2WO_4-Mn/SiO_2 catalyst was described elsewhere[4] and the reaction system was also presented in our previous work[7,8].

3. RESULTS AND DISCUSSION

3.1 Oxidative dehydrogenation of C_2H_6 with O_2

Considering the positive effect of temperature on the homogeneous reaction of ethane oxidation, prior to the performance of the catalytic reactions, the effect of temperature on the blank reaction of ethane oxidation was investigated. As may be seen in Table 1, the conversion of ethane could reach 17.5%.at 750 ℃. This result indicates that temperature favor the gas phase reaction of ethane oxidation. The effect of ratio of C_2H_6 to O_2 on the oxidative dehydrogenation of ethane is depicted in Table 2. The conversion increased whereas the selectivity to ethylene decreased with the decreasing ratio. The better conversion and selectivity to ethylene(65.3% and 75.4%) was obtained at the ratio of 2.1:1. It should be noticed that the conversion and selectivity almost remained constant when the space velocity was increased to 1.1×10^5 h^{-1}, which is similar to the results of oxidative

Table 1
The effect of temperature on the blank reaction of ethane oxidation at a total flow rate of 150ml/min ,C_2H_6:O_2=2.1:1

T(℃)	C_2H_6 Conversion(%)	Selectivity (%)			
		C_2H_4	CH_4	CO	CO_2
600	0.6	66.5	0.5	17.3	15.7
650	3.8	51.8	2.1	23.5	22.6
700	11.6	48.5	5.8	17.4	18.3
750	17.5	45.7	13.5	19.0	21.8

Table 2
Effect of the ratio of C_2H_6 to O_2 on ethane oxidative dehydrogenation over Na_2WO_4-Mn/SiO_2 catalyst at GHSV: 49920h^{-1}.

C_2H_6:O_2	T(℃)	C_2H_6 Con.(%)	Selectivity (%)			
			C_2H_4	CH_4	CO	CO_2
2.5:1	750	60.7	77.2	8.7	15.2	4.1
2.1:1	750	70.5	72.2	9.0	14.4	4.4
1.8:1	750	77.4	66.7	9.3	19.4	4.5
2.1:1	700	65.3	75.4	6.8	11.5	6.6
2.1:1[a]	700	62.1	77.3	7.1	10.0	5.6
2.1:1[b]	700	52.9	83.8	6.2	5.7	4.3

[a]GHSV=1.1×10^5 h^{-1}, [b]GHSV=1.8×10^5 h^{-1}

coupling of methane over this catalyst[4,5]. When the space velocity increased to 1.5×10^5 h^{-1}, the conversion decreased to 52.9% and the selectivity to ethylene increased to 83.8%. The reason why ethane conversion was almost constant at the space velocity range of 5.0×10^4 h^{-1} to 1.1×10^5 h^{-1} is probably attributed to the nature of the catalyst. From our previous results[4,5], we consider that the surface lattice oxygen is possible responsible for ethane oxidation, and the Na_2WO_4-Mn/SiO_2 catalyst has strongly exchanging ability between gaseous oxygen and surface lattice oxygen, which allowed the ethane conversion to 'remain constant at a wide range of high space velocity. Now, further experiments are being carried out to confirm the above suggestion. So this catalyst could be used for OCM and ODE at high space velocity, which was economical in industrial application, since high space velocity means high productivity. The >70% ethane conversion and selectivity to C_2H_4 could be achieved at 700 ℃, with the space velocity higher than 30000h^{-1}. As shown in Figure 1, the stability test of the catalyst for oxidative dehydrogenation of ethane showed that ethane conversion and selectivity remained constant during 100 h.

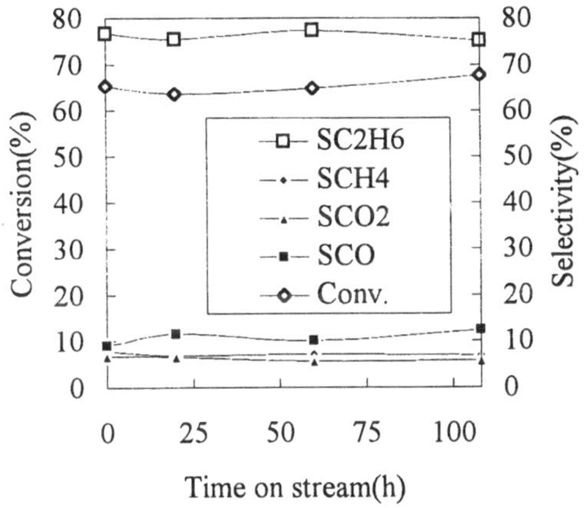

Figure 1 Ethane oxidative dehydrogenation over Na_2WO_4-Mn/SiO_2 catalyst as a function of time at C_2H_6:O_2=2.1:1, 700 ℃. GHSV: 55000h^{-1}.

3.2 Oxidative dehydrogenation of C_2H_6 with CO_2

Table 3 presents the results of conversion of C_2H_6 with CO_2 at different temperature. As may be seen, both C_2H_6 and CO_2 conversions increased with temperature. The conversion of C_2H_6 and selectivity to C_2H_4 could reach 53% and 95% respectively at 800 ℃. Moreover, it appeared that the selectivity to C_2H_4 decreased with increasing temperature, whereas the selectivity to CO and CH_4 increased. The increase in selectivity to CH_4 is attributed to the crack reaction of C_2H_6 at high temperature. The results further demonstrated that high temperature favored the activation of CO_2 on the surface of the catalyst. In addition, it is noticeable that the selectivity to CO increased from 0.5% at 800 ℃ to 3.4% at 850 ℃, whereas the selectivity to C_2H_4 decreased from 97.0% at 800 ℃ to

90.5% at 850 ℃. From our previous results[4,5,8], the surface lattice oxygen (which desorbed at 800 ℃ according to O_2-TPD profile) is the oxygen species for the production of C_2H_4, whereas the bulk lattice oxygen(which desorbed at 850 ℃) is the oxygen species to produce CO. Thus, the increase in selectivity to CO is possibly due to that the bulk lattice oxygen partially anticipated the oxidation of C_2H_6. This conclusion could be confirmed by ethane pulse reaction at 800 ℃ and 850 ℃ respectively. The results of C_2H_6 pulse reaction showed that only C_2H_4 and CH_4 were observed at 800, whereas, besides C_2H_4 and CH_4, the deep oxidation product-CO were also observed at 850 ℃ Based on the above results, we can concluded that the best selectivity to C_2H_4 in the reaction of C_2H_6 with CO_2 could be achieved below 800 ℃. In addition, the possible reaction mechanism for ethane oxidation to ethylene using CO_2 as an oxidant over Na_2WO_4-Mn/SiO_2 catalyst may speculated as follows: CO_2 may first be adsorbed on the catalyst. An oxygen atom in CO_2 dissociated to formed and active oxygen species, and CO is released. The active oxygen species(surface lattice oxygen) oxidize ethane to form ethylene. It should be note that a little amount of H_2 was also observed. However, the blank reaction of ethane with CO_2 at 850°C indicated that <20% ethane conversion was also observed and almost no CO was observed. This indicates that there is no contribution of stream reforming reactions. So the formation of hydrogen is attributed to the thermal dehydrogenation of ethane. Some ethylene is produced by the thermal dehydrogenation of ethane.

Table 3
Conversion of C_2H_6 with CO_2 over Na_2WO_4-Mn/SiO_2 catalyst at GHSV 3600h^{-1}, C_2H_6:CO_2=1:1.

T(℃)	Conversion(%)		Selectivity(%)		
	C_2H_6	CO_2	C_2H_4	CO	CH_4
700	18.6	9.2	99.6	-	0.4
750	25.8	16.3	97.8	0.5	1.7
800	53.3	43.8	97.0	0.5	2.5
850	69.5	49.2	90.5	3.4	6.1

4. CONCLUSIONS

Na_2WO_4-Wn/SiO_2 is also a better catalyst for oxidative dehydrogenation of ethane. The >70% ethane conversion and selectivity to C_2H_4 could be achieved at 700 ℃, with the space velocity higher than 30000h^{-1}. The stability test of the catalyst for ODE showed that ethane conversion and selectivity remained constant during 100 h.

In addition, CO_2 could be used as an oxidant instead of O_2 for oxidative dehydrogenation of ethane. 53.3% ethane conversion and 97% selectivity to ethylene could obtained at 800 ℃. The selective decreased when temperature was higher than 800 ℃. This is due to that the surface lattice oxygen is responsible for selective oxidative dehydrogenation of ethane, whereas the bulk lattice oxygen is responsible for deep oxidation of ethane.

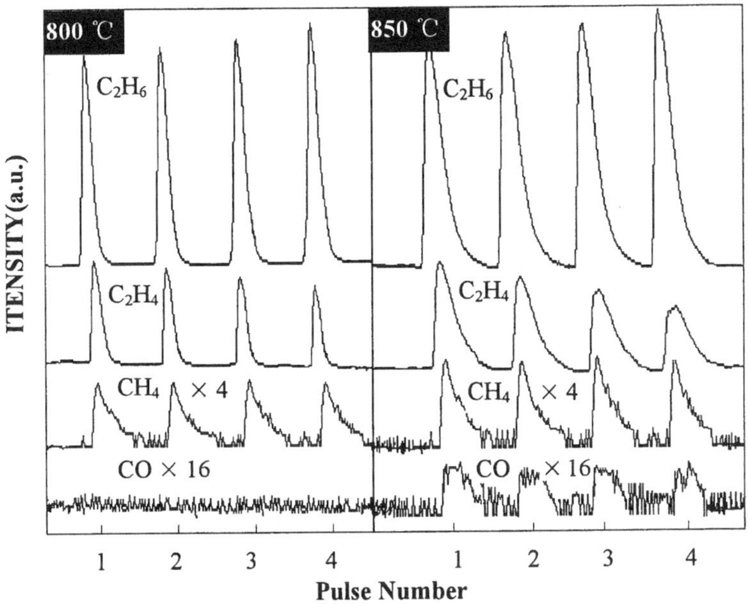

Figure 2 C_2H_6 pulse over Na_2WO_4–Mn/SiO_2 catalyst at 800 ℃ and 850 ℃.

REFERENCES

1. E.Morales, J.H.Lunsford, J.Catal.,118(1989)255.
2. S.J.Conway, J.H.Lunsfiord, J.Catal.,131(1991)513.
3. D.J.Wang, M.P.Rosynek, J.H.Lunsford, J.Catal.,151(1995)155.
4. Fang, S. Li, J. Lin, J.Molec. Catal.(China).,6(1992)255.
5. Fang, S. Li, J. Lin, J.Molec. Catal.(China).,6(1992)427.
6. D.J.Wang, M.P.Rosynek, J.H.Lunsford, J.Catal.,155(1995)523.
7. Y. Liu, S. Shen, Appl.Catal.,121(1995)57.
8. Y. Liu, X.-X Liu, J-Z Xue, S.S.Li, Proceedings of the 5th Natural Gas Conversion Symposium, Sicily, Italy, September 20-25, 1998, edited by A.Parmaliana et al.

NATURAL GAS CONVERSION V
Studies in Surface Science and Catalysis, Vol. 119
A. Parmaliana et al. (Editors)

Cofeeding of methane and ethane over Na_2WO_4-Mn/SiO_2 catalyst to produce ethylene

Y.-D. Zhang[a], S.-B. Li[a*], Y. Liu[a], J.-Z. Lin[a] G.-G. Lu[b], X.-Z. Yang[b], J. Zhang[b]

[a]State Key Laboratory for Oxo Synthesis and Selective Oxidation, Lanzhou Institute of Chemical Physics, The Chinese Academy of Sciences, Lanzhou, 730000, China

[b]Jinzhou Petrochemical Company, Jinzhou, 121001, China

Methane/ethane cofeeding experiments are conducted at 1093 K and atmospheric pressure over Na_2WO_4-Mn/SiO_2 catalyst. The addition of ethane to OCM feed at a fixed CH_4/O_2 ratio results in a decrease in methane conversion. But this decrease can be remedied by increasing oxygen flow rate. Ethane can be co-converted with methane. Methane/ethane cofeeding reaction can be simply regarded as two parallel reactions of oxidative coupling of the fed methane and oxidative dehydrogenation of the fed ethane. The selectivity of OCM reaction is *only* dependent on methane conversion and is not affected by the addition of ethane and the change of initial CH_4/O_2 ratio. The OXD reaction of ethane in the presence of methane can achieve very high conversion with high selectivity.

1. INTRODUCTION

The direct conversion of methane to higher hydrocarbons through oxidative coupling of methane (OCM) has attracted worldwide attention since the publication of Keller and Bhasin's work in 1982 [1]. Hundreds of catalysts have been found to be more or less active and selective for the coupling reaction. But even for the best OCM catalysts known at this time the sum of selectivity and conversion is about 100%, well below the criteria of 122% for OCM process to be economically attractive [2]. About one third of the total investment is in the cryo-separation section. The separation of ethylene from OCM product stream by ethylene reaction [3] or absorption [4], with ethane and methane being recycled to OCM reactor, can avoid cryo-separation and thus may improve the economics of OCM process.

Na_2WO_4-Mn/SiO_2 catalyst [5] is one of the best catalysts for OCM [6]. The catalyst is also very effective for the oxidative dehydrogenation (OXD) of ethane [7]. Methane and ethane can be expected to be co-converted over this catalyst with satisfactory results. $^{13}C_2H_6/CH_4$ or $^{13}C_2H_4/CH_4$ cofeeding experiments have been performed to determine the origin of CO_x [8-10]. And the effects of cofeeding ethane on OCM reaction have also been investigated without the use of isotope techniques [11-12]. But the reaction of the cofed ethane has not gained adequate attention. The present work reports methane/ethane cofeeding results over Na_2WO_4-Mn/SiO_2 catalyst in terms of conversion and selectivity for OCM reaction of methane as well as for OXD reaction of ethane.

* To whom correspondence should be addressed.

2. EXPERIMENTAL

5% Na_2WO_4-2%Mn/SiO_2 catalyst sample was prepared by incipient wetness impregnation, at 353 K, of silica gel (initial BET surface area of 350 $m^2.g^{-1}$ and particle size of 20-40 mesh) with aqueous solutions of $Mn(NO_3)_2$ and Na_2WO_4. The sample was then dried for 4h at 393 K and calcined for 2 h at 1093 K.

Experiments were conducted at 1093K in a quartz micro-reactor over 0.1g catalyst under atmospheric pressure. The total flow rate was maintained at 90ml.min^{-1}(STP). Reaction product analysis method was given in detail elsewhere [5]. GC analysis of the product gas composition was carried out on-line after reaction on the fresh catalyst had proceeded for 1 h. Results were reported on carbon atom basis except otherwise indicated. Cofeeding results at an exact methane conversion were obtained by interpolating the experimental results of two runs with methane conversion near the specified value (one being higher and the other lower, achieved by adjusting oxygen flow rate).

3. RESULTS AND DISCUSSION

The addition of ethane to OCM feed makes the analysis of the reaction results difficult because it is impossible to differentiate the origin of the products without the application of isotope technique. Methane conversion is the only value that can be calculated directly from the reaction results. The cracking of the fed ethane to methane, under methane/ethane cofeeding conditions, may be minimized by the presence of ample methane. Ekstrom [9] failed to observe the presence of $^{13}CH_4$ when $^{13}C_2H_6$ was added to OCM feed over Sm_2O_3 catalyst. Methane conversion *only* concerns the oxidative coupling reaction of the fed methane. The selectivity of OCM reaction of the fed methane and the conversion and selectivity of OXD reaction of the fed ethane can be derived indirectly from the mathematical treatment of the cofeeding reaction results.

3.1. OCM reaction in cofeeding mode

The effects of co-feeding ethane on methane conversion at given CH_4/O_2 ratios are shown in Figure 1. It can be seen that methane conversion decreases with the addition of ethane at a constant oxygen feed-rate and increases with the increase in oxygen feed-rate at a given

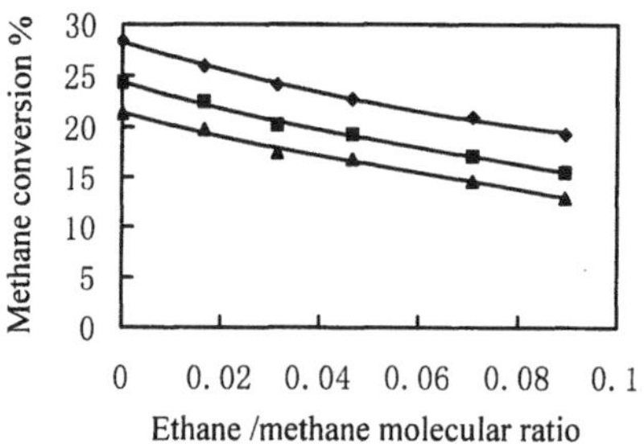

Figure 1 The effects of C_2H_6 addition on CH_4 conversion at different O_2 feed rates
◆ 5.58 ,■ 4.46 ,▲ 3.72 mmoles.min.$^{-1}$.g^{-1}
1093 K , CH_4 at 22.32 mmoles.min.$^{-1}$.g^{-1}

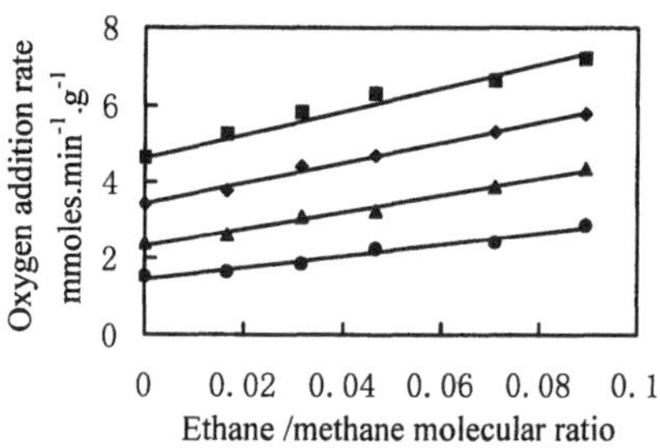

Figure 2 The relations between O_2 feed rates and C_2H_6/CH_4 to maintain CH_4 conversion at
■ 25% ,◆ 20% ,▲ 15% ,● 10%
1093 K , CH_4 at 22.32 mmoles.min.$^{-1}$.g^{-1}

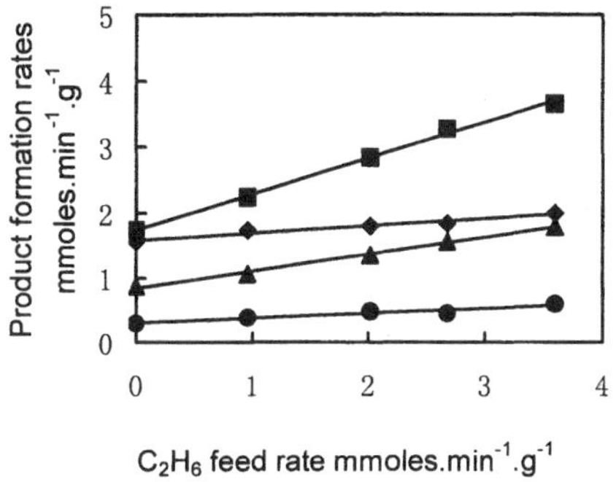

Figure 3. The dependence of product formation rates on C_2H_6 feed rates at 20%CH_4 conversion.
■ C_2H_4 ◆ C_2H_6 ▲ CO_x ● C_3^+
1097K, CH_4 at 22.32 mmole.min^{-1}.g^{-1}

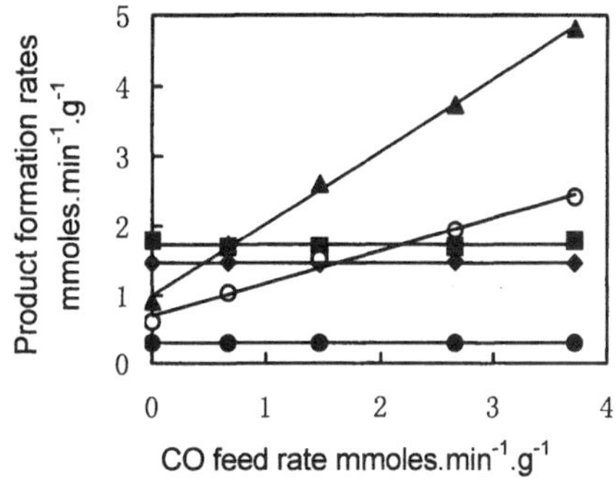

Figure 4. The dependence of product formation rates on CO feed rates at 20% CH_4 conversion.
▲ CO_x ○ CO_2 ■ C_2H_4 ◆ C_2H_6 ● C_3^+
1097K, CH_4 at 22.32 mmole.min^{-1}.g^{-1}

ethane feed rate. The decrease in methane conversion upon the addition of ethane was also observed by other authors [9-12]. But methane conversion can be maintained at a constant value after the addition of ethane by adjusting the flow-rate of oxygen as shown in Figure 2. The intercepts in the figure are the oxygen amounts needed for OCM reaction of pure methane to proceed to the stated conversions. It is well known that ethane is more active than methane and oxygen is the limiting reactant in OCM reaction (oxygen conversion is better than 99% in all the experiments). An extra amount of oxygen must be provided for the OXD reaction of the cofed ethane if methane conversion of OCM reaction is to be kept constant.

Methane/ethane cofeeding results at 20% methane conversion are shown in Figure 3. Product formation rates are plotted against ethane feed rate. The intercepts in Figure 3 are the product formation rates of OCM reaction of pure methane. Ethane in the cofeeding product stream is wholly treated as a product for comparison with OCM results (the intercepts) though unconverted feed ethane is included. Discussions on Figure 3 are difficult, since both methane and ethane contribute to the products. Carbon monoxide is a product of OCM reaction, no hydrocarbons would derive from carbon monoxide under OCM reaction conditions. Methane/carbon monoxide cofeeding results may be much easier to discuss. Ethylene can not be converted to ethane under OCM reaction conditions [10]. Methane/ethylene cofeeding results may also be easier to discuss. Discussions on the foregoing two reactions may be helpful to the understanding of methane/ethane reaction.

The addition of carbon monoxide or ethylene to OCM feed at constant oxygen flow-rate also causes a decrease in methane conversion. Oxygen flow-rate has to be adjusted to maintain methane conversion at a fixed level. Methane/carbon monoxide cofeeding results at 20% methane conversion are plotted in Figure 4 and methane/ethylene cofeeding results at 20% methane conversion are shown in Figure 5. Figure 4 shows that hydrocarbon (from methane only) formation rates do not vary with the addition of carbon monoxide and that more carbon dioxide is formed upon the addition of carbon monoxide. This increase in carbon dioxide formation rate accounts for the increase in oxygen flow-rate to maintain methane conversion. Since methane feed rate and conversion are kept unchanged, CO_x formation rate

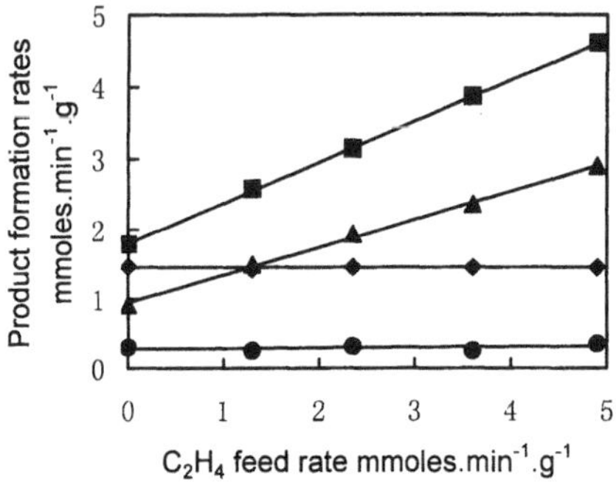

Figure 5. The dependence of product formation rates on C_2H_4 feed rates at 20%CH4 conversion.
■ C_2H_4 ◆ C_2H_6 ▲ CO_x ● C_3^+
1097K , CH_4 at 22.32 mmole.min^{-1}.g^{-1}

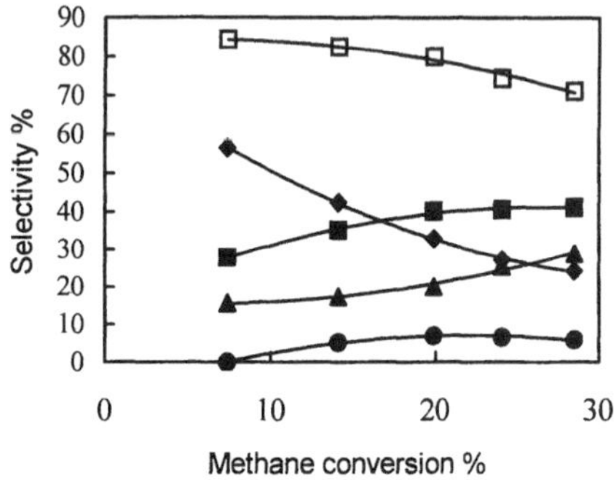

Figure 6 The relationship between selectivity and conversion of OCM reaction
□ C_2^+ ◆ C_2H_6 ■ C_2H_4▲ CO_x ● C_3^+
1097K , CH_4 at 22.32 mmole.min^{-1}.g^{-1}

from methane must remain unchanged. OCM product formation rates do not change at constant methane feed rate and conversion, though initial methane/oxygen ratios are adjusted and carbon monoxide is added. That is to say, the addition of carbon monoxide and the change of oxygen concentration do not affect the relationship between selectivity and conversion of OCM reaction.

Ethylene in methane/ethylene cofeeding product stream is also treated as a product. Figure 5 shows that ethane and C_3^+ formation rates do not change after the addition ethylene to OCM feed. Ethane and C_3^+ only come from the reaction of methane. Reaction of the cofed ethylene only produces CO_x. Since methane feed rate and conversion are kept constant, the sum of CO_x formation rate and ethylene formation rate from methane reaction must remain unchanged. This means that either CO_x formation rate and ethylene formation rate both remain unchanged or both change in the opposite directions and in equal magnitude. The latter case seldom occurs. Based on the above dicussions, it can be concluded that the product formation rates of OCM in Figure 5 do not change. CO_x formation rate can be divided into two parts, one part below the intercept coming from OCM reaction, and the other above the intercept coming from the oxidizing reaction of the cofed ethylene. Ethylene formation rate can also be divided into two parts, one part below the intercept coming from OCM reaction, and the other being the unconverted portion of the cofed ethylene. The slope of CO_x line in Figure 5 is the fraction of the fed ethylene converted to CO_x, and the slope of ethylene line is the fraction of the fed ethylene unconverted. The addition of ethylene and the change of oxygen concentration do not affect the relationship between selectivity and conversion of OCM reaction.

By analogy, it is easy to understand the methane/ethane cofeeding results in Figure 3. Each product formation rate can be divided into two parts as in the case of methane/ethylene cofeeding reaction. The product formation rates of OCM remain at the values of the intercepts. The relationship between selectivity and conversion of OCM reaction is not affected by the addition of ethane and the change of oxygen concentration. The relationship between selectivity and conversion of OCM reaction at different methane conversion is shown in Figure 6. It can be understood that the above relationship is not affected by the addition of an additive (ethane, ethylene or carbon monoxide) or by a change in oxygen concentration.

3.2. OXD reaction of the cofed ethane

Since both methane and ethane contribute to the cofeeding products, conversion and selectivity of the OXD reaction of the cofed ethane can not be calculated directly from analytical results of the reaction products. However, by arranging cofeeding results in the form of Figure 3, conversion and selectivity of OXD reaction of the cofed ethane can be easily calculated. The product formation rates from methane don not vary. The differences between cofeeding product formation rates and OCM product formation rates (the values of the intercepts) are the OXD product formation rates. The slope of C_2H_6 line (0.122) in Figure 3 is the fraction of the unconverted cofed ethane, and the slopes of CO_x line (0.256), C_2H_4 line (0.547) and C_3^+ line (0.07) are the fractions of the cofed ethane converted to the corresponding products. It can be calculated from the slopes that the conversion of OXD reaction is about 87% with a selectivity (C_2H_4 + C_3^+) of about 70%. The conversions and selectivities for OXD reaction of the cofed ethane at other methane conversions are shown in Figure 7.

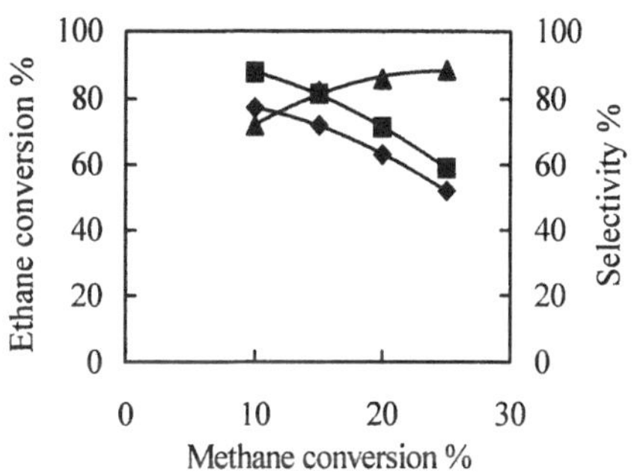

Figure 7 The conversion and selectivity of OXD reaction of the cofed ethane at different CH_4 conversion
▲ conversion, ■ selectivity to C_2H_4 + C_3^+
◆ selectivity to C_2H_4

It is well known that the activation of methane and ethane, under OCM conditions, involves the same kind of active sites. OCM and OXD reactions of the fed methane and ethane must proceed simultaneously. Methane conversion, which can be measured directly, may be used as an indication of the depth of the cofeeding reaction. It can be understood from Figure 6 and Figure 7 that once methane conversion in the cofeeding mode is controlled at a certain level by adjusting oxygen feed rate, the selectivity of OCM reaction and the conversion & selectivity of OXD reaction of the fed ethane are determined.

The reaction results of the cofed ethane in the cofeeding mode are better than those obtained on the same catalyst with pure ethane [7]. When methane conversion is kept below 20%, simultaneous high ethane conversion and high C_2H_4+C_3^+ selectivity can be obtained. Such results have not been achieved over a chloride-free catalyst ever reported for OXD of ethane [13]. It seems that the presence of ample methane has diluting effects on reactions of ethane, minimizing the unselective reactions. The presence of methane may also prevent the cracking of ethane to methane. As it is known, the oxidative dehydrogenation reactions are generally carried out under oxygen deficient conditions to maintain the catalyst surface in its selective form, on one hand, and to avoid the explosive limit, on the other. But in the cofeeding mode, the excessive methane (CH_4/O_2>3) keeps the catalyst surface in its selective form, at least, for the OCM reaction (comprising OXD of the formed ethane) and keeps the reaction mixture from the explosive limit. The OXD reaction of the cofed ethane can proceed selectively under oxygen sufficient conditions (C_2H_6/O_2 <1) at the onset of the reaction. It can be seen that the presence of methane favors the performance of the catalyst for OXD reaction. Cofeeding ethane to OCM feed is preferable to the separate feeding of methane to OCM and ethane to OXD.

4. CONCLUSION

Ethane can be co-activated with methane over $NaWO_4$-Mn/SiO_2 catalyst to produce ethylene. Methane/ethane cofeeding reaction can be regarded as two parallel reactions (OCM of methane and OXD of the cofed ethane). Both reactions consume oxygen.

The addition of ethane to OCM feed at a fixed CH_4/O_2 ratio causes a decrease in methane conversion. But this decrease can be remedied by increasing oxygen flow rate. This extra oxygen is needed by the OXD reaction of the cofed ethane.

The addition of ethane to OCM feed and the change in oxygen concentration does not affect the dependence of the selectivity of OCM reaction on methane conversion. The presence of methane can offer a favorable operational environment for OXD of the cofed ethane which otherwise could not be realized.

Cofeeding ethane in natural gas with methane or recycling ethane formed by OCM reaction back to the feed is preferable to the separate feeding of methane to OCM and ethane to OXD.

5. ACKNOWLEDGEMENTS

The authors are grateful to Jinzhou Petrochemical Company for financial support

REFERENCES

1. G. E. Keller and M. M. Bhasin, J. Catal., 73 (1982) 9
2. E.,E., Wolf (eds.), Methane Conversion by Oxidative Process: Fundamental and Engineering Aspects, Van Nostrand Rheinhold ,New York,1992, pp463-482
3. Y. Liu and S.-K Shen, Appl. Catal., 121 (1995) 57
4. Oil & Gas J., 76(52) (1978) 199
5. X.-P. Fang, S.-B. Li , J.-Z. Lin et al, J. Molecular Catalysis (China) , 6(4) (1992) 427
6. H. J. Lunsford , Angew. Chem. Int. Ed. Engl., 34 (1995) 970
7. Y. Liu, X.-X. Liu and S.-B. Li, J. Molecular Catalysis (China) , 10(6) (1996) 401
8. C. Shi, M. P. Rosynek and J. H. Lunsford, J. Phys. Chem., 98 (1994) 8371
9. E.,E., Wolf (eds.), Methane Conversion by Oxidative Process: Fundamental and Engineering Aspects, Van Nostrand Rheinhold ,New York,1992, pp99-137
10. P. F. Nelson and N. W. Cant, J. Phys. Chem., 94 (1990) 3756
11. Q. Chen, P. M. Couwenberg and G. B. Martin. , Catalysis Today, 21 (1994) 309
12. E.,E., Wolf (eds.), Methane Conversion by Oxidative Process: Fundamental and Engineering Aspects, Van Nostrand Rheinhold ,New York,1992, pp429-462
13. F. Cavani and F. Trifiro , Catalysis Today , 24 (1995) 307

NATURAL GAS CONVERSION V
Studies in Surface Science and Catalysis, Vol. 119
A. Parmaliana et al. (Editors)

A New Route for C_2H_4 Production by Reacting C_2H_6 with CO_2 over a Catalyst of Chromium Oxide Supported on Silicalite-2 Type Zeolite

Xu Longya*, Lin Liwu, Wang Qingxia, Yan Li, Wang Debao and Liu Weichen

State Key Laboratory of Catalysis,
Dalian Institute of Chemical Physics, Chinese Academy of Sciences,
Dalian P. O. Box 110, Dalian 116023, P. R. CHINA

A Cr/Si-2 catalyst was developed to possess good reactivity for the formation of ethylene from ethane oxidative dehydrogenation with CO_2. The addition of Mn and La oxide could improve the reactivity of the catalyst. It was also found that increasing the CO_2/C_2H_6 ratio of the feed gas could increase the ethane conversion and ethylene selectivity of the Cr/Si-2 catalyst. Besides, addition of steam to the feed gas could diminish coke formation during the reaction process. The main reactions for the dehydrogenation of ethane with CO_2 to ethylene have been deduced to consist of the cracking of ethane to ethylene and CO_2 hydrogenation to CO.

1. INTRODUCTION

Up to now, production of C_2H_4 from C_2H_6 is carried out by the conventional steam cracking process, in which a very high temperature (about 830℃) and large amount of steam are needed. In the present paper, a new process for the production of C_2H_4 from C_2H_6 is reported, which is realized by reacting C_2H_6 with CO_2. Some special features of this new process can be noted as following:

* CO_2 is acting as a mild oxidant for the oxidative dehydrogenation of C_2H_6 to yield C_2H_4.
* A significant lowering of the reaction temperature as comparing to the steam cracking process, resulting in lower coke formation on the catalyst.
* C_2H_6 conversion and C_2H_4 selectivity are higher in this process than in all other processes of C_2H_4 production from C_2H_6. Furthermore, no C_3^+ and C_2H_2 products are formed.

Thus, dehydrogenation of ethane by the unconventional oxidant CO_2 is an interesting process from both the scientific and practical points of view[1,2]

* Corresponding author.

Since the reaction of C_2H_6 with CO_2 to form ethylene has to be carried out at rather high temperatures(700-800℃), complicated reactions will unavoidably be occurred, during the dehydrogenation of ethane with CO_2 to ethylene, such as the cracking of ethane to ethylene, the reverse water-gas shift reaction, the hydrocracking reaction of ethane to methane and the reforming reaction of ethane with CO_2 to form syn-gas. After analyzing all these possibilities[3], an over-all reaction scheme for the reaction of ethane with CO_2 has been proposed as follows:

$$xC_2H_6 + y\,CO_2 \Leftrightarrow zC_2H_4 + wCO + pCH_4 + qH_2O + rH_2$$

It can be seen from this reaction scheme that in order to increase the selectivity to ethylene, the formation of methane must be suppressed and that the key problem is to develop a catalyst which can suppress the thermodynamically favorable side reactions.

2. EXPERIMENTAL

2.1. Preparation of the catalysts

Supported Cr/Si-2 catalysts were prepared by impregnating a solution with a given amount of $Cr(NO_3)_3$ and $KMnO_4$ or $Ni(NO_3)_3$ onto a Silicalite-2 zeolite under vacuum condition. After drying at 400 K for 6 h and calcined at 800 K for 8 h, the catalysts were cooled to room temperature, crushed and screened. Catalyst samples with a size of Φ1.2 mm x 2.5 mm were used for the C_2H_6 dehydrogenation with CO_2 as an oxidant.

2.2. C_2H_6 dehydrogenation with CO_2 as an oxidant

The catalysts were evaluated in a small fixed-bed flow reactor made of stainless steel. The catalyst charge was 6 ml. Reaction conditions were set as follows: reaction temperature 780-800℃, pressure 0.1 MPa, GHSV 1000 h^{-1}, and CO_2/C_2H_6=1/1(mole ratio). The products were analyzed on line by gas chromatography with a TCD detector.

3. RESULTS and DISCUSSION

3.1. C_2H_6 dehydrogenation with CO_2 as an oxidant over the Cr/Si-2 catalyst

Table 1
The performance of the Cr/Si-2 catalysts with different promoters for C_2H_4 production via the oxidative dehydrogenation of C_2H_6 with CO_2.

Catalyst	conversion (%)		Selectivity (wt%)		
	CO_2	C_2H_6	CH_4	C_2H_4	H_2/CO (mol. ratio)
Cr/Si-2	18.6	58.9	19.6	80.4	1.4
Cr-Mn/Si-2	22.2	62.4	18.4	81.6	1.4
Cr-Mn-Ni/Si-2	24.2	67.9	18.7	81.3	1.6
Cr-Mn-La/Si-2	20.5	64.2	13.8	86.2	1.4

*Reaction conditions: 1073K、0.1MPa、1000h^{-1}、CO_2/C_2H_6 (molar ratio) =1/1.

Table 1 lists the typical results of Cr/Si-2 catalysts with different promoters for C_2H_4 production from the oxidative dehydrogenation of C_2H_6 with CO_2. It can be seen that the Cr-Mn-Ni/Si-2 catalyst exhibited the highest activity, whilst the Cr-Mn-La/Si-2 catalyst gave the best selectivity. Furthermore, manganese oxide has been found to be an effective promoter for enhancing both the activity and the C_2H_4 selectivity. Ni, which is the active component for CH_4 or C_2H_6 reforming with CO_2 to syngas, is very favorable for the activation of C_2H_6 and CO_2 in the reaction of C_2H_6 with CO_2. On the other hand, the promotion effect of lanthanum oxide lies in the increasing of the C_2H_4 selectivity.

3.2. Analysis of C_2H_6 dehydrogenation with CO_2 over the Cr/Si-2 catalyst

From the product distribution of C_2H_6 dehydrogenation with CO_2, it is suggested that conversion of ethane takes place according to the following reactions:

$$16C_2H_6 + 9CO_2 \rightleftharpoons 14C_2H_4 + 12CO + 6H_2O + 12H_2 + CH_4$$

which can be separated as following basic reaction:

1. $C_2H_6 \rightleftharpoons C_2H_4 + H_2$ a
2. $CO_2 + H_2 \rightleftharpoons CO + H_2O$ b
3. $C_2H_6 + H_2 \rightleftharpoons 2CH_4$ c
4. $CH_4 + CO_2 \rightleftharpoons 2CO + 2H_2$ d
5. $C_2H_6 + 2CO_2 \rightleftharpoons 4CO + 3H_2$ e

practically, the reaction is made from (a)×14 + (b)×6 + (c) + (d) + (e). That is to say, the main reactions of ethane dehydrogenation with CO_2consist two consecutive coupling reactions:

$$C_2H_6 \Leftrightarrow C_2H_4 + H_2$$
$$CO_2 + H_2 \Leftrightarrow CO + H_2O$$

in which, ethane first dehydrogenates to form ethylene and hydrogen, and then hydrogen reacts with CO_2 according to the reverse water-gas shift reaction to form CO and water, thus resulting in a continual dehydrogenation of ethane to ethylene, and enhancing the conversion of ethane and the selectivity to ethylene.

3.3. Kinetic analysis of C_2H_6 dehydrogenation with CO_2 over the Cr/Si-2 catalyst

According to the results of the variation in the reaction temperature and space velocity for the reaction (listed in Table 2):

$$C_2H_6 \underset{k_{-1}}{\overset{k_1}{\rightleftharpoons}} C_2H_4 + H_2$$

The following relationship between the reaction velocity constant and the reaction temperature can be deduced.

$$\ln\frac{Xe}{Xe-X} = (k1+k\text{-}1)\times t = K_1\times t$$

Xe -- thermal-dynamic conversion of ethane; X -- actual conversion of ethane on catalyst
K_1 -- right reaction rate constant; K_{-1} -- left reaction rate constant; t -- contact time.

Table 2
The effects of reaction temperature and GHSV on the conversions of C_2H_6 and CO_2

Temp.720 ^{0}C	GHSV(h^{-1})	600	800	1000	1200	1500
Xe=48.7%	C_2H_6 conv.%	17.6	15.2	13.3	12.6	12.0
	CO_2 conv.%	14.8	14.0	12.2	9.7	8.3
	C_2H_4 sele.%	97.7	97.8	98.0	98.3	98.6
Temp.750 ^{0}C	GHSV(h^{-1})	600	800	1000	1200	1500
Xe=58.2%	C_2H_6 conv.%	32.6	31.0	26.2	23.1	19.8
	CO_2 conv.%	24.3	16.6	13.3	11.1	9.8
	C_2H_4 sele.%	96.5	96.9	97.3	97.6	97.8
Temp.780 ^{0}C	GHSV(h^{-1})	600	800	1000	1200	1500
Xe=67.3%	C_2H_6 conv.%	53.5	48.4	46.5	43.7	37.3
	CO_2 conv.%	37.3	29.5	26.4	21.1	15.5
	C_2H_4 sele.%	92.9	94.8	95.5	96.4	97.5
Temp.800 ^{0}C	GHSV(h^{-1})	800		1000	1200	1500
Xe=72.8%	C_2H_6 conv.%	69.8		68.6	57.6	53.3
	CO_2 conv.%	42.2		39.1	31.7	24.3
	C_2H_4 sele.%	90.8		92.3	92.9	93.3
Temp.820 ^{0}C	GHSV(h^{-1})			1000	1500	2000
Xe=77.6%	C_2H_6 conv.%			72.2	61.4	55.1
	CO_2 conv.%			55.7	40.1	29.2
	C_2H_4 sele.%			90.4	92.3	93.1

With investigation of the relationship between $\ln\frac{Xe}{Xe-X}$ and t, K_1 are achieved as following data.

Table 3
The relationship between K1 or lnK1 and reaction temperature

Temp.(K)	993	1023	1053	1073	1093
K_1	262.1	514.4	944.5	2763.5	2880.8
$\ln K_1$	5.57	6.24	6.85	7.72	7.97

From the relationship, as listed in Table 3, between K_1 ($\ln K_1$) and reaction temperature (1/T):

$$K_1 = K_1^0 \times e^{-\frac{E_1^0}{R.T}} \qquad \ln k1 = \ln k1^0 - \frac{E_1^0}{R.T}$$

the pre-exponential factor (K_1^0) of ethane dehydrogenation with CO_2 is 4.05×10^{14}, and the activation energy is 55.5 kcal/mole. Knowing that for the C_2H_6 steam cracking process, $K^0 = 1.274\times10^{13}$ and E^0 =61.99 kcal/mole, then the ratio of K_1 to K can be obtained as follows:

$$K_1/K = K_1^0 \times e^{-\frac{E_1^0}{R.T}} / K^0 \times e^{-\frac{E^0}{R.T}}.$$

The ratio of K_1 /K was found to be 675, 715 and 785 at the reaction temperature of 800℃, 780 ℃ and 750℃ respectively.

3.4. Dilute C_2H_6 in FCC tail-gas dehydrogenation with CO_2 as an oxidant over the Cr/Si-2 catalyst

Table 4
The performance of the Cr/Si-2 catalysts for the production of C_2H_4 from C_2H_6 in FCC tail gas dehydrogenation with CO_2 .

Catalyst	conversion (%)		Selectivity (wt%)		
	CO_2	C_2H_6	CH_4	C_2H_4	C_2H_4 content(V%)
Cr/Si-2	19.5	60.6	20.4	79.6	25.4
Cr-Mn/Si-2	23.5	63.1	18.9	81.1	26.1
Cr-Mn-Ni/Si-2	26.7	69.7	19.4	80.6	26.7
Cr-Mn-La/Si-2	20.7	63.6	14.2	85.8	27.2

*Reaction conditions: 1073K, 0.1MPa, $1000h^{-1}$, CO_2/C_2H_6(mole ratio)=1/1.
In FCC tail-gas, C_2H_6(V content)=18.8%, C_2H_4(V content)=19.2%.

Another objective of the present work is to make use of the results discussed above for the development of a new process for the utilization of the C_2H_6 component in the FCC dry gas to produce C_2H_4. We have reported previously on the development of a new process for the utilization of the dilute C_2H_4 in the FCC dry gas to produce ethylbenzene[4]. Now, by combining our new results with the ethylbenzene process, the C_2H_6 as well as the C_2H_4 components in the FCC dry gas can be utilized completely for the production of ethylbenzene.

Results in Table 4 show that almost the same result of oxidative dehydrogenation of C_2H_6 with CO_2 to produce C_2H_4 over the Cr/Si-2 catalysts were obtained by using the FCC tail-gas instead of pure C_2H_6 as the reaction feed gas. The Cr-Mn-La/Si-2 catalyst showed a stable reactivity with about 86% of C_2H_4 selectivity and 65% of C_2H_6 conversion, with a stable performance for more than 900h during the life test of the catalyst.

3.5. The effect of H_2O addition on the C_2H_6 dehydrogenation over the Cr/Si-2 catalyst

Table 5 demonstrates the effect of H_2O addition on the behavior of the Cr-Mn-La/Si-2 catalyst for C_2H_6 dehydrogenation with CO_2. With the addition of H_2O into the reaction feed-gas, both the CO_2 and C_2H_6 conversion decreased slightly, while the C_2H_4 selectivity increased. From the viewpoint of energy consumption, it is important to choose an appropriate proportion of $H_2O/CO_2+C_2H_6$ during the commercialization of the process.

The crucial point in the development of this process is to avoid carbon deposition over the catalyst surface. In the present paper, one of the effective means to reduce coke deposits on the catalyst is by adding H_2O into the reaction feed gas. The influence of H_2O on coke deposition was investigated by employing DTA and TG. From Table 6 it can be seen that, with the addition of water into the reaction feed gas, the coke deposition on the surface of Cr-Mn-La/Si-2 catalyst during C_2H_6 dehydrogenation with CO_2 was greatly diminished.

Table 5
The effect of H_2O on the performance of the Cr-Mn-La/Si-2 catalyst for C_2H_4 production from C_2H_6 dehydrogenation with CO_2.

$H_2O/CO_2+C_2H_6$	conversion (%)		Selectivity (wt%)		
mole ratio	CO_2	C_2H_6	CH_4	C_2H_4	H_2/CO mole ratio
0.0/2.0	20.5	64.2	13.8	86.2	1.4
0.5/2.0	20.3	63.2	14.7	85.3	1.3
1.0/2.0	20.0	62.7	13.3	86.7	1.4
2.0/2.0	19.2	60.1	13.0	87.0	1.4
3.0/2.0	19.1	60.3	12.7	87.3	1.4

Reaction conditions: 1073K, 0.1MPa, 1000/h, CO_2/C_2H_6(mole ratio)=1/1.

Table 6
The effect of H_2O addition on carbon deposition on the Cr-Mn-La/Si-2 catalyst during C_2H_6 dehydrogenation with CO_2

$H_2O/(CO_2+C_2H_6)$molar ratio	0.0/2.0	0.5/2.0	1.0/2.0	2.0/2.0**	3.0/2.0
depositing carbon (wt. %)	8.6	5.4	4.7	6.8	2.1

Reaction conditions: 1073K, 0.1MPa, 1000/h, CO_2/C_2H_6(mole ratio)=1/1.
The catalyst was sampled from the reactor after running for 40 h.
** The catalyst was sampled from the reactor after running for 240 h.

REFERENCES

[1], A. Kh. Mamedov , S. R. Mirzabekova and O. V. Krylov, << New Developments in Selective Oxidation>>, Elsevier Publishers(Amsterdam), 1990, pp. 477.

[2], A. Kh. Mamedov , S. R. Mirzabekova and O. V. Krylov, << New Developments in Selective Oxidation>>, Elsevier Publishers(Amsterdam), 1990, pp. 159.

[3], Xu Longya, Lin Liwu and Wang Qingxia, Chinese J. Catal., 18(1997), 353.

[4], Wang Qingxia, Zhang Shurong, Cai Guangyu, et al., CN 87105054.4, 1987.

NATURAL GAS CONVERSION V
Studies in Surface Science and Catalysis, Vol. 119
A. Parmaliana et al. (Editors)

Kinetics and mechanism of the selective oxidation and degradation of n-butane over nickel molybdate catalysts

L.M. Madeira* and M.F. Portela

GRECAT, Instituto Superior Técnico – Technical University of Lisbon,
Av. Rovisco Pais, 1096 Lisboa Codex, Portugal

The kinetics of the n-butane reactions in the presence of oxygen over α-$NiMoO_4$ has been investigated. The rates of dehydrogenation (C_4's) and degradation (CO_x) reactions as functions of the butane and oxygen partial pressures are described by kinetic models based on a generalized form of the Mars and van Krevelen mechanism.

1. INTRODUCTION

Unpromoted molybdate compounds have been widely used as oxidative dehydrogenation (ODH) catalysts [1]. Quite recently, Stern and Grasselli using several simple molybdates of the type $AMoO_4$ (with A = Ni, Co, Mg, Mn or Zn) found that the nickel one was the most active and selective in propane ODH [2]. The NiO-MoO_3 system also proved to be very promising for n-butane conversion [3-8]. Although much effort in the development of selective catalysts has appeared in the last few years, few kinetic studies have been carried out and no detailed kinetic studies with such catalytic system were found in the literature, except for the high temperature β phase of $NiMoO_4$ in propane conversion [9]. Therefore, for improving our knowledge of the factors that control the catalytic behavior and to understand some important details of the kinetics and mechanisms of the involved reactions, a systematic kinetic study of the n-butane ODH was undertaken with pure α-$NiMoO_4$ in a large range of experimental conditions.

2. EXPERIMENTAL

A stoichiometric α-$NiMoO_4$ catalyst was prepared by coprecipitation using $Ni(NO_3)_2.6H_2O$ and H_2MoO_4 solutions. The solid was carefully characterized by different techniques (BET, ICP, atomic absorption, HTXRD, FTIR, XPS, LRS, TPD, TPR and electric conductivity) and tested in a fixed-bed, continuous flow tubular reactor for n-butane ODH. Reactants (butane and O_2, diluted in N_2) and products (1-butene, *trans*-2-butene, *cis*-2-butene, butadiene and carbon oxides) were analyzed with an on line gas chromatograph with two columns. Temperature was varied between 500 and 560 °C and the influence of the reactants partial pressure was studied for a total pressure in the reactor of 1.10 bar. The catalyst charge and the contact time used (0.150 g and 4.45 $g.h/mol_{butane}$, respectively), allowed to obtain differential conversions. Experimental details and results can be found elsewhere [4-8].

* L.M. Madeira thanks PRAXIS XXI Program from JNICT for financial support.

3. RESULTS AND DISCUSSION

Characterization data evidenced that the prepared catalyst was pure α-$NiMoO_4$ and the structure is stable even at temperatures up to 600 °C without occurring the α to β phase transition. The surface area of the catalyst is 44.1 m^2/g.

The influence of temperature and reactants partial pressures in the dehydrogenation and degradation reaction rates (based on surface area) are shown in Figures 1 and 2, respectively. The general trends of the butenes and butadiene formation rates versus the reactants partial pressures are similar to those presented in Figure 1 for the total of C_4 products [8,10]. It is noteworthy that the O_2 partial pressure (P_{O2}) does not affect the rates of dehydrogenation processes but favors the carbon oxides formation. Moreover, all the products formation rates are positively influenced by the butane partial pressure (P_{butane}), but this effect is more significant in the case of CO and CO_2.

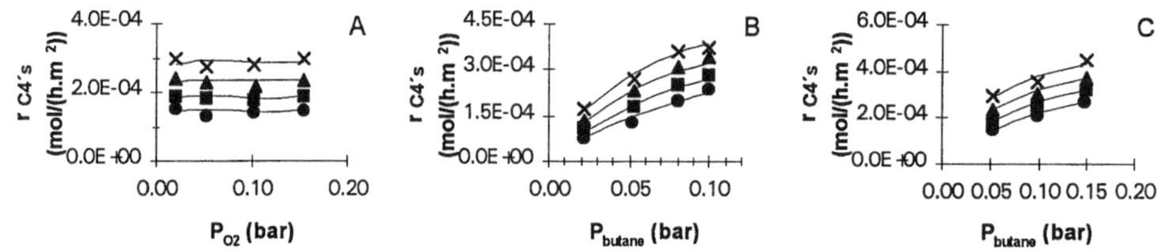

Figure 1. Influence of P_{O2} with P_{butane} = 0.05 bar (A) and of P_{butane} with P_{O2} = 0.05 bar (B) and 0.15 bar (C) in the total C_4 (butenes and butadiene) products formation rate. (●) 500 °C; (■) 520 °C; (▲) 540 °C; (×) 560 °C. Full lines obtained from eq. (4).

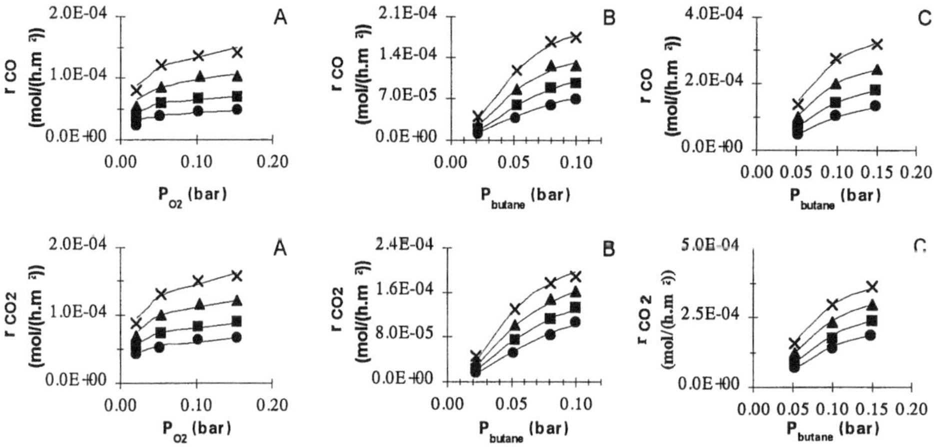

Figure 2. Influence of reactants partial pressures in the carbon oxides formation rates. Symbols and legend as in Figure 1. Full lines obtained from eq. (5).

In what concerns the products distribution, an increase of the oxygen concentration leads to a decrease of C_4's selectivities and an increase of CO_x formation at low partial pressures. Butane partial pressure does not affect the products distribution. The effect of temperature leads to an increase in butane conversion, decreasing the butenes selectivity and increasing the CO selectivity, the effects in butadiene and CO_2 being minor.

The results found in literature [10] suggest a Mars and van Krevelen mechanism [11] for the studied reaction. In this mechanism the agents responsible for oxidation are the oxygen ions of the oxide lattice which, in a first step, react with the hydrocarbon to give the products and a partially reduced catalyst. After that occurs the catalyst reoxidation by gaseous oxygen to restore the catalyst in its original state. The existence of a redox mechanism was also supported by previous data obtained in our laboratory: the relationship between catalytic activity and reducibility of different catalysts [8], the zero order in oxygen for the dehydrogenation process (see Figure 1), the experiments without O_2 in the reactor feed, which showed that butane can be converted to C_4 products with high selectivity even without gas phase oxygen [12], and finally the absence of oxygen adsorption in the O_2-TPD experiments [10]. Therefore, the reduction and oxidation rates of the catalyst surface can be described by the following equations considering a generalized Mars and van Krevelen mechanism:

$$r_r = k_r P_{butane}{}^{n} \theta_o \tag{1}$$

$$r_o = k_o P_{O_2}{}^{n_1} P_{butane}{}^{n_2} \theta_r \tag{2}$$

where k_r and k_o are kinetic constants, θ_o and θ_r represent the fractions of oxidized and reduced sites, respectively, P_i is the partial pressure of reactant and n_i the partial order. As described elsewhere [10], the combination of eqs. (1) and (2) with the condition of steady-state (r_r = r_o/α), where α is the stoichiometric coefficient of oxygen, leads to the general rate equation:

$$r = \frac{k_o k_r P_{butane}{}^{n} P_{O_2}{}^{n_1}}{k_o P_{O_2}{}^{n_1} + \alpha \cdot k_r P_{butane}{}^{(n-n_2)}} \tag{3}$$

When fitting eq. (3) to the experimental data of each dehydrogenation product formation rates, by optimization of a multiple correlation coefficient using non-linear regression analysis [10], a much simpler equation was obtained. It is noteworthy that the equation rate that best describes the formation of all dehydrogenation products is the same with $n = 1$ and $n_1 = n_2 = 0$. Therefore, eq. (3) can be simplified to:

$$r = \frac{k_o k_r P_{butane}}{k_o + \alpha \cdot k_r P_{butane}} \tag{4}$$

The kinetic constants k_o and k_r were obtained at different temperatures by fitting eq. (4) to the data of each dehydrogenation product (butenes and butadiene) formation rate [10] and to the total C_4 products formation rate presented in Figure 1. The fitting curves are included in such figure. The computed values of the kinetic constants agree with the Arrhenius law as evidenced in Figure 3 for the total of C_4's formation. From those plots, the pre-exponential factors k_i' and the activation energies Ea_i were computed [10].

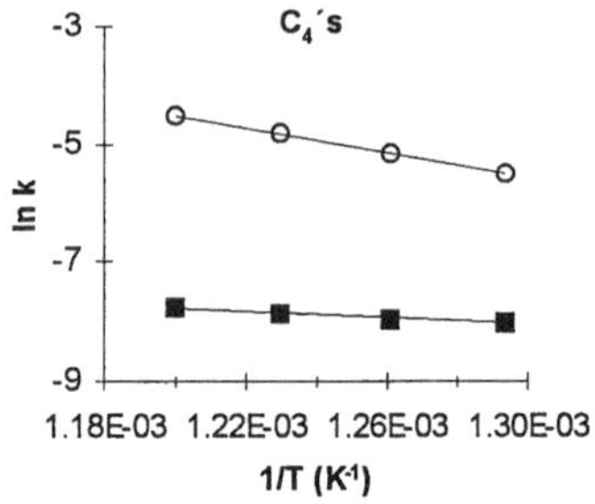

Figure 3. Arrhenius plots of the kinetic constants k_r (o) and k_o (■) for the total of dehydrogenation products.

It is well known that in the ODH reaction n-butane first reacts by cleavage of a secondary C–H bond to form an adsorbed alkyl species. After a second H–abstraction, olefinic intermediates are formed and desorbed as olefins or undergo another dehydrogenation to form butadiene [13]. To analyze the selective oxidation process, in opposition to total oxidation, the global C_4's formation was considered. The obtained parameters are presented in Table 1.

Table 1
Computed parameters for formation of dehydrogenation and degradation products.

	C_4's		CO	CO_2
k_o' (mol/(h.m^2))	9.6×10^{-3}	k_o' (mol/(h.m^2.bar$^{1/2}$))	1.0×10^{2}	5.0
k_r' (mol/(h.m^2.bar))	3.0×10^{3}	k_r' (mol/(h.m^2.bar^2))	2.6×10^{6}	1.1×10^{5}
Ea_o (J/mol)	2.2×10^{4}	Ea_o (J/mol)	7.0×10^{4}	4.5×10^{4}
Ea_r (J/mol)	8.7×10^{4}	Ea_r (J/mol)	1.2×10^{5}	9.7×10^{4}

When using high alkane conversion levels, alkenes are overoxidized and selectivity to carbon oxides increases due to the parallel-consecutive reaction network. However, at low contact times, dehydrogenation and combustion products are formed from alkane through parallel reactions [13]. The same procedure as above described for C_4 products was then applied to the carbon oxides. After a detailed optimization, we found that the equation that best describes the experimental data is the same for both CO and CO_2 with $n = 2$, $n_1 = 0.5$ and $n_2 = 0$. Therefore, eq. (3) can be simplified to:

$$r = \frac{k_o k_r P_{butane}^{\,2} P_{O_2}^{\,1/2}}{k_o P_{O_2}^{\,1/2} + \alpha \cdot k_r P_{butane}^{\,2}} \tag{5}$$

The kinetic constants were then computed at different temperatures. The corresponding fitting curves are included in Figure 2 and their Arrhenius behavior is evidenced in Figure 4. The computed pre-exponential factors and activation energies are presented in Table 1.

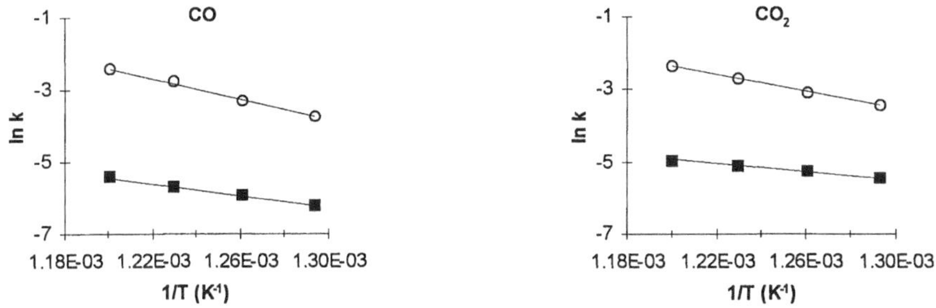

Figure 4. Arrhenius plots of the kinetic constants k_r (o) and k_o (■) for carbon oxides.

4. CONCLUSIONS

It was evidenced that the reaction rates of all the butane reactions in the presence of oxygen over α-$NiMoO_4$ (selective oxidation and degradation) can be described by a kinetic model based on a generalized form of the Mars and van Krevelen mechanism. No assumptions were done on the reaction orders in the reduction and reoxidation steps. Such orders were determined later in the fitting process leading to computation of kinetic parameters.

The influence of butane partial pressure in the reoxidation step, eq. (2), was investigated because a negative effect was previously found for 1-butene ODH over bismuth molybdates due to a competition of oxygen and 1-butene over the reduced site [14]. Nevertheless, such order was found to be null in both cases analyzed here. An interesting feature is that in the case of the dehydrogenation products P_{O2} does not affect the oxidation rate of the reduced active sites, i.e., $n_1 = 0$, which may be an evidence that the diffusion of lattice oxygen controls the global process of the catalyst reoxidation. For carbon oxides a partial order of 1/2 was obtained, suggesting that a dissociative adsorption of gaseous O_2 may determine the reoxidation rate.

In the dehydrogenation process the role of gaseous oxygen seems to be to replenish the consumed lattice oxygen, reoxidizing the solid. However, it must be remarked that such gaseous oxygen may have another role: the formation of carbon oxides by attacking the weakened C-C bonds of the adsorbed hydrocarbon molecules.

The higher partial order to oxygen for carbon oxides production suggests the convenience of avoiding the presence of gaseous oxygen in the reactor feed to obtain high C_4 selectivities, i.e., to operate in a redox mode using the lattice $NiMoO_4$ oxygen with periodical catalyst regeneration. This could be done in practice using a riser reactor. Another possibility is the use of membrane reactors where it is possible to control the oxygen distribution, maintaining its partial pressure sufficiently low in the catalytic bed.

Finally, it is important to point out that the effect of an increase in the reaction temperature is more pronounced in the solid reduction process since the activation energy for reduction is higher than the corresponding to reoxidation (Table 1).

REFERENCES

1. J.D. Burrington and R.K. Grasselli, J. Catal., 59 (1979) 79.
2. D.L. Stern and R.K. Grasselli, J. Catal., 167 (1997) 550.
3. C. Mazzocchia, R. Del Rosso and P. Centola, An. Quim., 79 (1983) 108.
4. R.M. Martín-Aranda, M.F. Portela, L.M. Madeira, F. Freire and M.M. Oliveira, Appl. Catal. A, 127 (1995) 201.
5. F.J. Maldonado-Hódar, L.M. Madeira, M.F. Portela, R.M. Martín-Aranda and F. Freire, J. Mol. Catal. A, 111 (1996) 313.
6. L.M. Madeira, J.M. Herrmann, F.G. Freire, M.F Portela and F.J. Maldonado, Appl. Catal. A, 158 (1997) 243.
7. L.M. Madeira, F.J. Maldonado-Hódar, M.F. Portela, F. Freire, R.M. Martín-Aranda and M. Oliveira. Appl. Catal. A, 135 (1996) 137.
8. L.M. Madeira, M.F. Portela, C. Mazzocchia, A. Kaddouri and R. Anouchinsky, Catal. Today, 40 (2-3) (1998) 91.
9. M. Sautel, G. Thomas, A. Kaddouri, C. Mazzocchia and R. Anouchinsky, Appl. Catal. A, 155 (1997) 217.
10. L.M. Madeira and M.F. Portela, in R.K. Grasselli, S.T. Oyama, A.M. Gaffney and J.E. Lyons (eds.), 3rd World Congress on Oxidation Catalysis, Elsevier, 1997, p. 797.
11. P. Mars and D.W. van Krevelen, Chem. Eng. Sci. (Spec. Suppl.), 3 (1954) 41.
12. F.J. Maldonado-Hódar, L.M. Madeira, M.F. Portela and R.M. Martín-Aranda, Proc. 15th Ibero-American Symp. Catal., Córdoba, Vol. 1, 1996, p. 251.
13. E.A. Mamedov and V.C. Corberán, Appl. Catal. A, 127 (1995) 1.
14. M.F. Portela, M.M. Oliveira, M.J. Pires, F.M.S. Lemos and L. Ferreira, Proc. 8th International Congress on Catalysis, Berlin, 1984, p. 533.

NATURAL GAS CONVERSION V
Studies in Surface Science and Catalysis, Vol. 119
A. Parmaliana et al. (Editors)

Application of the oscillating microbalance catalytic reactor: Kinetics and coke formation over Pt-Sn/Al_2O_3 in propane dehydrogenation.

H. P. Rebo, D. Chen, E. A. Blekkan and A. Holmen.

Department of Industrial Chemistry, Norwegian University of Science and Technology (NTNU), N-7034 Trondheim, Norway.

A Tapered Element Oscillating Microbalance (TEOM) has been used to study propane dehydrogenation over Pt-Sn/Al_2O_3 catalysts at 500 - 580°C. Focus has been on the deactivation kinetics, which is easily obtained from the TEOM. The fixed-bed reactor characteristics of the microbalance coupled with an on-line GC allow the rates and selectivities to be studied simultaneously with the mass changes of the catalyst bed due to coking. Both the initial reaction rate and the coking rate increased with temperature. The coking reaction deactivated faster than the main reaction which was explained by coke formation as a structure sensitive reaction which occurs over larger Pt ensembles than propane dehydrogenation. Hydrogen was found to decrease both the rate of coke formation and the deactivating effect of the coke formed.

1. INTRODUCTION

Light alkenes (C_2 - C_4) are mainly produced by steam cracking of various hydrocarbon feedstocks. Irrespective of feedstock, ethene is the major product from steam cracking. Additional C_3 and C_4 alkenes can be recovered from refinery streams or produced by catalytic dehydrogenation of the parent alkane. It is predicted [1] that the propene demand growth will outstrip that of ethene, suggesting an increased interest in propane dehydrogenation. Dehydrogenation reactions are limited by equilibrium and require high temperatures in order to achieve high conversions. Unfortunately, other catalyzed reactions become significant at higher temperatures, such as isomerization, hydrogenolysis and coking.

Supported Pt catalysts have been extensively studied for propane dehydrogenation. The high activity and the high selectivity to propene make such catalysts very suitable, but improvements regarding the selectivity and in particular the deactivation rate seem to be obtainable by catalyst modifications. It has become a common practice to incorporate a second metal (Sn, Ag, Ge, Pb) [2-7] and/or alkali metals (Li, Na, K) [4-6]. Padro et al. [4] found that the addition of K to a Pt-Sn/Al_2O_3-catalyst lowered the amount of coke formed on the metal as well on the support. The addition of K also produced coke with a lower degree of polymerisation. A second modification used by Padro et. al. [4] was the use of the non-acidic support $ZnAl_2O_4$ which resulted in less cracking and polymerisation reactions compared to an acidic support.

Sn increases the catalyst stability, but coke is still inevitably formed, leading to catalyst deactivation. Different mechanisms have been proposed for the enhanced catalyst lifetime as a result of Sn addition. Cortright et. al. [6] reports that Sn decreases the size of the surface platinum ensembles, and that hydrogenolysis and isomerization reactions are suppressed over the smaller platinum ensembles. Among a number of kinetic studies is the work of Biloen et al. [7] who reported a reaction order with respect to propane of about 1 for both platinum and dilute platinum-in-gold alloys, and a catalyst dependent reaction order in hydrogen. The variation was proposed to be due to differences in the steady-state coverage θ_H depending on the neighbouring atoms of Pt. Hydrogen is known to help maintain the catalyst activity in propane dehydrogenation, but the exact role of hydrogen for both the kinetics and deactivation is less clear. Propane dehydrogenation is industrially performed with recycled hydrogen and/or steam addition to reduce coking. This work deals with propane dehydrogenation on Pt-Sn/Al_2O_3 catalysts, and in particular with catalyst deactivation studied by a new type of microbalance.

2. EXPERIMENTAL

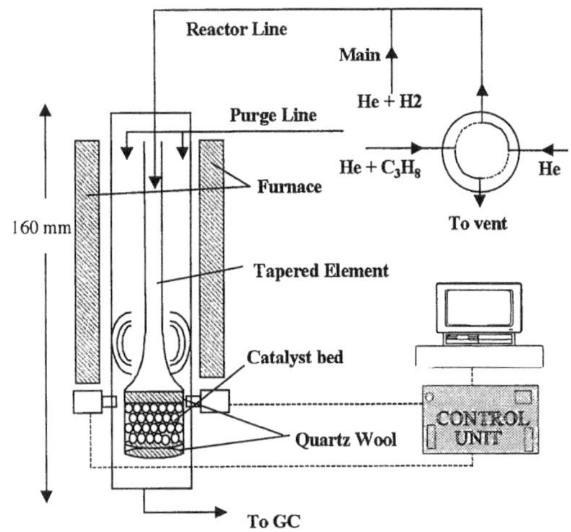

Figure 1. Schematic drawing of the TEOM apparatus.

Catalytic dehydrogenation of propane was carried out in an oscillating microbalance (TEOM) [8]. Originally developed for detecting particulates in the effluent of combustion systems [8], the TEOM has in recent years also been used to study catalytic conversion processes suffering from rapid deactivation such as methanol to olefins [9] and heptane reforming [10]. The main advantage of the TEOM reactor over a gravimetric microbalance is that all the gases in the reaction mixture are forced to flow through the catalyst bed as in a conventional fixed-bed reactor [11]. Coupled with on-line GC, the catalyst activity, the selectivities and the coking rate can be studied *in situ* simultaneously as a function of the coke content on the catalyst. The experimental set-up is shown schematically in Fig. 1, and it is designed for transient (pulses and step changes) as well as for continuous flow experiments [12]. All the results reported here were obtained in the continuous flow mode. The TEOM has been upgraded for use at temperatures up to 700°C, and a completely new tapered element (TE) was installed. A base line mass drift which gets significant above 600°C is observed, and must be accounted for. The base line has become more stable over time for our new TE as the number of repeated cycles at elevated temperatures has increased.

The catalysts studied were made according to [2] by impregnation to incipient wetness of a γ-Al_2O_3 (S_{BET} = 308 m^2/g) support with Sn using $SnCl_2{\cdot}2H_2O$ followed by drying (110°C for 12 hrs) and calcination (550°C for 2.5 hrs in humid air followed by 2 hrs in dry air). The purpose of the wet calcination (humid air) was to remove Cl originating from the metal salts. Impregnation to incipient wetness with Pt in the form of $H_2PtCl_6{\cdot}6H_2O$ was then performed followed by drying and calcination after the same procedure. Pt-Sn-catalysts with varying Pt contents were prepared and the compositions are given in Tab. 1.

The metal loadings reported are calculated from the amounts of metal salts added during the synthesis. Surface area measurements of the catalysts were performed by the BET method. The Pt dispersion was measured volumetrically by H_2 chemisorption at 25°C using a Micromeritics ASAP 2010 and assuming a H:Pt ratio of 1. Two isotherms were measured with evacuation of the sample in-between. Only the amount of H_2 irreversibly adsorbed calculated from the difference between the two isotherms is reported in Tab. 1.

Propane was dried over molecular sieves (4 Å), oxygen was removed by Alltech oxytraps and sulphur free operation was ensured by using a ZnO trap. Before the kinetic experiments, the catalyst (7 – 30 mg) was heated to 520°C at a heating rate of 4°C/min in hydrogen and kept at this temperature in a flow of hydrogen (17 ml/min) for 4 hrs. The temperature was then adjusted to the desired level and the dehydrogenation experiments were started when the mass signal was stable. A mixture of helium and propane was introduced by the four-way valve (Fig. 1), maintaining a constant flow rate during a switch. The space velocity was calculated as g propane/g cat·h and was varied to maintain low conversions.

Table 1.
Properties of the catalysts used in this study.

Catalyst	Pt (wt %)	Sn (wt %)	S_a (m^2/g)	H : Pt irr.
Pt(1.0)/Al_2O_3	1.0	0.0	267	0.68
Pt(0.15)-Sn/Al_2O_3	0.15	1.28	225	0.44
Pt(0.5)-Sn/Al_2O_3	0.50	1.28	230	0.41
Pt(1.0)-Sn/Al_2O_3	1.0	1.28	215	0.30

The reaction was followed by GC-analysis of the exit gas stream every 16 min. The GC was equipped with a FI-detector and a GS-Q column (30 m, 0.54 mm i.d.), allowing the separation of hydrocarbons up to C_3. The selectivities were calculated on a carbon-basis, and the reaction rates were calculated by assuming differential conditions in the catalyst bed. Initial reaction rates (r_0) were found by plotting the reaction rate r (mol propane/g cat·h) versus the coke content and extrapolating to zero coke content according to Eq (1):

$$r = r_0 \cdot \exp(-\alpha \cdot C) \tag{1}$$

where C is the coke content in wt % and α is a deactivation constant. To ensure the absence of wall effects or contributions from the support alone, a wet calcined γ-Al_2O_3 prepared by the same procedure was studied at 520°C. The activity of the γ-Al_2O_3 was negligible for all feed mixtures at the relatively low temperatures used in this study.

3. RESULTS

Experiments were performed at 500, 520, 540 and 580°C, propane partial pressures between 0.01 and 0.17 bar and hydrogen partial pressures below 0.14 bar. Helium was used as diluent, carrier gas and purge gas. A typical dehydrogenation experiment was stopped after 300 min. on stream. The selectivity to propene was better than 99 % for all the experiments below 520°C on all the Pt-Sn/Al_2O_3-catalysts. The selectivity to other products was found to depend on the feed composition for all temperatures. The high propene selectivity is probably a result of the elimination of acidic sites by wet calcination and reduction of the Pt ensemble size by Sn addition. At 520°C, P_{C3H8} = 0.044 and P_{H2}= 0.039 bar, only a very small amount of C_1 and C_2 was detected over the Pt(0.5)-Sn/Al_2O_3 compared to a Pt(1.0)/Al_2O_3-catalyst. This pattern suggests that Sn decreases the rate of coke formation and hydrogenolysis [6] by reducing the Pt ensemble size. Such pattern corresponds well with previous reports and with the observation of hydrogenolysis as a structure sensitive reaction.

Fig. 2 presents the effect of hydrogen on the deactivation and coke formation. The figure gives the coking and the propane conversion for two experiments carried out over Pt(1.0)-Sn/Al_2O_3 with and without hydrogen present in the feed at 540°C. The initial activity for the experiment without hydrogen present in the feed was higher, but the reaction suffered from faster deactivation as expected at these conditions. The weight increase after 150 min. on stream was about 1 wt % for the catalyst with hydrogen present in the feed, and 2.8 wt % when no hydrogen was added to the feed. Figs. 2 and 3 clearly indicate that the role of hydrogen in propane dehydrogenation is dual: The rate of coke formation is reduced, and the same holds for the toxicity of the coke, i.e. a relatively smaller drop in catalyst activity compared to the initial value at a certain coke content. Fig. 2 illustrates the main advantage of the TEOM: Rates, product selectivities and coke formation are obtained simultaneously at differential conditions.

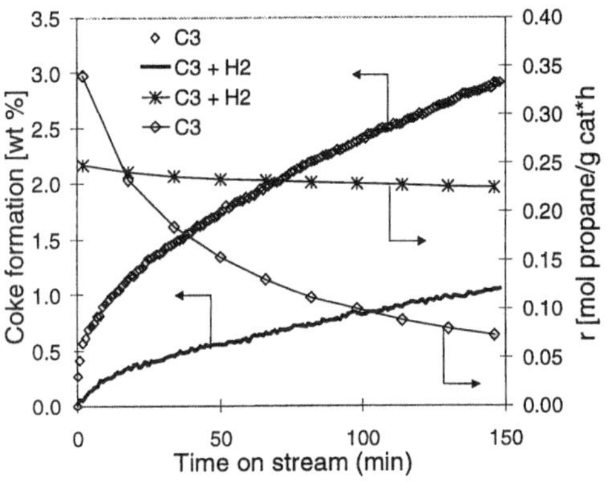

Figure 2. Reaction rates and coke formation at 540°C on Pt(1.0)-Sn/Al_2O_3. Conditions as given in Tab. 2.

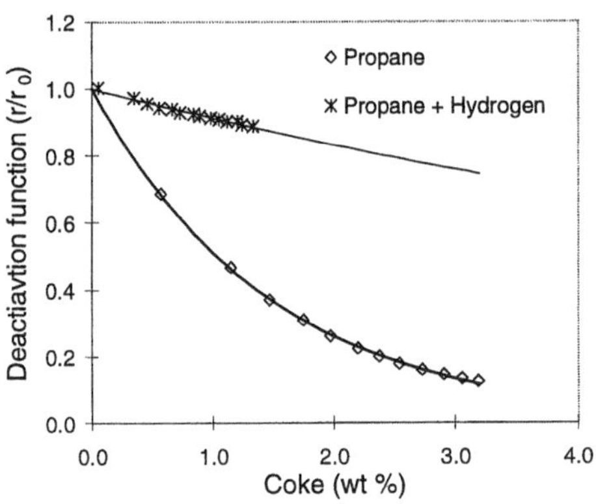

Figure 3. r/r_0 versus coke content. Experimental conditions as in Figure 2. Lines: Eq (1).

Fig. 3 gives the deactivation functions (r/r_0) for the experimental data in Fig. 2. At 1 wt % coke, the relative loss in reaction rate was about 10 % with hydrogen in the feed, whereas the reaction rate was found to be less than 50 % of the initial reaction rate without any hydrogen present in the feed stream. The data necessary to plot the deactivation functions in Fig. 3 is obtained directly from the TEOM. However, as shown in Fig. 3 it is not a straightforward relation between the coke content and the deactivation in propane dehydrogenation since the deactivation is dependent of the feed composition. An increased coke toxicity does probably mean that relatively more coke is formed on the metal compared to the support, and a differentiation between coke on the metal and coke on the support should be made. Lin et al.

[14] studied similar catalysts using TPO, and showed that the fraction of the total coke located on the support increased with the Sn/Pt-ratio.

Fig. 4 presents the deactivation functions versus coke contents at three different temperatures for the Pt(1.0)-Sn/Al_2O_3-catalyst. The partial pressure of propane was 0.091 bar and the space velocities between 39 and 42 h^{-1} for these experiments. Eq (1) illustrated by the lines in Fig. 4 represents the experimental data well. It can be seen that the coke is more toxic at lower temperatures. Different models for the relation between the coke formation and the time on stream have been developed earlier [13]. Fig. 5 shows the coke content obtained from experiments between 500 and 580°C versus time on stream together with the model according to Eq (2) which was the model that described our data best:

$$C = 1/\alpha_C \cdot \ln[1 + \alpha_C \cdot r_{0C} \cdot t] \qquad (2)$$

α_C is a deactivation constant for the coke formation, r_{0C} is the initial coking rate and t is time. The fitted parameters are shown in Tab. 2. Increased temperature resulted in the expected increase in coking rate. All these experiments were carried out without hydrogen in the feed, and the lack of hydrogen and high temperatures are conditions expected to increase the coke formation.

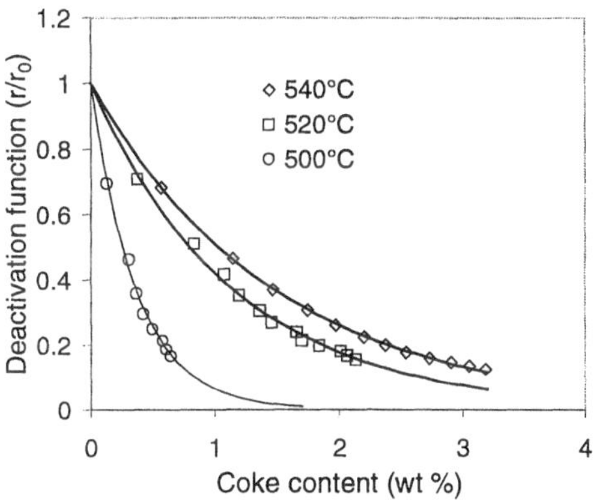

Figure 4. r/r_0 versus coke content for Pt(1.0)-Sn/Al_2O_3 at different temperatures. Line: Eq (1), Points: Experimental data.

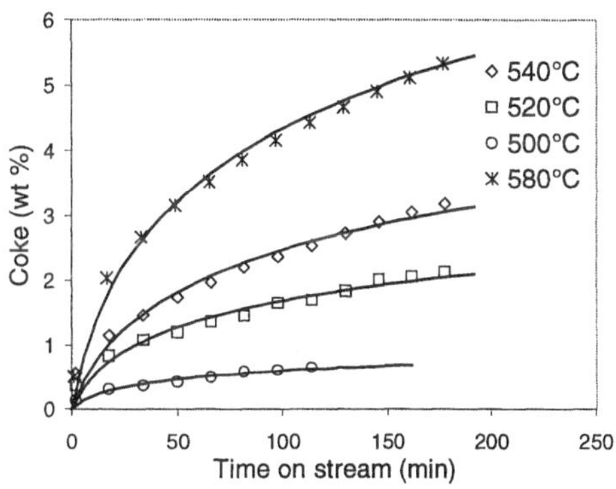

Figure 5. Coke formation versus TOS on Pt(1.0)-Sn/Al_2O_3 at different temperatures. Line: Eq (2), Points: Experimental data.

Tab. 2 shows that faster deactivation was found at lower temperatures represented by relatively higher α-values than at higher temperatures. Tab. 2 also displays an increase in initial coking rate and a decrease in α_C-values with increasing temperature. Interesting is the fact that the deactivation constant (α_C) for the coking reaction is higher than the deactivation constant for the main reaction (α) at the different temperatures. This trend is probably explained by coke formation as a structure sensitive reaction which needs larger Pt-ensembles [6]. As more coke is formed, the number of such ensembles available will decrease relatively faster than the number of single Pt atoms. Propane dehydrogenation has earlier been reported to occur over single Pt atoms [7]. Hydrogen in the feed resulted in a lower initial reaction rate, but this disadvantage is compensated for by the favourable effects of hydrogen on the initial coking rate

(r_{0C}) and the deactivation constant (α_C) for the coke formation.

Table 2.
Experimental conditions and parameters from Eq (1) and Eq (2).

T (°C)	SV (h^{-1})	P_{C3H8} (bar)	P_{H2} (bar)	r_0 mol/g cat·h	α g cat/g coke	r_{0C} g coke/g cat·h	α_C g cat/g coke
500	39	0.091	-	0.28	2.74	2.12	4.84
520	42	0.091	-	0.41	0.86	4.55	1.48
540	40	0.091	-	0.50	0.67	5.48	0.89
540	41	0.079	0.132	0.25	0.09	1.01	1.39

4. CONCLUSIONS

Propane dehydrogenation over Pt-Sn/Al_2O_3 catalysts have been studied using the TEOM reactor. The apparatus makes it possible to simultaneously obtain kinetics of the main reaction and the coking reaction. Hydrogen was found to help in maintaining the activity for longer time on stream and by reducing both the coking rate and the deactivating effect of the coke formed. The coke content as a function of time on stream could be described very well with a model of the form: $C = 1/\alpha_C * \ln[1+\alpha_C r_{0C} t]$. At each temperature, the coking reaction deactivated faster than the main reaction which is explained by the number of Pt ensembles available on the surface.

ACKNOWLEDGEMENTS
Financial support from the Norwegian Research Council is gratefully acknowledged.

REFERENCES

1. Anon., Eur. Chem. News., 67 (1997) 28.
2. O. A. Bariås, A. Holmen and E.A. Blekkan, J. Catal., 158 (1996) 1.
3. G. Corro, P. Marecot and J. Barbier, Stud. Surf. Sci. Catal., 111 (1997) 359.
4. C.L. Padró, S.R. de Miguel, A.A. Castro, A. Scelza, Stud. Surf. Sci. Catal., 111 (1997) 191.
5. G.J. Siri, M.L. Casella, G.F. Santori, O.A. Ferretti, Ind. Eng. Chem. Res., 36 (1997) 4821.
6. R.D. Cortright and J.A. Dumesic, J. Catal., 148 (1994) 771.
7. P. Biloen, F.M. Dautzenberg and W.M.H. Sachtler, J. Catal., 50 (1977) 77.
8. H. Patashnick, G. Rupprecht and J.C.F. Wang, Prepr. ACS Div. Petr. Chem., 25 (1980) 188.
9. D. Chen, H.P. Rebo, K. Moljord and A. Holmen, Ind. Eng. Chem. Res., 36 (1997) 3473.
10. K. Liu, S.C. Fung, T.C. Ho and D.S. Rumschitzki, Ind. Eng. Chem. Res., 36 (1997) 3264.
11. D. Chen, A. Grønvold, H.P. Rebo, K. Moljord and A. Holmen, Appl. Catal., 137 (1996) L1.
12. D. Chen, H.P. Rebo, K. Moljord and A. Holmen, Chem. Eng. Sci., 51 (1996) 2687.
13. G.F. Froment and K.B. Bischoff, Chemical Reactor Analysis and Design, John Wiley & Sons, New York, 1990.
14. H. Lin, T. Zao, J. Zang and Z. Xu, Appl. Catal., 67 (1990) 11.

NATURAL GAS CONVERSION V
Studies in Surface Science and Catalysis, Vol. 119
A. Parmaliana et al. (Editors)
1998 Elsevier Science B.V.

Conversion of n-butane over Pt-Me/Al_2O_3 catalysts

D. Nazimek and J. Ryczkowski

University of Maria Curie-Sklodowska, Faculty of Chemistry,
3 Maria Curie-Skłodowska Square, 20-031 Lublin, Poland
E-mail: Nazimek@hermes.umcs.lublin.pl

Platinum is an universal, versatile catalyst, used in many reactions, and therefore the literature concerning this metal and its alloys is very extensive: a large part of it deals with reactions of saturated hydrocarbons. Conversion of n-butane over Pt/Al_2O_3 catalysts modified by copper, gold or silver is an example of such investigations. It can be notice that Pt-Me system formed involves different active centres on the catalyst surface than pure platinum does.

1. INTRODUCTION

Bimetallic and multimetallic catalysts have been widely investigated in recent years [1-3]. Characterisation of highly dispersed, supported bimetallic catalysts is a subject of continuous research [3,4]. The most important alloy for catalytic process in the industry is Pt-Re. Platinum-tin catalysts are used quite widely in naphta reforming. From a fundamental point of view, the Pt-Cu, Pt-Au and Pt-Ag combinations are also interesting. The introducing of Cu, Au and Ag modifier into platinum catalyst can cause geometric changes such as site blocking and ensemble formation responsible for isomerization and hydrogenolysis reaction [5,6].

Sinfelt has studied the bimetallic systems of Ru and Os with Cu supported on silica, which were examined in the hydrogenolysis of ethane and the dehydrogenation of cyclohexane to benzene [7]. Copper atoms were found to be built into the Os or Ru phase, causing the formation of new active centres of different reactivity in the hydrogenolysis of ethane.

In the following paper Sinfelt has been reported the results two different effects when studying the Ni-Cu system in the hydrogenolysis of ethane and dehydrogenation of cyclohexane to benzene. The catalytic activity of Ni-Cu alloy in ethane hydrogenolysis decreased rapidly (three orders of magnitude), whereas for dehydrogenation of cyclohexane initial activity increased rapidly and remained on a stable level after reaching 80 wt. % of Cu in the bulk of the alloy [8].

The purpose of this paper was to study the influence of the Cu, Au and Ag additives to platinum catalysts on the course of the hydrogenolysis and isomerization of n-butane.

2. EXPERIMENTAL

2.1. Catalyst preparation

The used support was an γ-alumina (Pechiney, St.Gobain; specific surface: 82.4 m^2/g). The preparation of the 0.37 wt. % Pt/Al_2O_3 parent catalyst was carried out by double impregnation of the alumina support with an aqueous solution of ethylenediaminetetraacetic acid (EDTA), following by H_2PtCl_6 solution [9,10]. After drying, it was calcined in air (823 K, 1h). Bimetallic contacts were prepared by impregnation of the parent Pt/Al_2O_3 catalyst with an aqueous solution of $Cu(NO_3)_2$, $AgNO_3$ and $HAuCl_4$, respectively. All temperature treatments for obtained samples were the same as for the parent catalyst following by reduction with hydrogen (873 K, 1h). Metals loading were measured by X-ray fluorescence (XRF).

2.2. Catalyst testing

Mean crystallite size of metal was determined by X-ray diffraction (XRD) (accuracy ± 30%) and from chemisorption measurements. Total surface areas of the catalysts examined were determined from argon adsorption at liquid nitrogen temperature by the BET method in a volumetric apparatus ensuring a vacuum of at least 10^{-5} mm Hg. The same apparatus was used for the determination of active surface area of the reduced catalysts. The hydrogen adsorption was measured at room temperature, assuming the hydrogen binding stoichiometry to be 1:1 with a hydrogen atom occupying an area of 0.89 nm^2 [11]. The amount of irreversibly adsorbed hydrogen was determined by the method described elsewhere [12]. The mean size of platinum crystallites (d_{Pt}) was calculated from the equation [13]:

$$d_{Pt} = (12.56*10^4)/(S_{Pt} * \gamma_{Pt}) \quad [nm] \qquad (1)$$

where S_{Pt} is the surface area of 1g of platinum in the catalyst (taking into account 100% degree of metal reduction [13] and , γ_{Pt} is the specific density of Pt, assuming that a platinum crystallite touches the carrier with one plane of the cubo-octahedron.

The measurement of the rate of n-butane hydrogenolysis was carried out in an atmosphere of H_2/C_4H_{10} = 6:1, in a gradienless reactor of our design (523-623 K) [9]. The reaction products have been analysed chromatographically (Chrompack CP 9001). Temperature dependencies of the reaction rate as a function of the degree of n-butane conversion have been measured (for respective degree of conversion T = const.). From the functions of reaction rate obtained from the degree of n-butane conversion at given temperature, temperature dependencies of the reaction rate at a constant degree of n-butane conversion have been plotted for various catalyst. The reaction rates have been calculated using the formula applied for this type of reactor:

$$r = [(C_i - C_o)*F]/W_C \quad [mol/g_{cat}*h] \qquad (2)$$

where: r - reaction rate; C_i - final concentration of the product; C_o - initial concentration of the product; F - volumetric reactor effluent rate; W_C - catalyst weight.

3. RESULTS AND DISCUSSION

Some details of the catalyst used in this study are given in Table 1. Specific surface area of the support does not undergo changes, and in all catalyst samples is almost the same. Increasing amount of gold and copper introduced to the parent catalyst causes progressive decrease in an active surface area. In the case of Pt-Ag alumina supported catalysts independently on the amount of silver introduced (in the investigated range of metal concentration) changes in an active surface area are on the same level. Mean platinum crystallite size in the parent catalyst and bimetallic systems is in the rage of 2 nm.

Table 1. Characteristic of the catalysts used in this study (Me = Cu, Ag or Au).

Catalysts	Pt	Me	Active surface area (m^2/g_{cat})	Mean crystallite size (nm)	
	(wt. %)	(wt. %)	S_H	d_H	d_{XRD}
Pt/Al_2O_3		-	6.4	2.02	~ 2
		0.3×10^{-3}	5.8		~ 2.6
		0.8×10^{-3}	5.1		~ 2
$Pt\text{-}Au/Al_2O_3$		1.5×10^{-3}	2.4		> 2
		3.2×10^{-3}	0.9		~ 2
		7.7×10^{-3}	0.2		< 2
		16.2×10^{-3}	0.2		3.2 (± 20%)
		0.7	0.2		~ 2
		0.8	0.2		~ 2
		1.0	0.2		~ 2
		1.0	0.2		< 2
$Pt\text{-}Ag/Al_2O_3$	0.37	1.3	0.2		< 2
		1.3	0.2		> 2
		1.4	0.2		> 2
		1.5	0.2		~ 2
		1.7	0.1		~ 2
		0.2×10^{-3}	6.0		~ 2
		0.3×10^{-3}	5.3		~ 2
		0.8×10^{-3}	4.1		~ 2
$Pt\text{-}Cu/Al_2O_3$		1.6×10^{-3}	3.2		< 2
		3.2×10^{-3}	2.1		~ 2
		15.2×10^{-3}	1.1		< 2
		31.6×10^{-3}	0.3		2.8 (± 24%)

Figure 1 demonstrates the relationship between activity (relative reaction rate $r_{Pt\text{-}Me}/r_{Pt}$ catalysts) and the alloy composition. Activity is defined as the overall reaction rate ratio $r_{Pt\text{-}Me}/r_{Pt}$, $X_{C_4H_{10}}$ = 5% and T = 620 K. Obtained dependencies between relative reaction rate and volumetric composition of the alloys are different than foresee. It was expected a continuous decrease of the reaction rate connected with an increasing amount of the modifier in Pt-Me alloy. An increase of the relative reaction rate at 0.86 atomic ratio of modifier (Ag) to platinum is surprising (Figure 1). The observed phenomena could be explained as follows: in the mentioned range of the alloy composition new surface species are formed which enable the increase of the reaction rate. Based on the analysis of diagrams of the alloys examined it was found that Ag_3Pt phase could be responsible for the observed behaviour [14].

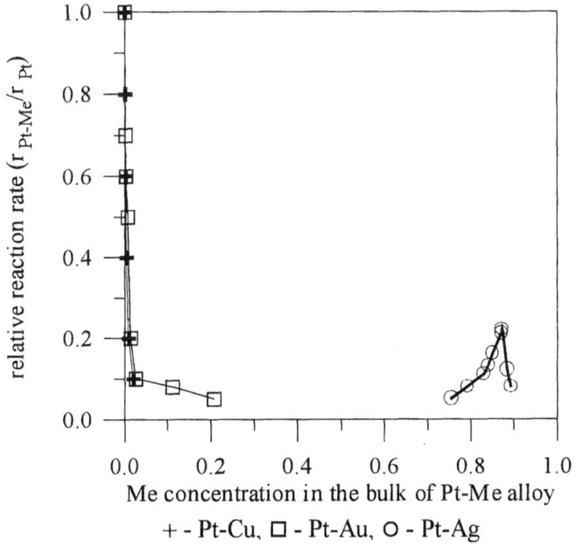

Figure 1. Relationship between activity and Pt-Me alloy composition.

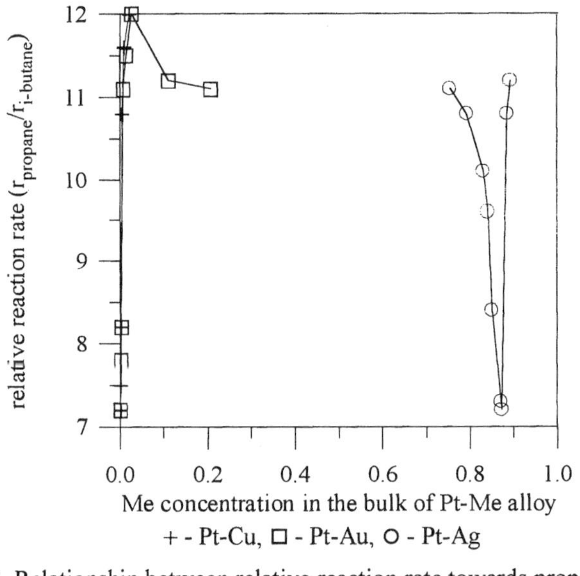

Figure 2. Relationship between relative reaction rate towards propane to i-butane and Pt-Me alloy composition.

Hydrogen activated by platinum can be accepted by the second metal through "spillover" effect. In this case platinum atom could be active in an isomerization reaction of hydrocarbon. Figure 2 shows the relationship between ratio of the reaction rate towards propane to i-butane on the bimetallic catalysts (Pt-Cu, Pt-Ag and Pt-Au) to Pt/Al_2O_3 parent catalyst. Relative reaction rate is defined as the reaction rates ratio towards propane to i-butane on the bimetallic catalysts to Pt/Al_2O_3 catalyst (r_{Pt-Me}/r_{Pt}), $X_{C4H10} = 5\%$ and $T = 620$ K.
Decreasing in the ratio of the reaction rate of propane towards i-butane (in studied range of modifiers concentration) suggests the above described presumption probable. There are literature data that isomerization of hydrocarbon can run over single platinum atom [15]. It seems that Ag_3Pt structure could be formed by the transformation of the existing B_5 centre (primary composite only with platinum atoms). Similar to B_5 centre this new structure consists with low coordinated atoms which could exhibit an increased activity in the "spillover" effect. The Pt-Me catalysts characterised by high selectivity towards propane exhibit also high selectivity towards i-butane and low selectivity towards ethane and methane. The analysis of data included in Table 1 and presented on Figures 1 and 2 leads to the conclusion that the Pt-Me system formed actually involves another system of active centres on the surface than pure platinum does. Slight decrease of the total reaction rate on those catalysts suggests that carbon deposit on the active surface is lower than on the others.

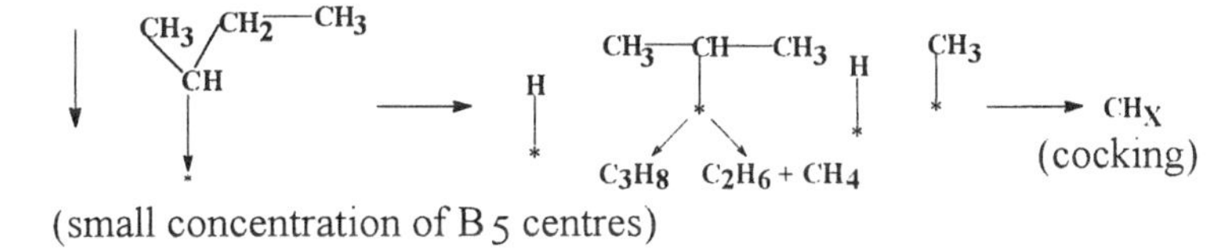

Low dispersion of metal or high concentration of the inert

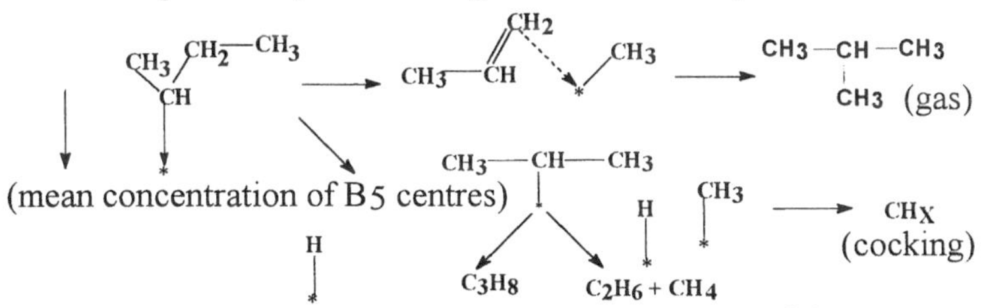

Mean dispersion of metal or medium concentration of the inert

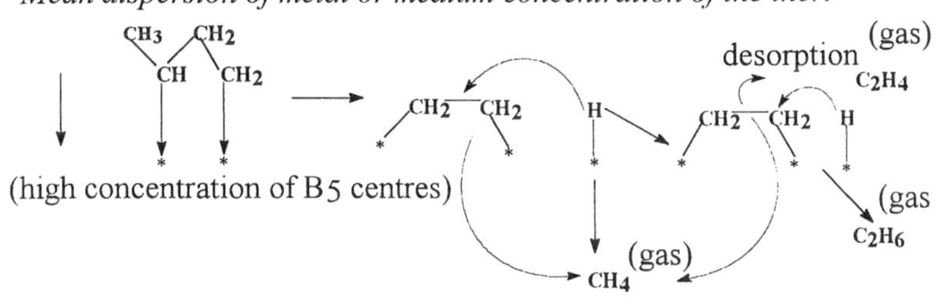

High dispersion of metal or very low concentration of the inert

where: * - *active sites*

From the proposed scheme is resulted that „dilution" of ensembles causes changes in n-butane chemisorption on the alloyed surface of catalysts, decrease of the total catalytic activity connected with an increase towards isomerization as well as carbon deposition.
The number of the new active centres formed (or modified existing sites) is regulated by the presence of Cu, Ag or Au contained in the catalysts, giving a constant change in quality, but different overall activities. These centres probably bind hydrocarbons differently from the surface of pure platinum.
The results obtained suggest the scheme of hydrogenolysis and isomerization of n-butane presented above.

4. REFERENCES

1. V. Ponec, Catal.Rev.-Sci.Eng., 11 (1975) 41.
2. V. Ponec and G.C. Bond (eds.), Catalysis by metals and alloys, Stud.Surf.Sci.Catal., Vol. 95, Elsevier, Amsterdam, 1995.
3. G.C. Bond, Chem.Soc.Rev., 20 (1991) 441.
4. M. Che and C.O. Bennett, Adv.Catal., 36 (1989) 55.
5. H.C. De Jongste, Ph.D. Thesis, Rijskuniversiteit te Leiden, 1980.
6. J.R. Anderson, Sci.Prog.Oxf., 69 (1985) 461.
7. J.H. Sinfelt, J.Catal., 29 (1973) 308.
8. J.H. Sinfelt, Adv.Catal., 23 (1973) 91.
9. D. Nazimek and J. Ryczkowski, Appl.Catal., 26 (1986) 47.
10. J. Ryczkowski, React.Kinet.Catal.Lett., 40 (1989) 189.
11. M.B. Palmer and M.A. Vannice, J.Chem.Tech.Biotech., 30 (1980) 205.
12. A. Daucher, F. Garin and G. Maire, J.Catal., 105 (1987) 233.
13. T.T.T. Wong and W.M.H. Sachtler, J.Catal., 141 (1993) 407.
14. M. Hansen, Constitution of binary alloys, McGraw Hill Book Company, Inc., New York-Toronto-London, 1958.
15. G. Leclercq, A. El Gharbi, L. Gengembre, T. Romero, L. Leclercq and S. Pietrzyk, J.Catal., 148 (1994) 550.

NATURAL GAS CONVERSION V
Studies in Surface Science and Catalysis, Vol. 119
A. Parmaliana et al. (Editors)

Propane Oxidative Dehydrogenation over Low Temperature Rare Earth Orthovanadate Catalysts Prepared by Peroxyl Method

Zhi Min FANG, Jing ZOU, Wei Zheng WENG and Hui Lin WAN*

Department of Chemistry, Institute of Physical Chemistry and State Key Laboratory for Physical Chemistry of the Solid Surface, Xiamen University, Xiamen 361005, CHINA

ABSTRACT

The high purity rare earth orthovanadates were prepared by the peroxyl method, and found to be a kind of the low temperature catalysts for propane oxydehydrogenation effectively at 320℃. The catalytic performances of $LnVO_4$ (Ln=Y, Ce~Yb) were much better than those of rare earth orthovanadate catalysts and also slightly exceeded that of magnesium orthovanadate $Mg_3(VO_4)_2$ reported in literatures. $LnVO_4$ (Ln=Y, Ce~Yb) were tetragonal active phases which could stabilize the existence of active sites for propane oxydehydrogenation. The catalysts with a small amount of $LnVO_3$ reduced from $LnVO_4$ (Ln=Dy~Yb) under reaction atmosphere exhibited better redox properties and catalytic performances possibly due to the existence of biphasic synergy catalysis. $LaVO_4$ was monoclinic unstable active phase, although its bulk structure did not change after reaction, it could not stabilize the existence of surface active sites, leading to the remarkable deactivation of $LaVO_4$ catalyst.

1. INTRODUCTION

The oxidative dehydrogenation of propane (ODP) to propene is one of the potentially important catalytic processes for the effective utilization of light alkanes. The VMgO catalysts which have better catalytic performances for the reaction have given rise to much interest and argument. Kung et al. [1] proposed that the active phase was magnesium orthovanadate ($Mg_3V_2O_8$) , but Volta et al. [2] suggested that magnesium pyrovanadate (α-$Mg_2V_2O_7$) was the active phase, in this phase, V^{4+} ions which was associated to the formation of oxygen vacancies could stably exist, and $Mg_3V_2O_8$ was responsible for the total oxidation due to nonexistence of V^{4+} ions. Based on the fact that the selectivity of $Mg_3V_2O_8$ could be promoted by the coexisting α-$Mg_2V_2O_7$, Delmon et al. [3] and Fang et al. [4-5] suggested that the biphasic synergy catalysis probably existed between $Mg_3V_2O_8$ and α-$Mg_2V_2O_7$ phases. In order to prepare catalysts with better catalytic performances than magnesium vanadates by replacing magnesium with rare earth, very few examples have been reported for the use of rare earth orthovanadates for this reaction, however, their catalytic performances were much lower than magnesium vanadates[6-10]. In this paper the preparation of rare earth orthovanadate catalysts by the peroxyl method is reported, together with characterization and good catalytic performances of the catalysts for the ODP to propene.

2. EXPERIMENTAL

2.1. Catalyst Preparation

The rare earth orthovanadate catalysts were prepared by the peroxyl method(-P). The rare earth oxide (AR) was dissolved in HNO_3 (AR) and H_2O_2 (AR) solution, then added into a transparent aqueous solution of NH_4VO_3 (AR) at 80 ℃. The resulting precipitate was evaporated to dryness while being stirred, and then dried at 120℃ for 1h. The resulting solid was ground into a fine powder and calcined in air at 600℃ for 6h.

2.2. Catalyst Characterization

XRD was performed on a Rigaku Rotaflex D/Max-C using CuKα radiation with λ =1.5406 Å. The target/filter (monochro) was graphite. The voltage was 40kV, and the current was 30mA.

FT-IR spectra were recorded on a PE Spectrum 2000 FTIR spectroscopy at a resolution of $4cm^{-1}$. The detector was MCT-B. The sample was diluted with KBr (5/100 by weight) and pressed into a thin wafer.

2.3. Catalyst Evaluation

The catalyst evaluation was performed in a fixed bed quartz microreactor in a flow system. Normally, the flow rate of feed was 60ml/min with the ratio of C_3H_8:O_2 being 2:1, the weight of catalyst was 0.30g with particle fractions between 60 to 100 meshes, and the reaction temperature was 320 and 500℃, respectively. The reaction products were analyzed by a on-line 102-GD gas chromatograph with squalane/active alumina column for C_3H_8, C_3H_6, C_2H_4,C_2H_6 analysis, and 601 carbon molecular sieve column for O_2, CO, CO_2, CH_4 analysis.

3. RESULTS AND DISCUSSION

3.1. Catalyst Characterization

3.1.1. XRD

$LaVO_4$ was monoclinic with trace V_2O_5 impurity. The other rare earth orthovanadates $LnVO_4$ (Ln=Y, Ce~Yb) was tetragonal with trace V_2O_5 impurity (for Ln=Y, Ce, Pr, Nd, Eu, Gd, Tb, Ho) and trace corresponding rare earth oxide impurity (for Ln=Y, Ce, Pr, Nd, Eu, Tb, Dy, Er, Yb), as shown in Fig.1. These indicated that high purity rare earth orthovanadates could be prepared by the peroxyl method. After reaction the trace V_2O_5 impurity vanished and the crystal structures of the most catalysts did not change except that a small amount of $LnVO_3$ phases reduced from $LnVO_4$ phases appeared in $LnVO_4$ (Ln=Dy, Ho, Er, Tm, Yb) catalysts and increased with the atomic number of rare earth, as shown in Fig.2.

3.1.2. FTIR

Six well resolved peaks at 781, 801, 819, 835, 850 and $880cm^{-1}$ appeared in the strongest VO_4 absorbance band of $LaVO_4$ with low monoclinic symmetry (Fig.3). $LaVO_4$ contained small V_2O_5 impurity at $1025cm^{-1}$. The strongest and broad unresolved VO_4 absorbance bands of the other rare earth orthovanadates with high tetragonal symmetry appeared around $791\sim816cm^{-1}$, and on the whole, their wavenumber increased with the atomic number of rare earth. Some $LnVO_4$ (Ln=Ce, Pr, Nd, Eu, Gd, Tb) contained trace V_2O_5 impurity at $1017\sim1025cm^{-1}$.

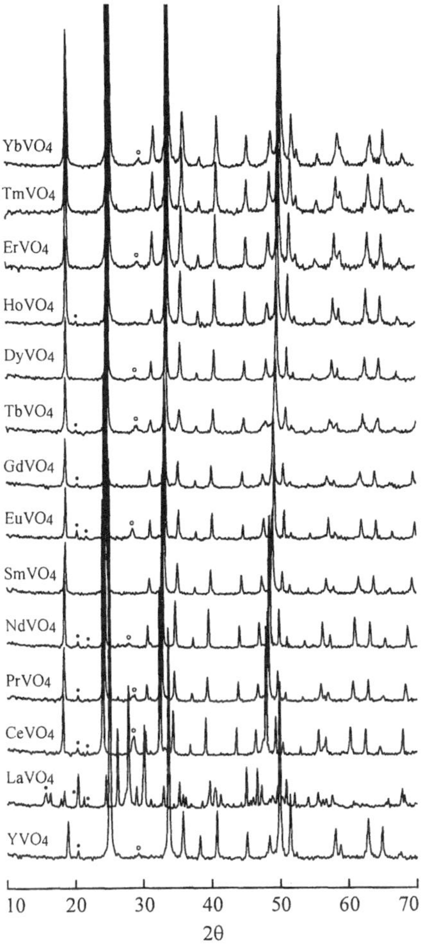

Fig.1 XRD spectra of fresh $LnVO_4$-P catalysts
(·) V_2O_5, (∘) Ln_xO_y

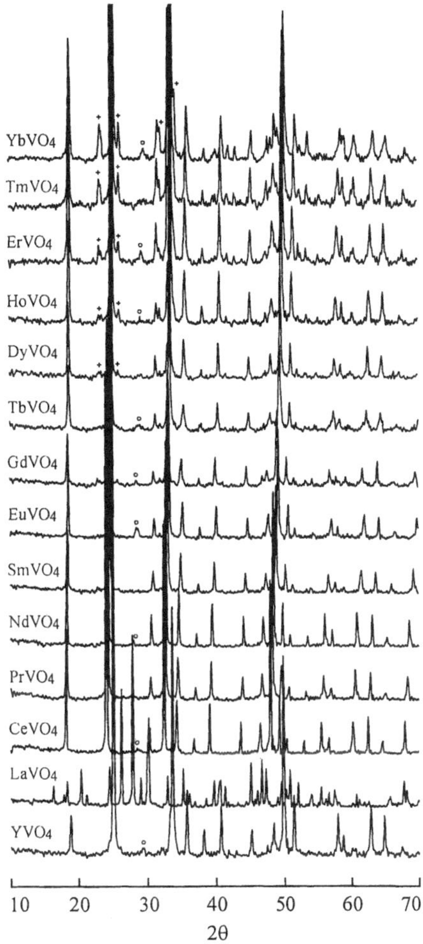

Fig.2 XRD spectra of used $LnVO_4$-P catalysts
(∘) Ln_xO_y, (+) $LnVO_3$

3.2. Catalytic evaluation

3.2.1. Catalytic performance

Table 1 lists the performances of propane oxydehydrogenation over rare earth orthovanadates. The catalytic performance of $LaVO_4$ decreased remarkably within 40min reaction with propene yield from 9.4 to 0.77% (Table 2). The other $LnVO_4$ (Ln=Y, Ce~Yb) catalysts were quite active, selective and stable for the propane oxydehydrogenation during 4h reaction (Table 3), and the propane conversions of 19.0~23.2% with the propene selectivities of 47.5~52.2% and the propene yields of 9.0~11.6% were achieved. These results were much better than those over rare earth orthovanadate catalysts reported in the literature [6-10]. Among the catalysts in Table 1, the propene yields over YVO_4, $SmVO_4$, $TbVO_4$, $HoVO_4$ and $TmVO_4$ exceeded the propene yield of 11.2% over magnesium orthovanadate $Mg_3(VO_4)_2$ with propane conversion and propene selectivity 24.0% and 46.6%, respectively [3].

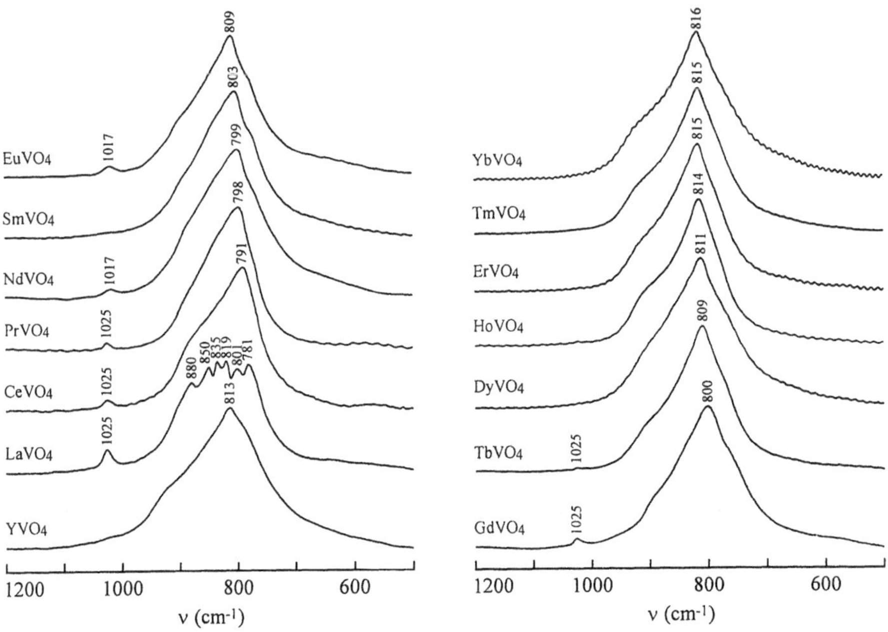

Fig.3 FTIR spectra of fresh $LnVO_4$-P catalysts

Table 1 Catalytic performances of rare earth orthovanadates for propane oxydehydrogenation

Catalyst	Conversion (%)	Selectivity (%)					Yield (%)
	C_3H_8	C_3H_6	Crack	CO	CO_2	CO_x	C_3H_6
EQR	0.4	85.0	3.2	4.0	7.8	11.8	0.3
V_2O_5	17.3	32.2	1.2	38.0	28.6	66.6	5.6
YVO_4	23.2	48.7	10.7	27.0	13.6	40.6	11.3
$LaVO_4$	0.9	80.5	4.5	4.8	10.2	15.0	0.7
$CeVO_4$	19.1	47.9	4.3	29.4	18.4	47.8	9.1
$PrVO_4$	19.6	47.7	4.8	27.4	20.1	47.5	9.3
$NdVO_4$	19.0	47.4	4.5	28.2	19.9	48.1	9.0
$SmVO_4$	22.3	52.2	8.5	26.0	13.3	39.3	11.6
$EuVO_4$	19.1	49.2	7.2	30.8	12.8	43.6	9.4
$GdVO_4$	21.3	47.5	10.2	27.9	14.4	42.3	10.1
$TbVO_4$	23.0	50.2	13.8	22.3	13.7	36.0	11.5
$DyVO_4$	21.9	50.6	9.8	26.7	12.9	39.6	11.1
$HoVO_4$	23.1	49.7	13.6	25.0	11.7	36.7	11.5
$ErVO_4$	22.7	48.4	13.2	22.2	16.2	38.4	11.0
$TmVO_4$	21.8	51.8	8.8	25.6	13.8	39.4	11.3
$YbVO_4$	21.8	49.1	10.5	24.2	16.2	40.4	10.7

500℃; SV=12000ml $g^{-1}h^{-1}$; C_3H_8:O_2=2:1; Crack=CH_4+C_2H_4+C_2H_6; CO_x=CO+CO_2;
EQR: Empty quartz reactor.

Table 2 Catalytic performance of $LaVO_4$ catalyst for propane oxydehydrogenation

Time on stream	Conversion (%)	Selectivity (%)					Yield (%)
(min)	C_3H_8	C_3H_6	Crack	CO	CO_2	CO_x	C_3H_6
20	22.2	42.4	15.9	24.4	17.3	41.7	9.4
40	0.95	81.4	2.6	5.3	10.7	16.0	0.77
60	0.93	80.8	3.5	5.0	10.7	15.7	0.75
120	0.90	80.6	4.1	4.9	10.4	15.3	0.73
240	0.89	80.5	4.5	4.8	10.2	15.0	0.72

500℃; SV=12000ml $g^{-1}h^{-1}$; C_3H_8:O_2=2:1; CO_x=CO+CO_2; Crack=CH_4+C_2H_4+C_2H_6.

Table 3 Catalytic performance of $SmVO_4$ catalyst for propane oxydehydrogenation

Time on stream	Conversion (%)	Selectivity (%)					Yield (%)
(min)	C_3H_8	C_3H_6	Crack	CO	CO_2	CO_x	C_3H_6
20	22.6	50.4	9.3	26.4	13.9	40.3	11.4
40	22.6	51.2	8.9	26.2	13.7	39.9	11.6
60	22.5	51.4	8.8	26.1	13.7	39.8	11.6
120	22.3	51.7	8.7	26.1	13.5	39.6	11.5
240	22.3	52.2	8.5	26.0	13.3	39.3	11.6

500℃; SV=12000ml $g^{-1}h^{-1}$; C_3H_8:O_2=2:1; CO_x=CO+CO_2; Crack=CH_4+C_2H_4+C_2H_6.

3.2.2. Low temperature catalytic performance

Table 4 lists the catalytic performances of rare earth orthovanadates for propane oxidative dehydrogenation at 320℃. The results indicated that rare earth orthovanadate catalysts could catalyze propane oxydehydrogenation effectively at 320 ℃, considerably lower than the temperature of 450 ℃ for VMgO catalyst [1-5]. Therefore, the rare earth orthovanadate catalysts prepared in this work were a kind of low temperature catalysts, and may be of practical significance. Moreover, it may be possible to extend the study of these catalysts to iso-butane oxydehydrogenation to iso-butene and propane selective oxidation to acrolein at lower temperature to avoid cracking and the total oxidation of acrolein over $LnVO_4$ based catalysts.

4. CONCLUSION

(1) The high purity rare earth orthovanadates could be prepared by the peroxyl method, and were a kind of low temperature catalysts. These catalysts could catalyze propane oxydehydrogenation effectively at 320℃, and may be of practical significance.

(2) The catalytic performances of $LnVO_4$ (Ln=Y, Ce~Yb) at 500℃ were much better than those of rare earth orthovanadate catalysts and slightly exceeded that of magnesium orthovanadate $Mg_3(VO_4)_2$ reported in the literatures.

(3) $LnVO_4$ (Ln=Y, Ce~Yb) was tetragonal active phases which could stabilize the existence of active sites for propane oxydehydrogenation. The catalysts with a small amount of $LnVO_3$ reduced from $LnVO_4$ (Ln=Dy~Yb) under reaction atmosphere presented better redox properties and catalytic performances probably due to the existence of biphasic synergy catalysis. $LaVO_4$ was monoclinic unstable active phase, although its crystal structure did not

change after reaction, it could not stabilize the surface active sites and leaded to the remarkable deactivation of $LaVO_4$ catalyst.

Table 4 Low temperature catalytic performances of rare earth orthovanadates

Catalyst	Conversion (%)	Selectivity (%)					Yield (%)
	C_3H_8	C_3H_6	Crack	CO	CO_2	CO_x	C_3H_6
YVO_4	21.9	46.6	9.6	28.3	15.5	43.8	10.2
$LaVO_4$	0.7	69.0	4.1	12.0	14.9	26.9	0.5
$CeVO_4$	18.0	46.3	3.9	29.8	20.0	49.8	8.3
$PrVO_4$	18.5	46.2	4.1	29.2	20.5	49.7	8.5
$NdVO_4$	18.0	46.1	4.0	29.6	20.3	49.9	8.3
$SmVO_4$	21.6	48.1	7.7	28.5	15.7	44.2	10.4
$EuVO_4$	18.3	47.0	6.6	31.2	15.2	46.4	8.6
$GdVO_4$	20.6	43.7	9.2	29.2	17.9	47.1	9.0
$TbVO_4$	22.1	46.6	12.7	24.3	16.4	40.7	10.3
$DyVO_4$	20.9	47.8	9.1	28.3	14.8	43.1	10.0
$HoVO_4$	22.0	46.8	12.4	27.0	13.8	40.8	10.3
$ErVO_4$	21.9	45.2	12.1	24.2	18.5	42.7	9.9
$TmVO_4$	20.8	48.6	8.0	27.7	15.7	43.4	10.1
$YbVO_4$	20.9	45.9	9.7	26.3	18.1	44.4	9.6

320℃; SV=12000ml $g^{-1}h^{-1}$; C_3H_8:O_2=2:1; CO_x=CO+CO_2; Crack=CH_4+C_2H_4+C_2H_6.

ACKNOWLEDGMENTS

The authors gratefully acknowledge the support of this work by the National Natural Science Foundation of China and the Doctoral Foundation of Education Committee of China.

REFERENCES

1. M. A. Chaar, D. Patel, H. H. Kung, J. Catal., 109 (1988) 463.
2. A. Guerrero-Ruiz, I. Rodriguez-Ramos, J. L. G. Fierro, V. Soenen, J. M. Herrmann, J. C. Volta, Stud. Surf. Sci. Catal., 72 (1992) 203.
3. X. T. Gao, P. Ruiz, Q. Xin, X. X. Guo, B. Delmon, J. Catal., 148 (1994) 56.
4. Z. M. Fang, W. Z. Weng, H. L. Wan and K. R. Tsai, J. Mol. Catal. (China), 9 (1995) 401.
5. Z. M. Fang, W. Z. Weng, H. L. Wan and K. R. Tsai, Chinese Sci. Bull., 42 (1997) 172.
6. A. Corma, J. M. Lopez-Nieto, N. Paredes, M. Perez, Y. Shen, H. Cao, S. L. Suib, Stud. Surf. Sci. Catal., 72 (1992) 213.
7. J. Castiglioni, P. Poix, R. Kieffer, Stud. Surf. Sci. Catal., 75 (1993) 2309.
8. B. Zhaorigetu, R. Kieffer, J. -P. Hindermann, Stud. Surf. Sci. Catal., 101 (1996) 1049.
9. C. T. Au, W. D. Zhang, H. L. Wan, Catal. Lett., 37 (1996) 241.
10. C. T. Au, W. D. Zhang, J. Chem. Soc. Faraday Trans., 93 (1997) 1195.

NATURAL GAS CONVERSION V
Studies in Surface Science and Catalysis, Vol. 119
A. Parmaliana et al. (Editors)

Propane catalytic oxidation and oxy-dehydrogenation over manganese-based metal oxides.

M. Baldi[a], V. Sanchez Escribano[b], J. M. Gallardo Amores[c], F. Milella[d], E. Finocchio[d], C. Pistarino[d] and G. Busca[d]

[a] Dipartimento di Ingegneria Idraulica e Ambientale, Università, I-27100 Pavia (Italy)
[b] Departamento de Quimica Inorgánica, Universidad, E-37008-Salamanca (Spain)
[c] Departamento de Quimica Inorgánica, Universidad Complutense, E-28040 Madrid (Spain)
[d] Istituto di Chimica, Facoltà di Ingegneria, Università, I-16129 Genova (Italy)

Pure Mn oxides (Mn_3O_4 and α-Mn_2O_3) and Mn-Fe and Mn-Al mixed oxides have been prepared by (co)precipitation methods, thoroughly characterized and tested as catalysts for propane oxy-dehydrogenation to propene using He-O_2-propane mixtures, using different flows, different contact times and different propane to oxygen molar ratios. These materials, due to their very high reactivities, allow propene production with yields not higher than 6-8 % but with productivities in excess of 3 kg_{C3H6} / kg_{CAT} h and in a quite low temperature range.

1. INTRODUCTION

The catalytic oxy-dehydrogenation of light alkanes to olefins represents a promising process for the utilization of natural gas and a concomitant limitation of the oil consumption. Total combustion to carbon oxides and also the production of oxygenates are main by-reactions affecting selectivity to oxydehydrogenation products. Some of such by-reactions likely arise from successive overoxidation of propene, which is more reactive than propane in oxidizing atmospheres. However, also competitive ways to products could also exist. Vanadium based catalysts such as Mg-vanadates, alkali-doped vanadia-titania and vanadia-alumina, vanadium-tin oxides and V-silicalite appear to be among potential catalyst canditates for propene synthesis from propane oxy-dehydrogenation [1,2]. In fact, vanadium oxides are among the more selective ones towards partial oxidation products although their activity is quite low [3].

To have more information on the selectivity behavior of transition metal oxides in propane oxidation catalysis and to test possible alternative approaches, we undertook a research on propane conversion over more easily reducible transition metal oxides, such as Mn oxides. Manganese oxides (MnO_2, Mn_2O_3 and Mn_3O_4) were found previously to be very active and stable catalysts for the combustion of organic compounds in the temperature range 373-773 K. On the other hand we observed that Mn_3O_4 also catalyzes selective oxidations, such as the oxydehydrogenation of alcohols to the corresponding carbonyl compounds [4] and gives rise to propene among the products of propane oxidation at low conversions [5]. These materials appeared to be good candidates to investigate the mechanisms involved in selective/unselective oxidations. Moreover, we undertook attempts to improve the selectivity of Mn oxides to partial oxidation products by mixing them with other oxides such as Fe and Al oxides.

2. EXPERIMENTAL

2.1. Catalyst preparation and characterization. The preparation of Mn-Al catalysts was carried out by dissolving $Al(NO_3)_3 \cdot 9H_2O$ and $(CH_3COO)_2Mn \cdot 4H_2O$ in water. Precipitation of pure compounds and coprecipitation of Mn-Al catalyst with atomic ratio 2:1 (M2A1), 1:1 (M1A1) e 1:2 (M1A2) were obtained rising the the pH up to 9.5 with NH_3 and ageing the mixture 24^h at 333 K. After separation by centrifugation, the catalysts were dried at 383 K and calcined at 723 K. The preparation of Mn_2O_3, Fe_2O_3 and Mn_2O_3-Fe_2O_3 mixtures was carried out by coprecipitation in ammonium carbonate water solution of manganese carbonate and iron nitrate, followed by ageing, filtering, drying and calcining at 1073 K for 3 h. The materials were characterized by means of DTA-TG, XRD, surface area, and elemental analyses by ICP.

Table 1: Physico-chemical characteristics of the catalysts tested in the present work.

Catalyst	Surface area m^2/g	Meas. atomic ratio Mn/Fe(Al)	XRD
Fe_2O_3	8	0	α-Fe_2O_3
M1F2	12	0,54	α-Fe_2O_3, α-Mn_2O_3
M1F1	11	1,17	α-Fe_2O_3, α-Mn_2O_3
M2F1	10	2,23	α-Fe_2O_3, α-Mn_2O_3
Mn_2O_3	9	∞	α-Mn_2O_3
Mn_3O_4	24	∞	α-Mn_3O_4
M2A1	64	1,85	α-Mn_3O_4
M1A1	106	0,89	α-Mn_3O_4, α-Mn_2O_3, γ-Al_2O_3
M1A2	92	0,51	α-Mn_2O_3, Mn_5O_8 ?, γ-Al_2O_3
Al_2O_3	70	0	γ-Al_2O_3

2.2. Flow reactor studies. Catalytic tests were carried out at atmospheric pressure in a continuous flow tubular glass fixed bed reactor. The catalyst (60-70 mesh) was mechanically mixed with calculated amounts of inert, low surface area material (quartz), to avoid preferential gas flow paths and hot spots in the catalytic bed. Different materials were tested comparing the activities of catalysts in the same amounts and in the same surface areas conditions (see table 1). The total gas flow was 350 cm^3/min and the partial pressures of C_3H_8, O_2 and He, in the above experimental runs, were 0.10, 0.05 and 0.85 atm, respectively, for the experiments performed with Mn-Al catalysts. The experiments on Mn-Fe catalysts were performed with 300 cm^3/min of 0.015, 0.09 and 0.905 atm of C_3H_8, O_2 and He, respectively.

Reactants and the reaction products were analyzed using two gas chromatographs (HP 5890) on line. Six-port valves, with a 5 ml loop, have been used as reproducible gas sampling injectors. Permanent gases (O_2, CO and CO_2) were separated using a Carbosieve S-II packed column (100-120 mesh) connected to a TCD detector. Other reaction products, as well as the organic reactant, were analyzed employing a PLOT fused silica column with Al_2O_3/KCl as stationary phase, connected to a FID detector. Only traces were sometimes found of the formation of alcohol or carbonyl compounds in a detectable amounts.

3. RESULTS AND DISCUSSION

3.1 Catalytic behavior of pure Mn oxides.

The spinel type compound Mn_3O_4 (hausmannite) is slightly more active than α-Mn_2O_3 (bixbyite) in propane oxidation. Conversion starts to be observable at 450 K and it is complete below 700 K. In both cases the oxidation is totally unselective in excess of oxygen and conversion is total, but high selectivities can be obtained when oxygen to hydrocarbon ratio is in the range 1-0.5 and conversion is low. Yields of about 6 % can be obtained in optimal conditions. In oxygen defect, propene selectivity ranges from 30 to 20 % on Mn_3O_4, and levels near 20 % even at total oxygen conversion; in excess oxygen, propene completely disappears at temperatures at which propane conversion is total (figure 1). Propane combustion gives rise to byproducts like ethylene and traces of methane (cracking products) and of 2-propanol, but no CO is observed, even at low temperatures and conversions.

3.2 Reaction scheme and kinetics of propane oxidation on Mn oxides.

The activation energies for propane oxidation on Mn_3O_4 and Mn_2O_3 were measured to be in the range 19-20 kcal/mol. They fall within the range observed for propane combustion on other catalysts and show that experiments are performed under chemical kinetic control. The reaction was found to be first order with respect to propane. Kinetic measures [6] can be interpreted assuming that the above activation energies are related to the first C-H bond breaking (assumed to be rate-determining) and that the reaction path splits into two ways. A common intermediate species has been identified (in agreement with IR spectroscopic data) to be a 2-propoxide species that can convert either to propene or to CO_2 via adsorbed acetone. Propene over-oxidation can also give rise to CO_2 and to CO (scheme I). The predominant propane combustion way is in defect oxygen through propene and in excess oxygen through adsorbed acetone.

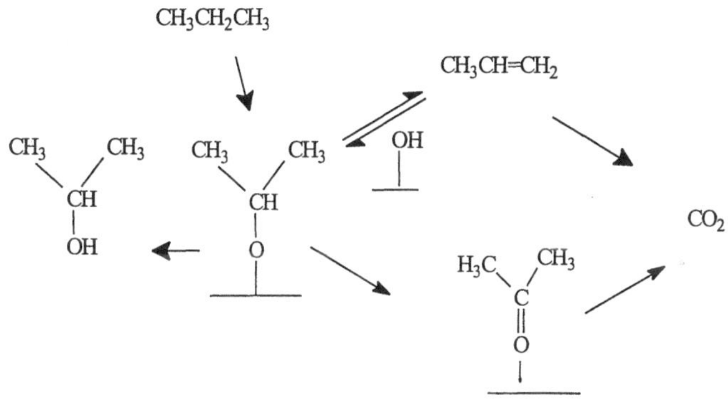

Scheme I. Proposed propane oxidation pathway.

3.3. Behavior of Mn-Fe complex oxides.

Catalysts composed by mixtures of solid solutions of α-Mn_2O_3 (bixbyite) and α-Fe_2O_3 (haematite) are more active than pure Mn oxides in propane conversion. These materials start to burn propane in the range 450-550 K showing some synergy occurring between iron and manganese oxide components. M2F1 appears to be the most active one. The addition of Fe tends to favor the way through propene, so increasing selectivity to propene. At high Fe content selectivity to propene increases very much even at low conversions but also conversion decreases. The activation energies measured on M2F1 and M1F1 are, within experimental error, the same as measured for pure Mn oxides. On the contrary, significantly higher values are measured for catalysts with higher Fe contents (22,9 kcal/mol for M1F2 and 23,5 kcal/mol for ferric oxide). In all cases CO_2 is the main reaction product.

3.4. Propane oxy-dehydrogenation over Mn-Al mixed oxides.

A series of experiments has been performed on MA samples, by varying both the Mn/Al atomic ratio and the contact times, as described in Table 2. In Figure 1 the selectivities to propene measured during the experimental runs A-L in the temperature range 450 - 700 K are reported. In all cases the propene selectivities decrease by increasing reaction temperatures (and propane conversion) down to a nearly stable value of 25-35%; in parallel, selectivities to CO_2

Table 2: Experimental conditions of the catalytic runs.

Run	Catalyst	Amount / g	Surface area / m^2	Contact time / h g_{cat} l^{-1}
A	Mn_3O_4	0,200	4,8	0,0095
B	M2A1	0.200	12.8	0,0066
C	M2A1	0.075	4.8	0,0036
D	M2A1	0,038	2,4	0,0018
E	M1A1	0.200	21.2	0,0066
F	M1A1	0.045	4.8	0,0021
G	M1A1	0.023	2.4	0,0011
H	M1A2	0,200	18,4	0,0095
I	M1A2	0,053	4,8	0,0025
L	M1A2	0,027	2,4	0,0013
M	Al_2O_3	0,200	14,0	0,0095

(the other main product), and to ethene and methane grow up. In the case of the high Al_2O_3 content catalysts a relevant production of CO is also found, and selectivity to C_2H_4 becomes significant (up to 11%). At temperatures above those needed to totally consume oxygen, the selectivities raise again, slightly. This is particularly evident in the case of Mn_3O_4, M2A1 e M1A1 at the highest contact times (runs A, B and E). In this range the selectivities to CO_2 begin to decrease and CO is formed. We can suppose that this is due to a reaction between propane and CO_2 to give propene, CO and water, where CO_2 acts as an oxidant with respect to propane.

In Fig. 2 the yield obtained are reported. The best yields have been obtained with the catalyst M1A1 and the lowest contact time, in agreement with our conclusion that propene selectivity is mainly limited (in defect oxygen) by overoxidation. We never measured higher yields than 7.5 % over our Mn-Al based catalysts. Our materials work in a lower temperature range than those based on vanadium [2], but give definitely lower propene yields..

In Fig.3 the selectivities to propene are plotted against propane conversions (with variable temperature conditions and contact times). As expected the selectivities decrease by increasing conversion. After oxygen total consumption all MA catalysts level to selectivities in the range 30-35 % except pure Mn oxide, that is the least selective.

In Fig. 4 the productivities obtained on our MA catalysts are reported. We observe values of above 3 kg_{C3H6}/ kg_{CAT} h with M1A1 and M1A2 and the lowest contact times. These values are definitely higher than the best ones reported in the review by Albonetti et al. [2]. This is in fact due to the high flows and the very low amounts of catalyst we used in our experiments, in spite of quite low yields. This, in agreement with the low temperature range we work with, underline the very high catalytic activity of Mn-based oxides, much higher than those of V-based oxides.

In Fig. 5 we plotted the selectivities to propene as a function of the Mn / Mn+Al atomic fraction in the catalyst, as measured in the conditions of highest conversions (12-17 %), using the same total catalyst surface areas. These data show that the addition of Al allows a progressive increase in selectivity to propene, while a maximum in productivity is likely existing at Mn / Mn + Al = 0,5 (M1A1). On the other hand, the propene productivity increases significantly by decreasing the total surface area and the contact time, as typical for an intermediate.

The effect of Al can be interpreted as due to its favoring effect on the way through propene in propane oxidation, with respect to the way to total oxidation via adsorbed acetone. This can be related to the increase of catalyst acidity that tends to favor the elimination of propene from 2-propoxides with respect to its further oxidation to acetone.

Figure 1:Propane selectivities vs. Temperature

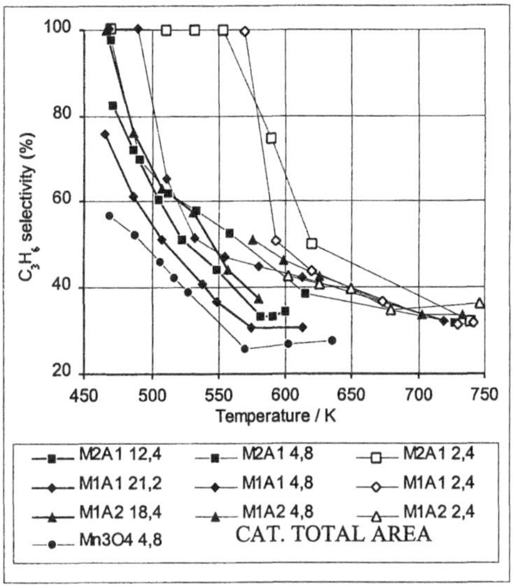

Figure 2: Propene yields vs. Temperature

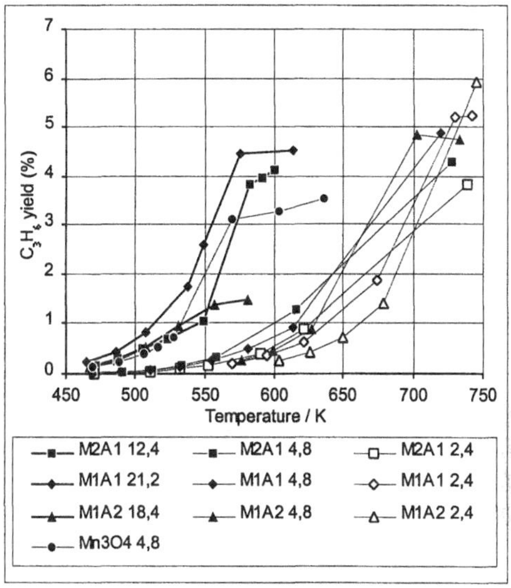

Figure 3: Propene selectivities vs. Propane Conversion

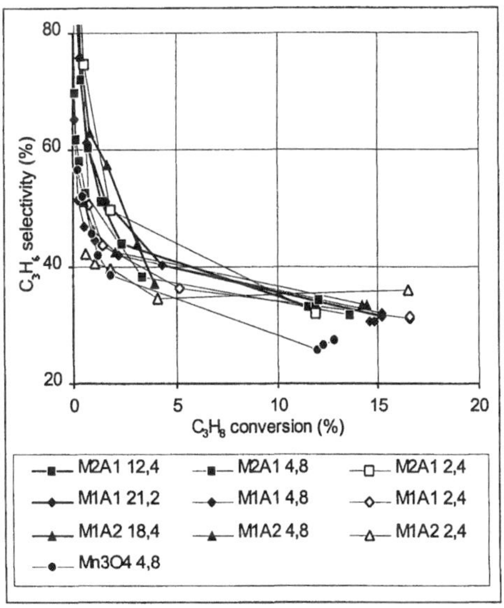

Figure 4: Propane productivities vs. temperature

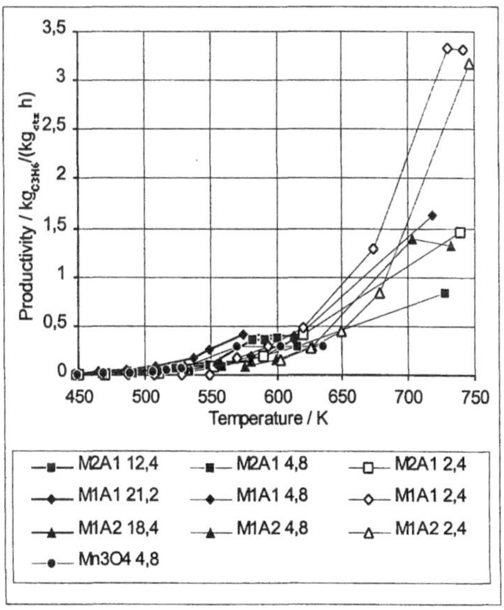

Figure 5: Selectivities and Productivities to propene at total oxygen consumption vs. catalyst composition.

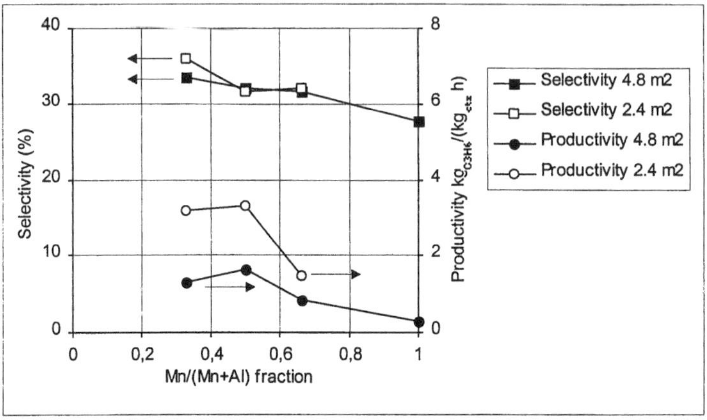

4. CONCLUSIONS

Mn- based oxides are very active materials in the oxidation of propane. Working with propane to oxygen ratios ratios near 2 : 1 (stoichiometric conditions for the oxidative dehydrogenation to propene) propene can actually be produced with yields in the range 6-8 %. However, the productivities found, above 3 Kg_{C3H6}/Kg_{CAT} h, appear to be interesting, in comparison with those reported for the best catalysts cited in the literature. Additionally, these results can be obtained at a quite low temperature, in the range 650-750 K.

ACKNOWLEDGEMENTS. This work has been supported in part by NATO (CRG grant n 960316). JMGA acknowledges the MEC (Spain) for a post-doctoral FPI grant.

REFERENCES

1 H.H. Kung, Advan Catal., 40 (1994) 1

2 S. Albonetti, F. Cavani and F. Trifirò, Catal. Rev. Sci. Eng. 38 (1996) 413

3 "Vanadia catalysts for selective oxidation of hydrocarbons and their derivatives", B. Grzybowska-Swierkosz, F. Trifirò and J.C. Vedrine eds., Appl. Catal. A General, 157 (1997) 1.

4 M. Baldi, F. Milella, G. Ramis, V. Sanchez Escribano and G. Busca, Appl. Catal. A General, 166 (1998) 75

5 M. Baldi, E. Finocchio, F. Milella, G. Busca, Appl. Catal. B Environ., 16, 43 (1998)

6 M. Baldi, E. Finocchio, C. Pistarino and G. Busca, Appl. Catal. A General, in press

NATURAL GAS CONVERSION V
Studies in Surface Science and Catalysis, Vol. 119
A. Parmaliana et al. (Editors)

Oxidative dehydrogenation of ethane over Pt and Pt/Rh gauze catalysts at very short contact times.

R. Lødeng[2], O.A. Lindvåg[2] , S. Kvisle[3] , H. Reier-Nielsen[4] , A. Holmen[1].

[1]Department of Industrial Chemistry, Norwegian University of Science and Technology (NTNU), N-7034 Trondheim, Norway. [2]SINTEF Applied Chemistry, N-7034 Trondheim, Norway.

[3]Norsk Hydro, Research Center, Box 2560, N-3901 Porsgrunn, Norway. [4]Norsk Hydro, Bygdøy alle 2, N-0257 Oslo, Norway.

The oxidative dehydrogenation of C_2H_6 has been studied at contact times of $5 \cdot 10^{-5}$ s using Pt or Pt/Rh gauze catalysts. The reaction was studied at gauze temperatures of 600-1050 °C and 1 bar pressure. The C_2H_6/O_2 ratio was varied between 1.3 and 2.5 and Ar and H_2O were used as diluents. The C_2H_4 selectivities obtained were always higher in the empty reactor (without any catalyst) or with only monoliths (as radiation shields), approaching the same values as in the presence of a gauze at conversion levels above 55-60 %. The gauze provides an extremely rapid heating of the reactants due to the increased importance of combustion, but the oxidative dehydrogenation reactions seem to take place mainly in the gas phase. The product distribution did not depend to any large extent on the composition of the gauze but minor differences in properties could be hidden due to the importance of the gas phase reactions close to the gauze. The byproduct synthesis gas was typically produced with a H_2/CO ratio close to 1.5 at severe conditions favouring the formation of C_2H_4.

1. INTRODUCTION

Lower olefins such as ethene and propene are mainly produced by steam cracking of hydrocarbons [1]. The formation of olefins is favoured by short contact times and in consequence by high temperatures. Propene and butenes are also produced commercially by catalytic dehydrogenation [2] and by catalytic cracking in oil refineries [3] .

It has recently been shown [4-6] that high selectivities of C_2-C_4 olefins can be obtained by oxidative dehydrogenation at very short contact times using noble metal impregnated monoliths or metal gauzes. The short contact time reactor holds the promise of greatly reducing the reactor volume as well as autothermal operation. It is proposed [5] that the catalytic reactor operates by rapid chemical heating followed by rapid quenching to avoid homogeneous reactions. It was also observed that a single Pt gauze produced very different selectivities from coated monoliths or multiple gauzes with parent olefin, formaldehyde and acetaldehyde being the most important partially oxidated products [5].

Previous experimental studies of methane partial oxidation on Pt and Pt-Rh gauzes [7,8] have indicated that the reactions in fact mainly take place on the metal surface. However, ethane and propane are much more reactive than methane and it is more difficult to avoid homogeneous reactions. The present work was initiated in order to study the potential of the short contact time reactor for oxidative dehydrogenation of ethane and in particular to study the relation between surface reactions and the possible gas phase reactions.

2. EXPERIMENTAL

The oxidative dehydrogenation was studied in different quartz reactors at atmospheric pressure and high temperatures. The experimental set-up has been described in detail elsewhere [7]. A circular gauze catalyst (diameter 15 mm) was held between two cylindrical pieces of ceramic monoliths (Cordierite from Corning, 400 cells/in^2). The catalyst consisted of only one layer of Pt or Pt-Rh gauze although one series of experiments with 5 layers of the gauze (diameter 4 mm) has also been carried out. The gauzes (1024 meshes/cm^2) were woven from wires of 60 μm diameter and the Pt-Rh gauze contained 10 wt-% Rh. The characteristics of the gauze catalyst are the same as for gauzes used in ammonia oxidation plants [9]. The contact time over the gauze as well as the gas phase residence time in the hot zone of the reactor have been varied by applying different reactor diameters (15 and 4 mm) and total flow rates. If nothing else is given the experiments were carried out using monoliths and a single gauze with a diameter of 15 mm.

Experiments were also performed without the two monoliths acting as radiation shields and with an empty reactor. The quartz reactor was heated in an electrical furnace and the axial temperature profile was measured by a moveable thermocouple (chromel-alumel protected by a quartz tube). Reliable temperature measurements at high temperatures are always very difficult. However, comparisons with an optical pyrometer [8] showed almost identical reactor (i.e. gauze) temperatures. Conversion and selectivities were obtained by on-line GC analysis of the reactor effluent [7].

The ratio of C_2H_6/O_2 in the feed was varied in the range 1.3-2.5. The feed was diluted with Ar and in some experiments also with H_2O. The experiments were performed in the temperature range 600-1050 °C. The residence time in the reactor is rather difficult to define since the reaction mixture is exposed to high temperatures in front as well as behind the gauze. Whenever a value is given for the residence time, it is calculated from the free volume of the gauze and the gas flow rate at a representative reaction temperature.

3. RESULTS AND DISCUSSION

The reaction was readily ignited by the gauze catalyst at an oven temperature of 650 °C. For the empty reactor at similar conditions an oven temperature of about 800 °C was required for obtaining even a C_2H_6 conversion in the order of 3-4 %.

The main products from the oxidative dehydrogenation of C_2H_6 at very short contact times were C_2H_4, CO, H_2 and H_2O with CH_4, C_2H_2, CO_2 and traces of C_3-C_6 hydrocarbons as

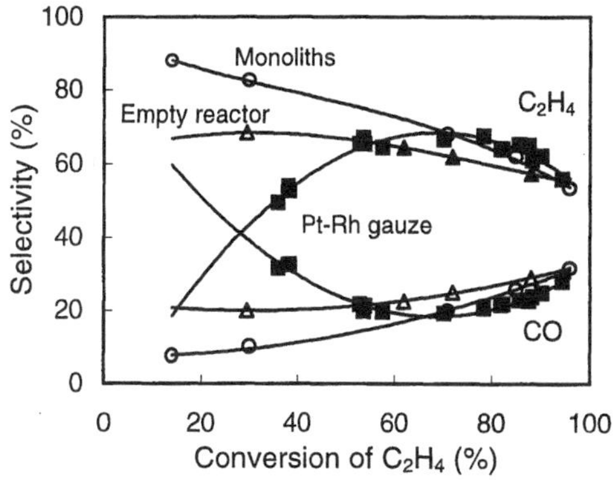

Figure 1. Selectivity to C_2H_4 and CO vs C_2H_6 conversion for different reactors. Feed: 268 C_2H_6, 158 O_2 and 1575 ml(STP) Ar/min.

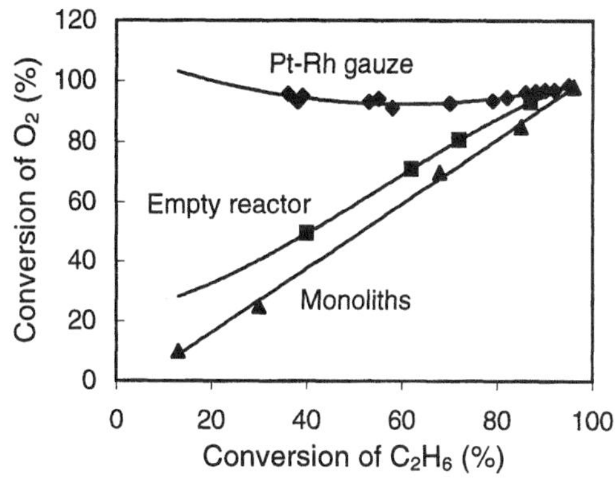

Figure 2. Conversion of O_2 vs. conversion of C_2H_6 for the reactors in Fig. 1. Conditions as in Fig. 1.

byproducts. The formation of C_2H_4, C_2H_2, CH_4 and the H_2/CO ratio were favoured by high temperatures. Lower temperatures resulting in low or moderate conversions, favoured strongly the formation of CO_2. The synthesis gas was typically produced with a H_2/CO ratio close to 1.5 at conditions also favouring the formation of C_2H_4. O_2 was never completely converted during these experiments. Oxygenates were not detected in the product gas although previous reports have indicated selectivities of oxygenates as high as 7 % on a carbon basis [5]. However, the condensate contained trace amounts of oxygenates.

As shown in Fig. 1 blank runs with an empty reactor or with only monoliths present demonstrated selectivities of C_2H_4 in the same order as in the presence of the gauze catalysts for C_2H_6 conversions above 55-60 %. In fact, almost identical results were obtained at the most severe conditions giving high conversions of C_2H_6. The reactor with monoliths showed the highest C_2H_4 selectivities at conversions of C_2H_6 below 40-50 %. C_2H_4 selectivities close to 70 % have been obtained with the Pt-Rh gauze at conversions of C_2H_6 in the range of 60-

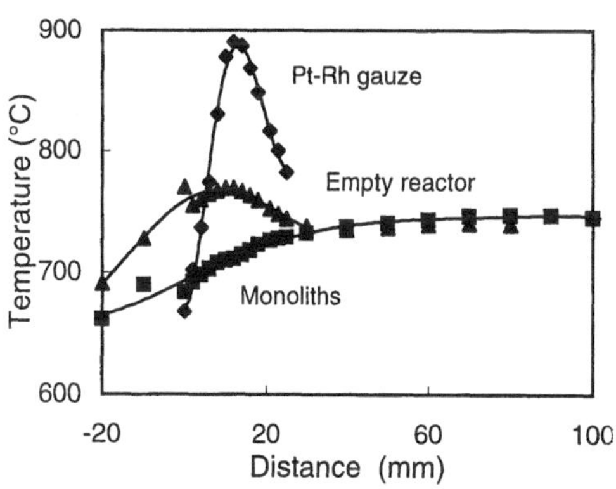

Figure 3. Temperature profiles for the reactors in Fig. 1. Conversion of C_2H_6: ~35 %.

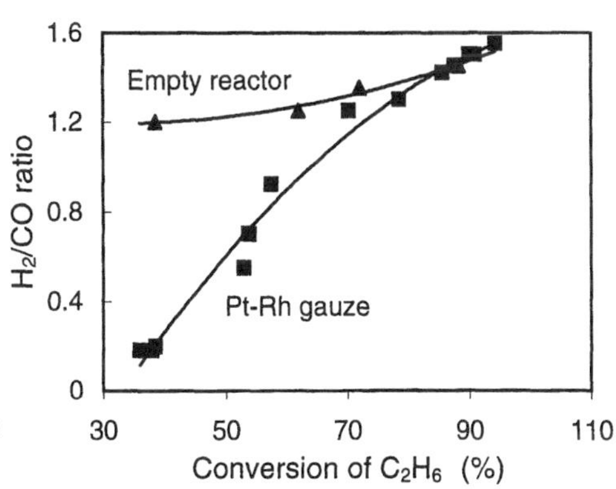

Figure 4. The H_2/CO ratio in the product gas Conditions as in Fig. 1.

80 %. It should be mentioned, however, that even higher yields of C_2H_4 have been obtained at conditions of increasing dilution and higher C_2H_6/O_2 ratios.

An explanation of the results given in Fig. 1 may be found in Fig. 2 which shows the conversion of O_2 at the same conditions as in Fig. 1. The data in Fig. 2 indicates that the noble metal gauze acts as a very efficient oxidation catalyst. The conversion of O_2 is well above 90 % for the Pt-Rh gauze at all conversions of C_2H_6. The results also show that more CO_2 is formed with the Pt-Rh gauze compared with the empty reactor at the same conversion of C_2H_6 and that the formation of CH_4 is favoured in the empty reactor. However, at high conversions of C_2H_6 the selectivity to CO_2 and CH_4 approaches the same value regardless of the reactor configuration. The superior selectivities with the monoliths are probably an effect of a reduction in the gas phase residence time.

The temperature characteristics are different between the reactor containing the gauze catalyst and the two other reactor configurations as shown in Fig. 3. The peak temperature is more than 100 °C higher on the gauze than measured in the empty reactor. More external heat is also required when the reactor is empty or containing only the monoliths. The reason being that the gauze proved to be very efficient for the internal production of heat through combustion reactions at or close to the gauze surface.

Of carbon containing products, the partial oxidation produces mainly C_2H_4 and CO. More than 90 % of the carbon in the feed is converted to C_2H_4 and CO. In Fig. 4 the H_2/CO ratio in the product gas from the empty reactor and the reactor with the Pt-Rh gauze catalyst is compared. Only small amounts of H_2 are leaving the gauze reactor at low conversion of C_2H_6 showing the efficiency of the gauze catalyst for combustion of H_2.

The C_2H_6/O_2 ratio is an important parameter for the product composition as well as for the amount of internal heat produced. Fig. 5 shows the selectivities to C_2H_4 at different C_2H_6/O_2 ratios in the presence of the gauze catalysts. The sum of the O_2 and C_2H_6 partial pressures is

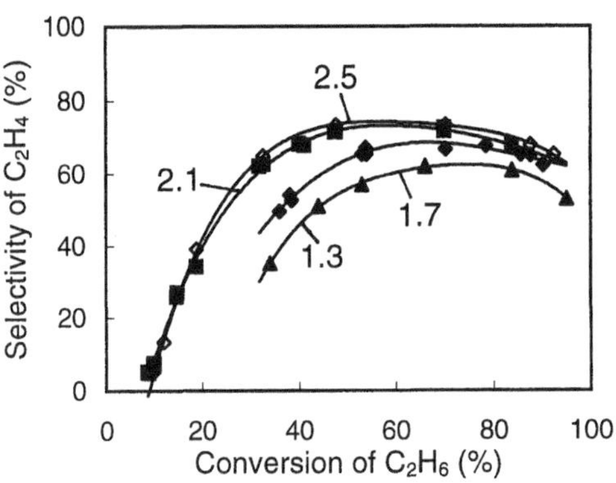

Figure 5. C_2H_4 selectivity at different C_2H_6/O_2 ratios for Pt and Pt-Rh gauzes. Pt-Rh gauze at C_2H_6/O_2 : 1.7.

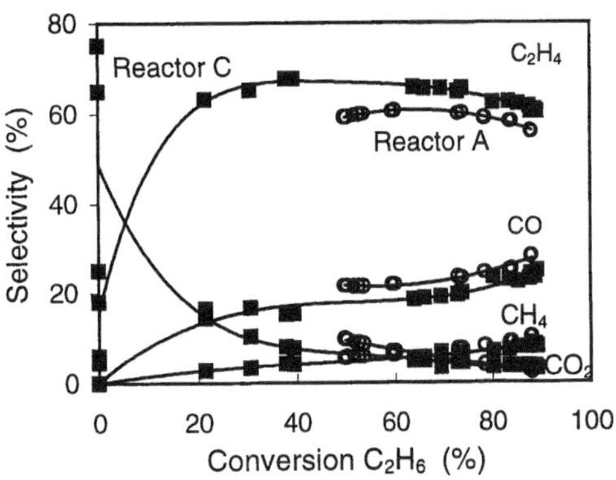

Figure 6. Effect of residence time. Reactor A (d: 15 mm, 1 Pt gauze). Reactor C (d: 4mm, 5 Pt gauzes). C_2H_6/O_2 : 1.7. Total flow rate: 2000 ml(STP)/min. Dilution: 32 % Ar.

kept constant. Reducing this ratio gave less C_2H_4, more CO and CO_2 whereas the selectivity to CH_4 remained almost unchanged.

In practical experimental work it is difficult to keep a high temperature only at or close to the gauze at very short contact times. The gas in front of and in particular behind the gauze will always be exposed to a comparable high temperature. Experiments have therefore also been performed in a small volume quartz reactor with an inside diameter of only 4 mm. In order to keep the same amount of catalyst as in the other reactors, five layers of Pt-Rh gauze were used. In this case it was necessary to increase the external heat supply in order to obtain comparable conversions of C_2H_6. The reason being that the conversion of O_2 was significantly less in the 4 mm reactor. It is evident from these results that the effect of a further reduction in the gas phase residence time is to give more C_2H_4 and less CO. The effect on CO_2 and CH_4 was less pronounced as shown in Fig. 6.

It is known that H_2 has a strong effect on the kinetics of methane pyrolysis [10]. Some experiments were therefore performed with a cofeed of H_2 (5.4% relative to C_2H_6). It was observed that H_2 had a positive effect on the formation of C_2H_4. However, the addition of H_2 reduced the net reactivity of the gas mixture which also indicates the dominance of the gas phase reactions relative to the catalytic reactions. Cofeeding of 5.6% CO relative to C_2H_6 did not have any effect on the reaction indicating that the oxidation of CO to CO_2 is not particular important at these conditions.

The high reaction temperatures in this study enables and favours formation of C_2H_2. C_2H_2 is unwanted in a process for making C_2H_4 since it has to be removed in a separate reactor. The results show that C_2H_2 is indeed formed at the highest temperatures. At about 1000 ^{0}C the ratio between C_2H_2 and C_2H_4 was as high as 1/20.

The product distribution did not depend significantly on the composition of the metal gauze although previous studies on the catalytic partial oxidation of methane showed large differences between Pt and Pt-10%Rh [8]. One reason could be that it is difficult to distinguish minor differences due to the dominant contribution from gas phase reactions.

In some runs the gauzes became very brittle in a very short time. In previous experiments of partial oxidation of methane it was observed that the Pt-10%Rh gauze was much more stable than the Pt gauze [8]. However, the opposite observation was made for the partial oxidation of ethane. The Pt gauze was more stable than the Pt-10%Rh gauze and in some runs the Pt gauze could be used for hours without being brittle. Of course, the gauze catalysts were heavily restructured as always after exposure to a high temperature oxidative atmosphere. Goetsch and Schmidt [6] did report that many hours of operation were required to achieve steady-state operation using a Pt-Rh gauze catalyst. Obviously, there must be some major differences between these two studies since we were not able to use the Pt-Rh gauze for a prolonged period of time. Goetsch and Schmidt [6] reported that the conversion of O_2 is 100 % with a single Pt-Rh gauze. In our case the conversion of O_2 was never complete even at severe conditions giving high conversions of C_2H_6.

The results from this study indicate that the gauze acts as an oxidation catalyst, effectively supplying internal heat to the endothermic cracking reactions mainly taking place in the gas phase in the post catalytic zone. In the presence of the gauze catalyst the conversion of C_2H_6 was almost completely decoupled from the always very high conversion of O_2. In contrast, in the absence of the gauze the extent of conversion of C_2H_6 and O_2 was strongly coupled.

Possible effects of the gauze could be to provide free radicals directly, or indirectly via formation of unstable products that could participate in a classical degenerate branching mechanism. Such products could be peroxides or oxygenates. This internal production of initiators would as observed lead to an increased rate of the gas phase dehydrogenation of C_2H_6. It would therefore be of interest to study the effect of different initiators on this reaction and such experiments are indeed in progress.

4. CONCLUSIONS

The high temperature oxidative dehydrogenation of C_2H_6 gives C_2H_4 and synthesis gas as the main products. A 70-75 % selectivity of C_2H_4 was obtained at high conversions of C_2H_6 (50 % or higher). More than 90 % of the carbon in the feed were converted to C_2H_4 and CO at high reactant conversions.

At severe conditions almost identical results were obtained with the gauze catalyst, with monoliths only or with the empty reactor. Less severe conditions caused the non-catalytic configurations to give superior selectivity to C_2H_4, due to less O_2 consumed. Combustion reactions at the Pt and Pt-Rh gauze catalysts proved to be very effective for the ignition of the reaction, for shortening the induction period and for rapid internal heating. No major differences could be distinguised between Pt and Pt-Rh gauzes due to the heterogeneous-homogeneous mechanism. In contrast to the empty reactor, the conversion of O_2 was always very high with the gauze. Runs at the shortest residence times demonstrated that both internal and external heat supply become limiting.

The stability of the gauzes is a problem. In some runs the gauze became brittle after very short time on stream. Surprisingly, the Pt gauze was more stable than the Pt-Rh gauze.

It was also demonstrated that H_2O is a suitable diluent for the above system.

ACKNOWLEDGEMENT

Financial support from Norsk Hydro is greatly acknowledged.

REFERENCES

1. L.F. Albright, B.L. Crynes and W.H. Corcoran (eds.), Pyrolysis. Theory and Industrial Practice, Academic Press, Inc., London, 1983.
2. D.E. Resasco and G.L. Haller, Catalysis (London), 11 (1994) 379.
3. P.B. Venuto and E.T. Habib, Fluid Catalytic Cracking with Zeolite Catalysts, Marcel Dekker, Inc., New York, 1979.
4. M. Huff and L.D. Schmidt, J. Phys. Chem., 97 (1993) 11815.
5. D.A. Goetsch and L.D. Schmidt, Science, 271 (1996) 1560.
6. P.M. Witt and L.D. Schmidt, J. Catal., 163 (1996) 465.
7. K. Heitnes Hofstad, O.A. Rokstad and A. Holmen, Catal. Lett., 36 (1996) 25.
8. K. Heitnes Hofstad, O.A. Rokstad and A. Holmen, Catal. Lett., 45 (1997) 97.
9. E. Bergene, O. Tronstad and A. Holmen, J. Catal., 160 (1996) 141.
10. A. Holmen, O. Olsvik and O. A. Rokstad, Fuel Proc. Technology, 42 (1995) 249.

NATURAL GAS CONVERSION V
Studies in Surface Science and Catalysis, Vol. 119
A. Parmaliana et al. (Editors)

Support effect on the n-hexane dehydrogenation reaction over platinum-tin catalysts

Jordi Llorca, Narcís Homs, Joaquim Sales and Pilar Ramírez de la Piscina

Departament de Química Inorgànica, Universitat de Barcelona, Diagonal 647, 08028 Barcelona, Spain*

The skeletal reactions of n-hexane have been studied over bimetallic Pt-Sn catalysts, which contain the PtSn phase supported on Al_2O_3, MgO or SiO_2, and compared to the monometallic catalysts. The bimetallic catalysts showed lower initial activity but also lower deactivation than the monometallic. A high selectivity to olefins (hexenes) has been related to the presence of the supported PtSn alloy. Differences in the catalytic behavior have been established as a function of the support. The MgO-supported catalyst was the most selective to olefins and SiO_2-supported was the most stable.

1. INTRODUCTION

Supported Pt-Sn catalysts are widely used in the study of hydrocarbon transformation reactions, mainly due to the observation that the steady-state activity of these bimetallic catalysts is superior to that of monometallic counterparts [1]. For a given support, their catalytic behavior and structural characteristics are determined by the Pt/Sn ratio and the preparation method used. We prepared silica- and alumina-supported Pt-Sn catalysts, in which only the presence of the PtSn phase was determined [2]. Recently, we studied the reaction of n-hexane over tailored silica-supported platinum-tin catalysts, in which well-defined Pt-Sn alloys were observed. Catalysts containing the PtSn alloy were more stable and showed high selectivity to dehydrogenation and low conversion to benzene and hydrogenolysis products [3]. Here we report the performance of alumina- and magnesia-supported catalysts that contain the PtSn phase [4] in the n-hexane dehydrogenation. Their catalytic behavior is compared to that of their monometallic and silica-supported counterparts.

* Financial support from spanish (CICYT MAT96-0859-C02) and catalan (1995SGR-00285) governments are gratefully acknowledged.

2. EXPERIMENTAL

Bimetallic catalysts with a nominal molar ratio Pt:Sn=1:1 were prepared by anchoring the *cis*-[PtCl($SnCl_3$)(PPh_3)$_2$] complex from a methylene chloride solution onto γ-Al_2O_3 (Girdler; surface area of 188 m^2 g^{-1}), MgO (obtained by adding ammonia to a $MgCl_2$ solution; 110 m^2 g^{-1}) and SiO_2 (Degussa aerosil; 200 m^2 g^{-1}) at room temperature followed by a vacuum treatment at 373 K overnight, as described elsewhere [2]. All supports were partially dehydrated by treatment under high vacuum at 473 K for 16 h. Samples were reduced in a glass reactor at atmospheric pressure in flowing hydrogen (40 ml min^{-1}) at 673 K for 16 h. The catalysts obtained are referred to as PtSn/Al_2O_3, PtSn/MgO and PtSn/SiO_2. For comparative purposes, platinum catalysts were also prepared in a similar manner by anchoring the *cis*-[$PtCl_2$(PPh_3)$_2$] complex onto the supports. The resulting samples are labelled Pt/Al_2O_3, Pt/MgO and Pt/SiO_2. All samples were prepared without contact with air. Table 1 shows the metal content of catalysts.

The n-hexane reaction was studied in a continuous-flow glass microreactor operating at atmospheric pressure. A feed of hydrogen saturated with n-hexane of varying composition was generated by bubbling hydrogen through a thermostatted saturator. 20-40 mg of catalyst diluted with inactive SiC was used for each reaction test. Products were separated in a TRB-1 capillary column on a gas chromatograph equipped with an automated gas sample valve. 3-hexenes were not quantified due to their low yield and proximity to the n-hexane peak. Prior to reaction studies, all catalysts were re-reduced *in situ* at 673 K for 1 h in flowing hydrogen. After reduction, the reactor temperature was lowered to 453 K and the catalyst was subsequently aged by increasing the temperature from 453 to 753 K at 5 K min^{-1} in a reaction mixture of H_2:n-hexane=4:1 (20 ml min^{-1}). The transformation of n-hexane was then monitored for 20-22 h in these reaction conditions for all catalysts. After the accelerated aging treatment, reaction studies were carried out at 753 K with H_2:n-hexane molar ratios ranging from 4 to 15 and reagent flow from 5 to 40 ml min^{-1} in order to study the performance and stability of catalysts.

3. RESULTS AND DISCUSSION

After aging, the initial catalytic activity at 753 K in the transformation of n-hexane (H_2:n-hexane=4:1) was determined for each catalyst (Table 1) and the reaction was monitored for 20 h in these conditions. In Figure 1 the deactivation percentage defined as the ratio [initial activity-activity (t) /initial activity] is represented as a function of time on stream.

Monometallic catalysts showed higher initial activity but also higher deactivation percentage than bimetallic. Moreover, significant differences between the PtSn catalysts as a function of support were found. The PtSn/SiO_2 catalyst was the most stable. No significant deactivation was observed during the study. These results confirm those obtained previously. Using a preparation method involving two impregnation steps with *cis*-[$PtCl_2$(PPh_3)$_2$] and $SnCl_2$ as precursors, we had obtained a catalyst in which only the PtSn phase was supported on silica [5], which had shown high stability in similar reaction conditions [3]. The catalyst used here, although it was prepared using a different method has similar structural characteristics to that previously studied.

Table 1
Composition and initial catalytic activity in the n-hexane transformations of platinum based catalysts.

Catalyst	Pt (%)	Sn (%)	Initial activity ($\mu mol.g^{-1}$ $Pt.h^{-1}$)[a]
Pt/Al_2O_3	2.37	--	1097
Pt/MgO	2.78	--	1071
Pt/SiO_2	1.96	--	1214
$PtSn/Al_2O_3$	2.79	1.81	799
PtSn/MgO	2.66	1.76	594
$PtSn/SiO_2$	2.52	1.55	278

[a]Reaction conditions: T=753 K, P=1 atm., GHSV=2400 h^{-1}, H_2/n-hexane=4.

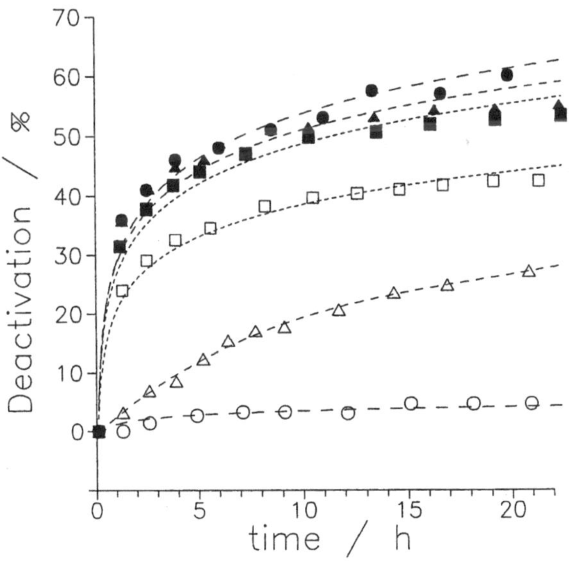

Figure 1. Deactivation percentage of catalysts as a function of reaction time: □ , Al_2O_3 supported; △ , MgO supported; O SiO_2 supported. Filled symbols Pt catalysts, empty symbols PtSn catalysts. Reaction conditions: T=753 K, P=1 atm., GHSV=2400 h^{-1}, H_2/n-hexane=4.

On the other hand, although a slight deactivation occurs for the PtSn/MgO catalyst, the curve profile differs strongly from that of the PtSn/Al_2O_3 sample, which in turn resembles that of monometallic catalysts in its deactivation behavior.

Table 2 shows the changes in conversion and selectivity values for all catalysts during the deactivation study (20 h). PtSn/MgO and PtSn/SiO_2 did not show significant change in the product distribution, and both showed high selectivity to hexenes from the beginning. When Pt is alloyed with Sn, reactions requiring ensembles with a large number of adjacent platinum atoms on the surface are expected to be suppressed. In this case, the catalytic dehydrogenation reaction occurs on single platinum entities, and the presence of Sn facilitate desorption of hexenes, which prevents the deactivation through the formation of carbonaceous residues from extensively dehydrogenated surface species [6].

Table 2
Initial and final conversion values and selectivity patterns of catalysts during the deactivation tests.

Catalyst	time	Conv.	Selectivity (%)[a]					
	(h)	(%)	C_1-C_5	o-C_6	mcp	ch	mp	bz
Pt/SiO_2	0.1	22.2	7.7	24.2	10.5	9.3	7.4	40.9
	19.3	8.5	3.1	50.3	6.9	11.6	6.2	21.9
PtSn/SiO_2	0.1	9.4	6.9	72.3	1.8	3.2	5.7	10.1
	19.0	8.9	6.3	74.8	1.7	3.1	4.2	9.9
Pt/Al_2O_3	0.1	17.0	8.4	42.0	13.8	11.8	9.5	14.5
	23.4	8.2	3.4	74.1	5.3	6.1	5.2	5.9
PtSn/Al_2O_3	0.1	14.6	16.0	36.4	16.2	9.1	16.8	5.5
	18.6	8.4	3.9	71.8	4.4	5.5	10.2	4.2
Pt/MgO	0.1	24.3	4.9	43.0	13.7	14.1	5.0	19.3
	19.2	11.4	1.3	76.5	4.8	7.9	2.6	6.9
PtSn/MgO	0.1	12.9	2.4	84.7	1.8	4.8	3.2	3.1
	20.8	9.0	0.7	90.8	1.3	3.4	1.6	2.2

[a]Reaction conditions: T=753 K, P=1 atm., GHSV=2400 h^{-1}, H_2/n-hexane=4. (o-C6=1- and 2-hexene, mcp=methylcyclopentane, ch=cyclohexane and cyclohexene, mp=2- and 3-methylpentane and bz=benzene).

Figures 2 and 3 show selectivity patterns of bimetallic catalysts in different reaction conditions. PtSn/MgO showed the lowest change on the selectivity in the different experimental conditions used. Moreover, the PtSn/MgO catalyst showed a high selectivity to hexenes and no significant change in its selectivity was observed when the H_2/n-hexane ratio was changed from 4 to 15 in all the conditions studied.

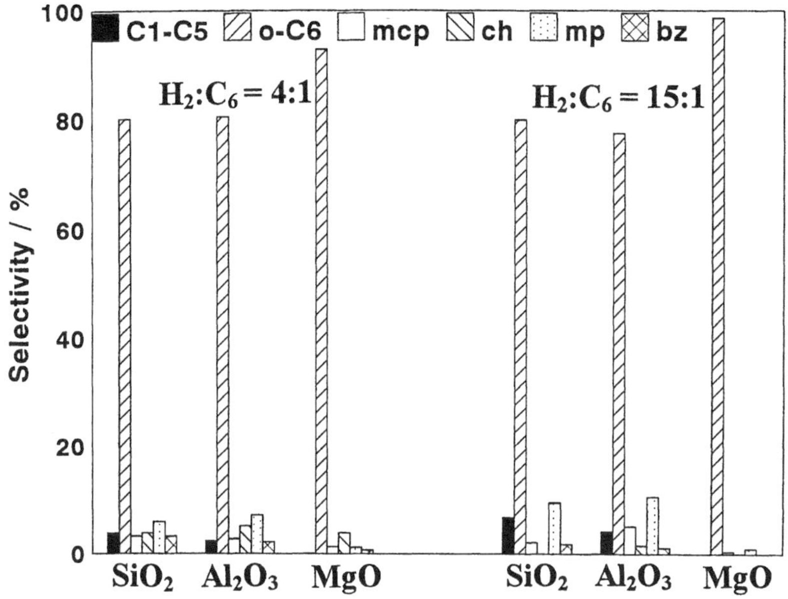

Figure 2. Product distribution at 753 K, GHSV=4800 h^{-1}, H_2/n-hexane=4, conversion values 5-7 %; H_2/n-hexane=15, conversion values 4-6 %.

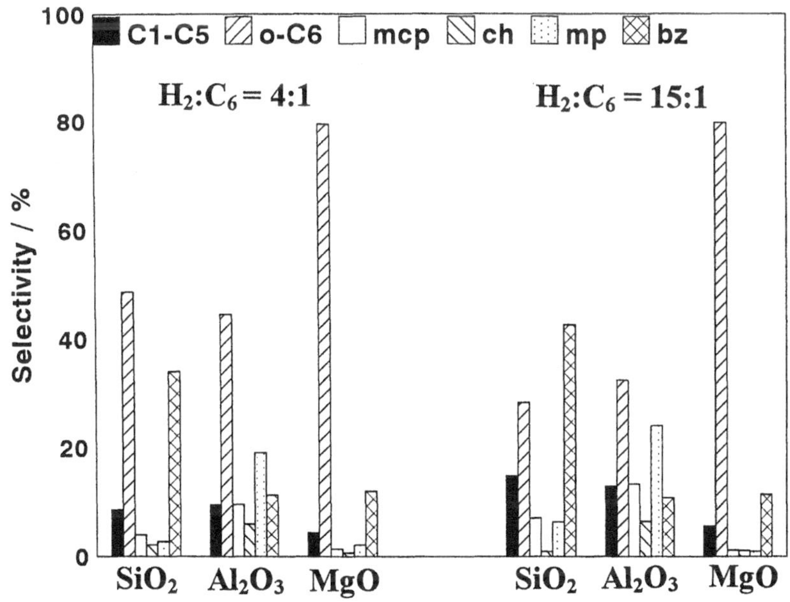

Figure 3. Product distribution at 753 K, GHSV=600 h^{-1}, H_2/n-hexane=4, conversion values 12-19 %; H_2/n-hexane=15, conversion values 12-14 %.

A decrease in the space velocity from 4800 h^{-1} to 600 h^{-1} in $PtSn/SiO_2$ and $PtSn/Al_2O_3$ catalysts resulted in a noticeable change in the selectivity. A decrease in the selectivity to hexenes and a high increase in the selectivity to benzene can be observed for $PtSn/SiO_2$ catalyst, this change was even higher for high H_2/n-hexane ratios. $PtSn/Al_2O_3$ catalyst also showed a decrease in the selectivity to hexenes when the space velocity was decreased. This catalyst showed the highest selectivity to methylpentanes, which increased at high H_2/n-hexane ratios. The main pathway of isomerization over metal centers involves cyclopentane-like surface intermediates; their desorption as methylpentanes requires the uptake of two hydrogen atoms, then a higher hydrogen excess should favour isomerization [7].

5. CONCLUSIONS

A clear effect of the support upon the catalytic performance of the PtSn alloy in the skeletal reactions of n-hexane is shown. SiO_2- and MgO- supported catalysts were more stable than Al_2O_3-supported samples. PtSn/MgO catalyst showed the highest selectivity to hexenes in all experimental conditions. The minimization of other reactions and the dehydrogenating properties of the isolated Pt atoms present in the PtSn phase, in addition to facilitating the desorption of hexenes, are responsible for this behavior.

REFERENCES

1. a) B. H. Davis, in "Selectivity in Catalysis" (M.E. Davis and S.L. Suib,Eds.), ACS Symposium Series 517, Washington DC, 1993 and references therein; b) V. Ponec and G.C. Bond in "Catalysis by Metals and Alloys" (B. Delmon and J.T. Yates, Eds.) Studies in Surface Science and Catalysis, Vol. 95, Chapter 13 and references therein, Elsevier Science B.V., Amsterdam, 1995.
2. J. Llorca, P. Ramírez de la Piscina, J.L.G. Fierro, J. Sales and N. Homs, J. Mol. Catal. A, 118 (1997) 101.
3. J. Llorca, N. Homs, J.L.G. Fierro, J. Sales and P. Ramírez de la Piscina, J. Catal., 166 (1997) 44.
4. Characterization of MgO-supported PtSn phase, J. Llorca, N. Homs, J. Sales and P. Ramírez de la Piscina, unpublished results.
5. J. Llorca, P. Ramírez de la Piscina, J.L.G. Fierro, J. Sales and N. Homs. J. Catal., 156 (1995) 139.
6. F.M. Dautzenberg, J.N. Helle, P. Biloen and W.M.H. Sachtler, J. Catal., 63 (1980) 119.
7. Z. Paál, Z. Zhan, Y. Manninger and W. M. H. Sachtler, J. Catal., 155 (1995) 43.

NATURAL GAS CONVERSION V
Studies in Surface Science and Catalysis, Vol. 119
A. Parmaliana et al. (Editors)
1998 Elsevier Science B.V.

Iron modified vanadyl phosphate as oxidation catalyst

G. Bagnasco[1], L. Benes[2], P. Galli[3], M. A. Massucci[3], P. Patrono[4], G. Russo[1], M. Turco[1]

[1] Dipartimento di Ingegneria Chimica, Università "Federico II" di Napoli, Naples, Italy
[2] Joint Laboratory of Solid State Chemistry, University of Pardubice, Czech Republic
[3]Dipartimento di Chimica Inorganica, Università "La Sapienza" di Roma, Rome, Italy
[4]IMAI-CNR, Area della Ricerca, Monterotondo Scalo, Rome, Italy

Iron modified vanadyl phosphate with formula $[Fe(H_2O)]_{0.2}(VO)_{0.8}PO_4 \cdot 2.25H_2O$ has been characterized by XRD analysis and NH_3 TPD technique. Catalytic oxidation properties were tested by NH_3 oxidation reaction. XRD analysis showed that the compound is isomorphous with layered tetragonal $VOPO_4 \cdot 2H_2O$ and is stable up to 600°C. NH_3 TPD measurements gave evidence of a wide distribution of adsorbing sites, the strongest ones being able to oxidize NH_3 mainly to N_2 , and to N_2O and NO in lower extent. NH_3 adsorbing properties were affected by the presence of water vapour. Catalytic tests showed appreciable activity, unlike the iron free compound that was found inactive. Catalytic activity was inhibited by water vapour. A reaction mechanism involving cooperation of Fe and VO groups was proposed.

1. INTRODUCTION

Vanadium phosphate catalysts are widely employed for several processes involving hydrocarbon oxidations, such as conversion of n-butane to maleic anhydride (1,2), propane to acetic or acrylic acid (3), or ammoxidation reactions (4). Such materials are generally prepared starting from vanadyl phosphates $VOPO_4$ or $VOHPO_4$ as precursors. These compounds are transformed under reaction conditions into V^{4+} pyrophosphate $(VO)_2P_2O_7$, which is the active phase for partial oxidation of hydrocarbons. However, there is good evidence that the presence of a $VOPO_4$ phase is essential for the catalytic activity (1,5). The role of different $VOPO_4$ phases in VPO catalysts has been discussed in recent studies, and the activity of α_I-$VOPO_4$ has been ascertained (2). It has been also reported that the catalytic activity of VPO systems can be influenced by the presence of water vapour in the feed (6). Moreover, some authors have reported that the addition of several metals can improve catalytic properties of VPO systems in oxidation reactions (7). Recently new materials, derived from layered $VOPO_4 \cdot 2H_2O$ by isomorphous substitution of some VO groups with trivalent metals, such as Fe, Cr, Mn, Al, were prepared and characterized (8,9). Such systems could be interesting catalysts due to oxidizing and dehydrogenating properties of the metals. However, the catalytic properties of these materials were scarcely investigated. In this work the catalytic activity of an iron modified vanadyl phosphate is studied, employing NH_3 oxidation as test reaction. The material is also characterized for structural, thermal and NH_3 adsorption properties.

2. EXPERIMENTAL

$[Fe(H_2O)]_{0.2}(VO)_{0.8}PO_4{\cdot}2.25H_2O$ (FeVOP), was obtained by refluxing V_2O_5 and $Fe(NO_3)_3.9H_2O$ (V/Fe molar ratio = 3) in 3M H_3PO_4 for 16 h. The suspension was hot filtered and the solid washed with cold water and dried in air (9). The catalyst was characterized by XRD, TG/DTA, NH_3 TPD techniques. A Philips PW 1100 model diffractometer was used for X-ray diffraction. BET surface areas were measured by N_2 adsorption at -196°C on a Quantachrom CHEMBET 300 instrument. NH_3 TPD measurements were carried out in a flow apparatus equipped with TCD and MS detectors. The measurements were carried out on samples treated at 450°C, by performing NH_3 adsorption at room temperature and thermodesorption at 10°C min^{-1}. A treatment with water vapour (0.6% by volume) could eventually be effected before NH_3 thermodesorption. NH_3 oxidation tests were carried out in a continuous flow laboratory plant with a fixed bed reactor (10), at T=200-450°C and contact time=$2.5x10^{-3}$ g h L^{-1}. Feed concentrations were: NH_3=400 or 700 ppm, O_2 =27000 ppm, H_2O= 0 or 1000 ppm. NH_3 was analyzed by an IR Radas Hartman & Braun spectrometer, NO by a chemiluminescence 955 Beckman analyzer, N_2 and N_2O by a 5890 HP gaschromatograph equipped with a TCD detector.

3. RESULTS AND DISCUSSION

FeVOP (S.A.= 5 m^2/g) is a well crystalline material, showing the XRD pattern characteristic of a layered tetragonal phase, as the parent material $VOPO_4{\cdot}H_2O$ (VOP), having interlayer distance of 7.16 Å against 7.41 Å of VOP. The TG/DTA measurements have shown that the material loses hydration water by steps in the range 25-250°C, forming first the monohydrated and then the anhydrous layered phase, with decrease of interlayer distance to 6.41 and 4.20 Å respectively (11). The XRD pattern of the material treated at 450°C (Figure 1b) shows the well evident signals of the anhydrous phase, which is less crystalline than the hydrated material (Figure 1a). After prolonged (24 h) exposition to moist atmosphere, only partial rehydration can be observed, as indicated by the appearance of a weak signal at d=7.16 Å. As previously reported (11), the crystalline structure is stable up to 650°C, while at higher temperatures the compound slowly decomposes. NH_3 TPD measurements were effected on FeVOP treated at 450°C, that corresponds to the temperature of the sample treatment before catalytic tests. Moreover NH_3 intercalation between the layers is excluded on FeVOP treated at this temperature (11), therefore only the external NH_3 adsorbing sites can be identified. The TPD spectrum, reported in Figure 2, curve a, shows a broad tailed peak with maximum at 180°C: the peak broadness suggests a wide distribution of the strength of NH_3 adsorption sites. The surface concentration of adsorbed NH_3 is $2.4x10^{14}$ molecules cm^{-2}, that is comparable with the theoretical surface concentration of V=O and Fe groups, as reported in (11). The TPD spectrum of VOP shows a similar shape as that of FeVOP, but with lower amount of NH_3, due to lower surface area, as already reported (11). In order to investigate on the oxidation properties of the material, mass spectrometric analysis of gases released during NH_3 desorption was carried out. The results of MS analysis reported in Figure 3, show that NH_3 is desorbed mainly unreacted and that some oxidation to N_2, and, in lower extent, to N_2O and NO occurs only at high temperature. N_2 is formed at temperature higher than 360°C, N_2O is produced in lower amounts at temperatures higher than 400°C, while NO is observed only above about 500°C. Therefore the strongest adsorbing sites are able to oxidize NH_3: the appearance of different oxidation products is probably related to different oxidation properties of such sites.

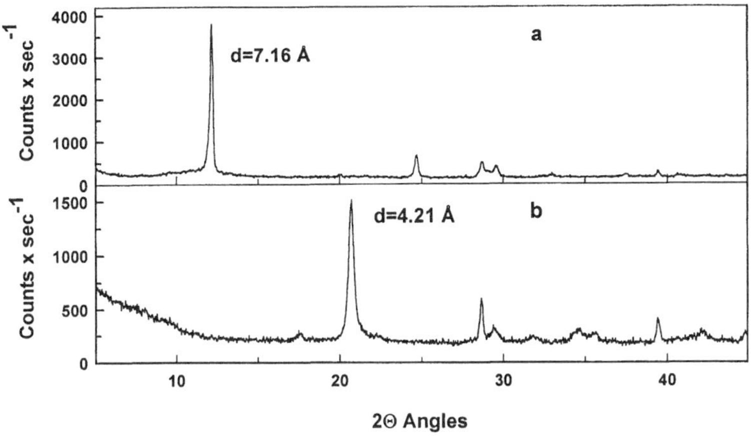

Figure. 1. XRD spectra of FeVOP: a) as prepared, b) after 12h treatment at 450°C

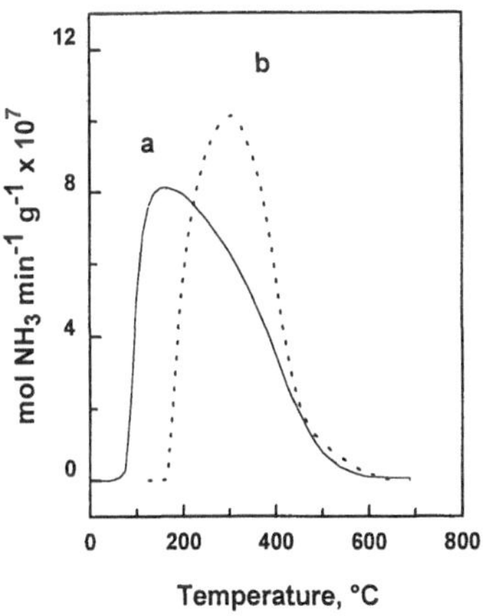

Figure 2. NH_3 TPD spectra of FeVOP. a) without b) after water vapour treatment

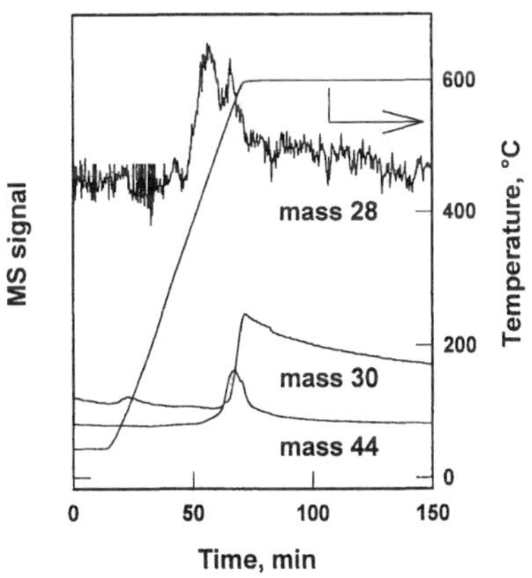

Figure 3. MS analysis of gases released during NH_3 TPD measurement. Conditions as curve a) of Figure 2

In order to investigate on the influence of water vapour on NH_3 adsorbing properties, TPD measurements were effected by treating the sample with water vapour / He mixture after NH_3 adsorption. The corresponding TPD spectrum is shown in Figure 2, curve b. As a consequence of the water vapour treatment, the peak maximum is shifted to higher temperature, although the amount of adsorbed NH_3 is practically unchanged ($2.2x10^{14}$

molecules cm^{-2}). This behaviour could be explained by supposing that the water vapour treatment produces strong adsorbing sites, so that NH_3, previously adsorbed on weaker sites, shifts onto such stronger sites. These sites could be protonic acid groups, such as $Fe(H_2O)^{3+}$ groups or V-OH species formed by hydrolysis of V=O groups. Nevertheless, it could be also hypothesized that water vapour causes some rehydration of the material, with consequent increase of interlayer distance, thus allowing shifting of surface adsorbed NH_3 into the interlayer space. It was reported that NH_3 intercalated in hydrated FeVOP is strongly retained, thus giving rise to a high temperature desorption peak (11).

NH_3 oxidation tests (Figures 4,5) show that FeVOP has appreciable activity only at temperatures higher than 300°C. The main product is N_2, while N_2O and NO are produced in lower extent and only at high temperature. Preliminary tests performed under the same conditions but without oxygen in the feed gave no N_2 product, so the occurrence of NH_3 decomposition can be excluded. NH_3 conversion appears quite independent of NH_3 feed concentration, suggesting a NH_3 reaction order close to 1. This would mean that NH_3 adsorption on active sites in the reaction conditions is far from saturation. As NH_3 adsorption could be affected by the presence of water vapour, some oxidation tests were effected by adding water vapour to the feed. A strong inhibiting effect of H_2O on NH_3 oxidation reactions is observed in Figueres. 4,5, mainly at high H_2O/NH_3 feed ratio. Anhydrous VOP (10), tested in the same conditions, gives no appreciable activity, suggesting that Fe plays an important role in catalytic oxidation properties. This can be attributed to the dehydrogenating properties

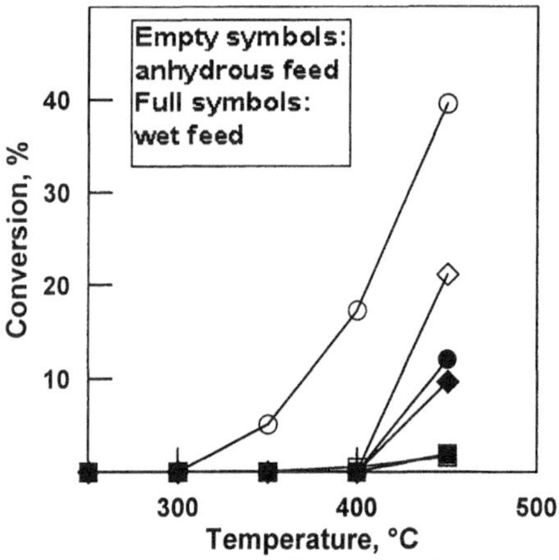

Figure 4. NH_3 conversion (●) and conversion to NO (■) and to N_2O (◆). (NH_3 feed concentration=400ppm)

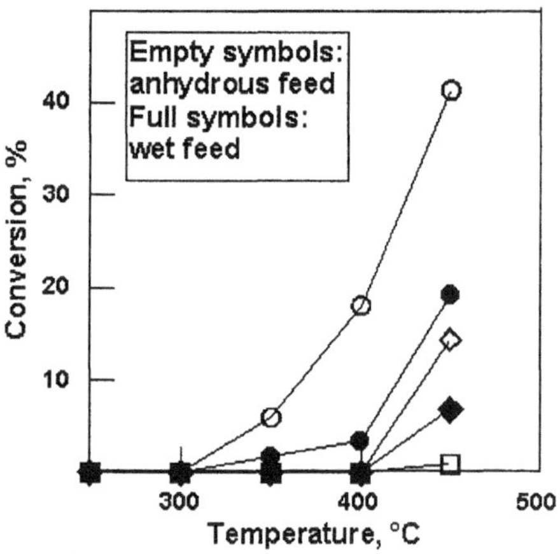

Figure 5. NH_3 conversion (●) and conversion to NO (■) and to N_2O (◆). (NH_3 feed concentration=700ppm)

of Fe, taking into account the proposed mechanism for NH_3 oxidation on vanadium based catalysts (10). According to this mechanism, ammonia, coordinatively adsorbed on VO groups, is dehydrogenated leading to an adsorbed NH_2 species. Then two NH_2 species form adsorbed hydrazine, which by subsequent dehydrogenation and oxidation gives rise to N_2. NH_3 dehydrogenation is probably the rate determining step. It can be supposed that Fe promotes dehydrogenation of adsorbed NH_3, so VO and Fe groups could cooperate as active sites. However it cannot be excluded NH_3 activation by adsorption on Fe^{3+} ions, since it is known that Fe_2O_3 phases can adsorb NH_3 forming amide species (12). TPD measurements give evidence of NH_3 oxidation products even in the absence of oxygen. Therefore it can be hypothesized that NH_3 dehydrogenation and oxidation steps occur by means of the oxygen from the catalyst, that is subsequently supplied by gaseous oxygen according to a Mars van Krevelen mechanism. The inhibiting effect of water vapour on catalytic activity could be due to coordination of water to Fe cations, thus reducing the active sites concentration. In fact, since water molecules are strongly bonded to Fe cations, as already observed, it is likely that free Fe^{3+} can take them again under reaction conditions.

4. CONCLUSIONS

A newly prepared iron modified vanadyl phosphate, isomorphous with $VOPO_4{\cdot}2H_2O$ shows noticeable catalytic oxidation properties, as evidenced by ammonia oxidation tests. By contrast no appreciable oxidation activity was observed on iron free material, thus indicating a key role of Fe in promoting oxidation activity. This also suggests that the substitution of VO groups with other trivalent metals can lead to the preparation of improved oxidation VPO catalysts.

REFERENCES

1. G. J. Hutchings, and R. Higgins, J. Catal., 162 (1996) 153
2. G. Centi, Catal. Today, 16 (1993) 5.
3. Y. Takita, S. Kikukawa, Y. Abe, Y. Mizukawa, and T. Shihara, Nippon Kagaku Kaishi 354 (1992).
4. A. Martin, B. Lücke, H. Seeboth and G. Ladwig, Appl. Catal., 49 (1989) 205.
5. G. J. Hutchings, A. Desmartin-Chomel, R. Olier and J. C. Volta, Nature, 368 (1994) 41
6. L. Savary, J. Saussey, G. Costentin, M. M. Bettahar. and J. C. Lavalley, Catalysis Letters, 38 (1996) 197
7. M. T. Sananés-Schulz, F. Ben Abdelouahab, G. J. Hutchings and J. C. Volta, J. Catal., 163, (1996) 346
8. K. Melànovà, J. Votinský, L. Beneš, and V. Zima, V., Mat. Res. Bull., 30 (1995) 1115
9. K. Richtrovà, J. Votinský, J. Kalousovà, L. Beneš, and V. Zima., J. Solid State Chem. 116 (1994) 400
10. G. Bagnasco, G. Peluso, M. Turco, G. Busca, and G. Ramis, in Studies in Surface Science and Catalysis, vol. 110, Grasselli, R.K., Oyama, S.T, Gaffney, A.M, Lyons, J.E., eds., Elsevier Science B. V. 1997, p. 643
11. G. Bagnasco, L. Benes, P. Galli, M. A. Massucci, P. Patrono, M. Turco, and V. Zima, J. Thermal Anal., in press
12. G. Ramis, L. Yi, G. Busca, M. Turco, E. Kotur, and R. J. Willey, J. Catal 157 (1995) 523

NATURAL GAS CONVERSION V
Studies in Surface Science and Catalysis, Vol. 119
A. Parmaliana et al. (Editors)
1998 Elsevier Science B.V.

Oxidative dehydrogenation of propane in annular reactor over a Pt/Al_2O_3 catalyst

A. Beretta[a], M. E. Gasperini[a], G. Trepiedi[a], L. Piovesan[b], P. Forzatti[a]

[a]Dipartimento di Chimica Industriale e Ingegneria Chimica "G. Natta", Politecnico di Milano, piazza L. da Vinci, 32, 20133 Milano, Italy

[b]Snamprogetti/Rice, via Maritano 26, 20097 San Donato Milanese (MI), Italy

Oxidative dehydrogenation of propane was tested inside a structured annular reactor with and without a Pt/Al_2O_3 catalyst. The Pt-containing catalyst was very active in the total combustion of propane, which was the prevailing reaction at temperatures lower than 500°C. At higher temperatures the formation of olefins was observed; this was very likely assisted by gas-phase reactions. Apparently, the presence of the catalyst shifted of 50°C towards lower T the onset of the ODH reaction with respect to the single homogeneous process.

1. INTRODUCTION

Among the novel routes for the chemical conversion of natural gas, oxidative dehydrogenation (ODH) of light alkanes represents a potential alternative to the traditional endothermic processes for the production of short-chain olefins. A recent review of the best results obtained in the oxidative dehydrogenation of ethane and propane has been provided by Baerns and Buyevskava [1]. They pointed out that, although a number of catalytic systems have been proved to be active in the ODH reactions including V-Mg-O systems, phosphates and molybdates of various transition metals, and structured noble metal-based systems, the development of a catalyst with economically feasible performance is still a "challenging task".

Very interesting data were obtained in the ODH of ethane and propane by Schmidt and coworkers; they observed almost complete conversion of the alkanes and selectivity as high as 60-70% of the olefins by using a Pt-coated foam monolith in an autothermal reactor at very short contact times [2]. However, due to the very high reaction temperature associated to the adiabatic reactor configuration (900-1000°C), possible contributions from homogeneous reactions were likely present; the effective role of the catalyst in the process seems thus unclear.

The present work addresses the study of the ODH of propane over a Pt/Al_2O_3 catalyst. Experiments were performed by using a novel annular reactor, wherein the catalyst is deposited as a thin and short layer onto a tubular ceramic support co-axially inserted inside a quartz tube. The gas stream flows across the annular chamber; given the absence of pressure drop, high flow rates can be guaranteed and, as very small amounts of catalyst can be deposited (10-100 mg), gas hourly space velocities as high as 10^6 L(STP)/kg_{cat}/h are easily realized [3, 4]. Also, the peculiar geometry of the system favors the dispersion of the heat produced by the reaction *via* radiation towards the oven internal wall; even in the presence of very exothermic processes, thus, a relatively good control of the catalyst temperature can be obtained and almost isothermal conditions arise along the catalyst bed. The catalyst temperature, then, can be read by exploiting the tubular support as a thermocouple-well.

Experiments were performed both in the presence and in the absence of catalyst in order to better understand the contribution of reactivity associated to gas phase reactions.

2. OXIDATIVE DEHYDROGENATION OF PROPANE OVER Pt/Al_2O_3

Few milligrams of a commercial 3% Pt/Al_2O_3 catalyst were deposited onto the ceramic support and a 13 mm long and about 50 μm thick catalyst layer was obtained. The activity tests were performed at a total flow rate of 120 Ncc/min, with feed composition $C_3H_8 : O_2 : N_2 = 1 : 1 : 4$, at atmospheric pressure and by increasing the temperature up to 700°C.

The flow rate to catalyst surface ratio herein used corresponds to the same value realized by Huff and Schmidt [2] in the foam monolith reactor at 9 STPL/min feed stream. In other words, the same "contact time" referred to the catalyst geometric surface was reproduced as a reference. Referred to the catalyst load, the flow rate corresponded to a gas hourly space velocity of 1.2 10^6 L(STP)/kg cat/h.

At a oven temperature as low as 150°C, the catalyst was already very active in the complete combustion of propane to CO_2 and water. The reaction rate was so high that the process was strongly limited by inter-phase diffusion resistances; as shown in Figure 1 in the range T>400°C, propane and oxygen conversions were in fact almost constant up to 500-550°C.

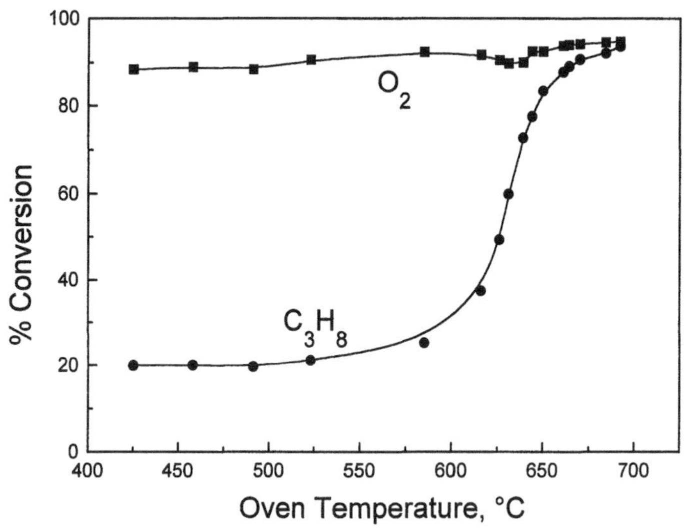

Figure 1 - Propane and oxygen conversions at increasing oven temperature over Pt/Al_2O_3

Beyond such temperature, propane conversion increased rapidly and was almost complete at 750°C. Oxygen conversion was always high in the whole range of temperature explored, so that the overall stoichiometry of the process changed from $C_3H_8 : O_2 = 1 : 5$ (stoichiometry of the combustion reaction) to $C_3H_8 : O_2 = 1 : 1$.

While at lower temperature CO_2 and water were the only reaction products, the increase of propane conversion was accompanied by a change in the product distribution. This is shown in Figure 2, where the % C-mole and % H-mole selectivities of the reaction products are plotted as functions of the oven temperature.

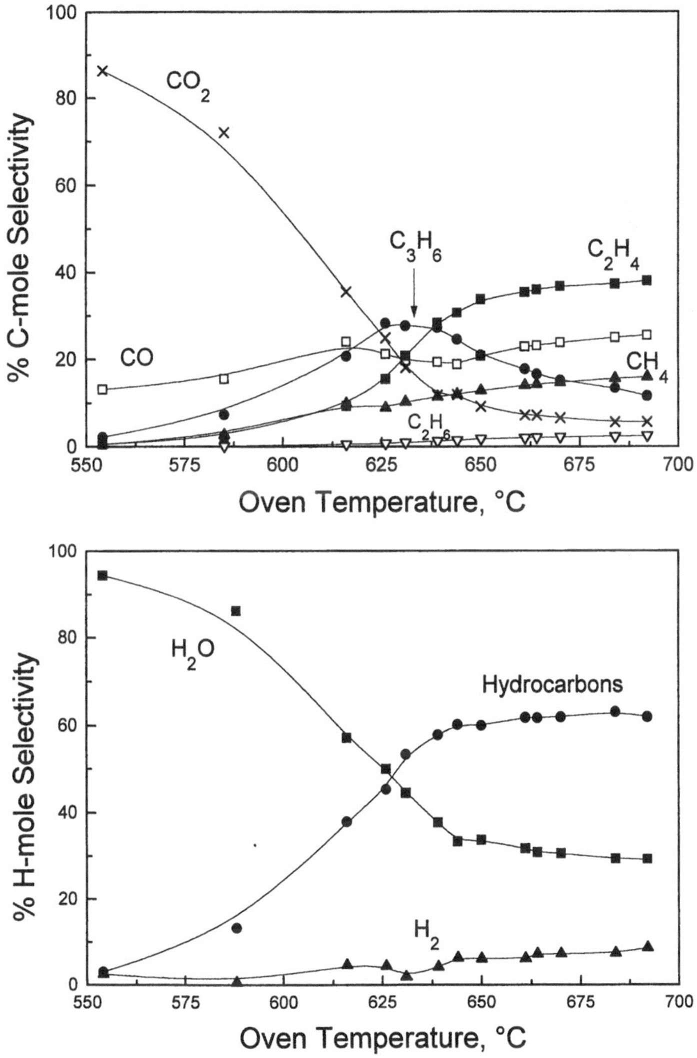

Figure 2 - Distribution of the C-containing and H-containing reaction products.

The presence of propylene traces in the product mixture was first observed at about 450°C oven temperature. The amount of the olefin became significant at 550°C; over this temperature also ethylene was observed, and the formation of both olefin increased remarkably. The selectivity of propylene increased up to a maximum (28-30%) at T_{oven} = 625°C. At higher temperatures its production decreased, probably due to the increasing relevance of secondary reactions, responsible for the formation of shorter C-containg species. The selectivity of ethylene and methane increased progressively. At high temperatures, while the formation of CO_2 and water had a dramatic drop, CO and H_2 concentrations in the product mixture increased.

The highest total yield to olefin was obtained at 640°C, with propylene + ethylene selectivity of 60% and propane conversion of 78%.

As mentioned in the Introduction, the mullite tube (coated with catalyst in a short central portion) was used as thermocouple-well, so that during each experiment the axial temperature profiles were measured. The axial temperature profile of the oven wall was also measured. While upstream and downstream from the catalytic portion mullite temperature and oven temperature were almost the same, the catalyst temperature was generally higher than the heating temperature. Figure 3 shows the relationship between oven and catalyst temperatures in the various experiments. At lower reaction T, in correspondence with the prevailing highly exothermic combustion route, the catalyst stabilized at nearly 150°C over the oven temperature. This gap decreased progressively at increasing oven T along with the lowering of CO_2 selectivity and the onset of the much less exothermic oxidative dehydrogenation.

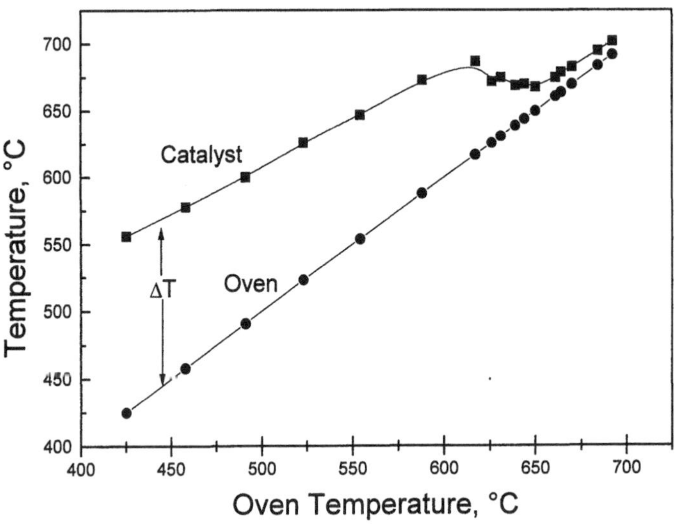

Figure 3 - Catalyst temperature vs. oven temperature during activity tests.

3. GAS-PHASE OXIDATIVE DEHYDROGENATION OF PROPANE

Figures 4 and 5 show the results of the experiments performed in the absence of catalyst under the same operating conditions of the catalytic tests. It was found that the oxidative dehydrogenation of propane was thermally activated at nearly 600°C, and a 90% conversion of the paraffin was obtained at the heating temperature of 710°C. Oxygen conversion was always slightly lower than propane conversion.

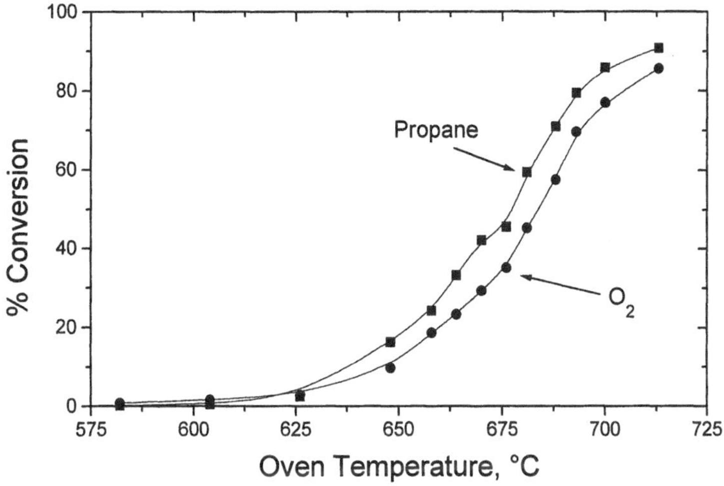

Figure 4 - Thermal activation of the $C_3H_8/O_2/N_2$=1/1/4 mixture. Conditions as in Fig.1

Propylene and ethylene were the most abundant products at lower propane conversions. At increasing temperature, propylene selectivity decreased while the selectivity to terminal products like CO and CH_4 increased. CO_2 was always present in very small amounts.

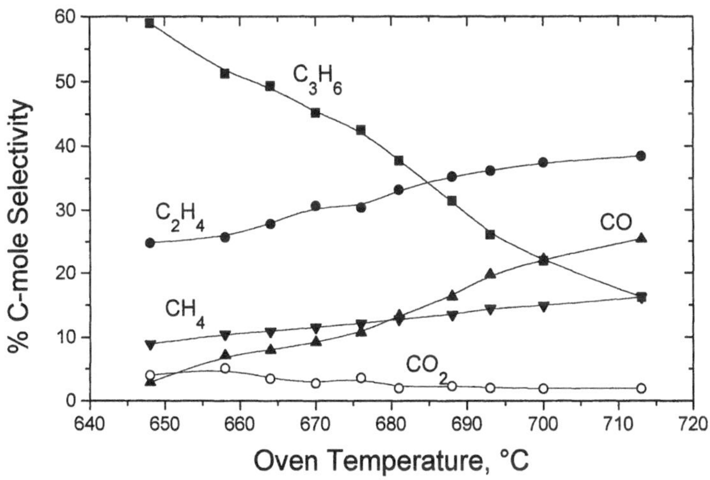

Figure 5 - Distribution of C-containing products of homogeneous propane ODH

Catalytic and homogeneous data are compared in Figure 6, where the outlet flow of propylene and ethylene are plotted vs. the reactor temperature. Both in the presence and in the absence of catalyst, the same reaction mechanism seemed to govern the formation of olefins; within this mechanism propylene behaved as intermediate product (going through a maximum productivity), while ethylene presented the feature of a terminal product. However, the presence of the small amount of catalyst (a negligible volume in comparison with the total gas-phase volume in the annular reactor) "anticipated" of about 50°C the activation of the propane/oxygen mixture in the production of olefins.

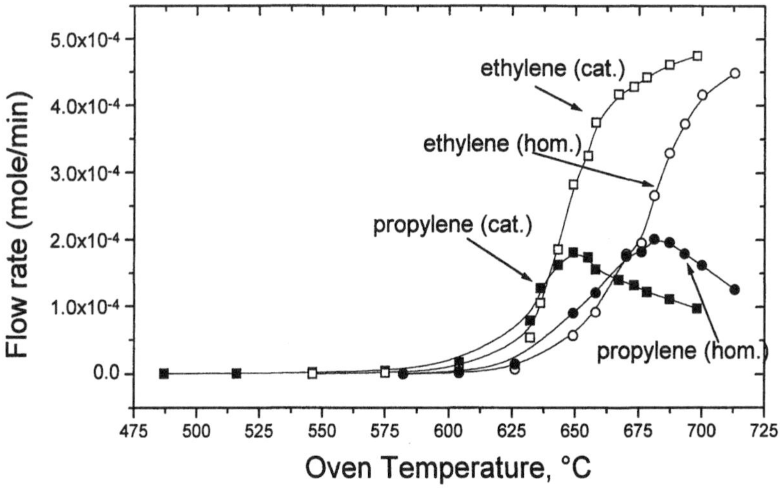

Figure 6 - Comparison of ethylene and propylene flow rates with and without catalyst.

4. CONCLUSIONS

The present data have clarified that: (1) the Pt-supported catalyst is active in the total oxidation of propane; this reaction prevails at temperatures lower than 500°C, maybe due to strongly adsorbed O-species which are very aggressive towards the paraffin; (2) the thermal activation of propane and oxygen mixtures largely contributed to the overall process at high temperatures (>625-650°C).

The effective contribution of the catalyst to the formation of olefins is still open. The presence of catalyst lowered the T-threshold for the production of propylene; further research is needed to better understand whether this effect resulted from catalytic and/or thermal factors.

REFERENCES

1. M. Baerns, O. Buyevskava, Proceedings of JECAT'97, p. 21, 1997.
2. M. Huff, L. D. Schmidt, J. Catal. 149 (1994) 127.
3. A. Beretta, Chem. Eng. Comm., (1998) in press.
4. A. Beretta, P. Baiardi, D. Prina, P. Forzatti, Chem. Engng. Sci, (1997)submitted

NATURAL GAS CONVERSION V
Studies in Surface Science and Catalysis, Vol. 119
A. Parmaliana et al. (Editors)
1998 Elsevier Science B.V.

Oxidative dehydrogenation of propane on supported V_2O_5 catalysts. The role of redox and acid-base properties

F. Arena [a], F. Frusteri [b], A. Parmaliana [a], G. Martra [c] and S. Coluccia [c]

a. Dipartimento di Chimica Industriale, Università degli Studi di Messina, Salita Sperone c.p. 29, I-98166 S. Agata (Messina), Italy

b. Istituto CNR-TAE, Salita S. Lucia 39, I-98126 S. Lucia (Messina), Italy

c. Dipartimento di Chimica Inorganica, Chimica Fisica e Chimica dei Materiali, Università degli Studi di Torino, via P. Giuria 7, I-10125 Torino, Italy

The influence of several oxide carriers on the activity of vanadia catalysts in the oxidative dehydrogenation of propane (POD) has been investigated. The effects of both redox and acid-base properties of the support on the surface structures and reactivity of vanadia have been assessed. The occurrence of competitive reaction paths on bare oxide carriers and supported vanadia catalysts has been discussed.

1. INTRODUCTION

The production of olefins by the oxidative dehydrogenation of light alkanes represents a very attractive alternative to conventional dehydrogenation processes since it could allow the use of natural gas as feedstock for production of large scale chemicals actually obtained from oil. Intensive research efforts recently focused on such topic have documented that supported vanadia systems are active and selective catalysts for attaining the oxidative dehydrogenation of light paraffins. The performance of such V_2O_5 systems is strongly affected by the nature of the oxide carrier and V_2O_5 loading /1-3/. Although the coordination of V^V-species mostly determines the redox properties and then the reactivity of the active phase /1/, also the influence of the acid-base features of supported vanadia catalysts on the oxidative dehydrogenation reactions has been claimed /1,2,4/.

Then, here we report on the effects of several oxide carriers (i.e., MgO, Al_2O_3, TiO_2, ZrO_2, SiO_2 and HY zeolite) characterised by different surface acid-base properties on the performance of same loaded (4.7-5.3 wt%) V_2O_5 catalysts in the POD reaction.

2. EXPERIMENTAL

2.1 Catalysts were prepared by incipient wetness of commercial oxide (i.e., MgO, Al_2O_3, TiO_2, ZrO_2, SiO_2 and HY zeolite) carriers with an aqueous solution (pH=11) of ammonium metavanadate. All samples were dried at 110°C and then calcined in air at 600°C. The list of samples along with their notation and physico-chemical properties is reported in Table 1.

2.2 Catalytic measurements in the POD were carried out in a flow quartz microreactor connected on line with a GC equipped with a three column analytical system /3/. The activity of the catalysts has been evaluated under the following conditions: T_R, 500°C; P_R, 1 bar; W_{cat}, 0.01-0.25 g (diluted with SiC (1/10, w/w)). The molar composition of the reaction mixture was C_3H_8:O_2:N_2:He=2:1:1:8 and a total flow rate of 100 STP $cm^3 \cdot min^{-1}$ was used.

2.3. Catalyst characterization

2.3.1. *TPR measurements* in the range 200-1000°C have been performed in a flow apparatus operating in both continuous and pulse modes using a 6% H_2/Ar mixture flowing at 60 STP cm^3/min and a heating rate of 20°C/min /5/.

2.3.2. *High Temperature Oxygen Chemisorption* (HTOC) tests /5,6/ were carried out at 367°C in a pulse mode after reduction of catalysts for 2h in H_2 at the same T. Vanadia dispersion (O/V) was calculated assuming a chemisorption stoichiometry of O_2:V=1:2.

Table 1. List of samples

Supports			Catalysts			
Code	Composition	S.A.$_{BET}$ ($m^2 \cdot g^{-1}$)	Code	V_2O_5 loading (wt%)	S.A.$_{BET}$ ($m^2 \cdot g^{-1}$)	Theor. surf. cov. ($V_{at.} \cdot nm^{-2}$)
M	MgO	24	VM-5	4.7	101	3.08
A	Al_2O_3	185	VA-5	4.8	209	1.52
T	TiO_2	51	VT-5	5.0	41	8.07
Z	ZrO_2	37	VZ-5	5.3	34	10.32
S	SiO_2	381	VS-5	5.0	286	1.16
H	HY zeolite	632	VH-5	5.3	350	1.00

3. RESULTS and DISCUSSION

3.1. Catalytic pattern of oxide supports and supported vanadia catalysts. The catalytic behaviour pattern in POD of the various oxide supports and vanadia catalysts has been evaluated operating with different contact times (τ) to attain a differential C_3H_8 conversion. The activity data of the various systems are summarised in Table 2 in terms of C_3H_8 conversion, reaction rate values, C_3H_6 and CO_x selectivity and C_3H_6 productivity (STY, $g_{C3H6} \cdot kg_{cat} \cdot h^{-1}$). C_3H_8 conversion never exceeds 6% while O_2 conversion is quite larger even if it keeps always less than 40%. All the catalysts produce essentially C_3H_6, CO_x and minor amounts of C_2 hydrocarbons (1-3%) and oxyproducts (1-7%).

Bare oxide supports exhibit rather different catalytic properties in POD, indeed differences in specific activity (SA, 0.4-10.1 $\mu mol_{C3H8} \cdot g^{-1} \cdot s^{-1}$), specific surface activity (SSA, 1.1-197 $nmol_{C3H8} \cdot s^{-1} \cdot m^{-2}$) and propylene selectivity (0-55%) up to ca. two orders of magnitude have been observed. In particular, titania exhibits the best functionality in POD with the highest values of SA (10.1 $\mu mol_{C3H8} \cdot g^{-1} \cdot s^{-1}$) and S_{C3H6} (55%) which account for a STY equal to 833 $g_{C3H6} \cdot kg_{cat} \cdot h^{-1}$. The HY zeolite bears SA (8.9 $\mu mol_{C3H8} \cdot g^{-1} \cdot s^{-1}$) and S_{C3H6} (44%) values close to those of titania and correspondingly a comparable STY (599 $g_{C3H6}\cdot kg_{cat}\cdot h^{-1}$). However, this system shows a fast deactivation likely due to a strong adsorption of reagents/products resulting in poisoning/coking phenomena. Zirconia and alumina show a comparable SA (2.4-3.5 $\mu mol_{C3H8}\cdot g^{-1}\cdot s^{-1}$) being less reactive than previous systems; the S_{C3H6} is 19 and 35% respectively accounting for much lower STY's (74-180 $g_{C3H6}\cdot kg_{cat}\cdot h^{-1}$). Magnesia and silica are the least active systems with SA values of 0.6 and 0.4 $\mu mol_{C3H8}\cdot g^{-1}\cdot s^{-1}$, respectively. Silica displays a S_{C3H6} (38%) comparable with that of the other oxides, while no propylene formation is detected on magnesia. However, taking into account that the above oxides are characterized by rather different surface area values, a more reliable comparison of their catalytic activity, made on the basis of the SSA, provides the following reactivity scale:

$$T \gg Z \gg M > A > H \gg S.$$

Table 2. Oxidative dehydrogenation of propane on V_2O_5-based catalysts

Sample	C_3H_8 conv. (%)	Reaction rate ($\mu mol_{C3H8}\cdot g^{-1}\cdot s^{-1}$)	Reaction rate ($nmol_{C3H8}\cdot m^{-2}\cdot s^{-1}$)	S_{C3H6} (%)	S_{COx} (%)	STY ($g_{C3H6}\cdot kg_{cat}\cdot h^{-1}$)
M	1.2	0.6	27.1	-	100	0
VM-5	4.3	4.7	46.4	31	68	222
A	6.2	3.4	18.2	35	62	180
VA-5	3.2	21.8	104	73	26	2407
T	3.7	10.1	197	55	48	833
VT-5	5.1	69.5	1,690	33	67	3200
Z	4.7	2.5	69.2	19	79	74
VZ-5	2.5	34.1	1,000	53	46	2356
S	0.8	0.4	1.1	38	32	25
VS-5	3.3	3.6	12.6	72	19	394
H	3.3	8.9	14.2	44	55	599
VH-5	4.4	11.9	34.2	38	60	690

Vanadia addition strongly promotes the functionality of all the oxide carriers giving rise to an enhancement of specific activity and STY_{C3H6}. The VT-5, paralleling the reactivity of titania support, is the most active system with a SA equal to 69.5 $\mu mol_{C3H8}\cdot g^{-1}\cdot s^{-1}$ which coupled with a S_{C3H6} of 33% leads on the whole to the highest STY (3200 $g_{C3H6}\cdot kg_{cat}\cdot h^{-1}$). VZ-5 sample shows a SA lower by ca. one half than that of the VT-5 catalyst and a higher selectivity to propylene (53%), much larger also than that of the relative carrier, accounting for a STY value of ca. 2400 $g_{C3H6}\cdot kg_{cat}\cdot h^{-1}$. VA-5 exhibits a SA still lower than that of previous systems even if the STY keeps equal to that of the VZ-5 sample mainly as a consequence of the highest S_{C3H6} (73%). All the other systems (VH-5, VS-5 and VM-5) are characterised by SA values lower than that of VA-5 catalyst. As the selectivity to propylene ranges from 72 (VS-5) to 31% (VM-%), on the whole such systems ensure much lower STY's (222-690 $g_{C3H6}\cdot kg_{cat}\cdot h^{-1}$). Analogously to the relative support, a fast deactivation of the VH-5 catalyst has been observed. Furthermore, taking into account the SSA of the various systems, the following reactivity scale is obtained:

$$VT\text{-}5 > VZ\text{-}5 \gg VA\text{-}5 > VM\text{-}5 > VH\text{-}5 > VS\text{-}5.$$

However, the fact that the strongly promoting effect induced by vanadia on the activity of supports is accompanied by a concomitant enhancement in the S_{C3H6} proves that a more effective reaction mechanism occurs on supported vanadia systems.

3.2. Reducibility, dispersion and surface structures of supported vanadia catalysts. The redox properties and dispersion of supported vanadia catalysts have been comparatively probed by TPR and HTOC to achieve basic insights into their surface structures /5,6/. The TPR pattern of vanadia catalysts and relative oxide support are shown in Figure 1. All such spectra monitor a complex pattern consisting of several peaks associated with the reduction of different surface species whose reducibility is controlled by the surface physico-chemical properties of the relative oxide carriers /1,2,5/.

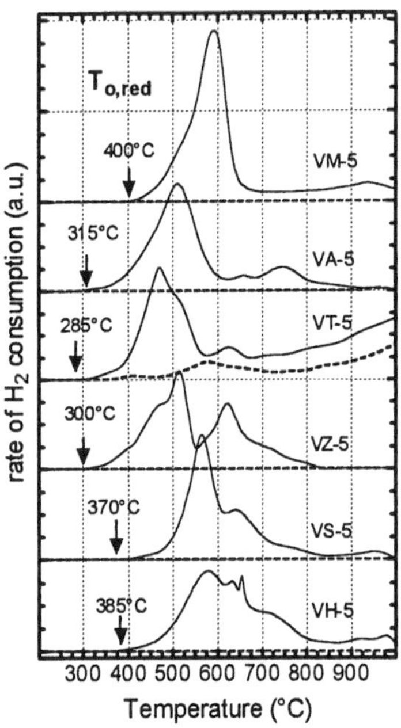

Figure 1. TPR pattern of supported vanadia catalysts (—) and relative support (—).

Namely, the promotional effect of the oxide carrier on the reduction of vanadia depends upon its *surface reducibility* as it determines the reactivity of the *bridging "V-O-support" functionalities* /1,5/. TiO_2 shows a reduction profile accounting for an enhanced reducibility both at the surface and in the bulk, whilst the flat TPR profiles of HY zeolite, SiO_2, ZrO_2, Al_2O_3, and MgO indicate that such systems are essentially not reducible. However, further insights into the surface reduction characteristics of metal oxide carriers can be achieved from the TPR of supported vanadia systems /5/. In fact, the reducibility scale of V_2O_5 catalysts:

VT-5 > VZ-5 > VA-5 > VS-5 > VH-5 > VM-5,

the same either based on the onset T of reduction (i.e., $T_{o,red}$) or T of the first peak maximum /5/, is in a very good agreement with literature data /1/, pointing to the reactivity of the surface oxygen layer of the support as factor governing the reducibility of vanadia catalysts. Then, while the above reducibility scale should account for an enhanced redox behaviour of TiO_2 and ZrO_2 explaining their highest SSA, it must be stressed that some other factor(s) control(s) the surface reactivity of other bare oxide carriers in POD.

Although surface redox properties of oxide carriers allow to explain the different reduction features of V_2O_5 catalysts, the above reducibility scale can be also rationalised in terms of acid-base properties of the carrier. Indeed, as shown in Figure 2, plotting the values of $T_{o,red}$ of the catalysts vs. the *pH of zero point charge* (i.e., pH_{zpc}) of the relative oxide carrier /5/ a hollow trend with the minimum for a pH_{zpc} close to 6 is found. Taking into account the easiest reduction of tetrahedrically coordinated (T_d) "isolated" species and that the support does not significantly affect their molecular structure and coordination symmetry /1,2,5/, these findings indicate that oxide carriers bearing either a strong surface acidity or basicity depress the reduction of the isolated vanadyl species. In particular, it can be inferred that the surface acidity of silica and HY zeolite exerts an electron-acceptor effect which hinders the reduction of monolayer species /5/.

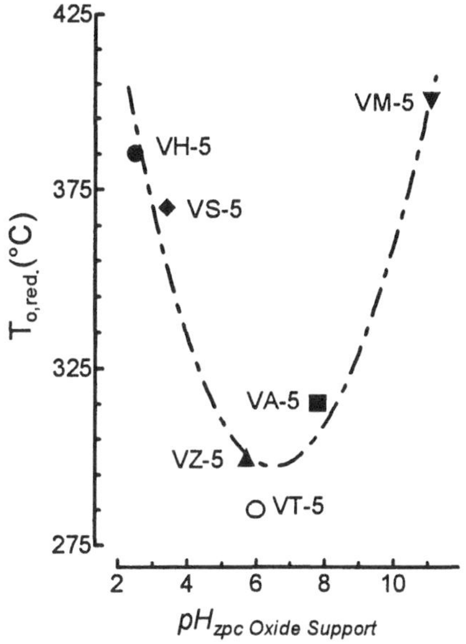

Figure 2. Relationship between the onset T of reduction ($T_{o,red}$) of vanadia catalysts and the point of zero charge (pH_{ZPC}) of the support.

Whereas, a strong interaction between basic MgO and acidic V_2O_5, resulting in the extensive formation of vanadates /1,2/, renders difficult the reduction of the VM-5 system.

On amphoteric oxides like TiO_2, ZrO_2 and Al_2O_3, the electron-donor positive effect of weakly basic sites should be prevalent enabling the prompt reduction of VO_4^{3-} units, while the intrinsic reducibility of TiO_2 further promotes the reduction of the VT-5 catalyst to lower T. Therefore, even if the structure of the monolayer species in the calcined catalysts is not affected by the nature of the support /1/, it appears that the surface acid-base features of the oxide carrier influence the electronic properties and reduction pattern of mono-oligomeric species besides to their maximum development /1,2/. Indeed, the HTOC data, shown in Table 3, reveal that vanadia dispersion on various catalysts ranges between 0.14 and 0.63. VA-5 catalyst is characterized by the highest vanadia dispersion (0.63), while similar oxygen uptakes account for analogous dispersion values on VZ-5 (0.46) and VS-5 (0.49) samples.

A dispersion value of 0.38 is found for the VT-5 system, even if this is an underestimated value /5/ likely as a consequence of the fact that the formation of a V-Ti-O solid solution /1/ renders $V^{(3 \text{ or } 4)+}$ stable against oxidation.

The VH-5 and VM-5 samples exhibit the lowest oxygen uptakes accounting for dispersion values equal to 0.14 and 0.17 respectively.

Table 3. HTOC of supported V_2O_5 catalysts: O_2 uptake and vanadia dispersion (O/V)

Sample	O_2 uptake ($\mu mol/g_{cat}$)	O/V	Sample	O_2 uptake ($\mu mol/g_{cat}$)	O/V
VM-5	43.3	0.17	VZ-5	133.8	0.46
VA-5	165.4	0.63	VS-5	134.9	0.49
VT-5	106.6	0.38	VH-5	40.5	0.14

As it has been previously ascertained that irrespective of the nature of the support /2,6/, oxygen chemisorption occurs on V^{3+} sites coming from the reduction of T_d mono-oligomeric species /5,6/, it is evident that the acid-base properties of the oxide support play a crucial role on the reducibility of vanadia because they control the development of the monolayer and thus the dispersion of the active phase which depends upon both vanadia-support interaction and hydroxyl population of the carrier /1,5/.

3.3. Redox and acid-base properties and catalytic pattern of oxide supports and supported vanadia catalysts. On the basis of catalytic data it is evident that no relationship between acid-base character of the oxide carrier and SA and SSA or S_{C3H6} exists. Moreover, vanadia acts as a promoter of both activity and selectivity pointing to its capability to enable a *surface concerted mechanism* involving the activation of propane molecule and gas-phase oxygen mostly over T_d mono-oligomeric sites /1,3/. This indicates the prevailing redox behaviour of vanadia catalysts as documented by Figure 3, showing the inverse relationships linking both SA and STY (A) with $T_{o,red}$ as well as SSA and Surface yield (B) with $T_{o,red}$.

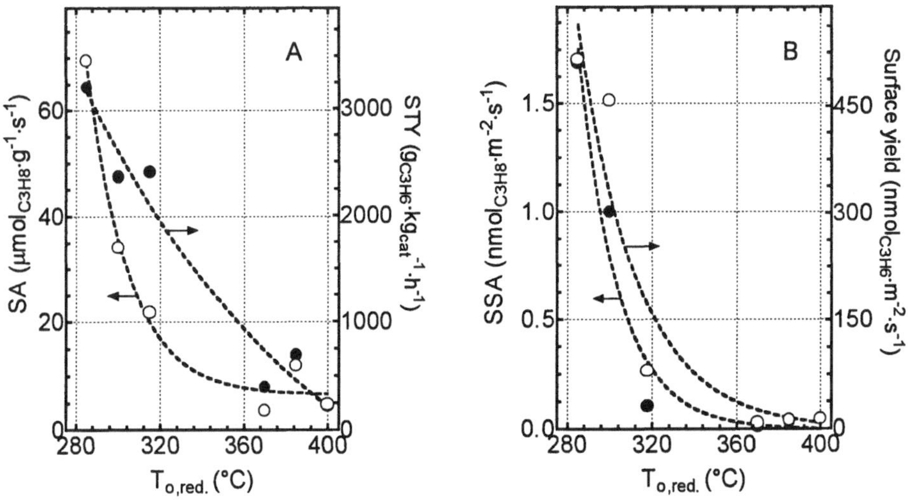

Figure 3. A: Relationships between SA, STY and $T_{o,red}$. B: Relationships between SSA, Surface yield and $T_{o,red}$

Such a mechanism is also prevalent on TiO_2 and could less effectively occur on ZrO_2. By contrast, on bare Al_2O_3, MgO and HY zeolite, possessing either basic or acidic Brönsted sites, a *surface-assisted* reaction mechanism, involving the primary stabilization of adsorbed carbocation and carbanion on acidic and basic sites respectively, should prevail. In particular, it is inferred that carbocation would react leading to the formation of C_3H_6, whereas the further reaction of the carbanion, formed on the basic sites of MgO, would mainly give the formation of carbon oxides. On silica /5/ and magnesia, however, a poor population of Brönsted sites should account for their very low surface reactivity in the POD reaction.

REFERENCES

1. I.E. Wachs and B.M. Wechuysen, *Appl. Catal.*, **157** (1997) 67
2. T. Blasco and J.M. López Nieto, *Appl. Catal.*, **157** (1997) 117
3. M. Puglisi, F. Arena, F. Frusteri, V. Sokolovskii, and A. Parmaliana, *Catal. Lett.*, **41**, (1996) 41
4. G. Busca, E. Finocchio, G. Ramis and G. Ricchiardi, *Catal. Today*, **32** (1996) 133
5. F. Arena, F. Frusteri and A. Parmaliana, *Appl. Catal.*, (1998) in press
6. F. Arena, F. Frusteri, G. Martra, S. Coluccia and A. Parmaliana, *J. Chem. Soc., Faraday Trans.*, **93** (1997) 3849

Selective Oxidative Conversion of Propane to Olefins and Oxygenates on Boria-Containing Catalysts

O. V. Buyevskaya, D. Müller, I. Pitsch and M. Baerns

Institute for Applied Chemistry Berlin-Adlershof
Rudower Chaussee 5, D-12484 Berlin, Germany

Abstract

Catalysts consisting of boria supported on Al_2O_3, TiO_2, ZrO_2 and MgO were studied in the partial oxidation of propane at 550°C. The highest propene yield amounted to 22 % (S = 45 - 48 %) on B_2O_3/Al_2O_3; in addition, yields to C_1-C_3 oxygenates of up to 8 % were achieved (S = 28 %). Surface and bulk composition of the B-Al-O catalysts were investigated using ^{11}B NMR, XRD and XPS to elucidate the factors determining propane conversion, product distribution and catalyst deactivation. It was shown that the coordination of boron influences catalyst activity; trigonal BO_3 species being present in both crystalline and amorphous phases are active in the dehydrogenation of propane to propene. The water formed during the reaction causes the loss of these species most probably as boric acid. The catalyst consisting of amorphous boria revealed high initial activity but resulted in faster volatilization of boria compared to a catalyst consisting of crystalline B_2O_3 and aluminium borates. Oxygenates formation was favoured in the presence of an $Al_6B_8O_{21}$ phase.

1. INTRODUCTION

Selective oxidative conversion of propane as a means of transforming natural gas to valuable chemicals is a challenging task [1-2]. Boria/alumina catalysts show good catalytic performance [3-4] but the catalysis is still not completely understood. In our recent work [4], we proposed that boria/alumina catalysts possess most likely Lewis acidity determining the activation of propane. A radical chain mechanism similar to that proposed for the partial oxidation of butane to acetic acid [5] with the formation of propyl hydroperoxide as a key intermediate was assumed. Less attention was, however, given to characterization of bulk and surface compositions and stability of the catalyst. The elucidation of the interplay between the initial catalyst composition and the formation of propene and oxygenates as well as of changes in catalyst compositions occurring under reaction conditions was the main objective of the present contribution. Since the catalyst stability is an important issue for B-containing catalysts special attention was given to the factors causing the loss of activity. Furthermore, the effect of different support materials on catalytic performance was studied.

2. EXPERIMENTAL

Catalyst preparation B_2O_3 (30 wt.%)/Al_2O_3 (A) catalyst was prepared by stirring of an aqueous solution of boric acid (3 wt.%) containing γ-Al_2O_3 (Merck) at room temperature; the water was then removed in vacuum at 70°C using a rotary evaporator. After drying at 120°C for 24 h the product was calcined at 600°C for 14 h. B_2O_3 (30 wt.%)/Al_2O_3 (B) was prepared by impregnating γ-Al_2O_3 (Riedel-de-Häen) with boric acid dissolved in warm water (5 wt. %

H_3BO_3). The excess of water was evaporated under stirring at *ca.* 100°C; the remaining solid was then dried at 110°C for 16 h and finally calcined at 600°C for 14 h. For preparing a B_2O_3 (30 wt.%)/Al_2O_3 (C) catalyst, boric acid was added to an aqueous solution of aluminium choride (Locron®, OH/Al = 2.5). The resulting gel (molar ratio B/Al = 0.63) was dried at 100°C and calcined at 600°C for 6 h. To synthesize the fourth B_2O_3 (13.2 wt.%)/Al_2O_3 (D) sample, aluminium isopropoxide was slowly added to an aqueous solution of boric acid. The resulting gel (molar ratio B/Al = 0.22) was transferred into an autoclave and kept at 150°C under autogenous pressure for 24 h. After drying at room temperature the sample was calcined at 1000°C for 2 h. Boria supported on TiO_2 (Degussa, P25), ZrO_2 (Alfa), MgO (Fluka) were prepared by the impregnation technique used for the preparation of a B_2O_3/Al_2O_3 (B) catalyst.

Catalyst characterization ^{11}B NMR spectra were recorded on a Unity-plus 500 spectrometer (Varian) at 160.4 MHz. The magic-angle spinning (MAS) technique was used with spinning rate of 10 kHz for the 5 mm rotor. For quantitative ^{11}B analysis short excitation pulses of 1 μs were applied. 100 - 200 scans were accumulated with repetition times of 10 s. B[4]/B[3] ratios were determined by a spectrum deconvolution using the WINFIT software (Bruker). Crystalline phases were characterized by XRD using STADIP transmission powder diffractometer (Stoe) with $CuK_{\alpha 1}$ radiation. The near-surface compositions were derived from XPS spectra recorded with an ESCALAB 220i-XL spectrometer (Fisons Instruments) using AlK_α radiation (1486.6 eV). BET surface areas were obtained by N_2 physisorption at -196°C.

Catalytic studies A quartz-made fixed-bed reactor ($\varnothing_{in}$ = 12 mm) with on-line GC analysis was operated at ambient pressure. A mixture of C_3H_8, O_2 and N_2 was passed through a bed of catalyst particles ($m_{cat.}$ = 1 g; d_p = 250 - 355 μm) at 550°C; total flow rates were varied from 20 to 100 ml/min. Partial pressures of propane amounted to 40 (C_3H_8/O_2 = 2) and 25 kPa (C_3H_8/O_2 = 0.5), respectively.

3. RESULTS AND DISCUSSION

3.1. Catalytic behaviour of various boria-containing catalysts

The degrees of propane conversion and the product selectivities obtained for all catalysts under similar reaction conditions are summarized in Table 1.

Table 1. Catalytic performance of different boron-containing catalysts

Catalyst	Conv. (%)		Selectivity (%)							
	C_3H_8	O_2	C_3H_6	C_2H_4	CH_4	C_1 oxi-genates	C_2 oxi-genates	C_3 oxi-genates	CO	CO_2
B_2O_3/MgO	20	80	27	12	3	0	<1$^{c)}$	2$^{e)}$	21	35
B_2O_3/TiO_2	8	15	74	11	3	0	1$^{c)}$	6$^{f)}$	4	1
B_2O_3/ZrO_2	15	26	68	14	3	0	2$^{d)}$	7$^{f)}$	5	<1
B_2O_3/Al_2O_3 (A)*	42	99	40	23	9	<1$^{a)}$	2$^{d)}$	1$^{e)}$	19	5
B_2O_3/Al_2O_3 (B)	30	46	42	13	3	16$^{b)}$	4$^{d)}$	8$^{g)}$	11	2
B_2O_3/Al_2O_3 (C)*	4	18	46	5	0	0	2$^{d)}$	2$^{f)}$	28	17
B_2O_3/Al_2O_3 (D)	28	100	41	12	5	0	0	0	27	14

Reaction conditions: 550°C, p(C_3H_8) = 40 kPa, p(O_2) = 20 kPa, τ = 2 g s ml^{-1}, 2 - 2.5 h of operation, *τ = 1 g s ml^{-1}; $^{a)}$ CHOOH; $^{b)}$ CH_3OH, CH_2O, HCOOH; $^{c)}$ CH_3CHO, CH_3COOH; $^{d)}$ C_2H_5OH, CH_3CHO, CH_3COOH; $CHOOCH_3$; $^{e)}$ CH_2CHCHO, CH_3CH_2CHO; $^{f)}$ CH_2CHCHO, CH_3CH_2CHO, CH_3COCH_3; $^{g)}$ CH_2CHCHO, CH_3CH_2CHO, CH_3COCH_3, C_3H_7OH.

Among supported boria catalysts, a low selectivity to propene compared to CO_2 was observed on B_2O_3/MgO. High propene selectivities were achieved when other support materials (TiO_2, ZrO_2, Al_2O_3) were used. Boria/alumina (A) was the most active catalyst revealing the highest propane conversion of 42 % at standard reaction conditions (cf. Table 1). The formation of oxygenates was favoured on the B_2O_3/Al_2O_3 (B) catalyst. The product distribution, in particular the oxygenates formation was affected by the propane-to-oxygen ratio applied. The dependency of product selectivities on propane conversion over sample (B) under oxygen-rich conditions ($C_3H_8/O_2 = 0.5$) is shown in Figure 1. Propene, ethylene, CO and CO_2 were the main products in this case, the total selectivity towards oxygenates did not exceed 8 %. A propene yield of 22 % was achieved at degrees of propane conversion of 45 - 48 %. Further increase in propane conversion led to a decrease in propene yield. The characterization of different B-Al-O catalysts was performed aiming on the elucidation of factors determining their activity and stability.

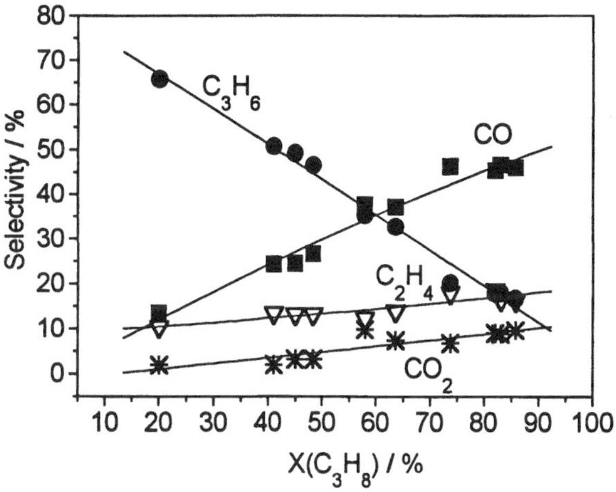

Figure 1. The dependence of product selectivities on propane conversion on B_2O_3(30 wt %)/Al_2O_3 (B); T = 550°C; $p(C_3H_8) = 25$ kPa; $p(O_2) = 50$ kPa.

3.2. Characterization of B_2O_3/Al_2O_3 catalysts

Results on bulk and surface composition of the four fresh B-Al-O catalysts are summarized in Table 2. Different phases were formed in the B_2O_3/Al_2O_3 system depending on the preparation procedure (cf. Experimental). The samples differ in their boron-containing crystalline phases. In sample (A), no crystalline boron-containing phases were present. Sample (B) consisted of B_2O_3 [JCPDS 6-297] and two aluminium borates $Al_6B_8O_{21}$ [JCPDS 23-759] and $Al_4B_2O_9$ [JCPDS 29-10]. The samples (C) and (D) consisted of $Al_4B_2O_9$ [JCPDS 29-10] and $Al_{18}B_4O_{33}$ [JCPDS 32-3], respectively.

Table 2. Results of bulk and surface characterization of B_2O_3/Al_2O_3 fresh catalysts

Catalyst	S_{BET} (m^2/g)	Crystalline boron-containing phases (XRD)	B[4]/B[3] (^{11}B NMR)	B/Al (XPS)
(A)	6	amorphous	13/87	1.3
(B)	52	B_2O_3, $Al_4B_2O_9$, $Al_6B_8O_{21}$	12/88	1.3
(C)	13	$Al_4B_2O_9$	21/79	0.8
(D)	89	$Al_{18}B_4O_{33}$	12/88	n.d.

The boron-to-aluminium ratio of 1.3 derived from XPS for samples (A) and (B) was significantly higher than the B/Al-ratio corresponding to the nominal loading (B/Al = 0.63). In sample (C), a lower boron concentration in the surface layers (B/Al = 0.8) was measured.

^{11}B NMR established as a powerful method for the determination of BO_4/BO_3 ratios (see e.g. [6]) was used to determine the oxygen coordination of boron in the catalysts consisting of crystalline and amorphous compounds. At the high magnetic field of 11.7 T used in this study the signals of tetrahedral BO_4 and trigonal BO_3 units are well separated in the ^{11}B MAS NMR spectra. The lowfield signal due to planar BO_3 units is dominated by effects of the quadrupole interaction of the ^{11}B nucleus (nuclear spin I = 3/2) with the electric field gradient at the nuclear site. Similar ratios of tetrahedral BO_4-to-trigonal BO_3 species were derived from ^{11}B NMR for samples (A) and (B). In sample (C), consisting of $Al_4B_2O_9$ as a main phase, the part of tetrahedral species was significantly higher compared to the other two samples. For sample (D) consisting of $Al_{18}B_4O_{33}$ as the only crystalline phase, the presence of a tetrahedral species (12 %) was derived although only BO_3 structural units had been expected according to the crystal structure of the pure phase [7].

From the comparison of the results on bulk composition with the catalytic behaviour it follows that an $Al_4B_2O_9$ phase revealing a high content of BO_4 species shows low activity and selectivity in the oxidative dehydrogenation of propane. Higher propane conversions were achieved on B-Al-O catalysts which reveal a high relative concentration (87 - 92%) of BO_3 species indepent of phase composition. The composition responsible for the preferential formation of oxygenates is still an open question. An involvement of a $Al_6B_8O_{21}$ phase in the formation of oxygenates might be assumed since the only catalyst revealing high oxygenates formation consists of this phase (cf Table 1 and 2); however, it should be considered, that oxygenates formation may also be due to secondary reactions of propyl radicals in the gas phase [4].

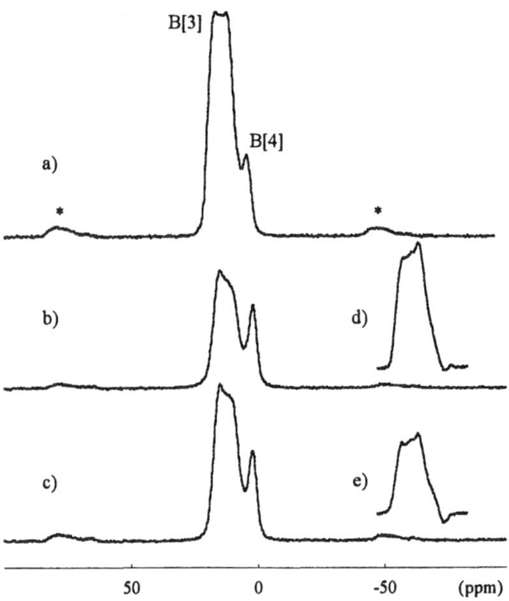

Figure 2. ^{11}B MAS NMR spectra of B_2O_3 /Al_2O_3 (A) catalyst: a) fresh; b) after 7 h at 550°C, C_3H_8/O_2 = 2; c) after treatment with 20 % H_2O, 5 % O_2 (in N_2) at 550°C for 3 h; d) difference spectrum (a)-(b), e) difference spectrum (a)-(c). Asterisks denote spinning sidebands.

3.3. Changes in catalyst compositions during catalytic reaction

The catalyst in which boron-containing crystalline phases were absent (sample A) showed a high initial activity towards propane dehydrogenation; the propene yield amounted to 18 % (X_{C3H8} = 47%, X_{O2} = 99 %, S_{C3H6} = 39 %, S_{C2H4} = 25 %, $S_{oxygenates}$ = 4 %, S_{CO} = 20 %, S_{CO2} = 1 %) after 0.5 h time-on-stream at 550°C, C_3H_8/O_2 =2 and τ = 1 g s ml^{-1}. After 7 h of operation the propene yield decreased to 11 % (X_{C3H8} = 30 %, X_{O2} = 99 %, S_{C3H6} = 36 %, S_{C2H4} = 17 %, $S_{oxygenates}$ < 1 %, S_{CO} = 24 %, S_{CO2} = 13 %). The decrease in propene yield was not only due to lower propane conversion but to a decrease in propene selectivity in favour to CO_2. To elucidate the reasons which caused a decrease of propene formation ^{11}B NMR spectra of fresh and used catalysts were compared (cf. Fig.2). From the difference in the peak areas of the spectra the absolute loss of boron could be determined. As seen from the difference spectrum (Fig. 2d), mainly a loss of BO_3 species (44 % from the boron amount in the fresh catalysts) occured within 7 h of time-on-stream on the catalyst consisting of only amorphous boron compounds. In Fig. 2c, the spectrum after treatment of the fresh catalyst with H_2O at 550°C for 3 h is given. The difference spectrum confirmed a similar loss of BO_3 (31 % from the initial boron amount) due to the presence of water in the feed gas. Thus, it may be safely concluded that the water formed during the reaction causes the loss of BO_3 species most probably as boric acid under reaction conditions. Another kind of surface species (BO_4) being present in the amorphous phase can be attributed according to Peil et al. [8] to the inclusion of tetrahedral boron into the alumina lattice. The amount of these species did not change during the reaction; thus, there is no evidence for the involvement of the four-coordinated boron being incorporated in the alumina lattice in the formation of the target products. BO_3 rather than BO_4 are obviously required for propane conversion to propene. The volatilization of the active component led to a boria-free alumina resulting in an increased formation of CO_2 after 7 h of time-on-stream.

The initial activity of catalyst (B) consisting of crystalline B_2O_3 and two aluminium borates was lower compared to sample (A); propane conversion amounted to 34 % after 0.5 h of operation ($p(C_3H_8) = 40$ kPa, $p(O_2) = 20$ kPa, $\tau = 2$ g s ml^{-1}). Partial oxidation products (e.g. propionaldehyde, acetone, acrolein, propanol, acetaldehyde, acetic acid, methanol, formaldehyde, formic acid, methylformate) were formed with selectivities of up to 30 %. Within 20 h of operation propane conversions amounted to 28 - 30 %. Contrary to sample (A), no increase in CO_x formation was observed with increasing time-on-stream; selectivities towards propene and oxygenates amounted to 82 - 85 %. After 60 h of operation propane conversion was 21 % (S_{C3H6} = 53 %, S_{C2H4} = 15 %, $S_{oxygenates}$ = 22 %).

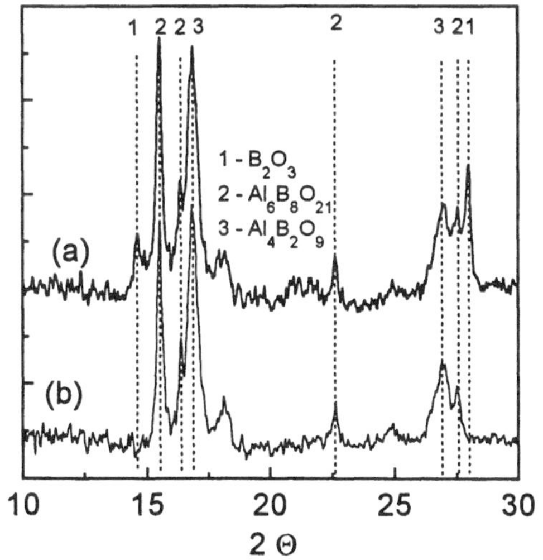

Figure 3. X-ray diagrams of B_2O_3/Al_2O_3 (B) catalyst; (a) -fresh; (b) after 60 h at 550°C, $C_3H_8/O_2 = 2$.

The characterization of the used catalyst (B) by XRD showed that the borate phases being present in the fresh catalyst remained unchanged but no crystalline B_2O_3 was present (cf. Fig. 3). No new phases compared to the fresh sample were formed during the reaction. The decrease of activity was due to a loss of boron (ca. 37 % from initial amount) as confirmed by ^{11}B NMR measurements; this caused also a decrease of the concentration of boron in the upper surface as derived from XPS. The B/Al ratio amounted to 0.7 after 60 h of reaction. It might be assumed that crystalline boria acted as a reservoir to replenish surface boria lost as proposed by Curtin et al [9] for similar catalysts used for the Beckmann rearrangement reaction. The boria-free surface of sample (B) consisting of borate phases had a less detrimental effect on selectivity compared to catalyst (A), on which the liberation of alumina surface resulted in an increased CO_2 formation.

As illustrated by the catalytic experiments, the loss of boria cannot be avoided especially when operating at high degrees of propane conversion due to the effect of water formed. The volatilization of boria seems to be slower if the initial catalyst consists of crystalline boria and aluminium borates but it cannot be completely avoided. Additional experiments were performed over an extended period of time at low propane conversions at which only minor amounts of water were formed. The XRD analysis of the used catalyst showed the presence of crystalline B_2O_3 and an increased content of the inactive $Al_4B_2O_9$ phase; a slight decrease of the B/Al ratio from 1.3 to 1.1 was derived from XPS measurements. Thus, the loss of active BO_3 species occurs under certain reaction conditions not only by volatilization but also by solid-state reaction with alumina forming inactive borate.

4. CONCLUSIONS

Catalysts prepared by impregnation of alumina with boric acid are suitable catalysts for the partial oxidation of propane; selectivities towards olefins and oxygenates of about 80 % can be obtained even at propane conversions above 20 %. Different boron-containing phases are formed in the B-Al-O system depending on the preparation procedure. It was shown that the trigonal BO_3 species being present in both crystalline and amorphous phases are active in the dehydrogenation of propane to propene. The water formed during the reaction causes a loss of these species most probably as boric acid. An $Al_4B_2O_9$ phase which is easily formed in the B-Al system under conditions applied cannot be considered as an active phase but the $Al_6B_8O_{21}$ phase seems to be involved in the oxygenates formation. In addition to volatilization of boria a solid state reaction resulting in inactive $Al_4B_2O_9$ have to be considered especially when operating at low conversions. The loss of boron can be avoided by using a $Al_{18}B_4O_{33}$ phase revealing high activity but due to increased CO_2 formation this phase cannot be considered as promising one.

Acknowledgements

This work was supported by the German Federal Ministry for Education, Science, Research and Technology (Contract No. 03C3012 0) and by the Department for Science, Research and Culture of the State of Berlin. Thanks are due to Dr. M. Schneider and Mrs. D.-C. Uecker for XRD analysis and to Dr. P. Druska for XPS measurements as well as to Mrs. R. Jentzsch for catalytic experiments.

REFERENCES

1. F. Cavani and F. Trifirò, Catal. Today 24 (1995) 307.
2. H.H. Kung, Adv. Catal 40 (1994) 1.
3. O.V. Buyevskaya, M. Kubik, M. Baerns, ACS Symp. Ser. 638 (1996) 155.
4. O.V. Buyevskaya and M. Baerns, Catal. Today, in print.
5. H. Pines, The Chemistry of Catalytic Hydrocarbon Conversions, Academic Press, New York, 1981, p. 215.
6. D. Müller, A.-R. Grimmer, U. Timper, G. Heller, M. Shakibaie-Moghadam, Z. anorg. Allg. Chem. 619 (1993) 1262.
7. M. Garsche, E. Tillmanns, H. Almen, H. Schneider and V. Kupcik, Eur. J. Mineral 3 (1991) 793.
8. K.P. Peil, L.G. Galya and G. Marcelin, J. Catal. 115 (1989) 441.
9. T. Curtin, J.B. McMonagle and B.K. Hodnett, Appl. Catal. A: General 93 (1992) 91.

NATURAL GAS CONVERSION V
Studies in Surface Science and Catalysis, Vol. 119
A. Parmaliana et al. (Editors)

NEW CATALYSTS AND CATALYTIC PROCESSES TO PRODUCE HYDROGEN AND SYNGAS FROM NATURAL GAS AND OTHER LIGHT HYDROCARBONS

V.N. Parmon, G.G. Kuvshinov, V.A. Sadykov, V.A. Sobyanin

Boreskov Institute of Catalysis
630090 Novosibirsk, Russia

The paper discusses new honeycomb catalysts for natural gas oxidation to syngas, a recently developed processes of moderate-temperature (450-650°C) catalytic pyrolysis of natural gas and/or light hydrocarbons into hydrogen and valuable carbonaceous materials, as well as partial oxidation of natural gas into syngas in high temperature solid oxide fuel-cell-like systems. The advantages of the new processes against conventional ones lie, e.g., in ability to produce hydrogen completely free of carbon monoxide, as well as to cogenerate hydrogen or syngas with other valuable materials, chemicals or even electricity.

1. INTRODUCTION

Natural gas and other available light hydrocarbons remain to be the main source for the cheapest hydrogen and syngas. The most conventional and cheap technology here is a catalytic steam or carbon dioxide reforming of hydrocarbons; such reforming being now among the largest scale processes of chemical industry. However, the technology advanced application meets several shortcomings. E.g., in hydrogen energetics, the reforming produces hydrogen mixed with carbon oxides, thus needing further very difficult and energy consuming extraction and cleaning of hydrogen from syngas obtained (see, e.g., [1]).

Thus, principal improvement or new concept in hydrogen and syngas production from light hydrocarbons and methane, especially, to avoid mentioned disadvantages, appear to be very important. Below we discuss some new approaches, which are under development at the Boreskov Institute of Catalysis.

2. NEW EFFICIENT HONEYCOMB CATALYSTS FOR PARTIAL OXIDATION OF METHANE TO SYNGAS

The catalytic partial oxidation of methane (POM) to syngas at short contact times is an alternative process to the conventional steam reforming of methane. POM has numerous advantages: this process is mildly exothemic and produces syngas with the H_2/CO ratio at about 2, which is ideal, e.g., for further methanol or Fischer-Tropsch synthesis. Also, the POM process requires much smaller reactors at lower investment costs.

Recently, a lot of catalysts for POM have been reported. High conversions and selectivities in the autothermal mode were obtained over rhodium (Pt-Rh gauze; 10-20% of Rh on alumina foam or cordierite extruded monoliths) catalysts [2-6]. Good performance was also demonstrated for catalysts likes pyrochlores La_2Ru_2(or Ir_2)O_7 [7-9] or perovskites ($LaRhO_3$) containing rare precious metals [10]. However, extremely high cost of such catalysts, caused by a high content of precious metals limits their future industrial application.

Sufficiently good performance was reported for oxide systems like NiO-CaO [11], NiO-MgO [12], CoO-MgO [13], NiO-rare earth oxides [14], Ni/Al_2O_3 [15], Ni-containing complex oxide systems [16]. However, these catalysts require, as a rule, longer contact times to achieve good activities and selectivities, and fail to keep autothermal mode. Their long-term stability in real operating conditions, especially at elevated pressures of reagents, remains to be an open problem due to recrystallization, coking, promoter loss, etc. The oxide catalysts performance appears to be extremely sensitive to the method of their synthesis. Moreover, most studies were conducted with powdered catalysts, while for practical application monolith catalysts with a low pressure drop are mostly suitable. Hence, the problem of cheap and efficient POM catalysts design is not yet solved.

In this work, we report on new very active, selective and thermally stable supported monolith POM catalysts based on mixed oxides of La, Co, Ni with perovskite structure containing only a small (< 0.5%) amount of noble metal (Rh, Ir). As a support, a proprietary thin-wall (wall thickness ca. 0.25 mm) straight-channel (channel size ca. 1 mm) corundum monolith with the specific surface area in the range of 1-10 m^2/g was used. The cumulative pore volume of the support was 0.6-0.7 m*l*/g, mean pore radius — ca. 30-40 μm. To increase its thermal shock resistance, a special coating by some refractory oxides or complex phosphates has been used. The catalytic tests in POM were conducted using an integral plug-flow reactor at contact times in a range of 0.1-0.02 s. In most cases, the reaction mixture CH_4: O_2: He = 2:1:12 was used. To check catalysts stability at higher concentrations of methane, the composition was changed for CH_4: O_2: He = 2:1:5.

Fig. 1 illustrates the developed catalyst performance compared to that of the supported pure Rh catalyst with a higher Rh loading. The promoted supported perovskite has a higher performance and selectivity to syngas, especially at short contact times. In terms of activity and selectivity, the promoted perovskites are superior of the catalysts with a high (5-10%) loadings of Rh on the alumina foam [5]. Even without precious metal additives some perovskites compositions have also demonstrated a good performance at comparable contact times, though at higher operating temperatures.

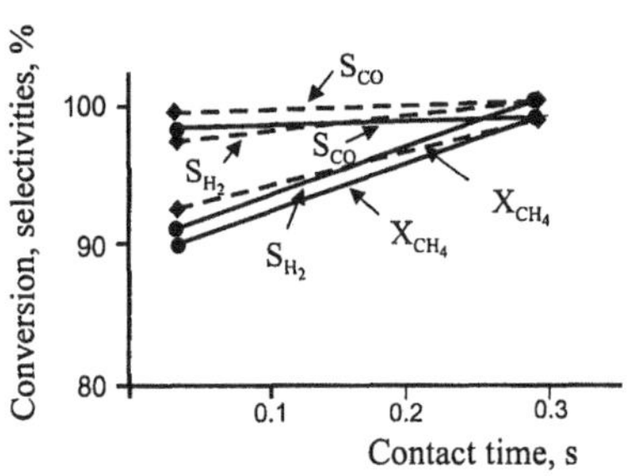

Figure 1. Selective oxidation of natural gas to syngas over honeycomb catalysts 2% Rh/α-Al_2O_3 (solid lines) and new catalyst BIC-6-10 with 0.2%Rh (dashed lines). X — CH_4 conversion, S — CO and H_2 selectivities. Temperature 850°C.

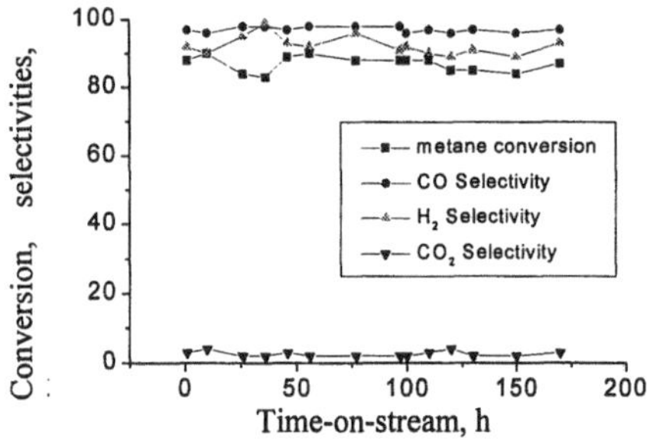

Figure 2. Stability tests for supported perovskite rhodium promoted catalysts at 800°C and CH_4: O_2:He = 2:1:5, contact time 0.03 s. After 110 h on stream, 5 vol% of ethane was added to methane.

Fig. 2 demonstrates a stability test of a $LaNi_{1-x}Rh_xO_y$ catalyst. The lifetime test was done using the feed gas of the composition close to that of natural gas containing ca. 10% of ethane, since higher hydrocarbons are known to be responsible for coking. The results obtained show that high activity and syngas selectivity of the catalyst tested were kept constant during the whole run. Moreover, coke formation on these catalysts was negligibly small. No cracking or active component spallation was observed after a set of thermal shocks from room temperature to 1000°C.

At present, only little is known about the state of the working surface of the studied catalysts in real conditions and on the mechanism of promotion by precious metals. However, one may assume a destruction of, at least, the surface layer of perovskite into lanthanum oxo(hydroxo)carbonate with partially oxidized clusters of transition metals being segregated. The precious metal additives can be incorporated into such clusters, forming mixed centers with improved performance and selectivity.

3. CARBON-OXIDES FREE CATALYTIC TRANSFORMATIONS OF METHANE AND LIGHT HYDROCARBONS INTO HYDROGEN AND VALUABLE CHEMICALS OR MATERIALS

The main idea of this approach is to transform hydrocarbons into hydrogen and valuable materials via direct and controllable pyrolitic processes. Examples of such processes are a high temperature catalytic pyrolysis of methane into H_2 and C_2-hydrocarbons

$$CH_4 \rightarrow H_2 + C_2H_4, C_2H_2, \qquad (1)$$

and a moderate temperature catalytic pyrolysis of CH_4 and higher hydrocarbons into hydrogen and a valuable carbonaceous material via reactions like

$$CH_4 \rightarrow 2H_2 + C. \qquad (2)$$

Highly endothermic process (1) is well known for many decades and is used as a base for either acetylene production in an electric arc or hydrogen production in nonequilibrium microwave plasma discharges. According to thermodynamics, predominant formation of H_2 and C_2-hydrocarbons should be achieved at temperatures above 1000-1100°C. The main reason of a yet not wide application of the mentioned plasmachemical technologies is an intensive coproduction of soot via rapid competing reaction (2). So, to obtain a high yield of C_2-hydrocarbons on the high-temperature (above 1100°C) pyrolysis of methane, one should overcome a very important problem of superrapid initial reagents heating and products quenching. Numerous attempts to reach the needed conditions of such quenching in plasmachemical and electric arc devices provide a moderate success due to a large enough geometric size of conventional types of plasma discharges, moreover being also nonpenetrative for the feeding gas.

A controversial approach to solve the problem of superrapid heating and quenching of the product mixture is to use specially designed catalytic reactors with very thin bed of a catalyst (or, more strictly, a catalyst-heat exchanger) heated by some sophisticated methods like microwave or IR-visible light irradiation. This unusual approach is under development [17], and could appear to be of practical interest.

Another approach to improve the parameters of the methane pyrolysis parameters is to follow reaction like (2), but producing a much more valuable carbon material rather than conventional soot. The respective technology can be based on pyrolysis of natural gas or other light hydrocarbons to carbon and hydrogen

$$C_mH_n \rightarrow C + H_2\ (+CH_4) \qquad (3)$$

at 450-650°C under atmospheric pressure over nickel or nickel/copper containing catalysts.

On the first glance, the above process is not suitable for hydrogen production, since thermodynamics limits hydrogen yeld at temperatures below 700°C, and the catalysts are deactivated by the carbon obtained. Indeed, usually processes (2) and (3) are considered only with regard to the problem how to prevent carbon deposition on the reforming catalysts [18].

The newly proposed concept of hydrogen production via a moderate temperature catalytic pyrolysis is based on a possibility to produce an additional valuable product — new filamentous carbonaceous materials [19-22]. Of importance is that the newly designed catalysts for the filamentous carbon growing [23] allow one to obtain the yield of carbon in amount of 200-400 times greater than the starting mass of the pyrolysis catalyst, and to enrich the treated methane 30-40 vol% hydrogen. Note, that it is much easier to use such methane-hydrogen containing mixture for, e.g., fueling the low temperature fuel cells than to use syngas with a large content of carbon monoxide inhibiting platinum electrodes, etc. Also, methane is much more easily separated from hydrogen than carbon oxides.

The growing carbonaceous filaments contain a nickel or nickel/copper particle on the top (see Fig.3 and [24,25]). The filament structure resembles graphite, its basal planes being not parallel to the filament axis. The edges of basal planes form the filament surface.

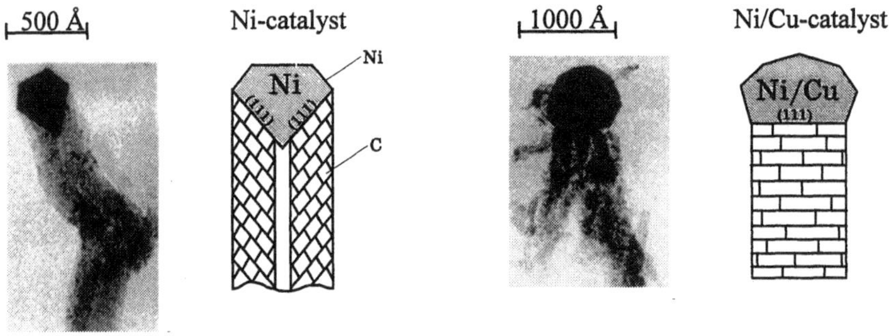

Figure 3. Examples of the catalytic filamentous carbon obtained together with hydrogen at the middle-temperature catalytic pyrolysis of methane and low hydrocarbons.

Pyrolysis filamentous carbon has some advanced properties to be used as promising sorbents, catalyst supports, etc. Also, it can be used to produce carbonaceous composite materials, etc. The properties of the latter can be easily controlled by the pyrolysis catalyst composition as well as by process conditions [26-28].

In the case of coproduction of hydrogen and the pyrolysis filamentous carbon, the main task is to obtain a maximum yield of the carbon material using a minimum catalyst amount. Studies on the kinetics of carbon formation from CH_4-H_2 mixtures over Ni-containing catalysts showed that hydrogen concentration is one of the main parameters of the process. The lower hydrogen concentration, the higher the rate of carbon deposition and catalyst deactivation. At low hydrogen concentration (< 2.5 vol%), the maximum amount of the carbonaceous material produced per unit catalyst mass until complete catalyst deactivation (ca. 50 g/g cat) does not depend on the temperature. When the hydrogen concentration is high (> 10%), the value can reach 150-400 g/g cat; an deactivation time increases from minutes to hours.

A principal scheme of the suggested process is shown in Fig.4. Basic elements are reactor 1 for hydrocarbons decomposition and hydrogen separator 2.

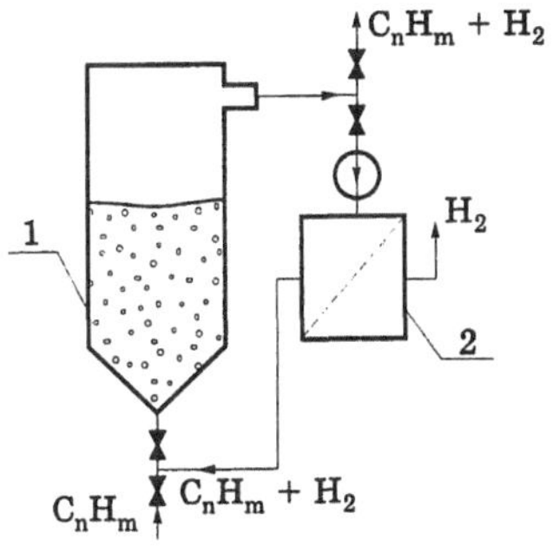

Figure 4. A tested technological scheme of the process of the moderate-temperature catalytic pyrolysis of methane and light hydrocarbons.

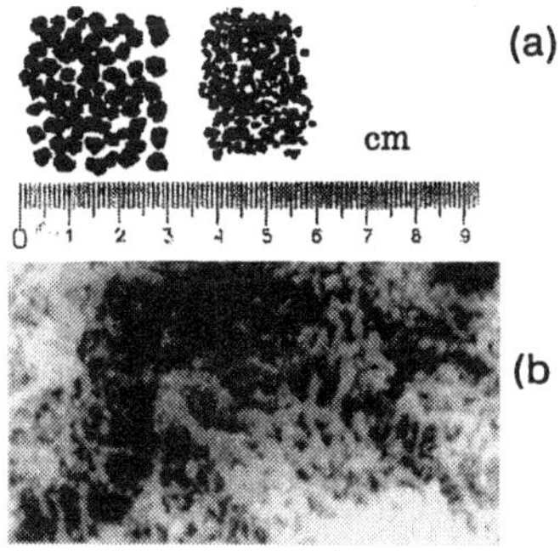

Figure 5. Outward appearance (a) and micrograph of the surface (b) of the granular catalytic carbon of filamentous structure obtained in the pilot setup.

A fresh gas feed mixed with the recycled flux of the gaseous reaction products, is directed to the reactor inlet to be partially decomposed into hydrogen and carbon. Carbon obtained on the catalyst stays in the reactor, while the gaseous reaction products go to a system for partial hydrogen separation and then return to the reactor inlet. As there are no gaseous or liquid byproducts, and regarding to hydrogen separation, nearly complete conversion of treated hydrocarbons into hydrogen and granular carbon can be obtained. In case some inert impurities are present in the treated gas, the conversion will be a little bit lower due to necessity to remove the inert gas accumulated in the system.

To prevent agglomeration of growing carbonaceous particles and to produce granulated carbon, we use the fluidized bed reactor which allows process performance in both continuous and periodic modes (see [1,27]). The yield of hydrogen in all modes was registered as high as 30-40%, in accordance to theoretical predictions. A method of an efficient supply of the heat energy into the reactor was elaborated as well.

To study the catalytic pyrolysis, both laboratory-scale and pilot (operating volume of the reactor 5 dm^3) setups were designed and tested. Carbon yield per catalyst unit mass until complete catalyst deactivation was regarded as the most important process parameter [1,26,27].

The main problems on upscaling are related to a tendency of the carbonaceous material to agglomerate, and thus decrease the carbon yield. However, it is possible to remove the agglomeration in the pilot reactor, adding some inert substrates, while a high yield of the carbonaceous material is retained via diminishing the methane flux through the reactor. The results obtained show a good practical prospects of the process scheme for the pilot reactor.

In all cases the carbonaceous material produced is nearly pure carbon with no more than 1 wt% admixtures. The carbonaceous granules, obtained in the pilot reactor, appear to be mesoporous particles 1-5 mm in the size composed of bounded graphite-like filaments (Fig.5). The experiments show that the properties of the carbonaceous material are well reproduced, and do not seriously depend on reactor size.

4. COGENERATION OF ELECTRICITY AND CHEAP SYNGAS VIA OXIDATIVE REFORMING OF METHANE IN SOLID OXIDE FUEL CELL SYSTEMS

The idea of this cogeneration is a possible use of high temperature fuel cells as both electricity generators and a catalytic reactors producing valuable chemicals.

An example is electrochemical oxidation of methane to syngas over a Pt electrode-catalyst in a high temperature cell with the yttria stabilized zirconia (YSZ) as a solid oxygen conducting electrolyte [29,30]. Schematically, such a cell is similar to the solid oxide fuel cell (SOFC):

CH_4 (+ N_2), Pt | YSZ | Pt, air.

At the electrochemical oxidation of methane in the SOFC system, the feeding methane flow does not contain oxygen at all. It is fed directly into reaction zone by passing an anodic current through the cell. In this case oxygen ions O^{2-} form at the air-blown electrode (cathode). Then, these ions transfer to the electrode-catalyst (anode) through the YSZ electrolyte. On the anode they can either discharge to produce O_2 or oxidize methane directly.

In the experiments, we used an electrochemical cell based on a YSZ tube, with two porous Pt electrodes (the anode and the cathode) containing 5 wt% of YSZ supported on the inner and outer surfaces of the tube. The geometrical surface area of the electrodes was 3 cm^2 and their thickness was ca. 10 μm. The experiments were carried out at atmospheric pressure within a temperature range of 660-800°C. The effects of temperature, residence time, methane concentration and current (or the electrode potential) on the process of methane conversion were studied. Typical experimental results [29,30] are presented on Figs. 6 and 7.

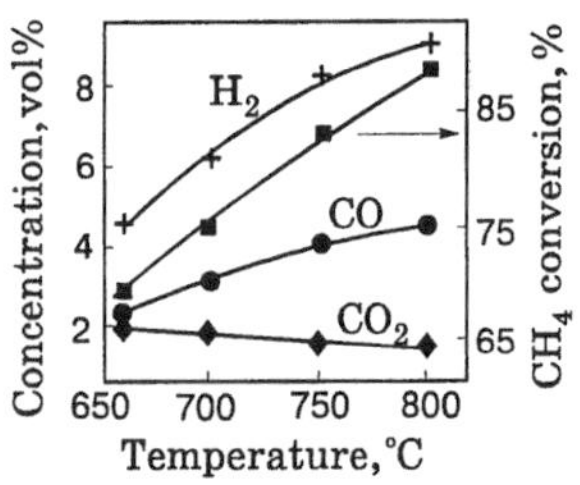

Figure 6. Effect of temperature on CH_4 conversion and H_2, CO and CO_2 product concentrations at electrochemical oxidation of methane. Flow rate 0.43 cm^3/s; inlet methane concentration 6.5 vol%; anodic current 300 mA.

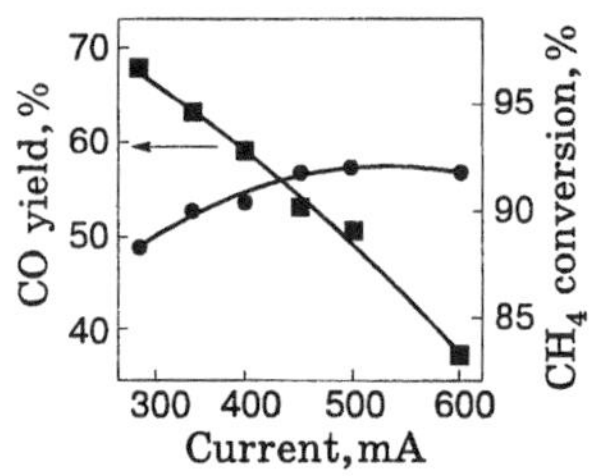

Figure 7. Effect of the anodic current on CH_4 conversion and the CO yield at electrochemical oxidation of methane. Temperature 800°C; flow rate 0.43 cm^3/s; inlet methane concentration 6.5 vol%.

In Fig.6 the methane conversion and outlet concentrations of H_2, CO and CO_2 are plotted against temperature at CH_4 electrochemical oxidation. Practically, a 100% conversion of the electrochemical oxygen pumped to reaction zone was observed, and only H_2, CO, CO_2 and H_2O were detected there as products. An increase in temperature results in a lower outlet concentration of carbon dioxide as well as in a higher outlet concentrations of hydrogen and carbon monoxide. At 800°C, the CO yield reaches 69%.

Fig.7 shows the CH_4 conversion and the yield of CO versus the anodic current. The conversion is equal to ca. 90% and practically does not change with the increase in anodic current, while CO yield decreases from 69 to 37%.

These results allow a conclusion that the Pt electrode is an active electrode-catalyst for the direct electrochemical conversion of methane to syngas in the SOFC system. Similar results were reported recently for the Ni [31,32] and Fe [33] electrode-catalysts as well. The recent experiments show a possibility to use not only methane-nitrogen mixture but also pure methane without essential coking of the catalyst-anode.

Table 1 presents the highest yields of CO produced by the electrocatalytic conversion of methane over Fe, Pt and Ni electrode-catalysts. These values were obtained at a practically complete conversion of electrochemically pumped oxygen, the inlet ratio of methane and oxygen flows being close to stoichiometry ($CH_4/O_2 = 2 \pm 0.2$). Table 1 shows considerable yields of CO ($\approx$ 65-97 %) over all tabulated electrode-catalysts, thus indicating their high electrocatalytic activity for syngas production from methane. The results are quite natural since both Fe, Pt and Ni are known as active catalysts for a common catalytic partial oxidation of methane to syngas.

Table 1
CO yields at electrocatalytic conversion of methane to syngas in the SOFC reactor

Electrode-catalyst	Temperature, °C	CO yield, %	Ref.
Fe	950	73	[33]
Pt	1100	97	[34]
	800	65	[29,30]
Ni	800-850	88	[31,32]

Note also that electricity is cogenerated together with syngas generation in the above mentioned fuel-cell-like systems.

5. CONCLUSIONS

The presented examples of the recent studies on new catalysts and some innovative processes for hydrogen and syngas production from natural gas and other light hydrocarbons evidence for good perspectives of principally new technologies, which will be able to compete with well elaborated and nowadays huge industrial scale technologies. The technologies under elaboration allow one either to produce hydrogen free of heavily extractable carbon oxides or to cogenerate simultaneously hydrogen and some valuable materials or even electricity.

Evidently, a simplicity and environmental harmless of the moderate-temperature catalytic pyrolysis of natural gas, as well as its ability to produce hydrogen simultaneously with a valuable carbonaceous material show this process as one of the most effective and promising processes to produce hydrogen from hydrocarbons in future. An evident advantage of this process for the nowadays needs is possibility of its direct application for fueling the low-temperature hydrogen fuel cells without a reaction mixture separation. Moreover, this process may be regarded as an alternative for even industrial steam reforming, since it does not pollute atmosphere with carbon dioxide.

Some expected advantages has also the electrochemical oxidation of methane to syngas in the SOFC systems in comparison to conventional partial catalytic oxidation. Thus, (i) it is possible to produce not only syngas alone but electricity as well; (ii) the probability of the reaction mixture to explode decreases, since feeded methane and air are everywhere separated by the solid electrolyte membrane, and it is easy to control the oxygen amount in the reaction zone by regulating electric current passing through the SOFC system; (iii) when air is used as an oxidant, no additional nitrogen accumulates in the reaction zone, since oxygen supplied for the methane partial oxidation is separated from other components of air just at the stage of its electrochemical supply to the reaction zone. Note, that oxygen separation from air is typically the most energy expensive stage of any oxidative conversion of light hydrocarbons into hydrogen, carbon oxides and other oxygenates.

For electrochemical oxidation of methane to syngas in the fuel-cell systems, there is also a possibility to use water vapor and carbon dioxide instead of air, e.g., by arranging systems like "CH_4, electrode | YSZ | electrode, H_2O" or "CH_4, electrode | YSZ | electrode, CO_2". The summed-up electrode semireactions expected in such systems represent the overall chemical reactions of steam and carbon dioxide reforming of methane. However, unlike conventional catalysis, their electrochemical performance allows arrangement of the required ratio CO/H_2 in the syngas via mixing the outlet gas flows from the anodic and cathodic compartments of the SOFC system.

One should expect that the electrochemical oxidation of CH_4 to syngas in the SOFC systems will be a subject for extensive and deep studies for the future hydrogen technologies.

REFERENCES

1. V.N.Parmon, G.G.Kuvshinov, V.A.Sobyanin, in: Hydrogen Energy Progress XI, Proc. of the 11th World Hydrogen Energy Conference, Stuttgart, 23-28 June, 1996, 1, vol. 3, 2439.
2. D.A.Hickman and L.D.Schmidt. J. Catal., 138 (1992) 267.
3. D.A.Hickman and L.D.Schmidt, in: Catalytic Selective Oxidation, ACS Symp. Ser., (1993) 416.
4. A.Hickman, E. A.Haupfear and L.D.Schmidt, Catal. Lett., 17 (1993) 223.
5. P.M.Torniainen, X.Chu and L.D.Schmidt, J.Catal., 146 (1994) 1.
6. L.D.Schmidt and C.T.Goralski. Studies in Surf. Sci. Catal., 110 (1967) 491.
7. R.H.Jones, A.T.Ashcroft, D.Waller, A.K.Cheetham and J.M.Thomas, Catal. Lett., 8 (1991) 169.
8. T.Ashcroft, A.K.Cheetham, J.S. Food, M.L. H. Green, at al. Nature, 344 (1990) 319.
9. GB Patent No 2239406.
10. Slagtern and V.Oslbye. Appl. Catal. A, 110 (1994) 99.
11. V.R.Choudhary, A.M.Rajput and B.Prabhakar, Catal. Lett., 15 (1992) 363.
12. V.R.Choudhary, A.S.Mamman and S.D.Sansare. Angew. Chem. Int. Ed., 31 (1992) 1189.
13. Y.-F.Chang and H.Heinemann, Catal. Lett., 21 (1993) 215.
14. V.R.Choudhary, V.H.Rane and A.M.Rajput, Catal. Lett., 22 (1993) 289.
15. D.Dissanayake, M.P.Rosynek, K.C.C.Kharas and J.H.Lunsford, J. Catal. 132 (1992) 117.
16. Miao, G.Xiong, Sh.Sheng, W.Cui, L.Xu and X.Guo. Appl. Catal. A: 154 (1997) 17.
17. V.I.Fedoseev, Yu.Yu.Tanashev, Yu.I.Aristov and V.N.Parmon, Kinet. Catal., 37 (1996) 869.
18. J.R.Rostrup-Nielsen. Catalysis Sci. Technol., Berlin: Springer, 5 (1984) 73.
19. R.T.K.Baker, Carbon, 27 (1989) 315.
20. V.B.Fenelonov, L.B.Avdeeva, V.I.Zheivot, L.G.Okkel, at al. Kinet. Catal., 34 (1993) 483.
21. L.B.Avdeeva, O.V.Goncharova, G.G.Kuvshinov, at al., Rus. Patent No 2064889, 1993.
22. V.A.Likholobov, V.B.Fenelonov, L.G.Okkel, et al., React. Kinet. Catal. Lett., 54 (1995) 381.
23. M.A.Ermakova, D.Yu.Ermakov, G.G.Kuvshinov, Kinet. Catal., 39 (1998) No 4 (in press).
24. V.N.Parmon, Catal. Lett., 42 (1996) 195.
25. V.N.Parmon, Catal. Today (1998) (in press).
26. L.B.Avdeeva, G.G.Kuvshinov, O.V.Goncharova, Yu.I.Mogilnikh and V.A.Likholobov, Proc. of the 1st World Congress Envir. Catalysis, Italy, Pisa, (1995) 459.
27. G.G.Kuvshinov, Yu.I.Mogilnykh, D.G.Kuvshinov, S.G.Zavarukhin, V.N.Parmon, in: Hydrogen Energy Progress XI, Proc. 11th World Hydrogen Energy Conference, 1996, vol.1, 655.
28. . G.G.Kuvshinov, Yu.I.Mogilnykh, D.G.Kuvshinov, V.I.Zaikovskii at al., Carbon, 39 (1998) 87.
29. V.V.Gal'vita, V.D.Belyaev, V.N.Parmon and V.A.Sobyanin, Catal.Lett., 39 (1995) 209.
30. V.V.Gal'vita, V.D.Belyaev, V.A.Sobyanin, Kinet. Catal., 38 (1997) 736.
31. V.V.Gal'vita, V.D.Belyaev, A.K.Demin and V.A.Sobyanin, Appl. Catal. A, 165 (1997) 301.
32. V.A.Sobyanin, V.D.Belyaev and V.V.Gal'vita, Catal. Today (1998) (in press).
33. H.Alquahtany, D.Eng and M.Stoukides, J. Electrochem. Soc., 140 (1993) 1677.
34. T.J.Mazanec, T.L.Cable and J.G.Frege, Solid State Ionics, 53-56 (1992) 111.

NATURAL GAS CONVERSION V
Studies in Surface Science and Catalysis, Vol. 119
A. Parmaliana et al. (Editors)

Modeling the Partial Oxidation of Methane to Syngas at Millisecond Contact Times

L.D. Schmidt, O. Deutschmann, and C.T. Goralski, Jr.

Department of Chemical Engineering and Materials Science, University of Minnesota, Minneapolis, Minnesota 55455

Synthesis gas can be produced by the autothermal direct oxidation of methane over rhodium coated monoliths at contact times on the order of several milliseconds. Modeling these reactors is useful both to better understand the system as well as to predict operation in regimes where experiments are either difficult or dangerous. These systems are difficult to model because the gradients are very large, parameters are strong functions of temperature and concentration, and the processes are very sensitive to catalyst geometry. We have simulated these processes using both a plug flow model to investigate the detailed gas phase and surface chemistry within the monolith as well as a 2-dimensional model to investigate the role of temperature and composition dependent transport with spatial gradients in temperature, concentration, and velocity.

1. PARTIAL OXIDATION OF METHANE AT MILLISECOND CONTACT TIMES

Synthesis gas is a very important chemical intermediate for many relevant processes including the production of methanol and Fischer-Tropsch synthesis of synthetic fuels. Syngas is currently produced by the endothermic steam reforming of methane or by homogeneous oxidation in autothermal reforming. There has recently been significant interest in alternative routes to syngas production, and one of the most promising technologies has been the development of a high temperature, short contact time partial oxidation reactors where methane is converted to syngas via direct partial oxidation with oxygen over rhodium catalysts [1]. Using this type of reactor, selectivities to syngas in excess of 90% has been achieved at methane conversions over 90% [2]. The symbols in Figure 1 show typical selectivities, conversions, and temperatures for the partial oxidation of methane in oxygen at various fuel to oxygen ratios.

The reactor used in these experiments was a well insulated quartz tube reactor 18 mm in diameter using an α-alumina ceramic foam monolith coated with ~5% rhodium as a catalyst. The catalyst had a void fraction of 0.79 and a nominal porosity of 45 pores per linear inch (ppi). Gases were fed to the reactor at a total flow rate of 5 slpm, resulting in residence times on the order of 4 ms at reaction temperatures [2]. The catalyst was insulated on each side with an inert monolith to act as a radiation shield. The temperature of the upstream face of the catalyst was in each case within 50 °C of the temperature of the downstream face. The temperature of the downstream face was within 25 °C of the calculated adiabatic reaction temperature for the exhaust composition for each case examined.

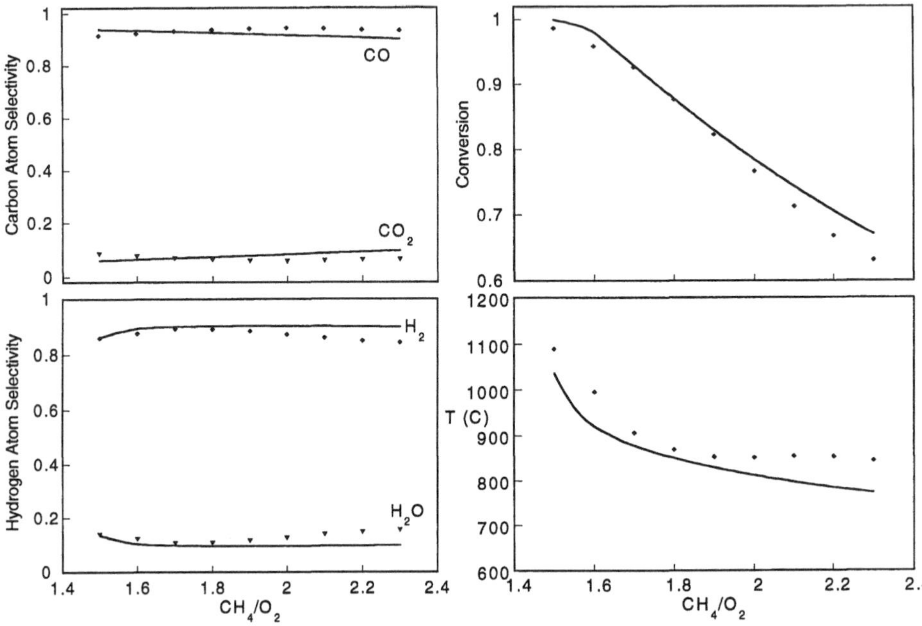

Figure 1. Syngas production at atmospheric pressure. Symbols represent experimental data [2]. Solid lines are PFTR modeling results. 45 ppi Rh catalyst, 30% N_2 diluent, 4 ms residence time.

2. MODELING PARTIAL OXIDATION REACTORS

Detailed modeling of high temperature oxidation reactors is a difficult problem because the very fast reaction and mass transfer rates generate equations that are stiff both in space and time. The temporal stiffness can be avoided by solving steady state problems, but detailed steady state models often require so much computing time to solve that parameter studies are impractical. We have approached this problem by tackling individual pieces of the problem separately to try to understand the global importance of all parameters in the system. To this end, we have developed two separate computer models of millisecond contact time partial oxidation reactors. The first is a plug flow tubular reactor (PFTR) model to study the interaction of heterogeneous and homogeneous chemistry in the partial oxidation of methane over noble metal coated monoliths. The second is a 2-dimensional model that includes only heterogeneous chemistry to study the importance of heat and mass transfer and radial and axial gradients in high temperature catalytic monolith reactors.

3. PLUG FLOW MODEL

Most downstream industrial processes utilizing syngas require feed pressures in excess of 20 atm. It is desirable to operate commercial syngas production reactors at these same pressures in order to avoid expensive compression costs and maximize throughput in the reactor. However, high pressure experiments can be expensive and dangerous. Therefore modeling these reactors at high pressures can help to elucidate the chemistry under these conditions and reduce the number of experiments that must be performed.

The competition between homogeneous and heterogeneous chemistry is very important. The heterogeneous chemistry is much more selective for the formation of syngas, whereas the homogeneous chemistry tends to produce more complete oxidation products and drives up the catalyst temperature. The contribution of homogeneous chemistry is very small at atmospheric pressure. In the absence of heterogeneous chemistry, the homogeneous chemistry takes place on time scales of ~50 ms which is much longer than the 4 ms residence times for the results shown in Figure 1. Unfortunately, the rates of the homogeneous reactions tend to be second order in pressure, whereas the heterogeneous reactions rates are essentially first order in pressure. Therefore, as the pressure in the reactor is increased, the role of homogeneous chemistry becomes more important.

The PFTR model was developed for simulating the detailed heterogeneous and homogeneous chemistry that takes place in a short contact time partial oxidation reactor. The flow in a single channel pore of the catalyst was modeled as a plug flow reactor so that the mass and energy balances for the reactor are reduced to ordinary differential equations that can be solved as an initial value problem. The equations were solved by integrating down the length of the monolith using the ordinary differential equation solver VODE [3]. The gas phase reactions rates were calculated using CHEMKIN [4]. The net flux of species to the surface was calculated by first solving the steady state equations describing the fractional coverage of each surface species and then calculating the net flux from the calculated coverage at each time step in the integration. The net reaction rate for each species at each time step was then calculated as the sum of the net homogeneous reaction rate plus the net surface flux times the surface area to volume ratio for the catalyst. The temperature is assumed to be uniform along the length of the monolith and was calculated iteratively using Newton's method by guessing a temperature and comparing the adiabatic temperature for the resulting exhaust stream to the temperature used in the calculation.

The mechanism used for the homogeneous chemistry was GRI-mech 2.11 [5], an elementary step mechanism for methane oxidation which contains 49 reaction species and 277 reactions. The heterogeneous mechanism used was published by Hickman and Schmidt for the partial oxidation of methane on rhodium catalysts [6]. This mechanism was derived primarily from published experimental data for elementary surface reactions on polycrystalline rhodium surfaces and contains 6 gas phase species, 8 surface species and 19 surface reactions.

The model was verified by comparing model predictions with experimental results. Figure 1 shows both experimental data (symbols) as well as model predictions (solid lines) for the partial oxidation of methane over a 45 pore per linear inch (ppi) catalyst at various fuel to oxygen ratios at atmospheric pressure and a residence time of 4 ms. The inlet stream in both the model and the experiments contained 30% nitrogen diluent. Overall, the model does an excellent job predicting the experimental results. The calculations predict slightly cooler temperatures than those observed experimentally, but this is primarily because the model tends to over predict the hydrogen selectivity by a few percent at high fuel to oxygen ratios and over predicts carbon monoxide by a few percent at low fuel to oxygen ratios. Negligible homogeneous reaction is predicted for these conditions.

Figure 2 shows plots of the mole fractions of the major and minor species as a function of position in the reactor at 20 atm pressure for a 45 ppi rhodium catalyst. The adiabatic reaction temperature for this case is 1365 °C. Most of the reaction takes place in less than 1 mm of catalyst because of this very high reaction temperature. The second panel in Figure 2 shows that although the oxygen is almost completely consumed, homogeneous reactions continue, most notably the reaction of ethane to ethylene and acetylene.

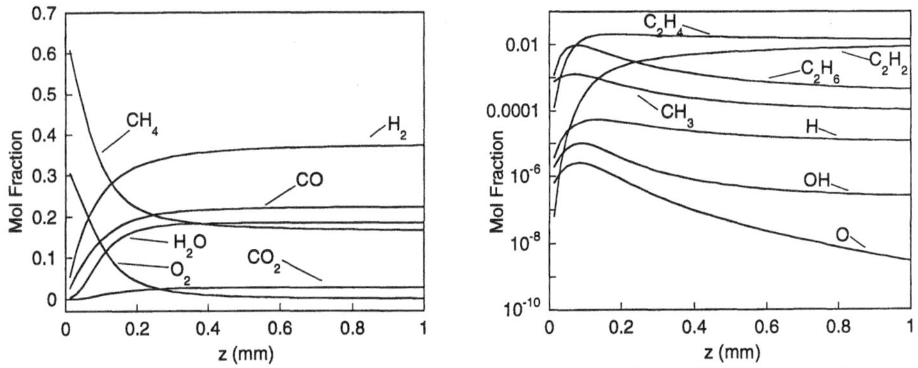

Figure 2. Species mole fractions as a function of axial position from PFTR model. 45 ppi Rh catalyst, 2/1 CH_4/O_2 ratio, $v_{inlet} = 2.5$m/s, T = 1365 °C, P = 20 atm.

Figure 2 shows an example of typical species profiles for a situation where homogeneous chemistry has become important. In general, this is typified by the production of water at the expense of hydrogen (leading to high temperatures) and the production of significant amounts of methane coupling products. The methane coupling reactions are driven by the high temperatures caused by the formation of water. This is illustrated in Figure 3 which shows selectivities to major products, conversion, and temperature for the partial oxidation of methane in oxygen as a function of pressure calculated using the PFTR model. The results for Figure 3 were calculated at a methane to oxygen ratio of 2.0 with no diluent, using both 45 and 80 ppi rhodium catalysts at a constant residence time of 5 ms.

The first panel of Figure 3 shows that at ~15 atm there is a sudden loss in CO selectivity. This is accompanied by only a slight increase in the CO_2 selectivity. At the same pressure, the upper right panel of Figure 3 shows a dramatic increase in the water selectivity and a corresponding decrease in the hydrogen selectivity. This increase in the selectivity to water drives up the temperature of the reactor, causing a dramatic increase in the selectivity to methane coupling products. This can be seen in the lower two panels of Figure 1. At the same time the temperature jumps, the methane conversion falls by more than 10%.

The rather dramatic change in the operation of the reactor at higher pressure is due to the onset of homogeneous chemistry. It is caused by oxygen in the gas phase stripping hydrogen atoms from methane molecules, resulting in the formation of gas phase methyl radicals and water. This increases the temperature of the reactor and allows the formation of methane coupling products by the combination of methyl radicals. The onset of homogeneous chemistry can be delayed by using a catalyst with smaller pores (more pores per inch). This is shown in Figure 3 where the dotted lines show that by using an 80 ppi catalyst, the onset of homogeneous chemistry can be delayed to up to ~23 atm. This occurs because smaller pores sizes allow for the oxygen to be consumed faster on the surface and thus not allowing the homogeneous chemistry to occur.

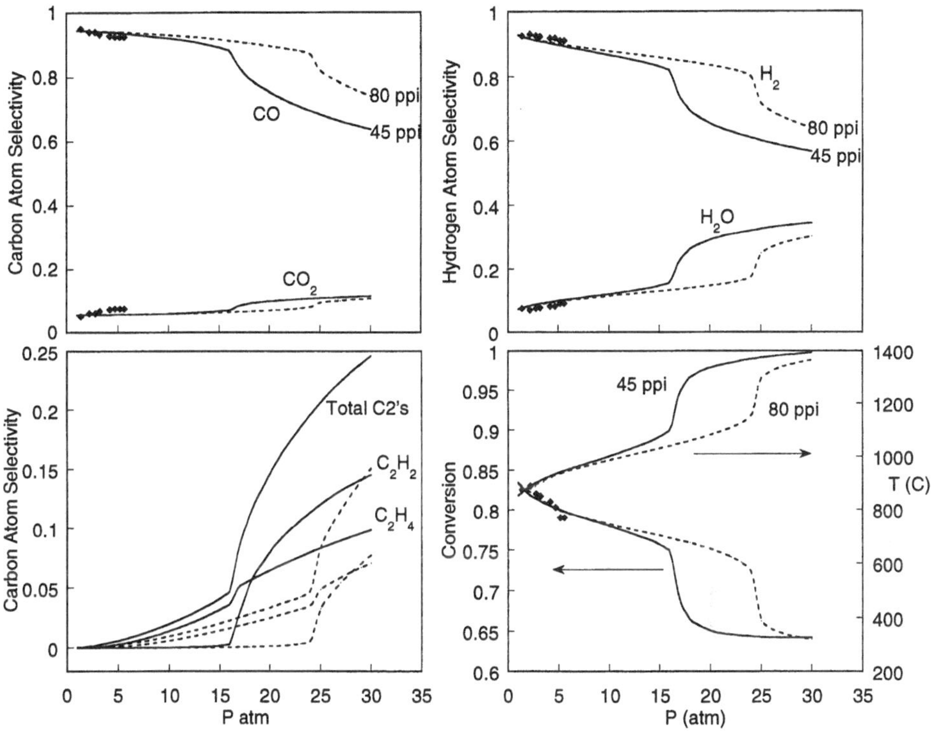

Figure 3. Syngas production at high pressure. Symbols = experimental results (80 ppi). Lines = PFTR model. Selectivities, conversion, and temperature as a function of pressure for 45 ppi (solid lines) and 80 ppi (dashed lines) Rh catalysts.

4. TWO-DIMENSIONAL MODEL

The short contact time reactor is characterized by an extremely fast variation of temperature, velocity, and transport properties of the inlet mixture at the catalyst entrance. These processes therefore involve a complex interaction of transport and kinetic effects which must be neglected in a PFTR model. We have developed computational tools which analyze the elementary chemical and thermal processes occurring at the catalytic surface and their coupling to the surrounding flow field in detail.

A tube wall catalytic reactor is a model of a single pore of the catalytic monolith. Flow in the pore is described by the two-dimensional governing equations for mass, momentum, and energy and for each species in cylindrical coordinates. The independent variables are the axial and the radial direction. Detailed surface chemistry and detailed transport models are used in the governing equations. Gas phase chemistry was neglected for this atmospheric pressure study . Mixing is modeled by a multi-component diffusion model, and all transport properties depend on composition and temperature. The problem is solved using the computational fluid dynamics code FLUENT which is coupled to external subroutines that model the detailed surface chemistry [7]. The newly developed subroutines calculate the surface coverage and the surface mass fluxes at each wall grid point using a point-implicit method.

In the simulations described here, the inlet gas has a velocity of 1 m/s and flows at 25 °C into a 3 cm long tube of 0.25 to 1 mm diameter. The inert heat shields placed in front of and behind the catalytic section (each 1 cm in length) are modeled as an adiabatic inert wall. Experiments indicate only small temperature gradients over the catalyst wall, so the catalytic wall is assumed to be isothermal. A wall temperature of 1000°C and a methane/oxygen ratio of 1.8 with 30% nitrogen diluent is chosen to replicate experimental conditions.

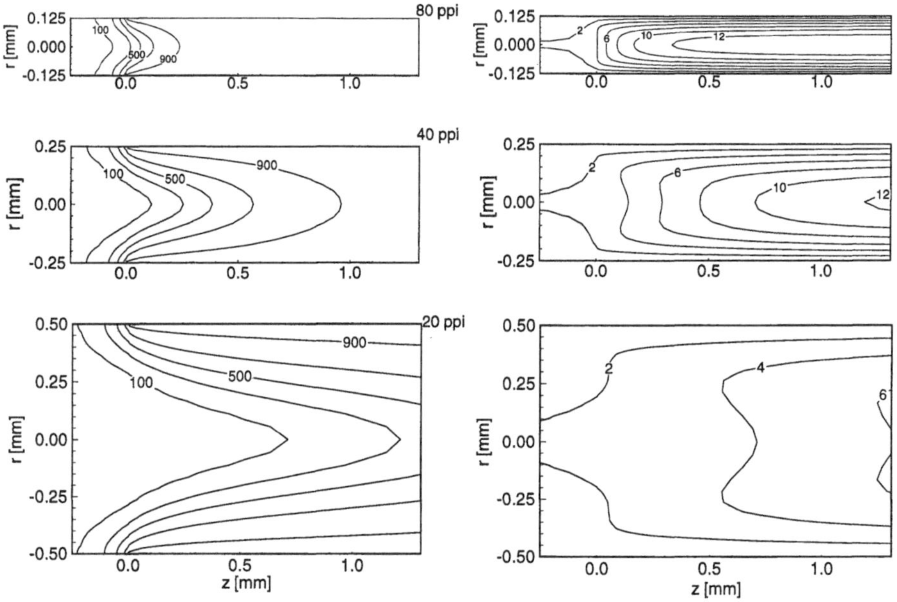

Figure 4. Two-dimensional simulation of temperature (°C, left) and axial velocity (m/s; right) around the catalyst entrance (z = 0.0) for different pore diameters (0.25, 0.5, and 1 mm).

Figure 4 illustrates the rapid variation of temperature and velocity at the catalyst entrance for three different pore diameters. The inlet mixture reaches the operating reactor temperature at a distance of only one tube diameter for the channel 0.25 mm diameter. In this region, the transport properties of the gases vary by approximately a factor of 10. Using averaged transport properties (instead of temperature and composition dependent values) would lead to a inaccurate prediction of conversion and selectivity. Furthermore, the incoming gases are preheated in the region immediately in front of the catalyst by upstream thermal conductivity. The fast increase in the velocity from 1 m/s to greater than 10 m/s at the catalyst entrance is not only a result of the density change due to the temperature increase but also caused by the

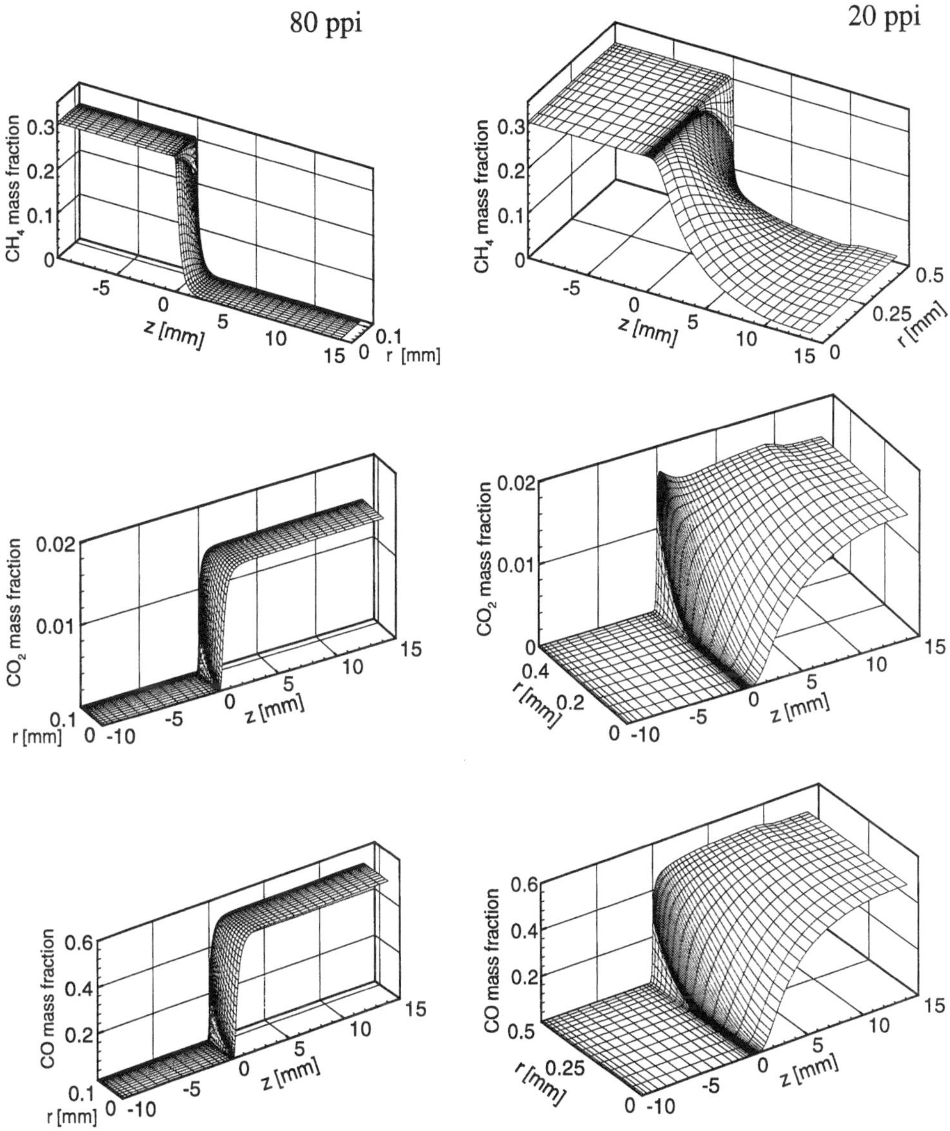

Figure 5. Mass fractions of CH_4, CO_2, and CO as function of the axial (z) and radial (r) position in a reactor pore of 0.25 mm diameter (left) and 1 mm diameter (right). The z-axis is zero at the catalyst entrance, and the r-axis is zero at the centerline of the channel.

change in composition due to reaction which results in almost two moles of products for each mole of reactant. The parabolic flow field which is quickly established in the first noncatalytic section of the reactor is strongly disturbed at the catalyst entrance, and two maxima occur.

The methane oxidation starts immediately at the catalyst entrance, as shown in Figure 5. Almost all methane conversion takes place in the very first part of the catalyst for the 0.25 mm diameter pore which corresponds to an 80 ppi monolith. Very strong radial and axial gradients are formed. In the case of a 1 mm diameter pore (20 ppi monolith), the reaction takes place over the whole catalyst length of 10 mm and it is not completed at the catalyst exit. At the catalyst entrance where the oxygen concentration is still high, the complete oxidation products CO_2 and H_2O are formed. Figure 5 illustrates the mixing of the CO_2 formed at the catalyst with the incoming flow. There is no CO_2 adsorption in the surface reaction mechanism, hence, the CO_2 concentration at the catalytic wall decreases only due to mass diffusion. There are still radial concentration gradients at the catalyst exit ($z = 10$ mm) because conversion is not yet complete. Therefore, mass diffusion leads to the small increase of the concentration of the reactants and decrease of the concentration of the products as shown in Figure 5.

5. CONCLUSIONS

Short contact time catalytic reactors show great promise as a future technology for the production of syngas. Modeling these reactors is a challenging problem because of the complex interaction of transport and surface chemistry. Additionally, detailed gas phase chemistry must be taken into account for the study of the reactor at higher pressures. The solution of the resulting system containing a large number of variables and nonlinearities in the governing equations requires the use of supercomputers and sophisticated numerical techniques.

We have applied a one-dimensional model with detailed gas and surface chemistry to study the effect of pressure on selectivity and conversion. A two-dimensional model was used to investigate the coupling between transport and surface kinetics at atmospheric pressure. The simulation results presented here clearly show that detailed modeling (reactor geometry, transport properties, and chemistry) can help to achieve a better understanding of and assist in the design of short contact time oxidation reactors.

6. REFERENCES

[1] D.A. Hickman and L.D. Schmidt, *Science*, **259** (1993), 343.

[2] A.S. Bodke and L.D. Schmidt, *The Effect of Ceramic Supports on Partial Oxidation of Hydrocarbons over Noble Metal Coated Monoliths*, (Accepted for publication in *Journal of Catalysis*).

[3] P.N. Brown and A.C. Hindmarsh, *J. Appl. Math*, **31** (1989), 40.

[4] R.J. Key, R.M. Rupley, and J.A. Miller, *CHEMKIN II: A Fortran Package for the Analysis of Gas Phase Chemical Kinetics*, Sandia National Laboratories, Livermore, Ca., Report SAND 89-8009 (1989).

[5] C.T. Bowman, R.K. Hanson, D.F. Davidson, W.C. Gardiner, Jr., V. Lissianski, G.P. Smith, D.M. Golden, M. Frenklach, and M. Goldenberg, *http://www.me.berkeley.edu/gri_mech/*.

[6] D.A. Hickman and L.D. Schmidt, *AIChE Journal*, **39** (1993), 1164.

[7] O. Deutschmann and L. D. Schmidt, *Two-Dimensional Modeling of the Partial Oxidation of Methane on Rhodium in a Short Contact Time Reactor*, 27th Symposium (International) on Combustion/The Combustion Institute, Pittsburgh, (Accepted for publication).

NATURAL GAS CONVERSION V
Studies in Surface Science and Catalysis, Vol. 119
A. Parmaliana et al. (Editors)

Catalytic Behaviour of Ni- and Rh-containing Catalysts in the Partial Oxidation of Methane at Short Residence Times

F. Basile[a], L. Basini[b], G. Fornasari[a], A. Guarinoni[b], F. Trifirò[a] and A. Vaccari[a]

[a]Dipartimento di Chimica Industriale e Materiali, Viale Risorgimento 4, 40136 Bologna, Italy
[b]Snamprogetti Research Lab.s, Via Maritano 26, 20097, San Donato Milanese MI, Italy

The temperature of the surface during CPO reaction on the Rh and Ni catalysts, obtained by HT precursors and previously selected as the best catalyst, has been investigated. A comparison of the activity of these two catalysts in CO_2 reforming and CPO reaction at short residence time has been carried out and extended to a Rh/Ni catalyst, which evidenced a possible synergic effect between the two metals.

1. INTRODUCTION

The catalytic partial oxidation (CPO) of natural gas has attracted much attention from scientists and technologists after indications were obtained that higher yields of CO and H_2 can be achieved by reducing the residence time (τ) to values between 10^{-4} and 10^{-2} s [1-4]. The majority of papers reported in the literature raise fundamental questions concerning relationships between flow velocity, transport phenomena and reaction kinetics. However many points remain to be investigated in order to understand short residence time CPO reactions such as: a) the role of the catalyst composition, b) the heat distribution along the catalytic bed and c) their correlation with the reaction kinetics.

The aim of this paper is to discuss these points providing experimental indications obtained with Ni, Rh and Rh/Ni catalysts prepared by decomposition of hydrotalcite-type (HT) precursors. Thanks to the homogeneity, and stability of the calcined samples obtained by this preparation method [5,7] it is possible to shed light on the role of the composition in the CPO reactions in terms of metal type and loading, as well as on the possible synergic effect of Rh and Ni. In order to obtain information on the mechanism of the CPO reaction CO_2 reforming was carried out using contact times and temperatures analogous to those of the CPO reaction. The gas and surface temperatures were determined simultaneously using a quartz reactor equipped with thermocouple and IR thermography.

2. EXPERIMENTAL

The HT precursors containing M/Mg/Al (M = Ni and/or Rh) and having the general formula $[M^{3+}{}_xM^{2+}{}_{1-x}(OH)_2]^{x+}(CO_3)^{2-}{}_{x/2}.mH_2O$ were prepared by coprecipitation at constant pH. A solution, containing the salts of the elements in the right atomic ratio was added to a

Table 1
Composition, present phases and surface areas of the investigated samples as a function of the thermal treatment

Sample	Composition	Dried at 90 °C		Calcined at 900 °C	
		Present phases	Surface area (m^2/g)	Present phases	Surface area (m^2/g)
Rh1	Rh/Mg/Al 1/71/28	HT	78	MgO, Spinel	97
Ni10	Ni/Mg/Al 10/61/29	HT	97	(Ni/Mg)O, Spinel	53
Rh/Ni-1	Rh/Ni/Mg/Al 0.1/5/66/29	HT	82	(Ni/Mg)O, Spinel	88
Rh/Ni-2	Rh/Ni/Mg/Al 0.1/6/59/35	HT	60	(Ni/Mg)O, Spinel	90

solution containing a slight excess of Na_2CO_3 and the pH was maintained constant by additions of NaOH. The precipitates were kept in suspension at 60°C for 40 min under stirring, then filtered and washed with distilled water till a Na_2O content lower than 0.02 wt% was obtained. The precipitates were dried overnight at 90°C and calcined at 900°C for 14h.

XRD powder analyses were carried out using a Philips PW1050/81 diffractometer equipped with a graphite monochromator and controlled by a PW1710 unit ($\lambda = 0.15418$ nm). A 2Θ range from 10° to 80° was investigated at a scanning speed of 70 °/h. The surface areas were determined by N_2 adsorption using a Carlo Erba Sorpty model 1700.

The catalytic tests were carried out at atmospheric pressure in a quartz microreactor with an oven temperature of 750°C using samples calcined at 900 °C (particle size 600-800 μm), previously reduced in situ at 750 °C in an equimolar H_2/N_2 mixture. The CPO tests were carried out at a contact time of 24 ms, using 0.075 g of catalyst and a gas mixture $CH_4/O_2/He = 2:1:4$ v/v. The CO_2 reforming tests were carried out at 12 ms using 0.025 g of catalyst and a gas mixture $CH_4/CO_2/He = 1:1:2$ v/v.

The thermography tests were carried out at atmospheric pressure using 1.6 g of catalyst and a residence time of 14 ms in autothermal conditions, after ignition in programmed temperature conditions. The surface temperatures were measured with IR thermography equipment (AGEMA) that collected emitted radiation in the 2-5 μm range. In all the tests the gas phase temperature inside the catalytic bed was measured by a moveable chromel-alumel thermocouple, inserted in a quartz wire inside the catalytic bed. The reaction products were analyzed on-line, after water condensation, using two gas-chromatographs equipped with HWD and Carbosieve SII columns, with He as the carrier gas for the analysis of CH_4, O_2, CO and CO_2 and N_2 for the analysis of H_2.

3. RESULTS AND DISCUSSION

3.1. Chemical-physical characterization

The chemical compositions of the dried samples, reported in Table 1, were analogous to

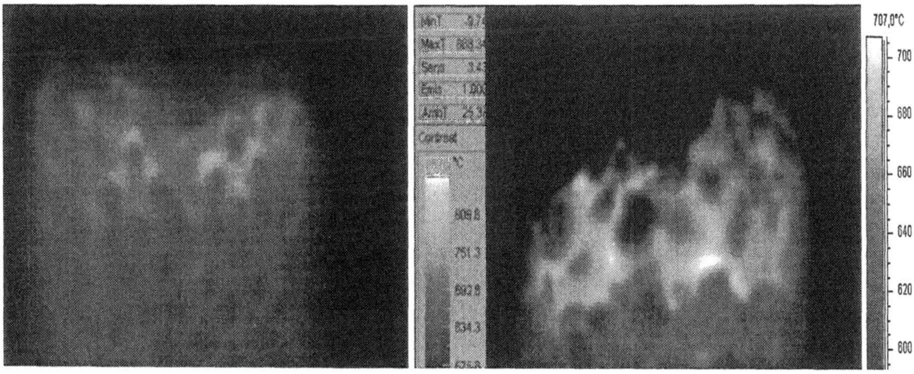

Figure 1. Temperature maps of the catalyst surface collected with IR thermography during CPO tests carried out at 14 ms using: a) Rh-1 catalyst; b) Ni-10 catalyst.

those of the precipitating solutions. The XRD powder patterns of the samples dried at 90 °C showed the presence only of an HT type phase confirming that all the cations coprecipitated in the well crystallized HT phase, in agreement with the amounts and nature of the ions present [5]. The surface areas of the precursors were not affected by the composition and the values are close to 100 m^2/g.

The HT precursors gave rise by calcination at 900 °C to a structural rearrangement, with formation of MgO and spinel type phases (Table 1). In some previous works [8, 9] a number of samples with different Ni or Rh loading were analyzed and it was found that Ni and Rh are dissolved inside the MgO type and spinel type phases respectively. These conclusions can also be extended to the samples containing both these metals considering that after calcination at 900°C only the MgO and spinel type phases are present without any other segregation. Notwithstanding the structural rearrangement, the surface area of the calcined samples is always very high, confirming the good stability at high temperature of the catalysts obtained from HT precursors.

3.2. Catalytic activity of Rh and Ni catalysts and temperature gradients in CPO reaction

Recently in our laboratories the influence of the metal loading on the catalytic performances of Ni and Rh catalysts obtained by HT precursors has been investigated [9,10]. For Ni containing catalysts, only the samples with the low Ni content ($\leq$ 20% as atomic ratio (a.r.)) were stable in the reaction conditions with a significant reduction of the parasitic carbon formation. However with decreasing Ni content the catalyst activation required more severe reducing conditions, since the Ni is more diluted and, as a consequence, more stabilized in the (Mg/Ni)O solid solution [10]. On the other hand the Rh catalysts with a low noble metal content (about 0.1% a.r.) showed poor catalytic performances. The catalytic performance increased as a function of the metal loading until a content of 1.0% a.r., over which the methane conversion and syngas selectivities did not significantly increase [9]. The rhodium catalysts showed a slightly lower activity than the nickel samples.

In order to evaluate whether the difference in activity of the two types of catalysts can be related to the difference in surface temperature and to analyze the temperature gradients, the surface and gas temperatures were monitored in autothermal tests, carried out by removing the oven after catalyst ignition. Furthermore the heat distribution and the temperature measurement are key parameters in developing the CPO reaction for use on an industrial scale. It is worth noting that although the CPO reaction is only slightly exothermic, the side total oxidation to CO_2 and H_2O is highly exothermic and the temperature of the reaction zone increases significantly causing problems of temperature control.

The experiments were carried out using the best Ni and Rh containing catalysts previously selected (i.e. Ni/Mg/Al 10/61/29 and Rh/Mg/Al 1/71/28 as atomic ratio). The Rh catalyst shows a high surface temperature at the beginning of the catalytic bed (Fig. 1a), much higher than that of the gas phase. The surface temperature decreases along the catalytic bed while the gas phase temperature increases, thus at the end of the catalytic bed the gas phase temperature is higher than that of the surface. These results suggest that in the first part of the catalytic bed exothermic reactions occur, while in the second part, endothermic reactions prevalently occur and decrease the surface temperature, even though the catalytic bed is warmed by the heat transported by the gas and irradiated by the solid.

On the other hand the Ni catalyst in autothermal conditions shows a cold zone at the beginning of the catalytic bed and the highest temperature in the second zone (Fig. 1b), where the highest gas phase temperature is also recorded. To analyze these data, it is useful to point out that the used catalyst was greenish in the first part and became dark with increasing distance along the catalytic bed [10]. It is clear then that in the first zone the catalyst is oxidized, probably because of the contact with fresh-oxygen during ignition, and its activity is negligible. The amount of reduced Ni increases along the catalytic bed, leading to an increase in the activity of the catalyst, thus in the last zone of the catalytic bed the temperature shows the highest value. Furthermore the maximum surface temperature of the Ni sample is lower, but reached in a larger zone, than that of the Rh catalyst. This can be explained considering the low activity of the Ni in the middle of the catalytic bed, where it is still partially oxidized. Indeed a certain amount of oxidized Ni was always observed during the catalytic tests and the extension and the degree of the oxidized zone depended on the ignition procedure and reactor shape.

Even though no catalyst deactivation was observed during these tests, in the experimental conditions employed, 10% graphitic carbon formation on the Ni containing catalyst was detected, while no-carbon formation was detected using the Rh containing catalyst.

3.3. Comparison of the Rh and Ni catalysts in CPO and CO_2-reforming

The comparison of the activity of Ni and Rh containing catalysts in CPO and CO_2 reforming reactions was carried out in a quartz microreactor in similar conditions, to make possible a correlation between the two reactions. In the CO_2 reforming of the methane, notwithstanding the low residence time, the two catalysts show a very high activity and a decrease in the temperature of the catalytic bed (ca. 680 °C) due to the endothermicity of the reaction. In particular the Rh catalyst shows higher methane conversion and H_2 selectivity, while the Ni catalyst exhibits higher CO_2 conversion (Fig. 2a). These data can be explained considering that Rh exhibits a higher efficiency in the methane activation and a lower efficiency in CO_2 activation [11]. These features also induce higher H_2 selectivities, because the amount of activated oxygen on the surface is lower and thus the oxidation of hydrogen species to OH, precursors of water, is lower.

To minimize the contact of the catalyst with fresh oxygen and the oxidation of the Ni catalyst, the tests in the CPO reaction were carried out by adding the oxygen to the CH_4 and He feed when the catalyst had already reached the desidred temperature. The catalysts show very high activity and an increase in the catalytic bed temperature at about 870°C. The Rh catalyst exhibits a lower methane conversion and CO selectivity, while the H_2 selectivities do

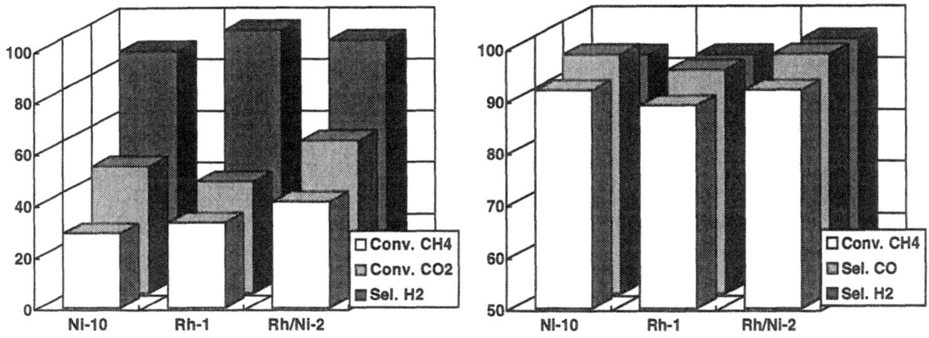

Figure 2. Comparison of the activity of Rh Ni and Rh/Ni catalysts in: a) CO_2 reforming of methane; oven temperature = 750 °C, contact time = 12 ms, reaction gas mixture $CH_4/CO_2/He$ = 1:1:2 (v/v); b) catalytic partial oxidation of methane; oven temperature = 750 °C, contact time = 24 ms, reaction gas mixture: $CH_4/O_2/He$ = 2:1:4 (v/v).

not show significant differences (Fig. 2b). The lower CO selectivity of the Rh catalyst can be related to the lower efficiency of CO_2 activation already observed during the CO_2 reforming tests, which diminishes the importance of the consecutive CO_2 reforming. The lower methane conversion also depends on the ratio of the total and partial oxidation reactions occurring during these process, taking into account that the oxygen conversion is complete. Therefore, the methane conversion is not easily related to the high efficiency of methane activation previously discussed.

3.4. Catalytic activity of a bimetallic Rh/Ni catalyst

The differences in the catalytic behavior of the Rh and Ni based catalysts and the flexibility of the HT and its calcined oxide suggested that it would be worthwhile to investigate the presence of possible synergic effects in a Ni/Rh catalyst. Furthermore the addition of a small amount of Rh could possibly lead to a decrease in the carbon formation in severe reaction conditions [11], occurring on the Ni based catalyst. An HT with reduced amounts of Rh and Ni, Rh/Ni/Mg/Al 0.1/5/66/29 (as atomic ratio), was prepared. However this catalyst showed low activity, due to the difficulty of the Ni reduction. In fact, the concentration of the Ni dissolved in the (Mg/Ni)O phase (Mg/Ni = 13) is very low, and therefore the Ni is extremely stabilized. For this reason an HT containing Rh/Ni/Mg/Al 0.1/6/59/35 (as a.r.) with a lower ratio of Mg^{2+}/Al^{3+} was prepared. In this way the Mg/Ni ratio is lower (Mg/Ni < 10) in the (Mg/Ni)O phase, making possible easy reduction of the Ni. The catalyst was successfully tested in the CO_2-reforming and CPO reaction at low residence time. The comparison with the Ni and Rh catalysts in the CO_2 reforming shows clearly that the Rh/Ni catalyst exhibits higher methane and CO_2 conversion and an intermediate H_2 selectivity (Fig. 2a). The high activity may be explained by the high efficiency in both CO_2 and CH_4 activation which increases the reaction rate. The high activity is confirmed by the temperature of the catalytic bed, which is lower than that of the Ni and Rh catalyst because of the endothermic CO_2 reforming reaction.

In the CPO tests the catalyst containing Ni and Rh also shows a very high activity. In the first zone of the catalytic bed no change in the oxidation state of Ni° is observed. The comparison with the Ni and Rh catalysts shows a methane conversion and CO selectivity higher than those of the Rh containing catalyst and similar to those of the Ni containing catalyst, while the H_2 selectivity is always higher than that obtained with the monometallic catalysts (Fig. 2b). The highest production of syngas reached in these conditions is surprising taking into account the low amount of Ni and Rh present in the catalyst and seems to be due to a synergic effect between Ni and Rh.

4. CONCLUSIONS

The Ni and Rh HT precursors can be profitably used in the preparation of metal catalysts by reduction of the calcined samples, giving rise to materials with high surface areas and high thermal stability.

The Rh and Ni catalysts having respectively 1 and 10% of metal as atomic ratio were selected as the best catalysts among the different metal loading samples. The measurements of the surface temperature in autothermal conditions evidence a hot zone in the first part of the catalytic bed of the Rh catalyst while with the Ni catalyst the temperature shows an increase along the bed, related to the increase in the amount of reduced Ni. The comparison of catalytic activity in CO_2 and CPO reactions shows high catalytic performances at very short contact times and a high efficiency of the Ni catalyst in CO_2 activation and a lower H_2 production. The Rh catalyst shows a high efficiency in methane conversion in the CO_2 reforming reaction.

The HT structure allows a bimetallic catalyst to be prepared by the insertion of Rh and Ni in the same structure. The metals are homogeneously distributed in the HT phase, giving the possibility of reducing the carbon formation occurring on Ni catalysts. The HT structure makes it possible to influence the Ni reducibility by changing the M^{2+}/M^{3+} ratio. The Rh/Ni catalyst showed the highest activity and selectivity in both reactions, notwithstanding the low amount of Rh and Ni present in the catalysts, supporting the possibility of a synergic effect.

REFERENCES

1. D.A. Hickman and L.D. Schmidt, J. Catal, 138 (1992) 267.
2. V.R. Choudhary, A.M. Rajput and V.H. Rane, J. Phys. Chem., 96 (1992) 8686.
3. G.A. Foulds and J.A Lapszewicz, in Catalysis (J.J. Spivey and S.K. Karval, Ed.s), Vol 11, The Royal Society of Chemistry, London 1994, p. 413.
4. S.C. Tsang, J.B. Claridge and M.L.H. Green, Catal. Today, 23 (1995) 3.
5. F. Cavani, F. Trifirò and A. Vaccari, Catal Today, 11 (1991) 173.
6. F. Trifirò and A. Vaccari, in Comprehensive Supramolecular Chemistry (J.L. Atwood, J.E.D. Davies, D.D. Mc Nicol and F. Vögtle Ed.s), Vol. 7, Pergamon, Oxford, 1996, ch.8.
7. J.R. Rostrup-Nielsen, Steam Reforming Catalysts, Teknisg Forlag, Copenaghen, 1975.
8. G. Fornasari, M. Gazzano D. Matteuzzi, F. Trifirò, A. Vaccari, Appl Clay Sc. 10 (1995) 69.
9. F. Basile, L. Basini, G. Fornasari, M. Gazzano, F. Trifirò and A. Vaccari, Chem. Commun., (1996) 2435.
10. F. Basile, L. Basini, M. D' Amore, G. Fornasari, A. Guarinoni, D. Matteuzzi, G. Del Piero, F. Trifirò and A. Vaccari J. Catal., 173 (1998) 247.
11. Y. Chen, K. Tomishige, K Yokoyama and K. Fujimoto, Appl. Catal. A: General, 165 (1997) 335.

NATURAL GAS CONVERSION V
Studies in Surface Science and Catalysis, Vol. 119
A. Parmaliana et al. (Editors)

Molecular aspects in short residence time catalytic partial oxidation reactions

L. Basini[a], A. Guarinoni[a], and K. Aasberg-Petersen[b]

[a] Snamprogetti S.p.A.,
Via Maritano 26, 20097, S. Donato. Mil. (MI), Italy

[b] Haldor Topsøe A/S,
Nymøllevej 55, DK-2800 Lyngby, Denmark

1. INTRODUCTION

The first step in manufacture of liquid products such as methanol, dimethyl ether (DME), or synthetic fuels from natural gas is the production of synthesis gas. Commercial technologies include tubular reforming, autothermal reforming or a combination [1-3], the optimal choice depending on desired end product and plant capacity [1].

A potential new technology is based on the catalytic partial oxidation (CPO) of natural gas into syngas as shown below (for methane):

$$CH4 + \frac{1}{2}O2 \Rightarrow CO + 2H2 \qquad \Delta H = H^{\circ}_{298K} = \quad -38 \quad kJ/mole$$

With a catalyst selective for CPO without further oxidation of the reaction products syngas could be produced at moderate temperature with limited heat production.

CPO has been the object of many studies since 1946. These have reported that Ni [4-6] or noble metals based catalysts are able to catalyze partial oxidation of methane into CO and H_2, and that Rh is unique to avoid parasitic reactions which lead to carbon formation [7-8].

The great part of the experimental work performed in the years before 1992 considered reaction conditions with residence time values between 50 and 0.1 s, but in 1992 Choudary et al.[9] and Schmidt et Al.[10] reported that the residence time (τ) conditions can be reduced by two orders of magnitude to values close to 10^{-3} s thus indicating a new way for obtaining syngas. Schmidt and coworkers are still very active on the issue which they have reviewed in 1995 [11].

Other authors have observed that reactivity at short residence time can be shifted from the equilibrium and have examined the possibility that conversions of methane and selectivities towards syngas could overcome equilibrium conditions [12-15]. However while it is clear the catalytic mechanisms operating at very short τ can generate non-equilibrium compositions, the comparisons between thermodynamics and experimental results remain difficult and questionable due to the occurrence of hot spot phenomena which render delicate the selection of "true" reaction temperature values.

The description of the molecular aspects of short residence time reactivity features and the discussion of the role of surface temperature is the object of this work which is also part of an extensive research program devoted to the study of surface chemistry of small Rh clusters [16-18]. The program is based on 14 experiments, schematized in figure 1, which have been performed in a reaction chamber equipped with diffuse reflectance Fourier transform IR spectroscopy (DRIFT) and mass spectrometry. The experiments here described are numbered 11, 12, 13 and 14 have been aimed to investigate stoichiometric and catalytic aspects of partial oxidation of methane. Some peculiar points of the last step of experiment 14, concerning differences between surface and gas temperatures have further been investigated with a quartz plug flow reactor equipped with both thermocouples and an IR camera.

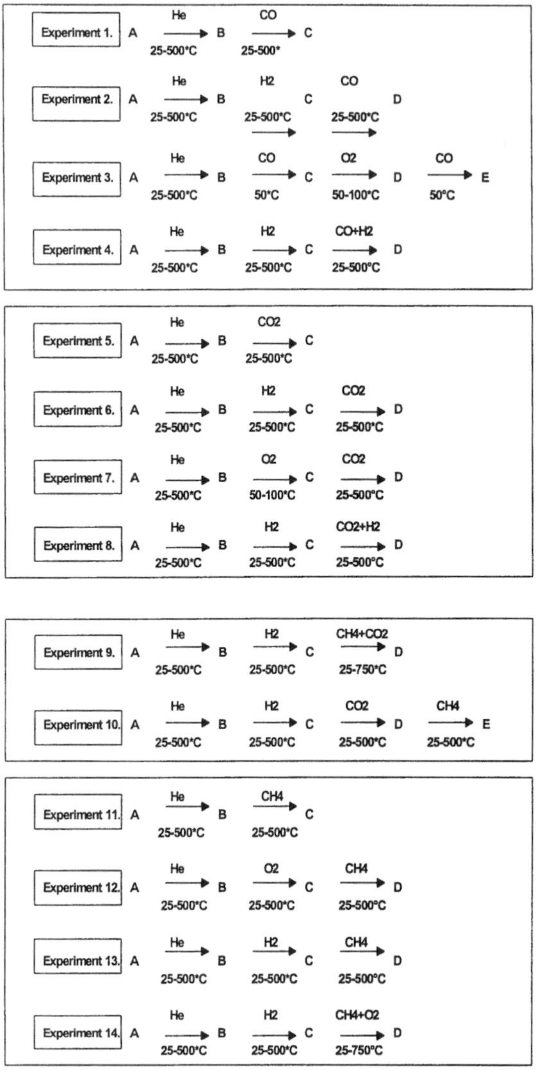

Figure 1. Schemes of the experimental sequences performed to investigate aspects of surface Rh cluster chemistry.

2. EXPERIMENTAL

2.1. Sample preparation

The preparation procedure and the characterization work of the cluster containing materials have been given in references [13-15]. Briefly it is here reported that preparation method involved a solid-liquid reaction between a n-hexane solution of small Rh clusters and the powdered oxides (α-Al_2O_3, MgO and CeO_2). Rh content has been varied between 0.1-0.8% wt. DRIFT spectra of the freshly prepared powdered materials showed, in all samples, carbonyl absorption bands (at ≈ 2090 and 2010 cm^{-1}), assigned to $Rh^I(CO)_2$ species formed through an oxidative disaggregation of the tetrametallic [20].

2.2. Apparatus for spectroscopic and spectrometric studies

DRIFT and mass spectra in flowing gaseous environments have been collected at 0.1 MPa and temperatures between 298 and 1023 K. The equipment description is reported in [16]. DRIFT spectra have been recorded at a resolution of 4 cm^{-1} and signal to noise ratios have been reduced collecting between 300 and 500 scans with a procedure which required between 2 and 3 minutes.

2.3. Apparatus for surface and gas temperature measurements

CPO reactions have been also studied with a cylindrical tubular quartz plug flow reactor (PFR) with an internal diameter of 15 mm and a wall thickness of 1 mm. The catalyst bed, filled with 1.5 g of regular α-Al_2O_3 spheres (∅ = 1 mm, Rh content = 0.32 % wt) was 10 mm high. Gas temperatures have been monitored by two movable thermocouples located at the inlet and outlet of the catalytic bed. An IR camera has collected the radiation emitted from the catalyst surfaces; temperatures maps had a spatial resolution of 250 μm. The reactor has been placed in a oven, which was switched off and lifted above the catalytic bed after the ignition of the reaction. Reactants (H_2, He, O_2, CH_4) were supplied from cylinders into different lines each equipped with mass flow meters and controllers. The output line composition was monitored with an HP gascromatograph equipped with two TCD detectors.

4. RESULTS

4.1. DRIFT and mass spectrometry measurements

Experiment 11

DRIFT and mass spectrometry measurements have been collected during two subsequent thermal cycles between 298 and 773 K while flowing pure He (step A→B) and a CH_4 + He mixture (CH_4/He = 1/4 v/v, step B→C). The first thermal cycle (common to all the experiments) has been performed to investigate thermal induced modification of the solid samples in inert environment and to obtain the same nude Rh clusters at point B of each experimental sequence. About the chemical features of step A→B described in [16] it is briefly mentioned that the identified chemical processes have been: a) a stoichiometric surface water gas shift reaction at T<473 K, followed by b) a complete decarbonylation of the clusters with desorption of CO and CO_2 and with the formation of a small amount of carbidic species.

The second step of experiment 11 (B→C) showed that the interaction with the nude Rh clusters dissociated CH_4 at 573 K (and higher temperatures too) in each of the three examined samples. Such dissociation has initially produced a partial recarbonylation of the nude Rh clusters and the formation of gaseous CO and H_2 with minor amounts of CO_2. The total amounts of desorbed H_2 and CO molecules during the heating step have determined a H_2/CO ratio equal to 3.8 v/v. Figure 2A shows the carbonyl stretching bands detected with DRIFT spectra on the Rh/Al_2O_3 sample.

Experiment 12

The experiment 12 has shown that the oxidizing treatment modified the DRIFT spectral features produced in experiment 11 and that such modification consisted in a sharpening and increased intensity of IR absorption bands (figure 2B). The changes produced in mass spectra also indicated that: a) CO and H_2 production initiated at temperatures 100 K lower than in experiment 11 and b) the molar ratio between produced H_2 and CO decreased to between 1-1.5 v/v.

Experiment 13

In this experiment the thermal decomposition of CH_4 was studied after a thermal treatment in He (A→B) and another one in flowing H_2 (B→C). Again the description of the reactions produced in H_2 [16] is omitted and we only mention that it was found that the reductive treatment originated small Rh hydrido clusters (with diameters smaller than 10 A). These clusters (particularly in the case of the Rh/MgO and Rh/Al_2O_3 samples) have been much less reactive towards CH_4 decomposition compared to the ones produced during experiments 11 and 12 and did not produce gaseous CO.

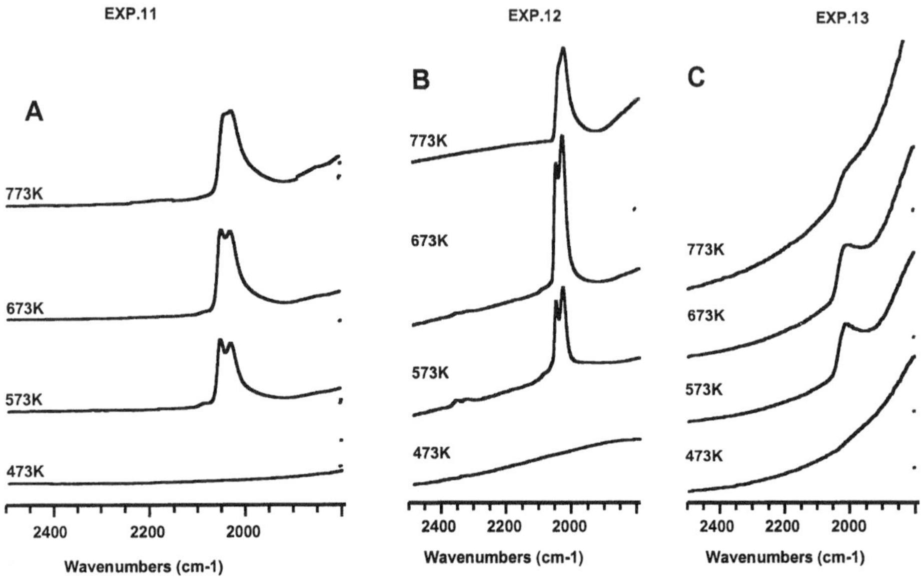

Figure 2. DRIFT spectra of surface species produced on Rh/Al_2O_3 during the interaction with gaseous CH_4 in experiments 11 (column A), 12 (column B) and 13 (column C).

Experiment 14

Two thermal treatments in He and H_2 flows have been performed as in experiment 13 and in conclusion a CH_4+O_2+He (CH_4/O_2/He = 2/1/4 v/v) flow has been admitted into the reaction chamber at two residence time (τ) conditions corresponding to 72 and 7.2 ms.. Moreover, different from the experiments 11, 12, and 13 where samples containing 0.1% wt of Rh have been used, in this case the experiment has been repeated with two set of materials with Rh loading of 0.1 and 0.8 wt %.

Rh = 0.1% wt cases. Methane conversion and selectivities towards CO and H_2 have increased monotonically with temperature between 298 and 1073 K during a thermal cycle performed at τ = 72 ms, while selectivities towards CO_2 and H_2O have reached a maximum value at ca. 673 K. Carbonyl cluster formation has also been detected between 673 and 1073 K but the IR bands appeared stronger and sharper during the heating cycles. These general features are common to all the three examined samples; although some difference have been determined in the intensity and positioning of the CO stretching bands of the surface carbonyl complexes. When the reactant mixture has been flowed with a shorter residence time (τ = 7.2 ms), the reactivity features have changed and the changes have been particularly different in the Rh/Al_2O_3 case. At τ = 7.2 the amount of product CO_2 and H_2O have increased as compared to CO and H_2 and in some cases also temporal oscillations in chemical reactivity have occurred. Surface carbonyl formation have been depressed compared to the τ = 72 ms case.

Rh = 0.8% wt cases.

The increase of metal loading improved the selectivity towards syngas at 7.2 ms and enhanced the intensity of the carbonyl stretching bands. It is underlined that this change in reactivity features apparently occurred at the same temperature and gaseous reactant composition.

4.2. Surface and gas temperature measurements

Experiment 14 was repeated in a quartz reactor measuring surface (with IR thermography) and gas temperatures (with thermocouples). The IR thermography measurements have shown that surface of the catalyst is rapidly heated from 298 K to about 1300 K in the initial 3 mm of catalytic bed while gas temperature has reached in the same area a maximum value of 810 K. Moreover, it has been observed that the surface temperature remained constant till the end of the bed while the gas temperature increased to about 930 K. IR thermography also showed that radial temperature variations were been relatively small (15-20 K). At 72 ms the same experiment could not be performed since the heat dispersion through the reactor walls become comparable with the chemical heat produced in the catalytic bed and reactions could not remain self sustained. The quantitative analysis of product composition showed that CPO produced a non-equilibrium mixture and that adiabatic exit temperature has been lower than average surface temperature.

5. DISCUSSION

The surface stoichiometric reaction performed in experiments 11 and 12, have shown that the CH_4 dissociation occurred at relatively low temperatures (lowest value 573 K in experiment 12) originating surface Rh carbonyl clusters, gaseous H_2, CO and traces of total oxidation products. The experiments 11-13 also showed that reactions have been enhanced by a pre-oxidizing treatment and almost entirely suppressed (particularly in cases of Rh/MgO and Rh/α-Al2O3) by a reductive pre-treatment. This chemistry occurred on surface sites containing reactive oxygen species and very small Rh clusters a conclusion also sustained by the observation that if the experiments 11-13 are repeated with a pure α-Al_2O_3 support only gaseous H_2 and surface carbon have been produced at temperatures higher than 723 K.

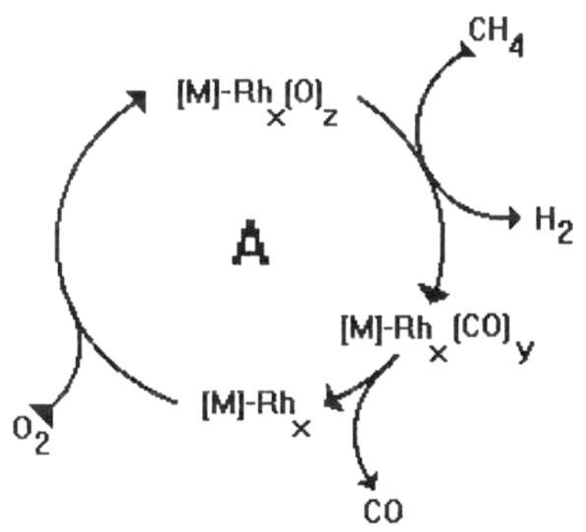

Figure 3. scheme of the repeating loop of the stoichiometric reactions.

The stoichiometric reactions produced in steps B→C and C→D during experiment 12, can be linked to produce a repeating loop which consumes CH_4 and O_2 producing CO and H_2. This cycle schematized in figure 3, has been followed three times for each examined sample without detecting any deactivation or modification of reactivity features. A cycle of stoichiometric reactions could also contribute to the reactivity features produced during CPO. However when CH_4 and O_2 are contemporaneously flowed through the reaction environment the complexity of the chemical system is largely increased and we have found that in this case, surface temperature is the main parameter which address the selectivity towards CO and H_2. IR thermography has shown that surfaces are much hotter that gases while CPO is occurring. Moreover measured surface temperatures have been found higher than estimated adiabatic gas temperatures. This is interpreted as a result of a combination of two main phenomena: a) heat recirculation by radiation through the solid particles, b) the occurrence in the initial layers of the catalytic bed of total combustion together with partial combustion reactions.

Surface temperature can also be increased by increasing the rate of heat generation per unit surface by augmenting the number of Rh containing active sites per unit surface. We believe that this is the reason why we have observed reactivity modifications in the DRIFT-mass spectrometry experiments performed with samples with two different Rh loading.

6. CONCLUSIONS

The study of stoichiometric surface chemistry allowed to generate separate reactions which periodically reproduced initial conditions according to the following steps: a) oxidative treatments originated highly reactive surface oxygen atoms, b) these reacted with CH_4 molecules selectively producing gaseous CO and H_2 species and minor CO_2 and surface carbonyl clusters, c) the carbonyl clusters thermally decomposed into nude Rh clusters which d) in flowing oxygen regenerated the oxygen species able to dissociate again CH_4.

IR thermography experiments allowed to measure large differences between surface and gas temperatures. Surface temperatures have been found much higher than experimental gas temperatures and also higher than estimated adiabatic gas temperatures indicating that reactions are largely confined in the heterogeneous solid-gas inter-phase regions whose conditions are largely responsible for the fast reaction rates of short residence time CPO.

7. REFERENCES

[1] Dybkjær, I. Presented at the 1996 Middle East Refining & PetrochemicalsConference and Exhibition, Bahrain June 10-12, 1996.

[2] Christensen, T.S. and Primdahl, I.I. Hydrocarbon Processing, 1994, p.39

[3] Rostrup-Nielsen, J.R. Catalysis Today, 18, p.305, 1993

[4] M. Prettre, C. Eichner, M. Perrin, Trans. Faraday Soc. **1946**, 43, 355

[5] D.Dissanayake, M.P. Rosinek, K.C.C. Kharas, J.H. Lunsford, *J.Catal.* **1991**, 132, 117

[6] G.R. Gavalas, C. Phichitcul, G.E. Voecks, *J.Catal.* **1984**, 88, 54

[7] J.K. Hockmuth Appl. Catal. B: Environmental **1992**, 1, 89

[8] A.T. Ashcroft, A.K. Cheetham, J.S. Foord, M.L.H. Green, C.P. Grey, A.J. Murrel, [9] P.D.F. Vernon, *Nature*, **1991**, 352, 225

[10] V.R. Choudary, A.S. Mamman, S.D. Sansare *Angew.Chem.Int.Ed.Engl.* **1992**, 31, 1189

[11] H. Hickman and Schmidt **J.Catal. 1992**, 138, 267

[12] S.S. Bharadwaj and L.D. Schmidt, *Fuel Processing Technology* **1995**, 42, 109

[13] D. Dissanake, M.P. Rosinek, J.H. Lunsford, *J.Phys. Chem.* **1993**, 97, 3644

[14] Y.F. Chang, H. Heinemann, *Catal.Lett.* **1993**, 21, 215

[15] Y. Matsumura and J.B. Moffat, *Catal.Lett.* **1994**, 24, 59

[16] F. Basile, L. Basini, M. D'Amore, G. Fornasari, A. Guarinoni, D. Matteuzzi, G. Del Piero, F. Trifirò and A. Vaccari, *J. Catal.* **1998**, 173, 247

[17] L. Basini, M. Marchionna and A. Aragno, *J.Phys.Chem.* **1992**, 23, 9431

[18] L. Basini and A. Aragno *J.Chem.Soc. Faraday Trans.* **1994**, 90(5), 787

[19] L. Basini and D. Sanfilippo, *J.Catal.* **1995**, 157, 162

[20] T. Basu, D. Panayotov, J.T. Yates Jr., *J.Am.Chem.Soc.* **1988**, 110, 2074

NATURAL GAS CONVERSION V
Studies in Surface Science and Catalysis, Vol. 119
A. Parmaliana et al. (Editors)

Effect of metal additives on deactivation of Ni/α-Al_2O_3 in the CO_2-reforming of methane

H.Halliche[a], R.Bouarab[a], O.Cherifi[a] and M.M.Bettahar[b]
[a]Laboratoire de Chimie du Gaz Naturel, Institut de Chimie, USTHB, BP 32 El-alia, 16111 Bab-Ezzouar, Algeria

[b]Laboratoire de Catalyse Hétérogène, Université Henri Poincaré, BP239, 54506 Vandoeuvre-Cedex, France

ABSTRACT
In the present paper we present physical characteristics, reducibility under H_2 atmosphere and catalytic performances in the CO_2 reforming of methane of Ni-M /α-Al_2O_3 (M = Fe, Co, Cu, Ce) catalysts. The metal additives, notably Co, decreased sensibly the reducibility of the nickel phase but increased its active surface. The order of activity, up to 730°C, was : Ni-Cu>Ni-Ce>Ni-Co>>Ni>>Ni-Fe. After aging treatment (700°C/12h under reactive atmosphere) the Ni catalyst became totally inactive whereas the activity of the Ni-Co was strongly enhanced. Conversly, the regeneration treatment (H_2/700°C/12h) strongly enhanced the activity of the Ni catalyst whereas it strongly decreased that of the Ni-Co catalyst. Medium behaviour was observed for the Ni-Ce catalyst.

1. INTRODUCTION

The process of carbon dioxide reforming of methane has received renewed interest in academia and industry since it produces syngas feedstock from cheap carbon-containing materials [1-8]. However, one of the main problem in this process is coking which deactivate the catalyst [4-6], notably in the case of nickel based catalysts [4]. It have been shown that the activity and stability of these catalysts depended on the nature of the support and the presence of metal addtives [6,7]. In a previous paper we have reported the influence of metal additives on the activity and coking of 1%Ni-1%M/ α-Al_2O_3 (M = Fe, Co, Cu, Ce) catalyst [9]. In the present study we report the results obtained on their reducibility under H_2 flow, aging behaviour and regeneration under H_2 atmosphere.

2. EXPERIMENTAL

2.1. Catalysts preparation, characterization, reducibility

The method of preparation of the catalysts was described previously [9]. The nickel composition, except for the Ni-Ce catalyst, but not that of the metal additives. The specific areas and XRD spectra were performed on a BET Coultronics 2100D device and Phillips 1710 diffractometer respectively. The obtained results are reported in Table 1.

The TPR experiments under H_2 were carried out in a pulse quartz reactor with 0.3g samples in the range temperature of 25°C-820°C.

2.2. Catalytic testings

The catalytic performances were carried out in a fixed-bed quartz tubular reactor with 0.1 g samples at atmospheric pressure at a total flow rate of $1.31.h^{-1}$. The reactant gas feed (Air Liquide) consisted of mixtures of CO_2/CH_4 (4/1) and the catalysts were in-situ reduced in a flow of H_2 at 600°C/12h before testing.

The reactants and products were analyzed on line using TCD gas chromatograph equiped with carbosieve column. Each reaction temperature was maintained constant until the corresponding steady-state was reached as indicated by the gas chromatograph analysis of the exit gases samples.

For aging and rejuvenation experiments see below.

3- RESULTS AND DISCUSSION

3-1 Characterization

The specific area of the Ni/α-Al_2O_3 référence catalyst was low (5.5 m^2g^{-1}) and sensibly decreased (divided by 4-8) in the presence of metal additive. As to the XRD spectra of the oxidized precursors, they exhibited the usual oxidation states for all metal components and, in addition, reduced states for Co, Cu and Ce additives.

Table 1 : BET specific areas and XRD spectra of the Ni-M*/ α-Al_2O_3 catalysts.

Catalyst	Ni / α-Al_2O_3	Ni-Co / α-Al_2O_3	Ni-Ce/ α-Al_2O_3	Ni-Cu / α-Al_2O_3	Ni-Fe / α-Al_2O_3
% Ni	1.19	0.86	-	1.17	1.16
BET area.(m^2/g)	5.5	1.1	1.0	0.7	1.3
XRD phases	α-Al_2O_3 NiO	α-Al_2O_3, NiO, CoO, Co_2O_3	αAl_2O_3, NiO CeO,CeO_2	αAl_2O_3, NiO CuO,Cu_2O	α-Al_2O_3, NiO,Fe_2O_3

* nominal composition : 1%.

3-2 Reducibility of the catalysts

The temperature of peaks of reduction and degree of reduction are reported in Table 2. Only one peak of reduction was observed for all catalysts, except the iron modified catalyst (for which two peaks were found) and the solids were uncompletly reduced (28%-90%).

Table 2 : Température of peaks of reduction (T_{max}) and percentage of reduction of the Ni-M / α-Al_2O_3 (M = Co,Ce,Cu,Fe) catalysts under H_2 atmosphere in the range temperature of 25°C-820°C .

Catalyseurs	Ni /α-Al_2O_3	Ni-Co /α-Al_2O_3	Ni-Ce/α-Al_2O_3	Ni-Cu /α-Al_2O_3	Ni-Fe / α-Al_2O_3
T_{max} (°C)	360	610	440	420	425, 815
Réduction (%)	28	90	59	69	38, 10

The metal additives decreased the reducibility of the nickel phase, specially the cobalt additive which shifted the temperature of peak of reduction from 360°C (Ni catalyst) to 610°C (Ni-Co catalyst). Moreover, the iron additive exhibited a second peak of reduction up to 815°C. The percentage of reduction was also increased in the presence of the metal additives (up to 90% for the Ni-Co catalyst against 28% for the Ni catalyst), indicating an increase of the nickel active surface of the catalysts. The uncomplete reduction was attributed to the existence of two NiO species of different size or/and differently attached to the support. In the case of the iron additive, the exitence of two peaks of reduction should correspond to small and easily reducible NiO particles and NiO particles more strongly interacting with the support respectively.

3.3 Catalytic performances

Testings showed that the reaction products were CO, CO_2, H_2, H_2O and carbon deposit (expressed as carbon balance). The obtained results are reported in Figure 1.

It can be seen that the rate of consumption of carbon dioxide is about twice that of methane and the order of activity is as following at 730°C: Ni-Cu > Ni-Ce > Ni-Co >> Ni >> Ni-Fe. Thus copper, cobalt and cerium had promoting effect on the nickel active phase whereas iron was inhibiting. The same order was obtained at the reaction temperature of 550°C, for which, moreover, the iron additive completly inhibited the nickel phase. The H_2/CO ratio was higher than expected (1/1) and depended both on the nature of the metal additive and the reaction temperature. Since the obtained values seemed roughly to parallel carbon deposit (not shown), the selectivity results were attributed to coking side-reactions. However, in the case of the iron additive, the H_2/CO ratio was also higher than the unit but low coking activity (carbon balance of 95.9%) was observed. This result should indicate that either the H_2/CO relative selectivity was independant of the nature of the metal additive [6] or, for the less reactive iron containing catalyst, water stemming from the reverse water gas shift reaction should play a similar rôle as in the mixed steam and CO_2 reforming of methane[10]: it decreased methane conversion and increased the H_2/CO ratio values.

3-3 Aging and regeneration of the catalysts

Aging/regeneration experiments were performed with the Ni, Ni-Co and Ni-Ce catalysts and the temperature heating cycle was as following :

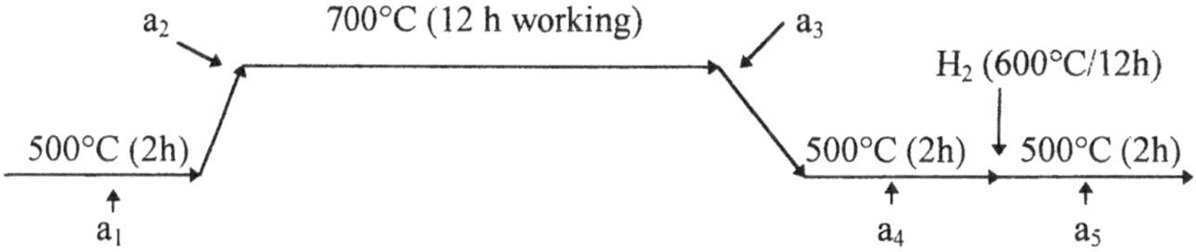

a_1, a_2, a_3, a_4 and a_5 represent the successive steady-state activities (according to the above testing procedure) at 500°C , 700°C (after heating), 700°C (after 12h working), 500°C/2h (after cooling) and 500°C/2h (after H_2/600°C/12h treatment). The obtained results are reported in Table 3.

1- At the reaction temperature of 700°C and after 12h/700°C treatment the order of activity was Ni-Co> Ni-Ce >> Ni as shown in Table 3. Indeed, the activity of the Ni catalyst was sensibly decreased (methane and carbon dioxide conversions and CO yield were divided by about 3, 2 and 4 respectively) whereas concomittently carbon deposit decreased as shown by the carbon balance values (94.0% against 96.0%). The Ni-Co catalyst was less sensitive to the thermal treatment since CH_4 conversion decreased only by about 7% whereas CO_2 conversion was not affected. However, the CO yield decreased by about 19% whereas carbon balance decreased from 85.8% to 77.3%. An intermediate behaviour was observed for the Ni-Ce catalyst (28% of decrease for CH_4 conversion, 23 % of decrease for CO yield).

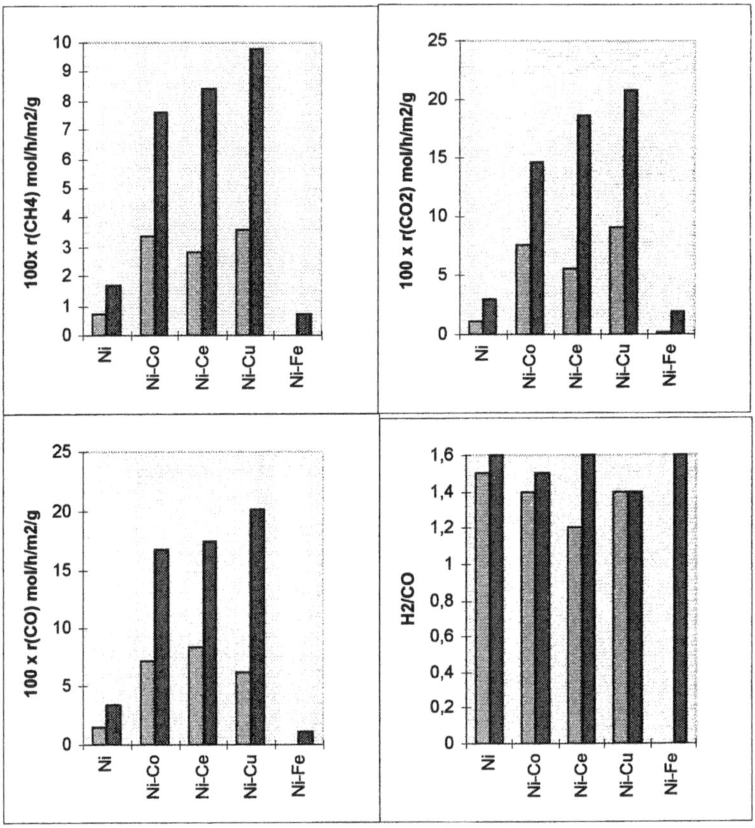

Figure 1 : Activity and H2/CO ratio in the CO2-reforming of methane over Ni-M/α-Al_2O_3 catalysts at 550°C and 730°C (dark symbols).

The important decrease of activity during aging treatment for the Ni catalyst should be attributed to sintering of the active phase rather than to coking since the carbon balance was better after (96.0%) than before (94.0%). In contrast, the decrease of the Ni-Co activity (see methane conversion changes) should be attributed, at least partly, to an increase of coking (see carbon balance changes). The Ni-Ce catalyst showed medium aging behaviour.

Moreover, in the case of the Ni-Co catalyst, the CO_2 conversion was quite not changed (a_3 /a_2 =104.1%) after aging treatment whereas CH_4 conversion, CO yield and coke deposition sensibly did (see above). It should be an indication that the CO_2 decomposition sites (s_2) were less prone or/and sensitive to coking than the CH_4 decomposition sites (s_1) :

$$CH_4(g) + s_1 \longrightarrow CH_4\text{-}s_1 \xrightarrow{-2\,H_2(g)} C\text{-}s_1 \xrightarrow{+\,s_2\text{-}O\,(-\,s_2)} CO(g) + s_1$$

$$C\text{-}s_1 \longrightarrow C\ \text{deposit-}s_1$$

$$CO_2(g) + s_2 \longrightarrow CO_2\text{-}s_2 \longrightarrow CO(g) + s_2\text{-}O$$

In such a case, the CO yield decrease should be related to that CO molecules originating from s_1 sites rather than s_2 sites.

Table 3 : Activity after aging (700°C/12h) and regeneration (H_2/600°C/12h) results in CO_2-reforming of methane for the Ni-M / α-Al_2O_3 (M = Co,Ce) catalysts .

operating conditions	catalyst	CH_4 activity variation	CO_2 activity variation	CO yield variation	carbon balance a_i-a_j (%)
a_3 / a_2	Ni /α-Al_2O_3	35.8	46.5	26.7	94.0-96.0
	Ni-Co/α-Al_2O_3	93.2	104.1	80.6	77.3 -85.8
(%)	Ni-Ce/α-Al_2O_3	72.2	82.3	77.2	89.2 -86.7
a_4 / a_1	Ni / α-Al_2O_3	0	0	0	- 99.0
	Ni-Co/α-Al_2O_3	243.0	435.0	30.0	55.6-95.4
(%)	Ni-Ce/α-Al_2O_3	24.5	15.0	31.8	99.2 -96.3
a_5 / a_1	Ni / α-Al_2O_3	620.0	104.3	16.0	91.0-99.0
	Ni-Co/α-Al_2O_3	28.0	55.1	34.0	95.8-95.4
(%)	Ni-Ce/α-Al_2O_3	182.0	16.7	64.0	97.6-96.3

Although to a lesser extent, it was also observed for the Ni and Ni-Ce catalysts, notably for the latter, that the conversion of CH_4 and yield of CO were more changed in the aging experiments than the CO_2 conversion did. These results should also be attributed to differences in the behaviour of the activation sites of methane and carbon dioxide respectively.

2- After aging at 700°C/12h and cooling the order of activity at 500°C was Ni-Co >> Ni-Ce >> Ni (Table 3). Indeed, the Ni catalyst became totally inactive whereas, in contrast, the Ni-Co catalyst seemed to be regenerated (a_4/a_1= 243.0% for CH_4 conversion) although this regeneration resulted in lower CO yield (a_4/a_1=30%) and higher carbon deposit (55.6% against 95.4% for carbon balance). Thus coking activity seemed to be enhanced by the aging treatment. On the other hand, high enhancement of CO_2 conversion (435.0%) was observed, indicating a better regeneration of the CO_2 decomposition active sites than that of CH_4 decomposition.

Medium results were obtained for the Ni-Ce catalyst and, in addition, regression of coking was observed (99.2% against 96.3 % for carbon balance). This diminution of carbon deposit at low reaction temperature should be attribuable to structural change of the active sites which probably arose during the aging treatment.

3- The results concerning the regeneration experiments under H_2/600°C/12h are also reported in Table 3. It can be seen that the Ni catalysts activity was not only restaured but also enhanced (a_5 / a_1= 620.0% for CH_4 conversion) but the yield of CO was strongly decreased (a_5

/ a_1= 16.0% and carbon balance passed from 99.0% to 91.0%). Thus the regeneration treatment led to the enhancement of coking activity for the Ni catalyst. In contrast, no variation was observed for CO_2 conversion (104.3%). Then, it can be concluded that the reductive treatment favoured the surface decomposition of methane and quite not that of carbon dioxide. This conclusion let us think that the former probably chemisorbed on metallic nickel surface sites whereas the latter rather chemisorbed on oxidized/basic surface sites.

Conversly, the activity of the Ni-Co catalyst was strongly decreased by the reductive treatment (a_5 / a_1= 34.0% for CO yield), the CO_2 decomposition sites being a little more resistent (a_5 / a_1= 55.1%) to the treament than the CH_4 decomposition sites (a_5 / a_1= 28.0 %). This deactivation cannot be attributed to coking since carbon balance was not changed but probably to structural changes on the surface of the solid. and and for CH_4

Medium results were obtained for the Ni-Ce catalyst and it can be seen in Table 3 that CH_4 decomposition was enhanced (a_5/a_1=182.0%) whereas both CO_2 decomposition (a_5/a_1 = 16.7%) and CO yield (a_5/a_1= 64.0%) were inhibited. Then it seemed that the surface sites for carbon dioxide decomposition were depleted by the reductive treatment. The decrease in the yield of CO cannot be attribuable to coke deposition (no change in carbon balance was observed) but most probably to structural changes on the surface of the solid.

4- CONCLUSIONS

Our study has shown that in the case of Ni-M /α-Al_2O_3 (M = Fe, Co, Cu, Ce) catalysts the metal additives, notably Co, decreased sensibly the reducibility of the nickel phase but seemed to increase its active surface. The order of activity up to 730°C was as following: Ni-Cu>Ni-Ce>Ni-Co>>Ni>>Ni-Fe. After aging treatment (700°C/12h under reactive atmosphere) the Ni catalyst became totally inactive whereas the activity of the Ni-Co was strongly enhanced (243% for methane conversion variation). Conversly, the regeneration treatment (H_2/700°C/12h) strongly enhanced the activity of the Ni catalyst whereas it strongly decreased that of the Ni-Co catalyst (620% against 28% for methane conversion variation). Medium behaviour was observed for the Ni-Ce catalyst. The changes of activity observed after these treatments were attibuted to either coking or structural changes of the surface of the solids. The turnover for methane and carbon dioxide consumption depended also on the nature of the metal additive and the nature of the treatment. Physical surface characterizations have to be undertaken in order to obtain more informations on the observed phenomena.

REFERENCES

1. O. Tokunaga and S. Ogasawara, React. Kinet. Catal. Lett., 39 (1989) 69.
2. A.M. Gadalla and M. Sommer, Chem. Engng. Sci., 44 (1989) 2825.
3. O. Yamasaki, T. Nozaki, K. Omata and K. Fujimoto, Chem. Lett. (The Chem. Soc. Japan), 1953 (1990).
4. J.R. Rostrup-Nielsen and J.H. Bak Hansen, J. Catal., 144 (1993) 38.
5. T. Sodesawa, A. Dobashi and F. Nozaki, React. Kinet. Catal. Lett., 12 (1979) 107.
6. W.H. Swan, V.K.H. Krill, G.A. Martin and C. Hirodatos, Catal. Today, 21 (1994) 571.
7. P. Turlier, E. Brum-Pereira and G.A. Martin, Proc. Intern. Congr. on CO_2 Utilization, Bari, 1990.
8. Z.Zhang, X.E.Vergkios, M.Susa,,J.Phys.Chem, 100(1996)744
9. D.Halliche, R.Bouarab, O.Cherifi and M.M.Bettahar, Catal.Today, 29(1996)373
10. D.Qin and J.Lapszewicz, Catal.Today, 21 (1994) 551-560

NATURAL GAS CONVERSION V
Studies in Surface Science and Catalysis, Vol. 119
A. Parmaliana et al. (Editors)
1998 Elsevier Science B.V.

The influence of promoters on the coking rate of nickel catalysts in the steam reforming of hydrocarbons

T.Borowiecki[a], A.Gołębiowski[b], J.Ryczkowski[a] and B.Stasińska[a]

[a]Faculty of Chemistry, Maria Curie-Skłodowska University, 20-031 Lublin, Poland

[b] Institute of Fertilisers, 24-100 Puławy, Poland

The effect of small amount of different promoters (K, Ba, Mo, W and Ce) on the properties of the nickel catalysts in steam reforming were investigated. The experimental results show that only the cerium promoter has very favourable influence on the nickel dispersion and its resistance for sintering. On the contrary the cerium promoter does not have the influence on the decrease of the rate of carbon formation in the steam reforming of n-butane.

1. INTRODUCTION

Steam reforming of hydrocarbons is one of the most important methods of commercial production of hydrogen or synthesis gas. Reaction is accompanied by very unfavourable and undesired formation of different kinds of carbon deposits [1,2], particularly filamentary carbonaceous deposits, which, by mechanically destroying the catalyst, make its deactivation irreversible. Therefore, one of the significant properties of a good catalyst for hydrocarbon steam reforming is its resistance to coking [3]. The simplest manner of improving the quality of the nickel catalysts is an introduction of certain additions into them, which increase their resistance to coking [3-6]. In the industrial practice, nickel catalysts with certain amount of potassium were applied (in spite of significant decrease of catalysts activity). Other promoters as U [5], La [6], Ce [7], rare earths [8] or Ba [9] can be applied, too. Borowiecki et al. [10,11] has shown that the introduction of small amounts (0.5-2 wt. %) of molybdenum or tungsten compounds to nickel catalysts greatly improved their resistance to coking.

At reforming temperature and at high pressure of steam and CO_2, potassium compounds decompose and they are slowly lost from the catalysts during use [3]. Cerium additions [7,8,12], increases resistance to coking and specific activity in the steam reforming of hydrocarbons. Other promoters also increase resistance to coking but lower specific nickel activity.

The aim of the present research was to compare the influence of some promoters (Mo, W, Ba, K and Ce) on the catalyst properties and their resistance to coking. It is known, that in the automobile catalysts ceria plays a key role in the oxygen storage capacity [13].

The second aim was to evaluate the thermostability of the promoting effect.

2. EXPERIMENTAL

2.1. Samples studied

Catalysts denoted henceforth as Ni-Me(-) (the number in brackets denotes the amount of MeO in wt. %) are samples of a commercial catalyst Ni/α-Al_2O_3 impregnated with a solution of appropriate promoter salts. The samples were dried at 105°C and calcined at 450°C for 4h.

The difficult conditions of performing the reaction cause that, in spite of the application of nickel on low-area supports (α-Al_2O_3 or $MgAl_2O_4$), one cannot prevent the phenomena of catalyst deactivation, such as sintering or nickel reaction with the support. The catalysts were heated for 24 h at 800°C, in mixture H_2O:H_2 = 1:1 under pressure 2.5 MPa to evaluate the thermostability of the promoting effect.

2.2. Methods

The methods used for measurements of the surface properties of the catalysts were described in detail elsewhere [10,11,14].

The total surface area of the samples was determined by physical adsorption of argon at the temperature of liquid nitrogen in a static - volumetric apparatus which ensured a vacuum better than 2 x 10^{-5} hPa.

The area surface of Ni was determined by hydrogen chemisorption at 20°C and 100 mm Hg pressure (after the reduction of catalysts), assuming that chemisorption stoichiometry was H: Ni = 1:1 and the surface area occupied by one hydrogen atom equalled 0.065 nm^2 [15].

Studies of the coking rates in the steam reforming of n-butane were carried out by the gravimetric method in a flow reactor, at constant temperature and partial pressure of n-butane (500°C and 6.1 kPa), and different ratios of reagent (more details can be found elsewhere [14]).

Temperature-programmed reduction (TPR) has been employed to study the reducibility of catalysts in the Altamira AMI-1 system.

Catalyst activity in steam reforming of methane was determined in the gradientless reactor under pressure of 0.1 MPa at various temperatures. The reaction rate constant was accepted as the activity measure (related to 1 g of the catalyst), assuming first order kinetics with respect to methane.

3. RESULTS

The properties of the catalysts have been set up in Table 1.

Figure 1 presents the effects of the promoters on the nickel surface area of the examined catalysts before and after pressurised heating. For four promoters (Mo, W, Ba and K) any noticeable changes of total and active surface areas cannot be observed. These promoters slightly increase the resistance of nickel for sintering.

In the contrary, catalysts promoted by cerium exhibit a significant increase in total and active surface areas, to 1.5 and 4 times, respectively, with maximum at ~1 wt. % Ce. Their treatment under pressure in the mixture of steam and hydrogen does not change surface development. The cerium promoter has very favourable influence on nickel dispersion and significantly increases its resistance for sintering (see Table 1).

Table 1
Main properties of the catalysts examined

Catalysts	Surface area ($m^2\ g^{-1}$)				Mean size of Ni crystallites (nm)				T_{max}*
	Total		Active		d_H		d_X		
	A	B	A	B	A	B	A	B	°C
Ni-Mo (0.5)	3.5	4.3	1.8	1.1	42.9	70.2	27	-	450
Ni-W (2.0)	3.3	3.9	1.8	1.1	42.9	70.2	31	-	465
Ni-Ba (2.0)	3.4	4.1	1.8	1.1	42.9	70.2	-	-	442
Ni-K (1.0)	3.7	3.7	1.8	0.9	42.9	85.8	-	-	430
Ni	3.8	3.8	2.0	0.9	38.6	85.8	42	90	445
Ni-Ce (0.2)	6.4	5.1	2.9	2.8	26.6	27.6	-	-	-
Ni-Ce (0.5)	6.2	4.9	3.4	3.4	22.7	22.7	52	56	-
Ni-Ce (1.0)	6.5	5.2	3.9	4.0	19.8	19.3	34	-	441
Ni-Ce (2.0)	5.5	5.0	3.3	3.5	23.4	22.1	39	54	430

A - samples after reduction at 800°C
B - samples after reduction and pressurised heating at 800°C in H_2O+H_2 mixture
d_H - from hydrogen chemisorption data at 25°C
d_X - from X-ray broadening data (CuK_α radiation)
T_{max} - temperatures at the maxima of the peak of reduction

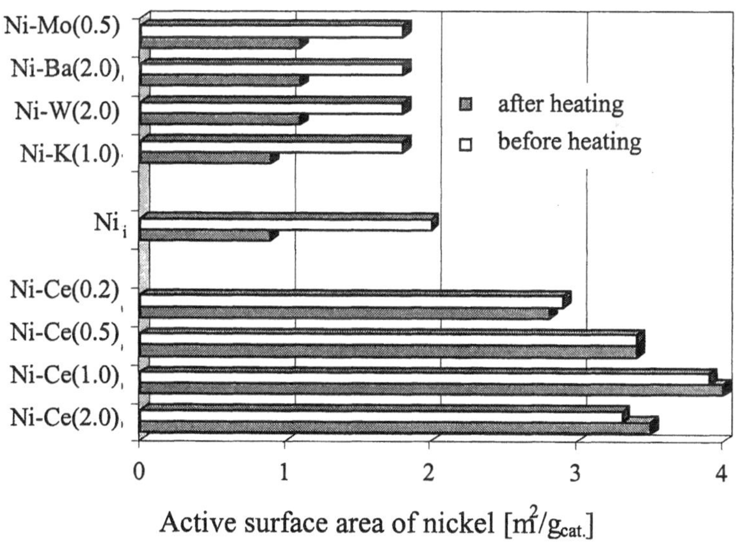

Fig. 1 - Active surface area of nickel in catalysts before and after pressurised heating

Cerium additions modify the course of NiO reduction. TPR spectra are shown in Figure 2. In comparison to the typical reduction course of $NiO/\alpha-Al_2O_3$ catalysts [16,17], TPR spectra of Ni-Ce catalysts with increasing Ce amounts show the progressive increase reducibility. Reduction moves towards lower temperature, but the effect is slight.

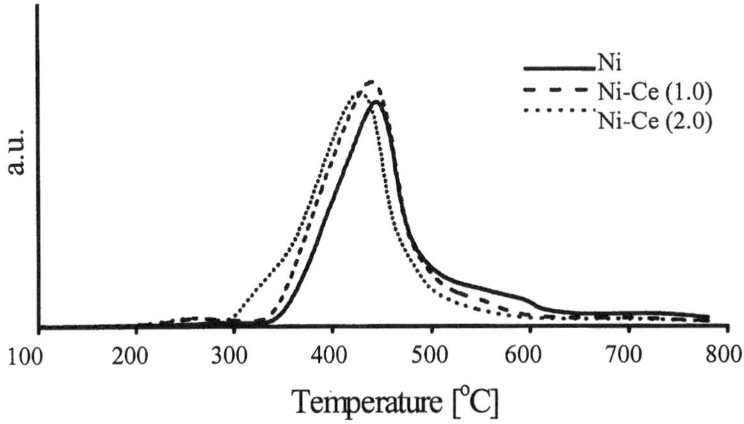

Fig. 2 - TPR spectra of Ni-Ce catalysts

Figure 3 presents carbon deposition on catalysts with different promoters in steam reforming of n-butane at temperature 500°C and at water-to-carbon ratio = 2.0.

Catalysts with the addition of Mo, K, W and Ba undergo coking much more slowly than the $Ni/\alpha-Al_2O_3$ catalyst. The greatest influence on the increase in resistance to coking has the addition of Mo or K.

The obtained results for the nickel-cerium catalysts do not confirm the literature data. In the range of the investigated promoter amounts (≤ 2 wt. %) we was found the rate increase of deposits formation (Fig. 3). For the impregnated nickel catalysts on industrial support $MgAl_2O_4$ with CeO_2 (up to 13.3 wt. % and the same sequence of preparation) [7,8] or indefinite oxide of rare earths (Re_2O_3) [12] there was found a slight promoter influence on the increase of nickel dispersion but a great decrease of the rate of deposit formation.

An increase of active surface area in Ni-Ce samples causes a decrease in the rate of catalyst coking (calculated per nickel surface area unit - Table 2).

For comparable and very small amounts of promoters (≤ 2 wt. %) the efficiency of their action in the limitation of nickel catalysts coking decreases as follows:

Mo, K >> W > Ba >Ce

Activities of catalysts at two temperatures after pressurised heating have been set up in Table 2. Other kinetic parameters for the steam reforming of n-butane are also presented in the Table 2.

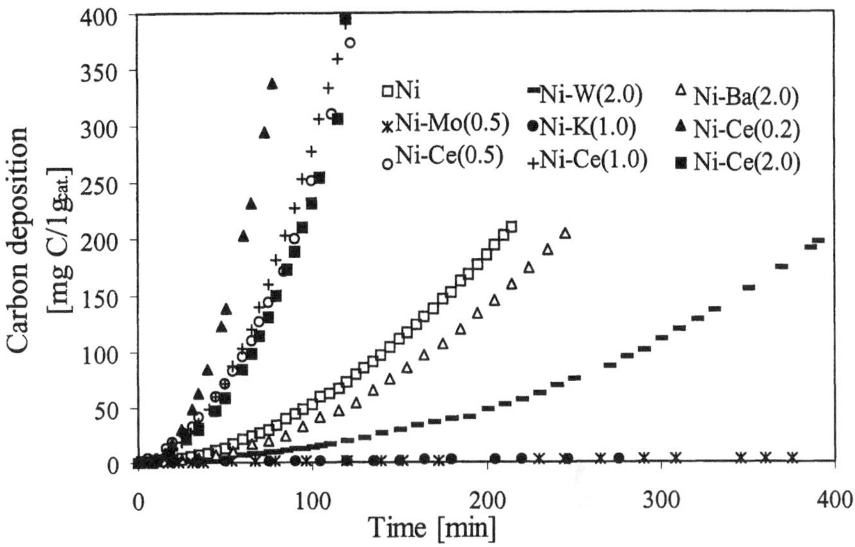

Fig. 3 - Carbon deposition on promoted nickel catalysts at reaction mixture H_2O:C = 2.0

Table 2
Catalytic properties of the catalysts examined

Catalysts	Activity*		E_{app}	lg A	Coking rate** ($\mu g\ g_m^{-2}\ min^{-1}$)	
	600°C	800°C	kJ mol^{-1}		H_2O:C = 0.5	H_2O:C = 2.0
Ni-Mo (0.5)	6.6	21.2	47.3	1.33	596	12
Ni-W (2.0)	11.9	28.9	42.7	1.29	5953	940
Ni-Ba (2.0)	8.7	24.4	45.7	1.39	7507	1351
Ni-K (1.0)	10.1	28.2	42.8	1.45	4916	13
Ni	12.2	38.9	45.2	1.59	12261	2299
Ni-Ce (0.2)	-	-	-	-	5415	2272
Ni-Ce (0.5)	13.2	48.0	51.8	1.13	5773	1348
Ni-Ce (1.0)	-	-	-	-	4752	1129
Ni-Ce (2.0)	15.8	47.8	48.2	1.47	-	1200

* - rate constant for steam reforming of methane [Nm^3/m^2 h at]
** - rate of coking at 20 wt. % C on the catalysts

Mo, W, Ba and K promoters cause a certain diminishing of nickel activity [18]. The lowering of this activity was about 10-30 %. Catalysts with cerium promoter exhibit slight increase of activity (~25 %) in the steam reforming of methane [18]. Activity of promoted catalysts in the steam reforming of methane decreases as follows:

Ni-Ce > Ni > Ni-W, Ni-K > Ni-Ba > Ni-Mo

4. CONCLUSIONS

Mo, W, Ba, K and Ce promoters of nickel catalysts for steam reforming of hydrocarbons influence on the catalytic activity and the resistance to coking. The effect of promoters presence does not disappear after high temperature and pressure treatments (these are modelling conditions for the catalyst at the beginning of reformer tube).

In the view of the literature data for the Ni-Ce catalysts with the same sequence of preparation [7,8], the results obtained for the catalysts promoted with Ce are very surprised. The great influence of cerium on the magnitude of the active surface area and good stability of the catalyst (in the conditions of heating at the elevated pressure) is connected with a small effect on the coking rate on nickel catalysts.

REFERENCES

1. J.R. Rostrup-Nielsen, in *Catalysis - Science and Technology*, (J.R. Anderson and M. Boudart, Eds.), Springer Verlag, Berlin, 1984, Vol.5, 1
2. E. Tracz, R. Scholz and T. Borowiecki, Appl.Catal., 66 (1990) 133
3. *Catalyst Handbook*, (M.V. Twigg, Ed.), Wolfe, London, 1989
4. S.P.S. Andrew, I&EC Prod.Res.Dev., 8 (1969) 321
5. K.S.M. Bhatta and G.M. Dixon, I&EC Prod.Res.Dev. 8 (1969) 324
6. K.B. Mok, J.R.H. Ross and R.M. Sambrook, in *Preparation of Catalysts III*, (G. Poncelet, P. Grange and P.A. Jacobs, Eds.), Elsevier 1983, 291
7. Q. Zhuang, Y. Qin and L. Chang, Appl.Catal., 70 (1991) 1
8. Z. Wu, J. Zhang and L. Chang, in *Catalyst Deactivation 1987*, (B. Delmon and G.F. Froment, Eds.), Elsevier, Amsterdam 1987, 209
9. T. Borowiecki, A. Gołębiowski and K. Stołecki, *Proc.7th Int.Symp.Heterog.Catalysis*, Bourgas, 1991, 199
10. T. Borowiecki and A. Gołębiowski, Catal.Lett., 25 (1994) 309
11. T. Borowiecki, A. Gołębiowski and B. Stasińska, Appl.Catal., A: General, 153 (1997) 141
12. R. Song, L. Zhang and S. Guo, in *Catalyst Deactivation 1991*, (C.H. Bartholomew and J.B. Butt, Eds.), Elsevier, Amsterdam 1991, 243
13. E. Rogemond, R. Frety, V. Perrichon, M. Primet, S. Salasc, M. Chevrier, C. Gauthier and F. Mathis, J.Catal., 169 (1997) 120
14. T. Borowiecki, Appl.Catal., 10 (1984) 273
15. C.H.Bartholomew and R.J.Farrauto, J.Catal., 45 (1975) 41
16. J.R. Richardson, M. Lei, B. Turk, K. Foster and M.V. Twigg, Appl.Catal.A: General, 110 (1994) 217
17. T. Borowiecki, J. Gryglicki, W. Gac, A. Denis and B. Stasińska, submitted to Appl.Catal.
18. A. Gołębiowski, K. Stołecki, T. Borowiecki and B.Stasińska, in preparation

NATURAL GAS CONVERSION V
Studies in Surface Science and Catalysis, Vol. 119
A. Parmaliana et al. (Editors)
1998 Elsevier Science B.V.

Reforming of methane with carbon dioxide over supported Ni catalysts.

R. Bouarab[a], S. Menad[a], D. Halliche[a], O. Cherifi[a] et M.M. Bettahar[b]

[a] Laboratoire de Chimie du Gaz Naturel, Institut de Chimie, USTHB, BP32, El Alia, 16111 Bab-Ezzouar, Alger, Algérie.

[b] Laboratoire de catalyse hétérogène, Université de Nancy I, Henry Poincarré 54506 - Nancy-Lés- Vandoeuvres Cedex France.

ABSTRACT

The reforming of methane with carbon dioxide was studied over nickel supported on αAl_2O_3, γAl_2O_3, $\alpha Al_2O_3 + SiO_2$, SiO_2 , ZrO_2, MgO and modified either by transition metallic adducts (Co, Cu and Fe) or by alkaline promoters Na, K. A strong influence of support nature on the catalytic activity was observed. The catalytic activity of $Ni/\alpha Al_2O_3$ was very similar to that of $Ni/\gamma Al_2O$; however, more coking was observed over the latter. Thus, the methane reforming activity decreases in the following order : αAl_2O_3 > γAl_2O_3 > SiO_2 > αAl_2O_3 + SiO_2 > ZrO_2, MgO. The nickel catalysts properties were also affected by metallic additives (Co, Cu and Fe) and by alkaline promoters Na, K .

1.INTRODUCTION

Carbon dioxide reforming of methane produces synthesis gas **($CO + H_2$) :**

$$\mathbf{CH_4 + CO_2 \rightleftharpoons 2CO + 2H_2}$$

This reaction has recently attracted much interest. One of the reasons of this interest is that reaction consumes both CH_4 and CO_2 which are greenhouse gases. Another advantage of CO_2/CH_4 reaction is its utilization with the reverse H_2/CO reaction as a means of converting solar energy to chemical energy for storage and transportation (SOLCHEM Process [1,2], CLEA Project [3]).

It has been proved in industry that the best metal for the catalysts of CO_2/CH_4 reforming is nickel. This is due to its large availability and lower cost in comparison with noble metals. However, one of the most important problems carrying out this reaction, is coke deposition on the surface which deactivates the catalysts [4]. The activity and the stability to coke deposition seems to depend strongly on the nature of support [5,6] and on the presence of additives [7].

The SPARG process [8] ameliorates the coking problem via sulfur passivation of the nickel catalysts.

Concerning CO_2 reforming of methane, over nickel-based catalysts, it has been suggested that the active phase consists of metal particles, as demonstrated for Ni / SiO_2 [9,10] and Ni/La_2O_3/Al_2O_3 [11]. Zhang et al [12-14] proposed a major role of the support over Ni/La_2O_3 catalysts for explaining their stability with time on stream. The catalytic stability of the same catalyst Ni/La_2O_3 is confirmed by Mirodatos et al [15] in their study combined with isotopic transient Drift and TAP experiments which reveal the existence of interface between a lanthanum carbonate phase and the nickel particles.

The present work reports the results of investigations about the effect of support (αAl_2O_3, γAl_2O_3, $\alpha Al_2O_3 + SiO_2$, SiO_2, ZrO_2, MgO), the influence of alkaline promoters Na, K and the effect of metallic additives Co, Cu and Fe.

2. EXPERIMENTAL

The supports used were ZrO_2 (Merck, 98%), MgO (Merck, 97%), γAl_2O_3 (Merck, 100 m^2/g), and SiO_2 (Degussa, 200 m^2/g). In order to obtain the αAl_2O_3 (3,7 m^2/g) support, the γAl_2O_3 was calcined under flowing air (1.3 $l.h^{-1}$, 5°C..min^{-1} heating rate) at 1200°C during 6 h. The supported catalysts with 1.2 wt- % loading of nickel were prepared by dried impregnation, followed by drying at 110°C for 12h and calcined 6h at 700°C under air flow (1.2 $l.h^{-1}$) with a heating rate of 5°C.min^{-1}. The Ni-Na/αAl_2O_3 , Ni-K/αAl_2O_3, Ni-Co/αAl_2O_3, Ni-Cu/αAl_2O_3 and Ni-Fe/αAl_2O_3 catalysts were elaborated by wetness impregnation of Ni/αAl_2O_3 catalyst using the appropriate metal salt solution and following the same procedure described above.

Catalysts were tested in a fixed bed quartz tubular reactor after reduction in situ overnight at 600°C under hydrogen (1.2 $l.h^{-1}$), temperature rise rate 4°C. min^{-1} using amount of catalyst 100 mg for all runs; particle size $\leq$ 160 μm. The reactants gas and products mixtures were periodically analyzed on line using TCD chromatograph which contained two 4 m Carbosieve B columns: 1/8 inch, 100 to 200 mesh. Prior to analyze, the effluent was passed through water-trap at 0°C in order to condense H_2O. H_2 concentration was not analyzed. In order to determine the metal loading of our catalysts, a spectrometer Perkin- Elmer was used. X-ray diffraction performed with a Philips PW 1710 diffractometer using filtered CuK_α radiation was used to identify the existing phases of fresh catalysts. Scanning electron microscopy (SEM) is used to determine the catalyst grain size. The results of characterization are summarized in Table 1.

3. RESULTS AND DISCUSSION

3.1. Results of characterization

The weight loading of all catalysts determined by spectroscopy are compared with nominal values. Except for potassium and copper loading, the agreement between analyzed metal and nominal loading is good.

X-ray diffraction of fresh Ni- alkaline/αAl_2O_3 samples shows sharp bands corresponding to well crystallized solids. In general, NiO and supports phases are present in all the catalysts. For the Ni- Co/αAl_2O_3 and Ni- Cu/αAl_2O_3 catalysts two oxidized phases appear for the cobalt (CoO - Co_2O_3) and copper (CuO - Cu_2O).

The particle size (metal + support) determined by SEM and TEM after calcination varies between 0.0019 and 5.3 μm.

Table 1
Characteristics of different catalysts.

Catalysts	metal content (%)		d (μm)
Ni(1,2) - Na(1)/αAl_2O_3	Ni 0,98	Na 0,93	2,4
Ni(1,2) - K(1)/αAl_2O_3	Ni 1,08	K 0,53	5,3
Ni(1,2) /αAl_2O_3	Ni 1,20		2,8
Ni(1,2)/γAl_2O_3	Ni 1,13		3,5
Ni(1,2) /αAl_2O_3 + SiO_2	Ni 0,82		
Ni(1,2) /ZrO_2	Ni 1,15		---
Ni(1,2) /MgO	Ni 1,23		---
Ni(1,2)/SiO_2	Ni 1,10		---
Ni(1)-Co(1)/ αAl_2O_3	Ni 0.86	Co 0.93	---
Ni(1)-Fe(1)/ αAl_2O_3	Ni 1.16	Fe 0.88	0,002*
Ni(1)-Cu(1)/ αAl_2O_3	Ni 1.17	Cu 0.03	0,0019*

* TEM measurement.

3.2. Catalytic measurements

In order to determine the catalytic properties of our catalysts, we have tested them for the methane dioxide reforming. The ratio CH_4/CO_2 = 1/4 of the feed was chosen with a total flow rate of 22 ml. min^{-1} . The catalytic performances were determined at atmospheric pressure by raising temperature from 450 to 730°C. The products formed are CO, H_2 and H_2O and carbon balance has shown that coke deposited. The conversions of CH_4 (%) and CO_2 (%) are expressed as the ratio of mols converted divides by the number of mols in the feed.

3.2.1 Effect of support

The effect of support on the conversions of methane and carbon dioxide over 1,2 wt -% nickel catalysts was studied at 600°C. The variation of the monoxide carbon yield as a function of time on stream is shown in Fig- 1. At this temperature, it seems like that the most active catalysts is Ni/αAl_2O_3 (CO yield = 15.5 $mol.h^{-1}.g^{-1}$), and the CO yield remains invariable with time on stream. This catalytic activity is similar to that of Ni/γAl_2O_3 (CO yield = 13.7 $mol.h^{-1}.g^{-1}$), but more coking is observed over this latter 84 % against 92,6 for Ni/αAl_2O_3 Table 2. This susceptibility to coking may be due to the formation of spinal $NiAl_2O_4$. Gadalla and Bower [16]observed thatγAl_2O_3 was converted into αAl_2O_3 and had also reacted with nickel forming $NiAl_2O_4$ .Ni/SiO_2 catalyst deactivates between 20 and 90 min of reaction, and than it tends to be moderately active and stable. The less activity of Ni/αAl_2O_3 + SiO_2 catalyst(CO yield = 1.5 $mol.h^{-1}.g^{-}1$). This feeble activity is due to the possible interaction between Nickel and the support (αAl_2O_3 + SiO_2) . Ni/MgO and Ni/ZrO_2 are inactive. The inertia of Ni/MgO is due to the weak weight loading of nickel [5]. Parmaliana et al [17,18] have shown that nickel loading of less about 11 wt-% for Ni/MgO catalyst are inactive for CH_4/H_2O reforming. This is due to the formation of an irreductible solid solution. The inactivity of Ni/ZrO_2 is probably due to the low temperature used for reduction (600°C). Swaan et al [19] have obtained moderately activity for Ni/ZrO_2 when reduced at 750°C. The order of activity observed for these different supported catalysts is : $Ni/\alpha Al_2O_3$ > $Ni/\gamma Al_2O_3$ > Ni/SiO_2 >$Ni/\alpha Al_2O_3 + SiO_2$ > Ni/ZrO_2, Ni/MgO.

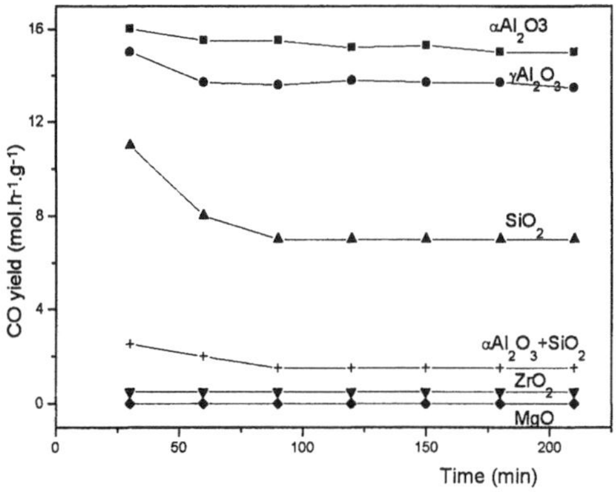

Figure 1. CO yield as a function of time on stream over $Ni/\alpha Al_2O_3$, $Ni/\gamma Al_2O_3$,Ni/SiO_2, Ni/ ($\alpha Al_2O_3 + SiO_2$) , Ni/ZrO_2 and Ni/MgO catalysts. Reaction Conditions : $CH_4/CO_2 = 1/4$, T = 600°C, metal loading = 1,2 wt-% and P = 1 atm.

Table 2
Performances of nickel supported catalysts Reaction Conditions : $CH_4/CO_2 = 1/4$, T = 600°C, metal loading = 1,2 wt-%, Flow rate = 22 ml. min^{-1} and P = 1 atm.

Support	αAl_2O_3	γAl_2O_3	SiO_2	MgO	ZrO_2
C[a] balance (%)	92.6	84.0	97.6	97.4	99.3
CH_4 conversion	62	63	22	3	3
CO_2 conversion	30	35	12	0,2	3

[a] C balance = ($X^{*}_{CH4} + X^{*}_{CO2} + X^{*}_{CO}$) Flow rate* / ($X^{**}_{CH4} + X^{**}_{CO2}$) Flow rate**
X_i = conversion of component i
* inlet reactor
** oulet reactor

3.2.2. Effect of metal additives

The influence of the metal additives on the catalytic performances depends on their nature (table 3), and the order of methane or carbon dioxide conversions are :

Ni-Co, Ni > Ni-Cu >> Ni-Fe

This difference in activity may be due to the possible interaction between nickel and the metal additives as have shown the XRD study for Co and Cu additives in the calcined catalysts.For the Fe additive, this interaction may be related to amorphous phases not detected by XRD.

Table 3
Effect of metallic adducts. Reaction Conditions : CH_4/CO_2 = 1/4, T = 600°C, metal loading = 1.2 wt-%, Flow rate = 22 ml. min^{-1} and P = 1 atm.

Catalyst	Ni-Co	Ni	Ni-Cu	Ni-Fe
CH_4 Conversion (%)	54.4	56.9	26.0	no active
CO_2 Conversion (%)	27.9	20.2	17	-

3.2.3 Influence of alkaline promoter

The influence of alkaline additive on the activity of Ni/αAl_2O_3 catalyst was investigated at atmospheric pressure. The obtained results at 600°C show that both alkaline additives decrease the CO yield of Ni(1,2)/αAl_2O_3 catalyst reference Fig- 2. The activities were also compared by CH_4 and CO_2 consumption and the order was found to be Ni(1,2)/αAl_2O_3 > Ni(1,2)-Na(1)/αAl_2O_3 > Ni(1,2)-K(1)/αAl_2O_3. Ni(1,2)/αAl_2O_3 catalyst is 5 times more active than Ni-K/αAl_2O_3 catalyst. The low activity of Ni-K/αAl_2O_3 results in blocking of nickel sites by potassium phase [20]. It appears also that potassium additive obstructs more sites than sodium.

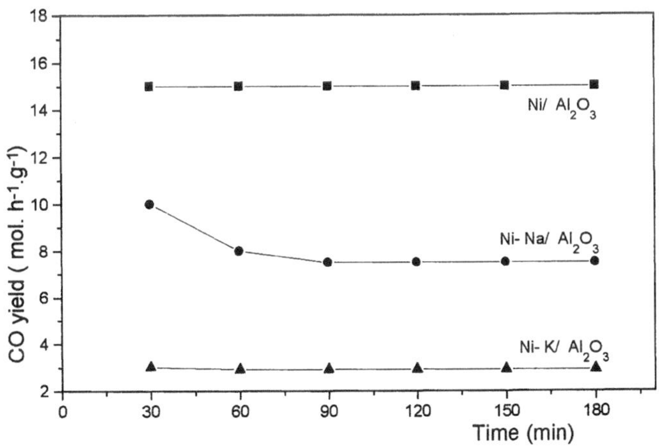

Figure 2. Effect of alkaline promoter on CO yield as a function of time on stream. Reaction Conditions : Nickel loading = 1,2 wt-% , CH_4/CO_2 = 1/4, T = 600°C, Flow rate = 22 ml. min^{-1} and P = 1 atm.

4. CONCLUSION

The present work describes the activity order in the reforming of methane with carbon dioxide over nickel supported catalysts : αAl_2O_3 > γAl_2O_3 > SiO_2 > (αAl_2O_3 + SiO_2)> ZrO_2, MgO.

It appears that the effect of alkaline promoter on the modified Ni/ αAl_2O_3 catalysts follows the order : Ni > Ni-Na > Ni-K.

The influence of the additives on the catalytic performances of Ni/ αAl_2O_3 modified catalyst depends on their nature and the observer order of activity is : Ni-Co, Ni > Ni-Cu >> Ni-Fe.

REFERENCES

1. T. A. Chubb, Sol. Energy, 24 (1980) 341.
2. J. McCrary, G. E. McCrary, T. A. Chubb, J. J. Nemecek and D. E. Simmons, Sol. Energy, 29 (1982) 141.
3. J. D. Fish and D. C. Hawn, J. Sol. Energy Eng., 109 (1987) 52.
4. J. R. Rostrup-Nielsen and J. H. Bak Hansen, J. Catal., 144 (1993) 38.
5. Michael C. J. Bradford, M. Albert Vannice, Appl. Catal., 142 (1996) 73-96.
6. T. Sodesawa, A. Dobashi and F. Nozaki, React. Kinet. Catal. Lett., 12 (1979) 107.
7. D. Halliche, R. Bouarab, O. Cherifi and M.M. Bettahar, Catal. Today, 29 (1996) 373-377.
8. N.R.Udengaard, J.H.B. Hansen and D.C. Hanson, Oil Gas J., 90 (1992) 265.
9. V.C.H. Kroll, H.M. Swaan and C. Mirodatos, J. Catal., 161 (1996) 409.
10. V.C.H. Kroll, H.M. Swaan, S. Lacombe and C. Mirodatos, J. Catal.,164 (1996) 387.
11. A. Slagtern, M. Olsbye, R. Blom, I.M. Dahl and H. Fjellrag , Appl. Catal. A : General 145 (1996) 375.
12. Z.L. Zhang and X.E. Verykios, Appl. Catal. 138 (1996) 109.
13. Z.L. Zhang , X.E. Verykios, S.M.MacDonald and S. Affrosnen, J. Phys. Chem.100, 744 (1996).
14. Z.L. Zhang and X.E. Verykios, J. Chem. Soc. Chem. Commun. 71 (1996).
15. A. Slagtern, Y. Schwrman, C. Leclerq, X. Verykios and C. Mirodatos, J. Catal., 172 (1997) 118.
16. A. M. Gadalla and B. Bower, Chem. Eng. Sci., 43 (1988) 3049.
17. F. Arena, B. A. Horrel, D. I. Locke, A. Parmaliana and N. Giordano, J. Catal., 132 (1991) 58.
18. A. Parmaliana, F.Arena, F. Frusteri, S. Collcera, L. Marchese, G. Martra and A.L. Chuvillin, J. Catal., 141 (1993) 34.
19. H.M. Swaan, V.C.H Kroll, G.A. Martin and C..Mirodatos, Catal. Today, 21 (1994) 571
20. G.A.Martin and H.Praliaud, Catal. Lett., 9 (1991) 151.

NATURAL GAS CONVERSION V
Studies in Surface Science and Catalysis, Vol. 119
A. Parmaliana et al. (Editors)

Ni/Al_2O_3 catalysts for syngas obtention via reforming with O_2 and/or CO_2

N.N.Nichio[1], M.L.Casella[1], E.N.Ponzi[1], O.A.Ferretti[1,2]

1.CINDECA (UNLP-CONICET), 47 N°257, (1900) La Plata, Argentina
2.Facultad de Ingeniería (UNLP), 47 N°257, (1900) La Plata, Argentina

The performance of Ni supported catalysts is analyzed for syngas obtention from methane. Catalysts were prepared employing two different precursor compounds (nickel nitrate and nickel acetylacetonate) and two supports ($\alpha-Al_2O_3$ and $\alpha-Al_2O_3$ modified by the addition of aluminium). The samples were tested in partial oxidation (POM), CO_2 reforming (R) and a mixed reforming (MR) reactions. It was observed a beneficial effect of the modification of the support upon the activity and stability. The catalyst prepared with the organometallic precursor and the modified support showed the highest stability. The presence of CO_2 as a reactant causes a diminution of the catalytic activity and the modification of H_2/CO ratio.

1. INTRODUCTION

Due to the great demand for H_2 and CO, both for petroleum refining processes (hydrotreating and hydrocracking) and petrochemical industry (production of methanol, ammonia and hydrocarbons via Fischer-Tropsch), its production by reforming of natural gas has achieved increased importance. While most syngas is obtained by steam reforming, some of the other routes like partial oxidation, CO_2 reforming and mixed reforming may be more attractive, depending on factors such as H_2/CO ratio, product purity, feedstock availability (purity and cost, including oxygen). Both steam reforming and CO_2 reforming reactions are highly endothermic, requiring large energy input (generally a combustion furnace), while partial oxidation reaction is an exothermic one.

Nickel and other transition metal-based catalysts have been reported to be active for these reactions [1-3]. This paper reports the use of nickel catalysts, prepared employing two different precursor compounds and two supports and the activity, selectivity and stability of the resulting systems are analyzed in the partial oxidation of methane (POM), CO_2 reforming of methane (R) and a mixed reforming (MR) reactions.

As partial oxidation of methane (POM) process is an exothermic reaction, local hot spots are usually formed which can irreversibly damage the active phase. CO_2 reforming is not only important in industry for CO production, but also is interesting as a CO_2 consuming reaction (greenhouse effect). Mixed reforming ($O_2 + CO_2$) appears as a technological alternative, in order to improve the control of temperature, critical in POM reaction, together with the possibility of obtaining H_2/CO ratios between 1 and 2.

2. EXPERIMENTAL

Two series of supported Ni catalysts, with a content of 2 wt% Ni, were prepared. One using $Ni(NO)_3 6H_2O$ (Mallinckrodt) as the metal precursor compound (Ni^I) and the other using Ni-acetylacetonate (Aldrich) (Ni^O). In one series, the support was α-Al_2O_3 (Ni^I_α and Ni^O_α), and in the other one, the α-Al_2O_3 support was modified by impregnation with $Al(NO_3)_3.9H_2O$ (Merck) so as to obtain 1% w/w Al in the solid ($Ni^I_{\alpha+Al}$ and $Ni^O_{\alpha+Al}$).

The catalysts were characterized by Temperature-Programmed Reduction (TPR), Transmission Electron Microscopy (TEM) (using a Jeol 2010 instrument), catalytic activity tests and Temperature Programmed Oxidation of carbon deposited by Thermogravimetric Analysis (TPO/TGA) (using a Shimadzu TGA50).

The experimental setup for the catalytic measurements in POM, R and MR reactions, consisted of a tubular quartz reactor (i.d.= 0.8 cm) and an axial thermowell containing a chromel-alumel thermocouple centered in the catalytic bed. The operating conditions were: atmospheric pressure, temperature range between 650 and 800°C, feed flow rate 65 cm^3min^{-1}, with a feed composition for POM: $N_2/CH_4/O_2$=11/2/1, for R: $N_2/CH_4/CO_2$ = 6/1/1, and MR: $N_2/CH_4/O_2/CO_2$ = 22/4/1.5/1. For all the studied reactions the ratio C/O = 1. The catalyst weight was 0.030 g, and the particle size between 0.12-0.15 mm. The catalyst was diluted with pure α-Al_2O_3, with a ratio diluent/catalyst of 8, in order to obtain isothermal conditions throughout the catalytic bed. These conditions were defined so as to eliminate intraphase and interphase transport resistances [4].

Stability tests were determined at constant temperature (750°C), for 24 h. Stability was evaluated in terms of an activity coefficient a_{CH4}, which is the ratio between the rate of consumption of CH_4 corresponding to t hours on stream and the initial rate value. The feed flow composition and the reaction products were analyzed by gas chromatography.

Carbon deposists produced during the catalytic tests were characterized by Temperature Programmed Oxidation (TPO/TGA), measuring weight variation as a function of temperature. Post-reaction samples of 0.010 g were tested, feeding with air at a flow rate of 40 cm^3min^{-1}, a He sealing gas system allowed the reacting zone of the electronic balance to be isolated. The heating programme was 10°C min^{-1} from room temperature to 850°C.

3. RESULTS AND DISCUSSION

As it has been previously report [5], TPR results for the catalysts studied show a stronger interaction between the precursor compound and the support when the α-Al_2O_3 support is modified by the addition of aluminium oxide; being this behaviour the same for both precursors employed, inorganic and organometallic. In the same paper, TEM results show a lower mean particle size and a more homogeneous particle size distribution for the catalysts prepared with the modified support, confirming the stronger interaction between Ni and α-Al_2O_3 + Al. Dispersion, estimated from TEM data using the Brunelle et al. equation [6], for Ni^I_α, Ni^O_α, $Ni^I_{\alpha+Al}$ and $Ni^O_{\alpha+Al}$ is 3.7, 3.5, 5.4 and 4.9%, respectively.

Table 1 present the results of catalytic activity tests for POM, R and MR reactions at 650 and 750°C. These results show that, despite the reaction, the most active catalysts are those prepared employing the modified support, a fact that is in agreement with the greater nickel dispersion of these samples.

Table 1. Catalytic results for POM, R and MR reaction.

Reaction	T (°C)	Ni^{I}_{α} X_{CH4} %	Ni^{I}_{α} H_2/CO	Ni^{O}_{α} X_{CH4} %	Ni^{O}_{α} H_2/CO	$Ni^{I}_{\alpha+Al}$ X_{CH4} %	$Ni^{I}_{\alpha+Al}$ H_2/CO	$Ni^{O}_{\alpha+Al}$ X_{CH4} %	$Ni^{O}_{\alpha+Al}$ H_2/CO
POM	650	45	2.5	54	2.3	73	2.4	78	2.6
POM	750	69	2.1	72	2.1	91	2.2	88	2.2
R	650	35	0.6	46	0.7	56	0.9	56	0.8
R	750	54	0.9	63	0.8	87	1.0	83	1.0
MR	650	36	1.5	41	1.7	50	1.8	42	1.3
MR	750	53	1.6	64	1.7	77	1.8	76	1.4

Concerning the mechanism of POM reaction, several authors [1-4] suggest that the reaction proceed via two-step mechanism in which there is first total oxidation of a part of methane to CO_2 and steam, and then the reforming of the remaining CH_4 with CO_2 and steam to syngas when employing catalysts such as Ni, Ir, Ru on different alumina supports. However, a direct partial oxidation mechanism (one-step mechanism) for Pt catalysts is also suggested [7]. Recently, D. Qin et al. [8] have proposed that POM reaction may proceed via both mechanisms simultaneously when carried out employing Rh, Ru, Pt, Pd on MgO. The contribution of each mechanism is dependent on the concentration and kinetics of adsorbed atomic oxygen and gaseous atomic oxygen. When comparing the behaviour of Ni^{I}_{α}, Ni^{O}_{α}, $Ni^{I}_{\alpha+Al}$ and $Ni^{O}_{\alpha+Al}$ catalysts for the three different reactions presented in Table 1, the lowest activity corresponds to R and MR reactions, a fact that would not agree with a two-step mechanism, for in that case, the addition of CO_2 should provoke an increment of the activity for MR reaction. Our results with these systems could be explained in terms of the coexistence of both mechanisms (one step and two steps mechanisms), being the lower rate of R and MR reactions due to the competition for the active sites for CH_4 activation by CO_2.

The values of H_2/CO ratio are close to those predicted theorically, i.e., 2 for POM, 1 for R and 1.5-1.8 for MR, mainly for higher levels of conversion.

For POM reaction, the samples showed a typical cycle on the methane conversion as a function of sequential changes in the reaction temperature, as it is well known for Ni and Rh supported catalysts [4,9]. This behaviour was explained in the following way: in a first step H_2 is generated by methane cracking reaction (at 800°C approximately) and then this H_2 provokes NiO reduction to metallic Ni which is the actual active phase. In R reaction O_2 is not present in the feed flow, the reduction of the oxidized Ni surface is achieved at a lower temperature and the cycle is not observed. With respect to the cycle, MR reaction has a behaviour similar to that of POM reaction, making evident that O_2 strongly inhibits the reduction of NiO and prevents the lower ignition temperature.

Under the operating conditions studied (750°C, O/C = 1, H/C = 2-4), carbon formation is highly favored and the high temperature and the presence of steam promote the sintering of the active phase [10]. So, the stability of these catalytic systems may be strongly affected by both factors. The catalysts were submitted to stability tests and the results shown that the stability varies slightly from one precursor to another, being more stable the $Ni^{O}_{\alpha+Al}$ catalyst. In

Figures 1 and 2 the evolution of the activity coeffient a_{CH4} are represented for Ni^{O}_{α},and $Ni^{O}_{\alpha+Al}$ catalysts as a function of the time, for POM, R and MR reactions. When analyzing the results for each reaction, it can be seen that the modification of the support provokes an important increment in the stability, and that the major degree of deactivation corresponds to R reaction.

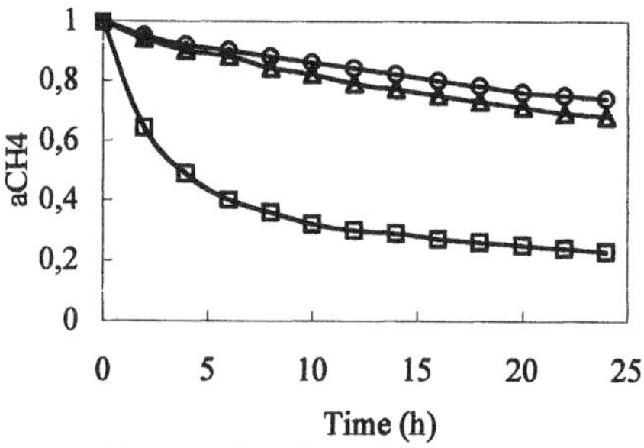

Figure 1. Stability of Nioα

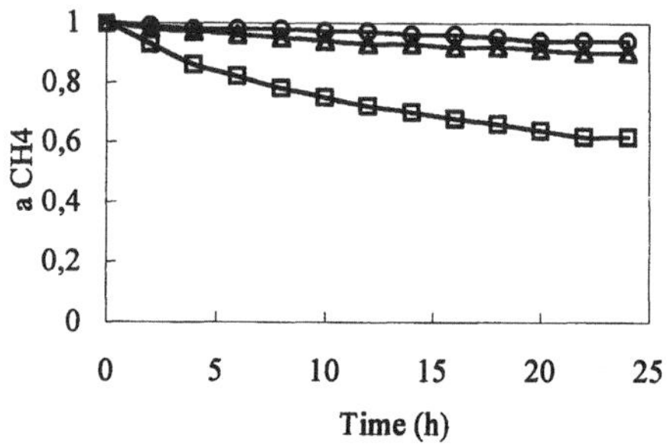

Figure 2. Stability of Nioα+Al

Particle size measured by TEM are presented in Table 2, showing the increment of mean particle size for the post reaction samples compared to the fresh ones. The samples supported on α-Al_2O_3 show greater sintering than those prepared on modified α-Al_2O_3, indicating the existence of a weak interaction between Ni and the support. The results for R reaction indicate that the deactivation of the catalysts should not be mainly attributed to a sintering process.

Table 2. Particle size measured by TEM.

		Ni^{I}_{α}	Ni^{O}_{α}	$Ni^{I}_{\alpha+Al}$	$Ni^{O}_{\alpha+Al}$
D_{av}(nm)	Fresh	22.4	23.5	14.6	16.7
	POM	29.5	48.5	24.6	20.2
	R	24.5	25.3	22.1	19.3
	MR	28.3	35.8	23.2	21.1

To further investigate carbon deposition that occurred with these reactions, it was examined the weight loss of these materials under heating in air with thermogravimetry. For all the analyzed catalysts, carbon deposition results greater for R post-reaction samples. Figure 3 shows this fact for $Ni^{I}_{\alpha+Al}$ sample, which could be a consequence of the slower rate of R reaction, what allows the development of intermediate species leading to carbon deposits. In Figure 4 it is represented a derivative change in weight as a function of temperature for all catalysts submitted to R reaction. As it can be seen, there is an important difference with respect to the amount of carbon deposited between the four samples analyzed. In relation to the support, catalysts prepared on modified alumina have a greater carbon content, which could be a consequence of their greater Ni dispersion and greater activity. Concerning the precursor, catalysts prepared from Ni-acetylacetonate lead to a lower carbon content. In the same figure it is observed a shift toward higher temperatures in the maximum of the peak

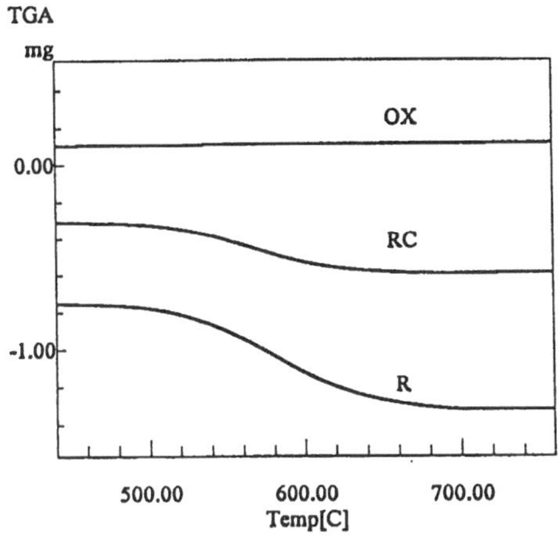

Figure 3. Variation of sample weight (mg) as a function of temperature (°C) for $Ni^{I}_{\alpha+Al}$ catalysts.

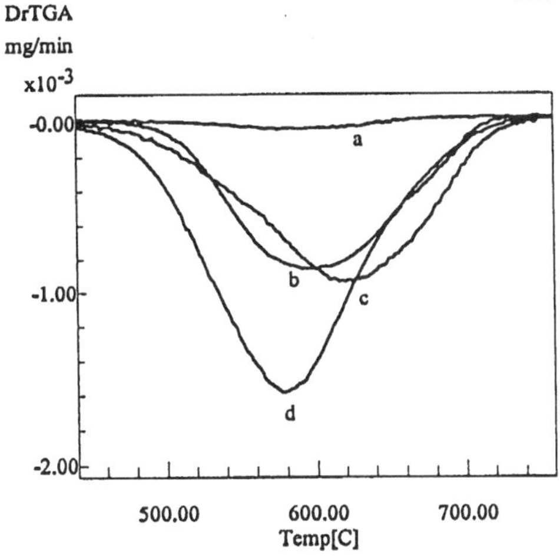

Figure 4. Derivative of weight variation ($mg.min^{-1}$) as a function of temperature (°C) for: (a) Ni^{O}_{α}; (b) Ni^{I}_{α}; (c) $Ni^{O}_{\alpha+Al}$;(d) $Ni^{I}_{\alpha+Al}$

corresponding to $Ni^{O}_{\alpha+Al}$ with respect to $Ni^{I}_{\alpha+Al}$ catalyst, that could be indicating the existence of different carbon species according to the precursor employed, in agreement with TEM micrographs (Figures 5 and 6), where it is observed that the contribution of filamentous carbon is more important in $Ni^{I}_{\alpha+Al}$. These results toghether with the high resistance to sintering, allow to explain why the $Ni^{O}_{\alpha+Al}$ catalyst presented the best stability performance for all the reaction studied.

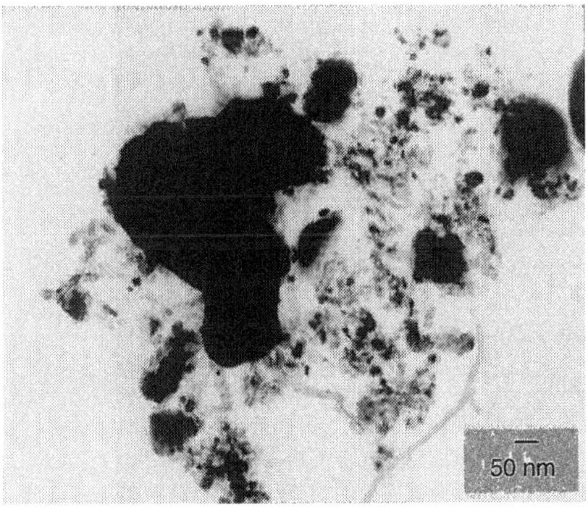

Figure 5. Electron micrograph (TEM) illustrating the morphology of carbon deposists on $Ni^{I}_{\alpha+Al}$ catalyst.

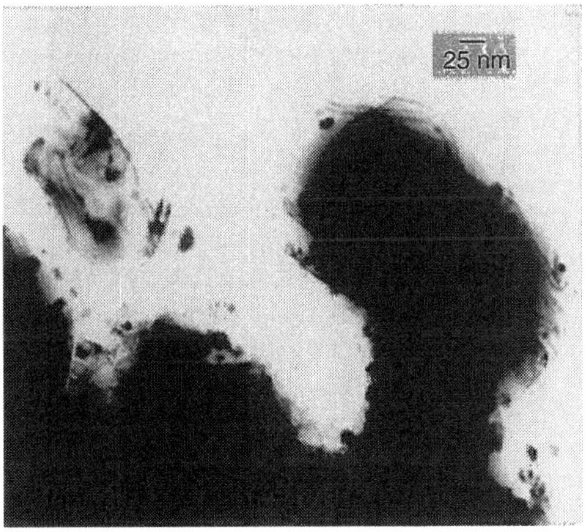

Figure 6. Electron micrograph (TEM) illustrating the morphology of carbon deposists on $Ni^{O}_{\alpha+Al}$ catalyst.

4. CONCLUSIONS

-Nickel catalysts prepared using α-Al_2O_3 support modified by aluminium oxide presented good dispersion and stability facing to sintering, for both precursors used.
-The lower activity of MR reaction compared to POM reaction would be compatible with the occurrence of two simultaneous (one step and two steps) mechanisms.
-MR reaction has a behaviour similar to that of POM reaction, with respect to the cycle on the CH_4 conversion as a function of sequential changes in the reaction temperature, making evident that O_2 strongly inhibits the reduction of NiO and prevents the lower ignition temperature.
-MR reaction leads to H_2/CO ratios between 1 and 2, a fact that represents an attractive alternative in order to obtain syngas for several processes.
-Carbon deposited not only depends on the nature of support, but also on the precursor employed. For samples prepared from nickel acetylacetonate, the carbon content is lower and the nature of carbon is different as shown TPO/TGA data and TEM results.
-Ni catalysts prepared from acetylacetonate precursor on α-Al_2O_3 modified by aluminium oxide resulted the best catalyst analyzed, in relation to the activity, selectivity and stability for all the reactions studied.

ACKNOWLEDGEMENT

The authors gratefully acknowledge Mr. Guillermo R. Bertolini for assistence with the experiments.

REFERENCES

1-Dissanayake, D., Rosynek, M., Kharas, K., and Lunsford, J., J.Catal. **132**, 117 (1991).
2-Vermeiren, W.J.M., Blomsma,E., and Jacobs, P.A., Catal Today **13**, 427 (1992).
3-Vernon, P.D., Green, M., Cheetham A.K., and Ashcroft, A.T, Catal. Today, **13**, 417 (1992)
4-Nichio,N., Casella, M,.Ferretti, O., González, M., Nicot, C., Moraweck, B., and Frety, R, Catal. Lett.**42**, 65 (1996).
5-Casella, M., Nichio, N., González, M., and Ferretti, O., Mat.Lett., in press.
6-Brunelle, J.P., Sugier, A., Le Page, J.F., J.Catal, **43**, 273 (1976).
7-Hickman,D., Haupfear, E., and Schmidt, L., Catal. Lett, **17**, 223 (1993).
8-Qin, D., Lapszewicz, J., Jiang, X., J.Catal, **159**, 140 (1996).
9-Boucouvalas, Y., Zhang, Z., Verykios, X., Catal.Lett. **27**, 131 (1994)
10-Rostrup-Nielsen, J.R. in "Catalysis Science and Technology" (J.Anderson and M.Boudart, Eds.), Vol. 5, Chap. 1. Springer-Verlag, Berlin, New York, 1984.

NATURAL GAS CONVERSION V
Studies in Surface Science and Catalysis, Vol. 119
A. Parmaliana et al. (Editors)

Characterization of Ni-honeycomb catalysts for high pressure methane partial oxidation

J. Lezaun[a], J.P. Gómez[b], M.D. Blanco[c], I. Cabrera[c], M.A. Peña[c] and J.L.G. Fierro[c]

[a]ENAGAS-GAS NATURAL, R&D Centre, 50080 Zaragoza, Spain
[b]REPSOL Petróleo S.A., Research Centre, 28045 Madrid, Spain
[c]Institute of Catalysis and Petrochemistry, CSIC, Cantoblanco, 28049 Madrid, Spain

Nickel-honeycomb catalysts prepared by different procedures have been characterized by temperature-programmed reduction and X-ray photoelectron spectroscopy. Prereduced catalysts were tested in the partial oxidation of methane under high pressure (20 bar). The influence of the method of incorporation of the catalytic component, the metal loading and the activation temperature on the catalytic performance were investigated. It appeared that the formation of a $NiAl_2O_4$ phase between the Ni precursor and the Al_2O_3 phase of the carrier, while maintaining high dispersion is a key parameter to maintain high activity and good stability during on-stream operation.

1. INTRODUCTION

Methane steam reforming (MSR) (Eq. 1) is the established process for converting natural gas into CO and H_2. This technology has been used for several decades and has been substantially improved over the years. However, as MSR is an energy demanding process, attention has moved to emerging technology such as partial oxidation (MPO) (Eq. 2).

$$CH_4 + H_2O \rightarrow CO + 3\,H_2 \qquad (\Delta H^\circ = +206\ \text{kJ/mol}) \qquad (1)$$
$$CH_4 + 1/2O_2 \rightarrow CO + 2\,H_2 \qquad (\Delta H^\circ = -\ 36\ \text{kJ/mol}) \qquad (2)$$

CH_4 conversions close to the equilibrium can be reached over supported Ni [1] or supported noble metals [2], although significant differences in product distribution with respect to equilibrium were reported for a few first row transition metal catalysts. As reported by Hickman and Schmidt [2-4], the use of preshaped honeycomb structures affords an extremely useful process for syngas production via the MPO process, working with high space velocities and very low pressure drop across the channels of the honeycomb. Hydrogen selectivity above 90% and CO selectivity in the order of 95% at CH_4 conversion somewhat above 90% were found on a Rh/honeycomb catalyst.

Reaction (2) has also been investigated on several group VIII metal coated honeycombs by Tornianen et al. [5]. Rh and Ni catalysts exhibited the best behavior, although Ni catalyst became deactivated by $NiAl_2O_4$ formation; Pt and Ir catalysts were found to be somewhat less active and selective but highly stable during on-stream operation; Pd and Co deactivated rapidly and Ru and Fe were inactive. According to these results, it appears that Rh and Ni catalysts are the most appropriate systems for syngas production according to reaction (2). Carefully reviewing the published results of the studies on reaction (2), the frequent presence of coke is reported, and specifically for Ni catalysts, suggesting the occurrence of coking reactions together with partial oxidation. In a previous contribution, Gómez et al. [6] disclosed the use of Ni/honeycomb catalysts for for reaction (2) working at temperatures above 973 K, CH_4/O_2 molar ratios of 2 in the feed and high pressures, typically of 20 bar. These conditions are even more favorable for the occurrence of carbon deposition on the catalysts surface.

The objective of the present work was to obtain additional information on the relevance of the method of incorporation of Ni to honeycomb structures in order to obtain active and stable catalysts for the aforementioned reaction. The nature of the nickel phase(s) presents in the calcined precursors and the temperature window at which the catalysts are activated may lead not only to an improved understanding of the generation of the metallic Ni particles and its role in partial oxidation. Hopefully, this information will provide a foundation for future research.

2. EXPERIMENTAL

Four different classes of nickel honeycombs were prepared. The first kind of honeycomb (NIK-21) was prepared by kneading a paste of γ-Al_2O_3 with 5% natural clay (55.0% SiO_2, 4.4% Al_2O_3, 25.4% MgO, 2.0% Fe_2O_3, CaO, K_2O and Na_2O minor impurities, and 8.3% loss ignition) and appropriate amounts of $Ni(NO_3)_2$ and rheological additives, followed by extruding under carefully controlled temperature and humidity environment. After extruding, the honeycombs were dried and finally calcined. The second type (NIKAI-12) was prepared by mixing γ-Al_2O_3 with 12 % natural clay, and after extruding, drying, and pre-calcination of this support at 523 K for 8 h, the resulting honeycomb was impregnated with a solution of $Ni(NO_3)_2$ of appropriate concentration, followed by drying and calcination. The third catalyst (NIKAI-30) was similar to the former prepared by impregnation, but with a higher pre-calcination temperature of the support (1273 K). The fourth catalyst (HOM-Ni) was prepared by impregnating a commercial silica honeycomb with a solution of $Ni(NO_3)_2$, drying and calcination. Finally, a cordierite honeycomb supported rhodium catalyst was prepared by impregnation of commercial cordierite honeycombs with a $RhCl_3$ solution, followed by drying and calcination. In every preparation, the final catalyst was calcined at 1273 K for 6 h. The honeycomb dimensions for both nickel and rhodium catalysts were: 16 mm diameter, 10 mm length, 52-64 cells/cm^2 and 0.3 mm wall thickness. The general characteristics of the catalysts are compiled in Table 1.

The reducibility of the catalysts was determined in a Micromeritics TPD/TPR 2900 apparatus, using a 5% H_2/Ar mixture with a flow rate of 50 mL(NTP)/min, 50 mg catalyst weight, and a temperature ramp of 10 K/min. XP spectra were acquired with a VG Escalab 200R spectrometer using a MgKα X-ray source. The calcined catalysts and samples used in reaction were directly transferred to the analysis chamber after evacuation at room temperature. The binding energy (BE) of C1s peak at 284.9 eV was taken as an internal standard.

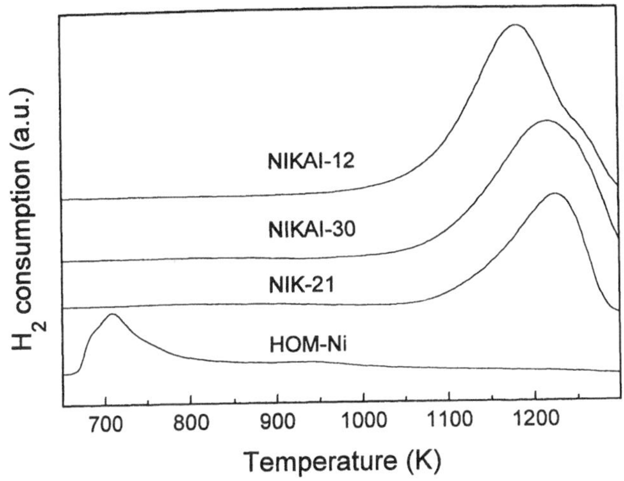

Figure 1. TPR profiles of nickel based honeycombs

The honeycomb catalysts were tested in a bench scale annular flow reactor system. The reactor consisted of a 25.4 mm O.D. stainless steel tube mounted vertically and covered internally with a 16 mm I.D. coaxially centred alumina tube. The honeycombs were pretreated in hydrogen flow for 1 h, at 473 K for rhodium catalyst, and at 1023-1223 K for nickel catalysts. The temperature of the reaction (in the range 973-1073 K) was controlled by an electrical furnace and measured at the bottom of the honeycomb. The reaction gases (methane and oxygen) were premixed with nitrogen and fed without preheating to the reactor, and their flow rates were controlled by mass flowmeters, with a total flow rate of 180 L(STP)/h (contact time 0.25 s) and a molar ratio N_2:CH_4:O_2 of 3:2:1. The pressure in the reaction system was maintained at 20 bar by means of a micrometric needle valve commanded by a servomotor drive. The effluent of the reactor was cooled for water separation, and the permanent gases were analyzed on-line by a HP 5890 GC equipped with a loop valve for automatic operation using two columns (Porapak Q and MS) and TCD.

Table 1
General characteristics of honeycomb catalysts

Catalyst	Support	Precalcination	Preparation	% Metal
NIK-21	γ-Al_2O_3 + 5% clay	-------	Kneading	9.4 % Ni
NIKAI-12	γ-Al_2O_3 + 12% clay	523 K, 8 h	Impregnation	16.0 % Ni
NIKAI-30	γ-Al_2O_3 + 12% clay	1273 K, 8 h	Impregnation	11.5 % Ni
HOM-Ni	Silica	-------	Impregnation	8.9 % Ni
RH-1	Cordierite	-------	Impregnation	0.9 % Rh

3. RESULTS AND DISCUSSION

Since the active phase in these catalysts for the MPO reaction is metallic nickel or rhodium, the catalyst should be pretreated in hydrogen before the reaction. Characterization of the catalysts by TPR yields information of their reducibility. The results for the nickel catalysts are presented in Fig. 1. For the honeycomb catalysts using alumina as the main support component, the hydrogen consumption occurs at relatively high temperature, in the range 1100-1300 K, associated with the reduction of very stable Ni^{2+}-containing phases. This kind of nickel is present in compounds formed by solid state reaction between NiO and Al_2O_3 during calcination, like $NiAl_2O_4$. The figure shows that TPR for NIKAI-30 and NIK-21 are very similar with an intense peak at ca. 1200 K. The main peak for the other Al_2O_3-based catalyst, NIKAI-12, is placed at a slightly lower temperature, 1170 K, with a shoulder at 1200 K. The lower precalcination temperature of the support used in the preparation of this catalyst should produce two different alumina phases and, consequently, two different modes of interaction with NiO, that produce the two nickel species observed. For this catalyst, no peak of reduction of the NiO phase is observed at low temperature (ca. 700 K). However, the silica-based catalyst HOM-Ni displays its main peak at ca. 700 K, characteristic of the reduction of free NiO without interaction with the support. This peak is very broad, with a long tail, indicating that nickel is also interacting heterogeneously with silica. This tail covers the range 720-850 K, a lower reduction temperature that in alumina-based catalysts, because nickel in the silica structure is reduced more easily than nickel interacting with alumina. The RH-1 catalyst, not shown in the figure, presents a narrow peak of reduction at 470 K, due to the reduction of Rh_2O_3, present on the surface of the honeycomb, into Rh°. The activation of rhodium-based honeycombs requires, therefore, much lower temperatures than nickel based catalyst.

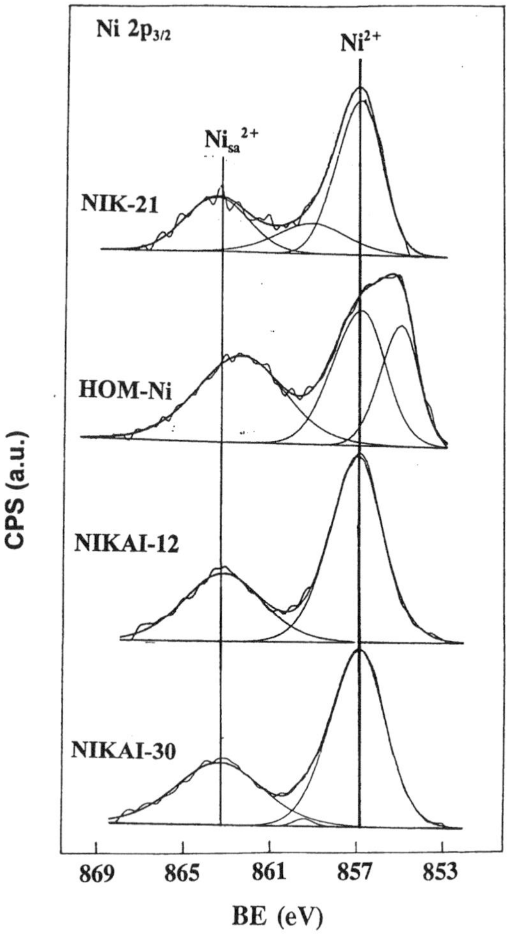

Figure 2. Ni $2p_{3/2}$ core level spectra for nickel based catalysts

In Fig. 2, Ni $2p_{3/2}$ core-level spectra for the nickel honeycomb catalysts are presented. The position of the peak at 855-857 eV is characteristic of Ni^{2+}, and this is also confirmed by the appearance of the satellite line at ca. 863 eV. The catalyst containing alumina displays a single peak at 856.8-857.0 eV, that can be assigned to nickel in the structure of alumina ($NiAl_2O_4$) [7], in agreement with the TPR data. For HOM-Ni catalyst, two components of this peak, at 855.0 (40%) and 856.9 (60%), can be resolved. The component at higher BE, as in the alumina honeycombs, is due to nickel interacting with the silica support, and the component at lower BE can be assigned to NiO. As it can be observed, the population of both species is roughly similar. This agrees with the TPR results, where the two reduction steps (700 K and 720-850 K, not resolved) produce similar peak area. The extent of the interaction of nickel with silica is lower than with alumina, and the resulting species are more reducible. XP spectrum of Rh $3d_{5/2}$ for RH-1 honeycomb displaysa single peak at 309.0 eV due to Rh^{3+} present in the surface of the calcined catalyst. When the catalysts are used in the reaction, some differences appear. For alumina-based catalysts, two components in the Ni $2p_{3/2}$ core-level can be detected: 856.8 eV (70) and 853.8 eV (30%) in NIK-21. The new second component is due to NiO produced by exposure of the metallic nickel present in the active catalyst to air, before it is transferred to the XPS analysis chamber. The fraction of 30% represents, therefore, the percentage of nickel reduced that is active in the reaction. A similar phenomenon occurs in the silica-based catalyst HOM-Ni, that presents the same components as the calcined catalysts, but with higher content of NiO: 854.7 eV (60%) and 857.0 (40%). Metal reduction can be detected in the rhodium catalyst, with a single peak at 307.3 eV in the Rh $3d_{5/2}$ core-level characteristic of Rh^{o}. The quantitative analysis of XP spectra shows that NIK-21 presents the lowest atomic ratio Ni/(Al+Si+Mg), since part of the nickel is in the bulk of the honeycomb due to the preparation method. The other catalysts present a similar nickel concentration, higher for NIKAl-30 after considering the differences in overal nickel content. The dispersion of nickel in alumina-based catalyst is maintained after reaction, but in the silica-based honeycomb decreases to half in the used catalyst.

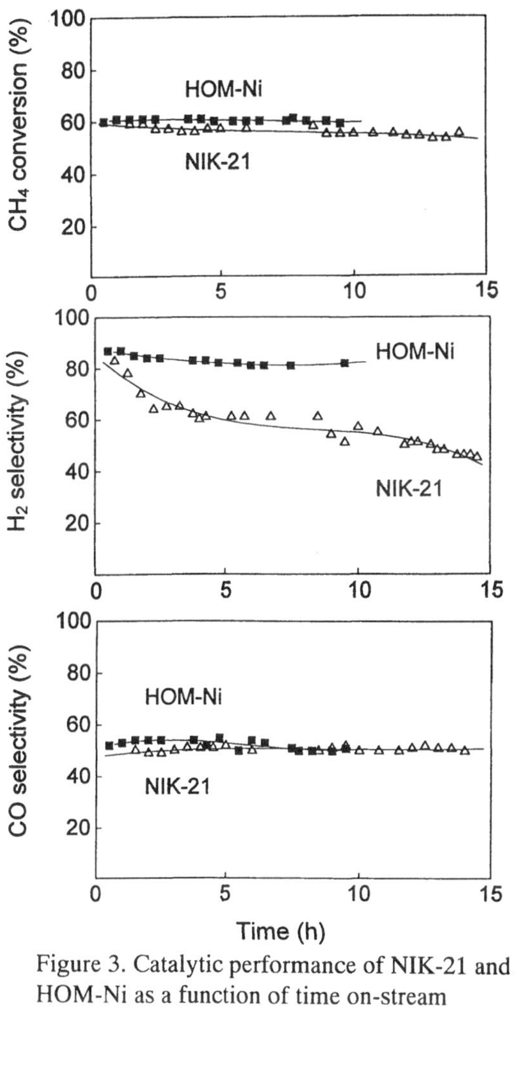

Figure 3. Catalytic performance of NIK-21 and HOM-Ni as a function of time on-stream

The honeycombs were pretreated in hydrogen at 1023 K for 1 h before reaction, in order to develop the active metal phase. The activity results at the reaction temperature of 1023 K are similar for the different nickel catalysts. The methane conversion was 54-57%, selectivity to hydrogen 75-78% and selectivity to CO 46-55%. The molar ratio H_2/CO was in the range 2.4-2.7, indicating that besides the partial oxidation reaction, there is an important contribution of the water gas shift reaction in the honeycomb, that modulates the final composition of the outlet gases, and produces a higher H_2/CO ratio somewhat higher than that expected from reaction (2). Nevertheless, increasing the reaction temperature up to 1073 K produces H_2/CO ratios close to 2. All the catalysts showed methane dimerization products with a selectivity of ca. 8% to C_{2+} hydrocarbons at 1023 K, suggesting that homogeneous reactions could also be involved. The RH-1 catalyst yields lower methane conversion (49%) and lower selectivities (H_2 36% and CO 50%), showing that its behavior under high pressure conditions is worse than that for nickel catalysts. However, the higher selectivity to C_{2+} hydrocarbons (ca. 20% at 1023 K) for RH-1 indicates that most of the dimerization products could be formed on the catalyst surface under these conditions. The reduction temperature is also important in the performance of the catalysts and increasing it from 1023 to 1223 K produces an increase in selectivity to H_2 and CO. Figure 3 shows the behavior of NIK-21 and HOM-Ni honeycombs at 1073 K. Both catalyst maintain the methane conversion stable close to 60%. However, selectivities to H_2 and CO do not change to a significant extent although a clear decline in H_2 is observed for NIK-21. For the other Ni-impregnated catalysts, H_2 selectivity remains stable.

In the light of these results, it can be concluded that Ni-honeycombs prepared by kneading and impregnation are active and selective for syngas production via methane partial oxidation when working under high pressure conditions. XPS has been used to reveal the nature of Ni species and their dispersion and TPR to define the activation temperatures of the catalysts. The higher Ni exposure and the slightly lower reduction temperatures of the impregnated catalysts favor stable operation.

ACKNOWLEDGEMENTS

The authors gratefully acknowledge financial support from ENAGAS-GAS NATURAL and REPSOL, and would like to thank the Management of the Technology Departments of both companies for offering them the opportunity to publish these results.

REFERENCES

1. D. Dissanayake, M.P. Rosynek, K.C.C. Kharas ans J.H. Lunsford, J. Catal., 132 (1991) 117.
2. D.A. Hickman and L.D. Schmidt, J. Catal., 138 (1992) 267.
3. D.A. Hickman and L.D. Schmidt, Science, 259 (1993) 343.
4. S.S. Bharadwaj and L.D. Schmidt, J. Catal., 146 (1994) 11.
5. P. Torniainen, X. Chu and L.D. Schmidt, J. Catal., 146 (1994) 1.
6. J.P. Gómez, J.M. Jiménez, S. Vic, J. Lezaun, P. Terreros, I. Cabrera, M.A. Peña and J.L.G. Fierro, Stud. Surf. Sci. Catal., 107 (1997) 397.
7. S. Damyanova, L. Daza and J.L.G. Fierro, J. Catal., 159 (1996) 150.

NATURAL GAS CONVERSION V
Studies in Surface Science and Catalysis, Vol. 119
A. Parmaliana et al. (Editors)

Transient reactions in CO_2 reforming of methane.

P. Gronchi, P. Centola, A.Kaddouri, and R. Del Rosso

Industrial Chemistry and Chemical Engineering Department, CIIC, Politecnico di Milano, P.za Leonardo da Vinci 32, 20133 Milano, Italy. Fax: +39-2-70638173.
E-mail: Paolo.Gronchi@polimi.it

The activity of Ni catalysts supported on SiO_2, La_2O_3-SiO_2 and La_2O_3 oxides were analysed in in the transient period using a single pulse of the pure reagents alone or a pulse of the 1/1 CH_4/CO_2 gas mixture (single-PSRA). The data have been compared with those obtained using the supports alone. The reduced and the unreduced form of both the species were investigated. The activity depends on the support and on the oxidation degree. The conversion of each reagent is ascribed to a redox mechanism between the metal and the support. The shape of the curve that represents the data of each test and the time of the peak maximum agree with the proposed pathway carbon surface specie.

1. INTRODUCTION

Due to the large natural gas reserve (the same order of magnitude as crude oil), methane appears the cheapest and most available carbon source for the petrochemical industry [1]. The most frequently industrial route to exploit methane is actually the carbon oxidation to synthesis gas by a catalytic steam reforming unit. The carbon and thermal efficiency of the reforming however deeply depends on the CO_2 content [2] formed both by the hydrocarbon oxidation and the CO disproportionation, or present in the natural gas (some reserves have till 40% of CO_2).

As a consequence the reactivity of CO_2 in the reforming conditions must be firstly studied to enhance the activity of the reforming catalyst towards the CO_2 conversion and to optimise the reactor design. Besides, the study helps to develop the industrial reforming to obtain synthesis gas with customised CO/H_2 ratio for specific organic hydrocarbon synthesis by adding CO_2. Moreover, the dry reforming appears a ready available process for the CO_2 recycling [3].

We recently reported some catalytic data, referred to the early stages of catalytic tests, performed on Ni metal catalyst supported on silica, on silica modified with La_2O_3 and on pure La_2O_3 which put in evidence a specific role of the rare earth oxide [4].

In this work we investigated the same catalysts and supports by single pulses of the reactants (CO_2, CH_4, CO_2+CH_4). Our aim is to better elucidate the catalytic activity by identifying surface species formed in a well defined oxidation state of the catalyst surface such those produced in the initial steps of the reaction. This study of the CO_2 reactivity can be helpful to design new catalyst for the dry reforming.

2. EXPERIMENTAL

Ni was loaded on three different supports: pure SiO_2, SiO_2 modified with La_2O_3 and pure La_2O_3. The catalysts (4% wt. Ni) were prepared by wet impregnation with $Ni(NO_3)_2$ and $La(NO_3)_3$ salts of SiO_2 (Merck7754, 70-230 mesh; 550 m^2/g) or La_2O_3 (Fluka) and following the same procedures for the drying and oxidation as previuosly reported [5]. The silica supported catalysts showed high BET area (Ni (4%)/SiO_2= 460 m^2/g, Ni (4%)//La_2O_3 (4%)-SiO_2= 220 m^2/g) while the lanthana supported catalyst showed a surface area < 20 m^2/g.

The catalyst (0.2 g; 120/70 mesh or 0.12-0.2 mm) was mixed with SiC of the same particle size, at 1/10 wt. ratio and charged into a vertical quartz reactor (10 mm i.d.) heated in a ventilated oven. The H_2 reduction was conducted, where necessary, in situ at 873K for 1 hour with a H_2+He mixture, 40ml/min and 20ml/min respectively. A He flow of 30 ml/min was maintained before the reduction and afterwards during the reaction. CO_2, CH_4 and CO_2+CH_4 single pulses (1 ml; 85% reactants and 15% of N_2 used as internal reference) were injected 873K, at controlled rate in the He stream. After the single pulse, samples (1 ml each) of the effluent gas were collected every 6 s with a Valco 16 multiport heated valve. The samples were successively analysed by a gas chromatograph. The gas chromatographic data give a medium value of the reacted pulse composition over each 6 s time. The activity tests are performed and repeated with both unreduced and reduced catalyst.

3. RESULTS

The chromatographic data of each test are represented by a graph (see Fig.1 as an example) having a time scale on the *X*-axis and the moles of the substance on the *Y*-axis. The time origin is arbitrary placed at the highest concentration of the reference gas in order to compare the different pulses. The total amount of a substance in a single pulse is obtained by time integration applying a correction factor based on the N_2 feeded/N_2 measured ratio. Single pulses of CO_2, CH_4 and CO_2+CH_4, are separately injected into the He stream.

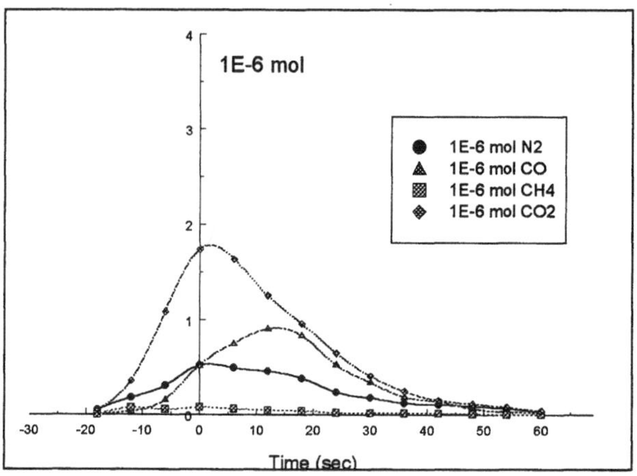

Fig.1. Single pulse gas chromatographic data of CH_4+CO_2 over H_2 reduced Ni/SiO_2.

The CO_2/CH_4 molar ratio in CH_4+CO_2 pulse was of ¼.
The results reported in Tables 1, 2 and 3 refer to the tree types of supports: pure SiO_2, SiO_2 modified with La_2O_3 and pure La_2O_3 respectively.
The selectivity data have been calculated from the relation:

$$Sel = \frac{\sum_{1}^{15} COmol}{CO_2 consumed + CH_4 consumed} \times 100$$

where the whole CO is calculated adding the CO amounts sampled by the 15 valve loops.
Some values are reported as not detectable, meaning that they are near the zero within the experimental error.

3.1 Pure CO_2 and CH_4 pulses

Comparing the activity of the supports without Ni, silica converts low amounts of both CO_2 and CH_4. A selectivity increase of the CO from CO_2 is observed when reduced lanthana is present, which can be entirely ascribed to the La_2O_3 by observing the high CO selectivity on the pure reduced rare earth oxide (Table 3). No clear effect is referred to silica after H_2-reduction. On the contrary the easy reducible lanthana converts high quantity of carbon dioxide and produces more CO when reduced than in the oxidised form.

Table 1. Pulse conversion and selectivity on SiO_2 supports and catalysts.

	CO_2 pulse		CH_4 pulse			$CO_2 + CH_4$ pulse		
	Conv. (%)	CO sel. (%)	Conv. (%)	CO sel. (%)	CO_2 sel. (%)	CO_2 Conv.	CH_4 Conv.	CO sel. (%)
Support								
SiO_2	3.7	traces	n.d.	n.d.	n.d.	4.8	n.d.	6.7
SiO_2red	0.5	traces	1.6	n.d.	n.d.	0.4	n.d.	n.d.
Catalyst								
Ni/ SiO_2	n.d.	n.d.	42.2	66.4	33.5	n.d.	4.1	8.5
Ni/ SiO_2 red	4.6	0.7	86.8	39.2	11.7	negative	94.7	34.0

n.d. = not detectable

Comparing the metal catalysts, CO_2 is converted at significant level only in presence of lanthana both as additive of the silica catalyst after reduction and as pure unreduced support. With this type of catalyst CO is produced with high selectivity after reduction (see Table 3)
On the other side, CH_4 is easily converted in presence of Ni. The conversion does not depend on the type of support and increases by 1.2 ÷ 2.0 times with the reduced forms.
CO and CO_2 are the gaseous products: the former being major product with the reduced metal on SiO_2, SiO_2-La_2O_3 and La_2O_3 oxide

Table 2. Pulse conversion and selectivity on La_2O_3 modified SiO_2 supports and catalysts.

	CO_2 pulse		CH_4 pulse			$CO_2 + CH_4$ pulse		
	Conv. (%)	CO sel. (%)	Conv. (%)	CO sel. (%)	CO_2 sel. (%)	CO_2 Conv.	CH_4 Conv.	CO sel. (%)
Support								
SiO_2-La_2O_3	1.4	68.6	traces	n.d.	n.d.	0.5	3.3	11.5
SiO_2-La_2O_3 re	0.3	90.0	n.d.	n.d.	n.d.	3.7	n.d.	46.7
Catalyst								
Ni/SiO_2-La_2O_3	n.d.	n.d.	47.6	78.9	21.0	3.1	n.d.	28.4
Ni/SiO_2-La_2O_3 red	8.7	90.5	57.8	16.3	2.6	59.5	88.7	51.3

Contrary to the other two catalysts a greater amount of CO_2 than CO has been observed with pure unreduced lanthana. With the other Ni catalysts, on the contrary the CO_2 selectivity is ever less than CO.

Table 3. Pulse conversion and selectivity on La_2O_3 containing supports and catalysts.

	CO_2 pulse		CH_4 pulse			$CO_2 + CH_4$ pulse		
	Conv. (%)	CO sel. (%)	Conv. (%)	CO sel. (%)	CO_2 sel. (%)	CO_2 Conv.	CH_4 Conv.	CO sel. (%)
Support								
La_2O_3	30.7	38.8	n.d.	traces	n.d	24.6	n.d.	2.0
La_2O_3 red	25.8	93.0	n.d.	traces	n.d.	14.8	n.d.	18.9
Catalyst								
Ni/La_2O_3	37.6	0.8	47.3	9.7	57.3	52.8	17.3	2.7
Ni/La_2O_3 red	7.9	62.3	75.0	56.3	12.4	58.8	85.6	50.1

3.2 Mixed CO_2/CH_4 pulses

The conversion of the two reagents is separately reported in the tables 1, 2 and 3. The conversion and selectivity features previously observed for the activity of the supports can again be evidenced.

Moreover on the unreduced Ni catalysts CH_4 is converted less than in the pure CH_4 pulse. Its conversion reachs the same values after the reduction of the catalyst. The methane reactivity seems strictly related on the CO_2 conversion. This is enhanced in presence of lanthana comparing the CO_2 pure pulse at the same conditions, and, inversely, blocked on Ni/SiO_2. With silica we observed a low negative value confirmed after several repeated tests which could signify a CO_2 formation by successive reaction.

4. DISCUSSION

Many papers appeared in the last 5 years on the reforming of CH_4 with CO_2 describing the influence of the support on the catalyst activity, on the carbon deposition and on the CO/H_2 ratio.

MgO, TiO_2, Al_2O_3, ZrO_2, La_2O_3 have been extensively compared with high temperature stable oxide like SiO_2 [6, 7]. The results of single pulse surface reaction analysis (single PSRA), reported in the Table 1, 2 and 3 confirm the role of the support. In the transient time, lanthana and silica show different selectivity at comparable conversion and activity: generally the presence of La_2O_3 promotes the CO_2 conversion and increases the CO selectivity. The differences appear both on the pure support and in presence of Ni.

The simultaneous injection of CO_2 and CH_4 in the inert gas stream highlights the different behaviours of the reduced and unreduced catalyst.

On the unreduced ones carbon dioxide appears to decrease the conversion of methane with respect to the one of the pure CH_4 pulse. The data confirm a competitive inhibition mechanism of CO_2 of the CH_4 adsorption and dissociation. For instance, comparing the CH_4 conversion values reported in Table 1 for Ni/SiO_2, we observe that from 42.2% it decreases to 4.1% in the presence of CO_2. The conversion is very low with SiO_2 and SiO_2-La_2O_3 supports, and reaches 17% with lanthana alone.

Moreover on Ni/La_2O_3, the CO_2 conversion increases becoming greater than that observed with the pure CO_2 pulse (from 37.6 % vs. 52.8 %).

In our opinion lanthana can be reduced from CH_4 producing O-vacancies. These promote the CO_2 conversion that causes the re-oxidation of the support itself forming CO.

A similar redox cycle has been observed on CeO_2 that is reduced by CH_4 and successively re-oxidate by CO_2 or by H_2O [8].

Thus the data obtained with Ni on SiO_2 and SiO_2-La_2O_3 indicate that in the unreduced oxidation state, the metal does not activate the CH_4 and justify the assumption that La_2O_3 can be partially reduced by hydrocarbons.

The reaction begins on the La_2O_3 support and successively takes place following the previously accepted way characterised by the CH_4 adsorption and reaction on the metal [9]. The strong activity of a reduced metal surface towards CH_4 catalytic conversion to CH_x species and H_2, has been investigated by many researchers and generally reported as easy.

On the reduced catalysts this mechanism takes place *ab initio* and our results agree with a relevant conversion of CH_4 together with the greater disappearance of CO_2 than the one observed with the CO_2 pure pulse in presence of La_2O_3.

It is noteworthy to observe that CO_2 can be differently adsorbed on lanthana pure or dispersed on silica: if unreduced (Table 2 and 3) forming carbonate or, on the reduced oxide, dissociatively, producing CO.

As a consequence the previously observed CO_2 conversion in presence of CH_4 on Ni/La_2O_3 is probably due to:

a) the formation of carbonate species on the unreduced oxide

b) the dissociative adsorption on the extensively H_2-reduced support.

In our previous works on dry reforming we observed a larger conversion of CO_2 than CH_4 at the steady state condition with Rh and Ni [10]. At the initial reaction time (5 min), which also means a deep reduction state of catalyst, we observed instead a greater conversion of CH_4 than CO_2 on the catalysts after reduction [11]. The last result is confirmed by the data obtained with single pulse of reagents all over the three type of catalysts.

The results suggest that in the transients the higher conversion of CH_4 than CO_2 on lanthana is due to the reduced Ni form which is successively oxidated by the CO_2 itself or the H_2O. Approaching the conditions of the unreduced catalyst the CH_4 conversion becomes less than that of CO_2 (table 3).

Some further indication on the mechanism can be drawn from the analysis of the time of peak maxima and of the shape of the curve. The maximum of the curve can be reasonably associated to the pathway of the reaction. A detailed analysis of time data must be developed, even if it is evident that the carbon oxide appearance depends on the support and on the oxidation degree.
With CH_4+CO_2 pulse, on reduced Ni/SiO_2 and Ni/La_2O_3 the CO maximum appears after the one of CO_2 and at the same time respectively. Probably it means that the low converted amount of CO_2 reacts with the CHx species from methane on the Ni/SiO_2 to give CO while on Ni/La_2O_3 the formation of CO is simultaneous to the CO_2 dissociative adsorption.

5. CONCLUSIONS

The behaviour of Ni and Ni-O species has been studied on SiO_2, SiO_2-La_2O_3 and La_2O_3 during the initial steps of the dry reforming. The activity of La_2O_3 has been studied in the unreduced form and in presence of O-vacancies. It is confirmed that on Ni-O, the CO_2 adsorption competes with the CH_4 dissociation according to the activity reduction registered in steam reforming catalysts at increasing CO_2 amount.
On La_2O_3 CO_2 adsorbs forming carbonate-formiate species and then decomposes to CO reducing the O-vacances. The oxygen transfer from La_2O_3 to Ni constitutes the O resources for the CO formation from methane. CH_4 reduces the Ni-O species and also produces vacancies on La_2O_3.

REFERENCES

[1] D. Sanfilippo, ECCE-1, Florence (Italy), May 4-7, 1997, Proceedings Vol.IV, pag.2219-2220
[2] Rostrup-Nielsen ECCE-1, Florence (Italy), May 4-7, 1997, Proceedings Vol. I , pag.327.
[3] H. Arakawa, ICCDU IV, Kyoto, September 1997, closing remarks.
[4] P. Gronchi, P. Céntola, and R. Del Rosso, AIDIC Conf.Series, vol. **2**, p.87-93, (I.Pasquon et al., Eds.), ERIS, Milano, 1997.
[5] P. Gronchi, R. Del Rosso, and P. Céntola, Applied Catalysis A, **152** (1997) 83-92
[6] M.C.J. Bradford and M.A. Vannice, Appl.Catal., A, **142** (1996) 73-96.
[7] Z. Zhang and X. Verykios, Catal.Lett., **38** (1996) 175-179.
[8] K. Otsuka, E. Sunada, T. Ushiyama, and I. Yamanaka, in "Natural Gas Conversion IV"(M.de Pontes, R.L.Espinoza, C.P.Nicolaides, J.H.Scholtz, M.S.Scurrell, Eds.) p. 531, Elsevier, Amsterdam, 1997.
[9] F. Solymosi and J. Cserény,Catal.Today, **21** (1994) 561-569
[10] P. Gronchi, P. Centola, and R. Del Rosso, Appl.Catal., A,**152** (1997)83-92
[11] P. Gronchi, P. Céntola, and R. Del Rosso, *in* "Advances in Chemical conversion for Mitigating Carbon Dioxide", (T.Inui, K.Anpo, K.Izui, S.Yanagida, T.Yamaguchi, Eds.) Elsevier, Amsterdam, 1998

NATURAL GAS CONVERSION V
Studies in Surface Science and Catalysis, Vol. 119
A. Parmaliana et al. (Editors)

Dry reforming of methane. Interest of La-Ni-Fe solid solutions compared to $LaNiO_3$ and $LaFeO_3$.

H. Provendier[1], C. Petit[1], C. Estournes[2] and A. Kiennemann[1]

1 LERCSI, ECPM, UMR CNRS 7515, 1 rue B. Pascal, 67008 Strasbourg, France

2 GMI, UMR CNRS 046, 23 rue du Loess, 67037 Strasbourg, France

$LaNi_xFe_{(1-x)}O_3$ perovskites ($0 \leq x \leq 1$) are highly efficient catalysts in dry reforming of methane for syngas production. During the catalytic tests, the starting mixed structures are destroyed, but for $x \leq 0.5$, the catalysts can be regenerated by calcination. For $0.3 \leq x \leq 0.8$ Ni-Fe alloys are evidenced, with Ni/Fe ratios depending on x. It is proposed that this alloy prevents surface poisoning by carbon through a dilution effect on Ni particles.

1. INTRODUCTION

Methane dry reforming to produce synthesis gas with a low H_2/CO ratio, is of particular industrial interest in the oxo- and Fischer-Tropsch syntheses. Moreover, this reaction is a valorization pathway of CO_2 contained in methane-fields [1]. The conventional catalytic systems in methane dry reforming are based on numerous supported metals: Ni, Co, Fe and noble metals (Pt, Pd, Rh) [1,2]. The fast deactivation by excess carbon deposits is the most important drawback of these catalytic systems, especially for supported nickel. However, recent results claimed that Ni/La_2O_3 [3] and $Ni/La\text{-}Al_2O_3$ [4,5] exhibit good activities and excellent stabilities. In a previous work, it has been shown the mixed $LaNi_xFe_{(1-x)}O_3$ perovskites are of high interest in oxyreforming of methane, given the good ageing properties, the lack of surface carbon formation, the control of nickel loss from the structure and the easy regeneration possibility for the catalyst [6,7]. The aim of the present study is to discuss the results obtained in dry reforming and the limitation of coke formation by controlling the loss of nickel from the structure during the catalytic tests. This control has been obtained by addition of iron in various amounts into the initial $LaNiO_3$ perovskite structure, in order to stabilize the structure and to limit the sintering of the active metal species.

2. EXPERIMENTAL PART

2.1. Preparation

The mixed $LaNi_xFe_{(1-x)}O_3$ perovskites with x values varying from 0 to 1 with a 0.1 step were prepared from $La(NO_3)_3$, $6H_2O$; $Ni(NO_3)_2$, $6H_2O$ and $Fe(NO_3)_3$, $9H_2O$ via a sol-gel related method [8]. The La, Ni and Fe nitrate salts were separately dissolved in hot propionic acid under stirring and mixed. After 30 minutes stirring, the resulting solution was evaporated until formation of a resin or of a foam. This resin or foam was calcined at increasing temperature, with an increasing slope of 3°C/min and maintained at 750°C for 4 hours.

2.2. Characterisation

The catalysts were characterized before and after catalytic test by X-Ray Diffraction (XRD) on a Siemens D5000TT diffractometer using a Cu Kα radiation, by BET surface measurements, by Scanning Electron Microscopy (SEM), using a JEOL JSM 840 device, by High Resolution Transmission Electron Microscopy (HRTEM) coupled to Electron Dispersive X-Ray Spectroscopy (EDS), using a EM-002B microscope. The statistical distribution of heavy elements (La, Ni, Fe), observed by microanalysis using a 200 nm focused beam on one area and using a 14 nm focused beam on different areas of the sample shows a good homogeneity of the samples before test. The XRD diagrams confirm the formation of the mixed $LaNi_xFe_{(1-x)}O_3$ perovskites all over the 0 to 1 range for x [7,10]. The total amount of carbon deposited on the catalysts during the test was measured by elemental and thermogravimetric analyses under hydrogen flow using a SETARAM 92 12.

2.3. Reaction conditions

The operating conditions were: 200 mg of catalyst placed in a fixed bed quartz reactor (6.6 mm ID), atmospheric pressure, inlet temperature: 400-800°C, feed flow rate: 0.3 $L.h^{-1}$ of CH_4 and 0.3 $L.h^{-1}$ of CO_2 with dilution in argon 2.4 $L.h^{-1}$. The outlet gas was analyzed by two on-line gas chromatographs: one giving the amount of remaining CH_4 and of produced CO and CO_2; the other quantifying CO and H_2 production. The catalytic systems became active under the gas mixture and under heating, without previous reduction, because the reduction occurred *in situ* during the test. The temperature program for this reaction was composed of three cycles. The first cycle was formed by a continuous temperature increase to 800°C with a slope of 3°C/min, during which chromatographic analyses were carried out every 60°C, followed by a step of one hour at 800°C and by a stepwise decrease every 50°C, down to 400°C. This first cycle was carried out to activate the catalyst. The second temperature cycle was performed like the first one, and the characteristic catalytic behaviour was obtained in the second temperature increase. A third temperature increase confirmed the results obtained in the second one. The overall duration of one test was 30 hours.

3. RESULTS AND DISCUSSION

Preliminary results on these catalysts have shown by XRD and HRTEM that nitrate precursors gave a highly homogeneous solid solution of $LaFeO_3$ and $LaNiO_3$ [10]. These systems, $La(Ni_xFe_{1-x})O_3$ with $0 \leq x \leq 1$, have been tested in the CH_4 + CO_2 reaction and the CO yields obtained at 800°C in the first and second cycles are given in Figure 1, except for $LaNiO_3$ because the carbon formation was too high and plugged completely the reactor during the second heating ramp. For the other catalysts, the results obtained in the second and third cycles are similar. Figure 2 gives the CO yields for all x values at different temperatures in the second cycle. For catalytic systems with $0.3 \leq x \leq 0.9$, the CO yields observed at 800°C are higher than 84% and reach 95% for x=0.3. At 800°C, the H_2/CO yield takes values between 0.81 (x=0.3) and 0.94 (x=0.8) and does not depend on the nature of the catalyst. This value is lower than 1 (stoichiometry of the reaction), because the water-gas shift reaction occurs [4]. The H_2/CO ratio increases with increasing temperature, as well as with the CH_4 and CO_2 conversions, e.g. for $LaNi_{0.3}Fe_{0.7}O_3$, it varies from 0.3 at 500°C (CO_2 conversion of 16% and CH_4 conversion of 10%), to 0.88 at 800°C (CO_2 conversion of 97.5%, CH_4 conversion of 96.7%). The results for the catalysts with $x \leq 0.1$ show that iron alone is not very active in dry reforming of methane (CO yield of 16%, CO_2 conversion of 23%, CH_4 conversion of 11% for $LaFeO_3$ at 800°C), due to the low reduction of the iron in these testing conditions. For $LaNiO_3$, the initial activity is very high (CO yield of 97.1% and CH_4 conversion of 97.5% at 800°C), but the catalyst is rapidly deactivated by carbon formation, whereas for $0.2 \leq x \leq 0.8$, the lifetime of the catalyst exceeds 200 hours.

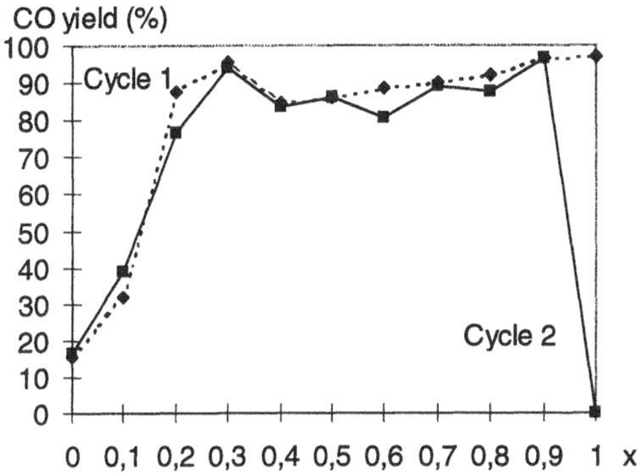

Figure 1. CO yields at 800°C versus x for $LaNi_xFe_{1-x}O_3$ systems in the first and second cycles.

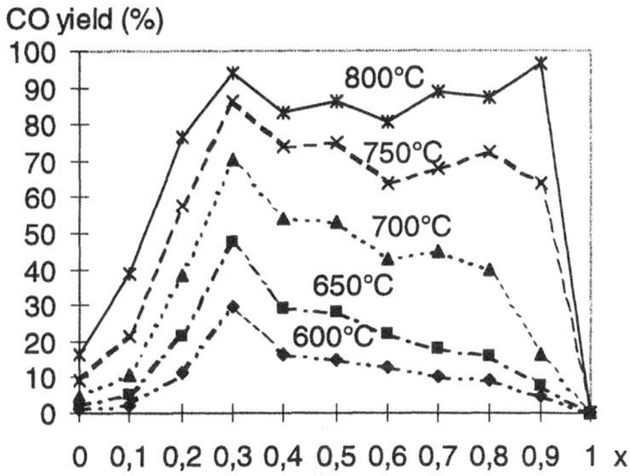

Figure 2. CO yields at different temperatures versus x In the second cycle.

The characterization of the catalysts after test shows that nickel has left the perovskite structure and is totally reduced for $x > 0.1$. Ni and NiO deposited on La_2O_3 are evidenced by XRD after test for the initially Ni rich perovskites (0.9 and 1. For $0.2 \leq x \leq 0.7$, part of the structure is present as a $LaFeO_3$ perovskite, and La_2O_3 and metallic nickel are also observed. However, in this case, a slight shift of the Ni peak suggests the formation of a Ni-Fe alloy. The destruction of the La-Ni-Fe structure is confirmed by TEM and EDS analyses in different areas (2-9) of the catalyst (Figure 3).

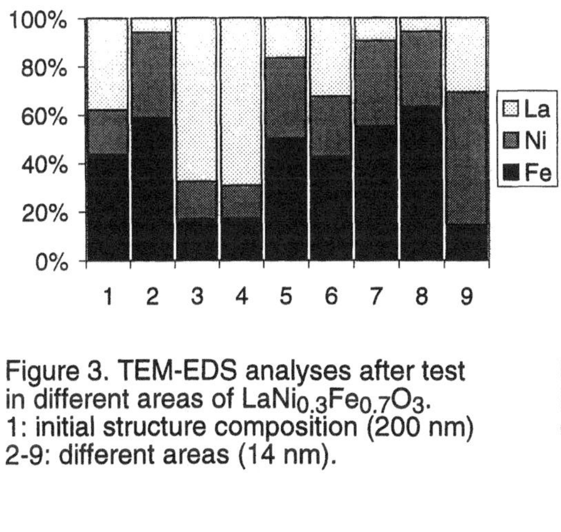

Figure 3. TEM-EDS analyses after test in different areas of $LaNi_{0.3}Fe_{0.7}O_3$.
1: initial structure composition (200 nm)
2-9: different areas (14 nm).

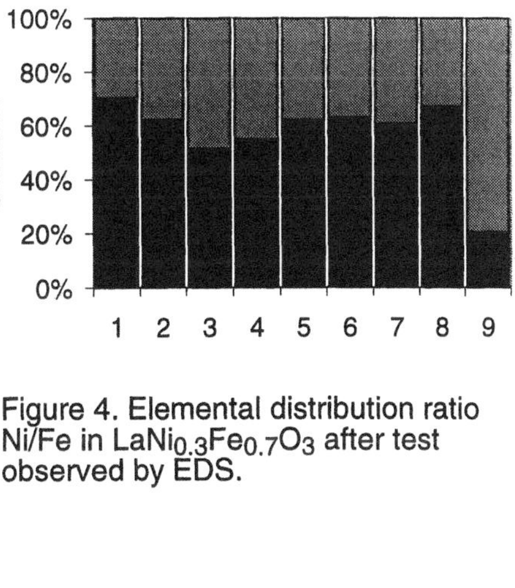

Figure 4. Elemental distribution ratio Ni/Fe in $LaNi_{0.3}Fe_{0.7}O_3$ after test observed by EDS.

The Ni and Fe proportions in $LaNi_{0.3}Fe_{0.7}O_3$ after test (Figure 4) appear almost constant in the sample, which supports the hypothesis of a Ni-Fe alloy formation. This alloy has been confirmed by Mössbauer measurements. Indeed Mössbauer analyses have shown the presence of metallic Fe, but the peak corresponding to metallic Fe presented a shift of the hyperfine field from 33T for α-Fe to 32.5T for $LaNi_{0.3}Fe_{0.7}O_3$ and to 29.4T for $LaNi_{0.7}Fe_{0.3}O_3$. This leads to conclude to the formation of an alloy presenting different Ni/Fe ratios depending on x and especially an enrichment in Ni content with increasing x.

However, whereas the catalyst has been deeply modified by the dry reforming, its regeneration is possible from Ni-Fe alloy deposited on $LaFeO_3$ and La_2O_3 for $0.2 \leq x \leq 0.5$ by a simple 24 hour calcination under air at 750°C. For $0.5 \leq x \leq 0.9$, the initial mixed perovskite structure is definitively lost because the regeneration test by calcination leads to the separation of $LaFeO_3$ and $LaNiO_3$ phases. In the case of $LaNi_{0.3}Fe_{0.7}O_3$, the XRD measurements, for which a zoom in the area of 2Θ between 32.0 and 33.0 can be seen in Figure 5, and the TEM-EDS analyses (Figure 6) evidence the recombination of the elements, leading to the formation of a mixed structure close to the initial perovskite and also the good homogeneity of the regenerated system. However, the x value (0.15) determined from the XRD, remains lower than the initial value (0.3).

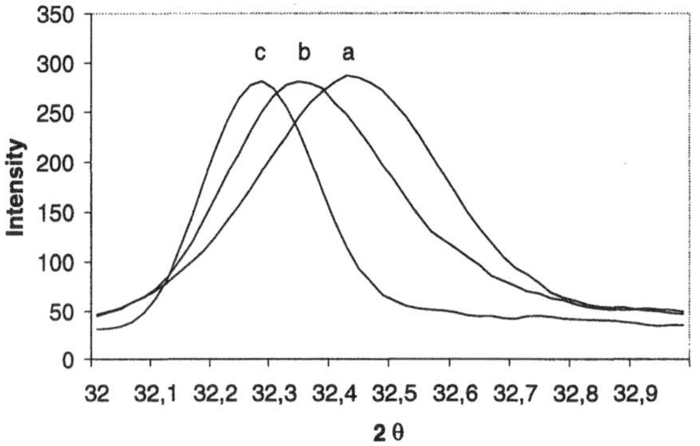

Figure 5. XRD of $LaNi_{0.3}Fe_{0.7}O_3$ in the area $32.0 \leq 2\Theta \leq 33.0$
a: before test, b: after test then recalcination, c: directly after test.

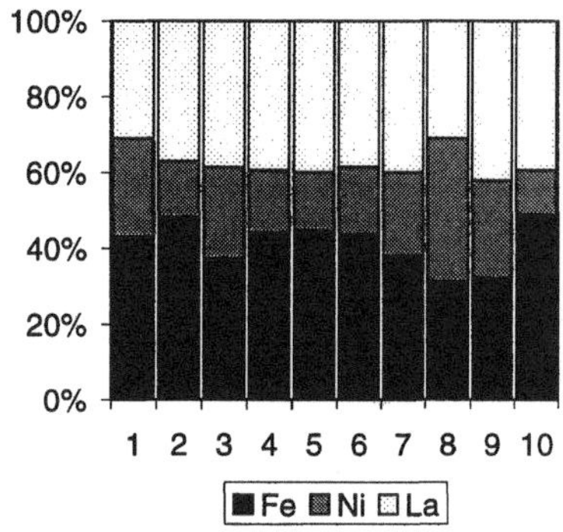

Figure 6. EDS elemental distribution for $LaNi_{0.3}Fe_{0.7}O_3$ after test then regeneration.

The amount of carbon deposited on the catalyst after test was determined by elemental analysis after the three reaction cycles performed during test and is given versus x (Figure 7). This curve shows clearly, that the $LaNi_xFe_{(1-x)}O_3$ mixed perovskite structures can resist to coke formation.

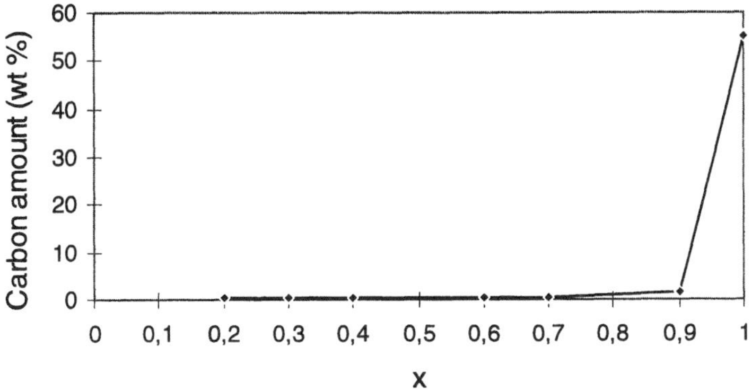

Figure 7. Amount of carbon (weight %) deposited on the $LaNi_xFe_{1-x}O_3$ catalysts after test versus x.

On the catalysts containing nickel and lanthanum, numerous arguments have been given to explain the decrease of carbon formation. It has been reported [11], that supports presenting a high affinity for CO_2 (like La_2O_3) prevent accumulation of carbon by reaction of metal with coke precursors from methane. It has also been suggested that carbon formation was linked to the ensemble effect [12, 13], 16 neighbouring sites are necessary for coke formation and 12, for syngas production. The formation of a definite structure (Ni-La-Al) leads to a more difficult reduction of nickel, preventing metal from sintering on the surface and from coke formation [4, 5, 14]. This effect depends on the catalyst pre-treatment and on the La_2O_3 content, hence possibly on the nature of the formed definite structures. Using Ni/La_2O_3 (17 wt% Ni), Zhang et al. [3] indicate no carbon formation, even on Ni particles up to 1000 Å. They explain this phenomenon by a covering of metallic nickel particles by La_2O_3, as shown in syngas chemistry. *A priori*, the $LaNi_xFe_{(1-x)}O_3$ catalysts can combine at least partially all the effects cited hereabove: basicity of La_2O_3, small metal particle size, coverage of the nickel particles by lanthanum oxide and a harder reduction of the nickel inserted in the mixed perovskite structure. Compared to the catalyst containing 17 wt % of Ni on La_2O_3, $LaNiO_3$ with 24 wt % of Ni, leads to a large coke formation and nickel particles of a lower size (100 Å to 200 Å) were measured by TEM. In this case, the covering effect of La_2O_3 can not be evoked, since decrease in Ni particle size should unfavour carbon formation, as proposed previously [12, 13]. On the opposite, the initial La-Ni-Fe perovskite permits to delay significantly the nickel reduction, especially when the amount of nickel in the structure is low [7, 10], and allows to decrease the size of the formed metal particles. For $LaNi_{0.7}Fe_{0.3}O_3$, the mean particle size measured by TEM is 300 Å (diameter), which is low, given the percentage of nickel in the structure (17 wt %). A CO_2 fixation on La_2O_3 could also be proposed. However, the $LaNi_xFe_{(1-x)}O_3$ catalysts show a particular Ni-Fe alloy formation during test, which is not observed in the catalysts proposed in literature. The presence of this alloy with a variable Ni-Fe composition depending on the initial perovskite composition, but a relatively steady composition for a given catalyst, modifies the nickel particles at the surface of the catalyst and seems to create a dilution of nickel, as already seen in Fischer-Tropsch synthesis for Ni-Cu catalysts [15]. It can be suggested that this phenomenon of alloying constitutes a new research direction for active catalysts in dry reforming of methane preventing carbon deposition.

4. CONCLUSION

The $LaNi_xFe_{(1-x)}O_3$ catalysts with $0.3 \leq x \leq 0.8$ performed well in dry reforming of methane. They are modified during the catalytic tests by formation of a Ni-Fe alloy deposited on La_2O_3 and $LaFeO_3$. The coke formation remains negligible for $0.2 \leq x \leq 0.8$, whereas the reactivity of $LaNiO_3$ is blocked because of carbon deposition. It is suggested that the presence of Ni-Fe alloy, besides the positive effects proposed by others, is an important point in preventing coke formation, because the iron atoms dilute nickel leading to lower ensembles.

ACKNOWLEDGEMENTS

The authors are grateful to the European Community for its financial support under Joule contract JOR3-CT95-0037.

REFERENCES

1. S.C. Tsang, J.B. Claridge and M.L.H. Green, Catal. Today, 23 (1995) 3.
2. J.H. Edwards and A.M. Maitra, Fuel Proces. Technol., 42 (1995) 269.
3. Z. Zhang, X.E. Verykios, S.M. Mac Donald and S. Affrossman, J. Phys. Chem., 100 (1996) 744.
4. R. Blom, I.M. Dahl, A. Slagtern, B. Sortland, A. Spjelkavik and E. Tangstad, Catal. Today, 21 (1994) 535.
5. A. Slagtern, U. Olsbye, R. Blom, I.M. Dahl and H. Fjellvag, App. Catal. A, General 145 (1996) 375.
6. C. Petit, A. Kiennemann, P. Chaumette and O. Clause, US Patent No. 5 447 705 (1995).
7. H. Provendier, C. Petit, C. Estournes, S. Libs and A. Kiennemann, Proceedings of The First European Congress on Chemical Engineering, Florence, Italy, 4-7 mai 1997, 1 (1997) 363.
8. A.C. Roger, C. Petit and A. Kiennemann, J. Catal. 167 (1997) 447.
9. M. Houalla, F. Delannay, I. Matsumura and B. Delmon, J. Chem. Soc. Faraday Trans., 76 (1980) 2128.
10. H. Provendier, C. Petit, G. Ehret and A. Kiennemann, J. Chimie Physique (under press).
11. O. Yamazuki, T. Nozaki, K. Omata and K. Fujimoto, Chem. Lett., (1992) 1953.
12. A. Slagtern and U. Olsbye, Appl Catal. A General 110 (1994) 99.
13. J.R. Rostrup-Nielsen, Catalysis, Science and Technology, 5 (1984) 1, J.R. Anderson and M. Boudart (eds.), Springer, Berlin.
14. A. Slagtern, U. Olsbye, R. Blom and I.M. Dahl, Stud. Surf. Sci. Catal. 107 (1997) 497.
15. M. Araki and V. Ponec, J. Catal., 44 (1976) 439.

NATURAL GAS CONVERSION V
Studies in Surface Science and Catalysis, Vol. 119
A. Parmaliana et al. (Editors)

Effects of Alkali and Rare Earth Metal Oxides on the Thermal Stability and the Carbon-deposition over Nickel Based Catalyst

Shenglin Liu Guoxing Xiong* Shishan Sheng Qing Miao Weishen Yang

State Key Laboratory of Catalysis, Dalian Institute of Chemical Physics, Chinese Academy of Sciences, P.O. Box 110, Dalian 116023, P.R.China

ABSTRACT

Effects of alkali and rare earth metal oxides on the thermal stability and the carbon-deposition over nickel based catalyst for partial oxidation of methane to syngas were investigated by a series of characterization techniques including flow-reaction, TG, TPO, XPS, XRD and BET. The results indicated that the introduction of Li and La oxides could suppress the carbon-deposition on the nickel based catalyst. LiNiLaO/γ-Al_2O_3 catalyst not only possessed excellent reaction performance (CH_4 conversion $\geq$ 96%, CO selectivity $\geq$ 98%), carbon-deposition resistance and improved the thermal stability of the nickel based catalyst, but also had comparatively stable porous structure and stable crystallity during the 200h life test experiment under the conditions of reaction temperature at 1123K, CH_4/O_2 ratio of 1.96 and space velocity of 2.7×10^4 l/kg.h.

1. INTRODUCTION

Partial oxidation of methane to syngas (POM) has received intense attention since 1990[1,2]. The catalysts used in POM reaction can be classified into two types: (1) noble metal (Pd, Rh, Ru, Pt, Ir) catalysts; (2) transition metal(Ni, Co, Fe) catalysts. The nickel based catalysts emerged as the most practical one because of their high turnover rates and inexpensive cost. However, nickel based catalysts supported on γ-Al_2O_3 are usually unstable at high temperature[3]. Preventing the alumina support from thermal deterioration was studied by many authors [4 - 6] and a large number of additives to the γ-Al_2O_3 support have been shown to inhibit sintering and phase transformation, such as SiO_2, BaO, CeO_2, La_2O_3, etc. The other reason causing the deactivation of NiO/Al_2O_3 catalyst is the loss and sintering of nickel that results in certain decreasing of active surface and some change of the interaction between active compound and support. In addition, the carbon-deposition over NiO/Al_2O_3 is also often considered as a possible cause of deactivation for the catalyst. Schmidt found the loss of nickel on the 3wt% Ni/Al_2O_3 catalyst during the POM reaction [7]. The serious carbon-deposition over Ni/Al_2O_3 that filled the catalyst pores and caused the granules to disinergrate into a fine

* Author to whom all correspondence should be addressed

powder at CH_4/O_2 ratio ≥ 2 was reported by Lunsford [8]. Hence, the stability of NiO/Al_2O_3 catalyst is one of the important considerations before this process can be used on an industrial scale.

Previously, we reported that the $ABNiO/\gamma\text{-}Al_2O_3$ (A=Li, Na, K; B=La, Sm, Ce, Y) catalysts were excellent POM reaction catalysts[9,10]. Meanwhile, we discussed the oxidative transformation of methane over the nickel based catalysts modified by alkali and rare earth metal oxides[11]. In the present paper, effects of alkali and rare earth metal oxides on the thermal stability and the carbon-deposition over nickel based catalyst and the 200h life test experiment of $LiNiLaO/\gamma\text{-}Al_2O_3$ for POM reaction were investigated by a series of characterization techniques including flow-reaction, TG, TPO, XPS XRD and BET.

2. EXPERIMENT

2.1. Preparation of catalysts

The catalysts were prepared by impregnating appropriate amounts of $LiNO_3$, $Ni(NO_3)_2$ and $La(NO_3)_3$ (for $LiNiLaO/\gamma\text{-}Al_2O_3$, Ni:2.7wt%; Li_2O: 1.25 wt%; La_2O_3:7.0wt%) and $Ni(NO_3)_2$ (for $NiO/\gamma\text{-}Al_2O_3$, Ni:2.7wt%), respectively, on $\gamma\text{-}Al_2O_3$ support for 24 h, followed by drying at 393K and calcination in air at 823-1073K for 4 h.

2.2. Test of Catalytic Performance

Catalysts were tested in atmosphere pressure fixed-bed microreactors. The carbon-deposition over catalysts was tested using a microreactor with an internal diameter of 4 mm with 100mg catalyst employed. The 80h and 200h life test experiments were performed using a microreactor with an internal diameter of 8 mm with 500mg catalyst employed. The analysis of reaction products was provided elsewhere[11].

2.3. Characterizations of Catalysts

TG test was performed with a Perkin-Elmer TGS-2 instrument. The TG profiles were recorded and treated by Perkin-Elmer 3600 work station at a programmed temperature rate of 10K/min in the air with the flow rate 25ml/min. TPO profiles were tested by on-line Mass Sepectroscopy (m/e: 44(CO_2), 28(CO or CO_2), 18(H_2O)) at a programmed temperature rate of 16K/min in a 5vol% O_2/He flow.

X-ray photoelectron spectoscopy (XPS) characterization was performed using a VG ESCA LABMK¢ò spectrometer. The sample chamber was pumped to a pressure of $1x10^{-9}$ Torr and monochromatised Mg exciting radiation was used. The C_{1s} peak at 284.6ev due to adventitious carbon was used as an internal standard. The surface relative atomic ratio of catalyst was calculated according to the spectra line peak intensity of XPS.

X-ray diffraction(XRD) characterization of the catalysts was performed with a Riguku D/Max-RB X-ray diffractometer using a copper target at 40KV x 100mA and a scanning speed of 8 degree/min. The specific surface area and pore volume of samples were determined by the BET nitrogen method in a volumetric equipment OMNISORP-100CX.

3. RESULTS AND DISCUSSION

3.1. Comparison of carbon deposition between NiO/γ-Al_2O_3 and LiNiLaO/γ-Al_2O_3

The introduction of Li and La improves the ability of carbon-deposition resistance, besides the activity of the nickel based catalyst[10]. The deposition of surface carbon over the NiO/Al_2O_3 catalyst during the POM reaction results in the deactivation of NiO/Al_2O_3[7,8,12]. It is well known that the acidity of catalyst surface favors carbon-deposition and the basity of catalyst surface prevents carbon-deposition[13]. Hence, the addition of Li and La oxides is reasonable to prevent carbon-deposition over catalyst surface[14]. The TG results of the samples after the POM reaction for10h ($CH_4/O_2 = 2$; GHSV = 27000 l/kg.h; 1123K; 1atm) indicated that the carbon-deposition resistant ability of LiNiLaO/γ-Al_2O_3 was much better than that of NiO/γ-Al_2O_3. During the period of 10 h of POM reaction, there was almost no carbon deposition over LiNiLaO/γ-Al_2O_3 and the amount of carbon-deposition over LiNiLaO/Al_2O_3 was only 0.24% of its net weight, but that over NiO/γ-Al_2O_3 was about 12% of its net weight. It was shown that the introduction of Li and La oxides could suppress the carbon-deposition of the nickel based catalyst.

TPO tests were also performed to demonstrate the above results. According to the TPO results (see Fig. 1), the only gaseous product was CO_2 during the TPO tests of NiO/γ-Al_2O_3 and LiNiLaO/γ-Al_2O_3 catalysts after POM reaction for 10h under the same reaction conditions, and no CO and H_2O were detected. Two CO_2 desorption peaks at 813K and 1013K for NiO/γ-Al_2O_3 and two at 793K and 933K for LiNiLaO/γ-Al_2O_3 appeared in the profiles, which indicates there were two different kinds of carbon produced during the POM reaction. The amount of carbon over NiO/γ-Al_2O_3 was far more than that over LiNiLaO/γ-Al_2O_3 and it was more difficult to be depleted the carbon by oxygen.

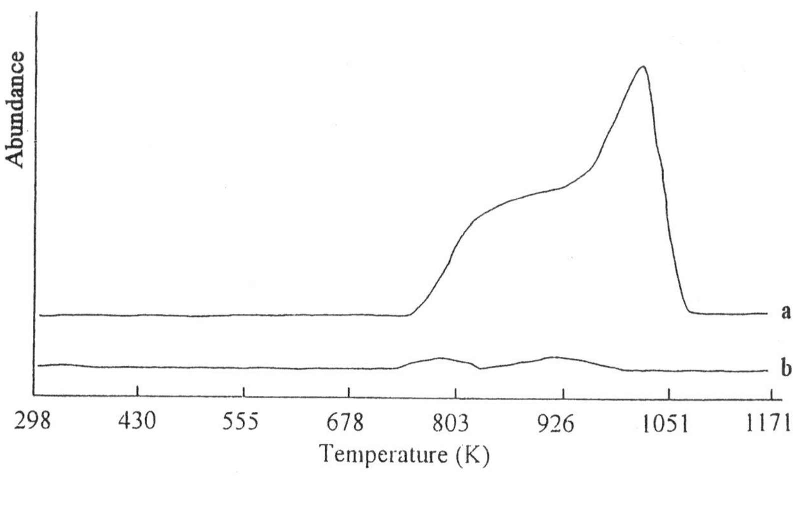

Figure 1. TPO profiles of catalysts after POM reaction for 10h (m/e = 44)
a: NiO/γ-Al_2O_3 b: LiNiLaO/γ-Al_2O_3

Another interesting phenomenon was that no H_2O was detected, and it was inferred that the carbon species produced during the POM reaction was carbon but no other carbon species with hydrogen atom.

3.2. Life test of LiNiLaO/γ-Al_2O_3 catalyst

3.2.1 The catalytic performance and stability

Previously, we investigated the 50h life test of LiNiLaO/γ-Al_2O_3 catalyst for POM reaction and found that CH_4 conversion kept exceeding 96% and CO selectivity 95% during the 50h life-test[10]. In the present work, we performed the 200h life test of LiNiLaO/γ-Al_2O_3 catalyst for POM reaction. The 200h life test experiment was performed using a microreactor with an internal diameter of 8 mm instead of 4 mm (for 50h life test experiment), and the weight of catalyst was 500 mg instead of 100 mg (for 50h life test experiment), while the stability of LiNiLaO/γ-Al_2O_3 catalyst was studied under the conditions of reaction temperature at 1123K, CH_4/O_2 ratio of 1.96 and spaces velocity of 2.7 x10^4 l/kg.h (see Fig. 2). During the 200 h life test experiment, CH_4 conversion kept exceeding 96.00%, H_2 selectivity and CO selectivity kept exceeding 98.00% and 96.00%, respectively. Those values approached that of the thermodynamic equilibrium. It suggested that the LiNiLaO/γ-Al_2O_3 catalyst was quite stable during high temperature reaction and the introduction of Li and La could improve the thermal stability of the nickel based catalyst.

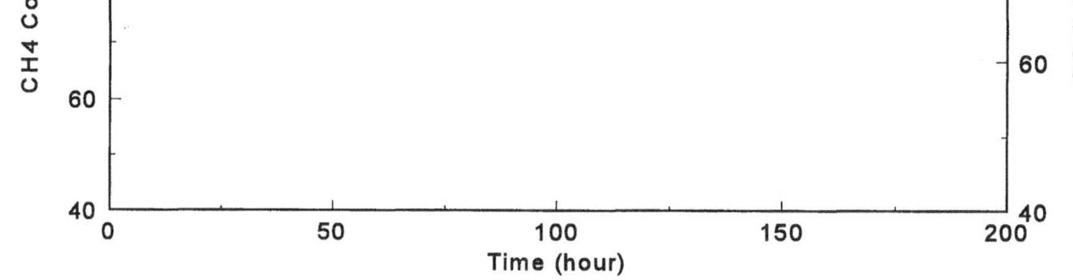

Figure 2. Performance of catalyst as a function of time (CH_4/O_2=1.96; GHSV =2.7 x 10^4 l/(kg.h); 1123K ; 1 atm)

3.2.2. The catalyst structure and crystallity analysis

After the 80h or 200 h life test experiment, the reactor was cooled to room temperature in N_2 gas flow. The catalyst was then investigated by BET method(see table 1). The results of BET showed that the catalyst pore volume changed little. In comparison with the specific surface area of fresh catalyst(Cat-0h), that of Cat-80h decreased by~20%, but the specific surface area of Cat-200h decreased only by ~3% (Compared with that of Cat-80h).Those results indicated the catalyst pore structure was comparatively stable during high temperature reaction.

The XRD tests were performed to determine the crystallity of those catalysts(see Fig.3). The XRD spectra of Cat-0h (reduced) was similar to that of Cat-200h. Only γ-Al_2O_3 and reduced nickel appeared, NiO and $NiAl_2O_4$ did not appear, namely, after the 200h life test experiment, the support γ-Al_2O_3 did not undergo crystallity transformation into α-Al_2O_3. The results indicated that the reduced nickel was the active center of LiNiLaO/γ-Al_2O_3 for POM reaction (It was consistent with other paper [11]) and the catalyst could kept the acidic property (that favors keeping the reduced nickel[11]) during high temperature reaction. It can be concluded that LiNiLaO/γ-Al_2O_3 catalyst had stable crystallity structure during high temperature reaction.

Table 1
Comparison of Porous volume, Specific surface area and Surface composition of catalysts

Sample*	Porous volume	Specific surface area	Surface relative atomic ratio (XPS)					
	ml/g	m^2/g	C	O	Li	La	Ni	Al
Cat-0h	0.60	120	0.45	2.11	0.084	0.0177	0.011	1.00
Cat-80h	0.62	96						
Cat-200h	0.58	93	0.40	1.90	0.050	0.0198	0.004	1.00

* Cat - 0h: The fresh LiNiLaO/γ-Al_2O_3 catalyst
Cat - 80h: The catalyst that has reacted for 80h, then was cooled to room temperature in N_2 gas flow
Cat - 200h: The catalyst that has reacted for 200h, then was cooled to room temperature in N_2 gas flow

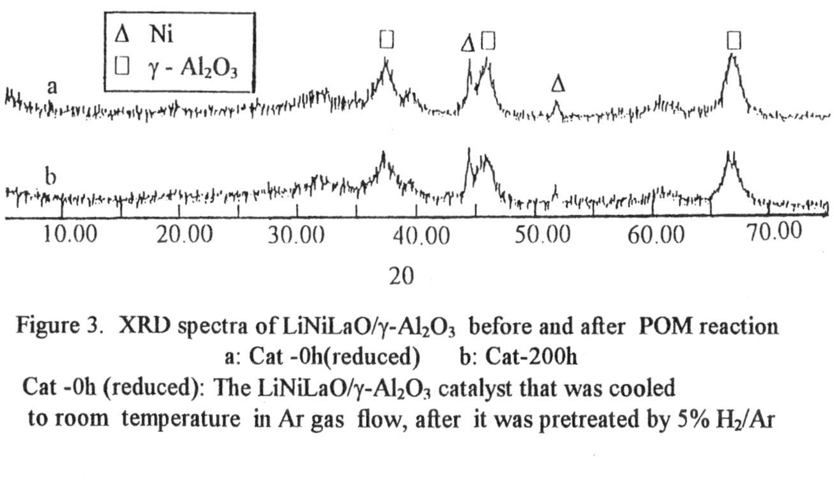

Figure 3. XRD spectra of LiNiLaO/γ-Al_2O_3 before and after POM reaction
a: Cat -0h(reduced) b: Cat-200h
Cat -0h (reduced): The LiNiLaO/γ-Al_2O_3 catalyst that was cooled to room temperature in Ar gas flow, after it was pretreated by 5% H_2/Ar

3.2.3 The carbon deposition and surface composition of the catalysts

After the 200 h life test experiment, the reactor was cooled to room temperature in N_2 gas flow. The catalyst was then investigated by TG (no shown) and XPS(see Table 1), the results of TG and XPS all indicated that the used catalyst had almost no carbon-deposition. The results demonstrated that the LiNiLaO/γ-Al_2O_3 catalyst had excellent resistance to carbon deposition during high temperature reaction

The surface relative atomic ratio of catalyst was investigated by XPS(see Table 1). The results indicated that the loss of Li on the surface of the catalyst was not serious and the intensity of La changed slightly. And the surface nickel lost obviously, but the reaction performance did not decrease during 200h life test experiment, which suggested that this amount of lost nickel was not enough to cause any change of reaction performance at low space velocity such as 2.7 x10^4 l/kg.h. Previously, the influence of space velocity on the performance of LiNiLaO/γ-Al_2O_3 (Ni:2.7wt%) for POM reaction was reported. The results indicated that CH_4 conversion decreased at space velocity $\geq 9 \times 10^4$ l/kg.h[10]. In the present work, the effect of the amount of nickel loaded on the reaction performance was discussed. The similar results were obtained , that was, CH_4 conversion decreased at space velocity $\geq 9 \times 10^4$ l/kg.h when the amount of nickel over γ-Al_2O_3 was 2.7wt%. However, when the amount of nickel increased to 10.0 wt%, the reaction performance did not decreased (CH_4 conversion 96%), even though the space velocity was increased to 5 x10^5 l/kg.h. So it could be concluded that the appropriate amount of nickel over γ-Al_2O_3 must be preserved in order to keep high performance for long period at high space velocity and temperature. The work will be published in other articles.

ACKNOWLEDGEMENT

The financial support of the National Natural Science Foundation of China is gratefully acknowledged.

REFFRENCES

1. A.T. Ashcroft, A. K. Cheetham, and P. D. F. Vernon, Nature, 344(1990)319.
2. S. C.Tsang, J. B. Claridge, M. L. H. Green, Catal.Today, 23(1995)3.
3. W. G. Schlaffer, C. Z. Morgan and J. N.Wilson, J. Phys. Chem., 61(1957)714.
4. S. Blonski, S. H. Garofalim, Catal. Lett., 25(1994) 324.
5. M. Pijolat, M. Dauzat, and M. Soustelle, Solid State Ionics, 50(1992)31.
6. Z. R. Ismagilov, R. A. Shkrabina, N. A. Koryabkina, Catal. Today, 24(1995)269.
7. P. M.Torniainen, X. Chu, L. D. Schmidt, J. Catal., 146(1994)1.
8. D. Dissanayake, M. P. Rosynek, J. H. Lunsford, J. Catal., 132(1991) 117.
9. G. X. Xiong, Q. Miao, China Patent, Appl. No. 95110071.8.
10. Q. Miao, G. X. Xiong, and X. X. Guo, Appl. Catal. A, 154(1997) 17.
11. Q. Miao, G. X. Xiong, and X. X. Guo, Stud. Surf. Sci. Catal., 101(1996)453.
12. J. B. Claridge, M. L. H. Green, and P. D. Battle, Catal. Lett., 22(1993)299.
13. D. L. Trimm, Design of Industrial Catalyst, Elsevier Scientific Publishing Company, Amsterdam-Oxford-New York, 1980.
14. C. H. Bartholomew, Catal. Rev. -Sci. Eng., 24(1982) 67.

NATURAL GAS CONVERSION V
Studies in Surface Science and Catalysis, Vol. 119
A. Parmaliana et al. (Editors)

Kinetics of methane reforming over Ni/SiO2 catalysts based on a step-wise mechanistic model

V.C.H. Kroll[a], G.J. Tjatjopoulos[b], C. Mirodatos[a*]

[a] Institut de Recherches sur la Catalyse, 2, avenue Albert Einstein, F-69626 Villeurbanne Cédex, France
[b] Chemical Process Engineering Research Institute, PO Box 361, GR-57001 Thermi, Thessaloniki, Greece

1. INTRODUCTION

Until now only few kinetic studies on the CO_2 reforming reaction have been reported in literature. Most of them concern simple power-law models, e.g., [1-4]. The advantage of these models is their simplicity in determination and application for engineering means. However, they are rather limited with respect to the comprehension of the reactions taking place on the catalyst surface. Richardson and Paripatyadar [5] report a more detailed kinetic study based on a Rh/Al_2O_3 catalyst. The forward rate of the carbon dioxide reforming reaction is described by a Langmuir-Hinshelwood model with competitive adsorption of carbon dioxide and methane. This paper aims at developing a kinetic scheme for the carbon dioxide reforming reaction for a Ni/SiO_2 catalyst. The detailed mechanistic scheme proposed by Kroll *et al.* [6] was the starting point for this development. The experiments were carried out in a recycle reactor run under perfectly mixed conditions.

2. EXPERIMENTAL

The Ni/SiO_2 catalyst was the same as described in [7]. After calcination at 750°C for 8 h under flowing oxygen the catalyst pellet size was brought to 0.2-0.3 mm. The catalyst deactivation was negligible using this pretreatment [8]. The catalyst was loaded in a tubular quartz reactor (i.d. 9 mm) with a thermocouple placed in a thermo-well (e.d. 3 mm) located in the catalyst bed. The typical amount of catalyst was 30 mg diluted with 270 mg quartz. The exhaust gases were recycled externally to the reactor entrance by a heated membrane pump. A recycle ratio of at least 20 was needed to attain perfectly mixed conditions [9]. All species were detected quantitatively by gas chromatography except water which was calculated using the oxygen balance. Argon was used as an inert tracer to measure the expansion factor. Since the reactor is operated under differential conditions the reaction rates can directly be determined from the inlet and the outlet flow rates. All the experiments were carried out at atmospheric pressure and most at 700°C. Up to 800°C both the empty reactor and the quartz diluent did not show any catalytic activity under all reaction conditions tested here.

* Corresponding author

3. RESULTS

3.1 Global influence of concentration on reaction rate. Figures 1 to 4 display the methane disappearance rate as a function of the species concentration measured at the reactor outlet.

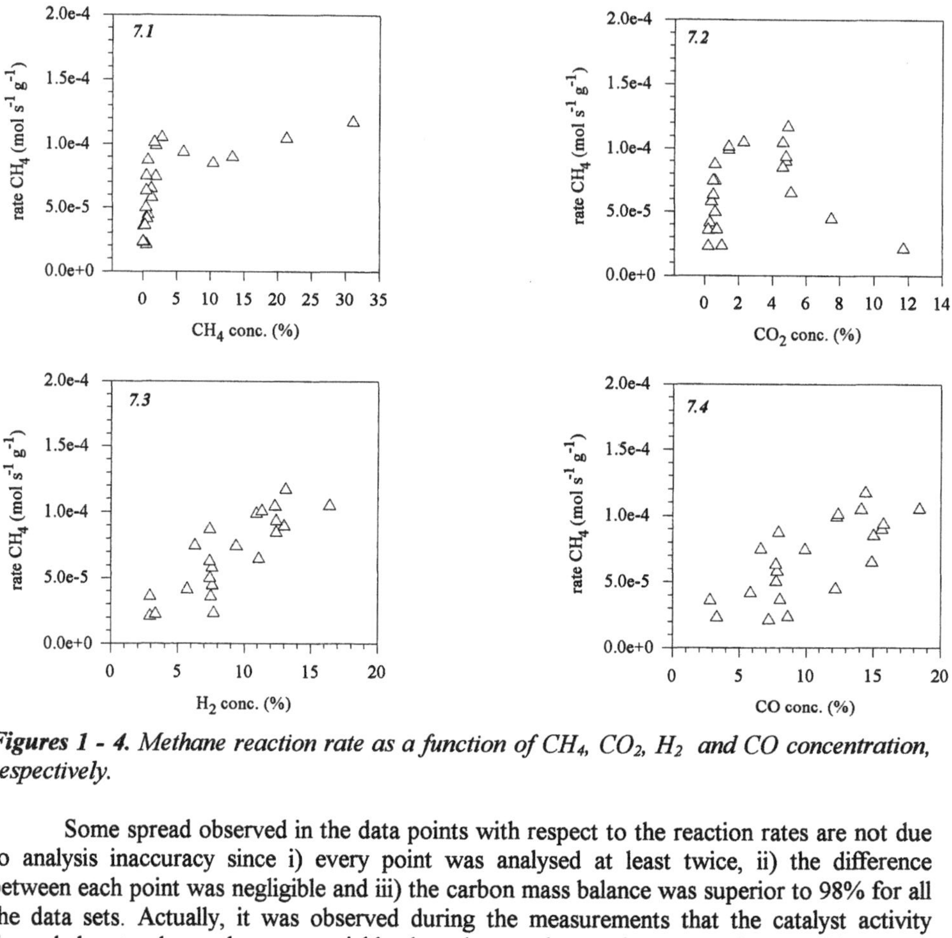

Figures 1 - 4. *Methane reaction rate as a function of CH_4, CO_2, H_2 and CO concentration, respectively.*

Some spread observed in the data points with respect to the reaction rates are not due to analysis inaccuracy since i) every point was analysed at least twice, ii) the difference between each point was negligible and iii) the carbon mass balance was superior to 98% for all the data sets. Actually, it was observed during the measurements that the catalyst activity depended not only on the state variables but also on the previous operating conditions. For example, the activity at a CH_4/CO_2 ratio equal to 1 was not necessarily the same if one approached this point from a higher or from a lower CH_4/CO_2 ratio. Moreover, this phenomenon was only observed when the reactor was operated differentially and not under integral conditions. These variations cannot be ascribed to irreversible deactivation phenomenon like nickel sintering. In contrast, the chemical state of the active surface and sub-surface, i.e., the relative degree of reduction/carbidization/oxidation, which was shown to be sensitive to the history of the catalyst [6], could originate this slight data spreading.

Figure 1 shows the reaction rate of methane first increases but then stabilises to zero order dependency as a function of CH_4 concentration. The reaction rate as a function of the CO_2 concentration (Figure 2) and shows a similar trend, except that the rate tends to decrease at higher CO_2 concentrations. Both figures suggest a competitive adsorption on the catalyst surface of CH_4 and CO_2 with a negative order for CO_2 at higher concentrations. Figures 3 and 4 show the CH_4 disappearance rate increases linearly with the H_2 and CO concentration.

As observed previously [6], the water gas shift reaction is always in equilibrium during the experiments performed in this kinetic study, which determines the concentration of both CO and H_2. At low CO_2/CH_4 ratios a high CO selectivity is observed that decreases only slightly and remains approximately 73%. The H_2 selectivity increases rapidly up to 71% but then slowly decreases. At low CO_2/CH_4 ratios the CO/H_2 ratio is nearly 1.6 but drops quickly to unity at stoichiometric mixtures. With further increasing the CO_2/CH_4 ratio the CO/H_2 ratio decreases again. The CO/H_2 ratio was always larger than one.

3.2 Activation energy. The apparent activation energy of the reforming reaction, E_{act}, was measured within the temperature range 680-720°C taking a point every 10°C. In order to check the sensitivity of E_{act} to CH_4 and CO_2 conversion, various contact times (W/F ratio) or reactants concentrations were chosen, but maintaining the CH_4/CO_2 ratio = 1. Straight lines were obtained in all cases, attesting the absence of temperature dependent diffusion limitation. As shown in Figure 5 the activation energy for both the reactants and the products is found to decrease when increasing the methane (or carbon dioxide) conversion. The kinetic scheme to be developed must be able to explain these variations.

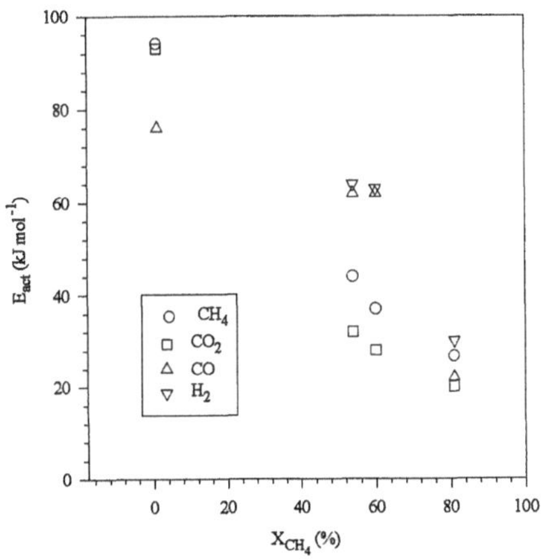

Figure 5. The activation energy of CH_4, CO_2, H_2, and CO as a function of CH_4 conversion.

4. KINETIC MODEL AND PARAMETER OPTIMISATION

4.1 Kinetic model. The following mechanistic features proposed by Kroll *et al.* [6] was taken as a starting point for the kinetic model :

i) CH_4 undergoes dehydrogenation on a free active site leaving a carbon adsorbed on the surface while the produced H_2 is immediately released into the gas phase (Figure 6, step 1). It has been shown from isotopic tracing experiments that there is a partial reversibility for this reaction with respect to carbon and a complete reversibility with respect to H_2.

ii) CO_2 is activated on a free site leaving an adsorbed oxygen and producing gaseous CO (Figure 6, step 2). This reaction was shown to be completely reversible [6].

iii) the rate determining step is the surface reaction between adsorbed carbon and adsorbed oxygen (Figure 6, step 3). This step is considered being the slowest step in the complete reaction scheme.

iv) while the water gas shift reaction was shown to be in equilibrium, the decomposition of H_2O on a free active site leaving an adsorbed oxygen and releasing H_2 in the gas phase is also assumed to be in equilibrium (Figure 6, step 4).

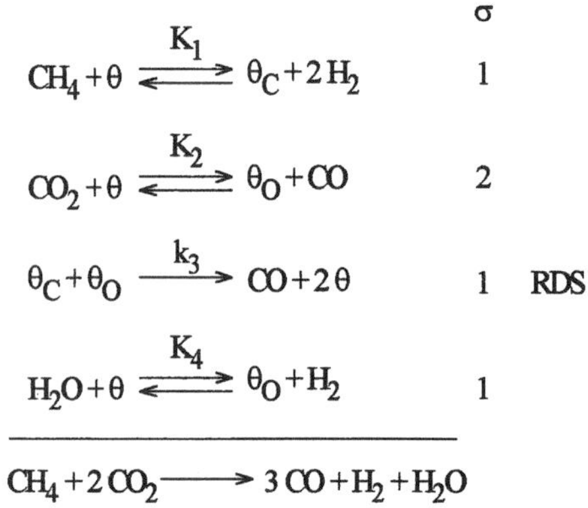

Figure 6. Kinetic scheme for CO_2 reforming of methane: one active site.

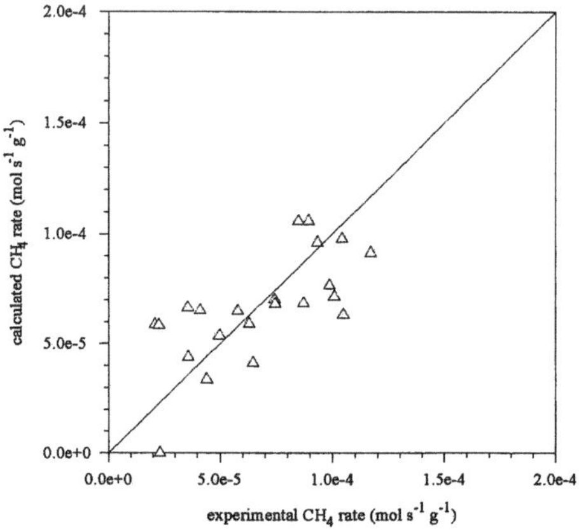

Figure 7. Calculated CH_4 reaction rate as a function of the experimental CH_4 reaction rate.

The above kinetic scheme can mathematically be written as :

$$K_1 = \frac{p_{H_2}^2 \cdot \theta_C}{p_{CH_4} \cdot \theta} \qquad K_2 = \frac{p_{CO} \cdot \theta_O}{p_{CO_2} \cdot \theta} \qquad K_4 = \frac{p_{H_2} \cdot \theta_O}{p_{H_2O} \cdot \theta} \tag{1}$$

in which p_i is the partial pressure of component i and θ_C and θ_O the surface coverage of adsorbed oxygen and carbon, respectively, θ the free site density and, K_i the adsorption equilibrium constants. The balance of the active surface sites is given by:

$$\theta + \theta_O + \theta_C = 1 \tag{2}$$

The rate determining step

$$r = k_3 \cdot \theta_C \cdot \theta_O \tag{3}$$

can be considered as the rate of methane consumption r_{CH4} since it measures the irreversible oxidation of surface carbon arising from methane into CO. It can be transformed using equations (1)-(3) as a function of directly measurable variables into:

$$r_{CH_4} = \frac{k_3 \cdot K_1 \cdot K_2 \cdot p_{CH_4} \cdot p_{CO_2} \cdot p_{CO} \cdot p_{H_2}^2}{\left(p_{H_2}^2 \cdot p_{CO} + K_1 \cdot p_{CH_4} \cdot p_{CO} + K_2 \cdot p_{CO_2} \cdot p_{H_2}^2\right)^2} \tag{4}$$

The water concentration is determined by equilibrium K_2/K_4 which constitutes the water gas shift reaction. The equilibrium constant was calculated as 1.6 [10]. This model can describe the dependency of the methane rate between an order of -1 and 1 as a function of

CH_4, CO_2, and CO concentrations, depending on the conversion of the various species. The order with respect to H_2 is in the range -2 and +2.

4.2 Parameter estimation. The model parameters K_1, K_2 and k_3 were calculated using non-linear least squares optimisation [11] applying the Levenberg-Marquardt search method. The estimated parameters were tested statistically following [12].

Table 1. *Estimated parameters.*

parameter	value	standard deviation	97.5% confidence interval
K_1 (%)	108.3	42.6	89.5
K_2	22.66	10.3	21.7
k_3 (mol s^{-1} g_{cat}^{-1})	$4.850*10^{-4}$	$6.63*10^{-5}$	$1.39*10^{-4}$

Table 2. *Variance-covariance and correlation matrix, respectively, of the estimated parameters.*

	K_1	K_2	k_3	K_1	K_2	k_3
K_1	1813	353	$-2.3*10^{-3}$	1.000	0.804	-0.809
K_2		106	$-5.4*10^{-4}$		1.000	-0.795
k_3			$4.4*10^{-9}$			1.000

Reaction rate coefficients. For the modelling procedure a set of 22 data obtained at 700°C were considered. Figure 7 shows both the experimental and calculated data points of the methane reaction rate. The estimated parameters are given in Table 1. It can be seen that K_1 and K_2 have a rather large confidence interval probably because they both appear in the nominator as well as in the denominator. Their direct effect on the reaction rate will be limited under the reaction conditions considered here. Although the correlation coefficients given in Table 2 are not very small (i.e. without correlation), they are neither close to unity (i.e. correlated). The calculated F_c value following [13] is 65 which is much larger than the tabulated $F(3,19;0.95) = 3.13$. So, the model is not rejected because of lack of fit.

The validity of the model was also checked by calculating the surface coverages θ_C and θ_O from K_1 and K_2 values in order to compare them with the values obtained on the same catalyst and for a similar set of partial pressures from isotopic transient kinetic analysis (SSITKA), as reported in [6, Table 4].

Table 3. *Comparison between surface coverages obtained by isotopic transient kinetics and by the present kinetic model.*

surface coverage	θ_C	θ_O
SSITKA [6]	0.43	0.10
Kinetic model	0.48	0.28

As can be seen a satisfying agreement between the two independent measurements is obtained, confirming that the working surface is largely covered with surface carbon atoms (most likely under a carbide-like structure) but also with a significant pool of surface oxygen atoms. Such a quantitative description of the working surface is also consistent with the mechanistic steps proposed in Figure 7 for the present kinetic study.

Activation energies. The activation energies were fitted using the reparametrisation of the Arrhenius and the van 't Hoff equation which improves the convergence during optimisation

[12]. However, with the available experimental data it was not possible to find statistically independent estimates for E_i parameters. Even the independent analysis per temperature did not provide realistic information. Equation (4) does not permit the simplification by decoupling parameters K_1, K_2 and k_3 under selected experimental conditions. However, the model does explain qualitatively the fact that the overall apparent activation energy depends on the conversion of the components. This difficulty to model a large range of kinetic data could arise from mechanistic changes with temperature. Thus, a shift of the RDS from the reaction between adsorbed C and O to the CH_4 activation into adsorbed C could occur at high temperature. As a matter of fact, the latter step was found to determine the overall rate for noble metals such as Ru [14].

5. CONCLUSIONS

The CH_4 reforming by CO_2 was studied in an external recycle reactor under differential conditions using a Ni/SiO_2 catalyst. The CH_4 reaction rate as a function of CH_4 concentration first increased rapidly and then levelled off. As a function of CO_2 concentration the rate first increased but then decreased again. The rate was a first order function of CO_2 and H_2 concentration. The water gas shift reaction was found to be at equilibrium in the range of investigated experimental conditions. The H_2 concentration at the reactor outlet was always lower than the CO concentration.

A kinetic model was proposed and validated statistically for the experimental data obtained at 700°C. The rate determining step is the surface reaction of adsorbed carbon and oxygen, both adspecies occupying a large fraction of the working surface. However, due to a lack of experimental data and/or to possible changes in the mechanism with temperature, the activation energies and the adsorption heats could not be determined from the model.

REFERENCES

1. Y. Sakai, H. Saito, T. Sodesawa, and F. Nozaki, React. Kinet. Catal. Lett., **24**, 3-4, (1984) 253-257.
2. A. Erdöhelyi, J. Cserényi and F. Solymosi, J. Catal., **141**, (1993) 287-299.
3. A. Erdöhelyi, J. Cserényi, E. Papp, and F. Solymosi, Appl. Catal. A: General **108**, (1994) 205-219.
4. O. Tokunaga, and S. Ogasawara, React. Kinet. Catal. Lett., **39**, 1, (1989) 69-74.
5. J.T. Richardson and S.A. Paripatyadar, Appl. Catal., **61**, 2, (1990) 293-309.
6. V.C.H. Kroll, H.M. Swaan, S. Lacombe, and C. Mirodatos, J. Catal., **164**, (1996) 387-398.
7. H.M. Swaan, V.C.H. Kroll, G.-A. Martin and C. Mirodatos, Catalysis Today, **21**, (1994) 571-578.
8. V.C.H. Kroll, H.M. Swaan, and C. Mirodatos, J. Catal., **161**, (1996) 409-422.
9. V.C.H. Kroll, Université Claude Bernard Lyon, Ph.D. Thesis (1996).
10. R.J. Kee, F.M. Rupley, and J.A. Miller, The Chemkin Thermodynamic Data Base, Sandia National Laboratories, Livermore, 1992.
11. P.T. Boggs, R.H. Byrd, Rogers, J.E., and R.B. Schnabel, ODRPACK, version 2.01, National Institute of Standards, Gaithersburg, 1992.
12. G.F. Froment, and L.H. Hosten, Catalysis, Science and Technology, (1981) 98-170.
13. G.B. Froment, and K.B. Bischoff, Chemical reactor analysis and design, 2nd edition, Wiley & Sons, New York, 1990.
14. Y. Schuurman, V.C.H. Kroll, P. Ferreira-Aparicio, and C. Mirodatos, Catalysis Today, **38**(1) (1997) 129-135.

NATURAL GAS CONVERSION V
Studies in Surface Science and Catalysis, Vol. 119
A. Parmaliana et al. (Editors)

Catalysts based on zirconium phosphates for selective methane oxidation to synthesis gas.

S. N. Pavlova[a], V. A. Sadykov[a], E. A. Paukshtis[a], E. B. Burgina[a], S. P. Degtyarev[a], D. I. Kochubei[a], A. V. Kalinkin[a], N. F. Saputina[a], R. I. Maximovskaya[a], V. I. Zaikovskii[a], R. Roy[b], D. Agrawal[b].

[a]Boreskov Institute of Catalysis SD RAN,
630090 Novosibirsk, pr. Lavrentieva, 5, Russia

[b]Materials Reseach Laboratory, Pennsylvania State University,
University Park, PA, 16802-4801

1. INTRODUCTION

The catalytic partial oxidation of methane (POM) to syngas conducted at high pressures and temperatures demands active catalysts with high thermal stability [1-2]. Complex zirconium phosphates of a NASICON type structure have a higher sintering resistance and a lower thermal expansion coefficient than conventional supports such as cordierite, α-Al_2O_3, etc. Further, a flexible structure of these framework phosphates allows to incorporate various substituting cations into the lattice [3-5], and, hence, to regulate their acid/base properties important at least for side reactions (coking, methane dimerization etc). Hence, such systems appear to be promising as supports for POM catalysts. In this work, powdered framework zirconium phosphates $MeZr_4P_6O_{24}$ loaded with $LaNiO_3$, $LaNi_{(1-x)}Rh_xO_y$, $LaNi_{(1-x)}Ir_xO_y$, $LaCo_{(1-x)}Ir_xO_y$ have been studied as catalysts for POM. To elucidate the factors determining their catalytic properties, chemical composition, real structure of the bulk and surface properties of the catalysts and supports have been investigated.

2. EXPERIMENTAL

2.1. Catalyst preparation.

Samples of crystalline and amorphous zirconium phosphates with nominal compositions $MeZr_4P_6O_{24}$. nH_2O (Me- Ca, Sr, Co or their mixture), $Me_{2/3}Zr_4P_6O_{24}$ nH_2O (Me = La) were synthesized as in [5]. All samples were calcined at 500 °C for 3 h except for a sample with the $Ca_{0.5}Sr_{0.5}Zr_4P_6O_{24}$ composition, which was sintered at 1300 °C, ball milled and annealed at 1000 °C. To obtain the catalysts, the powders of supports were impregnated by appropriate solutions of La, Ni, Co, nitrates and Rh, Ir chlorides, then dried and calcined at 900 °C.

2.2. Catalyst characterization.

Bulk structure of parent zirconium phosphates was characterized by EXAFS (spectra were acquired at the EXAFS Station of the Siberian Center of Synchrotron Radiation,

Novosibirsk using the same procedure as in [6] and fitted with the help of EXCURVE-92 program) and IR spectroscopy of the lattice modes (a BOMEM MB-102 IR-Fourier spectrometer). Crystalline samples were also studied by TEM (Jeol 200 C, 200 kv). ^{31}P NMR MAS (a Bruker CXP-300 spectrometer) was used to characterize the composition and local environment of the phosphate groups in amorphous and crystalline samples. Surface properties were probed by the IR spectroscopy of surface hydroxyls and CO test molecule using a IFS-113 v Bruker spectrometer.

The concentration of elements in the parent zirconium phosphates and catalysts was determined with the help of the atomic emission spectrometry (a BAIRD spectrometer). The surface composition of catalysts before and after reaction was identified by XPS (a VG ESCA-3 spectrometer with Al K_α radiation). Relative atomic concentrations of elements were calculated using empirical sensitivity factors [7].

2.3. Catalytic tests.

Catalytic tests were conducted using an integral U-shaped 4 mm inner diameter plug-flow quartz reactor. A sample of catalyst (0.05 g) was placed near the reactor bottom, and 2 cm^3 of a quartz sand were added upstream and downstream of the catalyst layer. Prior to reaction, the catalysts were pretreated at 850 °C in the 0.5 cm^3/sec He flow. The experiments were carried out at 800-850 °C with the feed gas of constant (CH_4:O_2:He = 2:1:25) composition.

3. Results and discussion

3.1. The properties of dispersed zirconium phosphates .

Bulk structure. For high temperature $SrZr_4P_6O_{24}$ sample annealed at 1000 °C, EXAFS data are in agreement with the ideal structure [4, 8]. Each Zr is surrounded by 6 oxygen atoms at a distance of ca 2.08 Å, and by six phosphorous atoms at a distance of ca 3.6 Å. The mean Zr-Sr distance ~3.2 Å with a coordination number ~3 also agrees with the ideal model. A set of well-resolved atomic planes with distances corresponding to a basic structure were observed in the high resolution TEM images (Fig. 1a).

Fig. 1. Typical TEM micrographs of $SrZr_4P_6O_{24}$ (a) and $Ca_{0,5}Sr_{0,5}$ $Zr_4P_6O_{24}$ (b) samples. <010> zone. Magnification; $5 \cdot 10^6$ Interplane distances for (104) planes: 4.65 Å for (a) and 4.5 Å for (b).

For high- temperature samples containing both Sr and Ca, some structure modification was introduced due to ordering of both cations in separate planes along the C axes. As a result, separate distances Zr- Sr (3.21Å) and Zr- Ca (3.07 Å) were revealed by EXAFS. Such ordering appears to be reflected in the complex pattern of the TEM images of planes other than those of the (003) type (Fig. 1b). However, this phenomenon does not cause any detectable changes in the phosphate groups properties studies by MAS NMR: only narrow band corresponding to a rather symmetric environment was observed (Fig. 2).

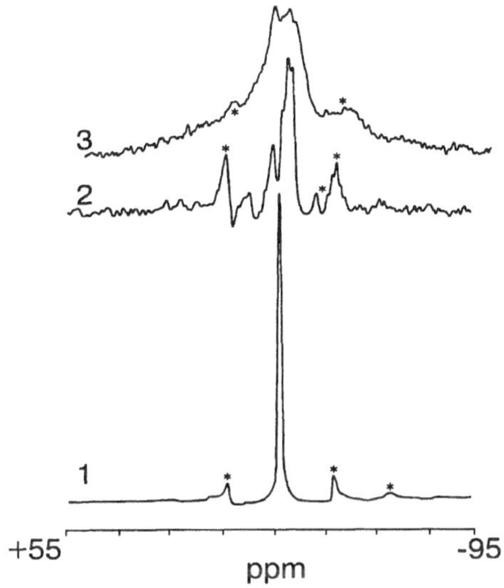

Fig.2. ^{31}P MAS spectra of $Ca_{0,5}Sr_{0,5}Zr_4P_6O_{24}$ (1), crystalline (2) and amorphous (3) $La_{2/3}Zr_4P_6O_{24}$. * - rotation side bands.

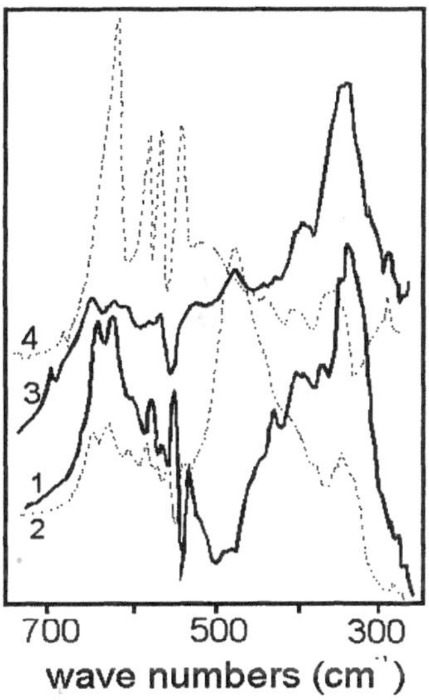

Fig.3. Differential (catalyst minus support) spectra of the lattices modes for the catalysts before (1,3) and after (2,4) reaction. 1.2 -sample 1, 3,4 - sample 4.

For crystalline samples calcined at temperatures around 500 °C, their structure is formed by bands (layers) with a rather perfect ordering along the C direction of the basic units of NASICON-type structure -two zirconium-oxygen octahedra [9]. However, the layers stacking is not perfect forming a great number of stacking faults and surface steps. Up to 5-6 water molecules per $MeZr_4P_6O_{24}$ formula unit are accommodated in the lattice of low-temperature samples, being in part retained here up to 900 °C. As a result, a mean Zr-O coordination number increases up to 8. NMR data (Fig.2) are in favor of the absence of any acidic phosphate groups in the lattice: for dispersed samples, superposition of several peaks situated in the range of δ -30 – -21 and corresponding to phosphate groups in a different environment [10] were revealed. Similarly, a broad IR band situated at ~ 1000 cm^{-1} can be assigned only to ν_3 stretching vibration of the PO_4^{3-} group.

For amorphous samples, the main structural features were the same as for the crystalline samples, though disordering was accompanied by decline of the Zr-P and Zr-Me coordination numbers.

Surface structure. For all samples, the surface was found to be mainly covered by weakly acidic surface P-OH groups (absorption bands at ~3660 cm^{-1}). More acidic hydroxyls bound with zirconium cations were manifested by less intense bands at ~ 3745- 3765 cm^{-1}.

Lewis acid centers were absent for samples containing lanthanum. Addition of the alkaline-earth cations was accompanied by appearance of the surface cus Zr^{4+} cations (adsorbed CO bands at~ 2200 cm^{-1} for Sr- containing sample) or Me^{2+} cations (CO band at ~ 2170 cm^{-1} for Ca-containing sample). The intensity of bands corresponding to Zr^{4+} - CO complexes for the Sr-containing sample is low. Amorphous samples usually have a lower density of the acid centers as compared with the crystalline ones of the same composition. It can be explained by easier relaxation of the surface structure in the former case leading to disappearance of the coordinatively unsaturated centers.

Hence, disordering of the dispersed framework zirconium phosphates structure, presence of residual water molecules in their lattice and exposed phosphate groups at the surface imply ability of these supports to strongly interact with the oxidic active components.

3.2. Bulk and surface composition of the catalysts.

In agreement with the IR data, XPS shows the surface of supports to be enriched by phosphate groups (Table 1). It implies possibility of their chemical interaction with the active

Table 1. Supports and catalysts composition.

Catalyst	Support	S m^2/g	Content and composition of the active component	Catalyst composition (relative to Zr)					
				Bulk (AES)		Surface (XPS)**			
				Ni (La)	Rh (Ir)	Ni (La)	Rh (Ir)	La	P
1	$Ca_{0,5}\ Sr_{0,5}$	2	16% $LaNiO_3$	0.2				0,9	5
2	$Sr_{0,5}\ La_{1/3}$	4	14% $LaNiO_3$	0.19		0.9		2.6	4.1
2*						0.18		1.3	2.9
3	$Ca_{0,5}\ Sr_{0,5}$	2	13% $LaNi_{0,95}Rh_{0,05}O_3$	0.16	0.008	0.27	0.04	0.93	4.5
3*						0.12		0.5	4.2
4	$La_{2/3}Zr_4$	3,3	29% $LaNi_{0,9}Rh_{0,1}O_3$	0.43	0.04	2.3	0.86	21	3.7
5	$CoZr_4$	3,5	21% $LaCo_{0,95}Ir_{0,05}O_3$	0.26	0.013	1.6	0.62	2.6	2.7

* - Catalysts after reaction

** - La $3d_{5/2}$, Ni $2p_{1/2}$, Co $2p_{3/2}$, Ir $4f_{7/2}$, Rh $3d_{5/2}$, P $2p_{3/2}$, Zr $3d_{5/2}$

component during impregnation and calcination steps depending upon the support nature. Indeed, for systems loaded onto Ca-Sr-NZP, (samples 1 and 3), a higher surface concentration of phosphorus correlates with a lower concentrations of Ni and Rh. A further drop in the La and Ni surface concentrations was also observed after catalysts contact with the reaction mixture testifying their incorporation into the bulk of support under reaction conditions. Hence, framework zirconium phosphates interact with the active components due to specific features of their real structure (vide supra). This conclusion is confirmed by analysis of the spectra of the lattice modes before and after testing in POM reaction. For all catalysts, in addition to bands of supports, new bands at 600-640 and 330-400 cm^{-1} assigned to (La-Ni)-O vibrations in the perovskite structure appear (Fig.3). Their exact position varies from sample to sample indicating a different degree of interaction with support. Such an interaction is also reflected in the variation of the framework PO_4^{3-}- group stretching vibration at 900-1100 cm^{-1} not shown here for brevity. In the first approximation, for catalysts with strong interaction between the active component and support (samples 1 and 3), reaction media has no effect on the perovskite bands, though triggering further changes in the support structure. Thus, after reaction, for sample 1, a band at 466 cm^{-1} due to deformation of the Zr-O polyhedra emerges (Fig.3). This effect is nearly absent for sample 3, where Rh is added into the active component. Since for the latter sample a degree of oxygen conversion in the reaction mixture is higher (Table 2), this difference can be assigned to more reducing reaction conditions favorable for Ni segregation in the surface layer of the catalysts. For samples 2 and 4 with a less pronounced interaction between the active component and support, after reaction, the changes in the bands assigned to perovskite were clearly observed (Fig. 3).

3.3. Catalytic activity.

Methane conversion and selectivities were found to depend upon the nature of both active component and support (Table 2). Thus, La -containing amorphous support clearly favor C_2 formation (cf. samples 1 and 2). It can be explained by decline of the density of

Table 2. Catalytic properties of zirconium phosphates supported catalysts in POM.

Cata-lyst	Contact time, sec.	X_{CH4}	X_{O2}	S_{CO} %	S_{CO2}	S_{C2H4}	S_{C2H6}
1	0,144	28	73	10	84	1,5	4,5
	0,03	10	31	25	68	-	7
2	0,144	32	92	13	71	8	8
	0,03	14	24	25	52	6	17
3	0,144	38	93	45	54,6	-	0,4
	0,03	24	74	13	86	-	1
4	0,144	95	98	99	1	-	-
	0,03	70	99	85	15	-	-
5	0,144	98	98	96	4	-	-
	0.03	76	99	84	16	-	-

Lewis acid sites for amorphous La-containing zirconium phosphate (vide supra). Indeed, for methane dimerization, not acid but basic centers are required [11]. Addition of Rh or Ir causes the activity and S_{CO} to increase and S_{C2} to fall. The highest performance and CO selectivities were observed for samples 4 and 5 with the lowest degree of the active component -support interaction. It implies that strong interaction of lanthanum nickelate with the framework phosphate matrix somewhat suppresses formation of Ni^{o} species required for POM [12-13], and, hence, should be avoided. Even Rh addition to the active component is not able to counteract this effect (sample 3). However, for the case of moderate interaction with support (samples 4 and 5), precious metal addition makes catalysts very active and promising for the practical application. For the last two samples, the increase of CO selectivities with the contact time typical for precious metal performance [14], suggests formation in the reaction conditions of surface Rh(Ir)-Ni^{o} active species responsible for activity. The opposite trend typical for oxidic catalysts was revealed for unpromoted supported perovskite samples. For all catalysts tested, activity was stable, and no coking was observed.

CONCLUSIONS.

Dispersed framework zirconium phosphates of optimum composition can be used as supports for POM catalysts. To ensure good catalytic performance, the degree of the active component -support interaction must be tuned by varying the chemical composition and real structure of support.

RFERENCES

1. S. C. Tsang, J. B. Claridge, M. L. Green, Catal. Today, 23 (1995) 3.
2. R. Burch, S. C. Tsang, Current Opinion in Solid State and Materials Science, 2 (1997) 90.
3. R. Roy, D. K. Agrawal, and H. A. McKinstry, Annu. Rev. Mater. Sci., 19 (1989) 59.
4. J. Alamo and R. Roy, J. Mater. Sci., **21** (1986) 444.
5. J. F. Clover, D. K. Agrawal and H. A. McKinstry, J. Mater. Sci. Lett., **7** (1988) 422.
6. D. I. Kochubei, S. N. Pavlova, B. N. Novgorodov, G. N. Kryukova and V. A. Sadykov, J. Catal., **162** (1996) 500.
7. D. Briggs, M. P. Seah, Auger and X- Ray Photoelectron Spectroscopy for Analysis of Solid Surfaces, Mir, Moscow, 1987.
8. JCPDS, 1983, No. 33-321; No. 33-1360.
9. V. A. Sadykov, D. I. Kochubei, S. P. Degtyarev, E. A. Paukshtis, E. B. Burgina, S. N. Pavlova, R. I. Maximovskaya, G. S. Litvak, V. I. Zaikovskii, S. V. Tsybulya, V. F. Tretyakov, D. Agrawal, R. Roy, In: Recent Advances in Catalytic Materials, N. M. Rodrigues, S. L. Soled and J. Hrbek (eds.), Mater. Res. Soc. Symp. Ser., v. 497, MRS, Warrendale, PA USA, 1998, (in press)
10. H. Nakayama, T. Eguchi, N. Nakamura, Sh. Yamaguchi, M. Danjyo and M. Tsuhako, J. Mater. Chem., **7** (1997) 1063.
11. V. D. Sokolovskii and E. A. Mamedov, Cat. Today, 14 (1992) 415.
12. V. R. Choudhary, B. S. Uphade, and A. A. Belhekar, J. Catal., 163 (1996) 312.
13. D. Dissanayake, M. P. Rosynek, K. C. C. Kharas and J. H. Lunsford, J. Catal., 132 (1991) 117.
14. J. S. Slaa, R. J. Berger and G. B. Marin, Cat. Lett., 43 (1994) 63.

NATURAL GAS CONVERSION V
Studies in Surface Science and Catalysis, Vol. 119
A. Parmaliana et al. (Editors)

Mechanistic Study of Partial Oxidation of Methane to Syngas over a Ni/Al_2O_3 Catalyst*

Shikong Shen, Chunyi Li, Changchun Yu
(Lab. for Catalytic Conversion of Natural Gas, University of Petroleum, Beijing 102200)

Abstract: The mechanism of methane partial oxidation to syngas over an 8 mol% Ni/Al_2O_3 catalyst has been studied by means of transient response technique at atmospheric pressure and 973K. The results show that the metallic Ni sites are responsible for dissociation of methane to form H_2 and carbide (Ni_xC), and dissociation of oxygen to form NiO. CO is probably formed via the carbide reaction with surface NiO. Both H_2 and CO are the primary products of methane partial oxidation. In combination with pulse reaction experiments, it could be derived that methane partial oxidation to syngas follows the direct oxidation mechanism.

Keywords: methane, partial oxidation, syngas, Ni/Al_2O_3 catalyst, transient response, reaction mechanism

1. INTRODUCTION

In recent years, partial oxidization of methane (POM) to syngas has attracted much interest[1-10]. POM is a mild exothermic reaction and the reaction contact time is less than 10^{-2}s. The selectivities of both CO and H_2 are higher than 95%. The ratio of H_2/CO is about 2, which is suitable for methanol and F-T synthesis. Comparing with traditional steam-reforming and combined reforming, POM has the advantages of high efficiency and low energy consumption. It was reported that Exxon had finished the development of a new process of POM with fluidized bed reactor using Ni-based catalyst [1].

There are two main different POM reaction mechanisms on Ni-based catalysts. Dissanayake *et al* investigated POM over Ni/Al_2O_3 catalyst by means of XPS and suggested that CH_4 combusts with O_2 to CO_2 and H_2O first, then both CO_2 and H_2O reform with CH_4 to syngas[4]. Looij *et al* held similar viewpoints by XPS and HTXRD studies on Ni/SiO_2 catalyst[5]. However, Hu *et al* inferred that CO is the primary product, and CO_2 is generated from CO according to the delay of transient response of CO_2 in comparison to that of CO over Ni/La_2O_3[6]. They also thought that the reaction between surface oxygen species and surface carbon species is the rate-controlling step.

In the current work, we obtained further evidence for direct oxidation mechanism of POM over an 8mol %Ni/Al_2O_3 catalyst by means of transient response technique combining with pulse reactions at atmospheric pressure and 973K.

* Supported by the National Natural Science Foundation of China, Issue No. 29673027

2. EXPERIMENTAL

2.1. Catalyst preparation

The 8mol% Ni/Al_2O_3 catalyst was prepared by co-precipitating 1mol/L mixture solution of $Ni(NO_3)_2$ and $Al(NO_3)_3$ using a 1 mol/L NH_4OH solution as precipitant. The pH of the solution was maintained at 7.5. The precipitate was aged for 6 hours, then filtrated, washed with distilled water, and dried at 373K for 10 hours, calcined in air at 673K for 10 hours, pulverized and sieved to 40-60 mesh for experimental usage.

2.2. Experimental method of transient response

The experimental apparatus for transient response and pulse reaction is shown in fig.1, which includes three main parts: a gas flow system, a reactor with temperature controller and an online-detection mass spectrometer (AMTEK QuadLink 1000). The reactor, which is an i.d. ϕ5.5mm and 400mm in length quartz tube, is placed in an electric furnace. The temperature of the furnace is controlled by an AI-FUZZY temperature controller using a K type thermal couple. 30mg Ni/Al_2O_3 catalyst is packed in the middle of the reactor and the free space of the reactor is filled with 20-40 mesh quartz particles. The quadrupole mass spectrometer can detect 8 mass channels simultaneously with minimum dwell time of 3 milli-second.

The oxidization treatment of the catalyst was performed in 17mol% O_2/He flow of 20ml/min at 973K for 30 minutes. The reduction treatment of the catalyst was performed in 17mol% H_2/He flow of 20ml/min at 973K until the intensity of H_2O down to its baseline. Both transient responses and pulse reactions were carried out in a 20ml/min total gas flow rate, atmospheric pressure and 973K.

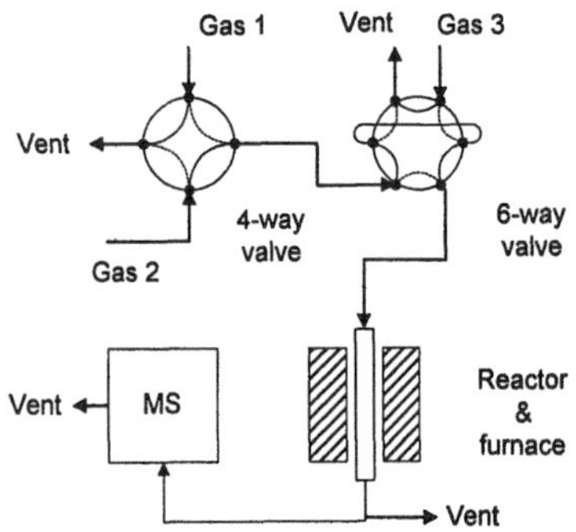

Fig.1 Schematic of the transient response kinetic analysis system

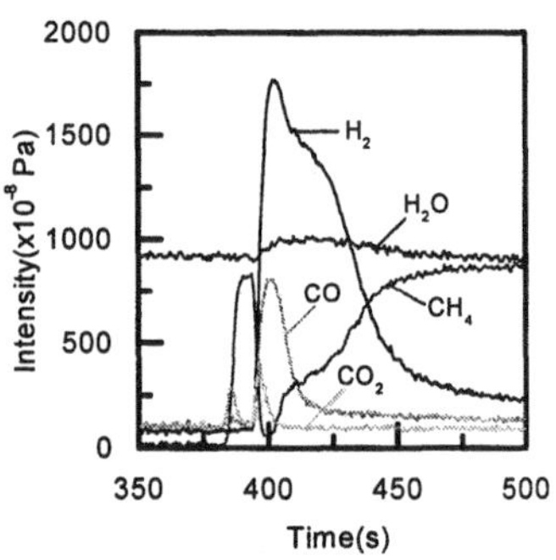

Fig.2. Responses after the switch: O_2/He → CH_4/He.

3. RESULT AND DISCUSSION

3.1. Methane reaction on oxidized and reduced surface of the catalyst

The responses of a step switch from 17mol% O_2/He flow to 17mol %CH_4/He flow are shown in fig.2. Methane intensity increases rapidly and reaches a steady level after switch, and maintains this level for 5-6 seconds which is an inducing period. The treating time of the catalyst by 17mol% O_2/He flow at 973K does not have any effect on the inducing period.

However, when the catalyst has been swept by He flow after the oxidation, no inducing period can be observed. The phenomenon may be due to the desorption of a little surface oxygen weakly bonded with nickel leading to the formation of metallic nickel sites for the dissociation of methane.

The methane intensity falls abruptly down to the baseline and keeps for 4-5 seconds after the period, then increases slowly to a steady level. Meanwhile, a large amount of H_2, CO, and a small amount of CO_2 and H_2O appear together with the methane intensity decrease. According to the previous XPS researches, nickel in the oxidized surface was existed as NiO[11,12]. Methane not dissociation immediately after the switch indicates that CH_4 direct reaction with NiO is difficult. However, once a little NiO is reduced to metallic nickel, methane would dissociate rapidly on Ni sites and form H_2 and surface carbon species, which have been proved to be carbide species (Ni_xC, x=1-3) by XPS[11,12]. The carbide species can react quickly with surface NiO to CO and/or CO_2, the ratio of CO to CO_2 is determined by the concentration of NiO around Ni_xC on the catalyst surface. The higher concentration of NiO, the more CO_2 is produced; otherwise, the more CO is produced. The surface reactions after the switch occur as follows:

$CH_4 + xNi \longrightarrow Ni_xC + 2H_2$ (main reaction) (1)
$H_2(2H) + NiO \longrightarrow Ni + H_2O$ (minor reaction) (2)
$Ni_xC + NiO \longrightarrow CO + (x+1)\,Ni$ (main reaction) (3)
$Ni_xC + 2NiO \longrightarrow CO_2 + (x+2)Ni$ (minor reaction) (4)

During the first stage of methane dissociation, the carbide carbon from CH_4 dissociation does not accumulate on the surface because Ni sites can be restored by the reaction between carbide and NiO.

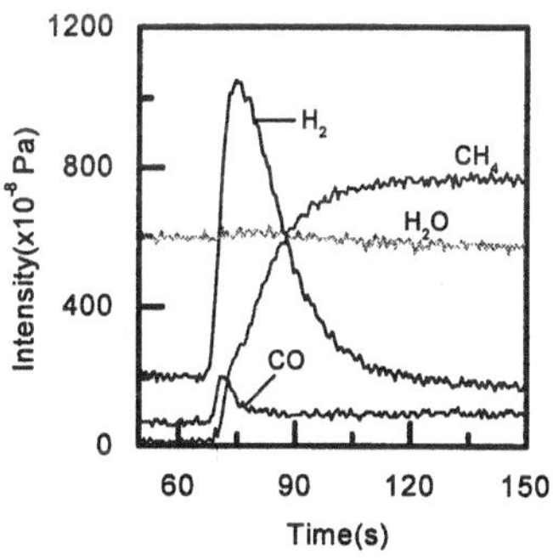

Fig.3. Responses after the switch: He → CH_4/He over the catalyst pre-reduced by H_2/He at 973K.

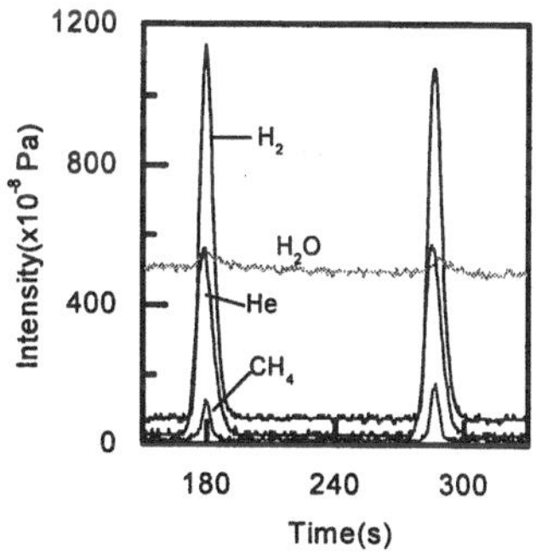

Fig.4. Pulse reactions of CH_4/He in N_2 gas flow over the catalyst pre-reduced by H_2/He at 973K.

The result of transient response from He flow to 17mol% CH_4/He flow over the catalyst pre-reduced by 17mol% H_2/He flow at 973K is shown in fig.3. After the switch, a large amount of H_2 and a small amount of CO are produced immediately, and the methane intensity increases slowly to a steady level. The results show that methane dissociates easily to H_2 and surface carbide species on Ni sites, and a small CO peak is probably come from the reaction between surface carbide and surface NiO which is still not reduced by H_2.

Pulse experiments of 17mol% CH_4/He in N_2 flow on the reduced catalyst is shown in fig.4. A great deal of H_2 appears immediately together with the inert reference gas He. It provided a further evidence for methane dissociating easily to H_2 and surface carbide on the Ni sites.

Certainly, it can not be excluded that methane dissociation is very fast through the consecutive dehydrogenation steps: $CH_4 \rightarrow CH_3 \rightarrow CH_2 \rightarrow CH \rightarrow C$. Hall *et al* reported that methane dissociates and methyl C-H breaks on Ni(111) and Ni(100) at around 220K[13]. Because in our experiment 0.2-second time resolution of the mass spectrometer can not observe the consecutive dehydrogenation steps of methane even if methane dissociation is step by step. The above results just show that methane dissociation is fast at 973K on Ni sites.

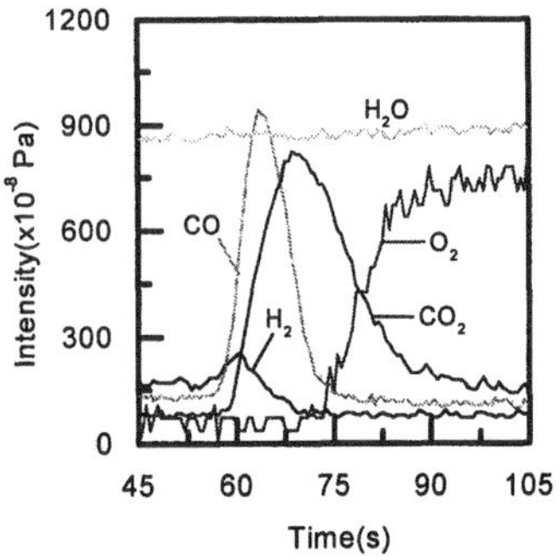

Fig.5. Responses after the switch: CH_4/He $\rightarrow$ O_2/He.

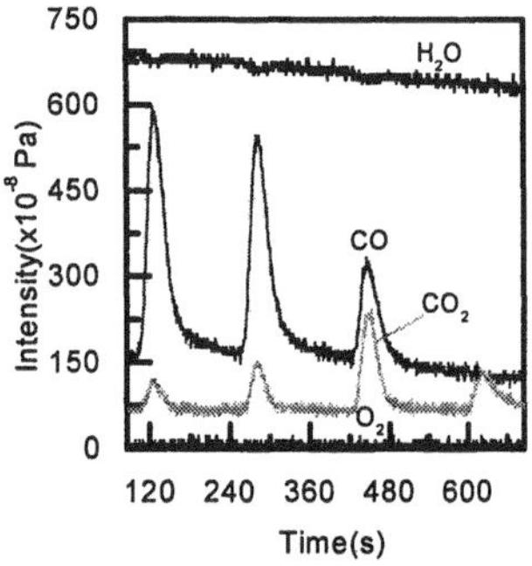

Fig.6. Pulse reactions of 17mol%O_2 /He in He flow over the catalyst pre-treated by CH_4/He for 20min at 973K.

3.2. Surface carbide species reaction with O_2 and CO_2

To investigate the reaction between surface carbide and O_2, a step switch from 17mol% CH_4/He flow to 17mol% O_2/He flow was carried out (fig.5). A large amount of CO and CO_2 appear immediately after the switch, but O_2 intensity gradually goes up about 10 seconds later. The intensity of CO reaches its maximum earlier than that of CO_2. It seems that CO and CO_2 are not from the reaction between surface carbide and gas phase oxygen, but from the reaction between surface carbide and NiO. The ratio of CO to CO_2 is determined by the relative concentrations of surface carbide to its nearby NiO. In order to further demonstrate this viewpoint, 17mol% O_2/He pulse reactions in He flow were performed (fig.6). The result shows that CO is the main product of the first two pulses of 17mol O_2/He in He flow over 17mol% CH_4/He treated catalyst at 973K. However, the concentration of surface carbide decreases with the increase of pulse number, while the corresponding NiO concentration increases. Thus, the ratio of CO/CO_2 and the total quantity of CO+CO_2 decrease with the pulse number.

Fig.7 shows that a large amount of CO was produced when switching 17mol% CH_4/He flow to 17mol% CO_2/He flow. It indicates that CO_2 can also react with surface carbide and produce CO.

$$CO_2 + Ni_xC \longrightarrow 2CO + xNi \quad (5)$$

However, it is not sure whether adsorbed or gas phase CO_2 reacts with surface carbide according the present results.

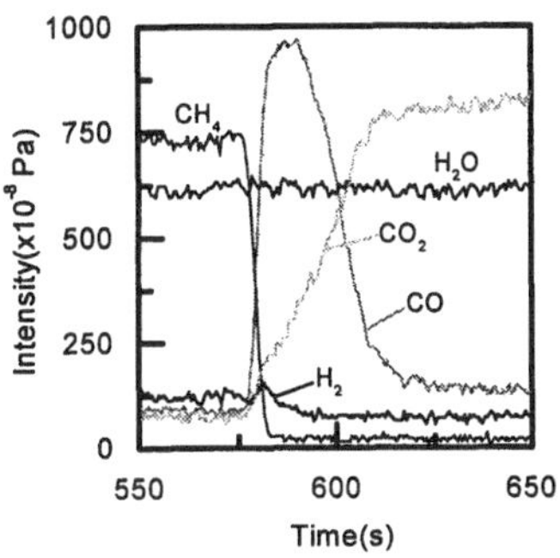

Fig.7. Responses after the switch: CH_4/He → CO_2/He.

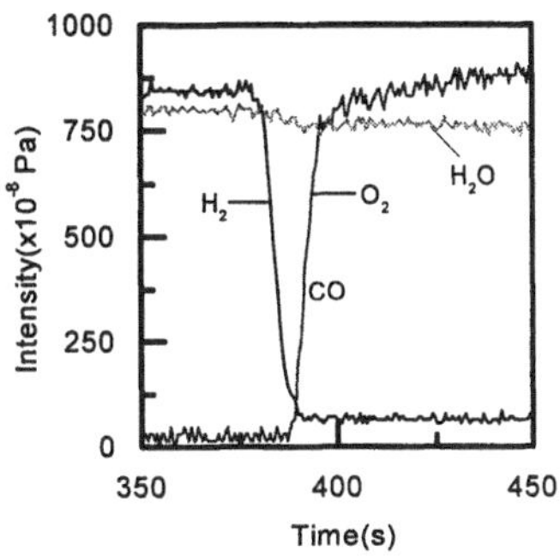

Fig.8. Responses after the switch: H_2/He → O_2/He.

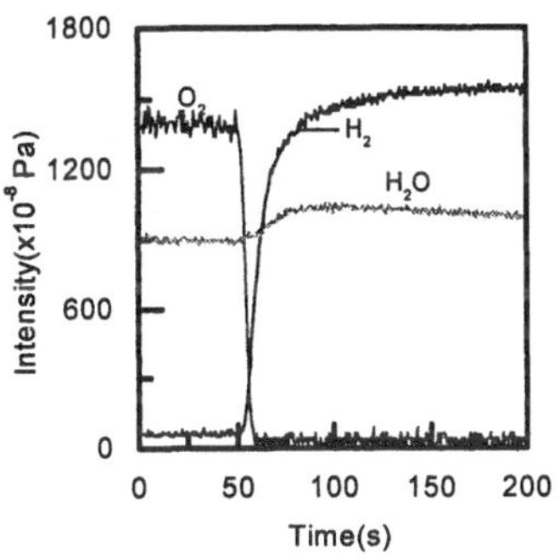

Fig.9. Responses after the switch: O_2/He →H_2/He.

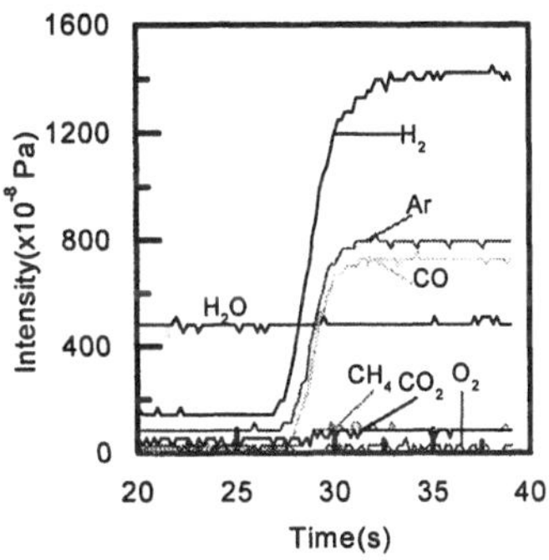

Fig.10. Responses after the switch: He→ CH_4/O_2/He/Ar over the catalyst pre-reduced at 973K.

3.3. H_2 reaction with NiO on catalyst surface

Fig.8 shows the switch from 17 mol%H_2/He flow to 17 mol%O_2/He flow, in which O_2 appears after H_2 coming down to the baseline, no obvious water can be observed. The results demonstrate that H_2 desorbed from the catalyst surface is very fast at high temperature and oxidation reaction of surface Ni occurs in 10 seconds after the switch.

$$O_2 + 2Ni \longrightarrow 2NiO \quad (6)$$

Fig.9 shows the responses of switching 17 mol%O_2/He flow to 17 mol%H_2/He flow. It indicates that most of NiO is easily reduced by H_2, whereas a small amount of nickel oxide is difficultly reduced, which may be a little spinel nickel aluminate in the catalyst. This opinion has been further proved by the result of pulse of 17 mol% H_2/He in He flow over the 17mol% O_2/He oxidized catalyst at 973K, because only trace water appears corresponding to each H_2 pulse.

3.4. CH_4/O_2 reaction over the reduced catalyst

The responses of the switch from He flow to CH_4/O_2/He/Ar (10/5/84/1, mol) flow (fig.10) shows that H_2 and CO appear together with inert tracer gas argon. It indicates that both H_2 and CO are primary products of partial oxidation of methane. Comparing with CO, the a small amount of corresponding CO_2 was observed and appeared a little delay, because CO_2 formation requires higher surface concentration of NiO.

4. CONCLUSION

In summary, the mechanism of partial oxidation of methane to syngas over the Ni/Al_2O_3 catalyst is illustrated in fig.11.Oxygen reacts with surface metallic nickel sites to form NiO, and CH_4 dissociates on the surface metallic nickel sites to form surface carbide and liberate H_2 immediately. Then surface carbide reacts with its nearby NiO to produce CO and CO_2. The reason of high selectivity of CO is duo to the fact that concentration of surface carbide is higher than concentration of surface NiO under the co-feed of stoichiometric CH_4/O_2 ratio at 973K and atmospheric pressure.

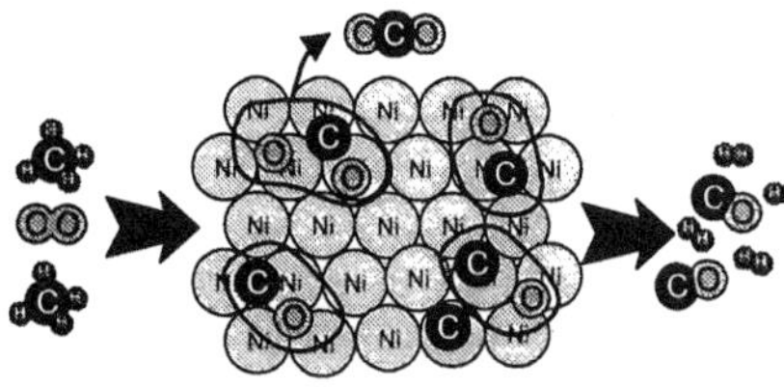

Fig.11 The mechanism schematic of partial oxidation of methane

REFERENCES

1. A.G. Duane, R.S. Geoffrey, Exxon Research and Engineering Company, US Patent No. 4877550 (1989)
2. A.T. Ashcroft, A.K. Cheetham, J.S. Foord, et al, Nature, 344 (1990) 319
3. P.D.F. Vernon, M.L.H. Green, A.K. Cheetham, A.T. Ashcroft, Catal. Lett., 6 (1990) 181
4. D. Dissanayake, M.P. Rosynek, J.H. Lunsford, J. Catal., 132 (1991) 117
5. F. van Looij, J.W. Geus, *J. Catal* 168 (1997) 154.
6. Y.H. Hu, E. Ruckenstein, *J. Catal* 158 (1996) 158.
7. D.A. Hickman, E.A. Haupfear, L.D. Schmidt, Catal. Lett., 17 (1993) 223
8. D.A. Hickman, L.D. Schmidt, AIChE J., 39 (1993) 1164
9. S.S. Bharadwaj, L.D. Schmidt, J. Catal., 146 (1994) 11
10. A.G. Dietz, L.D. Schmidt, Catal. Lett., 33 (1995) 15
11. C.C. Yu, S.K. Shen, J. Chem. Phys. (Chinese), 10 (1997) 233
12. C.C. Yu, Y. Lu, X.J. Ding, S.K. Shen, Stud. Surf. Sci. Catal., 107 (1997) 503
13. R. B. Hall, M. Castro, C. M. Kim, C. A. Mims, Stud. Surf. Sci. Catal., 101 (1996)327

NATURAL GAS CONVERSION V
Studies in Surface Science and Catalysis, Vol. 119
A. Parmaliana et al. (Editors)

Improved Stability of Nickel-Alumina Aerogel Catalysts for Carbon Dioxide Reforming of Methane

Jin-Hong KIM[a], Dong Jin SUH[b]*, Tae-Jin PARK[b], and Kyung-Lim KIM[a]

[a]Department of Chemical Engineering, Yonsei University, 134 Shinchon-dong, Sudaemoon-ku, Seoul 120-749, Korea

[b]Clean Technology Research Center, Korea Institute of Science and Technology, P.O.Box 131, Cheongryang, Seoul 136-791, Korea

CO_2 reforming of methane was carried out over nickel-alumina aerogel catalysts prepared with various Ni loadings. Compared with alumina-supported Ni catalyst prepared by conventional impregnation method, nickel-alumina aerogel catalysts exhibited remarkably low coking rate due to highly dispersed metal particles. From TEM studies, it was observed that the metal particle size significantly influenced the formation of filamentous carbon. Addition of potassium to the nickel-alumina aerogel catalysts appeared to improve the stability of the catalyst at 923 K while the effect was not clear at higher temperatures.

1. INTRODUCTION

CO_2 reforming of methane to produce synthesis gas with low H_2/CO ratio recently attracted a lot of interest. Environmentally, this reaction is attractive due to the reduction of both greenhouse gases which may be converted into useful feedstock for further chemical processes. However, the major obstacle to the successful industrial application is the formation of coke which deactivates the catalyst. Although the catalysts based on noble metals are reported to be less susceptible to coking [1], it is still worthwhile, from the industrial point of view, to develop stable nickel-based catalysts. Recently, we proposed a fast synthetic route for obtaining high-surface-area alumina aerogels and extended it successfully to nickel-alumina aerogel catalysts, which showed good activity and excellent stability in CO_2 reforming of methane [2,3]. In this work, we have tried to optimize catalytic properties of nickel-alumina aerogel catalysts by varying Ni/Al mole ratios in the preparation step and by the addition of potassium as promoter.

2. EXPERIMENTAL

Nickel-alumina aerogel catalysts were prepared by the sol-gel processing of nickel acetate and aluminum *sec*-butoxide (ASB) in ethanol and subsequent supercritical drying with carbon dioxide at 333 K and 24 MPa (designated as SAA*x*, *x* means the Ni/Al mole ratio multiplied by 100). The dried aerogel was subjected to a standard calcination procedure which consisted of heating in helium at 573 K and in oxygen at 773 K [4]. Further details of the procedure have been described previously [2,3]. For comparison purposes, alumina-supported 10 wt.% nickel catalysts were prepared by impregnating a commercial alumina (supplied by Strem Chemicals, specific surface area 362 m^2/g) with nickel nitrate solution (designated as ICN). Potassium-doped nickel-alumina aerogel catalysts were prepared by impregnating SAA25 with potassium acetylacetonate in ethanol.

The catalysts were characterized by XRD (Rigaku D/MAX-Ⅲ A diffractometer, CuKα radiation), XPS (SSI 2803-S), TEM (Philips CM 30), and N_2 adsorption-desorption isotherm (ASAP 2000, Micromeritics) at 77 K. Total carbon contents were determined with a Perkin-Elmer 240DS elemental analyzer in order to quantify the amount of carbonaceous deposits on the catalysts after reaction.

The reforming reaction was carried out in the temperature range of 873 to 1073 K at atmospheric pressure using a flow-through quartz reactor. The catalyst (50mg) was activated in hydrogen flow for 2 h at 973 K prior to the catalytic measurements. The feed composition was $CH_4/CO_2/N_2$ = 1/1/2 and the total flow rate 40 ml/min.

3. RESULTS AND DISCUSSION

3-1. Nickel-alumina aerogel catalysts

Table 1. Properties of impregnated nickel/alumina catalyst and nickel-alumina aerogel catalysts after H_2 treatment at 973 K for 2 h

Catalysts	Textural properties			Bulk phase [d]	Composition R = Ni/(Ni + Al)	
	S_{BET} [a]	V_P [b]	D_P [c]		Bulk [e] (R_B)	Surface [f] (R_S)
ICN	145	0.25	6.9	Ni (fcc)	-	0.064
SAA10	436	3.37	30.9	Amorphous	0.09	0.029
SAA15	448	3.21	30.4	Amorphous	0.13	0.062
SAA20	439	3.02	27.5	Poorly crystallized	0.17	0.067
SAA25	406	2.33	23.0	Poorly crystallized	0.20	0.063

[a] BET surface area (m^2/g), [b] total pore volume (cc/g), [c] average pore diameter (nm).
[d] bulk phase identified by XRD, [e] theoretical value, [f] surface composition calculated by XPS analysis.

Table 1 shows textural properties, bulk phases and surface Ni/Al mole ratios of the nickel-alumina aerogel catalysts pretreated in hydrogen at 973 K. Compared with the ICN catalyst, the SAA series had much favorable textural properties for catalytic use and were X-ray amorphous at relatively low metal loadings. This result indicated that the nickel particles contained in the nickel-alumina aerogel catalysts were well dispersed even after hydrogen treatment at 973 K. According to our TEM studies, the preparation of alumina-supported Ni catalysts via sol-gel synthesis and subsequent supercritical drying led to the formation of very small Ni particles which were homogeneously distributed over the support. Compared with ICN which had nickel particles distributed in the broad range of 5 to 30 nm, very small nickel particles of about 4 nm in diameter were homogeneously formed on the SAA20 catalyst even though its surface composition was higher than ICN.

The catalytic reaction data were summarized in Table 2. The ICN catalyst deactivated immediately with the coke formation and in a few hours no further reaction could be carried out by operational problems such as the blockage of the pore mouth and eventually blockage of the reactor tube. On the other hand, the nickel-alumina aerogel catalysts, regardless of different nickel contents, exhibited good activity and excellent stability at least during 30 h of reaction. Moreover, the SAA25 catalysts showed slightly higher CO_2 conversion than a 5 wt.% Ru/alumina catalyst (ESCAT44 from Engelhard). This indicates that highly dispersed nickel particles prepared by sol-gel process could help to enhance activity and decrease coking rate.

Table 2. Activities and coke forming rates of SAA series catalysts [a]

Catalysts	CH_4 conversion (%)		CO_2 conversion (%)		Rate of coke formation [b]	Amount of filamentous carbon
	1 h	30 h	1 h	30 h		
SAA10	61	47	66	54	1.53×10	Not detected
SAA15	64	55	69	62	6.81×10	Very small
SAA20	69	66	73	67	1.07×10^2	Small
SAA25	70	66	75	71	2.50×10^2	Moderate
ICN [c]	66	-	68	-	3.11×10^3	Serious
ESCAT44	72	71	74	72	6.84	Not detected

[a] Reaction conditions; 973 K, 1 atm, $CH_4/CO_2/N_2 = 1/1/2$, $W/F = 1.25 \times 10^{-3}$ g-cat · s/ml.
[b] calculated from total carbon contents in the used catalysts (μmol C/g-cat · h),
[c] reaction was terminated within 4–5 h.

From our TEM studies, it was clear that the type of surface carbon formed during the reaction depended on metal particle size. In the case of ICN, whisker carbon formed seriously on the large metal particles during the reaction, but no evidence of whisker carbon formation was observed at the surface of the SAA10 catalyst. The SAA15 catalyst with a mean particle size of 3.3 nm could form very small amount of filamentous carbon, but formation of carbon filament was accelerated with a Ni/Al mole ratio in excess of 0.2. Therefore, it is likely that a minimum diameter is required for nickel particles to generate carbon filaments as reported by Duprez et al [5].

Sintering seemed to be another important cause of filamentous carbon formation. After H_2 treatment, the SAA20 catalyst had very small nickel particles, but some particles sintered during the reaction and the sintered metal particles are likely to offer the sites for carbon filament growth. From this result, formation of large metal particles in preparation or pretreatment step and sintering during the reaction may be two important factors to induce filamentous carbon formation. If seriously formed, filamentous carbon could affect the pore size distributions of the used catalysts.

Figure 1 shows typical changes of pore size distribution. For the ICN catalyst, new pores, which were not present before reaction, were observed after reaction and these newly formed pores of above 30 nm were caused by formation of fabric consisted of whisker carbon and fractured nickel particles. Although we could also observe filamentous carbon in the SAA25 catalyst after reaction, they were not densely formed and had the radius of 5–6 nm, which were about half of those on the ICN catalyst. Because of this difference, the SAA25 catalyst did not show any noticeable changes of pore size distribution after reaction as the other nickel-alumina aerogel catalysts.

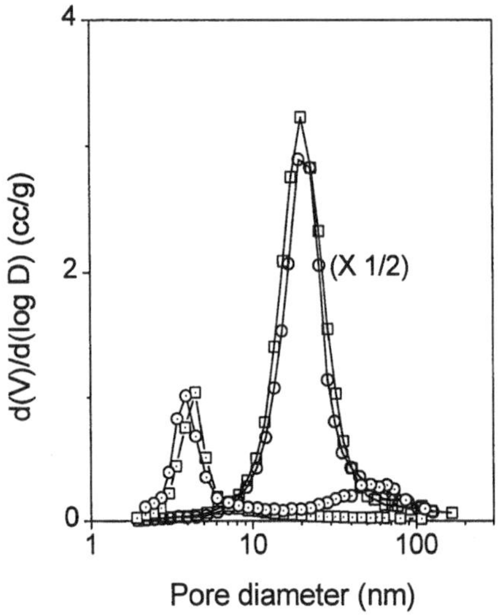

Fig. 1. Effect of filamentous carbon formation on pore size distribution: (□) before reaction; (○) after reaction; open symbol, SAA25; dot-centered symbol, ICN.

Figure 2 shows changes of CO yields during the reforming reaction. At 873 K, with the increase in nickel content, CO yield markedly decreased during the reaction. This deactivation can be explained by rapid coke formation on the catalyst surface at low temperature which is the thermodynamically favorable condition for the coke formation, especially at high nickel content. However, at above 973 K, CO yield decreased dramatically as the nickel loading decreased. As shown in Table 2, at low nickel content, carbon deposition may not fully explain this substantial loss of activity. This deactivation may be caused by thermal degradation of active sites at high temperature, in relation to sintering or the interaction of small nickel particles with neighboring Al atoms.

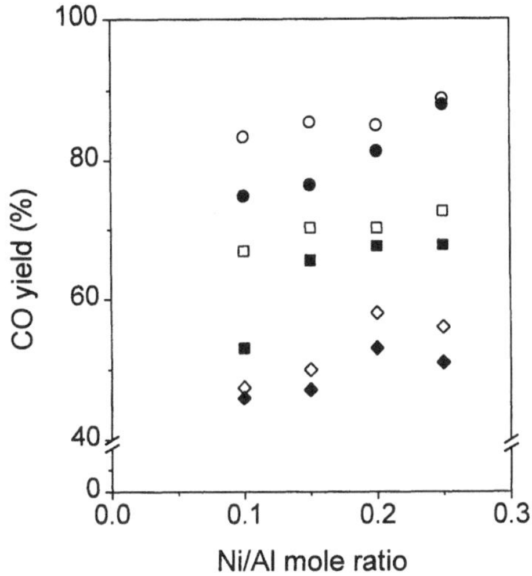

Fig. 2. Changes of CO yield during the reaction at different temperatures: (○) at 1073 K; (□) at 973 K; (◇) at 873 K; open symbol, after 1 h; closed symbol, after 30 h (at 873 K, after 5h).

3-2. Potassium-doped nickel-alumina aerogel catalysts

Nickel-alumina aerogel catalysts showed good activity and excellent stability during CO_2 reforming reaction. However, carbon formation at low temperatures still remained as an obstacle to be overcome, especially in the case of the SAA25 catalyst with high Ni loading. The effects of potassium addition on the catalytic activity and the rate of coke formation were summarized in Table 3. As shown in Table 3, at 973 K, potassium-doped on SAA25 catalysts did not show the promotional effect clearly. The highest CH_4 conversion was obtained for the 5 wt.% potassium-doped catalyst while no further improvement in CO_2 conversion and carbon formation rate was observed. However, promotional effect of potassium became clear in the reaction at lower temperatures. At 923 K, the SAA25 catalyst, without potassium-doping, showed rapid deactivation due to coke formation within 10 h, and no further reaction could be carried out due to the clogging of the reactor tube as in the case of the ICN catalyst at 973 K.

Table 3. Effect of potassium on activities and carbon deposits of nickel-alumina aerogel catalysts (SAA25) [a]

K content (wt. %)	CH_4 conversion (%)		CO_2 conversion (%)		H_2/CO	Rate of Coke formation [b]
	1 h	30 h	1 h	30 h		
0	70	66	75	71	1.03	2.50×10^2
3	70	68	73	71	1.05	2.79×10^2
5	74	71	75	73	1.03	3.94×10^2
10	71	68	73	70	1.05	2.56×10^2

[a] Reaction conditions; 973 K, 1 atm, $CH_4/CO_2/N_2 = 1/1/2$, $W/F = 1.25 \times 10^{-3}$ g-cat · s/ml.
[b] calculated from total carbon contents in the used catalysts (μmol C/g-cat · h).

On the other hand, SAA25 catalysts containing 5–10 wt. % potassium showed higher activity and stability than the raw SAA25 catalyst. Such temperature-depending behavior of potassium-doped SAA25 catalysts indicates that the role of potassium has a certain relationship with changing reducibility or metal-support interactions. Hence, it is likely that the promotional effect caused by potassium is related to the increased chance of CH_4 to access to nickel particles and the role of potassium is to trap CO_2 and to activate it. This consideration is somewhat in line with other studies on the promotional effect of alkali metal or basic oxides [6,7].

In conclusion, further optimization of nickel-alumina aerogel catalysts may offer interesting possibility to reduce catalyst coking and allow to increase the economic viability of CO_2 reforming of methane.

REFERENCES

1. A. T. Ashcroft, A.K. Cheetham, M. L. H. Green, and P. D. F. Vernon, *Nature*, **352** (1991) 225.
2. D. J. Suh, T.-J. Park, J.-H. Kim, and K.-L. Kim, *Chem. Mater.*, **9** (1997) 1903.
3. D. J. Suh, T.-J. Park, J.-H. Kim, and K.-L. Kim, *J. Non-Cryst. Solids.*, in press.
4. D. J. Suh and T.-J. Park, *Chem. Mater.*, **8** (1996) 509.
5. D. Duprez, M. C. Demicheli, P. Marecot, J. Barbier, O. A. Ferretti, and E. N. Ponzi, *J. Catal.* **124** (1990) 324.
6. F. Solymosi, *J. Mol. Catal.*, **65** (1991) 337.
7. A. Erdohelyi, J. Cserenyi, and F. Solymosi, *J. Catal.* **141** (1993) 287.

NATURAL GAS CONVERSION V
Studies in Surface Science and Catalysis, Vol. 119
A. Parmaliana et al. (Editors)

Influence of molybdenum and tungsten dopants on nickel catalysts for the dry reforming of methane with carbon dioxide to synthesis gas

Andrew P. E. York, Thomas Suhartanto and Malcolm L. H. Green*

The Catalysis Centre, Inorganic Chemistry Laboratory, University of Oxford, South Parks Road, Oxford, OX1 3QR, U. K.

The reforming of methane with carbon dioxide using molybdenum or tungsten oxide doped nickel catalysts has been studied. The nickel catalyst doped with a small amount of Mo deactivated, although no deposited carbon was detected at low dopant levels. In contrast, no deactivation was seen for the W doped sample, and a significant decrease in deposited carbon, compared with the undoped samples, was observed. The inhibition of carbon formation by introducing oxides of Mo or W is explained by showing that the carbides of molybdenum and tungsten are also active for methane dry reforming. It is proposed that carbide sites are formed in close proximity to nickel sites during the dry reforming over Mo/W doped Ni/Al_2O_3, and that these sites enhance the rate of carbon dioxide dissociation and thus provide more surface oxygen for reaction with the carbon formed over nickel.

1. INTRODUCTION

The dry reforming reaction to produce synthesis gas from methane and carbon dioxide *(1)* was first reported by Fischer and Tropsch in 1928 [1], but has recently enjoyed a renaissance due to the potential future importance of synthetic fuels produced from synthesis gas [2-5].

$$CH_4 + CO_2 \rightarrow 2\,CO + 2\,H_2 \qquad \Delta H^{\circ}_{298} = +247\ \text{kJ mol}^{-1} \qquad (1)$$

Nickel based catalysts are reported to be suitable for this reaction and commercially available. The major drawback is that these catalysts also catalyse carbon deposition by methane decomposition *(2)* or the Boudouard reaction *(3)*, leading to potential catalyst breakdown and deactivation, and reactor clogging [6].

$$CH_4 \rightarrow C_{(s)} + 2\,H_2 \qquad (2)$$
$$2\,CO \rightarrow C_{(s)} + CO_2 \qquad (3)$$

A number of methods for reducing the amount of carbon formed on nickel catalysts during reforming reactions have been found. For example, the addition of alkali and alkaline-earth

* We would like to thank CANMET for supporting this work and for funding to APEY. TS is grateful to Pertamina for a grant.

metal dopants [7-10], such as K, Na, MgO, CaO, or lanthanide/actinide dopants, such as UO_2 and U_3O_8 [10], and La_2O_3 [11], all lead to a more carbon resistant catalyst, in most cases due to the promotion of CO_2 dissociation resulting in more surface oxygen for reaction with carbon. It is also possible to inhibit the size of the nickel ensembles, since large ensembles have been implicated in carbon formation [12,13]; the most common method for ensemble size control is sulphur passivation, which is used in the industrial SPARG process, but the catalyst activity is rather low and sulphur must be consistently introduced to the reactor [14]. Other poisons and alloy forming components have also been investigated [12,15-17], e.g. Sn, Pb, Sb, Bi, Cu, Pt and Co, though in all cases activity was affected detrimentally. Another approach is to synthesise nickel catalysts with very small highly dispersed nickel particles [18]. Alternatively, it is possible to use supported group VIII metals of the second and third transition series for the methane reforming reactions, giving high yields of synthesis gas, without carbon formation [2,3,19,20].

Recently, Borowiecki and Golebiowski showed that the addition of molybdenum or tungsten oxide to a supported nickel material lead to an improvement in the catalytic activity and stability for steam reforming [21]; no explanation for this phenomenon was given. In this paper we report the effect of molybdenum or tungsten oxide dopants on a nickel catalyst used for the CO_2 reforming of methane. In addition, some results for the use of molybdenum and tungsten carbide for methane dry reforming are presented, and a possible connection between the Ni-MoO_3/WO_3 and Mo_2C/WC systems is outlined.

2. EXPERIMENTAL

2.1. Catalyst preparation and characterisation

The Al_2O_3 supported nickel catalysts were prepared by wet impregnation: $NiCl_2 \bullet 6H_2O$ (BDH) was added to Al_2O_3 (Akzo, high purity, 250-355μm), dried at 170°C for 24 h and calcined at 500°C for 3h in air. The Mo and W doped catalysts were prepared by adding the appropriate amount of $(NH_4)_6Mo_7O_{24}.4H_2O$ (Fisons) or $(NH_4)_{10}W_{12}O_{41}.5H_2O$ (Johnson-Matthey) to the calcined Ni/Al_2O_3 material, followed by drying at 170°C for 24 h and calcining at 500°C for 3h in air. Before reaction all catalysts were dried at 200°C overnight under N_2, then reduced at 700°C in H_2 for 1h. Mo_2C and WC were prepared by temperature programmed reduction (TPR) of the corresponding oxides, under flowing 20% v/v CH_4/H_2 (150 ml min^{-1}), at a heating rate of 1°C min^{-1} from room temperature to 750°C and 880°C, respectively. Normally the catalysts were prepared *in situ* and tested immediately.

N_2 BET isotherms were obtained using an all glass high vacuum line. Temperature programmed oxidation (TPO) was carried out in flowing O_2, ramping the temperature from 150-850°C at 10°C min^{-1}. The oxidation products were then analysed using a Hewlett Packard mass spectrometer, calibrated for carbon dioxide by $CaCO_3$ decomposition.

2.2. Apparatus

Catalysts were tested for dry reforming in microreactors built using 1/8" and 1/16" o.d. 316 stainless steel tubing and 316 stainless steel Swagelok fittings throughout. The catalyst sample was placed between two quartz wool plugs in the centre of a 4 mm i.d. silica tube, and inserted into a vertical Severn Science tube furnace. This was heated to the required reaction temperature using a Eurotherm 905 temperature controller. For safety reasons, in experiments carried out at elevated pressures the silica tube was placed inside a steel tube. Inlet gas flow rates were controlled using Brooks 5850TR mass flow controllers. The exit gas stream from the reactor passed through a Tescom two stage back-pressure regulator to allow elevated pressure experiments to be carried out. All the pipework was heated to prevent condensation of the products.

Methane (Union Carbide, >99.95%), hydrogen (BOC, C.P. grade), argon (BOC, C.P. grade), and carbon dioxide (BOC, C.P. grade) were used as received without further purification.

2.3. Product Analysis

Product analysis was carried out using Hewlett-Packard 5890II and 5880A gas chromatographs, fitted with both a thermal conductivity detector, and a methanator/flame ionisation detector. Separation of the products was achieved using 3m Porapak Q, and 3m Porapak T and Hayesep R packed columns, respectively, with argon carrier gas allowing accurate analysis of hydrogen. Reference data and pure component injections were used to identify the major peaks, and response factors for the products and reactants were determined and taken into account in the calculation of the conversion and product distribution. In all cases stoichiometric gas mixtures were used and carbon balances were better than 97%.

3. RESULTS AND DISCUSSION

3.1. Dry reforming over MoO_3 and WO_3 doped alumina supported nickel catalysts

The results obtained for the dry reforming of methane over nickel supported on alumina, and the molybdenum and tungsten doped supported nickel catalysts are presented in Table 1. In all cases the results are those obtained at the start of the experiments. The 2% Ni/Al_2O_3 catalyst was found to be the most active, and the most stable, of a number of supported nickel catalysts tested, e.g. TiO_2, SiO_2, ZrO_2. This catalyst gave high CH_4 and CO_2 conversions, in close agreement with those expected from thermodynamic equilibrium calculations, however, after 150h some catalyst deactivation had occurred, the conversions dropping to around 60% and 70% for CH_4 and CO_2, respectively. TPO of the post-catalytic sample showed that a large amount of carbon was had been deposited on the catalyst during reaction, and HRTEM analysis confirmed that whisker and encapsulate carbon was present on the catalyst, in agreement with previous studies on carbon formation during methane reforming [14,19].

The addition of MoO_3 or WO_3 to the 2% Ni/Al_2O_3 catalyst resulted in markedly different changes in the catalytic behaviour for the dry reforming of methane; previously it was reported that their effects were similar for methane steam reforming [21]. The addition of a small amount of Mo resulted in a decrease in carbon formation, and no change in the initial CH_4 and CO_2 conversion. However, deactivation was observed in the higher doped samples, perhaps due to spreading of MoO_3 and coating of the Ni particles leading to a less active catalyst; this phenomenon is not surprising since the melting point of MoO_3 is only 795°C [22]. The presence of excess Mo (Mo/Ni = 3) resulted in a small increase in carbon deposition, compared with some of the lesser doped samples, but in this case extremely rapid deactivation was observed. When W was added to the catalysts, no deactivation was observed (m.p. = 1473°C [22]), and only a small decrease in initial CH_4 and CO_2 conversion. In addition, a significant decrease in deposition of carbon during the dry reforming reaction was seen.

Table 1
Influence of Mo or W on a 2% Ni/Al_2O_3 catalyst for methane dry reforming (T = 700°C, p = 1 bar, 30ml min^{-1}, CH_4/CO_2 = 1:1.1).

Catalyst	Dopant % M:Ni	Conversion / %		Yield / %	S_g	mg C/
		CO_2	CH_4	CO	m^2g^{-1}	mg Ni
2% Ni/Al_2O_3	None	81.0	74.1	77.7	219.8	178.2
"	33.3% W	81.6	75.4	78.7	214.4	28.2
"	100% W	77.9	71.0	74.6	206.0	U/D
"	300% W	69.8	61.5	65.9	194.8	U/D
"	33.3% Mo	77.8	70.7	74.4	N/A	U/D
"	100% Mo	76.8	69.6	73.4	N/A	U/D
"	300% Mo	79.2	71.8	75.7	N/A	27.2

N/A = not available; U/D = undetected

Currently there is no literature available on the advantageous effects of the addition of group VI compounds to Ni catalysts for the dry reforming of methane, however, the trends described above for these Ni-Mo/W catalysts are mirrored by those described by Mauchausse *et al.* [23] in a study of the hydrotreatment of coal.

A number of plausible reasons for the role of tungsten oxide in the inhibition of carbon formation can be advanced. For example, i) the tungsten oxide may change the solubility of carbon in nickel by electronic effects; ii) it could limit the size of nickel particle ensembles; or iii) it could change the rate of dissociation of carbon dioxide, leading to more surface oxygen for reaction with surface carbon. Here we forward the latter explanation, and propose that tungsten carbide sites are formed *in situ* in the reaction between oxide and methane, since recently it has been shown that the group VI transition metal carbides are active catalysts for methane dry reforming themselves; this was unknown when Borowiecki and Golebiowski [21] published their paper on the inhibition of carbon formation in steam reforming by adding MoO_3/WO_3. The formation of low oxidation state Mo/W compounds (MO_2/M_xC) in the reactor is not surprising since the atmosphere is very reducing. In the next section some results obtained for methane dry reforming over metal carbides will be presented, and the possible implications of these results to the current system will be discussed.

3.2. Dry reforming over Mo_2C and WC catalysts

The results obtained for the dry reforming of methane over Mo_2C (90 m^2g^{-1}) and WC (40 m^2g^{-1}) are presented in Table 2. These materials initially catalysed the dry reforming reaction to close to thermodynamic equilibrium, however, HRTEM analysis of the post-catalysis samples indicated that no bulk carbon deposition had occurred. At ambient pressure catalyst deactivation occurred, and this was clearly observable after 8 hours (Figure 1), but under certain conditions (e.g. elevated pressure) high activity was maintained for the duration of the experiments (> 3 days). XRD studies of the deactivated catalysts (Figure 2) showed that bulk carbide was still present, as well as MO_2. No oxide was found for the catalysts tested at elevated pressure [24].

Table 2
Dry reforming over carbide catalysts at 850°C.

	P / bar	$C_{(CH4)}$	$C_{(CO2)}$	$Y_{(CO)}$	H_2/CO
Mo_2C	1.0	92.5	92.4	92.5	0.93
WC	1.0	92.0	93.1	92.6	0.94
Mo_2C	8.3	63.5	75.9	70.0	0.76
WC	8.3	62.7	75.4	68.6	0.79

$CH_4/CO_2 = 1$, GHSV = 2.87×10^3 h^{-1}.

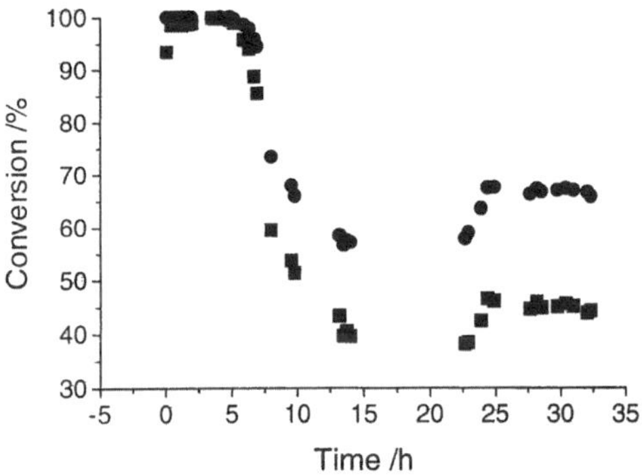

Figure 1. Conversion of CH_4 (■) and CO_2 (●) over bulk WC, 950°C, 1 bar, $CH_4/CO_2 = 1$, GHSV = 410 h^{-1}.

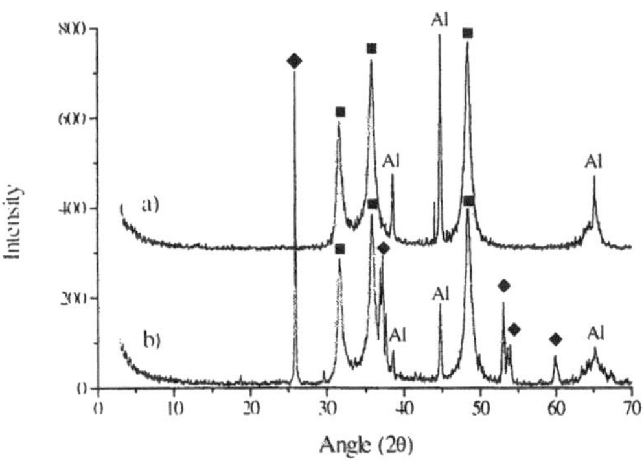

Figure 2. XRD of WC a) as prepared, and b) after dry reforming at 1 bar: (◆) WO_2; (■) WC (850°C, GHSV =2.87×10^3 h^{-1}, CH_4/CO_2 =1).

It has been found that the carbide catalysts can be cycled between carbide-oxide, depending on the conditions, and it is, therefore, possible that this is occurring in the Ni/MO_3 system. Studies of the dry reforming reaction mechanism over the group VI metal carbides have indicated that the dissociation of CO_2 is very fast, and CH_4 dissociation is rate limiting [25]; this is in contrast to that found for nickel catalysts [26]. Under the reaction conditions used, the formation of some carbide sites would not be surprising. If these sites are in close proximity to the nickel sites, then the surface oxygen atoms ($O_{(s)}$) formed on the carbide could react with surface carbon ($C_{(s)}$) formed on the nickel, by a spillover mechanism, and therefore remove the carbon before it can dissolve into the nickel crystals, and polymerise to form 'whisker' carbon.

Carbide: $CO_2 \rightarrow CO + O_{(s)}$
Nickel: $CH_4 \rightarrow C_{(s)} + 2H_2$

After spillover: $C_{(s)} + O_{(s)} \rightarrow CO$

This proposed 'carbide effect' is similar to that suggested for alkali and alkaline-earth metals, where the increased basicity of the catalyst after doping with, for example CaO, also results in an increase in the rate of dissociation of CO_2. However, at this time the role of MoO_3/WO_3 as an ensemble size growth inhibitor cannot be ruled out. Further work is currently ongoing.

4. CONCLUSION

In conclusion, it has been shown that the addition of MoO_3 or WO_3 to an alumina supported Ni catalyst results in a dramatic decrease in carbon formation during methane dry reforming; no deactivation was observed for the WO_3 doped samples. This is potentially of great interest to industry as an alternative to conventional nickel catalysts.

REFERENCES

1. F. Fischer and H. Tropsch, Brennst. Chem., 9 (1928) 39.
2. A.T. Ashcroft, A.K. Cheetham, M.L.H. Green and P.D.F. Vernon, Nature, 352 (1991) 225.
3. P.D.F. Vernon, M.L.H. Green, A.K. Cheetham and A.T. Ashcroft, Catal. Today, 13 (1992) 417.
4. Y.H. Hu and E. Ruckenstein, Catal. Lett., 36 (1996) 145.
5. A. Erdohelyi, J. Cserenyi, E. Papp and F. Solymosi, Appl. Catal. A-General, 108 (1994) 205.
6. T. Osaki, T. Horiuchi, K. Suzuki and T. Mori, Catal. Lett., 35 (1995) 39.
7. A.M. Gadalla and M.E. Sommer, J. Am. Ceram. Soc., 72 (1989) 683.
8. A.M. Gadalla and M.E. Sommer, Chem. Eng. Sci., 44 (1989) 2825.
9. O. Yamazaki, T. Nozaki, K. Omata and K. Fujimoto, Chem. Lett., (1992) 1953.
10. K.S.M. Bhatta and C.M. Dixon, Ind. Eng. Chem. Prod. Res. Dev., 8 (1969) 324.
11. Q. Zhuang, Y. Qin and L. Chang, Appl. Catal., 70 (1991) 1.
12. C.A. Bernado, I. Alstrup and J.R. Rostrup-Nielsen, J. Catal., 96 (1985) 517.
13. T. Borowiecki, Appl. Catal., 4 (1982) 223.
14. J.R. Rostrup-Nielsen, in "Catalysis Science and Technology", J.R. Andersen and M. Boudart (eds.), Springer, Berlin, 5 (1984) 1.
15. I. Haque, Ph.D. Thesis; University of New South Wales, Sydney, 1990.
16. C. Padeste, D.L. Trimm and R.N. Lamb, Catal. Lett., 17 (1993) 333.
17. J.R. Rostrup-Nielsen, J. Catal., 33 (1974) 184.
18. T. Hayakawa, H. Harihara, A.G. Andersen, A.P.E. York, K. Suzuki, H. Yasuda and K. Takehira, Angew. Chem. Int. Ed. Engl., 35 (1996) 192.
19. J.B. Claridge, M.L.H. Green, S.C. Tsang, A.P.E. York, A.T. Ashcroft and P.D. Battle, Catal. Lett., 22 (1993) 299.
20. J.R. Rostrup-Nielsen and J.-H. Bak Hansen, J. Catal., 144 (1993) 38.
21. T. Borowiecki and A. Golebiowski, Catal. Lett., 25 (1994) 309.
22. CRC Handbook of Chemistry and Physics, R.C. Weast and M.J. Astle (eds.), CRC Press Inc., Florida, 1980.
23. C. Mauchausse, E. Kural, D.L. Trimm and N.W. Cant, Fuel, 71 (1992) 203.
24. A.P.E. York, J.B. Claridge, A.J. Brungs, S.C. Tsang and M.L.H. Green, J. Chem. Soc., Chem. Commun., (1997) 39.
25. A.P.E. York, unpublished results.
26. J.R. Rostrup-Nielsen and D.L. Trimm, J. Catal., 48 (1977) 155.

NATURAL GAS CONVERSION V
Studies in Surface Science and Catalysis, Vol. 119
A. Parmaliana et al. (Editors)
1998 Elsevier Science B.V.

Sustainable Ni Catalysts Prepared by SPC Method for CO_2 Reforming of CH_4

Shu Suzuki,[1] Takashi Hayakawa,[1] Satoshi Hamakawa,[1] Kunio Suzuki,[1] Tetsuya Shishido[2] and Katsuomi Takehira[2]*

[1]*National Institute of Materials and Chemical Research, Tsukuba Research Center, AIST, Higashi 1-1, Tsukuba, Ibaraki 305, Japan.*
[2]*Department of Applied Chemistry, Hiroshima University, Kagamiyama 1-4-1, Higashi-hiroshima 739, Japan*

$Ni/Ca_{0.8}Sr_{0.2}TiO_3$ and $Ni/BaTiO_3$ catalysts have been prepared by *solid phase crystallization* (*spc*) method and used successfully for CO_2 reforming of CH_4 into synthesis gas at 850℃. The Ni catalysts were *in situ* obtained during the reaction from starting materials, where nickel species was homogeneously incorporated in the perovskite structure. The high activity as well as the high sustainability against coke formation may be due to well dispersed and stable Ni metal particles over the perovskite, where the nickel species thermally evolve from the cations homogeneously distributed inside an inert perovskite matrix as the precursor. Oxygen mobility in the perovskite was also important for removing coke from nickel surface, resulting in the effective sustainability.

1. Introduction

The CO_2 reforming of CH_4 has been intensively studied for the purpose of its use in industry for the production of synthesis gas.[1] The conversion of CH_4 to synthesis gas is usually carried out by its H_2O reforming.[2] Since the replacement of H_2O by CO_2 results in a lower H_2/CO ratio in the product gas, the combination of these two reforming reactions widens the utility of synthesis gas, *i.e.*, in methanol or acetic acid synthesis. Recently, this process has also received attention from an environmental view point because the emission of CH_4 and CO_2 in the atmosphere brings about global warming by the greenhouse effect and these harmful gases can simultaneously be converted to useful synthesis gas. Ni or precious metals are reported to be active as the catalyst for the reaction, however the reaction is frequently accompanied by coke formation, especially on Ni catalysts, leading to catalyst deactivation.

High dispersion of metal species over catalyst may reduce coke formation.[3] Metal-supported catalysts are conventionally prepared by wet impregnation of different supports. This method is not fully reproducible and may give rise to some inhomogeneity in the distribution of the metal on the surface. A new concept of the catalyst preparation, therefore, may be required. Use of the precursors containing homogeneously distributed metal inside in the structure, which on further calcination and reduction, may result in the formation of well dispersed and stable metal particles on the surface. We have proposed a new method of the preparation of well dispersed and stable metal supported catalyst, *i.e.*, "solid phase crystallization." This method was successfully applied to the preparation of Ni-supported catalyst for the partial oxidation of CH_4 to synthesis gas.[4-6] By using $CaTiO_3$ or $BaTiO_3$ perovskite containing small amount of Ni in the Ti site as the precursor, finely dispersed and stable Ni metals were formed over the catalyst, resulting in the high activity and sustainability against coke formation during the reaction. Here we report high activity and sustainability of $Ni/Ca_{0.8}Sr_{0.2}TiO_3$ and $Ni/BaTiO_3$ catalysts prepared by the *spc* method in the CO_2 reforming of CH_4.

2. Experimental

The catalysts $Ni/Ca_{0.8}Sr_{0.2}TiO_{3(spc)}$, $Ni/BaTiO_{3(spc)}$ and $Ni/TiO_{2(spc)}$ were obtained by the solid phase crystallization method from the precursors prepared by the citrate method.[4-6] An aqueous solution of reagent grade nickel nitrate, alkaline earth carbonates and titanium isopropoxide was treated with an excess amount of citric acid and ethylene glycol, evaporated at 80-90℃ to make a sol of organic metal complex. This was followed by two-step decomposition by heating at 200℃ for 5 h and 500℃ for 5 h, and finally calcining at 900℃ in air for 10 h. Moreover, three catalysts $Ni/Ca_{0.8}Sr_{0.2}TiO_{3(imp)}$, $Ni/BaTiO_{3(imp)}$ and $Ni/TiO_{2(imp)}$ were prepared by impregnation (*imp*) method. $Ca_{0.8}Sr_{0.2}TiO_3$, $BaTiO_3$ and TiO_2 were separately prepared by the citrate method, impregnated with aqueous solution of nickel nitrate with an equimolar amount of citric acid and ethylene glycol, and calcined at 900℃ in air for 5 h. In the both cases of *spc* and *imp*, atomic ratio of Ni/Ti was fixed at 0.2/1.0. The $Ni/\alpha\text{-}Al_2O_3$, Ni/MgO, Ni/ZrO_2 and Ni/SiO_2 catalysts were prepared by the similar impregnation method and finally calcined at 900℃ in air for 5 h. The amount of Ni was 10.3 wt% on each support. The structure of the catalysts was studied by using XRD, XPS, BET, SEM and TEM as reported in the previous paper.[4-6]

All the catalysts have been tested as prepared and *in situ* reduced during the reaction by using a fixed bed catalyst in a mixture of CO_2 ($1.0\ l\cdot h^{-1}$), CH_4 ($1.0\ l\cdot h^{-1}$) and N_2 ($1.4\ l\cdot h^{-1}$) at 850℃. The catalytic activities were also tested by changing the space velocity from 20,000 to 70,000 ($ml\cdot h^{-1}\cdot g\text{-}cat^{-1}$). Moreover, the reactions for testing the catalyst life were carried out at 850℃ for 30 h under the same conditions. A U-shaped quartz reactor was used with 150 mg of catalyst dispersed in 2 ml of quartz wool. The thermocouple was placed in the middle of the catalyst bed. Product gases were sampled immediately after the reactor and injected into a gas chromatograph for analysis. The amount of coke formed over the catalyst after the reaction for 6 h was estimated from the amount of CO_2 formed during the TPO reaction from 250 to 850℃ at the rate of $2.5℃\cdot min^{-1}$ under air flow ($1.4\ l\cdot h^{-1}$).

150 mg of the supports was treated by H_2 (1 ml x 5) pulses at 850℃ and then CO_2 was pulsed (1 ml x 15) over the supports. The amount of mobile oxygen in the supports was estimated from the amount of CO formed from CO_2. Coke was deposited over 10 mg of the catalyst by CH_4 (1 ml x 10) pulses at 850℃ and the catalyst was further treated for 4 or 19 h

separately under the atmosphere of He. Coke can be converted to CO_2 by the mobile oxygen species during this treatment, resulting in self-decoking. The amount of coke was measured by oxygen (1ml x 10) pulses over the catalyst.

3. Resuts and discussions

3.1. Activities of the catalysts

Activities of the various Ni catalysts in the CH_4 reforming with CO_2 (1) were tested for 30

$$CH_4 + CO_2 \longrightarrow 2CO + 2H_2 \quad (1)$$

h at the space velocity of 22700 $ml \cdot h^{-1} \cdot g\text{-}cat^{-1}$ and typical results are shown in Figures 1, 2 and 3. Among the catalysts prepared by the *imp* method using conventional support (Fig. 1), Ni/MgO showed the highest activity, followed by Ni/α-Al_2O_3. It was reported that NiO-MgO system forms 'ideal' solid solutions over the whole molecular fraction range and was succesfully used as the catalyst for the CH_4 steam-reforming.[7] Ni/TiO_2 showed a clear decrease in the activity during the reaction in the both cases of *spc* and *imp*. Ni/TiO_2 showed a similar change in the activity in the partial oxidation of CH_4 to synthesis gas.[6] This may be due to the partial oxidation of Ni surface by mobile oxygen species from TiO_2 surface in the both reactions. Among the catalysts supported on perovskite by the *imp* method, Ni/$BaTiO_{3(imp)}$ was the most active, followed by Ni/$Ca_{0.8}Sr_{0.2}TiO_{3(imp)}$, while Ni/$SrTiO_{3(imp)}$ showed a low activity, which increased gradually during the reaction (Fig. 2).

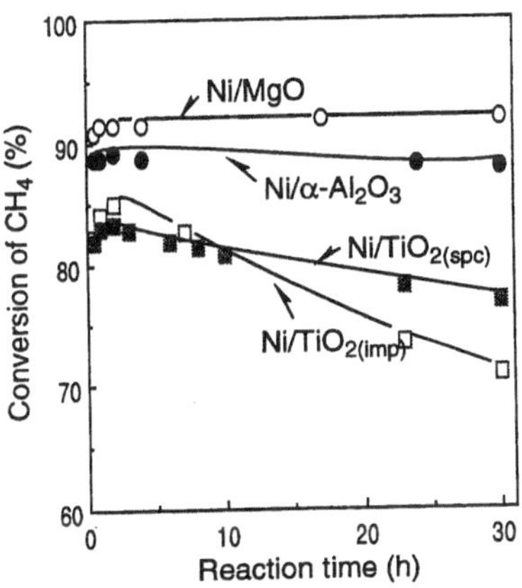

Fig.1 CH_4 reforming with CO_2 over the Ni supported catalyst.

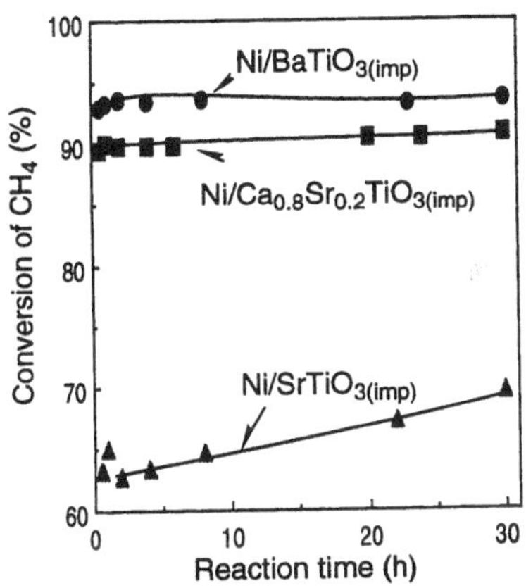

Fig.2 CH_4 reforming with CO_2 over the Ni supported catalyst.

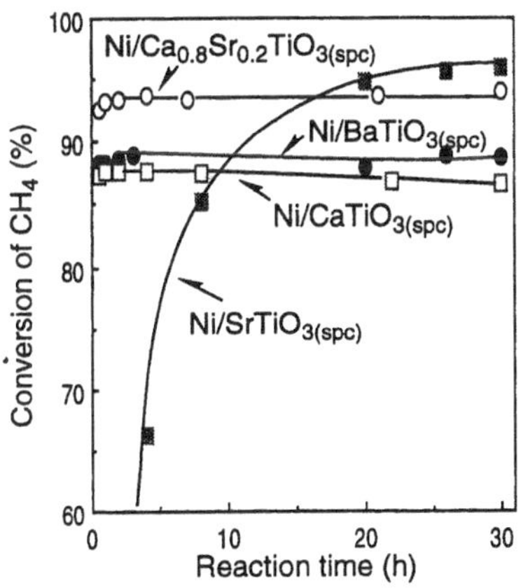

Fig. 3 CH_4 reforming with CO_2 over the Ni supported catalyst.

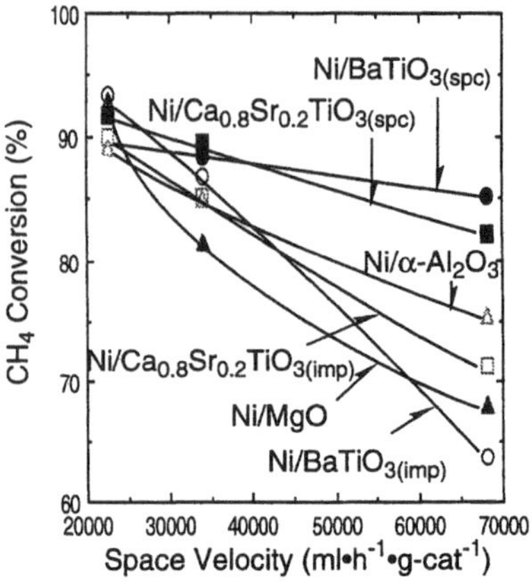

Fig. 4 CH_4 reforming with CO_2

When the catalyst was prepared by the *spc* method (Fig. 3), the highest activity was obtained by $Ni/Ca_{0.8}Sr_{0.2}TiO_{3(spc)}$, followed by $Ni/BaTiO_{3(spc)}$ and $Ni/CaTiO_{3(spc)}$. $Ni/SrTiO_{3(spc)}$ showed again a low activity at the beginning of the reaction, while the activity slowly increased and reached the highest value among the catalysts tested. XRD analyses suggest that $SrTiO_3$ forms a stable cubic crystal structure and contains much amount of Ni in the Ti site compared to $CaTiO_3$ and $BaTiO_3$.[5] This may cause a slow formation of highly dispersed nickel particle, resulting in the high activity after the long induction period.

In the partial oxidation of CH_4, the highest activity was observed with $Ni/Ca_{0.8}Sr_{0.2}TiO_{3(spc)}$.[5] Also in the present reaction, the replacement of 20% of Ca with Sr in $Ni/CaTiO_{3(spc)}$ resulted in an increase as well as in a stabilization in the activity, the value of which was much higher than Ni/MgO. This may be due to the formation of active structure during phase transition from orthorhombic $CaTiO_3$[7] to cubic $SrTiO_3$.[8] No clear difference was observed in the activities of the effective catalysts, *i.e.*, $Ni/BaTiO_{3(spc)}$, $Ni/Ca_{0.8}Sr_{0.2}TiO_{3(spc)}$, $Ni/\alpha\text{-}Al_2O_3$, $Ni/Ca_{0.8}Sr_{0.2}TiO_{3(imp)}$, Ni/MgO and $Ni/BaTiO_{3(imp)}$, selected by the catalytic screening at 850℃ for 30 h at the space velocity of 22700 $ml\cdot h^{-1}\cdot g\text{-}cat^{-1}$ (Figs. 1, 2 & 3). This may be due to the fact that thermodynamic equilibrium of the reaction (1) was attained in these conditions. The activity was further tested by increasing space velocity (Figure 4). At the higher space velocity, the reaction (1) must be far from thermodynamic equilibrium on the catalyst surface and, therefore, the catalytic activity may be compared more precisely. The catalysts were clearly devided into two groups depending on the preparation method, *i.e.*, the *spc* method and the *imp* method. Upon increasing the space velosity, those prepared by the *spc* method, $Ni/BaTiO_{3(spc)}$ and $Ni/Ca_{0.8}Sr_{0.2}TiO_{3(spc)}$, were still enough active, while those prepared by the *imp* method, $Ni/BaTiO_{3(imp)}$, $Ni/Ca_{0.8}Sr_{0.2}TiO_{3(imp)}$, $Ni/\alpha\text{-}Al_2O_3$ and Ni/MgO, quickly lost their activities. It is thus very likely that the *spc* method is much more effective for the catalyst preparation. The high activity of the catalyst prepared by the *spc* method may be due to the formation of highly dispersed and stable fine Ni particles.[4-6]

3.2. Coking over the catalysts

Amount of coke formed over the catalysts during the CO_2 reforming of CH_4 for 6 h was measured as follows: $Ni/CaTiO_{3(spc)}$, 2.0; $Ni/SrTiO_{3(spc)}$, 1.3; $Ni/Ca_{0.8}Sr_{0.2}TiO_{3(spc)}$, 3.4; $Ni/BaTiO_{3(spc)}$, 1.3; $Ni/\alpha\text{-}Al_2O_3$, 21.3; Ni/MgO, 5.2; Ni/ZrO_2, 41.8; Ni/SiO_2, 11.5; $Ni/TiO_{2(spc)}$, 2.3; $Ni/TiO_{2(imp)}$, 0.2 and $Ca_{0.8}Sr_{0.2}TiO_3$, 4.0 wt%. Generally, Ni catalyst supported on single metal oxide showed the higher amount of coke than those supported on perovskite-type mixed oxide. The low coking observed over Ni/MgO may be due to the basicity of MgO as well as the high dispersion of Ni particles produced from the NiO-MgO solid solutions.[9&10] The *spc* method was definitively effective for decreasing the coke formation in the partial oxidation of CH_4,[4-6] while this was not the case in the present study. This may be due to much easier formation of coke in the reaction (1). The low amount of coke, while the activity decrease, observed over $Ni/TiO_{2(spc)}$ and $Ni/TiO_{2(imp)}$ may be due to the presence of mobile oxygen over TiO_2 as seen in the strong metal-support interaction (SMSI).[10] Mobile oxygen species can also exist in the perovskite materials used as the supports in the present study.

The amount of mobile oxygen in the support was measured by the pulse reactions of CO_2. The mobile oxygen in the supports, $Ca_{0.8}Sr_{0.2}TiO_{3(spc)}$, $BaTiO_{3(spc)}$ and α-Al_2O_3, first reacted with H_2 to form H_2O. The oxygen vacancies thus formed then reacted with CO_2 to form CO, from which the amount of mobile oxygen was estimated (Figure 5). Both $Ca_{0.8}Sr_{0.2}TiO_{3(spc)}$ and $BaTiO_{3(spc)}$ showed quite high values of the oxygen mobility compared to that of α-Al_2O_3. During the pulse reaction, $Ca_{0.8}Sr_{0.2}TiO_{3(spc)}$ trapped quickly oxygen from CO_2 to form CO within 10 pulses of CO_2, while $BaTiO_{3(spc)}$ slowly did resulting in the long term formation of CO from many pulses of CO_2. This suggests a slow mobility of oxygen in $BaTiO_{3(spc)}$ compared to that of $Ca_{0.8}Sr_{0.2}TiO_{3(spc)}$.

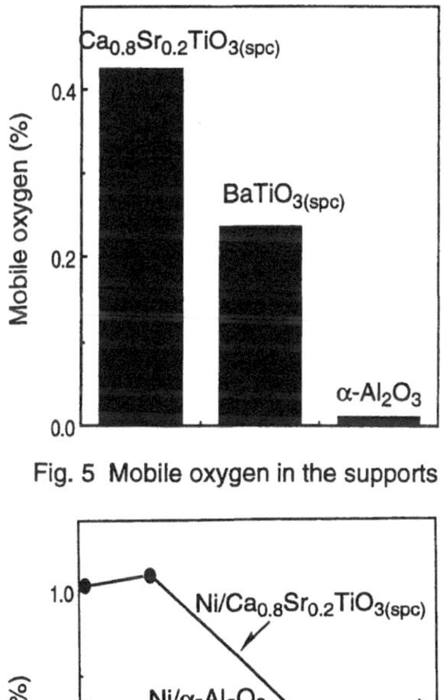

Fig. 5 Mobile oxygen in the supports

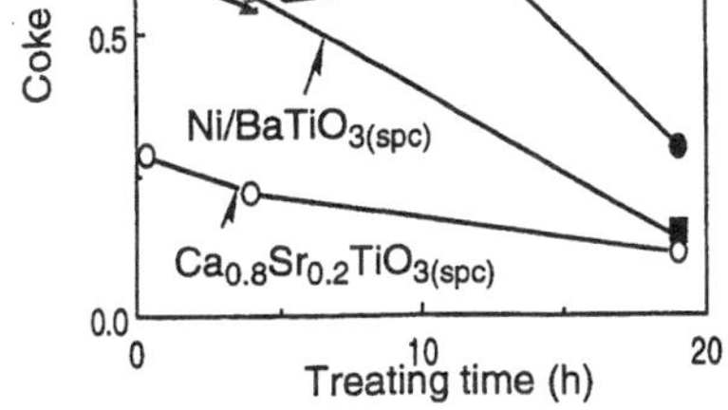

Fig. 6 Amount of coke after treatment

Most likely such mobility of oxygen can affect a self-decoking over the catalysts. The catalysts were first coked by CH_4 pulses at 850℃ and were then treated in the He atmosphere for a certain period (4 or 19 h) at 850℃. During the treatement in He, part of coke over the catalysts can be eliminated by the mobile oxygen species from the supports. The remaining coke was finally converted to CO_2 by oxygen pulse. The results are shown in Figure 6 together with the amount of coke first deposited. Coke was formed over the support $Ca_{0.8}Sr_{0.2}TiO_{3(spc)}$ itself by the CH_4 pulses and was eliminated by the mobile oxygen during the He treatment. Much amount of coke was formed over the Ni catalysts and was also eliminated over the perovskite supported catalysts. Ni/α-Al_2O_3 alone showed no elimination of coke during the He treatment, coinciding well to small amount of mobile oxygen in the α-Al_2O_3 structure in Figure 5.

The ability of decoking may also depend on the size of nickel particle formed on the catalyst surface. On nickel particle of large size, approach of mobile oxygen from the support to decoking point must be long, resulting in slow decoking. This is the case of the catalyst prepared by impregnation. TEM observation of the catalyst reveals that the size of nickel particle reaches 50-100 nm in diamter on the *imp* catalyst, while the size of substantial part of the particles is less than 1 nm on the *spc* catalyst. In the latter case, decoking may take place quickly by a easy approach of the mobile oxygen from the support. It is thus likely that both Ni/$Ca_{0.8}Sr_{0.2}TiO_{3(spc)}$ and Ni/$BaTiO_{3(spc)}$ are effective catalyst for CO_2 reforming of CH_4 also from the view point of sustainability against coking.

4. Conclusion

The Ni/perovskite catalyst was prepared by the *spc* method and used for the CO_2 reforming of CH_4, resulting in the high activity as well as the high sustainability against coke formation. Highly dispersed and stable Ni metal particles was formed over the perovskite by thermal evolution of the Ni cations homogeneously distributed inside perovskite matrix. Oxygen mobility in the perovskite was also effective for removing coke from nickel surface.

References

1) A. T. Ashcroft, A. K. Cheetham, J. S. Foord, M. L. H. Green and P. D. F. Vernon, Nature, 352 (1991) 225.
2) J. R. Rostrup-Nielsen, in Catalysis, Science and Technology, ed. J. R. Andersen and M. Boudart, Springer, Berlin, 1984, vol. 5, p. 1.
3) C. H. Bartholomew, Catal. Rev.-Sci. Eng., 24 (1982) 67.
4) T. Hayakawa, H. Harihara, A. G. Andersen, A. P. E. York, K. Suzuki, H. Yasuda and K. Takehira, Angew. Chem., Int. Ed., 35 (1996) 92.
5) T. Hayakawa, H. Harihara, A. G. Andersen, K. Suzuki, H. Yasuda, T. Tsunoda, S. Hamakawa, A. P. E. York, Y. S. Yoon, M. Shimizu and K. Takehira, Appl. Catal., A: General, 149 (1997) 391.
6) R. Shiozaki, A. G. Andersen, T. Hayakawa, S. Hamakawa, K. Suzuki, M. Shimizu and K. Takehira, J. Chem. Soc., Faraday Trans., 93 (1997) 3225.
7) H. F. Kay and P. C. Bailey, Acta. Cryst., 10 (1957) 219.
8) M. Ceh, D. Kolar and L. Golic, J. Solid State Chem., 68 (1987) 68.
9) F. Arena, F. Frusteri, A. Parmaliana, L. Plyasova and A. N. Shmakov, J. Chem. Soc., Faraday Trans., 92 (1996) 469.
10) T. Osaki, J. Chem. Soc., Faraday Trans., 93 (1997) 643.

NATURAL GAS CONVERSION V
Studies in Surface Science and Catalysis, Vol. 119
A. Parmaliana et al. (Editors)

Metal-Support Interactions In Steam Reforming

N. P. Siswana[a], D. L. Trimm[a] & N. W. Cant[b]

[a]School of Chemical Engineering & Industrial Chemistry,
The University of New South Wales, Sydney NSW 2052

[b]Department of Chemistry, Macquarie University, Sydney NSW 2109

Abstract

Reactions used to model the steam reforming of natural gas show that the influence of the support on the performance of the metal is minimal. On the other hand, hydrogen produced during the reactions adsorbs on the metal and can spill over to the support. This spilt over hydrogen could be responsible in part for the gasification of carbonaceous residues.

Introduction

The production of hydrogen by the steam reforming of natural gas is an important industrial reaction (1, 2). Both methane and smaller amounts of light hydrocarbons in natural gas react with steam at high temperatures to give hydrogen and carbon oxides.

$$C_nH_m + nH_2O = nCO + (\frac{2n+m}{2})H_2 \qquad 1$$

$$CO + H_2O = CO_2 + H_2 \qquad 2$$

Industrial catalysts consist of nickel supported on an alkaline support (1, 2, 3). Although nickel is most cost-effective, Rh, Ru, Pd and Pt are at least as active for the reaction (1, 2). Thermodynamics require operation with steam at high temperatures (1, 2), which results in several unwanted side reactions such as coke formation (3). As a result, the composition of the catalyst and the operating conditions are adjusted to favour coke gasification. Steam: carbon ratios are kept high to favour gasification, and alkali is added to the catalyst to promote the reaction (1, 2, 3).

Steam reforming is fast, and all of the functions of the catalyst support have yet to be established. Efficient dispersion of catalyst with high temperature stability in the presence of steam is essential (4), as is the capacity to promote gasification of coke (5). However the importance of metal-support interactions in the system has yet to be determined.

Previous studies of platinum supported on a range of transition metal oxides has revealed metal support interactions in hydrogenolysis and oxidation (6).

The present studies were carried out to explore the approach in the context of the steam reforming of natural gas. The reactions of methane proved too fast for the study, and propene steam reforming was used as the test reaction.

Experimental

Transition metal oxides (TMO) supported on γ-alumina were prepared by the incipient wetness technique (7) using metal nitrate solutions. γ-alumina was ground to 210-300 micron size and mixed in a slurry containing nitrates calculated to give a 10% oxide loading. After 2h the solids were dried for 16h at 383 K and calcined (673K: 4h). Supported vanadium oxide was prepared from vanadium oxalate using the same method.

A portion of each calcined catalyst was impregnated with chloroplatinic acid solution intended to give 3% loading of platinum. The catalysts were dried and calcined using the procedure described above.

All catalysts were characterised. Total surface area was determined using a flow method and applying the BET equation. Metal surface area of the reduced catalyst (H_2: 673K: 4h) was determined from carbon monoxide adsorption measurements assuming a stoichiometry of one. The results are summarised in Table 1.

Table 1: Metal and total surface areas

Dopant on γ-Al_2O_3	Total surface area (m^2g^{-1})	Catalyst on γ-Al_2O_3	Total surface area (m^2g^{-1})	Metal surface area ($m^2(Pt)g^{-1}$)
nil	200	Pt alone	190	5
Co_3O_4	170	Pt/Co_3O_4	173	4
Fe_2O_3	183	Pt/Fe_2O_3	174	2.4
MnO_2	170	Pt/MnO_2	181	3
Cr_2O_3	181	Pt/Cr_2O_3	182	3.5
V_2O_5	184	Pt/V_2O_5	178	3
CuO	198	Pt/CuO	189	4
NiO	180	Pt/NiO	170	5

Propylene steam reforming was studied using a flow system. The flows of propylene and nitrogen to the reactor were controlled by Brooks 5850E mass flow controllers. Water was pumped to an evaporator using a ISCO syringe pump. The steam: propylene ratio was maintained at a ratio of 100:3 in order to achieve measurably slow rates.

Tests were carried out using 0.3g samples mixed with quartz beads and placed in a 4 mm I. D. reactor held in the constant temperature zone of an electrically heated furnace. The samples were reduced in hydrogen at 673K (2h) before flushing with nitrogen and introducing the reaction mixture. Product analysis was carried out after trapping water. A Gow-Mac 580 TCD gas chromatograph was used to analyse reactants & products. Hydrogen, carbon monoxide and methane were separated using a CRT column (Alltech) at 313K. Carbon dioxide, ethylene, ethane and propylene were separated on a 2.5 m Porapak N column (343K).

Initial experiments established the absence of mass transfer effects (10) and determined the initial rate period. All catalysts were tested at a molar flow rate (dry) of 1.86 x 10^{-3} mol min^{-1} and 101 kPa pressure.

Results & Discussion

Initial experiments showed that steam reforming was promoted over transition metal oxides (TMO) containing platinum. The reaction was initiated at 873 K over TMO, a temperature that reduced to 673 K when Pt was present. Only carbon dioxide and hydrogen were produced over TMO while Pt-TMO also produced ethane, ethylene and traces of methane. These additional products are expected to arise from cracking and hydrogenation reactions

$$C_3H_6 = C_2H_4 + C + H_2 \qquad 3$$

$$C_2H_4 + H_2 = C_2H_6 \qquad 4$$

Thermodynamic calculations (1) showed that carbon monoxide would not be expected as a major product.

The disappearance of propylene over both TMO and Pt-TMO was found to show a clear maximum at nickel (figure 1) with a minimum at Cr/Mn. The H_2:CO_2 ratios observed in the products obtained over TMO were close to the 3:1 ratios expected on the basis of reactions 2 & 3 (figure 2) with the exception of manganese where the hydrogen yield was low.

The activity pattern was more complex over Pt/TMO (figure 3) but similar H_2:CO_2 ratios could be calculated once allowance had been made for the production of ethylene and ethane (reactions 7 & 8, figure 4). In the Pt/TMO series, cobalt dopant gave lower hydrogen values than expected

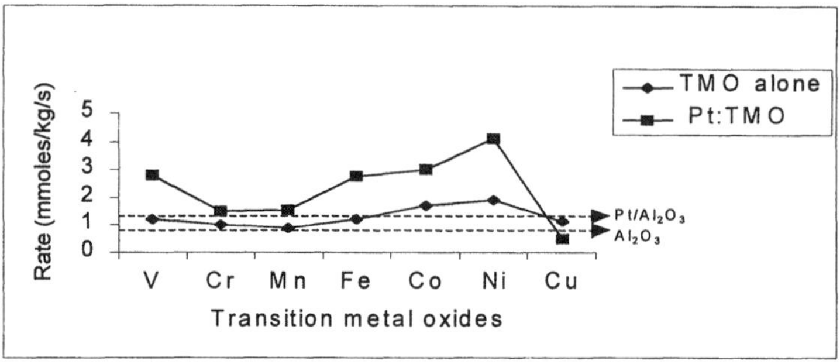

Figure 1:Activity patterns for the disappearance of propylene. Temperature TMO tests = 873 K. Temperature Pt/TMO tests = 673 K. Other variables described in the experimental section.

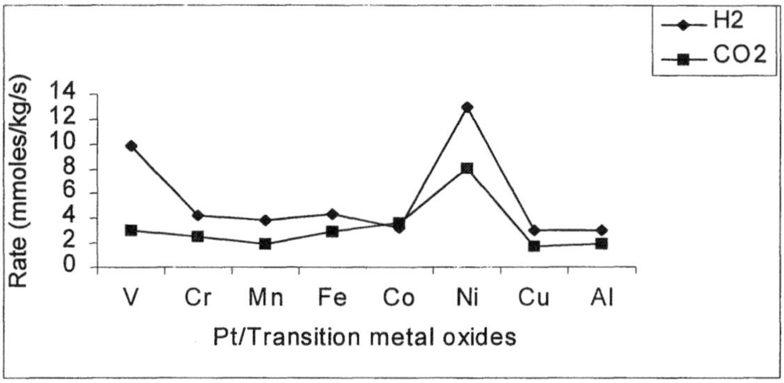

Figure 2:
Activity patterns for on the appearance of hydrogen and carbon dioxide over TMO.
T = 873 K.

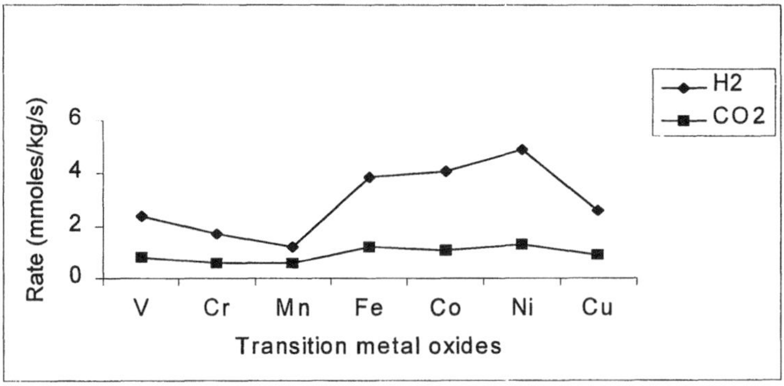

Figure 3: Activity patterns for the appearance of hydrogen and carbon dioxide over Pt/TMO.
T = 673 K

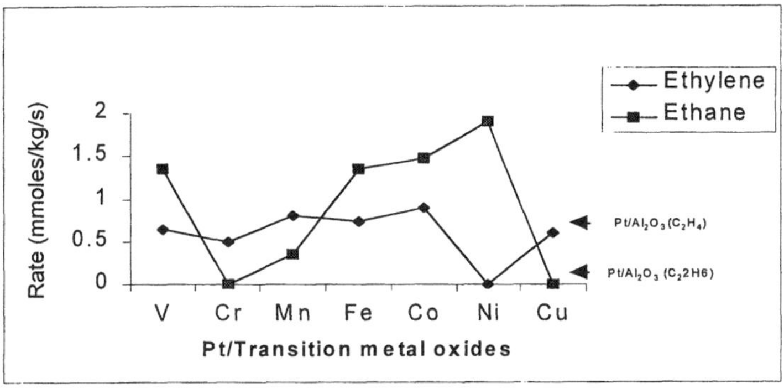

Figure 4: Ethane:Ethylene production over Pt/TMO catalysts. T = 673 K.

The results allow some estimation of activity patterns in steam reforming (carbon dioxide production), hydrocarbon cracking (total C_2 production) and hydrogenation (ethylene: ethane rations).

Dowden (9) has suggested that activity patterns for metal oxide catalysts can be explained in terms of the contribution of crystal field stabilisation energies in the oxides. Theoretical calculations and experimental measurements (9, 10) have shown that twin peak patterns are expected in the systems, with maxima at cobalt and manganese oxides. If the metal oxide supports are influencing the catalytic activity for platinum for the test reactions, then similar twin peak patterns should be observed (6). At first sight, it would seem that such patterns are observed but detailed examination shows the peaks in activity to be in the wrong positions.

Some differences can be expected as a result of the fact that hydrogen, produced during steam reforming, may reduce some of the metal oxides to a lower valency state. Thermodynamic calculations (11) carried out for the conditions of reaction show that the preferred form of the oxide/metal may be different from that added. Copper oxide is expected to be reduced to copper while most oxides may be reduced to a lower valency state. The steam:hydrogen ratio (ca 11:1) is such that nickel may remain as the oxide, even though, under normal reducing conditions, nickel metal would be expected (12).

It is seen from figures 2 & 3 that considerably more hydrogen can be produced at lower temperatures in the presence of platinum, and that this is particularly obvious in the case of added vanadia and nickel oxide. Nickel is known efficiently to catalyse steam reforming (1, 2) and one obvious inference is that Ni is being formed in the presence of Pt. Similar effects have been observed previously with hydrogen adsorbing on Pt and spilling over to the surface to reduce NiO to Ni (12, 13). The result in the present system is a more efficient catalyst in which both Pt and Ni catalyse steam reforming.

The presence of Ni metal is also indicated by the results in figures 4. Some C_2 production occurs over Pt/Al_2O_3, but considerably more is seen in the presence of vanadia, iron, cobalt and nickel oxides. A clear minimum in the activity pattern is seen with chromia, which indicates that cracking is not acid catalysed.

Metal catalysed cracking and hydrogenation would be consistent with the presence of nickel, cobalt and iron on the surface of the dopant. However, the role of vanadia in accelerating this process is uncertain.

Thus it is clear that involvement of the support in steam reforming, metal catalysed cracking and olefin hydrogenation is unlikely. However, hydrogen - adsorbed on Pt and spilt over to the support - is accelerating the reduction of metal oxides and improving the efficiency of the catalyst. For the steam reforming process itself, the inference is that hydrogen, spilt over from the active metal, may play a role in removing carbonaceous intermediates from the support surface.

Acknowledgements

NPS would like to express her thanks to the Australian International Development Assistance Bureau for financial support.

References

1. Catalysts Handbook, ed Twigg, M.V., 2nd Edn. Manson Publishing (1996).

2. Bridger, G. W., Specialist Periodical Reports of the Chemical Society: Catalysis Vol. 3, p 39 (1980).

3. Trimm, D. L., Catal. Rev. Sci. Eng. 16 155 (1977).

4. Trimm, D. L., Design of Industrial Catalysts, Elsevier Amsterdam (1980).

5. Trimm, D. L., & Bernardo, C. A, Carbon 17 115 (1979).

6. Siswana, N. P. & Trimm, D. L., Catal. Letters 46 27 (1997).

7. Geus, J. W., Stud. Surf. Sci. Catal. 16 1 (1983).

8. Thomas, J. M. & Thomas, W. J., Principles & Practice of Heterogeneous Catalysis, VCH Germany (1997).

9. Dowden, D. A. & Wells, D., Actes. Congr. Int. Catal 2nd, p 1499 (1960).

10. Dowden, D. A., Chem. Eng. Progr. Symp. Ser. 63 90 (1967).

11. Richardson, J. T., Turk, B., Twigg, M. V., App. Cat. A 148 97 (1996).

12. Ponec, V. & Bond, G. C., Stud. Surf. Sci. Catal. 95 (1995).

13. Cookson, D. J., Percival, G. G., White, N., Jiang, C. & Trimm, D. L., Arabian J. Sci. Eng. 21 211 (1996).

NATURAL GAS CONVERSION V
Studies in Surface Science and Catalysis, Vol. 119
A. Parmaliana et al. (Editors)

Kinetic study of the catalytic partial oxidation of methane to synthesis gas over Ni/La_2O_3 catalyst

V.A. Tsipouriari and X.E. Verykios*

Department of Chemical Engineering, University of Patras,
GR-26500 Patras, Greece Tel: +3061 997826, Fax: +3061 991527

The kinetic behavior of the Ni/La_2O_3 catalyst in the partial oxidation of methane was investigated as a function of temperature and of partial pressures of CH_4 and O_2. It has been found that CPO reaction over the Ni/La_2O_3 catalyst follows mainly the sequence of total oxidation of CH_4 to CO_2 and H_2O, followed by reforming reactions to synthesis gas, while CO formation via the direct route is possible at low oxygen partial pressures, and depends on the oxidation state of the Ni crystallites. Hysteresis phenomena, which were also observed, suggest that multiple states of the nickel surface can exist depending on the history of the catalyst. It is proposed that the reaction mechanism over the Ni/La_2O_3 catalyst involves adsorption of methane and oxygen on different sites, while the decomposition of CH_4 into C and H and the reaction of adsorbed C and O towards CO are suggested as the rate determining steps (RDS).

1. INTRODUCTION

The catalytic partial oxidation of methane (CPO), constitutes an interesting alternative process for the conversion of natural gas to synthesis gas. This reaction is slightly exothermic and leads to the production of syngas with an H_2/CO ratio of about 2, which is suitable for methanol and Fischer-Tropsch synthesis. In comparison, the steam reforming process over Ni-based catalysts, is less energy efficient and produces less desirable H_2/CO ratio. Thus, the CPO process has attracted the attention of many research groups and has been investigated over various metal-based catalysts. Group VIII metals such as Rh, Ru, Pt, Ir, Pd and Ni are reported to be active and selective towards partial oxidation of methane to synthesis gas[1].

The performance of various Ni-based catalysts under condition of CPO of methane, including Ni/La_2O_3 catalysts, has been studied by various investigators.

Hu and Ruckenstein [2] investigated the kinetics of methane oxidation over Ni/La_2O_3 catalysts in the temperature range of 450-700°C, by the pulse transient analysis method. They showed that CO may be the primary product of CPO reaction, while CO_2 is subsequently generated from combustion of CO. Au et al. [3, 4] performed pulse studies of the reaction over Ni/La_2O_3 and Ni/SiO_2 catalysts. They have found that direct formation of CO via a pyrolysis mechanism takes place over reduced Ni sites, whereas combustion of methane is favored in an oxidized (NiO) surface. CO selectivity is, therefore, determined by the ratio of Ni^o/Ni^{2+} on the surface, which is expected to increase with increasing reaction temperature.

Recent studies conducted in this laboratory [5,6] shown that high activity for CO_2 reforming of methane to synthesis gas, combined with excellent stability, can be achieved over a Ni catalyst consisting of Ni crystallites appropriately dispersed on a La_2O_3 support. This catalyst was also studied in the CPO process [7] and it has been shown that the enhanced stability of the Ni/La_2O_3 catalyst towards the CPO reaction is related to the the decoration of the Ni particles by LaO_x species which favor removal of excess carbon deposited on Ni crystallites.

In the present work, the kinetic behavior of the Ni/La_2O_3 catalyst has been investigated under steady-state conditions as a function of temperature and the partial pressures of CH_4 and O_2. A reaction mechanism is proposed, which explains satisfactorily the experimental data.

2. EXPERIMENTAL

2.1 Catalyst preparation

The methods of preparation of the Ni/La_2O_3 catalyst have been described earlier [5,6].

2.2 Kinetic apparatus and measurements

The apparatus employed for kinetic measurements consists of a flow measuring and control system, a heated quartz tube reactor and an on line analytical system. Feed flow rates were measured and controlled by thermal mass flow meters. Weighed amounts of catalyst (5-10 mg) were placed in the middle of the reactor supported by quartz wool. A chromel-alumel thermocouple, inside a thermowell which run through the catalyst bed was used for temperature measurement and control. Analysis of the feed and of the reaction mixtures was performed by a gas chromatograph connected on-line to the reactor apparatus via a gas sampling valve. The TC detector was used to analyse H_2, O_2, CO, CH_4, CO_2, and H_2O, separated by a carbonsieve S-II 100/120 mesh column. Kinetic measurements were performed under differential conditions in the temperature range of 500-650°C, employing dilute CH_4/O_2/He feeds. The total feed flow rate was 500 ml/min, while the catalyst particle size was in the range of 0.12-0.18 mm.

3. RESULTS AND DISCUSSION

3.1 Mass and heat transport effects

Reliable kinetic data can be only be obtained in the absence of mass and heat transfer resistances. The effects of interface and intraparticle heat and mass transport resistances on kinetic parameters were determined employing theoritical and experimental procedures. In the theoritical estimations, the feed mixture was assumed to consist of 4% CH_4, 2% O_2, 94% He. For a reaction temperature of 800°C, it was estimated that external mass transport resistances are marginally significant, resulting in an O_2 partial pressure difference between the bulk gas phase and the catalyst surface of less than 7%. However, temperature differences were estimated to be in the neighborhood of 12K. these estimates imply that, neglected any internal transport effects, the observed reaction rate will be higher than the intrinsic rate by 5-10%, which may be considered negligible. Based on the theoretical analysis, it was determined that in order to eliminate transport resistances, kinetic experiments must be conducted at temperatures lower than 800°C, the catalyst particles must be very small, of the order of 0.12-0.18 mm in diameter, and the feed flow rate must be in excess of 300 ml/min. These conditions were verified by appropriate experimental procedures and were used in the kinetic mechanism in the present study.

3.2 Kinetic results

The influence of the partial pressures of CH_4 and O_2 on the rate of the catalytic partial oxidation of methane was studied over the Ni/La_2O_3 catalyst at atmospheric pressure in the temperature range of 500-650°C in the absence of mass and heat transfer limitations. The total flow rate was controlled at 500 ml/min. In all experiments methane and oxygen conversion were less than approximately 10%. A constant partial pressure of CH_4 equal to 4 kPa was used when the O_2 partial pressure was varied and a constant partial pressure of O_2 equal to 2 kPa was used when the CH_4 partial pressure was varied.

The effect of the variation of CH_4 partial pressure on the rate of methane combustion in the temperature range of 500-600°C is presented in Fig.1, while the effect of the variation of O_2 partial pressure on the rate of methane conversion in the same temperature region is

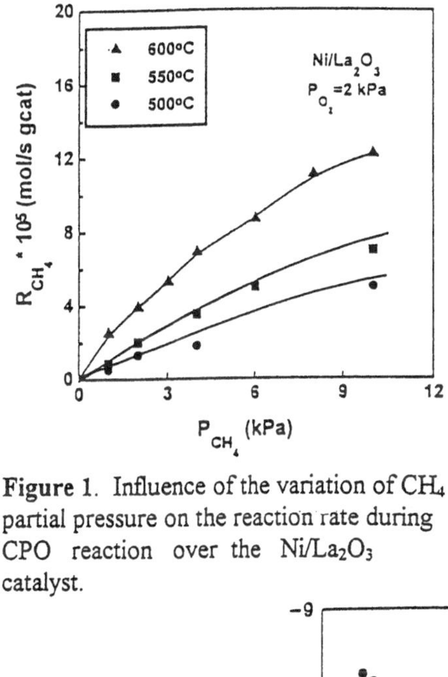

Figure 1. Influence of the variation of CH_4 partial pressure on the reaction rate during CPO reaction over the Ni/La_2O_3 catalyst.

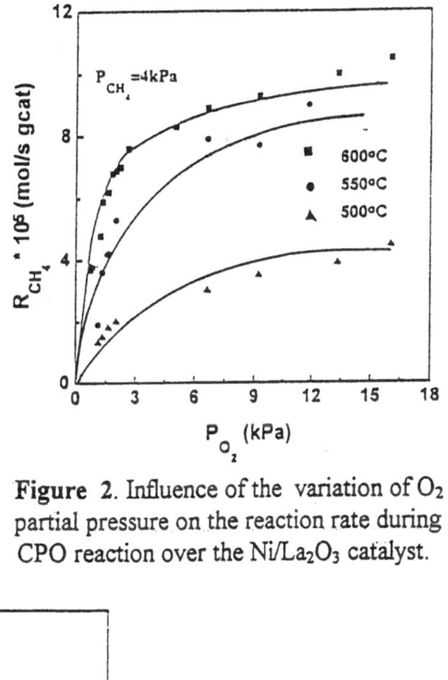

Figure 2. Influence of the variation of O_2 partial pressure on the reaction rate during CPO reaction over the Ni/La_2O_3 catalyst.

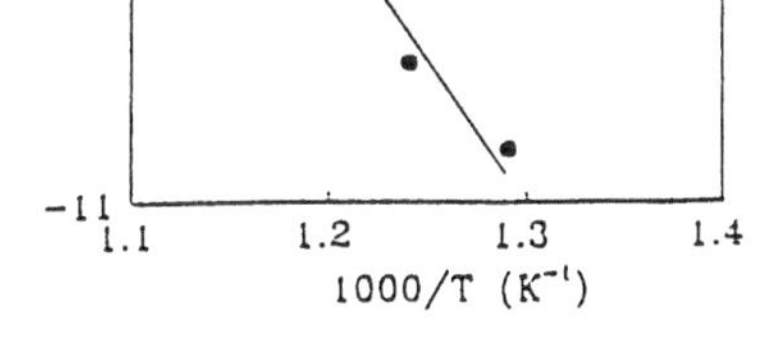

Figure 3. Influence of the reaction temperature on the rate of methane combustion. Experimental conditions: $CH_4/O_2/He$=4/2/94 vol%, Q=500 ml/min.

presented in Fig. 2. It is shown that the rate of reaction is strongly affected by the partial pressure of O_2 at low O_2 partial pressures, while a smaller influence seems to exist when partial pressure of O_2 exceeds approximately 3 kPa. However, the influence of CH4 partial pressure on the reaction rate is not so significant at low CH_4 partial pressures, as in the case of O_2 partial pressures. Indeed, no break in the kinetic curves is observed within the partial pressure range shown in Fig.1. The activation energy of the CH_4/O_2 reaction was determined by employing a feed containing 4 kPa of methane and 2 kPa of oxygen. Initial reaction rates were determined in the temperature range of 500-650°C. The corresponding Arrhenius plot is shown in Fig.3. Under these conditions, CO_2 is produced exclusively with an activation energy of 19 kcal/mol.

Although nickel is generally believed to be an indirect CPO catalyst, kinetic studies performed at low oxygen partial pressures show that this is not the case under specific conditions. Hysteresis phenomena, which were observed, suggest that multiple states of the nickel surface can exist depending on the history of the catalyst. This behavior is depicted in Figs. 4a and 4b, where the steady state rate of methane consumption and CO selectivity at 600°C are plotted as a function of methane pressure at a constant oxygen pressure of 0.6 kPa. As is shown, in Fig. 4 (a), when the methane partial pressure is varied in ascending and descending orders, a large hysteresis loop develops in the rate of methane oxidation, the rate being significantly higher when the experiment is started at high methane pressures (or high P_{CH4}/P_{O2}).

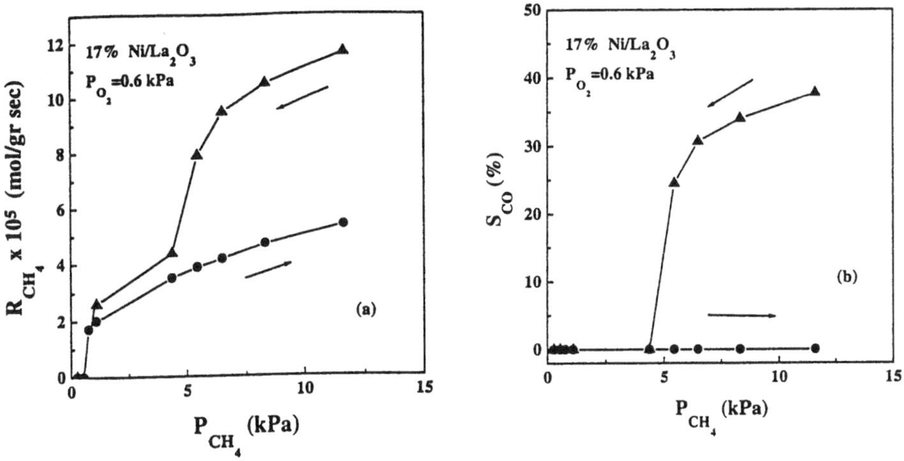

Figure 4. Rate of CH_4 consumption (a) and CO selectivity (b) as a function of CH_4 partial pressure under CPO at 600°C and constant oxygen pressure of 0.6 kPa. Methane partial pressure is varied both in an ascending and a descending sequence.

Further more, as illustrated in Fig. 4(b), selectivity towards CO formation is zero when the experiment is initiated at low methane pressures (low P_{CH4}/P_{O2}), but it is significant (30-40 %) when the experiment is initiated at high methane pressures. At a critical methane pressure (~5 kPa) CO selectivity drops to zero while the methane conversion rate is drastically reduced.

This phenomenon was observed to be sensitive to the temperature. A narrower loop and a higher selectivity towards CO formation was detected at higher temperatures.

More detailed studies of the hysteresis phenomenon reveal that there is a critical value of the P_{CH4}/P_{O2} ratio above which the catalyst is very active and is also selective towards formation of CO in the presence of oxygen. This critical ratio depends strongly on the absolute value of the oxygen partial pressure. As the oxygen pressure increases, the critical P_{CH4}/P_{O2} ratio also increases and, when the oxygen pressure exceeds 1 kPa, (at 600°C) direct CO formation in the presence of O_2 is not observed.

The results presented in the previous paragraphs may be interpreted by considering most realistic values of the P_{CH4}/P_{O2} ratio, the nickel surface is in an "oxidized" state. This state is active for combustion of methane to CO_2. However, under ultra lean oxygen conditions, or very high P_{cH4}/P_{O2} ratios, the nickel surface may exist in a "partially reduced" state. This state is more active and is also selective towards the direct formation of CO. At 600°C, the "reduced" state cannot existd at oxygen pressures higher than 1 kPa. The window in which nickel can exist in a "reduced" state becomes wider at higher reaction temperatures. These results are in agreement with the results obtained over Ni/La_2O_3 catalyst by Hu and Ruckenstein [2]. Their results indicate that CO is a primary product of the reaction and CO_2 is produced by subsequent oxidation of CO. Oxidation of CO was found to be strongly favored over oxidized Ni. Au et al. [3,4] have also proposed that direct production of CO is determined by the ratio of Ni^{o}/Ni^{2+} on the surface, which is expected to increase with increasing reaction temperature.

3.3 Kinetic model

It is apparent from the results presented above that the surface state and kinetic behavior of the Ni/La_2O_3 catalyst under conditions of CPO of methane is affected by the partial pressure of oxygen and the fuel-to-oxygen ratio. However, under realistic conditions of the variables, the catalyst seems to be in the "oxidized" state and its kinetic behavior may be described following a regular approach and procedure. Thus, based on the kinetic results which have been obtained, a kinetic model has been developed, which predicts with satisfactory accuracy the kinetics of the methane combustion reaction. The model is based on the following mechanism: methane adsorbs on particular sites (S_1) on the surface of Ni via an intermediate species, (rate determing step) which decomposes to adsorbed C and H. Oxygen adsorbs on different sites, S_2, via a molecular precursor and this step is at equilibrium. Adsorbed carbon and adsorbed oxygen react to form adsorbed CO, a step which is also rate-determining. The CO thus produced reacts very rapidly with adsorbed oxygen to produce CO_2. The same is true of adsorbed hydrogen which also reacts very rapidly to produce H_2O. The mechanistic steps are illustrated as follows:

$CH_4 + S_1 \rightarrow \text{intermediate I} \rightarrow C\text{-}S_1 + 4H\text{-}S_1$	RDS	(1)
$O_2 + S_2 \rightleftarrows O_2\text{-}S_2$	Equilibrium	(2)
$O_2\text{-}S_2 + S_2 \rightarrow 2O\text{-}S_2$	Fast	(3)
$C\text{-}S_1 + O\text{-}S_2 \rightarrow CO\text{-}S_1 + S_2$	RDS	(4)
$CO\text{-}S_1 + O\text{-}S_2 \rightarrow CO_2 + S_1 + S_2$	Fast	(5)
$H\text{-}S_1 + O\text{-}S_2 \rightarrow OH\text{-}S_1 + S_2$	Very fast	(6)
$OH\text{-}S_1 + H\text{-}S_1 \rightarrow H_2O + 2S_1$	Very fast	(7)

A rate expression for the combustion of methane is developed, based on the mechanism illustrated above:

$$R = \frac{k_1 K_2 k_3 P_{CH_4} P_{O_2}}{k_1 K_2 P_{CH_4} P_{O_2} + k_1 P_{CH_4} + K_2 k_3 P_{O_2}}$$

The predictions of the model, i.e. the rate of methane combustion as a function of methane and oxygen partial pressures, is compared in Figures 1 and 2 with the experimental data (the solid lines are the model predictions). It is apparent that the kinetic model is in good agreement with the experimental data indicating that the basic features of the reaction mechanism have been correctly incorporated into the model. Other kinetic mechanism were also analysed but their kinetic predictions were significalty different than the experimental observations.

4. CONCLUSIONS

The following conclusions can be drawn from the results of the kinetic study of the CH_4/O_2 reaction over the Ni/La_2O_3 catalyst:

- The CPO reaction follows mainly the indirect reaction scheme over Ni/La_2O_3 catalyst, while the CO formation via the direct route is observed at low oxygen partial pressures and high P_{CH4}/P_{O2} ratios.
- Hysteresis phenomena observed during kinetic measurements suggest that multiple states of the nickel surface can exist depending on the history of the catalyst.
- The kinetic model based on adsorption of CH_4 and O_2 on different surface sites and rate determining steps, the adsorption of CH_4 and the formation of CO, predicts satisfactory the experimental results.

ACKNOWLEDGEMENT

Financial support by the Commission of the European Union (Contract no. JOU2-0073 and JOF3-CT95-0026) is gratefully acknowledged.

REFERENCES

1. M.A Pena, J. P. Comez and J. L. G. Fierro, Appl. Catal. A 144 (1996) 7.
2. Y.H. Hu and E. Ruckenstein, J. Catal. 158 (1996) 260.
3. C.T. Au, Y.H. Hu and H.L. Wan, Catal. Lett., 27 (1994) 199.
4. C.T. Au, H.Y. Wang and H.L. Wan, J. Catal., 158 (1996) 343.
5. Z.L. Zhang and X. E. Verykios, Appl. Catal., 138 (1996) 109.
6. Z.L. Zhang and X. E. Verykios, S.MacDonald and S.Affrossman, J. Phys. Chem.,100 (1996) 744.
7. V.A. Tsipouriari, Z.L. Zhang and X.E. Verykios, J. Catal., submitted.

NATURAL GAS CONVERSION V
Studies in Surface Science and Catalysis, Vol. 119
A. Parmaliana et al. (Editors)

Kinetic behaviour of the Ru/TiO_2 catalyst in the reaction of partial oxidation of methane

C. Elmasides, T. Ioannides and X. E. Verykios

Department of Chemical Engineering, University of Patras
GR-265 00 Patras, HELLAS

The kinetic behavior of the Ru/TiO_2 catalyst in the partial oxidation of methane was investigated as a function of temperature and partial pressures of CH_4, O_2, CO, CO_2, H_2 and H_2O. It has been found that the catalyst promotes, to a large extent, the direct formation of syn-gas from the reaction of methane and oxygen, while the reforming and water-gas shift reactions are negligible under these conditions. It is proposed that the reaction mechanism over the Ru/TiO_2 catalyst involves formation of CO-H_2 and CO_2-H_2O in two in series reaction routes, possibly on different active sites of the Ru surface. Part of H_2, and, to a lesser extent, of CO is consecutively oxidized by adsorbed oxygen to H_2O and CO_2 giving the final product distribution.

1. INTRODUCTION

Formation of synthesis gas via catalytic partial oxidation of methane has received significant attention during the past several years. Concerning the reaction pathway of partial oxidation of methane to synthesis gas, two main reaction schemes have been proposed: one is the sequence of total oxidation (CO combustion) followed by reforming reactions (indirect scheme), and the other is the direct conversion of methane to synthesis gas without the experience of CO_2 and H_2O as reaction intermediates (direct scheme).

The indirect scheme is supported by experimental findings of several investigators [1-3]. It has been found that the direct reaction scheme for synthesis gas formation is followed to a large extent when the reaction is carried out over Ru/TiO_2 catalysts [4,5]. More specifically, it has been found that, in the absence of mass and heat transfer resistances, high selectivity to synthesis gas (>65%) is obtained over Ru/TiO_2 catalysts in the low methane conversion range (oxygen conversion < 100%), whereas small selectivity to synthesis gas is observed over Ru catalysts supported on carriers other than TiO_2 or over other metal catalysts under the same conditions. The extent of the direct scheme was also found to be sensitive to modifications of the TiO_2 carrier [5]. Doping of the TiO_2 carrier with cations of lower valence, such as Ca^{2+}, was found to promote further the direct formation of synthesis gas over Ru catalysts. Direct formation of synthesis gas over metal-loaded monoliths and ceramic foams operating at high temperatures (~1000°C) and short contact times (~1 ms) has been reported by Schmidt and co-workers [6,7].

In the present work, the kinetic behaviour of a Ru/TiO_2 catalyst has been investigated under steady-state conditions as a function of temperature and partial pressures of CH_4, O_2, CO, CO_2, H_2 and H_2O. The carrier of the present catalyst contains Ca^{2+} cations as dopant and La_2O_3 as stabilizer.

2. EXPERIMENTAL

The parent titania carrier used in the present study was obtained from Degussa(P-25). The preparation of the doped TiO_2 carrier and of the Ru catalyst is described in Ref. [5].

Kinetic measurements were performed under differential conditions in the temperature range of 680-800°C employing dilute $CH_4/O_2/N_2$ feeds. Nitrogen was used as a diluent instead of helium in order to be able to analyze produced hydrogen by gas chromatography. The quantity of catalyst used was 30 mg. The catalyst particle size was in the range of 0.12-0.18 mm. A type K thermocouple was positioned inside the catalyst bed for accurate measurement of catalyst temperature.

Two gas chomatographs were used for the analysis of reactants and products. CH_4 and CO_2 were analyzed in a Carboxen 1000 column and O_2, N_2 and CO in a Molecular Sieve 5A column with He as a carrier gas, while hydrogen was analyzed in the second chromatograph employing N_2 as the carrier gas.

3. RESULTS AND DISCUSSION

3.1. Mass and heat transport effects

The experimental conditions of the kinetic measurements of the present work, were carefully chosen in order to minimize the effect of mass and heat transfer resistances. Thus, experiments were conducted at rather low reaction temperatures (~700°C) with very dilute CH_4/O_2 feeds and high feed flow rates (~800-1000 cm^3/min). It has been calculated theoretically and verified experimentally [4] that mass and heat transfer resistances do not influence the kinetic results of the present study.

3.2. Effect of CH_4 and O_2 partial pressure on catalyst behaviour

The effect of the partial pressures of CH_4 and O_2 on the activity and selectivity of the catalyst was investigated at reaction temperatures of 680, 700 and 720°C. The rates of CO and CO_2 production and of CH_4 consumption are presented in Fig. 1 as a function of CH_4 partial pressure at a reaction temperature of 700°C and at constant partial pressure of oxygen of 2.7 kPa. Two different working states of the catalyst can be observed. In the first state, which is observed at CH_4/O_2 ratios lower than approximately 2, the catalyst is essentially inactive for the reaction (only traces of CO_2 are detected in the reactor effluent). At CH_4/O_2 ratios higher than 2, the catalyst exhibits high activity combined with good selectivity for CO and H_2 formation. Similar results were also obtained during kinetic experiments at reaction temperatures of 680 and 720°C.

The effect of oxygen partial pressure on the activity and selectivity of the catalyst is presented in Fig. 2, where the rates of CO and CO_2 production and of CH_4 consumption are plotted as a function of oxygen partial pressure at a reaction temperature of 700°C and at constant partial pressure of methane of 6.4 kPa. It can be seen that the reaction order with respect to oxygen pressure is positive at low oxygen pressures. At higher partial presures of oxygen (2.5-3.0kPa) a plateau in catalyst activity is reached and the catalyst transforms to its inactive state at an oxygen pressure of 3.0-3.2 kPa ($P_{CH_4}/P_{O_2}<2$).

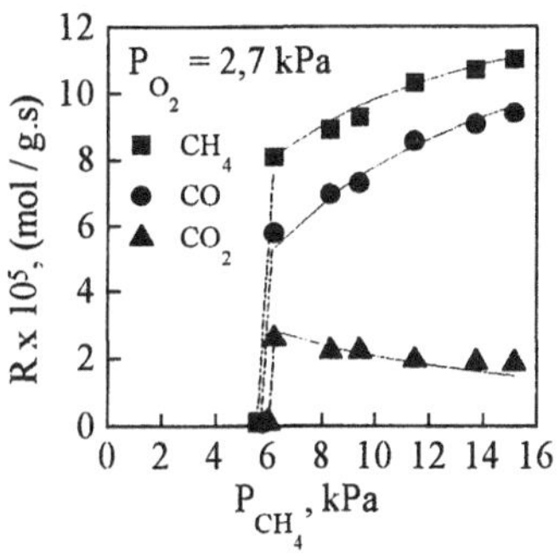

Figure 1. Influence of methane partial pressure on the activity of the catalyst Solid lines: model predictions

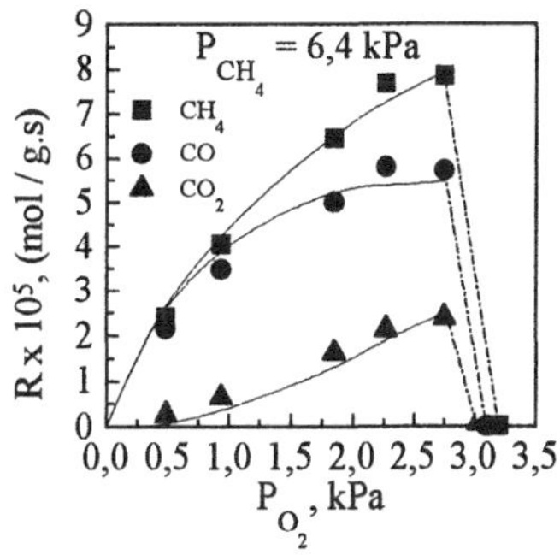

Figure 2. Influence of oxygen partial pressure on the activity of the catalyst Solid lines: model predictions

Selectivity towards CO and H_2 formation observed during the experiments presented in Figs. 1 and 2 is plotted in Fig. 3 as a function of the CH_4/O_2 ratio. It can be seen that the data from all experiments appear to lie in a single curve (one for CO and one for H_2). This result implies that, in the range of partial pressures which was examined, selectivity to CO and H_2 is determined primarily by the CH_4/O_2 ratio rather than by the individual values of the partial pressures of methane and oxygen. Both selectivities increase with increasing CH_4/O_2 ratio and they appear to be sensitive to this ratio when it acquires values close to two, which corresponds to the transition of the catalyst to the extinguished state.

The effect of temperature on the activity and selectivity of the catalyst was examined for a feed containing 4.2 kPa of methane and 1.9 kPa of oxygen at temperatures in the range of 680-800°C. The corresponding Arrhenius plots for the rates of CO, H_2 and CO_2 formation and methane consumption are presented in Fig. 4. The rate of H_2O formation is not shown for reasons of clarity. The activation energy of CO_2 and H_2O production was found to be quite small (8-9 kJ/mol), as the rate of CO_2 and H_2O production increases only slightly with temperature. The activation energies of CO and H_2 production were found to be 72 and 80 kJ/mol, respectively, which implies that selectivity towards synthesis gas formation increases with increasing reaction temperature.

3.3. Effect of addition of reaction products in the CH_4/O_2 feed

The addition of steam in the feed (Fig. 5) results in the decrease of the rates of CH_4 consumption and CO and H_2 production, while the rate of CO_2 production is not affected.

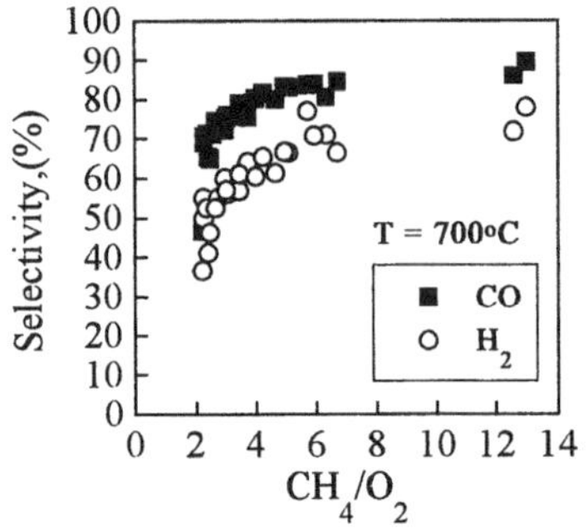

Figure 3. Selectivity of the catalyst for CO and H_2 formation as a function of CH_4/O_2 ratio

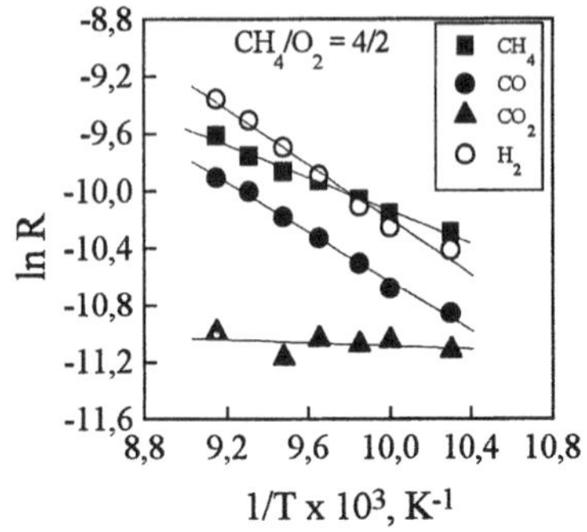

Figure 4. Arrhenius plots of the rates of CO, H_2 and CO_2 production and CH_4 consumption under CPO conditions

This result suggests that water acts as an inhibitor of the reaction of partial oxidation of methane towards CO and H_2. This is, probably, due to adsorption of H_2O on the surface of the catalyst. The observed behaviour shows also that the steam reforming and water-gas shift (WGS) reactions do not take place at a rate high enough to influence the product distribution.

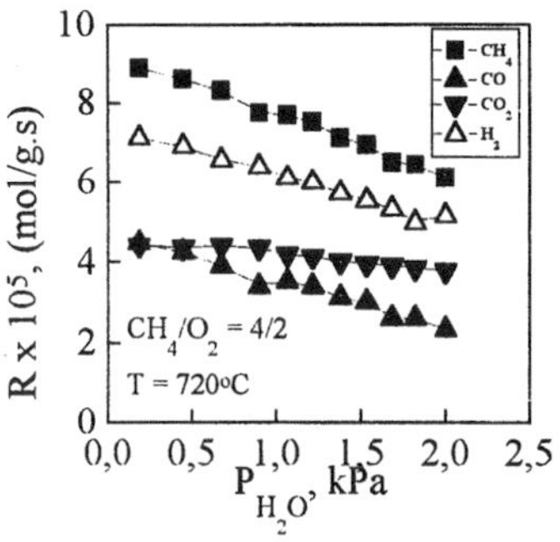

Figure 5. Effects of steam partial pressure on the rates of CPO of methane

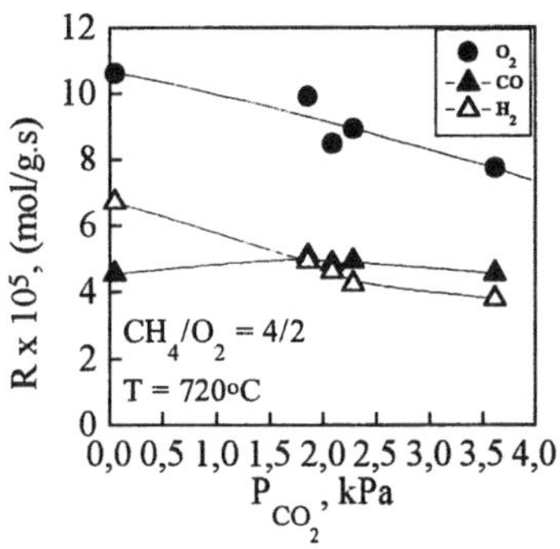

Figure 6. Effects of partial pressure ofCO_2 on the rates of CPO of methane

The same conclusion arises in the case of CO_2 addition in the CH_4/O_2 feed (Fig. 6). The almost constant rate of CO production with CO_2 addition indicates that the CO_2 reforming does not take place at an appreciable rate under these conditions.

An increase in the rate of CO_2 production, by up to 70% was observed upon addition of CO in the feed (Fig. 7). The enhancement of the CO_2 production rate is not due to the WGS reaction, because, in this case, the rate of H_2 production should also increase, which is not the case. The observed increase in CO_2 formation rate is most probably due to oxidation of the CO added in the feed. However, it should be pointed out that the rate of partial oxidation of methane to CO is at least 75% higher than the rate of oxidation of CO under these conditions.

The effect of H_2 partial pressure on the rates of CO and CO_2 production and the rates of O_2 and CH_4 consumption is presented in Fig. 8. It can be observed that addition of H_2 has a positive effect on the activity of the catalyst in the partial oxidation of methane. The dashed line in Fig. 8 (denoted as $R^*_{O_2}$) refers to the estimated rate of oxygen consumption during the reaction of partial oxidation of methane. The difference between the dashed line and the true rate of O_2 consumption corresponds to the rate of O_2 consumption for hydrogen oxidation. The rate of hydrogen oxidation at a hydrogen pressure of 1.8 kPa is $16x10^{-5}$ mol/g.s. Comparison of this value with the rate of CO oxidation, when CO is added in the feed($2.1x10^{-5}$ mol/g.s) shows that the rate of oxidation of hydrogen added to the feed is approximately 8 times higher than the rate of oxidation of CO added to the feed. This explain the higher selectivity of the catalyst toward CO formation than H_2 formation (Fig. 3).

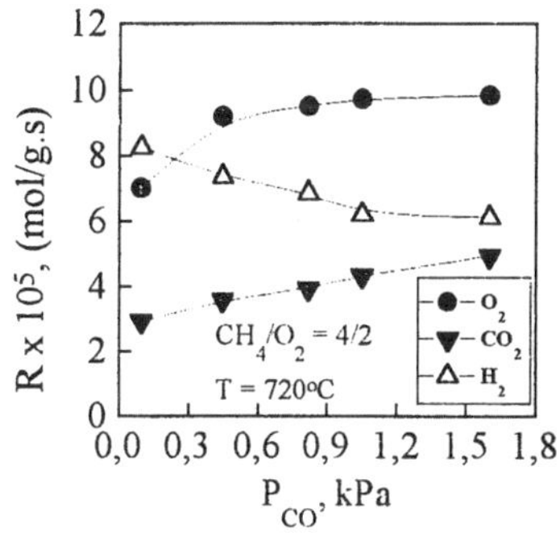

Figure 7. Effects of partial pressure of CO on the rates of CPO of methane

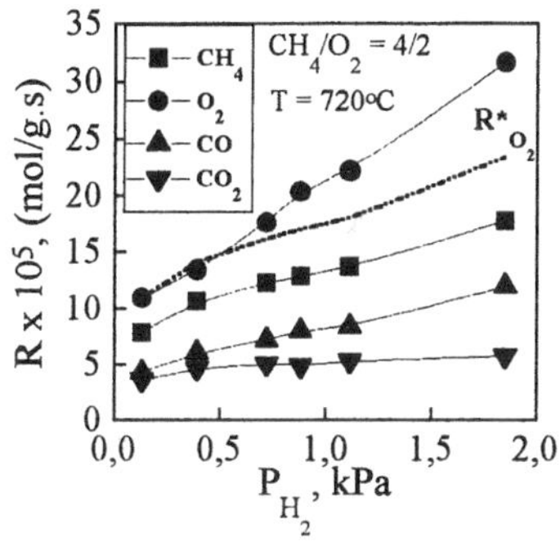

Figure 8. Effects of partial pressure of H_2 on the rates of CPO of methane

3.4. Kinetic model

A model which describes the experimental results assumes CH_4 cracking via an adsorbed precursor and the reaction of the surface carbon with atomically adsorbed oxygen (Langmuir-Hinshelwood) to adsorbed CO which either desorbs or reacts further toward CO_2. Based on the Langmuir-Hinshelwood model the following expressions for the rates of CH_4 consumption and CO and CO_2 production are obtained:

Methane consumption

$$R_{CH_4} = \frac{k_1 K_2 k_3 P_{CH_4} P_{O_2}}{K_2 k_3 P_{O_2} + k_1 P_{CH_4} + k_1 K_2 P_{O_2} P_{CH_4}} \tag{1}$$

CO_2 production

$$R_{CO_2} = \frac{K_2^2 k_3 k_4 P_{O_2}^2 P_{CO}}{\left(1 + K_2 P_{O_2}\right)\left(K_2 k_3 P_{O_2} + k_1 P_{CH_4} + k_1 K_2 P_{O_2} P_{CH_4}\right)} \tag{2}$$

CO production
$R_{CO} = R_{CH_4} - R_{CO_2}$ (3)

The model prediction is presented in Figures 1 and 2, in the form of a continuous line, for the rates of methane consumption and CO and CO_2 production under conditions of variable partial pressure of methane or variable partial pressure of oxygen. The prediction of the model sutisfies well the experimental observations.

4. CONCLUSIONS

The following conclusions can be drawn from the results of the kinetic study of the CH_4/O_2 reaction over the Ru/TiO_2 catalyst:

- Two different states of the working catalyst surface exist: a state at low P_{CH_4}/P_{O_2} ratios, which is almost inactive and non-selective and a state at high P_{CH_4}/P_{O_2}, which is active and highly selective for CO and H_2 formation. The transition between these two states occurs at a CH_4/O_2 ratio of approximately 2.
- The initial selectivity for CO and H_2 formation can be as high as 90% and increases with increasing CH_4/O_2 ratio and with temperature.
- The high CO selectivity observed under conditions of partial oxidation of methane can be attributed to the comparably low activity of the catalyst to oxidize CO in the presence of oxygen.
- A kinetic model based on the mechanism of CH_4 cracking, where surface C reacts with atomicaly adsorbed oxygen toward CO while CO_2 is the product of the CO oxidation, describes satisfactory the experimental results.

ACKNOWLEDGEMENT

Financial support by the Commission of the European Union (Contract no. JOF3-CT95-0026) is gratefully acknowledged.

REFERENCES

1. D. Dissanayake, M. P. Rosynek, K. C. C. Kharas and J. H. Lunsford, J. Catal., 132 (1991) 117.
2. D. Dissanayake, M. P. Rosynek and J. H. Lunsford, J. Phys. Chem., 97 (1993) 3644.
3. M. Prettre, C. Eichner and M. Perrin, Trans. Faraday Soc., 43 (1946) 335.
4. Y. Boucouvalas, Z. L. Zhang and X. E. Verykios, Catal. Lett., 40 (1996) 189.
5. Y. Boucouvalas, Z. L. Zhang, A. M. Efstathiou and X. E. Verykios, Studies in Surface science and Catalysis, Vol. 101, Elsevier, Amsterdam, 1996, p. 443.
6. D. A. Hickman and L. D. Scmidt, Science, 259 (1993) 343.
7. P. M. Torniainen, X. Chu and L. D. Schmidt, J. Catal., 146 (1994) 1.

NATURAL GAS CONVERSION V
Studies in Surface Science and Catalysis, Vol. 119
A. Parmaliana et al. (Editors)

Catalytic Partial Oxidation of Methane in a Spouted Bed Reactor. Design of a Pilot Plant Unit and Optimization of Operating Conditions.

S.S. Voutetakis [a], G. J Tjatjopoulos [a], I.A. Vasalos [a] and U. Olsbye [b]

[a]Chemical Process Engineering Research Institute, P.O. Box 361, 57001 Thermi, Thessaloniki, Greece
[b]SINTEF Applied Chemistry, P.O. Box 124 Blidern, N-0314 Oslo, Norway

In order to investigate the feasible operation of a spouted bed reactor in catalytic partial oxidation a pilot plant was constructed in CPERI using a high degree of automation. The results imply that the spouted bed reactor concept is promising for application in the CPO process, on the condition that the adverse fountain effects are eliminated or minimized.

1. INTRODUCTION

Catalytic partial oxidation of Methane can be an attractive process for converting natural gas to synthesis gas (SYNGAS), because it gives high methane conversions (over 90% per pass) combined with high selectivities to carbon monoxide [1-3]. However, the successive application of such a reaction at industrial scale depends on the feasibility of safe operation even at high pressures and large flows. Based on results reported in a British Petroleum patent on oxidative pyrolysis [4], analysis shows that oxidative pyrolysis is a combination of methane partial oxidation to carbon monoxide which produces heat and of methane pyrolysis which consumes heat [5]. The methane pyrolysis reaction is thermodynamically limited and requires temperatures above 1000 °C for significant conversion [6]. In order to investigate the feasible operation of a spouted bed reactor in catalytic partial oxidation a pilot plant was constructed in CPERI using a high degree of automation. The results imply that the spouted bed reactor concept is promising for application in the CPO process, on the condition that the adverse fountain effects are eliminated or minimized.

2. DESCRIPTION OF THE SYNGAS PILOT PLANT UNIT

Spouted bed is a promising reactor configuration in oxidative partial oxidation process, as it combines short residence times for the gas phase along with almost perfect mixing of the catalytic particles, which enhances the heat transfer inside the bed. The concept of the spouted beds is that gas enters from a small nozzle at the base of the catalytic bed at high velocity, creating a central dilute phase core. The particles which rise inside the core rain back onto the bed in the form of a fountain, and they slowly travel downwards inside the annular dense phase. The gas flows mainly inside the core, while a small percentage of flow is distributed to

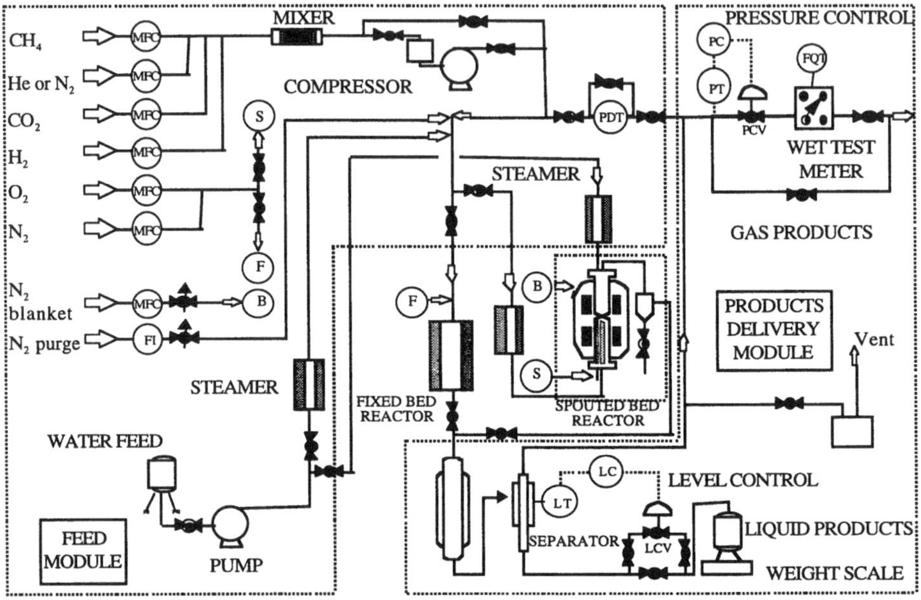

Fig. 1. Simplified flow diagram of the Spouted Bed reactor pilot plant

the annulus. The system is characterized by cyclic pattern of solids movement and high gas velocity [7]. The SYNGAS pilot plant unit was designed and constructed by Chemical Process Engineering Research Institute (CPERI). The unit supports the spout bed type reactor and a fixed bed reactor used alternatively for preliminary activity measurements. The construction material for the reactors is heat resistant stainless steel and high alloy while the rest of the unit is made of SS-316. The two reactors share the same utilities and feed systems. The pilot plant is fully computerized and automated. Graphical sheets can represent the alternative flowsheet of either reactor unit through a simple command. A simplified process flowchart is given in Figure 1. The pilot plant is constructed module wise, and can be distinguished to the following discrete sections:

2.1. The feed module

This section consists of 8 independent gas feed lines and one liquid feed line with all the necessary safety and automation equipment, supplying the following substances: Methane (or natural gas), Nitrogen (or Helium), Carbon Dioxide, Hydrogen, Oxygen and Nitrogen. The feed lines of the first four substances are mixed together prior to feeding into the reactor. The pressure of this "reducing" gaseous mixture can be elevated (in case of using natural gas), by means of a gas compressor. The two other feed lines (Oxygen and Nitrogen mixture), consisting the "oxidizing" feed, are mixed separately and fed to the reactor by a separate route. The above feeds are controlled by mass flow controllers. Two more Nitrogen feed lines are included: a Nitrogen line for purging the system controlled by a rotameter, and another line of Nitrogen for blanketing the spout reactor shell which is measured by mass flow indicator. Liquid water is delivered at the system pressure from a liquid feed vessel by a precise syringe pump. Water is converted into steam, after passing through a steaming unit before entering into the process. The mass flow controllers, the water pump and the air operated valves are connected to the computer, allowing remote operation of the feed system.

Table 1
Specifications of spouted bed reactor

Residence time	<0.5 sec
Internal Diameter	54 mm
Maximum Reactor Temperature	1100 °C
Maximum Reactor Pressure	25 barg
Maximum Gas Flow	45 Slpm
Quench Steam Feed Rate	3-5 gr/min
Solids Catalyst Load	70-200 gr

2.2. The spouted bed reactor

The specifications of the reactor are reported in Table 1. The reactor consists of a pipe which is installed inside a high pressure shell (figure 1). A nitrogen flow maintains a slight positive pressure differential between the reactor shell and the reactor wall (figure 1, line B). By using this concept, the stress at the reactor wall along with the potential danger in case of material failure are minimized. Continuous recording of the above pressure difference is essential for the safe operation of the reactor. The whole construction is included inside an air tight cell with heavy walls, which is continuously purged with fresh air. The reactor, includes a pre-heating section and a reaction section. The reducing and oxidizing streams are preheated separately by an electric furnace at the preheating section and they are mixed together just at the entrance of the reaction section where they pass through a 1.6 mm nozzle at an extremely high speed. The reaction section is cylindrical, with an inverted conical base. The catalytic bed has a height of 1 to 10 cm above the bottom of the inverted cone. The gaseous feed creates a spout in the core of the bed and a fountain of particles above the bed. Observation of spouting state of the bed inside the reactor is accomplished by monitoring the differential pressures around the reactor. The bed is initially heated by another furnace. After the reaction has started, the system is producing enough heat to maintain autothermal conditions. Above the fountain region, the gas is cooled down by quench steam which is fed concurrently inside the reactor, and also by a cooling coil located at the external surface of the reactor exit. A wire filter after the exit of the reactor recovers any fine particles that are formed due to attrition of the catalyst, which may be entrained by the upcoming gas.

2.3. The product delivery and analysis module

The reactor effluent is cooled down to 3-7 °C, passing through a heat exchanger. The liquid product (water) is recovered at the bottom of a cooled condenser-separator at 2 °C, and it is collected into a liquid product vessel, while its production rate is measured by means of an electronic scale. The total gas flow-out is recorded by means of a wet test meter with a digital interface, while a sub-stream is led to the analysis system, for on-line determination of the obtained composition of the gas products. Pressure regulators and filters installed in the sampling line, provide humid-free gas sample to the connected Gas Chromatograph, which integrates the analysis within 15 minutes.

3. EXPERIMENTAL CONDITIONS

The performance of the spouted bed reactor in the catalytic partial oxidation of methane was extensively tested in a series of experiments using a catalyst 0.15wt%Ni/1.7wt%La/Al_2O_3 with a mean particle diameter of 150 and 670 microns [8]. The operating conditions covered in these experiments were: pressure of 2, 5 and 8 bar, dense bed temperature of 850-1000°C, and methane-to-oxygen feed ratios in the range of 1.6-3.0 mol/mol with 0-40% nitrogen dilution. A typical catalyst load amounted to 0.145 kg (or 100 ml) with a particle size distribution in the range of 150-250 microns. The nominal feed flow rates were 10 lit/min at 2 bar, 20 lit/min at 5 bar and 32 lit/min at 8 bar, leading to a catalyst contact time (W_{cat}/F_{CH4}) of 10-40 $kg_{cat}.s/mol_{CH4,o}$. The mass balance for Carbon was closed with an approximation of 1%, while the error for the balances of oxygen and hydrogen reached up to +/-5% in some experiments. The product distribution was measured at the reactor outlet while the axial temperature profile was recorded below the bed surface and inside the freeboard region up to middle zone of the particle fountain.

4. DISCUSSION OF SYNGAS PILOT PLANT RESULTS

The performance of several catalysts was tested for various experimental conditions. The conversion and selectivity were always increasing with temperature and decreasing with pressure as expected from equilibrium trends. A percentage of diluent nitrogen in the gaseous feed seems to have a significant effect on reaction condition. By increasing nitrogen up to a value of 40% the conversion and selectivity show considerable improvement. When using large particles leading to shorter fountains an improved performance was observed. However, prohibitively large flow rates that are needed with large particles at high pressures for the described system. The catalyst remained stable in all experiments (2-4 h duration) with small attrition which amounted to a loss of 0.3-0.4wt% in the form of fines. A comparison of the experimental results with the theoretically calculated chemical equilibrium is made in figure 2 (experimental a) for methane conversion versus temperature and figure 3 (experimental a) for CO selectivity. The experimental curve for conversion has a considerable distance from equilibrium, especially at low temperatures. This distance is significantly decreased for the measured selectivity. Thermodynamic equilibrium has not yet been achieved for the real spouted bed operation. The same behavior is observed in figures 4 and 5 where the effect of pressure is investigated. Even though the overall trend is consistent with theoretical expectations a significant distance from equilibrium is observed. Catalyst evaluation and optimization of reaction conditions or reactor configuration were investigated out in order to provide even better results than the already presented promising yields. Therefore, it was observed that a temperature drop of 50-100 °C was found between the lower (dense bed) and

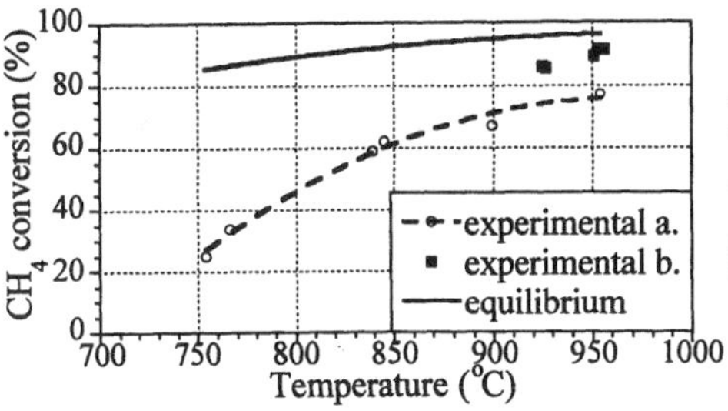

Figure 2. Conversion vs. Temperature under previous (a) and new (b) reactor designs ($CH_4 : O_2 : N_2 = 40:20:40$, P = 2 barg)

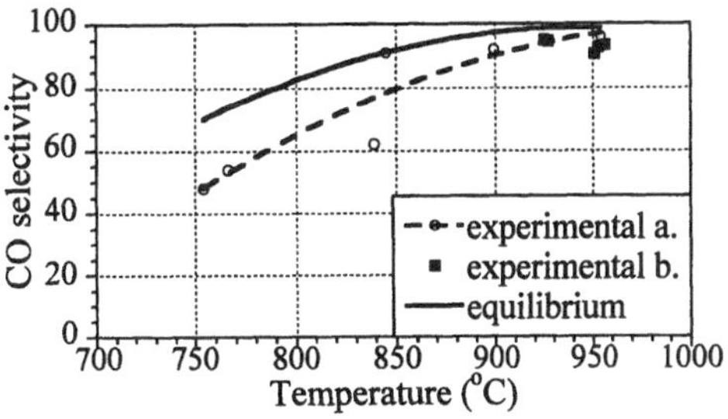

Figure 3. CO Selectivity vs. Temperature under previous (a) and new (b) reactor designs ($CH_4 : O_2 : N_2 = 40:20:40$, P = 2 barg)

upper (mid-fountain) thermocouple positions. Additionally, it was observed that in all experiments, the methane conversion and syngas selectivity at the reactor outlet were consistent with the equilibrium product distribution calculated at the specified feed composition and operating pressure, but at a reaction temperature about 150 °C lower than the temperature recorded at the mid-fountain position. This result can be explained by noting that an additional temperature drop of as much as 200 °C occurs from the middle to summit of the fountain due to the design requirement that the temperature at the reactor outlet be maintained lower than 150 °C because of material limitations.

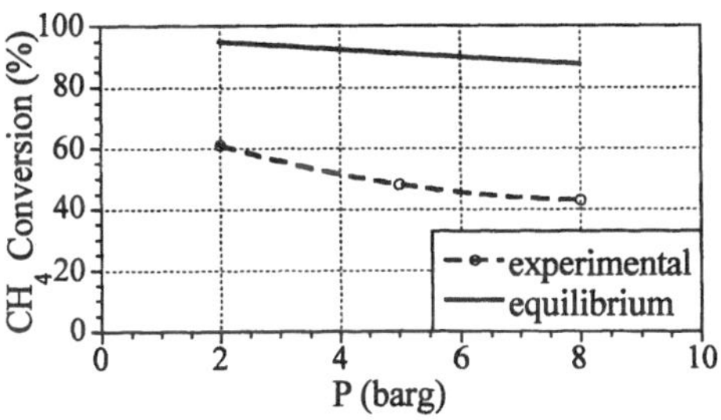

Figure 4. Conversion vs. Pressure ($CH_4 : O_2 : N_2 = 40:20:40$, T = 908° C)

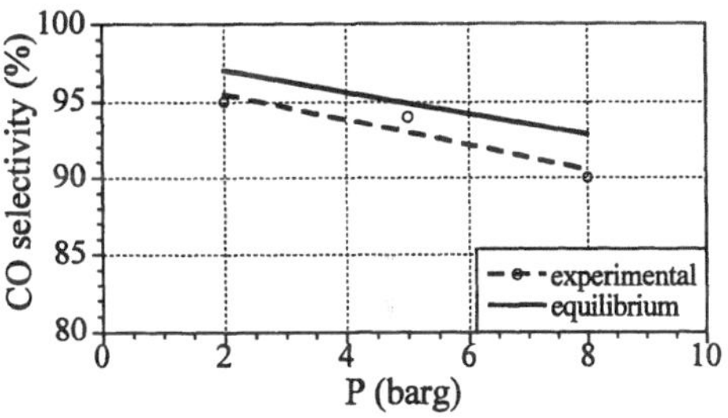

Figure 5. CO Selectivity vs. Pressure ($CH_4 : O_2 : N_2 = 40:20:40$, T = 908° C)

4. REACTOR DESIGN OPTIMIZATION

Based on those experiments and the explanations given above, it was concluded that with the original reactor design (freeboard height and cooling of the exit zone) it was difficult to complete the second stage of the catalytic partial oxidation reaction, i.e. endothermic reforming of the unreacted methane with produced steam and carbon dioxide. For optimizing the reactor operation for the specific process, the design was modified and the upper section of the reactor was reconstructed in order to minimize the influence of the exit boundary conditions on the reactor performance (figure 6). This design modification was proven successful during a series of comparative experiments. Specifically the temperature span observed between the 1st and 4th thermocouples of the previous design which correspond to the locations of 1st and 2nd of the new design was reduced from 200 °C to 40 °C, leading to conversions closer to the reaction equilibrium temperatures than before as seen in figure 2 (experimental b.) without adverse effects in CO selectivity (figure 3, experimental b.) The results of the optimized reactor design are more profound when combined with the large diameter particles which form shorter fountains and therefore more confined boundary conditions for the completion of the reaction near the main reactor bed.

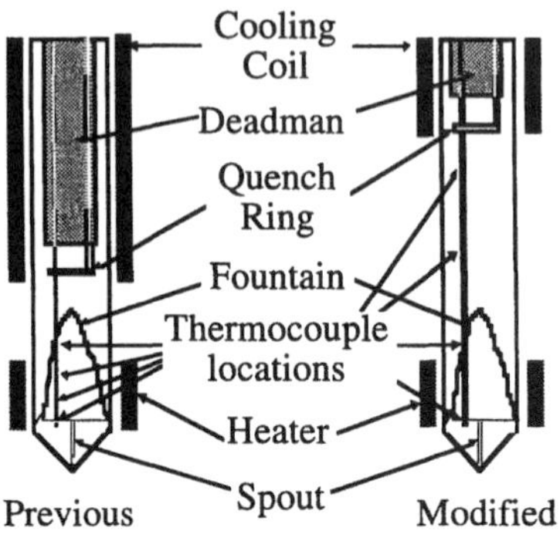

Figure 6. Comparison of Previous and Modified Reactor Designs

5. CONCLUSIONS

As shown in the presented results the unit is performing satisfactorily based on the theoretical expectations and the results are repeatable. The control system helps to assure the effective operation and safety of the unit. The perfect integration of state of the art computer control with a novel type reactor led to practical understanding of the specific requirements of the partial oxidation of methane reaction. Thus, it was made possible to alter the initial reactor design in order to achieve even better process results. Further effort that is currently carried out in the modeling of such a reactor will prove that this process path is successful. Especially when high pressures of synthesis gas are required for further processing, the ability of spouted bed reactor to operate at such conditions is of significant importance when one regards the improvement in overall plant economics that is achieved by eliminating a costly compression.

REFERENCES

1. A.A. Lemonidou, A.E. Stambouli, G.J. Tjatjopoulos and I.A. Vasalos, Catal. Letters, 43 (1997) 235
2. D. Dissanayake, M.P. Rosynek, K.C.C. Kharas and J.H. Lunsford, J. Ctal., 132 (1991)117
3. A.T. Ashcroft, A.K. Cheetman J.S. Foord, M.L.H. Green, C.P. Grey, A.J. Murrel and P.D.F. Vernon, Nature 344 (1990) 319
4. J.H. Brophy, R.P. Manning, B.P., Patent No. WO 85/05094
5. G.A. Huff, I.A Vasalos, (1997) Proceedings of 5th European Workshop on Methane Activation, 8-10 June, Limerick, Ireland.
6. F.J. Weinberg, T.G. Bartleet, F.B. Carleton, P. Rimbotti, J.H. Brophy, R.P. Manning, Combust. Flame, 72 (1988) 235.
7. K.B. Mathur and N. Epstein, Spouted Beds, Academic Press, New York, 1974.
8. A. Slagtern, U. Olsbye, R Blom, I. M. Dahl, H. Fjellvag, Appl. Catal. A, General 165 (1997) 379

NATURAL GAS CONVERSION V
Studies in Surface Science and Catalysis, Vol. 119
A. Parmaliana et al. (Editors)

Effect of promoters on supported Pt catalysts for CO_2 reforming of CH_4

S. M. Stagg, and D. E. Resasco*

School of Chemical Engineering and Material Science,
University of Oklahoma, Norman, OK 73019

The addition of promoters to Pt/ZrO_2 catalysts for the dry reforming of CH_4 at 800°C has been investigated. It has been previously proposed that the ZrO_2 aids in the dissociation of CO_2, which helps to remove carbon that is deposited on the metal during the decomposition of CH_4. The results of this study support this idea and show that the addition of promoters such as cerium and lanthanum improves catalyst stability. It is believed that the promoters have multiple effects. They stabilize the surface area for high temperature operation, increase the density of CO_2 adsorption sites near the metal particle, and reduce particle growth under reaction conditions.

1. INTRODUCTION

In recent years, the utilization of CO_2 for the reforming of CH_4 has attracted interest due to environmental and industrial advantages over the conventional steam reforming [1-3]. The greatest obstacle preventing commercialization is the lack of a catalyst capable of operating at the high temperatures and pressures required by industry. Recent studies [4-7] have shown that Pt/ZrO_2 catalysts exhibit high activity and stability at moderate temperatures, 650°C. It has been proposed that on this catalyst, the decomposition of CH_4 and the dissociation of CO_2 occur via two independent pathways. Methane decomposition occurs on the metal, resulting in the production of hydrogen and the formation of carbonaceous deposits. The role of the support is to adsorb CO_2 and facilitate dissociation at the metal-support interface. The stability of the ZrO_2-supported Pt catalyst can be ascribed to the ability of the support to adsorb large quantities of CO_2 and provide a cleaning mechanism for the carbon deposited from CH_4 decomposition.

One possible mechanism is that the CO_2 that adsorbs near the metal-support interface dissociates to form CO and adsorbed O. This adsorbed O then reacts with carbon deposits on the metal particle to form an additional CO and clean the metal surface. Another possibility for the cleaning mechanism is that the O that reacts with the carbon on the metal comes from the support itself. Pulse experiments using $^{13}CH_4$ have shown that, in a reducing atmosphere,

* This work was partially supported by the Donors of the Petroleum Research Fund administered by the American Chemical Society. We gratefully acknowledge the National Science Foundation for a graduate traineeship for one of us (SMS).

a partial reduction of the ZrO_2 support can occur [8]. The O formed during the dissociation could then re-oxidize the support, providing a reduction/oxidation mechanism for continuous cleaning.

Recently, we have shown that, when operating at high temperatures, Pt/ZrO_2 catalysts deactivate due to carbon deposition, particularly at CH_4:CO_2 ratios greater than 1 [8]. We believe that under such severely deactivating conditions, the rate of CH_4 decomposition exceeds the rate of CO_2 dissociation and carbon accumulation occurs on the metal particle. In this study, we have primarily used CH_4:CO_2 ratios of 2:1 to accelerate deactivation and to be able to investigate the effects of the promoters in relatively short reaction times.

In previous work [8,9], we have investigated the addition of Sn as a promoter, in an attempt to reduce the rate of CH_4 decomposition and carbon formation. We have shown that under oxidizing conditions, or in the presence of excess Sn, the performance of the catalyst is not improved. In these cases, the presence of Sn disrupts the metal-support interaction, inhibiting the dissociation of CO_2 and eliminating the cleaning mechanism. However, we have also demonstrated that the controlled addition of Sn to Pt/ZrO_2, using surface reduction deposition techniques, can result in increased activity and stability at high temperatures and high CH_4:CO_2 ratios. It is believed that these bimetallic catalysts reduce the amount of carbon deposited from CH_4 decomposition without significantly altering the effect of the support.

In this contribution, we are focussing on the promotion of the ZrO_2 support. It is well established that adding cations such as Y^{3+} or La^{3+} enhances the surface area and thermal stability of ZrO_2 [10]. Similarly, recent studies [11] performed on automotive oxidation catalysts have shown that doping cerium oxide with Zr^{4+} resulted in a material that, after calcination at 800°C, had more than six times the surface area of the unpromoted CeO_2 catalyst. Furthermore, these materials were shown to exhibit good oxygen storage capacities and high thermal stability. We have prepared catalysts supported on both Ce^{4+}- and La^{3+}-doped ZrO_2 and compared their activity, stability, CO_2 adsorption capacity and BET surface area with an unpromoted Pt/ZrO_2 catalyst, under severely deactivating conditions.

2. EXPERIMENTAL

The zirconium hydroxide and lanthanum (5 wt %)-doped ZrO_2 were obtained from Magnesium Elektron Inc. The cerium-doped ZrO_2 was made by aqueous impregnation of cerium nitrate to the zirconium hydroxide. The incipient wetness of the $Zr(OH)_4$ was 0.2 cm^3/g and the weight percent of Ce in the final material 5%. Two Ce-doped supports were prepared. The first support was prepared by drying $Zr(OH)_4$ at 110°C overnight, prior to impregnation of the cerium nitrate. The second preparation involved adding the Ce^{4+} to the $Zr(OH)_4$ without a prior drying step. The materials were dried overnight at 110°C, and then calcined at 800°C for 4 h in stagnant air, prior to the impregnation of the metals. The Pt catalysts were prepared by incipient wetness impregnation of an aqueous solution of $H_2PtCl_6.6H_2O$ with a Pt loading of 1.5 wt %. The nomenclature for the catalysts prepared can be found in Table 1. The Ce-doped support prepared without a prior drying step was used in the catalyst preparation as that preparation resulted in the higher surface area. The Pt/Z catalyst was dried overnight at 110°C, calcined in air (30cm^3/min) for 2 h at 400°C, and then reduced *in-situ*, in H_2 (30 cm^3/min) at 500°C for 1 h, prior to reaction. The Pt/CeZ and Pt/LaZ catalysts were dried overnight at 110°C and then reduced, *in-situ*, in H_2 (30 cm^3/min) at 500°C for 1 h, prior to reaction.

The CH_4 reforming activity was measured in a quartz flow reactor at 800°C using a CH_4:CO_2 ratio of 2:1, and a GHSV of 180,000 h^{-1}. Temperature-programmed desorption (TPD) experiments of CO_2 were performed on all samples. In each run, the samples were heated to 300°C in air, cooled to room temperature in He, and then exposed to ten 50μL pulses of CO_2. After flushing in He (15 cc/min) at room temperature for 30 min., the sample was then heated in He (15 cc/min) to 800°C at a rate of 8°C/min.

Extended X-ray Absorption Fine Structure (EXAFS) was performed on beamline X-23A2 at the National Synchrotron Light Source at Brookhaven National Laboratory, Upton, New York. For each sample, 0.3 g of catalyst was diluted with 0.2 g of SiO_2 and pressed into pellets. The pellets were then exposed to reaction at 800°C and a 3:1 ratio of CH_4:CO_2 for 2 h. After reaction, the pellets were transferred to a glove bag where they were placed into holders and sealed in Kapton tape, without exposure to air. EXAFS spectra were recorded at liquid nitrogen temperatures on the reduced samples and samples exposed to reaction. All of the EXAFS spectra were taken in fluorescence mode near the Pt L_{III} edge (11,564 eV) with the scans ranging from 150 eV below the edge to at least 1000 eV past the edge. The EXAFS data was analyzed using the program BAN from Tolmar Instruments.

3. RESULTS AND DISCUSSION

Table 1 shows the BET surface area and CO_2 desorption capacity for the unpromoted and promoted ZrO_2 supports. The impregnation of the Ce^{4+} to a pre-dried $Zr(OH)_4$ did not result in the stabilization of the surface area after calcination at 800°C. More importantly, it resulted in a reduction in the CO_2 adsorption capacity. However, the addition of Ce^{4+} to the $Zr(OH)_4$, prior to drying and calcination, did result in a 20% increase in the surface area. An even greater effect was observed when La^{3+} was added as a promoter, with more than a 50% increase in surface area, and 6 times the CO_2 desorption capacity. X-ray diffraction studies have shown that the structure of the zirconia is very different for the three catalysts after calcination at 800°C for 4 h. The zirconia in the unpromoted sample was completely monoclinic, while only tetragonal zirconia was observed in the La^{3+} promoted sample. The addition of Ce^{4+} resulted in the partial stabilization of the tetragonal form with a small fraction of monoclinic zirconia present. These results show that the addition of promoters to the ZrO_2 not only increases the surface area by stabilizing the tetragonal form of zirconia, but also increases the density of CO_2 adsorption sites. Both of these promotional effects are believed to have a large impact on the cleaning ability and subsequent stability of the catalysts under reaction conditions.

Table 1.
Characteristics of the catalysts investigated

Support	Pretreatment of support prior to impregnation of Pt	BET (m^2/g)	TPD of CO_2* (moles/g)	Catalyst 1.5 wt% Pt
ZrO_2	Dried at 110C, calcined at 800C for 4 hours	35	$1x10^4$	Pt/Z
La-ZrO_2	Dried at 110C, calcined at 800C for 4 hours	55	$6x10^4$	Pt/LaZ
Ce-ZrO_2	Dried at 110C, calcined at 800C for 4 hours	41	$7x10^5$	Pt/CeZ
Ce-ZrO_2	Dried at 110C before and after impregnation of Ce, calcined at 800C for 4 hours	36	$5x10^5$	--------

* TPD of CO_2 on support prior to impregnation of Pt

In agreement with these expectations, Fig. 1 shows the fractional conversion of CO_2 and the product ratio. It demonstrates that, when exposed to reaction at 800°C and a 2:1 ratio of CH_4:CO_2, both promoted catalysts were much more stable than the unpromoted catalyst. The initial CO_2 conversion and product ratio for all three catalysts were the same. However, the rate of deactivation for the unpromoted Pt/ZrO_2 catalyst was much faster than the promoted samples, and within 1 h the unpromoted catalyst had experienced a significant drop in activity.

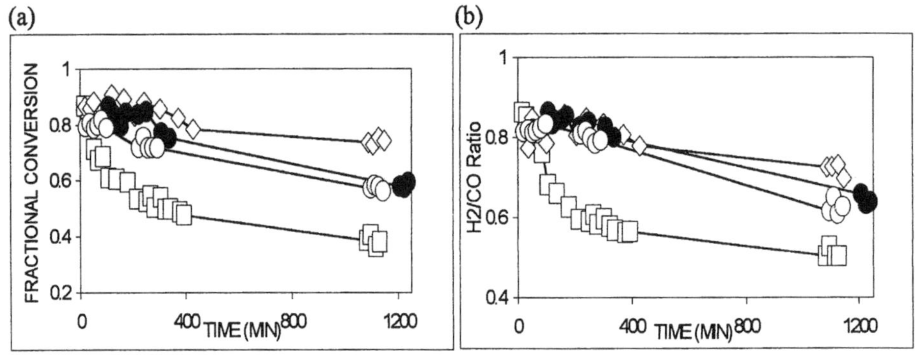

Figure 1 – (a) CO_2 fractional conversion and (b) H_2:CO Ratio for a 2:1 feed ratio of CH_4:CO_2 at 800°C. Pt/Z calcined (❑), Pt/CeZ uncalcined (•), Pt/CeZ calcined (o), Pt/LaZ uncalcined (◇).

After approximately 4 h, the rate of deactivation for the Pt/Z catalyst decreased and the drop in activity was much less over the next 15 h of reaction. The promoted catalysts were significantly more stable, with the La-promoted sample exhibiting almost 75% conversion of CO_2 after 19 h on stream. Similar behavior was observed for the CH_4 conversion. Fig. 1a also shows the CO_2 fractional conversion for a Pt/CeZ catalyst exposed to reaction after calcination in air (30 cm^3/min) at 400°C for 2 h. The results of this experiment show that calcination at low temperatures after impregnation of the metal results in no significant changes in activity or stability.

As mentioned above, one explanation for the higher stability of the promoted catalysts is that they have higher surface area and CO_2 adsorption capacity, which result in an enhancement of the dissociation and cleaning ability. This explanation is supported not only by the TPD and BET data in Table 1, but also by the product ratio shown in Fig. 1b. The unpromoted Pt/Z catalyst exhibited an initial H_2:CO product ratio of approximately 0.85. However, the ratio decreased rapidly within the first 4 h on stream, in parallel to the rapid drop in CO_2 and CH_4 conversion, explained previously. Both drops can be ascribed to the formation of carbon deposits on the metal, which decreases the initial CH_4 decomposition capacity and subsequently, the H_2:CO product ratio. In contrast, the promoted catalysts maintained a high (0.7-0.75) product ratio throughout the entire experiment. This suggests that either the rate of carbon formation has decreased, or the cleaning ability of the promoted catalysts was greater than the unpromoted catalyst. In light of the TPD data and the higher surface area, the latter explanation appears to be more reasonable.

The final effect of the addition of the promoter can be seen in the X-ray absorption studies. Fig. 2a shows the Fourier transform of the EXAFS data corresponding to the reduced Pt/Z, Pt/CeZ, and Pt/LaZ catalysts, along with that for a Pt foil. After reduction at 500°C for 1 h, the magnitude of the Fourier transform for both of the promoted catalysts was larger than that for the unpromoted Pt/Z. The increased magnitude indicates that after reduction the promoted samples have larger metallic clusters, or larger particles, than the unpromoted Pt/Z catalyst. However, after 2 h of reaction at 800°C and a 3:1 CH_4:CO_2 ratio, the Pt particles on the unpromoted catalyst were much larger than those on the Ce- or La-promoted samples (Fig. 2b). In fact, the promoted catalysts exhibited only a modest particle growth. Similar results were also observed by Transmission Electron Microscopy. The degree of particle growth observed follows the same order as the rate of deactivation. That is, the Pt/Z had the greatest growth and the fastest rate of deactivation, while the Pt/LaZ had minimal particle growth and exhibited the lowest rate of deactivation.

It is then possible that sintering is a major contributor to the deactivation of the Pt/Z catalyst at 800 C. Sintering not only would reduce the area of active metal, but more importantly, it also decreases the metal-support interaction, limiting the removal of carbon from the metal. Furthermore, the amount of carbon required to deactivate the catalyst would be much smaller, and consequently the activity of the catalyst would decrease rapidly. The promoted catalysts do not exhibit significant sintering. Therefore, another benefit of the cerium and lanthanum promoters is that their presence retards the Pt particle growth and the role of the support is not hindered at high temperatures, leading to increased stability.

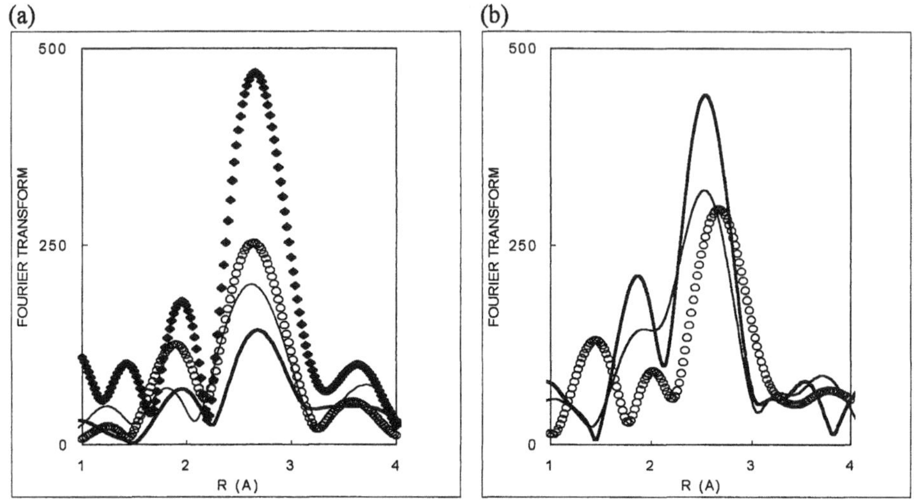

Figure 2 – EXAFS data. a) after reduction (500°C) for 1 hour and, b) after 2 hours of reaction at 800°C and CH_4:CO_2 ratio 3:1. Pt foil (♦), Pt/Z (━), Pt/CeZ (—), and Pt/LaZ (o).

The results of this contribution suggest that both the reduction of the particle growth, as well as the enhanced cleaning ability of the promoted catalysts, are responsible for the improved stability when compared to the unpromoted Pt/Z catalyst. However, at this time we can not rule out that the addition of the promoters may have additional promotional effects. As stated previously, promotion of ZrO_2 with Ce has been shown to improve the oxygen storage capacities of the support. It is possible that increasing the oxygen storage capacity can result in enhanced cleaning ability and higher catalyst stability.

REFERENCES

1. Erdohelyi, A., Cserenyi, J., and Solymosi, F., J. Catal. 141(1993) 287.
2. van Keulen, A. N. J., Seshan, K., Hoebink, J. H. B. J., and Ross, J. R. H., J. Catal. 166 (1997) 306.
3. Rostrup-Nielsen, J. R., Stud. Surf. Sci. Catal. 81 (1994) 25.
4. Ross, J. R. H., van Keulen, A. N. J., Hegarty, M. E. S., and Seshan, K., Catal. Today 30 (1996) 193.
5. Lercher, J. A., Bitter, J. H., Hally, W., Niessen, W., and Seshan, K., Stud. Surf. Sci. Catal. 101 (1996) 463.
6. Bitter, J. H., Seshan, K., and Lercher, J. A., J. Catal. 171 (1997) 279.
7. Bradford, M., Vannice, M. A., J. Catal. (in press).
8. Stagg, S. M., Romeo, E., Padro, C., Resasco, D. E., J. Catal (in press).
9. Stagg, S. M., and Resasco, D. E., Stud. Surf. Sci. Catal. 111 (1997) 543.
10. Duchet, J. C., Tilliette M. J., Cornet D., Catal. Today 10 (1991) 507.
11. A. Trovarelli, C. de Leitenburg, and G. Dolcetti, Chemtech 27 (6) (1997) 32.

NATURAL GAS CONVERSION V
Studies in Surface Science and Catalysis, Vol. 119
A. Parmaliana et al. (Editors)

An In-situ DRIFTS Study of the Mechanism of the CO_2 Reforming of CH_4 over a Pt/ZrO_2 Catalyst

Aisling M. O'Connor, Frederic C. Meunier and Julian R.H. Ross

Centre for Environmental Research, University of Limerick, Limerick, Ireland

Abstract

In-situ DRIFTS experiments have been carried out to investigate the mechanism of the CO_2 reforming of methane at 600°C over a Pt/ZrO_2 catalyst and these have been used to postulate a mechanism for the reaction. CH_4 dissociation occurs on the Pt and the resultant carbon species react with lattice oxygen species from the ZrO_2. The CO_2 then replenishes the lattice oxygen, a slow step in the reaction.

1. INTRODUCTION

Supported noble metals are effective catalysts for the CO_2 reforming of CH_4 [1-4]:

$$CH_4 + CO_2 \Leftrightarrow 2H_2 + 2CO \qquad \Delta H_{298K} = +261 kJmol^{-1} \tag{1}$$

Various related mechanisms for the reaction over noble metal catalysts have been proposed [2, 5-9]. Rostrup-Nielsen and Hansen [5], Erdohelyi et al. [2, 6, 7] and Efstathiou et al. [8] have all reported that both the CH_4 and CO_2 adsorb dissociatively:

$$CH_{4(g)} \Rightarrow CH_{x(s)} + (4\text{-}x)H_{(s)} \tag{2}$$

$$CO_{2(g)} \Leftrightarrow CO_{(s)} + O_{(s)} \tag{3}$$

Erdohelyi et al. [2, 6, 7] and Basini and Sanfilippo [9] found that the dissociation of the CO_2 was aided by the presence of hydrogen species on the catalyst surface (equation 4):

$$CO_{2(s)} + H_{(s)} \Leftrightarrow CO_{(s)} + OH_{(s)} \tag{4}$$

However, Efstathiou et al. [8] found no evidence for this.

Two main routes have been proposed for the formation of CO: dissociation of CO_2 by Reaction 3 [2, 5-8] and reaction of adsorbed CH_x with adsorbed oxygen species [2, 5-9]:

$$CH_{x(s)} + O_{(s)} \Leftrightarrow CO_{(g)} + xH_{(s)} \tag{5}$$

Erodehelyi et al. [2, 6, 7] have also reported that adsorbed CH_x species can react with gaseous CO_2 to form CO:

$$CH_{x(s)} + CO_{2(g)} \Leftrightarrow 2CO_{(g)} + xH_{(s)} \tag{6}$$

Our own research and that of the Twente group has shown that Pt/ZrO_2 is an active and stable catalyst for the CO_2 reforming [10-14]. The aim of this paper is to propose a mechanism for the reaction mechanism over this catalyst based on DRIFTS (Diffuse Reflectance Infrared Fourier Transform Spectroscopy) and kinetic [15] measurements.

2. EXPERIMENTAL

Details of the preparation of the Pt/ZrO_2 catalyst used here are given elsewhere [10, 11]. A Nicolet Magna 550 FTIR Spectrometer was used for the DRIFTS experiments. The catalyst sample was crushed to a particle size of <106 μm in order to produce a smooth matt surface in the sample holder of the DRIFTS cell. Ar was passed through the cell for at least 12 h to flush out any contaminants and the catalyst was then reduced *in-situ* at 400°C in a flowing 20% H_2/Ar mixture (100 cm^3min^{-1}). After reduction, the catalyst was brought to the reaction temperature in a flow of Ar and a background spectrum was recorded at this temperature. Various gas mixtures were then introduced and spectra were recorded every 5 min; the background was subtracted using OMNIC FTIR Windows - based software.

3. RESULTS AND DISCUSSION

3.1. Interaction of CO_2 with Pt/ZrO_2 and ZrO_2

Fig. 1 shows a series of DRIFTS spectra obtained when CO_2 was passed over the reduced catalyst at 600°C. After 1 min, two IR bands were observed, one in the region of 2040cm^{-1} and a second broader band stretching from 1600 to 1300cm^{-1}. The small band at 2040cm^{-1}, probably due to linearly adsorbed CO [8, 16], disappeared after 10 min. This disappearance may be linked to re-oxidation by the CO_2 of the ZrO_2 support, this having been reduced partially during the reduction step [17-19]; in other words, exposure of the freshly reduced catalyst to CO_2 gives rise to CO and O_{ads}, the latter replenishing the lattice oxygen of the partially reduced ZrO_2 and the CO species desorbing relatively slowly to give gaseous CO.

The broad band observed in the region from 1600 to 1300cm^{-1} is most likely due to several overlapping bands of carbonate species; the intensity of this band initially increased with time on stream for the first 10 min exposure of the catalyst to CO_2 and remained constant thereafter. The maximum of this broad band was at 1435 cm^{-1} and part of the band can be assigned tentatively to an ionic carbonate species weakly adsorbed on the catalyst surface [20,21]. As the intensity of the IR band of the ionic carbonate is quite high, it is difficult to ascertain whether or not other types of carbonate species are produced from the interaction of CO_2 with the catalyst. An equivalent experiment using the ZrO_2 support alone (not shown) gave a band in the 1600 to 1300cm^{-1} region of the spectrum with a similar shape and intensity to that found with the Pt/ZrO_2 catalyst. We therefore conclude that the carbonate species is adsorbed predominantly on the ZrO_2 and not the on Pt of the catalyst.

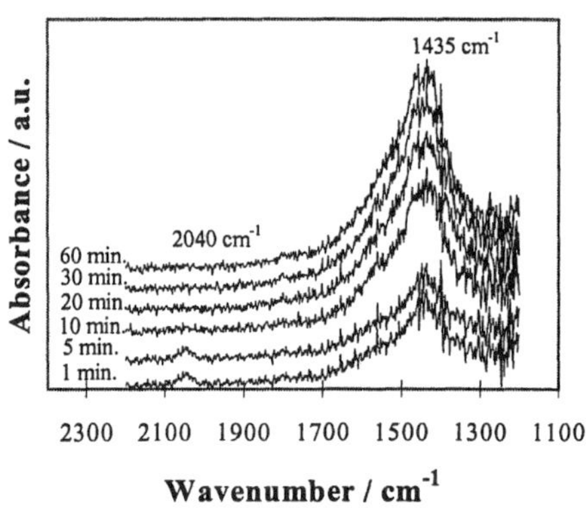

Figure 1. In-situ DRIFTS spectra recorded at 600°C while passing CO_2 over the catalyst.

3.2. Interaction of CH_4 and CO_2 on Pt/ZrO_2

A sample spectrum taken after 20 min during the reaction of CH_4 and CO_2 over the Pt/ZrO_2 catalyst is shown in Fig. 2; similar spectra resulted at intervals from 1 to 60 min. The characteristic complex bands due to gas phase CH_4, CO_2 and CO have not been subtracted from the spectra of Fig. 2 as the subtraction procedure caused some of the bands due to adsorbed species to become obscured. The band appearing at 2040cm^{-1} in Fig. 2 corresponds to linearly adsorbed CO; the intensity of this band is much greater than those obtained in the absence of CH_4 (Fig. 1) and this indicates that the CH_4 enhances the dissociation of CO_2 and also confirms the suggestion made above that the desorption of adsorbed CO is a relatively slow step. The band at 1435 cm^{-1}, attributable to ionic carbonate species, was lower in intensity under reaction conditions than in the absence of CH_4 and this suggests that the carbonate species must be consumed during the reforming reaction. A band at 1570cm^{-1} is also clearly visible (compare the broader unresolved bands of Fig. 1) and this is evidence for the presence of bidentate carbonate species; the symmetric stretching of this species would also be expected to occur in the 1300 to 1360cm^{-1} region of the spectrum [20-22] but the appropriate band may be obscured by the bands due to ionic carbonate and gas phase methane species. In order to show that there were no adsorbed formate species, which would give rise to bands due to C-H bonds in the 3000 to 2800cm^{-1} region of the spectrum [20], a subtraction for the gas phase CH_4 was carried out. Since no bands were visible in this area of the corrected spectrum, it was unlikely that formate species were present.

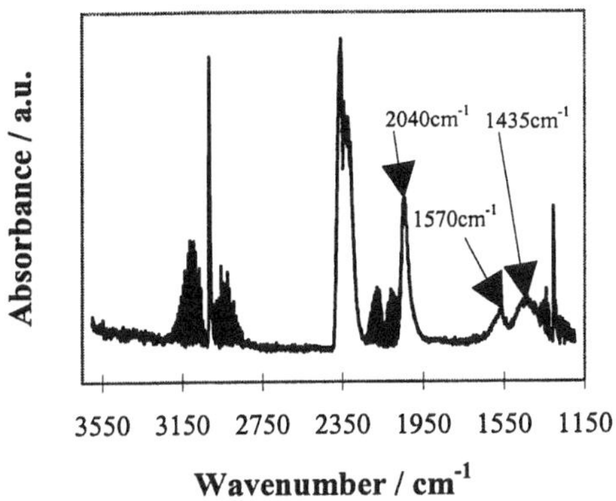

Figure 2. In-situ DRIFTS spectrum recorded after 20 minutes under reaction conditions at 600°C.

Data obtained in the 4000 to 3200cm^{-1} region of the spectrum under reaction conditions (corresponding to those used to obtain the data of Fig. 2) after various times on stream are shown in Fig. 3. The band in the 3800 - 3500cm^{-1} region can be attributed to hydroxyl groups on the ZrO_2 [20-22]. The intensity of this band remained constant with time on stream. These hydroxyl groups are probably intermediates in the formation of H_2O and H_2. This result indicates that the catalyst has a significant storage capacity for hydrogen, in agreement with TAP results to be reported elsewhere [23]. Hoang and Lieske [24] have observed that ZrO_2 can bind hydrogen reversibly at high temperatures and

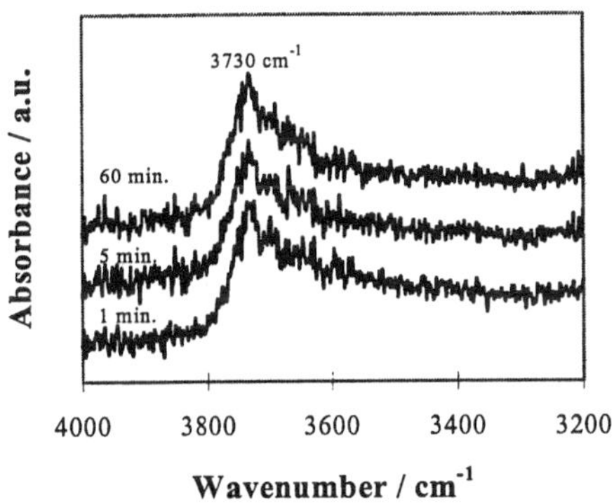

Figure 3. In-situ DRIFTS spectra recorded in the 4000 - 3200cm^{-1} region of the spectrum.

that the presence of Pt enhances the hydrogen uptake. This phenomenon was also observed by He and Ekerdt [25] and by Dall'Agnol *et al.* [26]; both groups attributed the increase in the concentration of surface hydroxyl groups to the uptake of hydrogen by ZrO_2.

3.4. Switching experiments

The following gas-delivery sequence was carried out over the Pt/ZrO_2 catalyst: CH_4/Ar (10min) → CO_2/Ar (10min) → CH_4/Ar (10min). Fig.4 shows DRIFTS spectra taken 1 min and 10 min after the introduction of each gas mixture. Two IR bands, in addition to the bands due to gas phase CH_4, were observed after 1 min exposure of the catalyst to CH_4; the band at 2040 cm^{-1} is due to linearly adsorbed CO while the doublet centring at 2150cm^{-1} corresponds to gas-phase CO. After 10 minutes exposure to CH_4, only bands due to gas-phase CH_4 were visible. The presence of gas phase and adsorbed CO after the initial exposure to CH_4 indicates that methane reacts with the lattice oxygen of the ZrO_2 to produce adsorbed CO species which then desorb as gaseous CO; once the pool of easily removable lattice oxygen has been depleted, no more CO can be produced. When CO_2 was subsequently passed over the catalyst, a band at 2040cm^{-1}, together with the doublet centred at 2150cm^{-1}, appeared after 1 min, as was also observed for the reaction of CO_2 with the fresh catalyst (see Fig. 1). The presence of linearly adsorbed CO indicates that replenishment of the lattice oxygen (which had been depleted by reaction with CH_4 in the last step) is brought about by the CO_2. After exposure to CO_2 for 10 min, the bands due to linearly adsorbed and gas phase CO had disappeared and a band at 1435cm^{-1}, which could be assigned to an ionic carbonate species, had developed. This indicates that the adsorbed CO desorbed once all the lattice oxygen had been replenished and that a very weakly adsorbed carbonate species then developed on the surface. When CH_4 was subsequently passed over the catalyst once again (top spectra), the band due to ionic carbonate species decreased significantly in intensity. It is therefore probable that this species reacts in the presence of CH_4 and is not merely a spectator species. Gas phase and linearly adsorbed CO as well as small amounts of bidentate carbonate species were also formed. A portion of the CO formed is probably derived from the reaction of CH4 (or surface species derived from methane) with the lattice oxygen replenished during the last step by CO_2. However, since the intensity of the band due to ionic carbonate species decreased in intensity in this step, the remainder of the CO formed, (and possibly also the bi-dentate carbonate species) was probably derived from this species. Two possibilities exist for the formation of CO from the ionic carbonate

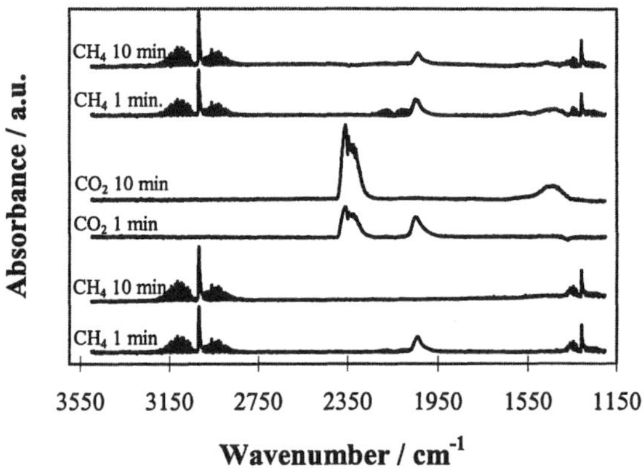

Figure 4. In-situ DRIFTS spectra obtained by performing the following switches over the reduced catalyst at 600°C: CH_4 (10min) → CO_2 (10min)→ CH_4 (10min). The sequence is from the bottom upwards.

species: (i) it replenished the lattice oxygen once the latter had been depleted by the CH_4; or (ii) it reacted directly with the CH_4 or with CH_4-derived species.

3.5. A model for the reaction

A model for the CO_2 reforming reaction based on the observations described above is shown schematically in Fig. 5 and some of the steps are set out in Eqs.7-13 below. The CH_4 first adsorbs on the Pt and then dissociates further to give a CH_x species which then dissociates further to form atomic carbon and hydrogen (Eqs. 7 and 8). These reactions appear to occur rapidly since no C-H species could be detected on the catalyst surface by DRIFTS under reaction conditions. This suggestion is in accordance with the

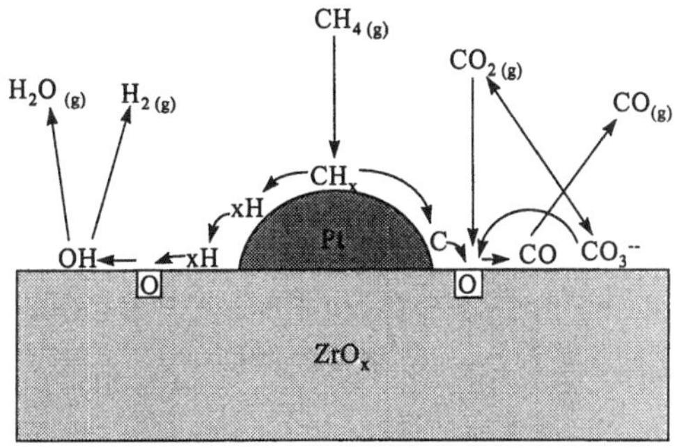

Figure 5. A model for the reaction mechanism of the CO_2 reforming of CH_4 over the Pt/ZrO_2

$$CH_{4(g)} \Rightarrow CH_{x(Pt^*)} \quad (7) \qquad CH_{x(Pt^*)} \Rightarrow C_{(Pt^*)} + xH_{(Pt^*)} \quad (8)$$

$$C_{(Pt^*)} + O_{(ZrO_x^*)} \Rightarrow CO_{(ZrO_x^*)} \quad (9) \qquad CO_{2(g)} + \square^{O-} \Rightarrow CO_{(ZrO_x^*)} + O_{(ZrO_x^*)} \quad (10)$$

$$2CO_{2(g)} \Rightarrow CO^{2-}_{3(ZrO_x^*)} + CO_{(ZrO_x^*)} \quad (11) \qquad CO^{2-}_{3(ZrO_x^*)} \Leftrightarrow CO_{(ZrO_x^*)} + 2O_{(ZrO_x^*)} \quad (12)$$

$$CO_{(ZrO_x^*)} \Rightarrow CO_{(g)} \quad (13)$$

(where Pt^* represents a Pt surface site and ZrO_x^* represents a ZrO_x surface site)

observation that the reaction order in CH_4 was found to be close to zero [15]. The majority of the surface carbon then reacts with the lattice oxygen of the ZrO_2 in the vicinity of the Pt-ZrO_2 interface to form adsorbed CO (Eq. 9). From the results of DRIFTS experiments in which adsorbed CO was observed after performing the $CH_4 \rightarrow CO_2$ switch (Fig. 4), it appears that the lattice oxygen removed by the reaction with $C_{(Pt*)}$ can be replenished by CO_2. Furthermore, the observation that the reaction was first order in CO_2 [15] indicates the involvement of CO_2 in the rate-determining step. The role of the CO_3^{2-} species remains unclear; however; from the DRIFTS experiments, it is evident that this species is formed once the lattice oxygen has been replenished by the CO_2. Since this species is consumed on switching from CO_2 to CH_4, it is probably not merely a spectator in the reaction and is most likely involved in the replenishment of the lattice oxygen during reaction (Eq. 12). The replenishment of the lattice oxygen of the ZrO_2 of a Rh/ZrO_2 catalyst by CO_2 under reaction conditions has been demonstrated by Efstathiou *et al.*[8] using transient isotopic techniques. TAP experiments carried out by Van Keulen *et al.* [27] over the same Pt/ZrO_2 catalyst used in this study also pointed to the presence of an oxygen pool, the CO_2 acting as a the supplier of oxygen to this pool.

Acknowledgements

We thank the EC (JOU2-CT92-0073) and KTI BV for each supporting part of this work. We also thank MEL Chemicals for supplying the ZrO_2 used.

REFERENCES

1. A.T. Ashcroft, A.K. Cheetham, M.L.H. Green and P.D.F. Vernon, Nature, 352 (1991) 225.
2. F. Solymosi, Gy. Kutsan and A. Erdohelyi, Catal. Lett., 11 (1991) 149.
3. D. Qin and J. Lapszewicz, Catal. Today, 21 (1994) 551.
4. D. Dissanayake, M.P. Rosynek, K.C.C. Kharas and J.H. Lunsford, J. Catal., 132 (1991) 117.
5. J.R. Rostrup-Nielsen and J.H.B. Hansen, J. Catal. 144 (1993) 38.
6. A. Erdohelyi, J. Cserenyi and F. Solymosi, J. Catal., 141 (1993) 287.
7. A. Erdohelyi, J. Cserenyi, E. Papp and F. Solymosi, Appl. Catal. A: Gen., 108 (1994) 205.
8. A.M. Efstathiou, A. Kladi, V.A. Tsipouriari and X.E. Verykios, J. Catal., 158 1 (1996) 64.
9. L. Basini and D. Sanfilippo, J. Catal., 157 (1995) 162.
10. A.M. O'Connor and J.R.H. Ross, Catal. Today, in press.
11. A.N.J. van Keulen, M.E.S. Hegarty, J.R.H. Ross and P.F. van den Oosterkamp, Stud. Surf. Sci. Catal., 107 (1997) 537
12. K. Seshan, H.W. ten Barge, W. Hally, A.N.J. van Keulen and J.R.H. Ross, Stud. Surf. Sci. Catal., 81 (1994) 285.
13. J.H. Bitter, W. Hally, K. Seshan, J.G. van Ommen and J.A. Lercher, Catal. Today, 29 (1996) 732.
14. K. Seshan, J.R.H. Ross, P.D. Mercera and E. Xue, German Patent DE 94 00 513, 1994.
15. A.M. O'Connor and J.R.H. Ross, to be published.
16. J.H. Bitter, Ph.D. Thesis, University of Twente, The Netherlands, 1997.
17. A.N.J. van Keulen, Ph.D. Thesis, University of Twente, The Netherlands, 1996.
18. A.L. Borer, C. Bronnimann, R. Prins, J. Catal., 145 (1994) 516.
19. D.L. Hoang, H. Lieske, Catal. Lett., 27 (1994) 33.
20. I.A. Fischer and A.T. Bell, J. Catal., 172 (1997) 222.
21. D. Bianchi, T. Chafik, M. Khalfallah, and S.J. Teicher, Appl. Catal. A: Gen., 112 (1994) 219.
22. J. Kondo, H. Abe, Y. Sakata, K. Maruya, K. Domen and T. Onishi, J. Chem. Soc., Faraday Trans., 84 (1988) 511.
23. A.M. O'Connor, Y. Schuurman, C. Mirodatos and J.R.H. Ross, to be published..
24. D.L. Hoang and H. Lieske, Catal. Lett., 27 (1994) 33.
25. M.Y. He and J.G. Ekerdt, J. Catal., 87 (1984) 238.
26. C. Dall'Agnol, A. Gervasini, F. Morazzoni, F. Pinna, G. Strukul and L. Zanderighi, J. Catal., 96 (1985) 106.
27. A.N.J. van Keulen, K. Seshan, J.H.B.J. Hoebink and J.R.H. Ross, J. Catal., 166 (1997) 306.

NATURAL GAS CONVERSION V
Studies in Surface Science and Catalysis, Vol. 119
A. Parmaliana et al. (Editors)

Kinetic Modelling of the Partial Oxidation of Methane to Syn-Gas at High Temperatures

C.R.H. de Smet*, R.J. Berger, J.C. Slaa and G.B. Marin+

Laboratorium voor Chemische Technologie, Schuit Institute of Catalysis,
Eindhoven University of Technology, P.O. Box 513, 5600 MB Eindhoven, The Netherlands

ABSTRACT

The partial oxidation of methane to syn-gas was investigated at conditions as close as possible to those industrially relevant: pressures up to 800 kPa and temperatures above 1200 K. Both the kinetics of the gas-phase reactions and the catalytic reactions, using a $Rh/\alpha-Al_2O_3$ catalyst, were studied. The gas-phase reactions were modelled successfully with a reaction network consisting of 40 elementary radical reactions. The Rh catalyst showed two active states: a low activity oxidized state and a high activity reduced state. Simulations with the gas-phase model indicated that gas-phase reactions are important in the catalytic oxidation reactions. In addition, quench of radicals on the inert, the catalyst pellets and the reactor wall was found to suppress the reactant conversions. Detailed kinetic modelling of the reactions on the catalyst in the oxidized state requires the simultaneous treatment of catalytic reactions, (surface initiated) gas-phase reactions and quench reactions.

1. INTRODUCTION

The catalytic partial oxidation of methane with oxygen to synthesis gas has recently received increasing attention as an alternative for the steam reforming of methane. Reactor design and heat management require knowledge of the reaction network and the reaction kinetics. The kinetic investigation was performed at high temperatures (above 1200 K) and high pressures (up to 800 kPa). High temperatures enhance the selectivities to CO and H_2, as reported by Schmidt et al. [1], and decrease the risk of carbon formation. At such conditions, however, gas-phase reactions are likely to play a role besides the catalytic reactions. Therefore, both the kinetics of the gas-phase reactions and the catalytic reactions using a typical partial oxidation catalyst were investigated. The interaction between the homogeneous and heterogeneous reactions will be discussed, in order to derive a kinetic model describing the performance of the Rh catalyst.

---

*Corresponding author
+Present address: Laboratorium voor Petrochemische Techniek, University of Ghent, Ghent, Belgium

2. KINETIC INVESTIGATION OF THE GAS-PHASE REACTIONS

The kinetic investigation of the gas-phase reactions was carried out with undiluted CH_4/O_2 mixtures ($2.1 < CH_4/O_2|_0 < 3.1$) at 300-800 kPa and 1000-1330 K in a tubular reactor, shown schematically in Figure 1. The dense sintered aluminium-oxide reactor is heated by a fluidized quartz sand bed, which enables a fast heat exchange with the reactor tube, thus promoting the isothermicity of the reactor. Nevertheless, it was difficult to control the reactor temperature and to prevent runaway during the experiments.

The influence of the residence time on the conversion and the selectivity is shown in Figure 2. The conversions of methane and oxygen increase more than proportionally with the residence time, indicating the occurrence of an ignition delay. At the conditions investigated, the gas-phase reactions mainly produce CO and H_2O. In addition, significant amounts of C_2-products (20-50%) are found, especially at low residence times.

Modelling of the experimental results resulted in a kinetic network consisting of 40 elementary radical reactions [2].

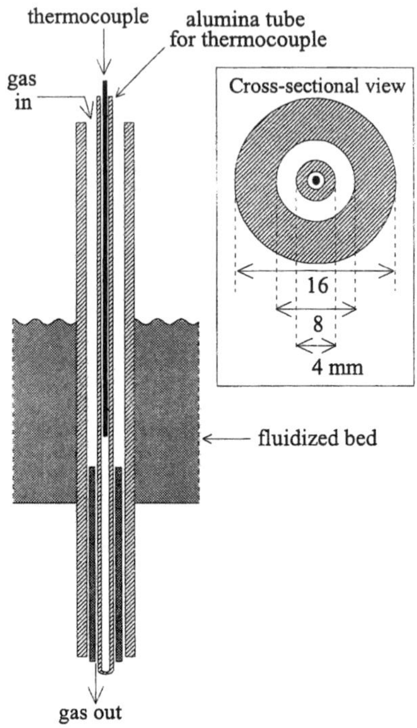

Figure 1. Schematic representation of the tubular reactor

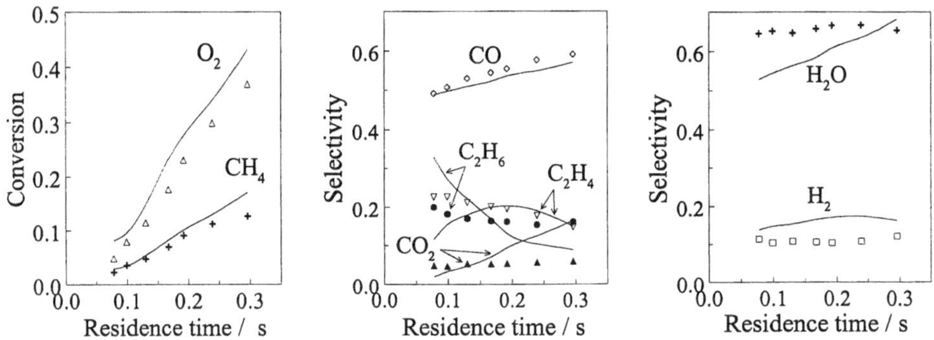

Figure 2. Conversions and selectivities as a function of residence time. Symbols: experimental results. Lines: results obtained with the gas-phase model. Conditions: T=1110 K, p=500 kPa, $CH_4/O_2/He|_0 = 67/33/0$.

3. KINETIC INVESTIGATION OF THE CATALYTIC REACTIONS

The catalytic reactions were investigated in the same sintered alumina reactor using a low-area 0.3 wt-% Rh/α-Al_2O_3 catalyst (80-105 μm), diluted with non-porous alumina pellets (106-150 μm). In order to minimize the empty volume of the reactor, non-porous alumina spheres (0.5-1.0 mm) and sintered alumina tubes were placed above and below the catalyst bed. Intrinsic kinetic data were obtained when the feed was diluted with 70-90% helium. The measurements were carried out at 1060-1400 K, 140-800 kPa, $1<CH_4/O_2|_0<4$ and $0.2<W_{cat}/F_{CH4,0}<3.0$ kg s mol^{-1}. The performance of the catalyst was stable for the duration of the experiments (15 days). The conversions and selectivities as a function of the space time are shown in Figure 3.

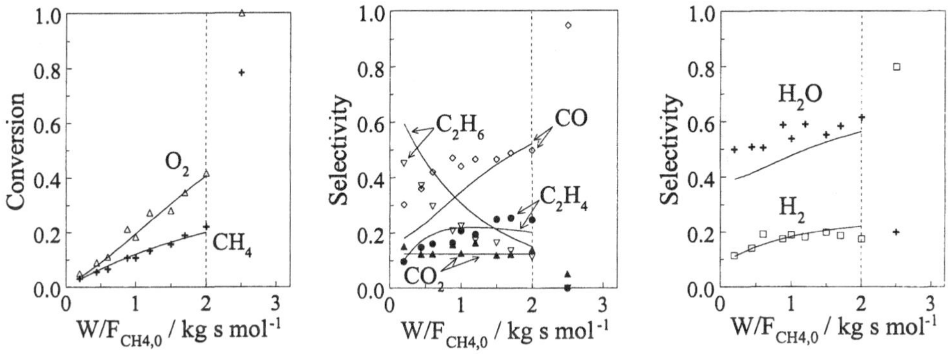

Figure 3. Conversions and selectivities on Rh/α-Al_2O_3 as a function of space time. Symbols: experimental results. Lines: results obtained with the global reaction network. Conditions: T=1298 K, p=400 kPa, $CH_4/O_2/He|_0 = 6.7/3.3/90$.

CO_2, C_2H_6 and H_2O are the primary products. C_2H_4 is formed consecutively by oxidative dehydrogenation of C_2H_6. At an oxygen conversion of 0.4, the behaviour of the catalyst changes significantly; the oxidation reactions are strongly accelerated leading to complete O_2 conversion followed by rapid reforming of the remaining CH_4 with H_2O and CO_2 to CO and H_2. This transition is probably due to reduction of the Rh [3].

Simulation of the catalytic experiments with the gas-phase model showed that the measured conversions were at least 3 times higher than the calculated conversions. This indicates that (1) catalytic reactions are important at the conditions investigated, and (2) that gas-phase reactions may also have a significant influence.

The kinetic model listed in Table 1 provided an accurate description of the behaviour of the catalyst in the oxidized state (i.e. at oxygen conversions below 40%), see Figure 3. However, more detailed kinetic modelling using Langmuir-Hinshelwood-Hougen-Watson type models revealed that the formation of C_2H_6, C_2H_4 and H_2 could not be described adequately. An alternative route for the formation of the C_2 products is a heterogeneous-homogeneous reaction pathway, in which methyl radicals formed at the catalyst surface couple in the gas phase to give ethane [4]. In addition, C_2 formation can occur via quench of methyl radicals, produced in the gas phase, by the inert alumina and catalyst pellets and the reactor wall.

Table 1: Set of global reactions describing the behaviour of the Rh/α-Al_2O_3 catalyst at $X(O_2) < 40\%$. (p_i expressed in Pa)

Reaction	Rate [mol s^{-1} m^{-3}]
$CH_4 + 2\,O_2 \Rightarrow CO_2 + 2\,H_2O$	$1.96\ 10^{-8}\ p_{O2}\ p_{CH4}$
$CH_4 + O_2 \Rightarrow CO + H_2 + H_2O$	$1.54\ 10^{-8}\ p_{O2}\ p_{CH4}$
$4\,CH_4 + O_2 \Rightarrow 2\,C_2H_6 + 2\,H_2O$	$1.87\ 10^{-8}\ p_{O2}\ p_{CH4}$
$C_2H_6 \Rightarrow C_2H_4 + H_2$	$3.02\ 10^{-2}\ p_{C2H6}$
$C_2H_4 + 2\,O_2 \Rightarrow 2\,CO + 2\,H_2O$	$3.69\ 10^{-6}\ p_{O2}\ p_{C2H4}$

4. INVESTIGATION OF THE EFFECT OF QUENCHING REACTIONS

The effect of quenching of gas-phase radicals at the reactor wall and the non-porous alumina spheres was investigated by performing experiments in the absence of the Rh catalyst and comparing the results with those obtained from simulations with the gas-phase model, see Table 2.

Table 2: Influence of radical quenching

Test	p [kPa]		Experi-ment	Gas-phase model	Quenching included
Empty tube	400	$X(O_2)$	6.0	5.9	5.0
		$S(C_2H_6)$	43	32	34
Tube filled with alumina	400	$X(O_2)$	0.5	2.3	1.2
		$S(C_2H_6)$	57	42	48
spheres	800	$X(O_2)$	9.1	13	6.8
(0.5-1 mm)		$S(C_2H_6)$	32	21	30

Conditions: T = 1273 K, $CH_4/O_2/He = 6.7/3.3/90$, $F_{tot,0} = 3.4\ 10^{-3}$ mol s^{-1}

From Table 2 it is clear that the oxygen conversion decreases and the C_2H_6 selectivity increases when the empty reactor is filled with non-porous alumina spheres. This is caused primarily by a reduction of the gas-phase volume, resulting in a decreased residence time. In case of the empty tube experiment, the experimental conversion agrees well with the gas-phase model. In case of the filled tube, however, the calculated oxygen conversions are too high and the C_2H_6 selectivities are too low. This indicates that in addition to the reduction of the gas-phase volume, quench reactions at the alumina spheres also suppress the oxygen conversion.

The effect of quenching of radicals at the alumina spheres and reactor wall, was investigated by performing simulations with a heterogeneous reactor model [4], taking into account both axial convection and diffusion of molecules and radicals towards the external surface of the pellets. In addition to the gas-phase reaction network, the quench reaction of methyl radicals towards ethane on the alumina spheres and reactor wall was considered. The quench reaction of the most abundant gas phase radical HO_2 did not influence the conversions and selectivities to a significant extent and was therefore not taken into account in the simulations.

The experimental results were used to optimize the quench reaction rate constant. Table 2 depicts that the heterogeneous model, with incorporation of this effect of radical quench, gives a better description of the experimental results: the oxygen conversion decreases and the ethane selectivity increases. The heterogeneous reactor model will be used to develop a detailed kinetic model to describe the performance of the Rh catalyst in the oxidised state.

5. CONCLUSIONS

The partial oxidation of methane to syn-gas was studied at industrially relevant conditions: pressures up to 800 kPa and temperatures above 1200 K. The kinetics of the homogeneous gas-phase reactions were investigated experimentally and modelled successfully with a kinetic network consisting of 40 elementary reactions. In addition, catalytic experiments were performed using a 0.3 wt-% Rh/α-Al_2O_3 catalyst. The experimental results indicated that the Rh catalyst can exist in two states: (1) a low activity oxidic state, promoting oxidation reactions only, and (2) a high activity reduced state, where only reforming reactions take place. Simulations with the gas-phase model showed that gas-phase reactions play an important role in the catalytic partial oxidation of methane to syngas. Radicals produced via gas-phase reactions quench at the surfaces of the non-porous alumina and catalyst pellets and reactor wall, and decrease the oxygen conversion. In order to develop a detailed kinetic model describing the performance of the Rh catalyst in the oxidised state, knowledge of the interplay between the surface reactions, the gas-phase reactions and the quench reactions is crucial.

REFERENCES

1. D.A. Hickman and L.D. Schmidt, AIChE J., 39 (1993) 1164.
2. R.J. Berger and G.B. Marin, to be published.
3. J.C. Slaa, R.J. Berger and G.B. Marin, Catal. Lett., 43 (1997) 63.
4. P.M. Couwenberg, Q. Chen and G.B. Marin, Ind. Eng. Chem. Res., 35 (1996) 415.

NATURAL GAS CONVERSION V
Studies in Surface Science and Catalysis, Vol. 119
A. Parmaliana et al. (Editors)
1998 Elsevier Science B.V.

Partial Oxidation of Methane to Synthesis Gas over Titania and Yttria/Zirconia Catalysts

A.G. Steghuis, J.G. van Ommen and J.A. Lercher

Catalytic Processes and Materials, Faculty of Chemical Technology,
University of Twente, PO Box 217, Enschede, The Netherlands

The oxidation of methane on titania and yttria/zirconia occurs *via* two parallel oxidation pathways, i.e., partial oxidation to $CO + H_2 + H_2O$ and full oxidation to $CO_2 + 2H_2O$. Both reactions proceed *via* a Mars van Krevelen mechanism. Methane is homolitically dissociated over $O^-_{(s)}$ sites and reacts *via* the formation and decomposition of a formaldehyde precursor to CO, H_2 and H_2O. CO_2 is obtained in parallel from decomposition of bicarbonate that is formed *via* further oxidation of dioxymethylene.

1. INTRODUCTION

Over the past years a large number of supported transition metal and noble metal catalyst along with a number of partially reduced mixed metal oxides were reported to be suitable for partial oxidation of methane [1,2,3,4]. In general, these catalyst are found to produce synthesis gas *via* the so called indirect partial oxidation mechanism that was first reported by Prettre [5] in 1946 for a supported Ni catalyst. Indirect partial oxidation first gives full oxidation of methane to carbon dioxide and water (ΔH°_{298K} = -802 kJ/mol) and subsequently reforming the remaining methane with the produced water (ΔH°_{298K} = 206 kJ/mol) and carbon dioxide (ΔH°_{298K} = 247 kJ/mol. Though high conversions to synthesis gas can be obtained in this way the primary full oxidation is problematic, it easily results in runaways due to the heat production from the highly exothermic full combustion [6].

In contrast direct partial oxidation has been claimed to occur over Rh and Pt coated monoliths in ultra short contact time reactors above 1000°C [7,8]. Direct partial oxidation was also reported over Ru/TiO_2 [9] and over $Ni/CaO{\cdot}2Al_2O_3$ [10]. The oxidized metal surface is suggested to be crucial for direct partial oxidation

In this paper, we report on the oxidative conversion of methane over Titania catalysts suitable for direct partial oxidation in comparison with earlier work on yttria/zirconia [11,12]. Steady state and transient kinetics measurements are used to elucidate the possible mechanistic steps over this catalyst.

2. EXPERIMENTAL

Titania (Degussa P-25) was pressed and subsequently crushed and sieved to a particle size of 0.3-0.6 mm. The oxide was calcined at 900°C for 15 hours; BET area measured is 7.6 m^2/g. Catalytic experiments were carried out in a quartz tubular flow reactor with a 5-mm internal diameter. A gas flow of 170 ml/min containing He, CH_4, O_2 and N_2 in the ratio of 13:2:1:1 with a total pressure of 1 bar was used. The catalyst bed was diluted with 50 wt% quartz particles with the same diameter as the catalyst The product gases were analysed with a Varian

3300 GC equipped with TCD and a 3m. 1/8" Carboxan 1000 column with He as a carrier gas. Conversions (X) and yields (Y) were calculated on the basis of CH_4 in the feed.

Pulse experiments were conducted at atmospheric conditions in a quartz tubular reactor with 4-mm internal diameter. Prior to the pulses the catalyst was heated in flowing oxygen up to 800°C. after 30 minutes at 800°C, oxygen was replaced by helium for 30 minutes. Subsequently, methane pulses were admitted. The products H_2, H_2O, CO and CO_2 were measured by mass spectrometer using masses 2, 18, 28 and 44.

Formaldehyde decomposition was measured, with a formaldehyde flow, generated by thermal decomposition of para-formaldehyde at 120°C in a 20 ml/min helium flow, which was lead through a second heated tubular quartz reactor at 1 atmosphere. The product stream is qualitatively analysed by mass spectroscopy.

3. RESULTS

3.1. Steady state performance

Figure 1 presents the results of conversions and yields obtained with 300 mg of titania between 600 and 900°C. Oxygen conversion was measurable at temperatures of 600°C and higher and was completed at 900°C. In this temperature range CO, CO_2, H_2 and H_2O were formed. The yields of all products increased with temperature. Contrary to the yttria/zirconia catalyst ethane or ethene were not formed.

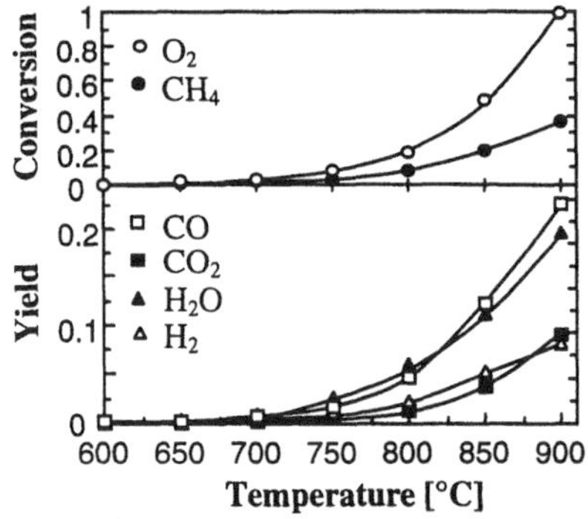

Figure 1: *Conversion and yield for partial oxidation of CH_4 with 300 mg TiO_2 versus temperature.*

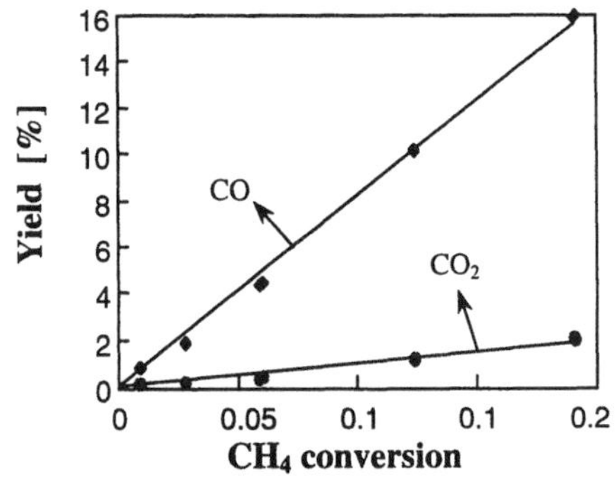

Figure 2: *CO and CO_2 yield as a function of methane conversion measured at 850°C.*

To distinguish between primary and secondary reaction products the CO and CO_2 yields are plotted in Figure 2 as a function of methane conversion for a series of experiments with a constant gas flow and different catalyst loadings at 850°C. The results show that formation of CO and CO_2 starts from the origin, indicating that both are primary products.

The influence of secondary reactions is shown in Figure 3, where the CO and H_2 selectivities are related to the methane conversion. The CO selectivity was hardly influenced by the conversion while the H_2 selectivity started at 45% and strongly decreased with increasing conversion. This indicates that over Titania H_2 is consumed in a secondary reaction.

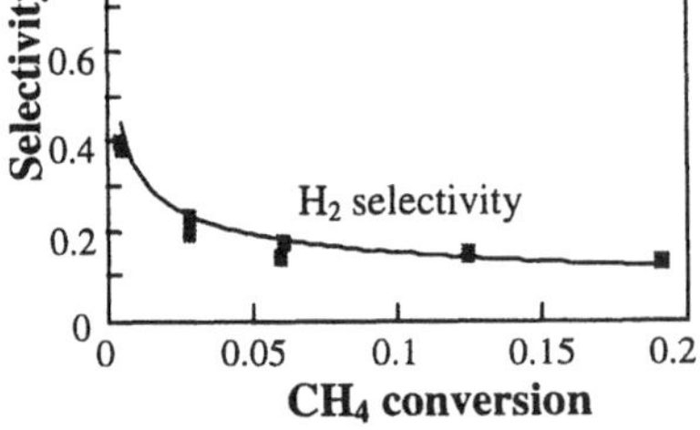

Figure 3: *Selectivity of CO and H_2 as a function of methane conversion at 850°C.*

3.2. Methane pulse experiments

The methane conversion in absence of gas phase oxygen was subsequently tested with pulse experiments. Methane was pulsed over 500 mg of titania at 800°C. The catalyst was pre-treated with oxygen at 800°C for 30 minutes followed by flushing with helium at the same temperature for a further 30 minutes to remove gas phase and adsorbed oxygen. The result is shown as pulse (a) in Figure 4. CO and CO_2 were formed from lattice oxygen with a CO selectivity of 0.9. Compared to the results for yttria/zirconia a smaller hydrogen peak and sharper water peak was seen. The H_2:CO ratio was 0.8.

Pulse (b) from Figure 4 is characteristic of the conversion of methane after including a 30 minute reduction in hydrogen before the helium flush in the catalyst pretreatment. Compared to pulse (a), the activity of the catalyst was lower. Hydrogen, CO and water but no carbon dioxide was formed from the pulse. In this case the H_2:CO ratio was 1.3.

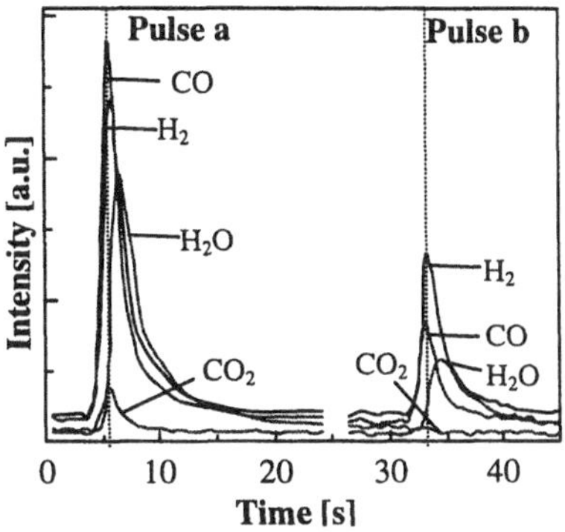

Figure 4: *CH_4 pulses over 500 mg TiO_2. Pulse (**a**): CH_4 pulse after pretreatment of 30 min. in O_2 followed by 30 min. in He, both at 800°C. Pulse (**b**): CH_4 pulse after pretreatment of 30 min. in O_2 followed by 30 min. in H_2 followed by 30 min. in He, all at 800°C.*

3.3. Formaldehyde decomposition

The measured product distribution as a function of temperature for thermal decomposition of formaldehyde in an empty tubular reactor is shown in Figure 5. At 550°C the formaldehyde conversion is starting. The conversion reaches about 50% at 700°C and is close to 80 % at 800°C. Over the entire temperature range the main products are CO and H_2. Small amounts of water were also found. When the same reactor is loaded with 300 mg Titania the product distribution as a function of temperature becomes significantly different (Figure 6). Already at 200°C a significant amount of formaldehyde is converted. The main products are CH_3OH, CO and CO_2. At 550°C the conversion of formaldehyde is already close to 80%, while the gas phase conversion was still close to zero at this temperature. At higher temperatures the main products are CO and H_2.

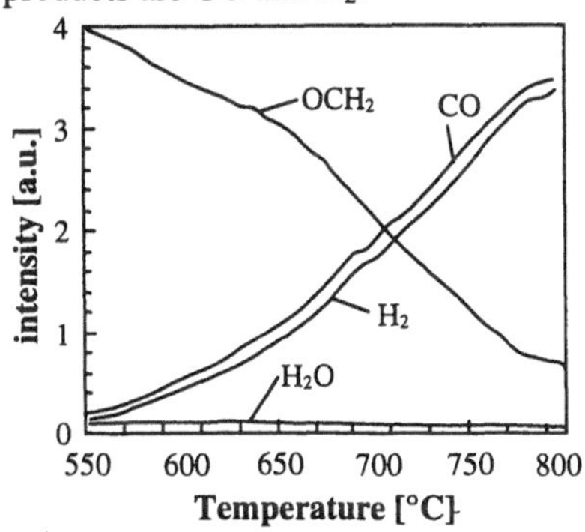

Figure 5: *Gas phase decomposition of formaldehyde in a 4 mm i.d. tubular quartz reactor.*

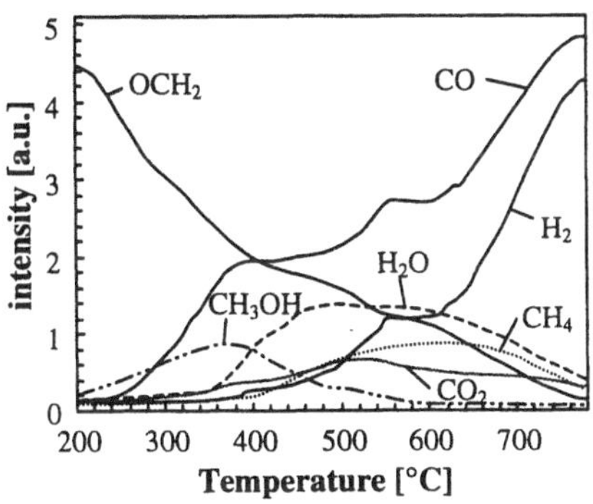

Figure 6: *Formaldehyde reaction over 300 mg titania. (H_2, CH_4, H_2O, CO, CH_2O, CH_3OH and CO_2 from m/e 2,16,18,28,29,32,44, respectively)*

4. DISCUSSION

As outlined above partial oxidation of methane may occur via two different mechanisms, i.e., direct partial oxidation or total oxidation followed by reforming reactions [5]. In figure 2 the straight lines for CO and CO_2 indicate that CO and CO_2 are primary products and are not interconverted. The same results were found earlier for yttria/zirconia [11,12]. Characteristic for the indirect route is that all oxygen is converted, before significant amounts of CO and H_2 are formed, which is not the case in figure 1. Note that we have excluded the possibility of oxygen slip through the catalyst bed by varying the bed height.

As expected from figure 2, addition of CO_2 in the reactant flow did not significantly influence the product distribution. This excludes the possibility of CO formation from fast CO_2 reforming and simultaneously excludes the possibility that CO is formed *via* reverse water gas shift reaction ($CO_2 + H_2 \rightarrow CO + H_2O$). Like yttria/zirconia, titania has no activity for steam-reforming.

As is shown in Figure 3 the hydrogen selectivity is decresing with increasing CH_4 conversion, indicating that on TiO_2 a secondary reaction of hydrogen oxidation occurs.

TiO_2, like yttria/zirconia does catalyze direct partial oxidation of methane, although not with a 100 % selectivity. Taking into account the reaction to CO_2 and water, the initial water yield is too high and the hydrogen yield is too low to account for a partial oxidation reaction according to $CH_4 + 1/2O_2 \rightarrow CO + 2H_2$. This suggests that water is also a primary product in the direct partial oxidation according to $CH_4 + O_2 \rightarrow CO + H_2 + H_2O$ (1) [11,12]

In Figure 4 the evolution of products stems from exposing the oxidized catalyst, where only lattice oxygen is present, to a pulse of methane. The results are in agreement with the reaction proposed for partial oxidation (reaction 1). Approximately equal amounts of H_2, CO and H_2O were formed together with a smaller peak of CO_2 from the complete oxidation. Water has a somewhat longer residence time than H_2 and CO, indicating that water is stronger adsorbed on titania. The total product composition agrees nicely with the steady state results. A 'Mars- van Krevelen' type mechanism is the foundation for methane oxidation on these oxides, so lattice oxygen is active for forming CO, CO_2 and H_2O.

The interaction of methane with the oxidic surface will start with the dissociative adsorption of methane to form a hydroxyl with lattice oxygen and a methyl group. The active site for the CH_4 can be an acid-base pair where the Lewis acid site (coordinatively unsaturated cation) accepts the CH_3^- and the adjactant lattice oxygen (base) accepts the H^+ after a heterolytic rupture of the C-H bond. A fast oxygen insertion or methyl migration will result in the methoxy formation [13,14,15]. The active sites for CH_4 activation may however, also be O^- sites, in case homolytic CH_4 dissociation and direct formation of methoxy and hydroxyl species occurs [16].Thus, the interaction of methane with the oxide surface in both cases leads to the formation of a methoxy and a hydroxyl group.

Recombination and desorption of hydroxyl groups energetically favors the formation of water and not hydrogen, especially in the presence of oxygen. However, comparison of Figure 5 and 6 shows that above 500 °C the catalyst is very active for the decomposition of formaldehyde to CO and H_2. An adsorbed formaldehyde precursor could therefore be the source of H_2 formation.

As is mentioned in the work of Finocchio *et al.*[17], in case of a stable molecule like methane the dissociation of the first C-H bond requires considerably more energy then the dissociation of a C-H bond in the formed surface species. Consequently a C-H bond in the methoxy is dissociated to form a second OH group and the remaining OCH_2 forms a dioxymethylene. The dioxymethylene can be considered the precursor of formaldehyde that easily decomposes

The dioxymethylene can be considered the precursor of formaldehyde that easily decomposes to CO and H_2 at the temperatures applied for methane oxidation. This is supported by measurements of Busca[18] and Feil[19], who showed that the primary adsorbate from formaldehyde over zirconia and titania is dioxymethylene. Furthermore the formation of CO and H_2 from sorbed formaldehyde was proposed by Wada *et al.* on the basis of a kinetic study over lanthanum-boron oxides [20, 21].

The two OH groups formed can leave the surface forming water, leaving a oxygen vacancy and a free lattice oxygen site behind. The oxygen vacancies (also those from oxymethylene decomposition) are replenished by gas phase oxygen, closing the catalytic cycle.

Infrared studies of CO_2 interaction with zirconia and titania show that bicarbonate is the precursor for CO_2 formation [22]. The studies of Busca and Feil also showed that upon heating above 300K the dioxymethylene species react to form methoxy and formate. This means that dioxymethylene can also disproportionate to formate and methoxy species (Figure7). With a free oxygen site the formate can easily form a bicarbonate and decompose to form CO_2 and a hydroxyl group. Note that while forming CO_2 all hydrogen from methane ended up as hydroxyl groups who will desorb to form two water molecules. When the oxygen vacancies are replenished with gas phase oxygen again the catalytic cycle for the full oxidation path closes.

Comparing the routes to CO and CO_2 formation one can conclude that to increase the CO selectivity the concentration of free oxygen sites should be lowered. Finally a summary of the proposed mechanism is presented in Figure 7.

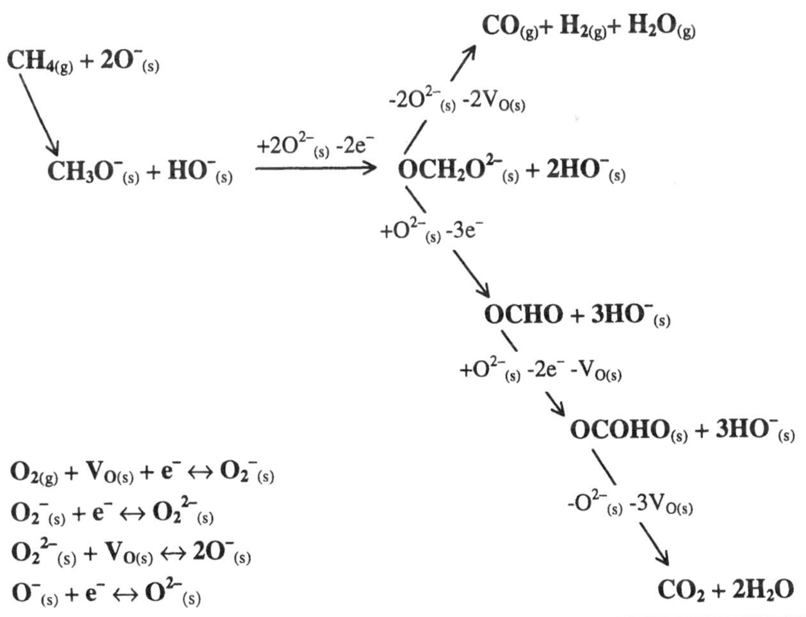

Figure 7: Proposed mechanism for CO and CO_2 formation over titania.

5. CONCLUSIONS

TiO_2, like yttria/zirconia is able to catalyse the direct partial oxidation of methane *via* $CH_4 + O_2 \rightarrow CO + H_2 + H_2O$. Next to this selective oxidation reaction the full oxidation of methane to CO_2 and H_2O occurs as a primary reaction. Both reactions involve a Mars- van Krevelen oxidation reduction route with lattice oxygen sites. Contribution of secondary reactions over TiO_2 is small, except for hydrogen oxidation. The mechanism of the selective oxidation is proposed to proceed *via* decomposition of a formaldehyde intermediate while carbon dioxide is formed via decomposition of a bicarbonate. With the activation of methane over $O^-_{(s)}$ sites as the rate determining step the mechanism supports the first order reaction kinetics in both oxygen and methane for the partial and the full oxidation.

ACKNOWLEDGEMENTS
Financial support of the E.U. under contract JOU2-0073 is gratefully acknowledged.

REFERENCES

[1] Dissanayake, D., Rosynek, M.P., Kharas, K.C.C., Lunsford, J.H., J. Catal., 132 (1991) 117.
[2] Kunimori, K., Umeda, S., Nakamura, J., Uchijima, T., Bull. Chem. Soc. Jpn., 65 (1992) 2562.
[3] Vermeiren, W.J.M.; Blomsma,E.; Jacobs, P.A. Catal. Today, 13 (1992) 427.
[4] Ashcroft, A.T.; Cheetham, A.K.; Foord, J.S.; Green, M.L.H.; Grey, C.P.; Murrell, A.J.; Vernon, D.F., Nature, 344 (1990) 319.
[5] Prettre, M.; Eichner, Ch.; Perrin, M., Trans. Faraday Soc., 43 (1946) 335.
[6] Chang, Y.-F., Heinemann, H., Catal. Lett., 21 (1993) 215.
[7] D.A.Hickman, L.D. Schmidt, J. Catal., 138 (1992) 267.
[8] D.A. Hickman, L.D. Schmidt, Science, 259 (1993) 343.
[9] Y. Boucouvalas, Z. Zhang, X.E. Verykios, Catal. Lett., 40 (1996) 189.
[10] A.A. Lemonidou, A.E. Stambouli, G.J. Tjatjopoulos, I.A. Vasalos, Catal. Lett., 43 (1997) 235.
[11] A.G. Steghuis, J.G. van Ommen, K. Seshan, J.A. Lercher, Stud. Surf. Sci. Catal., 107 (1997) 403.
[12] A.G. Steghuis, J.G. van Ommen and J.A. Lercher, Catal Today (1998) in press
[13] V.D. Sokolovskii, S.m. Aliev, O.V. Buyevskaya, A.A. Davydov, Catal. Today, 4 (1989) 293.
[14] T. Ito, T. Tashiro, M. Kawasaki, T. Watanabe, K. Toi, H. Kobayashi, J. Phys. Chem., 95(11) (1991) 4476.
[15] T. Ito, T. Watanabe , T. Tashiro, M. Kawasaki, K. Toi, H. Kobayashi in, "Acid-Base Catalysis", eds, K.Tanabe, H. Hattori, T. Yamaguchi, T. Tanaka, Weinheim,Tokyo, (1989) p483.
[16] A. Bielanski, J. Haber. "Oxygen in Catalysis", Marcel Dekker Inc., New York, (1991) p423.
[17] E. Finocchio, G. Busca, V. Lorenzelli, R.J. Willey, J. Catal., 151, (1995) 204.
[18] G. Busca, J. Lamotte, J.C. Lavalley, V. Lorenzelli, J. Am. Chem. Soc., 109 (1987) 5197.
[19] F.S.Feil, J.G. van Ommen, J.R.H. Ross, Langmuir, 3 (1987) 668.
[20] S. Wada, T. Tagawa, H. Imai, Appl. Catal., 47 (1989) 277.
[21] S. Wada, H. Imai, Catal. Lett., 8 (1991) 131.
[22] M-Y. He, J.G. Ekerdt, J. Catal., 87 (1984) 381.

NATURAL GAS CONVERSION V
Studies in Surface Science and Catalysis, Vol. 119
A. Parmaliana et al. (Editors)
1998 Elsevier Science B.V.

MORPHOLOGICAL CHANGES OF Ca PROMOTED Ni/SiO$_2$ CATALYSTS AND CARBON DEPOSITION DURING CO$_2$REFORMING OF METHANE

C.E.Quincoces, S.Perez de Vargas, A.Diaz[1], M.Montes[1], M.G.González

CINDECA (UNLP - CONICET). Calle 47 N^0 257- (1900) La Plata - Argentina..
[1]Grupo de Ing.Química, Univ.del País Vasco, 20080 San Sebastián, España

SUMMARY

Carbon dioxide reforming of methane on Ni/ CaO-SiO$_2$ catalysts was investigated at 923 K and atmospheric pressure using a CH$_4$/CO$_2$ ratio of 1:3, to analyze the effects of support composition on catalyst activity and stability. The catalysts were prepared by impregnating nickel on a SiO$_2$ support, modified by calcium oxide. This modification decreased the textural properties and reduced dispersion of the active phase.On the other hand, the increase of the Ca contents enhanced the reforming activity of the Ni/SiO$_2$ catalyst. Carbon characterization by TPH and TPO revealed the formation of filaments which did not deactivate the reaction in the conditions tested.

1 . INTRODUCTION:

There is a growing interest in the production of synthesis gas by methane reforming with CO$_2$. This process produces CO and H$_2$ in a suitable ratio for the further production of oxygenated derivatives, methanol synthesis and Fisher-Tropsch processes. Besides, this reaction is of great significance for environmental protection, for it enables the transformation of methane and CO$_2$ (responsible for the "greenhouse effect") into valuable compounds.

Reaction conditions (temperature, CH$_4$/CO$_2$ ratio) used in the methane reforming are the main cause for the Ni support deactivation process. These catalysts are frequently used in industry, and the deactivation caused by carbon and metallic phase sintering affect their catalytic properties. Therefore, with the purpose of increasing the useful life and the stability of the Ni support catalysts, they were added with elements that modify the support or the active phase. The addition of alkali and alkaline-earth promoters to those catalysts may result in significant changes in activity and stability. Zhang and Verykios (1) have reported that the support modification by CaO improves the stability of the Ni/γ-Al$_2$O$_3$ catalyst. Published works (2,3) have shown that alumina modification with alkaline-earth metals causes a significant effect on the reforming hidrocarbon reaction when promoted with nickel.

The present work analyzes the influence of support modifications of Ni/SiO$_2$ catalysts on the activity, stability and carbon deposition during the CO$_2$ reforming of methane.

2 . EXPERIMENTAL

Unpromoted nickel-silica catalysts were prepared by impregnating the support (Kali-Chemie AF 125 silica) with an aqueous solution of nickel nitrate. The promoted nickel catalysts were prepared by consecutive impregnation of the support with aqueous solutions of calcium and nickel nitrate (Merck,p.a.) at concentrations of 15 wt% Ni, and between 1 and 5% wt Ca salt, respectively. After each impregnation , the catalysts were oven - dried at 393 K and calcined at 723 K. The catalyst were reduced in a hydrogen stream (100 cm^3/min), at a heating rate of 8 K/min until reaching 923 K, a final temperature kept for subsequent two hours.

The catalysts were characterized by H_2 chemisorption, temperature programmed reduction (TPR) and X-ray diffraction (XRD).

The metallic surface areas (A_M) were measured by hydrogen adsorption and oxygen uptake using the dynamic pulse method with a Pulse Chemisorb 2700 Micromeritics. The Ni surface area was determined assuming Ni/H=1 and 6.33 A^2 for the area occupied per nickel atom. The mean crystal size was calculated from H_2 adsorption(dp_{Ad}) and by XRD according to the Scherrer equation (dp_{XRD}). The BET surface area and pore volume distribution were measured by nitrogen adsorption in an Accusorb 2100E Micromeritics analyzer

To identify the crystalline phase of supported nickel catalysts a Philips PW 1720 X-ray diffractrometer was used, which was equipped with a Cu Kα source at 40kV and 20 mA.

The TPR experiments were carried out in a conventional equipment. The samples (0.04 g) were heated from room temperature to 1073 K at a rate of 10 K/min, using a 10% H_2/N_2 gas flow at a rate of 20 ml/min.

Catalytic activity in fresh and spent samples was determined in a continuous flow reaction system at 923 K and atmospheric pressure using a reaction mixture of p_{CH4} =0.03 atm , CH_4/CO_2= 1.3, the total flow rate was of 200 cm^3/ min. A mass of 0.01 g of catalysts was loaded into the fixed bed reactor.

Reactants were analyzed by gas chromatography (Perkin-Elmer Sigma 1) at 40 K through a column packed with Porapak Q, using a thermal conductivity detector.

After the reaction, the catalytic bed was cooled in reaction mixture from 923 to 348 K, so as to avoid carbon gasification. The reactor was then cooled down to room temperature and purged in hydrogen for 12 h. In a subsequent stage, the reactor was heated at a linear rate of 10 K/min from room temperature to 1073 K in an hydrogen flow rate of 20 cm^3/min while monitoring the effluent gas composition.

The reactivity of the carbon species was analyzed by temperature programmed oxidation (TPO). The experiments were carried out in a Shimadzu TGA-50H under air stream (20 cm^3/min) by increasing the temperature from room temperature to 1073 K at heating rate of 10 K/min.

3 . RESULTS AND DISCUSSIONS

Morphological propierties:

Table 1 summarizes the characteristics of unpromoted (Ni/Si) and Ca-promoted (Ni/$SiCa_x$) catalysts. In it, we can observed that the Ca addition results in a smaller BET

Table 1. Morphological characteristics of the catalysts.

Catalysts	Ca content (%)	S_{BET} ($m^2.g^{-1}$)	pore vol. ($cm^3.g^{-1}$)	Ni surface area ($m^2.g^{-1}$)	dp_{ABS} (nm)	dp_{XRD} (nm)
Ni/Si	0	209	0.51	34	16	12
Ni/SiCa$_1$	1	195	0,45	22	25	18
Ni/SiCa$_3$	3	155	0,42	22	25	17
Ni/SiCa$_5$	5	143	0,36	28	19	11

surface area and pore volumes. These parameters decrease for increasing promoter amounts revealing that the addition of Ca to the silica produces significant textural changes in the support. Pereira and Martin (4) have related silica sintering to surface silicate formation in alkali-promoted Ni/SiO_2 catalysts. Nickel crystal particle size calculated from H_2 adsorption and XRD presented a good agreement. Table 1 also reveals that Ca addition to Ni/SiO_2 decreases the Ni surface area, suggesting that support sintering is related to Ni dispersion.

In order to determine the crystalline species and the metal-support interaction, we characterized differents samples by X-ray diffraction and temperature programmed reduction

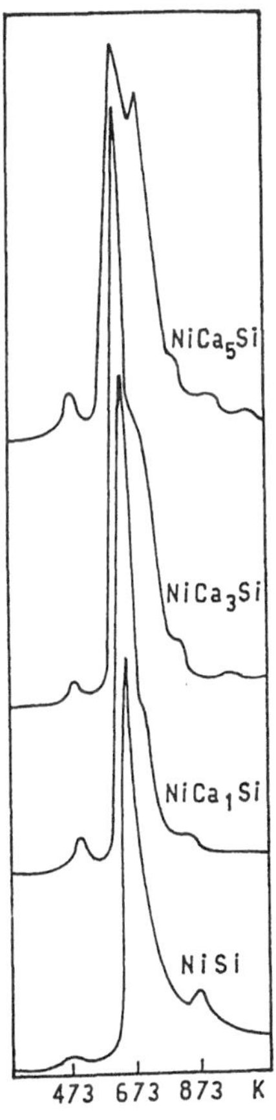

Fig 1. TPR patterns of Ca promoted and unpromoted catalysts.

Temperature Programmed Reduction

The reducibility of the catalysts was characterized by TPR experiments. Fig 1 shows the TPR profiles of the different samples. The unpromoted Ni catalyst presents three peaks whose maxima are observed at 473, 663 y 873 K. Compared with pure NiO, the peak at 663 K can be assigned to the NiO reduction with a low support interaction. The occurence of the peak at 873 K is attributed to the reduction of the nickel oxide that has interacted with the support. According to the literature, it is due to formation of chemical interaccion between the nickel hydroxide and silica. Once reduced, this compound shows a strong resistance to sintering and, as a consequence, allows a smaller particle size to be obtained

In the Ca-modified catalysts a shoulder on the higher temperature side is observed, which becomes more pronounced as the Ca content increases.

For Ni/SiCa$_5$, the signal at

about 663 K is overlapped by a peak at 713 K.These results suggest that the addition of Ca may influence the propierties of the active site.

X-ray Diffraction Analysis.

In order to determine crystalline species and crystal size, the differents samples were studied by XRD. Fig 2 shows the XRD patterns obtained for calcined and reduced samples of Ca-promoted catalysts. For the calcined Ni/SiCa$_x$ (spectra a, b and c), the diagrams exhibits the NiO and silica signals. In the Ni/SiCa$_5$ sample (spectrum c) a broadening of the NiO signal is observed in correspondence with a decreases in the metallic crystal size. In the reduced Ni/SiCa$_5$ catalyst at 1073 K (spectrum d) we observed peaks of metallic Ni, calcium silicate hydrate (Ca_2 SiO_4 H_2O) and silica. It is evident that the reduction process leads to a chemical interaction of Ca with the support while decreasing the metallic dispersion.

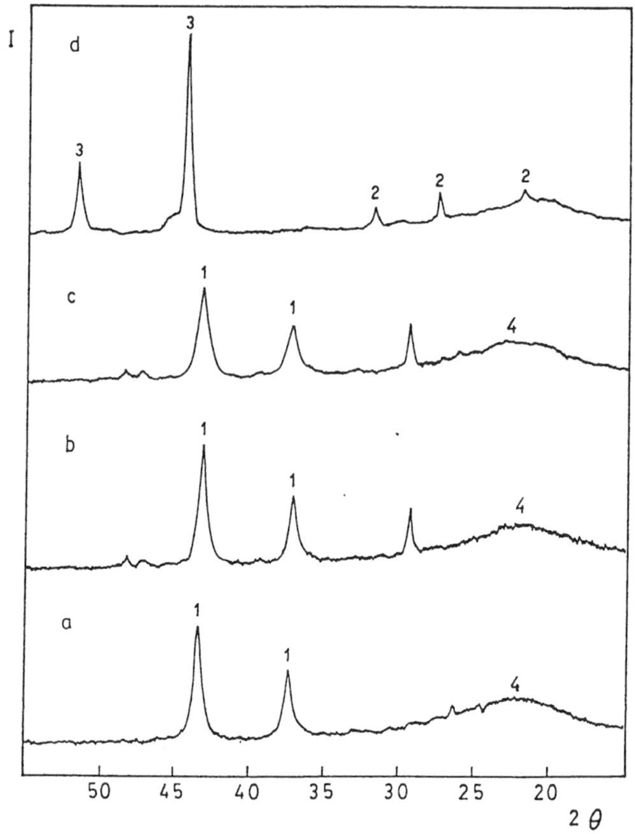

Fig 2 :X-ray diffraction patterns of promoted catalysts calcined at 773 K. (a) Ni/Si Ca$_1$, (b) Ni/Si Ca$_3$, (c) Ni/SiCa$_5$.(b) Ni/SiCa$_5$ catalyst reduced at 1073 K. References: 1= NiO, 2= $Ca_2.SiO_4.H_2O$, 3=Ni, 4=silica

Activity and stability of the catalysts.

Table 2 shows the activity measured during CO_2 reforming of CH_4 at atmospheric pressure, 923 K and a 1.3 CH_4/CO_2 ratio, the data indicates that the specific activity (TON) of the Ni/SiCa$_x$ increases with Ca content. This may be due to the fact that the Ca -modified silica improves the interaction between active component and support, enhancing the activity of the reaction. Choudhary (5) has reported that supported nickel catalysts using the supports coated with CaO, MgO and rare-earth show higher activity, selectivity and productivity in methane to syngas conversion reactions, than the non-precoated version.

After the reaction, the promoted and unpromoted catalysts were subjected to "in situ" TPH treatment, and carbon formation was suggested. Catalysts deactivation was not observed during reaction.

Activity catalyst measurements after hydrogenation revealed reaction deactivation and decreased catalysts stability.

Table 2. Activity of catalysts for CO_2 reforming of CH_4.

Catalysts	Ca content (%)	TON 10^{16}(cm^3/at$_{Ni}$.min)	x_1	x_2
Ni/Si	0	1.877	53.5	48.7
Ni/SiCa$_1$	1	1.783	32.5	23.9
Ni/SiCa$_3$	3	2.574	47.9	31.7
Ni/SiCa$_5$	5	2.778	58.2	51.1

x_1:methane conversion % over fresh catalysts . x_2: methane conversion % after carbon hydrogenation

Carbon species characterization by Temperature Programmed Reduction

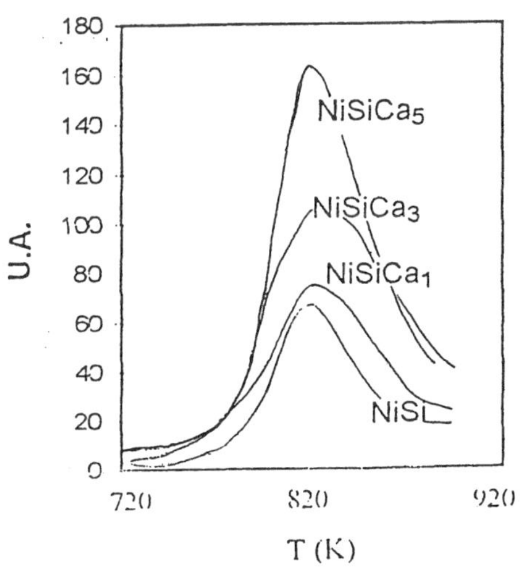

Fig.3.-TPH spectra of carbon deposited on unpromoted and promoted catalysts

In order to characterize the carbon deposited on the catalysts during the reforming reaction, TPH experiments were carried out after reaction at 923 K for 150 min. It can be observed that only methane was detected as hydrogenation product and that the Ca-based promotion increases carbon deposits. In all cases, a signal between 743 and 973 K is observed (fig3), which can be assigned to hydrogenation of whisker-like filamentous carbon (6). According to the literature, this species is produced by adsorbed carbon atoms derived from methane decomposition and CO dissociation. Adsorbed carbon atoms can first dissolve in the Ni crystallite and then diffuse through the metal to form a filament. In most cases, the nickel crystallites are attached to the filaments tips.

As the major type of deposit observed in this study was filamentous carbon, it is possible to ascribe the catalyst stability during the reforming reaction to the fact that the nickel placed on the tip of the filaments is active for the reaction. On the other hand, after carbon hydrogenation we observed that the catalyst stability decreases. The literature (7) has indicated that this is caused by the nickel loasing interaction with the support. This mechanism may describe the decrease of catalyst activity after the TPH experiments.

4 . CONCLUSIONS

Modification of the support after CaO addition to Ni/SiO_2 catalyst produces morphological changes and decreases the metallic dispersion of Ni.

The promotor content increases the specific activity (TON) for the methane reforming with carbon dioxide

Carbon formation is promoted by the content of CaO added to the support

In the reaction conditions the carbon deposited, characterized by TPH and TPO, was of the whisker-like filamentous type. Catalysts deactivation is not produced during reaction by formation of carbon.

Carbon hydrogenation decreases the catalysts stability. This behaviour may be attributed to metal- support interaction breaking off by gasification of carbon filaments.

5 . ACKNOWLEDGEMENTS

The authors acknowledge the technical assistence of Mrs Silvia Garcia and the financial support of CONICET and Universidad Nacional de La Plata.

6 . REFERENCES

1.- Zhang, Z.L.,Verykios,X.E., Catal.Today **21** (1994) 589

2.-Lemonidou,A.A. and Vasalos,I.A., Appl.Catal. **54** (1989) 119

3.-Tomita,T. and Kawamura,S., Proc. 7th Int.Congr.on Catalysis (1980).p.804

4 - Pereira, E.B and Martin,G., Appl. Catal. A **115** (1994) 135

6.- Bartholomew, C.H., Catal.Rev. Sci. Eng. **24** (1982) 67

7.- Chen,I. and Chen, F., Ind. Eng. Chem. Res. **29** (1990) 550

8.- Bartholomew, C.H.,Strasburg,M.V., Yen-Hsieh,H., Appl. Catal. **36** (1988) 147

NATURAL GAS CONVERSION V
Studies in Surface Science and Catalysis, Vol. 119
A. Parmaliana et al. (Editors)

Methane Dry Reforming over Well-dispersed Ni Catalyst Prepared from Perovskite-type Mixed Oxides

Jun Woo Nam,[a] Hoon Chae,[a,+] Seong Ho Lee,[a] Heon Jung,[b] and Kwan-Young Lee[a,*]

[a] Department of Chemical Engineering, Korea University, Anam-dong, Sungbuk-ku, Seoul 136-701, Korea

[b] Energy Conversion Research Department, Korea Institute of Energy Research, P.O.Box 103, Yusong, Taejon 305-600, Korea

Methane dry reforming with carbon dioxide was performed using perovskite-type mixed oxide catalysts, $La_{1-x}Sr_xNiO_3$ (x = 0, 0.1) and $La_{2-x}Sr_xNiO_4$ (x = 0-1) at 600-900°C and atmospheric pressure. Catalytic activity and resistance against coke formation were varied with the type of perovskite and the amount of Sr-substitution. $LaNiO_3$ showed high activity without coke formation while La_2NiO_4 did not show any activity. Among Sr-substituted catalysts, $La_{0.9}Sr_{0.1}NiO_3$ and $La_{1.8}Sr_{0.2}NiO_4$ had the highest catalytic activity without coke formation. For these catalysts, initially the catalytic activity increased with time and then reached a steady state. The XRD spectra of used catalysts showed that catalysts were transformed to mixed phases of $La_2O_2CO_3$ and $SrCO_3$ with highly dispersed Ni metal during the reaction. This transformation could be caused by the removal of lattice oxygen by Sr-substitution, which was promoted by the reducing atmosphere of the reaction (CH_4/CO_2 = 1). The enhanced catalytic activity of Sr-substituted perovskite could come from dual active sites, La_2O_3 for CO_2 adsorption and Ni for CH_4 activation.

1. INTRODUCTION

As a measure to reduce green house gases, dry reforming of methane using CO_2 serves the purpose well in that the reaction not only produces syn-gas (CO + H_2) which can be used for commercial Fischer-Tropsch process but also eliminates green house gases, CO_2 and CH_4, simultaneously.

Research on the catalyst of dry reforming of CH_4 has been concentrated on supported noble metal catalysts [1]. Especially, Ru, Rh, and Ni have been reported as promising catalysts. The type of the support changed the activity of these catalysts to a great extent and deactivation by coke formation was subject to be solved. Ni catalyst has been studied

---

* To whom correspondence should be addressed.

+ Present address : Central Research & Development, SKC Limited, #500, Chongja-dong, Changan-ku, Suwon, Kyonggi-do 440-300, Korea.

widely since it is cheap and initial activity was high. However, it tends to lose its activity because of coke formation. It has been reported that coke formation over Ni catalyst depends on the particle size (dispersion). Highly dispersed Ni catalyst is favorable for lowering coke formation [2,3]. In addition, the formation of coke has been reported to be suppressed by the addition of oxides of alkaline earth metals with high basicity [4-6].

In these points of view, perovskite-type catalysts, $LaNiO_3$ and La_2NiO_4, in which Ni ions are uniformly distributed with basic component, La, are expected to be a good catalyst itself or a precursor for an effective catalyst. Perovskite-type catalysts have well-defined structure and show various redox properties with substitution of their components [7]. In this work, Ni-containing perovskites were investigated as catalysts for methane dry reforming which is closely related to the redox property of the catalyst.

2. EXPERIMENTAL

The Ni-based perovskite mixed oxides tested in this study were prepared by Ohbayashi's method, in which an aqueous mixed solution of metal acetates, $La(CH_3COO)_3 \cdot 1.4H_2O$ (Aldrich, 99.9%), $Sr(CH_3COO)_2$ (Aldrich), $Ni(CH_3COO)_2 \cdot 4H_2O$ (Aldrich, 98%), was used [8,9]. The mixed solution was evaporated under vacuum at 90°C for 12 h and solid powder was obtained. And then, it was decomposed in the air at 350°C for 1 h. For $La_{2-x}Sr_xNiO_4$, it was calcined at 1000°C for 10 h, while in the case of $La_{1-x}Sr_xNiO_3$, it was mixed with the same amount of Na_2CO_3 and calcined at 900°C for 5 h. Na_2CO_3 was added as a flux in order to prepare single phase of $LaNiO_3$ [9]. After calcination, the catalyst was washed with distilled water repeatedly to eliminate remaining Na.

Ni-supported catalysts were prepared to compare with Ni-containing perovskite catalysts. Ni was supported to Al_2O_3 and SiO_2 by an incipient wetness method using nitrate solution. The supported catalyst was dried at 120°C for 24 h and then used for reaction after reduction by H_2 at 500°C for 5 h.

The prepared catalysts were characterized by XRD (Rigaku MAX IIA), ICP, BET (Micromeritics ASAP 2010), and TGA/DTA (Mac Science).

Methane dry reforming was carried out using a continuous flow fixed bed reactor with total flow rate of 40 ml/min (CH_4/CO_2=1) and catalyst loading of 0.03-1g under atmospheric pressure at the temperature range of 600–900°C. The reactants and products were analyzed by on-line GC. The amount of coke formed on the surface during reaction was measured by analyzing CO_2 production by repeated O_2 pulse injections.

3. RESULTS AND DISCUSSION

Structure of the prepared catalysts was examined by the X-ray diffraction, which patterns are shown in Fig. 1(A). $LaNiO_3$ showed only a single phase of perovskite structure where Ni exists as Ni^{3+}, while $La_{0.9}Sr_{0.1}NiO_3$ was composed of the mixed phases of La_2NiO_4 and NiO where Ni exists as Ni^{2+}. As stated previously, a flux such as Na_2CO_3 is necessary to keep Ni^{3+} for preparation of $LaNiO_3$. In case of $La_{0.9}Sr_{0.1}NiO_3$, it is considered that Ni^{3+} is more easily reduced by the substitution of very small amount of Sr with the loss of lattice oxygen. On the other hand, $La_{2-x}Sr_xNiO_4$ where Ni exists as Ni^{2+} was relatively stable even though some Ni_2O_3 appeared in the large substitution of Sr (x=1) as shown in Fig. 1(A).

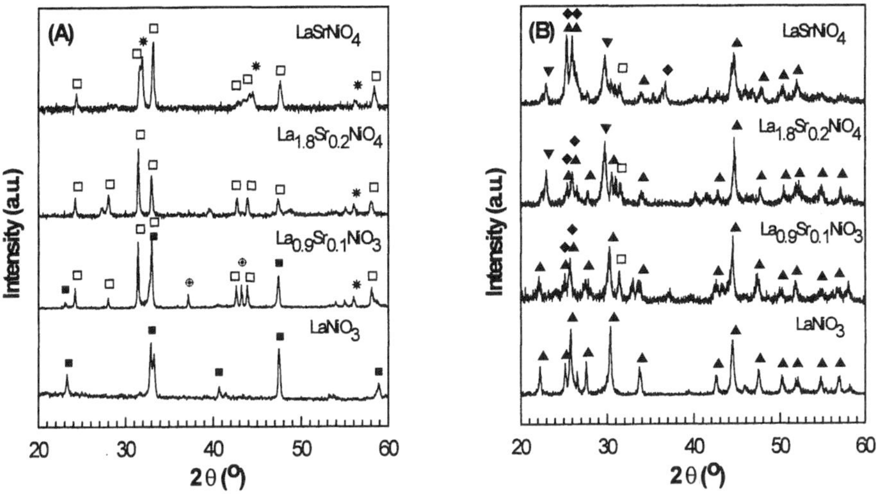

Figure 1. XRD patterns of $La_{1-x}Sr_xNiO_3$ (x=0, 0.1) and $La_{2-x}Sr_xNiO_4$ (x=0.2, 1) (A) before and (B) after reaction at 700°C. ■: $LaNiO_3$, □: La_2NiO_4, ▲: $La_2O_2CO_3$ (hexagonal), ▼: $La_2O_2CO_3$ (monoclinic), ⊕: NiO, ✱: Ni_2O_3, ◆: $SrCO_3$

The catalytic activity and the stability of perovskite-type $LaNiO_3$ were compared with those of conventional Ni-supported catalysts, 5wt%-Ni/SiO_2 and 5wt%-Ni/Al_2O_3 at 700 °C (Fig. 2 and Table 1). The stability was evaluated by the amount of surface coke formed during the reaction. Although initial activity of $LaNiO_3$ at 700°C was lower than those of the Ni-impregnated catalysts, its activity increased noticeably with time before reaching a steady state. The increase of activity with reaction time had been reported by Lee et al. in case of $La_{1-x}Sr_xCoO_3$ [10], which was explained by the formation of highly dispersed active sites resulted from the reducing atmosphere caused by the reaction condition (CH_4/CO_2 = 1). Surface coke of $LaNiO_3$ was hardly detected after the reaction. Considering that coke formation can be suppressed at low coordinated Ni sites as stated previously [2,3], it can be concluded that deactivation by coke was prevented due to good dispersion of Ni particles which was realized by the proper reduction of $LaNiO_3$. $La_{0.9}Sr_{0.1}NiO_3$ also revealed the similar activity changes with time having the higher activity of 38% with only 0.03 g of catalyst after 6 h of reaction at 700 °C as shown in Fig. 2. Coke formation was not observed with this catalyst, either.

The catalytic activity and stability against coke formation were investigated at 700°C for $La_{2-X}Sr_XNiO_4$ (x=0-1), a K_2NiF_4-type catalyst which has similar structure to perovskite. Figure 3 shows dependency of activity and the amount of surface coke on Sr content (x). La_2NiO_4 showed no activity but the catalysts with Sr-substitution revealed high activities, which increased with the amount of Sr-substitution. However, for the catalysts of $x \geq 0.4$, a considerable pressure drop due to coke formation was observed and the amount of surface coke measured after reaction increased with x as shown in Fig. 3. Therefore, $La_{1.8}Sr_{0.2}NiO_4$ was the most highly active catalyst without coke formation among $La_{2-X}Sr_XNiO_4$.

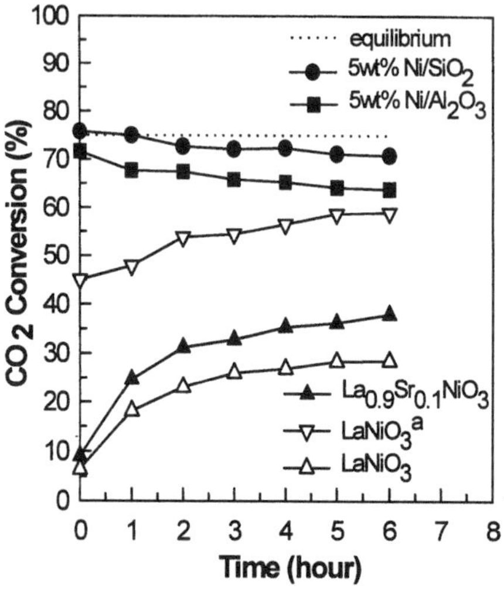

Figure 2. The change of CO_2 conversion over $LaNiO_3$, $La_{0.9}Sr_{0.1}NiO_3$, and Ni-supported catalysts with time on stream at 700 °C. [a] 0.1 g of catalyst was used, while 0.03 g was used for others.

Table 1
CO_2 conversion, the amount of coke formation, and BET surface area of $LaNiO_3$ and Ni-supported catalysts

	5wt%-Ni/SiO_2	5wt%-Ni/Al_2O_3	$LaNiO_3$	$LaNiO_3$
Catalyst loading (g)	0.03	0.03	0.03	0.1
BET surface area (m^2/g)	215	123	2	2
CO_2 conversion (%)[a]	71	64	28	59
Coke (μ mol)[b]	132.4	26.9	0	2.2

[a] CO_2 conversion after 6 h of reaction at 700 °C.
[b] The amount of coke formed during the reaction at 700 °C for 6 h.

XRD spectra of the catalysts after the reaction are shown in Fig. 1(B). XRD analysis of the used catalysts were performed for $LaNiO_3$, $La_{0.9}Sr_{0.1}NiO_3$, and $La_{1.8}Sr_{0.2}NiO_4$ which showed high resistance to the coke formation with high activity, and for $LaSrNiO_4$ with the most amount of coke formed among the catalysts tested. For all the catalysts, perovskite structure was not maintained after reaction. The used catalysts consist of $La_2O_2CO_3$, with

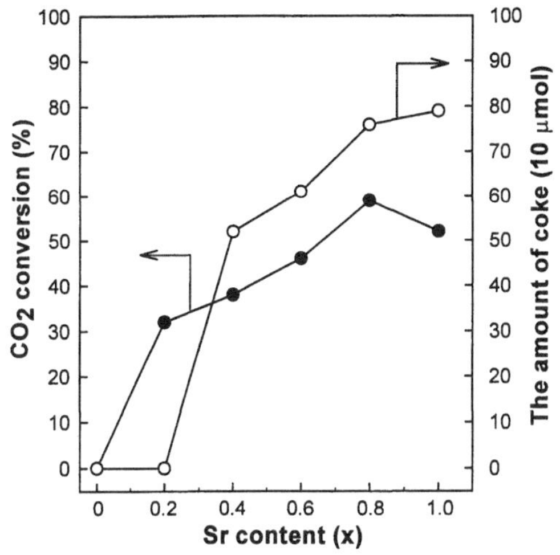

Figure 3. Dependency of CO_2 conversion and the amount of coke on Sr content (x) for $La_{2-x}Sr_xNiO_4$ at 700°C.

$SrCO_3$ and a small amount of La_2NiO_4. The $La_2O_2CO_3$ phase is formed due to the adsorption of CO_2 on La_2O_3 which has been previously reported as the active site for CO_2 adsorption [11,12]. It should be pointed out that neither Ni metal nor Ni oxides were observed. Ni oxides might be present in an amorphous form or particles of Ni are highly dispersed. Considering that Ni ions are highly dispersed in the crystal lattice of perovskite at the beginning and the reaction condition is in reducing atmosphere, Ni could be well dispersed as metal on the mixed phases even after the reaction. $LaMO_3$ (M = Co, Cr, Ni, Rh) has been reported to be transformed to M/La_2O_3 with high dispersion of metal by the pretreatment under reducing atmosphere [13,14], which supports the present result well. However, it should be noted that the well-dispersed Ni catalyst in the present study was prepared not by intentional reduction but by the natural reducing reaction condition, from which the optimum dispersion state was obtained. Ni is well known as an active site for methane in steam reforming of methane. It has been reported that Ni site with 12 neighboring sites is necessary for the activation of methane, while Ni site with 16 neighboring sites acts for the formation of coke [3], suggesting that highly dispersed Ni catalysts are required to suppress the coke formation. Thus, it can be concluded that the high activity and the stability against coke formation resulted from the high dispersion of Ni sites for CH_4 activation and the formation of La_2O_3 for CO_2 adsorption. The reason for the initial increase of the catalytic activity in Fig. 2 can also be explained that Ni and La_2O_3 active sites were formed by the gradual reduction caused by the appropriate reducing reaction condition.

As shown in Fig. 3, both catalytic activity and the coke formation increased with Sr contents. $SrCO_3$ formed after the reaction as shown in Fig. 1(B) is the product of CO_2 adsorption on SrO and is distinguished in used $LaSrNiO_4$ where the large amount of coke was formed during the reaction. Unlike La_2O_3, the basicity of SrO seems to be too strong for CO_2 to be decomposed from $SrCO_3$. The reason for the high activity of catalysts with high Sr contents despite of large coke formation is explained as the transformation of perovskite structure to the active one of La_2O_3 and well-dispersed Ni is also promoted by Sr-substitution as stated previously.

3. CONCLUSION

Perovskite-type mixed oxides, $La_{1-x}Sr_xNiO_3$ (x = 0, 0.1) and $La_{2-x}Sr_xNiO_4$ (x = 0-1) were transformed during methane dry reforming to a stable and highly effective catalytic system of well-dispersed Ni over La_2O_3. The transformation was subjected to the type of perovskite and the amount of Sr-substitution, which resulted from the removal of lattice oxygen by Sr-substitution promoted by the reducing atmosphere of the reaction (CH_4/CO_2 = 1). When $La_{0.9}Sr_{0.1}NiO_3$ and $La_{1.8}Sr_{0.2}NiO_4$ were used as starting catalyst, the highest catalytic activity was obtained without coke formation. The reaction over these catalysts is considered to proceed by the catalytic reaction of CO_2 adsorbed on La_2O_3 and activated CH_4 over well-dispersed Ni sites.

ACKNOWLEDGEMENT

This study has been supported by Korea Science and Engineering Foundation through Research Fund and authors appreciate its financial support.

REFERENCES

1. J. Nakamura and T. Uchijima, *Shokubai*, **35**, 478 (1993).
2. T. Borowiecky, *Appl. Catal.*, **10**, 273 (1984).
3. J. R. Rostrup-Nielsen, *"Catalysis,"* Vol. 5, Springer-Verlag, (1984) p. 1; J. R. Rostrup-Nielsen, J. H. B. Hansen, *J. Catal.*, **144**, 38 (1993).
4. E. Ruckenstein and Y. H. Hu, *Appl. Catal. A: General*, **133**, 149 (1995).
5. S. B. Tang, F. L. Qiu, and S. J. Lu, *Catal. Today*, **24**, 253 (1995).
6. T. Horiuchi, K. Sakuma, T. Fukui, Y. Kubo, T. Osaki, and T. Mori, *Appl. Catal. A: General*, **144**, 111 (1996).
7. e.g., L. G. Tejuca and J. L. G. Fierro, *"Properties and Applications of Perovskite-type Oxides,"* Marcel Dekker (1993).
8. H. Ohbayashi, T. Kudo, T. Geju, *Jpn. J. Appl. Phys.*, **13**, 1 (1974).
9. T. Nakamura, M. Misono, T. Uchijima, and Y. Yoneda, *Nippon Kagaku Kaishi*, **11**, 1697 (1980).
10. H. Chea, J. W. Nam, and K.-Y. Lee, *J. Eng. Sci. & Tech., Korea Univ.*, **32**, 1 (1996).
11. T. Inui, M. Funabiki, Y. Takegami, *J. Chem. Soc., Faraday Trans. 1*, **76**, 2237 (1980).
12. M. P. Rosynek, K. T. Magnuson, *J. Catal.*, **48**, 417 (1977).
13. Å. Slatern and U. Olsbye, *Appl. Catal. A: General*, **110**, 99 (1994).
14. R. Shiozaki, A. G. Anderson, T. Hayakawa, S. Hamakawa, K. Suzuki, M. Shimizu, and K. Takehira, *Stud. Surf. Sci. Catal.*, **110**, 701 (1997).

NATURAL GAS CONVERSION V
Studies in Surface Science and Catalysis, Vol. 119
A. Parmaliana et al. (Editors)

Rare earth promoted nickel catalysts for the selective oxidation of natural gas to syngas*

Wei Chu[a], Qiangu Yan[a], Xi Liu[a], Qing Li[a], Zuolong Yu[a†], Guoxing Xiong[b]

[a] Chengdu Institute of Organic Chemistry, Academia Sinica,
P.O.Box 415, Chengdu 610041, China

[b] Dalian Institute of Chemical Physics, Academia Sinica,
P.O.Box 110, Dalian 116023, China

Abstract: Selective oxidation of natural gas to yielding syngas was investigated over rare earth promoted nickel catalysts. It was found that the rare earth oxides like Y_2O_3, CeO_2 were good promoters for better product yields and selectivities. The operating parameters could have an important influence on the catalyst performances, and the syngas content (CO/H_2) could arrive to 97.68% in the reaction with a space velocity of 1.5×10^5 h^{-1} CH_4 and at 870°C. The promoted catalysts were studied by XRD, TPR and XPS techniques, the ceria could suppress the growth of nickel crystallites, enhancing the catalyst stability and the catalytic performances.

Keywords: Natural gas, Selective oxidation, Syngas production, Nickel catalyst, Yttria promoter, Ceria effect

1. INTRODUCTION

Natural gas was one of the most important resources like oil and coal. Considerable efforts have bee devoted to its conversion into easily transportable and/or value added products[1,2]. The synthesis gas production was the first step in the indirect conversions.

The traditional stream reforming suffered from several disadvantages[3]. However, the newly reported direct catalytic oxidation ($CH_4+1/2O_2 \rightarrow CO+2H_2$) was a real challenge. This new process was mildly exothermic and could be operated at very high space velocity, and presenting also some other profits[4].

Recently, a few studies have been presented over the noble catalysts[5,6] and the transitional

* This work was supported by National Natural Science Foundation of China, Project No. 29573128
† To whom correspondence should be addressed

metal based catalysts[7,8]. The sintering of active sites and the carbon deposition could cause the catalyst deactivation. The promoters were then introduced into the catalysts and some improving could be achieved[9]. The operating conditions could also have important influences on the catalytic activities[10]. And different propositions of reaction mechanism were proposed and remained to be elucidated[11].

In this paper, the influences of rare earth oxides like Y_2O_3 and CeO_2 on nickel catalysts were studied in the partial oxidation of methane to synthesis gas. The XRD, TPR, XPS techniques were used for the studies of active sites and the catalyst characterizations.

2. EXPERIMENTAL

The rare earth promoted catalysts were prepared by a wet impregnation technique[9]. An aqueous solution of nickel nitrate and rare earth compounds were added to 2 g of 20~40 mesh support, and the samples were dried and subsequently calcined in air at high temperature, then sieved before use.

Catalyst testing was performed in a continuous flow micro-reactor made up of quartz, using 300 mg catalyst, pretreated in hydrogen[9,10]. The reaction conditions were controlled at a space velocity of 1.0×10^5~4.5×10^5 h^{-1}, 700~900°C, and a total pressure of 1.5 atm. The feed and product gases were analyzed by an on-line gas chromatogrph with thermal conductivity detector, and the two columns were Porapak Q and 13X molecular sieve respectively.

The temperature programmed reduction (TPR) tests were operated on a homemade apparatus. The sample was pretreated in N_2 flow and saturated with N_2+H_2, the temperature was then augmented to 930°C with a rate of 12°C/min, the TPR profiles were recorded on line. The X-ray diffraction (XRD) and X-ray photoelectron spectroscopy (XPS) were also used for the catalyst characterizations.

3. RESULTS AND DISCUSSION

3.1. Influence of rare earth oxide promoters

The nickel based catalysts were investigated in the partial oxidation of natural gas to the feedstock CO/H_2, showing a fairly good performance[12]. The promoted catalysts were prepared by a wet impregnation method, with the addition of rare earth oxide like Y_2O_3, CeO_2, *etc*. The catalytic performances were compared for the ceria promoted nickel catalysts. It was observed that the H_2 selectivity was enhanced gradually with CeO_2 addition, a better result of 93.4% was obtained when the catalyst had a content of 4% CeO_2 (reaction at 800°C). Similar phenomena were observed for CO selectivity. The conversion of methane was slightly decreased at first (when 1% CeO_2 was added); the better effect was achieved when 2% CeO_2

promoter was added, the methane conversion reached 93.2%. For the product yield, the better results were also obtained with 2% CeO_2 promoter (see Fig.1); the yields of CO and H_2 were respectively 88.2% and 85.7%.

The natural gas was used as raw material which contains 92.25% CH_4, 3.24% C_2H_6 and some other hydrocarbons with a carbon number of more than 2. For 100 mole natural gas, there would be 103.14 mole carbon and 399.08 mole hydrogen, and the H/C ratio was equal to 3.8693, (the H_2/C = 1.93465). In addition, during the reaction, the products were CO and H_2 with small amount of CO_2 and H_2O, then the product ratio H_2/CO would not be unity. In fact, it was 1.93 for the $NiCeX_2$ catalyst at 870°C. In the other part, when the temperature was raised from 700°C to 870°C, the H_2/CO ratio varied from 2.03 to 1.93. This could be also explained as that the partial oxidation to CO and H_2 was operated at a different degree compared with the total oxidation to CO_2 and H_2O.

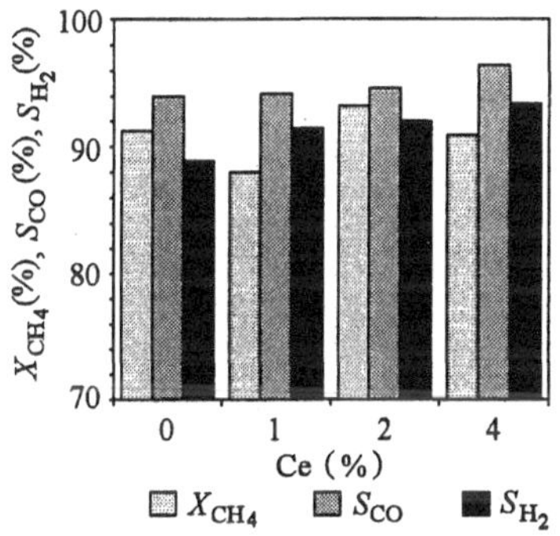

Fig.1 Effect of ceria promoter on product yields

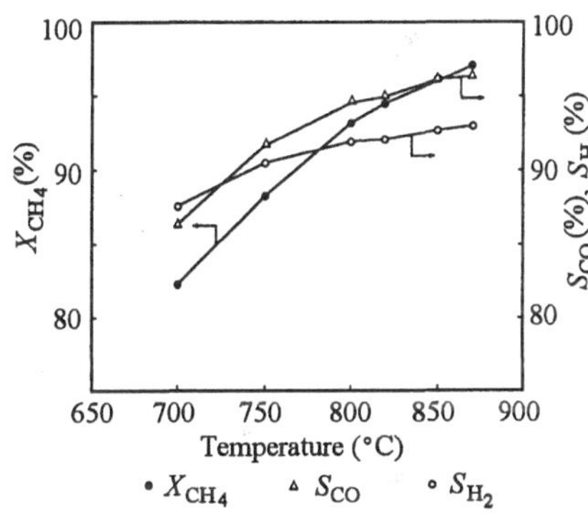

Fig.2 Catalytic activities on function of temperature

The influence of reaction temperature on the methane conversion and the selectivities was demonstrated in Fig.2 for the $NiCeX_2$ catalyst. When the temperature was changed from 700° C to 870°C, the methane conversion was augmented gradually; so did the selectivities of product CO and H_2. The methane was more active to be dissociated to $C_{(s)}$ and $H_{(s)}$ when the temperature was higher, so did the desorption of $CO_{(s)}$ and $H_{2(s)}$. These could be the reason of the observed phenomena. At 870°C with a space velocity of 1.5×10^5 h^{-1}, a methane conversion of 97.1% was obtained with a hydrogen selectivity of 93.0%, the ratio of H_2/CO was 1.93.

There were always several propositions of reaction pathway, the CO and H_2 were supposed to be the initial products of the selective partial oxidation, the CO_2 and H_2O could be the products of total oxidation of methane. The CO_2 could also be obtained from the oxidation of

CO while the H_2 oxidized to H_2O. In our reaction condition, the selectivity to CO and that to H_2 were both high, and $S_{CO} > S_{H_2}$, this could be explained as that the rate of H_2 oxidation to H_2O was somewhat greater than that of CO oxidation to CO_2.

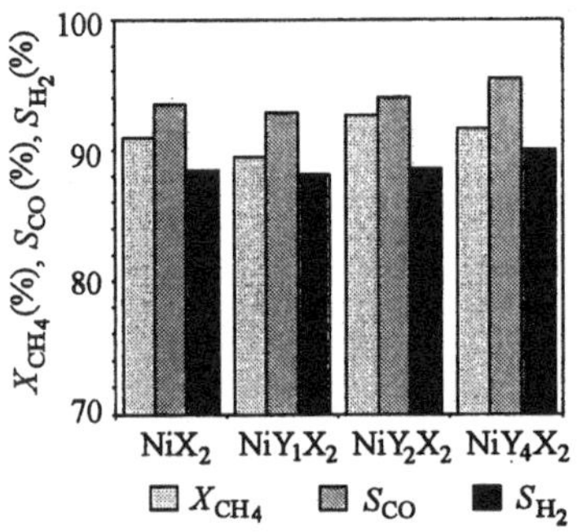

Fig.3 Influence of yttria additive on performance

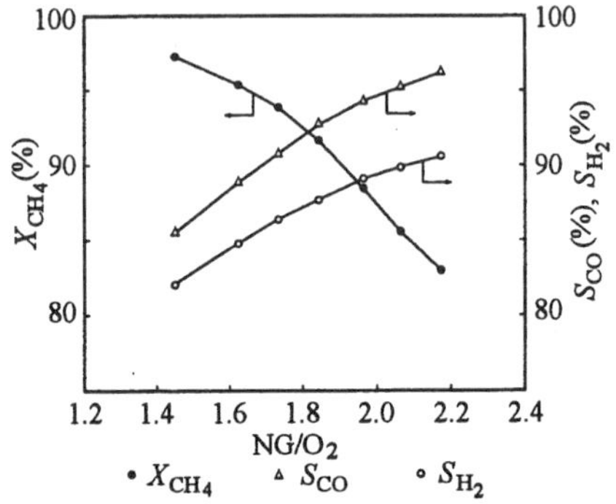

Fig.4 Influence of inlet gas composition on activities

In the case of yttria promoted catalysts, the comparison of catalytic performance was illustrated in Fig.3. When an amount of 1%Y_2O_3 was added to the metal catalyst, the activity decreased slightly. With the augmentation of yttria content, the selectivity of CO rose gradually, so did the H_2 selectivity. The better result of the methane conversion was obtained with 2% yttria addition, and reached 92.8% under the reaction at 800°C. For the product yields, the catalyst promoted with 4% yttria gave the better, a CO yield of 87.8% was achieved with 82.8% H_2 yield. Then, it could be drawn that the rare earth oxides like CeO_2 and Y_2O_3 were good promoters for the desired selective reaction.

3.2. Effect of different parameters

In the new process to produce synthesis gas, the operating conditions could play an important role in the catalytic selective oxidation[10]. The influence of inlet gas composition on activities was indicated in Fig.4.

When the NG/O_2 ratio was increased from 1.46 to 2.17 in the reaction at 780°C, the methane conversion changed gradually from 97.3% to 83.0%. However, the CO selectivity was enhanced from 85.5% to 96.2%; so did the H_2 selectivity, from 82.1% to 90.6%.

In the case of product yields, the best result was obtained when the NG/O_2 ratio was equal to 1.73. A CO yield of 85.3% was reached with a H_2 yield of 81.1% at 780°C.

The product gas composition of this selective oxidation could be varied with the reaction temperature. When the center temperature was changed from 700°C to 870°C over the $NiCe_2X_2$ catalyst, the remained content of methane diminished, from 7.38% to 1.04%.

However, the amounts of CO and H_2 was enhanced gradually, the total amount of CO and H_2 was augmented to 97.68% from 87.93%; while the content of CO_2 decreased. When the reaction took place at 870°C with a space velocity of 1.5×10^5 h^{-1} CH_4, a content of 63.59% for hydrogen with 34.09% of CO could be obtained.

3.3. Physico-chemical characterization of catalysts

The temperature programmed reduction (TPR) tests were operated at a homemade apparatus. For the catalyst NiX_2, there were three reduction regions, the main one was from 620°C to 840°C with a peak at 760°C, an obvious reduction position at more than 900°C was also observed. The broad and low peak at 500°C was corresponding to the reduction of three-dimensional NiO aggregates at minor amount; the sharp one at 760°C was the reduction of highly dispersed two-dimensional NiO; and the higher one at more than 900°C suggested a part of slowly and hardly reducible $NiAl_2O_4$.

For the catalyst promoted with ceria, the reduction behavior was quite different, only a broad and middle peak around 701°C was observed. This phenomenon demonstrated that there was an synergistic effect between the active nickel component, ceria and support, which was favorable for the catalyst stability and the performance[9]. In the case of yttria promoted catalysts, the TPR graphs were compared. There were also three reduction ranges for NiY_2X_2 and NiY_4X_2 samples respectively. The peak temperature of the main one diminished to 672°C for the promoted NiY_2X_2 catalyst from 760°C, a higher reduction peak at 843°C was also observed. These two new reduction peaks were lower and broader than those of the unpromoted NiX_2 sample. For the NiY_4X_2 catalyst, the main one was shifted even more to 666°C, and the higher was more than 900°C; these two were also lower and broader. The modifications of reduction behaviors could play an important role in the formation of active sites, then giving an improved catalytic activity. Although ceria and yttria were both rare earth oxides, the mechanism of their functions could be at a different way according to their different reduction characters.

The X-ray diffraction (XRD) measurements were used for the structure phase determination. For the NiX_2 catalyst, several peaks of NiO crystallises could be detected (*d* value equals to 2.09, 2.411, 1.475). In the case of ceria promoted $NiCe_2X_2$ sample, the intensities of NiO diffraction peaks decreased; and this suggested that the ceria suppressed the growth of nickel crystallite and enhanced its dispersion.

The X-ray photoelectron spectroscopy (XPS) technique was used for the surface study of the ceria promoted catalysts. For the unpromoted NiX_2 catalyst, a severe surface enrichment of nickel was observed, and the nickel surface amount was equal to 60.8 wt%. When ceria was added, there was less amount of nickel on the surface (42.5%), and the alumina content was augmented to 32.0% from 12.0%.

The above observations suggested that there were several beneficial effects of ceria addition: (1) for the unpromoted catalyst, the nickel should be covering on the support surface and aggregated; however, the addition of ceria could diminish the strong interaction between the nickel and alumina, enable the nickel to enter into the lattice of Al_2O_3, then the surface content of alumina was enhanced and that of nickel decreased. (2) the ceria presence suppressed the growth of nickel crystallites, and they kept staying in smaller particles. The carbon deposition reaction was abated, and the catalyst stability was improved.

ACKNOWLEDGMENT

The authors would like to thank Mr. Kang Xingwu, Mr. Lin Shihu, Mr. Sheng Shishan, Dr. Gao Lizhen, Mr. Chen Dongliang for their helps in techniques and experiments, as well as useful discussions.

REFERENCES

1. M.J.Brown, M.D.Parkyns, *Catal.Today*, 8, (1991), 305
2. J.R.Anderson, *Applied Catalysis*, 47, (1989), 177
3. E.R.Kirk, D.F.Othmer, eds., *Encyclopedia of Chemical Technology*, Vol.12, 3rd Ed., (Wiley-Interscience), New York, (1990), 938p
4. D.A.Hickman, L.D.Schmidt, Science, 259, 243, (1993)
5. R.H.Jones, A.T.Ashcroft, D.Waller, A.K.Cheetham, J.M.Thomas, *Catal.Lett.*, 8, (1991) 169
6. A.T.Ashcroft. A.K.Cheetham, J.S.Foord, M.L.H.Green, P.D.F.Vemon, *et al.*, *Nature*, 344, 319 (1990)
7. V.R.Choudhary, V.H.Rane, A.M.Rajput, *Catalysis Lett.*, 22, (1993), 289
8. P.D.Battle, J.B.Claridge, F.A.Copplestone, S.C.Tsang, *et al.*, *Appl.Catalysis*, 118, 217, (1994)
9. Chu Wei, Yan Qiangu, Li Qing, Yu Zuolong, Liu Xi, *et al.*, *Catalysis Today*, accepted, (1998)
10. Yu J. S., Zhang L.Q., Yu Z.L., *Nat.Gas Chem. Industry*, 21(4), 5, (1996)
11. F.Van Looij, J.C.Van Giezen, E.R.Stobbe, J.W.Geus, *Catal.Today*, 21, 495, (1994)
12. Chu Wei, Liu Xi, Yu Zuolong, Li Qing, Luo Chunrong, *et al.*, *Nat.Gas Chem.Industry*, accepted, (1998)

NATURAL GAS CONVERSION V
Studies in Surface Science and Catalysis, Vol. 119
A. Parmaliana et al. (Editors)

Reactivity of Pt/Al_2O_3 and $Pt/CeO_2/Al_2O_3$ catalysts for partial oxidation of methane to syngas*

Qian-Gu Yan[a] Wei Chu[a] Li-Zhen Gao[a] Zuo-Long Yu[a†] Song-Yue Yuan[b]

[a]Chengdu Institute of Organic Chemistry, The Chinese Academy of Sciences
Chengdu, Sichuan 610041 P.R.China

[b]Changchun Institute of Applied Chemistry, The Chinese Academy of Sciences
Changchun, Jilin 130022 P.R.China

Abstract: The ceria modified $Pt/CeO_2/Al_2O_3$ and Pt/Al_2O_3 catalysts were studied in the partial oxidation of methane to syngas. The SEM, XRD, TPR and TPD techniques were used for the catalyst characterization. The addition of ceria could enhance the Pt dispersion and decrease the Pt crystallise size; the activity and selectivity of catalyst for partial oxidation were improved significantly, and the methane total oxidation was suppressed sharply. The ceria effect was also discussed in a detailed way.

Keywords: Partial oxidation of methane (POM), Syngas, Pt catalyst, Ceria promoter, Characterization

INTRODUCTION

Partial oxidation of methane (POX) to produce syngas has more potential than the steam reforming (SR) process. In POX process, supported metal catalysts, including noble metals and transition metal catalysts are often used[1,2]. Under the operation conditions, supported Pt catalyst showed great stability and carbon deposition resistance, however, it showed lower conversion and selectivities than Rh and Ni, especially for H_2 selectivity.

Cerium oxide and CeO_2-containing materials have been used as catalysts and as structural and electronic promoters for heterogeneous catalytic reactions since the past decades. Ceria is added as a promoter to noble metals in many reactions. Several possible mechanisms of these promoting effects have been proposed, such as ceria affecting the dispersion of supported metals, promoting the WGS reaction and SR reaction, increasing the thermal stability of the support, and preventing the sinter of the metal particles[3~5]. The potential of CeO_2 to oxide hydrocarbons has been known for several years[6]. Total oxidation of hydrocarbons on ceria

* This project is supported by the National Natural Science Foundation of China, No. 29573128.
† To whom correspondence should be addressed.

have been examined. Several studies on methane oxidation have been conducted using CeO_2 supported on Al_2O_3 or SiO_2 with noble metals. This study reports the effect of ceria on Pt/Al_2O_3 catalyst for partial oxidation of methane to syngas. The TPR and H_2-TPD techniques were used to elucidate the effect of ceria on Pt/Al_2O_3 catalyst and the possible mechanism.

EXPERIMENTAL

Preparation of catalysts

The support used in this study was alpha alumina (5 m^2/g) with the particle size of 40~50 mesh. Six catalysts were prepared with different amounts of ceria of 0, 0.5, 1.0, 2.0, 4.0 and 8.0%(wt). The incipient wetness impregnation method was used. The support was first impregnated with a solution of cerium nitrate, dried at 383 K for 5 h, calcined at 873 K for 10 h and cooled, then it was impregnated with a solution of $Pt(NH_3)_4(NO_3)_2$ and dried at 423 K for 5 h. The platinum amounts are all 1.0 %(wt).

Test of catalyst

The catalytic performance tests were carried out in a fixed quartz microreactor (8 mm i.d.). The apparatus has been described previously[7]. The experimental conditions for catalyst performance testing were as follows: 250 mg catalyst, total space velocity of 2×10^5 h^{-1}, mole ratio of CH_4/O_2=2, reaction system pressure of 0.05 MPa, temperature range 1023~1073 K.

Catalyst characterization

In the TPR and TPD of this study, 100 mg catalyst sample was placed into a quartz tube (3 mm o.d.). The sample was preheated *in situ* with He at 423 K for 1 h and then cooled to room temperature. A gas stream of 8%(V) H_2 in argon at a total flow rate of 40 mL/min passed through the catalyst. The temperature was increased at a rate of 10 K/min from 303 K to 1023 K. The water produced by reduction was trapped into a column of silica gel. The amount of H_2 consumption was detected with a thermal conductivity detector (TCD). The reduced sample was cleaned at 1023 K in He for 2 h, then a pure hydrogen flow of 30 mL/min (purified before using) passed through the catalyst for 0.5 h. The catalyst sample was cooled to room temperature in pure hydrogen stream, then the catalyst was blown with He to remove the excess gas. Desorption experiments were carried out with He flowing at 30 mL/min by increasing catalyst bed temperature from 303 K to 1023 K at the rate of 20 K/min.

SEM was carried out on a Jeol TXA-800 scanning electron microscopy. The Pt crystal sizes of catalysts were measured on a powder X-ray diffractometer (D/max-rB).

RESULTS AND DISCUSSION

TPR and TPD experimentals

The TPR profiles of catalysts are shown in Fig.1. A blank test of CeO_2/Al_2O_3 sample was performed for the TPR analysis, there were two reduction peaks: one peak was around 773 K, the higher was at 923 K. The peak area was bigger when the ceria content was higher. For the Pt/Al_2O_3 catalyst, there was only one peak at 458 K; for the $Pt/CeO_2/Al_2O_3$ catalyst with different content of ceria, the reduction profiles were illustrated in Fig.1. It was shown that the presence of Pt decreased the low temperature peak of CeO_2 on alumina to a range between 483 K and 623 K; and this peak became broader and more multiple when the ceria content augmented; that of Pt was almost unchanged. As it was knows, the platinum could activate hydrogen and this activation could be enough to decrease the reduction temperature of ceria through a spillover mechanism. In other aspect, some of the reduced ceria could cover a part of platinum.

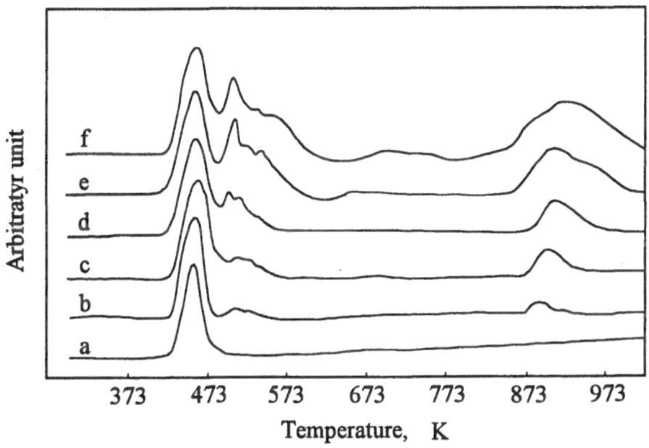

Fig.1 TPR spectra of Pt/Al_2O_3 and $Pt/CeO_2/Al_2O_3$ catalysts
(a) 0%CeO_2, (b) 0.5%CeO_2, (c) 1.0% CeO_2
(d) 2.0%CeO_2, (e) 4.0%CeO_2, (f) 8.0%CeO_2

The TPD profiles of catalysts with H_2 were discussed.

In the TPD experiment, a blank test of CeO_2/Al_2O_3 sample was operated and there was only one peak at 823 K for H_2 desorption, and it was explained as that the ceria could behave as a container of hydrogen which desorbed at 823 K. The profiles of Pt/Al_2O_3 and $Pt/CeO_2/Al_2O_3$ catalysts were detected. Three broad peaks were shown for the Pt/Al_2O_3 catalysts, at 393 k, 623 K and 803K respectively. When the ceria was added, there were some obvious changes: (a) a much broader region between 373 K and 723 K was obtained rather than two peaks at 393 K and 623 K, this new desorption could be from Pt surface with a ceria environment. Another explanation could be that during the reduction the partially reduced CeO_2-Ce_2O_3 could move and cover the surface of platinum and this could diminish the desorption amount of hydrogen at low temperature. (b) however, the H_2 desorption amount increased with the content of ceria at high temperature of 823 K.

SEM and XRD analyses

Table 1

Pt crystallite sizes of catalysts before and after reaction

	Pt particle size /Å	
CeO_2 / % (wt)	Before reaction	After reaction
0	210	282
0.5	115	121
1.0	90	92
2.0	85	86
4.0	83	87
8.0	80	83

The dispersion of Pt over catalyst was studied using SEM-EDX and it was found that the Pt on $Pt/CeO_2/Al_2O_3$ is well-distributed but it is bad-distributed on Pt/Al_2O_3. The Pt crystallite size was calculated from the half-width of the Pt (111) peak in XRD curve of the catalysts (Table 1). From Table 1, we find that the Pt crystallite size on $Pt/CeO_2/Al_2O_3$ is smaller than that on Pt/Al_2O_3, which indicates that ceria can decrease the Pt crystallite size and inhibit Pt crystallite growth during reaction. This may be derived from the high temperature reduction, partially reduced ceria species are able to migrate through the Pt particles to separate them to smaller ones or to cover part of the Pt particle surface.

Methane partial oxidation reaction

The results of catalytic performance testing are shown in Table 2.

From Table 2, we can find the improvement both in catalytic activity and in selectivity of products, especially in H_2 selectivity over $Pt/CeO_2/Al_2O_3$ catalyst with increasing of the amount of Ce. The effect was obvious at the beginning and became slight after the addition of 1% CeO_2.

The measured axial temperatures along catalyst bed (center temperature was all at 1023 K) were compared. Over the CeO_2/Al_2O_3 sample, the combustion activity of methane oxidation was very high and the center temperature could increase sharply to more than 1273 K. For the unpromoted Pt/Al_2O_3 catalyst, there was the selective oxidation of methane to CO and H_2, together with a part of total oxidation to CO_2 and H_2O. And the temperature at the inlet could arrive to 1171 K. When the ceria was added, there were both dispersion effect and electronic effect; the activity of total oxidation was suppressed and the temperature at the inlet arrived only at 1109 K when 1% CeO_2 was added.

Table 2
The conversion and selectivity (at 1023 K and 1073 K) of catalyst in the partial oxidation of methane to syngas

	1023 K			1073 K		
CeO_2 %(wt)	$C(CH_4)$ %	$S(CO)$ %	$S(H_2)$ %	$C(CH_4)$ %	$S(CO)$ %	$S(H_2)$ %
0	75.1	86.5	80.3	83.8	90.2	84.5
0.5	80.5	89.0	84.6	88.2	93.3	90.6
1.0	82.0	90.5	86.0	89.5	93.4	91.2
2.0	82.8	91.0	87.2	90.2	93.0	92.8
4.0	83.0	91.2	87.1	90.6	93.3	91.8
8.0	82.5	91.5	87.0	90.4	93.0	92.2

L.D.Schmidt[1] pointed out that the monolith-supported Pt catalyst exhibited less activity and H_2 selectivity than Rh and Ni catalysts for partial oxidation reaction. It was shown the total oxidation (combustion reaction) activity of methane over Pt/Al_2O_3 catalyst decreased significantly as ceria was added, CH_4 conversion and the selectivity of partial oxidation products were increased. S.H.OH[6] found that the total oxidation of CH_4 over $Pt/CeO_2/Al_2O_3$ was suppressed.

Discussion of ceria effect

According to the results given above, the following discussions are proposed. The strong interaction between Pt and ceria may result in a larger extent of boundary area between cerium oxide and platinum. We believe that the active sites involve both Pt and ceria. It was found that during the SR process or under the reductive atmosphere, some of the cerium were in the state of Ce(III)[8]. Ce^{3+} sites formed after reduction of Pt based catalyst, and those sites are at the metal -ceria interface, the lower-valence-state cerium may dissociatively adsorb oxygen or water, and the resulting adspecies -O or -OH may be transferred to adjacent platinum and reacted with surface carbon species (CH_x, x=0~3) to give out CO, CO_2, H_2 and H_2O. So the activity of $Pt/CeO_2/Al_2O_3$ is higher than Pt/Al_2O_3.

Otsuka[9] reported that cerium oxide itself can oxidize methane to syngas. Though its activity is several orders lower than other transition metal catalysts, however, when noble metals coexist with cerium oxide, the reaction may be much accelerated. The mechanism of redox cycle of CeO_2 is proposed as follows:

$$CeO_2 + nCH_4 \longrightarrow CeO_{2-n} + nCO + 2nH_2 \quad (1)$$

$$CeO_{2-n} + n/2\ O_2 \longrightarrow CeO_2 \quad (2)$$

When the amount of ceria was increased on the surface of catalyst, and following high-temperature reduction, CeO_x species are able to migrate through the Pt particles to separate them to smaller ones or to cover part of the particle surface, the activation energy of CH_4 on Pt would increase, or the chemisorption sites of CH_4 on Pt decrease, therefore, in some extent, the dissociative adsorption of CH_4 onto Pt would be suppressed. So, the reaction over $Pt/CeO_2/Al_2O_3$ is not as violent as that over Pt/Al_2O_3, this depressed the combustion reaction of methane over the catalyst, avoided high temperature for local catalyst bed and decreased the temperature gradient of catalyst bed (decreased more than 100°C). The selectivity increase for $Pt/CeO_2/Al_2O_3$ catalyst should be resulted from the inhibition of the total oxidation activity of methane over catalyst. The suppressing of total oxidation to avoid high temperature benefits also to the stability of catalyst.

Moreover, another role of ceria in this reaction is promoting the water-gas-shifted (WGS) reaction by a redox cycle mechanism, which benefits the increase in selectivity to hydrogen.

REFERENCES

1. P.M.Torniainen, X.Chu, and L.D.Schmidt, *J.Catal.*, **146** (1994) 1
2. D.Dissanayake, M.P.Rosynek, K.C.C.Kharas, and J.H.Lunsford, *J.Catal.*, **132** (1991) 117
3. J.C.Summers, and S.A.Ausen, *J.Catal.*, **58** (1979) 131
4. B.Harrison, A.F.Diwell, and C.Hallett, *Platinum Met.Rev.*, **32** (1988) 73
5. Hattori, Inoko, and Murakami, *J.Catal.*, **79** (1983) 493
6. S.H.OH, P.J.Mitchell, and R.M.Siewert, *J.Catal.*, **132** (1991) 287
7. Q.G.Yan, Z.L.Yu, and S.Y.Yuan, *J.Nat.Gas Chem.*, **6** (1997) 93
8. J.M.Schwartz, and L.D.Schmidt, *J.Catal.*, **138** (1992) 238
9. K.Otsuka, T.Ushiyama, and I.Yamanaka, *Chem.Lett.*, (1993) 1517

NATURAL GAS CONVERSION V
Studies in Surface Science and Catalysis, Vol. 119
A. Parmaliana et al. (Editors)

Carbon-free CH_4-CO_2 and CH_4-H_2O Reforming Catalysts —Structure and Mechanism

K. Tomishige, Y. Chen, O. Yamazaki, Y. Himeno, Y. Koganezawa and K. Fujimoto

Department of Applied Chemistry, School of Engineering, The University of Tokyo, Hongo, Bunkyo-ku, Tokyo 113-8656, Japan

Nickel-magnesia solid solution catalyst with low Ni content had excellent stability for CO_2 reforming of methane and steam reforming under low steam to methane ratio at 1123 K without deposited carbon. The deactivation due to oxidation of active nickel species observed under some reaction condition. Nickel-magnesia solid solution and magnesia supported nickel catalysts were characterized by means of FTIR, temperature programmed reaction of CO_2. It was found that small nickel particles formed on $Ni_{0.03}Mg_{0.97}O$ solid solution catalyst have high ability to reduce CO_2. Though this ability causes the deactivation due to oxidation, this also plays an important role in inhibiting carbon deposition.

1. INTRODUCTION

CO_2 and steam reforming of methane (equations 1 and 2) have been proposed as one of the most promising technologies for effective utilization of methane, and this synthesis gas can be used as liquid fuel by Fischer-Tropsch synthesis. Especially the key point in the utilization of syngas is to control the ratio of H_2 to CO. Generally this is performed by mixing H_2O and CO_2. A serious problem in this reaction is carbon deposition via methane decomposition and CO disproportionation.

$$CH_4+CO_2 \rightarrow 2CO+2H_2 \qquad \Delta H=+247\ \text{kJ/mol} \qquad (1)$$
$$CH_4+H_2O \rightarrow CO+3H_2 \qquad \Delta H=+206\ \text{kJ/mol} \qquad (2)$$

Carbon deposition has been reported to cause the catalyst deactivation, plugging the reactor. This has also been observed in steam reforming, but much more serious in CO_2 reforming of methane as expected by thermodynamic calculations [1]. Recently we have reported that nickel-magnesia solid solution $Ni_{0.03}Mg_{0.97}O$ has exhibited high and stable activity without carbon deposition in CO_2 and steam reforming of methane [2-7]. But deactivation due to oxidation of active nickel species has been observed on this catalyst [6]. In this article, we investigated the relation between inhibition of carbon deposition and deactivation due to oxidation through characterization of catalyst structure and surface reactivity.

2. EXPERIMENTAL

Nickel-magnesia solid solution catalyst was prepared by coprecipitating nickel acetate and magnesium nitrate aqueous solution with potassium carbonate. After being filtered and washed with hot water, the precipitate was dried overnight at 393 K, and then calcined in air at 1223 K for 10 h. Supported nickel catalyst (Ni/MgO) was prepared by impregnating MgO with $Ni(C_5H_7O_2)_2$ acetone solution. MgO was prepared by the same method as solid solution catalyst. The loading was denoted as the molar ratio Ni/(Ni+Mg). CH_4-H_2O and CH_4-CO_2 reactions were performed in a fixed bed flow reaction system equipped with gas chromatograph. Catalysts were usually reduced with H_2 at 1123 K before the reaction for 0.5 h. Reaction condition was CH_4/(H_2O or CO_2) =1/1, 773 K-1123 K, W/F=0.1-1.2 gh/mol. Carbon deposition was characterized by temperature programmed hydrogenation (TPH) method, in which pure hydrogen was introduced, and then temperature was raised from room temperature to 1123 K at a heating rate of 20 K/min [7]. The signal corresponding to CH_4 formation was recorded continuously by FID detector without separating column. Temperature programmed reaction of CO_2 with reduced catalysts was preformed in a fixed bed flow reactor. CO formation was measured by FID-GC. Pure CO_2 was introduced to the sample and the heating rate was 10 K/ min. FTIR spectra were obtained in a transmission mode using *in-situ* IR cell connected to closed circulating system. The sample was pressed into a self-supporting disk and put into a slit of the holder in this IR cell. The sample was reduced with hydrogen at 1123 K for 0.5 h. CO and CO_2 was introduced to IR cell through the closed circulating system at room temperature.

3. RESULTS AND DISCUSSION

Figure 1 shows dependence of methane conversion in CH_4-CO_2 and CH_4-H_2O reaction over $Ni_{0.03}Mg_{0.97}O$ solid solution catalyst on reduction temperature at 873 K. At reduction temperature higher than 1073 K, catalytic activity of CH_4-CO_2 reaction is almost the same over $Ni_{0.03}Mg_{0.97}O$, while that of CH_4-H_2O reaction is a little higher than that of CH_4-CO_2 reaction. This corresponded to the ability of CO_2 and H_2O as the oxidizing reagent. On the other hand, at low reaction temperature 1023 K, methane conversion at 10-min time on stream of CH_4-H_2O reaction is much lower than that of CH_4-CO_2 reaction. Apparently this is strange, but these results are probably due to the deactivation of oxidation. On the sample reduced at 1023 K, the deactivation was observed in both CH_4-CO_2 and CH_4-H_2O reaction. The rate of deactivation was much more rapid in CH_4-H_2O reaction. The deactivation was not observed on the sample reduced at higher temperature under this reaction condition.

Figure 2 shows the effect of the addition of H_2 on the catalytic stability of CH_4-CO_2 reaction on $Ni_{0.03}Mg_{0.97}O$ and 3 mol% Ni/MgO catalysts. The rapid deactivation was observed on $Ni_{0.03}Mg_{0.97}O$ catalyst in CH_4-CO_2 reaction under this small W/F condition. The addition of 5% H_2 to the reactant gas improved drastically catalyst stability. This indicated that the deactivation on $Ni_{0.03}Mg_{0.97}O$ catalyst is due to the oxidation, and this supports the results shown in Figure 1. On 3 mol% Ni/MgO catalyst, by the addition of hydrogen, the activity was decreased and the stability was not so improved. This indicated that the deactivation is not due to the oxidation but probably due to carbon deposition.

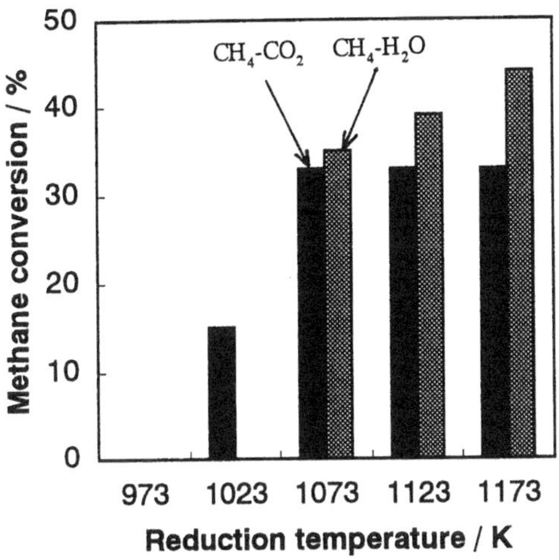

Figure 1. Dependence of methane conversion in CH_4-CO_2 and CH_4-H_2O reaction over $Ni_{0.03}Mg_{0.97}O$ solid solution catalyst on reduction temperature.
Reaction condition: CH_4/CO_2 or CH_4/H_2O=1/1, total pressure 0.1 MPa, W/F=1.2 gh/mol, reaction temperature 873 K, methane conversion after 10 min.

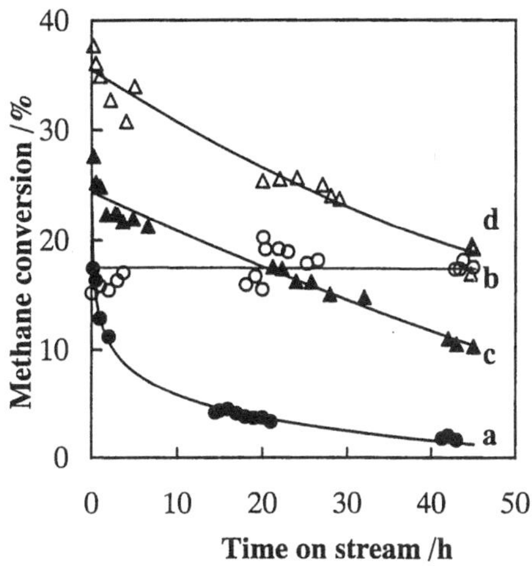

Figure 2. The effect of the addition of H_2 gas to CH_4+CO_2 reactant gas on catalytic stability in CH_4-CO_2 reaction over $Ni_{0.03}Mg_{0.97}O$ (a, b) and 3 mol% Ni/MgO (c, d).
Reaction condition: reaction temperature 1023 K, W/F=0.5 gh/mol for $Ni_{0.03}Mg_{0.97}O$ and W/F=0.2 gh/mol for 3 mol% Ni/MgO, total pressure 0.1 MPa, $CH_4/CO_2/N_2$=50/25/25 (a, c) or $CH_4/CO_2/H_2/N_2$=50/25/5/20 (b, d).

Catalyst properties of $Ni_{0.03}Mg_{0.97}O$ and 3 mol% Ni/MgO catalysts are listed in Table 1. Carbon deposition was observed on 3 mol% Ni/MgO, but that was not observed at all on $Ni_{0.03}Mg_{0.97}O$ even at low reaction temperature (~773 K). This indicated that $Ni_{0.03}Mg_{0.97}O$ has high resistance to carbon deposition. This is consistent with the suggestion that the deactivation is due to oxidation. And it was found that Ni particles are highly dispersed on $Ni_{0.03}Mg_{0.97}O$ [5], but Ni particles formed on 3 mol% Ni/MgO have much larger particle size.

Figure 3 shows FTIR spectra of CO adsorption on nickel magnesia catalysts. On 3 mol% Ni/MgO catalyst, linear (2100-2000 cm^{-1}), bridge (2000-1850 cm^{-1}) and physisorbed $Ni(CO)_4$ (2057 cm^{-1}) were mainly observed. In contrast, on $Ni_{0.03}Mg_{0.97}O$ nickel monomer and dimer carbonyl species which are interacted with MgO were mainly observed as previously reported[8]. These species were increased with the CO pressure, therefore they are found to be formed via CO induced structural change. This difference in FTIR spectra corresponded to the dispersion of Ni metal particles.

Table 1. Catalyst properties of nickel-magnesia solid solution and supported catalysts.

Catalyst	BET	O_2 [a)]	H_2 [b)]	Ni^0/Ni_{total} [c)]	D [d)]	R_{CO} [e)]	carbon deposition rate [f)]	sel. [g)]
	/m^2/g	/μmol/g-cat		/%	/%	/μmol $g^{-1}s^{-1}$	/μ C-mol $g^{-1}s^{-1}$	/%
$Ni_{0.03}Mg_{0.97}O$	22	10.5	3.1	2.9	29.5	31	0.00	0.00
3 mol%Ni/MgO	25	226.5	3.9	62.4	1.7	138	0.15	0.109

a: adsorption temperature 873 K, b: adsorption temperature 298 K, c: reduction degree of Ni was estimated by 2x(amount of O_2 adsorption)/(total amount of Ni), d: dispersion of Ni was estimated by (amount of H_2 adsorption)/(amount of O_2 adsorption), e: CO formation rate in the reforming of CH_4 with CO_2 under 773 K, 0.1 MPa, CH_4/CO_2=1/1, W/F=0.1 gh/mol, catalyst weight: 0.05 g, f:carbon deposition rate under the same reaction condition, g: carbon deposition selectivity is estimated by (carbon deposition rate)/(carbon deposition rate + CO formation rate).

Figure 4 shows FTIR spectra of CO_2 adsorption on these samples. Mainly three kinds of adsorbed CO_2 were observed. They are two kinds of bidentate carbonates (1670 cm^{-1}+1305 cm^{-1}, 1625 cm^{-1}+1272 cm^{-1}) and one bicarbonate (1652 cm^{-1}+1405 cm^{-1})[9]. $Ni_{0.03}Mg_{0.97}O$ catalyst had similar spectra to MgO, bidentate carbonates were main. On the other hand, on 3 mol% Ni/MgO catalyst, the peak intensity was smaller than that of $Ni_{0.03}Mg_{0.97}O$ and bicarbonate species was main.

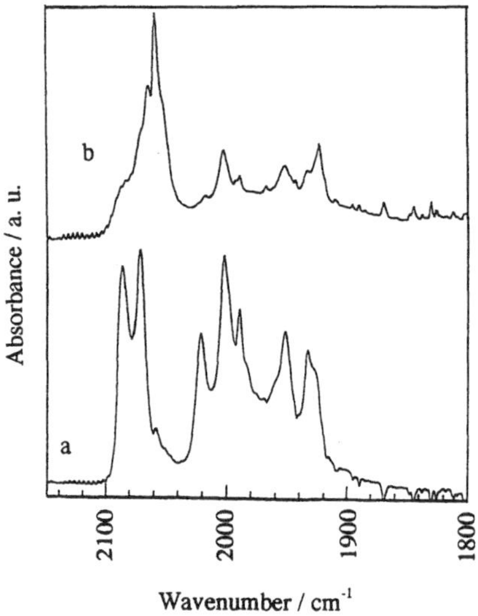

Figure 3. FTIR spectra of CO adsorption on $Ni_{0.03}Mg_{0.97}O$ (a) and 3 mol% Ni/MgO (b) catalysts. P_{CO}=13.3 kPa, 298 K.

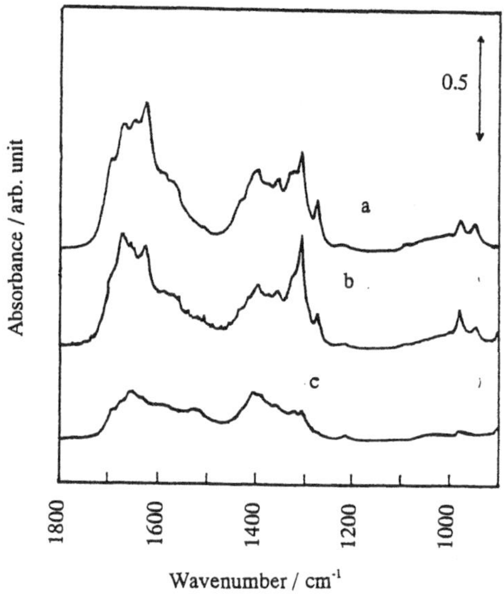

Figure 4. FTIR spectra of CO_2 adsorption on MgO (a), $Ni_{0.03}Mg_{0.97}O$ (b) and 3 mol% Ni/MgO (c) catalysts. P_{CO2}=1.3 kPa, 298 K, after evacuation.

This supports that not only Ni particle size but also structure of support surface is different from these catalysts. According to reduction degree of Ni listed in Table 1, about 40% Ni was not reduced on 3 mol% Ni/MgO catalyst. It is suggested that these Ni species were present at near-surface layer of MgO to form solid solution with high Ni content. The difference between FTIR spectra of CO_2 adsorption seems to be reflected by Ni content of support surface.

Reaction order of CH_4-CO_2 and CH_4-H_2O reaction on $Ni_{0.03}Mg_{0.97}O$ and 3 mol% Ni/MgO catalysts were listed in Table 2. Reaction mechanism of CH_4-CO_2 and CH_4-H_2O reaction is described simply as equations 3 and 4.

$$CH_4 \rightarrow C+2H_2 \quad (3)$$

$$C+CO_2 \rightarrow 2CO \text{ (CO}_2\text{ reforming)} \qquad C+H_2O \rightarrow CO+H_2 \text{ (steam reforming)} \quad (4)$$

In CH_4-H_2O reaction, reaction order of CH_4 was close to 1, and that of H_2O was close to zero. On both catalysts, methane dissociation (eq. 3) is rate-determining step. In CH_4-CO_2 reaction on 3 mol% Ni/MgO catalyst, reaction order of methane and CO_2 was much less than one. This indicated that rate-determining step switches to CO_2 dissociation or the reaction between C and O formed from CO_2. This switching is probably corresponding to the ability of H_2O and CO_2 as an oxidizing reagent. On the other hand, in CH_4-CO_2 reaction on $Ni_{0.03}Mg_{0.97}O$, reaction order of methane was also close to one. This suggests that the rate of eq. 4 is much higher than that of eq. 3 like in CH_4-H_2O reaction.

Table 2. Reaction order of CH_4-CO_2 and CH_4-H_2O reaction.

Catalyst	Reaction	Reactant	
		CH_4	CO_2 or H_2O
$Ni_{0.03}Mg_{0.97}O$	CH_4-CO_2	0.9	0.1
	CH_4-H_2O	0.8	0.1
3 mol% Ni/MgO	CH_4-CO_2	0.4	0.2
	CH_4-H_2O	1.0	0.0

CH_4-CO_2 reaction: 773 K, W/F=0.5 gh/mol($Ni_{0.03}Mg_{0.97}O$) and 0.2 gh/mol (3 mol% Ni/MgO), CH_4-H_2O reaction: 1023 K, W/F=0.3 gh./mol.

Figure 5 shows the profiles of CO_2 TPR on $Ni_{0.03}Mg_{0.97}O$ and 3 mol% Ni/MgO reduced at 1123 K. CO_2 reduction started at 453 K and 513 K on $Ni_{0.03}Mg_{0.97}O$ and 3 mol% Ni/MgO catalysts, respectively. This indicated that $Ni_{0.03}Mg_{0.97}O$ has much higher ability to reduce CO_2 than 3 mol% Ni/MgO. This is consistent with the result that $Ni_{0.03}Mg_{0.97}O$ catalyst tends to deactivate by oxidation of active site. From the comparison between these two catalysts, it is suggested that the reaction of nickel metal surface with CO_2 from gas phase does not seem to proceed at low temperature, but CO_2 is adsorbed on support surface and adsorbed CO_2 interacted with nickel metal at the interface between metal and support. $Ni_{0.03}Mg_{0.97}O$ has much larger interface than 3 mol% Ni/MgO since metal particles on $Ni_{0.03}Mg_{0.97}O$ is much more highly dispersed. In addition, since $Ni_{0.03}Mg_{0.97}O$ is similar to MgO in terms of CO_2 adsorption, the interaction between CO_2 and $Ni_{0.03}Mg_{0.97}O$ is suggested to be stronger than 3 mol% Ni/MgO. These two factors are suggested to promote the activation of CO_2 on $Ni_{0.03}Mg_{0.97}O$. This promotion is strongly related to the characteristic inhibition of carbon deposition in CH_4-CO_2 reaction on $Ni_{0.03}Mg_{0.97}O$.

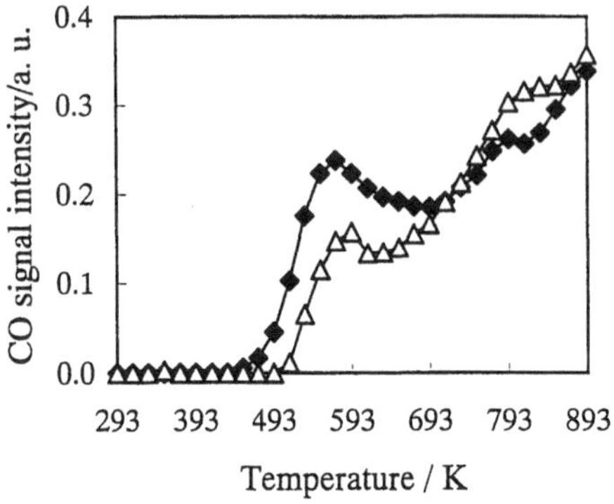

Figure 5 Profiles of CO_2 TPR on $Ni_{0.03}Mg_{0.97}O$ (◆)and
3 mol% Ni/MgO (△)reduced at 1123 K.
TPR condition: heating rate 10 K/min, 100% CO_2 was used.
Reduction condition: 1123 K, 30 min, 100% H_2 was used.

4. CONCLUSION

$Ni_{0.03}Mg_{0.97}O$ solid solution catalyst has high resistance to carbon deposition in CO_2 reforming of methane, but the deactivation due to oxidation of active nickel species was observed under some reaction condition. This deactivation was able to be avoidable by the addition of hydrogen to reactant gas. On $Ni_{0.03}Mg_{0.97}O$ catalyst, nickel particles were much more highly dispersed than that on 3 mol% Ni/MgO. And CO_2 was adsorbed on $Ni_{0.03}Mg_{0.97}O$ similarly to MgO. In addition, CO_2 was activated on $Ni_{0.03}Mg_{0.97}O$ at 60 K lower temperature than on 3 mol% Ni/MgO. This is suggested to be promoted by large interface between metal and support , and strong interaction of CO_2 with support surface. Combined with the results of reaction order, it is suggested that high resistance to carbon deposition in methane reforming is closely related to high ability of CO_2 activation and this causes rapid oxidation of carbon species on nickel before converting to deposited carbon.

REFERENCES

1. J. R. Rostrup-Nielsen, Catalysis Science and Technology, J. R. Anderson, M. Boudart (eds), Germany, Berlin, Springer, 5 (1984) 3.
2. O. Yamazaki, K. Omata, T. Nozaki and K. Fujimoto, Chem. Lett. (1992) 1953.
3. O. Yamazaki, K. Tomishige and K. Fujimoto, Appl. Catal. A:General 136 (1996) 49.
4. K. Tomishige and K. Fujimoto, Catalysis surveys from Japan, in press.
5. Y. Chen, K. Tomishige and K. Fujimoto, Appl. Catal. A:General, 165 (1997) 335.
6. Y. Chen, O. Yamazaki, K. Tomishige and K. Fujimoto, Catal. Lett., 39 (1996) 91.
7. Y. Chen, K. Tomishige and K. Fujimoto, Appl. Catal., A:General, 161 (1997) L11.
8. A. Zecchina, G. Spoto, S. Coluccia and E. Guglielminotti, J. Chem. Soc., Faraday Trans. 1, 80 (1984) 1891.
9. R. Phillip and K. Fujimoto, J. Phys. Chem., 96 (1992) 9035.

NATURAL GAS CONVERSION V
Studies in Surface Science and Catalysis, Vol. 119
A. Parmaliana et al. (Editors)

Advanced Gas-to-Liquids Processes for Syngas and Liquid-Phase Conversion

E. P. Foster, P. J. A. Tijm, D. L. Bennett
Air Products and Chemicals, Inc.
7201 Hamilton Boulevard
Allentown, PA 18195-1501

INTRODUCTION

Conventional technology options may lead to commercially viable gas-to-liquids (GTL) projects which are very large, have favorable site specific factors or very low natural gas costs. New and lower cost technology will be required to enable GTL to be broadly useful for remote gas monetization as a liquid fuel. Air Products and Chemicals, Inc. is currently developing two separate technologies which would result in a significant reduction in the capital investment required for GTL product plants. ITM Syngas is one of Air Products proprietary syngas technologies. It is in the early stages of development, but has the potential for very significant reductions in the cost of syngas, an important intermediate for GTL production. Air Products, along with its partners, have recently been selected by the U.S. Department of Energy (DOE) for an $85MM, three phase program to develop this ITM Syngas technology. The program will take eight years and culminate in a 15,000,000 SCFD pre-commercial syngas demonstration plant. In addition to ITM Syngas, in April 1997 Air Products started up a commercial scale Liquid Phase Methanol (LPMEOH) plant which converts coal derived syngas to methanol using a slurry bubble column reactor. This technology is expected to reduce the cost of liquid synthesis. It also produces an environmentally superior alternative fuel and/or chemical feedstocks.

1. CONVENTIONAL TECHNOLOGY OPTIONS AND OBJECTIVES FOR COST REDUCTION

GTL plants typically require synthesis gas generation, liquid product synthesis, and depending upon the liquid product synthesis route, possibly separation and upgrading. The major portion of the GTL plant capital investment is required for the generation of the synthesis gas, a mixture of hydrogen and carbon monoxide. Synthesis gas can be generated in a variety of processes. For liquid fuel products, the H_2/CO ratio of the synthesis gas should be about 2. The more economical processes, when an H_2/CO ratio of 2 is required, are typically: Partial Oxidation (POX) or Autothermal Reforming (ATR).

A 25% reduction in the cost of GTL products would be a significant improvement, should allow GTL products to compete with oil at $20/bbl or less, and is probably sufficient to bring about a substantial increase in GTL production. Overall GTL cost reduction will have to come from a variety of sources. Air Products has extensive development programs to decrease the cost of both the syngas and the liquid synthesis portions of the GTL plant. Together, these two technologies have the potential for a 20 to 35% reduction in the capital cost for a GTL facility.

2. AIR PRODUCTS' ITM SYNGAS PROGRAM

Some cost reduction for conventional syngas production techniques will occur, but these technologies are quite mature and a significant breakthrough is not likely. If we could combine the functions of both the cryogenic air separation plant and the high temperature syngas generation steps into a single process step, a significant breakthrough would result. Air Products and its development partners, in conjunction with the U.S. DOE, are developing Ionic Transport Membranes (ITM) technology for syngas production (ITM Syngas) which has the potential for a significant reduction in GTL production costs (1,2). ITMs are non-porous ceramic membranes which operate at high temperature. The formulation of the ceramic materials is relatively complex, and incorporates oxygen ion vacancies within the crystalline structure. When heated to a high temperature (typically higher than 700°C), ions of oxygen can flow through these membranes by sequentially occupying these oxygen ion vacancies. ITMs can be used to generate many industrial gas products and Air Products has been developing ITMs for over a decade. Air Products holds about forty patents in the field. A few representative patents are cited in the reference list.

Certain types of ceramic membrane materials can be used in partial pressure driven processes, known as ITM Oxygen processes, which make tonnage quantities of oxygen with the cogeneration of electricity (11). Other types of mixed conducting ceramic membrane materials can be used in processes in which a driving force for oxygen transport is developed by depleting the oxygen partial pressure through reaction (7).

A general characteristic of ITM technology is the very high flux of oxygen transfer which is possible. The net result is that ceramic membrane reactors are very compact and low weight. This, in particular, will make ITM Syngas very attractive for applications requiring compact design and a very small footprint, such as oil platforms, barges, ships or where there is a high cost associated with construction, e.g., arctic areas.

ITM Syngas Process Mechanism

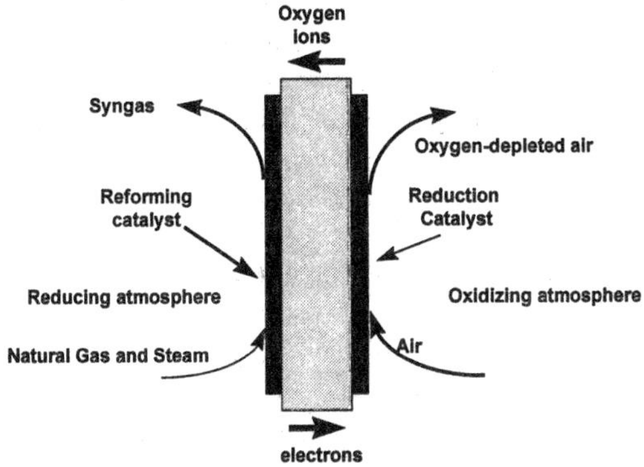

The membrane structure is complex. As illustrated above, a simplified membrane has three layers: the non-porous membrane which allows only oxygen ion and electron flow, the reduction catalyst layer, and the reforming catalyst layer. The membrane material must show long-term stability in reducing and oxidizing atmospheres and long-term compatibility with the reduction and reforming catalysts. Intimate contact of the reforming catalyst with the ceramic membrane is critical to accomplish depletion of the oxygen transported through the membrane by reaction at the ceramic membrane/catalyst interface. This is a necessary requirement to control the partial oxidation reaction exotherm and provide a stable thermal operating mode for the membrane reactor.

Air Products has been working with the DOE on syngas technology since 1995. In 1997, Air Products and our partners won a competitive bid for an $85 million, eight year program to develop the ITM Syngas process and related technology. Air Products has assembled a highly qualified team from industry, national labs, and universities. The team comprises leading ionic ceramic material scientists, experienced ceramic fabrication experts, process and project engineers, and commercial experts. Our team includes; Air Products and Chemicals, Inc., ARCO Exploration and Production Company, Argonne National Laboratory, Ceramatec, Inc., Chevron, Eltron Research, Inc., McDermott Technology Incorporated, Pacific Northwest National Laboratory, Norsk Hydro, Pennsylvania State University, University of Alaska and University of Pennsylvania.

3. MAJOR TECHNICAL GOALS

Material selection is a highly iterative process requiring consideration of many factors. The material must have an acceptable oxygen flux, stability, strength and fabricability. The membrane system must be able to tolerate the pressure differentials, reasonable ramp up and ramp down rates and emergency shutdown transients. The reactor design must address seals, differential expansion, and safety. Ceramic membranes require complex, multi-step ceramic processing. Feedstock purity, control of potential contamination during processing, powder preparation, part forming, firing and assembly are among the many considerations during the development of the ceramic manufacturing processes. The fundamental performance of the membrane system must be evaluated under many operating cases. A full understanding of performance and stability must be developed. In addition, tests are needed at a variety of scale sizes so that developed models for flux, conversion, and stability can be applied with confidence. Scale-up of chemical reactors and processing equipment is a concern for all technology developers. In addition to conventional gas production process scale-up concerns, scale-up of the ceramic manufacturing processes for ceramic components is a critical issue which will be addressed.

4. RISK MANAGEMENT

The Oil and Gas Industry is skeptical and cautious when it comes to new technology promises. A critical aspect of our development program is the close cooperation with key end-users so that their input directly impacts the development approach. Air Products and its partners have chosen to invest heavily to understand scale-up. Initial testing is at the bench scale. An engineering Process Development Unit will test sub-scale membranes under full operating conditions at 12,000 SCFD rates. The next scale for development is a 500,000 SCFD Sub-scale Engineering Prototype. This large facility will be integrated into the Alternate Fuels Development Unit facility operated by Air Products at DOE's LaPorte, Texas site. The plant shall demonstrate the suitability of ITM Syngas technology for further scale-up. During the final stage of ITM Syngas development, a 15,000,000 SCFD Pre-Commercial Technology Demonstration Unit will be tested.

5. AIR PRODUCTS' LIQUID PHASE SYNTHESIS PROCESS

In the early seventies, Chem Systems and Air Products initiated efforts to achieve a cost reduction in the GTL synthesis step by improving the heat transfer characteristics in the reactor through the use of a slurry. This started the development of the liquid phase methanol (LPMEOH™) technology.

The liquid medium is the feature that differentiates the Liquid Phase™ process from conventional technology. Conventional methanol reactors use fixed beds of catalyst pellets and operate in the gas phase. The Liquid Phase™ reactor uses catalyst in powder form, slurried in an inert mineral oil. The mineral oil acts as a temperature moderator and a heat removal medium, transferring the heat of reaction from the catalyst surface via the liquid slurry to boiling water in a simple internal tubular heat exchanger. The heat exchanger occupies only a small fraction of the cross-sectional area of the reactor. The slurry reactor can achieve high syngas conversion per pass due to its capability to remove heat while maintaining a constant, highly uniform temperature through the entire length of the reactor. In essence the Liquid Phase™ process enables operating a highly exothermic reaction in an isothermal process. These differentiating features result in a cost reduction compared to conventional gas phase technology. A key for Air Products success in the development of Liquid Phase™ technology is its understanding of the scale-up factors of this liquid phase reactor technology.

The Liquid Phase™ reactor's unique temperature control capabilities permits direct processing of syngas which is rich in either hydrogen or in carbon oxides (carbon monoxide and carbon dioxide). As a result, the syngas production process does not influence the liquid product loop. Gas phase production of methanol would require such a feedstock to undergo the water gas shift reaction (to increase the hydrogen content) and to have the carbon dioxide (CO_2) removed (to reduce the excess carbon oxides). In a gas phase reactor, temperature moderation is only achieved by recycling large (typically 6/1) amounts of hydrogen (H_2)-rich gas which is also required to stay away from carbon deposition on the catalysts, while using high gas velocities to enhance the heat transfer. Typically, a gas phase reactor is limited to about 16% CO gas in the inlet to the reactor in order to limit the conversion per pass to avoid excess heating. However, with the LPMEOH™ reactor, CO gas concentrations in excess of 50% have been routinely tested without any adverse effect on the catalyst activity.

A second differentiating feature of the Liquid Phase™ reactor is its robust character. The slurry reactor, which is a simple vessel with an internal heat exchanger, is suitable for rapid ramping, idling, and extreme stop/start actions. The thermal moderation provided by the liquid inventory in the reactor acts to buffer sharp transient operations that would not normally be tolerable in a gas phase methanol synthesis reactor. The ($CO + H_2$) syngas stream produced by the gasification of oil, coke or coal can be used as both a clean turbine fuel and a chemical synthesis feedstock. Thus, an Integrated Gas Turbine Combined Cycle (IGCC) unit can be configured as a highly flexible facility adaptable on a real-time basis to changing conditions of demand for power, methanol fuel, and/or chemicals.

The key to economically exploiting this flexibility is to efficiently maintain high, constant levels of utilization of the most capital intensive portions of the facility, i.e., the gasification island and baseload combined cycle power production unit. Thus, a hybrid IGCC-LPMEOH™ facility can maintain base-loaded IGCC's power cost and superior environmental performance, while adding the ability for peaking power supply and chemical co-products for further economic improvement.

A third differentiating feature of the Liquid Phase™ process is that a high quality product is produced directly from carbon dioxide-rich syngas. This is directly related to the isothermal character of the process. In many processes, undesirable by-products are produced under extreme operating conditions. Gas phase methanol synthesis, which relies on hydrogen-rich syngas, results in a crude methanol product with 4 to 20% water by weight. The product from the LPMEOH™ process, under CO-rich conditions, typically contains only 1% water by weight. This methanol product is suitable for many applications, and at a substantial savings in purification costs. The steam produced in the Liquid Phase™ reactor is suitable for purification of the product (for upgrading to a higher quality) or for use in an IGCC power generation cycle.

Another unique feature of the Liquid Phase™ technology is the ability to add fresh catalyst on-line. Liquid Phase™ catalysts deactivate at a very slow rate. With the Liquid Phase™ reactor, spent catalyst slurry can be withdrawn and fresh catalyst slurry pumped in during operation whenever desired. This allows continuous and uninterrupted operation, a high number of stream days, and the maintenance of a high productivity level in the reactor. Furthermore, choice of replacement rate permits on-line optimization of productivity versus catalyst replacement cost.

The Liquid Phase reactor technology has also been used to make Fischer-Tropsch liquids, dimethyl ether (DME), and other oxygenates.

The LPMEOH™ technology was developed during the 1980's, with the support of the U. S. DOE. The concept was proven in over 7,400 hours of test operation, and thousands of gallons were produced at LaPorte, Texas, to test the liquid product. The LaPorte facility provided the scale-up rules for the first commercial plant. The first commercial-scale plant is now operating at Eastman Chemical Company's coal gasification facility in Kingsport, Tennessee under the DOE's Clean Coal Technology Program. Construction began in October 1995 and was completed in December 1996.

Eastman Chemical and Air Products completed commissioning by April 1997. The LPMEOH™ plant came on-stream on April 2, 1997. On April 6, design-capacity was met, and on April 10, 1997 120% of design capacity was achieved. The success of the Kingsport plant, which produces at least 80,000 gallons of methanol per day, has confirmed our ability to scale-up our technology from the LaPorte facility by a size ratio of 25 to 1. The product methanol is being refined by distillation to chemical grade (99.85 wt % purity) for feedstock to Eastman's chemicals production. Some of the raw methanol product (98% purity) will be used for off-site product-use tests.

6. CONCLUSIONS

Air Products and Chemicals is developing two significant technologies which would have significant impact on GTL capital costs and overall economics. The ITM Syngas process, by combining the air separation and the high-temperature syngas generation processes into a single compact ceramic membrane reactor, has the potential for substantially reducing the capital investment for syngas generation. Air Products, with its partners and the U.S. DOE, are developing this technology through a co-funded $85 million, eight year program. This program is being managed with critical attention to: 1) gas process and reactor scale-up, 2) development of ceramic materials with properties required for a commercial large-scale process, 3) ceramic membrane manufacturing scale-up, and 4) risk management so that technical success along with successful execution of this program will lead to timely and rapid commercialization.

Air Products liquid phase reactor technology is expected to significantly reduce the capital cost of the liquid synthesis portion of the GTL plant. The reactor technology is applicable to methanol, FT liquids, dimethyl ether and other oxygenates production This liquid phase technology has been demonstrated in the Kingsport plant which produces at least 80,000 gallons of methanol per day. The stability and flexibility of the reactor system were also confirmed. While still in an early stage of performance, the plant achieved over 5,000 hours of stable operation at varying feed rates, showing an impressive availability of over 93%. A four year operating demonstration period is ongoing to demonstrate the commercial advantages of the liquid phase technology.

REFERENCES

1. L. J. Frost, E. P. Ted Foster, S. L. Russek, D. R. Rowley, Use of Ceramic Membranes for Oxygen Separation and Syngas Production; Syngas for Fuel Cells, International Business Communications (IBC) Floating Platform Situation Offshore (FPSO)/Remote Gas Utilisation Conference, 1-3 December 1997, London, England.
2. D. L. Bennett, P. N. Dyer, Ionic Transport Membrane Technology for Gas-to-Liquids Processing, Intertech Gas-to-Liquids Processing Conference, 18-20 March 1998, San Antonio, Texas.
3. M. F. Carolan, et al., Inorganic Membranes, US Patent No. 5,683,797 (1997).
4. M. F. Carolan, et al., Ion Transport Membranes with Catalyzed Dense Layer US Patent No. 5,569,633 (1996).
5. E. A. Hazbun, Ceramic Membrane and Use Thereof for Hydrocarbon Conversion, US Patent No. 4,827,071 (1989).
6. D. Kang, et al., Integrated Production of Oxygen and Electric Power, US Patent No 5,657,624 (1997).
7. J. Kilner, et al., Ceramic Ion Conducting Membranes for Oxygen Separation, Chemistry & Industry, (17 November 1997), pages 907-911.

NATURAL GAS CONVERSION V
Studies in Surface Science and Catalysis, Vol. 119
A. Parmaliana et al. (Editors)

High Pressure Autothermal Reforming (HP ATR)

Ola Olsvik and Roger Hansen

R&D Centre STATOIL, N-7005 Trondheim, Norway

ABSTRACT

The High Pressure Autothermal Reformer (HP ATR) for production of synthesis gas has been evaluated. The aim of using HP ATR is to lower the plant investment costs by converting more gas per unit volume and eliminating the synthesis gas compression step, bringing the synthesis gas from 20-40 up to 70-100 bars, needed for the methanol and DME synthesis. The energy efficiency of the process may also be improved. Thermodynamic calculations have been performed showing the possibility to obtain a low methane slip at high pressures. Fluid flow simulations and combustion kinetic calculations are carried out to study the performance of the ATR in the high pressure range (70-100 bar). Detailed reaction mechanisms were used to study the pressure effect on the tendency to form higher hydrocarbons in the combustion chamber.

1. INTRODUCTION

World wide reserves of natural gas have risen exponentially over the past decade. Over the past two or three years, oil and gas companies have started to look more closely into the economics and tchnically feasibility of building huge gas conversion plants to take fully advantage of the economics of scale.

Exxon has announced that they have studied a 100,000 BPD Fischer-Tropsch plant in the Middle East.

There are speculations that the methanol marked may develop into a fuel marked in the future. These speculations have been based on mainly two scenarios for the future energy marked: 1) Daimler Benz and Toyota have announced that methanol is at the moment their preferred hydrogen source in their future automobiles based on fuel cells. 2) Further, Japan considers methanol as an interesting alternative fuel source for power generation (alternative to oil and LNG import).

A mega methanol plant and a Fisher Tropsch plant both scaled to the same Natural Gas consumption, would roughly have the same economics when methanol is priced at fuel value (80 $/T).

1.1 Why HP ATR ?

When ICI announced their new methanol synthesis catalyst in 1966 and the LP Methanol process which allowed the methanol synthesis to operate at 50 barg compared to 300 barg for the old and less active catalyst, this represented the most important step in cutting the costs of making methanol. However later in the early 70's, ICI modified the catalyst because new and larger plants (> 2000 MTPD) turned out to have an economical optimum when the methanol synthesis operated at 75 - 100 barg.

In general , the optimum operating pressure rises with plant capacity. Higher pressure reduces equipment size. A higher pressure in the methanol synthesis, increases the costs and energy consumption in the synthesis gas compression step. Rising the pressure in the reforming section reduces the compressor duty but since almost all methanol plants have a steam reformer this cannot be justified. Higher reformer pressure means higher outlet temperatures and as a consequence higher reformer tube temperatures. By generating all synthesis gas in an ATR the problem of expensive reformer tube materials is no longer relevant. This is the main reason why ATR is considered the technology of choice for mega methanol plants.

If we consider a single train 7,500 MTPD methanol plant, the optimum operating pressure for the methanol synthesis could be around 120 barg. In order to reduce synthesis gas compressor costs and energy consumption, the ATR should operate at as a high pressure as technically possible. This is the background why we have started to pay attention to ATR at around 80 barg. Recent studies have indicated that the capital costs of making methanol can be reduced by approximately 30% compared to a conventional 2500 MTPD [1].

1.2 The Autothermal Reformer (ATR)

The most important part in a gas conversion plant, both in terms of investments and in operating costs is the synthesis gas section. The technologies most relevant for production of synthesis gas are:

1. conventional steam reforming (tubular fired reforming)
2. autothermal reforming (ATR)
3. different variations of gas heated reforming (GHR)
4. different combinations of 1. and 2.

The ATR concept is the most favourable for large - scale manufacturing of synthesis gas [2]. An ATR (see Fig. 1) is a refractory lined vessel containing catalyst and an injector. The catalyst bed is located in the lower part of the ATR to ensure complete conversion of methane. The feedstock and steam is fed to the ATR and mixed directly with oxygen or air from an injector located near the top of the vessel.

The main net reactions in the combustion zone are a combination of combustion, steam reforming and pyrolysis, where the pyrolysis reactions are not shown here:

$$CH_4 + 1.5\,O_2 \Rightarrow CO + 2H_2O \qquad -\Delta H_{298} = +519\ \text{kJ/mol} \qquad (1)$$

$$CH_4 + H_2O = CO + 3H_2 \qquad -\Delta H_{298} = -206\ \text{kJ/mol} \qquad (2)$$

$$CO + H_2O = CO_2 + H_2 \qquad -\Delta H_{298} = +41\ \text{kJ/mol} \qquad (3)$$

$$H_2 + 0.5O_2 \Rightarrow H_2O \qquad -\Delta H_{298} = +242\ \text{kJ/mol} \qquad (4)$$

The present paper will mainly focus on the performance of an HP ATR: fluid flow, combustion kinetics and soot formation.

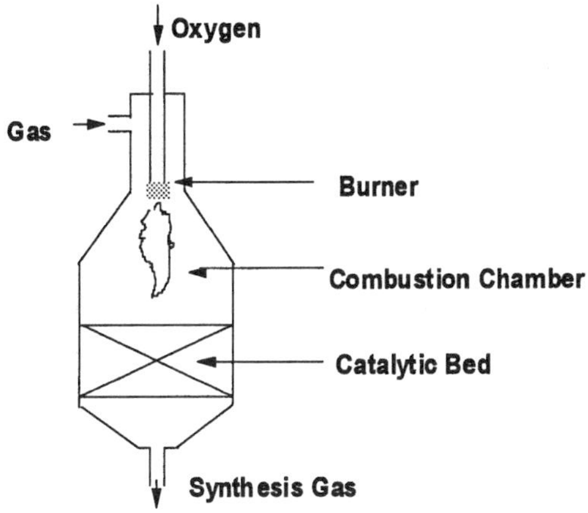

Figure 1. A schematic sketch of the HP-ATR.

2. NUMERICAL MODEL FOR THE FLUID FLOW CALCULATIONS

A numerical model of this secondary reformer has been developed using PHOENICS [3], a general purpose CFD package. The catalytic bed is included in the model to account for the pressure drop and flow through a porous media. A separate model of the flow field through the burner has been developed. The output from the burner model serve as accurate boundary conditions in the model of the combustion chamber.

2.1 Turbulence

The influence of turbulence is captured by a two-equation k-ε model. In the standard version the two transport equations for the turbulent fluctuations k, and its dissipation rate ε are given as:

$$\frac{Dk}{Dt} = \frac{\partial}{\partial x_i}\left(\frac{\nu_T}{\sigma_k}\frac{\partial k}{\partial x_i}\right) + P - \varepsilon \tag{5}$$

$$\frac{D\varepsilon}{Dt} = \frac{\partial}{\partial x_i}\left(\frac{\nu_T}{\sigma_\varepsilon}\frac{\partial \varepsilon}{\partial x_i}\right) + c_{\varepsilon 1}\frac{\varepsilon}{k}P - c_{\varepsilon 2}\frac{\varepsilon^2}{k} \tag{6}$$

where the production term (P) and the turbulent viscosity (ν_T) are given by:

$$P = \nu_T\left(\frac{\partial U_i}{\partial x_j} + \frac{\partial U_j}{\partial x_i}\right)\frac{\partial U_i}{\partial x_j} \tag{7}$$

$$\nu_T = c_\mu \frac{k^2}{\varepsilon} \tag{8}$$

here D/Dt denotes the material derivative and c_μ, σ_k, σ_ε, $c_{\varepsilon 1}$ and $c_{\varepsilon 2}$ are all model constants.

However, it is shown [4] that the standard version, with its built in assumption of isotropic turbulence, fails in mimicing the behaviour of a swirling flowfield. Therefore two versions of the k-ε model, the so-called *two-scale* model [5] was used in the present work. In the *two-scale* model an additional production term, $c_{\varepsilon 3}\, P^2/k$, has been added to the transport equation for ε. The new production term is proportional to a timescale k/P which is the time scale where energy from large turbulence is transmitted to smaller eddies, whereas in the standard k-ε model the only timescale present is the dissipative, k/ε.

2.2 Combustion

Four different turbulent combustion model were implemented and tested: (1) Mixed Is Burned (MIB), 2) Eddy Dissipation Concept (EDC), 3) Eddy Break Up (EBU), 4) Double Delta Probability Function (DDEL) in a previous work [4]. The comparisment carried out in the work by Olsvik et al.[4] showed that the Eddy Break Up (EBU) model did best represent the partial oxidation of methane which takes place in an autothermal reformer

EBU may be used in cases where the chemistry is fast and the flow is turbulent. The model takes into account that oxidizer and fuel may occur at the same location at the same time:

$$R = -C\ \mathrm{MIN}(\mathrm{MFU}, \mathrm{MOX}/S)\ \rho\ \mathrm{Volum}\ \varepsilon/k \tag{9}$$

where MFU is the mass fraction of fuel, MOX is the mass fraction of oxygen, S is the stoichometric factor, MIN is an argument that implies that R is dependent on the species in shortes supply, ρ is the density, and C is the Eddy Break Up constant.

3.1 Soot calculations

The effect of pressure on the soot (C_2H_2) formation was carried out in a premixed laminary flame [6] to reduce the simulations time when you include both non-premixed diffusion flame, detailed chemical kinetics and a complex geometry. A detailed reaction mechanism with 49 species and 277 elementary reactions [7], including C_2 chemistry, was used in the study. The program computes species and temperature profiles in a freely propagating premixed laminar flame. The program accounts for finite rate chemical kinetics and molecular transport.

3. RESULTS

3.1 Equilibrium calculations

Table 1 shows results from equilibrium calculations carried out at different pressures, temperatures and steam to carbon ratio (S/C). A low methane slip (leakage) is important for the efficiency of the overall process. The calculations show that a low methane slip can be obtained at high pressures when the HP ATR exit temperature and/or the S/C are high. The composition of the feed to the ATR will also influence the methane leakage.

The table shows a methane slip of 2.3 % at 75 bars, S/C equal to 1.2 and a outlet temperature of 1050 °C. To obtain even a lower methane slip, at the same S/C, the temperature out of the ATR have to be increased by adding more oxygen into the burning chamber. The disadvantage by increasing the temperature is sintring of the catalyst and plugging of the bed which will influnce the flow pattern in the combustion chamber.

Table 1
Equilibrium calculations at different pressures, temperatures and S/C

CH4 slip, mole % dry basis	1.0	2.3	1.2	3.5	2.0	1.9
Reforming pressure, bar	40	75	75	100	100	75
HP ATR exit temp. °C	1000	1050	1150	1050	1150	1050
Steam to carbon ratio (S/C)	1.2	1.2	1.2	1.2	1.2	1.8

The coke formation reactions:

$$CH_4 \quad = \quad H_2 + C \qquad (10)$$

$$2CO \quad = \quad CO_2 + C \qquad (11)$$

$$H_2 + CO \quad = \quad H_2O + C \qquad (12)$$

are not taken into account in the calculations. Reaction (11), known as the Boudouard reaction, and reaction (12) are favourized at higher pressures, according to the le Chatelier principle.

3.2 Fluid flow simulations and combustion kinetic

Figure 2 and 3 show fluid flow calculations carried out at 30 and 70 bars respectively. The fluid flow pattern are simular in the two cases (see Figures 2a and 3a), and figures 2b and 3b show that the flame core is closer to the burner tip for the 70 bars case. This may be explained by the fact that when the partial pressure increases the reaction rate increases:

$$R \quad = \quad k\ P^{n}_{CH4}\, P^{m}_{O2} \qquad (13)$$

The temperature at the burner tip is critical for the performance of ATR because when the temperature gets high enough the material in the burner tip will start melting. To avoid melting of the burner tip a new design and/or other flow rates of the natural gas and/or oxidizer streams are necessary

The calculations did also show that the flame core temperature increased with increasing pressure.

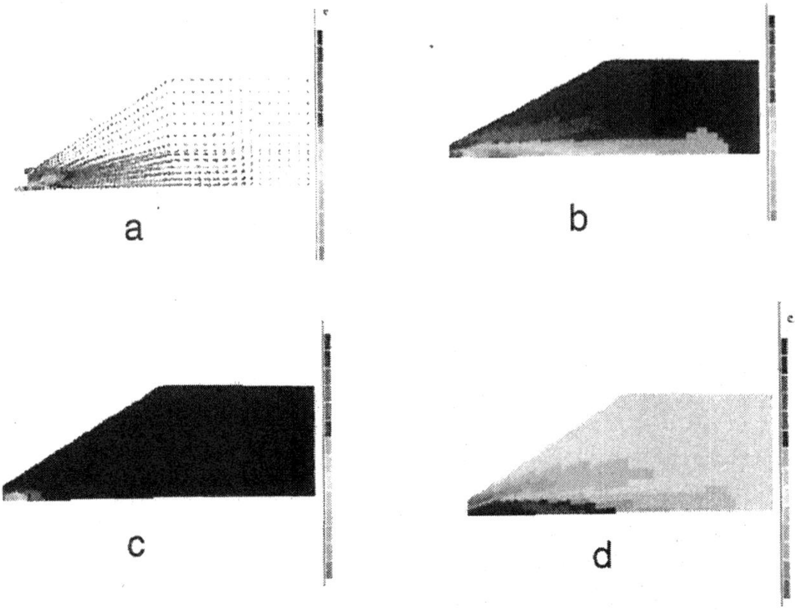

Figure 2. 2-D plots of velocity vectors (a), temperature (b), and the concentration of oxygen (c) and methane (d) for calculations carried out at 30 bar. Red colour indicates high values, and blue colour indicates low values.

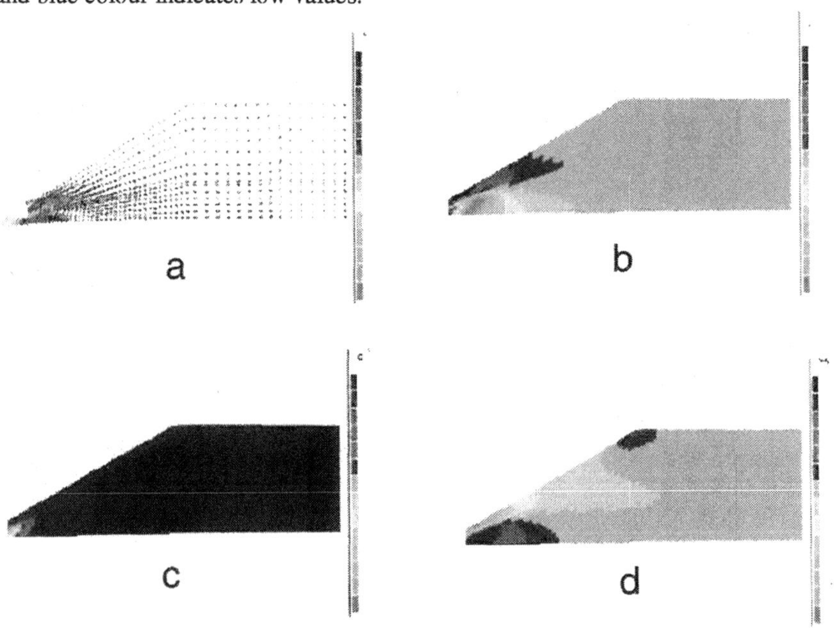

Figure 3. 2-D plots of velocity vectors (a), temperature (b), and the concentration of oxygen (c) and methane (d) for calculations carried out at 70 bar. Red colour indicates high values, and blue colour indicates low values.

3.3 Formation of higher hydrocarbons ($C2^+$)

Figure 4 shows that the formation of C_2H_2 decreases as a function of pressure, from 2.9 mol % at 30 bar down to 2.3 mol % at 100 bar. This may be explained by the following soot forming mechanisms [8]:

CH4 -> C2H6 + H2 -> C2H4 +H2 -> C2H2 +H2 -> aromatics + H2 -> soot (13)

According to the le Chatelier principle reaction (13) is not favourd at higher pressure.

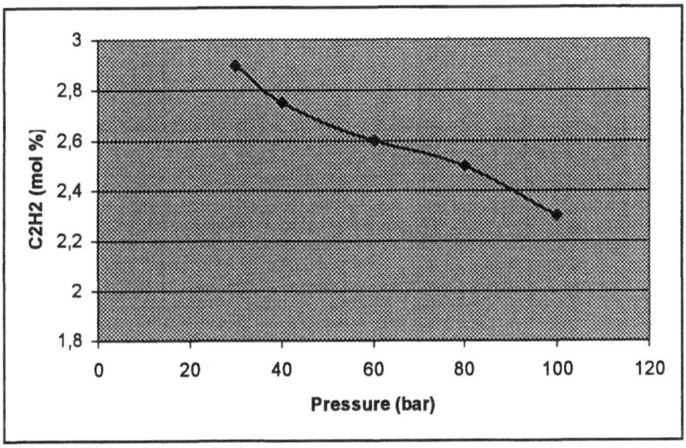

Figure 4. C_2H_2 formation as a function of pressure.

One of the most critical issues in a synthesis gas production unit is the high temperature section downstream of the ATR because of material corrosion (metal dusting). The main reactions causing metal dusting are reactions (11) and (12) mentioned earlier.

Results from metal dusting studies at higher pressure are not found in the literature, and the authors think that very little is done so far, but according to reactions (11) and (12) metal dusting may be even a larger problem at higher pressures.

CONCLUSION

Our interest in HP ATR is the fact that we consider this technology to be the optimum choice for mega methanol or DME plants.

The results showed that when the reforming is carried out at higher pressure the outlet temperature has to be increased. Increasing the outlet temperature may cause sintring of the catalyst and plugging of the bed which will influence the flow pattern in the combustion chamber.

Results from turbulent combustion simulations showed that the burner concept and/or flow rates used in present study have to be modified to obtain better combustion performance when the pressure is increased.

Calculations carried out with detailed chemistry, 49 species and 277 elementary reactions including C_2 chemistry, showed that the formation of soot decreased with increasing pressure, and that metal dusting, in the heat recovery section downstream the ATR, may be favoured at higher pressure.

REFERENCES

1 Helge H. Larsen, "Selection of Technology for Large Methanol Plants", presented on World Methanol Conference (1994).

2 Dybkjær, I., and Hansen, J.B., "Large-scale Production of Alternative Synthetic Fuels from Natural Gas", Presented at the Fourth International Gas Conversion Symposium, Kruger National Park, South Africa, 19-23 November, 1995

3 Phoenics User Manual (1997)

4 Olsvik, O., Fredriksen, G.R., and Gravdahl, A.R., "Mathematical Modeling of Turbulent Combustion in a Chemical Reactor", Presented at The First European Congress on Chemical Engineering, Florence, Italy, 4-7 May, 1997.

5 Kim,S.W. and Chen,C.P., A multi-time-scale turbulence model based on variable partitioning of the turbulent kinetic energy spectrum", Numerical Heat Transfer, 16,193-211 (1989).

6 R.J. Kee, J.F.Grcar,M.D.Smoke and J.A. Miller, A Fortran Program for Modeling Steady Laminar One-Dimensional Premixed Flames, Sandia Report, SAND85-8240 (1992)

7. C.T. Bowman, R.K. Hanson, D.F. Davidson, W.C. Gardiner, Jr., V. Lissianski, G.P. Smith, D.M. Golden, M. Frenklach and M. Goldenberg, http://www.me.berkeley.edu/gri_mech/ (1998)

8. O. Olsvik, Pyrolysis of Methane, PhD thesis, NTH Trondheim (1993)

NATURAL GAS CONVERSION V
Studies in Surface Science and Catalysis, Vol. 119
A. Parmaliana et al. (Editors)

Developments in Autothermal Reforming

Thomas S. Christensen, Peter S. Christensen, Ib Dybkjær, Jens-Henrik Bak Hansen and Ivar I. Primdahl

Haldor Topsøe A/S, Nymøllevej 55, DK-2800 Lyngby, Denmark

1. INTRODUCTION

The interest in conversion of natural gas to liquid products (Gas to Liquids, GTL) such as MeOH, DME, and liquid hydrocarbons, has grown significantly over the last decade, and the relevant technology is in demand. GTL plants generally consist of syngas preparation, synthesis, and product upgrade/purification. The syngas preparation is the most capital-intensive part of the process, requiring one half to two thirds of plant investment. Obviously, the choice of syngas technology and optimization of the chosen technology are very important for the overall economy of GTL projects. Autothermal Reforming (ATR) has been identified as an attractive technology for syngas units for very large MeOH and DME plants [1, 2]. Also for Fischer-Tropsch plants and in general for production of CO-rich syngas ATR has been found to be attractive [3].

2. NEW DEVELOPMENTS

The ATR technology was described in some detail in [3], which also contains a summary of the results of a pilot plant program, carried out in 1990-1992 in order to demonstrate new burner developments and operation at overall steam to carbon (S/C) ratio down to 0.6. Burner design and development were especially addressed in [4], and the two publications [3, 4] may together be seen as a description of the state-of-the-art of the ATR technology in 1994.

Engineering studies of syngas production have been carried out for optimal selection of operating conditions and design. Based on these engineering studies it was decided to carry out tests to demonstrate the feasibility of the process lay-outs. The pilot plant used in 1990-1992 [3] was reused after certain modifications, mainly to facilitate operation at very low S/C ratio. Numerous tests with variation of operating parameters were performed in order to establish safe design background at very low steam to carbon ratio. This also included the influence of variations in S/C ratio, temperatures, and pressure as well as the effect of feed composition such as content of higher hydrocarbons, CO_2, and H_2 on the limits for soot-free operation. In Table 1 results of some of these explorative tests are shown. All data in Table 1 represent operating conditions without soot formation, but they do not represent the limits of the technology. Results and operating conditions of tests of longer duration are shown in Table 2. The syngas composition was measured daily by GC-analysis, and the composition was found to be stable. The measured approach to equilibrium for reforming and shift is in the range of 0-10°C. During the tests the condensate was checked frequently for soot formation with gravimetric and spectrophtometric methods. The condensate was clear in all tests with a content of suspended solids less than 0.5 mg/l. Due to heat loss from the pilot unit, the measured exit temperature given in Table 1 and 2 is lower than the adiabatic exit temperature.

In large industrial units the heat loss is relatively much smaller, and the exit temperature will be very close to the adiabatic temperature. For comparison, the calculated adiabatic equilibrium temperature is shown in Table 2.

Table 1 – ATR Pilot Plant Explorative Tests

Feed ratios (mole/mole)	**A**	**B**	**C**	**D**	**E**	**F**
S/C	0.59	0.21	0.19	0.46	0.46	0.27
CO_2/C	0.01	0.01	0.20	0.01	0.01	0.01
C_2+/C 1)	0.06	0.06	0.07	0.22	0.21	0.06
C_5 +/C 2)	0.00	0.00	0.00	0.00	0.15	0.00
O_2 /C	0.62	0.59	0.61	0.60	0.58	0.59
Product gas						
Temperature, °C	1055	1065	1055	1050	1030	1065
Pressure, bar	24.5	24.5	24.5	24.5	24.5	24.5
H_2/CO, mole/mole	2.27	1.96	1.61	2.05	2.08	2.00
CO/CO_2, mole/mole	4.77	9.93	6.46	6.45	6.06	8.93
CH_4 leakage, dry mole %	0.49	1.15	0.77	0.81	1.13	0.95
M = (H_2-CO_2)/(CO+CO_2)	1.70	1.69	1.26	1.64	1.64	1.70

1) Carbon in hydrocarbons higher than C_1 divided by total hydrocarbon carbon
2) Carbon in hydrocarbons higher than C_4 divided by total hydrocarbon carbon

Table 2 – ATR Pilot Plant Runs With and Without CO_2 Addition

Feed ratios, mole/mole	**G**	**H**
S/C	0.51	0.36
CO_2 /C	0.19	0.01
C_2+/C	0.09	0.05
O_2/C	0.62	0.57
Product gas		
Temperature, °C	1025	1025
Pressure, bar	27.5	28.5
H_2/CO, mole/mole	1.80	2.14
CO/CO_2, mole/mole	4.44	6.71
CH_4 leakage, dry mole %	0.92	1.80
M = (H_2-CO_2)/(CO+CO_2)	1.29	1.74
Adiabatic exit temp.,°C	1055	1065
Run time, hours	350	370

3. SYNGAS PROPERTIES

The synthetic fuels, MeOH, DME, synthetic gasoline and diesel, can all be considered as MeOH or derivatives of MeOH obtained by dehydration [5]. However, the required properties of the syngas are different for the different syntheses. One important aspect is that in the synthesis of MeOH, DME and in the High Temperature Fischer-Tropsch process [6] the water gas shift is active in addition to the synthesis process. This means that CO_2 is a reactant in the synthesis process, and that the syngas ideally has the same stoichiometry as the final product.

The composition is normally characterized by the "Module" $M = (H_2 - CO_2)/(CO + CO_2)$, which should be close to 2. Furthermore, the content of CO_2 is important. Catalysts for production of MeOH require a certain minimum content of CO_2 to remain active and selective. In DME synthesis, a catalyst is used with activity for both MeOH synthesis and dehydration of MeOH. Sufficient water is produced by the dehydration to ensure, via the shift equilibrium, the presence of enough CO_2 to maintain the MeOH synthesis activity. In the Low Temperature Fischer-Tropsch process [6], which is used for production of synthetic diesel (via production of waxy hydrocarbons, which are, in a final process step, hydrocracked to diesel), the shift equilibrium is not established. The main reactants are H_2 and CO, while CO_2 is not reacting. The desirable composition of syngas for Low Temperature Fischer-Tropsch synthesis is best described by the ratio between the reactants H_2 and CO, and the best results are obtained when the ratio is around 2.

The composition of the raw syngas produced by ATR can be adjusted within limits by proper choice of the operating conditions. Critical parameters are the S/C ratio, the pressure, the inlet and outlet temperatures, and obviously the compositions of the feed streams. The syngas composition can be calculated by a heat and mass balance over the adiabatic reactor, assuming complete reforming of higher hydrocarbons, complete consumption of O_2, and establishment of the CH_4 reforming and the shift equilibria. Qualitatively, the influence of the critical operating parameters is:

- S/C Ratio: Low S/C ratio yields a gas with a low H_2/CO ratio and a high CO/CO_2 ratio. This is discussed further below. Increasing the S/C ratio reduces the CH_4 slip and increases the distance to the limit for soot formation.

- Inlet Temperatures: High preheat temperatures reduce the O_2 consumption, thereby reducing the CO_2 content in the product gas and increasing the H_2/CO ratio. Installation of a prereformer with reheat will have an even stronger effect in this direction.

- Outlet Temperature: High outlet temperature increases the O_2 consumption, decreases the H_2/CO ratio, increases the CO/CO_2 ratio, and reduces the CH_4 slip.

- Hydrocarbon Feed Composition: Increasing content of higher hydrocarbons and/or increasing content of CO_2 reduces the H_2/CO ratio. CO_2 addition or recycle can be used to adjust the H_2/CO ratio within wide limits.

Figure 1 shows the H_2/CO ratio and the module M obtained from lean natural gas by ATR as a function of the S/C ratio and the exit temperature (feed preheat temperature of 550°C, no prereformer). High preheat temperature and especially addition of prereformer with reheat increases both M and the H_2/CO ratio. However, although certain adjustments are possible, direct production of a gas with a module of 2 is impossible. Such gases are best produced by a combination of steam reforming and O_2-fired secondary reforming or, for very large capacities, by ATR followed by adjustment of the gas composition by addition of H_2 and/or removal of CO_2.

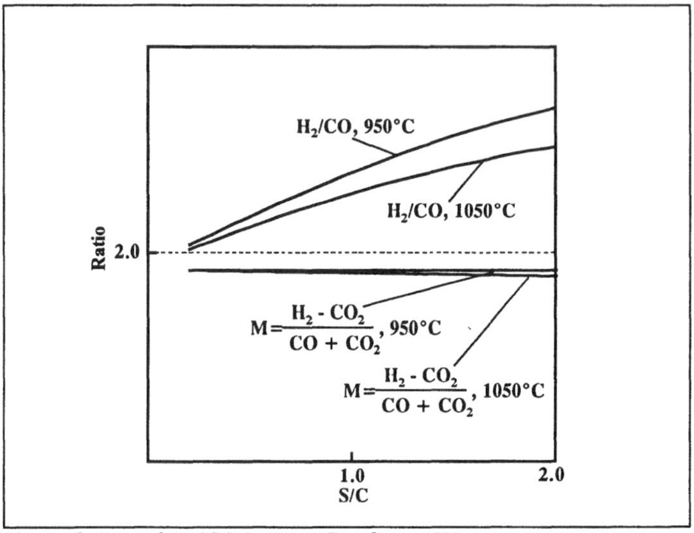

Figure 1: M and H_2/CO in Raw Gas from ATR

Direct production of a gas with $H_2/CO = 2$ is achieved only at very low S/C ratio and high exit temperature. At higher S/C ratio, the value of 2 is best obtained by partial recycle of CO_2 or CO_2-rich tail gas. However, it can be combined with removal of H_2 from the syngas. This could be an attractive way of producing H_2 for the hydrocracking of Fischer-Tropsch wax to diesel. Operation at low S/C ratio is in all cases attractive since it reduces the formation of CO_2 and thus the amount of gas to be recycled.

Low preheat temperature and/or a content of higher hydrocarbons and/or CO_2 facilitate production of syngas with a low H_2/CO ratio.

4. ENGINEERING STUDIES

Several studies have been performed on optimization of syngas production for large GTL plants. Early studies relating to MeOH and DME production were reported in [1, 2]. The conclusion was that for small plants, single-step tubular reforming is preferred. For intermediate capacities (corresponding to the largest MeOH plants built today) two-step reforming is most competitive, while ATR with adjustment of the gas composition is attractive for very large plants. Further studies have shown that operation at low S/C ratio significantly improves the economics of plants based on ATR. Table 3 shows the effect of S/C ratio in cases where the gas composition is adjusted by removal of CO_2. Table 4 shows similar results with adjustment by addition of H_2. In both cases the feed is a lean natural gas (95% CH_4). A prereformer with reheat to 650°C is included.

Table 3 – Production of Syngas for MeOH/DME - Adjustment by CO_2-removal; M = 2.06

S/C	0.2	0.4	0.6	1.0	2.0
Feed, Nm^3/h	84652	80996	79305	77973	77813
Fuel, Nm^3/h	(5023)	(712)	1525	4114	7796
Total NG, Nm^3/h	79629	80284	80830	82087	85609
Oxygen, Nm^3/h	42022	42058	42217	42796	44873
CO_2 export, Nm^3/h	5610	5782	5820	6085	7135
Syngas: CO/CO_2	∞	25.9	12.74	6.42	2.93
CH_4, dry vol%	3.94	2.26	1.48	0.76	0.24
Capacity 2500 MTPD $MeOH_{eq}$ (assuming 100% C efficiency in loop + dist.). P = 30 bar					

Table 4 – Production of Syngas for MeOH/DME - Adjustment by H_2 Addition

S/C	0.2	0.4	0.6	1.0	2.0
Feed, Nm^3/h	78967	75434	73777	72339	71309
Fuel, Nm^3/h	(4711)	(694)	1379	3778	7110
Total NG, Nm^3/h	74256	74740	75156	76117	78419
Oxygen, Nm^3/h	39224	39192	39295	39723	41143
H_2 import, Nm^3/h	15815	15514	15587	16214	18712
Syngas: CO/CO_2	13.75	8.60	6.25	4.05	2.15
CH_4, dry vol%	3.59	2.06	1.35	0.69	0.21
Capacity 2500 MTPD $MeOH_{eq}$ (assuming 100% C efficiency in loop + dist.). P = 30 bar					

It will be seen that the advantage of going below S/C = 0.6 in both cases is limited; only above this level, a significant increase in consumption of natural gas and O_2 is seen. Furthermore, the amount of CO_2 to be removed (Table 3) or H_2 to be added (Table 4) is rather constant. It may be mentioned that the amount of H_2 required corresponds well to what can be recovered from the purge gas from a synthesis loop with a "reasonable" efficiency. This is true for lean natural gas. For heavy feed gas or gas containing CO_2, the required amount of H_2 increases, and CO_2-removal becomes necessary.

Table 5 shows the effect of S/C on production of gas with H_2/CO = 2.0. In this case, a prereformer is not included. The feed preheat temperature is 550°C. Adjustment of the composition is achieved by recovery and recycle of CO_2 from the syngas. It will be seen that the effect of S/C in this case is much more pronounced, especially on the amount of CO_2 to be recycled; at very low S/C, the ratio H_2/CO = 2.0 can be achieved without CO_2 recycle.

Table 5 – Production of Syngas with $H_2/CO = 2.0$ - Adjustment by CO_2 Recycle

S/C	0.2	0.4	0.6	0.8	1.0
Feed, Nm^3/h	38374	36931	36510	36431	36508
Fuel, Nm^3/h	(1835)	(5)	870	1424	1847
Total NG, Nm^3/h	36539	36926	37380	37855	38355
Oxygen, Nm^3/h	20279	20683	21188	21732	22302
Syngas: CO_2, dry vol%	2.28	2.48	2.71	2.97	3.23
CH_4, dry vol%	3.18	1.65	1.01	0.68	0.48
CO_2-removal/recovery, Nm^3/h	489	2504	4485	6464	8478
Capacity: 100,000 Nm^3/h H_2 + CO. P = 30 bar.					

Specific studies have been made recently on the use of ATR for syngas production for very large MeOH, DME and Fischer-Tropsch plants. It was found that in no case does the ATR technology set the limit for the size of single-stream units. Capacities of at least 10,000 MTPD of MeOH equivalents or 35,000 bbl/day Fischer-Tropsch products can be accommodated in one ATR reactor (but not necessarily by other parts of the plant, e.g. boilers, compressors, etc.). In production of syngas for MeOH and DME the optimum S/C ratio is in the range 0.5-1.0, lowest for DME. For low temperature Fischer-Tropsch plants, the lowest possible S/C ratio in the ATR unit is preferred, since it reduces or eliminates the costly removal and recycle of CO_2 or CO_2-rich gas.

5. CONCLUSIONS

It has been shown in an extensive development program that the limits for safe and soot-free operation in ATR units can be extended to very low S/C ratio, close to what is relevant for POX units. In parallel, engineering studies have identified interesting process concepts, which take advantage of the newly identified operating conditions and allow competitive design of large GTL plants.

REFERENCES

1. Dybkjær, I., Bøgild Hansen, J., "Large-scale Production of Alternative Synthetic Fuels from Natural Gas", Stud. Surf. Sci. Catal. 107, (1997) 99.
2. Holm-Larsen, H., "Selection of Technology for Large Methanol Plants", 1994 World Methanol Conference, November 30-December 1, 1994, Geneva, Switzerland.
3. Christensen, T.S., Primdahl, I.I., "Improve syngas production using autothermal reforming", Hydroc. Process. 73 (1994) 39.
4. Christensen, T.S., Dybkjær, I., Hansen, L., Primdahl, I.I., "Burners for Secondary and Autothermal Reforming – Design and Industrial Performance", Ammonia Plant Saf. 35, (1994) 205.
5. Dybkjær, I., Sandholm Hansen, J., "Syngas Production Technology for Conversion of CO_2 Rich Hydrocarbon Feedstocks", Indonesian Conference on Natural Gas for Petrochemicals, June 26-27, 1996, Jakarta, Indonesia.
6. Jager, B., "Developments in Fischer-Tropsch Technology", Stud. Surf. Sci. Catal. 107 (1997) 219.

NATURAL GAS CONVERSION V
Studies in Surface Science and Catalysis, Vol. 119
A. Parmaliana et al. (Editors)

Synthesis Gas Production: Comparison of Gasification with Steam Reforming for Direct Reduced Iron Production.

Gary A Foulds*, Geoff R Rigby and W Leung; BHP Research, Newcastle Laboratories, PO Box 188, Wallsend, NSW, 2287, Australia.

Jim Falsetti and Fred Jahnke; Texaco Inc., Global Gas and Power, White Plains, New York, 10650, USA.

1. INTRODUCTION

Synthesis gas (syngas) can be produced from a number of carbon feedstocks. Traditionally, it has been produced from natural gas or coal, depending on the availability and cost of the natural gas or coal and the end use of the syngas. Generally, it has been found that steam reforming is favoured, especially when natural gas is available and syngas with a higher H_2/CO ratio is required. Gasification has been used where natural gas is unavailable or expensive and when syngas with a high CO content is required. However, recent developments in gasification technology have made it more competitive with steam reforming. These developments include reduction in capital cost and the ability to utilise very inexpensive "opportunity" fuels such as petroleum coke (petcoke), refinery wastes (vacuum residuals, asphalt) high ash/sulphur coal and Orimulsion®, which are being produced at an increasing rate.

In this paper, we compare gasification with steam reforming, not in isolation, but coupled to the production of direct reduced iron (DRI). This is done because there are significant process integration benefits to be gained by using the gasifier in this application. However, it should be noted that the benefits of integration will vary, depending on the particular downstream application. The end result of the study is that an "equivalent cost" of natural gas and gasifier feedstock is obtained, giving some idea as to how much of a threat gasification of "opportunity" fuels is to natural gas steam reforming.

2. DIRECT REDUCED IRON AND HOT BRIQUETTED IRON PRODUCTION

Fuelled by the scenario of a shortage of scrap iron and the increase in steel mini mills as alternatives to integrated steel plants, the production of direct reduced iron (DRI) in the form of briquettes (hot briquetted iron - HBI) is currently undergoing a major expansion. This is manifested by persistent news of new plants both in

* Author to whom correspondence should be addressed. Tel: +61 02 4979 2531; Fax +61 02 4979 2023; email foulds.gary.ga@bhp.com.au

Australia and internationally, including BHP's Port Hedland HBI plant. DRI technology can essentially be divided into two broad technologies according to the iron ore feed; lump- and/or pellet-based or fines-based. Further subdivision can be made according to hydrocarbon feedstock (coal or gas). Table 1 contains a summary of the major DRI processes [1]. At present the lump/pellet-based technologies of Midrex and HYL dominate commercial DRI production with 65% and 26% of world production respectively, while the only commercialised fines technologies to date are the Fior and Nucor Iron Carbide processes. The soon-to-be-commissioned BHP plant in Port Hedland is the first Finmet plant, the technology of which is based on a variant of the Fior process. All of these processes are natural gas-based, with the coal-based technologies essentially still waiting for the right economic situation or at the research and development stage.

Table 1 - Summary of Major DRI Processes

Feed	Natural Gas-based	Coal-Based
Lump/Pellets	• Midrex[a] • HYL[b] • Arex	• Rotary Kiln
Fines	• Fior[c] • Finmet • Iron Carbide • Circored	• Fastmet • Circofer

a - 65% of current world DRI production; b - 26% of current world production; c - less than 1% of current world production.

Given that most commercial DRI plants to date (Midrex, HYL III, Fior) are natural gas-based, the most favourable situation for this enterprise is where inexpensive gas and high grade iron ore are located in close proximity, since this obviates excessive transportation costs. In the absence of natural gas, one alternative is to produce syngas from alternative hydrocarbon feedstocks, and then to integrate this within an existing DRI/HBI process. BHP and Texaco have developed and optimised such a process concept which can utilise a range of hydrocarbon feedstocks [2].

2. NATURAL GAS-BASED DRI/HBI PROCESSES

Table 2 lists the process parameters for the natural gas-based Midrex, HYL III and Fior/Finmet processes. Essentially, all of the commercial processes involve the conversion of natural gas to synthesis gas (syngas), which is adjusted to the required H_2/CO ratio. The syngas is then usually mixed with recycled, unreacted syngas, heated, and fed into the DRI reactor(s). The HYL and Finmet processes use conventional steam reforming of natural gas to produce syngas (CO_2 is removed via scrubbers), while the Midrex process employs a combination of steam and CO_2 reforming of natural gas in a proprietary reformer. Other differences can be observed in the H_2/CO ratio (ranging from 1.5 to over 6), temperature (800 to 900°C) and pressure (1 to 10 bar) at which iron ore reduction takes place [2-4].

Table 2 - Process parameters for Midrex, HYL III and Fior/Finmet.

Process	Midrex	HYL III	Fior/Finmet
Metallic Charge	pellet/lump	pellet/lump	fines
Capacity	0.25 to 1.2 mtpa	0.25 to 1.5 mtpa	0.5 to 1 mtpa
Reactor type	counter current shaft	counter current shaft	fluidised bed
Residence Time	3 to 5 hours	3 to 5 hours	1.5 to 2 hours
Operating Pressure	1 atm	5 atm	10 atm
Reducing gas temperature	800 to 885°C	900 to 960°C	850°C
H_2/CO ratio	1.5 to 1.8	0.25 to 5	5 to 6
Fuel	Natural gas, Corex gas	Natural gas, Corex gas	Natural gas
Reformer	CO_2 and steam	Steam	Steam
% metallisation	88-94%	92-94%	93%
Product carbon	1-3%	1.8-4.5%	1.5%
Product type	DRI,HBI	DRI,HBI, Hytemp Fe	DRI, HBI

3. GASIFICATION-BASED DRI/HBI PRODUCTION

Figure 1 is a process flow sheet for a gasification-based HBI process. In this description, we have used petcoke as the chosen carbon feedstock. Petcoke is ground and slurried with water before being pumped into the down-flow Texaco quench gasifier. The quench gasifier was chosen because of its lower capital cost and ease of operation. If desired, a full heat recovery or partial heat recovery gasifier configuration could also be used. Oxygen, produced at an air separation plant (ASP), is fed co-currently with the slurry into the gasifier. The entrained flow gasifier operates at 50 atm pressure, and the reaction temperature in the gasification zone reaches about 1350°C. Generally, the gasifier pressure is chosen to match the requirements of the downstream process. However, in this case we have consciously chosen a much higher pressure so that power can be recovered when expanded to the DRI loop pressure. The hot gas exiting the gasifier is then quenched with water to remove slag and to partially clean the syngas. The saturated syngas, which is composed primarily of H_2 and CO, along with steam, CO_2 and traces of methane and H_2S, exits the quench zone at about 270°C and 50 atm pressure. The syngas is then water-scrubbed in a soot scrubber which essentially removes all traces of entrained particulates, alkali metals, heavy metals, ammonia and chlorides. After scrubbing, the syngas enters a water gas shift reactor where the H_2/CO ratio is adjusted from about 0.8 to the desired composition, before passing to the acid gas removal system where H_2S and CO_2 are removed in a high pressure scrubber. A

Claus unit is used to recover the sulphur in its elemental form. The sweet syngas then passes through an expander to generate power and let the pressure down from 50 atm to the DRI reactor operating pressure. This syngas is then mixed with part of the recycled syngas from the DRI reactor, heated rapidly to 850-900°C in a gas heater, and fed to the DRI reactor. The hot reducing gas enters the reactor and flows upward, reducing the counter-flowing iron oxide feed. The reduced iron descends to the bottom of the reactor, where it is cooled, carburised, continuously discharged and briquetted. The top gas leaving the DRI reactor is split into a recycle stream and a fuel stream for the gas heater. The recycle top gas contains additional CO_2 and H_2O, generated from reduction of the iron ore, which must be removed prior to recycling. Water and entrained particulates are removed via the wet scrubber, while CO_2 is removed in the low pressure CO_2 removal system. A novel CO_2 removal system has also been proposed and patented which uses a common steam stripper for both the high and low pressure scrubbers [2].

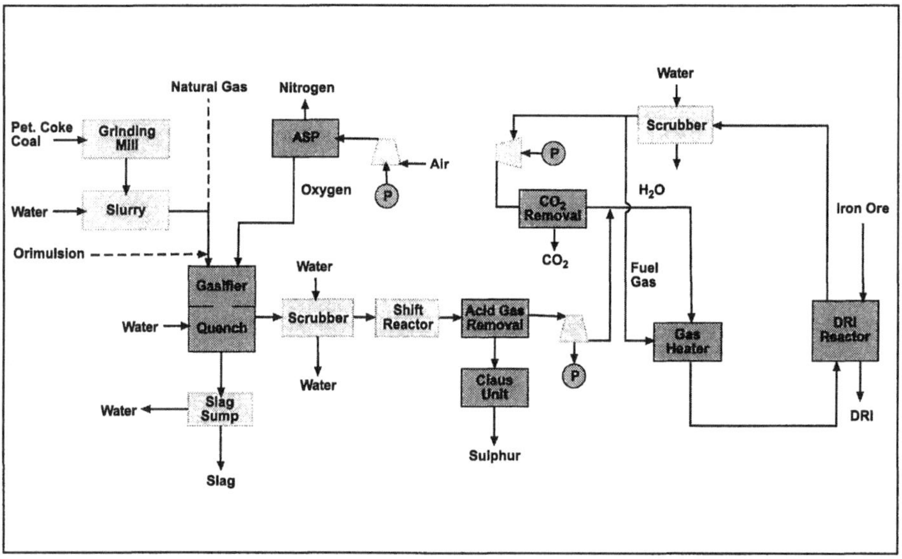

Figure 1 - Gasification based DRI process flow sheet.

4. PROCESS ECONOMICS

Process economics will vary considerably depending on location, sourcing and cost of feedstocks, and financing. The aim of this analysis is to determine the equivalent cost of natural gas in the gas-based processes versus a low-cost opportunity fuel such as petcoke in the gasification-based system. The cost basis has been defined as follows:

- Plant capacity 1.2-1.25 million tonnes of HBI per annum
- Pellet-based plant. Pellets supplied at US$37/t (67% Fe)
- Petcoke delivered at US$ 20.00/t (average price for delivered coke to US utilities)
- Oxygen supplied at US$ 30/t

- Power supplied at US$ 0.04/kWh
- Water supplied at US$ 0.1/m^3
- Labour supplied at US$ 15/mh
- Capital recovery factor of 0.1993 (15% over 10 years)
- Capital cost for gas-based plant US$ 200 - 225 million (US Gulf)
- Capital cost for gasification-based plant US$ 245 - 290 million (US Gulf)

Costs have been divided into operating cost and a capital cost component. We have determined the parity gas price by determining the operating cost using petcoke and then calculating backwards in the gas-based process using this operating cost value. Table 3 lists the average operating costs for gas-fired pellet-based HBI processes and the analogous gasification-based systems. As can be seen, on a cash operating basis, the gas-price equivalent to a petcoke cost of US$ 20/t varies between US$ 1.91 and US$ 2.22/GJ.

Table 3. Operating costs for gas-based and gasification based HBI processes.

Raw Materials	Consumption/ t HBI	Gas-based ($/t HBI)	Gasification-based ($/t HBI)
Iron Ore Pellets	1.45-1.5 t	53.65-55.50	53.65-55.50
Natural gas	10.5-11.0 GJ	***20.10-24.41***	
Petcoke	0.45-0.55 t		9.00-11.00
Oxygen	0.4-0.5 t		12.00-15.00
Power (gas-based)	80-130 kWh	3.20-5.20	
Power (gasification-based)	0-40 kWh		0.00-1.60
Water	1.5-1.75 m^3	0.15-0.175	0.15-0.175
Labour	0.20 mh	3.00	
Labour	0.22-0.24 mh		3.30-3.50
O&M		5.00-6.00	7.00-7.50
Administration		1.40-5.00	1.40-5.00
Total Operating Costs		**95.47-112.25**	**86.50-99.28**
Natural gas	10.5-11.0 GJ	***29.07-37.37***	
Capital Recover Factor		39.86-44.84	48.83-57.80
Total		**135.33-157.08**	**135.33-157.08**

However, this does not tell the full story, as the capital costs are not equivalent, and a cost of capital component needs to be included. Using the capital cost and capital recovery factors listed earlier, it turns out that the equivalent gas price for petcoke at US$ 20/t ranges between US$ 2.77 and US$ 3.40/GJ. This agrees well with the value (US$ 3.50-US$ 4.00/MMBtu) determined by HYL for a coal-based (US$ 32/t) system [7].

5. CONCLUSIONS

Integrating gasification-produced syngas with DRI/HBI in an optimised way offers a technically viable alternative to natural-gas based processes. The process allows a range of feedstocks to be used giving it considerable fuel flexibility. "Opportunity" fuels such as petcoke are being produced at an increasing rate, thereby ensuring a cheap, abundant supply. Gasification technology offers a way to process these feedstocks while at the same time meeting the most stringent environmental requirements. The process is especially adept at handling high sulphur fuels, with sulphur being recovered in its elemental form or as sulphuric acid. A marketable slag is produced. Carbon dioxide is captured as part of the process and can be used if a suitable use is available

Cost comparisons are highly site specific. In the case studied here, it was found that operating costs are quite competitive with the gas-based processes, with costs being equivalent at ≈ US$ 2.00/GJ for gas versus US$ 20/t for petcoke. Notwithstanding the great advances made in capital cost reduction in the development of the quench design, this is still a significant factor. For the recovery factors selected, the cost of capital contribution lifts the equivalent gas price to ≈ US$ 3.00/GJ. This contribution is highly sensitive to the local interest rate and can be reduced by appropriate financing, but efforts to reduce capital cost should still be a major thrust of future work.

6. REFERENCES

1. Battelle Europe/ITM Iron and Steel Industry multiclient: Follow up 1994-1995
2. F.C. Jahnke, J.S. Falsetti, G.A. Foulds and G.R. Rigby, US patent Application, February 6, 1998.
3. M. Thompson, 5 part series reprinted from DIRECT FROM MIDREX, 4th Quarter 1991 to 4th Quarter1992, Midrex Corporation.
4. C. Dominguez and P.E. Duarte, SEAISI Quarterly, July 1994, 36.
5. R. Quintero, Gorham/Intertech Conference on Iron & Steel Scap, Scrap Substitutes and Direct Steelmaking Conference, March 5-7, 1996, Atlanta, Georgia, USA.
6. A. Hassan and R. Whipp, Iron and Steel Scrap, Scrap Substitutes and Direct Steelmaking Conference, March 21-23, 1995, Atlanta, Georgia, USA; A. Hassan and R. Whipp, MPT International, 3, 1996, 50.
7. P.E. Duarte, E.O. Gerstbrein and H. Smegal, 3rd Annual Asian Steel Summit '97, 17-19 June, Hong Kong, 1997.

NATURAL GAS CONVERSION V
Studies in Surface Science and Catalysis, Vol. 119
A. Parmaliana et al. (Editors)

Research progress and pilot plant test on SDTO process

Z. M. Liu, G. Y. Cai, C. L. Sun, C. Q. He, Y. J. Chang, L. X. Yang, R. M. Shi and J. Liang

State Key Laboratory for Catalysis, Dalian Institute of Chemical Physics, Chinese Academy of Sciences, Dalian 116023, China

In the earlier of 1990's, we have proposed a new process for the synthesis of light olefins from syngas, which was designated as SDTO method. That is to convert syngas to dimethylether and than to convert dimethylether to light olefins. The present work deals with the R&D research progress of this process. The catalysts were developed for the two reactions, scale-up preparation of the catalysts and pilot plant test were performed. Evaluation of the pilot plant data showed that 190~200 grams of DME were yielded by single-pass for each standard cubic meter of syngas. For the second reaction, material-balance data indicated that 1.880 tons of DME or 2.615 tons of methanol can produce one ton of light olefins, which constituting 0.533 ton of ethylene, 0.349 ton of propylene and 0.118 ton of butenes.

1. INTRODUCTION

Synthesis of light olefins from syngas has attracted many attentions in the past two decades because it may provide a chance for the production of light olefins from natural gas other than from the conventional oil route. Direct conversion of syngas to light olefins [1] is particularly expected. However, it is difficult since the reaction is thermodynamically unfavorable and the products are normally Schulz-Flory distributed. Many efforts have been focused on indirect methods [2-5]. Most recently, an important progress was reported on the well-known methanol to olefin (MTO) process[6]. This indicates that indirect conversion method is still more efficiency than direct conversion. In the earlier of 1990's, we proposed a new indirect method designated as SDTO [7, 8] (*Syngas via Dimethylether to Olefin*) based on our understanding of GTO and MTO processes. That is to convert synthesis gas to dimethylether(DME), and then to convert DME to light olefins. SDTO method may exhibit greater efficiency than the method through methanol, since it is well known that the reaction of syngas to DME is far more favorable in thermodynamics than in the case of methanol synthesis. The present paper reports the progress of our SDTO research and some results from pilot plant test that was finished in 1995.

2. SYNTHESIS GAS TO DME

2.1. Catalyst

For synthesis gas to DME, metal-acid bifunctional catalyst was applied to the reaction[9-11]. The metallic component of the catalyst can be similar to that of methanol synthesis catalyst, while solid acids, such as zeolites, γ-Al_2O_3 and SiO_2-Al_2O_3, can be used as acid component. The fundamental aspect about the incorporation between the two components has been studied. It has been found that when γ-Al_2O_3 used as acid component, the catalyst exhibits rather low DME selectivity and single-pass CO conversion, both are in the range of 70%. Furthermore, the initial reaction temperature is also rather high(250~270 ℃), which is unfavorable for the stability of the catalysts as well as for the thermodynamic equilibrium of the reaction, since this reaction is strongly exothermic, and CO conversion will be diminished when the reaction temperatures are higher than 240°C. Zeolites that have much stronger acidic strength than γ-Al_2O_3 can greatly enhance the activity and selectivity of the catalyst. Also, the Cu-Zn components of the conventional methanol synthesis catalysts were modified with transition metal promoters, so that the initial reaction temperature can be effectively lowered. Accordingly, three generations of the catalyst have been developed using Cu-Zn components and ZSM-5 zeolite[9-11]. Typical performance data of these catalysts are listed in table 1. It is worth to point out that besides their superior activity and selectivity, the SD219 catalysts also show activities for the hydrogenation of CO_2. As a result of this special feature of the SD219 catalysts, the carbon utilization efficiency (carbon selectivity) of the DME synthesis process in single-pass can exceed the theoretical value of 66.67% when the feed gas contains CO_2. Long term (2200 hours) test of 219-1 catalyst under conditions of 3.5MPa and GHSV=1000h^{-1} (H_2/CO=2) shows that the catalysts possess good activity and stability (CO conversion > 75%).

Table 1
Properties of SD219 catalysts in synthesis gas to DME reaction (CO/H_2=1/2, P=3.0MPa).

Catalyst	Reac. Temp. (℃)	GHSV(h^{-1})	CO Conv. (%)	DME Sele. (wt.%)*
SD219-1	240	1000	90.0	95.8
SD219-2	230	1000	95.1	93.8
SD219-3	220	1000	94.4	95.5
	240	1000	96.1	91.4
	240	3000	90.1	94.0

* DME selectivity = [DME/(DME + MeOH)]

2.2. Pilot plant test

The pilot plant test of synthesis gas to DME reaction was carried out on a ϕ40×4000mm fixed bed reactor with catalyst SD219-2 (pellet size: ϕ5×6-7mm). Under reaction conditions of 3.4-3.7MPa, 240±5°C, GHSV=1000h^{-1} and with semi water gas (H_2 42.43%, CO 23.21%, CO_2 11.59%, N_2 21.54%) as reaction feed, the pilot plant test had been operated continuous and smoothly for 1000 hours, with a CO single-pass conversion of 75~78% and a DME selectivity of 95% (Fig. 1). No evident decrease has been found for the activity and selectivity of the catalyst during the 1000 hours tests. Evaluation of the pilot plant data showed that 190~200 grams of DME were yielded for each standard cubic meter of syngas. Also, it can be seen from Fig. 2 that the stability of the SD219 catalyst was very good, and the results of the pilot plant tests proved that this process has shown a concrete potentiality for further scaling-up to become a commercial process.

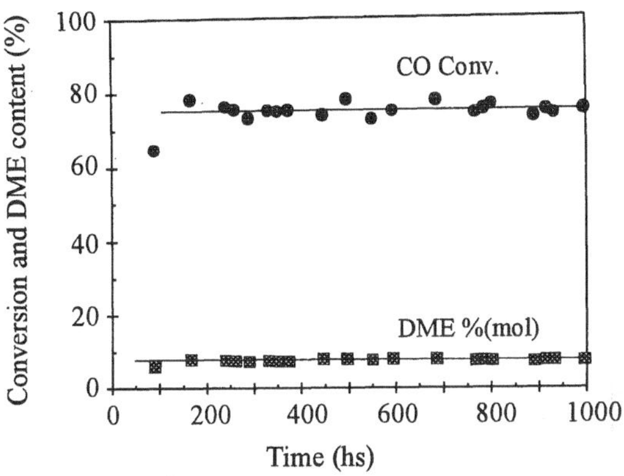

Fig. 1. Pilot plant performance over a SD219-2 catalyst for the direct synthesis of DME from syngas

3. DME TO LIGHT OLEFINS

3.1. DO123 catalyst

In the SDTO process, a SAPO-34 [12, 13] type catalyst, designated as the DO123 type catalyst, has been developed for the conversion of DME to lower olefins with excellent performance in a fluidized-bed reactor system. One of the features of the DO123 catalyst lies in that the molecular sieve is synthesized with a new method different from the conventional procedure. Usually, SAPO-34 molecular sieve is synthesized by using tetraethyl-hydroxyl-amine(TEAOH) as the template agent [5, 14]. However, TEAOH is an expensive reagent, and this will cause a high cost for the production of the catalysts. Many efforts were paid to the synthesis of SAPO-34 to develop a cheap synthesis method. Triethylamine was successfully applied as template in the synthesis [15, 16] instead of TEAOH. A dual template method was invented to control the crystal size of the molecular sieve [17]. It has been estimated that the SAPO type molecular sieve synthesized with templates of these cheap amines and with optional procedures can reduce the cost to 20% of the molecular sieve prepared with the TEAOH template.

In the synthesis of the SAPO-34 type molecular sieve, the conditions of crystallization were monitored by means of IR., NH_3-TPD, SEM and MASNMR techniques. The properties of the catalysts, such as the number and strength of the acid sites, the contents and distributions of the Si, Al and P elements, the crystal size and the pore size, etc., can be optimized to give the best performances of the catalysts for DME conversion to lower olefins. Consequently, the selectivity for ethylene formation can be remarkably enhanced, and the formation of alkanes can be suppressed.

By incorporating appropriate binders to the SAPO-34 molecular sieve, the DO123 catalyst has been prepared by spray drying method into microspheres with excellent mechanical as well as abrasive-resistant properties. This catalyst was tested with a bench-scale fluidized-bed reactor of the size of ϕ20×500mm, and under the reaction conditions of ambient pressure, 550°C, WHSV of 5 ~7 h^{-1}, and linear velocity of 12~15cm/s, the conversion of DME was 100%, and the selectivities to ethylene, $C_2^=$—$C_3^=$ olefins and $C_2^=$—$C_4^=$ olefins were 50~60%, 85% and 90% respectively. Methanol was also used as the feedstock for the reaction test, and nearly identical results as the DME feedstock was obtained(Table 2). This implied that both DME and methanol can be used as the feedstock for the production of lower olefins over the DO123 catalyst. It has also been found from the bench-scale operations that coke depositions on the DO123 catalyst could be regenerated easily by burning in air at 600°C within 10 minutes (Fig. 2). More than 100 cycles of regeneration have been carried out on the catalyst during the bench-scale tests, and the reactivity of the catalyst did not change very obviously. Because large amount of water exists in the products and the catalyst has to be regenerated frequently at high temperature (>550°C), high stability of the catalyst is important. The performance of the catalyst was also tested after thermal and hydrothermal treatments under severe conditions. The activity and selectivity of the catalyst were almost the same as those of the fresh catalyst after 300 hours calcination in air or 80 hours stream treatment at 800°C [8].The X-ray diffraction results show that only a little decrease of the relative crystallinity of the treated samples comparing with that of the fresh catalysts. All these results indicate that the DO123 catalyst not only has very good activity and selectivity for the conversion of DME or methanol to lower olefins, but also possesses the advantages of easiness in regeneration, excellent thermal and hydrothermal stability, and no need of water addition during the reaction operations.

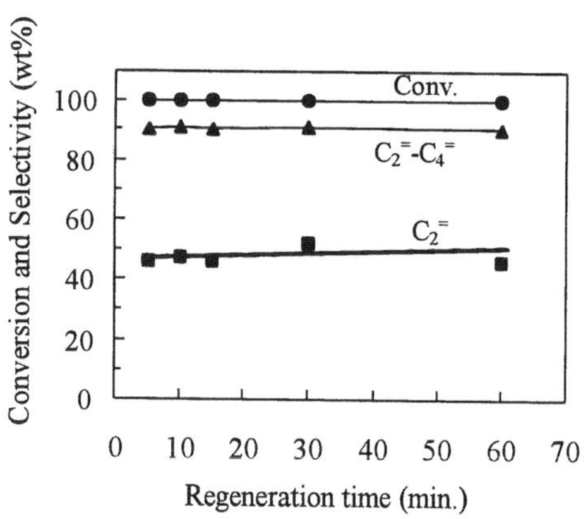

Fig. 2. Influence of regeneration time on DO123 catalyst performance (regeneration condition: 600°C in air; catalytic evaluation: fluidized reactor, 500°C, MeOH line velocity = 14.2cm/s, reaction time = 2 min.)

Table 2
Performance of the DO123 catalyst with different feed components in a fluid bed bench-scale reactor (550°C, reaction time = 10 min. conversion = 100%).

	Products (wt.%)			
Feed	C_2H_4	C_3H_6	$C_2^=$—$C_3^=$	$C_2^=$—$C_4^=$
Methanol [a]	62.79	22.34	85.13	89.57
DME + H_2O [b]	62.80	22.65	85.45	90.23
DME [c]	59.35	24.22	83.57	88.32

[a] WHSV of methanol 6.45 h^{-1}, linear velocity of the reactants 15.21cm/s, [b] WHSV of DME 4.64 h^{-1}, linear velocity of reactants 15.21cm/s. [c] WHSV of DME 7.16 h^{-1}, linear velocity of reactants 11.75cm/s.

3.2. Pilot plant test

The SAPO molecular sieve was then prepared in autoclaves of $1M^3$ capacity under commercial production conditions, and the DO123 catalyst so produced was put into scale-up tests in a pilot plant fluidized reactor system that was connected in series with the "syngas to DME" fixed-bed reactor mentioned above. Pilot plant test of DME to light olefin was carried out on a fluid bed reaction system (reactor ϕ100mm) with continuous regeneration of the catalyst. The fluidized reactor was an up-flow dense bed type, with a capacity of 15~25 tons/year. The DME feed charge for the second stage reaction came from the concentrated product of the first stage, with a concentration of greater than 98%. The DO123 catalyst was operated in this pilot plant under different reaction conditions for process studies, and a total of 1500 cycles of regeneration was carried out on the same batch of catalyst. Characterization investigations have been done on the catalyst after the pilot plant operations, and no remarkable changes on its physic-chemical properties had been detected. Data from the pilot plant performances(Fig.3) showed that there was a good duplication of the pilot plant results

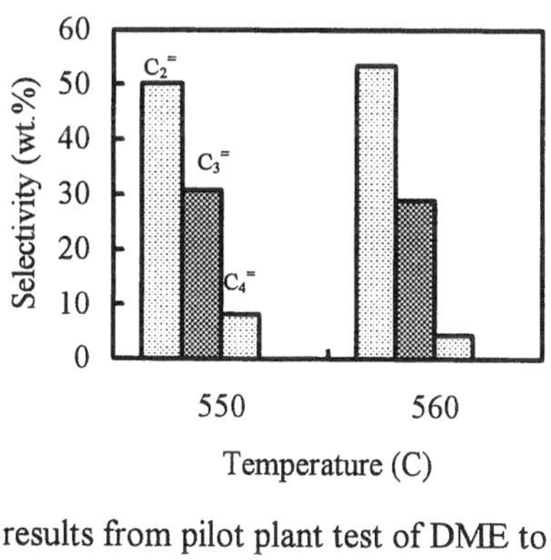

Fig. 3. Typical reaction results from pilot plant test of DME to light olefins.

with that of the bench-scale tests. Estimations from material-balance data indicated that for the production of a ton of lower olefins(constituting 0.533 ton of ethylene, 0.349 ton of propylene and 0.118 ton of butenes), 1.880 tons of DME or 2.615 tons of methanol have to be consumed.

4. CONCLUSION

The proposed STDO method was developed and demonstrated on pilot plant scale. The data showed that 190~200 grams of DME were yielded by single-pass for each standard cubic meter of syngas, and 1.880 tons of DME or 2.615 tons of methanol can produce one ton of light olefins which constituting 0.533 ton of ethylene, 0.349 ton of propylene and 0.118 ton of butenes. This new method may provide an alternative way for light olefin production from synthesis gas.

REFERENCES

1. B. Bussemeier, C.D. Forhning and B. Cornids, Hydroc. Proc., 35(1), 105(Nov. 1976).
2. M.B. Scherwin, Hydroc. Proc., 60(3), 79(March 1981).
3. C.D. Chang, Catal. Rev. -Sci. Eng., 25(1981).
4. G.Q.Chen and J. Liang, Stud. Surf. Sci. Catal., 28, 907(1986).
5. T. Inui, S. Phatanasri and H. Matsuda, J. Chem. Soc. Chem. Commun., 205(1990).
6. B. V. Vora, R. A. Lentz and T. L. Marker, Petrochemical Review, Mar., 19-21(1996).
7. G.Y. Cai, Q.X. Wang, G. Q. Chen, C.L.Sun, Z.M. L, Z.Z.Wang, L.Y. Xu, and Z.Q. Z, A New Method for the Production of Light Olefins from Synthesis Gas, Chinese Patent No. 11061576(1991).
8. G. Cai, Z. Liu, R. Shi, C. He, L. Yang, C. Sun and Y. Chang, Appl. Catal. A, 125, 29(1995).
9. R.M. Shi, G.Y. Cai and Z.M. Liu, J. Natural Gas Chem., 5(4), 287(1996).
10. G.Y. Cai, C.L.Sun and Z.M. Liu, Catalyst for Dimethylether Production from Syngas, Chinese Patent No. CN1087033A.
11. G.Y. Cai, R.M. Shi and Z.M. Liu, Manufacture of Dimethylether from Syngas, Chinese Patent No. CN1085824A.
12. S.T. Wilson, B.M. Lok and E.M. Flanigen, US Patent No. 4310440(1982).
13. B.M. Lok, C.A. Messina and P.R. Lyle, J. Am. Chem. Soc., 106, 6092(1984).
14. J. Liang, H.Y. Li, S.Q. Zhao, Proceedings of 7th Int. Zeolite Conference, p231, 1986.
15. H.Y. Li and J. Liang, Shiyouhuagong, 16(5), 340(1987).
16. C.Q. He, Z.M. Liu, G.Y. Cai, J. Fuel Chem. Tech. (Chinese), 23(3), 306(1995).
17. C.Q. He, Z.M. Liu, G.Y. Cai, Synthesis of Molecular Sieve by Duel-template agent, Chinese Patent No. 94110059.6.

NATURAL GAS CONVERSION V
Studies in Surface Science and Catalysis, Vol. 119
A. Parmaliana et al. (Editors)

Short-stage technology of synthetic diesel and jet fuels production from natural gas for small-scale installations of low-pressure

V.M.Batenin[a], D.N.Kagan[a], A.L.Lapidus[b], F.N.Pekhota[c], M.N.Radchenko[d], A.D.Sedykh[d], E.E.Shpilrain[a]

[a]Institute for High Temperatures of the Russian Academy of Sciences
13/19 Izhorskaya, Moscow 127412, Russia

[b]Institute of Organic Chemistry of Russian Academy of Sciences
47 Leninsky pr., Moscow 117913, Russia

[c]Ministry of Science and Technologies of Russia
11 Tverskaya, Moscow 103905, Russia

[d]Russion Joint-Stock Company "GAZPROM"
16 Nametkina, Moscow 117884, Russia

1. CONCEPTION

Analysis of the state of the fuel and power engineering system of Russia showed that liquid synthetic fuels from natural gas and other source gases should be produced locally by small-scale plants. Using such plants, potential consumers, which can afford only limited capital investments, would in a short space of time make up the deficit of motor fuels (including most scarce diesel and jet fuels) at the sites of production and transportation of hydrocarbon gases, where liquid fuel supply with motor transport or product pipelines is uneconomical. Besides there are a great number of low-pressure wells which are comprising a significant part (from 13 up to 17%) of total gas reserves but closed now (so named underbalanced wells).

Both factors are extremely important for significant part of regions of Russia and perhaps other countries with large territory.

Hence *the working out process is intended for*:

- autonomous provision of motor fuel to above consumers;
- significant extension of the natural gas resource base through using units (with no preliminary gas comprimating) on low-pressure underbalanced fields.

The above purpose defines *the following requirements to the process*:

- minimum number of stages and, consequently, maximum conversion at each stage;
- smooth physicochemical conditions of the process to effect it at natural gas initial pressure corresponding that at the output of low-pressure and underbalanced wells, and in local networks after gas distribution stations;
- production of fuel with maximum calorific value, high motor and environmental characteristics;

- possibility to use existing fueling network and motor construction existing infrastructure.

2. THE PROCESS

The analysis of the state of modern art of technologies [1-3] has revealed absence of ready-made patterns in the world practice which could be applied to develop compact units to resolve the task, first of all because of multi-stage processes intended for large-scale production.

The above requirements to the process are realized by means of the environmentally friendly low-stage process being developed, which is adequate to be implemented in a compact form [1,2,4].

The process involves consequently two main stages (Fig.1):

- combined (H_2O/CO_2) conversion of natural gas or other gaseous hydrocarbons (P=1-2 MPa, t=650-750 °C) which allows to produce synthesis gas of a regulated composition to reach a strictly optimal H_2/CO ratio and provides a high degree of conversion in single pass;
- highly selective, with respect to diesel and jet fractions, Fischer-Tropsch synthesis (P=1-2 MPa, t=180-200 °C) allowing to obtain high output of the motor fuels (C_5-C_{18}) directly in a single-pass process;

  and a standard fractionating stage.

As the second stage of the process (HSLF synthesis from synthesis gas) in the process suggested is the highly selective Fischer-Tropsch synthesis in the reaction (in obtaining paraffin hydrocarbons)

$$nCO + (2n+1)H_2 \rightarrow C_nH_{2n+2} + nH_2O + n(\sim 165 kJ), \tag{1}$$

optimum composition of synthesis gas (target product of the first stage, conversion of natural gas into synthesis gas) is

$$H_2/CO \approx 2 \tag{2}$$

An analysis of methane conversion into synthesis gas [1] has revealed that:

- from the viewpoint of meeting condition (2);
- by the degree of methane single-pass conversion (up to 80-85%);
- be the level of process development and acceptability for remote dispersed consumers,

steam-carbon dioxide conversion of methane

$$3CH_4 + 2H_2O + CO_2 + 659\,kJ \rightarrow 4(CO + 2H_2) \tag{3}$$

is optimum.

As the Fig.1 reveals, a synthesis product in reactor 3 acts as a source of H_2O in steam-carbon dioxide conversion, which significantly determines the lack of discharges in the process, i.e. its environmental purity. Nickel acts as a catalyst basis in the conversion, among others well-tested catalysts GIAP-3,8,16,21 are used.

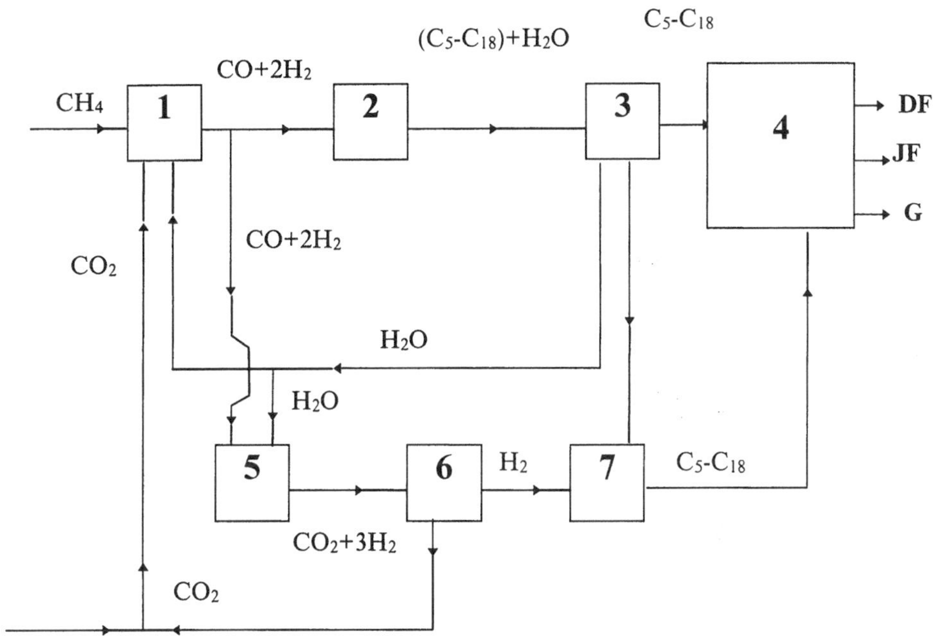

Figure 1. Principle diagram of SLF (DF and JF) production from CH_4

Main technology (1-4):

1 – combined (H_2O/CO_2) conversion of CH_4 into synthesis gas ($CO+2H_2$);
2 – selective Fischer-Tropsch synthesis;
3 – liquid product separation (fractions C_5-C_{18}, H_2O; high molecular paraffins <3%, ROH: C_1-C_6<2%);
4 – distillation of obtained liquid hydrocarbons: DF - diesel fuel, JF - jet fuel, G - gasoline.

Possible additional stages 5-7 (if fuel is to meet special requirements):

5 - shift-conversion (obtaining H_2, CO_2);
6 - absorption-desorption of CO_2;
7 - hydrorefining (hydrocracking, hydrostabilization, hydroisomerization)

The main stage of the process – Fischer-Tropsch synthesis of liquid hydrocarbons from synthesis gas – is effected in a fixed-bed Arge reactor which features sufficient reliability and allows to thoroughly regulate thermodynamic parameters and heat removal in the process. The latter factor is very significant due to the fact that Fischer-Tropsch synthesis is a highly exothermic reaction (1).

Optimized have been:

- thermodynamic characteristics of the process in order to reveal modes with maximum thermodynamic probability to obtain fuel of required composition [4];
- and than (within this optimized range of thermodynamic parameters) kinetic characteristics and catalytic systems in order to regulate the process with respect to target product output, selectivity, hydrocarbons composition and structure [1,4];

which serves as a basis for low number of stages in the process (with high degrees of conversion and target product output in a single-pass mode) and obtaining fuel with high motor and environmental qualities.

3. THERMODYNAMICS

Thermodynamic data for reaction balancing constant calculation for typical hydrocarbon synthesis reactions in a real temperature range are given in Table 1 (in column 5: E+n means 10^{+n}). The Table reveals (with C_6 hydrocarbons as an example) that T=400K is thermodynamically favorable for obtaining hydrocarbons of any structure. At T=500K, the balancing constant of the reaction producing paraffin hydrocarbons is 3 order higher than that for producing naphtenes and, which is especially important, is 5 orders higher than that form aromatic hydrocarbons. Meanwhile, it is at T=600K that the constants differ by one order only, whereas at T=700K it is thermodynamically more probable to obtain aromatic hydrocarbons. Thus, in the temperature range of 400K≤T<500K the synthesis is thermodynamically not selective one, while in the temperature range of 500<T≤700K there occurs selectivity unfavorable for production of diesel fuels.

Table 1
Reaction Thermodynamic Characteristics

T, K	ΔH_T^0, kJ/mol	S_T^0, J/mol·K	ΔG_T^0, kJ/mol	K_T^0
	$6CO+9H_2 \rightarrow C_6H_6+6H_2O$ (aromatic)			
400	-718.7	-1001.5	-330.3	>E+50
500	-729.4	-1025.1	-216.6	4.4E+22
600	-737.7	-1040.3	-113.6	7.8E+0.9
700	-743.7	-1049.9	-8.9	4.61
	$6CO+12H_2 \rightarrow C_6H_{12}+6H_2O$ (naphtene)			
400	-930.5	-1380.7	-390.3	>E+50
500	-945.4	-1434.2	-234.3	3.1E+24
600	-956.4	-1434.1	-96.0	2.3E+08
700	-963.9	-1445.9	48.1	2.5E-04
	$6CO+13H_2 \rightarrow C_6H_{14}+6H_2O$ (paraffin)			
400	-973.9	-1419.5	-418.3	>E+50
500	-988.9	-1452.6	-262.4	2.6E+27
600	-1000.5	-1473.7	-116.2	1.3E+10
700	-1008.9	-1487.1	32.0	4.1E-3
	$12CO+25H_2 \rightarrow C_{12}H_{26}+12H_2O$ (paraffin)			

Table 1
Reaction Thermodynamic Characteristics

400	-1902.1	2856.2	-793.6	>E+50
500	-1929.9	-2917.7	-470.6	E+50
600	-1951.3	-2956.7	-177.3	2.7E+15
700	-1966.5	-2981.0	120.0	1.1E-09

Therefore, a fine regulation of temperature with selection of modes in the area of maximum thermodynamic probability of obtaining paraffin hydrocarbons and, consequently, minimum probability of obtaining aromatic hydrocarbons (T≈500K) opens a way to directed synthesis of diesel fuels with high cetane ratings (n-paraffins) and favorable environmental characteristics (lack of aromatics).

As to the thermodynamic probability of obtaining hydrocarbons of similar structure but different molecular weight, Table 1 demonstrates (with C_6-C_{12} hydrocarbons as an example) a prevailing probability of obtaining diesel fractions in the temperature range selected.

4. CATALYSTS AND KINETICS

Within the process thermodynamic parameter optimum range considered above, composition and structure of the hydrocarbons obtained are determined by catalytic system kinetic characteristics. Catalysts developed for the process provide:

- high activity with hydrocarbon output of 80-85% which excludes recycling;
- maximum selectivity in the area of chain growth probability (CGP) of Schultz-Florie distribution a = 0.82-0.86 that corresponds to the highest relative outputs in diesel-reactive fractions C_5-C_{18} (over 80% of obtained liquid hydrocarbons and 70% of entire hydrocarbon fraction) which excludes necessity of hydrocracking unit (unlike the Shell process where high outputs are obtained in heavy paraffins that require an additional hydrocracking to transform them into motor fractions);
- high cetane ratings of diesel fuel due to low olefin content (the main component is n-paraffins) which allows to exclude synthesis product hydrostabilization unit or implement it as a very small modular attachment in soft conditions (P≤2 MPa, t=200°C);
- possibility to regulate isoparaffin content for significant decrease of crystallization temperature and fuel "cold" viscosity, which allows to obtain winter kinds of without additional hydroisomerization unit;
- environmental purity through the lack of aromatic hydrocarbons in the synthetic fuel obtained;
- high thermal stability, retaining the characteristics for a year under operating conditions in a fixed-bed Arge reactor.

5. PHYSICAL-CHEMICAL PROPERTIES AND TECHNICAL CHARACTERISTICS OF SYNTHETIC DIESEL FUEL

Table 2 gives main characteristics of synthetic diesel fuel produced from natural gas in a single-pass mode with no hydrorefining whatsoever; the fuel quality exceeds that of the summer diesel fuel produced from oil (GOST 305-82) approaching values of winter fuels.

Table 2
Synthetic Diesel Fuel (Fraction 150-300 °C) Quality Indices

Index	Norm according to GOST 305-82		Actual values
	summer	winter	
1. Fractional composition			
50% distilled at t, °C, not exceeding	280	280	200-230
96% distilled at t, °C, not exceeding	360	340	290-310
2. Solidification temperature, °C, not exceeding	-10	-35	complies -15 to -20
3. Turbidity temperature, °C, not exceeding	-5	-25	complies
4. Acidity, *mg* KOH per 100 cm^3 product, not exceeding	5	5	0.5-1.5
5. Cetane rating, not less than	45	45	45-47
6. Kinematic viscosity at 20 °C, mm^2/s (sSt)	3-6	1.8-5	2.5-2.8
7. Flash temperature in closed crucible, °C, not lower than	40	35	50
8. Sulfur mass fraction, %, not more than	0.5	0.5	0.0015
9. Copper plate test	Passes		Passes
10. Actual resin concentration, mg per 100 cm^3 product	40	30	
11. Iodine number, *g* iodine per 100 *g* fuel, not more than	6	6	
12. Ash content, %, not more than	0.01	0.01	
13. Coking capacity 10% residue, %, not more than	0.2	0.1	
14. Mechanical impurity content	absent		absent
15. Water content	absent		absent
16. Density at 20 °C, kg/m^3, not more than	860	840	770

REFERENCES

1. V.M.Batenin, D.N.Kagan, A.L.Lapidus, F.N.Pekhota, M.N.Radchenko, A.D. Sedykh, E.E.Shpilrain, The I-st Intern. Congress "New High Technologies for Oil and Gas Industry" (Tyumen, June 4-6, 1996), Moscow: Ed. by GASPROM (1996) 111.
2. E.E.Shpilrain., M.D.Navalikhina, D.N.Kagan, 3-rd USSR-Australia Workshop on Advanced Fossil Fuel Based Energy Technologies (Sydney, June 20-21, 1989). Sidney: Ed. by School of Electrical Engineering, University of Sidney (1989) A4.1.
3. Yu.I.Bokserman, V.R.Grunvald, O.B.Braginsky, Prospects for development, production and application of motor fuels and chemical products from natural gas in Russia up to the year of 2010, Moscow: Ed. by GASPROM, 1996.
4. M.N.Radchenko., D.N.Kagan, G.A.Krechetova, Preprints IVTAN No. 8-385, No. 8-387, No. 8-389, Moscow, 1995.

NATURAL GAS CONVERSION V
Studies in Surface Science and Catalysis, Vol. 119
A. Parmaliana et al. (Editors)

Advanced Catalytic Converter System for Natural Gas Powered Diesel Engines

Vadim O. Strots[a], Grigorii A. Bunimovich[a], Yurii Sh. Matros[a], Ming Zheng[b], and Edward A. Mirosh[b]

[a]Matros Technologies, Inc., 14963 Green Circle Drive, Chesterfield, Missouri 63017, USA

[b]Alternative Fuel Systems, Inc., #420, 107 - 11th Avenue S. W., Calgary, Alberta T3C 0M5, Canada

The paper discusses the development of catalytic converter for aftertreatment of exhaust gas from diesel engines powered with natural gas. The converter, operated with periodical reversals of the flow, ensures destruction of CO and hydrocarbons, including methane. Both computer simulation and engine testing results are presented.

1. INTRODUCTION

Numerous diesel engines (over 100 million, according to some estimates [1]) supply motive and stationary power worldwide and make an important pollution source, emitting hydrocarbons, CO, NO_x, particulates and CO_2. Pollution concerns, along with anticipated depletion of oil reserves, warrant research and development efforts in the area of alternative fuels.

Natural gas is considered to provide for an economical and cleaner alternative fuel. Its high hydrogen to carbon ratio, ~ 3.8:1, results in the lowest, after pure hydrogen, CO_2 emissions per unit of energy produced. Combustion of natural gas can reduce emissions of hydrocarbons, CO and oxides of nitrogen and sulfur. Combination of these advantages gas with fuel efficiency of diesel engine creates a potentially superior power source.

A diesel dual-fuel (DDF) engine uses both of diesel fuel and natural gas [1]. Pilot diesel charge helps to achieve stable ignition of natural gas mixed with air. A microprocessor-based fuel management system is used to meter and time separate diesel and natural gas injections. This results in fuel efficiency, power and torque similar to those of pure diesel fuel operation.

Depending on engine type and operating conditions, up to 80 % of diesel fuel can be substituted by natural gas in a DDF engine. This results in about 15 % reduction of CO_2 emission. Sulfur content in natural gas is lower than in diesel fuel, which results in SO_2 emissions reduction by ~ 30 % [2]. Emission of particulates can also be reduced because premixing air with natural gas produced better homogeneity of the air-fuel mixture.

However, using natural gas as a fuel creates exhaust treatment problems. Incomplete combustion results in high concentration of methane in the engine exhaust, which has to be reduced substantially to meet stringent emission regulations in Western Europe, Japan and other countries. Methane molecule is exceptionally stable and requires much higher catalyst temperature than those in diesel oxidation catalysts (Fig. 1). However, such temperatures are not always available because of low exhaust temperature. For example, low torque operation of Isuzu 4BE1 diesel engine retrofitted to dual fuel operation produces 4,000 - 6,000 ppm of CH_4 at under 250°C, which is lower than needed for effective methane destruction.

Solving the problem of destruction of methane from diesel dual fuel engine exhaust can dramatically increase acceptance of natural gas as an alternative to oil-derived diesel fuel.

Reverse flow converter (RFC) concept was suggested as a potential solution. The concept, pioneered by Matros [3], is used for oxidation of SO_2 to SO_3, oxidation of volatile organic compounds, and selective catalytic NO_x reduction [4]. Evaluation of the concept included development of dynamic process model that incorporated reaction kinetics of methane oxidation and processes of heat and mass storage and transport in a catalyst monolith, extensive process simulation, and prototype testing. The results are discussed in this paper.

2. REVERSE FLOW CONVERTER

The Matros reactor (Fig. 2) operates with periodical reversals of the gas flow through catalyst bed or monolith. The gas flow entering the catalyst picks up heat from the solid material. When the gas temperature rises to the catalyst ignition point, catalytic reaction occurs with heat release. After the reaction zone, the gas enters the cooling zone where it gives up its heat energy to cooler solid material. Then it leaves the reactor.

After the input portion of the monolith becomes cooled and the output portion becomes heated, the flow is reversed using the diverting valve. This causes re-heating of the formerly cooled end. Periodical reversals create an oscillating profile of temperature with low temperatures at the ends and high temperature in the center of the bed.

Heat energy is effectively trapped inside the monolith, thus ensuring effective oxidation of CO and hydrocarbons at low temperature of incoming gas. In addition, the converter was found to be insensitive to, or is barely affected by short-term changes in the input gas

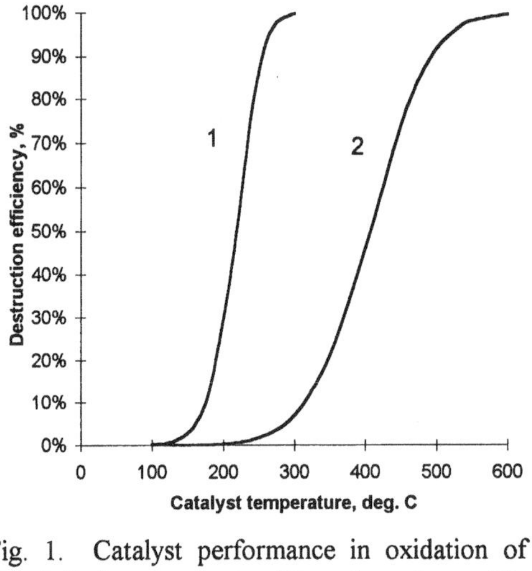

Fig. 1. Catalyst performance in oxidation of diesel fuel components (1) and methane (2). Residence time = 0.1 s.

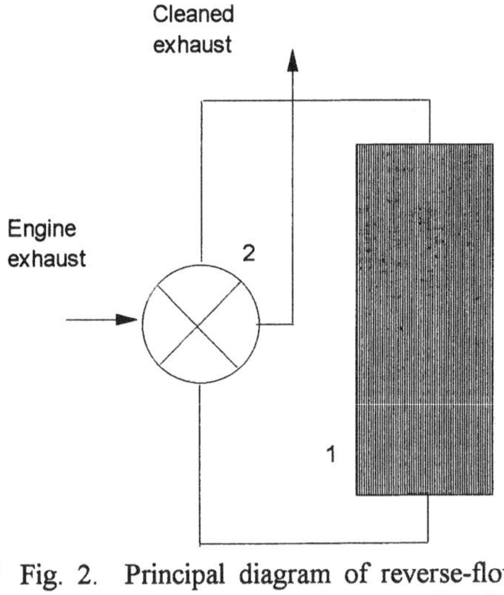

Fig. 2. Principal diagram of reverse-flow converter. 1: catalyst canister, 2: diverting valve

temperature, flow rate or composition changes.

Applications of Matros reactor to methane oxidation have been investigated by a number of researchers. Matros *et. al.* [5] suggested using the reactor to utilize the energy of methane combustion of low-concentrated gases. Both theoretical and experimental results obtained with a $CuO\text{-}Cr_2O_3$ / $\gamma\text{-}Al_2O_3$ catalyst demonstrated self-sustainable process operation with high catalyst temperature, 800 - 1,000 °C, while inlet temperature and adiabatic temperature rise of combustion were low, 25 to 122.5 °C, respectively. Sapundzhiev *et. al.* [6] conducted extensive simulation study of a system comprising a combination of reverse flow reactor with heat exchanger installed between two halves of perovskite catalyst bed. Chaouki *et. al.* [7] investigated methane oxidation on 0.2 % Pd/Al_2O_3 catalyst in a reverse flow reactor as a method for NO_x-free combustion. In these studies, beds of pelleted catalyst were flanked by beds of ceramic packing to facilitat thermal energy regeneration.

3. PROCESS SIMULATION

Earlier studies of Matros reactor [3,4] lead to a system of model equations that incorporates essential factors governing the dynamic behavior of the converter, yet remains simple enough to obtain stabilized periodic time domain solutions with reasonable computing time. This study was conducted with a one-dimensional, two-phase model of plug-flow reactor which assumed that both gas phase and surface concentrations of reactants, and the gas flow temperature are in quasi-steady state related to the temperature of solid material. The model equations and references to using similar models for simulation of automotive converters, are given in [8].

An essential assumption made in the model is that profiles of temperature, composition, and velocity in the converter are considered uniform throughout the cross-section of the monolith. Sometimes this assumption may cause dramatic differences between model predicted and actual operation of Matros reactors. For example, small heat losses can result in formation of cold zones near the walls of large-scale reactors, accompanied by low conversion of reactants (see [4] for references). It turns out, however, that high velocity of the gas flow through the channels of the catalyst monolith and high hydrocarbon concentrations result in relatively uniform temperature distribution throughout the cross-section areas of automotive converter.

An example of the converter simulation is given in Figs. 3 and 4. In these examples, rate equations obtained from literature data were used (see [8] for references). The incoming gas flow (engine exhaust) has the temperature of 178 °C and contains 4090 ppm of CH_4 and 2470 ppm of CO. The data were obtained from actual engine test results at moderate engine speed of 1400 rpm. At this speed, the residence time in the converter was 0.22 s. The resultant steady state operation of conventional converter is characterized by 93.4 % destruction of CO but only 5.3 % destruction of methane (Fig. 4, lines 4 and 3, respectively). There is small temperature rise along the catalyst monolith due to CO oxidation heat release.

Stabilized periodical reverse-flow operation presents dramatic contrast to conventional operation. While both ends of the monolith are cold, the temperature in the central part of the monolith is as high as 660°C. This temperature results in complete destruction of both CO and methane. It should be noted that the high-temperature regime is possible only if periodic operation begins with preheated catalyst. If the start-up temperature is too low, the heat relesed by oxidation reactions will not drive the temperature up, and the performance will be

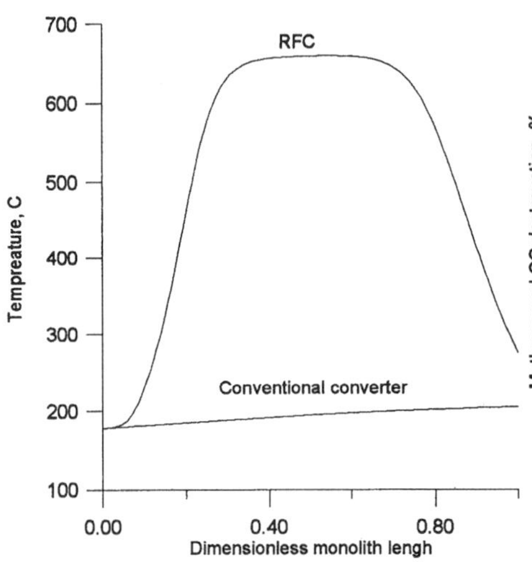

Figure 3. Simulated stabilized axial profiles of temperature in conventional and reverse-flow converters.

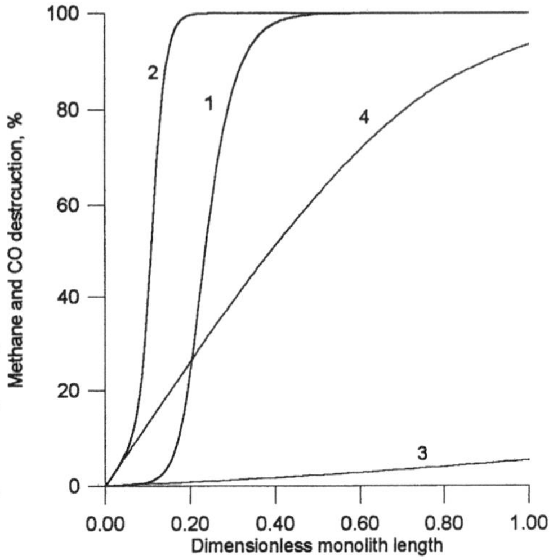

Figure 4. Simulated stabilized axial profiles of CH_4 (1,3) and CO (2,4) conversion in reverse-flow (1,2) and conventional (3,4) converters.

close to that of conventional converter.

Apparently, stabilized periodic regime may not be achieved in an actual city truck or bus because their operation is typically transient, including frequent accelerations, braking, and idling. In an Isuzu 4BE1 engine, such operation may cause oscillations of exhaust temperature from 100 to 700°C, flow rate from 9.3 to 75 l/s, and hydrocarbons concentration from 82 to 7700 ppm. RFC simulation showed that it is ideally suited for this sort of operation. For example, engine idling caused reduction of methane conversion from > 90 % to < 10 % during about 4 min, while it took over an hour for the RFC to experience similar transition.

As an illustration of dynamic operation of the RFC, Fig. 5 presents the results of simulation of 13-mode Japanese test. Passing this or similar regional, national and local tests is a necessary condition for commercial use of the converter. Rate constants for this simulation were estimated based on actual catalyst performance during engine tests.

During the test, engine exhaust temperature (line 4, Fig. 5), flow rate and composition vary widely, following complex pattern of engine operation. The test begins with the catalyst preheated, however, the temperature in the conventional converter falls quickly during first several modes that include two periods of idling and some low-torque operation (line 3). The output methane concentration (line 6) is low during idling but reaches over 5,000 ppm at low-torque operation. Only after about 20 min of test, when the catalyst gets heated by hot engine exhaust (high-torque operation), substantial conversion of methane is obtained.

In reverse-flow operating mode, the converter keeps high temperature in its central part in all engine operation modes (line 1, Fig. 5). Accordingly, the methane concentration remains low during all the test (line 5). Temperature in the converter ends is generally lower and oscillates as a result of flow reversals (line 2).

4. ENGINE TESTING

The RFC was tested for treatment of exhaust from 3.9 l Isuzu 4BE1 diesel engine retrofitted to dual fuel operation with AFS conversion kit. The converter, designed according to the scheme in Fig. 2, included a catalyst can and a four-way diverting valve. When the valve is not actuated, the unit works as a common catalytic converter. Thus the performance with and without flow reversals can be compared. The catalyst temperature was monitored by six thermocouples. One more thermocouple was used for engine exhaust temperature monitoring. The pipes between engine and converter were thermally insulated. Total concentration of hydrocarbons was measured using an FID detector, samples being taken from upstream and downstream of the diverting valve.

Each test run began with conventional operation, i.e. without flow reversals, until the time catalyst activity stabilized. Periodic operation was tested in a wide range of engine speed and

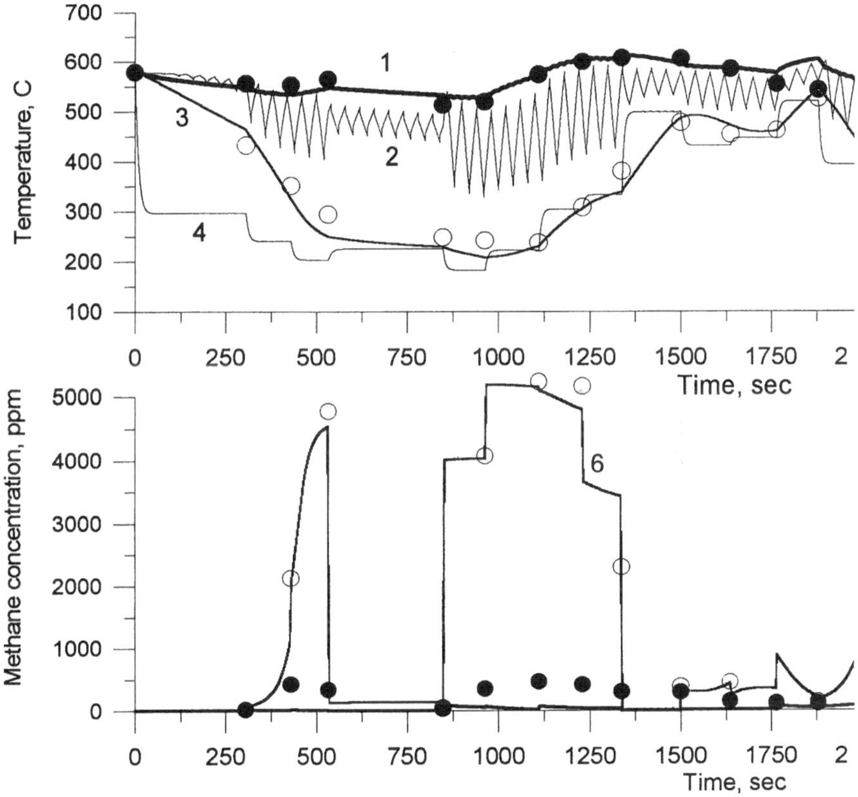

Fig. 5. Converter performance in 13-mode Japanese emission test. Lines - simulation, circles - experiment. 1: average catalyst temperature, RFC operation, 2: boundary catalyst temperature, RFC operation, 3: average catalyst temperature, conventional operation, 4: inlet (engine exhaust) temperature, 5: outlet methane concentration, RFC operation, 6: outlet methane concentration, conventional operation.

torque values. After each reverse-flow run the catalyst performance was tested in conventional mode. The tests showed that the reverse-flow operation did not cause deactivation.

Fig. 5 presents the results obtained during 13-mode Japanese test. It shows that the converter operates in a good agreement with model predictions. In conventional mode, the catalyst temperature decreased during first half of the test, then rose. Methane destruction was observed only in the second half of the test. Integration of inlet and outlet hydrocarbon concentrations over the test duration showed that the average methane destruction was only 27 %. Reverse-flow operation resulted in high temperature in the catalyst, 500 - 600°C, during the entire test, and in high efficiency of hydrocarbons destruction, over 85 % at the average.

5. CONCLUSIONS

Both simulation and experimental results presented here prove that the reverse-flow catalytic converter can effectively treat exhausts of natural gas fueled diesel engines. Highly efficient destruction of methane, other hydrocarbons and CO was achieved with currently available catalysts, despite the temperature of engine exhaust falling far below the catalyst ignition temperature. The converter was found to be able to handle engine exhaust gas flow with widely variable temperature, flow rate and composition, intrinsic to diesel engines. In addition, the RFC possesses a unique capability to retain heat for during long time engine idling, when the exhaust gas is cold and contains very little concentrations of CO and hydrocarbons. The superiority of RFC was demonstrated during 13-mode Japanese test where 85 % destruction of hydrocarbons was achieved, vs. 27 % with conventional converter.

The results show that the reverse-flow converter presents a viable solution for the problem of methane removal from the exhaust of diesel dual fuel engines. It is anticipated that, upon testing completion and approval of government agencies, DDF engines coupled with novel converters will gain acceptance and thus provide a route for increasing replacement of oil derived fuels by natural gas.

6. REFERENCES

1. A. Badakhshan, J. Beck, P. Beck, and E. A. Mirosh. Autotech 95 - European Perspective and Opportunities, Birmingham, Nov., 1995.
2. Dual-Fuel Conversion System Developed for Isuzu Truck Engine; Tests Show Significant Reductions in NO_x, CO, Smoke. Diesel Progress, Feb. 1997, 44.
3. Yu. Sh. Matros. JVKhO im. Mendeleeva, 22 (1977), 567.
4. Yu. Sh. Matros and G. A. Bunimovich., Catal. Rev. - Sci. Eng., 38 (1996), 1.
5. Yu. Sh. Matros, A. G. Ivanov, and L. L. Gogin. Teoreticheskie Osnovy Khimicheskoi Tekhnologii, 22 (1988), 481.
6. C. Sapundzhiev, J. Chaouki, C. Guy, and D. Klvana. Chem. Eng. Comm., 125 (1993), 171.
7. J. Chaouki, C. Guy, C. Sapundzhiev, D. Kusohorsky, and D. Klvana. Ind. Eng. Chem. Res., 33 (1994), 2957.
8. V. O. Strots, G. A. Bunimovich, Yu. Sh. Matros, M. Zheng, and E. A. Mirosh. SAE Paper 980194, 1998.

NATURAL GAS CONVERSION V
Studies in Surface Science and Catalysis, Vol. 119
A. Parmaliana et al. (Editors)
1998 Elsevier Science B.V.

Overview of U.S. DOE's Natural Gas-to-Liquids RD&D Program and Commercialization Strategy

V.K. Venkataraman[a], H.D. Guthrie[a], R.A. Avellanet[b] and D.J. Driscoll[a]

[a]U.S. Department of Energy, Federal Energy Technology Center, Morgantown, WV, USA

[b]U.S. Department of Energy, Office of Fossil Energy, Oil and Gas Programs, Washington, DC, USA

1. INTRODUCTION

Natural gas, which is comprised mostly of methane, is one of our most abundant natural resources, both in the U.S. and world wide. In the United States alone, recoverable natural gas resources are several times its current estimate of reserves, 166 trillion cubic feet (TCF) [1]. Unfortunately, many of these resources are located offshore or in remote areas. High transportation costs, or complete lack of any transportation mechanism, prohibits extensive use of this "stranded" natural resource. To overcome this limitation, the U.S. Department of Energy's (DOE) Federal Energy Technology Center (FETC) has developed a highly diversified gas-to-liquids research program to evaluate , promote and develop processes that convert natural gas into higher value products (*i.e.*, liquid fuels) which will offset the high transportation costs and allow use of this untapped, environmentally friendly resource.

By advancing technologies to convert unmarketable gas resources into valuable products, cooperative efforts between DOE and industry could yield the following benefits by 2010. (1) Our shortfall in domestic production of oil will be reduced by 200,000 to 500,000 barrels per day of high quality transportation fuel made from Alaska's North Slope gas resources; (2) Advanced gas-to-liquids conversion technology that yields ultra clean burning diesel fuels that meet the most stringent emissions requirements, at costs competative with those of comparable fuels made from crude oils, will be utilized; and (3) Small-scale gas-to-liquids technology for both natural gas liquefaction and chemical conversion to higher hydrocarbon liquids will provide an economic and environmentally sound option for utilization of the associated gas of remote offshore oil reservoirs, and also for onshore gas reservoirs without pipeline access. In summary, development of efficient gas conversion technologies will enhance U.S. energy security, reduce dependence on oil imports and strengthen the economic competitiveness of the U.S. oil and gas industry.

2. NATURAL GAS CONVERSION RESEARCH

Three potential routes for the conversion of natural gas have emerged: physical, direct and indirect conversion. Physical conversion techniques center on the conversion of natural gas to liquefied natural gas (LNG). Direct conversion focuses on the chemical transformation of natural gas to ethane, ethylene, acetylene or methanol, precursors of liquid fuels. During indirect conversion, the natural gas is first converted into synthesis gas (a mixture of carbon monoxide and hydrogen) which is then converted into the desired end product, such as Fischer-Tropsch (F-T) liquids, methanol, MTBE, or mixed alcohols. All three approaches are currently under investigation under the gas-to-liquids conversion program at FETC. A graphical summary of the program is illustrated in Figure 1.

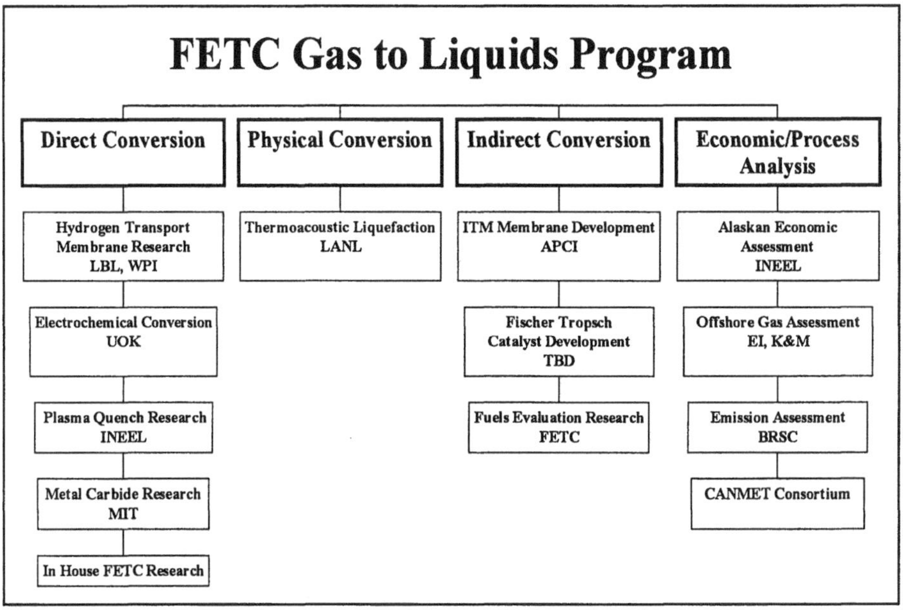

Figure 1. Federal Energy Technology Center's Emerging Processing Technology Applications Research Program

2.1. Physical Conversion

Physical conversion is the simplest of the gas-to-liquids concepts, involving only changing the physical state of the natural gas. This usually means converting the natural gas to LNG. LNG occupies a much smaller volume than the original gas and can be maintained at ambient pressure, although it does have to be kept at cryogenic temperatures. One difficulty with LNG has been the cost of liquefaction. One novel approach being explored is known as Thermoacoustically Driven Orifice Pulse Tube Refrigeration (TADOPTR). This technology, which is well suited for liquefaction capacities in the range of 500-10,000 gallons per day, has the unique capability of producing refrigeration at cryogenic temperatures with no moving parts. Research on this concept is being carried out under a Cooperative Research and Development Agreement (CRADA) between Los Alamos National Laboratory (LANL) and Cryenco, a small business located in Denver, CO. TADOPTR has already been demonstrated at a production rate of 100 gal/day, and scale up to 500 gal/day is underway.

2.2. Direct Conversion

The direct conversion of methane to higher hydrocarbons has been extensively investigated over the past 15 years [2]. However, the inherent stability of methane makes its direct conversion difficult, since most of the desired products are more reactive than methane itself and, once formed, tend to react further to undesirable by-products (CO and CO_2). With conventional catalytic systems, yields have remained low. In the case of C_2 production, yields have been limited to 30% or less. Yields of oxygenated products, such as methanol and formaldehyde, have been even lower, on the order of 5-6%. To overcome these limitations, FETC has funded relatively novel research projects, including the use of electric fields, plasma torches, and hydrogen transport membranes.

As an example of this approach, hydrogen transport membrane technology operates as follows. Methane is allowed to react in the absence of oxygen (pyrolytically) over a catalyst on one side of the membrane. Although conversion to higher hydrocarbons, in particular C_2, is highly equilibrium limited, the hydrogen produced during the reaction is selectively removed via transport through the membrane, thus removing the equilibrium constraint and allowing the reaction to proceed further. On the other side of the membrane, pure hydrogen can be recovered or the hydrogen can react with oxygen to produce water and generate heat needed for the pyrolysis reaction. In this latter case, the overall reaction is: $2CH_4 + O_2 \rightarrow C_2H_4 + 2H_2O$, which is simply the oxidative coupling of methane.

2.3. Indirect Conversion

Indirect conversion of natural gas requires the intermediate production of synthesis gas (syngas) which is then converted into the desired end product. Syngas can be produced by a variety of processes, two of the most important being steam reforming and partial oxidation. In steam reforming, the hydrocarbons in the natural gas react with steam at high temperature to form a mixture of CO and H_2. In partial oxidation, the natural gas reacts with oxygen and, optionally, steam to form syngas. The H_2/CO ratio is higher from steam reforming than from partial oxidation. Both of these processes are energy and capital intensive, and syngas production remains the most expensive step in the indirect conversion of natural gas.

Since much of the cost of indirect conversion is associated with syngas production, any improvement which reduces this cost will have a significant beneficial effect on indirect conversion economics. In turn, any reduction in the cost of oxygen manufacture would translate into a reduction in the cost of syngas by partial oxidation. One technology which shows considerable promise is the use of ceramic membranes for oxygen production. Air is passed over the outside of the membrane, which is a dense, nonporous ceramic material capable of withstanding high temperatures. Because of the nature of the ceramic material, the oxygen partial pressure differential across the membrane causes oxygen from the air to diffuse through the membrane in the form of oxide ions (O^{-2}). Thus, air separation is achieved without the use of relatively expensive cryogenics or compression.

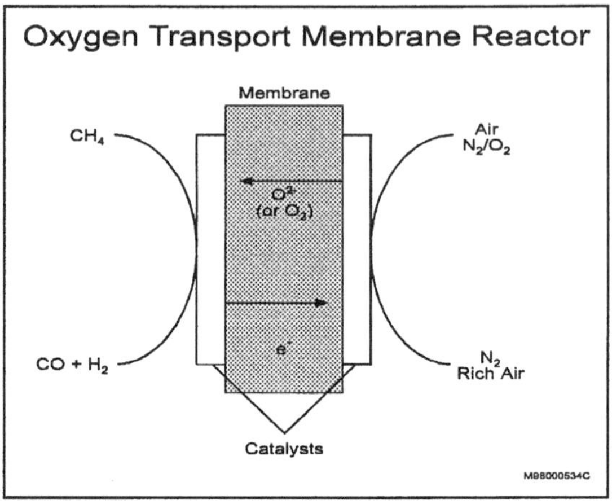

Figure 2. Oxygen Ion Transport/Methane Reforming Ceramic Membrane Reactor

In addition to functioning as an oxygen separation unit, the membrane can also serve as the synthesis gas reactor. The process is depicted in Figure 2. If natural gas is passed over the inside of the membrane, it will react with the oxygen diffusing through the membrane, resulting in the formation of synthesis gas. When fully developed, this approach should substantially reduce both capital and operating costs for syngas production, thus allowing indirect conversion of natural gas to become an alternative, cost-effective source of liquid fuels and value added chemicals.

Table 1
Emission Performance of SynGas Derived Diesel Fuels

	Emission Reduction vs Petroleum Diesel	
	Low S	Low S/Aromatics
Hydrocarbons	41-46%	25-31%
CO	45-47%	34-38%
NO_x	9%	5%
Particulate	27-32%	23-29%

The use of natural gas derived syngas for the production of Fischer-Tropsch products, *i.e.*, diesel fuels, has recently become one of the hottest research areas in the field of gas-to-liquids conversion. The products produced show considerably improved cetane values and, equally important, reduced emissions when compared with conventional petroleum diesel fuels. Typical reductions in several types of emissions are illustrated in Table 1.

3. ECONOMIC AND PROCESS ANALYSIS

Continued economic and process analysis of existing and potential natural gas conversion technologies is a necessary and integral part of the gas-to-liquids program. Recent studies include the potential and economics for offshore gas-to-liquids conversion and an economic assessment of Alaskan North Slope gas utilization options. Of particular importance in the latter work was the identification of the window of opportunity (Figure 3) to extend the life of the Trans Alaskan Pipeline System (TAPS) [3]. Identification of a viable technology to convert gas to pipeline quality liquids by 2009-2016 could extend the life of TAPS by 20 or more years. Continued operation of TAPS is vitally important to Alaska's economy; therefore, there is considerable incentive to develop and commercialize new gas-to-liquids conversion technologies, capable of a 200,000-500,000 bbl/day production rate, by early in the 21st century.

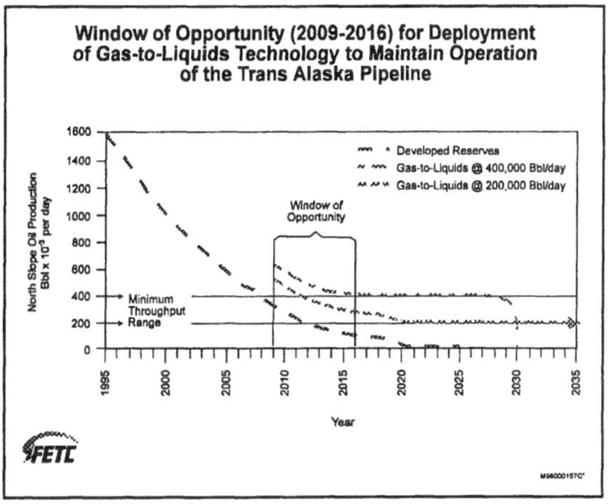

Figure 3. TAPS Window of Opportunity

Recently, Bechtel conducted an economic study of a full scale F-T facility, using both natural gas and coal as feedstocks [4]. Estimated total capital cost for a plant feeding 410 million SCFD of natural gas and producing about 45,000 BPD of F-T liquid products is $1,840 million. At a southern Illinois location, the estimated Crude Oil Equivalent (COE) price is $32.8/bbl (1994 dollars) if natural gas costs $2/million Btu [5]. However, the COE price is strongly dependent on gas cost. If natural gas is available at $0.50/million Btu, the COE price drops to $19.7/bbl. From this analysis, it is clear that a low natural gas cost is necessary if F-T technology is to be competitive domestically. One place with potentially low cost natural gas is the North Slope of Alaska. The low value of North Slope natural gas is dictated by the high infrastructure cost required to bring this gas to market. However, if the gas were first liquefied using F-T technology, the F-T liquids could be blended in with North Slope crude and transported through TAPS, not only potentially extending the life of the existing pipeline but also avoiding the cost of a new natural gas pipeline. This scenario appears to offer a significant incentive for utilizing F-T technology on the North Slope.

If a F-T facility were to be built on the North Slope, its configuration would probably be somewhat different from that in Bechtel's baseline design. It is very unlikely that finished products would be produced on the North Slope. Rather, a so-called syncrude would be produced, which would consist essentially of the C_4+ product directly from the F-T reactors. Therefore, the product upgrading section of the baseline plant would be eliminated, thus reducing capital costs by 10-15%. However, construction costs will be higher on the North Slope than in the lower 48 states, and this may offset the savings from eliminating product upgrading. If some of the improvements discussed above, such as use of membranes for syngas production, can be incorporated, F-T technology for conversion of remote gas could be very attractive.

Recently, Energy International Corporation completed a DOE funded study evaluating the conversion of associated natural gas to liquid hydrocarbons [6]. Of particular interest was their analysis of the potential of employing F-T conversion as a means of utilizing offshore associated gas. The report indicted that an offshore F-T could best be accommodated by a Floating Production, Storage, Offloading vessel (FPSO) based on a converted surplus tanker, such as have been frequently used around the world recently. Estimated cost for a facility of this type was approximately $525MM. Other structure types used in deep water (platforms) are more expensive and cannot handle the required load. The FPSO is a mobile, small footprint unit and provides an alternative to fixed platform GTL conversion units.

ACKNOWLEDGMENT

All of the work presented above was funded by the U.S. Department of Energy through the Federal Energy Technology Center's Emerging Processing Technologies Applications Program.

REFERENCES

1. U.S. Crude Oil, Natural Gas, and Natural Gas Liquids Reserves 1966 Annual Report, U.S. Department of Energy/Energy Information Administration, November 1997.
2. Driscoll, D.J., Direct Methane Conversion, Proceedings of Gas Conversion: Strategies, Technologies & Activities, Energy Frontiers International, Girdwood, AK, July, 1997.
3. Final Report, "Economics of Alaskan North Slope Gas Utilization Options", Contract No. DE-AC07-94ID13223, August 1996.
4. Topical Report VI, Natural Gas Fischer-Tropsch Case, Volume II, Plant Design and Aspen Process Simulation Model, Baseline Design/Economics for Advanced Fischer-Tropsch Technology, Contract No. DE-AC22-91PC90027, August 1996.
5. Choi, G.N., Kramer, S.J., Tam, S.S., Srivastava, R.D., and Stiegel, G.J., "Natural Gas Based Fischer-Tropsch to Liquid Fuels: Economics," Society of Petroleum Engineers 66th Western Regional Meeting, Anchorage, Alaska, May 22-24, 1996.
6. Singleton, A.H. and Cooper, P.G. (Energy International Corporation), Final Report, "Conversion of Associated Natural Gas to Liquid Hydrocarbons", Contract No. DE-AC21-95MC32079, November 1997.

NATURAL GAS CONVERSION V
Studies in Surface Science and Catalysis, Vol. 119
A. Parmaliana et al. (Editors)

Paraffins activation through Fluidized Bed Dehydrogenation: the answer to light olefins demand increase

Domenico Sanfilippo, Franco Buonomo, Giorgio Fusco, Ivano Miracca
Snamprogetti SpA, Via Maritano, 26 - 20097 San Donato Milanese, Italy
Georgii R. Kotelnikov - Yarsintez - Pr. Oktyabrya, 88 - 150040 Yaroslavl, Russia

1. INTRODUCTION

Light olefins (C_2-C_5) are the starting material for manufacturing some of the most important chemical products as polymers, synthetic rubbers and fuel components: ethylene and propylene derivatives form the core of the commodities chemical industry and C_4–C_5 olefins are transformed into valuable components for reformulated fuels.

Unfortunately light olefins are not available as a natural resource and their production requires expensive technologies: steam cracking (SC) of ethane produces mainly ethylene, while the use of heavier feedstocks (Natural Gas Liquids, naphtha, heavier liquids) implies the co-production of a huge variety of intermediates (with limited flexibility between them) requiring the downstream integration of a complex network of chemical productions. C_3–C_5 olefins are also available from FCC (Fluid Catalytic Cracking) plants in oil refineries.

Isobutylene both from SC or FCC is mainly used to produce methyl tert-buthyl ether (MTBE). World MTBE production capacity grew up to about 20 million MTPY [1], exceeding the traditional sources of isobutylene, and requiring the massive use of field butanes via the n-butane isomerization to isobutane and its dehydrogenation to isobutylene.

The growth rate of world consumption of light olefins is foreseen to continue at a differentiated rate ($C_3 > C_2$) [2]. Paraffins dehydrogenation, with its specificity respect to the olefins, gives the chance of answering selectively to the market demand of olefins. Dehydrogenation provides furthermore an interesting way to locally exploit Natural Gas Liquids (NGL) sources by converting the relatively low value paraffins into valuable intermediates of fuel and petrochemical industries.

These considerations have led in the last decade to the development and commercialization of some processes able to dehydrogenate C_3-C_5 paraffins to the respective mono-olefins with high yields.

2. Thermodynamic and kinetic considerations

The reactions of dehydrogenation can be schematized as:

$$\text{Paraffin} \Leftrightarrow \text{Mono-olefin} + H_2$$

are characterized by the following features [3]:

- *conversion is limited by thermodynamics*, as reported in figure 1. Commercially viable conversions require reaction temperature exceeding 800 °K, higher for lower paraffins.

- *side reactions occur in the required operating conditions*. The use of a catalyst is necessary in order to keep a suitable conversion per pass while obtaining high selectivity toward the desired olefin. *Only alkali promoted chromia/alumina and platinum/alumina* have been industrially developed.
- *coke is a fatal by-product* formed on the catalyst. A periodical regeneration of the catalyst is mandatory.
- *dehydrogenation is strongly endothermic*, resulting in a high adiabatic temperature decrease; the way how the heat is supplied characterizes the industrial processes presently commercialized.
- reactions happen with an increase of the number of moles: *high pressure adversely affects the reactions* which must be performed in the gas phase.

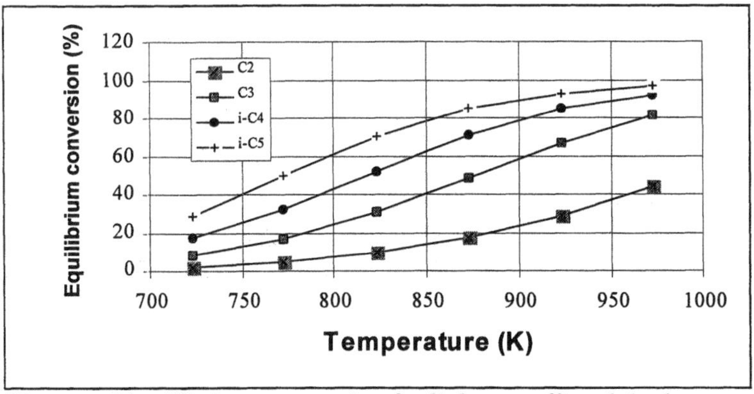

Figure 1: Equilibrium conversion for light paraffins dehydrogenation

3. HEAT SUPPLY AND CATALYST REGENERATION

The main problem to be solved in the development of a suitable reactor for an industrial dehydrogenation process is how to supply the huge quantity of heat needed by the reaction at temperatures well above 800°K, maintaining in the meantime a careful control of temperature to minimize the formation of by-products, and to find how to perform periodical regenerations to remove the small amount of coke on the surface of the catalyst.

These constraints have stimulated the fantasy of engineers in finding suitable solutions. Available technologies are characterized by different choices as concerns the system catalyst / heat supply mode / catalyst regeneration mode. Typical examples are [4]:

- *Adiabatic fixed bed reactors* (fig. 2a) in cyclic operation, quickly alternating reaction and regeneration periods. Reaction heat is stored on the catalyst during the regeneration step and released by the catalyst during the reaction period. Since each reactor is operated batchwise, to achieve continuous operation more reactors in parallel are used.
- *Adiabatic moving bed reactors* in series (fig. 2b) with a slow downflow of the catalyst. Heat of reaction is supplied to the gas by intermediate heating in furnaces between reactors. The catalyst is conveyed to the regenerator and then back to the first reactor.

Fig. 2: Commercial reactors conceptual design

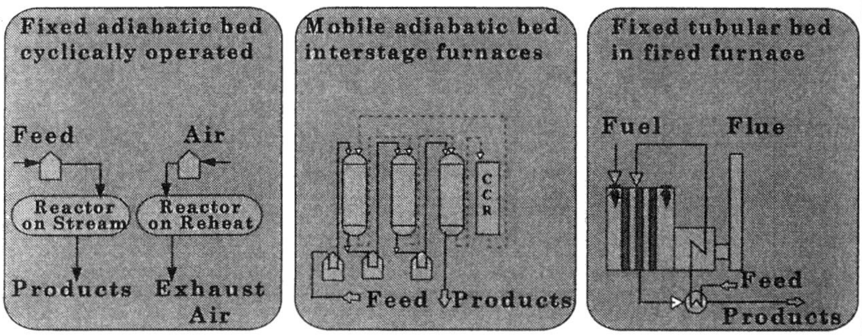

Packed tubes in furnace fireboxes (fig. 2c) for heat supply similar to steam-reforming processes. Several reactors in parallel are used, one of these being alternatively in regeneration.

- The *Snamprogetti–Yarsintez fluidized bed reactor* below described.

4. THE FBD TECHNOLOGY.

Originally a dehydrogenation technology employing a fluidized bed reactor was developed in the former Soviet Union in the 1960s, and over 30 units have been supplying C_4-C_5 olefins for the production of synthetic rubbers. In the last decade, Snamprogetti and Yarsintez (Russia) have further developed this process improving catalyst and engineering design [5]. The result is the FBD (Fluidized Bed Dehydrogenation) technology, commercialized by Snamprogetti, for the dehydrogenation of C_3, C_4, C_5 paraffins (FBD-3, FBD-4 and FBD-5 processes) or their mixtures. A commercial unit for the production of 450,000 MTPY of isobutylene, as a feedstock to a 700,000 MTPY MTBE unit, is in operation for SADAF in Al-Jubail (Saudi Arabia) since 1997. Other projects are in progress.

5. FLUIDIZED BED CONCEPTUAL DESIGN

The heart of the process is the conceptual design of the reactor/regenerator. The reaction is carried out in a catalytic bubbling fluidized bed operating at 800-870 °K and 120-150 kPa. Reaction heat is supplied by the heat capacity of the solid catalyst continuously circulating from the bottom of the reactor to the top of the regenerator and viceversa through a U- or J-bend piping system, originating a countercurrent flow of gas and solid either in the reactor and in the regenerator. Catalyst is pneumatically conveyed from reactor to regenerator and viceversa using feed, nitrogen or air as lift gas.

In the regenerator, the catalyst restores its initial performances by burning the small quantity of coke formed on it during the reaction and restructuring the active sites. Additionally some fuel is burned directly on the catalyst to satisfy the overall thermal balance [6]. Before being conveyed to the regenerator the catalyst is stripped with nitrogen to avoid loss of adsorbed products.

The achievement of optimal yields and reactor volumes would be favoured by a plug-flow movement of the gas phase, while a fluidized bed approaches an ideal

CSTR (Continuous Stirred Tank Reactor). Also a sharp distribution of the catalyst particles residence times in the reactor/regenerator sections is essential to preserve its life and performances.

Figure 3: Reactor - regenerator cycle

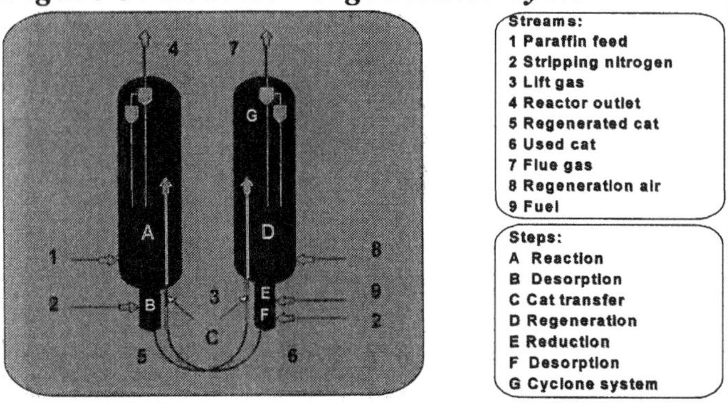

The insertion of horizontal baffles in a fluidized bed hinders gas backmixing (roughly exponentially with the inverse of their fractional open area). A low open area is necessary to approach closely a plug-flow of the gas phase. For instance, an open area of 50% results in halving the degree of gas backmixing compared to a freely bubbling bed, while an open area of 20% decreases gas backmixing by a factor of 20 [7]. On the other hand restrictions of the free solid and gas flows determines the approach to flooding. Therefore the final choice of number and fractional open area of baffles must be a compromise.
The control of backmixing affects also the temperature profile: Figure 4 shows equilibrium conversion versus temperature for the dehydrogenation of isobutane. The other curves show conceptual profiles of conversion vs. temperature that can be established in the reactor following the path of the reacting gases.

Figure 4: Conceptual temperature profiles

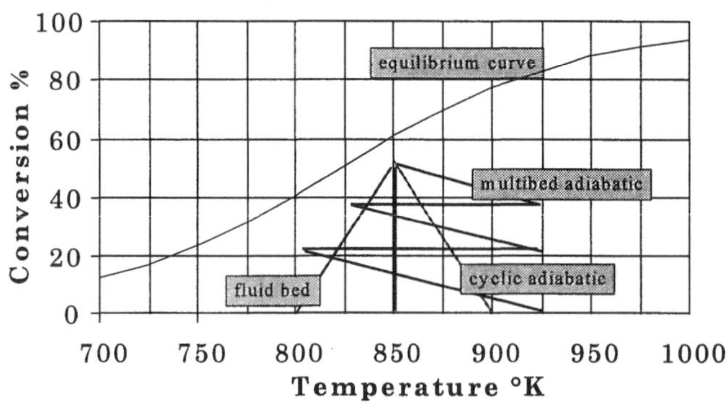

A profile with gas temperature increasing along the reactor is the most desirable, because it allows to approach the equilibrium at higher temperature maximizing the yield, while the mean reaction temperature is kept acceptably low to minimize the formation of by-products. To reach the same conversion to isobutylene with an isothermal or, even worse, decreasing temperature profile, would require a much higher mean reaction temperature, increasing the rate of thermal cracking reactions. The fluid bed operation, with a higher exit temperature vs the inlet one, allows the use of a feed effluent heat exchanger, avoiding feed preheating furnaces (cost, production of NO_x, risk of coking).

6. THE PROCESS SCHEME.

Fresh feed of hydrocarbons mixed with a recycle from an olefins user unit (P-P splitter for C_3, MTBE for C_4 e.g.) is vaporized, preheated in a gas-gas feed-effluent heat exchanger and fed to the reactor on the bottom of the catalyst bed through a distributor whose pressure drop is able to ensure a perfect distribution of the gas in the whole cross section.

Paraffins are dehydrogenated according to thermodynamic equilibrium with very high selectivity (>90% mol.) to the corresponding olefins and leave the reactor after separation of the entrained catalyst. Reaction products are purified from the entrained catalyst powder (cyclones and scrubber), it is sent to compression and separation sections to recover the C_4 stream from hydrogen and by-products.

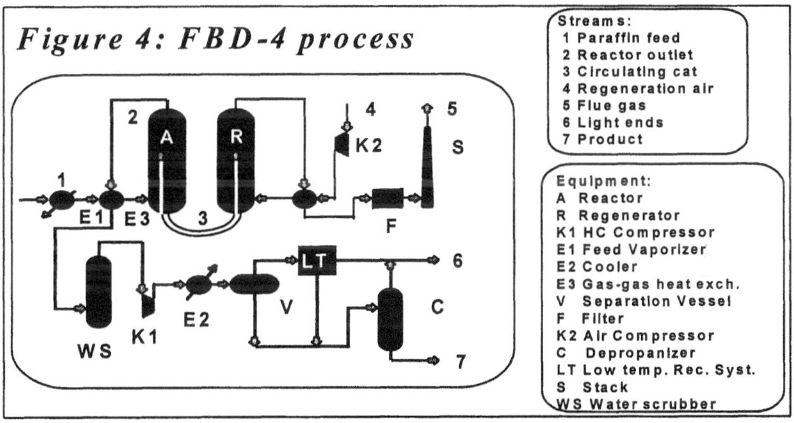

Figure 4: FBD-4 process

Very high efficiency cyclones and filters, the best available technology, clean from the dust the flue gas upstream of the stack up to a maximum increase of chromium (active component) concentration at ground level lower than the values of Canadian Arctic [8].

7. THE CATALYST

The FBD processes use a chromia/alumina catalyst with alkaline promoter [9]. The catalyst is produced in a form suitable to be used in a fluidized bed system. It has a microspheroidal shape with average diameter less than 10^{-4} m and density less than 2000 kg/m^3, typical of 'A' group according to Geldart [10], and it is very similar to a FCC catalyst. Conversion of isobutane around 50 %, with molar selectivity to isobutylene higher than 90 % are obtained.

The dehydrogenation capability of chromia-alumina systems is well known since a long time [11]. The active sites have been assumed to be Cr^{3+} [12-14] , both Cr^{2+} and Cr^{3+} [15], or coordinatively unsaturated Cr^{2+} [16] by expansion of the

coordination number to six by chemisorbing a paraffin molecule. The redox atmospheres and the relevant residence time that the catalyst meets in its cycle become therefore very important. Mechanisms probably occur to an extent that depends on the operative conditions and the detailed catalyst formulation.
The catalyst design takes into consideration other noticeable roles played by the carrier (transition alumina mainly) and promoters: mechanical properties regulation, control of the Cr oxidation state during catalyst regeneration, reduction of the tendency of chromium to enter into the alumina framework (chemical death of the catalyst), tuning of the catalyst surface acidity to inhibit cracking, skeletal isomerization and oligomerization reactions. Alkali affect the oxidation state of chromium as well; while a low alkaline content improves formation of mixed-valence chromium oxide, which are the precursors of the active center, higher contents lead to formation of well distinct alkali chromates and crystalline chromia, thus reducing the dehydrogenating activity [17].

8. CONCLUSIONS

In all chemical processes it is of course very important to closely integrate the catalytic system and the reactor engineering and this is particularly true for the dehydrogenation reaction, where several choices are possible, but the thermodynamic limitations force on the one hand the catalyst design in a very narrow window of characteristics and performances and on the other the reactor design to answer to very demanding operating conditions. The fluidized bed choice of FBD, in which catalyst and reactor design benefit from reciprocal advantages, has found the most adequate technological solutions.

REFERENCES

1. MTBE and Oxygenates, DeWitt & Co. Inc. (1997)
2. Parpinelli-TECNON: World Petrochemical Industry 1997. Olefins (1997)
3. D. Sanfilippo, F. Buonomo, F. Trifirò, in G. Ertl, H. Knotzinger, J. Weitkamp edts. Handbook of heterogeneous catalysis, 5, chap. 4.3.1, 2140-51, VCH (1997)
4. F. Trifirò, F. Cavani, Catalytica Studies Div., Study # 4192 OD, (1993)
5. D. Sanfilippo, Chemtech 35-9, Aug. 1993
6. D. Sanfilippo, F. Buonomo, G. Fusco, M. Lupieri, I. Miracca, Chem. Eng. Sci., 47 (9-11) 2313 (1992)
7. Overcashier, Todd, Olney, AIChE J. 5, (1), 54-60 (1959)
8. Syracuse Res. Corp. Contract 68-C8-0004 ATSDR Public Health Service U.S. EPA, Oak Ridge National Laboratory, (1989)
9. F. Buonomo, R. Iezzi, B. Notari, G. Kotelnikov, R. Michailov, A. Patanov, USP 4,746,643, to Snamprogetti (1985)
10. D. Geldart, Powder Technol., 1, 285, (1973)
11. P.H. Emmett Catalysis, 3 chap. 10, Reinhold Pub. Corp. (1955).
12 W. Grunert, W. Saffert, R. Feldhaus, K. Anders, J. Catal. 99 149 (1986)
13. C.H. Marcilly, B. Delmon, J. Catal. 24, 336 (1972).
14. P. Konig, P. Tetenyi, Acta Chim. Acad. Sci. Hung. 88, 123 (1976)
15. F.M. Ashmawy, J. Chem. Soc. Faraday Trans., 76, 2096 (1980)
16. H.J. Lugo, J.H.Lunsford, J.Catal., 91, 155 (1985)
17. F. Cavani et al. J. Catal. 158, 236-250 (1996)

NATURAL GAS CONVERSION V
Studies in Surface Science and Catalysis, Vol. 119
A. Parmaliana et al. (Editors)

Oxidation of n-butane and n-pentane over V/P/O-based catalysts: comparison between fresh and "equilibrated" catalysts

C. Cabello[1], F. Cavani, S. Ligi and F. Trifirò
Dipartimento di Chimica Industriale e dei Materiali, Viale Risorgimento 4, 40136 Bologna, Italy
[1]On leave from CINDECA, La Plata, Argentina

The catalytic performance of V/P/O-based catalysts in the selective oxidation of n-butane to maleic anhydride and of n-pentane to maleic and phthalic anhydrides is examined as a function of the time-on-stream. The fresh catalyst is made of a largely amorphous compound, containing relatively small amounts of $(VO)_2P_2O_7$ and large amounts of V^{III} ions, formed as a consequence of the calcination treatment. With increasing time-on-stream, the formation of well crystallized $(VO)_2P_2O_7$ is accompanied by considerable modifications in the catalytic performance.

Introduction

The oxidation of n-butane to maleic anhydride is catalyzed by a V/P mixed oxide. The same compound also efficiently catalyzes the oxidation of n-pentane to maleic and phthalic anhydride [1]. The main component of the catalyst composition is vanadyl pyrophosphate, $(VO)_2P_2O_7$, but the catalytic performance both in terms of activity and of selectivity is considerably affected by the chemical physical properties of the active phase. In particular, the following characteristics are considered fundamental in influencing the initial reactivity towards n-butane and n-pentane: i) the presence of additional oxidized or reduced phases under reaction conditions, namely V^V/P/O (typically $VOPO_4$), or V^{III}/P/O, either crystalline or amorphous; and ii) the crystallinity of the vanadyl pyrophosphate, which in turn affects the specific surface area and the preferential exposure of particular crystallographic planes. Even though the initial properties of such "nonequilibrated" catalysts can differ considerably, under reaction conditions all of them tend to evolve towards the formation of an "equilibrated" catalyst, which exclusively contains well crystallized $(VO)_2P_2O_7$ (a V^{IV} phase).

The objective of the work reported here was to compare the catalytic properties of nonequilibrated and equilibrated V/P/O-catalysts in the oxidation of n-butane and n-pentane, in order to obtain information on how surface and bulk properties may affect the catalytic performance. In particular, a catalyst was prepared characterized by the presence of both $(VO)_2P_2O_7$ and an amorphous V^{III} phase. Under reaction conditions typical of laboratory reactors this compound undergoes a slow transformation towards the formation of a well-crystallized vanadyl pyrophosphate, and the catalytic activity is considerably modified with time-on-stream.

Results and discussion

Catalysts were prepared using the classic procedure in an organic medium. V_2O_5 was reduced by means of isobutanol in the presence of H_3PO_4 and the solid obtained was then filtered and dried, with formation of the catalyst precursor, $VOHPO_4{\cdot}0.5H_2O$. The precursor was heated at T > 400°C in a nitrogen flow. This treatment leads to a compound (*calcined catalyst*) which is in part amorphous (see XRD pattern in Figure 1), and contains $(VO)_2P_2O_7$

as well as an additional nonidentified crystalline phase. The catalyst obtained was then sieved into 0.5 mm particles, and loaded (3 g) into a laboratory flow microreactor, operating at atmospheric pressure. The catalyst was left to age in a flow of 1.7 % n-butane in air, at 340°C (W/F 3 g s / ml). From time to time the activity of the catalyst was tested in n-butane and in n-pentane oxidation. After 100 h of reaction the catalyst has been characterized (*sample 1*, nonequilibrated catalyst), and left again in the reaction environment up to 400 h of rèaction (*sample 2*, nonequilibrated catalyst), and then again up to 800 h of reaction (*sample 3*, almost equilibrated catalyst), and finally up to 1200 h (*sample 4*, equilibrated catalyst), after which the catalyst showed a stable catalytic behaviour. The specific surface area of the samples was 24 m^2/g (calcined sample), 24 m^2/g (sample 1), 17 m^2/g (sample 2), 20 m^2/g (sample 3) and 23 m^2/g (sample 4). The XRD patterns of samples 1-4 (after reaction) are reported in Figure 1. Sample 1 is still identical to the catalyst before reaction, while samples 2, 3 and 4 exhibit the XRD pattern typical of $(VO)_2P_2O_7$. A slight progressive increase in crystallinity is observed in samples 2, 3 and 4, in correspondence with an increase in the time-on-stream.

UV-Vis DRS spectra of the four samples are shown in Figure 2; all samples exhibit absorptions typical of octahedral V^{IV}, and moreover a band at around 480-500 nm which corresponds to a d-d transition of octahedral V^{III}. No bands are observed which can be attributed to V^{V} ions. The intensity of the band relative to V^{III} decreases progressively from the calcined sample to sample 4. In sample 4 this band is very weak but still visible, and remains unchanged for times-on-stream longer than 1200 h.

These results indicate that the ageing of the catalyst in the reaction medium leads to the following modifications: i) transformation of amorphous phases and of crystalline phases other than $(VO)_2P_2O_7$ to vanadyl pyrophosphate, and increase in the crystallinity of the latter, and ii) increase in the average oxidation state of vanadium, with transformation of V^{III} ions to V^{IV}.

The absence of V^{V} ions in the unloaded catalysts, as evidenced by both chemical analysis and UV-Vis DRS, does not exclude the possibility that V^{V} species develop on the catalyst surface during the redox cycles involved in paraffin oxidation.

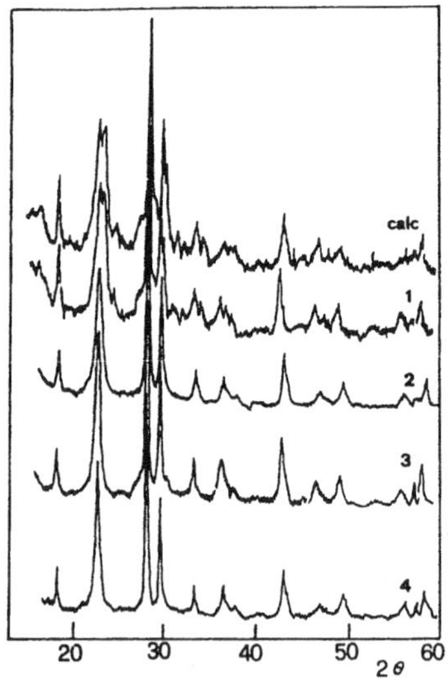

Figure 1. XRD patterns of calcined catalyst, and of samples 1 to 4.

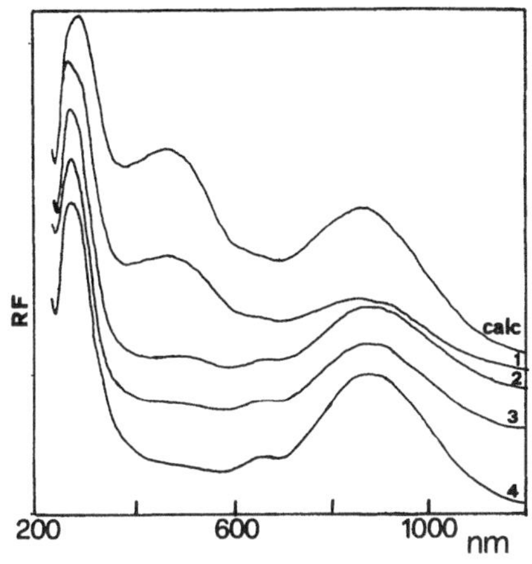

Figure 2. UV-Vis-DRS spectra of calcined catalyst and of samples 1 to 4

Figure 3 reports the conversion of n-butane as a function of temperature for samples 1 to 4, and the selectivity to maleic anhydride as a function of temperature (other conditions: W/F 3 g s/ml, feedstock composition 1.7 % n-butane in air). Sample 1 was very active; however, after a few hundred hours of reaction (sample 2) the activity fell considerably, and then increased with further increasing time-on-stream (samples 3 and 4). The selectivity trend of samples show differences which however are likely due to the very different activity of the catalysts. At low temperature (320°C) all catalysts had similar selectivity to maleic anhydride (around 70%). On increasing the temperature, the selectivity was initially constant for all samples, and then it decreased. However, the temperature at which the fall of selectivity begins was the lowest for the most active sample (already at 320°C for sample 1, at 30% n-butane conversion), while was the highest for the least active sample (approx. 390°C for sample 2, at 45-50% n-butane conversion).

The results obtained in the oxidation of n-pentane are shown in Figure 4, which reports n-pentane conversion and selectivity to maleic anhydride and to phthalic anhydride as functions of the residence time. Once again, an initial fall in activity was observed with increasing time-on-stream (samples 2 was much less active than sample 1), which was then followed by a progressive increase in conversion. The distribution of the products was affected considerably by the modification of the catalyst properties. In sample 1 the prevailing product was maleic anhydride, while the selectivity to phthalic anhydride was low; in sample 2 the two anhydrides were formed in comparable amounts, and in samples 3 and 4 phthalic anhydride was formed with higher selectivity than maleic anhydride. Another difference concerns the stability of products. In sample 1 both the phthalic and maleic anhydride selectivities were substantially unaffected by n-pentane conversion (up to almost total paraffin conversion). In samples 2, 3 and 4 the selectivity to maleic anhydride was stable (in the range of conversions examined), while the selectivity to phthalic anhydride decreased with increasing conversion: the more active was the catalyst (sample 4), the more relevant was the decrease in selectivity.

The results for n-butane and n-pentane oxidation are summarized in Figure 5. Figure 5a reports the n-butane conversion at 340°C and W/F 3 s g/ml and the selectivity at 30% n-butane conversion plotted vs. time-on-stream. Figure 5b reports the n-pentane conversion at 340°C and W/F 3 s g/ml and the selectivity to maleic and phthalic anhydrides (at approximately 20% n-pentane conversion) as functions of time-on-stream. The selectivity to maleic anhydride from n-butane at fixed conversion was not much affected by time-on-stream. This means that variations in chemical physical features do not considerably affect the ratio between the rates of the parallel reactions of n-butane transformation to maleic anhydride and to CO_x, (which determines the initial selectivity), while they do affect the catalyst activity and the rate of the consecutive reaction of maleic anhydride combustion. From n-pentane instead the initial selectivity was greatly affected by time-on-stream. This means that modifications in chemical physical properties considerably affect the ratio between the parallel reactions of maleic anhydride and phthalic anhydride formation from n-pentane. The overall selectivity to maleic + phthalic anhydride was the highest for sample 2, that is for the sample characterized by the presence of crystalline vanadyl pyrophosphate as well as by a considerable amount of surface V^{III}.

The data reported indicate that the catalytic performance in n-butane and in n-pentane oxidation is affected by modifications which occur in the catalyst during activation inside the reactor. During the first 100 h reaction the catalyst (sample 1) does not undergo considerable changes of the crystallinity (as evidenced by XRD patterns) with respect to the calcined catalyst. The main modification concerns the surface average oxidation degree, which is considerably increased as a consequence of the decrease of the surface concentration of V^{III} ions.

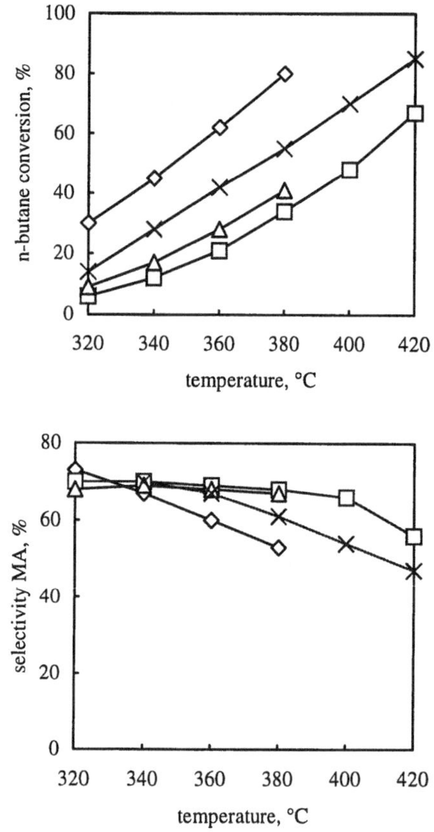

Figure 3. n-butane conversion (top) and selectivity to maleic anhydride (bottom) as functions of the reaction temperature. Catalysts: sample 1(◇), 2 (□), 3 (△) and 4(✕).

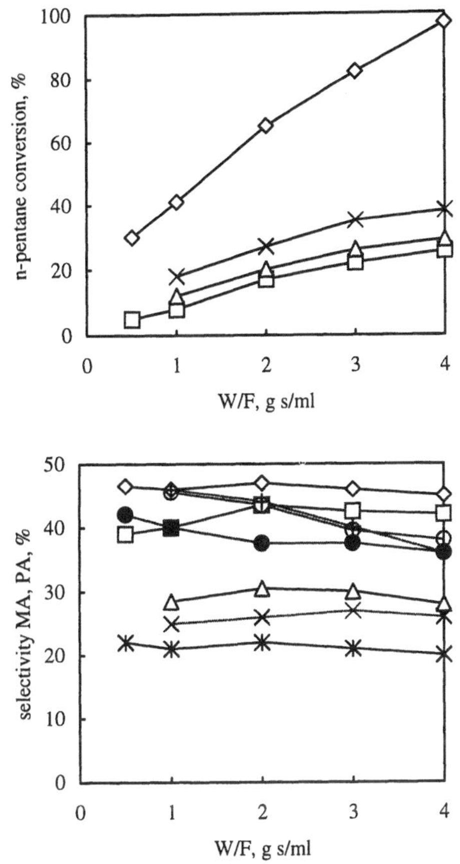

Figure 4. n-pentane conversion (top) and selectivity to maleic and phthalic anhydrides (bottom) for samples 1 (◊,◊,✱), 2 (□,□,●), 3 (△,△,O) and 4(✖,✖,+).

After 400 h time-on-stream (sample 2) changes involve i) the oxidation of a large part of surface V^{III} ions, as evidenced by the comparison of UV-Vis DRS spectra, ii) and the transformation of amorphous and crystalline phases to well-crystallized $(VO)_2P_2O_7$. All this occurs with a contemporaneous fall in activity; this suggests that i) the amorphous V^{III}/P/O phase is unstable in the reaction environment, and that ii) the presence of V^{III} species creates very active sites for paraffin oxidation. The disappearance of these sites causes the decrease of activity. In a previous study an equilibrated catalyst was subjected to a reducing treatment with n-butane [3]. In agreement with the results reported here a temporary increase in activity was observed, and the selectivity to maleic anhydride exhibited a trend analogous to that of sample 1. This confirms the role of V^{III} ions developed on the surface of $(VO)_2P_2O_7$. This species generates centres which are very active, but the catalyst becomes nonselective with increasing reaction temperature.

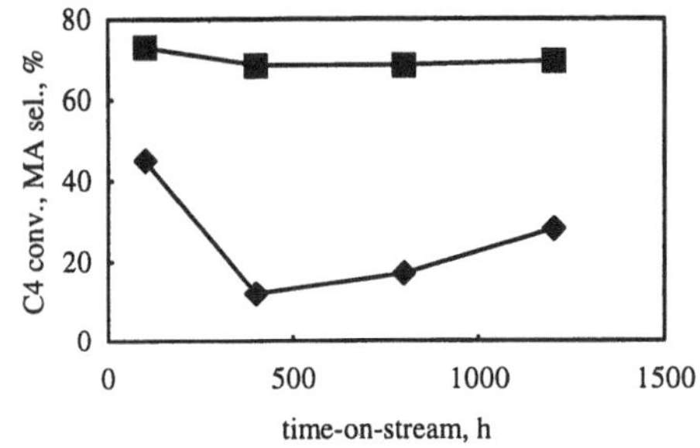

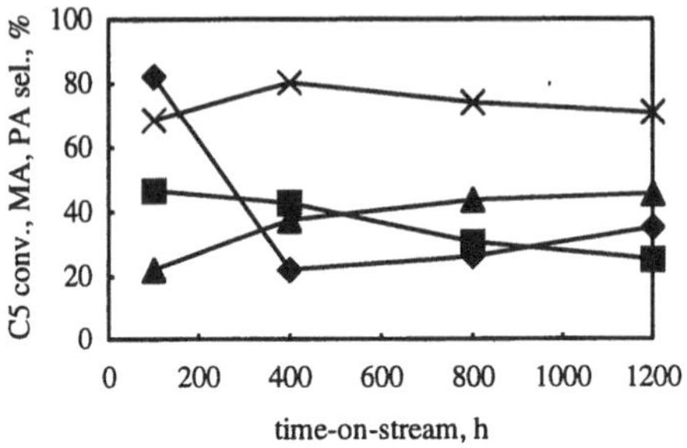

Figure 5. (a: left): n-butane conversion at 340°C (◆) and selectivity to maleic anhydride (■) (at 30% n-butane conversion) as functions of time-on-stream. (b: right): n-pentane conversion (◆) and selectivity to maleic (■), to phthalic (▲) anhydrides and to maleic + phthalic anhydride (✕) (at 20% conversion) as functions of time-on-stream.

From n-pentane, maleic anhydride was formed rather than phthalic anhydride on sample 1. At 340°C these two products were stable, and did not undergo successive reactions of combustion. Since the two anhydrides are formed via parallel reactions starting from a common intermediate (adsorbed pentadiene), and since the formation of phthalic anhydride possibly involves the contribution of surface acid centres [2], the results indicate that sample 1 possesses a low number of sites capable of catalyzing this transformation. The oxidation of V^{III} to V^{IV} (sample 2) leads to a modification of surface properties (availability of acid centres and of O^{2-}-insertion centres), with a modification of the distribution of the products obtained from n-pentane. On further prolonging time-on-stream (samples 3 and 4) further changes include a slow increase in the crystallinity of $(VO)_2P_2O_7$, a further decrease in the surface concentration of V^{III} species, and an increase in the specific surface area. An increase in the phthalic-to-maleic anhydride ratio from n-pentane is observed. The equilibrated catalyst is a well crystallized vanadyl pyrophosphate, containing exclusively V^{IV} and V^{III} ions at the surface. These results confirm the importance of having well crystallized $(VO)_2P_2O_7$ to achieve as high as possible selectivity to phthalic anhydride [4,5].

ACKNOWLEDGEMENTS. Lonza is gratefully acknowledged for financial support.

REFERENCES

1) F.Cavani and F. Trifirò, in "Catalysis Vol. 11", Royal Society of Chemistry, p. 246
2) F. Cavani, A. Colombo, F. Giuntoli, E. Gobbi, F. Trifirò and P. Vazquez, Catal. Today, 32 (1996) 125
3) F. Cavani, A. Colombo, F. Giuntoli, F. Trifirò, P. Vazquez and P. Venturoli, in "Advanced Catalysts and Nanostructured Materials", W.R. Moser (Ed.), Academic Press, 1996, p. 43
4) G. Calestani, F. Cavani, A. Duran, G. Mazzoni, G. Stefani, F. Trifirò and P. Venturoli, in "Science and Technology in Catalysis, 1994", (Y. Izumi, H. Arai, M. Iwamoto, Eds.), Kodansha Ltd., Tokyo, Elsevier, Amsterdam, 1995, p. 179
5) Z. Sobalik, S. Gonzalez, P. Ruiz and B. Delmon, in "3rd World Congress on Oxidation Catalysis", (R.K. Grasselli, S.T. Oyama, A.M. Gaffeny and J.E. Lyons, Eds.), Elsevier, Amsterdam, 1997, p. 1213

NATURAL GAS CONVERSION V
Studies in Surface Science and Catalysis, Vol. 119
A. Parmaliana et al. (Editors)

Economic Gas to Liquids: A New Tool for the Energy Industry

Mark A. Agee

President and Chief Operating Officer, Syntroleum Corporation, 1350 South Boulder, Tulsa, OK 74119, USA

Recent technological advances within the energy industry have made dramatic improvements in lowering the cost of finding, producing and refining oil. As wonderful as these advances have been, the industry has left behind a vast storehouse of energy in the form of remote and substandard gas reserves. These stranded reserves are largely unmarketable because of the prohibitive costs of transportation to market. And they account for between half and two-thirds of all known gas in the world. By some estimates, as much as 900 trillion cubic feet of gas (25 trillion cubic meters) are beyond the economic reach of commercial markets as gas or LNG.[1] Frequently they also trap underlying oil reserves.

With this perspective, consider the litany of problems currently facing the energy industry: environmental mandates for cleaner fuels; raw materials getting heavier and dirtier; process upgrading investments more difficult to justify; tightening supplies for middle distillates and naphtha; surging demand for clean fuels for power generation; trouble in financing LNG mega-projects; and, as noted above, most of the reserves of the cleanest fuel—natural gas—too remote for economic delivery to markets.

GTL— THE NEW TOOL

Against this background, economical gas to liquids (GTL) technology is an event that has been waiting to happen. GTL is currently one of the more interesting topics in the energy industry and it is easy to understand why. GTL turns unmarketable reserves of natural gas into a form that overcomes prohibitive transportation costs, avoids long-term, high-risk, take-or-pay contracts, and helps satisfy the demand for cleaner liquid products. Economic GTL is tantamount to the discovery of several hundred billion barrels of new oil. No technology in the last 50 years will affect as many segments of the petroleum industry as GTL will. It offers a welcome new source of super-clean fuels and feedstocks for refiners, petrochemical companies, and power companies—all facing more stringent environmental requirements.

Transformed into liquids, stranded gas resources could have access to world commodity markets through the existing conventional oil transportation infrastructure. The old barriers of treatment and transportation costs would be neatly erased. At the same time, GTL would liberate remote gas reserves from the limitations of gas markets, which cannot begin to absorb the huge volume of stranded reserves in many regions. In the form of liquids, they could flow instead into the much larger oil market via a 100% compatible infrastructure. This will provide a welcome, cost-effective supplement to many problems of product mix, quality, and the environment with which industry now struggles.[2,3]

SAME CHEMISTRY - DIFFERENT APPROACHES

All current GTL processes start from the same point, the chemistry discovered in 1923 by the German scientists Franz Fischer and Hans Tropsch. However, the execution of that chemistry to create an economic process varies widely.

Shell and Exxon, two pioneers in GTL, took the conventional approach, depending on incremental improvements and economies of scale to achieve acceptable economics. Currently, this approach limits applications to large plants of 50,000 BPD and larger. Sasol has focused on larger sizes as well, but claims economic viability at capacities down to 10,000 BPD.

Syntroleum takes a somewhat different approach, in executing the chemistry, as well as how to distribute, support and improve the technology. As a technology provider to the industry, we view the needed solution as one that can be widely applied to the problems of oil companies with stranded reserves, large and small. In order for GTL technology to live up to its potential, however, it must overcome several barriers to extensive commercialization.

Cost is the most obvious barrier. It is generally accepted that capital costs must be below $30,000 per barrel of daily capacity for GTL to be economical. Although several companies, including Syntroleum, claim ability to break this barrier, it has yet to be demonstrated in any commercial scale plant. We believe that it will be demonstrated by the end of 1999. The ultimate goal is to lower capital costs to $12,000 per barrel of daily capacity, roughly the same cost as a conventional world scale refinery.

Size is the next barrier to widespread application (Figure #1). Shut-in fields come in all sizes, from small but worthwhile to super-giant. The range of economic plant sizes available will determine the extent GTL will be used. A 50,000 BPD plant requires roughly 500 MMCFD of gas (180 BCF a year) or 5.4 trillion cubic feet over a 30-year life. That size plant could tap a significant amount of reserves but is limited to a small number of possible fields—only 2% of the 4,448 identified gas fields outside North America.[4] A plant size of 10,000 BPD, requiring 100 MMCFD of gas can be supported for 30 years in roughly 7% of the world's known gas fields. At 2,000 BPD, more than half the world's gas fields, containing more than 95% of the world's gas, could be monetized.

Portability is the third barrier and a crucial one. Gas pipelines can be built only where reserves are adequate to amortize their cost over the projected life of a field. Once installed, a pipeline cannot be moved when the reserves have been depleted. The limitations inherent with pipelines thus leave enormous reserves of oil and gas out in the cold, i.e., remote oil fields with associated gas that cannot be economically marketed, and remote gas fields that are too risky for pipeline investment. Thus, for extensive application in isolated areas, especially involving associated gas that cannot be flared and smaller gas reserves, portability brings a new factor into the economics equation. GTL plants that can be mounted on barges or ships have this critical capability. They can be moved from one site to another as needed, giving operators the option of faster depletion rates, as well as repeat use of the asset. This flexibility for GTL technology opens up myriad applications without the financial and reserve risks associated with fixed, stationary pipelines. It enables the monetization of many smaller offshore fields without the need for long-term reserves. This factor is particularly important when you consider that roughly half of the world's stranded gas is located in water.

POSSIBILITIES FOR GTL

Economic GTL offers a new option for the energy industry with potential application to a range of situations where current technology falls short. Industry and producing countries can now re-evaluate investment strategies in exploration, production, refining and marketing, with GTL creating an entirely new set of opportunities:

Monetization of known stranded gas is the first and most obvious opportunity (Table #1). Exploitation of remote and substandard gas deposits, now shut in by the prohibitive costs of transportation and treatment, will add decades to the supply of clean fuels. It will help to fuel economic growth while meeting environmental requirements and, at the same time, transfer huge unrecovered exploration costs to booked assets of great value. This will be a big boost to the companies and countries that had the misfortune to find gas in remote locations where previously only oil could be produced commercially.

Table 1
Stranded Gas Fields Outside North America (Undeveloped, Non-Associated, Larger than 0.5 Trillion Cubic Feet)

Size of Reserves (TCF)	Number of Fields
5+	58
1 - 5	165
0.5 - 1	175
Total	398

Source: Petroconsultants S.A.

Early development of remote gas discoveries will now be possible, even routine. Previously, remote gas discoveries too distant from markets for economic reach by pipelines or LNG tanker were simply shut in. Absent construction of a local power plant or other gas-using facilities, they had zero market opportunity and zero value. Development was postponed indefinitely. Where resources fit GTL requirements, development will now move ahead similar to commercial-size oil discoveries. New oil deposits are developed routinely with the certain knowledge that commodity status guarantees market opportunity. With the advent of GTL, many remote gas deposits will enjoy similar advantages.

Early development of remote oil discoveries with associated gas will be an important benefit of GTL technology. Frequently these discoveries are shut in until a market can be found for the associated gas. If no market is available, the cost of reinjection is excessive, and flaring is prohibited, the oil remains undeveloped. In many instances, GTL will be the answer.

A new opportunity for exploiting large gas deposits already in production, with only a portion of their reserves dedicated to LNG or pipeline projects will be early candidates for GTL plants, just as remote, stranded reserves are. These large deposits—in the Middle East, Africa, Asia-Pacific, and the former Soviet Union—comprise the bulk of the world's current proved reserves. Most of these resources do not currently have firm additional markets as gas or LNG. But when the GTL option is introduced, the picture changes. First of all, these large fields have the reserves to support one or more synfuel plants in addition to existing LNG or pipeline projects. And there will be an immediate market for the high-quality syncrude and synfuels produced. Furthermore, GTL is an option that may be more attractive than an additional LNG plant. LNG has the edge in energy efficiency. GTL, however, offers significant offsets that will be attractive to reserves owners and customers alike. For example, front-end capital costs will be greatly reduced compared to LNG, which requires costly tankers and terminals. Also, financing will be easier since the liquid output can be readily sold in world commodity markets via conventional transportation, without rigid long-term contracts that are required for LNG. These factors will force consideration of GTL as a possibly superior method of monetizing natural gas compared to a greenfields LNG project. The LNG market, projected to double over the next 20 years, is still small in absolute terms. Its growth, though rapid, translates to only about 50,000 BPD annually of oil equivalent. In fact, the entire combined alternative gas markets—LNG, ammonia, and methanol—are only slightly above 2 million BPD, equating to no more than one and a half year's demand growth for oil products.

Re-evaluation of exploration strategy will be another result of the new technology. Explorationists may now consider gas prospects in locations that have been ruled out heretofore for lack of economic access to markets with previous technology. Good prospects in remote locations will now be upgraded in accordance with geologic merit as well as new economic considerations. This should lead to lower finding costs, higher overall success rates, and increased booked reserves.

GTL presents new investment options to countries with shut-in gas and heavy oil or tar sand reserves. Petroleum ministries and companies may now consider whether to exploit the gas instead of the heavy oil or tar sands. They may also decide to exploit the advantages of synergism by doing both together. Instead of making the required large processing investment to upgrade heavy crude to refinery-acceptable standards, they may be able to accomplish the same purpose by blending with high-quality syncrude made from natural gas. This approach will monetize more total resources. In many situations it will be more capital efficient and yield greater total returns.

GTL offers the opportunity for preemptive investment in advance of increasingly stringent environmental regulations to end or avoid flaring of natural gas. Where flaring is being phased out, or is likely to be proscribed, GTL provides a chance to comply with or get in ahead of new regulations, generating additional income at the same time. Some governments are already examining GTL technology within this context.

GTL promises to find extensive applications offshore and adjacent to coastal reserves. With half the world's gas reserves in water and significant additional amounts near water, offshore will be a major sphere of GTL operations. Offshore, GTL will have its usual big advantage—ready access to commodity markets for its output—*plus portability*. Where reserves are small or rapid depletion is indicated, a gas pipeline presents a big financial risk. When reserves are exhausted, you cannot pick up the pipeline and move it to the next location. But with a barge or ship-mounted GTL plant, you can do just that. This fact will create a large role for portable and platform-mounted GTL plants in a variety of situations. These will range from small plants servicing small reserves of gas-well gas or associated gas, to very large fields offshore and adjacent to coastlines. Portability and easy adaptability to offshore sites may turn out to be one of the most valuable features of this new technology. And the value will increase as exploration moves farther from shore and deeper into the oceans.

GTL presents multiple options for refiners facing heavier, higher-sulfur crude slates, rising middle distillate and naphtha demand, and more stringent product specifications. Demand for middle distillates is projected to grow by 3.1 million BPD over the next decade. In the case of naphtha, petrochemical demand for this basic feedstock is already outpacing byproduct supply from refiners. Projected growth already has led petrochemical companies to conclude that they will be forced to produce their own raw material. With existing and projected capacity, refiners are ill prepared to meet prospective future requirements, either as to volume or quality. Additional processing investment and higher crude runs could achieve compliance and meet demand—but at a cost the market may not justify.

GTL offers two options that may be superior. The first is to use GTL-based syncrude, with its high API gravity and highly paraffinic and contaminant-free nature, for upgrading the crude charge. This approach will yield lighter, lower-sulfur products across the board, aiding compliance with clean-air rules for diesel fuel and other products. The second is to use GTL-based synfuels to blend with existing product to stretch supplies and upgrade quality of finished products and feedstocks. These options are a function of the superior quality of GTL products, which are free of sulfur, metals, and aromatics.

THE OUTLOOK

These are just a few of the numerous new opportunities opened up by economical GTL. Oil, gas and petrochemical companies, examining their own project lists and strategies, will discover many others. In this context, GTL is a new tool whose innumerable applications will influence investment decisions from exploration all the way downstream.

Like most new technologies, GTL will undergo rapid improvement. Those responsible for the new processes will see to that. In order to facilitate rapid development, Syntroleum has striven to create an environment that encourages joint development and rapid implementation and sharing of improvements. For several years now, Syntroleum has cultivated relationships with companies who share our vision of how GTL technology should be developed and used. There are "grant-back" provisions in our license agreements with Texaco, ARCO, Marathon, YPF, Kerr-McGee and Enron, giving each licensee access to the improvements of other licensees. Syntroleum also has alliances and/or joint development agreements with companies such as Texaco, ARCO, Marathon, Enron, Criterion Catalyst, Lyondell Petrochemicals, Catalytica and ABB, Brown & Root, Bateman Engineering and AMEC—all focused on specific improvements to the Syntroleum Process. The power of this "*technology pool*" continues to grow. As a result of these growing relationships, the Syntroleum Process is the most widely supported GTL process available today. We believe that this strategy, along with a more flexible, lower cost process, insures that the Syntroleum Process will continue to improve rapidly and that our users will lead the GTL race now and in the future.

Momentum continues to grow. GTL has made a quantum leap in the short year since the energy industry heard the first announcements of commercial breakthrough. It has gone from curiosity and skepticism to serious consideration for a number of industry projects, which are now turning into actions. We anticipate that at least two GTL plants will be under construction within the next 12 months, plus the announcement of several others. The first plant should be up and running by late 1999. Given the traditionally conservative approach to new technology in the energy industry, that will be a truly significant achievement.

On the other hand, consider that emerging GTL technology provides a cost-effective technology that can be installed anywhere, on the scale required, and is portable too. Its products are 100% compatible with existing infrastructure, and are a perfect fit for the demand mix, quality and environmental requirements of a growing world commodity market. This picture is either too good to be true, or it is truly the stuff of which megatrends are made.[5] We believe the latter. And it appears that the industry does too.

---

[1] *Oil & Gas Journal,* June 19, 1995, p. 37.

[2] *Oil & Gas Journal Newsletter,* July 8, 1996.

[3] *Oil & Gas Journal Newsletter,* July 22, 1996.

[4] Ivanhoe, L. F., and George H. Leckie, *Oil & Gas Journal,* Feb. 15, 1993, p. 87.

[5] East-West Center Energy Advisory, Honolulu, "Economical Conversion of Natural Gas to Liquid Synthetic Fuels - the Next Megatrend?" January 1997

Figure #1

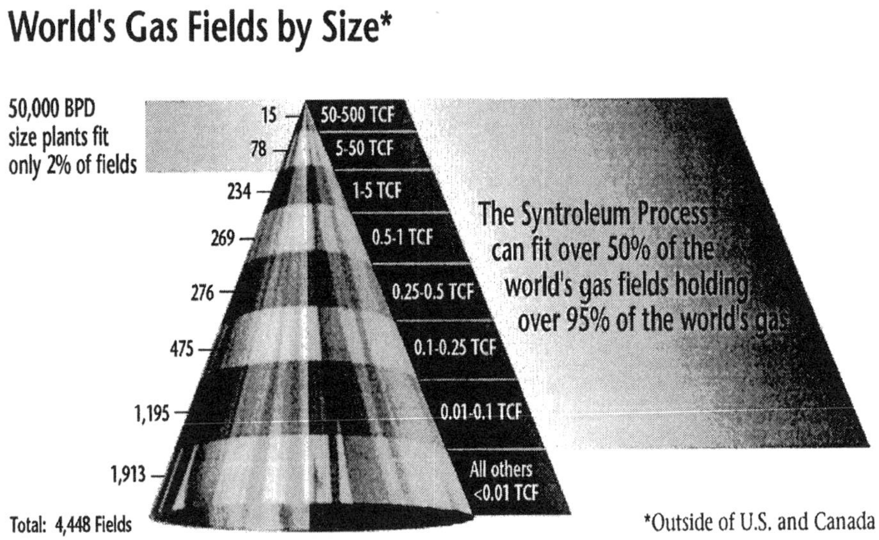

NATURAL GAS CONVERSION V
Studies in Surface Science and Catalysis, Vol. 119
A. Parmaliana et al. (Editors)

KTI's innovative reformer design for Hydrogen and Syngas Plant

F. Giacobbe and O. Loiacono, KTI SpA - Rome, Italy

1. Introduction

The hydrogen and syngas market is in a continuous grow due to the increasing demand by the chemical, petrochemical, metallurgical and the electronic industries and by the petroleum refineries.

More recently there has been a surge of interest in the conversion of natural gas to high value liquid products ranging from heavy lubes and waxes to premium transportation fuels.

The technologies available for hydrogen and syngas production are :

- steam reforming of hydrocarbons
- non catalytic partial oxidation
- catalytic partial oxidation or autothermal reforming

Partial oxidation and autothermal reforming are based on the use of oxygen which could be replaced by the air in future when the ceramic membrane reactors will be technically and economically viables. However at the present the use of oxygen poses safety problems in addition to its cost.

The partial oxidation and the autothermal reforming are actractive especially for syngas, when oxygen can be available at low cost or when are used in combination with steam reforming.

Steam reforming is the most mature and predominant technology and the 95% of current hydrogen and syngas production is based on the steam reforming of hydrocarbons.

In response to market increased demand, KTI is introducing technology innovations which permit to decrease operating and capital cost of the hydrogen/syngas production units.

These innovations, which will be illustrated and discussed in this paper, resulted in the development of the HER (Heat Exchanger Reformer) and of the ART (Advanced Reforming Technology) and are focused in the reformer area which is the heart of the hydrogen and syngas plant.

2. Limits of current Steam Reforming Technology

The steam reforming technology process for hydrogen and syngas production is a threshold problem in terms of pinch analysis. In other words in the steam reforming process is possible to close the heat balance by only supplying heat to the process and a process sink is not required.

However because of a design constraints it is not possible to use the reformer effluent stream to supply part of the heat required for the steam reforming reaction. This calls for an extra heat input to the process that then must be rejected to an external sink generally represented by the steam export from the plant and by cooling water consumption. In this situation the steam export must be maximized to increase the efficiency of the plant.

The above mentioned constraint has the following consequences on plant design:

– extra fuel is fired in the steam reforming furnace increasing the waste heat to be recovered from the flue gases;

– the extra waste heat recovery calls for additional investment on the steam generation facilities.

KTI has developed and patented innovative designs for steam reforming units which remove the above design constrains and decrease significantly the production cost of hydrogen and syngas. These innovative designs are based on the use of regenerative catalysts tubes. The concept of regenerative catalyst tubes is illustrated in figure N.1.

3. The Innovative KTI Exchanger Reformer

The KTI Heat Exchanger Reformer (H.E.R.) is a regenerative reformer patented in Italy, Europe, Germany, Netherlands, France, Belgium, UK, Japan and USA. The KTI Heat Exchanger Reformer differs from the other regenerative reformers because of the presence of a metal sleeve surrounding the catalyst tubes as illustrated in figure 2.

The flue gas flows downward from the burners to the bottom of the metal sleeve, then it flows upwards inside the sleeve and in contact with the catalyst tube.

The heat is transferred from the flue gas outside the sleeve to the flue gas inside the sleeve and from the flue gas inside the sleeve to the catalyst tubes.

This arrangement permits:

- To control the maximum skin tube temperature by controlling the temperature of the flue gas entering the sleeve (inversion temperature).
- To maintain the skin tube temperature, along the tube immerged in the sleeve, close to the bottom maximum value with consequent increase of the heat flux and of the H2 production.

Additional features include:
- absence of flame impingement on the catalyst tubes
- the high temperature outlet system is eliminated.
- the size of the radiant box is decreased.

4. The operation of the demo unit and the design of the commercial unit.

In 1993 a prototype unit has been put in operation in an Air Products facility located at Rozemburg - Netherlands (see Figure 3).

The prototype unit was sized for a net hydrogen production of 100 Nm3/hr, which corresponds to the smallest industrial size.

A test program has been conducted from March 1993 since December 1993 which has proved the viability and reliability of the Heat Exchanger Reformer (HER) as a stand-alone unit.

Two plants incorporating the H.E.R. technology are presently under construction. The plant is illustrated in the Figure 4.

The process flow includes the following design features:

The expensive quench boiler installed in the standard H2 plant is replaced by a kettle design to work as a steam drum connected to the convection section of the Heat Exchanger Reformer.

The LP steam required to deaerate the boiler feed water is generated in the deaerator itself making use of the heat supplied by the CO shift converter effluent. This arrangement permits to decrease the consumption of cooling water.

In figure 5 a 3D picture of the plant layout is shown.

The plant is modularized and it can be totally prefabricated in the shop and to be transported to the site on skids which can be assembled in a couple of weeks.

5. The KTI ART (Advanced Reforming Technology) Reformer

The Heat Exchanger Reformer makes use of a large diameter catalyst tubes with an internal diameter of 6-8 inches. Because of this large diameter the hydrogen or the syngas production pressure is practically limited to 18 Barg. This pressure is quite satisfactory for most of the small-medium size hydrogen users but for a large hydrogen plant for refineries it is mandatory to produce the hyrogen at pressure higher than 18 Barg. This calls for the use of the standard 4 inches catalyst tubes.

KTI has patented a new riser with a large heat exchange surface and a small cross sectional area which can be installed in the standard 4 inches catalyst tubes.

This innovation offer the possibility to revamp existing units or to design new units with lower capital expenditures and higher energy efficiency.

One of these unique regenerative catalyst tubes will be installed shortly in an existing reformer furnace for testing its performance.

This new regenerative catalyst tube has been patented in Italy, Europe, Germany, Netherlands, France, Belgium, UK, Japan and USA.

The use of KTI's ART regenerative catalyst tubes introduces step-out changes in the hydrogen/syngas process flow scheme for the following reasons:

- the ART regenerative catalyst tubes incorporates also the reforming feed preheater and the reformer effluent quench cooler;
- the reformer inlet and the reformer outlet nozzles are close and at the same elevation as for the catalyst tubes of the H.E.R.;
- the high temperature bottom transfer line, the high temperature pigtails,
- the reformer feed preheater and the quench boiler are eliminated;
- the hydrogen production of the reformer is increased with consequent decrease of the reformer box size;
- the high reformer efficiency permit a saving of fuel
- the excess steam production is drastically reduced.

6. Conclusions

KTI's know-how and market position in hydrogen/syngas technology represents the result of an extensive development effort spent in the last ten years. The innovative concepts introduced by KTI combine technical know-how and operational experience in H2/Syngas production.

These innovations have been conceived to produce hydrogen/syngas at low cost over a wide range of plant capacities.

FIGURE 1 - THE REGENERATIVE CATALYST TUBE

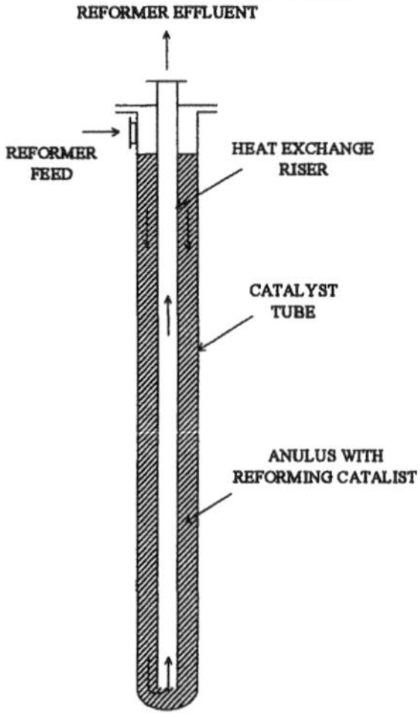

FIGURE 4

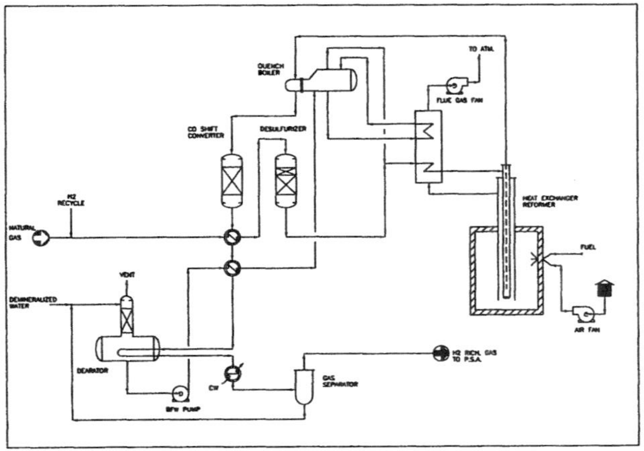

Process Flow Scheme incorporating the H.E.R.

FIGURE 2

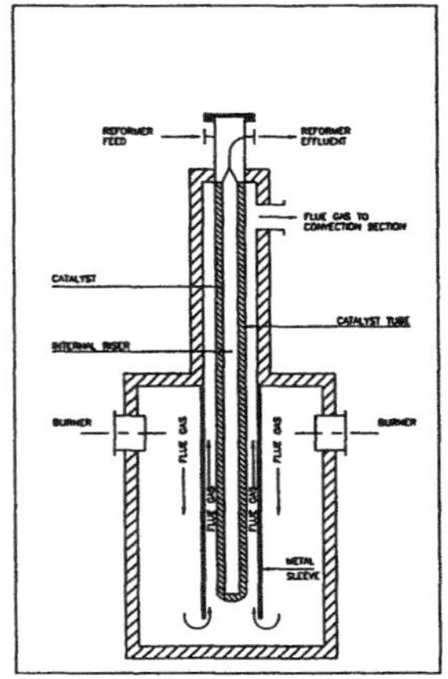

Heat Exchanger Reformer General Assembly

ATTACHMENT N°:
OF REQUISITION N°

INTERNAL REVISION
A

GBE-0520-521003-0 NUUU

FIGURE 3 - THE DEMO UNIT

FIGURE 5 - 3D PICTURE FOR THE PLANT USING H.E.R.

NATURAL GAS CONVERSION V
Studies in Surface Science and Catalysis, Vol. 119
A. Parmaliana et al. (Editors)

EXXON'S ADVANCED GAS-TO-LIQUIDS TECHNOLOGY

B. Eisenberg and R. A. Fiato *
Exxon Research and Engineering Company
P.O. Box 101, Florham Park, New Jersey 07932

C. H. Mauldin and G. R. Say
Exxon Research and Development Laboratories
P. 0. Box 2226, Baton Rouge, Louisiana 70821

S. L. Soled
Exxon Research and Engineering Company
Rt. 22 East, Clinton Township, Annandale, New Jersey 08801

Technology for conversion of natural gas to liquid hydrocarbons is emerging as an attractive alternative for bringing large, static gas resources to market. Exxon's AGC-21** advanced gas conversion technology utilizes new reactors and novel catalyst systems to produce high quality, highly paraffinic water-clear liquids, free of sulfur, metals, nitrogen, asphaltenes, and salt. A high capacity, fluid bed synthesis gas generation system and high productivity, cobalt-based, three-phase slurry hydrocarbon synthesis have been demonstrated in large process development units. These processes link with a flexible product upgrading technology in a three-step configuration for high capacity gas-to-liquids conversion. Protected by extensive patents, this unique, proprietary technology is being studied for a range of commercial applications.

INTRODUCTION

Exxon's Advanced Gas-to-Liquids process AGC-21 is the result of a $300 million research and development effort that has been underway for almost 20 years. Over the last several years, Exxon has established a strong proprietary position and holds over 1500 patents worldwide on various aspects of the process. Recent studies have shown that Exxon's fully integrated three-step process is the clear industry leader with process-costs less per barrel than those of the closest competitors. Indeed, improvements to catalysts and reactor designs have reduced the company's cost of gas conversion by 60%, down to levels that are now competitive with crude oil.

---

* Author to whom correspondence should be addressed.
** AGC-21 is a proprietary mark of Exxon Corporation.

The AGC-21 process consists of three distinct steps for transforming natural gas and other light hydrocarbons into high quality liquids. In the first step, synthesis gas is generated by contacting methane with limited amounts of steam and oxygen in a catalyzed fluid-bed reactor.

Hydrocarbons are synthesized in the second step operating in a novel slurry reactor using new, high productivity cobalt catalysts operating at high syngas conversion levels for wax synthesis. The full range, primarily normal-paraffin, product contains substantial levels of a 650oF+ boiling material which is solid at ambient conditions, has a melting point above 250oF, and is unsuitable for transporting via conventional pipeline of crude carriers.

The final step in AGC-21 is accomplished with proprietary catalysts in a fixed-bed or trickle-bed reactor with mild isomerization of the normal paraffins to slightly branched analogs that make excellent feeds for refineries of chemical plants.

Advances that have been developed in AGC-21 include high performance catalysts in all process steps, high capacity reactor systems, heat and steam integration enhancements, and tailored chemistry for high quality synthetic crude, fuels and specialties production.

RESULTS AND DISCUSSION

The first step in AGC-21 involves syngas production in single-step fluid-bed reactor to produce the optimum H2/CO ratio for hydrocarbon synthesis. The novel reactor system is the result of Exxon's decades long experience with fluid-solids technology. The single-step FBSG system avoids the extreme temperatures of autothermal or partial oxidation reactor systems, with temperatures well within the experience range of a flexicoker gasifier reactor. Exxon's FBSG process has been demonstrated in over three years of PDU operation. Fluid-bed reactors were chosen because they are especially suited to scaleup to large efficient sizes, and they are amenable to the combined partial oxidation and reforming chemistries required to convert methane to syngas.

Confidence for the commercial scale-up of FBSG comes from Exxon's extensive experience in worldwide refining applications, including our proprietary Fluid Cat Cracking and Flexicoking facilities. The Amuay Venezuela Flexicoker, the world's largest reactor with a diameter of 70 feet, was designed by Exxon. Exxon's FBSG reactors will be about one-half that size.

The second step involves hydrocarbon synthesis (HCS) for production of wax from syngas in a three-phase slurry reactor. The cobalt based catalysts and the slurry reactor system are at the heart of AGC-21. Our research leadership in this area continues, resulting in key scientific and technological breakthroughs for advancing catalyst performance. Our catalyst systems have been specifically designed for slurry operation, now recognized by others as being the most efficient way to convert synthesis gas to wax. Exxon has spent over a decade optimizing the cobalt slurry reactor, including several years of fully integrated FBSG - HCS operations.

Figure 1a
EXXON AGC-21 HYDROCARBON SYNTHESIS (HCS) PROCESS

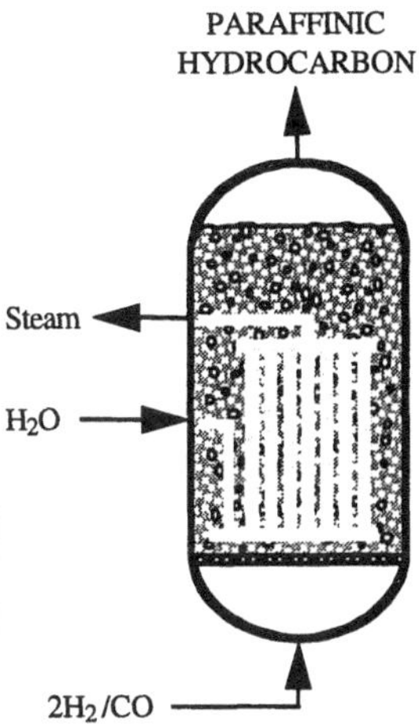

<u>Revolutionary, Patented Catalyst Family</u>

- Proprietary Multimetallic Cobalt-based Catalysts
- Patented 'Thin-Layer' Concept
- Tailored for Slurry Reactor

<u>Slurry Reactor - Route to Viable GTL plants</u>

- Significantly More Economic than Fixed Bed
- Broad, Patented Hydrodynamic Operating Regime
- Proprietary Catalyst Activation and Maintenance
- Two Years of Integrated Operation w/FBSG

Figure 1b
AGC-21 SLURRY REACTOR EXHIBITS HIGH PRODUCTIVITY

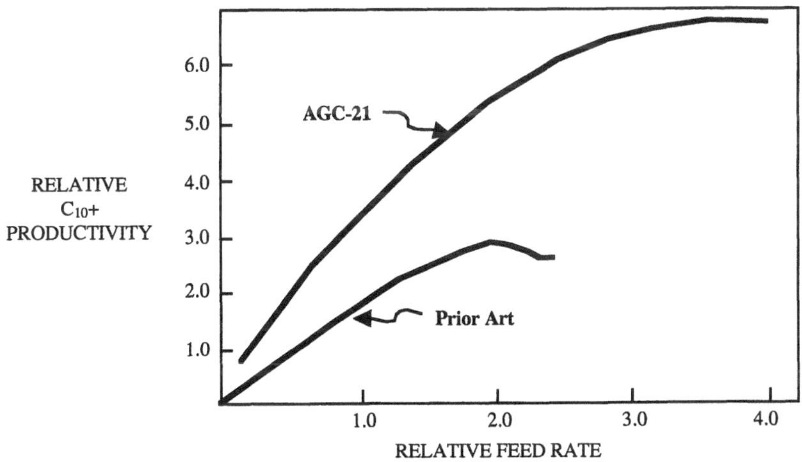

- Slurry Productivity 2-3 Times Higher Than Prior Art
- Major Improvement Achieved by Controlling Interplay of
  - Three Phase Hydrodynamics
  - Process Conditions
  - Catalyst Properties

As noted there are a number of technological breakthroughs incorporated into AGC-21, including new state-of-the-art cobalt HCS catalysts. Our studies have shown that intraparticle mass transfer can be important, especially when operating at high productivities. We have found that by maintaining cobalt in a dispersed RIM or thin layer on the outside of the catalyst, it is possible to achieve very high productivity with minimum gas make. Our RIM catalysts and advanced reactor engineering concepts have demonstrated the highest wax productivities available to date.

Figure 2
RIM CATALYSTS PROVIDE HIGH PRODUCTIVITY AND SELECTIVITY TO HEAVY HYDROCARBONS

ELECTRON MICROPROBE REVEALS CATALYTIC "RIM"

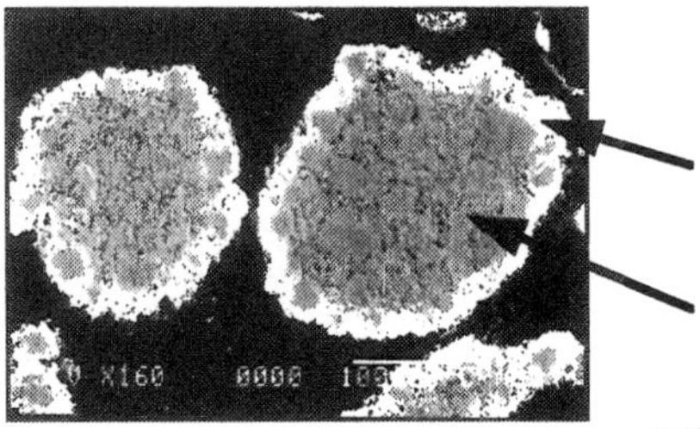

- Peak Performance Provided by Highly Active Cobalt Sites Present in Thin Outer Layer of Catalyst Particles

Reference: US 5,545,674, US 5,128,377 and EP 434,284

Figure 3
EXXON HYDROCONVERSION UNIT IN KAWASAKI, JAPAN

- 25,000 BPD
  Multi-Phase Slurry
  Hydroconversion Unit

High catalyst productivity, combined with our commercial experience with large-scale three-phase reactors is one of the keys to AGC-21. Exxon has operated a large-scale multi-phase reactor at its Kawasaki Japan refinery, to convert heavy crude into clean products. A single unit with a capacity of 25,000 B/D has provided valuable experience and confidence for scale-up of slurry HCS to high capacities.

The final step in AGC-21 involves wax upgrading to a range of liquid products. Exxon's significant Hydroprocessing experience coupled with recent advances in Hydroisomerization catalyst technology, have led to a number of upgrading options, depending on the target product slate. The upgrading reactor converts HCS wax to a liquid suitable with a pour point below 35oF that is suitable for transport via pipeline or tanker. Coupled with proprietary catalyst with a balance of hydroisomerization and hydrocracking activities, the upgrading step is able to produce a mixture of diesel and naphtha in varying yield. Diesel which can approach 80% of the total plant output is an excellent fuel or blendstock.

Further analysis of AGC-21 diesel shows the high quality of the clear, colorless paraffin liquid.

Figure 4
HIGH QUALITY DIESEL PROPERTIES

	Conventional	Pseudo-CARB	AGC-21™
Cetane	45	50	74
Sulfur (ppm)	330	345	<10
Specific Gravity	0.84	0.84	0.78
Flash Point (°C)	71	82	81
Cloud Point (°C)	-17	-16	-12

The absence of high-density aromatics along with the abundance of paraffins lead to a somewhat lighter diesel with gravity below that of conventional fuel. While this could reduce driving range if used neat, it opens attractive possibilities for blending with denser refinery streams to produce a fuel of high quality.

Figure 5
F-T DIESEL HIGHER CETANE THAN TYPICAL REFINERY STREAMS

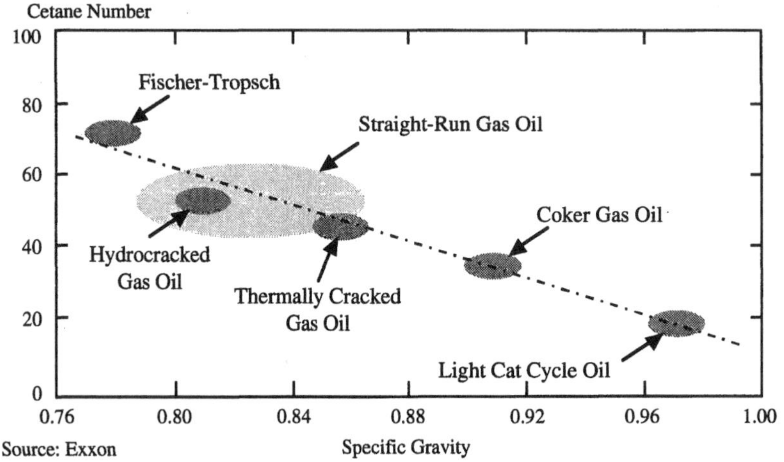

Source: Exxon

REFERENCES

J. M. Hochman, G. C. Lahn, R. F. Bauman, and B. Eisenberg, "Development of Advanced Gas Conversion Technology", Proceedings from the European Applied Research Conference on Natural Gas, Eurogas '92, A. Holmen, B. F. Magnussen and F. Steinke, Eds., Trondheim, May 1-3, 1992.

B. Eisenberg, G. C. Lahn, R. A. Fiato and G. R. Say, "Exxon AGC-21 Advanced Gas Conversion Technology", presented at Alternate Energy '93, Council on Alternate Fuels, April 27-30, 1993.

C. M. Eidt, Jr., R. F. Bauman, B. Eisenberg, J. M. Hochman and G. C. Lahn, "Current Developments in Natural Gas Conversion Technology", Natural Gas Processing, 13, 5, 1-8, 1994.

B. Eisenberg, L. L. Ansell, R. A. Fiato and R. F. Bauman, Advanced Gas Conversion Technology for Remote Natural Gas Utilization", presented at the 73rd Annual Gas Processors Association Convention, March 7-9,1994.

B. M. Everett, B. Eisenberg and R. F. Bauman, "Advanced Gas Conversion Technology: A New Option for Natural Gas Development", presented at the First Doha Conference on Natural Gas, March 14, 1995.

B. Eisenberg, R. A. Fiato, C. H. Mauldin, G. R. Say, S. L. Soled, "Exxon's Advanced Gas-To-Liquids Technology In Natural Gas Development", presented at AIC Conference on Gas-To-Liquids Conversion, Singapore, March 5, 1998.

NATURAL GAS CONVERSION V
Studies in Surface Science and Catalysis, Vol. 119
A. Parmaliana et al. (Editors)

Generation of Synthesis Gas Off-Shore : Oxygen Supply and Opportunities for Integration with GTL Technologies

Dennis M. Brown*, Derek Miller, Rodney J. Allam
Air Products PLC, Walton-on-Thames, Engl
and Peter J. A. Tijm
Air Products and Chemicals, Inc., Allentown, USA

SUMMARY

Oxygen is frequently used in generating the synthesis gas required for converting natural gas to liquid products (GTL). With the increasing need to convert gas associated with deposits of crude oil, and as a large part of such deposits are in remote areas and frequently off-shore, there is a need to design and build oxygen plants capable of operating at sea.

This paper gives an overview of Air Products' proposals for oxygen generation suitable for Floating Production, Storage and Off-Loading (FPSO) applications. It pays particular attention to safety, size and weight issues, and integration of the various unit operations. Air Products' considerable experience in methanol, DME and Fischer-Tropsch available from its syngas-to-liquids pilot plant, makes it particularly well placed to address integration of the exothermic downstream GTL unit.

ASSOCIATED GAS - THE GTL OPTION

The conversion of natural gas to synthetic fuels such as Fischer-Tropsch "white crude" (1,2,3) offers an alternative to flaring, re-injection, or LNG production. New Fischer-Tropsch technology developments have reduced the costs of producing the sulphur-free white crude, and the increasing environmental constraints in the developed world are leading to close study of this application. In particular, the operation of a Fischer-Tropsch unit on board a ship (FPSO) is a very interesting and productive way of converting usually-flared gas into a clean, synthetic crude which can be mixed with the regular crude oil in the ships' tanks. This recent application has become classed as the gas-to-liquids (GTL) option. Although primarily aimed at Fischer-Tropsch, other liquid products include methanol (4,5,6), for either chemical or fuel use, and possibly, Dimethyl Ether (DME) which is proposed as a clean diesel alternative (7).

SYNGAS GENERATION AND GTL TECHNOLOGIES

In each case of a GTL technology, the natural gas is first converted to a synthesis gas (syngas) prior to conversion of that syngas to the ultimate product.

Conventional Steam-Methane Reforming (SMR) is now often combined with oxygen blown secondary reforming - usually an autothermal reformer (ATR). Recent developments in reformer technology include the Advanced Gas Heated Reformer offered by ICI and the proposed BP-Kvaerner reformer. Partial oxidation processes (POX) offer an alternative to endothermic reforming by reacting methane directly with oxygen. This reaction is highly exothermic and the heat can be used for steam/power generation.The choice of syngas generating technology can be complicated. The H_2:CO ratio required for Fischer-Tropsch and methanol synthesis is essentially 2:1 which can be conveniently provided directly by oxygen-blown ATR units. However, conventional SMR technology can be preferred in add-on situations or where the "extra" H_2 offered from SMR's can be utilised in hydro-treating the Fischer-Tropsch white crude to clean jet or diesel fuels.

Grass-roots applications usually favour POX, particularly as the heat generated can be used for site power requirements. From a CO_2 emission perspective, and this is becoming increasingly important in those parts of the world where such emissions are taxed, a combination of conventional SMR with an ATR secondary unit may be preferred.

Off-shore, a whole new set of additional rules apply. On a ship, the size - particularly the height - of the units is important as are weight and footprint considerations. Efficiency may be of secondary importance to convenience of size, ease of operation and safety issues. The oxygen plant takes on a wholly new perspective on a ship that is rocking, rolling, heaving and yawing in heavy seas. These operating and safety considerations have a major influence on the choice of syngas generation off-shore.

OXYGEN SUPPLY AT SEA - ADDRESSING THE ISSUES

Oxygen at the tonnage scale can be supplied by several technologies. The traditional process for high purity (99%) large (>300TPD) plants uses cryogenic distillation of air. Vacuum Swing Adsorption (VSA) technology is also offered commercially generally in modules up to 150 TPD and where high O2 purity is not required (90-98% purity). Polymeric membranes can also be harnessed to produce oxygen, although many multi-modular units are usually required for any significant tonnage. The issue of weight, size or footprint area, is important at sea and for GTL processes currently being considered, which use upwards of 500 TPD oxygen, regular membrane production of O2 is probably not viable. VSA units are also large, although stackable, and may offer an option where purity is not required. However, use of VSA will require compression of the gaseous oxygen product to serve in the syngas/GTL trains. Compressors do leak, and gaseous oxygen compression is not recommended for ship-board applications.

The traditional cryogenic ASU can avoid O2 compression by producing liquid oxygen at low pressure and pumping this liquid to the pressure required for downstream service. This cycle (with vaporisation) is the recommended route for providing large quantities of oxygen at sea. The main challenges for large ASU's on ships relate to the design and operation of such plants to maintain production under the swaying motion experienced on board a ship located in open sea, especially the special safety features required for an application involving fuel, oxidant, ignition sources and personnel in close proximity.

A. Process Design

In a cryogenic ASU, air is compressed, cleaned and cooled to cryogenic temperatures, where it is distilled in a double column arrangement to produce a low pressure O2 product. There are then fundamentally two ways in which pressurised O2 can be produced : compressor or pump. The compressor route takes cold gas O2 from the column, warms it to ambient and compresses it. The pump route, takes liquid O2 (LOX) from the column, pumps it to delivery pressure and vaporises it against a high pressure air stream. For a shipboard application, a pumped LOX design is preferred. Oxygen compressors always have some risk of leaks and fire, which is handled on-shore by the use of safety barriers. This is inappropriate for a ship application, where an 'inherently safe' option is preferred.In order to prevent operational and safety problems on an ASU, it is important that all contaminants in the air are adequately dealt with. Most impurities are removed by the front end molecular sieve units- taking out CO2, water, and most hydrocarbons. Some hydrocarbons get through the front end and naturally accumulate in the LOX. With a pumped LOX cycle there is effectively a very large draw-off or purge, which minimises the chances for impurity build up.

Another 'inherently safe' feature recommended for ship applications is the use of a copper-wound coil heat exchanger for the vaporisation of the pumped LOX. For on-shore application, the LOX is typically vaporised in an aluminium plate fin exchanger. However, due to the potential for a fire (aluminium will burn in pressurised O2) a blast shield is provided. Again, this is inappropriate on a ship. Since copper will not burn as readily in O2, a copper exchanger is preferred off-shore. This choice has the added advantage that it can vaporise LOX at any pressure (aluminium plate fin exchangers have limited design pressures) and is therefore equally viable for Fischer-Tropsch or methanol production.

Distillation can be performed using either trays or packing. On-shore, structured packing is the technology of choice due to its lower power requirements resulting from low pressure drop. In off-shore GTL applications, power saving is probably not a significant issue, but packing is still the preferred choice - the reason being the potential difficulties associated with column movement on a ship (8). Trays have free liquids which can be severely maldistributed by the swaying motion of the ship. Packing only has thin films of liquid, which will be affected much less - this subject will be discussed in more detail later on.

Another safety feature is the selection of copper packing in sections of the plant where O2 purities are sufficiently high that aluminium or stainless steel packing could burn. In practice this means the bottom section of the LP column.

The reboiler/condenser which provides reflux for the HP column and boil-up for the LP column needs careful consideration. Two technology choices exist : the traditional thermosyphon reboiler and the more efficient downflow reboiler. The main issue of importance in the equipment selection is safety. If the liquid O2 is allowed to dry boil, build up of potentially hazardous impurities can occur. On land, both types can operate successfully. Currently, however, the thermosyphon is the preferred choice for off-shore applications since it is easier to maintain good liquid flow distribution and avoid dry-boiling.

One of the major technical challenges of building an ASU on a ship is in designing distillation column systems which are tolerant of wave motion. The most significant motions in terms of impact on the distillation column are swaying and permanent tilt. Typical values for North Sea / West of Shetland locations are :- Swaying : up to 16 degrees from vertical; Swaying Period : 10 - 20 seconds, with lower movements (say 2 degrees) up to 40 seconds; Permanent Tilts : up to 4 degrees.

To put these numbers in perspective, a shipboard ASU making O2 might have a cold box height of 35m and be located 15m above the centre of rotation of the ship - with a 16 degree sway, the top of the cold box will move 14m away from the vertical position. A typical land-based, high efficiency cold box might be 50m high and will be installed with an allowable deviation away from vertical of less than 25mm.

B. Packed Column Performance under Swaying and Offset Conditions

Performance of packed columns, particularly on ASU's, can be very sensitive to the ratio of liquid to vapour (L/V). Column tilt will adversely affect performance in two ways. Part of the column will be over-irrigated with liquid, with some liquid running down the wall and bypassing the packing. Other parts of the packing will have too little vapour, perhaps with some vapour not seeing any liquid at all and also bypassing. In addition, high liquid loading provides more resistance to vapour flow, which will tend to force vapour to drier areas of the packing. There may also be problems with initial distribution of liquid into the packing, when the distributors are not horizontal.

C. Effect of Bypassing on Oxygen Plant Performance

In practice, the distillation columns will be designed to avoid complete liquid/vapour bypassing. However, in order to understand the consequences of these phenomena taking place, a sensitivity study indicates the effect of either gas or liquid bypassing. The study is based on an ASU designed to make 99.6% purity O2 at a recovery of 95% of the oxygen in the feed air. Bypassing is assumed to take place equally and simultaneously in all distillation sections in the HP and LP columns.

At a product purity of 99.6% it can be seen that even relatively small amounts of liquid bypassing will cause a dramatic reduction in the O2 recovery. Above 3% bypassing, purity cannot be maintained at any recovery. Vapour bypassing is not as severe, for example 10% bypassing causes a loss in recovery of about 20%.

It is interesting to note that if we allow the O2 purity to drop a little, the sensitivity to bypassing is greatly reduced (e.g. even 10% liquid bypassing only reduces recovery of O2 by 10%).

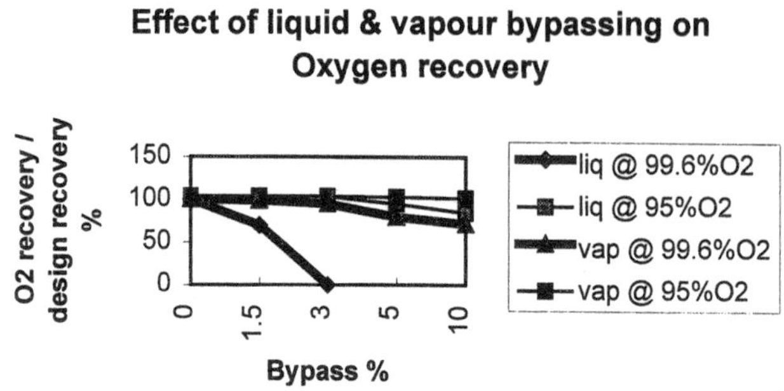

The following can be concluded from this exercise:
(i) distillation performance is potentially very sensitive to column motion, if this results in vapour or, in particular, liquid bypassing. The column internals will need to be designed to minimise the effects of column motion;
(ii) impact on the downstream GTL process of varying O2 purity needs to be understood. For example in particularly bad weather with a lot of column movement, if column performance starts to drop off, a lower O2 purity will permit significantly higher O2 production;(iii) another way of maintaining O2 production, even under adverse conditions, is to provide an oversized Main Air Compressor, such that even if O2 recovery starts to fall, O2 production can be maintained - this might well be more economical than installing extra packing to keep O2 recovery up. In other words there is an optimisation between investment in the distillation column and investment in the Main Air Compressor.

D. Dynamic Study of Swaying Column Performance

Under ship movement, it is quite likely that column performance will vary as a function of time. Liquid inventories in the columns will tend to buffer out this effect. A study was carried out to understand the impact of varying distillation performance by creating a full dynamic simulation model of the plant (complete with liquid/vapour inventories process control loops, etc.). The simulation modelled the plant running at steady state and then oscillated the HETP.

The simulation assumed a constant air feed to the plant and constant O2 product, and the O2 purity was varied. The HETP was varied between 70% and 100% over a 20 second period to simulate performance as the column passes through the vertical position (good performance) through to the tilted condition (poor performance).

The purities in the packing were seen to start oscillating and that O2 purities near the bottom of the LP column tend to get worse, before stabilising at a lower purity. The O2 product, which is based on the purity of the liquid in the sump of the LP column drops as expected, but does not oscillate. It also takes about 2 hours for the purity to drop from 99.6% to about 99.4%. There is a lot of dampening of the product purity due to the liquid inventory in the sump and the purities in the plant as a whole only move slowly due to the total liquid hold up in the columns.

One of the important design considerations relates to the amounts of liquid in the plant. More liquid helps column stability, but could adversely affect safety. The dynamic simulation can be used to investigate this in order to determine the optimum capacities. The largest liquid inventory is in the sump of the LP column - for a thermosyphon reboiler between 30-50% of the total plant inventory is located here. If we reduce this inventory to 10% of its normal value and rerun the simulator, we find that the O2 purity changes more quickly (it now takes about 1 hour to change) but is still quite stable.

E. Packed Column Design for Shipboard Applications

There are fundamentally two problems associated with designing packed columns for operation under swaying conditions : how to minimise liquid maldistribution feeding into the packing, and how to minimise liquid maldistribution within the packing (9,10). A novel design developed by Air Products, uses a mildly pressurised distributor to give a coarse distribution, feeding into a second segmental "egg box" distributor. This distributor consists of individual irrigation cells which do not connect with each other - each cell will typically have one distribution hole. In between each cell is a space for the vapour to pass through. The purpose of the egg box distributor is to prevent liquid slopping from one side of the column to the other, and to provide a liquid buffer which dampens down any variations in the feed to each cell.

The combination of the two distributors can cope with all expected oscillating movements (typically 8-12 degrees) and also permanent tilts (typically 2-3 degrees) without any significant additional complexity or cost.

Structured packing consists of vertical corrugated perforated sheets which are arranged in layers. The sheets in each layer are aligned. Usually, each layer is rotated in relation to the one above and below, in order to promote mixing, which helps the distillation process. In a column which is tilting, there will be little tendency for liquid to maldistribute across the sheets, but there will be a tendency for sideways movement along the sheets. In order to minimise this tendency as the column moves, the basic structure of the packing can be used to advantage. By having the sheets arranged in individual cells within a layer, with some sheets being orientated perpendicularly to others, the degree to which liquid can move is severely restricted to the size of each cell. In this way, although there will still be some local maldistribution, there will be no gross effects such as large amounts of liquid running down the walls of the column.

COLD BOX /PLANT LAYOUTS

A variety of different cold box layouts have been investigated. A number of parameters have been considered such as cost/height/plot area/safety. Fundamentally height is an issue for several reasons: taller columns will sway more and will be more likely to be affected by ship motion; there are height limitations (typically 35m) on ships due to helicopters, etc.

Conventionally, the LP column is located on top of the HP column - this can be done within the 35m limit. Different arrangements for the columns and heat exchangers have been investigated. It is believed that the best option will be to have all of the cryogenic equipment within one large box or can. Primarily for safety reasons, it will also give a smaller plot area at similar cost to other options. Options to reduce cold box height include installing the LP and HP columns side by side, with transfer pumps. These arrangements can also be accommodated within a single cold box, within a larger plot area. Overall, a conventional stacked column arrangement is simplest, cheapest and will perform adequately under all required operating conditions.

There are many permutations for overall ship layouts - the turret can be at the front, middle or rear of the ship. This study has taken a typical layout and, in discussion with shipping experts W.S. Atkins, has come up with a conceptual layout based on the principle of keeping all of the equipment containing hydrocarbons in a separate area to the ASU/control room (incorporate the philosophy of an "invisible line" across the ship - fuel on one side, non-fuel on the other).

SAFETY : COLD BOX

On land, cold boxes are designed to provide a containment for the cryogenic equipment and insulation. It is an accepted risk that very occasionally there will be process leaks within the box. The risk is manageable because personnel are generally well away from the box and there is plenty of space for dispersion of leaking fluids. On a ship such incidents could lead to major problems both to personnel and to surrounding equipment, and even the ship itself. Such incidents are totally unacceptable for off-shore application.

Air Products has developed a contained system which addresses all of the common potential hazards associated with ASU cold boxes. The process equipment is located within a container having a central void surrounded by perimeter insulation. In on-shore plants, the whole free space in the box is packed with insulation. If a gas leak should occur, there may be a restriction to gas flow that panel buckling can occur. On the perimeter insulated box recommended for off-shore, there is free passage to the top of the cold box, where a relief system will direct the cold vapours to a safe location without over pressuring the cold box structure.

This containment will hold all of the process equipment containing cryogenic fluids, including the LOX pumps, such that any leak will be contained. The bottom of the box contains a sealed membrane, sized to contain the full liquid inventory held within the process. The sides and top of the perimeter are internally clad with metal sheeting designed to deflect any spraying liquid down into the sump. If there is a leak, any spilt liquid can remain in the sump indefinitely.

In practice, in order to effect repairs, the liquid will have to be removed and a special "dip tube" arrangement for removal of any spilt liquid is incorporated using a venturi run on compressed dry air. The concept of perimeter insulation and liquid sump, whilst being unusual in the ASU industry, is somewhat similar to the methods already employed in LNG transportation - a proven technique for bulk cryogenic liquid containment on ships.

SAFETY : AIR IMPURITIES

There are a variety of different air contaminants of concern when designing an ASU. A shipboard ASU will need particular care due to the close proximity of hydrocarbon sources and the potential for poor air circulation patterns due to confined equipment layouts. Most contaminants are removed in the front end clean-up adsorbers, which are particularly designed for this purpose. However, if design limits are exceeded, breakthrough into the plant can occur. The main area of concern is the build-up of hydrocarbons in the liquid oxygen. Unfortunately, most contaminants which are not totally adsorbed on the molecular sieves end up in the LOX. The build-up of these contaminants will be limited by the LOX draw-off or purge. With a pumped LOX cycle, this purge is effectively about 20% of the air so contaminants can only build up by a factor of 5.

CONCLUSIONS

We can conclude by saying that today's cryogenic ASU's, with certain modifications, are technically viable for off-shore production of O2 and that the safety considerations, whilst critical, can be adequately dealt with. It is clear that there are a number of unique features required for a shipboard ASU which will make the design different than a land-based plant. These key features are covered by technology developments specifically tailored to this application.

SHIPBOARD OXYGEN PLANT PFD: SPECIAL FEATURES

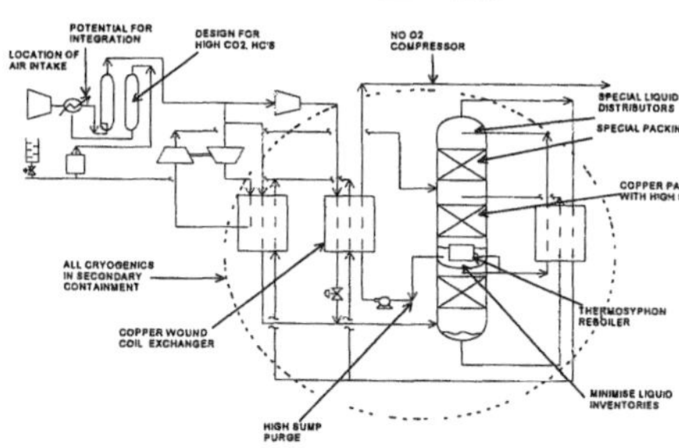

References:

1 M.J.Gradassi; N.W.Greene, " Economics of Natural Gas Conversion Processes", Fuel Processing Technology, 42, 65-83, 1995.

2 Choi et.al, " Design/Economics of Natural Gas Based Fischer-Tropsch Plant", AIChE Spring General Meeting, Houston, TX, 2-3 March 1997.

3 B.Jager, R.Espinoza, " Adv. in Low Temperature Fischer-Tropsch Synthesis", Catal..Today, 23, 17-28, 1995

4 R.Hymas, "The Impact of New Methanol Technology on Global Technology", World Methanol Conference, Tampa, FL, 89-101, 8-10 Dec 1997.

5 J.P.Lange, "Perspective for Manufacturing Methanol at Fuel Value",Ind.Eng.Chem.Res., Vol.36, No.10, 1997.

6 H.Gohna, "Concepts for Modern Methanol Plants", World Methanol Confer., Tampa, FL, 139, 8 Dec 1997.

7 Fleisch et.al, "A New Clean Diesel Technol:, SAE, Intern. Congress & Exposition, Detroit, MC, Feb, 1995.

8 D.Dunne, "Distillation Performance Improvement", Proceedings of ICI Methanol Technology Operators Forum, San Diego, California, 1997.

9 Tanner et.al, "Modelling the Performance of Packed Column Subjected to Tilt", Trans ICHemE, 74, partA ,March 1996.

10 Tanner et.al, " Effects of Tilt and Motion on a Packed Deaeration Column", IChemE Sym Ser No 128, B111-B119

NATURAL GAS CONVERSION V
Studies in Surface Science and Catalysis, Vol. 119
A. Parmaliana et al. (Editors)

Conversion of Natural Gas to Ethylene and Propylene: The Most-Profitable Option

B. V. Vora,[a] T. Marker,[a] E. C. Arnold,[a] H. Nilsen,[b] S. Kvisle,[b] and T. Fuglerud[b]
[a] UOP LLC, Des Plaines, Illinois 60017, U.S.A.[b] Norsk Hydro a.s. Bygdøy allé 2, Oslo, Norway

INTRODUCTION

In 1996, proven gas reserves were estimated to be at 150 x 10^{12} (trillion) cubic meters.[1] At an annual production of about 2.2 trillion cubic meters, the current reserves will last more than 65 years. To date, the use of natural gas for chemical production has been mostly limited to ammonia and methanol. The gas-conversion options that producers should consider are converting gas to liquids (transportation fuel), to methanol and ammonia, and to petrochemical intermediates or final products. Niche liquefied natural gas (LNG) applications should also be evaluated. This paper examines these options and shows that the conversion of natural gas to ethylene and propylene and their derivatives is a most-attractive option.

NATURAL GAS CONVERSION TECHNOLOGIES

Direct Conversion

A number of direct natural gas conversions to higher hydrocarbons have been studied for more than 50 years.[2-4] Of these methods, oxidative coupling has received significant attention. The best result to date approaches 30% conversion of methane and about an 80% selectivity to a C_2^+ hydrocarbon product, which is economically not attractive.

Indirect Conversion

Indirect conversion of methane via synthesis gas shows many promising options. The technology for the production of synthesis gas is widely used and well established. Steam reforming to partial oxidation and most recently autothermal reactor systems have been used commercially. The natural gas conversion options via synthesis gas that are of interest are gas to liquids (GTL), gas to methanol and ammonia, and gas to olefins (GTO).

Synthesis Gas to Liquids

During World War II, several small plants built in Germany to convert coal gas to liquid fuels used a Fischer-Tropsch (F-T) process to support the war efforts. In 1955, Sasol, in South Africa, commissioned Sasol-I, the world's first large-scale GTL plant and expanded it during the 1980s with Sasol-II and Sasol-III for a total capacity of around 125,000 barrels per day (BPD) of syncrude production[5]. 1993, Shell began operating its 12,000 BPD GTL plant in Malaysia to primarily produce wax. More recently, Exxon and Syntroleum have announced their version of an improved F-T process.[6] During the 1970s, Mobil developed a process for converting methanol to gasoline and built a 10,000 BPD plant in New Zealand. The economics of this plant were based on continuously rising fuel prices following the oil embargo of 1973. However, stable and lower oil prices and moderate inflation made the project uneconomical, and the MTG plant has been shut down for many years.

Methanol and Ammonia

The conversion of natural gas to methanol and ammonia via synthesis gas is used in many parts of the world, and several commercial processes are available for their production. Today, more than 50 operating plants range in capacity from 100,000 to about 1.0 million metric tons per year of methanol production. Recent developments suggest that a single-train methanol plant with a capacity in the range of 5,000 to 10,000 metric tons per day (MTD) is possible.[7-10]

Natural Gas to Olefins (GTO)

No *direct* process for converting natural gas to olefins (GTO) currently exists. An *indirect* two-step process involves first the production of methanol and then its selective conversion in a second step to ethylene and propylene. The UOP/HYDRO MTO process converts crude methanol to ethylene and propylene. Thus, significant savings are realized in methanol synthesis by not requiring the methanol purification-distillation section. The catalyst used in this step, which was first discovered and disclosed by Union Carbide[11] in 1988, has been scaled up jointly by UOP and two Norsk Hydro. The UOP/HYDRO MTO process has a wide range of operating flexibility depending on the demand for ethylene or propylene, as illustrated by the material balances of two cases in Table 1. Over the last two years this technology has been disclosed in numerous publications.[12-14]

Table 1. Material Balance for UOP/HYDRO MTO Process

Component	High-Ethylene Case, KMTA	High-Propylene Case, KMTA	Market Value, $/MT
Methanol feed	2,502	3,370	Natural gas feed @ 0.75$ / MM Btu
Products:			
Ethylene	500	500	500
Propylene	325	660	400
Mixed C_4's	105	185	126
Light Ends	62	30	100
C_5^+	60	75	37
Total	1,052	1,450	—

The key benefits of the UOP/HYDRO MTO process are the following:

- High ethylene and propylene yields improve profitability.
- Product slate flexibility allows high-ethylene or high-propylene production.
- Simple fluid-bed reactor and regenerator design enables steady-state operation at high efficiency.
- Low production of diolefins and acetylenes makes product purification easier.
- Low production of H_2 and methane results in a less-expensive simplified cold-box design.
- Low production of ethane and propane allows the production of chemical-grade ethylene and propylene without $C_2^=$ and C_2 or $C_3^=$ and C_3 splitter columns.
- Low emissions make the process environmentally friendly.

GAS UTILIZATION MARKET

A quantity of natural gas equivalent to 1 million British thermal units (MM Btu) is valued at $0.75 for a Middle East location. The $0.75 paid for the 1 MM Btu of natural gas is converted to $9 for low-density polyethylene or polypropylene. By contrast, the other products, such as LNG, distillate, and gasoline, have three to four times less value. The profitability issue must also be coupled with market size.

For the most part, LNG projects are viable only in large gas fields. The number of LNG producers is limited, and only eight LNG importers are active.[15] Currently, about 103 billions of cubic meters per year

(BCM/yr) of natural gas is processed to make 75 MM MTA of LNG. The capital cost for LNG is substantial: total costs can be in the $3 to $5 billion range for large-scale units, taking into account the entire LNG supply train.[16]

The theoretical consumption of 20 BCM/yr of natural gas to meet current methanol demand of 24 MM MTA is the smallest by far, accounting for only 1% of the total natural gas produced. The 1996 demand for ammonia, which is used mostly for fertilizers such as urea, ammonium phosphates, and nitrates, is about 120 MM MTA. To meet this ammonia production, 108 BCM of natural gas is required. During the 1950s to 1970s, most of the methanol or ammonia plants were located in industrial nations and close to demand centers. However, with discoveries of new gas fields in many developing nations, the ammonia and methanol industry has shifted to areas where natural gas is available at a low price. If all current demand for ethylene and propylene were to be met by the GTO technology, then about 340 BCM/yr, which is less than GTL but higher than LNG or ammonia or methanol, is required. Certainly, an ethylene and propylene capacity of 120 MM MTA is appreciable and could efficiently use natural gas as a feed source. Because the LNG plants rely on large reserves, the number of opportunities is relatively slim. A plant producing 4 MM MTA of LNG requires about 6 BCM/yr. Over a life of 20 years, this requirement amounts to 120 BCM, but the number of gas fields with this capacity is limited.

The GTL processes, such as those from Exxon or Shell, also rely on large gas reserves: a GTL plant producing 50,000 BPD requires about 5 BCM/yr of natural gas, or 100 BCM over 20 years. As with LNG, this amount limits the number of opportunities to less than 5% of the existing fields. Even if the minimum economic size becomes 10,000 BPD, less than 10% of the existing fields are available to meet a 20-year production span.

A world-scale UOP/HYDRO MTO unit producing 500,000 MTA of ethylene and 325,000 MTA of propylene needs about 2 BCM/yr of natural gas feed. Further, the UOP/HYDRO MTO process is ideal for revamp situations: when only 50,000 MTA of ethylene may be needed, only 0.2 BCM/yr of natural gas is required. Roughly 30 to 40% of the world's gas fields, which accounts for more than 90% of the available natural gas, will meet the requirement for ethylene and propylene production by using the UOP/HYDRO MTO process.

ECONOMICS OF GAS-CONVERSION PROCESSES

The following cases were chosen for economic comparison:

- 50,000 BPD GTL complex
- 500,000 MTA ethylene and 325,000 MTA propylene GTO complex
- 4 MM MTA LNG plant
- 500,000 MTA ethylene and 228,000 MTA propylene naphtha cracker

The battery limit capital cost (BLCC) for the UOP/HYDRO MTO plant is estimated at $270 million. The total GTO investment is estimated at $1.1 billion, which includes offsites at 30% of BLCC investment and an allowance of $240 million for storage, terminal, and contingency allowance. On a similar basis, the total investment for naphtha cracking is estimated to be $740 MM.

The average values for the GTL and LNG plants came from some recent publications.[14,16-18] As seen from the economic comparison in Table 5, a GTO complex producing 825,000 MTA of ethylene and propylene uses 2.0 BCM/yr of natural gas, less than half the amount of gas required for a 50,000 BPD GTL complex. Although the GTO plant uses less than half to a one-third of the natural gas feed used by the GTL and LNG options, it has significantly better economics. As shown in Table 2, the payback period on capital for the GTO option is 5.1 years compared to the 7.7 years required for the 50,000 BPD GTL case.

The LNG production option is also no bargain. For a natural gas utilization of 6.5 BCM/yr at an estimated capital investment of $2.8 billion, the payback, at 7.9 years, is similar to that for the GTL option. The comparison shows that GTO economics are significantly more attractive compared to the GTL or LNG options.

Table 2. Economic Comparison of GTL, LNG, GTO, and Naphtha-Cracking Options

	GTL	LNG	GTO	Naphtha Cracking
Product rate, BPD	50,000	—	—	—
Ethylene + propylene, KMTA	—	4250	825	728
Gas utilization, BCM/yr	5	6.5	2.0	—
Naphtha feed, KMTA	—	—	—	1610
Investment, $MM	1500	2800	1100	740
Gas cost, @ $.75/MM Btu	139	181	56	—
Naphtha cost @ $150/MT	—	—	—	241
Other op. cost, $MM/yr	70	120	90	87
Total op. cost, $MM/yr	209	301	146	328
Revenues, $MM/yr	428	917	402	431
Transportation, $MM/yr	25	260	41	40
Revenues, FOB	403	657	361	391
Cash flow excluding capital	194	356	215	63
Simple payback, yr	7.7	7.9	5.1	11.8

Table 2 also shows the economics of ethylene-propylene production via GTO and naphtha cracking. These economics are based on a Middle East location and use the same product values shown in Table 1. The naphtha feed was valued at 150 $/MT. The breakdown of the investment requirement for these two options is shown in Table 2. The basis for the naphtha cracking plant is 500,000 MTA ethylene production and 228,000 MTA propylene coproduction. The payback period of 11.8 years for naphtha cracking is also poor compared to the 5.1 years for the GTO technology. A sensitivity analysis was performed by varying the price of natural gas from $0.20 to $2/MM Btu. Natural gas prices much above $1.25/MM Btu result in a nearly exponential rise in payback time for LNG and GTL options. By contrast, the GTO payback is nearly linear with gas price. This result points out that the GTO option looks even more attractive than alternative processes as fuel prices sink or rise.

CONCLUSION

Methanol and ammonia plants formerly sited in industrial nations are now being built close to low-cost natural gas sources. Now with the availability of an economical route for the GTO conversion, the olefin industry should show a similar shift over time.

The GTO option is more attractive than the GTL or LNG options. The GTL or LNG option is limited to large gas fields. The GTO option provides wider application to 30% of the gas fields. The GTO is also more attractive than naphtha cracking for ethylene and propylene production.

The UOP/HYDRO MTO process provides the most-attractive option for upgrading natural gas to higher-value ethylene and propylene. The process is flexible over a wide range of ethylene-to-propylene ratios and has a combined ethylene and propylene yield of 80%. The process has been demonstrated in a fluid-bed reactor-regenerator demonstration unit at Norsk Hydro. UOP's MTO-100™ catalyst has been commercially manufactured with good reproducibility. UOP and Norsk Hydro are confident about the scale-up of the process to a commercial unit and are actively licensing the process.

REFERENCES

1. Department of Energy, "Natural Gas," Energy Information Administration (May 4, 1996).
2. N. D. Parkyns, C. I. Warburton, and J. D. Wilson, "Natural Gas Conversion to Liquefied Fuels and Chemicals: Where Does It Stand?" *Catal. Today*, 18(4), 1993, 385-442.
3. J. M. Fox, T. P. Chen, and B. Degen, "An Evaluation of Direct Methane Conversion Processes," *Chem. Eng. Prog.*, 42, Apr. 1990.
4. M. J. Gradassi and N. W. Green, "Economics of Natural Gas Conversion Processes," *Fuel Processing Technology, 42, 65-83, 1995*
5. B. Jager, "Developments in Fischer-Tropsch Technology," Natural Gas Conversion ConferenceKruger National Park, South Africa, Nov. 19-23, 1995
6. A..K. Rhodes,"Downstream Catalyst Companies Ally with Gas-to-Liquids Process Developer," *Oil & Gas J., 85-86, Dec. 30, 1996.*
7. H. H. Larsen, "A Selection of Technology for Large Methanol Plants," 1994 World Methanol Conference, Geneva, Switzerland, Nov.-Dec. 1994.
8. H. Göhna, "Concepts for Modern Methanol Plants," 1997 World Methanol Conference, Tampa, FL, Dec. 8-10, 1997.
9. R. Hymas, "The Impact of New Methanol Technology on Global Markets," 1997 World Methanol Conference, Tampa, FL, Dec. 8-10, 1997.
10. I. Rees, "Methanol Technology: The Next Generation," presented at 1997 World Methanol Conference, Tampa, FL, Dec. 8-10, 1997.
11. S. W. Kaiser, U.S. Patent 3,677,243.
12. 12.N. Nilsen, S. Kvisle, T. Fuglerud, and A. Slattsveen, "The UOP/HYDRO MTO Process: The New Way to Convert Natural Gas into Petrochemicals," European Applied Research Conference on Natural Gas, Trondheim, Norway, June 3-5, 1996.
13. B. V. Vora, T. L. Marker, P. T. Barger, H. R. Nilsen, S. Kvisle, and T. Fuglerud, "Economic Route for Natural Gas Conversion to Ethylene and Propylene," *Studies in Surface Science and Catalysis*, 107, 87-98, 1997.
14. B. V. Vora, E. C. Arnold, T. L. Marker, and C. Eng, "The UOP/HYDRO MTO Process: The Critical Link in Upgrading Natural Gas to Olefins," 47th Canadian Chemical Engineering Conference, Edmonton, Canada, Oct. 5-8, 1997.
15. "Financing LNG," *Petroleum Economist*, Aug. 1997
16. Chem Systems, "Liquefied Natural Gas," Process Evaluation Research Planning (PERP), Nov. 1997.
17. Chem Systems, "Advances in Fischer-Tropsch Technology," Process Evaluation Research Planning (PERP), July 1994.
18. Gerald N. Choi and others, "Design/Economics of a Natural Gas Based Fischer-Tropsch Plant," AIChE Spring National Meeting, Houston, TX, Mar. 9-13, 1997.

NATURAL GAS CONVERSION V
Studies in Surface Science and Catalysis, Vol. 119
A. Parmaliana et al. (Editors)

Gas-To-Liquids Processes: Current Status & Future Prospects

M.M.G. Senden, A.D. Punt, A. Hoek

Shell International Oil Products B.V., Shell Global Solutions / OGTS
P.O. Box 38 000, 1030 BN Amsterdam, The Netherlands

Most of the world's Natural Gas resources are remote from the market and their exploitation is constrained by factors such as transportation costs and market outlets. To increase the economic utilisation of Natural Gas, techniques other than pipeline transmission or LNG shipment have been developed. Chemical conversion (liquefaction) of gas to make gas transportable as a liquid and add value to the products is now a proven technology.

The Shell Middle Distillate Synthesis (SMDS) process is well known as Shell's contribution to the novel techniques of gas utilisation. The technology has been applied now for several years by Shell MDS (Malaysia) Sendirian Berhad on a commercial scale of 12,000 bbl/d. The plant produces automotive fuels, special chemicals and waxes of exceptional quality. Commercial experience in the production and marketing of these products has been obtained over four years of successful operation.

On-going research has led to promising developments with the synthesis catalyst, leading to further cost savings as well as scope for improved thermal efficiency. These have greatly enhanced the robustness of the SMDS economics. However, the economic viability of Gas-to-Liquid processes today and in the future still depends on key factors such as low-cost gas, capital costs, site-specific conditions and possible synergy with other industrial activities, e.g. power generation. The SMDS process has developed into a commercially proven technology, a position Shell intends to hold and strengthen.

1. INTRODUCTION

In the not so distant future (some 20 years from now), it is expected that crude oil production will peak and so the supply of oil will thereafter start to decline. This fact, in conjunction with the growing acceptance of Natural Gas fuelled power generation and the environmental advantages of Natural Gas, will increase the importance of Natural Gas as energy source in coming decades.
Today Natural Gas is mainly used as a heating medium and to power-generation fuel. To a limited extent, due to market constraints, it is also used to produce chemicals such as Methanol and Ammonia.

The option of exporting gas through pipeline grids is normally the preferred route for regional markets. For large transportation distances, the "physical" conversion of Natural Gas to Liquefied Natural Gas (LNG) by cooling is applied. LNG projects are constrained by the need to have large gas reserves (typically larger than 7 TCF), require high capital investment and also entail high (cryogenic) transportation costs. The world-wide LNG market is a large one, now accounting for more than 4 % of world gas consumption.

Reason enough to look for alternative Natural Gas commercialisation routes which are suitable for smaller gas reserves and can transport the energy content over long distances economically.

The "chemical" conversion of Natural Gas by catalytic Fischer-Tropsch synthesis is such an alternative route. The process produces Middle Distillates of an extraordinary quality. Its products can be targeted on the oil products market, which is huge in size and has existing (conventional) distribution / marketing systems in place.

At present Shell operates an integrated gas to Middle Distillates plant MDS(M) Sdn Bhd in Bintulu, Malaysia with a capacity of 12,000 bbl/day. This paper will focus on the four years of operation of this plant, the quality of its products and the prospects of reducing its specific capital investment costs so that it can become an attractive alternative to gas firing for the commercialisation of gas reserves.

2. SHELL'S COMMERCIAL EXPERIENCE

In 1989 the first commercial project for SMDS Malaysia was approved, the plant was then built and in 1993 production started, twenty years after the first research effort was initiated.

Instead of the original plan to produce only Naphtha, Kerosene and Gasoil with storage and loading facilities, the Bintulu project was extended to include the production of speciality chemicals. This addition supported the economics of a relatively small pioneer project. It took advantage of the high quality of the raw wax and in particular its very high content of straight chain molecules. 12,000 bbl/d is a very small capacity in terms of oil products, but in terms of chemical products, the 470,000 Mt/annum 'chemical products' is regarded as a mega-chemical project.

MDS(Malaysia) Sdn Bhd is a joint venture of Shell, Petronas, Mitsubishi and the State of Sarawak. The plant converts 100 MMSCFD of Natural Gas into 12,000 bbl/d of products ranging from Middle Distillates to Solvents and Paraffinic Waxes. The process scheme is shown in Fig. 1

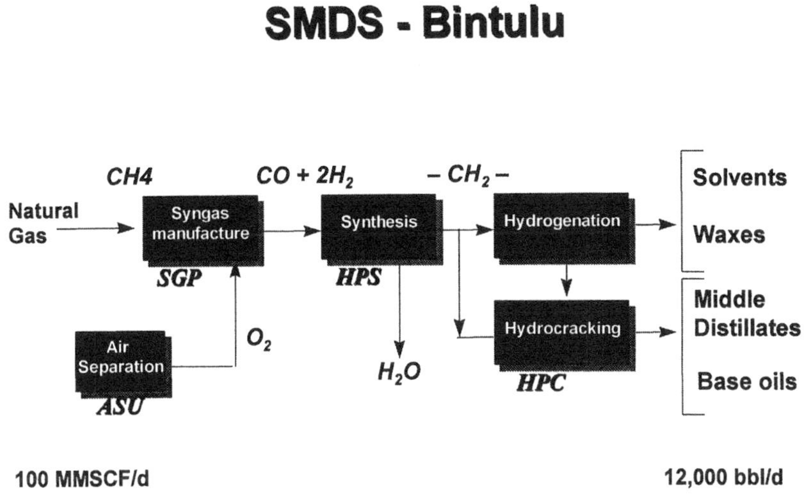

Figure 1. Simplified process flow scheme SMDS Bintulu

The raw Fischer Tropsch wax produced in the HPS step is firstly hydrogenated to saturate olefins present in the raw wax. Solvents and detergent feed stock grades are subsequently distilled off in various distillation columns. Surplus wax is again mixed with the surplus distillates and then hydrocracked to the desired middle distillate products with the use of a proprietary hydrocracking catalyst. Simultaneously the product is isomerised to improve cold flow properties. A new catalyst was developed for this service. The product from the hydrocracker is a "synthetic crude oil" which is refined in a conventional crude distillation unit. Water co-produced in the synthesis reaction is purified, treated to specifications in a biotreater and subsequently discharged.

MDS(Malaysia) Sdn Bhd now has experience with four years of operation in a fully integrated complex and has gathered a wealth of experience such as with the online regeneration of the FT catalyst, the behaviour and confirmed yield figures of the proprietary developed hydrocracker catalyst, and also with the operational distillation of wax grades and bio-treating of effluent water.

3. QUALITY OF THE PRODUCTS

By their nature, products from carbon monoxide and hydrogen are extremely clean. They contain no sulphur, no nitrogen and no (detectable) aromatics. The SMDS products have impurities that are several orders of magnitude lower than highly refined crude oil derived products. Here we focus on the middle distillate fuel qualities (see table 1).

Naphtha

The ability to use the SMDS naphtha as chemical feed stock for petrochemicals instead of converting it into gasoline, presents an advantageous opportunity. The tops/naphtha fraction is completely paraffinic and therefore makes an excellent ethylene cracker feed stock. The absence of aromatic structures in the cracker feed gives up to 10 percent higher yield of ethylene and propylene than from petroleum-derived naphtha feed stock.

Kerosene

SMDS kerosene is a good clean-burning product for domestic heating. It can also be used to upgrade kerosene fractions which have a low smoke point and high aromatics and would otherwise be unsuitable for use in jet fuel.

Moreover there are attractive solvent/chemical applications for SMDS kerosene. It can be tailored to a solvent of high purity, which together with its low odour and water-clear appearance make it particularly attractive in applications such as dry cleaning and other "speciality solvent" applications.

SMDS kerosene is also suitable as a trimming agent for heavy gas oils that need to be upgraded to specification diesel, for example winter diesel for automotive use in cold climates.

Gas oil

SMDS gas oil has excellent properties, far better than the minimum specifications in terms of cetane number. Given its exceptional quality, SMDS gas oil is an ideal, high-value blending component for upgrading lower-quality stock derived from catalytic cracking operations, for example cycle oils. However, also the forthcoming EU specifications for Diesel which require e.g. a minimum cetane number of 58, a maximum density of 825 kg/m3 and a maximum sulphur content of 50 ppm mass, could ideally use SMDS gas oils as blending component. Alternatively the products could do well on a market where premium specifications are desired to meet local requirements.

Table 1
Typical Middle Distillate Fuel properties

Property	Unit	Naphtha	Kerosene	Gas oil	Method
Density @ 15 °C	kg/m3	690	738	780	ASTM D1298
Saybolt colour		+29	+30	n/a	ASTM D156
ASTM colour		n/a	n/a	0	ASTM D1500
Distillation range					ASTM D86
IBP	°C	43	155	201	
FBP	°C	166	191	358	
Sulphur 1)	ppm	<3	<10	<15	ASTM D1266
Cetane index		n/a	58	75	ASTM D976
Smoke point	°C	n/a	>50	n/a	ASTM D93
Flash point	°C	n/a	42	88	ASTM D93
Aromatics	%v	0	<0.1	<0.1	ASTM D5186

One example of this is the Californian diesel market, where the Californian Air Resources Board (CARB) requires commercial fuels to give lower emissions than a reference fuel which has a minimum cetane number, low sulphur and low aromatics. As it has zero aromatics, zero sulphur (below detection limits) and 76 cetane index, SMDS gas oil can be used in California, USA, in blends that meet these severe CARB regulations.
An other attractive characteristic of SMDS Diesel is that this material is fully bio-degradable. For certain applications this is of utmost importance (e.g. where spills into environment are feasible), but this feature will become more and more important in the near future.

4. FUTURE PROSPECTS

The economic viability of Gas-to-Liquid projects today and in the near future depends on several key economic factors: the availability of low-cost gas, crude prices, capital/operating costs, site specific factors and the fiscal regime of the host country.
Since crude prices remain at a persistently low level, low-cost gas as well as fiscal friendly regimes are of the utmost importance to make a Gas-to-Liquids project viable in the next few years.
If natural gas is priced at 0.50 US$/MMBtu, then the feedstock cost element in the product is about US$ 5/bbl. The total selling price in US$/bbl further includes a capital charge which depends on numerous factors, including fiscal regimes, local incentives, debt/equity ratio, type of loans and corporate return requirements.

Another important factor is whether the products are for domestic use or export. For countries with sufficient gas but needing to import oil or oil products to meet local demand, SMDS products manufactured in that country should realise at least import parity values. For such countries, therefore, the national benefit of the SMDS process may be substantial. In addition to these factors, the capacity of the plant is of great importance. Especially for remote locations, where self-sufficiency of the plant is essential, larger plants in the range of 50,000 bbl/d benefit from the economy of scale.

MDS(Malaysia) Sdn Bhd has demonstrated the commercial viability of the SMDS process.
Meanwhile further developments have reduced the specific capital cost such as:

- equipment scale-up, notably in the synthesis gas manufacturing plant, which accounts for more than 50 % of the total process capital cost;
- further catalyst improvements. A second-generation catalyst which yields significantly more liquids than the catalyst used in Bintulu, has been developed and is ready for application in the next plant;
- general process integration within the project.

Operational experience, coupled with technological improvements, has resulted in specific capital costs to well below 30,000 US$/bbl. This capex figure includes process units, utility systems, common facilities e.g. flare system, effluent water treating, fire fighting, storage, and general facilities such as buildings, roads, control room, substation, telecommunication etc.

The successful application of SMDS conversion technology at Bintulu represents an important advance in the commercialisation of that technology and is an asset in Shell's portfolio of technologies for making natural gas transportable. It provides exciting opportunities in terms of marketing hydrocarbon products of a quality ideally suited for a business environment requiring increasingly high performance standards. Shell intends to hold that position and will further strengthen it by actively seeking commercial opportunities for a project in the 50,000 bbl/d range.

Author Index

STUDIES IN SURFACE SCIENCE AND CATALYSIS

Advisory Editors:
B. Delmon, Université Catholique de Louvain, Louvain-la-Neuve, Belgium
J.T. Yates, University of Pittsburgh, Pittsburgh, PA, U.S.A.

Volume 1 **Preparation of Catalysts I.** Scientific Bases for the Preparation of Heterogeneous Catalysts. Proceedings of the First International Symposium, Brussels, October 14–17, 1975
edited by **B. Delmon, P.A. Jacobs and G. Poncelet**

Volume 2 **The Control of the Reactivity of Solids.** A Critical Survey of the Factors that Influence the Reactivity of Solids, with Special Emphasis on the Control of the Chemical Processes in Relation to Practical Applications
by **V.V. Boldyrev, M. Bulens and B. Delmon**

Volume 3 **Preparation of Catalysts II.** Scientific Bases for the Preparation of Heterogeneous Catalysts. Proceedings of the Second International Symposium, Louvain-la-Neuve, September 4–7, 1978
edited by **B. Delmon, P. Grange, P. Jacobs and G. Poncelet**

Volume 4 **Growth and Properties of Metal Clusters.** Applications to Catalysis and the Photographic Process. Proceedings of the 32nd International Meeting of the Société de Chimie Physique, Villeurbanne, September 24–28, 1979
edited by **J. Bourdon**

Volume 5 **Catalysis by Zeolites.** Proceedings of an International Symposium, Ecully (Lyon), September 9–11, 1980
edited by **B. Imelik, C. Naccache, Y. Ben Taarit, J.C. Vedrine, G. Coudurier and H. Praliaud**

Volume 6 **Catalyst Deactivation.** Proceedings of an International Symposium, Antwerp, October 13–15, 1980
edited by **B. Delmon and G.F. Froment**

Volume 7 **New Horizons in Catalysis.** Proceedings of the 7th International Congress on Catalysis, Tokyo, June 30–July 4, 1980. Parts A and B
edited by **T. Seiyama and K. Tanabe**

Volume 8 **Catalysis by Supported Complexes**
by **Yu.I. Yermakov, B.N. Kuznetsov and V.A. Zakharov**

Volume 9 **Physics of Solid Surfaces.** Proceedings of a Symposium, Bechyňe, September 29–October 3, 1980
edited by **M. Láznička**

Volume 10 **Adsorption at the Gas–Solid and Liquid–Solid Interface.** Proceedings of an International Symposium, Aix-en-Provence, September 21–23, 1981
edited by **J. Rouquerol and K.S.W. Sing**

Volume 11 **Metal-Support and Metal-Additive Effects in Catalysis.** Proceedings of an International Symposium, Ecully (Lyon), September 14–16, 1982
edited by **B. Imelik, C. Naccache, G. Coudurier, H. Praliaud, P. Meriaudeau, P. Gallezot, G.A. Martin and J.C. Vedrine**

Volume 12 **Metal Microstructures in Zeolites.** Preparation - Properties – Applications. Proceedings of a Workshop, Bremen, September 22–24, 1982
edited by **P.A. Jacobs, N.I. Jaeger, P. Jirů and G. Schulz-Ekloff**

Volume 13 **Adsorption on Metal Surfaces.** An Integrated Approach
edited by **J. Bénard**

Volume 14 **Vibrations at Surfaces.** Proceedings of the Third International Conference, Asilomar, CA, September 1–4, 1982
edited by **C.R. Brundle and H. Morawitz**

Volume 15 **Heterogeneous Catalytic Reactions Involving Molecular Oxygen**
by **G.I. Golodets**

Volume 106 **Hydrotreatment and Hydrocracking of Oil Fractions**
Proceedings of the 1st International Symposium / 6th European Workshop, Oostende, Belgium, February 17-19, 1997
edited by **G.F. Froment, B. Delmon and P. Grange**

Volume 107 **Natural Gas Conversion IV**
Proceedings of the 4th International Natural Gas Conversion Symposium, Kruger Park, South Africa, November 19-23, 1995
edited by **M. de Pontes, R.L. Espinoza, C.P. Nicolaides, J.H. Scholtz and M.S. Scurrell**

Volume 108 **Heterogeneous Catalysis and Fine Chemicals IV**
Proceedings of the 4th International Symposium on Heterogeneous Catalysis and Fine Chemicals, Basel, Switzerland, September 8-12, 1996
edited by **H.U. Blaser, A. Baiker and R. Prins**

Volume 109 **Dynamics of Surfaces and Reaction Kinetics in Heterogeneous Catalysis.**
Proceedings of the International Symposium, Antwerp, Belgium, September 15-17, 1997
edited by **G.F. Froment and K.C. Waugh**

Volume 110 **Third World Congress on Oxidation Catalysis.**
Proceedings of the Third World Congress on Oxidation Catalysis, San Diego, CA, U.S.A., 21-26 September 1997
edited by **R.K. Grasselli, S.T. Oyama, A.M. Gaffney and J.E. Lyons**

Volume 111 **Catalyst Deactivation 1997.**
Proceedings of the 7th International Symposium, Cancun, Mexico, October 5-8, 1997
edited by **C.H. Bartholomew and G.A. Fuentes**

Volume 112 **Spillover and Migration of Surface Species on Catalysts.**
Proceedings of the 4th International Conference on Spillover, Dalian, China, September 15-18, 1997
edited by **Can Li and Qin Xin**

Volume 113 **Recent Advances in Basic and Applied Aspects of Industrial Catalysis.**
Proceedings of the 13th National Symposium and Silver Jubilee Symposium of Catalysis of India, Dehradun, India, April 2-4, 1997
edited by **T.S.R. Prasada Rao and G. Murali Dhar**

Volume 114 **Advances in Chemical Conversions for Mitigating Carbon Dioxide.**
Proceedings of the 4th International Conference on Carbon Dioxide Utilization, Kyoto, Japan, September 7-11, 1997
edited by **T. Inui, M. Anpo, K. Izui, S. Yanagida and T. Yamaguchi**

Volume 115 **Methods for Monitoring and Diagnosing the Efficiency of Catalytic Converters.**
A patent-oriented survey
by **M. Sideris**

Volume 116 **Catalysis and Automotive Pollution Control IV.**
Proceedings of the 4th International Symposium (CAPoC4), Brussels, Belgium, April 9-11, 1997
edited by **N. Kruse, A. Frennet and J.-M. Bastin**

Volume 117 **Mesoporous Molecular Sieves 1998**
Proceedings of the 1st International Symposium, Baltimore, MD, U.S.A., July 10-12, 1998
edited by **L.Bonneviot, F. Béland, C. Danumah, S. Giasson and S. Kaliaguine**

Volume 118 **Preparation of Catalysts VII**
Proceedings of the 7th International Symposium on Scientific Bases for the Preparation of Heterogeneous Catalysts, Louvain-la-Neuve, Belgium, September 1-4, 1998
edited by **B. Delmon, P.A. Jacobs, R. Maggi, J.A. Martens, P. Grange and G. Poncelet**

Volume 119 **Natural Gas Conversion V**
Proceedings of the 5th International Gas Conversion Symposium, Giardini-Naxos, Taormina, Italy, September 20-25, 1998
edited by **A. Parmaliana, D. Sanfilippo, F. Frusteri, A. Vaccari and F. Arena**

FSC
www.fsc.org
MIX
Papier | Fördert gute Waldnutzung
FSC® C083411

Kolibri 360 GmbH
Ferdinand-Jühlke-Straße 7
99095 Erfurt, Deutschland
produktsicherheit@kolibri360.de